P9-AOS-881

Your steps to success.

STEP 1: Register

All you need to get started is a valid email address and the access code below. To register, simply:

1. Go to www.mybiology.com
2. Click the appropriate book cover.
   Cover must match the textbook edition being used for your class.
3. Click "**Register**" under "**First-Time User?**"
4. Leave "**No, I Am a New User**" selected.
5. Using a coin, scratch off the silver coating below to reveal your access code.
   Do not use a knife or other sharp object, which can damage the code.
6. Follow the on-screen instructions to complete registration. You may want to click Log In Now on your registration confirmation screen and bookmark that page.

During registration, you will establish a personal login name and password to use for logging into the website. You will also be sent a registration confirmation email that contains your login name and password.

Your Access Code is:

Note: If there is no silver foil covering the access code, it may already have been redeemed, and therefore may no longer be valid. In that case, you can purchase access online using a major credit card. To do so, go to www.mybiology.com and click "Buy Access" and follow the on-screen instructions.

STEP 2: Log in

1. Go to www.mybiology.com and "**Log In.**"
2. Click your book cover and under enter the login name and password that you created during registration. Cover must match the textbook edition being used for your class. If unsure of this information, refer to your registration confirmation email.
3. Click "**Log In.**"

STEP 3: (Optional) Join a class

Instructors have the option of creating an online class for you to use with this website. If your instructor decides to do this, you'll need to complete the following steps using the Class ID your instructor provides you. By "joining a class," you enable your instructor to view the scored results of your work on the website in his or her online gradebook.

To join a class:

1. Log into the website. For instructions, see "STEP 2: Log in."
2. Click "**Join a Class**" near the top right.
3. Enter your instructor's "**Class ID**" and then click "**Next.**"
4. At the Confirm Class page you will see your instructor's name and class information. If this information is correct, click Next.
5. Click "**Enter Class Now**" from the Class Confirmation page.
- *To confirm your enrollment in the class, check for your instructor and class name at the top right of the page. You will be sent a class enrollment confirmation email.*
- *As you complete quizzes on the website from now through the class end date, your results will post to your instructor's gradebook, in addition to appearing in your personal view of the Results Reporter.*

To log into the class later, follow the instructions under "STEP 2: Log in."

Got technical questions?

SITE REQUIREMENTS

For the latest updates on Site Requirements, go to www.mybiology.com, choose your text cover, and click Site Reqs.

WINDOWS
OS: Windows XP
Resolution: 1024 x 768
Plugins: Latest version of Flash/QuickTime/Shockwave (as needed)
Browsers: Internet Explorer 6.0 (XP only); Internet Explorer 7.0; Firefox 2.0.
Internet connection: 56K minimum

MACINTOSH
OS: 10.3.x; 10.4
Resolution: 1024 x 768
Plugins: Latest version of Flash/QuickTime/Shockwave (as needed)
Browsers: Safari 1.3 (10.3.x only); Safari 2.0 (10.4.x only); Firefox 2.0
Internet connection: 56K minimum

Register and log in

Join a class

Important: Please read the Subscription and End-User License agreement, accessible from the book website's login page, before using the *myBiology* website. By using the website or CD-ROM, you indicate that you have read, understood, and accepted the terms of this agreement.

Features of the Website for *Life on Earth*, **Fifth Edition**

Student Home

The Student Home organizes media under central concepts that correlate directly to each chapter in the text.

Animations

Explore approximately 75 animations with interactive review exercises and videos.

Quizzes

Assess your understanding with almost 2000 multiple-choice questions. Each chapter includes a Pre-Test to diagnose your current knowledge and provide you with a study plan and a comprehensive Post Test (40 questions per chapter on average). Each quizz has hints, immediate feedback, grading, and e-mailable results.

Cumulative Test

Create customized tests on multiple chapters at once by choosing the chapters and number of questions. Each question provides a hint and reference to the relevant sections in the text.

Word Study Tools

Use electronic Flash Cards to test your knowledge of the key terms and definitions for each chapter or multiple chapters. Includes audio pronunciations for selected terms. Word roots are also provided to improve vocabulary skills. The Glossary provides definitions of boldface terms and selected audio pronunciations.

Discovery Channel Videos

View a collection of 85 videos.

Web Links and References

Organized by chapter, Web Links allow you to extend your knowledge through links to relevant websites, access news links on recent developments in biology, and consult an archive of relevent biology news articles.

LIFE ON EARTH

FIFTH EDITION

Teresa Audesirk
Gerald Audesirk
Professors Emeriti, University of Colorado, Denver

Bruce E. Byers
University of Massachusetts, Amherst

PEARSON

Benjamin Cummings

San Francisco   Boston   New York
Cape Town   Hong Kong   London   Madrid   Mexico City
Montreal   Munich   Paris   Singapore   Sydney   Tokyo   Toronto

Editor-in-Chief: Beth Wilbur
Executive Director of Development: Deborah Gale
Acquisitions Editor: Star MacKenzie
Senior Project Editor: Mary Ann Murray
Media Producer: Katherine Brayton
Supplements Project Editor: Blythe Robbins
Editorial Assistant: Erin Mann
Executive Marketing Manager: Lauren Harp
Marketing Manager: Jay Jenkins
Executive Managing Editor: Erin Gregg
Managing Editor: Mike Early
Senior Production Supervisor: Shannon Tozier
Media Production Supervisor: Liz Winer

Production Service: S4Carlisle Publishing Services
Illustrations: Dartmouth Publishing, Inc.
Text and Cover Design: Hespenhiede Design
Manufacturing Manager: Michael Penne
Director, Image Resource Center: Melinda Patelli
Manager, Rights and Permissions: Zina Arabia
Image Permission Coordinator: Debbie Latronica
Photo Research: Yvonne Gerin
Cover Printer: Phoenix Color
Printer and Binder: QuebecorWorld Versailles
Cover Image: Marguerite Smits Van Oyen/ npl / MINDEN PICTURES
About Cover Image: Meerkat (Suricata suricatta) family standing on guard, Tswalu Kalahari Reserve, South Africa

Library of Congress Cataloging-in-Publication Data
Audesirk, Teresa.
    Life on earth / Teresa Audesirk, Gerald Audesirk, Bruce E. Byers.—5th ed.
        p. cm.
    Includes bibliographical references and index.
    ISBN 978-0-13-175535-2
    1.   Biology.   I. Audesirk, Gerald.   II. Byers, Bruce E.   III. Title.
    QH308.2.A936 2008
    570—dc22
                                                            2007052445

ISBN 10-digit 0-13-175535-8; 13-digit 978-0-13-175535-2 (Student edition)
        10-digit 0-321-55895-2; 13-digit 978-0-321-55895-4 (Professional copy)

Printed on recycled paper. Printer used soy-based ink.

1 2 3 4 5 6 7 8 9 10—QWV—11 10 09 08 07
www.pearsonhighered.com

About the Authors

Terry and Gerry Audesirk grew up in New Jersey, where they met as undergraduates. After marrying in 1970, they moved to California, where Terry earned her doctorate in marine ecology at the University of Southern California, and Gerry earned his doctorate in neurobiology at the California Institute of Technology. As postdoctoral fellows at the University of Washington's marine laboratories, they worked together on the neural bases of behavior, using a marine mollusk as a model system.

The Audesirks are now professors emeriti of biology at the University of Colorado Denver, where they taught introductory biology and neurobiology, and researched mechanisms of heavy metal toxicity from 1982 until their retirement in 2006.

Terry and Gerry share a deep appreciation of nature and of the outdoors. After retirement, they moved to Northern Colorado, where they enjoy hiking, horseback riding, and snowshoeing. Long-time members of many conservation organizations, they are looking forward in retirement to volunteer work with the Nature Conservatory and the Colorado Division of Wildlife.

Bruce E. Byers, a midwesterner transplanted to the hills of western Massachusetts, is a professor in the biology department at the University of Massachusetts, Amherst. He has been a member of the faculty at UMass (where he also completed his doctoral degree) since 1993. Bruce teaches introductory biology courses for both nonmajors and majors; he also teaches courses in ornithology and animal behavior.

A lifelong fascination with birds ultimately led Bruce to scientific exploration of avian biology. His current research focuses on the behavioral ecology of birds, especially on the function and evolution of the vocal signals that birds use to communicate. The pursuit of vocalizations often takes Bruce outdoors, where he can be found before dawn, tape recorder in hand, awaiting the first songs of a new day.

To Heather, Jack, and Lori, and in memory of Eve and Joe

T. A. and G. A.

To Maija

B. E. B.

Preface

Should scientists be allowed to clone humans? Are genetically engineered crops safe? Are people causing climate change? Will avian flu become epidemic? Will physicians soon be growing artificial organs for transplantation into people? Why are antibiotic medicines becoming less effective? Many of today's most important and controversial social, medical, environmental, and ethical issues are related to biology. The need for voters, jurors, and citizens in general to understand the basic concepts of biology has never been more urgent.

Many of the students who will use this text are enrolled in a course that will provide both their first and final exposure to biology before they leave formal education. We hope that they will emerge from the course prepared to ask intelligent questions, make informed choices, and scrutinize science articles in the popular press with an educated and critical eye. We also hope that students will better understand and appreciate their own bodies, the other organisms with which we share Earth, the evolutionary forces that mold all life-forms, and how complex interactions within ecosystems sustain us and all other life on our planet. Perhaps most of all, we hope that students will develop a fascination with life that will inspire them to keep learning science. To help instructors achieve these teaching goals, we offer this revision of *Life on Earth*. Now in its fifth edition, *Life on Earth* helps students effectively manage a wealth of scientific information and motivates their learning.

Helping Students Manage Information and Get Motivated

The fifth edition of *Life on Earth* has been revised with three goals in mind:

- To help instructors effectively manage their presentation of biological concepts and processes.

- To help students understand how biology informs issues that are relevant to their lives and concerns; we focus in particular on issues related to health and the environment.

- To craft a presentation that reflects the latest scientific findings, up-to-date statistics, and topical examples.

Life on Earth
. . . Is Organized Clearly and Consistently

Throughout each chapter, students will find aids that help them navigate through the large amount of information they face in this course. We have organized this text to help students see the forest and the trees, without getting lost in thickets of detail.

- "At a Glance" at the start of each chapter brings together the chapter's major subheadings and now includes the titles of essays as well. Instructors can easily assign—and students can easily locate—key topics within each chapter.

- Major sections are introduced as questions to which the student will find answers in the section, while minor subheadings are presented as summary statements that reflect content. A key consequence of this organizational scheme is that it imparts an understanding of biology as a hierarchy of interrelated concepts, rather than a set of isolated, independent subjects. It also reminds students of the importance in science of asking questions.

- The "Summary of Key Concepts" section at the end of each chapter pulls together the important concepts. To be clear and consistent, we use the same major subheadings that also appear in both the "At a Glance" and the chapters themselves, allowing instructors and students to move efficiently among the different components within a chapter.

- Throughout each chapter are references to the text's accompanying media resources, which can be found on the companion Web site to the text. These callouts appear where the most challenging topics are covered, referencing Web Animations and, new to this edition, BioFlix.

. . . Actively Engages and Motivates Students

Scientific literacy cannot be imposed on students; they must actively participate in acquiring both core concepts and the habit of thinking like a scientist. The likelihood of this active participation is increased if students recognize that biology is connected to their personal lives and to issues that they care about. To help engage and motivate students, this new edition incorporates the following:

- **Case Study Approach.** Each chapter opens with a strikingly illustrated brief essay. These chapter-opening case studies are based on recent news items, on situations in which students might find themselves, or on particularly fascinating biological topics. For example, students will investigate blood doping by elite athletes (p. 98); contemplate the use of DNA to absolve people wrongly convicted of crimes (p. 188); explore why a flu virus that infects birds might be a serious threat to people (p. 418); and learn about a species that has been brought back from the brink of extinction (p. 625). Each chapter-opening case study ends with a cliff-hanger, enticing students to read more. At the end of each chapter, we revisit the case study, allowing students to explore the topic a bit further in light of what they have just learned and, often, to find answers to questions raised in the initial piece. The "revisited" essays conclude with a "Consider This" segment that poses an open-ended question to encourage deeper thinking about the topic. If you want to encourage students to read the openers, you can assign these questions.

- **Case Study Continued Feature.** The chapter-opening case studies are separated from their concluding "revisited" sections by a substantial stretch of text, so student readers may lose track of the opener topic and how it is linked to the chapter material. To counter this tendency and to better integrate chapter-opening themes and chapter content, we have added a new feature to the fifth edition. Each chapter contains several "Case Study *Continued*" segments, placed in the margin at key junctures, that briefly remind the reader of the case study topic, add a bit of new information, and tie the opener theme to the concept under discussion in the adjacent text.

- **Figure-Based Questions.** Selected figure captions in each chapter include questions designed to encourage readers to review and extend their new knowledge of the pictured structure or process. These help link the graphics to the content, a helpful exercise for today's more visual learners.

- **Feature Box Essays.** We have retained, updated, and added new essays to our three essay libraries: "Earth Watch" environmental essays, which explore issues such as the loss of biodiversity, the growing ozone hole, and invasions of exotic species; our medically related "Health Watch" essays that investigate topics such as sexually transmitted diseases, muscular dystrophy, and how smoking damages the lungs; and "Scientific Inquiry" essays that illuminate the process of discovery by describing how application of the scientific method leads to outcomes, such as the discovery of plant hormones or the structure of DNA.

- **Is This Science?** "Is This Science?" questions are among the questions in the "Applying the Concepts" critical-thinking sections at the end of each chapter. These questions are designed to help students practice their scientific reasoning and critical-thinking skills.

- **NEW Conservation Biology Chapter.** Environmental concerns are increasingly prominent, and *Life on Earth* has a long tradition of making connections between the environment and many areas of biological inquiry. In this edition, we expand this tradition by devoting a new chapter to the emerging discipline of conservation biology and the biodiversity crisis it addresses—Chapter 31: Conserving Earth's Biodiversity.

. . . Contains Superior Illustrations

Based on the advice of reviewers and careful scrutiny by the authors and developmental editor, we have revised many illustrations. For the fifth edition, we have:

- **Expanded the consistent use of color.** We have been vigilant in tracking the use of color to provide consistency in illustrating specific atoms, structures, and processes. We have also made the colors more vibrant to better distinguish individual parts of a figure, to help engage the modern readers' interest, and to focus attention on the most important aspects of the illustration.

- **Improved overall quality.** We have redrawn the more diagrammatic figures for greater interest and accuracy.

- **Enhanced label clarity.** We have revised the size, placement, and font of figure labels for more consistency and readability.

- **Refined our use of "talking boxes."** We have incorporated additional explanatory statements within figures, and have improved the talking boxes for greater clarity and ease of reading.

. . . Provides Print and Media Resources That Aid User Exploration

We understand the challenges presented to students in learning biology and to instructors in teaching it. As such, we've designed a supplements package to help both instructors and students to make the most of the course.

Instructor Supplements

Instructor Guide
by Michael Wenzel
ISBN: 0-13-175536-6 • 978-0-13175536-9
This helpful resource provides chapter outlines, answers to text questions, and innovative lecture activities with accompanying handouts that can be torn out, copied, and distributed to the class.

Test Item File
by Stephen T. Kilpatrick and Gregory Pryor
ISBN: 0-13-175537-4 • 978-0-13175537-6
The Test Item File offers hundreds of multiple-choice questions to use on tests and quizzes, all rated for difficulty and tied to a chapter learning objective. New to this edition are figure and scenario-based questions to test students' critical-thinking abilities.

Instructor Resource CD/DVD
ISBN: 0-13-175549-8 • 978-0-13175549-9
This powerful media package includes all of the teaching resources you'll need in one convenient location. Enhance your lectures with BioFlix and BLAST animations, PowerPoint presentations, active lecture questions to facilitate class discussions (for use with or without clickers), and an image library with all of the photos, illustrations, and tables from the text. Also included is an electronic version of the Test Item File. The files are categorized by chapter objective for instructor ease in searching for question types.

Transparency Pack
ISBN: 0-13-175538-2 • 978-0-13175538-3
More than 300 full-color acetates include select key figures and tables from the text.

OneKey CourseCompass™
ISBN: 0-13-175542-0 • 978-0-13175542-0
This nationally hosted, dynamic, interactive online course management system is powered by Blackboard, the leading platform for Internet-based learning tools. This easy-to-use and customizable program enables professors to tailor content to meet individual course needs! Every CourseCompass™ course includes preloaded content such as testing and assessment question pools. CourseCompass is also available with an e-book.
http://www.aw-bc.com/coursecompass/

OneKey WebCT
ISBN: 0-13-175541-2 • 978-0-13175541-3
This open-access course management system contains preloaded content such as testing and assessment question pools.
http://www.aw-bc.com/webct

OneKey Blackboard
ISBN: 0-13-175540-4 • 978-0-13175540-6
This open-access course management system contains preloaded content such as testing and assessment question pools.
http://www.aw-bc.com/blackboard

Student Supplements

Life on Earth Premium Website
This powerful study tool includes an e-book, interactive tutorials, and animations—including BioFlix and Web Animation activities. Each chapter includes learning objectives, quizzes with grade book functionality, flash cards, links to relevant Web sites, and more.
http://www.aw-bc.com/audesirk

Student Access Card for Companion Website
ISBN: 0-13-175551-X • 978-0-13175551-2

Current Issues in Biology, Vol. 5
ISBN: 0-321-54187-1 • 978-0-321-54187-1
Give your students the best of both worlds—accessible, dynamic, relevant articles from *Scientific American* magazine that present key issues in biology, paired with the authority, reliability, and clarity of Benjamin Cummings' nonmajors biology texts. Articles cover current topics in biology such as obesity, stem cell research, performance-enhancing drugs, global warming, the genetics of race, and more. A host of questions at the end of each article help students check their comprehension and see the connections between science and society. Each volume is available at no additional charge when packaged with a new text.

Student Study Guide
By Sonja Stampfler and Ruthanne Thompson
ISBN: 0-13-175539-0 • 978-0-13175539-0
The Student Study Guide for this edition is more useful than ever, with 50 test questions per chapter to help students prepare for exams. New to this edition are flash cards with each chapter's key terms, as well as word roots sections that help students to make connections between a word's root and its definition.

Study Card for *Life on Earth*
ISBN: 0-321-55914-2 • 978-0-321-55914-2
New to this edition, this helpful study tool summarizes each chapter's key concepts and directs students to informative illustrations in the text.

Acknowledgments

This edition of *Life on Earth* is our first endeavor under the auspices of Benjamin Cummings. We are pleased to be aboard, and thank Editor-in-Chief Beth Wilbur for welcoming us to our new publisher, easing our transition, and assembling a skilled and talented team to perform the dauntingly complex task of producing a text and supplement package. We are fortunate that the team included Senior Project Editor Mary Ann Murray, who carefully monitored our work, keeping us on track and providing copious, terrifically valuable editorial feedback and advice. Mary Ann helped us keep the text clear, cogent, and student friendly. She was ably assisted by Editorial Assistant Erin Mann. Melissa Parkin also provided helpful developmental editing and coordination, especially during the initial phase of the revision.

Working under the capable guidance of Managing Editor Michael Early, Senior Production Supervisor Shannon Tozier oversaw production, skillfully coordinating the multiple threads of the production process. Shannon brought art, photos, and manuscript together into a seamless whole. Her challenging task was made easier by the excellence of those whose efforts she oversaw: Production Project Manager Mary Tindle managed the complex flow of art and manuscript and somehow made sure that everything reached its various destinations on time. Photo Researcher Yvonne Gerin tracked down a wide selection of excellent photos, and Photo Production Manager Travis Amos ably made the arrangements necessary to make the photos publishable. Lorretta Palagi performed her copyediting duties with exceptional skill and care. Editorial Proofreader Art Bush caught as many of our errors as is humanly possible. Carlisle Communication completed a vast volume of composition accurately and with dispatch; the artists at Dartmouth Publishing brought a deft touch and fine aesthetic sense to the illustration program.

We also thank Marilyn Perry, Design Manager, for coordinating text and cover designs, and Gary Hespenheide for developing those lovely and effective designs.

Supplements Project Editor Blythe Robbins managed development of our diverse and useful array of supplemental instructional materials. Katherine Brayton, Media Producer, and Brienn Buchanan, Media Project Editor, did great work on the text's Web site, as did Liz Winer on the Instructor's Resource CD-ROM.

Jay Jenkins, Senior Marketing Manager, provided inspired marketing concepts and oversaw a large and dedicated sales force with energy, talent, and enthusiasm. This book would not be in your hands if not for Jay's tireless efforts.

Finally, Acquisitions Editor Star MacKenzie has brought kindness, enthusiasm, and deep experience to the task of leading our team. We thank her for her commitment to the project, her organizational ability, and her sensitivity to all of the people involved.

So here we acknowledge, with deep appreciation, our "coach" and all of our teammates!

Terry and Gerry Audesirk
Bruce E. Byers

Comments and questions are welcome from students and faculty. Please e-mail us at star.mackenzie@pearson.com.

Fifth Edition Reviewers

Cindy Jo Arrigo, *New Jersey City University*
Marilyn C. Baguinon, *Kutztown University of Pennsylvania*
Sarah F. Barlow, *Middle Tennessee State University*
Karen S. Borgstrom, *Moraine Valley Community College*
Francie S. Cuffney, *Meredith College*
Carl Estrella, *Merced College*
Diane W. Fritz, *Northern Kentucky University*
Sushmita Ghosh, *Heidelberg College*
Tom Knoedler, *James A. Rhodes State College*
Kathleen H. Lavoie, *State University of New York–Plattsburgh*
Stephen Lebsack, *Linn-Benton Community College*
Mary O'Sullivan, *Elgin Community College*
Kim Cleary Sadler, *Middle Tennessee State University*
Fayla Schwartz, *Everett Community College*
Howard Singer, *New Jersey City University*
Mark Smith, *Chaffey College*
Anna Bess Sorin, *University of Memphis*
Sonja Stampfler, *Kellogg Community College*
Diana Wheat, *Linn-Benton Community College*
Michelle Zurawski, *Moraine Valley Community College*

Fifth Edition Contributors

Test Bank
Stephen T. Kilpatrick, *University of Pittsburgh at Johnstown*
Gregory Pryor, *Francis Marion University*

Study Guide
Sonja Stampfler, *Kellogg Community College*
Ruthanne Thompson, *University of North Texas*

Instructor's Guide
Michael Wenzel, *California State University Sacramento*

Companion Website
Susan Lustick, *San Jacinto College*
Nancy Pencoe, *University of West Georgia*
Ethan Prosen, *New Jersey City University*
Michael Wenzel, *California State University Sacramento*

Previous Edition Reviewers

W. Sylvester Allred, *Northern Arizona University*
Judith Keller Amand, *Delaware County Community College*
William Anderson, *Abraham Baldwin Agriculture College*
Steve Arch, *Reed College*
Kerri Lynn Armstrong, *Community College of Philadelphia*
G. D. Aumann, *University of Houston*

Vernon Avila, *San Diego State University*
Marilyn C. Baguinon, *Kutztown University of Pennsylvania*
J. Wesley Bahorik, *Kutztown University of Pennsylvania*
Gail F. Baker, *LaGuardia Community College*
Michelle Baker, *Utah State University*
Neil Baker, *Ohio State University*
Bill Barstow, *University of Georgia, Athens*
Colleen Belk, *University of Minnesota, Duluth*
Michael C. Bell, *Richland College*
Gerald Bergtrom, *University of Wisconsin*
Arlene Billock, *University of Southwestern Louisiana*
Kathleen L. Bishop, *University of North Texas*
Brenda C. Blackwelder, *Central Piedmont Community College*
Raymond Bower, *University of Arkansas*
Robert Boyd, *Auburn University*
Marilyn Brady, *Centennial College of Applied Arts & Technology*
Edward M. Brecker, *Palm Beach Community College*
Neil Buckley, *State University of New York, Plattsburgh*
Virginia Buckner, *Johnson County Community College*
Arthur L. Buikema, Jr., *Virginia Polytechnic Institute*
Sharon K. Bullock, *Virginia Commonwealth University*
William F. Burke, *University of Hawaii*
Robert Burkholter, *Louisiana State University*
Kathleen Burt-Utley, *University of New Orleans*
Linda Butler, *University of Texas, Austin*
W. Barkley Butler, *Indiana University of Pennsylvania*
Bruce E. Byers, *University of Massachusetts, Amherst*
Sara Chambers, *Long Island University*
Nora L. Chee, *Chaminade University*
Joseph P. Chinnici, *Virginia Commonwealth University*
Dan Chiras, *University of Colorado, Denver*
Bob Coburn, *Middlesex Community College*
Martin Cohen, *University of Hartford*
Mary U. Connell, *Appalachian State University*
Joyce Corban, *Wright State University*
Ethel Cornforth, *San Jacinto College, South*
David J. Cotter, *Georgia College*
Lee Couch, *Albuquerque Technical Vocational Institute*
Dave Cox, *Lincoln Land Community College*
Donald C. Cox, *Miami University of Ohio*
Patricia B. Cox, *University of Tennessee*
Peter Crowcroft, *University of Texas, Austin*
Carol Crowder, *North Harris Montgomery College*
Donald E. Culwell, *University of Central Arkansas*
Robert A. Cunningham, *Erie Community College, North*
Karen Dalton, *Community College of Baltimore County– Catonsville Campus*
David H. Davis, *Asheville-Buncombe Technical Community College*
Jerry Davis, *University of Wisconsin, LaCrosse*
Douglas M. Deardon, *University of Minnesota*
Lewis Deaton, *University of Southwestern Louisiana*
Fred Delcomyn, *University of Illinois, Urbana*
Lorren Denney, *Southwest Missouri State University*
Katherine J. Denniston, *Towson State University*
Charles F. Denny, *University of South Carolina, Sumter*
Jean DeSaix, *University of North Carolina, Chapel Hill*
Ed DeWalt, *Louisiana State University*
Daniel F. Doak, *University of California, Santa Cruz*
Matthew M. Douglas, *University of Kansas*
Ronald J. Downey, *Ohio University*
Ernest Dubrul, *University of Toledo*
Michael Dufresne, *University of Windsor*
Susan A. Dunford, *University of Cincinnati*

Mary Durant, *North Harris College*
Ronald Edwards, *University of Florida*
Rosemarie Elizondo, *Reedley College*
George Ellmore, *Tufts University*
Joanne T. Ellzey, *University of Texas, El Paso*
Wayne Elmore, *Marshall University*
Carl Estrella, *Merced College*
Nancy Eyster-Smith, *Bentley College*
Deborah A. Fahey, *Wheaton College*
Gerald Farr, *Southwest Texas State University*
Rita Farrar, *Louisiana State University*
Marianne Feaver, *North Carolina State University*
Linnea Fletcher, *Austin Community College, Northridge*
Charles V. Foltz, *Rhode Island College*
Matthew Fountain, *State University of New York, Fredonia*
Douglas Fratianne, *Ohio State University*
Scott Freeman, *University of Washington*
Donald P. French, *Oklahoma State University*
Don Fritsch, *Virginia Commonwealth University*
Teresa Lane Fulcher, *Pellissippi State Technical Community College*
Michael Gaines, *University of Kansas*
Irja Galvan, *Western Oregon University*
Sandi Gardner, *Triton College*
Gail E. Gasparich, *Towson University*
Farooka Gauhari, *University of Nebraska, Omaha*
George W. Gilchrist, *University of Washington*
David Glenn-Lewin, *Iowa State University*
Elmer Gless, *Montana College of Mineral Sciences*
Charles W. Good, *Ohio State University, Lima*
Margaret Green, *Broward Community College*
Carole Griffiths, *Long Island University*
Lonnie J. Guralnick, *Western Oregon University*
Martin E. Hahn, *William Paterson College*
Madeline Hall, *Cleveland State University*
Georgia Ann Hammond, *Radford University*
Blanche C. Haning, *North Carolina State University*
Helen B. Hanten, *University of Minnesota*
John P. Harley, *Eastern Kentucky University*
Stephen Hedman, *University of Minnesota*
Jean Helgeson, *Collins County Community College*
Alexander Henderson, *Millersville University*
James Hewlett, *Finger Lakes Community College*
Alison G. Hoffman, *University of Tennessee, Chattanooga*
Dale Holen, *Pennsylvania State University, Worthington Scranton Campus*
Leland N. Holland, *Paso-Hernando Community College*
Laura Mays Hoopes, *Occidental College*
Michael D. Hudgins, *Alabama State University*
Donald A. Ingold, *East Texas State University*
Jon W. Jacklet, *State University of New York, Albany*
Rebecca M. Jessen, *Bowling Green State University*
Florence Juillerat, *Indiana University–Purdue University at Indianapolis*
Thomas W. Jurik, *Iowa State University*
Arnold Karpoff, *University of Louisville*
L. Kavaljian, *California State University*
Hendrick J. Ketellapper, *University of California, Davis*
Kate Lajtha, *Oregon State University*
Tom Langen, *Clarkson University*
Elizabeth E. LeClair, *DePaul University*
Patricia Lee-Robinson, *Chaminade University of Honolulu*
William H. Leonard, *Clemson University*
Edward Levri, *Indiana University of Pennsylvania*
Graeme Lindbeck, *University of Central Florida*
Jerri K. Lindsey, *Tarrant County Junior College, Northeast*

John Logue, *University of South Carolina, Sumter*
William Lowen, *Suffolk Community College*
Ann S. Lumsden, *Florida State University*
Steele R. Lunt, *University of Nebraska, Omaha*
Kimberly G. Lyle-Ippolito, *Anderson University*
Douglas Lyng, *Indiana University–Purdue University, Fort Wayne*
Daniel D. Magoulick, *The University of Central Arkansas*
Paul Mangum, *Midland College*
Michael Martin, *University of Michigan*
Linda Martin-Morris, *University of Washington*
Kenneth A. Mason, *University of Kansas*
Margaret May, *Virginia Commonwealth University*
Amy McMillan, *Buffalo State College*
D. J. McWhinnie, *DePaul University*
Gary L. Meeker, *California State University, Sacramento*
Thoyd Melton, *North Carolina State University*
Joseph R. Mendelson III, *Utah State University*
Karen E. Messley, *Rockvalley College*
Timothy Metz, *Campbell University*
Glendon R. Miller, *Wichita State University*
Neil Miller, *Memphis State University*
Jack E. Mobley, *University of Central Arkansas*
John W. Moon, *Harding University*
Richard Mortenson, *Albion College*
Gisele Muller-Parker, *Western Washington University*
James Murphy, *Monroe Community College*
Kathleen Murray, *University of Maine*
Lance Myler, *State University of New York, Canton*
Robert Neill, *University of Texas*
Harry Nickla, *Creighton University*
Daniel Nickrent, *Southern Illinois University*
Jane Noble-Harvey, *University of Delaware*
David J. O'Neill, *Community College of Baltimore County, Dundalk Campus*
James T. Oris, *Miami University, Ohio*
Marcy Osgood, *University of Michigan*
C. O. Patterson, *Texas A&M University*
Fred Peabody, *University of South Dakota*
Harry Peery, *Tompkins-Cortland Community College*
Rhoda E. Perozzi, *Virginia Commonwealth University*
Gary Pettibone, *Buffalo State College*
Bill Pfitsch, *Hamilton College*
Ronald Pfohl, *Miami University, Ohio*
Bernard Possident, *Skidmore College*
Elsa C. Price, *Wallace State Community College*
James A. Raines, *North Harris College*
Karen Raines, *Colorado State University*
Mark Richter, *University of Kansas*
Todd Rimkus, *Marymount University*
Robert Robbins, *Michigan State University*
William D. Rogers, *Ball State University*
Paul Rosenbloom, *Southwest Texas State University*
K. Ross, *University of Delaware*
Mary Lou Rottman, *University of Colorado, Denver*
Albert Ruesink, *Indiana University*
Christopher F. Sacchi, *Kutztown University of Pennsylvania*
Kim Cleary Sadler, *Middle Tennessee State University*

Mark Sandheinrich, *University of Wisconsin–La Crosse*
Alan Schoenherr, *Fullerton College*
Edna Seaman, *University of Massachusetts, Boston*
Linda Simpson, *University of North Carolina, Charlotte*
Anu Singh-Cundy, *Western Washington University*
Russel V. Skavaril, *Ohio State University*
John Smarelli, *Loyola University*
Shari Snitovsky, *Skyline College*
Jim Sorenson, *Radford University*
Ruth Sporer, *Rutgers University*
Mary Spratt, *University of Missouri, Kansas City*
Benjamin Stark, *Illinois Institute of Technology*
William Stark, *Saint Louis University*
Kathleen M. Steinert, *Bellevue Community College*
Barbara Stotler, *Southern Illinois University*
Gerald Summers, *University of Missouri, Columbia*
Marshall Sundberg, *Louisiana State University*
Bill Surver, *Clemson University*
Eldon Sutton, *University of Texas, Austin*
Dan Tallman, *Northern State University*
David Thorndill, *Essex Community College*
William Thwaites, *San Diego State University*
Professor Tobiessen, *Union College*
Richard Tolman, *Brigham Young University*
Jeff Travis, *University of Albany*
Dennis Trelka, *Washington & Jefferson College*
Sharon Tucker, *University of Delaware*
Gail Turner, *Virginia Commonwealth University*
Glyn Turnipseed, *Arkansas Technical University*
Lloyd W. Turtinen, *University of Wisconsin, Eau Claire*
Robert Tyser, *University of Wisconsin, La Crosse*
Robin W. Tyser, *University of Wisconsin, LaCrosse*
Kristin Uthus, *Virginia Commonwealth University*
F. Daniel Vogt, *State University of New York, Plattsburgh*
Nancy Wade, *Old Dominion University*
Susan M. Wadkowski, *Lakeland Community College*
Jyoti R. Wagle, *Houston Community College, Central*
Michael Weis, *University of Windsor*
DeLoris Wenzel, *University of Georgia*
Jerry Wermuth, *Purdue University, Calumet*
Jacob Wiebers, *Purdue University*
Carolyn Wilczynski, *Binghamton University*
P. Kelly Williams, *University of Dayton*
Roberta Williams, *University of Nevada, Las Vegas*
Sonya J. Williams, *Langston University*
Sandra Winicur, *Indiana University, South Bend*
Bill Wischusen, *Louisiana State University*
Chris Wolfe, *North Virginia Community College*
Colleen Wong, *Wilbur Wright College*
Jennifer Woodhead, *Brunswick Community College*
Wade Worthen, *Furman University*
Robin Wright, *University of Washington*
Mark L. Wygoda, *McNeese State University*
Brenda L. Young, *Daemen College*
Cal Young, *Fullerton College*
Tim Young, *Mercer University*
Samuel J. Zeakes, *Radford University*

Real-world applications
and human interest case studies engage students

Case Study — Muscles, Mutations, and Myostatin

No, the bull shown at the top of the chapter opening photo hasn't been pumping iron—he's a Belgian Blue, and they always have bulging muscles. What makes a Belgian Blue look like a bodybuilder, compared to an ordinary bull, such as the Hereford below it?

When a mammal develops, its cells divide many times, enlarge, and become specialized for a specific function. The size, shape, and cell types in any organism are precisely regulated during development, so you don't wind up with a head the size of a basketball, or have hair growing on your liver. Muscle development is no exception. When you were very young, cells destined to form your muscles multiplied, fused together to form long, relatively thick cells with multiple nuclei, and synthesized the specialized proteins that cause muscles to contract and thereby move your skeleton. A protein called *myostatin*, found in all mammals, puts the brakes on this process. The word "myostatin" literally means "to make muscles stay the same," and that is exactly what myostatin does. As muscles develop, myostatin slows down—and eventually stops—the multiplication of these pre-muscle cells. A bodybuilder can bulk up by lifting weights, which *enlarges* the muscle cells, but doesn't usually add many *more* cells.

Belgian Blues have more muscle cells than ordinary cattle do. Why? You may have already guessed: They don't produce normal myostatin. And why not? Proteins such as myostatin are synthesized from the genetic directions contained in deoxyribonucleic acid, or DNA. The DNA of a Belgian Blue is very slightly different from the DNA of normal cattle—it has a change, or mutation, in the DNA of its myostatin gene. As a result, it produces defective myostatin. Belgian Blue pre-muscle cells multiply more than normal, producing remarkably buff cattle. Other animals, including some breeds of dogs (photo to the left), may also have myostatin mutations, with similar effects.

How does DNA contain the instructions for traits such as muscle size, flower color, or gender? How are these instructions usually passed unchanged from generation to generation? And why do the instructions sometimes change? The answers to these questions lie in the structure and function of DNA. ■

Like Belgian Blue cattle, "bully" whippets have defective myostatin, causing them to have enormous muscles.

Muscles, Mutations, and Myostatin
Continued

In addition to Belgian Blues, several other breeds of cattle, including Maine Anjou, Piedmontese, Limousine, Charolais, and Blonde d'Aquitaine, have excessive muscle development caused by mutated myostatin genes. From what you've learned about the "language" of DNA and the mechanisms of DNA replication, would you expect that all of these cattle would have the same mutation? Check your reasoning in the "Muscles, Mutations, and Myostatin, *Revisited*" section at the end of the chapter.

Muscles, Mutations, and Myostatin
Continued

All "normal" mammals have a DNA sequence that encodes a functional myostatin protein, which limits their muscle growth. Belgian Blue cattle have a mutation that changes a "friendly" gene to a nonsensical "fliendly" one that no longer codes for a functional protein, so they have excessive muscle development.

Muscles, Mutations, and Myostatin Revisited

Belgian Blue cattle have a mutation in their myostatin gene, causing their cells to stop synthesizing the myostatin protein about halfway through the process. Several other breeds of "double-muscled" cattle have the same mutation, but some Maine Anjou, Piedmontese, Limousine, and Charolais cattle have totally different mutations. What they all have in common is that their myostatin proteins are nonfunctional. This is an important feature of the "language" of DNA: The nucleotide words must be spelled just right, or at least really close, for the resulting proteins to function. Any one of an enormous number of possible mistakes will render the proteins useless.

Humans have myostatin, too, and, not surprisingly, mutations can occur in the human myostatin gene. As you learned in Chapter 9, a child inherits two copies of most genes, one from each parent. In 1999, a child was born in Germany who inherited a mutated myostatin gene from both parents. Although the mutation is different from the one in Belgian Blue cattle, it also results in short, inactive, myostatin proteins. Even at 7 months, this boy had well-developed calf, thigh, and buttock muscles. At 4 years, he could hold a 7-pound dumbbell in each hand, with his arms fully extended horizontally out to his sides (try it—it's not that easy for many adults).

You know that mutations may be neutral, harmful, or beneficial. Into which category do myostatin mutations fall? Belgian Blue cattle are born so muscular, and consequently so large, that they usually must be delivered by cesarean section. A few become so muscular that their muscles get in the way, and they can hardly walk. So far, the German boy appears to be healthy. What will happen as he grows up? Will he become a super-athlete, or suffer debilitating health effects as he ages, or both? Only time will tell.

Consider This
If a person becomes a super athlete as a result of a known mutation, such in as the myostatin gene, it is "fair" to allow him or her to compete against people who don't have this mutated allele?

◄ A case study approach
The text's unique approach engages students with human-interest case studies that run throughout each chapter.

▲ Expanded human interest case study coverage

Each chapter begins and ends with a case study relevant to the chapter topic and, new to this edition, includes two to three *Case Study Continued* sections throughout each chapter, relating the case study to the chapter topic at hand. The *Case Study Revisited* section at the end of the chapter allows students to further reflect on what they have learned in the chapter, and ends with a *Consider This* section that encourages even deeper thinking.

Section headings ► help students to make connections

Every major section's heading is written as a question. As students read the material, the answers to the question are revealed. This format helps to stimulate analytical thinking and reminds the student of the importance of asking questions in scientific exploration.

22.3 How Does the Immune System Recognize Invaders?

To understand how the immune system recognizes invaders and initiates a response, we must answer three related questions: How do immune cells recognize foreign cells and molecules? How can immune cells produce specific responses to so many different types of cells and molecules? How do immune cells avoid mistaking the body's own cells and molecules for invaders?

Immune Cells Recognize Invaders' Complex Molecules

From the perspective of the immune system, hepatitis viruses and humans differ from one another because each contains specific, complex molecules that the others don't. The immune system can also distinguish intravenous (IV) solutions from rattlesnake venom because the venom contains specific, complex molecules that IV solutions lack. (Most common IV solutions are principally water, salts, and glucose.) Finally, *your* immune cells distinguish *your* body's cells and molecules from those of all the other organisms on Earth because some of your complex molecules are unique to you (unless you have an identical twin), and some of the complex molecules of all other organisms are unique to them. These large, complex molecules, usually proteins or polysaccharides, are called **antigens,** because they can be "*antibody generating*" molecules; that is, they can provoke an immune response that includes the production of antibodies (see the next section). Antigens may be individual molecules dissolved in your body fluids (for example, in the venom injected by a rattlesnake) or they may be located on the surfaces of cells (for example, on invading microbes).

Antibodies and T-Cell Receptors Recognize and Bind to Antigens

Cells of the immune system generate two types of proteins that recognize, bind, and then help to destroy specific antigens. **Antibody** proteins are produced by B cells and **T-cell receptor** proteins are produced by T cells. Antibodies may remain attached to the surfaces of the B cells that produced them or they may be released into the blood plasma. In contrast, T-cell receptors always remain attached to the surfaces of the T cells that produced them; they are never secreted into the plasma.

Antibodies Both Recognize and Help to Destroy Invaders

Antibodies are Y-shaped molecules composed of two pairs of peptide chains: one pair of identical large (heavy) chains and one pair of identical small (light) chains. Both heavy and light chains consist of a **constant region,** which is similar in all antibodies of the same type, and a **variable region,** which differs among individual antibodies (**Fig. 22-5**). The combination of light and heavy chains results in antibodies with two functional parts: the two "arms" of the Y and the "stem" of the Y. The variable regions (at the tips of the arms) form binding sites for specific antigens. These binding sites are a lot like the active sites of enzymes (see pp. 79–80). Each binding site has a particular size, shape and electrical charge, so only certain molecules can fit in and bind. The binding sites are so specific that each antibody can bind only a few, very similar, types of antigen molecules.

Antibodies may act as *receptors*, binding to specific antigens and eliciting a response to them, or as *effectors*, helping to destroy the cells or molecules that bear the antigens. As a receptor, the stem of an antibody anchors it to the plasma membrane of the B cell that produced it, while its two arms stick out from the B cell,

HELP STUDENTS BECOME INFORMED CITIZENS

Health Watch ▶

Health Watch boxes focus on human health-related topics that will interest today's students and help them apply the concepts to real-life examples.

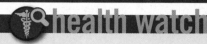

health watch

Golden Rice

Rice is the principal food for about two-thirds of the people on Earth (**Fig. E12-2**). Rice provides carbohydrates and some protein, but is a poor source of many vitamins, including vitamin A. Unless people eat enough fruits and vegetables, they often lack sufficient vitamin A, and may
...mmune system defects,
...atory, digestive, and
...to the World Health
...100 million children
...iciency; as a result
...0,000 children
...in Asia, Africa, and
...deficiency typically
...rice may be all they
...9 biotechnology
...dy: rice genetically
...ta-carotene, a pigment
...ht yellow, and that the
...rts into vitamin A.

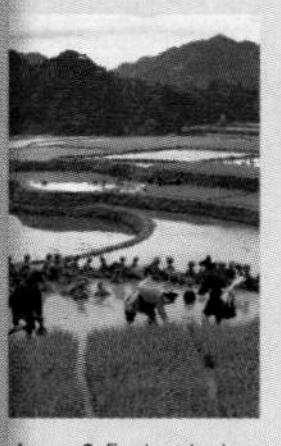

...dreams? For hundreds
...provides the major
...enough vitamins and
...gy improve the quality of
...of life for these people?

Creating a rice strain with high levels of beta-carotene wasn't simple. However, funding from the Rockefeller Institute, the European Community Biotech Program, and the Swiss Federal Office for Education and Science enabled molecular biologists Ingo Potrykus and Peter Beyer to tackle the task. They inserted three genes into the rice genome, two from daffodils and one from a bacterium. As a result, "Golden Rice" grains synthesize beta-carotene (**Fig. E12-3**, upper right).

The trouble was, the original Golden Rice didn't make enough beta-carotene, so people would have had to eat enormous amounts to get enough vitamin A. The Golden Rice community didn't give up. Researchers at the biotech company Syngenta set about increasing the beta-carotene levels. It turns out that daffodils aren't the best source for genes that direct beta-carotene synthesis. Golden Rice 2, with genes from corn, produces over 20 times more beta-carotene than the original Golden Rice (compare the rice in the upper right and left-hand sections of **Fig. E12-3**). About 3 cups of cooked Golden Rice 2 should provide enough beta-carotene to equal the full recommended daily amount of vitamin A.

In addition, Syngenta and multiple patent holders have given the technology—free—to research centers in the Philippines, India, China, and Vietnam, with the hope that they will modify native rice varieties for local use. Further, any individual farmer who produces less than about $10,000 worth of Golden Rice doesn't have to pay any fees to Syngenta or the other patent holders. Finally, Syngenta has donated Golden Rice 2 to the Humanitarian Rice Board for experiments and planting in Southeast Asia.

Is Golden Rice 2 the best way, or the only way, to solve the problems of malnutrition in poor people? Perhaps not. For one thing, many poor people's diets are deficient in many nutrients, not just vitamin A. To help solve that problem, the Bill and Melinda Gates Foundation is funding research by Peter Beyer to increase the levels of vitamin E, iron, and zinc in rice. Further, not all poor people eat mostly rice. In parts of Africa, sweet potatoes are the main source of starches. Eating orange, instead of white, sweet potatoes, has dramatically increased vitamin A intake for many of these people. Finally, in many parts of the world, governments and humanitarian organizations have started vitamin A supplementation programs. In some parts of Africa and Asia, as many as 80% of the children receive large doses of vitamin A a few times when they are very young. Some day, the combination of these efforts may result in a world in which no children suffer blindness from the lack of a simple nutrient in their diets.

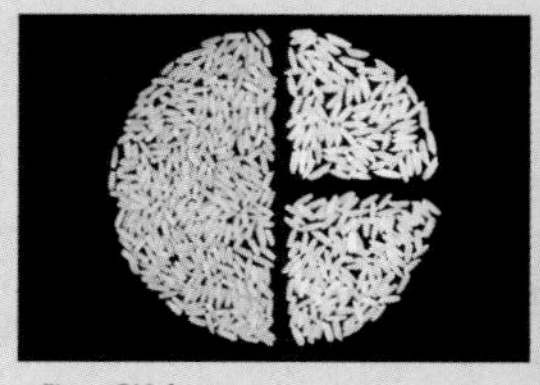

▲ **Figure E12-3 Golden Rice** Conventional milled rice is white or very pale tan (lower right). The original Golden Rice (upper right) was pale golden-yellow because of its increased beta-carotene content. Second-generation Golden Rice 2 (left) is much deeper yellow, because it contains about 20 times more beta-carotene than original Golden Rice.

earth watch

People Promote High-Speed Evolution

You probably don't think of yourself as a major engine of evolution. Nonetheless, as you go about the routines of your daily life, you are contributing to what is perhaps today's most significant cause of rapid evolutionary change. Human activity has changed Earth's environments tremendously, and when environments change, populations adapt or perish. The biological logic of natural selection, spelled out so clearly by Darwin, tells us that environmental change leads inevitably to evolutionary change. Thus, by changing the environment, humans have become a major agent of natural selection.

Unfortunately, many of the evolutionary changes we have caused have turned out to be bad news for us. Our liberal use of pesticides has selected for resistant pests that frustrate efforts to protect our food supply. By overmedicating ourselves with antibiotics and other drugs, we have selected for resistant "supergerms" and diseases that are ever more difficult to treat. Heavy fishing in the world's oceans has favored smaller fish that can slip through nets more easily, thereby selecting for slow-growing individuals that remain small even as mature adults. As a result, fish of many commercially important species are now so small that our ability to extract food from the sea is compromised.

Our use of pesticides, antibiotics, and fishing technology has caused evolutionary changes that threaten our health and welfare, but the scope of these changes may be dwarfed by those that will arise from human-caused modification of Earth's climate. Human activities, especially activities that use energy derived from fossil fuels, modify the climate by contributing to global warming. In coming years, species' evolution will be influenced by environmental changes associated with a warming climate, such as reduced ice and snow, longer growing seasons, and shifts in the life cycles of other species that provide food or shelter.

There is growing evidence that global warming is already causing evolutionary change. Warming-related evolution has been found in populations as diverse as mosquitoes, birds, and squirrels. For example, researchers have discovered that, in northern populations of a mosquito species, the insects' genetically programmed response to changing day length has shifted during the past four decades. (Mosquitoes use day length as a cue to tell them what time of year it is.) As a result, shorter days are now required to stimulate mosquito larvae to enter their overwintering pupal stage, and the transition takes place much later in the autumn. The delay allows the larvae to take advantage of the longer feeding and growing season produced by global warming.

Climate change is also affecting the evolution of bird migration, for example by causing European blackcap birds to evolve into genetically distinct populations with different migration patterns. Historically, these birds have bred in Austria and Germany and migrated south into Spain and Morocco for the winter. Since the early 1960s, however, increasing numbers of blackcaps have instead spent the nonbreeding season in southern England, where winters have become milder and food more abundant. Individuals in this population have a selective advantage over birds that spend the winter farther away from the breeding grounds; they reach their nesting area earlier in the spring, gain better territories, and produce more offspring. Further, laboratory studies have demonstrated that blackcaps that overwinter in England prefer to mate with each other, and that their offspring inherit the tendency to migrate to England. Thus, the proportion of blackcaps wintering in England is increasing, and this group of birds is evolving to become increasingly distinct from other blackcaps.

In Canada, red squirrels in a colony closely monitored by scientists now produce litters 18 days earlier, on average, than they did 10 years ago (**Fig. E13-3**). The change is tied to a warming climate, because spring now arrives earlier and spruce trees produce earlier crops of seeds, the squirrels' only food. Squirrels that breed earlier gain a competitive edge by better exploiting the warmer weather and more abundant food. And, because the time at which a squirrel gives birth is influenced by the animal's genetic makeup, early-breeding squirrels pass to their offspring the genes th...
result, the ge...
population is...
squirrels are...
breeding one...
The availa...
climate char...
evolutionary...
evolution of...
evolutionary...
ecosystems...
is not readily...
evolution is...
however, tha...
species and...
evolutionary...
appropriate...
well-being as...

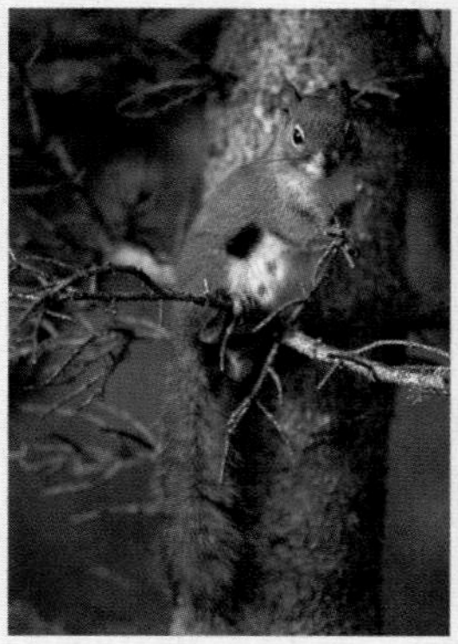

▲ **Figure E13-3 Red squirrels have evolved in response to global warming**

▲

Earth Watch

Earth Watch boxes focus on environment-related issues that students will recognize from current news stories.

scientific inquiry

Neuroimaging—Observing The Brain In Action

For most of human history, the brain has been a mystery. Now, however, imaging techniques provide exciting insights into brain function. These techniques include *PET* (positron emission tomography) and *fMRI* (functional magnetic resonance imaging). PET and fMRI are both based on the fact that regions of the brain that are most active have higher energy demands; they use more glucose and attract a greater flow of oxygenated blood than do less active areas. In PET scans, scientists inject the subject with a radioactive substance, such as a radioactive form of glucose, and then monitor levels of radioactivity that reflect differences in metabolic rate. These are translated by computer into colors on images of the brain. By monitoring radioactivity while a specific task is performed, scientists can identify which parts of the brain are most active during that task. In contrast, fMRI detects differences in levels of oxygenated and deoxygenated blood among different brain regions. The presence of oxygen bound to the hemoglobin in blood causes it to respond differently to the powerful magnetic field and pulses of radio waves generated by an fMRI machine. Active brain regions can be distinguished with fMRI without using radioactivity and over much shorter time spans than those required by PET.

Using fMRI or PET, researchers can observe changes as the brain performs a specific reasoning task or responds to an odor or a visual or auditory stimulus. Through PET and fMRI, scientists have confirmed that different aspects of the processing of language occur in distinct areas of the cerebral cortex. Using fMRI, other researchers analyzed the frontal lobe areas used in generating words in individuals who spoke two languages. In contrast, subjects who had grown up speaking two languages, the same region of the frontal lobe was used in speaking both languages. In contrast, subjects who had learned a second language later in life used different but adjacent frontal lobe areas for the two languages.

Functional MRI scans have also been used to determine the parts of the brain that are most active during various emotional states. For example, when a person is frightened, the amygdala lights up (**Fig. E24-2a**). When people in love view photos of their lovers, other areas in the brain are activated (**Fig. E24-2b**). Interestingly, most of these same areas are also activated by drugs of abuse, such as cocaine.

By the way, never let anyone tell you that you use only a small fraction of your brain! Although PET and fMRI images sometimes make it appear that only a small area of the brain is active, this is because activity of other regions is subtracted out during the imaging process, to show where brain activity has *changed* as a result of stimulation. The images shown here merely highlight areas where activity is more intense than in a resting state.

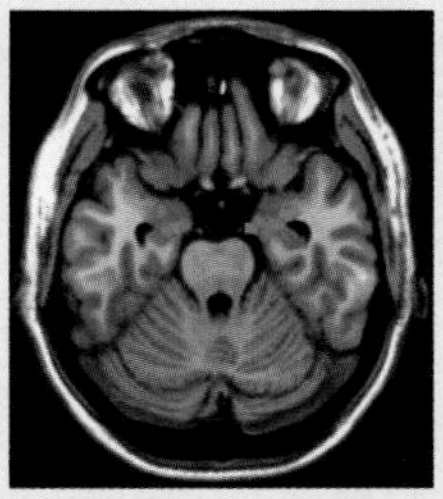

(a) Activation of the amygdala (red) by a frightening stimulus

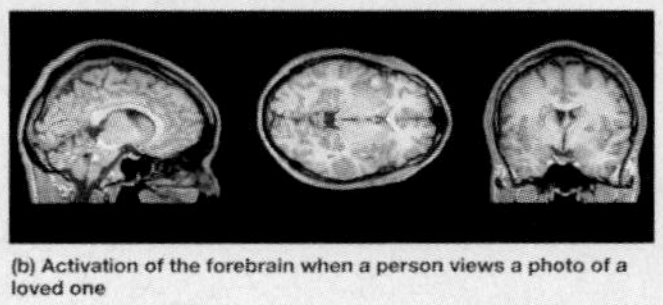

(b) Activation of the forebrain when a person views a photo of a loved one

▲ **Figure E24-2 Localization of emotions in the human brain** ***(a)*** A frightening experience activates the amygdala, a part of the forebrain that apparently produces emotions such as fear and rage. ***(b)*** Looking at photos of lovers activates multiple parts of the brain, including areas in the cerebral cortex (left) and structures deep within the forebrain (right).

Scientific Inquiry ▶

Scientific Inquiry boxes illuminate how we know what we know, stepping the student through the scientific process.

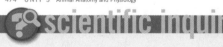

Web tools

engage students and assist instructors

BioFlix Mitosis

Web Animation Mitosis

8.5 How Does Mitotic Cell Division Produce Genetically Identical Daughter Cells?

As we described earlier, mitotic cell division consists of mitosis (nuclear division) and cytokinesis (cytoplasmic division) (Fig. 8-9). For convenience, biologists divide mitosis into four phases, based on the appearance and behavior of the chromosomes: (1) prophase, (2) metaphase, (3) anaphase, and (4) telophase. Cytokinesis usually occurs during telophase. However, mitosis sometimes occurs without cytokinesis, producing cells with multiple nuclei.

During Prophase, the Chromosomes Condense and Are Captured by the Spindle Microtubules

During **prophase,** three major events occur: (1) the duplicated chromosomes condense, (2) the spindle microtubules form, and (3) the chromosomes are captured by the spindle.

Easy-to-locate web ▶ icons direct students to online animations

BioFlix™ and Web Animation icons are placed throughout each chapter and direct students to **mybiology.com** to view informative animations on the topic that they are reading about.

▼ **Figure 8-9 Mitotic cell division in an animal cell** In the fluorescent micrographs, chromosomes are blue and microtubules are green. **QUESTION:** What would be the genetic consequences for the daughter cells if one set of sister chromatids failed to separate at anaphase?

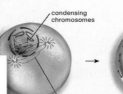

INTERPHASE **MITOSIS**

nuclear envelope — chromatin — nucleolus

condensing chromosomes — spindle pole — spindle microtubules

beginning of spindle formation — kinetochore — spindle pole

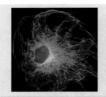

Prophase

Kinetochore

Mitotic Spindle

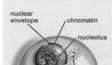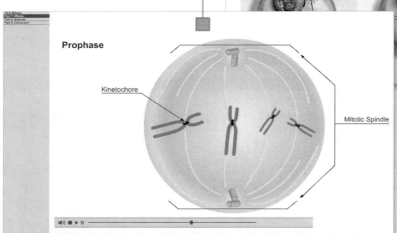

◀ Web Animations

Web Activities help students better understand difficult concepts with animations, interactive exercises, and quizzes.

BioFlix Animations ▶

BioFlix cover the toughest topics in biology with high-quality, 3-D animations.

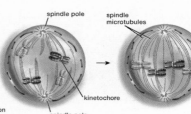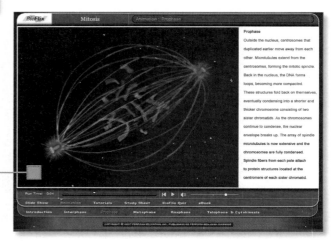

Prophase

Outside the nucleus, centrosomes that duplicated earlier move away from each other. Microtubules extend from the centrosomes, forming the mitotic spindle. Back in the nucleus, the DNA forms loops, becoming more compacted. These structures fold back on themselves, eventually condensing into a shorter and thicker chromosome consisting of two sister chromatids. As the chromosomes continue to condense, the nuclear envelope breaks up. The array of spindle microtubules is now extensive and the chromosomes are fully condensed. Spindle fibers from each pole attach to protein structures located at the centromere of each sister chromatid.

mybiology *promotes efficient studying*

mybiology.com is a comprehensive study resource that provides all of the study tools students need, including quizzes, an integrated eBook, and BioFlix™—movie-quality animations that tackle the toughest topics in biology.

The four-step structure of mybiology.com allows students to identify what they don't know, select the appropriate activity to match their learning style, and practice outside of the lectures.

1 **Focus Your Effort** uses a pre-test as a road map to identify a student's study needs.

2 **Direct Your Learning** utilizes the resources that support the various learning styles, from BioFlix for visual learners to an eBook for those who prefer to read.

3 **Test Yourself** offers a post-test to check students' understanding and prepare for exams.

4 **Extend Your Knowledge** allows students to go beyond the chapter material to explore case studies and the process of science with Discovery Channel videos and other tools.

The easy-to-navigate **eBook** is available on each new **mybiology.com** website and allows students to view the complete textbook online. Links to media activities are embedded in the **eBook**, and a helpful Google-based search tool allows students to find the topics they are looking for. Students can also annotate pages using a note feature.

SEE SCIENCE IN ACTION WITH DISCOVERY CHANNEL VIDEO CLIPS

These brief, 3–5 minute Discovery Channel video clips cover topics from fighting cancer to antibiotic resistance and introduced species. The video clips are located on **mybiology.com** and are accompanied with easily assignable quiz questions that report to a gradebook.

Encourage students
to take a more active role in their learning

NEW! **BioFlix™ animations** and student tools invigorate classroom lectures and cover the most difficult biology topics with 3-D, movie-quality animations, labeled slide shows, carefully constructed student tutorials, study sheets, and quizzes.

BioFlix Topics

- Tour of an Animal Cell
- Tour of a Plant Cell
- Cellular Respiration
- Photosynthesis
- Mitosis
- Meiosis
- Protein Synthesis
- Water Transport in Plants
- Muscle Contraction
- How Neurons Work

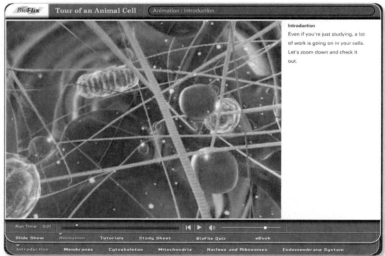

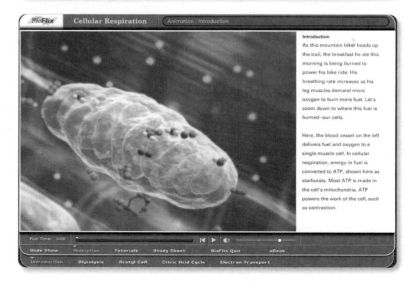

STUDY TOOLS HELP STUDENTS INTERNALIZE BIOLOGY

Student Quizzes ▶

Keep track of students' progress with these assignable, online quizzes. Scores can be recorded in an online gradebook.

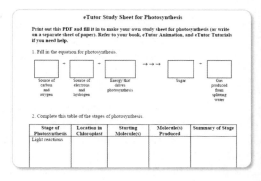

Tutorials ▶

Carefully constructed student tutorials help students learn about important biological processes.

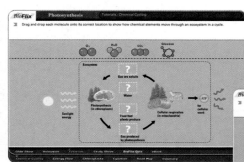

▲ e-Study Sheets

Students can study for tests by completing printable study sheets.

PRESENTATION AND COURSE MANAGEMENT TOOLS HELP YOU TEACH

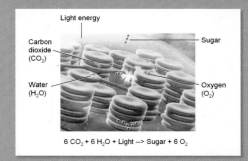

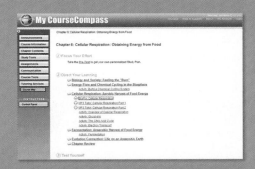

PowerPoint® Lecture Slides

PowerPoint slides supplement instructors' lectures with chapter outlines and summaries of key concepts.

Flexible Presentation

Instructors have the flexibility to add their own narration to their BioFlix™ animations by disabling the BioFlix sounds.

Course Management

Open-access course management systems—BlackBoard, WebCT, and CourseCompass™—contain pre-loaded content, such as testing and assessment question pools.

Dynamic resources

help you teach biology

REVISED!
INSTRUCTOR RESOURCE CD/DVD-ROM

The Instructor Resource CD/DVD-ROM package combines all of the instructor media for **Life on Earth**, **Fifth Edition**, *in one convenient location, organized chapter-by-chapter.*

978-0-131-75537-6 • 0-131-75549-8

Expanded Test Bank questions are available as downloadable Microsoft® Word® files.

Customizable **PowerPoint®** ▶ **Lectures** outline the major concepts presented in each chapter and include key figures to get the points across.

NEW! BioFlix™ animations invigorate classroom lectures with 3-D, movie-quality graphics. Each animation is typically three minutes long.

▼

BLAST Animations are ▶ scientifically clear and accurate animations that instructors can step through in PowerPoint for lecture presentations.

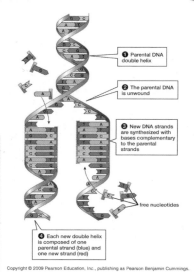

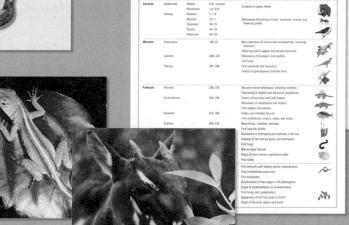

▲

All of the art, tables, and photos are available as JPEGs and in PowerPoint format. Stepped-art and PowerPoint art files allow instructors to customize labels to best fit their teaching needs.

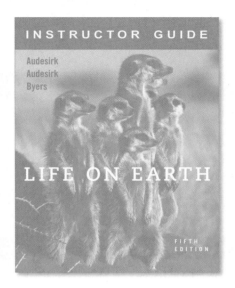

The Instructor Guide

This helpful resource includes chapter outlines, answers to text questions, and innovative lecture activities with accompanying handouts that can be torn out, copied, and distributed to the class. The Instructor Guide is available in print and in downloadable Microsoft® Word® format.

978-0-1317-5536-9 • 0-131-75536-6

SCIENTIFIC AMERICAN CURRENT ISSUES IN BIOLOGY

Scientific American Current Issues in Biology gives your students the best of both worlds—accessible, dynamic, relevant articles from *Scientific American* magazine that present key issues in biology, paired with the authority, reliability, and clarity of Benjamin Cummings' non-majors biology texts. Articles include questions to help students check their comprehension and make connections to science and society. *Scientific American* is available in Volumes 1–5. This resource is available at no additional charge when packaged with a new text.

Volume 1: 978-0-8053-7507-7 • 0-8053-7507-4
Volume 2: 978-0-8053-7108-6 • 0-8053-7108-7
Volume 3: 978-0-8053-7527-5 • 0-8053-7527-9
Volume 4: 978-0-8053-3566-8 • 0-8053-3566-8
Volume 5: 978-0-321-54187-1 • 0-321-54187-1

Brief Contents

Contents

unit two
Inheritance 109

chapter 10
DNA: The Molecule of Heredity 158

chapter 11
Gene Expression and Regulation 169

chapter 12
Biotechnology   187

unit three
Evolution   209

chapter 13
Principles of Evolution   210

chapter 16
The Diversity of Life   279

unit four
Plant Anatomy and Physiology   317

chapter 17
Plant Form and Function   318

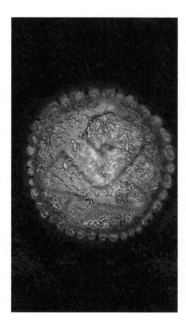

chapter 22
Defenses Against Disease 417

Case Study Avian Flu 418

chapter 23
Chemical Control of the Animal Body: The Endocrine System 440

Case Study Anabolic Steroids—Fool's Gold? 440

chapter 26
Animal Behavior   513

unit six
Ecology   535

chapter 27
Population Growth   536

chapter 31
Conserving Earth's Biodiversity   624

BioFlix

BioFlix™ 3-D animations bring difficult biological topics to life. Instructors can use the animations in lectures with or without narration. Each animation is accompanied by a tutorial, a study sheet, a quiz, and a PowerPoint® slide show with labels that identify the animation's main components. BioFlix animations are ideal for study groups and can be accessed by just logging in to www.mybiology.com.

In *Life on Earth,* Fifth Edition, BioFlix are denoted at appropriate locations in each chapter with icons that look like this:

BioFlix

Here is a list of the BioFlix animations produced to date, the chapters they are linked to, and the page number on which each BioFlix logo can be found within the chapters.

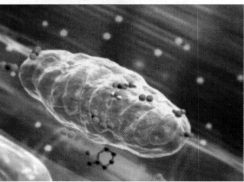

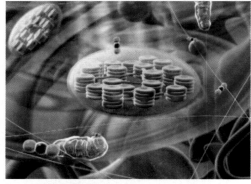

chapter 1

An Introduction to Life on Earth

Seen from space, Earth provides few hints of the abundant life on its surface.

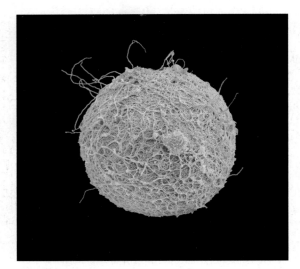

Seen through a microscope, Earth's life offers a view of spectacular beauty and diversity. Here, sperm cells (the hair-like structures) gather on the surface of a human egg. The first sperm cell to penetrate the surface will fertilize the egg.

1.1 Why Study Biology?

Why study biology, the science of life? Perhaps you are already fascinated by the study of life and need no further encouragement, but even if you have not yet discovered the thrill of biological inquiry, there are still good reasons to explore biology.

Biology Helps You Understand Your Body

Much of biology is devoted to understanding how bodies work and how disruptions and diseases can prevent them from working properly. Studying biology will help you understand more about the inner workings of your own body, how your behavior and diet affect your health, and how different diseases do their damage. This kind of knowledge can be empowering. When you discuss your health with your physician, you will have a much better understanding of the subject matter. Biological knowledge helps you become an informed consumer of health care, and helps you take charge of your own health.

Biology Helps You Become an Informed Citizen

Our society currently faces a host of complex social and ethical issues, many of which will require difficult decisions. As a member of society, you have a voice in those decisions and a responsibility to make your views known (and to elect representatives who share your views). Many of today's most controversial problems involve biology: Do we need to worry about the impact of pollution on our health? Should we do anything to halt the extinction of endangered species? Are genetically modified foods helpful or harmful? Should human cloning be legal? Should scientists be allowed to use cells from human embryos to develop cures for diseases? Is population control a good idea? The more you know about the basic biology that underlies these and other issues, the more likely you will be to have an informed opinion and to be an effective advocate for your point of view. And the more you know about how science works, the better equipped you will be to make your own, independent evaluation, based on evidence, of the claims you encounter in the media or in discussions with other people.

Biology Can Open Career Opportunities

Biology is relevant to many of today's careers. Are you considering a career in biotechnology, in the pharmaceutical industry, or in the medical equipment business? Do you plan to be a nurse, physician's assistant, physical therapist, medical technologist, or some other type of health care professional? Perhaps your plans include a job as a wildlife manager, forester, environmental consultant, veterinarian, or zookeeper. Or maybe you have set your sights on becoming a lawyer, with a specialization in environmental law or patent law. Perhaps a career in biomedical research, bioinformatics, or genetic counseling appeals to you. And the nation always needs qualified science teachers for its schools, colleges, and universities.

A working knowledge of biology will increase your chances of entering any of these careers and many more. Biological knowledge is an important prerequisite for a host of jobs, including many of today's fastest-growing professions.

Biology Can Enrich Your Appreciation of the World

Some people feel that science promotes a cold, clinical view of life, and that scientific explanations of the natural world rob us of a sense of wonder and awe. Nothing could be further from the truth. Biological knowledge only deepens our appreciation of nature's majesty.

Imagine walking across your school's campus on a warm spring day, past a green lawn in which some honeybees buzz around a patch of sunflowers, a rabbit nibbles at some dandelion leaves, and grasshoppers move among the blades of grass. Before you began studying biology, you might have paid little attention to the lawn as you made your way to class. But, as a biology student, you pause to contemplate the intertwined lives of the lawn's inhabitants. You understand that the soil beneath the grass blades is home to an invisible world of recyclers: fungi, bacteria, and tiny animals that feed on and decompose dead plants and animals, returning nutrients to the soil. You recall that grass gets its green color from a unique molecule, chlorophyll, that traps solar energy. Photosynthesis in the grass plant's cells uses the captured energy to manufacture sugar. The sugar feeds the plant, as well as the rabbit at the lawn's edge and the multitude of grasshoppers that move through the grass.

The buzzing of the bees reminds you that the sunflowers use some of their sugar to sweeten the nectar in their brightly colored flowers. You recognize that showy flowers evolved to entice insects to the energy-rich nectar. Looking carefully at a bee that is probing the sunflowers, you see yellow pollen clinging to its legs and to the hairs on its body (**Fig. 1-1**). The pollen grains carry sperm that can fertilize the eggs that lie within other sunflower flowers, beginning the production of new seeds. The sunflower plants "use" the insects to ensure fertilization, and both plants and insects benefit.

As your knowledge of biology grows, you may find yourself making closer and more frequent observations of the living world and reflecting more deeply on your observations. Perhaps you will also feel the increased sense of wonder that can come with greater understanding. Perhaps you *won't* think of biology as just another course to take, just another set of facts to memorize. Biology can be much more than that. It can be a pathway to a new appreciation of life on Earth.

1.2 How Do Biologists Study Life?

A biologist's job is to answer questions about life. What causes cancer? Why are frog populations shrinking worldwide? What happens when a sperm and egg meet? How does HIV cause AIDS? When did the earliest mammals appear? How do bees fly? The list of questions is both endless and endlessly fascinating.

Anyone can ask an interesting question about life. A biologist, however, is distinguished by the manner in which he or she goes about finding answers. A biologist is a scientist and accepts only answers that are supported by evidence, and then only by a certain kind of evidence. Scientific evidence consists of observations or measurements that are easily shared with others and that can be repeated by anyone who has the appropriate tools. The process by which this kind of evidence is gathered is known as the **scientific method.**

The Scientific Method Is the Basis for Scientific Inquiry

The scientific method proceeds step-by-step (**Fig. 1-2**). It begins when someone makes an **observation** of an interesting pattern or phenomenon. The observation, in turn, stimulates the observer to ask a **question** about what was observed. Then, after a period of contemplation (that perhaps also includes reflecting on the scientific work of others who have considered related questions), the person proposes an answer to the question, an explanation for the observation. This proposed explanation is a **hypothesis.** A good hypothesis leads to a **prediction,** typically expressed in "if . . . then" language. The prediction is tested with further observations or with **experiments.** These experiments produce results that either support or refute the hypothesis, and a **conclusion** is drawn about it. A single experiment is never an adequate basis for a conclusion; the experiment must be repeated not only by the original experimenter but also by others.

▲ **Figure 1-1 Bees and flowering plants are interdependent** Bees depend on the nectar in flowers for food, and flowering plants depend on the bees' inadvertent transfer of pollen from one plant to another for reproduction.

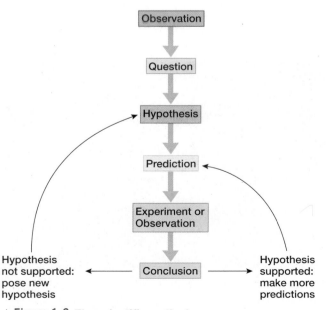

▲ Figure 1-2 **The scientific method**

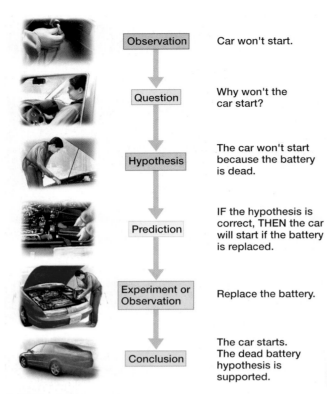

	Observation	Car won't start.
	Question	Why won't the car start?
	Hypothesis	The car won't start because the battery is dead.
	Prediction	IF the hypothesis is correct, THEN the car will start if the battery is replaced.
	Experiment or Observation	Replace the battery.
	Conclusion	The car starts. The dead battery hypothesis is supported.

▲ **Figure 1-3 The scientific method in everyday life**

Web Animation Experimental Design

▲ **Figure 1-4 A feathered dinosaur** Note the impressions of feathers on the forelimbs and tail of this fossil *Caudipteryx* dinosaur.

The Scientific Method Is Useful in Everyday Life

You may find it easier to visualize the steps of the scientific method through an example from everyday life (**Fig. 1-3**). Imagine this scenario: Late to an appointment, you rush to your car and make the *observation* that it won't start. This observation leads directly to a *question*: Why won't the car start? You quickly form a *hypothesis*: The battery is dead. Your hypothesis leads in turn to an if-then *prediction*: If the battery is dead, then a new battery will cause the car to start. Next, you design an *experiment* to test your prediction: You replace your battery with the battery from your roommate's new car and try to start your car again. Your car starts immediately, and you reach the *conclusion* that your dead battery hypothesis is correct.

Well-Designed Experiments Incorporate Controls

Simple experiments (like our car battery experiment) test the hypothesis that a single factor, or **variable,** is the cause of an observed phenomenon. The most straightforward test is usually an experiment in which only a single variable is changed. To be scientifically valid, however, the experiment must also rule out *other* possible variables that might have caused the phenomenon. So experimenters usually also include **controls,** parts of the experiment in which no variable is changed. The results from the controls can then be compared with results from the experimental sections. For two classic examples of well-designed experiments, see "Scientific Inquiry: Controlled Experiments, Then and Now" on pp. 6–7.

With the importance of controls in mind, let's revisit the imaginary car battery experiment. Can you really be confident that the old battery was dead? Perhaps the battery was fine all along, and you just needed to try to start the car again. Or perhaps the battery cable was loose and simply needed to be tightened. To control for these other variables, you might reinstall your old battery, making sure the cables are secured tightly, and attempt to restart the car. If your car repeatedly refuses to start with the old battery, but repeatedly starts immediately with your roommate's new battery, you have isolated a single variable, the battery. And (although you may have missed your appointment) you can now safely draw the conclusion that your old battery is dead.

The scientific method is powerful, but it is important to recognize its limitations. In particular, scientists can seldom be sure that they have controlled *all* the variables other than the one they are trying to study. For this reason, scientific conclusions remain tentative and are subject to revision if new observations or experiments demand it. In any case, nearly every conclusion immediately raises further questions that lead to further hypotheses and more experiments. Science is a never-ending quest for knowledge.

Experiments Are Not Always Possible

A well-designed experiment is usually the most convincing way to test a hypothesis, but biology includes many hypotheses that are not suited to experimental tests. For example, evolutionary biologists often ask questions about events from the prehistoric past. Consider, for example, the hypothesis that the ancestors of today's birds were dinosaurs. These hypothesized ancestors became extinct long ago, of course, and there is no experiment that can demonstrate how they evolved millions of years ago. Nonetheless, the biologists who study this question do use the other parts of the scientific method by recognizing the testable predictions implicit in their hypotheses. For example, if dinosaurs were the ancestors of birds, then we predict the discovery of fossils of dinosaurs with feathers. Such fossils have indeed been found, providing evidence consistent with the conclusion that the hypothesis is correct (**Fig. 1-4**).

In some cases, an experiment would be theoretically possible but is impractical or unethical. For example, consider the hypothesis that smoking causes lung cancer in people. In principle, we could test this hypothesis with an experiment that divided a large sample of people who had never smoked into two groups. The members of one group would be required to smoke a pack of cigarettes each day, and the members of the other group would serve as controls and would not be allowed to smoke. After, say, 20 years, we could count the number of cases of lung cancer in each group. Such an experiment would provide a powerful test of the hypothesis but would, needless to say, be highly unethical.

Again, however, an inability to experiment does not mean that the scientific method must be abandoned. The hypothesis generates predictions that can be tested by careful observation. For example, if smoking causes lung cancer, then we predict that a random sample of smokers should contain more lung cancer victims than a comparable sample of nonsmokers. Many studies have indeed found an association between smoking and lung cancer, evidence that supports the hypothesis that smoking causes cancer.

Science Requires Communication

Finally, note that *communication* is an important part of science. Even the best experiment is useless if it is not communicated. If the results of experiments are not communicated to other scientists in enough detail to be repeated, conclusions cannot be verified. Without verification, scientific findings cannot be safely used as the basis for new hypotheses and further experiments. Scientific effort is wasted if its results are not reported in clear and accurate detail.

Life Can Be Studied at Different Levels of Organization

As they investigate questions about life, biologists often view the living world as a series of *levels of organization*, with each level providing the building blocks for the next level (Fig. 1-5). For example, the **cell** is the smallest unit of life (Fig. 1-6), but in many life-forms, cells of similar type combine to form structures known as **tissues.** Different tissues, in turn, can combine to form **organs** (for example, a heart or a kidney), and a collection of different organs forms the basis of an **organism.** A collection of organisms of the same species constitutes a **population,** and a collection of different populations makes up a **community.** Note that each level of organization incorporates multiple members of the previous level; a community contains many populations, a population contains many organisms, and so on. (Also note that we could describe many more levels of organization than we have mentioned here.)

One of the first decisions a biologist must make when designing an experiment is to choose an appropriate level of organization at which to study a problem. This decision is ordinarily based on the question to be answered. For example, if you want to know how frogs make croaking sounds, you would study frog organs (structures within the frog body). The question of how frogs croak would be impossible to answer if you focused on frog cells or frog communities. On the other hand, if you want to know whether frog numbers are declining, it would do you no good to study frog organs. To answer that question, you would have to study frog populations. It is important for scientists to recognize and choose the level of organization that is most appropriate to the question at hand.

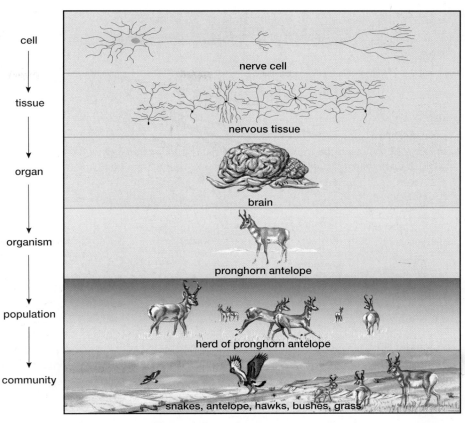

▲ **Figure 1-5 Levels of organization of life** Life is organized in levels, some of which are represented here. Items at each level are the building blocks of the next level. **EXERCISE:** Think of a scientific question that can be answered by conducting an investigation at the cell level, but that would be impossible to answer at the tissue level. Then think of one answerable at the tissue level but not at the cell level. Repeat the process for two other pairs of adjacent levels of organization.

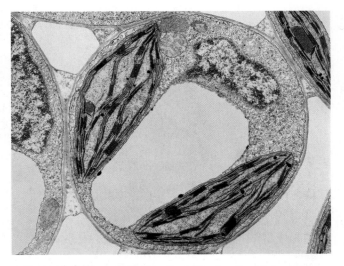

▲ **Figure 1-6 The smallest unit of life** A cell wall forms the boundary of this plant cell, separating the cell's contents from its surroundings. Portions of adjacent cells are also visible in this colorized micrograph.

scientific inquiry

Controlled Experiments, Then and Now

A classic experiment by the Italian physician Francesco Redi (1621–1697) beautifully demonstrates the scientific method. Redi investigated why maggots appear on spoiled meat. When he began this work, the appearance of maggots was considered to be evidence of *spontaneous generation*, the production of living things from nonliving matter.

Redi *observed* that flies swarm around fresh meat and that maggots appear on meat left out for a few days. He formed a testable *hypothesis*: The flies produce the maggots. In his *experiment*, Redi wanted to test just one variable: the access of flies to the meat. Therefore, he took two clean jars and filled them with similar pieces of meat. He left one jar open (the *control* jar) and covered the other with gauze to keep out flies (the *experimental* jar). He did his best to keep all the other variables the same (for example, the type of jar, the type of meat, and the temperature). After a few days, he observed maggots on the meat in the open jar, but saw none on the meat in the covered jar. Redi *concluded* that his hypothesis was correct and that maggots are produced by flies, not by the nonliving meat (Fig. E1-1). Only through controlled experiments could the age-old hypothesis of spontaneous generation be laid to rest.

More than 300 years after Redi's experiment, today's scientists still use the same approach to design their experiments. Consider the experiment that Malte Andersson designed to investigate the long tails of male widowbirds. Andersson *observed* that male, but not female, widowbirds have extravagantly long tails, which they display while flying across African grasslands (Fig. E1-2). This observation led Andersson to ask the *question*:

Observation:	Flies swarm around meat left in the open; maggots appear on the meat.
Question:	Why do maggots appear on the meat?
Hypothesis:	Flies produce the maggots.
Prediction:	If flies are kept away from the meat, then no maggots will appear.

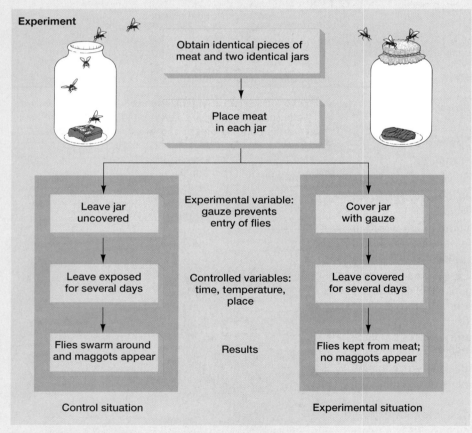

Experiment

Obtain identical pieces of meat and two identical jars

Place meat in each jar

Leave jar uncovered	Experimental variable: gauze prevents entry of flies	Cover jar with gauze
Leave exposed for several days	Controlled variables: time, temperature, place	Leave covered for several days
Flies swarm around and maggots appear	Results	Flies kept from meat; no maggots appear

Control situation   Experimental situation

Conclusion:	Spontaneous generation of maggots from meat does not occur; flies are probably the source of maggots.

▲ **Figure E1-1 The experiments of Francesco Redi** **QUESTION:** Redi showed that maggots don't appear by spontaneous generation, but did his experiment conclusively demonstrate that flies cause maggots? What kind of follow-up questions might be asked, and what kind of experiment would be necessary if Redi really wanted to determine the source of maggots?

▲ **Figure E1-2** **A male widowbird**

do the males, and only the males, have such tails? His *hypothesis* was that males have tails because females prefer to mate with -tailed males, which therefore have more pring than shorter-tailed males. From this othesis, Andersson *predicted* that if his othesis were true, then more females would d nests on the territories of males with icially lengthened tails than would build nests he territories of males with artificially tened tails. He then captured some males,

trimmed their tails to about half their original length, and released them (*experimental* group 1). Another group of males had the tail feathers that had been removed from the first group glued on as tail extensions (*experimental* group 2). Finally, Andersson had two *control* groups. In one, the tail was cut and then glued back in place (to control for the effects of capturing the birds and manipulating their feathers). In the other, the birds were simply captured and released. The experimenter was doing his best to make sure that

tail length was the only variable that was changed. After a few days, Andersson counted the number of nests that females had built on each male's territory. He found that males with lengthened tails had the most nests on their territories, males with shortened tails had the fewest, and control males (with normal-length tails) had an intermediate number (**Fig. E1-3**). Andersson *concluded* that his hypothesis was correct, and that female widowbirds prefer to mate with males that have long tails.

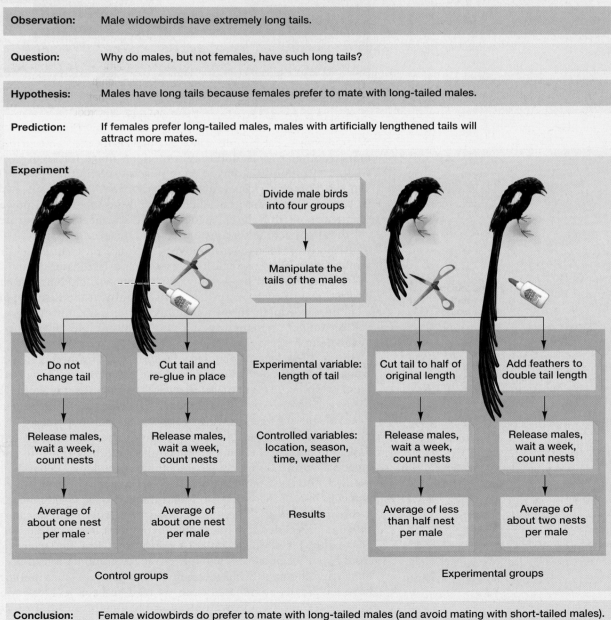

Observation:	Male widowbirds have extremely long tails.
Question:	Why do males, but not females, have such long tails?
Hypothesis:	Males have long tails because females prefer to mate with long-tailed males.
Prediction:	If females prefer long-tailed males, males with artificially lengthened tails will attract more mates.

Experiment

Divide male birds into four groups
↓
Manipulate the tails of the males

Do not change tail	Cut tail and re-glue in place	Experimental variable: length of tail	Cut tail to half of original length	Add feathers to double tail length
↓	↓		↓	↓
Release males, wait a week, count nests	Release males, wait a week, count nests	Controlled variables: location, season, time, weather	Release males, wait a week, count nests	Release males, wait a week, count nests
↓	↓	Results	↓	↓
Average of about one nest per male	Average of about one nest per male		Average of less than half nest per male	Average of about two nests per male

Control groups   Experimental groups

Conclusion:	Female widowbirds do prefer to mate with long-tailed males (and avoid mating with short-tailed males).

▲ **Figure E1-3 The experiments of Malte Andersson**

▶ **Figure 1-7 Penicillin kills bacteria** **QUESTION:** Why do some molds produce substances that are toxic to bacteria?

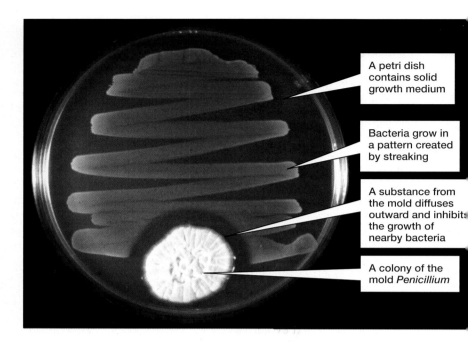

A petri dish contains solid growth medium

Bacteria grow in a pattern created by streaking

A substance from the mold diffuses outward and inhibits the growth of nearby bacteria

A colony of the mold *Penicillium*

Science Is a Human Endeavor

Scientists are real people. They are driven by the same pride, ambitions, a fears as other people, and they sometimes make mistakes. As you will rea Chapter 10, ambition played an important role in the discovery by James W son and Francis Crick of the structure of DNA. Accidents, lucky guesses, c troversies with competing scientists, and the intellectual powers of individ scientists contribute greatly to scientific advances. To illustrate what we mi call "real science," let's consider an actual case.

To study bacteria, microbiologists must use pure *cultures*—that is, plates c taining colonies of a single type of bacteria free from contamination by ot bacteria, molds, and so on. Only by studying a single type at a time can we le about that particular bacterium. Consequently, at the first sign of contaminati a culture is normally thrown out. On one occasion, however, Scottish bacteric gist Alexander Fleming turned a ruined culture into one of the greatest med advances in history.

In the 1920s, one of Fleming's bacterial cultures became contaminated wi patch of a mold called *Penicillium*. As he was about to throw out the culture d Fleming observed that no bacteria were growing near the mold (**Fig. 1-7**). Why n Fleming hypothesized that perhaps *Penicillium* had released a substance t killed off the bacteria growing nearby. To test this hypothesis, Fleming grew sc pure *Penicillium* in a liquid nutrient broth. He then filtered out the *Penicilli* mold and applied the liquid in which the mold had grown to an uncontamina bacterial culture. Sure enough, something in the liquid killed the bacteria. Furt research into these mold extracts resulted in the production of the first *antibioti* penicillin, a bacteria-killing substance that has since saved millions of lives.

Fleming used the scientific method. His experiment began with an obser tion and proceeded to a question, a hypothesis, and a prediction, followed by perimental tests, which led to a conclusion. But the scientific method alc would have been useless without the lucky combination of accident and a b liant scientific mind. Had Fleming been a "perfect" microbiologist, he woulc have had any contaminated cultures. Had he been less observant and less cr ous, he would have thrown out the spoiled culture dish. Instead, he turned contents of the contaminated dish into the beginning of antibiotic therapy bacterial diseases. As French microbiologist Louis Pasteur said, "Chance fav the prepared mind."

Scientific Theories Have Been Thoroughly Tested

Scientists use the word "theory" in a way that is different from its everyday usage. If the nonscientist Dr. Watson were to ask Sherlock Holmes, "Do you have a theory as to the perpetrator of this foul deed?" he would be asking for what a scientist would call a hypothesis—an educated guess based on observable evidence or clues. A scientific theory is far more general and more reliable than a hypothesis. Far from being an educated guess, a **scientific theory** is a general explanation of important natural phenomena, developed through extensive and reproducible tests. For example, the theory of gravity, which states that objects exert attraction for one another, is fundamental to the science of physics. Similarly, the *cell theory*, which states that all living things are composed of cells, is fundamental to the study of biology. Scientific theories have been repeatedly supported and never refuted by sound scientific evidence.

Perhaps the most important theory in biology is the *theory of evolution*. Since its formulation in the mid-1800s by Charles Darwin and Alfred Russel Wallace, the theory of evolution has been supported by an overwhelming accumulation of evidence, including fossil finds, geological studies, radioactive dating of rocks, breeding experiments, and research results in genetics, molecular biology, and biochemistry. People who say that evolution is "just a theory" profoundly misunderstand what scientists mean by the word "theory."

1.3 What Is Life?

What is life? This short, simple question does not have a short, simple answer. Although each of us has an intuitive understanding of what it means to say that something is alive, that intuition cannot be easily translated to a precise definition. Life is so diverse and complex that it has proved impossible to devise a definition that neatly divides the living from the nonliving. Standard dictionary definitions are of little help, because they typically use phrases such as "the quality that distinguishes living organisms from dead organisms," without providing much insight as to what that mysterious "quality" might be.

Because we cannot define life precisely, we must instead build our definition bit by bit, by describing a series of different features of living things. In fact, this entire textbook is really an extended attempt to define life. As you read the book and attend your biology classes, you will learn about many different aspects of the living world. Our hope is that, as you proceed, a picture of life will emerge, in much the same way that a painted image gradually takes shape from the patches of color an artist applies to a canvas. We begin with a brief discussion of some properties that are shared by living things. Taken together, these properties form a combination of characteristics that is not found in nonliving objects.

Living Things Are Both Complex and Organized

Compared with nonliving matter of similar size, living things are highly complex and organized. A nonliving crystal of table salt consists of just two chemical elements, sodium and chlorine, arranged in a precise way; the salt crystal is *organized* but simple (**Fig. 1-8a**). The nonliving water of an ocean contains atoms of all the naturally occurring elements, but these atoms are randomly distributed; seawater is *complex* but not organized (**Fig. 1-8b**). In contrast, even the simplest organisms contain dozens of different elements linked together in thousands of specific combinations to form cells. Cells are both complex *and* organized, and every living thing consists of at least one cell. In many organisms, cells are further organized into larger and more complex assemblies such as eyes, legs, digestive tracts, and brains (**Fig. 1-8c**).

(a) Organized

(b) Complex

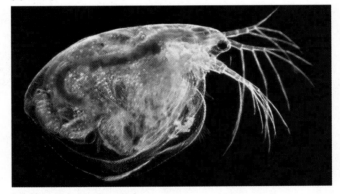

(c) Organized and complex

▲ **Figure 1-8 Life is both complex and organized** *(a)* Each crystal of table salt, sodium chloride, is a cube, showing great organization but minimal complexity. *(b)* The water and dissolved materials in the ocean have complexity but very little organization. *(c)* The waterflea is only 1 millimeter long (1/1,000 meter; smaller than the letter *i*), yet it has legs, a mouth, a digestive tract, reproductive organs, light-sensing eyes, and even a rather impressive brain in relation to its size.

▶ Figure 1-9 **Living things reproduce** As they grow, these polar bear cubs will resemble, but not be identical to, their parents. The similarity and variability of offspring are crucial to the evolution of life.

Web Animation What Is Life?

Living Things Grow and Reproduce

At some time in its life cycle, every organism becomes larger—that is, it *grows.* This characteristic is most obvious in plants and animals, which tend to start out very small and grow tremendously during their lives. Even single-celled bacteria, however, grow to about double their original size before they divide.

Organisms are also able to reproduce, giving rise to offspring of the same type (**Fig. 1-9**). Methods for producing offspring vary quite a bit among different kinds of organisms, but the result is always the same—the production of new individuals.

Living Things Respond to Stimuli

Organisms detect and respond to stimuli in their environments. For example, plants grow toward a source of light. Some kinds of bacteria can move to get away from a poisonous chemical. Animals have sensory organs and muscular systems that allow them to detect and respond to light, sound, chemicals, and many other stimuli in their surroundings. Animals can also respond to stimuli inside their bodies. For example, when you feel hungry, you sense the contractions of your empty stomach and the low levels of sugars and fats in your blood. You then respond by finding some food and eating.

Living Things Acquire and Use Materials and Energy

To grow, reproduce, and maintain their high level of organized complexity, organisms must obtain materials and energy from the environment (**Fig. 1-10**). The materials and energy power the chemical reactions needed to sustain life.

All organisms must obtain **energy**—the ability to do work, such as carrying out chemical reactions, growing leaves, or contracting a muscle. Plants and some single-celled organisms capture the energy of sunlight and store it in sugar molecules, a process called **photosynthesis.** Organisms that are not capable of photosynthesis obtain energy by extracting it from energy-containing molecules. In most cases, these energy-rich molecules are obtained by consuming the bodies of other organisms. Some kinds of single-celled organisms, however, extract energy from molecules found in the surrounding environment.

▲ Figure 1-10 **Living things acquire energy and materials from the environment** The green plants on which the toad stands capture energy from the sun and materials from the air, water, and soil. Small animals gain energy and materials by eating the plants, and the toad gains energy and materials by eating some of those small animals.

Living Things Use DNA to Store Information

All known forms of life use a molecule called **deoxyribonucleic acid,** or **DNA,** to store information (**Fig.** 1-11). Each organism's DNA acts as an instruction manual, a blueprint that guides the construction and operation of the organism's body. Each segment of DNA that contains a set of instructions is a **gene.**

When an organism reproduces, it passes a copy of its DNA to its offspring. The resulting transmission of genes to offspring is the basis of *heredity*—the means by which offspring inherit their parents' characteristics.

1.4 Why Is Life So Diverse?

One of the most striking characteristics of living things is that they all contain DNA. The DNA molecule is a rather complex structure, consisting of a distinctive arrangement of a specific set of parts. How could this particular unique structure have ended up in the bodies of every living thing from the smallest bacterium to the largest whale? The answer to this question lies in the common ancestry of all life. That is, every organism has descended from the same ancestor, and that common ancestor used DNA to store information. DNA has been passed down from generation to generation. (Remember that when an organism reproduces, it passes a copy of its DNA to its offspring.) Every organism living today carries the legacy of a shared ancient heritage.

If modern organisms descended from earlier organisms, and if all life therefore shares such fundamental qualities as the presence of DNA, why are living things so diverse? Because living things have been evolving for billions of years. Over eons, organisms on different branches of the tree of life have changed in different ways, increasing the differences between them even as all of them retained some key features that were present in the earliest living things.

Evolution Accounts for Both Life's Unity and Its Diversity

Over long periods of time, organisms change. This kind of change is not the kind that occurs within an individual organism during its lifetime, the kind that makes you look different now than you did when you were a small child. Instead, we're talking about the change that takes place within groups of organisms, from generation to generation. Each generation is slightly different from the one before it, so, after many generations, a typical member of the group may bear little resemblance to the typical group members of earlier times. This kind of accumulating generation-to-generation change within a group is known as **evolution.**

Evolutionary change has led to the amazing variety of life-forms on Earth. At the same time, evolution accounts for the presence of features that demonstrate life's unity, because evolutionary change has taken place along the branches of a single, gigantic "family tree" that formed as ancestors gave rise to new groups of modified descendants. All of the biological structures, mechanisms, systems, and interactions that we will discuss in this book arose through evolution. The idea that evolution gave rise to all of life's key features and characteristics is the guiding principle that unifies the study of biology.

Natural Selection Causes Evolution

The most important process by which evolution occurs is **natural selection.** Natural selection occurs because the characteristics of the different individuals in a population vary, and some individuals possess characteristics that help them survive and reproduce more successfully than do others that lack those traits. The individuals with these favorable traits tend to have a greater number of offspring, and those offspring tend to inherit the favorable traits from their parents. Those traits thus become more common in the group.

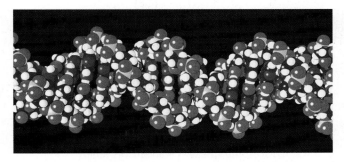

▲ **Figure 1-11 DNA** A computer-generated model of DNA, the molecule of heredity.

earth watch

Why Preserve Biodiversity?

The loss of species is the folly our descendants are least likely to forgive us.

E. O. Wilson, Professor, Harvard University

Scientists have estimated that there are 8 million to 10 million species on Earth today. The vast majority of these species live in the tropics. Unfortunately, tropical habitats are being rapidly destroyed and disrupted by human activities. For example, a recent analysis of thousands of satellite photos concluded that worldwide tropical rain-forest cover has decreased by about 23,000 square miles per year for the past decade. Most of lost forest has been destroyed by logging or to clear land for agriculture. Similarly, a worldwide survey of coral reefs revealed that about 20% of Earth's reef area has already been destroyed and an additional 25% is severely damaged, again mostly as the result of human influences such as pollution.

The rapid destruction of habitats in the tropics is causing many species to go extinct, as their homes disappear (**Fig. E1-4**). The biologist E. O. Wilson has calculated that current rates of habitat loss result in the extinction each year of as many as 25,000 species, most of them disappearing before they are even discovered and named. If Wilson's estimate is reasonably accurate, organisms are being driven to extinction

at a staggering rate, and the Earth of the future will have many fewer species than are currently present. Does it matter? Is there any reason for us to try to slow the loss of biodiversity?

One possible reason to protect Earth's biodiversity is an ethical one. In light of our power to destroy other species, perhaps we have an ethical obligation to protect them from extinction. But even if you do not agree with this ethical argument, there may be compelling practical reasons to save other species from destruction. Our ecological self-interest may be at stake.

For example, Earth's species form communities, highly complex webs of interdependent life-forms whose interactions sustain one another. These communities play a crucial role in processes that purify the air we breathe and the water we drink, build the rich topsoil in which we grow our crops, provide the bounty of food that we harvest from the oceans, and decompose and detoxify our waste. We depend entirely on these "ecosystem services." When our activities cause species to disappear from communities, we take a big risk. If we remove too many species, or remove some especially crucial species, we may disrupt the finely tuned processes of the community and undermine its ability to sustain us.

▲ **Figure E1-4 Biodiversity threatened** Destruction of tropical rain forests by indiscriminant logging threatens Earth's greatest storehouse of biological diversity.

For example, consider how natural selection might have influenced the evolution of beaver teeth. Beavers with larger teeth might have been able to chew down trees more efficiently, build bigger dams and lodges, and eat more bark than "ordinary" beavers. Because these big-toothed beavers obtained more food and better shelter than their smaller-toothed relatives, they raised more offspring. The offspring inherited their parents' genes for larger teeth. Over time, less-successful, smaller-toothed beavers became increasingly scarce. After many generations, all beavers had large teeth.

Life's Diversity Is Threatened

One outcome of evolution is that Earth is now home to a tremendous variety of **species,** or types of organisms. In many places, these species have evolved complex interrelationships with one another and with their surroundings. The word **biodiversity** is often used to describe this wealth of species and the complex interrelationships that sustain them. In recent decades, a single species, *Homo sapiens* (modern humans), has drastically increased the rate at which the environment is changing. Unfortunately, many other species have been unable to cope with this rapid change. In the habitats most affected by humans, many species are being driven to extinction. This problem is further explored in "Earth Watch: Why Preserve Biodiversity?"

Chapter Review

Summary of Key Concepts

For additional study help and activities, go to www.mybiology.com.

1.1 Why Study Biology?
The practical benefits of studying biology include greater understanding of health and disease, new career options, and more informed opinions on environmental and bioethical controversies. Most important, the more you know about living things, the more fascinating they become.

1.2 How Do Biologists Study Life?
Biology is the science of life. Knowledge in biology is acquired through the scientific method. First, an observation is made, which leads to a question. Then a hypothesis is formulated that suggests a possible answer to the question. The hypothesis is used to predict the outcome of further observations or experiments. A conclusion is then drawn about the hypothesis. Conclusions are based only on results that can be shared, verified, and repeated. A scientific theory is a general explanation of natural phenomena, developed through extensive and reproducible experiments and observations.

Web Animation   Experimental Design

1.3 What Is Life?
Organisms possess the following characteristics: Their structure is complex and organized; they consist of cells; they grow; they reproduce; they respond to stimuli; they acquire energy and materials from the environment; and they use DNA to store information.

Web Animation What Is Life?

1.4 Why Is Life So Diverse?
Evolution by natural selection has brought about both life's unity (shared features) and its diversity (a huge number of incredibly different forms). Evolution is the theory that modern organisms descended, with modification, from preexisting life forms. Species evolve as a consequence of (1) variation among members of a population, (2) inheritance of those variations by offspring, and (3) natural selection of the variations that best help an organism to survive and reproduce in its environment.

Key Terms

biodiversity *p. 12*
cell *p. 5*
community *p. 5*
conclusion *p. 3*
control *p. 4*
deoxyribonucleic acid (DNA)
  p. 11

energy *p. 10*
evolution *p. 11*
experiment *p. 3*
gene *p. 11*
hypothesis *p. 3*
natural selection *p. 11*

observation *p. 3*
organ *p. 5*
organism *p. 5*
photosynthesis *p. 10*
population *p. 5*
prediction *p. 3*

question *p. 3*
scientific method *p. 3*
scientific theory *p. 9*
species *p. 12*
tissue *p. 5*
variable *p. 4*

Thinking Through the Concepts

Suggested answers to end-of-chapter and figure-based questions can be found at the end of the text.

Fill-in-the-Blank

1. List the six levels of organization of life described in this chapter, from the simplest to the most complex: _____; _____; _____; _____; _____; _____.

2. A(n) _____ is a general explanation of natural phenomena supported by extensive, reproducible tests and observations. In contrast, a(n) _____ is a proposed explanation for observed events. To answer specific questions about life, biologists use a general process called the _____.

3. Fill in the remaining five steps of the scientific method in the order in which they are performed:

Observation, _____, _____, _____, _____, _____.

4. The scientific method relies on isolating a single factor that is allowed to change; this factor is the _____. Other factors are kept constant; these are _____.

5. An important scientific theory that explains why organisms are at once so similar and also so diverse is the theory of _____. This theory includes an explanation of the process by which the characteristics of organisms change over time so that favorable traits become more common; this process is called _____.

6. The molecule that guides the construction and operation of an organism's body is called (complete term) _____, abbreviated as _____. This large molecule contains discrete segments with specific instructions; these segments are called _____.

7. Fill in the five important qualities that, taken together, distinguish living from nonliving things: (1) Living things are both complex and _____; (2) living things grow and _____;

(3) living things respond to _____; (4) living things acquire and use _____ and _____; and (5) living things use _____ to store information.

Review Questions

1. Describe the scientific method. In what ways do you use the scientific method in everyday life?

2. Starting with the cell, list the levels of organization of life, briefly explaining each level.

3. What is the difference between a scientific theory and a hypothesis? Explain how each is used by scientists.

4. What are the differences between a salt crystal and a tree? Which is living? How do you know?

5. What is evolution? Briefly describe how evolution occurs.

Applying the Concepts

1. Design an experiment to test the effects of a new dog food, "Super Dog," on the thickness and water-shedding properties of the coats of golden retrievers. Include all six parts of a scientific experiment. Be sure that your experiment assesses both coat thickness and water-shedding ability.

2. Think of two different types of organisms that you have seen interacting, for example, a monarch butterfly caterpillar on a milkweed plant or a beetle in a flower. What questions do you have about this interaction? Choose one of your questions and devise a single, simple hypothesis about its answer. Use the scientific method and your imagination to design an experiment that tests this hypothesis. Be sure to identify variables and control for them.

For additional resources, go to www.mybiology.com.

unit one

The Life of a Cell

Single cells can be complex, independent organisms, such as these two protists. A large *Euplotes* prepares to eat a much smaller *Paramecium*.

chapter 2

Atoms, Molecules, and Life

Many of today's most popular foods contain artificially modified molecules. Olestra, a new molecule that mimics fat, has proved to be an especially controversial addition to our diet.

Case Study   Improving on Nature?

Obesity is a major public health problem, and today's consumers are increasingly aware that they should limit their consumption of high-calorie foods. Many people, however, would like to reduce calories without giving up the pleasure of eating fatty and sweet foods. This widespread desire has long been apparent to the food industry, which has responded by developing artificial substitutes for fats and sweeteners that mimic the taste of real fats and sweeteners but lack the calories.

The most widely used artificial fat is known as olestra. Like real fat, olestra adds a pleasing texture and taste to foods. Our bodies, however, are not able to extract the energy stored in olestra's molecules. Olestra is completely indigestible and passes through the body unchanged. This indigestibility ensures that potato chips, for example, made with olestra have far fewer calories than normal chips.

Indigestibility is also a key feature of the popular artificial sweetener sucralose (also known as Splenda). Sucralose adds no calories to foods and beverages that contain it, but our sense of taste perceives it to be 600 times sweeter than normal table sugar. Another widely used artificial sweetener, aspartame (also known as NutraSweet), is not indigestible, but the amount of aspartame present in a typical artificially sweetened food or beverage is so small that aspartame is effectively calorie free.

Olestra, sucralose, and aspartame could be valuable allies in the battle against obesity, but are they safe to eat? Consumer advocates have raised concerns about the safety of these manufactured food additives. These critics cite researchers who have suggested a link between artificial sweeteners and cancer, and consumer complaints that blame artificial fats and sweeteners for a host of ailments including headaches, diarrhea, arthritis, and chest pains. In response to these concerns, the health effects of olestra, sucralose, and aspartame have been examined in hundreds of scientific studies. Policy makers at the U.S. Food and Drug Administration (FDA) have reviewed this research and have concluded that these food additives are safe for human consumption. The FDA's declaration, however, has not settled the question to everyone's satisfaction. Skeptics continue to maintain that the FDA is downplaying the importance of studies that show negative health effects. They believe extra caution is warranted in assessing substances that, like olestra, sucralose, and aspartame, are found in thousands of food products and consumed by millions of people.

A scientifically inclined observer might wonder how these controversial faux foods achieve their ability to deceive the senses. What properties enable these molecules to mimic food but provide no nutrition? Before answering this question, we will need to explore some background information about biological molecules. ■

Can fake foods help treat the obesity epidemic? Or are they unsafe to eat?

Web Animation Interactive Atoms

2.1 What Are Atoms?

An **element** is a substance that cannot be broken down or converted to other substances by ordinary chemical means. For example, carbon is an element, so if you took a diamond (a form of carbon) and cut it into pieces, each piece would still be carbon. If you could make finer and finer divisions, you would eventually produce a pile of carbon **atoms.** Atoms are the basic structural units of matter.

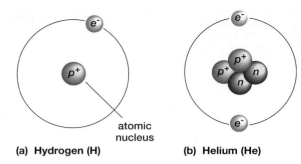

(a) Hydrogen (H)   **(b) Helium (He)**

▲ **Figure 2-1 Atomic models** The two smallest atoms, **(a)** hydrogen and **(b)** helium. In these simplified models, the electrons are represented as miniature planets, circling around a nucleus. The nucleus of hydrogen consists of a proton. The nucleus of helium consists of two protons and two neutrons.

Atoms Are Composed of Even Smaller Particles

Atoms are made up of even smaller units called subatomic particles. Physicists have discovered dozens of different subatomic particles, but we will discuss only three important types: protons, neutrons, and electrons. Positively charged **protons** are found in the **atomic nucleus,** which is the central core of an atom. An atomic nucleus also contains uncharged **neutrons** (with the exception of the nucleus of a hydrogen atom, which contains a single proton but no neutrons). **Electrons,** which are lighter than neutrons and protons and are negatively charged, orbit the atomic nucleus (**Fig. 2-1**). An individual atom has an equal number of electrons and protons and is therefore electrically neutral.

Atoms are the basic units of elements, and the atoms that make up different elements vary in the number of subatomic particles they contain. In particular, each element is composed of atoms that have a unique **atomic number,** which is the number of protons in its nucleus. For example, a hydrogen atom has one proton in its nucleus and therefore has an atomic number of 1, a helium atom has two protons and an atomic number of 2, and so on, for all of the elements. Each element also has particular chemical properties. For example, some elements, such as oxygen and hydrogen, are gases at room temperature; others, such as lead and iron, are solids.

Atoms of the same element may have different numbers of neutrons; when they do, the differing atoms are called **isotopes** of the element. Some, but not all, isotopes are **radioactive;** that is, they spontaneously break apart, forming different types of atoms and releasing energy in the process. For example, radioactive isotopes of uranium decay to form lead. Some scientists use radioactive isotopes to determine the age of fossils (see "Scientific Inquiry: How Do We Know How Old a Fossil Is?" on page 000 in Chapter 15).

Electrons Orbit the Nucleus, Forming Electron Shells

As you may know from experimenting with magnets, like poles repel each other and opposite poles attract each other. In a similar way, negatively charged electrons repel one another but are drawn to the positively charged protons of the nucleus. The attraction between electrons and protons holds an atom together as its electrons orbit around the nucleus. But because the electrons repel each other, only a limited number of them can be at the same distance from the nucleus.

Electrons orbit at only seven different distances from the nucleus, and these different distances are called **electron shells** (**Fig. 2-2** illustrates the first four shells). Electrons in different shells have different amounts of energy. Electrons in the innermost shell have the lowest energy, electrons in the second shell have

▶ **Figure 2-2 Electron shells in atoms** Most biologically important atoms have at least two shells of electrons. The first shell, closest to the nucleus, can hold only two electrons, and the next shell can hold a maximum of eight electrons. More distant shells can also hold eight electrons each.

QUESTION: Why do biologically active atoms tend to be ones whose outer shells are not full?

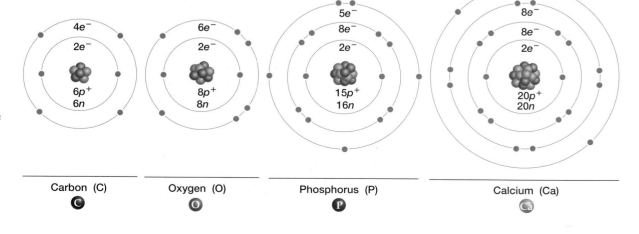

Carbon (C)   Oxygen (O)   Phosphorus (P)   Calcium (Ca)

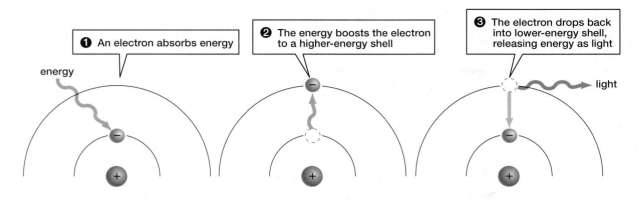

more energy, and so on. In some circumstances, electrons can move from one shell to another, with corresponding changes in energy level. An electron must absorb energy to move from an inner to an outer shell, and an electron moving from an outer to an inner shell releases energy (**Fig. 2-3**).

The electrons in an atom normally first fill the shell closest to the nucleus and then begin to occupy the next shell. The electron shell closest to the atomic nucleus can hold only two electrons, and the second shell can hold up to eight electrons. Thus, a carbon atom, with six electrons, has two electrons in the first shell, closest to the nucleus, and four electrons in its second shell (see Fig. 2-2).

2.2 How Do Atoms Form Molecules?

A **molecule** consists of two or more atoms, which can be of the same or different elements. The atoms are held together by interactions among their outermost electron shells.

Atoms Interact When There Are Vacancies in Their Outermost Electron Shells

Atoms interact with one another according to two basic principles:

- An atom will not react with other atoms when its outermost electron shell is completely full or empty. Such an atom is described as being *inert*.

- An atom will react with other atoms when its outermost electron shell is only partially full. Such atoms are described as *reactive*. Reactive atoms are unstable and tend to interact with other atoms in ways that increase their stability.

An atom with an outermost electron shell that is partially full can gain stability in three ways. It can lose electrons (to empty the shell), gain electrons (to fill the shell), or share electrons with another atom (with both atoms behaving as though they had full outer shells, as in the case of water). The results of losing, gaining, and sharing electrons are **chemical bonds,** attractive forces that hold atoms together in molecules. Each element has chemical bonding properties that arise from the arrangement of electrons in its outer shell.

Chemical reactions make and break chemical bonds to form new substances. Such reactions are essential for the maintenance of life. Whether chemical reactions occur in a plant cell as it captures solar energy, in your brain as it forms new memories, or in your car's engine as it guzzles gas, the reactions make new chemical bonds, or break existing ones, or both.

H—C—C—C—C—O—H

(with H atoms above and below each carbon)

(a) All bonds shown

$CH_3—CH_2—CH_2—CH_2—OH$

(b) Bonds within common groups omitted

OH

(c) Carbons and their attached hydrogens omitted

(d) Overall shape depicted

▲ **Figure 2-4 Different ways to represent a molecule**
Each of the diagrams illustrates a molecule of butanol, $C_4H_{10}O$. **(a)** In one common method of illustrating molecules, each bond in a molecule is represented by a line connecting two atoms. In some depictions **(b)**, the bonds in commonly occurring groups of atoms are not shown, but viewers familiar with this convention will understand that the molecule depicted this diagram is identical to the one portrayed in (a). **(c)** In another common space-saving convention, carbon atoms and the hydrogen atoms attached to them are not shown. **(d)** To illustrate a molecule's shape, a more complex depiction is necessary.

A Molecule Can Be Depicted in Different Ways

When describing and discussing molecules and the chemical reactions that they undergo, scientists use certain standards and conventions that govern how molecules are depicted. The simplest way to represent a molecule is to simply add up the number of each type of atom in the molecule and then write the totals as subscripts in a molecular formula. For example, the molecular formula of butanol (an industrial solvent) is $C_4H_{10}O$, indicating that a molecule of butanol consists of 4 carbon atoms, 10 hydrogen atoms, and 1 oxygen atom.

Molecular formulas are useful for concise summaries of reactions, but they have a major drawback: They do not show how the atoms in a molecule are bonded to one another. The particular bonds in a molecule help determine its properties and how it will react with other molecules, so it is important to have molecular representations that include depictions of bonds. In most standard depictions, bonds are represented by straight lines connecting atoms, which are represented by their atomic symbols, as in molecular formulas (**Fig. 2-4a**).

You will encounter a few other conventions in depictions of molecules in this text and elsewhere:

- Certain small groupings of atoms occur repeatedly in many different biological molecules. To save space, these common groupings may be shown as formulas (for example CH_3 or OH), without showing lines to represent the bonds within the grouping (**Fig. 2-4b**).

- Carbon atoms are very common in biological molecules, so they are often not labeled in depictions. Instead, the position of a carbon atom is indicated by a junction between two bond lines or by a bond line's empty tip (**Fig. 2-4c**). The "don't label carbon atoms" convention is especially common for representations of molecules in which carbon atoms are linked in a ring shape.

- The carbon atoms in biological molecules are typically bonded to one or more hydrogen atoms. In depictions that do not label carbon atoms, the hydrogen atoms bonded to the unlabeled carbons may also be omitted from a depiction (see Fig. 2-4c). The viewer is intended to assume their presence.

Although diagrams of the types described above contain information about the bonds present in a molecule, they do not include information about the molecule's shape. Because a molecule's shape is often important to its function, you will also encounter depictions of molecules that are designed to help a viewer visualize their shape (**Fig. 2-4d**).

Charged Atoms Interact to Form Ionic Bonds

The stability that atoms gain by emptying or filling their outermost shells is demonstrated by the formation of table salt (sodium chloride). Sodium (Na) has only one electron in its outermost electron shell, and chlorine (Cl) has seven electrons in its outer shell—one electron short of being full (**Fig. 2-5a**). Sodium, therefore, can become stable by losing an electron to chlorine (leaving its outer shell empty), and chlorine can fill its outer shell by gaining that electron. Once sodium loses an electron, the protons in the atom outnumber the electrons, so the sodium atom becomes positively charged (Na^+). Similarly, when chlorine picks up an electron, it becomes negatively charged (Cl^-). Positively or negatively charged atoms are called **ions** (**Fig. 2-5b**).

Opposite charges attract, so sodium ions and chloride ions tend to stay near one another. They form crystals that contain repeating orderly arrangements of the two ions (**Fig. 2-5c**). The electrical attraction between oppositely charged ions that holds them together in crystals is called an **ionic bond.** Ionic bonds are easily broken, as occurs when salt is dissolved in water.

Uncharged Atoms Share Electrons to Form Covalent Bonds

An atom with a partially full outermost electron shell can also become stable by sharing electrons with another atom, forming a **covalent bond.** Consider the hydrogen atom, which has one electron in a shell built for two. A hydrogen atom can become stable if it shares its single electron with another hydrogen atom, forming a molecule of hydrogen gas, H_2. Because the two hydrogen atoms are identical, neither nucleus can exert more attraction and capture the other's electron. So each electron's orbiting time is divided equally between the two nuclei, forming a single covalent bond. Each hydrogen atom behaves almost as if it had two electrons in its shell.

Most Biological Molecules Use Covalent Bonding

The molecules in proteins, sugars, carbohydrates, fats, and virtually every other biological molecule are formed of atoms held together by covalent bonds. The atoms most commonly found in biological molecules (hydrogen, carbon, oxygen, nitrogen, phosphorus, and sulfur) all need at least two electrons to fill their outermost electron shell and can share electrons with one or more other atoms. Hydrogen can form a covalent bond with one other atom; oxygen and sulfur with two other atoms; nitrogen with three; and phosphorus and carbon with up to four. This diversity of bonding possibilities permits biological molecules to be constructed in almost infinite variety and complexity.

Covalent Bonds Are Either Nonpolar or Polar

In hydrogen gas, the two nuclei are identical, and the shared electrons spend equal time near each nucleus. Therefore, not only is the molecule as a whole electrically neutral, but each end, or *pole*, of the molecule is also electrically neutral. Such an electrically symmetrical bond is called a *nonpolar* covalent bond, and a molecule (such as H_2) formed with such bonds is a **nonpolar molecule** (Fig. 2-6a).

Electron sharing in covalent bonds, however, is not always equal. In many molecules, one nucleus may have a larger positive charge (due to more protons) and therefore attract the electrons more strongly than does the other nucleus. This situation produces a *polar* covalent bond. Although the polar molecule as a whole is electrically neutral, it has charged parts. The atom that attracts the electrons more strongly picks up a slightly negative charge and is thus the negative pole of the molecule. The other atom has a slightly positive charge and is the positive pole. In water, for example, oxygen attracts electrons more strongly than does hydrogen, so the oxygen end of a water molecule is negative and each hydrogen is positive (Fig. 2-6b). Water, with its charged poles, is a **polar molecule.**

Hydrogen Bonds Form Between Molecules with Polar Covalent Bonds

Because their covalent bonds are polar, water molecules attract one another. The partially negatively charged oxygen atoms of water molecules attract the partially positively charged hydrogen atoms of other nearby water molecules.

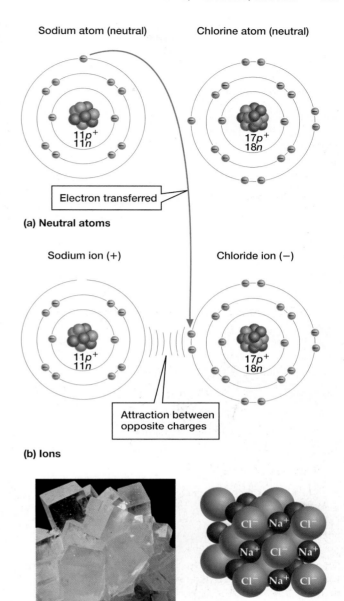

Sodium atom (neutral)   Chlorine atom (neutral)

$11p^+$
$11n$
$17p^+$
$18n$

Electron transferred

(a) Neutral atoms

Sodium ion (+)   Chloride ion (−)

$11p^+$
$11n$
$17p^+$
$18n$

Attraction between opposite charges

(b) Ions

Cl⁻ Na⁺ Cl⁻
Na⁺ Cl⁻ Na⁺
Cl⁻ Na⁺ Cl⁻

(c) An ionic compound: NaCl

▲ **Figure 2-5** The formation of ions and ionic bonds

► **Figure 2-6 Covalent bonds**
Shared electrons form covalent bonds. **(a)** In hydrogen gas, one electron from each hydrogen atom is shared, forming a single covalent bond. **(b)** Oxygen, lacking two electrons to fill its outer shell, makes one bond with each of two hydrogen atoms to form water. **QUESTION:** In water's polar bonds, why is oxygen's pull on electrons stronger than hydrogen's?

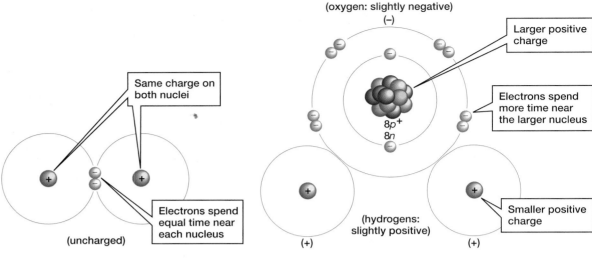

(a) Nonpolar covalent bonding in hydrogen **(b) Polar covalent bonding in water**

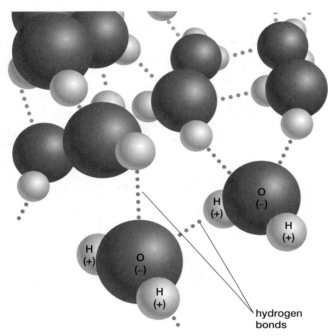

▲ **Figure 2-7 Hydrogen bonds** The partial charges on different parts of water molecules produce weak hydrogen bonds, with each bond joining the hydrogen of one water molecule to the oxygen of another water molecule.

Web Animation Water and Life

This electrical attraction is called a **hydrogen bond** (Fig. 2-7, Table 2-1). As we shall see shortly, hydrogen bonds give water several unusual properties that are essential to life on Earth. Hydrogen bonds are also important in many biological molecules.

2.3 Why Is Water So Important to Life?

Water is one of the most abundant substances on Earth and plays a crucial role in sustaining life. In fact, most scientists believe that life would be impossible without water—life on Earth is thought to have originated in water; all living things require water; and 60% to 90% of an organism consists of water. The importance of water stems from some properties of the water molecule that allow it to perform a unique role in support of life. Let's consider some of these properties.

Water Interacts with Many Other Molecules

Water enters into many of the chemical reactions in living cells. For example, water supplies the oxygen that green plants release into the air during photosynthesis. Water is also used when animals digest the molecules in food. Conversely, water is produced in the reactions that manufacture proteins, fats, and sugars. Whenever a chemical reaction occurs in a living cell, water is very likely to be involved.

Many Molecules Dissolve Easily in Water

Why is water so important in biological chemical reactions? One reason is that water is an extremely good **solvent;** that is, it is capable of dissolving a wide range of substances, including proteins, salts, and sugars. Because water can dissolve so

Table 2-1	**Chemical Bonds**
Type of Bond	**Bond Forms**
Ionic bond	Between positive and negative ions
Covalent bond	By sharing of electron pairs: equal sharing produces nonpolar covalent bonds; unequal sharing produces polar covalent bonds
Hydrogen bond	Between a hydrogen atom in a polar covalent bond and another atom in a polar covalent bond

health watch

Health Food?

Many of the routine chemical reactions that normally occur in cells give rise to molecules that contain atoms with unpaired electrons in their outer shells. This type of molecule, called a *free radical*, may also be generated in cells that absorb ultraviolet radiation from sunlight or that come into contact with contaminants such as tobacco smoke. A free radical is very unstable and reacts readily with nearby molecules, capturing an electron to complete its outer shell. But, by stealing an electron from another molecule, it creates a new free radical and begins a chain reaction that can lead to the destruction of biological molecules crucial to life. Cells have repair mechanisms that offset much of this harm, but the repair system is often not able to correct all of the damage.

Damage caused by free radicals contributes to a variety of human ailments, including heart disease, nervous system disorders such as Alzheimer's disease, and some forms of cancer. Fortunately, other molecules, called antioxidants, react with free radicals and render them harmless. Our bodies synthesize several antioxidants, and others can be obtained from a healthy diet. Vitamin C, for example, is an antioxidant found in many foods, including

oranges, red and green peppers, tomatoes, cantaloupe, and kiwi fruit.

Although it is difficult to do controlled studies on the effects of antioxidants in the human diet, there is some indirect evidence that a diet high in antioxidants is beneficial to health. For example, the low incidence of heart disease among the French (many of whom who eat a relatively high-fat diet) may be due in part to antioxidants in wine, which the French consume regularly. The French also eat considerably more fruits and vegetables, which are high in antioxidants, than do Americans.

Now, perhaps amazingly, it appears that chocolate, often a source of guilt for those who indulge in it, might contain antioxidants and thus be a type of health food (**Fig. E2-1**). Researchers have given us an excuse to eat chocolate and feel good about it. Cocoa powder (the dark, bitter powder made from the seeds inside the cacao pod) contains high concentrations of flavonoids, which are powerful antioxidants that are chemically related to those found in wine. So, does this finding mean that consumption of chocolate reduces the risk of cancer and heart disease? No studies have been done yet, but one

suspects there will be no shortage of volunteer subjects for the research. Although weight gained from eating too much chocolate candy could certainly counteract any positive effects of cocoa powder, chocoholics now have reason to relax and enjoy some modest indulgence.

▲ **Figure E2-1** **Chocolate**

---

many molecules, the watery environment inside a cell provides an excellent setting for the countless chemical reactions essential to life.

Water is such an excellent solvent because it is a polar molecule, with positive and negative poles. Thus, if a crystal of table salt (sodium chloride) is dropped into water, the positively charged ends of water molecules will be attracted to and will surround the negatively charged chloride ions in the salt crystal. At the same time, the negatively charged poles of water molecules will surround the positively charged sodium ions. As water molecules enclose the sodium and chloride ions and shield them from interacting with each other, the ions separate from the crystal and drift away in the water; the ionic bonds are broken and the salt dissolves (**Fig. 2-8**).

Water also dissolves molecules that are held together by polar covalent bonds. Its positive and negative poles are attracted to oppositely charged regions of dissolving molecules. Ions and polar molecules that dissolve readily in water are termed *hydrophilic* (Greek for "water-loving") because of their electrical attraction for water molecules. (Molecules that are uncharged and nonpolar and do not dissolve in water are called *hydrophobic*. Fats and oils are examples.) Many biological molecules, including sugars and amino acids, are hydrophilic and dissolve in water. In addition, many gases, such as oxygen and carbon dioxide, also dissolve in water. Thus, fish swimming in a lake use oxygen that is dissolved in the water and release carbon dioxide into the water.

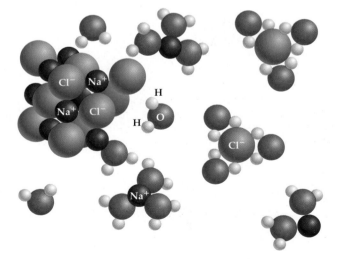

▲ **Figure 2-8** **Water as a solvent** When a salt crystal is dropped into water, the oppositely charged poles of the water molecules surround the sodium and chloride ions. Notice how the water molecules insulate the ions from the attractiveness of other molecules of salt. The ions disperse, and the whole crystal gradually dissolves. **QUESTION:** Why is it important to human physiology that sugars dissolve easily in water?

►**Figure 2-9 Cohesion among water molecules** *(a)* With webbed feet bearing specialized scales, the basilisk lizard of South America makes use of surface tension, caused by cohesion, to support its weight as it races across the surface of a pond. *(b)* Within a redwood tree, cohesion holds water molecules together in continuous strands from the roots to the highest leaves, which may be more than 300 feet above the ground.

(a) Cohesion causes surface tension

(b) Cohesion helps water to reach treetops

Water Molecules Tend to Stick Together

In addition to interacting with other molecules, water molecules interact with each other. Because hydrogen bonds link individual water molecules, liquid water has high **cohesion;** that is, water molecules have a tendency to stick together. Cohesion among water molecules at the water's surface produces **surface tension,** the tendency for the water surface to resist being broken. If you've ever experienced the slap and sting of a belly flop into a swimming pool, you've discovered firsthand the power of surface tension. Surface tension can support fallen leaves, some spiders and water insects, and even a running lizard (**Fig. 2-9a**).

Cohesion plays an important role in land plants. A plant absorbs water through its roots, but is then faced with the big problem of moving the water to its leaves, which can be more than 300 feet up in a tall tree (**Fig. 2-9b**). The problem is solved by cohesion. The tiny tubes that connect the leaves, stem, and roots are filled with water, and when water molecules evaporate from the leaves, water is pulled up the tubes from below. The system works because the hydrogen bonds between water molecules are stronger than the weight of the water in the tubes. Even in a 300-foot-tall tree, the water "chain" doesn't break.

Another property of water that helps plants get water from their roots to their leaves is adhesion. *Adhesion* describes water's tendency to stick to surfaces that have a slight charge that attracts polar water molecules. Adhesion helps water move within small spaces, such as the thin tubes in plants that carry water from roots to leaves. If you stick the end of a narrow glass tube into water, the water will move a short distance up the tube. Put some water in a narrow glass bud vase or test tube and you will see that the upper surface is curved. The water is pulling itself up the sides of the glass by adhesion to the surface of the glass and cohesion among water molecules.

Water Can Form Ions

Although water is generally regarded as a stable compound, individual water molecules constantly gain, lose, and swap hydrogen atoms. As a result, at any given time about two of every billion water molecules are *ionized*—that is, broken apart into hydroxide ions (OH^-) and hydrogen ions (H^+) (**Fig. 2-10**).

A hydroxide ion has gained an electron from the hydrogen atom in a water molecule, and it has a negative charge. The hydrogen ion, which has lost its elec-

water
(H_2O)   hydroxide ion
(OH^-)   hydrogen ion
(H^+)

▲ **Figure 2-10 A water molecule is ionized** The molecule breaks apart into a hydroxide ion and a hydrogen ion. The reaction is reversible; the ions can recombine to form water.

tron, now has a positive charge. Pure water contains equal concentrations of hydrogen ions and hydroxide ions.

In many solutions, however, the concentrations of H+ and OH− are not the same. If the concentration of H+ is greater than the concentration of OH−, the solution is *acidic*. An **acid** is a substance that releases hydrogen ions when it is dissolved in water. For example, when hydrochloric acid (HCl) is added to pure water, almost all of the HCl molecules separate into H+ and Cl−. Now the concentration of H+ is much greater than the concentration of OH−, and the solution is acidic. (Many acidic substances, such as lemon juice and vinegar, have a sour taste because the sour-taste receptors on your tongue are specialized to respond to the excess of H+.)

If the concentration of OH− is greater than that of H+, the solution is *basic*. A **base** is a substance that combines with hydrogen ions, reducing their number. For instance, if sodium hydroxide (NaOH) is added to water, the NaOH molecules separate into Na+ and OH−. The OH− combines with H+, reducing the number of H+ ions. The solution then contains an excess of OH− and is basic. (The tongue's bitter-taste receptors are stimulated by many bases, and basic substances thus tend to have a bitter taste. Most people find bitter tastes unpleasant, so few foods are basic.)

pH Measures Acidity

The degree of acidity is expressed on the **pH scale,** in which neutrality (equal numbers of H+ and OH−) is assigned the number 7. Acids have a pH below 7; bases have a pH above 7 (**Fig. 2-11**). Pure water, with equal concentrations of H+ and OH−, has a pH of 7.0. Each unit on the pH scale represents a tenfold change in the concentration of H+. Thus, beer (pH 4.1) has a concentration of H+ that is about 10 times greater than that of coffee (pH 5.0), and a carbonated soft drink (pH 3.0) has a concentration of H+ that is 10,000 times higher than that of water (pH 7.0).

A Buffer Maintains a Solution at a Constant pH

In most mammals, including humans, both the cell interior and the fluids that bathe the cells are nearly neutral (pH about 7.3 to 7.4). Small increases or decreases in pH

▼ **Figure 2-11 The pH scale** The pH scale expresses the concentration of hydrogen ions in a solution on a scale of 0 (very acidic) to 14 (very basic).   **QUESTION:** How would the concentration of hydrogen ions in a cup of tea change if you added lemon juice to it?

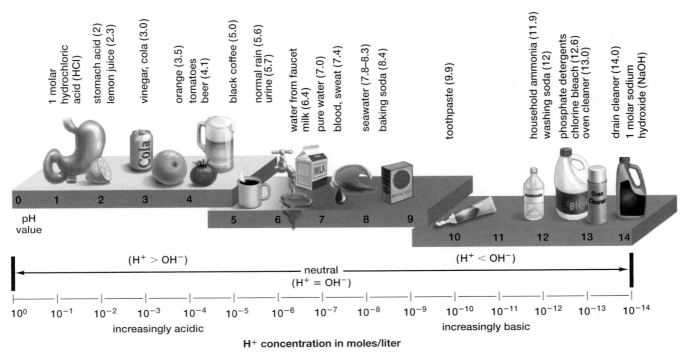

may cause drastic changes in both the structure and function of biological molecules, leading to the death of cells or entire organisms. Nevertheless, living cells seethe with chemical reactions that take up or give off H^+. How, then, does the pH remain constant overall?

The answer lies in the many buffers found in organisms. A **buffer** is a substance that tends to maintain a solution at a constant pH by accepting or releasing H^+ in response to small changes in H^+ concentration. If the H^+ concentration rises, buffers combine with them. If the H^+ concentration falls, buffers release H^+. The result in both cases is that the concentration of H^+ is restored to its original level.

2.4 Why Is Carbon So Important to Life?

Organisms contain a tremendous variety of substances, but a great many of these diverse molecules share a common feature: They contain carbon atoms. Carbon is widespread in living things because the carbon atom is extraordinarily versatile. Carbon can combine with other atoms in many different ways to form a huge number of different molecules.

This vast array of combinations is possible because a carbon atom has four electrons in its outermost shell, leaving room for four more. Therefore, carbon atoms are able to form many bonds. They become stable by sharing four electrons with other atoms, forming up to four covalent bonds. Carbon-containing molecules can contain many carbon atoms and can assume a variety of complex shapes, including chains, branches, and rings. Carbon thus forms the basis for an amazing diversity of molecules and makes it possible for living things to construct the many different substances required to sustain life.

In chemistry, molecules that have a carbon skeleton and also contain some hydrogen atoms are known as **organic molecules.** The carbon skeletons of organic molecules can be quite complex, but carbon alone does not account for the diversity of organic molecules. Instead, groups of atoms, called **functional groups,** attach to the carbon backbone and determine the characteristics and chemical reactivity of the molecules. These functional groups are far less stable than the carbon backbone and are more likely to participate in chemical reactions. The common functional groups found in organic molecules are shown in Table 2-2.

2.5 How Are Biological Molecules Joined Together or Broken Apart?

Organic molecules participate in a multitude of chemical reactions inside organisms. Many of these reactions serve to build or break down the complex molecules that are essential to life.

Construction of Large Molecules Yields Water

In living organisms, large, complex molecules are constructed by first assembling smaller molecules and then hooking them together. Small organic molecules (for example, *sugars*) are used as subunits to construct longer molecules (for example, *starches*), just as railcars are joined to form a train.

The chemical reactions that link small subunits together to make large biological molecules often release water as a by-product. (A *by-product* is an incidental product of a chemical reaction whose primary function is to produce a different product.) These water-releasing reactions are collectively known as **dehydration synthesis** reactions (Fig. 2-12a).

In a dehydration synthesis reaction that links two subunits, a hydrogen atom (—H) is removed from one subunit and a hydroxyl group (—OH) is removed

Web Animation Structure of Biological Molecules

Table 2-2	Important Functional Groups in Biological Molecules		
Group	**Structure**	**Properties**	**Types of Molecules**
Hydrogen (—H)	—H	Polar or nonpolar, depending on which atom hydrogen is bonded to; involved in dehydration synthesis and hydrolysis	Almost all organic molecules
Hydroxyl (—OH)	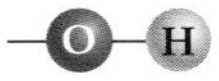 —O—H	Polar; involved in dehydration synthesis and hydrolysis	Carbohydrates, nucleic acids, alcohols, some acids, and steroids
Carboxylic acid (—COOH)	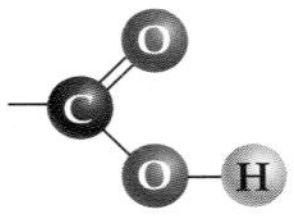 —C(=O)—O—H	Acidic; negatively charged when H+ separates from it; involved in peptide bonds	Amino acids, fatty acids
Amino (—NH₂)	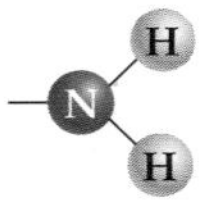 —N(H)(H)	Basic; may bond an additional H+ becoming positively charged; involved in peptide bonds	Amino acids, nucleic acids
Phosphate (—H₂PO₄)	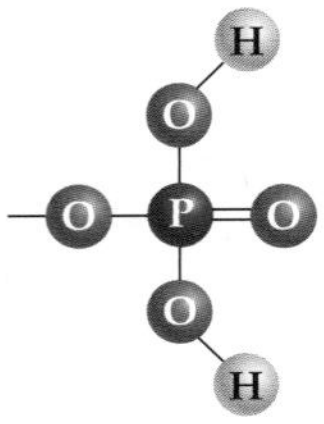 —O—P(=O)(O—H)(O—H)	Acidic; up to two negative charges when H+ separates from it; links nucleotides in nucleic acids; energy-carrier group in ATP	Nucleic acids, phospholipids
Methyl (—CH₃)	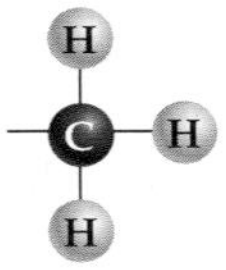 —C(H)(H)(H)	Nonpolar; tends to make molecules hydrophobic	Many organic molecules; especially common in lipids

from a second subunit, creating openings in the outer electron shells of the two subunits. These openings are then filled by electrons shared between the subunits, creating a covalent bond that links them. The previously removed free hydrogen and hydroxyl ions combine to form a molecule of water. This by-product water molecule might later be used in another chemical reaction in the cell or be released to the surrounding environment.

Breakdown of Large Molecules Uses Water

Chemical reactions in organisms not only build biological molecules, but also break them down. For example, large molecules that serve as food must be broken down into subunits that the organism can use. In these breakdown reactions, the covalent bonds that link the subunits of a large molecule are broken, separating the subunits and creating openings in their outer electron shells. The openings are filled by formation of bonds between each subunit and either a hydrogen ion or a hydroxyl ion; these ions are drawn from water molecules that participate in the reaction. The overall breakdown reaction, then, consumes water. Such water-consuming breakdown reactions are collectively known as **hydrolysis** reactions (**Fig. 2-12b**). Hydrolysis can be viewed as reversing the process of dehydration synthesis. Dehydration synthesis yields water as it links two molecules; hydrolysis consumes water as it breaks the links.

Considering how complicated living things are, it might surprise you to learn that nearly all biological molecules fall into one of only four general categories: carbohydrates, lipids, proteins, or nucleic acids (**Table 2-3**).

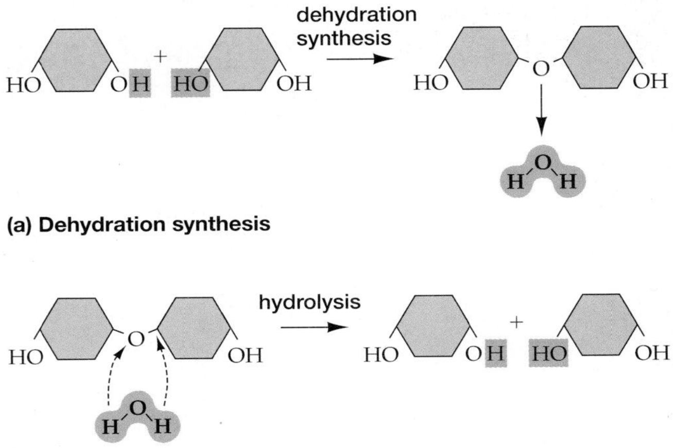

(a) Dehydration synthesis

(b) Hydrolysis

▲ **Figure 2-12 Dehydration synthesis and hydrolysis**

Table 2-3	The Principal Biological Molecules		
Class of Molecule	**Principal Subtypes**	**Example**	**Function**
Carbohydrate: Usually contains carbon, oxygen, and hydrogen, in the approximate formula $(CH_2O)_n$	**Monosaccharide:** Simple sugar	Glucose	Important energy source for cells; subunit of most polysaccharides
	Disaccharide: Two monosaccharides bonded together	Sucrose	Principal sugar transported throughout bodies of land plants
	Polysaccharide: Many monosaccharides (usually glucose) bonded together	Starch	Energy storage in plants
		Glycogen	Energy storage in animals
		Cellulose	Structural material in plants
Lipid: Contains high proportion of carbon and hydrogen; usually nonpolar and insoluble in water	**Triglyceride:** Three fatty acids bonded to glycerol	Oil, fat	Energy storage in animals, some plants
	Wax: Variable numbers of fatty acids bonded to long-chain alcohol	Waxes in plant cuticle (surface covering)	Waterproof covering on leaves and stems of land plants
	Phospholipid: Polar phosphate group and two fatty acids bonded to glycerol	Phosphatidylcholine	Component of cell membranes
	Steroid: Four fused rings of carbon atoms with functional groups attached	Cholesterol	Component of eukaryotic cell membranes; precursor of other steroids such as testosterone, bile salts
Protein: Chain of amino acids; contains carbon, hydrogen, oxygen, nitrogen, and sulfur		Keratin	Principal component of hair
		Silk	Principal component of silk moth cocoons and spider webs
		Hemoglobin	Transport of oxygen in vertebrate blood
Nucleic acid: Made of nucleotide subunits; may consist of a single nucleotide or long chain of nucleotides		Deoxyribonucleic acid (DNA)	Genetic material of all living cells
		Ribonucleic acid (RNA)	Genetic material of some viruses; in living cells, essential in transfer of genetic information from DNA to protein
		Adenosine triphosphate (ATP)	Principal short-term energy carrier molecule in cells
		Cyclic adenosine monophosphate (cyclic AMP)	Intracellular messenger

2.6 What Are Carbohydrates?

Carbohydrates are molecules composed of carbon, hydrogen, and oxygen in the approximate ratio of 1:2:1. Carbohydrates can be small, water-soluble sugars or long chains that are made by stringing sugar subunits together. A carbohydrate consisting of just one sugar molecule is called a simple sugar, or *monosaccharide*. When two monosaccharides are linked, they form a *disaccharide*. Three or more form a *polysaccharide*.

Carbohydrates are important energy sources for most organisms. Consider a breakfast that includes blueberry pancakes, syrup, and orange juice. The pancakes consist mainly of carbohydrates that were originally stored in the seeds of wheat or other grains. The sugar that sweetens the syrup, blueberries, and orange juice was also stored by plants as an energy source. The carbohydrate molecules that once served the plants that manufactured them now provide energy to the humans who consume them. Other carbohydrates, such as cellulose and similar molecules, provide structural support for individual cells or even for the entire bodies of organisms as diverse as plants, fungi, bacteria, and insects.

Most simple sugars have a carbon backbone of three to seven linked carbon atoms (**Fig. 2-13a**). Each carbon atom in the backbone generally has both a hydrogen (—H) and a hydroxyl group (—OH) attached to it, so carbohydrates have the general chemical formula $(CH_2O)_n$ where n is the number of carbons in the backbone. This formula explains the origin of the name "carbohydrate," which literally means "carbon plus water." When dissolved in water, the carbon backbone of a simple sugar usually "circles up" to form a ring (**Fig. 2-13b**). Simple sugars also assume the ring form when they link together to form disaccharides and polysaccharides.

A Variety of Simple Sugars Occurs in Organisms

Glucose is the most common simple sugar in organisms and is the subunit of which most polysaccharides are made. Glucose has six carbons, so its chemical formula is $C_6H_{12}O_6$. Many organisms synthesize other simple sugars that have the same chemical formula as glucose but have slightly different structures. These include fructose (the "corn sugar" found in corn syrup and also the molecule that makes fruits taste sweet). Some other common simple sugars, such as ribose and deoxyribose, have five carbons.

Disaccharides Store Energy

Simple sugars, especially glucose and its relatives, have a short life span in a cell. Most are either quickly broken down, so that their chemical energy is freed to fuel various cellular activities, or are linked together by dehydration synthesis to form disaccharides or polysaccharides (**Fig. 2-14**). Disaccharides are often used for short-term energy storage, especially in plants. Common disaccharides include sucrose (table sugar: glucose plus fructose) and lactose (milk sugar: glucose plus galactose). When cells require energy, disaccharides are broken apart into their monosaccharide subunits to release energy.

Polysaccharides Store Energy and Provide Support

Try chewing a cracker for a long time. Does it taste sweeter the longer you chew? It should, because the cracker contains starch, a polysaccharide that breaks down into its sweet-tasting glucose subunits as you chew. Certain polysaccharides, such as starch (in plants) and glycogen (in animals), are used mainly for long-term energy storage. Starch is commonly formed in roots and seeds, usually as huge, branched chains of up to half a million glucose subunits. Glycogen molecules, which are stored as an energy source in the liver and muscles of animals such as humans, are generally much smaller than starch molecules.

Many organisms use polysaccharides as structural materials. For example, the polysaccharide *chitin* is the main component of the hard body coverings of insects, spiders, crabs, and lobsters. One of the most important structural polysaccharides is *cellulose*, which provides support for plant cells and makes up about half the bulk of a typical tree trunk (**Fig. 2-15**). Cellulose is by far the most abundant organic molecule on Earth (which is not that surprising when you consider the vast fields and forests that blanket much of our planet). Ecologists estimate that about a trillion tons of cellulose are made each year.

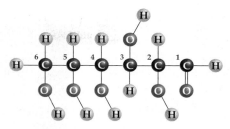
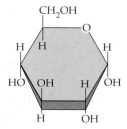

(a) Glucose, linear form

(b) Glucose, ring form

▲ **Figure 2-13 A simple sugar** Two diagrams of glucose, a six-carbon monosaccharide ($C_6H_{12}O_6$). Glucose molecules may assume **(a)** a linear (straight) form or **(b)** a cyclic (ring) form. Note that the diagram of the ring form follows the convention of not labeling the carbons that form the ring.

Improving on Nature?

Continued

Some artificial sweeteners, including sucralose, are produced by chemical reactions that modify naturally occurring disaccharide molecules.

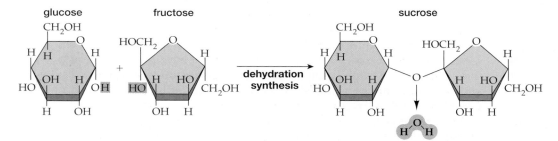

◀ **Figure 2-14 Manufacture of a disaccharide** The disaccharide sucrose is made by a dehydration synthesis reaction in which a hydrogen (—H) is removed from glucose, and a hydroxyl group (—OH) is removed from fructose, another simple sugar. The two simple sugars join by means of a covalent bond, and a water molecule forms in the process.

▶ **Figure 2-15 Cellulose structure** Cellulose, like starch, is composed of glucose subunits. Unlike starch, however, cellulose has great structural strength, due partly to the arrangement of parallel molecules of cellulose into long, cross-linked fibers. **QUESTION:** Many types of plastic are composed of molecules derived from cellulose, but engineers are working hard to develop plastics based on starch molecules. Why might starch-based plastics be an improvement over existing types of plastic?

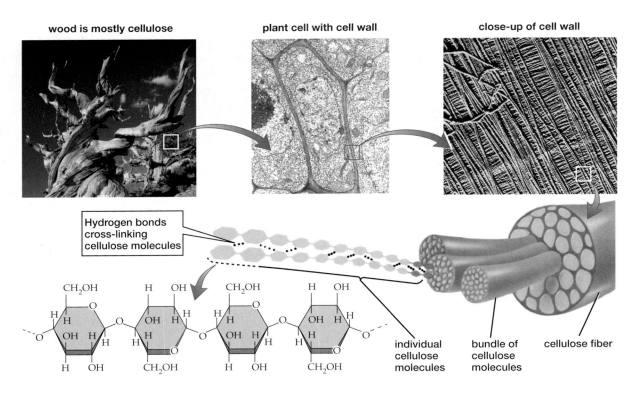

Like starch, cellulose consists of long chains of glucose subunits. Cellulose's digestibility, however, differs sharply from that of starch. Most animals can readily digest starch, but they cannot digest cellulose at all. For most animals, cellulose is roughage or fiber, material that passes undigested through the digestive tract. The only exceptions are animals such as cows or termites, which have special one-celled organisms living in their digestive tracts. These microscopic organisms are among the few that are able to break down cellulose, so the animals in whose guts they live can eat leaves, wood, and other foods that contain a lot of cellulose. The microbes digest the cellulose, and the animal hosts can absorb the nutrient molecules that are freed.

2.7 What Are Lipids?

Lipids are molecules that share two important features. First, they contain large regions composed almost entirely of hydrogen and carbon, with nonpolar carbon–carbon or carbon–hydrogen bonds. Second, these nonpolar regions make lipids hydrophobic and insoluble in water. Lipids are classified into three major groups: (1) oils, fats, and waxes; (2) phospholipids; and (3) steroids.

Oils, Fats, and Waxes Contain Only Carbon, Hydrogen, and Oxygen

Oils, fats, and waxes have several features in common. First, they contain only carbon, hydrogen, and oxygen. They also include one or more *fatty acid* subunits, which are long chains of carbon and hydrogen with a *carboxylic acid group* (—COOH) at one end. Finally, they usually do not have ring structures.

Fats and oils form by dehydration synthesis from three fatty acid subunits and one molecule of glycerol, a short, three-carbon molecule with one hydroxyl group (—OH) attached to each carbon atom (**Fig. 2-16**). The resulting molecules, known collectively as *triglycerides*, are used for long-term energy storage in both plants and animals. For example, during summer and fall, bears

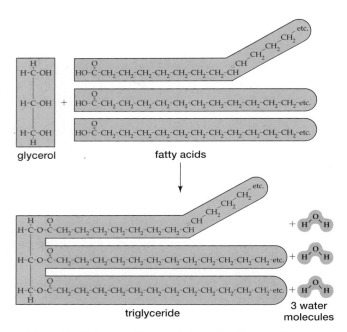

▲ **Figure 2-16 Synthesis of a triglyceride** Dehydration synthesis links a single glycerol molecule with three fatty acids to form a triglyceride and three water molecules.

(a) Fat

(b) Wax

◀ **Figure 2-17** **Lipids** *(a)* Fat is an efficient way to store energy. If this bear stored the same amount of energy in carbohydrates instead of fat, it probably would be unable to walk! *(b)* Wax is a highly saturated lipid that remains very firm at normal outdoor temperatures. Its rigidity serves well in the strong but thin-walled hexagons of this honeycomb.

consume more energy than they spend and store fat on their bodies, which tides them over during their winter hibernation (**Fig. 2-17a**). Because fats store the same energy with less weight than do carbohydrates, fat is an efficient way for animals to store energy.

The difference between a fat (such as beef fat), which is a solid at room temperature, and an oil (such as peanut oil), which is liquid at room temperature, lies in their fatty acids. In fats, all carbons in the fatty acid chains are joined with single covalent bonds (one pair of electrons shared between the two atoms). The remaining bond positions on the carbons are occupied by hydrogens. Such fatty acids are said to be saturated, because they are "saturated" with hydrogens; they have as many hydrogens as possible. Saturated fatty acids tend to be very straight, and they nestle closely together, forming solid lumps at room temperature (**Fig. 2-18a**).

In oils, some of the carbons in the fatty acid chains are joined by double covalent bonds (two pairs of electrons shared between the two atoms). Consequently, there are fewer attached hydrogens, and the fatty acid is said to be unsaturated. Unsaturated fatty acids tend to have bends and kinks in the fatty acid chains (**Fig. 2-18b**). The kinks keep oil molecules apart; as a result, an oil is liquid at room temperature.

Improving on Nature?

Continued

Fats and oils have a high concentration of chemical energy; a gram of fat or oil stores about twice as much energy as a gram of sugar. Because fats are so high in calories, fat substitutes such as olestra may be especially appealing to dieters.

▼ **Figure 2-18** Saturated fat and unsaturated oil

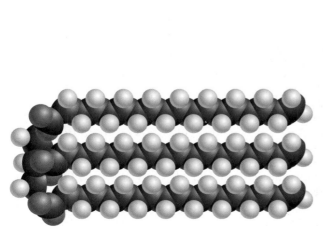

(a) Beef fat (saturated)

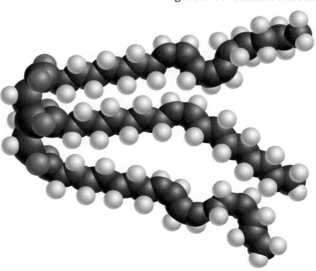

(b) Peanut oil (unsaturated)

❓ health watch

Cholesterol—Friend and Foe

Cholesterol is a steroid with a bad reputation. Why are so many products now advertising themselves as "cholesterol free" or "low in cholesterol"? After all, cholesterol is a crucial component of cell membranes. It is also the raw material for the production of bile (which helps us digest fats), vitamin D, and both male and female sex hormones.

Although cholesterol is crucial to life, medical researchers have found that people with excessively high levels of cholesterol in their blood are at increased risk for heart attacks and strokes. Unfortunately, the cholesterol builds up "silently" and gives no warning signs. A person may not know that anything is wrong until he or she actually suffers a heart attack. Cholesterol contributes to the formation of obstructions in arteries, called *plaques*, which in turn can promote the formation of blood clots (Fig. E2-2). These clots can break loose and block an artery carrying blood to the heart, causing a heart attack, or to the brain, causing a stroke.

Have you heard of "good cholesterol" and "bad cholesterol"? Because cholesterol molecules are nonpolar, they do not dissolve in blood (which

is mostly water). Instead, the cholesterol molecules are transported through blood in packets surrounded by special carrier molecules called *lipoproteins* (phospholipids plus proteins). Cholesterol in high-density lipoprotein packets ("HDL cholesterol," which has more protein and less lipid) is the good kind. These packets transport cholesterol to the liver, where it is removed from circulation. Cholesterol in low-density lipoprotein packets ("LDL cholesterol," with less protein and more lipid) is the bad kind. This is the form in which cholesterol circulates to cells throughout the body and can be deposited on artery walls. A high ratio of HDL (good) to LDL (bad) is correlated with reduced risk of heart disease. A complete cholesterol screening test will distinguish between these two forms in your blood.

Where does cholesterol come from? Cholesterol comes from animal-derived foods; it is essentially nonexistent in plants. Egg yolks are a particularly rich source, and meat, whole milk, and butter contain it as well. Another source of cholesterol is your own body, which synthesizes cholesterol from other lipids. Because of genetic differences, some

people's bodies manufacture more than others'. People with high cholesterol (about 25% of all adults in the United States) can often reduce their levels by eating a diet low in both cholesterol and saturated fats. For people with excessively high cholesterol who are unable to reduce it adequately by changing their diets, doctors often prescribe cholesterol-reducing drugs.

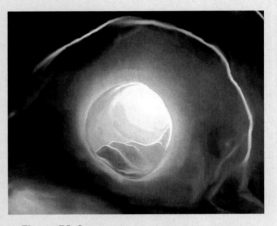

▲ **Figure E2-2 Plaque** A plaque deposit (ripply structure) partially blocks a carotid artery.

Although **waxes** are chemically similar to fats, they are not a food source. We and most other animals cannot digest them. They are highly saturated, making them solid at normal outdoor temperatures. Waxes form a waterproof coating over the leaves and stems of land plants. Some animals synthesize waxes as waterproofing (for mammalian fur, for example), and a few others, such as bees, use waxes to build elaborate structures (see Fig. 2-17b).

Phospholipids Have Water-Soluble Heads and Water-Insoluble Tails

The membrane that separates the inside of a cell from the outside world contains several types of *phospholipids*. A phospholipid is similar to an oil, but instead of having three fatty acid subunits attached to its glycerol subunit, one of the fatty acids is replaced by a smaller, phosphate-containing subunit. This subunit sits at one end of the phospholipid molecule, where it forms a polar head that is water soluble. The two fatty acids form tails that are not soluble in water. Thus, a phospholipid has two dissimilar ends: a hydrophilic head attached to hydrophobic tails (Fig. 2-19). As you will see in Chapter 3, this dual nature of phospholipids is crucial to the structure and function of cell membranes.

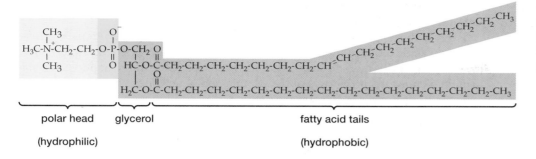

◀ **Figure 2-19 Phospholipids** Phospholipids are similar to fats and oils, except that only two fatty acid tails are attached to the glycerol subunit. The glycerol is bonded to a polar head composed of a phosphate group plus another functional group.

polar head   glycerol   fatty acid tails

(hydrophilic)   (hydrophobic)

Steroids Consist of Four Carbon Rings Fused Together

Steroids are structurally different from the other lipids. All steroids are composed of four rings of carbon fused together with various functional groups protruding from them (**Fig. 2-20**; note the basic steroid "skeleton" in color). One type of steroid is cholesterol. Cholesterol is a vital component of the membranes of animal cells and is also used by cells to synthesize other steroids. Steroids made from cholesterol include male and female sex hormones, the hormones that regulate salt levels, and bile, which assists in fat digestion. Why, then, has cholesterol gotten so much bad publicity? Find out in "Health Watch: Cholesterol—Friend and Foe."

2.8 What Are Proteins?

Proteins perform many different functions in organisms. An especially important role is played by *enzymes*, proteins that guide almost all of the chemical reactions that occur inside cells. Because each enzyme assists only one or a few specific reactions, most cells contain hundreds of different enzymes. Other types of proteins are used for energy storage or structure (**Fig. 2-21**). Proteins may also function in transport or movement (for example, a protein carries oxygen in the blood, and others help muscle cells move).

Proteins Are Formed from Chains of Amino Acids

A molecule of **protein** consists of one or more chains of amino acids. All **amino acids** have the same fundamental structure, in which a central carbon is bonded to four different functional groups: an amino group ($-NH_2$); a carboxylic acid group

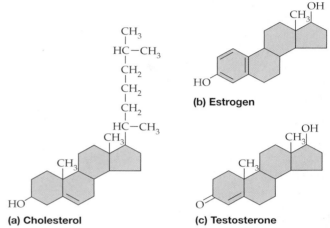

(a) Cholesterol   **(b) Estrogen**   **(c) Testosterone**

▲ **Figure 2-20 Steroids** Some steroids are synthesized from *(a)* cholesterol. All steroids have almost the same molecular structure (colored rings). Great differences in steroid function result from the differences in functional groups attached to the rings. Notice the similarity in structure between *(b)* the female sex hormone estradiol (a type of estrogen) and *(c)* the male sex hormone testosterone.

▼ **Figure 2-21 Structural proteins**

(a) Hair   **(b) Horn**   **(c) Silk**

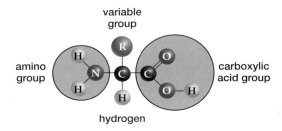

▲ Figure 2-22 Amino acid structure

(—COOH); a hydrogen; and a variable group (represented by the letter R and therefore often called an *R group*) (**Fig. 2-22**).

Each of the 20 amino acids that are found in proteins has a different R group, so amino acids differ in their chemical and physical properties—size, water solubility, electrical charge, and so on. Therefore, the sequence of the amino acids in a protein dictates the properties of the protein and determines whether it is an enzyme or a hormone or a structural protein.

Amino Acids Join by Dehydration Synthesis to Form Chains

Like lipids and polysaccharides, proteins form by dehydration synthesis. In the synthesis reaction, two amino acids are joined by the formation of a *peptide bond*— a particular kind of bond in which the nitrogen of the amino group (—NH$_2$) of one amino acid is joined by a single covalent bond to the carbon of the carboxylic acid group (—COOH) of another amino acid. The resulting chain of two amino acids is called a *peptide* (**Fig. 2-23**). More amino acids are added, one by one, until the protein is complete. Amino acid chains in living cells vary in length from three to thousands of amino acids. Often, the word "protein," or "polypeptide," is reserved for long chains—say, 50 or more amino acids in length—and "peptide" is used for shorter chains.

Three-Dimensional Shapes Give Proteins Their Functions

The phrase "amino acid chains" may evoke images of proteins as floppy, featureless structures, but they are not. Instead, proteins are highly organized molecules that fold themselves into complex, three-dimensional shapes (**Fig. 2-24**). Each different type of protein has a different shape, because each one has a different sequence of amino acids, and this sequence determines the protein's shape.

A protein's *primary structure* is simply the unique sequence of its long amino acid chain, but most chains also have a *secondary structure*. The secondary structure is caused by hydrogen bonds that bend the chain in a characteristic pattern, which in most proteins is either a spiral shape or a pleated shape. The hydrogen bonds result from attraction between the weakly positive hydrogen atom adjacent to a peptide bond and the weakly negative oxygen atom adjacent to another, nearby peptide bond. Thus, hydrogen bonds form between the carboxylic acid groups and the amino groups in peptide bonds all along the length of the chain, causing the chain to fold up or form a spiral.

A protein's secondary structure may be further modified into complex *tertiary structures*, caused by interactions among the different R groups along the chain. Because each amino acid has a distinctive R group with distinctive chemical characteristics, the different amino acids in a protein may attract or repel one another, causing further bending and folding into complex, irregular shapes. Finally, protein shape can be still further modified when two or more folded polypeptides join to form even more complicated *quaternary structures*.

A protein's shape is what enables it to perform its function. If its shape is disrupted, the protein may no longer be able to function correctly, even if the peptide bonds between amino acids remain intact. Proteins whose shape has been

▼ **Figure 2-23 Protein synthesis** In protein synthesis, dehydration synthesis joins the carbon of the carboxyl acid group of one amino acid to the nitrogen of the amino group of a second amino acid. (The blue oval identifies the atoms that ultimately form the water by-product of the reaction.)

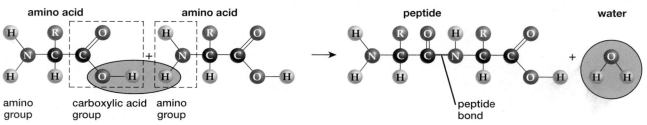

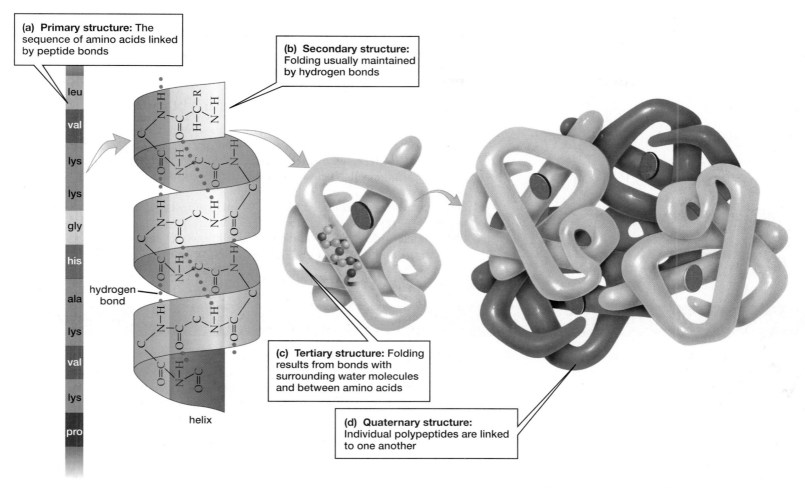

(a) Primary structure: The sequence of amino acids linked by peptide bonds

(b) Secondary structure: Folding usually maintained by hydrogen bonds

(c) Tertiary structure: Folding results from bonds with surrounding water molecules and between amino acids

(d) Quaternary structure: Individual polypeptides are linked to one another

hydrogen bond

helix

▲ Figure 2-24 **Protein chains fold into complex shapes** Each kind of protein has **(a)** a unique sequence of amino acids and a distinctive shape. A protein's shape is influenced by **(b)** hydrogen bonds between carboxylic acid groups and amino groups and **(c)** interactions among the protein's component amino acids. Many proteins consist of **(d)** a combination of multiple peptide subunits.   **QUESTION:** Why do most proteins, when heated, lose their ability to function?

disrupted are said to be **denatured.** There are many ways to denature a protein. For example, when an egg is fried, the heat of the frying pan denatures the albumin protein in the egg white, causing the egg white to change from clear to white and from liquid to solid. Sterilization using heat or ultraviolet rays kills bacteria and viruses by denaturing the proteins they need to live. Salty or acidic solutions also denature proteins—dill pickles are preserved in this way.

2.9 What Are Nucleic Acids?

Nucleic acids are long chains of similar but not identical subunits called *nucleotides*. All nucleotides have a three-part structure: (1) a five-carbon sugar (ribose or deoxyribose), (2) a phosphate group, and (3) a nitrogen-containing molecule called a *base* that differs among nucleotides (**Fig. 2-25**).

There are two types of nucleotides, the ribose nucleotides (containing the sugar ribose) and the deoxyribose nucleotides (containing the sugar deoxyribose). Nucleotides string together in long chains to form nucleic acids, with the phosphate group of one nucleotide covalently bonded to the sugar of another (**Fig. 2-26**).

DNA and RNA, the Molecules of Heredity, Are Nucleic Acids

There are two types of nucleic acids: *deoxyribonucleic acid*, or *DNA*, and *ribonucleic acid*, or *RNA*. DNA consists of chains of deoxyribose nucleotides millions of units long. DNA is found in all living things, and its sequence of nucleotides,

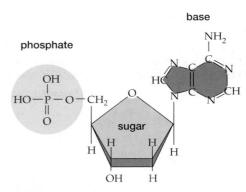

▲ Figure 2-25 **Deoxyribose nucleotide**

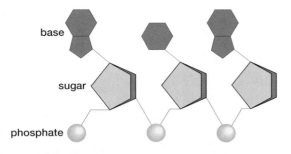

▲ **Figure 2-26** **Nucleotide chain**

Web Animation Functions of Macromolecules

like the dots and dashes of a biological Morse code, spells out the genetic information needed to construct the proteins of each organism. Chains of RNA are copied from the DNA in the nucleus of each cell. These RNA copies carry the message of DNA's genetic code and direct the manufacture of proteins.

Other Nucleotides Perform Other Functions

Not all nucleotides are part of nucleic acids. Some exist singly in the cell or as parts of other molecules. Cyclic nucleotides, such as *cyclic adenosine monophosphate (cyclic AMP)*, act as messengers within cells, carrying information to other molecules in the cell. Nucleotides with extra phosphate groups, such as *adenosine triphosphate (ATP)*, carry energy from place to place within a cell. They pick up energy where it is produced (during photosynthesis, for example) and deliver it to drive energy-demanding reactions elsewhere (say, to manufacture a protein). Other nucleotides (NAD$^+$ and FAD) are known as electron carriers and transport energy in the form of high-energy electrons.

Improving on Nature?   Revisited

Now that you know more about biological molecules, let's revisit the topic of artificial foods and see how scientists have modified molecules to eliminate their caloric content. Let's first consider the fake fat, olestra. Earlier fat substitutes used protein, carbohydrate, and fat molecules to build new molecules that imitated the molecular shape of fat—one molecule of glycerol attached to three molecules of fatty acids. In olestra, however, the glycerol molecule is replaced with sucrose and has six, seven, or eight fatty acids attached. The large number of fatty acid chains prevents digestive enzymes from reaching the digestible sucrose at the center of the olestra molecule. The molecule thus never breaks down into fragments that can be absorbed by the body and, hence, is excreted unchanged.

The super-sweet sucralose molecule is built by modifying a molecule of sucrose. That is, as its marketers like to say, sucralose is "made from sugar, so it tastes like sugar." In the synthesis of sucralose, three hydroxyl groups on a sucrose molecule are replaced by chlorine atoms (see Fig. 2-14). This modification has a twofold effect: It both increases the molecule's tendency to bind to the sweet receptor on a human tongue and renders the molecule unrecognizable to any of the enzymes that normally digest carbohydrates. Like olestra molecules, sucralose molecules remain unchanged during their journey through the digestive tract.

Aspartame is also very sweet but it is not made from sugar. Instead, it is synthesized by combining methanol with two amino acids (phenylalanine and aspartic acid). In the digestive system, aspartame breaks down into its constituent molecules, which, in the amounts present in a glass of diet soda or other aspartame-sweetened product, have negligible energy content.

Consider This
Some experts argue that we should encourage people to use artificial sweeteners and fake fats, because obesity will decline if people eat appealing food while limiting fat and sugar consumption. Critics of fake foods, however, contend that people should avoid any possible risks associated with artificial food additives and, instead, consume nutritious foods that are naturally low in sugar and fats. What do you think?

Chapter Review

Summary of Key Concepts

For additional study help and activities, go to www.mybiology.com.

2.1 What Are Atoms?
An element is a substance that can neither be broken down nor converted to different substances by ordinary chemical means. The smallest possible particle of an element is the atom, which is composed of a central nucleus, containing protons and neutrons, and electrons which orbit outside the nucleus. All atoms of a given element have the same number of protons, which is different from the number of protons in the atoms of every other element. Electrons orbit the nucleus in electron shells, at specific distances from the nucleus. Each shell can contain a fixed maximum number of electrons.

Web Animation **Interactive Atoms**

2.2 How Do Atoms Form Molecules?

The chemical reactivity of an atom depends on the number of electrons in its outermost electron shell. An atom is most stable, and therefore least reactive, when its outermost shell is either completely full or empty. Atoms combine to form molecules, which are held together by chemical bonds. Oppositely charged ions may be held together by ionic bonds. When two atoms share electrons, covalent bonds form. In a nonpolar covalent bond, the two atoms share electrons equally. In a polar covalent bond, one atom may attract the electron more strongly than the other atom does; in this case, the strongly attracting atom bears a slightly negative charge, and the weakly attracting atom bears a slightly positive charge. Some polar covalent bonds give rise to hydrogen bonding, the attraction between charged regions of individual polar molecules.

2.3 Why Is Water So Important to Life?

The water molecule is important to organisms because of its ability to interact with many other molecules and to dissolve many polar and charged substances, to participate in chemical reactions, and to cohere to itself.

Web Animation Water and Life

2.4 Why Is Carbon So Important to Life?

Organisms depend on carbon's ability to form a huge variety of different molecules. Carbon-containing molecules are so diverse because the carbon atom is able to form many types of bonds. This ability, in turn, allows organic molecules (molecules with a backbone of carbon and hydrogen atoms) to form many complex shapes, including chains, branches, and rings.

2.5 How Are Biological Molecules Joined Together or Broken Apart?

Most large biological molecules are synthesized by linking together many smaller subunits. Chains of subunits are connected by covalent bonds through dehydration synthesis. These chains may be broken apart by hydrolysis reactions. The most important organic molecules fall into four classes: carbohydrates, lipids, proteins, and nucleic acids. Their major characteristics are summarized in Table 2-3.

Web Animation Structure of Biological Molecules

2.6 What Are Carbohydrates?

Carbohydrate molecules are generally composed of carbon, hydrogen, and oxygen in the ratio 1 carbon: 2 hydrogens: 1 oxygen. Carbohydrates include sugars, starches, and cellulose. Sugars (monosaccharides and disaccharides) are used for temporary storage of energy and for the construction of other molecules. Starches and glycogen are polysaccharides that serve for longer-term energy storage in plants and animals, respectively. Cellulose and related polysaccharides form the cell walls of fungi, plants, bacteria, and some other microorganisms.

2.7 What Are Lipids?

Lipids are nonpolar, water-insoluble molecules with diverse chemical structures. They include oils, fats, waxes, phospholipids, and steroids. Lipids are used for energy storage (oils and fats), as waterproofing for the outside of plants and animals (waxes), as the principal component of cellular membranes (phospholipids), and as hormones (steroids).

2.8 What Are Proteins?

Proteins are chains of amino acids. The sequence of amino acids in the chain determines the structure of a protein. A protein is functional when folded into its characteristic three-dimensional shape. Proteins include enzymes (which guide chemical reactions), structural molecules, and transport molecules.

2.9 What Are Nucleic Acids?

Nucleic acid molecules are chains of nucleotides. Each nucleotide is composed of a phosphate group, a sugar group, and a nitrogen-containing base. The two types of nucleic acids are deoxyribonucleic acid (DNA) and ribonucleic acid (RNA). Nucleotides that function singly include intracellular messengers (cyclic AMP) and energy-carrier molecules (ATP).

Web Animation Functions of Macromolecules

Key Terms

acid *p. 25*

amino acid *p. 33*

atom *p. 17*

atomic nucleus *p. 18*

atomic number *p. 18*

base *p. 25*

buffer *p. 26*

carbohydrate *p. 28*

chemical bond *p. 19*

chemical reaction *p. 19*

cohesion *p. 24*

covalent bond *p. 21*

dehydration synthesis *p. 26*

denature *p. 35*

electron *p. 18*

electron shell *p. 18*

element *p. 17*

functional group *p. 26*

hydrogen bond *p. 22*

hydrolysis *p. 27*

ion *p. 20*

ionic bond *p. 21*

isotope *p. 18*

lipid *p. 30*

molecule *p. 19*

neutron *p. 18*

nonpolar molecule *p. 21*

nucleic acid *p. 35*

organic molecule *p. 26*

pH scale *p. 25*

polar molecule *p. 21*

protein *p. 33*

proton *p. 18*

radioactive *p. 18*

solvent *p. 22*

surface tension *p. 24*

wax *p. 32*

Thinking Through the Concepts

Suggested answers to end-of-chapter and figure-based questions can be found at the end of the text.

Fill-in-the-Blank

1. An atom that has lost or gained one or more electrons is called a(n) _____. If an atom loses an electron it takes on a(n) _____ charge. If it gains an electron it takes on a(n) _____ charge. Atoms with opposite charges attract one another, forming _____ bonds.

2. In addition to protons, all atoms except hydrogen have _____ in their nuclei. Atoms of the same element that differ in the number of neutrons in their nuclei are called _____ of one another. Some of these atoms spontaneously break apart, releasing _____ and forming different atoms in the process. Such atoms are described as _____.

3. An atom with an outermost electron shell that is either completely full or completely empty is described as _____. Atoms with partially full outer electron shells are _____ and may gain, lose, or share electrons, forming _____.

4. Water is described as _____ because each water molecule has partial negative and positive charges. This property of water allows water molecules to form _____ bonds with one another. The bonds between water molecules give water a high _____ and result in surface tension.

5. Large biological molecules are often formed from a series of similar smaller molecules that are linked together. The chemical reaction that creates bonds between the smaller subunits by removing water is called _____. The reverse reaction that breaks down these molecules by adding water is called _____. Starch and cellulose are both formed from subunit molecules of _____, and proteins are formed from _____.

6. The four general classes of biological molecules are _____, _____, _____, _____. After each molecule, identify the general class to which it belongs: cellulose: _____; steroid: _____; enzyme: _____; fat: _____; disaccharide: _____; DNA: _____; glycogen: _____.

Review Questions

1. Distinguish atoms from molecules; elements from compounds; and protons, neutrons, and electrons from each other.

2. Compare and contrast covalent bonds, ionic bonds, and hydrogen bonds.

3. Describe how water dissolves a salt.

4. Define *acid*, *base*, and *buffer*. How do buffers reduce changes in pH when hydrogen ions or hydroxide ions are added to a solution? Why is this phenomenon important in organisms?

5. Which elements are common components of biological molecules?

6. List the four principal types of biological molecules, and give an example of each.

7. What roles do nucleotides play in organisms?

8. Distinguish among the following: monosaccharide, disaccharide, and polysaccharide. Give two examples of each and their functions.

9. Describe the manufacture of a protein from amino acids.

Applying the Concepts

1. **IS THIS SCIENCE?** Headlines on a magazine cover proclaim: "Turn fat into muscle!" Evaluate this claim from a scientific standpoint.

2. Drugstores sell many different brands of "antacid" remedies, which are intended to bring relief from "acid stomach." Each brand claims to eliminate symptoms faster than its competitors. How do these compounds work? Use your knowledge of acids, bases, and buffers to design an experiment to determine which brand of antacid works best.

3. Many of water's unique properties are the result of its polar covalent bonds, which allow water molecules to form hydrogen bonds with each other. What if water molecules instead had nonpolar covalent bonds? Using information from this chapter, make a list of hypotheses about the ways in which this change might affect the properties of water. Describe how each change would affect living things. Design an experiment to test one of your hypotheses.

For additional resources, go to www.mybiology.com.

chapter 3

Cell Membrane Structure and Function

A rattlesnake
prepares to strike.

Case Study   Vicious Venoms

Like many college freshmen, Karl and Mark were eager to explore their new environs. On the first weekend of the fall semester, they left their Southern California campus and drove to a trailhead in the Mojave Desert. Shortly after beginning their hike, they spotted a rocky bluff and decided to climb it. Scrambling upward, Karl reached for a handhold on a rock outcropping. But rather than feeling solid rock beneath his hand, he was instead shocked to feel a writhing, scaly, animal body. He heard an unmistakable rattling sound, and almost immediately felt an intense burning pain in his hand. As he yanked his arm back, he got a glimpse of a big snake slithering into a crevice.

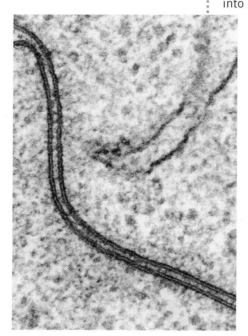

Many animal venoms attack cell membranes. This micrograph shows portions of the cell membranes of two adjacent cells.

Fortunately for Karl, Mark had his cell phone along and quickly dialed 911. In less than an hour, the two hikers were aboard a medical evacuation helicopter. As they flew toward the hospital, a paramedic listened to their story and radioed the hospital to alert the emergency room staff to the impending arrival of a patient who had been bitten by a western diamondback rattlesnake. This was sobering news, because the western diamondback is responsible for most of the 10 to 15 snakebite deaths in the United States each year. By the time Karl reached the hospital, a large bruise was spreading over his hand. His blood pressure had dropped, and the paramedics were administering oxygen because he was gasping for air.

Meanwhile, a couple of thousand miles to the east, in Kentucky, another person was about to have an unfortunate encounter with a venomous animal. Melissa was in her cellar, restacking some old cartons. As she worked, she was unaware that a brown recluse spider clung to one of the cartons. When she inadvertently pressed her arm against the spider, she did not even feel its bite. Hours later, however, her arm began to sting and she noticed a small, red swollen area on it. That night, Melissa had trouble sleeping as the pain in her arm increased. The next morning, alarmed to see a spreading, purplish welt, Melissa sought medical help.

After a series of tests to rule out other causes, the doctor said she suspected a brown recluse bite. In many cases, the doctor warned, such bites destroy both the surrounding skin and underlying tissue, resulting in an open wound that can become quite large and might take months to heal.

Many of the harmful consequences of rattlesnake and brown recluse spider venoms stem from their effects on cell membranes. How do venoms attack cell membranes? Why does damage to membranes have such destructive results? Understanding the answers to these questions will be easier if we first review some of what is known about the structure and function of cell membranes. ■

3.1 What Does the Plasma Membrane Do?

The cell is the smallest unit of life. Each cell is surrounded by a thin **plasma membrane,** which isolates the cell's contents from the external environment. The membrane also acts as a gatekeeper, controlling which substances are allowed to pass in or out and transferring chemical messages from the external environment to the cell's interior. These are formidable tasks for a structure so thin that 10,000 plasma membranes stacked on one another would scarcely equal the thickness of this page.

At first glance, a plasma membrane might appear to be a simple film surrounding a cell, similar to a soap bubble. Membranes, however, are complex structures that contain many different components, each of which performs a particular function.

3.2 What Is the Structure of the Plasma Membrane?

The plasma membrane's ability to do its jobs is closely tied to the way membranes are structured. The overall organization of membranes can be described as proteins floating in a double layer of lipids. The lipids are responsible for the isolating function of membranes (separating the cell's interior from its surroundings). The proteins are responsible for the gatekeeper function (controlling the exchange of substances across the membrane and responding to chemical signals).

Membranes Are "Fluid Mosaics"

One way to picture the structure of a membrane is to think of it as a **fluid mosaic.** A fluid is a liquid or a gas—that is, a substance that changes shape without breaking apart and whose molecules move freely relative to one another. The word "mosaic" refers to anything that resembles a mosaic artwork made by setting small, colored pieces of a material, such as tile, in mortar. A membrane, then, when viewed at high magnification, looks something like a lumpy, constantly shifting mosaic of tiles. A double layer of phospholipids forms a thick but still liquid "mortar" for the mosaic, and various proteins are the "tiles," which can move about within the phospholipid layers (**Fig. 3-1**). This movement gives a dynamic, ever-changing quality to membranes (even though the actual ingredients of the membrane remain relatively constant).

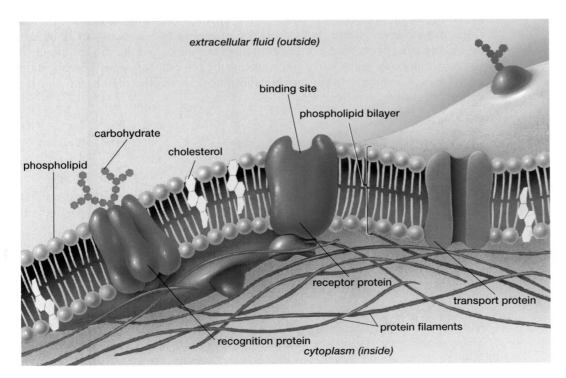

extracellular fluid (outside)

binding site

phospholipid bilayer

carbohydrate

cholesterol

phospholipid

receptor protein

transport protein

protein filaments

recognition protein

cytoplasm (inside)

◀ Figure 3-1 **The plasma membrane is a fluid mosaic** The plasma membrane is a bilayer of phospholipids in which various proteins are embedded. Many proteins have carbohydrates attached to them. The various types of membrane proteins fall into three categories: recognition proteins, receptor proteins, and transport proteins.

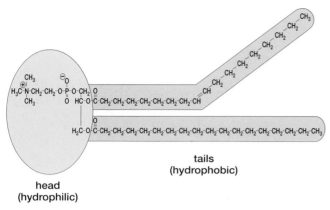

tails
(hydrophobic)

head
(hydrophilic)

▲ Figure 3-2 A phospholipid molecule

Web Animation Membrane Structure and Transport

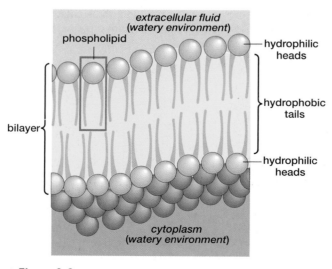

▲ Figure 3-3 A phospholipid bilayer

The Phospholipid Bilayer Is the Fluid Portion of the Membrane

Both the external and internal environments of a cell are watery. Externally, cells are surrounded by water. For example, single-celled organisms may live in fresh water, in the ocean, or in the moisture that clings to soil particles and other surfaces. Each cell in a multicelled organism is similarly bathed in water, because the spaces between the cells in an organism's body are filled with a fluid that consists largely of water. Internally, the *cytoplasm*, which includes all of a cell's contents except the nucleus, is mostly water. So, plasma membranes separate a cell's watery cytoplasm from its watery external environment.

Under these wet conditions, phospholipids spontaneously arrange themselves into a double layer called a **phospholipid bilayer.** As you learned in Chapter 2, a phospholipid molecule consists of a polar head and a pair of nonpolar tails (**Fig. 3-2**). The head is **hydrophilic,** which means that it is attracted to water molecules. The attraction is due to the formation of hydrogen bonds between water and the phospholipid heads. The phospholipid tails are **hydrophobic,** which means that they repel water molecules. As a result of the differing properties of the heads and tails, phospholipids in water organize themselves into a bilayer in which the hydrophilic heads form the outer, water-facing surfaces, with each hydrophilic head facing either the cytoplasm or the extracellular fluid. In contrast, the hydrophobic tails repel water and "hide" inside the membrane (**Fig. 3-3**). Because the phospholipid molecules in the membrane are not bonded to one another, individual phospholipids can move about easily within each layer.

In most cells, the phospholipid bilayer of membranes also contains cholesterol molecules. Cholesterol makes the bilayer stronger, more flexible, and less permeable to water-soluble substances such as ions or simple sugars. (You may want to review Fig. 2-20 on p. 33 and "Health Watch: Cholesterol—Friend and Foe" on p. 32.)

A Mosaic of Proteins Is Embedded in the Membrane

Thousands of proteins are embedded within or attached to the surface of a membrane's phospholipid bilayer. Some of the proteins can move about within the fluid phospholipid bilayer, but others are anchored to a network of filaments within the cytoplasm. Many of the proteins have carbohydrate groups attached to them, especially to the parts that stick outside the cell. Together, this collection of proteins is responsible for moving substances across the membrane and for communicating with other cells.

3.3 How Does the Plasma Membrane Play Its Gatekeeper Role?

One of the main functions of the plasma membrane is to control the movement of substances into and out of cells. Both the phospholipid bilayer and the embedded proteins play roles in this gatekeeper job.

The Phospholipid Bilayer Blocks the Passage of Most Molecules

Most biological molecules, including salts, amino acids, and sugars, are polar and water soluble. These substances cannot easily pass through the nonpolar, hydrophobic tails within the phospholipid bilayer. The phospholipid bilayer thus plays the main role in isolating the cell's contents from the external environment. The isolation is not complete, however. Very small molecules, such as water, and uncharged, lipid-soluble molecules can pass relatively freely through the phospholipid bilayer.

The Embedded Proteins Selectively Transport, Respond to, and Recognize Molecules

Most molecules that cross the membrane do so with the assistance of membrane proteins. In addition to providing transportation services, membrane proteins also play an important role in a cell's responses to the substances in its environment. Membrane proteins fall into three major categories—transport proteins, receptor proteins, and recognition proteins—each of which serves a different function (see Fig. 3-1).

Transport proteins allow the movement of hydrophilic (water-soluble) molecules through the plasma membrane. They do so either by forming channels through which molecules pass or by grabbing onto molecules and carrying them across the membrane. For example, glucose molecules, from which cells extract energy, can enter a cell only if carried across the membrane by molecules of specialized glucose-transport proteins.

Receptor proteins deliver chemical messages to the cell. They trigger responses inside the cell when specific molecules outside the cell, such as hormones or nutrients, bind to them. For example, when molecules of the hormone insulin bind to receptor proteins on the surface of one of the cells in your body, the cell responds by activating the transport proteins that move glucose molecules across the plasma membrane and into the cell.

Recognition proteins serve as identification tags and cell-surface attachment sites. For example, the cells of your immune system use recognition proteins to distinguish your own cells from those of disease-causing invaders. Thus, a bacterium is recognized as a foreign invader to be destroyed while your blood cells are ignored, because the two types of cells have different recognition proteins on their surfaces.

Each of the three broad categories of membrane proteins includes many different proteins, each of which has a distinctive shape and does a very specific job. Thus, each type of transport protein carries only a particular type of molecule across the plasma membrane, each kind of receptor protein binds only to one particular kind of molecule, and different cells bear distinctive collections of recognition proteins.

Vicious Venoms

Continued

To survive, a cell must maintain the appropriate chemical composition of its cytoplasm by allowing some molecules to move across its plasma membrane while preventing others from doing so. A break in a cell's plasma membrane can disrupt this function and kill the cell. Some body cells, such as those in animal muscles and other tissues that frequently move or change shape, are subject to a lot of bumping and jostling and are at especially high risk of breaking. Usually, though, a cell damaged in this way is able to quickly repair the broken membrane. If, however, a plasma membrane comes into contact with a substance that harms the phospholipid molecules from which membranes are constructed, the repair mechanisms may be overwhelmed. Many venoms, including components of rattlesnake venom and brown recluse spider venom, do their damage in just this fashion, by a direct attack on membrane molecules.

3.4 What Is Diffusion?

Cell membranes function in a watery environment, so before continuing our discussion of the transport of substances across the cell's plasma membrane, let's look at some characteristics of molecules in fluids. We begin with some definitions:

- The *concentration* of a substance in a fluid is the number of molecules of that substance in a given unit of volume.

- A *gradient* is a physical difference between two regions of space. A gradient can cause molecules to move from one region to another. Molecules frequently encounter gradients of concentration, pressure, and electrical charge.

Molecules in Fluids Move in Response to Gradients

Concentration gradients are important because they influence the movement of molecules or ions within a fluid. For example, consider perfume molecules moving from an open bottle into the air (remember that both gases and liquids

▶ **Figure 3-4** Diffusion of a dye in water

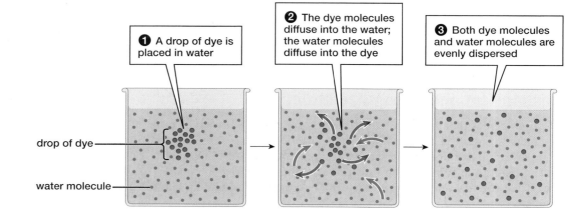

❶ A drop of dye is placed in water

❷ The dye molecules diffuse into the water; the water molecules diffuse into the dye

❸ Both dye molecules and water molecules are evenly dispersed

drop of dye

water molecule

are fluids). The perfume is moving from an area of high perfume concentration (the air inside the bottle) to an area of low perfume concentration (the air outside the bottle). In other words, the perfume molecules move in response to a **concentration gradient,** a difference in concentration between one region and another. The movement of molecules from regions of higher concentration to regions of lower concentration is called **diffusion.** We refer to such movements as going "down" the concentration gradient (from higher to lower). Thus, the reason that the air is fragrant in a room containing an open perfume bottle is that perfume molecules have diffused down their concentration gradient. If there are no other factors opposing diffusion, it will continue until the diffusing substance is evenly dispersed throughout the fluid.

A Drop of Dye in Water Illustrates Diffusion

To watch diffusion in action, place a drop of food coloring in a glass of water, and check its progress every few minutes. With time, the drop will seem to spread out and become paler, until eventually, even without stirring, the entire glass of water will be uniformly faintly colored. Molecules of dye move from the region of higher dye concentration into the surrounding water where the dye concentration is lower (**Fig. 3-4**). Simultaneously, some water molecules enter the dye droplet, and the net movement of water is from the higher water concentration outside the drop into the lower water concentration inside the drop.

At first, there is a very steep concentration gradient, and the dye diffuses rapidly. As the concentration differences lessen, the dye diffuses more and more slowly. In other words, the greater the concentration gradient is, the faster the rate of diffusion will be. However, as long as the concentration of dye within the expanding drop is greater than the concentration of dye in the rest of the glass, the net movement of dye will continue until the dye becomes uniformly dispersed in the water. At that point, the concentration gradient has been eliminated, and the rate of diffusion has therefore dropped to zero. Water and dye molecules continue to move, of course, but they move at random in all directions, with no net movement in any particular direction.

SUMMING UP: The Principles of Diffusion - - - - - - - - - -

- Diffusion is the movement of molecules down a gradient from higher concentration to lower concentration.
- The greater the concentration gradient, the faster the rate of diffusion.
- If no other processes intervene, diffusion will continue until the concentration gradient is eliminated.

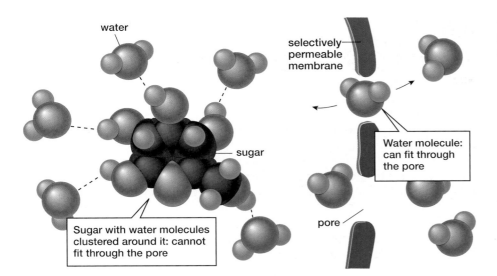

◀ **Figure 3-5 A selectively permeable membrane** Water molecules move freely across the membrane but larger dissolved molecules cannot.

3.5 What Is Osmosis?

Like other molecules, water diffuses from regions of higher water concentration to regions of lower water concentration. What do we mean when we describe a solution as having a "higher water concentration" or a "lower water concentration"? The answer is simple: Pure water has the highest water concentration. Any substance added to pure water displaces some of the water molecules, and the resulting solution will have a lower water content than pure water. The higher the concentration of dissolved substances, the lower the concentration of water.

The movement of water across membranes from areas of higher water concentration to areas of lower water concentration has such dramatic and important effects that we refer to it by a special name: **osmosis.** Osmosis moves water across **selectively permeable** membranes, which are so called because they allow the passage of some molecules but prevent the passage of other molecules. A very simple selectively permeable membrane might have pores just large enough for water to pass through but small enough to prevent the passage of sugar molecules (**Fig. 3-5**). Consider a bag made of a special plastic that is permeable to water but not to sugar. What will happen if we place a sugar solution in the bag and then immerse the sealed bag in pure water? The principles of osmosis tell us that the bag will swell. If it is weak enough, it will eventually burst (**Fig. 3-6**).

Web Animation Osmosis

SUMMING UP: The Principles of Osmosis

- Osmosis is the diffusion of water across a selectively permeable membrane.
- Dissolved substances reduce the concentration of water molecules in a solution.
- Water moves across a membrane down its concentration gradient from a higher concentration of water molecules to a lower concentration of water molecules.

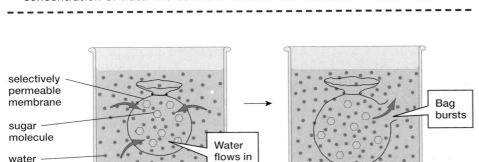

◀ **Figure 3-6 Osmosis** Water molecules diffuse down their concentration gradient across a selectively permeable membrane, from a region of higher water concentration to a region of lower water concentration. **QUESTION:** Imagine a container of glucose solution, divided into two compartments (A and B) by a membrane that is permeable to water and glucose but not to sucrose. If some sucrose is added to compartment A, how will the contents of compartment B change?

3.6 How Do Diffusion and Osmosis Affect Transport Across the Plasma Membrane?

As shown in **Table 3-1**, there are two main types of transport across membranes: passive transport and energy-requiring transport. We describe passive transport below; transport that requires energy is described in section 3.7.

During **passive transport,** substances move into or out of cells down concentration gradients. Cells have concentration gradients across their plasma membranes because the composition of the cytoplasm is very different from that of the extracellular fluid. The direction of movement across membranes by passive transport is determined entirely by concentration gradients, but whether a particular substance can cross depends on how the substance interacts with the phospholipid bilayer and transport proteins of the membrane. Passive transport requires no expenditure of energy. It may be accomplished by simple diffusion, facilitated diffusion, or osmosis. These processes are described in the following sections.

Plasma Membranes Are Selectively Permeable

Many molecules cross plasma membranes by diffusion, driven by differences between their concentration in the cytoplasm and in the external environment. Not all molecules, however, are equally likely to be able to diffuse across a membrane, because plasma membranes are selectively permeable. Some molecules can pass through; other molecules cannot.

Some Molecules Move Across Membranes by Simple Diffusion

Molecules that dissolve in lipids, such as ethyl alcohol and vitamin A, easily diffuse across the phospholipid bilayer, as do very small molecules, including water and

Table 3-1	Transport Across Membranes
Passive transport	Movement of substances across a membrane down a concentration, pressure, or electrical charge gradient. Does not require the cell to expend energy.
Simple diffusion	Diffusion of water, dissolved gases, or lipid-soluble molecules through the phospholipid bilayer.
Facilitated diffusion	Diffusion of molecules through a channel or carrier protein.
Osmosis	Diffusion of water across a selectively permeable membrane.
Energy-requiring transport	Movement across a membrane of substances that travel against a concentration gradient. Requires the cell to expend energy.
Active transport	Movement of individual small molecules or ions through membrane-spanning proteins, using cellular energy, usually ATP.
Endocytosis	Movement into a cell of large particles, which are engulfed as the plasma membrane forms vesicles that enter the cytoplasm.
Exocytosis	Movement out of a cell of materials that are enclosed in a membranous vesicle. The vesicle moves to the cell surface, fuses with the plasma membrane, and opens to the outside, allowing its contents to diffuse out.

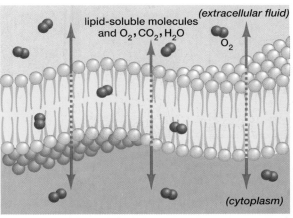

lipid-soluble molecules
and O_2, CO_2, H_2O

(extracellular fluid)

O_2

(cytoplasm)

(a) Simple diffusion through the phospholipid bilayer

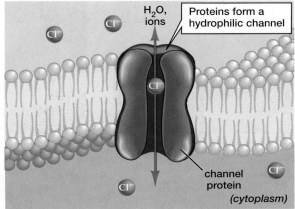

H_2O, ions

Proteins form a hydrophilic channel

Cl^-

Cl^-

Cl^-

channel protein

Cl^-

Cl^-

(cytoplasm)

(b) Facilitated diffusion through a channel protein

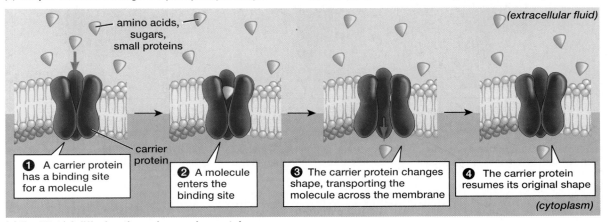

amino acids, sugars, small proteins

(extracellular fluid)

carrier protein

❶ A carrier protein has a binding site for a molecule

❷ A molecule enters the binding site

❸ The carrier protein changes shape, transporting the molecule across the membrane

❹ The carrier protein resumes its original shape

(cytoplasm)

(c) Facilitated diffusion through a carrier protein

◄ **Figure 3-7 Diffusion through the plasma membrane**
(a) Lipid-soluble molecules and gases such as oxygen, water, and carbon dioxide can pass by simple diffusion directly through the phospholipids. *(b)* Water and some water-soluble molecules enter or exit the cell by facilitated diffusion through a channel protein. *(c)* Certain molecules cross a membrane by facilitated diffusion through a carrier protein that changes shape to allow the passage. **EXERCISE:** Imagine an experiment that measures the initial rate of diffusion into cells placed in sucrose solutions of various concentrations. Sketch one graph (initial diffusion rate versus solution concentration) that shows the result expected if diffusion is simple and another graph that shows the result expected for facilitated diffusion.

dissolved gases such as oxygen and carbon dioxide. This type of passive transport is called **simple diffusion** (Fig. 3-7a). Generally, the rate of simple diffusion depends on the concentration gradient across the membrane, the size of the molecule, and how easily it dissolves in lipids. Large concentration gradients, small molecule size, and high lipid solubility all increase the rate of simple diffusion.

Other Molecules Cross the Membrane by Facilitated Diffusion

Most water-soluble molecules, such as ions (K^+, Na^+, Ca^{2+}), amino acids, and simple sugars, cannot move through the phospholipid bilayer on their own. These molecules can diffuse across only by **facilitated diffusion,** a type of passive transport in which diffusion is aided by either of two types of transport proteins: channel proteins or carrier proteins.

Channel proteins form pores, or channels, in the lipid bilayer through which certain ions can cross the membrane (Fig. 3-7b). In general, each type of channel protein is specialized and allows only particular ions to pass through. Nerve cells, for example, have separate channels for potassium ions (K^+), sodium ions (Na^+), and calcium ions (Ca^{2+}). Although water can diffuse directly through the phospholipid bilayer in all cells, many cells have specialized water channels called *aquaporins*. Aquaporins allow water to cross membranes by facilitated diffusion, which is faster than simple diffusion.

A **carrier protein** grabs onto a specific molecule on one side of a membrane and carries it to the other side. Each different type of carrier protein is able to bind to a specific molecule (typically an amino acid, sugar, or small protein). Binding triggers a change in the shape of the carrier that allows the bound molecule to pass through the protein and across the plasma membrane (Fig. 3-7c).

▶ **Figure 3-8 The effects of osmosis** Red blood cells are normally suspended in the fluid environment of the blood. *(a)* If red blood cells are immersed in an isotonic salt solution, which has the same concentration of dissolved substances as the blood cells do, there is no net movement of water across the plasma membrane. The red blood cells keep their characteristic dimpled disk shape. *(b)* A hypertonic solution, with more salt than is in the cells, causes water to leave the cells, shriveling them. *(c)* A hypotonic solution, with less salt than is in the cells, causes water to enter, and the cells swell and eventually burst.   **QUESTION:** A freshwater fish swims in a solution that is hypotonic compared with the fluid inside its body. Why don't freshwater fish swell up and burst?

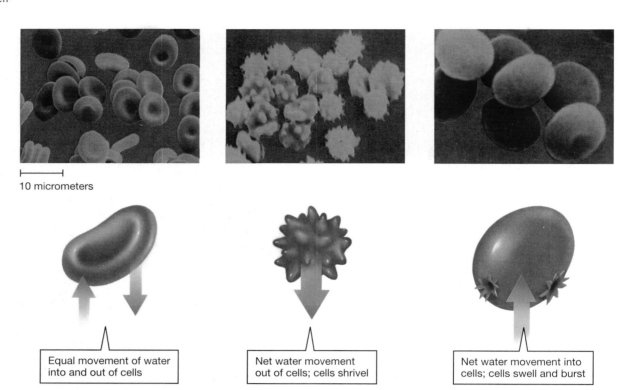

10 micrometers

Equal movement of water into and out of cells

Net water movement out of cells; cells shrivel

Net water movement into cells; cells swell and burst

(a) Isotonic solution has the same salt concentration as the cytoplasm

(b) Hypertonic solution has a higher salt concentration than the cytoplasm

(c) Hypotonic solution has a lower salt concentration than the cytoplasm

Water Can Move Across Plasma Membranes by Osmosis

Most plasma membranes are highly permeable to water. The flow of water across them depends on whether water is more highly concentrated inside or outside the cell. Ordinarily, the extracellular fluid outside the cell has the same water concentration as the cytoplasm inside the cell. In that case, the extracellular fluid is said to be *isotonic* to the cytoplasm, and there will be no net movement of water into or out of the cell (**Fig. 3-8a**). Note that the *types* of dissolved particles are seldom the same inside and outside the cells, but the *total concentration* of all dissolved particles is equal, with the result that the water concentration inside is equal to that outside of the cells.

If red blood cells are taken out of the body and immersed in salt solutions of varying concentrations, the effects of osmosis become dramatically apparent. If the solution has a higher salt concentration than the cytoplasm of the red blood cell (that is, if the solution has a lower water concentration), water will leave the cells by osmosis. The cells in such a *hypertonic* solution will shrivel up until the concentrations of water inside and outside become equal (**Fig. 3-8b**). If, on the other hand, the solution has little salt (is *hypotonic*), water will enter the cells, causing them to swell (**Fig. 3-8c**). This process explains why your fingers wrinkle up after a long bath. It may seem as if your fingers are shrinking, but they're not. Instead, water is moving into the outer skin cells of your fingers, swelling them more rapidly than the cells underneath and causing the wrinkling.

3.7 How Do Molecules Move Against a Concentration Gradient?

Passive transport, though undeniably efficient, could never by itself meet all of a cell's needs for moving molecules across membranes. Passive transport can carry substances only down concentration gradients, but all cells also need to move some materials "uphill" across their plasma membranes, *against* concentration gradients. For example, it is often necessary to move nutrients from the environment, where they are less concentrated, into the cell's cytoplasm, where they are more concentrated. Passive transport would instead move nutrient molecules out of the cell, so a different type of transport—one that requires energy—is necessary (see Table 3-1).

Active Transport Uses Energy to Move Molecules Against Their Concentration Gradients

During **active transport,** the cell uses energy to move substances *against* a concentration gradient. A helpful analogy for understanding the difference between passive and active transports is to consider what happens when you ride a bike. If you don't pedal, you can go only downhill, as in passive transport. However, if you put enough energy into pedaling, you can go uphill as well, as in active transport.

Membrane Proteins Regulate Active Transport

In active transport, membrane proteins use energy to move individual molecules across the plasma membrane (**Fig. 3-9**). Active-transport proteins span the width of the membrane and have two active sites where other molecules can bind. One active site binds a particular molecule, say, a calcium ion. The second site binds the energy-carrier molecule adenosine triphosphate (ATP). The ATP donates energy to the active-transport protein, causing it to change shape and move the calcium ion across the membrane. Active-transport proteins are often called *pumps*, in analogy to water pumps, because they use energy to move molecules "uphill" against a concentration gradient.

Cells Engulf Particles or Fluids by Endocytosis

Sometimes cells must acquire particles that are too large to move across a membrane by either passive transport or active transport. One way in which cells can accomplish this task is by an energy-consuming process called

Vicious Venoms

Continued

Some disease-causing bacteria, such as those that cause pneumonia in humans and those that cause stomach ulcers, make people sick in part by producing toxic proteins that attack cell membranes. These toxins bind to cell membranes and function rather like out-of-control transport proteins, forming open pores that allow any kind of molecule to pass unimpeded across the membrane. A cell under attack from these protein toxins is deprived of its ability to keep its contents isolated from the surrounding environment, and it quickly dies.

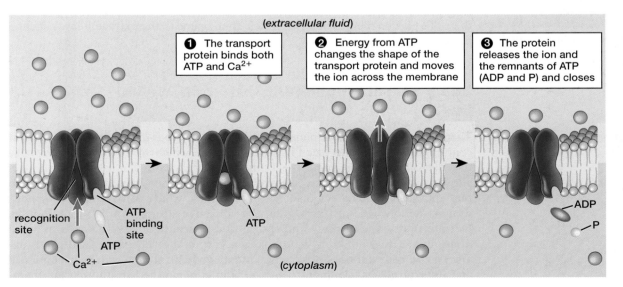

◀ **Figure 3-9 Active transport** Active transport uses cellular energy to move molecules across the plasma membrane, often against a concentration gradient. An active-transport protein binds the molecule to be transported and the energy-carrier ATP, and then changes shape to move the ion across the membrane. Notice that when ATP donates its energy, it loses one of its phosphate groups and becomes ADP.

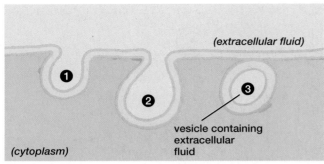

❶ A dimple forms in the plasma membrane, which **❷** deepens and surrounds the extracellular fluid. **❸** The membrane encloses the extracellular fluid, forming a vesicle.

(a) Pinocytosis

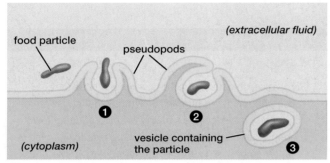

❶ The plasma membrane extends pseudopods toward an extracellular particle (food, for example). **❷** The ends of the pseudopods fuse, encircling the particle. **❸** A vesicle that contains the engulfed particle is formed.

(b) Phagocytosis

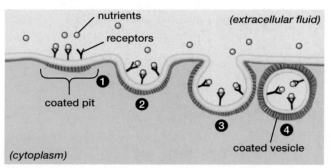

❶ Receptor proteins for specific molecules or complexes of molecules are localized at coated pit sites. **❷** The receptors bind the molecules and the membrane dimples inward. **❸** The coated pit region of the membrane encloses the receptor-bound molecules. **❹** A vesicle ("coated vesicle") containing the bound molecules is released into the cytosol.

(c) Receptor-mediated endocytosis

▲ **Figure 3-10** Three types of endocytosis

endocytosis. During endocytosis, the plasma membrane engulfs an extracellular molecule, particle, or droplet of fluid and pinches off a membranous sac called a **vesicle,** with the particle or fluid inside, into the cytoplasm.

Endocytosis can proceed in several different ways, depending on the size and nature of the material to be captured. When a drop of liquid must be moved into a cell, a very small patch of plasma membrane dimples inward as it surrounds extracellular fluid and buds off into the cytoplasm as a tiny vesicle (**Fig. 3-10a**), a process called *pinocytosis*.

Another process, *phagocytosis*, is used to pick up larger particles, such as when an amoeba engulfs food or when a white blood cell destroys bacteria that have invaded a human body. In phagocytosis, the cell first extends parts of its surface membrane. These extensions are called *pseudopods*. The tips of the pseudopods fuse around the particle and carry it to the interior of the cell inside a vesicle (**Fig. 3-10b**).

Pinocytosis and phagocytosis are somewhat indiscriminate; these processes transport into the cell whatever is dissolved in the drop of liquid captured by pinocytosis or whatever is in or near the particle engulfed by phagocytosis. However, another type of endocytosis, called *receptor-mediated endocytosis*, transports only specific molecules (**Fig. 3-10c**). This process depends on the many receptor proteins that cells have on their outside surfaces; each type of receptor protein has a binding site for a particular molecule. In some cases, the receptors accumulate in depressions on the plasma membrane called *coated pits*. If the appropriate molecule contacts a receptor protein in a coated pit, the molecule attaches to the binding site. The pit deepens into a U-shaped pocket that eventually pinches off and becomes a vesicle that contains the receptor, its bound molecules, and a bit of extracellular fluid. The vesicle moves into the cytoplasm where it ultimately breaks up, releasing the bound molecules. Among the substances transported by receptor-mediated endocytosis in humans is LDL cholesterol, which is the so-called "bad cholesterol" whose presence at high levels in the blood is associated with risk of cardiovascular disease. One way in which LDL cholesterol is removed from blood is through uptake into cells by receptor-mediated endocytosis.

Exocytosis Moves Material Out of the Cell

Cells often use the reverse of endocytosis, a process called **exocytosis,** to dispose of unwanted materials, such as the waste products of digestion, or to secrete materials, such as hormones (**Fig. 3-11**). During exocytosis, a vesicle carrying material to be expelled moves to the cell surface, where the vesicle's membrane fuses with the cell's plasma membrane. The vesicle then opens to the extracellular fluid, and its contents diffuse out.

Some Plasma Membranes Are Surrounded by Cell Walls

The plasma membranes of plants, fungi, and many types of bacteria lie inside stiff coatings called **cell walls** (see Fig. 4-4). Cell walls, which are produced by the cells they surround, support and protect otherwise fragile cells. For example, cell walls allow plants and mushrooms to resist the forces of gravity and blowing winds so that they can stand erect on land. Tree trunks, which can support enormous weight, demonstrate just how great the collective strength of cell walls can be. Despite their strength, cell walls are porous, permitting easy passage of small molecules such as minerals, water, oxygen, carbon dioxide, amino acids, and sugars. Thus, a cell wall does not control the movement of materials between a cell and its external environment. This function is performed by the plasma membrane, just as in cells that lack cell walls.

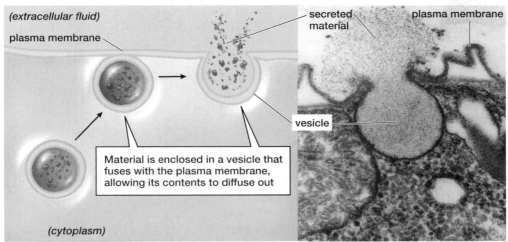

(extracellular fluid)

plasma membrane

secreted material

plasma membrane

Material is enclosed in a vesicle that fuses with the plasma membrane, allowing its contents to diffuse out

vesicle

(cytoplasm)

0.2 micrometer

◄ Figure 3-11 **Exocytosis** Exocytosis is functionally the reverse of endocytosis.

Vicious Venoms   Revisited

Many venomous snakes and spiders have venom that acts, at least in part, by destroying the plasma membranes of victims' cells. These venoms are rich in enzymes called *phospholipases*. (As you will learn in Chapter 5, enzymes are proteins that speed up the rate of chemical reactions such as the breakdown of biological molecules.) Phospholipases break down the phospholipids of plasma membranes, causing cells to rupture and die.

Cell death caused by phospholipases can destroy the tissue around a rattlesnake or brown recluse spider bite (**Fig. 3-12**). If the venom reaches the victim's bloodstream, phospholipases attack the membranes of red blood cells (which carry oxygen throughout the body). The resulting loss of blood cells reduces the blood's ability to carry oxygen and may cause the victim to become short of breath. The rattlesnake can inject far more venom far more deeply than the spider can, so snake venom is more likely than spider venom to enter the bloodstream and cause breathing problems. Both venoms, however, break down the membranes of cells that form the tiny blood vessels called *capillaries*, causing bleeding under the skin around the bite and, in severe cases, in internal organs as well.

Although snake and spider bites can have serious consequences, it is important to realize that only a tiny fraction of Earth's large number of spider and snake species are dangerous to people. The best defense is to learn which, if any, venomous snakes and spiders live in your area and where they prefer to "hang out." If your activities bring you to such places, wear protective clothing and always look before you reach. With appropriate caution, we can comfortably coexist with spiders and snakes, avoid their bites, and keep our cell membranes intact.

Consider This

Phospholipases and other digestive enzymes are found in animal (including human, snake, and spider) digestive tracts as well as in snake and spider venom. What function might these seemingly dangerous enzymes serve in digestive systems?

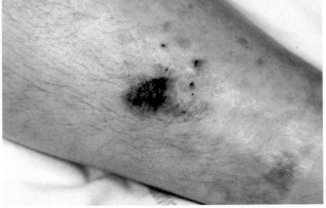

(a) **Brown recluse spider bite**

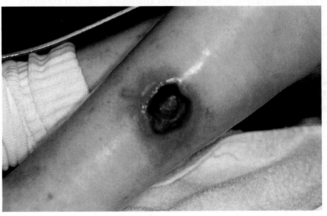

(b) **Rattlesnake bite**

▲ Figure 3-12 **Phospholipases in venoms can destroy cells**

Chapter Review

For additional study help and activities, go to www.mybiology.com.

3.1 What Does the Plasma Membrane Do?

The plasma membrane isolates the cytoplasm from the external environment. It also regulates the flow of materials into and out of the cell and the flow of chemical messages into the cell.

3.2 What Is the Structure of the Plasma Membrane?

The plasma membrane consists of a bilayer of phospholipids with proteins embedded in it.

> **Web Animation** Membrane Structure and Transport

3.3 How Does the Plasma Membrane Play Its Gatekeeper Role?

The phospholipid bilayer blocks most molecules from passing through the membrane (although some small molecules and lipid-soluble molecules do pass through). Larger molecules can cross the membrane only with the help of membrane proteins. Transport proteins regulate the movement of water-soluble substances through the membrane. Receptor proteins bind molecules in the external environment, triggering changes in the cell. Recognition proteins serve as identification tags and attachment sites.

3.4 What Is Diffusion?

Diffusion is the movement of molecules of a substance from regions of higher concentration of that substance to regions of lower concentration.

3.5 What Is Osmosis?

Osmosis is the diffusion of water across a selectively permeable membrane and down its concentration gradient. Dissolved substances decrease the concentration of water molecules.

> **Web Animation** Osmosis

3.6 How Do Diffusion and Osmosis Affect Transport Across the Plasma Membrane?

In simple diffusion, water, dissolved gases, and lipid-soluble molecules diffuse through the phospholipid bilayer. In facilitated diffusion, water-soluble molecules cross the membrane through protein channels or with the assistance of carrier proteins. In all these types of passive transport, molecules move down their concentration gradients, and cellular energy is not required. Water moves across plasma membranes by osmosis. Osmosis does not require cellular energy.

3.7 How Do Molecules Move Against a Concentration Gradient?

In active transport, proteins spanning the membrane use cellular energy (ATP) to drive the movement of molecules across the plasma membrane, usually against concentration gradients. Large molecules (for example, proteins), particles of food, microorganisms, and extracellular fluid may be acquired by endocytosis. The secretion of substances such as hormones and the excretion of wastes from a cell are accomplished by exocytosis.

Key Terms

active transport *p. 49*
carrier protein *p. 47*
cell wall *p. 50*
channel protein *p. 47*
concentration gradient *p. 44*
diffusion *p. 44*

endocytosis *p. 50*
exocytosis *p. 50*
facilitated diffusion *p. 47*
fluid mosaic *p. 41*
hydrophilic *p. 42*
hydrophobic *p. 42*

osmosis *p. 45*
passive transport *p. 46*
phospholipid bilayer *p. 42*
plasma membrane *p. 40*
receptor protein *p. 43*

recognition protein *p. 43*
selectively permeable *p. 45*
simple diffusion *p. 47*
transport protein *p. 43*
vesicle *p. 50*

Thinking Through the Concepts

Suggested answers to end-of-chapter and figure-based questions can be found at the end of the text.

Fill-in-the-Blank

1. The general term for the movement of substances through the plasma membrane from an area of high concentration to an area of lower concentration of that substance is _____. If movement of a substance through the membrane requires energy, the process is called _____.

2. The plasma membrane is often described as a _____. It consists of a double layer of _____ in which _____ are embedded. Which portion of the plasma membrane is primarily responsible for isolating the cell from its surroundings? _____ Which portion allows the cell to exchange materials with its environment? _____

3. A membrane that is permeable to some substances but impermeable to others is described as being _____. Assume you have a thin sac made of membrane permeable to water but not to sugar. You fill it with a concentrated sugar solution and place it in a container of pure water. The sac will *(choose one: shrink/swell/ remain unchanged)*.

4. Cell membranes include a bilayer of molecules called _____. These molecules have parts called tails that are hydrophobic, meaning _____. These tails face the *(choose one: outside/inside)* of the membrane.

5. After each molecule, place the term that most specifically describes the process by which it moves through a plasma membrane: carbon dioxide: _____; ethyl alcohol: _____; a sodium ion: _____; an amino acid: _____; water: _____; glucose: _____.

6. The general process by which fluids or particles are transported into cells is called _____. Does this process require energy?

_____ The specific term for engulfing fluid is _____, and the term for engulfing solid particles is _____. The engulfed substances are taken into the cell in membrane-lined sacs called _____.

Review Questions

1. Describe and diagram the structure of a plasma membrane. What are the two principal types of molecules in plasma membranes?

2. What are the three main categories of proteins commonly found in plasma membranes, and what is the function of each?

3. Define *diffusion*, and compare that process with osmosis.

4. Define *hypotonic*, *hypertonic*, and *isotonic*. What would happen to an animal cell immersed in each of the three types of solution?

5. Describe the following types of transport processes: simple diffusion, facilitated diffusion, active transport, pinocytosis, phagocytosis, and exocytosis.

Applying the Concepts

1. IS THIS SCIENCE? In humans, cells obtain the cholesterol they need by transporting it across the cell membrane from the surrounding blood or extracellular fluid. In a healthy person, this movement of cholesterol into cells prevents cholesterol levels in the blood from reaching harmful levels. But in many people, blood cholesterol levels are too high. The inventors of the drug ezetimibe demonstrated that the drug prevents cholesterol from moving across the membranes of the cells that line the intestine. On the basis of this finding, the inventors hypothesized that taking ezetimibe would reduce blood cholesterol. Do you agree? Provide a scientifically sound explanation for your answer.

2. Different cells have somewhat different plasma membranes. For example, the plasma membrane of a *Paramecium*, a single-celled organism that lives in ponds, is only about 1% as permeable to water as the plasma membrane of a human red blood cell. Why do you think that *Paramecium* has evolved a membrane with such low water permeability? What molecular differences do you think might account for this low water permeability? On the basis of your answers to these questions, develop a hypothesis

about how habitat affects the evolution of plasma membrane structure. On the basis of your hypothesis, what testable predictions can you make about how the plasma membranes of different species will differ?

3. A preview question for Chapter 22: The integrity of the plasma membrane is essential for cellular survival. Could the immune system make use of this fact to destroy foreign cells that have invaded the body? How might cells of the immune system disrupt membranes of foreign cells?

4. A preview question for Chapter 17: Plant roots take up minerals (inorganic ions such as potassium) that are dissolved in the water of the soil. The concentration of such ions is usually much lower in the soil water than in the cytoplasm of root cells. Design the plasma membrane of a hypothetical mineral-absorbing cell, with special reference to mineral-permeable channel proteins and mineral-transporting active-transport proteins. Justify your choice of channel proteins and active-transport proteins.

For additional resources, go to www.mybiology.com.

chapter 4

Cell Structure and Function

Scientists studying animal limb regeneration hope that their findings may one day allow humans to regrow lost limbs.

Case Study   Can Lost Limbs Grow Back?

When an antitank mine exploded beneath his Humvee, Captain David Rozelle lost his right leg. As did many U.S. soldiers wounded in Iraq, Rozelle survived injuries that might have killed him in an earlier war, before recent advances in the treatment of traumatic injuries. These advances have increased the number of people who, like Rozelle, are living without limbs. The growing number of amputees has increased scientific interest in a long-standing question: Can humans regrow lost body parts?

If a person loses a finger, or a hand, or a whole limb, thick scar tissue forms at the site of the wound, and the missing structure does not regrow. This inability to regrow stands in stark contrast to the regenerative powers of many other animals. A starfish that loses an arm quickly regrows the lost part, as does a crab that loses a claw, a lizard that loses its tail, or a salamander that loses a leg.

The difference between an animal that regenerates lost parts and one that does not is, in large measure, a difference in how cells at the site of a wound behave. In humans and other mammals, the cells in the scar tissue at a wound site do not reproduce or reproduce only in a limited way that does not regrow the lost body part. In species that can regenerate lost parts, the cells at wound sites behave quite differently—they reproduce steadily and replace the missing tissues. Biologists hypothesize that the cells of regenerating animals gain this ability from exposure to proteins or other molecules, produced by the animal itself, that prevent scar formation and cause cells at a wound site to reproduce.

Researchers are working hard to identify the molecules that allow regeneration in salamanders, starfish, and other species. Recent progress in this endeavor delights proponents of the emerging field of *regenerative medicine*. This field seeks to develop medical treatments to stimulate regeneration in humans. Regeneration-inducing substances discovered by scientists studying salamanders and other animals might someday be applied to human wounds to hinder scar tissue formation and promote regeneration. What changes might such substances make to the way cells behave? ∎

If a salamander such as this axolotl loses a limb, it grows back.

4.1 What Features Are Shared by All Cells?

Cells are the smallest units that retain the properties of life. All living things, from the tiniest bacterium to the largest whale, are composed of cells. The smallest organisms consist of only a single cell, but larger organisms can contain trillions of cells, each specialized to perform a specific function. Because cells can be specialized for a variety of tasks, and because there are many different species of single-celled organisms, there are many different types of cells. Despite their diversity, cells share certain features, which are described below.

Cells Are Enclosed by a Plasma Membrane

All cells are surrounded by a **plasma membrane.** The plasma membrane consists of a phospholipid bilayer in which proteins are embedded, and it controls the movement of substances into and out of the cell. A description of plasma membrane structure and function can be found in Chapter 3.

▶ Figure 4-1 **Relative sizes**
Dimensions commonly encountered in biology range from about 100 meters (the height of the tallest trees) to a few micrometers (the diameter of the smallest cells) to a few nanometers (the diameter of many large molecules, such as proteins).

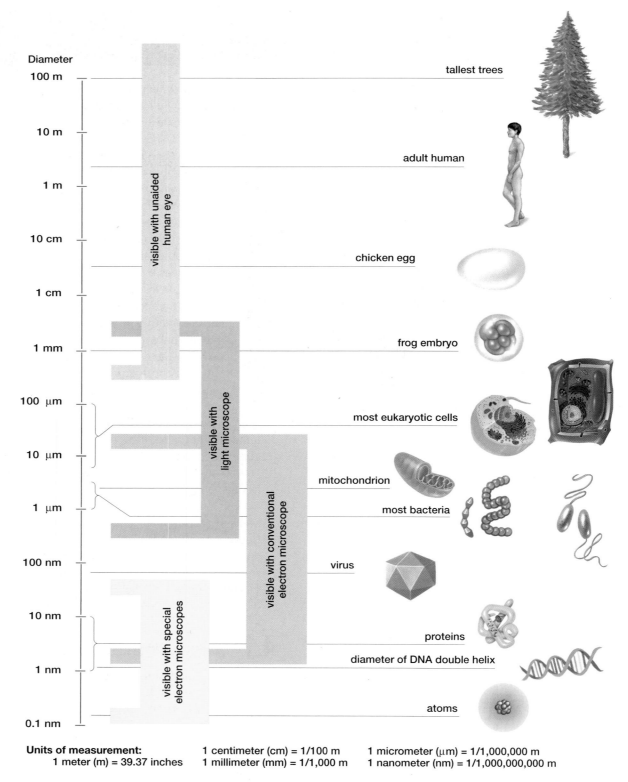

Diameter

100 m — tallest trees

10 m — adult human

1 m —

10 cm — chicken egg

1 cm —

1 mm — frog embryo

100 μm — most eukaryotic cells

10 μm —

1 μm — mitochondrion
most bacteria

100 nm — virus

10 nm — proteins

1 nm — diameter of DNA double helix

0.1 nm — atoms

visible with unaided human eye
visible with light microscope
visible with conventional electron microscope
visible with special electron microscopes

Units of measurement:
1 meter (m) = 39.37 inches

1 centimeter (cm) = 1/100 m
1 millimeter (mm) = 1/1,000 m

1 micrometer (μm) = 1/1,000,000 m
1 nanometer (nm) = 1/1,000,000,000 m

Cells Use DNA as a Hereditary Blueprint

All cells contain **deoxyribonucleic acid (DNA),** genetic material that stores the instructions for making all the parts of a cell and for producing new cells. Each cell inherits its genetic material from the cells that give rise to it, so all cells in a single body generally contain identical DNA molecules.

Cells Contain Cytoplasm

The **cytoplasm** consists of all the material inside the plasma membrane and outside the DNA-containing region. The fluid portion of the cytoplasm contains water, salts, and an assortment of organic molecules. It forms a thick soup of proteins, lipids, carbohydrates, salts, sugars, amino acids, and nucleotides (see Chapter 2). Most of the cell's *metabolic activities*—that is, the sum of all the biochemical reactions that underlie life—occur in the cell's cytoplasm. For example, the manufacture of proteins takes place in the cytoplasm on special structures called *ribosomes.*

Cells Obtain Energy and Nutrients from Their Environment

To maintain themselves, all cells must continuously acquire and expend energy. The ultimate source of this energy is sunlight. Cells in some organisms, such as plants, can harness the sun's energy directly and store it in high-energy molecules. Once solar energy has been captured in this way, some of the energy-containing molecules become available to cells that are not capable of capturing solar energy on their own. For example, an animal may consume plant tissue, thereby making the energy captured by the plant cells available to its own cells.

In addition to energy, cells need nutrients. Carbon, nitrogen, oxygen, minerals, and other building blocks of biological molecules are found in the environment—the air, water, rocks, and living things. As with energy, cells must obtain these materials from their environment.

Table 4-1 summarizes some features shared by all cells.

Cell Function Limits Cell Size

Most cells are small, ranging from about 1 to 100 micrometers (millionths of a meter) in diameter (**Fig. 4-1**). Why are cells so small? The answer lies in the method by which cells move nutrients and wastes across the plasma membrane. As you learned in Chapter 3, many nutrients and wastes move into and out of cells by *diffusion,* the movement of molecules from places of higher concentration to places of lower concentration. Diffusion, however, is a slow process; oxygen molecules take more than 200 days to diffuse over a distance of 4 inches in cytoplasm. At such speeds, diffusion is effective only over short distances. If cells were any larger than they are, vital materials could not diffuse from the plasma membrane to the innermost portion of the cell (or vice versa) quickly enough to be useful.

Further, as a cell enlarges, its volume increases more rapidly than its surface area does. For example, a cell that doubles in radius becomes eight times greater in volume but only four times greater in surface area (**Fig. 4-2**). So, as a cell grows, its volume of cytoplasm (and thus the amount of nutrients and wastes that must be exchanged with the environment) grows much faster than the area of plasma membrane through which the exchanges are made. If cells get too big, metabolic needs will overwhelm the available membrane surface area.

Table 4-1	Features Common to All Cells
Molecular components	Proteins, amino acids, lipids, carbohydrates, sugars, nucleotides, DNA, RNA
Structural components	Plasma membrane, cytoplasm, ribosomes
Metabolism	Extracts energy and nutrients from the environment; uses energy and nutrients to build, repair, and replace cellular parts

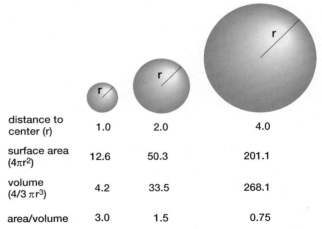

distance to center (r)	1.0	2.0	4.0
surface area (4πr²)	12.6	50.3	201.1
volume (4/3 πr³)	4.2	33.5	268.1
area/volume	3.0	1.5	0.75

▲ **Figure 4-2 As radius increases, volume grows faster than surface area**

scientific inquiry

The Search for the Cell

Human understanding of life's cellular nature came slowly. In 1665, English scientist and inventor Robert Hooke reported observations that he made through a primitive microscope. He aimed his instrument at a thin piece of cork (which comes from the bark of an oak tree) and saw "a great many little Boxes" (**Fig. E4-1a**). Hooke called the boxes "cells" because he thought they resembled the tiny rooms, or cells, occupied by monks. He wrote that in the living oak and other plants, "These cells [are] fill'd with juices."

By the 1670s, Dutch microscopist Anton van Leeuwenhoek was observing a previously unknown world through simple microscopes of his own construction (**Fig. E4-1b**). His descriptions of myriad "animalcules" (his term for single-celled organisms) in rain, pond water, and well water caused quite an uproar, because water was consumed without treatment in those days. Leeuwenhoek made careful observations of an enormous range of microscopic specimens, including red blood cells, sperm, and the eggs of small insects.

More than a century passed before biologists began to understand more about the role of cells in life on Earth. Microscopists first noted that plants consist of cells. The thick wall surrounding all plant cells, first observed by Hooke, made their observations easier. Animal

▶ **Figure E4-1 Microscopes yesterday and today** *(a)* Robert Hooke's drawings of the cells of cork, as he viewed them with an early light microscope similar to the one shown here. Only the cell walls remain. *(b)* One of Leeuwenhoek's microscopes and a photograph of blood cells taken through a Leeuwenhoek microscope. The specimen is viewed through a tiny hole just underneath the lens. *(c)* A modern electron microscope. This instrument is capable of performing both scanning and transmission electron microscopy.

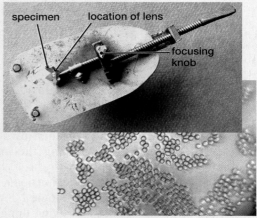

specimen   location of lens

focusing knob

blood cells photographed through Leeuwenhoek's microscope

(b) Leeuwenhoek's microscope

(a) 17th century microscope and cork cells

(c) Electron microscope

4.2 How Do Prokaryotic and Eukaryotic Cells Differ?

Despite the features that are shared by all cells, cells can be grouped into two fundamentally different types. The first type, the **prokaryotic cell,** is found only in two groups of single-celled organisms, the bacteria and the archaea. Biolo-

cells, however, escaped notice until the 1830s, when German zoologist Theodor Schwann saw that cartilage contains cells that "exactly resemble [the cells of] plants." In 1839, after studying cells for years, Schwann was confident enough to publish his *cell theory*, calling cells the elementary particles of both plants and animals. By the mid-1800s, German botanist Matthias Schleiden had further refined science's view of cells when he wrote: "It is . . . easy to perceive that the vital process of the individual cells must form the first, absolutely indispensable fundamental basis" of life.

Ever since the pioneering efforts of Hooke and Leeuwenhoek, biologists, physicists, and engineers have collaborated to improve the capabilities of microscopes. Today's microscopes fall into two basic categories: *light microscopes* and *electron microscopes* (**Fig. E4-1c**).

Light microscopes use lenses, usually made of glass, to focus and magnify light rays that either pass through or bounce off a specimen. Light microscopes provide a wide range of images, depending on how the specimen is illuminated and whether it has been stained (**Fig. E4-2a**). The resolving power of light microscopes—that is, the smallest structure that can be seen—is about 1 micrometer (1 μm, a millionth of a meter).

Electron microscopes use beams of electrons instead of light. The electrons are focused by magnetic fields rather than by lenses. Some types of electron microscopes can resolve structures as small as a few nanometers (billionths of a meter). *Transmission electron microscopes* (TEMs) pass electrons through a thin specimen and can reveal the details of interior cell structure, including organelles and plasma membranes (**Fig. E4-2b**). *Scanning electron microscopes* (SEMs; see Fig. E4-1c) bounce electrons off specimens that have been coated with metals and provide three-dimensional images. SEMs can be used to view the surface details of structures that range in size from entire insects down to cells and even organelles (**Figs. E4-2c,d**). Photographs made through electron microscopes are often called **electron micrographs**.

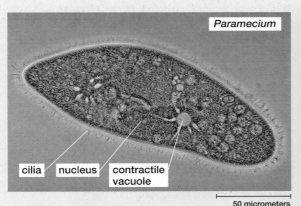

(a) Light microscope

50 micrometers

Paramecium

cilia | nucleus | contractile vacuole

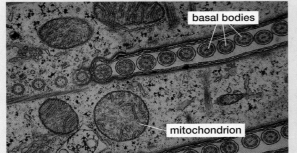

(b) Transmission electron microscope

basal bodies

mitochondrion

0.5 micrometers

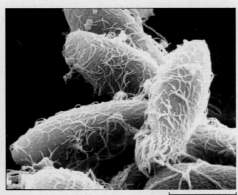

(c) Scanning electron microscope

10 micrometers

(d) Scanning electron microscope

mitochondria

5 micrometers

◄ **Figure E4-2 A comparison of microscope images** *(a)* A living *Paramecium* (a single-celled freshwater organism) photographed through a light microscope. *(b)* A transmission electron microscope (TEM) photo showing the basal bodies at the bases of the cilia that cover *Paramecium*. Mitochondria are also visible. *(c)* A false-color scanning electron microscope (SEM) photo of *Paramecium*. *(d)* An SEM photo at much higher magnification, showing mitochondria (many of which are sliced open) within the cytoplasm.

gists often refer to bacteria and archaea as *prokaryotes*. The second type of cell, which makes up the bodies of all other living things, is called the **eukaryotic cell,** and organisms with these cells are known as *eukaryotes*. One striking difference between the two cell types is that the DNA of eukaryotic cells is contained within a membrane-enclosed structure inside the cell that is called the **nucleus.** In contrast, the genetic material of prokaryotic cells is not enclosed within a membrane.

The earliest cells on Earth were probably prokaryotic. Eukaryotic cells arose much later, most likely as the result of a partnership among different kinds of prokaryotic cells. This scenario for the origin of eukaryotic cells is described in more detail in Chapter 15.

4.3 What Are the Main Features of Eukaryotic Cells?

The nucleus is perhaps the most distinctive feature of eukaryotic cells, but eukaryotic cells also differ from prokaryotic cells in many other ways. For one thing, they are usually larger than prokaryotic cells—typically more than 10 micrometers in diameter (see Fig. 4-1). Also, the cytoplasm of eukaryotic cells houses membrane-enclosed structures called **organelles** that perform specific functions within the cell. A network of protein fibers, the **cytoskeleton,** gives shape and organization to the cytoplasm of eukaryotic cells. Many of the organelles are attached to the cytoskeleton.

Different types of eukaryotic cells may contain different kinds of organelles. For example, animal cells (**Fig. 4-3**) have a few organelles that are not found in plant cells (**Fig. 4-4**), and vice versa. Even within a single organism, the organelle content of cells can vary considerably. For example, in plants, leaf cells contain many chloroplasts (an organelle needed for photosynthesis); root cells contain no chloroplasts but may instead contain organelles that store starch. In general, multicelled eukaryotes contain many different types of

BioFlix   Tour of an Animal Cell

BioFlix   Tour of a Plant Cell

Web Animation   Cell Structure

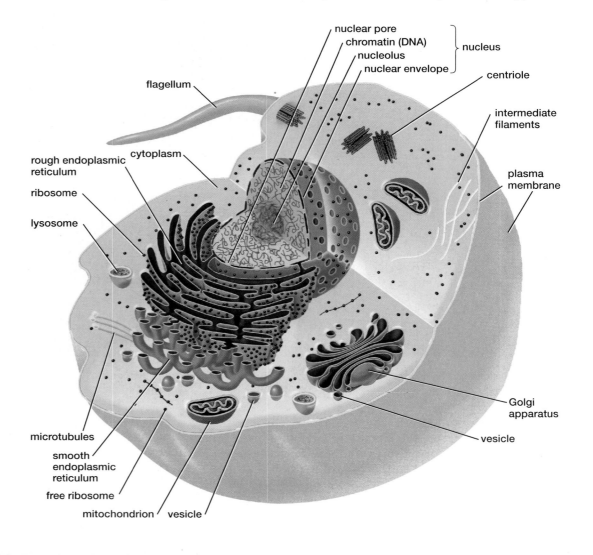

▶ **Figure 4-3 A generalized animal cell**

cells, each with a distinctive structure suited to its function. In vertebrate animals, for example, the cells that make up muscles differ from those that make up skin; each tissue includes distinctive cell types that are not present in other tissues. (For information on how cells of different types are used to build replacement parts for bodies, see "Health Watch: Spare Parts.")

As you read the following sections, which describe the structures of the cell in more detail, you may wish to refer to the illustrations in Figures 4-3 and 4-4. Bear in mind that these illustrations represent generic, "typical" animal and plant cells. As discussed earlier, the appearance of real cells varies greatly, depending on a cell's function and the kind of organism in which it occurs.

Table 4-2 provides an overview of the main structures present in eukaryotic cells.

4.4 What Role Does the Nucleus Play?

The nucleus is usually the largest organelle in a cell (**Fig. 4-5**). It is bounded by a *nuclear envelope* that separates the interior of the nucleus from the cytoplasm. Inside the nuclear envelope, the nucleus contains a granular-looking material called chromatin. The nucleus also contains a region called the *nucleolus*, which, in cells that have been stained in preparation for viewing under a microscope, appears as a darker-colored area.

Can Lost Limbs Grow Back?

Continued

When an animal cell divides, the usual result is two cells just like the one that divided. Thus, liver cells generally give rise to more liver cells, bone cells to more bone cells, skin cells to skin cells, and so on. This kind of cell proliferation works well for maintaining bodies, but is not suitable for regenerating lost limbs, a process that requires growth of many tissue types from whatever cells are present at the wound site.

Some cells have a special ability that would be helpful for limb regeneration: They can give rise to many different cell types. These cells are called *stem cells*. When a stem cell divides, each resulting new cell may either remain a stem cell or become a cell of some other type. Animal embryos, which begin with only a few cells but ultimately give rise to all of the body's many different cell types, are rich in stem cells. But in adult humans and other adult mammals, stem cells are rare.

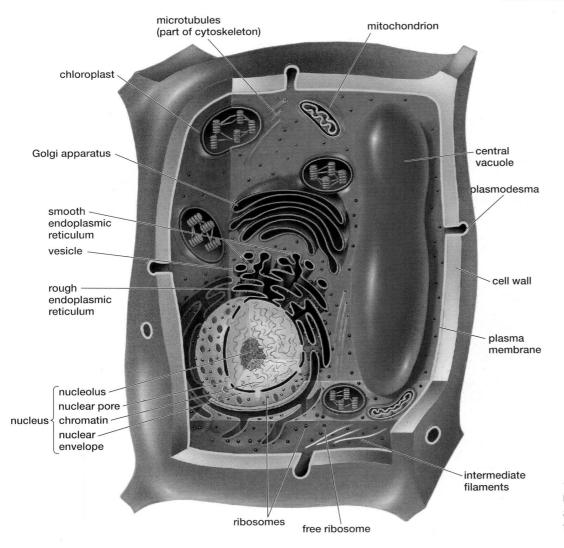

◀ Figure 4-4 **A generalized plant cell**
QUESTION: Of the nucleus, ribosome, chloroplast, and mitochondrion, which appeared earliest in the history of life?

Table 4-2	Structures Found in Eukaryotic Cells	
Structure	**Description**	**Function**
Cell surface		
Cell wall	Stiff outermost layer of plant and fungal cells	Protects and supports the cell
Plasma membrane	Phospholipid bilayer that surrounds and contains cell contents	Isolates cell contents from the environment; regulates movement of materials into and out of the cell; communicates with other cells
Organization of genetic material		
Chromosomes	Long threads composed of DNA and proteins; located inside the nucleus	Contain and control use of DNA; DNA encodes the information needed to construct the cell and control cellular activity
Nucleus	Spherical, membrane-bound structure; usually a cell's largest organelle	Container for chromosomes
Nuclear envelope	Double membrane that encloses the nucleus	Regulates movement of materials into and out of the nucleus
Nucleolus	Roughly spherical structure located inside the nucleus; consists of DNA, RNA, and proteins	Synthesizes components of ribosomes
Cytoplasmic structures		
Mitochondria	Membrane-bound sacs in the cytoplasm	Produce energy by aerobic metabolism
Chloroplasts	Membrane-bound sacs in the cytoplasm; contain the pigment chlorophyll; not in animal cells	Perform photosynthesis
Ribosomes	Complex assemblies of protein and RNA in the cytoplasm; often bound to the endoplasmic reticulum	Site of protein synthesis
Endoplasmic reticulum	Network of membranes inside the cell	Synthesizes membrane components and lipids
Golgi apparatus	Set of membranes shaped like stacks of flattened sacs	Modifies and packages proteins and lipids; synthesizes carbohydrates
Lysosomes	Vesicles containing digestive enzymes	Digest excess membrane, worn-out organelles, and biological molecules
Central vacuole	Fluid-filled sac in the cytoplasm; mainly in plant cells	Contains water and wastes
Cytoskeleton	Network of protein filaments	Gives shape and support to the cell; positions and moves cell parts
Centrioles	Barrel-shaped structures made of microtubules	Synthesize and organize microtubules
Cilia and flagella	Thin, microtubule-filled extensions of the cell membrane	Move cell through fluid or move fluid past the cell surface

▶ **Figure 4-5 The nucleus** *(a)* The nucleus is bounded by a nuclear envelope. Inside are chromatin (DNA and associated proteins) and a nucleolus. *(b)* An electron micrograph of a yeast cell that was frozen and broken open to reveal its internal structures. The large nucleus, with nuclear pores penetrating its nuclear envelope, is clearly visible.

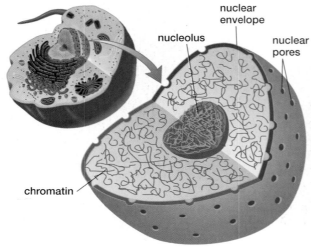

(a) Structure of the nucleus

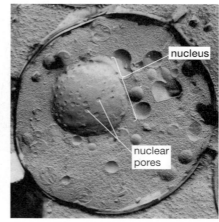

(b) Yeast cell

health watch

Spare Parts

Despite its great potential, regenerative medicine has yet to yield treatments that have actually been used on people. An alternative treatment approach that is currently available is *tissue engineering.* Tissue engineers build replacements for damaged body tissues by applying cells to "scaffolds" made of special biodegradable plastics (**Fig. E4-3**). The cells divide and spread over the scaffold to form an artificial tissue that can be surgically implanted.

Tissue engineers have successfully built artificial skin and cartilage (the stiff tissue that supports your nose, ears, and knees). Artificial skin has been especially useful for treating burn victims. Artificial cartilage has been used to rebuild damaged knees and as a bone substitute in replacements for amputated fingers.

Another recent application of tissue engineering is in artificial corneas (the outer layer of the eye). These are built by growing cornea cells on the surface of a very thin plastic membrane. Engineered corneas have been implanted in several patients with severe eye damage, and the results have been promising, with most of the recipients experiencing improved vision. The days of long, often futile, waits for cornea donors may be coming to an end.

The scientists who now shape cells into living cartilage, skin, and corneas might someday be

▲ **Figure E4-3 An artificial ear grafted onto the back of a mouse** In an early demonstration of the potential of tissue engineering, scientists painted cartilage cells onto an ear-shaped scaffold. The resulting artificial ear was grafted onto the back of a mouse so that the cartilage cells could be nourished by the mouse's circulatory system.

able to sculpt cells into working organs, such as livers, kidneys, and lungs. The path to engineered organs recently reached a milestone with the transplantation of laboratory-grown bladders to seven patients with bladder disease (**Fig. E4-4**). These transplants represented the first successful human transplants of an engineered organ. A bladder, however, is a relatively simple organ, and it will be more difficult to produce complex organs

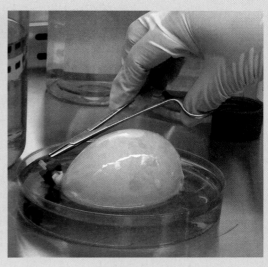

▲ **Figure E4-4 A human bladder, grown in a laboratory and ready for transplantation**

that are thicker, include multiple tissue types, and contain intricate networks of blood vessels. Nonetheless, researchers have already managed to grow bone, heart valves, heart muscle, teeth, tendons, intestines, blood vessels, and breast tissue on plastic scaffolds and to implant some of these tissues into animals. With each passing day, our ability to control the behavior of cells gets a little better.

The Nuclear Envelope Controls the Passage of Materials into and out of the Nucleus

The nucleus is isolated from the rest of the cell by a **nuclear envelope** that consists of a double membrane. The membrane is perforated with tiny membrane-lined channels called *nuclear pores.* Water, ions, and small molecules can pass freely through the nuclear pores, but the passage of large molecules, such as proteins, pieces of ribosomes, and RNA, is regulated by special "gatekeeper proteins" that line each nuclear pore. These gatekeepers permit the passage of certain molecules and prevent the passage of others. DNA does not cross the nuclear membrane—it remains within the nucleus for the life of the cell.

The Nucleus Contains Chromosomes

The DNA molecules contained inside the nucleus are closely associated with certain kinds of protein molecules, and this DNA-protein complex is known as **chromatin.** A eukaryotic cell's chromatin is arranged in a set of long, thread-shaped structures called **chromosomes.** When cells divide, each chromosome coils upon itself, becoming thicker and shorter. The resulting "condensed" chromosomes are easily visible under even relatively low-power microscopes (**Fig. 4-6**).

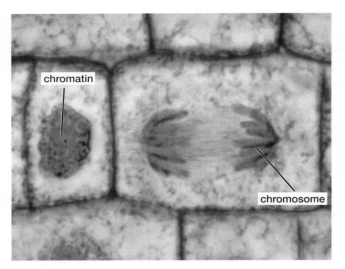

▲ Figure 4-6 **Chromosomes** The chromosomes in this light micrograph of a dividing cell in an onion root tip are made of the same material (DNA and proteins) as the chromatin seen in the nondividing cell adjacent to it but are in a more compact state. **QUESTION:** Why doesn't chromatin remain in its condensed form in nondividing cells?

Ribosome Components Are Made at the Nucleolus

Most eukaryotic nuclei have at least one darkly staining region called a **nucleolus** (see Fig. 4-5a). The nucleolus consists of DNA, RNA, proteins, and ribosomes in various stages of construction.

Nucleoli are the sites where the components of **ribosomes** are manufactured. Finished components leave the nucleus and move to the cytoplasm, where they come together to form complete ribosomes. A ribosome is composed of RNA and proteins and serves as a kind of "workbench" for the manufacture of proteins. Just as a workbench can be used to construct many different objects, a ribosome can be used to synthesize any of the thousands of proteins made by a cell. In electron micrographs, ribosomes appear as dark granules, either distributed in the cytoplasm or clustered along membranes (see Figs. 4-3 and 4-4). The functions of ribosomes are described in greater detail in Chapter 11.

4.5 What Roles Do Membranes Play in Eukaryotic Cells?

Eukaryotic cells have an elaborate system of membranes that not only enclose the cell but also create internal compartments within the cytoplasm. This membrane system is composed of the plasma membrane and several organelles, including the endoplasmic reticulum, nuclear envelope, and Golgi apparatus, and membrane-enclosed sacs such as lysosomes (see Figs. 4-3 and 4-4).

The Plasma Membrane Isolates the Cell and Helps It Interact with Its Environment

As described in Chapter 3, the plasma membrane forms the outer boundary of a cell, enclosing the cytoplasm. It is a complex structure that performs the seemingly contradictory functions of separating the cytoplasm of the cell from the outside environment and providing for the transport of selected substances into or out of the cell. In addition to a plasma membrane, the cells of plants, fungi, and some single-celled organisms have a stiff *cell wall* that forms an outer protective coating.

The Endoplasmic Reticulum Manufactures and Processes Proteins and Lipids

The endoplasmic reticulum (ER) is a series of interconnected, membrane-enclosed channels in the cytoplasm (**Fig. 4-7**). A cell's ER is continuous with its nuclear envelope. Eukaryotic cells have two forms of ER: rough and smooth. Numerous ribosomes stud the outside of the **rough endoplasmic reticulum;** in contrast, **smooth endoplasmic reticulum** lacks ribosomes.

The rough ER and smooth ER have different functions. Rough ER, with its embedded ribosomes, provides sites where proteins are manufactured. As the proteins are made, many of them move through the ER membrane into the ER interior. The proteins then move through channels in the ER and accumulate in pockets. The pockets bud off, forming membrane-bound sacs called **vesicles.** The vesicles move about the cell, carrying their protein cargo to its destination.

Smooth ER provides sites for the manufacture of lipids and other molecules that, like the proteins made in the rough ER, can be packaged for shipment elsewhere in the cell. The particular substances produced by smooth ER vary among different types of cells. In cells of the reproductive organs of vertebrates, for example, smooth ER manufactures sex hormones. Smooth ER can also perform other functions. For example, enzymes bound to the smooth ER in liver cells help detoxify waste products, drugs, and alcohol. In muscle cells, smooth ER stores the large amounts of calcium that are required for muscle contraction.

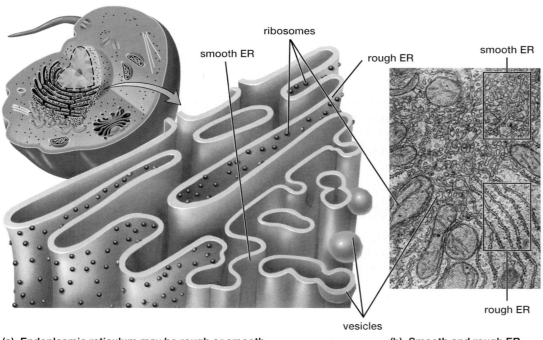

ribosomes

smooth ER

rough ER

smooth ER

rough ER

vesicles

(a) Endoplasmic reticulum may be rough or smooth

(b) Smooth and rough ER

◀ **Figure 4-7 Endoplasmic reticulum** *(a)* The endoplasmic reticulum is connected to the nuclear envelope. There are two types of endoplasmic reticulum: rough ER, coated with ribosomes, and smooth ER, without ribosomes. *(b)* Although in electron micrographs the ER looks like a series of tubes and sacs, it is actually a maze of folded sheets and interlocking channels.

Among the molecules made by both the smooth and rough ER are phospholipids and cholesterol, which are key components of membranes. Rough ER can also produce membrane proteins. Thus, ER produces all of the components of membranes and is the cell's site for membrane construction. Most of this new membrane is used for new or replacement ER, but some of it moves toward the nucleus to replace nuclear membrane or into the cytoplasm to maintain other membranes of the cell.

Web Animation Membrane Traffic

The Golgi Apparatus Sorts, Chemically Alters, and Packages Important Molecules

The **Golgi apparatus** is a specialized set of membranes derived from the endoplasmic reticulum (**Fig. 4-8**). It looks very much like a stack of flattened sacs and functions as a transfer station for the substances that are transported around the cell.

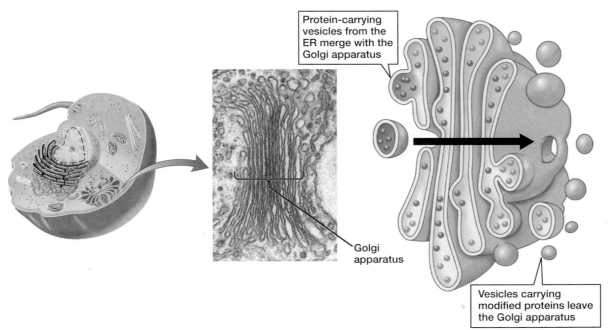

Protein-carrying vesicles from the ER merge with the Golgi apparatus

Golgi apparatus

Vesicles carrying modified proteins leave the Golgi apparatus

◀ **Figure 4-8 The Golgi apparatus** The Golgi apparatus is a stack of flat membranous sacs. Vesicles transport material from the endoplasmic reticulum (ER) to the Golgi apparatus (and vice versa) and from the Golgi apparatus to the plasma membrane, lysosomes, and vesicles. Arriving vesicles join the Golgi apparatus on one face, and departing vesicles bud off from the opposite face.

Vesicles arrive from the ER and fuse with one face of the Golgi apparatus, adding their membrane to the Golgi apparatus and emptying their contents into the Golgi sacs. At the same time, other vesicles bud off the Golgi apparatus on the opposite face of the stack, carrying away proteins, lipids, and other complex molecules. These vesicles then move to other parts of the cell or to the plasma membrane for export.

As molecules pass through the Golgi apparatus, they are sorted and sometimes undergo further processing. Proteins and lipids received from the ER are sorted according to their destinations. For example, proteins that will remain inside the cell are separated from those that the cell will secrete. The Golgi apparatus modifies some molecules. For instance, it adds sugars to proteins to make glycoproteins, which serve many functions in cells. For example, many of the recognition proteins on cell membranes (see Chapter 3) are glycoproteins.

We Can Follow the Travels of a Secreted Protein

To understand how the membranous organelles work together, let's look at the secretion of an antibody within your body. An antibody is a protein, secreted by a type of white blood cell, that binds to foreign invaders (such as bacteria) and helps destroy them. The antibody protein is manufactured on ribosomes of the rough ER and then packaged into vesicles formed from the ER membrane. These vesicles travel to the Golgi apparatus, where the membranes fuse, releasing the protein into the Golgi apparatus. Here, carbohydrates are attached to the protein, which is then repackaged into vesicles formed from Golgi membrane. The vesicle containing the completed antibody travels to the plasma membrane, where the antibody is released outside the cell by exocytosis. Once outside the cell, the antibody will make its way into the bloodstream to help defend your body against infection.

Lysosomes Serve as a Cell's Digestive System

Some of the proteins manufactured in the ER and sent to the Golgi apparatus are digestive enzymes that can break down proteins, fats, and carbohydrates into their component subunits. In the Golgi apparatus, these enzymes are packaged in vesicles called **lysosomes.** One function of lysosomes is to digest excess cellular membranes or defective organelles. After identifying these organelles, the cell encloses them in vesicles made of membrane from the ER. These vesicles fuse with lysosomes, and digestive enzymes within the lysosome enable the cell to recycle valuable materials from the defunct organelles.

4.6 Which Other Structures Play Key Roles in Eukaryotic Cells?

Eukaryotic cells include a number of other important structures. Among them are vacuoles, mitochondria, chloroplasts, and the elements of the cytoskeleton. Some eukaryotic cells also have structures called cilia and flagella that help with locomotion.

Vacuoles Regulate Water and Store Substances

Most cells contain one or more vacuoles, which are fluid-filled sacs surrounded by a single membrane. For example, many plant cells contain a large **central vacuole** (see Fig. 4-4). Filled mostly with water, the central vacuole helps keep the right amount of water in the cell and helps support the cell. This support function may be compromised if a plant cannot obtain enough water to maintain the proper amount in its cells' central vacuoles; this is why a houseplant that has not been watered wilts.

The central vacuole also provides a dump site for hazardous wastes, which plant cells often cannot excrete. Some plant cells store extremely poisonous substances, such as sulfuric acid, in their vacuoles. These substances deter animals from feeding on the otherwise nutritious leaves. Vacuoles may also store sugars and amino acids not immediately needed by the cell.

Can Lost Limbs Grow Back?

Continued

Animals that can regenerate limbs probably have stem cells at wound sites, whereas animals, such as humans, that cannot regenerate limbs do not. Biologists hypothesize that, in response to wounds, cells of limb-regenerating species secrete proteins (using a process much like the one described here for antibodies) that cause other nearby cells to become stem cells.

Mitochondria Extract Energy from Food Molecules

The energy that cells need to survive, grow, and reproduce is harvested in the **mitochondrion** (plural, mitochondria). The mitochondrion is sometimes called the powerhouse of the cell because it is the organelle in which energy is extracted from sugar molecules and stored in the high-energy bonds of ATP. Once the energy is stored in ATP molecules, it is available to power the many energy-consuming reactions that occur in every cell. Mitochondria are round, oval, or tubular sacs made of a pair of membranes. The outer mitochondrial membrane is smooth, but the inner membrane loops back and forth to form deep folds (see Figs. 4-3 and 4-4). The structure of the mitochondrion is described in greater detail in Chapter 7, as is the mitochondrion's role in energy production.

Chloroplasts Capture Solar Energy

Plant cells contain **chloroplasts,** organelles that capture energy directly from sunlight and store it in sugar molecules. Chloroplasts are the site of photosynthesis, the energy-capturing process on which all life ultimately depends. Each chloroplast is surrounded by a double membrane and contains stacks of hollow membranous sacs (see Fig. 4-4). The pigment chlorophyll gives chloroplasts a green color (plants are green because they contain chloroplasts). The structure of chloroplasts and the reactions of photosynthesis are described in more detail in Chapter 6.

The Cytoskeleton Provides Shape, Support, and Movement

Organelles do not drift haphazardly about the cytoplasm. Most are attached to a network of protein fibers called the *cytoskeleton* (**Fig. 4-9**). Several types of protein fibers, including thin *microfilaments*, medium-sized *intermediate filaments*, and thicker, hollow, cylindrical tubes called **microtubules,** make up the cytoskeleton.

The cytoskeleton performs the following important functions:

- **Cell shape.** In cells without cell walls, the cytoskeleton determines the shape of the cell.

- **Cell movement.** Many types of cells move about. Much of this cell movement is generated by the movement of microfilaments and microtubules inside the cell.

- **Organelle movement.** Microtubules and microfilaments move organelles from place to place within a cell. For example, the vesicles that bud off of the ER and Golgi apparatus are guided to their destinations by the cytoskeleton.

- **Cell division.** Microtubules and microfilaments are essential to cell division in eukaryotic cells. When eukaryotic nuclei divide, microtubules move the chromosomes into the daughter nuclei. In animal cells, microtubules also form the building blocks of *centrioles*, barrel-shaped structures that play a role in dividing up the genetic material during cell division. In addition, the final stage of animal cell division involves microfilaments, which form a ring that divides the cytoplasm by pinching the parent cell around its middle.

Cilia and Flagella Move the Cell or Move Fluid Past the Cell

Both **cilia** and **flagella** (singular, cilium and flagellum) are slender, movable extensions of the plasma membrane. They typically contain microtubules that extend along their length. Some single-celled eukaryotic organisms use cilia or flagella to move about. Most animal sperm rely on flagella for movement. Some small animals use cilia for movement, swimming by the coordinated beating of rows of cilia. More commonly, however, cilia are used to move fluids and suspended particles past a surface. Cells with cilia line such diverse structures as the gills of oysters (moving food- and oxygen-rich water), the oviducts of female

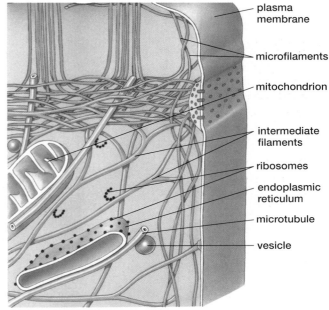

plasma membrane

microfilaments

mitochondrion

intermediate filaments

ribosomes

endoplasmic reticulum

microtubule

vesicle

(a) Components of the cytoskeleton

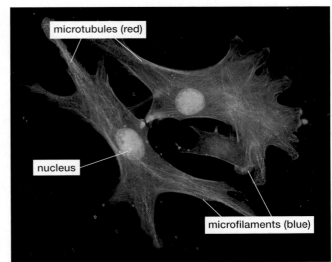

microtubules (red)

nucleus

microfilaments (blue)

(b) Cell with stained cytoskeleton

▲ **Figure 4-9 The cytoskeleton (a)** The cytoskeleton gives shape and organization to eukaryotic cells. It consists of three types of proteins: microtubules, intermediate filaments, and microfilaments. **(b)** These cells from the lining of a cow artery have been treated with fluorescent stains to reveal microtubules and microfilaments, as well as the nucleus.

▶ **Figure 4-10 How cilia and flagella move** *(a)* (Left) Cilia usually "row," providing a force of movement parallel to the plasma membrane. Their movement resembles the arms of a swimmer doing the breast stroke. (Right) Scanning electron micrograph of cilia lining the trachea (which conducts air to the lungs); these cilia sweep out mucus and trapped particles. *(b)* (Left) Flagella move in a wavelike motion, providing continuous propulsion perpendicular to the plasma membrane. In this way, a flagellum attached to a sperm can move the sperm straight ahead. (Right) A human sperm on the surface of a human egg cell.

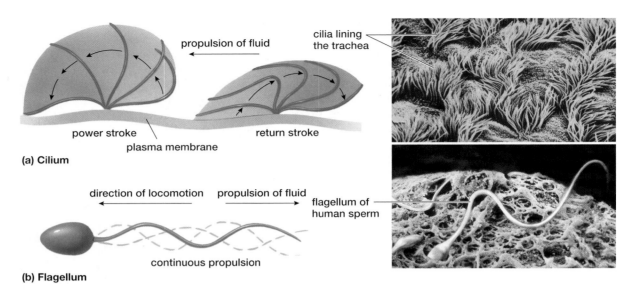

propulsion of fluid

cilia lining the trachea

power stroke

return stroke

plasma membrane

(a) Cilium

direction of locomotion   propulsion of fluid

flagellum of human sperm

continuous propulsion

(b) Flagellum

mammals (moving the eggs along from the ovary to the uterus), and the respiratory tracts of most land vertebrates (clearing mucus that carries debris and microorganisms from the windpipe and lungs).

The main differences between cilia and flagella lie in their length, number, and the direction of the force they generate. In general, cilia are shorter (about 10 to 25 micrometers long) and more numerous than flagella. They provide force in a direction parallel to the plasma membrane, like the oars in a rowboat. This force is accomplished through a "rowing" motion (**Fig. 4-10a**). Flagella are longer (50 to 75 micrometers), usually are fewer in number, and provide force perpendicular to the plasma membrane, like the engine on a motorboat (**Fig. 4-10b**).

4.7 What Are the Features of Prokaryotic Cells?

Although Earth's larger, more conspicuous organisms are made up of eukaryotic cells, prokaryotic cells are extremely abundant. The number of prokaryotes in a single handful of soil is greater than the total number of humans that ever lived. Invisible to us, prokaryotic bacteria and archaea inhabit virtually every nook and cranny of the planet, including on and inside our own bodies. You may be tempted to view the comparatively simple structure of prokaryotic cells as an indication that prokaryotes are primitive and inferior to eukaryotes. That simple structure, however, has formed the basis of enormous biological success.

Prokaryotic cells are usually described as lacking organelles, but recent research has shown that some prokaryotes contain membrane-bound, organelle-like structures that help maintain the correct level of acidity in the cell. Nonetheless, prokaryotic cells lack nuclei, mitochondria, choloroplasts, Golgi apparatus, ER, and the other complex organelles that characterize eukaryotic cells. Prokaryote cytoplasm does, however, contain ribosomes, the workbenches on which proteins are manufactured in both prokaryotes and eukaryotes. The cytoplasm also may contain food granules that store energy-rich materials, such as glycogen.

Most prokaryotic cells are very small (less than 5 micrometers in diameter) and have a relatively simple internal structure (compare **Fig. 4-11** with Figs. 4-3 and 4-4). In general, they are surrounded by a stiff cell wall, which provides shape and protects the cell. Some prokaryotes can move, propelled by whiplike protrusions that undulate or spin.

The cytoplasm of most prokaryotic cells is relatively uniform in appearance. Each prokaryotic cell has a single, circular strand of DNA. It is usually coiled, attached to the plasma membrane, and concentrated in a region of the cell called

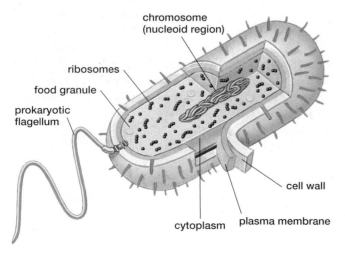

chromosome (nucleoid region)

ribosomes

food granule

prokaryotic flagellum

cell wall

cytoplasm   plasma membrane

▲ **Figure 4-11 A generalized prokaryotic cell**

the **nucleoid.** The nucleoid is not, however, separated from the rest of the cytoplasm by a membrane.

Prokaryotic cells are compared with the eukaryotic cells of plants and animals in Table 4-3.

Table 4-3	Comparison of Structures in Prokaryote, Plant, and Animal Cells		
Structure	**Prokaryotes**	**Plants**	**Animals**
Cell surface			
Cell wall	Present	Present	Absent
Plasma membrane	Present	Present	Present
Organization of genetic material			
Genetic material	DNA	DNA	DNA
Chromosomes	Single, circular, no proteins	Many, linear, with proteins	Many, linear, with proteins
Nucleus	Absent	Present	Present
Nuclear envelope	Absent	Present	Present
Nucleolus	Absent	Present	Present
Cytoplasmic structures			
Mitochondria	Absent	Present	Present
Chloroplasts	Absent	Present	Absent
Ribosomes	Present	Present	Present
Endoplasmic reticulum	Absent	Present	Present
Golgi apparatus	Absent	Present	Present
Lysosomes	Absent	Present	Present
Central vacuole	Absent	Present	Absent
Cytoskeleton	Absent	Present	Present
Centrioles	Absent	Absent (in most)	Present
Cilia and flagella	Present[a]	Absent (in most)	Present

[a] Many prokaryotes have structures called flagella, but these are not made of microtubules and move in a fundamentally different way than eukaryotic cilia or flagella do.

Can Lost Limbs Grow Back? Revisited

The ability of some animals to regenerate lost parts depends on their cells' ability to turn back the clock. Regeneration of lost parts requires stem cells. As a body grows and develops, however, most of its stem cells lose their ability to give rise to any cell type. Thus, if a human loses a leg, the stem cells needed to replace it don't exist, and new ones are not created. Instead, the body's resources are devoted to healing the wound. If a salamander loses a leg, however, something very different happens. Cells near the wound undergo a transformation that turns them into stem cells and a new leg begins to grow.

There is increasing evidence that, in salamanders and other limb-regenerating species, the transformation of normal cells into stem cells is initiated by proteins secreted by other body cells. There is also some evidence that these proteins might have the same effect on the cells of species that do not normally regenerate lost parts. For example, in one recent experiment, researchers made an extract of tissue from a regenerating salamander limb and applied it to cultures of mouse muscle cells. Mice cannot regenerate lost parts, and their mature muscle cells normally form ropy, nondividing muscle fibers. Under the influence of the salamander extract, however, some mature mouse muscle cells reverted to an earlier state, dividing over and over again as if still part of a young, growing mouse. The researchers concluded that a protein in the salamander's regenerating tissue converted the mouse cells into a growing state.

In recent years, researchers have identified some of the proteins involved in controlling regeneration in salamanders, flatworms, zebrafish, and other regenerators. Will these proteins prove effective at promoting regeneration of damaged human body parts? No one knows for sure, but this approach is promising enough that the U.S. Defense Department has committed millions of dollars to research limb regeneration, in hopes of one day restoring lost limbs to amputees.

Consider This

In addition to the limb regeneration effort, a far larger program of research is aimed at medical use of stem cells grown outside the body. Scientists hypothesize that such stem cells could promote regrowth of damaged tissues. Compared with regenerative medicine, what might be the advantages and disadvantages of the "add new stem cells" approach?

Chapter Review

Summary of Key Concepts

For additional study help and activities, go to www.mybiology.com.

4.1 What Features Are Shared by All Cells?

The smallest units of life are single cells. Cells are the functional units of multicellular organisms, and every living organism is made up of one or more cells. Cells are limited in size because they must exchange materials with their surroundings by diffusion. Diffusion is relatively slow, so the interior of the cell must never be too far from the plasma membrane, and the plasma membrane must have a large surface area (relative to the volume of its cytoplasm) for diffusion. Both of these constraints limit the size of cells.

4.2 How Do Prokaryotic and Eukaryotic Cells Differ?

All cells are either prokaryotic or eukaryotic. Prokaryotic cells are small and relatively simple in structure, lack a cell nucleus, and are found only in bacteria and archaea. More complex eukaryotic cells make up all other forms of life.

4.3 What Are the Main Features of Eukaryotic Cells?

Eukaryotic cells are generally much larger than prokaryotic cells. Eukaryotic cells also have organelles (including a nucleus) and a cytoskeleton. Structures present in eukaryotic cells are listed in Table 4-2.

Web Animation   Cell Structure

4.4 What Role Does the Nucleus Play?

The nucleus is bounded by the double membrane of the nuclear envelope and contains the cell's genetic material (DNA). Pores in the nuclear envelope regulate the movement of molecules between the nucleus and the cytoplasm. The genetic material of eukaryotic cells is organized into chromosomes, which consist of DNA and proteins. Ribosomes are particles that are the sites of protein synthesis. The nucleolus is the site at which the components of ribosomes are manufactured.

4.5 What Roles Do Membranes Play in Eukaryotic Cells?

The membrane system of a cell consists of the plasma membrane, endoplasmic reticulum (ER), Golgi apparatus, and vesicles derived from these membranes. Rough ER manufactures many cellular proteins. Smooth ER manufactures phospholipids. The ER is the site of all membrane synthesis within the cell. The Golgi apparatus is a series of membranous sacs derived from the ER. The Golgi apparatus processes and modifies materials synthesized in the rough and smooth ER. Some substances in the Golgi apparatus are packaged into vesicles for transport elsewhere in the cell. Lysosomes are vesicles that contain enzymes that digest food particles and defective organelles.

Web Animation   Membrane Traffic

4.6 Which Other Structures Play Key Roles in Eukaryotic Cells?

All eukaryotic cells contain mitochondria, organelles that use oxygen to complete the breakdown of food molecules, capturing much of their energy as ATP. Plant cells contain chloroplasts, which capture the energy of sunlight during photosynthesis, enabling the cells to manufacture organic molecules, particularly sugars, from simple inorganic molecules. Many eukaryotic cells contain vacuoles that are bounded by a single membrane and that store food or wastes.

The cytoskeleton organizes and gives shape to eukaryotic cells and moves and anchors organelles. The cytoskeleton is composed of microfilaments, intermediate filaments, and microtubules. Cilia and flagella are whiplike extensions of the plasma membrane that contain microtubules. These structures move fluids past the cell or move the cell through its fluid environment.

4.7 What Are the Features of Prokaryotic Cells?

Prokaryotic cells are generally very small and have a relatively simple internal structure. Most are surrounded by a relatively stiff cell wall. The cytoplasm of prokaryotic cells lacks membrane-enclosed organelles and contains a single, circular strand of DNA.

Study Note: Figures 4-3 and 4-4 illustrate the overall structure of animal and plant cells, respectively. Table 4-2 lists the principal organelles and their functions. Table 4-3 compares structures in prokaryotic cells, plant cells, and animal cells.

Key Terms

cell *p. 55*
central vacuole *p. 66*
chloroplast *p. 67*
chromatin *p. 63*
chromosome *p. 63*
cilia *p. 67*
cytoplasm *p. 57*
cytoskeleton *p. 60*

deoxyribonucleic acid (DNA) *p. 57*
electron micrograph *p. 59*
eukaryotic cell *p. 59*
flagella *p. 67*
Golgi apparatus *p. 65*
lysosome *p. 66*
microtubule *p. 67*

mitochondrion *p. 67*
nuclear envelope *p. 63*
nucleoid *p. 69*
nucleolus *p. 64*
nucleus *p. 59*
organelle *p. 60*
plasma membrane *p. 55*

prokaryotic cell *p. 58*
ribosome *p. 64*
rough endoplasmic reticulum *p. 64*
smooth endoplasmic reticulum *p. 64*
vesicle *p. 64*

Thinking Through the Concepts

Suggested answers to end-of-chapter and figure-based questions can be found at the end of the text.

Fill-in-the-Blank

1. Fill in the structure that each statement describes: Provides external support for each plant cell: _____; membrane-enclosed storage sac important in plant cells: _____; energy-capturing organelle found in plant but not animal cells: _____.

2. The nucleus is surrounded by a double membrane called the _____, which is perforated by membrane-lined _____ that allow passage of molecules such as _____, _____, and _____.

3. List four components of the nucleolus: _____, _____, _____, _____.

4. A secreted protein, such as an antibody, is synthesized on a "workbench" called a _____ that is attached to the rough _____. The protein is then packaged into sacs called _____, which travel to the _____, where they are modified.

5. The cell's "digestive system" consists of vesicles called _____ produced by the _____. These vesicles contain _____ and their function is to digest and recycle the molecules from defective cell parts such as _____ and _____.

6. List four major functions of the cytoskeleton: _____, _____, _____, _____.

7. Cilia and flagella both contain cytoskeletal elements called _____. List three ways in which cilia differ from flagella: _____, _____, _____.

Review Questions

1. Review the section entitled "Can Lost Limbs Grow Back?" at the beginning of this chapter, focusing especially on the investigation described in the third paragraph. Then, using Figure E1-1 on p. 6, as a model, describe the observation, question, hypothesis, prediction, experiment, and conclusion that make up the investigation.

2. Diagram "typical" prokaryotic and eukaryotic cells, and describe their important similarities and differences.

3. Which organelles are common to both plant and animal cells, and which are unique to each?

4. Describe the nucleus, including the nuclear envelope, chromatin, chromosomes, DNA, and the nucleolus.

5. What are the functions of mitochondria and chloroplasts?

6. What is the function of ribosomes? Where in the cell are they typically found?

7. Describe the structure and function of the endoplasmic reticulum and Golgi apparatus.

Applying the Concepts

1. IS THIS SCIENCE? Researchers studying a meteorite made of rock from Mars found small, tubular structures that are very similar to fossil bacteria found on Earth. The Martian rock also contained certain organic molecules that on Earth are found in decomposing cells. Some scientists claimed that these findings are evidence of past life on Mars. Do you agree that science has demonstrated the past existence of living cells on Mars? If so, explain why the evidence is scientifically persuasive. If not, describe the kind of scientific evidence that would persuade you.

2. If muscle biopsies (samples of tissue) were taken from the legs of a world-class marathon runner and a typical couch potato, which would you expect to have a higher density of mitochondria? Why? How would the density of mitochondria in a muscle biopsy from the biceps of a weight lifter compare with those of the runner and the couch potato?

3. One of the functions of the cytoskeleton in animal cells is to give shape to the cell. Plant cells have a fairly rigid cell wall surrounding the plasma membrane. Does the rigid cell wall make a cytoskeleton unnecessary for a plant cell? Defend your answer in terms of other functions of the cytoskeleton.

4. Most cells are very small. What physical and metabolic constraints limit cell size? What problems would an enormous cell encounter? What adaptations might help a very large cell survive?

For additional resources, go to www.mybiology.com.

chapter 5

Energy Flow in the Life of a Cell

In a healthy person's body, enzymes help convert energy stored in fats and carbohydrates to the energy of movement.

Case Study A Missing Molecule Makes Mischief

Imagine that you could never eat pizza, hamburgers, ice cream, or bagels with cream cheese. For one out of every 10,000 children born in the United States, eating such foods can lead directly to neurological disorders and severe mental retardation. These children, who have a syndrome known as phenylketonuria (PKU), cannot fully digest proteins. In particular, their bodies cannot break down the amino acid phenylalanine. As a result, phenylalanine builds up in their blood. High levels of phenylalanine are toxic and have devastating effects on developing nervous systems.

When enzymes fail to work properly, the simple act of eating a hamburger can have dangerous consequences.

PKU was not discovered until 1934, when a woman brought her two severely retarded children to Asbjørn Følling, a Norwegian physician. The distraught mother sought an explanation for her children's condition. As part of his examination, Følling performed routine urine tests, including one in which he added the chemical ferric chloride to the urine. Ferric chloride turns purple when it reacts with ketones (molecules whose presence in urine can indicate diabetes), but in samples from the two children, it turned bright green. No substance was known to cause this change, and a puzzled Dr. Følling began a long series of chemical analyses to identify the mysterious substance in the urine. Eventually, he determined that it was phenylpyruvic acid, an indicator of high levels of phenylalanine. Følling hypothesized that the children's retardation was caused by excess phenylalanine and predicted that other individuals with mental disabilities would also have phenylpyruvic acid in their urine. To test his prediction, he obtained urine samples from 430 mentally retarded people and found that eight of the samples did indeed turn bright green when ferric chloride was added.

Why can't PKU victims process phenylalanine? It turns out that people with PKU have a defect in a single type of molecule. What kind of molecule could have such a dramatic effect, turning ordinary foods into poisons? ■

5.1 What Is Energy?

Energy is the capacity to do work. In this definition, "work" refers to any of a wide range of actions, including manufacturing molecules, moving objects, and generating heat and light. Energy can be either *kinetic* or *potential*, depending on whether it is in use or stored. **Kinetic energy**, or energy of movement, includes light (movement of photons), heat (movement of molecules), electricity (movement of electrically charged particles), and movement of large objects. **Potential energy,** or stored energy, includes chemical energy stored in the bonds that hold atoms together in molecules, electrical energy stored in a battery, and positional energy stored in a diver poised to spring.

Under the right conditions, kinetic energy can be transformed into potential energy, and vice versa. For example, a penguin climbing up onto an ice floe converts kinetic energy of movement into potential energy. When the penguin dives off, the potential energy is converted back into kinetic energy (**Fig. 5-1**).

Energy Cannot Be Created or Destroyed

The laws of thermodynamics describe the basic properties and behavior of energy. The **first law of thermodynamics,** often called the law of conservation of energy, states that energy can neither be created nor destroyed, so the total amount of energy in the universe remains constant.

▲ **Figure 5-1 From potential energy to kinetic energy** A penguin on an ice floe has potential energy, because the heights of the ice and the ocean are different. As it dives, the potential energy is converted to the kinetic energy of the motion of its body. Finally, some of this kinetic energy is transferred to the water, which is set in motion.

Although energy cannot be created or destroyed, it can change its form. For example, a car converts the chemical potential energy of gasoline into the kinetic energy of movement and heat. Similarly, a runner converts the chemical potential energy of food into the kinetic energy of movement and heat. In a coal-burning power plant, the chemical potential energy of coal is converted to the kinetic energy of heat, which is then converted to the kinetic energy of electricity. In your home, the kinetic energy of electricity is converted to the kinetic energy of light and heat, and perhaps to the potential energy stored in rechargeable batteries. In all of these cases, energy changes form, but the total amount of energy remains unchanged.

Energy Tends to Become Less Useful

Although energy is never destroyed, energy that is in the forms most useful for accomplishing work tends to be converted to forms that are less useful for accomplishing work. This tendency is described by the **second law of thermodynamics,** which states that whenever energy is converted from one form to another, the amount of energy that is in useful forms decreases. The second law tells us that all conversions of energy bring about a loss of usefulness—that is, no process is 100% efficient.

To illustrate the second law, let's examine a car engine burning gasoline. In a moving vehicle, the kinetic energy of movement is much smaller than the amount of chemical energy in the gasoline that was burned to cause the movement. The first law tells us that energy cannot be destroyed, so where is the "missing" energy? It has been transferred to the car's materials and surroundings. The burning gas not only moved the car but also heated up the engine, the exhaust system, and the air around the car. In addition, the friction of tires on the pavement slightly heated the road. No energy is missing. However, the energy that was released as heat has been converted to a less useful form. It merely heated the engine, the exhaust system, the air, and the road.

Matter Tends to Become Less Organized

Regions in which energy is concentrated tend to be more organized than regions in which energy is widely dispersed. For example, the eight carbon atoms in a single molecule of gasoline have a much more orderly arrangement than do the eight carbon atoms in the eight separate, randomly moving molecules of carbon dioxide that are formed when the gasoline molecule burns. Therefore, we can also phrase the second law in terms of the organization of matter: Processes that proceed spontaneously lead to increasing randomness and disorder. This randomness and disorder is called **entropy.** Entropy always tends to increase, a tendency that can be overcome only by adding energy.

We all experience the tendency toward entropy in our homes. Frequent inputs of energy are required to keep debris confined to the trash can, newspapers to folded stacks, books to their shelves, and clothes to drawers and closets. Without our energetic cleaning and organizing efforts, these items tend to end up in their lowest-energy state—a state of disorder. When the ecologist G. Evelyn Hutchinson said, "Disorder spreads through the universe, and life alone battles against it," he was making an eloquent reference to entropy and the second law of thermodynamics.

Living Things Use the Energy of Sunlight to Create Low-Entropy Conditions

If you think about the second law of thermodynamics, you may wonder how life can exist at all. All chemical reactions, including those inside living cells, cause the amount of usable energy to decrease. Matter tends toward increasing randomness and disorder. Given these facts, how can organisms accumulate the concentrated energy and precisely ordered molecules that characterize living things?

The answer is that living things use a continuous input of solar energy to construct complex molecules and maintain orderly structures—to "battle against disorder." This solar energy arrives on Earth in the form of sunlight, which is produced by nu-

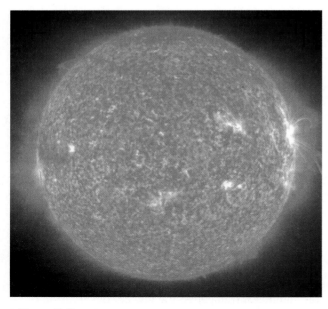

▲ **Figure 5-2 The sun** Energy-releasing reactions in the sun provide almost all of the energy used by life on Earth.

clear reactions in the sun (**Fig. 5-2**). These reactions produce huge increases in entropy, so the highly organized, low-entropy systems of life do not violate the second law. The increased order of living things is achieved at the expense of an enormous loss of order in the sun. The entropy of the solar system as a whole constantly increases.

5.2 How Does Energy Flow in Chemical Reactions?

A **chemical reaction** converts one set of chemical substances, the **reactants,** into another set, the **products.** Some chemical reactions release energy, and others consume it. A reaction that releases energy is classified as an **exergonic reaction,** and its products contain less energy than the reactants (**Fig. 5-3a**). Conversely, an **endergonic reaction** requires a net input of energy from some outside source. The products of an endergonic reaction contain more energy than the reactants (**Fig. 5-3b**).

Let's look at two processes that illustrate these types of reactions: burning sugar and photosynthesis.

Exergonic Reactions Release Energy

When sugar is burned by a flame, it reacts with oxygen (O_2) to produce carbon dioxide (CO_2), and water (H_2O). This reaction is exergonic; the reactants contain more energy than the products. That is, the molecules of sugar and oxygen contain much more energy than the molecules of carbon dioxide and water. The extra energy is released as heat (**Fig. 5-4**).

Once started, exergonic reactions proceed without an input of energy. For example, a spoonful of sugar, once ignited, continues to burn spontaneously until all sugar molecules are consumed. It's as if exergonic reactions run "downhill," from high energy to low energy, just as a rock pushed from the top of a hill rolls to the bottom.

Endergonic Reactions Require an Input of Energy

Unlike the energy-releasing reaction that takes place when sugar burns, many reactions in living things create products that contain *more* energy than the reactants. For example, the sugar produced by photosynthesis contains far more energy than the carbon dioxide and water from which it is formed (**Fig. 5-5**). Similarly, a protein molecule contains more energy than the individual amino acids that were joined together to build it.

The reactions that form complex biological molecules require an overall input of energy. The energy for photosynthesis, for example, comes from sunlight. We can view endergonic reactions such as photosynthesis as "uphill" reactions, going from low energy to high energy, like pushing a rock to the top of a hill.

All Reactions Require an Initial Input of Energy

All chemical reactions, even exergonic ones that release energy overall, require an initial input of energy to get started. For example, even though burning sugar releases energy, a spoonful of sugar can't burst into flames by itself. The fire doesn't begin until some energy is added. This initial energy input to a chemical reaction is called the **activation energy** and is something like the push you might give a rock poised at the top of a hill to start it rolling down (**Fig. 5-6**).

Chemical reactions require activation energy to get started because a shell of negatively charged electrons surrounds atoms and molecules. For two molecules to react with each other, their electron shells must be forced together, despite their mutual electrical repulsion. That force requires energy.

The usual source of activation energy is the kinetic energy of movement. Molecules moving with sufficient speed collide hard enough to force their electron shells to mingle and react. Because molecules move faster as the temperature increases, most chemical reactions occur more readily at high temperatures than at low temperatures.

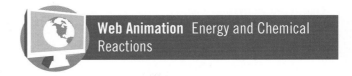

Web Animation Energy and Chemical Reactions

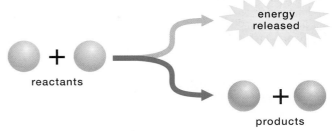

(a) Exergonic reaction

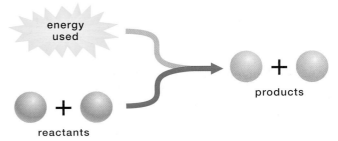

(b) Endergonic reaction

▲ **Figure 5-3 Exergonic and endergonic reactions** *(a)* An exergonic reaction releases energy, but *(b)* an endergonic reaction consumes energy.

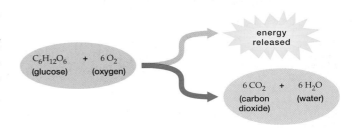

▲ **Figure 5-4 Burning glucose releases energy**

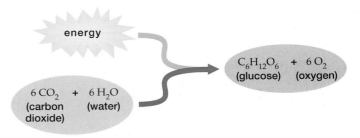

▲ **Figure 5-5 Photosynthesis requires energy**

▶ **Figure 5-6 Activation energy**
(a) An exergonic ("downhill") reaction, such as the burning of glucose, proceeds from high-energy reactants (here, glucose) to low-energy products (CO_2 and H_2O). Starting the reaction, however, requires an initial input of energy—the activation energy. *(b)* An endergonic ("uphill") reaction, such as photosynthesis, proceeds from low-energy reactants (CO_2 and H_2O) to high-energy products (glucose) and therefore requires a large activation energy, in this case provided by sunlight. **QUESTION:** In addition to heat and sunlight, what are some other sources of activation energy?

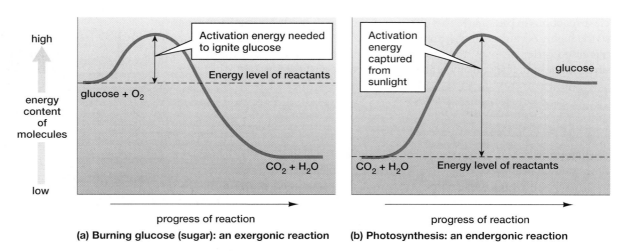

(a) Burning glucose (sugar): an exergonic reaction   (b) Photosynthesis: an endergonic reaction

Exergonic Reactions May Be Linked with Endergonic Reactions

Endergonic reactions require energy from other sources, and they obtain this energy from exergonic, energy-releasing reactions. In a **coupled reaction,** an exergonic reaction provides the energy needed to drive an endergonic reaction. For example, when you drive a car, the exergonic reaction of burning gasoline provides the energy for the endergonic reaction of starting a stationary car into motion and keeping it moving. Photosynthesis is another coupled reaction. In photosynthesis, an exergonic reaction in the sun releases light energy that drives endergonic sugar-making reactions in plants. Because some energy is always lost as heat when energy is transferred between reactions (remember the second law of thermodynamics), the energy provided by an exergonic reaction must be greater than that needed to drive the endergonic reaction.

5.3 How Is Energy Carried Between Coupled Reactions?

Coupled reactions are common in organisms. The energy released by exergonic reactions in cells is used to drive endergonic, energy-consuming reactions. The exergonic and endergonic parts of coupled reactions often occur in different parts of the cell, so there must be some way to transfer energy from the exergonic reaction that releases energy to the endergonic reaction that consumes it. The job of transferring energy from place to place is done by **energy-carrier molecules.**

Energy carriers work something like rechargeable batteries, picking up an energy charge at an exergonic reaction, moving to another location within the cell, and releasing the energy to drive an endergonic reaction. Energy-carrier molecules are unstable, so they are used only for energy transfer within cells. They are not used to transfer energy from cell to cell, nor are they used for long-term energy storage.

Web Animation Energy and Life

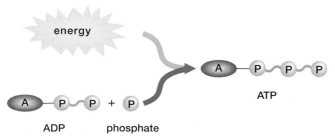

▲ **Figure 5-7 ATP synthesis stores energy in ATP**

ATP Is the Principal Energy Carrier in Cells

The most common energy-carrier molecule in cells is the nucleotide **adenosine triphosphate,** or **ATP.** ATP provides energy for a wide variety of endergonic reactions, serving as a common currency of energy transfer. ATP is made in a reaction that adds a phosphate molecule to an **adenosine diphosphate (ADP)** molecule (**Fig. 5-7**). The energy required for the reaction comes from exergonic reactions such as the breakdown of glucose.

ATP stores energy in its chemical bonds and can carry the energy to sites in the cell that perform energy-requiring reactions. Then, the bond to one of ATP's phosphate molecules breaks, yielding ADP and phosphate (which are later recycled to make more ATP) and energy (**Fig. 5-8**). The energy released from the broken bond is transferred to the energy-requiring reaction.

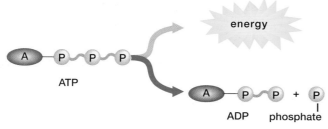

▲ **Figure 5-8 Breakdown of ATP releases energy**

In summary, exergonic reactions (such as glucose breakdown) drive the endergonic reaction that converts ADP to ATP. The ATP molecule moves to a different part of the cell, where the ATP breaks down and liberates some of its energy to drive an endergonic reaction (Fig. 5-9). For example, the endergonic reaction of muscle cell contraction is powered by energy transferred from the exergonic breakdown of ATP (Fig. 5-10). During the energy transfers in coupled reactions, heat is given off and there is an overall loss of usable energy.

The life span of an ATP molecule in a living cell is very short, because this energy carrier is continuously formed, broken down, and remade. If the ATP molecules that you use just sitting at your desk all day could be captured (instead of recycled), they would weigh nearly 90 pounds. A person running a marathon may use a pound of ATP every minute. (ADP must be quickly converted back to ATP, or it would be a very brief run.) As you can see, ATP is not a long-term energy-storage molecule. Long-term storage is the job of more stable molecules, such as glycogen, fat, starch, or sucrose, which can store energy for hours, days, or months.

Electron Carriers Also Transport Energy Within Cells

In addition to ATP, other carrier molecules also transport energy within a cell. **Electron carriers** capture the energetic electrons to which energy is transferred in some exergonic reactions. The loaded electron carriers then donate the electrons, along with their energy, to molecules participating in endergonic reactions (Fig. 5-11). Common electron carriers include nicotinamide adenine dinucleotide (NAD$^+$) and its relative, flavin adenine dinucleotide (FAD). You will learn more about electron carriers and their role in cells in Chapters 6 and 7.

5.4 How Do Cells Control Their Metabolic Reactions?

A cell is a miniature chemical factory, and the multitude of different chemical reactions that take place in a cell together constitute the cell's **metabolism.** Many of these reactions are linked in sequences called **metabolic pathways** (Fig. 5-12). Photosynthesis (Chapter 6) is one such pathway. Glycolysis, the series of reactions that begins the digestion of glucose (Chapter 7), is another.

▲ Figure 5-9 **Coupled reactions** Energy from exergonic reactions such as the breakdown of glucose powers the endergonic synthesis of ATP. In turn, the exergonic breakdown of ATP powers endergonic metabolic reactions such as the synthesis of proteins from amino acids. **QUESTION:** Why does breakdown of ATP release energy for cellular work?

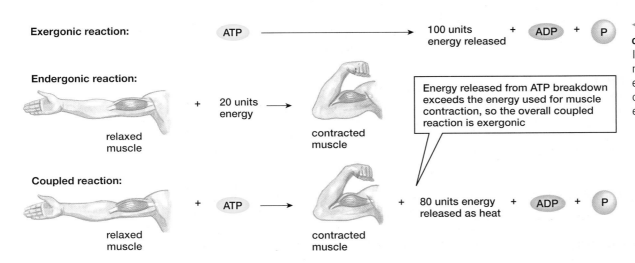

Exergonic reaction:

ATP → 100 units energy released + ADP + P

Endergonic reaction:

relaxed muscle + 20 units energy → contracted muscle

Coupled reaction:

relaxed muscle + ATP → contracted muscle + 80 units energy released as heat + ADP + P

Energy released from ATP breakdown exceeds the energy used for muscle contraction, so the overall coupled reaction is exergonic

◄ Figure 5-10 **ATP breakdown is coupled with muscle contraction** In the coupled system as a whole, more energy is produced by the exergonic reaction than is used to drive the endergonic one. The extra energy is released as heat.

▶ Figure 5-11 **Electron carriers** An electron-carrier molecule such as NAD+ picks up an electron generated by an exergonic reaction and holds it in a high-energy outer electron shell. The electron is then deposited, energy and all, with another molecule to drive an endergonic reaction, typically the synthesis of ATP.

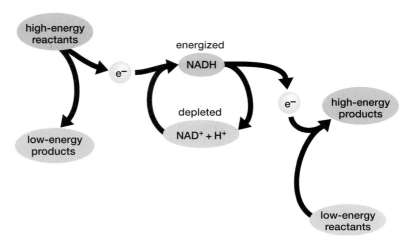

Orderly metabolic pathways are essential to proper cell function, but the environment inside a cell presents some possible obstacles to effective initiation and completion of the pathways. In particular, a cell contains a mixture of many different molecules, and this diversity of potentially reactive molecules could cause a chaotic free-for-all of uncontrolled reactions. Therefore, a cell must control reactions so that they proceed in an orderly manner.

At Body Temperatures, Many Spontaneous Reactions Proceed Too Slowly to Sustain Life

One way to control a reaction is to control the speed at which it occurs by influencing its activation energy (the amount of energy required to start the reaction; see Fig. 5-6). In general, the speed at which a reaction occurs is determined by its activation energy. At a given temperature, reactions with low activation energies proceed more rapidly than do reactions with higher activation energies.

The activation energies of many metabolic reactions are high enough that, at the temperatures found in the bodies of organisms, the reactions proceed extremely slowly. The reaction rate is so slow that the reactions effectively do not occur spontaneously. For example, sugar molecules in organisms almost never spontaneously break down and give up their energy. Nonetheless, the breakdown of sugar is one of the most important energy sources for living cells. If sugar molecules do not break down on their own, how is their energy released? Cells are able to gain access to the energy stored in sugar molecules through the action of catalysts.

Catalysts Reduce Activation Energy

Catalysts are molecules that speed up a reaction without themselves being used up or permanently altered. They speed up reactions by reducing the activation energy (**Fig. 5-13**). In the absence of a catalyst, a reaction with high activation

A Missing Molecule Makes Mischief

Continued

The molecule that is missing from the cells of PKU victims is an enzyme, phenylalanine hydroxylase. In normal cells, this enzyme catalyzes a key reaction in the metabolic pathway of the amino acid phenylalanine, namely its conversion to another amino acid, tyrosine. When the hydroxylase enzyme is absent, the reactant phenylalanine is not converted to the product tyrosine. The reactant instead accumulates in the body, reaching toxic levels that can cause a range of ill effects, including mental retardation. (For an example of a missing enzyme whose effects are less devastating, see "Health Watch: Is Milk Just for Babies?")

▶ Figure 5-12 **Simplified view of two metabolic pathways** In pathway 1, a reactant molecule (A) undergoes a series of reactions to yield a final product (E). Each reaction along the pathway from A to E yields an intermediate product (B, C, or D) that becomes a reactant in the next reaction of the pathway. Intermediate products of one pathway may become reactants in another pathway. For example, intermediate C of pathway 1 can also serve as the initial reactant in pathway 2, which yields the final product G.

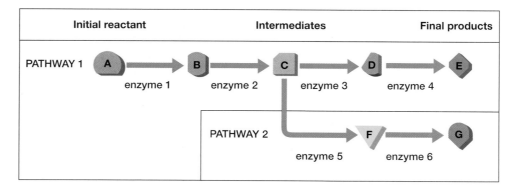

health watch

Is Milk Just for Babies?

Infants thrive on an all-milk diet, but most adults cannot digest milk. About 75% of people worldwide, including 25% of those in the United States, lose the ability to digest lactose, or "milk sugar," in early childhood, and they become *lactose intolerant*. Roughly 75% of African Americans, Hispanics, and Native Americans, as well as 90% of Asian Americans, are lactose intolerant. Only a relatively small proportion of people, primarily those of northern European descent, retain the ability to digest lactose into adulthood (**Fig. E5-1**).

Lactose intolerance arises when the body stops producing the enzyme lactase, which catalyses the breakdown of lactose. When someone who lacks this enzyme consumes milk products, undigested lactose draws water into the intestines by osmosis and also feeds intestinal bacteria that produce gas. The excess water and gas lead to abdominal pain, bloating, diarrhea, and flatulence—a rather high price to pay for indulging in ice cream. But compared with the consequences of other enzyme deficiencies, such as PKU, the inability to tolerate milk is a relatively minor inconvenience.

Most people who are lactose intolerant do not need to avoid milk products altogether. Some such people produce enough lactase to tolerate a few servings daily, and most can eat aged cheeses (such as cheddar) and yogurt with live bacteria, which contains relatively little lactose because the bacteria break it down. Those unwilling to forgo ice cream can consume lactase supplements along with dairy products.

▲ **Figure E5-1 Risky behavior** For most adults, drinking a glass of milk invites unpleasant consequences.

---

energy will proceed only slowly, because few molecules collide hard enough to react. But when the activation energy is lowered by a catalyst, a much higher proportion of molecules react when they collide. The reaction proceeds much more rapidly.

Note three important characteristics of all catalysts:

- Catalysts speed up reactions.

- Catalysts can speed up only those reactions that would occur spontaneously anyway, if their activation energy could be surmounted.

- Catalysts are not consumed in the reactions they promote. No matter how many reactions they participate in, the catalysts themselves are not permanently changed.

Enzymes Are Biological Catalysts

The catalysts that are made by living organisms are called **enzymes.** Almost all enzymes are proteins. Enzymes share with all other kinds of catalysts the three characteristics that we just described, but they also have additional attributes that set them apart from nonbiological catalysts. One of the most important of these is that each enzyme is usually very specialized, catalyzing a single reaction or at most a few different reactions. In metabolic pathways such as those in Figure 5-12, each reaction is catalyzed by a different enzyme.

The Structure of Enzymes Allows Them to Catalyze Specific Reactions

The function of enzymes is closely related to their structure. Each enzyme has a complex three-dimensional shape that includes a cleft or notch, called the **active site,** into which reactant molecules can enter. Reactants on which an

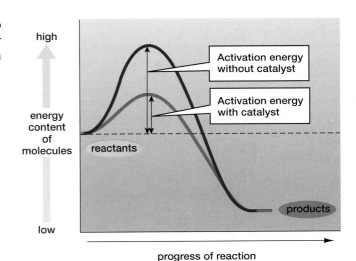

▲ **Figure 5-13 Catalysts reduce activation energy**
QUESTION: Can a catalyst make a nonspontaneous reaction occur spontaneously?

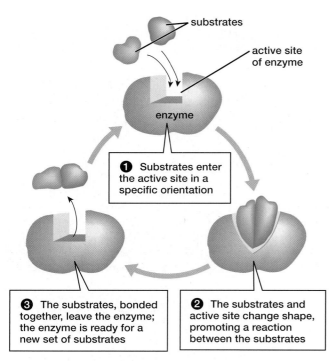

▲ Figure 5-14 The cycle of enzyme–substrate interactions
QUESTION: How would you change reaction conditions if you wanted to increase the rate at which an enzyme-catalyzed reaction produced its product?

① Substrates enter the active site in a specific orientation

③ The substrates, bonded together, leave the enzyme; the enzyme is ready for a new set of substrates

② The substrates and active site change shape, promoting a reaction between the substrates

substrates

active site of enzyme

enzyme

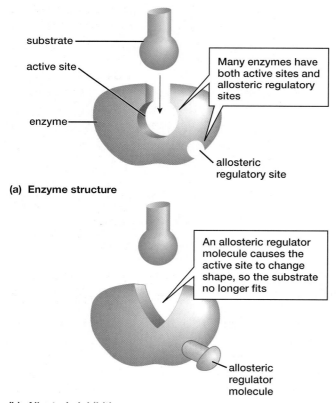

substrate

active site

enzyme

Many enzymes have both active sites and allosteric regulatory sites

allosteric regulatory site

(a) Enzyme structure

An allosteric regulator molecule causes the active site to change shape, so the substrate no longer fits

allosteric regulator molecule

(b) Allosteric inhibition

▲ Figure 5-15 Allosteric regulation (a) Many enzymes have an allosteric regulatory site at which **(b)** a regulator molecule may bind, thereby altering the enzyme's active site.

enzyme acts—that is, reactants that enter an enzyme's active site—are called **substrates.**

Why does each enzyme catalyze only one particular reaction? The active site of each enzyme has a distinctive shape and distribution of electrical charge. Because the enzyme and its substrate must fit together, only certain substrate molecules can enter the active site of a given enzyme, much as only certain keys can fit into a lock. As a result of this lock-and-key connection between enzyme and substrate, a given enzyme can interact with only one or a small number of substrates. For example, several enzymes are required to completely digest all of the proteins we eat, because each enzyme breaks apart only a specific sequence of amino acids.

Although it can be helpful to think of a substrate as a "key" that fits tightly into the "lock" formed by an enzyme's active site, this *lock-and-key model* is not completely accurate. Unlike rigid keys and locks, substrates and enzymes are flexible and elastic. When substrate molecules enter an enzyme's active site, both substrate and enzyme change shape to fit more tightly together (**Fig. 5-14**). These shape changes are comparable to the adjustments two hands make to more firmly grasp one another in a handshake.

After the substrates bind to the enzyme, chemical interactions between the active site and the substrate molecules may alter the chemical bonds within the substrates. These changes promote the particular chemical reaction catalyzed by the enzyme. When the reaction between the substrates is finished, the products no longer fit properly into the active site and are expelled. The enzyme is then ready to accept more substrate molecules.

The breakdown or production of a molecule within a cell usually occurs in many separate steps. Just as carving a staircase into a cliff allows the cliff to be climbed one small step at a time, breaking a reaction into small steps (each with a low activation energy) allows the overall reaction to surmount its high total activation energy. Each step is catalyzed by a different enzyme, and each enzyme lowers the activation energy for its particular step (see Fig. 5-6). For example, our cells break down sugar in many small steps, with each step liberating a small amount of energy. This step-by-step procedure is the reason that the reaction can proceed at body temperature, and we are not roasted by the sugar that is "burning" in our bodies.

Cells Regulate Metabolism by Controlling Enzymes

To be useful, the metabolic reactions in cells must be controlled. Reaction products must be produced at the proper speed, in a suitable location, and at the appropriate time. The necessary regulation of metabolic reactions is accomplished by managing the action of enzymes. Cells control the action of enzymes in a number of ways, including allosteric regulation and competitive inhibition.

Allosteric Regulation Can Increase or Decrease Enzyme Activity

In **allosteric regulation,** an enzyme's activity is modified by a regulator molecule. The regulator molecule binds temporarily to a special regulatory site on the enzyme; the site is separate from the enzyme's active site (**Fig. 5-15a**). The binding of the regulator molecule modifies the active site of the enzyme, causing the enzyme to become either more or less able to bind its substrate (**Fig. 5-15b**). Thus, allosteric regulation can either promote or inhibit enzyme activity, depending on the particular enzyme and regulator molecules involved.

In some cases of allosteric regulation, the end product of a metabolic pathway binds to the regulatory site and acts as an inhibitor of an enzyme that helped produce it. This arrangement ensures that the pathway will stop producing its product when the amount produced reaches a certain level, just as a thermostat automatically turns off a heater when a room becomes warm enough.

Competitive Inhibition Can Be Temporary or Permanent

Some regulatory molecules temporarily bind directly to an enzyme's active site, thereby preventing substrate molecules from binding (**Fig. 5-16**). Such molecules in effect compete with the enzyme's normal substrate for access to the active site, so they are said to control enzyme activity by **competitive inhibition.**

Competitive inhibition is part of a cell's normal system for regulating metabolic reactions, but it is also the means by which some poisons act. For example, methanol (a highly toxic form of alcohol) competes with ethanol (the form of alcohol found in alcoholic beverages) for the active site of the enzyme alcohol dehydrogenase. Alcohol dehydrogenase can break down methanol, but the process produces formaldehyde, a toxin that makes people ill and can cause blindness. Doctors treat victims of methanol poisoning by administering ethanol. Ethanol blocks formaldehyde production by competing with methanol for the active site of ethanol dehydrogenase.

The example of methanol and ethanol illustrates an important property of competitive inhibition: In the competition for the active site, the winner is usually the molecule with the higher concentration. The normal substrate and the inhibitor can each displace the other. Some poisons, however, bind irreversibly to an enzyme. For example, some nerve gases and insecticides permanently block the active site of acetylcholinesterase, the enzyme that breaks down acetylcholine (a substance that nerve cells release to activate muscles). The resulting inactivation of acetylcholinesterase allows acetylcholine to build up and overstimulate muscles, causing paralysis. Death follows, because victims are unable to breathe.

The Activity of Enzymes Is Influenced by Their Environment

The complex three-dimensional structure of an enzyme is sensitive to changes in environmental conditions. The hydrogen bonds that play a key role in determining enzyme structure can be altered by the enzyme's chemical and physical surroundings. Each enzyme functions optimally at a particular pH, salt concentration, and temperature, and deviations from these optimal conditions can change an enzyme's shape and destroy its effectiveness. For example, one reason that dill pickles stay well preserved in a vinegar-salt solution is that the salty, acidic conditions in the solution distort the shape of enzymes used in bacterial metabolism. With their enzymes inoperative, the bacteria that ordinarily cause food to spoil cannot survive.

When temperatures rise too high, the hydrogen bonds that determine enzyme shape may be broken apart and the enzyme rendered useless. Smaller changes in temperature, however, may simply change the rate at which enzyme-catalyzed reactions proceed. At higher temperatures, substrate molecules move more rapidly, and their random movements are more likely to bring them into contact with the active site of an appropriate enzyme. Thus, many reactions are accelerated by moderately higher temperatures and slowed by lower temperatures.

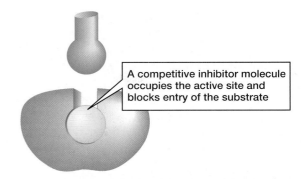

A competitive inhibitor molecule occupies the active site and blocks entry of the substrate

▲ **Figure 5-16 Competitive inhibition** A nonsubstrate molecule reversibly blocks the active site, competing with the normal substrate.

A Missing Molecule Makes Mischief

Continued

People with PKU lack working phenylalanine hydroxylase enzyme molecules, but the cells of most victims actually do produce the protein. Unfortunately, however, the molecule fails to perform its normal catalytic function. Why? A thoughtful biologist might speculate that the problem is likely to be caused by a malformed active site, because an enzyme's active site is what enables it to catalyze a reaction. But researchers have discovered that most people with PKU produce hydroxylase protein with a normal active site. However, changes in the molecule's amino acid sequence cause other parts of the protein to fold incorrectly and yield a misshapen molecule. (For a description of how a protein's sequence affects its shape, see Chapter 2, pp. 34–35.) The misshapen protein is not recognized by the cell as a normal cell component, so the cell's defenses attack and destroy it shortly after it is produced.

Web Animation Enzymes

A Missing Molecule Makes Mischief Revisited

The only current treatment available to people with PKU, whose bodies lack the enzyme needed to break down phenylalanine, is a diet that excludes proteins (almost all of which contain at least some phenylalanine). So people with PKU cannot eat many foods that the rest of us take for granted, including meat, cheese, and fish. They must also avoid the widespread artificial sweetener aspartame, which contains phenylalanine.

Because sticking to the special diet can be difficult, other treatments would be welcome, and scientists at several labs are working to develop such treatments. One research group is testing a treatment in which young adults with PKU eat a much less restrictive diet but also eat a mixture of amino acids that slows the movement of phenylalanine from the blood into the brain. Because these patients eat protein, the level of phenylalanine in their blood increases, but the researchers believe that the higher levels won't reach and damage the brain.

(continued)

Another promising potential treatment involves phenylalanine ammonia-lyase (PAL), an enzyme that degrades phenylalanine. Researchers have shown that PAL, administered orally to patients, can break down phenylalanine in the intestine, before it is absorbed into the bloodstream. Digestive enzymes, however, quickly degrade PAL, so it will not be a viable treatment option until researchers find a way to protect it from the body's digestive enzymes.

Consider This

Newborns in all developed nations are screened for PKU, and affected individuals are identified before any neurological damage can occur. Infants are also screened for some other ailments, but there are many potentially detectable conditions for which they are not tested. In your view, what criteria should be used to decide which tests are mandatory for all newborns?

Chapter Review

Summary of Key Concepts

For additional study help and activities, go to www.mybiology.com.

5.1 What Is Energy?

Energy is the capacity to do work. Kinetic energy is the energy of movement (light, heat, electricity, movement of large particles). Potential energy is stored energy (chemical energy, positional energy). The first law of thermodynamics states that energy cannot be created or destroyed, so the total amount of energy remains constant, although it may change in form. The second law of thermodynamics states that any use of energy within a system causes a decrease in the quantity of concentrated, useful energy and an increase in the randomness and disorder of matter (entropy).

5.2 How Does Energy Flow in Chemical Reactions?

Chemical reactions fall into two categories. In exergonic reactions, the product molecules have less energy than do the reactant molecules, so the reaction releases energy. In endergonic reactions, the products have more energy than do the reactants, so the reaction requires an input of energy. Exergonic reactions can occur spontaneously, but all reactions, including exergonic ones, require an input of activation energy to overcome electrical repulsions between reactant molecules. In a coupled reaction, the energy liberated by an exergonic reaction drives the endergonic reaction. Organisms couple exergonic reactions, such as light-energy capture or sugar metabolism, with endergonic reactions, such as the synthesis of organic molecules.

Web Animation Energy and Chemical Reactions

5.3 How Is Energy Carried Between Coupled Reactions?

Energy released by chemical reactions within a cell is captured and transported within the cell by energy-carrier molecules, such as ATP and electron carriers. These molecules are the major means by which cells couple exergonic and endergonic reactions that occur at different places in the cell.

Web Animation Energy and Life

5.4 How Do Cells Control Their Metabolic Reactions?

Cellular reactions are linked in metabolic pathways and are regulated in large measure by enzymes. High activation energies cause many reactions, even exergonic ones, to proceed very slowly, or not at all, under normal environmental conditions. Catalysts lower the activation energy and thereby speed up chemical reactions without being permanently changed themselves. Organisms synthesize enzymes that catalyze one or a few reactions. The reactants temporarily bind to the active site of the enzyme, making it easier to form the new chemical bonds of the products. Cells regulate enzyme-catalyzed reactions through processes that include allosteric regulation and competitive inhibition. Environmental conditions including pH, salt concentration, and temperature can promote or inhibit the function of enzymes, by altering their three-dimensional structure.

Web Animation Enzymes

Key Terms

activation energy *p. 75*
active site *p. 79*
adenosine diphosphate (ADP) *p. 76*
adenosine triphosphate (ATP) *p. 76*
allosteric regulation *p. 80*
catalyst *p. 78*

chemical reaction *p. 75*
competitive inhibition *p. 81*
coupled reaction *p. 76*
electron carrier *p. 77*
endergonic reaction *p. 75*
energy *p. 73*
energy-carrier molecule *p. 76*

entropy *p. 74*
enzyme *p. 79*
exergonic reaction *p. 75*
first law of thermodynamics *p. 73*
kinetic energy *p. 73*
metabolic pathway *p. 77*
metabolism *p. 77*

potential energy *p. 73*
product *p. 75*
reactant *p. 75*
second law of thermodynamics *p. 74*
substrate *p. 80*

Thinking Through the Concepts

Suggested answers to end-of-chapter and figure-based questions can be found at the end of the text.

Fill-in-the-Blank

1. According to the first law of thermodynamics, the total amount of energy in the universe is always _____. Energy occurs in two major forms: _____, the energy of movement, and _____, or stored energy.

2. According to the second law of thermodynamics, when energy changes forms it tends to be converted from _____ useful to _____ useful forms. This leads to the conclusion that matter spontaneously tends to become less _____. This tendency is called _____.

3. Energy needed to get any chemical reaction started is called _____ energy. This energy is required to force the _____ of the reactants together. This energy is usually supplied by _____.

4. Once started, some reactions release heat and are called _____ reactions. Others require a net input of energy and are called _____ reactions. Which type will continue spontaneously once it starts? _____. Which type allows the formation of complex biological molecules from simpler molecules (for example, proteins from amino acids)? _____. When these two types of reactions are linked, with the energy released from one providing the energy required by the other, the two reactions are described as _____ reactions.

5. The abbreviation "ATP" stands for _____. This molecule is the principal _____ molecule in living cells. The molecule is synthesized by cells from _____ and _____. This synthesis requires _____, which is then stored in the molecule of ATP.

6. Enzymes are (type of biological molecule) _____. Enzymes promote reactions in cells by lowering the _____. Each enzyme possesses a specialized region called a(n) _____ into which the reactant molecules fit. Each of these specialized regions has a distinctive _____ and a distinctive distribution of _____ that make it quite specific for its substrate molecules.

Review Questions

1. Explain why organisms do not violate the second law of thermodynamics. What is the ultimate energy source for most forms of life on Earth?

2. Define metabolism, and explain how reactions can be coupled to one another.

3. What is activation energy? How do catalysts affect activation energy? How does this change the rate of reactions?

4. Describe some exergonic and endergonic reactions that occur in plants and animals very regularly.

5. Describe the structure and function of enzymes.

Applying the Concepts

1. IS THIS SCIENCE? Creationists sometimes critique the theory of evolution with the following argument: "According to evolutionary theory, organisms have increased in complexity through time. However, evolution of increased biological complexity contradicts the second law of thermodynamics. Therefore, evolution is impossible." Develop a scientific response to this argument.

2. IS THIS SCIENCE? Some practitioners of alternative medicine believe that people can improve their health by swallowing capsules of various plant and animal enzymes. This "enzyme therapy" is based on the hypothesis that the enzymes will act in the stomach to predigest food, so that the body will have to use fewer of its own enzymes for digestion and can instead use them to "maintain metabolic harmony" and improve overall health. Evaluate this hypothesis from a scientific viewpoint and suggest an experiment to test it.

3. A preview question for ecology: When a brown bear eats a salmon, does the bear acquire all the energy contained in the body of the fish? Why or why not? What implications do you think this answer would have for the relative abundance (by weight) of predators and their prey? Does the second law of thermodynamics help explain the title of the book *Why Big, Fierce Animals Are Rare*?

4. As you learned in Chapter 2, the subunits of virtually all organic molecules are joined by dehydration synthesis reactions and can be broken apart by hydrolysis reactions. Why, then, does your digestive system produce separate enzymes to digest proteins, fats, and carbohydrates—in fact, several of each type?

For additional resources, go to www.mybiology.com.

chapter 6

Capturing Solar Energy: Photosynthesis

A giant meteorite may have ended the reign of *Tyrannosaurus* and the other dinosaurs.

Case Study   Did the Dinosaurs Die from Lack of Sunlight?

About 65 million years ago, life on Earth suffered a catastrophic blow. Within a short period, most of the species on the planet were destroyed. This devastating mass extinction eliminated more than 70% of the species then existing, including the dinosaurs. *Triceratops*, *Tyrannosaurus*, and all of the other dinosaur species disappeared forever. Land and sea were nearly emptied of life, and many millions of years passed before new species arose to take the place of those that had disappeared.

Most scientists believe that this devastation began when a gigantic meteorite, 6 miles in diameter, entered the atmosphere and crashed into Earth. As the meteorite plowed into the ocean at the tip of the Yucatan Peninsula, it dug a crater a mile deep and 120 miles wide. Any organism in the immediate area was, of course, killed by the blast wave from the impact. This kind of direct destruction, however, must have been limited to a relatively small area. How, then, did the meteorite impact eliminate thousands of species in all corners of the globe? In all likelihood, the most serious damage was done not by the meteorite itself, but by the lingering effects of its sudden arrival. In particular, many scientists have concluded that the meteorite's most damaging long-term effect was disruption of the most important chemical reaction on Earth: photosynthesis.

What exactly does photosynthesis do? What makes it so important that interrupting it brought down the mighty dinosaurs? To find out, read on. ∎

The meteorite's impact darkened the skies and started forest fires worldwide.

6.1 What Is Photosynthesis?

Organisms need energy to live. But, as you learned in Chapter 5, the first law of thermodynamics tells us that energy cannot be created, so living things cannot manufacture their own energy. Some organisms can, however, capture energy from sunlight and convert it to chemical energy. The process by which this conversion is accomplished is known as **photosynthesis.**

Life on Earth Depends on Photosynthesis

The solar energy captured by photosynthesis is stored in sugar and other organic molecules. This stored energy is absolutely essential for the continuation of life on Earth. It provides nourishment not only for the photosynthetic organisms themselves but also for the organisms that eat them and, ultimately, for almost every organism that eats another organism.

In addition to capturing the energy that nourishes life, photosynthesis also makes a crucial, life-sustaining contribution to the atmosphere. As it stores solar energy in chemical bonds, photosynthesis consumes carbon dioxide (CO_2) and water (H_2O) and releases oxygen gas (O_2) as a by-product. As a result, photosynthesis removes carbon dioxide from the atmosphere and adds oxygen to it. Both of these actions help maintain the atmosphere, but the formation of oxygen gas is especially important. Most organisms need oxygen to live and they continually consume it. If atmospheric oxygen were not constantly replenished by photosynthesis, it would soon dwindle to levels too low to sustain life.

Photosynthesis Converts Carbon Dioxide and Water to Glucose

The energy captured by photosynthesis is stored in the bonds of glucose ($C_6H_{12}O_6$). When written in its simplest form, the overall chemical reaction for photosynthesis is:

$$6\ CO_2 + 6\ H_2O + \text{light energy} \rightarrow C_6H_{12}O_6 + 6\ O_2$$

This basic photosynthetic reaction, in which carbon dioxide and water are converted to glucose and oxygen, takes place in photosynthetic organisms including plants, seaweeds, and various types of single-celled organisms. Each group of photosynthetic organisms exhibits distinctive variations on the basic process. In this chapter, we focus on photosynthesis in plants.

Plant Photosynthesis Takes Place in Leaves

Plant photosynthesis takes place mainly in leaves (**Fig. 6-1a**). Leaves have several features that make them especially well suited for this role. For example, the leaves of most land plants are only a few cells thick. The thinness of leaves ensures that sunlight can easily penetrate them. In addition, the flattened shape of leaves exposes a large surface area to the sun.

A leaf obtains CO_2 for photosynthesis from the air, through adjustable pores in the leaf surface called **stomata** (singular, stoma). Stomata can open and close, and they open at appropriate times to admit CO_2. Inside the leaf are a few layers of cells collectively called **mesophyll**; photosynthesis occurs mainly in these cells (**Fig. 6-1b**). A system of veins supplies water and minerals to the mesophyll cells and carries the sugars they produce to other parts of the plant. (The structures that transport materials around a plant's body are described in greater detail in Chapter 17.)

▶ Figure 6-1 **An overview of photosynthetic structures**

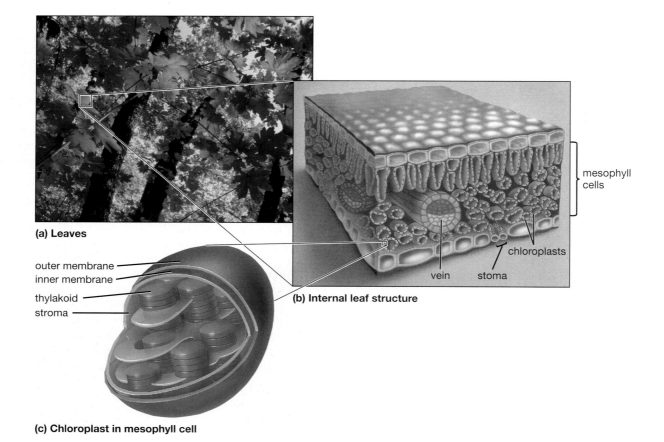

(a) Leaves

outer membrane
inner membrane
thylakoid
stroma

(b) Internal leaf structure

mesophyll cells

chloroplasts

vein   stoma

(c) Chloroplast in mesophyll cell

Leaf Cells Contain Chloroplasts

Leaf cells, especially mesophyll cells, contain large numbers of **chloroplasts,** the organelles in which photosynthesis occurs. Chloroplasts have a double outer membrane that encloses a semifluid substance, the **stroma** (Fig. 6-1c). Embedded in the stroma are disk-shaped, interconnected membranous sacs called **thylakoids.** The light-dependent reactions of photosynthesis occur within the membranes of the thylakoids.

Photosynthesis Consists of Light-Dependent and Light-Independent Reactions

The simple equation that summarizes the overall reaction of photosynthesis conceals the fact that photosynthesis actually requires dozens of steps, each catalyzed by a different enzyme. The steps can be divided into two groups, the light-dependent reactions and the light-independent reactions. The two groups of reactions occur at different locations within the chloroplast, but they are linked by the energy-carrier molecules **adenosine triphosphate (ATP)** and **nicotinamide adenine dinucleotide phosphate (NADPH).**

- In the *light-dependent* reactions, molecules in the membranes of the thylakoids capture sunlight energy and convert some of it into chemical energy stored, for the short term, in ATP and NADPH. The light-dependent reactions consume water and release oxygen gas as a by-product (Fig. 6-2).

- In the *light-independent* reactions, enzymes in the stroma use the chemical energy of the ATP and NADPH molecules produced by the light-dependent reactions to power the manufacture of glucose or other organic molecules, consuming carbon dioxide in the process. As energy is extracted from ATP and NADPH, these energy carriers are converted to ADP (adenosine diphosphate) and NADP+ respectively. These depleted energy carriers return to the light-dependent reactions to be recharged into ATP and NADPH.

6.2 How Is Light Energy Converted to Chemical Energy?

In the first stage of photosynthesis, the light-dependent reactions convert the energy of sunlight into chemical energy stored in carrier molecules. Sunlight consists of electromagnetic waves. The waves come in different sizes (wavelengths), ranging from short-wavelength gamma rays, through ultraviolet, visible, and infrared light, to very long wavelength radio waves. Sunlight includes only a portion of the full electromagnetic spectrum. Of the wavelengths emitted by the sun, photosynthesis uses only those that fall into the range of visible light. We perceive different wavelengths of visible light as different colors.

Light Energy Is First Captured by Pigments in Chloroplasts

The job of absorbing light energy for use in photosynthesis falls to certain molecules in the thylakoid membranes of chloroplasts. The membranes contain several types of *pigments* (light-absorbing molecules), each of which absorbs a different range of wavelengths. **Chlorophyll,** the key light-capturing molecule, strongly absorbs violet, blue, and red light but reflects green. The reflected wavelengths reach our eyes, so we see chlorophyll as green (Fig. 6-3). Most leaves appear green because they are rich in chlorophyll.

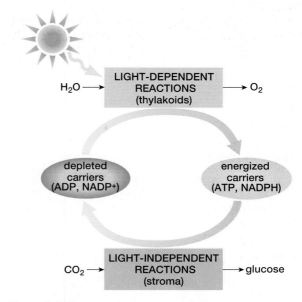

▲ **Figure 6-2 An overview of photosynthesis** The light-dependent reactions transfer solar energy to energy-carrier molecules. The light-independent reactions transfer energy from those molecules to longer-term storage in glucose molecules. The depleted energy carriers return to the light-dependent reactions to be reenergized.

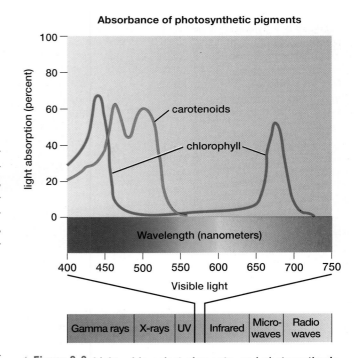

▲ **Figure 6-3 Light, chloroplast pigments, and photosynthesis** Visible light, a small part of the electromagnetic spectrum, consists of wavelengths that correspond to the colors of the rainbow. Chlorophyll (green curve) strongly absorbs violet, blue, and red light. Carotenoids (orange curve) absorb blue and green wavelengths. **QUESTION:** On the basis of the information in this graph, what color are carotenoids?

Web Animation Properties of Light

Thylakoids also contain other molecules, called *accessory pigments,* that capture light energy and transfer it to chlorophyll. The accessory pigments include carotenoids, which absorb blue and green light, but reflect yellow and orange. In leaves, the colors of carotenoids are usually masked by the presence of chlorophyll. When a leaf dies, however, its chlorophyll often breaks down before its carotenoids, and a bright yellow or orange color may be revealed.

The Light-Dependent Reactions Generate Energy-Carrier Molecules

The light-dependent reactions take place in special locations called photosystems, which are located in the thylakoid membranes. Each thylakoid contains thousands of **photosystems,** which are highly organized assemblages of proteins, chlorophyll, and accessory pigments. There are two types of photosystem: *photosystem I* (PS I) and *photosystem II* (PS II).

Each photosystem includes a **light-harvesting complex,** which contains many chlorophyll and accessory pigment molecules, including carotenoids. These molecules absorb light and pass the energy to a few specific chlorophyll molecules called the **reaction center.** The reaction center chlorophylls are located next to an **electron transport chain (ETC),** a series of electron-carrier molecules also embedded in the thylakoid membrane (**Fig. 6-4**). The electron transport chains associated with photosystems I and II do not have the same component molecules; each type of photosystem is coupled with a different kind of ETC.

Photosystem II Generates ATP

As you read the following sections about the photosystems, you may wish to consult **Figure 6-5** to help you follow the steps. In the first four steps shown, photosystem II captures light energy and produces ATP. ❶ The light-dependent reactions

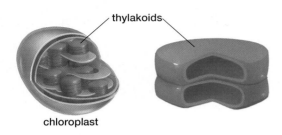

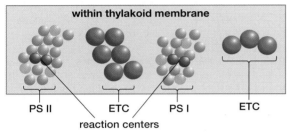

▲ **Figure 6-4 Structures associated with the light-dependent reactions**

▶ **Figure 6-5 The light-dependent reactions of photosynthesis**
QUESTION: If these reactions produce energy-carrier molecules (ATP and NADPH), then why do plant cells need mitochondria?

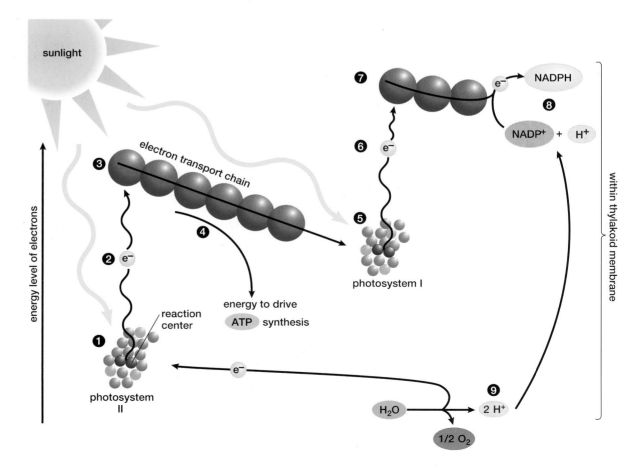

begin when the light-harvesting complex of photosystem II absorbs energy from light. The energy passes from molecule to molecule until it reaches the reaction center. ❷ When the reaction-center chlorophyll molecule receives the energy, electrons absorb the energy and jump from the chlorophyll molecule over to ❸ the adjacent electron transport chain. After arriving at the electron transport chain, these energetic electrons keep jumping, moving from one carrier molecule to the next. During the jumps, the electrons release energy. ❹ Some of this released energy powers reactions that synthesize the energy carrier ATP.

Photosystem I Generates NADPH

In the next steps of the light-dependent reactions, photosystem I captures light energy and produces the energy carrier NADPH. ❺ As photosystem II continues to generate ATP, light rays are also striking the light-harvesting complex of photosystem I. Just as in photosystem II, the energy from the light striking photosystem I is transmitted to the reaction-center chlorophyll, which ❻ ejects electrons. Almost immediately, these lost electrons are replaced by electrons that have reached the end of the electron transport chain associated with photosystem II and jumped over to photosystem I. ❼ At the same time, the ejected photosystem I electrons jump to the electron transport chain associated with photosystem I. The electrons move from molecule to molecule through the electron transport chain until they reach the electron-carrier molecule $NADP^+$. ❽ Each $NADP^+$ molecule picks up energized electrons from the electron transport chain and a hydrogen ion from water, forming NADPH.

Splitting Water Maintains the Flow of Electrons Through the Photosystems

Overall, electrons flow from the reaction center of photosystem II, through the electron transport chain associated with photosystem II, to the reaction center of photosystem I, through the electron transport chain associated with photosystem I, and on to form NADPH. To sustain this one-way flow of electrons, photosystem II's reaction center must be continuously supplied with new electrons to replace the ones it gives up.

The electrons that replenish photosystem II are generated by a water-splitting reaction that is catalyzed by an enzyme present in the photosystem. This reaction (step ❾ in Fig. 6-5) can be summarized as:

$$H_2O \rightarrow \tfrac{1}{2} O_2 + 2H^+ + 2\,e^-$$

Each molecule of water converted by this reaction provides two electrons to photosystem II. The hydrogen ions produced by the reaction may be used in the conversion of $NADP^+$ to NADPH (step ❽ in Fig. 6-5). The oxygen gas (O_2) produced by the reaction may be used directly by the plant in its own respiration or given off to the atmosphere.

SUMMING UP: Light-Dependent Reactions ----------

(The numbered steps in this list correspond to the numbers in Fig. 6-5.)

❶ The light-dependent reactions begin when light is absorbed by the light-harvesting complex of photosystem II.

❷ The absorbed energy causes electrons from the reaction center of the complex to be ejected.

❸ The ejected electrons are transferred to the electron transport chain adjacent to photosystem II. As the electrons pass through the transport system, they release energy.

❹ Some of the energy is used to manufacture ATP.

❺ Meanwhile, light is absorbed by the light-harvesting complex of photosystem I.

❻ The absorbed energy ejects electrons from the reaction center.

(continued)

BioFlix **Photosynthesis**

Web Animation Photosynthesis

Did the Dinosaurs Die from Lack of Sunlight?

Continued

The pigments in the light-harvesting complexes of photosystems I and II absorb light only of the appropriate wavelengths. If light energy at these wavelengths is prevented from reaching the photosystems, ATP and NADPH will not be produced, and will therefore not be available to power the manufacture of glucose. The meteorite impact of 65 million years ago may have drastically reduced the amount of sunlight reaching Earth's surface, thereby substantially reducing the amount of energy captured and stored by photosynthesis.

6.3 How Is Chemical Energy Stored in Glucose Molecules?

The ATP and NADPH molecules generated in the light-dependent reactions are used in the light-independent reactions to manufacture molecules for the long-term storage of energy. As their name suggests, the light-independent reactions can occur without light (as long as ATP and NADPH are available).

The Light-Independent Reactions Manufacture Glucose

The light-independent reactions take place in the fluid stroma that surrounds the thylakoids. There, the reactions make glucose from carbon dioxide and water, using enzymes that are present in the stroma.

The C₃ Cycle Captures Carbon Dioxide

The first step on the path to glucose is to capture carbon dioxide. It is captured in a set of reactions known as the **C₃ cycle** (three-carbon cycle). The C₃ cycle is so named because some of the important molecules in it contain three carbon atoms, and because the reaction pathway is cyclic (one of its outputs is also one of its key inputs). The cycle is also sometimes called the *Calvin-Benson cycle*.

The C₃ cycle requires the five-carbon sugar **ribulose bisphosphate (RuBP)**, CO_2 (usually from the air), enzymes to catalyze each of the reactions, and energy in the form of ATP and NADPH. The ATP and NADPH are usually supplied by the light-dependent reactions. As these ATP and NADPH molecules donate energy to power the reactions of the C₃ cycle, they are converted to ADP and $NADP^+$, which thereby become available for recharging to ATP and NADPH in the light-dependent reactions.

At the beginning of the C₃ cycle, CO_2 molecules are captured from the atmosphere, a process known as **carbon fixation.** During the carbon fixation step (step ❶ in **Fig. 6-6**) six molecules of carbon dioxide combine with six molecules of RuBP in a series of reactions that yields 12 three-carbon molecules of *phosphoglyceric acid* (PGA). Thus, carbon from gaseous CO_2 is "fixed" into a relatively stable organic molecule, PGA.

As the cycle proceeds, the 12 molecules of PGA are converted to 12 molecules of *glyceraldehyde-3-phosphate* (G3P; step ❷ in **Fig. 6-6**). The energy for this reaction is provided by the ATP and NADPH energy-carrier molecules that were generated by the light-dependent reactions. Then, through a series of reactions powered by ATP energy, 10 of the 12 G3P molecules are used to regenerate the 6 molecules of RuBP needed to restart the cycle (step ❹ in **Fig. 6-6**).

Carbon Fixed During the C₃ Cycle Is Used to Synthesize Glucose

Each turn of the C₃ cycle begins and ends with six molecules of RuBP but also captures additional carbon atoms from CO_2. These extra carbons are found in

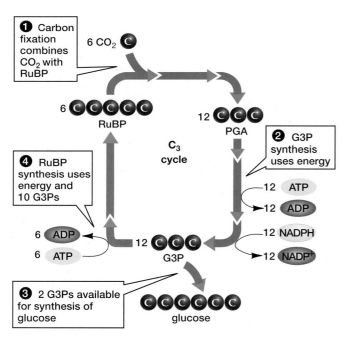

▲ **Figure 6-6 The C₃ cycle of carbon fixation** ❶ RuBP reacts with CO_2 to form PGA. ❷ The energy of ATP and the electrons and hydrogens of NADPH are used to convert PGA to G3P. ❸ Two G3P molecules are further processed into glucose. ❹ Energy from ATP is used to rearrange the remaining G3P into RuBP completing one turn of the C₃ cycle. The depleted energy carriers ADP and $NADP^+$ return to the light-dependent reactions for recharging.

earth watch

Are the Benefits of Biofuels Bogus?

In photosynthesis, plants use solar energy to extract CO_2 from the atmosphere and trap it in energy-rich organic molecules. Plants have been capturing atmospheric carbon in this fashion for hundreds of millions of years, and some of the captured carbon has been placed into long-term storage as fossil fuels. Over eons, heat and pressure have converted the bodies of dead plants—and their stored carbon—into coal, oil, and natural gas. So when you drive your car, turn up the thermostat, or turn on your desk lamp, you are actually unleashing the energy of prehistoric sunlight trapped by photosynthetic organisms. At the same time, you are freeing carbon from long-term storage and releasing it into the atmosphere as carbon dioxide. Since the industrial revolution of the mid-1800s, we have increased the CO_2 content of the atmosphere by about 36%, mostly by burning fossil fuels.

As a result of this increase in atmospheric CO_2, Earth's climate is growing warmer (a phenomenon described in more detail in Chapter 29). Although no one can say with certainty what the consequences of this warming trend will be, many biologists and climate scientists fear that a hotter future climate will place extraordinary stress on Earth's inhabitants, including humans.

Because of the threat posed by global warming, governments throughout the world are promoting the use of *biofuels* to replace fossil fuels. Biofuels are renewable energy sources derived from recently living organisms, mainly plants. The two main types of biofuel are bioethanol and biodiesel. Bioethanol is produced by fermenting sugar- or starch-rich plants, such as sugar cane or corn, to produce alcohol (fermentation is described in Chapter 7). Biodiesel is made from oils extracted from plants, especially soybean, canola, or palm oil.

Biofuels have the potential to reduce CO_2 emissions because the carbon stored in them was only recently extracted from the atmosphere by photosynthesis. Thus, burning biofuels simply returns to the atmosphere CO_2 that left it a short time earlier. And as more plants are grown to replace the ones used to manufacture biofuel, CO_2 is again removed from the atmosphere, recycling it. Is this a simple solution to global warming?

The environmental and social benefits of burning food crops in our gas tanks are hotly debated. In the United States, the emphasis has been on bioethanol made from corn. Estimates of the net reduction in CO_2 emissions achieved by burning ethanol instead of gasoline range from 15% to 40%. Achieving this reduction, though, would require major changes to the world's agricultural economies. For example, scientists in the United States estimate that to replace just 10% of the fuel used in American vehicles with bioethanol made from corn would require that 30% of our total agricultural land be converted to growing "fuel-corn." This shift would reduce the amount of corn available for export to food-poor countries and drive up its price as well. Demand for bioethanol has already caused world sugar prices to rise dramatically, spurring farmers in India to increase cultivation of this water-demanding crop. Increased sugar cultivation will further drain India's rapidly diminishing underground water stores, increasing the likelihood of crop failures. In Indonesia, plans for enormous new palm plantations to make biodiesel threaten the rain forests of Borneo, home to many endangered species. Brazil has been steadily increasing its reliance on biodiesel, and soybean plantations are replacing large expanses of Brazilian rain forest (**Fig. E6-1**). Ironically, clearing these forests for agriculture increases atmospheric CO_2 because rain forests trap more carbon than the crops that replace them.

Biofuels could become more environmentally friendly if new technologies are developed to produce them more efficiently. For example, if cellulose could be readily cleaved into its component sugars, ethanol could then be generated from cellulose-rich grasses or wood chips, or from material that is currently considered waste, such as corn stalks. Municipal wastes and manure from feedlots might also be transformed into fuels.

Although the benefits of most currently produced biofuels may not justify their environmental costs, the equation could change as we develop better methods for harnessing the energy captured by photosynthesis.

▲ **Figure E6-1 Converting forests to farmland** Increasing demand for biofuels provides an economic incentive to cut down ecologically important tropical forests, such as the forest that was cleared to make room for this soybean farm in Brazil.

the two extra molecules of G3P that are left over after RuBP is regenerated. As indicated in step ❸ of **Figure 6-6**, these two G3P molecules (three carbons each) combine to form one molecule of glucose (six carbons). (People harvest some of the energy stored in this glucose and use it in place of energy from fossil fuels; see "Earth Watch: Are the Benefits of Biofuels Bogus?")

(The numbered steps in this list correspond to the numbers in Fig. 6-6.)

❶ In the C_3 cycle, 6 molecules of RuBP capture 6 molecules of CO_2 to produce 12 molecules of PGA.

❷ A series of reactions driven by energy from ATP and NADPH uses the PGA to produce 12 molecules of G3P.

❸ Two of these molecules of G3P join to form 1 molecule of glucose.

❹ The other 10 G3P molecules are used to regenerate the 6 RuBP molecules that are needed to begin the cycle anew.

Each passage through the cycle yields one molecule of glucose and depleted energy carriers (ADP and $NADP^+$) that will be recharged during light-dependent reactions.

6.4 What Is the Relationship Between Light-Dependent and Light-Independent Reactions?

As discussed earlier, photosynthesis includes two separate sets of reactions, the light-dependent and light-independent reactions. The two sets of reactions are, however, closely linked. The light-dependent reactions capture solar energy, and the light-independent reactions use the captured energy to manufacture glucose (**Fig. 6-7**; see also Fig. 6-2). You can think of photosynthesis as a seamless combination of two components: the "photo" component of light-dependent energy capture and the "synthesis" component of light-independent glucose production.

▶ **Figure 6-7 Two sets of reactions are connected in photosynthesis**   **QUESTION:** Could a plant survive in an oxygen-free atmosphere?

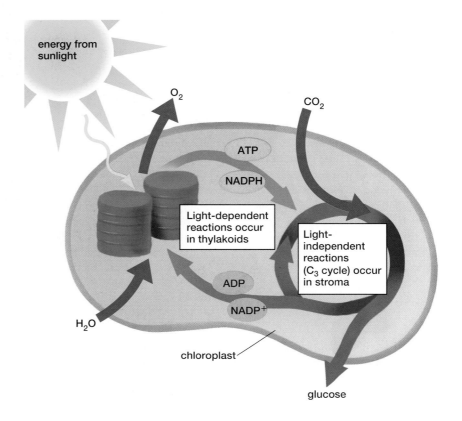

Energy-carrier molecules provide the link between the two sets of reactions. The light-dependent reactions in the membranes of the thylakoids use light energy to "charge up" molecules of ADP and NADP$^+$ to form ATP and NADPH. These energized carriers move to the stroma, where their energy is used to drive the light-independent reactions that synthesize glucose. The depleted carriers, ADP and NADP$^+$, then return to the light-dependent reactions for recharging to ATP and NADPH.

6.5 How Does the Need to Conserve Water Affect Photosynthesis?

Photosynthesis requires carbon dioxide. Therefore, you might think an ideal leaf would be very porous to allow lots of CO_2 to enter it from the air. For land plants, however, being porous to CO_2 also allows water to evaporate easily from the leaf. Loss of water from leaves can be stressful for land plants and may even be fatal. Leaf structure, therefore, must balance the need to obtain CO_2 with the need to retain water.

One solution to the problem of water loss has been the evolution of *stomata,* pores in leaves that can open and close. When water supplies are adequate, the stomata open, letting in CO_2 (**Fig. 6-8a**). If the plant is in danger of drying out, the stomata close (**Fig. 6-8b**). Closing of the stomata reduces evaporation, but, unfortunately, it also reduces CO_2 intake and restricts the release of the O_2 produced by photosynthesis.

When Stomata Are Closed to Conserve Water, Wasteful Photorespiration Occurs

In hot, dry conditions, stomata remain closed much of the time. As a result, the amount of CO_2 inside the leaf decreases and the amount of O_2 increases. Under these conditions, photosynthesis is hampered. The main problem is that RuBP, the molecule that combines with CO_2 during the normal C_3 cycle, can also combine with O_2. When RuBP combines with O_2 instead of CO_2 a wasteful process called **photorespiration** takes place. Photorespiration does not produce any useful cellular energy, and it prevents the C_3 cycle from synthesizing glucose. Thus, when photorespiration occurs, there is little or no photosynthesis and plants manufacture little or no glucose.

Alternative Pathways Reduce Photorespiration

Some plant species have evolved metabolic pathways that reduce photorespiration. Plants with these alternative pathways can produce sufficient glucose even under hot and dry conditions. The two most important alternative pathways are the C_4 **pathway** and **crassulacean acid metabolism (CAM).**

C_4 Plants Capture Carbon and Synthesize Glucose in Different Places

In typical plants, also known as C_3 plants, both carbon fixation and glucose synthesis are accomplished by the reactions of the C_3 cycle, which take place in mesophyll cells (**Fig. 6-9a**). When photorespiration blocks the carbon fixation step in these plants, the entire machinery of glucose production in the

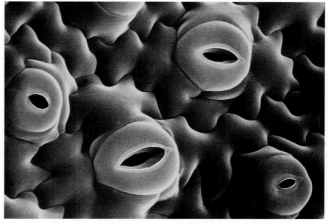

(a) Stomata open

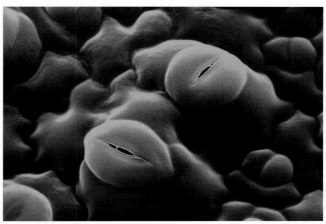

(b) Stomata closed

▲ **Figure 6-8 Stomata** Stomata can be **(a)** opened when the water supply is adequate or **(b)** closed when water is limited.

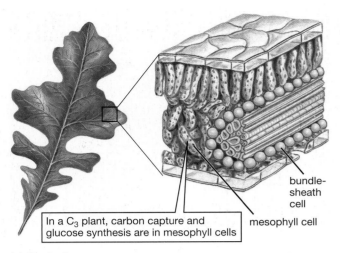

In a C_3 plant, carbon capture and glucose synthesis are in mesophyll cells

bundle-sheath cell

mesophyll cell

(a) C_3 plant

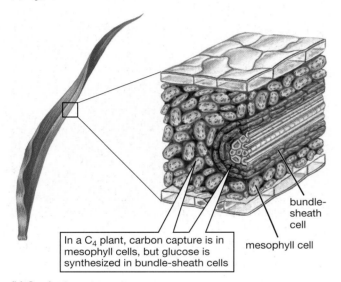

In a C_4 plant, carbon capture is in mesophyll cells, but glucose is synthesized in bundle-sheath cells

bundle-sheath cell

mesophyll cell

(b) C_4 plant

▲ **Figure 6-9** **Leaves of C_3 and C_4 plants** In C_3 plants, capture of atmospheric carbon dioxide and the C_3 cycle both take place in the chloroplasts of mesophyll cells. In C_4 plants, however, the two processes occur in different places: CO_2 is captured in mesophyll cells, but the C_3 cycle takes place in bundle-sheath cells.   **QUESTION:** Why do C_3 plants have an advantage over C_4 plants under conditions that are not hot and dry?

mesophyll cells shuts down. Plants that use the C_4 pathway, however, avoid this result by using a different reaction to capture carbon and then shuttling the captured carbon away from mesophyll cells for further processing elsewhere in the leaf.

The C_4 pathway thus includes two stages that take place in different parts of the leaf. In the first stage, which takes place in mesophyll cells, carbon dioxide is captured in a reaction that proceeds even when the stomata are closed and oxygen concentration is high. This carbon-fixing reaction in the mesophyll cells yields a four-carbon molecule (for which the C_4 cycle is named).

The four-carbon molecule is transferred from mesophyll cells to another group of cells known as *bundle-sheath cells* (**Fig. 6-9b**). In the bundle-sheath cells, the four-carbon molecule breaks down, releasing CO_2. This release creates a high CO_2 concentration in the bundle-sheath cells, where the second stage of the pathway—the regular C_3 cycle—can now proceed without much competition from oxygen (**Fig. 6-10a**).

The C_4 pathway is found in some plant species that experience strong sunlight and high daytime temperatures. Many C_4 species are grasses, including agriculturally important species such as sugar cane, maize (corn), and sorghum.

CAM Plants Capture Carbon and Synthesize Glucose at Different Times

C_4 plants counter the effects of photorespiration by physically separating the locations at which carbon capture and glucose synthesis take place, but some other plants have evolved a different method for separating the two processes. In plants that use CAM, photorespiration is reduced by fixing carbon in two stages that take place in the same cells but at different times of day. The first stage takes place at night, when CAM plants (unlike almost all other plants) keep their stomata open. During the night, reactions in mesophyll cells incorporate carbon dioxide into four-carbon molecules that are stored in vacuoles. The next day, the stomata close, conserving precious water during the heat of the day. The closed stomata, of course, also prevent carbon dioxide from reaching the mesophyll cells. But, during the second stage, the organic acids that had been stored there the previous night are broken down, releasing carbon dioxide. The regular C_3 cycle proceeds, using carbon dioxide captured the night before (**Fig. 6-10b**).

Species that use CAM tend to be ones that live in dry places and store water in fleshy leaves. For example, many cacti and other desert species are CAM plants.

▶ **Figure 6-10** **Two ways to reduce photorespiration** Photorespiration can be reduced by separating carbon fixation from the C_3 cycle. The two processes may take place **(a)** in different physical locations, as in C_4 plants, or **(b)** at different times, as in CAM plants.

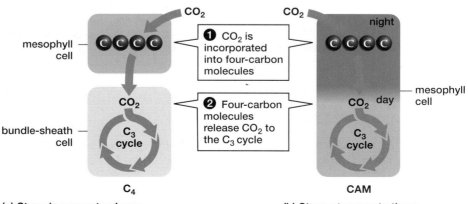

mesophyll cell

❶ CO_2 is incorporated into four-carbon molecules

CO_2

bundle-sheath cell

C_3 cycle

❷ Four-carbon molecules release CO_2 to the C_3 cycle

(a) Steps in separate places

C_4

CO_2

night

mesophyll cell

CO_2 day

C_3 cycle

(b) Steps at separate times

CAM

Did the Dinosaurs Die from Lack of Sunlight? Revisited

How did Earth's collision with a giant meteorite disrupt photosynthesis and exterminate the dinosaurs? Researchers hypothesize that the impact pulverized the meteorite and smashed Earth's crust, sending trillions of tons of debris rocketing into the atmosphere. The debris was flung so high that much of it went into orbit around the planet. Some of the orbiting material fell back to the ground, burning up as it plunged though the atmosphere and causing huge forest fires. The resulting smoke and ashes, combined with the orbiting dust cloud, obscured the sun and plunged Earth into a night that lasted for months.

In the perpetual darkness of the post-impact world, the temperature dropped dramatically. This sudden winter must have placed tremendous stress on organisms accustomed to the tropical conditions that then prevailed over large areas of the planet. Even more devastating, however, was the sudden loss of sunlight. With little sunlight reaching Earth's surface, the land plants that had survived the fires could not capture enough energy. In the sea, a huge number of photosynthetic algae perished. As the photosynthesizers declined, the organisms that depended on them for food also suffered. Large plant-eaters must have been especially vulnerable, and the predators that fed on them were soon without prey. Eventually, the stress of life on a photosynthesis-starved planet spelled extinction for most of Earth's species.

Consider This

Some paleontologists have concluded that the mass extinction of 65 million years ago was not instantaneous but was spread out over as many as 300,000 years. If this conclusion were correct, could the hypothesis that disrupted photosynthesis caused the extinction also be correct?

Chapter Review

Summary of Key Concepts

For additional study help and activities, go to www.mybiology.com.

6.1 What Is Photosynthesis?
Photosynthesis captures the energy of sunlight to convert the inorganic molecules of carbon dioxide and water into high-energy organic molecules such as glucose. In plants, photosynthesis takes place in the chloroplasts, in two major groups of steps: the light-dependent and the light-independent reactions.

6.2 How Is Light Energy Converted to Chemical Energy?
The light-dependent reactions occur in the thylakoids. Light excites electrons in chlorophyll molecules and transfers the energetic electrons to electron transport chains. The energy of these electrons drives three processes:

- *Photosystem II generates ATP*: Some of the energy from the electrons is used to drive ATP synthesis.
- *Photosystem I generates NADPH*: Some of the energy, in the form of energetic electrons, is added to electron-carrier molecules of NADP$^+$ to make the highly energetic carrier NADPH.
- *Splitting water maintains the flow of electrons through the photosystems*: Some of the energy is used to split water, generating the electrons needed by photosystem II and producing hydrogen ions and oxygen as by-products.

Web Animation Properties of Light
BioFlix Photosynthesis
Web Animation Photosynthesis

6.3 How Is Chemical Energy Stored in Glucose Molecules?
In the stroma of the chloroplasts, both ATP and NADPH provide the energy that drives the light-independent reactions that synthesize glucose from CO_2 and H_2O. The light-independent reactions include:

- **Carbon fixation:** Carbon dioxide and water combine with RuBP to form PGA.
- **Synthesis of G3P:** PGA is converted to G3P, using energy from ATP and NADPH.
- **Regeneration of RuBP:** Ten molecules of G3P are used to regenerate six molecules of RuBP, again using ATP energy.
- **Synthesis of glucose:** Two G3P molecules are used to manufacture one molecule of glucose.

6.4 What Is the Relationship Between Light-Dependent and Light-Independent Reactions?
The light-dependent reactions produce the energy carrier ATP and the electron carrier NADPH. Energy from these carriers is used in the synthesis of organic molecules during the light-independent reactions. The depleted carriers, ADP and NADP$^+$, return to the light-dependent reactions for recharging.

6.5 How Does the Need to Conserve Water Affect Photosynthesis?
When stomata close to conserve water, RuBP may combine with O_2 (rather than with CO_2). Such photorespiration prevents carbon fixation and does not generate ATP. Some plants have evolved an additional series of reactions for carbon fixation that minimizes photorespiration. The most common forms of this additional step are the C_4 pathway and CAM.

adenosine triphosphate (ATP) *p. 87*
C₃ cycle *p. 90*
C₄ pathway *p. 93*
carbon fixation *p. 90*
chlorophyll *p. 87*
chloroplast *p. 87*

crassulacean acid metabolism (CAM) *p. 93*
electron transport chain (ETC) *p. 88*
light-harvesting complex *p. 88*
mesophyll *p. 86*

nicotinamide adenine dinucleotide phosphate (NADPH) *p. 87*
photorespiration *p. 93*
photosynthesis *p. 85*
photosystem *p. 88*

reaction center *p. 88*
ribulose bisphosphate (RuBP) *p. 90*
stoma (plural, **stomata**) *p. 86*
stroma *p. 87*
thylakoid *p. 87*

Thinking Through the Concepts

Suggested answers to end-of-chapter and figure-based questions can be found at the end of the text.

Fill-in-the-Blank

1. Plant leaves contain special openings called _____ that allow _____ gas to enter and _____ gas to leave. The process of photosynthesis occurs in organelles called _____, which are concentrated in _____ cells within leaves.

2. Chlorophyll captures wavelengths of light corresponding to the following three colors: _____, _____, and _____. Which color does chlorophyll reflect? _____. Accessory pigments called _____ capture wavelengths of light that correspond to the following two colors: _____ and _____. What two colors do accessory pigments reflect? _____ and _____.

3. During the "first" stage of photosynthesis, sunlight is captured by pigments located in _____. Here, the energy is transferred to special _____ chlorophylls where it energizes an electron. This electron then travels through a series of electron-carrier molecules called the _____. As the electron travels, it loses _____, which is captured in molecules of _____.

4. The oxygen produced as a by-product of photosynthesis comes from _____, and the carbons used to make glucose come from _____. The raw material to make glucose is produced during the light-_____ reactions, which include carbon fixation during the _____.

5. C₃ plants have trouble in dry climates because leaving their stomata open to allow _____ to enter also allows water to evaporate. Under these conditions, RuBP combines with oxygen in a wasteful process called _____. Some types of plants, called _____ plants, avoid this by fixing carbon in two stages. Such plants are adapted to hot and _____ conditions.

6. Complete this summary statement about photosynthesis: Light-dependent reactions take in _____ and release _____. Light-dependent reactions provide energy-rich _____ and _____ molecules for use in the C₃ cycle. Light-independent reactions use _____ from the air and produce molecules that are used to make _____.

Review Questions

1. Write the overall equation for photosynthesis. Does the overall equation differ between C₃ and C₄ plants?

2. Draw a diagram of a chloroplast and label it. Explain specifically how chloroplast structure is related to its function.

3. Briefly describe the light-dependent and light-independent reactions. In what part of the chloroplast does each occur?

4. What is the difference between carbon fixation in C₃ and in C₄ plants? Under what conditions does each mechanism of carbon fixation work most effectively?

Applying the Concepts

1. IS THIS SCIENCE? Chlorophyll extracted from plants is often sold in capsules. One Web site that offers chlorophyll for sale states that "chlorophyll tablets reduce or eliminate offensive body and breath odors." Do you think this claim is likely to be true? (You might want to research the causes of breath and body odors before answering.) Devise an experiment to test the claim.

2. Suppose an experiment is performed in which plant 1 is supplied with normal carbon dioxide but with water that contains radioactive oxygen atoms. Plant 2 is supplied with normal water but with carbon dioxide that contains radioactive oxygen atoms. Each plant is allowed to perform photosynthesis, and the oxygen gas

and sugars produced are tested for radioactivity. Which plant would you expect to produce radioactive sugars, and which plant would you expect to produce radioactive oxygen gas? Why?

3. You are called before the Ways and Means Committee of the U.S. House of Representatives to explain why the U.S. Department of Agriculture should continue to fund photosynthesis research. How would you justify the expense of producing, by genetic engineering, an enzyme that catalyzes the reaction of RuBP with CO_2 but prevents RuBP from reacting with oxygen? What are the potential applied benefits of this research?

For additional resources, go to www.mybiology.com.

chapter 7

Harvesting Energy: Glycolysis and Cellular Respiration

Tyler Hamilton is among the professional cyclists who have been penalized for artificially boosting the oxygen supply to their cells.

Case Study

When Athletes Boost Their Blood Counts: Do Cheaters Prosper?

Thousands of spectators cheered wildly as the leaders of the 50-km cross-country ski race entered the home stretch at the 2002 Winter Olympics in Salt Lake City, Utah. As the grueling race drew to its conclusion, the skiers were clearly exhausted, struggling to find the energy for a final burst. One skier, however, came on strong. Johann Mühlegg, competing for Spain, raced to the front of the pack and pulled away, finishing almost 15 seconds ahead of the second-place skier to claim the gold medal. But Mühlegg's triumph was short lived. Shortly after the race, he was stripped of his medals and expelled from the Games.

Blood-doping athletes like Johann Mühlegg gain endurance but put their health at risk.

The American cyclist Tyler Hamilton is another Olympic winner whose moment in the sun was brief. At the 2004 Summer Olympics in Athens, Greece, Hamilton won a gold medal for his performance in the road time trials. Within months of his victory, however, he was banned from racing for 2 years.

Why were Mühlegg and Hamilton punished? Because, in both cases, post-race tests found evidence of blood doping by the athletes. Blood doping improves a person's physical endurance by increasing the blood's ability to carry oxygen. Hamilton gained this increase with a blood transfusion, adding someone else's blood to his own. The extra blood cells from the transfusion greatly increased the oxygen-carrying capacity of his blood. Mühlegg acquired extra blood cells by a different method. He injected the drug darbepoetin, which mimics the effect of another blood-doping substance, erythropoietin (Epo). Epo is present in normal human bodies, where it stimulates bone marrow to produce more red blood cells. A healthy body produces just enough Epo to ensure that red blood cells are replaced as they age and die. An injection of extra Epo (or darbepoetin), however, can stimulate the production of a huge number of extra oxygen-carrying red blood cells.

Why is endurance improved by extra oxygen molecules in the bloodstream? Think about this question as we examine oxygen's role in supplying energy to muscle cells. ■

7.1 What Is the Source of a Cell's Energy?

Cells do a lot of work. They manufacture and transport an enormous number and variety of molecules, initiate a multitude of chemical reactions, continually replace worn-out parts, dispose of waste, divide periodically, and (in many cases) move about. To do all this work, cells need energy (which we defined in Chapter 5 as the capacity to do work).

The energy that powers a cell's activities is stored in the bonds of molecules such as fats and carbohydrates. The energy in these storage molecules, however, must be transformed before it can be used to power a cell's metabolism. To be usable, stored energy must, in most cases, be transferred to the bonds of energy-carrier molecules, especially **adenosine triphosphate (ATP).** For this reason, some of the most important reactions in cells are the ones that transfer energy from energy-storage molecules to energy-carrier molecules. These reactions take place in all cells, because all cells perform work.

Glucose Is a Key Energy-Storage Molecule

Although most cells can harvest energy from a variety of organic molecules, this chapter's description of the harvesting process will focus on glucose. Almost all cells break down glucose for energy.

Photosynthesis Is the Ultimate Source of Cellular Energy

The energy released during glucose metabolism (glucose breakdown in cells) was originally acquired by photosynthesis. Photosynthetic cells capture and store the energy of sunlight, and this stored energy is later used by cells. Those cells may be in the photosynthetic organisms themselves, or they may be in other organisms that directly or indirectly consume photosynthesizers. For example, caterpillars eat plants, bluebirds eat caterpillars, and hawks eat bluebirds. The cells that make up the plants, caterpillars, bluebirds, and hawks are all powered by energy stored in glucose that was originally captured by photosynthesis in the plant's leaves.

Glucose Metabolism and Photosynthesis Are Complementary Processes

As photosynthesis captures solar energy for storage in cells, it consumes water and carbon dioxide and produces glucose and oxygen. As glucose metabolism releases stored energy, it consumes glucose and oxygen and produces water and carbon dioxide. Thus, the two processes are almost perfectly complementary—the products of each process provide reactants for the other. This symmetry is apparent in the chemical equations for glucose formation by photosynthesis and for the complete metabolism of glucose back to CO_2 and H_2O:

Photosynthesis:

$6\ CO_2 + 6\ H_2O + \text{sunlight energy} \rightarrow C_6H_{12}O_6 + 6\ O_2$

Complete Glucose Metabolism:

$C_6H_{12}O_6 + 6\ O_2 \rightarrow 6\ CO_2 + 6\ H_2O + \text{chemical energy and heat energy}$

This symmetry might lead you to hypothesize that a cell can convert all of the chemical energy in a glucose molecule to high-energy bonds of ATP. But remember the second law of thermodynamics, which tells us that conversion of energy into different forms always decreases the amount of useful energy (see Chapter 5 to review the laws of thermodynamics). Thus, most of the energy released during the breakdown of glucose is lost as heat rather than converted to the chemical energy of ATP. Nevertheless, a cell can extract and capture a great deal of energy when glucose is completely broken down.

7.2 How Do Cells Harvest Energy from Glucose?

Glucose metabolism proceeds in stages, as summarized below and in **Figure 7-1.**

- The first stage, **glycolysis,** takes place in the cytoplasmic fluid and splits a single glucose molecule (a six-carbon sugar) into two three-carbon molecules of **pyruvate.** The split releases a small

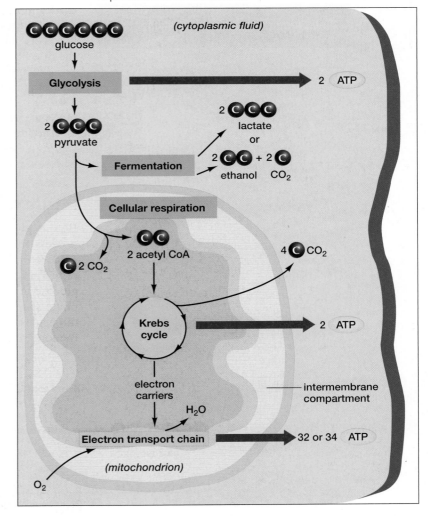

▼ Figure 7-1 **An overview of glucose metabolism** Refer to this diagram as we progress through the reactions of glycolysis (in the cytoplasmic fluid) and cellular respiration (in the mitochondrion). Glucose is broken down in stages, with energy captured in ATP along the way. Most of the ATP is produced in mitochondria.

amount of chemical energy, which is used to generate two ATP molecules. Glycolysis does not require oxygen and proceeds in exactly the same way under both *aerobic* (with oxygen) and *anaerobic* (without oxygen) conditions. Glycolysis is described in more detail in section 7.3.

■ When oxygen is available, the pyruvate produced by glycolysis is further processed in a second stage called **cellular respiration,** which uses oxygen to break the pyruvate down completely to carbon dioxide and water, generating an additional 34 or 36 ATP molecules (the number differs among cell types). Cellular respiration takes place in **mitochondria,** which are organelles specialized for the aerobic breakdown of pyruvate. Cellular respiration is described in more detail in section 7.4.

■ When oxygen is not available, the second stage of glucose metabolism is **fermentation.** In fermentation, the pyruvate produced by glycolysis does not enter the mitochondria, but instead remains in the cytoplasm and may be converted into lactate or ethanol. Fermentation does not produce additional ATP energy. We discuss fermentation in section 7.5.

7.3 What Happens During Glycolysis?

As cells do work, they consume ATP. A very active cell—say, a muscle cell in the leg of a running sprinter—consumes ATP rapidly. In a less active cell—say, a cell in the root of a pine tree during winter—ATP is consumed much more slowly. But regardless of the rate at which a cell consumes ATP, the cell's supply of this crucial energy-carrier molecule will sooner or later be depleted. Eventually, the concentration of ATP declines to a level low enough to activate an enzyme that is inhibited by high levels of ATP. The activated enzyme catalyzes the onset of glycolysis, thus initiating the first stage in the process of replenishing the cell's ATP.

Glycolysis is a series of reactions that, as previously mentioned, breaks a molecule of glucose into two molecules of pyruvate, and requires no oxygen to do so. The reactions of glycolysis produce relatively few energy carriers. Overall, each molecule of glucose yields just two molecules of ATP and two molecules of the electron carrier **nicotinamide adenine dinucleotide (NADH).** Nonetheless, glycolysis is an essential part of glucose metabolism. It consists of two major parts (each with several steps): *glucose activation* and *energy harvest* (**Fig. 7-2**).

Glucose Activation Consumes Energy

Before glucose is broken down to release its energy, it must be activated—a process that actually uses up energy. During glucose activation, a molecule of glucose undergoes two enzyme-catalyzed reactions, each of which uses ATP energy. These reactions use two phosphates from ATP to convert a relatively stable glucose molecule into a highly unstable, activated molecule of fructose bisphosphate (**Fig. 7-2**, step ❶). Forming fructose bisphosphate costs the cell two ATP molecules, an investment that is necessary to produce greater energy returns in the long run.

▶ Figure 7-2 The essentials of glycolysis

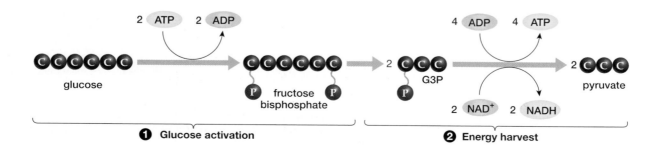

Energy Harvest Yields Energy-Carrier Molecules

In the energy-harvest steps, fructose bisphosphate splits apart into two three-carbon molecules of glyceraldehyde-3-phosphate (G3P). (In Chapter 6 we encountered G3P in the C_3 cycle of photosynthesis.) Each G3P molecule then goes through a series of reactions that converts it to pyruvate (**Fig. 7-2**, step ❷). During these reactions, two molecules of ATP are generated for each G3P, for a total of four ATPs. Because two ATPs were used to activate the glucose molecule in the first place, there is a net gain of only two ATPs per glucose molecule.

At another step along the way from G3P to pyruvate, the energized electron carrier NADH is produced (Fig. 7-2, step ❷). Each molecule of G3P yields one molecule of NADH, so two NADH carrier molecules are formed for each glucose molecule that is converted to pyruvate.

SUMMING UP:	Glycolysis

Each molecule of glucose is broken down to two molecules of pyruvate. During this process, ATP is both consumed and produced, but the process as a whole yields a net increase of two ATP molecules. In addition, two molecules of the electron carrier NADH are formed.

7.4 What Happens During Cellular Respiration?

If oxygen is present, glycolysis is followed by the second stage of glucose metabolism, cellular respiration. Cellular respiration is a series of reactions in which the pyruvate produced by glycolysis is broken down to carbon dioxide and water. Over the course of these reactions, large amounts of ATP are produced, and thus become available to power the activities of cells.

In eukaryotic cells, cellular respiration takes place in mitochondria. A mitochondrion has two membranes that produce two compartments. The inner membrane encloses a central compartment containing the fluid **matrix,** and the outer membrane surrounds the organelle, producing an **intermembrane compartment** between the membranes (**Fig. 7-3**, top). Active cells that consume large amounts of energy (and therefore require greater capacity to produce ATP) typically have more mitochondria than cells with lower energy needs.

For an overview of cellular respiration, review Figure 7-3 with the help of the summary below. A more detailed description follows the summary.

❶ The two molecules of pyruvate produced by glycolysis are transported into a mitochondrion, which is surrounded by two membranes, the inner membrane and the outer membrane. The pyruvate molecules cross both of them and pass into the fluid matrix that lies inside the inner membrane.

❷ Each three-carbon pyruvate is split into CO_2, which is released as a by-product, and a two-carbon molecule, acetyl CoA, which enters a series of reactions known as the **Krebs cycle.** The Krebs cycle produces one ATP from each pyruvate and donates energetic electrons to several molecules of the electron carriers NADH and **flavin adenine dinucleotide (FADH$_2$).** The Krebs cycle also releases CO_2.

❸ The electron carriers donate their energetic electrons to the electron transport chain, which is embedded in the inner membrane of the mitochondrion. After donating their electrons, the depleted carriers (NAD^+ and FAD) become available for recharging by the Krebs cycle.

❹ In the electron transport chain, the energy of the donated electrons is used to transport hydrogen ions from the matrix to the intermembrane compartment that lies between the inner and outer membranes.

When Athletes Boost Their Blood Counts: Do Cheaters Prosper?

Continued

An athlete's endurance will be improved if he or she can increase the amount of energy available to muscle cells. To be useful, the extra energy must be stored in the bonds of ATP molecules. One way to increase the amount of ATP in muscle cells is to increase the rate of glucose breakdown there. A blood-doping athlete increases the rate by using artificial methods to deliver extra oxygen to cells. Why does the extra oxygen increase glucose metabolism? Is it because the oxygen increases the rate at which glycolysis splits glucose molecules? No; we know that glycolysis does not require oxygen, so its rate cannot be affected by the concentration of oxygen. The reason for the effectiveness of blood doping, therefore, must lie not with its effect on glycolysis, but rather, with its effect on cellular respiration.

▶ **Figure 7-3 Cellular respiration**
Cellular respiration takes place in mitochondria. A mitochondrion's inner membrane separates the matrix (inner compartment) from the intermembrane compartment (between the inner and outer membranes).

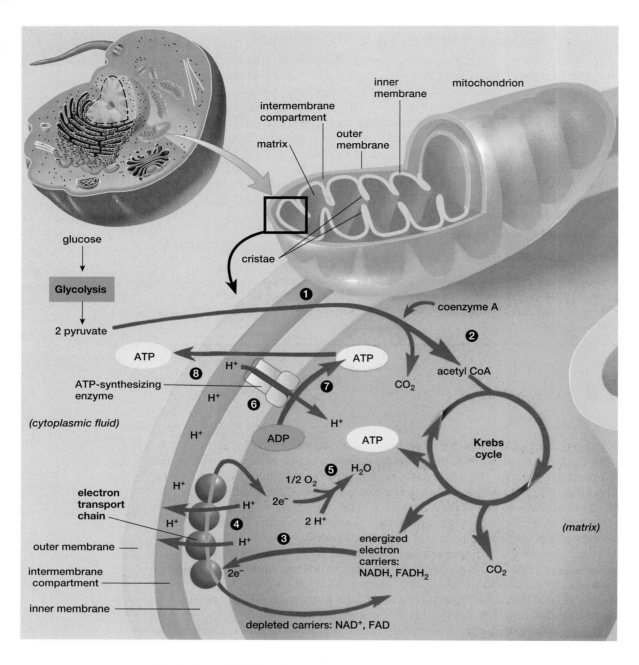

❺ At the end of the electron transport chain, the electrons combine with O_2 and H^+ to form H_2O.

❻ The hydrogen ions that were transported into the intermembrane compartment by the electron transport chain flow back across the inner membrane, down their concentration gradient.

❼ As the hydrogen ions travel back into the matrix, their flow provides the energy to produce ATP from ADP.

❽ ATP moves out of the mitochondrion into the fluid of the cytoplasm, where it provides energy for cellular activities.

The Krebs Cycle Breaks Down Pyruvate in the Mitochondrial Matrix

The fuel for the reactions of cellular respiration is the pyruvate produced by glycolysis in the cell cytoplasm. The pyruvate moves from the cytoplasm to the mito-

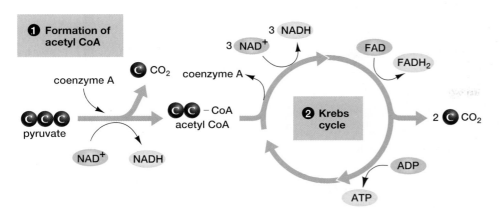

chondrial matrix, diffusing down its concentration gradient through pores in the two mitochondrial membranes.

After the pyruvate reaches the matrix, it reacts with a molecule called *coenzyme A* (**Fig. 7-4**, step ❶; see also Fig. 7-3, step ❶). In this reaction, each pyruvate molecule is split into CO_2 and a two-carbon molecule called an acetyl group. The acetyl group immediately attaches to coenzyme A, forming an *acetyl-coenzyme A complex* (acetyl CoA). During this reaction, two energetic electrons and a hydrogen ion are transferred to NAD^+, forming NADH.

The next group of reactions forms the cyclic pathway of the Krebs cycle (**Fig. 7-4**, step ❷), which is sometimes called the *citric-acid cycle* because citrate (the ionized form of citric acid) is the first molecule produced in the cycle. Citrate is formed when acetyl CoA combines with the four-carbon molecule oxaloacetate. As citrate forms, coenzyme A is released for reuse in further turns of the cycle. Each citrate then undergoes a series of reactions that regenerate oxaloacetate, give off two more CO_2 molecules, and capture most of the energy of the original acetyl group in one ATP and four electron carriers—one $FADH_2$ and three NADH.

SUMMING UP:	The Mitochondrial Matrix Reactions

Each pyruvate that enters the matrix reactions produces one CO_2 and one NADH during the synthesis of one acetyl CoA. Each acetyl CoA in turn yields two more CO_2, one ATP, three more NADH, and one $FADH_2$ via the Krebs cycle. Because each glucose feeds two pyruvates to the matrix reactions, the reactions yield a total of six CO_2 molecules, two ATPs, eight NADH electron carriers, and two $FADH_2$ electron carriers per glucose molecule.

Energetic Electrons Are Carried to Electron Transport Chains

At the end of the matrix reactions, only a small portion of the energy of glucose has been captured in ATP. The cell has thus far gained only four ATP molecules from each original glucose molecule: two during glycolysis and two during the Krebs cycle. The cell has, however, captured many energetic electrons in carrier molecules: 2 NADH during glycolysis plus 8 more NADH and 2 $FADH_2$ from the matrix reactions, for a total of 10 NADH and 2 $FADH_2$ (**Table 7-1**).

The carriers deposit their electrons in **electron transport chains** located in the inner mitochondrial membrane (**Fig. 7-5**, step ❶; see also Fig. 7-3, step ❹). After donating their electrons, the depleted carriers (NAD^+ and FAD) become available for recharging by the Krebs cycle.

Within the transport chains, the energetic electrons move systematically from molecule to molecule. During some of these transfers, energy is released and

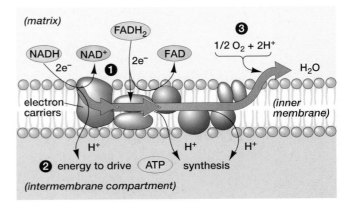

▲ **Figure 7-5 The electron transport chain in the inner mitochondrial membrane   QUESTION:** If no oxygen is present, how is the rate of ATP production affected?

Table 7-1   **Summary of Glycolysis and Cellular Respiration**

Process	Location	Reactions	Electron Carriers Formed (per glucose molecule)	ATP Yield (per glucose molecule)
Glycolysis	Fluid cytoplasm	Glucose breaks down to two pyruvates	2 NADH	2 ATP
Cellular Respiration				
Acetyl CoA formation	Matrix of mitochondrion	Each pyruvate combines with coenzyme A to form acetyl CoA and CO_2	2 NADH	none
Krebs cycle	Matrix of mitochondrion	Acetyl group of acetyl CoA is metabolized to two CO_2	6 NADH, 2 $FADH_2$	2 ATP
Electron transport	Inner membrane, intermembrane compartment	Energy of electrons from NADH and $FADH_2$ are used to pump H^+ into intermembrane compartment, H^+ gradient is used to synthesize ATP	n/a	32–34 ATP

When Athletes Boost Their Blood Counts: Do Cheaters Prosper?

Continued

We've seen that lack of oxygen interrupts the flow of electrons through the electron transport chain and thereby halts production of ATP (except for the small amount produced by glycolysis). This kind of interruption of ATP production often takes place in the muscle cells of exercising athletes, because cellular respiration in hard-working muscles consumes oxygen faster than it can be replaced by oxygen carried in the blood. In an endurance race, a great advantage accrues to a competitor who can delay the moment at which oxygen is exhausted and cellular respiration halted. Blood-doping athletes seek to gain this delay by artificially increasing the concentration of oxygen molecules in the blood that supplies their muscle cells.

used to pump hydrogen ions out of the matrix, across the inner mitochondrial membrane, and into the intermembrane compartment (**Fig. 7-5**, step ❷).

At the end of the electron transport chain, oxygen finally plays its part. An oxygen atom combines with two hydrogen ions and two energetically depleted electrons to form water (**Fig. 7-5**, step ❸). Oxygen thus accepts electrons after they run their course through the electron transport chain.

If oxygen is not present to pull spent electrons from the end of the transport chain, the electrons "pile up" in the chain, the process of pumping hydrogen ions out of the matrix comes to a halt, and cellular respiration ceases. When cellular respiration stops, little ATP is produced. ATP is required to supply the energy for cell metabolism, so most cells, and therefore most organisms, cannot survive for long without oxygen. (Certain kinds of bacteria are an exception to this generalization; see the discussion of fermentation in section 7.5.)

Energy from a Hydrogen-Ion Gradient Is Used to Produce ATP

The hydrogen ions that are pumped out of the matrix become a source of energy for the production of more ATP. The process of pumping hydrogen ions across the inner mitochondrial membrane generates a large H^+ concentration gradient; that is, the concentration of hydrogen ions in the intermembrane compartment becomes high and the concentration in the matrix becomes low. As dictated by the second law of thermodynamics, creating this nonuniform distribution of hydrogen ions requires an input of energy. It is the chemical equivalent of pumping water upward into an elevated storage tank. Energy is later released when the hydrogen ions are allowed to move down their concentration gradient—like opening the valves of the storage tank and allowing the water to rush out.

The energy released when the hydrogen ions move down their concentration gradient is captured and used to synthesize ATP molecule in a process called **chemiosmosis.** As hydrogen ions move from the intermembrane compartment to the matrix, they pass through protein channels that are part of ATP-synthesizing enzymes. The energy released as the ions flow through the channels is used to synthesize 32 to 34 molecules of ATP for each molecule of glucose that is broken down (see **Fig. 7-3**, steps ❻ and ❼). This ATP is transported from the matrix to the intermembrane compartment and then diffuses through the outer membrane of the mitochondrion to the surrounding cytoplasm, where it becomes available to power any of the cell's energy-consuming metabolic reactions.

Electrons from the electron carriers NADH and $FADH_2$ enter the electron transport chain of the inner mitochondrial membrane. Here their energy is used to generate a hydrogen-ion gradient across the inner membrane. The movement of hydrogen ions down the gradient through the pores of ATP-synthesizing enzymes drives the synthesis of 32 to 34 molecules of ATP. At the end of the electron transport chain, two electrons combine with one oxygen atom and two hydrogen ions to form water.

7.5 What Happens During Fermentation?

Some microorganisms live in places where oxygen is scarce or absent. Such oxygen-free environments can be found in sediments beneath lakes and oceans, deep beneath the surface of Earth, and in bogs and marshes. The anaerobic conditions in these places rule out cellular respiration, which requires oxygen as its final electron acceptor. So how do the inhabitants of anaerobic environments survive? They metabolize the products of glycolysis by fermentation.

Unlike the reactions of cellular respiration, the reactions of fermentation generate no ATP. Instead, they serve to regenerate the NAD^+ molecules that are needed for glycolysis. To understand why an alternative means of regenerating NAD^+ is needed in anaerobic conditions, recall that during cellular respiration when oxygen is present, the electron transport chains of the mitochondria accept electrons from NADH, transforming it into NAD^+ (see Fig. 7-5). But in an oxygen-free environment in which cellular respiration is impossible, the electron transport chains cannot accept electrons. If there were no alternative acceptor of electrons from NADH, the cell's supply of NAD^+ would quickly be used up, glycolysis would halt, and cells would lack even the meager amount of ATP produced by glycolysis.

In fermentation, the pyruvate produced by glycolysis acts as an electron acceptor for NADH. The pyruvate is converted to ethanol or lactate, and NADH is converted back to NAD^+. Thus, fermentation provides a way to regenerate NAD^+ so that glycolysis (and therefore modest ATP production) can continue.

There are two main types of fermentation: one converts pyruvate to ethanol and carbon dioxide and the other converts pyruvate to lactate.

Some Cells Ferment Pyruvate to Form Alcohol

The type of fermentation that yields ethanol and CO_2 is known as *alcoholic fermentation* and is the primary mode of metabolism in many microorganisms. The reactions of alcoholic fermentation use hydrogen ions and electrons from NADH, thus regenerating NAD^+(**Fig. 7-6**).

Alcoholic fermentation by single-celled fungi called yeasts is responsible for some gastronomic staples: wine, beer, and bread. Wine retains the ethanol product of the yeasts' fermentation of grape sugar. In most wines, the carbon dioxide is allowed to escape, but in sparkling wines and champagne the gas is retained. These bubbly wines are made by adding more yeast and sugar just before the wine is bottled. The added ingredients ensure that fermentation continues in the sealed bottle, trapping carbon dioxide inside (**Fig. 7-7**).

Like sparkling wines, beer retains both the alcohol and carbon dioxide (released as bubbles when the bottle is opened) produced by fermentation. In bread dough, however, the alcohol evaporates during baking, while the carbon dioxide is trapped in the dough and makes the bread rise.

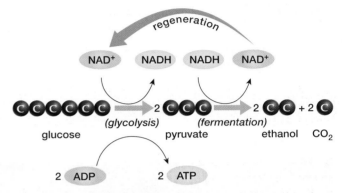

▲ **Figure 7-6** **Glycolysis followed by alcoholic fermentation**

▲ **Figure 7-7** **Alcoholic fermentation of grapes generates bubbles of carbon dioxide**   **QUESTION:** Some species of bacteria use aerobic respiration and other species use anaerobic (fermenting) respiration. In an oxygen-rich environment, would either type be at a competitive advantage? What about in an oxygen-poor environment?

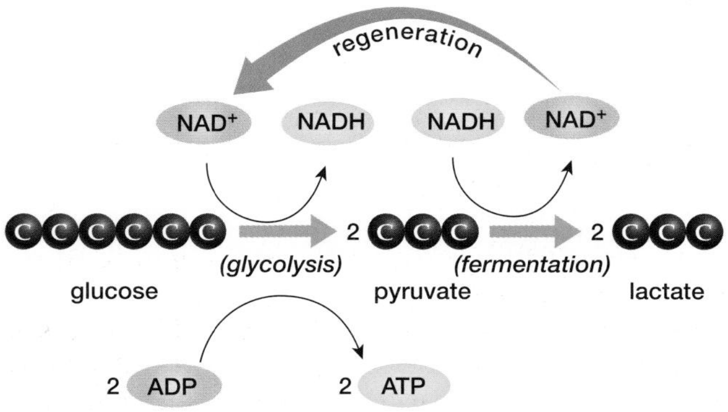

▲ Figure 7-8 Glycolysis followed by lactate fermentation

Other Cells Ferment Pyruvate to Lactate

Various microorganisms ferment pyruvate to lactate, including the bacteria that give yogurt, sour cream, and cheese their distinctive flavors. Lactate fermentation may also occur for short periods in the body cells of aerobic organisms when those cells are temporarily deprived of oxygen. For example, human muscle cells may use lactate fermentation during vigorous exercise. Working muscles need lots of ATP. They usually get the ATP from cellular respiration, which generates far more ATP than does glycolysis, but cellular respiration is limited by an organism's ability to capture oxygen (by breathing, for example). When you exercise vigorously, you may not be able to supply your muscles with enough oxygen to allow cellular respiration to meet the muscles' energy needs.

When deprived of adequate oxygen, your muscles do not immediately stop working. Instead, glycolysis continues for a while, providing its meager two ATP molecules per glucose and generating both pyruvate and NADH. Then, muscle cells ferment pyruvate molecules to lactate, a reaction that uses electrons and hydrogen ions from NADH and thereby regenerates NAD^+ (Fig. 7-8).

In high concentrations, however, lactate is toxic to your cells. It can cause intense discomfort and fatigue, forcing you to stop exercising or at least slow down. As you rest, breathing rapidly, oxygen once more becomes available and the lactate is converted back to pyruvate.

When Athletes Boost Their Blood Counts: Do Cheaters Prosper? Revisited

As you have seen, human cells most efficiently extract energy from glucose when an ample supply of oxygen is available to them. The aim of blood-doping athletes, then, is to extend as long as possible the period in which muscle cells have access to oxygen. During a difficult hill climb, a skier who has doped his blood with darbepoetin, or a cyclist who has doped his blood with a transfusion of red blood cells, has an unfair advantage. He continues to move efficiently, his muscle cells using cellular respiration to churn out abundant ATP. Meanwhile his "clean" competitors labor painfully, muscles laden with lactate from fermentation.

Blood doping is difficult to detect. Injected Epo (or its synthetic equivalent, darbepoetin) cannot be easily distinguished from the Epo that forms naturally in the human body. Sports officials assert that the difficulty of detecting Epo and darbepoetin have made them the drugs of choice among blood-doping skiers, cyclists, distance runners, and other competitive athletes. But as tests to detect the drugs improve, "old-fashioned" blood-doping methods such as transfusions are making a comeback. (Dopers who use transfusions typically use their own, previously stored blood, but even transfusions of others' blood are hard

to detect, because foreign blood cells are similar in almost all respects to a person's own blood cells.)

Contestants at the Olympics are now routinely tested for blood doping. The tests are mainly intended to deter athletes who seek to gain an unfair advantage, but the crackdown on doping also helps protect the athletes. The extra blood cells that are the goal of blood doping tend to thicken the blood and make it harder to pump through blood vessels, so those who inject Epo or darbepoetin suffer increased risk of heart attacks. A number of young professional cyclists have died of heart attacks in recent years, and many observers suspect that blood doping is to blame. It appears that some professional athletes are risking their lives to get more oxygen molecules to their muscles.

Consider This

Advances in gene therapy may one day make it possible to modify athletes' kidney cells so that they have extra copies of the genes that produce Epo. In your view, would such genetically modified individuals have an unfair advantage?

Chapter Review

For additional study help and activities, go to www.mybiology.com.

7.1 What Is the Source of a Cell's Energy?

Cells require energy to survive and function. To be usable by a cell, energy must be in the form of chemical energy in the bonds of the energy-carrier molecule ATP. A key component of cellular metabolism consists of reactions that harvest energy from food molecules such as glucose and convert it to ATP energy. The ultimate source of all cellular energy is solar energy that is captured during photosynthesis.

7.2 How Do Cells Harvest Energy from Glucose?

Cells produce usable energy by breaking down glucose to lower-energy compounds and capturing some of the released energy as ATP. In glycolysis, glucose is metabolized in the fluid portion of the cytoplasm to two molecules of pyruvate, generating two ATP molecules. If oxygen is available, the pyruvates are metabolized to CO_2 and H_2O through cellular respiration in the mitochondria, generating much additional ATP.

7.3 What Happens During Glycolysis?

During glycolysis, a molecule of glucose is activated to form fructose bisphosphate. In a series of reactions, the fructose bisphosphate is broken down into two molecules of pyruvate. These reactions produce four ATP molecules and two NADH electron carriers. Because two ATPs were used in the activation steps, the net yield from glycolysis is two ATPs and two NADHs.

BioFlix Cellular Respiration

Web Animation Glucose Metabolism

7.4 What Happens During Cellular Respiration?

If oxygen is available, cellular respiration can occur. The pyruvates are transported into the matrix of the mitochondria. In the matrix, each pyruvate reacts with coenzyme A to form acetyl CoA plus CO_2. One NADH is also formed at this step. The two-carbon acetyl group of acetyl CoA enters the Krebs cycle, which releases the two carbons as CO_2. One ATP, three NADHs, and one $FADH_2$ are also formed for each acetyl group that goes through the cycle. Thus, the Krebs cycle generates 2 ATPs for each molecule of glucose.

The NADHs and $FADH_2$s from glycolysis and the matrix reactions deliver their energetic electrons to the electron transport chain. The energy of the electrons is used to pump hydrogen ions across the inner membrane from the matrix to the intermembrane compartment. At the end of the electron transport chain, the depleted electrons combine with hydrogen ions and oxygen to form water. This is the oxygen-requiring step of cellular respiration. During chemiosmosis, the hydrogen-ion gradient created by the electron transport chain is used to produce ATP, as the hydrogen ions diffuse back across the inner membrane through channels in ATP-synthesizing enzymes. Electron transport and chemiosmosis yield 32 to 34 additional ATPs, for a net yield of 36 to 38 ATPs per glucose molecule.

Figure 7-1 and Table 7-1 summarize the locations, major processes, and overall energy harvest for the complete metabolism of glucose from glycolysis through cellular respiration.

7.5 What Happens During Fermentation?

In the absence of oxygen, the pyruvate produced by glycolysis cannot enter the reactions of cellular respiration. Instead, it is converted by fermentation to lactate or ethanol and CO_2. Fermentation regenerates NAD^+ so glycolysis may continue. However, no additional ATP is gained by fermentation.

Key Terms

adenosine triphosphate (ATP) *p. 98*

cellular respiration *p. 100*

chemiosmosis *p. 104*

electron transport chain *p. 103*

fermentation *p. 100*

flavin adenine dinucleotide (**FADH₂**) *p. 101*

glycolysis *p. 99*

intermembrane compartment *p. 101*

Krebs cycle *p. 101*

matrix *p. 101*

mitochondria *p. 100*

nicotinamide adenine dinucleotide (**NADH**) *p. 100*

pyruvate *p. 99*

Thinking Through the Concepts

Suggested answers to end-of-chapter and figure-based questions can be found at the end of the text.

Fill-in-the-Blank

1. The breakdown of glucose in the presence of oxygen occurs in two stages, first _____ and second _____. Where in the cell does the first stage occur? _____ Where does the second stage occur? _____ Which stage produces most of the ATP? _____.

2. In the absence of oxygen, many microorganisms break down glucose using the process of _____, which generates only _____ molecules of ATP. This process is followed by _____, in which no ATP is produced, but the high-energy electron-carrier molecule _____ is freed for reuse in further glucose breakdown.

3. Yeast in bread dough and alcoholic beverages use a type of fermentation that generates both _____ and _____.

4. During vigorous exercise, the energy demanded by an athlete's muscles may exceed the availability of _____ necessary to carry out cellular respiration. For a short time, muscles may be powered by the process of _____, followed by _____ fermentation, which causes the substance _____ to accumulate. As the athlete rests and breathes rapidly, this substance is converted back to _____ and used in cellular respiration.

5. During cellular respiration, the electron transport chain pumps hydrogen ions out of the mitochondrial _____ into the _____ of the mitochondrion, producing a large _____ of hydrogen ions. The ATP produced by cellular respiration is generated by a process called _____. During this process, hydrogen ions travel through membrane channels that are part of _____-synthesizing enzymes.

6. The cyclic pathway in cellular respiration is called the _____, or the _____. This cyclic pathway begins when _____ combines with oxaloacetate to form _____. The pathway liberates two molecules of CO_2 and "charges up" two different, high-energy electron carrier molecules: _____ and _____.

Review Questions

1. Starting with glucose ($C_6H_{12}O_6$), write the overall reactions for (a) glycolysis plus cellular respiration and (b) glycolysis plus fermentation in yeast.

2. Draw a labeled diagram of a mitochondrion, and explain how its structure is related to its function as the powerhouse of the cell.

3. What role do the following play in the complete metabolism of glucose: (a) glycolysis, (b) chemiosmosis, (c) fermentation, and (d) NAD^+?

4. Outline the major steps in the complete breakdown of glucose under (a) aerobic condtions and (b) anaerobic conditions. What is the overall energy harvest in ATP molecules generated per glucose molecule for each?

5. Describe the Krebs cycle. In what form is most of the energy harvested?

6. Describe the mitochondrial electron transport chain and the process of chemiosmosis.

7. Why is oxygen necessary for cellular respiration?

Applying the Concepts

1. IS THIS SCIENCE? Dinitrophenol (DNP) was a popular diet drug in the 1930s, but its use declined as potential users became aware of its dangerous toxicity. Today, the Food and Drug Administration (FDA) bans prescriptions for DNP, but it is nonetheless readily available for purchase on Web sites that offer it to desperate dieters and obsessive bodybuilders. DNP acts by breaking down the hydrogen ion gradient required for chemiosmosis and ATP production in mitochrondia. On the basis of your understanding of the reactions of cellular respiration, do you think that DNP really causes weight loss? Why or why not? Explain why DNP can kill those who take it.

2. More than a century ago, French biochemist Louis Pasteur described a phenomenon, now called the "Pasteur effect," in the wine-making process. He observed that in a sealed container of grape juice containing yeast, the yeast will consume the sugar very slowly as long as oxygen remains in the container. As soon as the oxygen is gone, however, the rate of sugar consumption by the yeast increases greatly and the alcohol content in the container rises. Discuss the Pasteur effect on the basis of what you know about cellular respiration and fermentation.

3. Imagine that a starving cell has reached the stage where every bit of its ATP has been depleted and converted to ADP plus phosphate. If that cell were placed in fresh nutrient broth at this point, would it recover and survive? Explain your answer on the basis of what you know about glucose breakdown.

For additional resources, go to www.mybiology.com.

unit two

Inheritance

Inheritance provides for both similarity and difference. All dogs share many similarities because their genes are nearly identical. The enormous variety of sizes, fur length and color, and bodily proportions result from tiny differences in their genes.

chapter 8

The Continuity of Life: How Cells Reproduce

Snuppy (pronounced Snoopy; center), the first cloned dog, poses for a family portrait with his genetic donor, the Afghan hound Tai (left), and his surrogate mother, a Labrador retriever (right).

Case Study   Send in the Clones

It seems that every few months the newspapers report yet another cloned animal—sheep (remember the famous Dolly?), mice, cattle, goats, cats, rats, horses, and now a dog. The first cloned dog, named Snuppy (short for "Seoul National University puppy," and pronounced Snoopy), is virtually identical to his "genetic parent," a male Afghan hound named Tai. Why do people keep cloning new species of animals? And why *is* a clone so similar to its "genetic parent"?

There are both scientific and practical reasons to clone animals. Because cloned animals are genetically identical to one another, they can provide important insights into normal development, responses to drugs, and susceptibility to environmental pollutants. For example, let's suppose that an environmental toxicologist used "normal," genetically diverse animals to determine what concentration of ozone, a common pollutant in city air, causes lung damage. She might take several groups of animals, raise one group under ozone-free conditions, and expose the other groups to various concentrations of ozone. Any differences in lung damage between control animals and ozone-exposed animals that she might observe will be caused both by ozone exposure and by the animals' genetic differences. If identical clones were used, the only differences would be caused by ozone exposure, making subtle health effects much easier to determine.

Cloning might also help to preserve endangered species. In 2003, African wildcats and bantengs (an endangered relative of the cow) were cloned. The banteng was cloned from cells that had been frozen for 23 years at the San Diego Zoo's Center for the Reproduction of Endangered Species, raising the possibility that endangered species could be brought back by cloning, even if there were no longer any living specimens. Of course, unless humans are willing to preserve suitable habitats, cloned bantengs or other endangered species will remain mere zoo specimens, and can scarcely be considered triumphs of conservation.

For emotional and economic reasons, domestic animals like Snuppy are favorite subjects for cloning. For example, in 2006, the ten-time world champion barrel-racing horse, Scamper, was cloned (photo at left). Why not just breed Scamper the old-fashioned way? By the time you've finished this chapter, you'll know why many famous horses don't sire offspring that are as athletic as they are. In any case, Scamper was a 29-year-old gelding—a castrated male—so he couldn't breed anyway.

Why is a clone nearly identical to its "genetic parent," while offspring are often so different from their biological parents and from each other? Keep that question in mind as we explore two types of cellular reproduction—mitotic and meiotic cell division—that provide for both constancy and variability in eukaryotic organisms. ■

Scamper's clone, named Clayton, already shows signs of Scamper's athleticism and temperament.

8.1 Why Do Cells Divide?

"All cells come from cells." This insight, first stated by the German physician Rudolf Virchow in the mid-1800s, captures the critical importance of cellular reproduction for all living organisms, from bacteria multiplying in an infected cut to a fertilized egg growing into an adult human. Cells reproduce by **cell division,** in which one cell gives rise to two or more cells, called **daughter cells.** In typical cell division, each daughter cell receives about half the cytoplasm and a complete set of hereditary information, usually identical to the hereditary information of the parent cell.

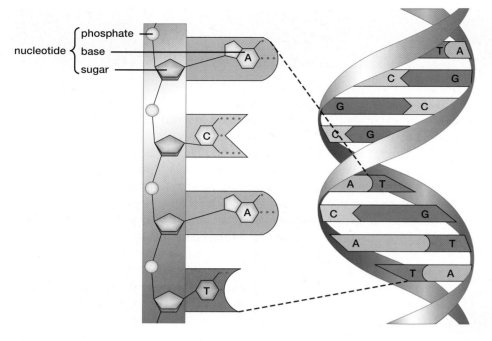

phosphate
nucleotide { base
sugar

(a) A single strand of DNA

(b) The double helix

▲ **Figure 8-1 The structure of DNA** *(a)* A nucleotide consists of a phosphate, a sugar, and one of four bases: adenine (A), thymine (T), guanine (G), or cytosine (C). A single strand of DNA consists of a long chain of nucleotides held together by bonds between the phosphate of one nucleotide and the sugar of the next. *(b)* Two DNA strands twist around one another to form a double helix.

Cell Division Transmits Hereditary Information to Each Daughter Cell

The hereditary information of all living cells is **deoxyribonucleic acid (DNA),** which is contained in one or more **chromosomes.** A single molecule of DNA consists of a long chain composed of smaller subunits called **nucleotides** (Fig. 8-1a). Each nucleotide consists of a phosphate, a sugar (deoxyribose), and one of four bases—adenine (A), thymine (T), guanine (G), or cytosine (C). In a chromosome, hydrogen bonds between the bases hold two strands of DNA together—adenine bonds with thymine, and guanine bonds with cytosine. The two DNA strands are wound around each other, as a ladder would look if it were twisted into a corkscrew shape. This structure is called a **double helix** (Fig. 8-1b). We will examine the structure of DNA in detail in Chapter 10.

Segments of DNA from a few hundred to many thousands of nucleotides long are the units of inheritance—the **genes**—that store genetic information in a cell. Like the letters of an alphabet, in a language with *very* long sentences, the specific sequences of nucleotides in genes spell out the instructions for making the proteins of a cell. We will see in Chapters 10 and 11 how DNA encodes genetic information and how a cell regulates which genes it uses at any given time.

For a cell to survive, it must have a complete set of genetic instructions. Therefore, when a cell divides, it cannot simply split its set of genes in half and give each daughter cell half a set. Rather, the cell must first replicate its DNA to make two identical copies, much like making a photocopy of an instruction manual. Each daughter cell then receives a complete "DNA manual" containing all of the genes.

Cell Division Is Required for Growth and Development

The familiar form of cell division in eukaryotic cells, in which each daughter cell is genetically identical to the parent cell, is called mitotic cell division (see sections 8.4 and 8.5). Since your conception as a single fertilized egg, mitotic cell division has produced all of the cells in your body, and continues every day in many organs, such as your skin and intestines. After cell division, the daughter cells may grow and divide again, or they may differentiate, becoming specialized for certain functions, such as contraction (muscle cells), fighting infections (white blood cells), or producing digestive enzymes (various cells of the pancreas, stomach, and intestine). This repeating pattern of divide, grow and (possibly) differentiate, then divide again is called the **cell cycle** (see sections 8.2 and 8.3).

Most multicellular organisms have three categories of cells, based on their abilities to divide and differentiate:

- **Stem cells.** Most of the daughter cells formed by the first few cell divisions of a fertilized egg, and a relatively small number of cells scattered throughout the bodies of adults, including in the heart, skin, intestines, brain, and bone marrow, are **stem cells.** Stem cells retain the ability to divide, perhaps for the entire life of the organism. In many cases, when a stem cell divides, one of its daughters remains a stem cell, replenishing the population of stem cells. The other daughter cell undergoes several additional rounds of cell division, but the resulting cells eventually differentiate. Depending on the combination of hormones and other "growth factors" that they encounter, these daughter cells can differentiate into a variety of cell types. Some stem cells in early embryos, in fact, can produce any of the cell types of the entire body.

- **Dividing cells with limited differentiation.** Many cells of the bodies of embryos, juveniles, and adults can divide, but typically differentiate only into one or two different cell types. Dividing cells in your liver, for example, can only become more liver cells.

- **Permanently differentiated cells.** Some cells differentiate and never divide again. For example, most of the cells in your heart and brain cannot divide.

Cell Division Is Required for Sexual and Asexual Reproduction

Sexual reproduction in eukaryotic organisms occurs when offspring are produced by the fusion of **gametes** (sperm and eggs) from two adults. Cells in the adult's reproductive system undergo a specialized type of cell division called meiotic cell division (described in section 8.6) to produce daughter cells with exactly half of the genetic information of their parent cells and of "ordinary" body cells. In animals, these cells become sperm or eggs. When a sperm fertilizes an egg, the resulting offspring once again contain the full complement of hereditary information typical of that species.

Asexual reproduction occurs by mitotic cell division and produces offspring that are genetically identical to the parent. During asexual reproduction, offspring are formed from a single parent, without having a sperm fertilize an egg. Bacteria (**Fig. 8-2a**) and single-celled eukaryotic organisms,

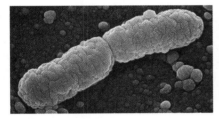

(a) Dividing bacteria

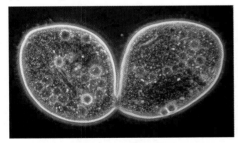

(b) Cell division in *Paramecium*

bud

(c) *Hydra* reproduces asexually by budding

The trees in this grove have already lost their leaves

The trees in this grove have begun to change color

The trees in this grove are still green

(d) A grove of aspens often consists of genetically identical trees produced by asexual reproduction

◄ **Figure 8-2 Cell division enables asexual reproduction** *(a)* Bacteria reproduce asexually by dividing in two. *(b)* In single-celled eukaryotic microorganisms, such as the protist *Paramecium,* cell division produces two new, independent organisms. *(c) Hydra,* a freshwater relative of the sea anemone, grows a miniature replica of itself (a bud) on its side. When fully developed, the bud breaks off and assumes independent life. *(d)* Trees in an aspen grove are often genetically identical. Each tree grows up from the roots of a single ancestral tree. This photo shows three separate groves near Aspen, Colorado. In fall, the timing of the changing color of their leaves shows the genetic identity within a grove and the genetic difference between groves.

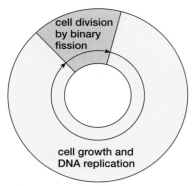

(a) The prokaryotic cell cycle

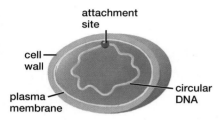

❶ The circular DNA double helix is attached to the plasma membrane at one point.

❷ The DNA replicates and the two DNA double helices attach to the plasma membrane at nearby points.

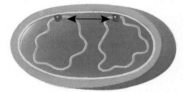

❸ New plasma membrane is added between the attachment points, pushing them further apart.

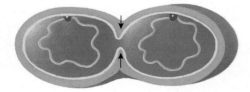

❹ The plasma membrane grows inward at the middle of the cell.

❺ The parent cell divides into two daughter cells.

(b) Binary fission

such as *Paramecium* commonly found in ponds (**Fig. 8-2b**), reproduce asexually by cell division, in which two new cells arise from each preexisting cell. Some multicellular organisms can also reproduce by asexual reproduction. A *Hydra* reproduces by growing a small replica of itself, called a bud, on its body (**Fig. 8-2c**). Eventually, the bud is able to live independently and separates from its parent. Many plants and fungi reproduce both asexually and sexually. The beautiful aspen groves of Colorado, Utah, and New Mexico (**Fig. 8-2d**) develop asexually from shoots growing up from the root system of a single parent tree. Although a grove looks like a population of separate trees, it is often a single individual whose multiple trunks are interconnected by a common root system. Aspen can also reproduce by seeds, which result from sexual reproduction.

Both prokaryotic and eukaryotic cells have cell cycles that include growth, metabolic activity, DNA replication, and cell division. However, major structural and functional differences exist between prokaryotic and eukaryotic cells, including differences in the organization of their DNA—the structure, size, number, and location of their chromosomes. Therefore, in this chapter we first describe the organization of the prokaryotic chromosome and the prokaryotic cell cycle. We then examine the eukaryotic chromosome and the essential processes of the eukaryotic cell cycle. Finally, we describe the two types of cell division in eukaryotes—mitotic cell division and meiotic cell division—and see how meiotic cell division produces genetic variation.

8.2 What Occurs During the Prokaryotic Cell Cycle?

A single, circular chromosome about a millimeter or two in circumference contains all of the DNA of a prokaryotic cell. A prokaryotic chromosome also contains proteins bound to the DNA. Unlike eukaryotic chromosomes, prokaryotic chromosomes are not contained in a membrane-bound nucleus (see Chapter 4).

The prokaryotic cell cycle consists of a relatively long period of growth—during which the cell also replicates its DNA—followed by a type of cell division called **binary fission,** which means "splitting in two" (**Fig. 8-3a**). The prokaryotic chromosome is usually attached at one point to the plasma membrane of the cell (**Fig. 8-3b**, step ❶). During the "growth phase" of the prokaryotic cell cycle, the DNA is replicated, producing two identical chromosomes that become attached to the plasma membrane at nearby, but separate, points (**Fig. 8-3b**, step ❷). As the cell grows, new plasma membrane is added between the attachment points of the chromosomes, pushing them apart (**Fig. 8-3b**, step ❸). When the cell has approximately doubled in size, the plasma membrane around the middle of the cell grows inward between the two DNA attachment sites (**Fig. 8-3b**, step ❹). Fusion of the plasma membrane through the middle of the cell completes binary fission, producing two daughter cells, each containing one of the chromosomes (**Fig. 8-3b**, step ❺). Because DNA replication produces two identical DNA molecules, the two daughter cells are genetically identical to one another and to the parent cell.

Under ideal conditions, binary fission in prokaryotes occurs rapidly. For example, the common intestinal bacterium *Escherichia coli* (usually called simply *E. coli*) can grow, replicate its DNA, and divide in about 20 minutes. Luckily, the environment in our intestines isn't ideal for bacteria growth; otherwise, the bacteria would soon outweigh the rest of our bodies!

◀ **Figure 8-3 The prokaryotic cell cycle (a)** The prokaryotic cell cycle consists of growth and DNA replication, followed by binary fission. **(b)** Binary fission in prokaryotic cells.

8.3 How Is the DNA in Eukaryotic Cells Organized?

Unlike prokaryotic chromosomes, eukaryotic chromosomes are separated from the cytoplasm in a membrane-bound nucleus. Further, eukaryotic cells always have multiple chromosomes. The smallest number, two, is found in the cells of females of a species of ant, but most animals have dozens and some ferns have more than 1,200! Each eukaryotic chromosome usually also contains more DNA than a prokaryotic chromosome does. Human chromosomes, for example, are about 10 to 80 times longer than the typical prokaryotic chromosome, and contain 10 to 50 times more DNA. The complex events of eukaryotic cell division are largely an evolutionary solution to the problem of sorting out a large number of long chromosomes. Therefore, to help you understand eukaryotic cell division better, we begin by taking a closer look at the structure of the eukaryotic chromosome.

The Eukaryotic Chromosome Consists of DNA Bound to Proteins

Fitting a huge amount of DNA into a nucleus only a few micrometers in diameter is no trivial task. For example, human chromosomes each contain a single DNA double helix, about 50 million to 250 million nucleotides long, and would be from 0.5 to 3 inches long (15 to 75 mm) if the DNA were completely relaxed and extended (Fig. 8-4a). However, for most of a cell's life, the DNA in each chromosome is coiled and looped around a variety of proteins, making the DNA almost 1,000 times shorter. When a cell needs to "read" some of its genetic information, it temporarily frees specific regions of DNA from these proteins, but the DNA is soon repacked. Nevertheless, even this enormous degree of compaction still leaves the chromosomes much too long to be sorted out and moved into the daughter nuclei during cell division. Just as thread is easier to organize when it is wound onto spools, sorting and transporting chromosomes is easier when they are condensed and shortened. During cell division, proteins fold up the DNA of each chromosome into compact structures that are about 10 times shorter than they are during the rest of the cell cycle (Fig. 8-4b).

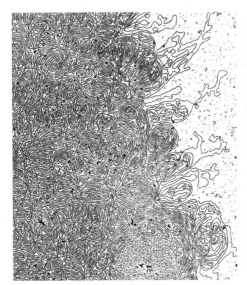

(a) Completely extended DNA

(b) A chromosome during cell division

◀ Figure 8-4 The eukaryotic chromosome (a) An electron micrograph of a small piece of completely extended DNA. At this magnification, a typical 2-inch long human chromosome would be about 3 miles long, so no photo can show an entire chromosome. (b) An electron micrograph of a human chromosome during cell division, when the chromosome has been condensed to about 4 micrometers in length. The "frayed" edges of the chromosome are loops of DNA and proteins, of about the same diameter that a chromosome would be when the cell is not dividing.

(a) A replicated chromosome consists of two sister chromatids

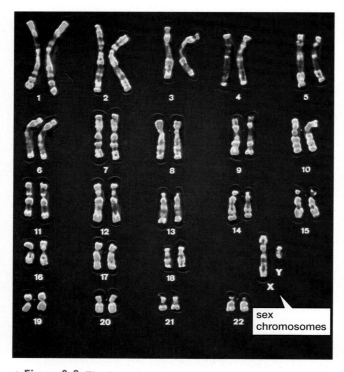

(b) Sister chromatids separate during cell division

▲ **Figure 8-5 Eukaryotic chromosomes during cell division** *(a)* Just before cell division, chromosomes coil up to form thick, short structures only about 2 to 20 micrometers long. Because the chromosomes have been replicated, they typically are "X-shaped" (see Fig. 8-4b), consisting of two sister chromatids joined at the centromere. *(b)* As cell division progresses, the sister chromatids separate, forming two independent chromosomes.

▲ **Figure 8-6 The karyotype of a human male** To make this picture, one man's chromosomes were stained and photographed. Individual chromosomes were then cut out of the photo and arranged in descending order of size. The two chromosomes above each number are a homologous pair that are similar in size and staining patterns and that have similar genetic material. Notice that the Y chromosome is much smaller than the X chromosome. If this karyotype were from a female, there would be two X chromosomes and no Y chromosome.

Duplicated Chromosomes Separate During Cell Division

Prior to cell division, the DNA within each chromosome is replicated, by a mechanism that we will describe in Chapter 10. At the end of DNA replication, the duplicated chromosome consists of two DNA double helices and associated proteins that remain attached to each other at the **centromere** (Fig. 8-5a). Although *centromere* literally means "middle body," a chromosome's centromere can be located almost anywhere along the DNA. While the two chromosomes are attached at their centromeres, we refer to each attached chromosome as a sister **chromatid.** Thus, DNA replication produces a duplicated chromosome with two identical sister chromatids. As you will see shortly, however, during mitotic cell division the two sister chromatids separate, and each chromatid becomes an independent chromosome that is delivered to one of the two daughter cells (Fig. 8-5b).

Eukaryotic Chromosomes Usually Occur in Pairs

When we view an entire set of stained chromosomes from a single cell (the *karyotype*), we see that the nonreproductive cells of many organisms, including humans, contain pairs of chromosomes (Fig. 8-6). With one exception that we will discuss shortly, both members of each pair are the same length and have the same staining pattern. This similarity in size, shape, and staining occurs because each chromosome in a pair carries the same genes arranged in the same order. Chromosomes that contain the same genes are called **homologous chromosomes,** or **homologues,** from Greek words that mean "to say the same thing." Cells with pairs of homologous chromosomes are called **diploid,** meaning "double."

Homologous Chromosomes Are Usually Not Identical

Despite their name, homologues usually don't say exactly the "same thing." Why not? Well, a cell might make a mistake when it copies the DNA of one homologue but not when it copies the DNA of the other homologue (see Chapter 10). Or a ray of ultraviolet light from the sun might zap the DNA of one homologue, damaging it. These changes in the sequence of nucleotides in DNA are called *mutations*, and will make one homologue "read" a little differently than its pair. A given mutation might have happened yesterday, or perhaps 10,000 years ago and have been inherited ever since. If we think of the DNA as an instruction manual for building a cell or an organism, then mutations are like misspelled words in the manual. Although some misspellings might not matter much, others might have serious consequences. As we'll see in the "Health Watch: Cancer-Mitotic Cell Division Run Amok" on p. 119, some mutations allow cell division to run rampant, causing life-threatening cancers.

As shown in Figure 8-6, a typical human cell has 23 pairs of chromosomes, for a total of 46. There are two copies of chromosome 1, two copies of chromosome 2, and so on, up through chromosome 22. These chromosomes—which have similar appearance, similar (but not identical) DNA sequences, and are paired in diploid cells of both sexes—are called **autosomes.** The cell also has two **sex chromosomes:** either two X chromosomes (in females) or an X and a Y chromosome (in males). The X and Y chromosomes are quite different in size (see Fig. 8-6) and in genetic composition. Thus, sex chromosomes are an exception to the rule that homologous chromosomes contain the same genes. However, the X and Y chromosomes behave as a pair during meiotic cell division.

Not All Cells Have Paired Chromosomes

Most cells within our bodies are diploid. However, during sexual reproduction, cells in the ovaries or testes undergo a special type of cell division, called meiotic cell division (see section 8.6) to produce gametes (sperm or eggs).

Gametes contain only one member of each pair of autosomes and one of the two sex chromosomes. Cells that contain only one of each type of chromosome are called **haploid** (meaning "half"). In humans, a haploid cell contains one each of the 22 autosomes, plus either an X or Y sex chromosome, for a total of 23 chromosomes. (Think of a haploid cell as one that contains half the diploid number of chromosomes, or one of each type of chromosome. A diploid cell contains two of each type of chromosome.) When a sperm fertilizes an egg, fusion of the two haploid cells produces a diploid cell with two copies of each type of chromosome. Because one member of each pair of homologues was inherited from the mother (in her egg), these are usually called maternal chromosomes. The chromosomes inherited from the father (in his sperm) are called paternal chromosomes.

In biological shorthand, the number of different types of chromosomes in a species is called the *haploid number* and is designated *n*. For humans, $n = 23$ because we have 23 different types of chromosomes (autosomes 1 to 22 plus one sex chromosome). Diploid cells contain $2n$ chromosomes. Thus, the body cells of humans have 46 (2×23) chromosomes. Every species has a specific number of chromosomes in its cells, from just a handful (e.g., 6 in mosquitoes) to hundreds (in shrimp and some plants).

Not all organisms are diploid. The bread mold *Neurospora*, for example, has haploid cells for most of its life cycle. Some plants, on the other hand, have more than two copies of each type of chromosome, with four, six, or even more copies of each chromosome per cell.

Web Animation Cell Division in Humans

8.4 What Occurs During the Eukaryotic Cell Cycle?

The eukaryotic cell cycle is divided into two major phases: interphase and mitotic cell division (Fig. 8-7). During **interphase,** the cell acquires nutrients from its environment, grows, and duplicates its chromosomes. During cell division, one copy of each chromosome and usually about half the cytoplasm (including mitochondria, ribosomes, and other organelles) are parceled out into each of the two daughter cells.

Most eukaryotic cells spend the majority of their time in interphase. For example, some cells in human skin, which divide about once a day, spend roughly 22 hours in interphase and only a couple of hours dividing. Immediately after cell division, a newly formed daughter cell enters interphase, where it grows and may differentiate. During this time, the cell is sensitive to internal and external signals that may cause it to continue through the cell cycle and divide, to differentiate into a particular cell type, or even to commit "cellular suicide."

If these signals trigger continuation of the cell cycle, the cell replicates its DNA. After DNA replication, the cell may grow some more before dividing. Often, if the signals trigger differentiation, the cell leaves the cell cycle. It may continue to grow in size, but it does not replicate its DNA or divide. Some differentiated cells, including most of those in your heart, eyes, and brain, never divide again. This is one reason why heart attacks and strokes are so devastating: the dead cells cannot be replaced. Other differentiated cells return to the cell cycle if they receive the necessary stimuli. Finally, some signals trigger cell death. For example, when you were an embryo, you had webbed fingers and toes, but the cells of the webbing received "death signals," thereby separating the fingers and toes. Far from replicating their DNA, cells undergoing programmed cell death usually produce enzymes that cut up their DNA into little pieces.

Sometimes cells divide regardless of the "stop signals" or "death signals" that they receive. You probably recognize this condition: cancer. In these cases, a single cell

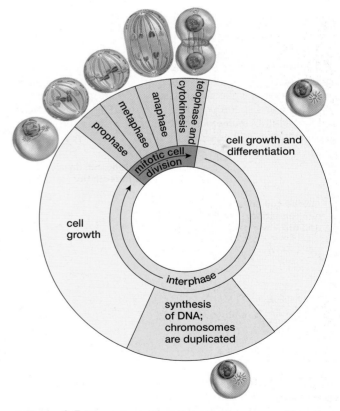

▲ **Figure 8-7 The eukaryotic cell cycle** The eukaryotic cell cycle consists of interphase (yellow) and mitotic cell division (blue). Some cells remain permanently in interphase and never divide again. As a prelude to sexual reproduction, meiotic cell division replaces mitotic cell division in the cell cycles of certain specialized cells, such as in the testes and ovaries of animals.

and its daughters divide repeatedly, using up vital nutrients, crowding out neighboring cells, and often killing the organism. The cellular changes that can cause cancer are explored in "Health Watch: Cancer—Mitotic Cell Division Run Amok."

There Are Two Types of Division in Eukaryotic Cells: Mitotic Cell Division and Meiotic Cell Division

Mitotic cell division may be thought of as "ordinary" cell division, the kind that occurs during development from a fertilized egg, during asexual reproduction in *Hydra* and aspens, and in your skin, liver, and digestive tract every day. Meiotic cell division is a specialized type of cell division required for sexual reproduction.

Mitotic Cell Division

Mitotic cell division consists of nuclear division (called **mitosis**) followed by cytoplasmic division (called **cytokinesis**). The word *mitosis* comes from the Greek word for "thread," because during mitosis the chromosomes condense into threadlike structures visible in a light microscope. Cytokinesis (from the Greek words for "cell movement") is the process by which the cytoplasm is divided between the two daughter cells. As we will see later in this chapter, mitosis gives each daughter nucleus one copy of the parent cell's replicated chromosomes, and cytokinesis usually places one of these nuclei into each daughter cell. Hence, mitotic cell division typically produces two daughter cells that are genetically identical to each other and to the parent cell, and usually contain about equal amounts of cytoplasm.

Meiotic Cell Division

Meiotic cell division is a prerequisite for sexual reproduction in all eukaryotic organisms. In animals, meiotic cell division occurs only in ovaries and testes. The process of meiotic cell division involves a specialized nuclear division called **meiosis** and two rounds of cytokinesis to produce four daughter cells that can become gametes (eggs or sperm). Gametes carry half of the DNA of the parent cell. Thus, the cells produced by meiotic cell division are not genetically identical to each other *or* to the original cell. During sexual reproduction, a sperm and an egg fuse, forming a fertilized egg with the full complement of DNA. The resulting offspring is genetically unique, similar to both parents, but identical to neither.

The Life Cycles of Eukaryotic Organisms Include Both Mitotic and Meiotic Cell Division

The life cycles of organisms can assume many different patterns, reflecting the great diversity of life on Earth. Generally, however, the life cycles of multicellular eukaryotic organisms include both meiotic and mitotic cell division (**Fig. 8-8**). A new generation begins when two gametes fuse, bringing together genes from two parents. Through mitotic cell division and differentiation of the resulting daughter cells, the fertilized egg grows and develops a multicellular body. At some point in the organism's life, meiotic cell division generates new gametes that may unite with other gametes to produce the next generation.

▼ **Figure 8-8 Mitotic and meiotic cell division in the human life cycle** Within ovaries, meiotic cell division produces eggs; within testes, meiotic cell division produces sperm. Fusion of the egg and sperm produces a fertilized egg that develops into an adult by numerous mitotic cell divisions and differentiation of the resulting cells.

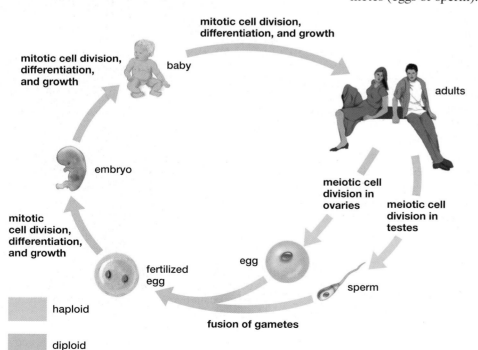

mitotic cell division, differentiation, and growth

mitotic cell division, differentiation, and growth

baby

adults

embryo

mitotic cell division, differentiation, and growth

meiotic cell division in ovaries

meiotic cell division in testes

egg

fertilized egg

sperm

fusion of gametes

haploid

diploid

health watch

Cancer—Mitotic Cell Division Run Amok

Mitotic cell division is essential for the development of multicellular organisms from single fertilized eggs, as well as for routine maintenance of body parts such as the skin and the lining of the digestive tract. Normally, mitotic cell division is tightly controlled. Cell division usually occurs only when a cell is stimulated by chemicals that are often called growth factors. For example, when a dog licks a cut on its paw, its saliva contains protein growth factors that stimulate cell division in skin cells, helping to heal the cut. However, even when a cell is stimulated by growth factors, progression through the cell cycle is regulated by a variety of proteins that check the condition of the cell (**Fig. E8-1a**): Is the DNA intact? Has it been completely and accurately replicated? Are the chromosomes lined up properly at metaphase? Under most circumstances, mitosis proceeds only if all of the answers are "yes."

Most cancers occur for two reasons. First, a virus may infect a cell and cause it to produce proteins that mimic the cell's response to growth factors. Second, a cell may suffer mutations in the genes for the cell cycle control proteins, so that the proteins tell mitotic cell division "go" instead of "no" even when the cell or its DNA have been damaged (**Fig. E8-1b**).

From Mutated Cell To Cancer

In most cases, a mutation is quickly fixed by enzymes that repair DNA. If the mutation is large or cannot be fixed, then "anti-tumor" proteins usually cause the cell to kill itself. If this original cancer cell dies, then of course no tumor will be formed. Even if the mutated cell evades its own internal controls, many mutations cause the cell's surface to "look different" to the cells of the immune system, which then kills the mutated cell. Occasionally, however, a renegade

cell survives and reproduces. Because mitotic cell division faithfully transmits genetic information from cell to cell, all daughter cells of the original cancerous cell will themselves be cancerous.

Why does medical science, which has conquered smallpox, measles, and a host of other diseases, have such a difficult time curing cancer? One reason is that both normal and cancerous cells use the same machinery for cell division, so most treatments that slow down the multiplication of cancer cells also inhibit the maintenance of essential body parts, such as the stomach, intestine, and blood cells. Truly effective and selective treatments for cancer must target cell division only in cancerous cells. Although great strides have been made in the fight to cure cancer, much remains to be done.

◄ **Figure E8-1**
Control of the cell cycle *(a)* The cell cycle is normally controlled by proteins that monitor the condition of the DNA, its replication, and the position of the chromosomes during metaphase. *(b)* Cancer is often caused by mutations that disable these control proteins. Mutations of different control proteins often cause different types of cancers.

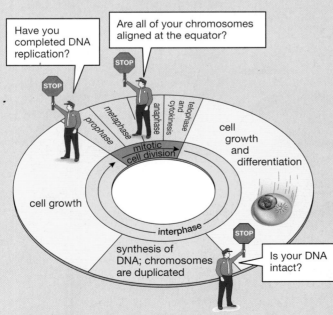

(a) Cell cycle control in a normal cell

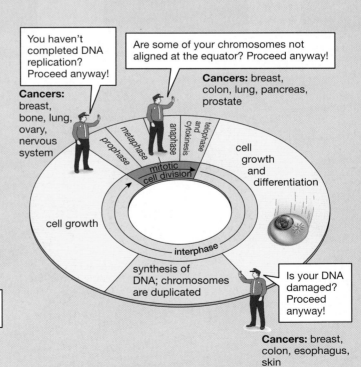

(b) Failure of cell cycle control leads to cancer

BioFlix **Mitosis**

Web Animation Mitosis

8.5 How Does Mitotic Cell Division Produce Genetically Identical Daughter Cells?

As we described earlier, mitotic cell division consists of mitosis (nuclear division) and cytokinesis (cytoplasmic division) (Fig. 8-9). For convenience, biologists divide mitosis into four phases, based on the appearance and behavior of the chromosomes: (1) prophase, (2) metaphase, (3) anaphase, and (4) telophase. Cytokinesis usually occurs during telophase. However, mitosis sometimes occurs without cytokinesis, producing cells with multiple nuclei.

During Prophase, the Chromosomes Condense and Are Captured by the Spindle Microtubules

During **prophase,** three major events occur: (1) the duplicated chromosomes condense, (2) the spindle microtubules form, and (3) the chromosomes are captured by the spindle.

▼ **Figure 8-9 Mitotic cell division in an animal cell** In the fluorescent micrographs, chromosomes are blue and microtubules are green. **QUESTION:** What would be the genetic consequences for the daughter cells if one set of sister chromatids failed to separate at anaphase?

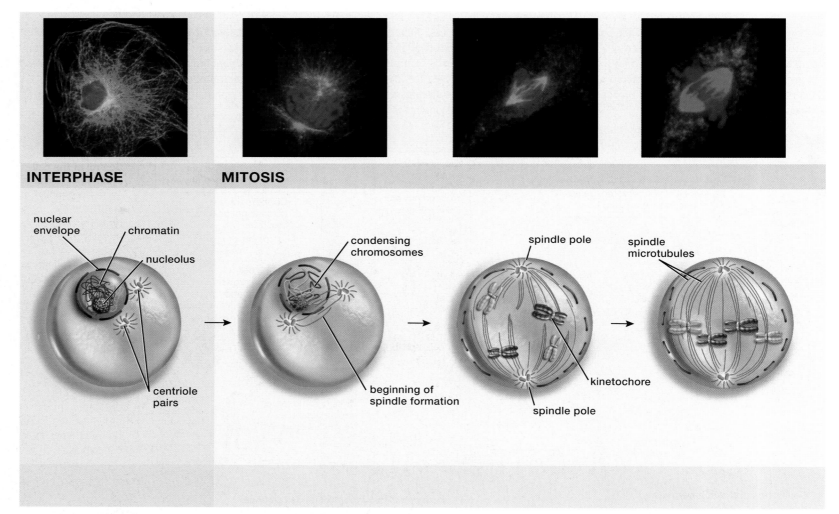

INTERPHASE   **MITOSIS**

nuclear envelope
chromatin
nucleolus
centriole pairs

condensing chromosomes
beginning of spindle formation

spindle pole
kinetochore
spindle pole

spindle microtubules

(a) Late Interphase
Duplicated chromosomes are in the relaxed uncondensed state; duplicated centrioles remain clustered.

(b) Early Prophase
Chromosomes condense and shorten; spindle microtubules begin to form between separating centriole pairs.

(c) Late Prophase
The nucleolus disappears; the nuclear envelope breaks down; spindle microtubules attach to the kinetochore of each sister chromatid.

(d) Metaphase
Kinetochores interact; spindle microtubules line up the chromosomes at the cell's equator.

Remember, chromosome duplication occurs during interphase. Therefore, when mitosis begins, each chromosome already consists of two sister chromatids attached to one another at the centromere. During prophase, the duplicated chromosomes coil up and condense. As the chromosomes condense, the nucleolus disappears.

Next, the spindle develops. In all eukaryotic cells, the proper movement of chromosomes during mitosis relies on **spindle microtubules.** The spindle forms near the nucleus as microtubules assemble, grow outward, and eventually surround the nucleus like a football-shaped basket. The tips of the football, from which the spindle microtubules radiate, are called the *spindle poles*.

Animal cells contain a pair of microtubule-containing organelles called **centrioles** near the nucleus. During interphase, a new pair of centrioles forms near the previously existing pair (**Fig. 8-9a**). During prophase, the centriole pairs migrate with the spindle poles to opposite sides of the nucleus (**Figs. 8-9b,c**). Therefore, when the cell divides, each daughter cell will receive a pair of centrioles. The centrioles *move with* the spindle poles but do not *produce* the spindle microtubules; the cells of plants and fungi, which do not contain centrioles, can make fully functional spindles.

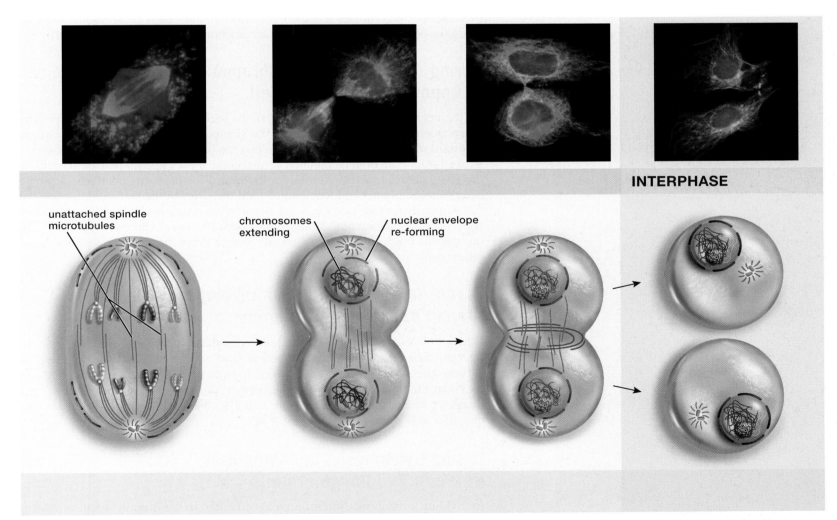

INTERPHASE

unattached spindle microtubules

chromosomes extending

nuclear envelope re-forming

(e) Anaphase
Sister chromatids separate and move to opposite poles of the cell; spindle microtubules that are not attached to the chromosomes push the poles apart.

(f) Telophase
One set of chromosomes reaches each pole and relaxes into the extended state; nuclear envelopes start to form around each set; spindle microtubules begin to disappear.

(g) Cytokinesis
The cell divides in two; each daughter cell receives one nucleus and about half of the cytoplasm.

(h) Interphase of daughter cells
Spindles disappear, intact nuclear envelopes form, chromosomes extend completely, and the nucleolus reappears.

As the spindle microtubules form into a complete basket around the nucleus, the nuclear envelope disintegrates, releasing the chromosomes. Every sister chromatid has a structure called a **kinetochore** located at its centromere. In each duplicated chromosome, the kinetochore of one sister chromatid binds to the ends of microtubules leading to one spindle pole, while the kinetochore of the other sister chromatid binds to microtubules leading to the opposite spindle pole (**Fig. 8-9c**). When the sister chromatids separate later in mitosis, the newly independent chromosomes will move along the spindle microtubules to opposite poles. Some spindle microtubules do not attach to chromosomes. Instead, they have free ends that overlap along the cell's equator. As we will see, these unattached spindle microtubules will push the two spindle poles apart during anaphase.

During Metaphase, the Chromosomes Line Up Along the Equator of the Cell

At the end of prophase, the two kinetochores of each duplicated chromosome are connected to spindle microtubules leading to opposite poles of the cell. As a result, each chromosome is connected to both spindle poles. During **metaphase,** the two kinetochores on a duplicated chromosome engage in a "tug of war." During this process, the microtubules lengthen and shorten, until each chromosome lines up along the equator of the cell, with one kinetochore facing each pole (**Fig. 8-9d**).

During Anaphase, Sister Chromatids Separate and Move to Opposite Poles of the Cell

At the beginning of **anaphase** (**Fig. 8-9e**), the sister chromatids separate, becoming independent daughter chromosomes. The kinetochores then pull the chromosomes poleward along the spindle microtubules. One of the two daughter chromosomes derived from each original parental chromosome moves to each pole of the cell. As the kinetochores tow their chromosomes toward the poles, the unattached spindle microtubules interact and lengthen to push the poles of the cell apart, forcing the cell into an oval shape (see Fig. 8-9e). Because the daughter chromosomes are identical copies of the parental chromosomes, both clusters of chromosomes that form on opposite poles of the cell contain one copy of every chromosome that was in the parent cell.

During Telophase, Nuclear Envelopes Form Around Both Groups of Chromosomes

When the chromosomes reach the poles, **telophase** begins (**Fig. 8-9f**). The spindle microtubules disintegrate, and a nuclear envelope forms around each group of chromosomes. The chromosomes revert to their extended state, and the nucleoli reappear. In most cells, cytokinesis occurs during telophase, enclosing each daughter nucleus within a separate cell (**Fig. 8-9g**). Following cytokinesis, the daughter cells enter interphase (**Fig. 8-9h**).

During Cytokinesis, the Cytoplasm Is Divided Between Two Daughter Cells

In dividing animal cells, microfilaments attached to the plasma membrane form a ring around the equator of the cell. During cytokinesis, the ring contracts and constricts the cell's equator, much as pulling the drawstring on a pair of sweatpants tightens the waist. Eventually the "waist" constricts completely, dividing the cytoplasm into two new daughter cells (**Fig. 8-10**).

Cytokinesis in plant cells is quite different, perhaps because their stiff cell walls make it impossible to divide one cell into two by pinching at the waist. In-

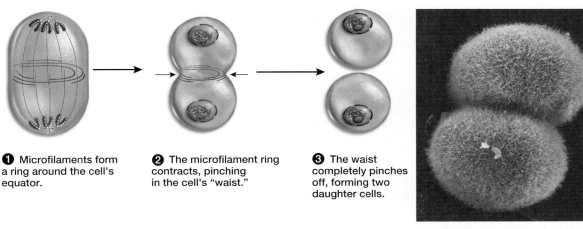

❶ Microfilaments form a ring around the cell's equator.

❷ The microfilament ring contracts, pinching in the cell's "waist."

❸ The waist completely pinches off, forming two daughter cells.

(a) Microfilaments contract, pinching the cell in two

(b) Scanning electron micrograph of cytokinesis

◀ Figure 8-10 Cytokinesis in an animal cell

stead, carbohydrate-filled vesicles bud off the Golgi apparatus and line up along the cell's equator between the two nuclei (Fig. 8-11). The vesicles fuse, producing a structure called the **cell plate,** which is shaped like a flattened sac, surrounded by plasma membrane, and filled with sticky carbohydrates. When enough vesicles have fused, the edges of the cell plate merge with the original plasma membrane of the cell. The carbohydrates contained in the vesicles become the cell wall between the two daughter cells.

8.6 How Does Meiotic Cell Division Produce Haploid Cells?

The key to sexual reproduction is meiotic cell division, the production of haploid cells with unpaired chromosomes from diploid parent cells with paired chromosomes. Each daughter cell receives one member from each pair of homologous chromosomes. For example, because each diploid cell in a human body contains 23 pairs of chromosomes, meiotic cell division produces gametes with 23 chromosomes, one from each pair. Cytokinesis in meiotic cell division is similar to cytokinesis in mitotic cell division. Therefore, we will describe only meiosis, the division of the nuclei, in the following sections.

BioFlix Meiosis

Web Animation Meiosis

cell plate forming a new cell wall

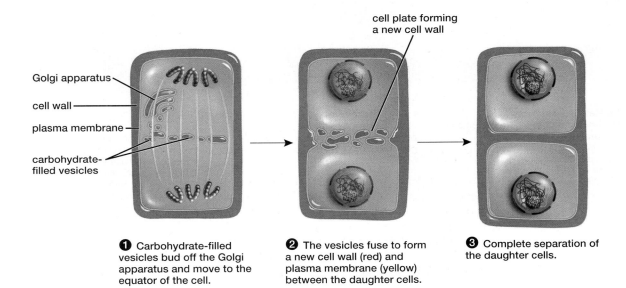

Golgi apparatus

cell wall

plasma membrane

carbohydrate-filled vesicles

❶ Carbohydrate-filled vesicles bud off the Golgi apparatus and move to the equator of the cell.

❷ The vesicles fuse to form a new cell wall (red) and plasma membrane (yellow) between the daughter cells.

❸ Complete separation of the daughter cells.

◀ Figure 8-11 Cytokinesis in a plant cell

Many of the events in meiosis are similar to those of mitosis. However, meiosis includes two nuclear divisions, so each cell that undergoes meiotic cell division yields four daughter cells. The two nuclear divisions are known as meiosis I and meiosis II. In meiosis I, homologous chromosomes pair up, but sister chromatids remain connected to each other (this differs from mitosis). In meiosis II, chromosomes behave as they do in mitosis: Sister chromatids separate and are pulled to opposite poles of the cell.

The phases of meiosis have the same names as the roughly equivalent phases in mitosis, followed by I or II to distinguish the two nuclear divisions.

Meiosis I Separates Homologous Chromosomes into Two Haploid Daughter Nuclei

The chromosomes are replicated in interphase before meiosis I. As we saw with mitosis, the sister chromatids of each duplicated chromosome are attached to one another at the centromere.

During Prophase I, Homologues Pair Up

Meiosis I differs considerably from mitosis. In mitosis, each chromosome moves independently, and the members of a pair of homologous chromosomes do not interact. In contrast, during *prophase I* of meiosis, pairs of homologues line up side by side (**Fig. 8-12a**). The two homologues in a pair often intertwine, forming crosses, or **chiasmata** (singular, chiasma). At some of the chiasmata, the homo-

▼ **Figure 8-12 Meiotic cell division in an animal cell** In these diagrams, two pairs of homologous chromosomes are shown, large and small. The yellow chromosomes are from one parent and the violet chromosomes are from the other parent. Note that each of the four daughter cells contains one member of each pair of parental homologous chromosomes. **QUESTION:** What would be the consequences for the gametes if one pair of homologues failed to separate at anaphase I?

MEIOSIS I

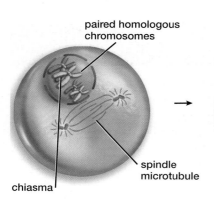

chiasma / spindle microtubule / paired homologous chromosomes

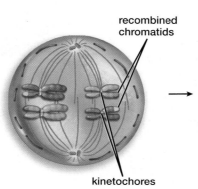

recombined chromatids / kinetochores

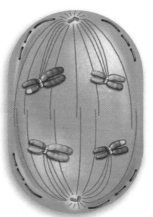

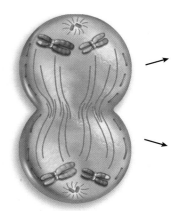

(a) Prophase I
Duplicated chromosomes condense. Homologous chromosomes pair up and chiasmata occur as chromatids of homologues exchange parts by crossing over. The nuclear envelope disintegrates, and spindle microtubules form.

(b) Metaphase I
Paired homologous chromosomes line up along the equator of the cell. One homologue of each pair faces each pole of the cell and attaches to the spindle microtubules via the kinetochore (blue).

(c) Anaphase I
Homologues separate, one member of each pair going to each pole of the cell. Sister chromatids do not separate.

(d) Telophase I
Spindle microtubules disappear. Two clusters of chromosomes have formed, each containing one member of each pair of homologues. The daughter nuclei are therefore haploid. Cytokinesis commonly occurs at this stage. There is little or no interphase between meiosis I and meiosis II.

logues exchange parts, in a process known as *crossing over* (see section 8.7). The chiasmata hold the homologues together in pairs, until anaphase I.

As prophase I continues, spindle microtubules begin to assemble outside the nucleus. Near the end of prophase I, the nuclear envelope breaks down and the chromosomes attach to spindle microtubules by their kinetochores. Unlike in mitosis, in which each sister chromatid has its own kinetochore, in meiosis I there is only one kinetochore for each replicated chromosome (one for each pair of sister chromatids).

During Metaphase I, Paired Homologues Line Up at the Equator of the Cell

During *metaphase I*, interactions between the kinetochores and the spindle microtubules move the paired homologues to the equator of the cell (**Fig. 8-12b**). The pairing of homologues in meiosis I makes the arrangement of chromosomes along the equator very different from the arrangement in mitosis. In mitosis, *individual* duplicated chromosomes line up along the equator, but during metaphase I of meiosis, *homologous pairs* of duplicated chromosomes line up along the equator.

During Anaphase I, Homologous Chromosomes Separate

In *anaphase I*, the spindle microtubules attached to each pair of homologues pull toward opposite poles. The chiasmata slide apart, allowing the homologues to separate from one another (**Fig. 8-12c**). One duplicated chromosome (still consisting of

MEIOSIS II

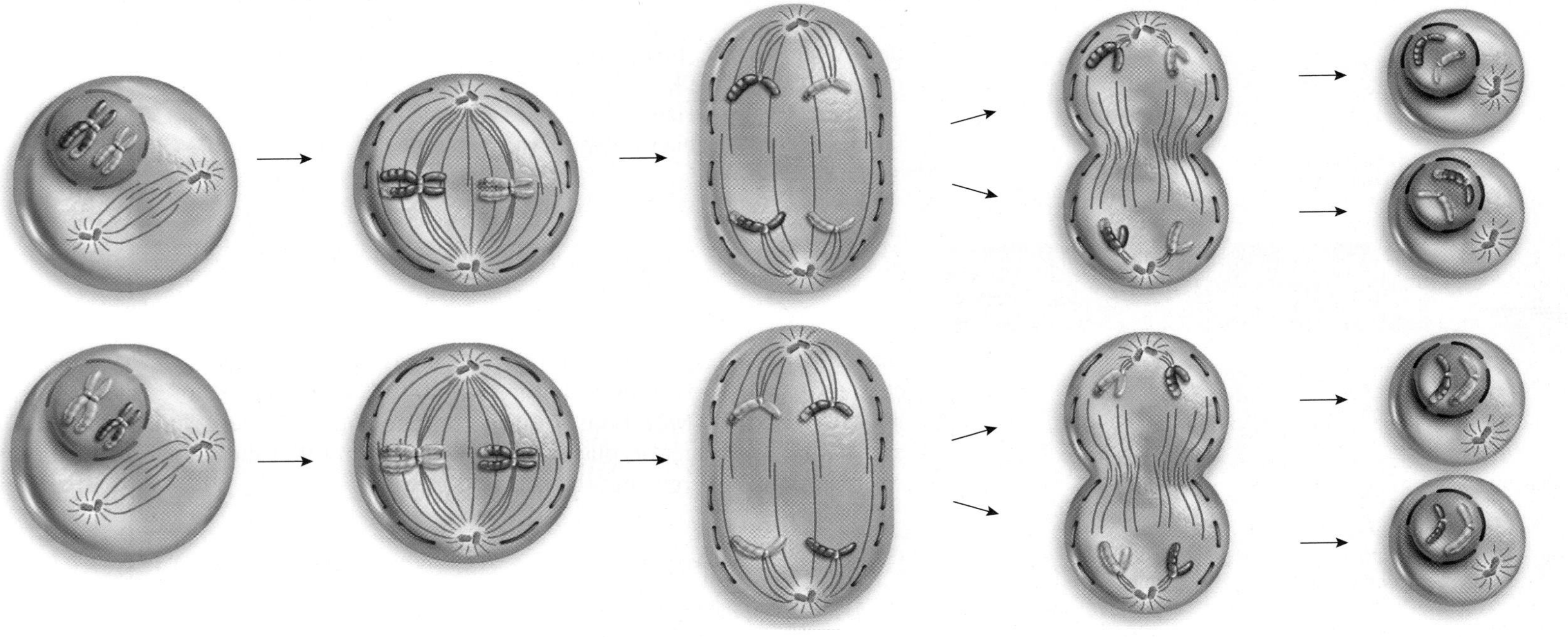

(e) Prophase II If the chromosomes have relaxed after telophase I, they recondense. Spindle microtubules re-form and attach to the sister chromatids.

(f) Metaphase II The chromosomes line up along the equator, with sister chromatids of each chromosome attached to spindle microtubules that lead to opposite poles.

(g) Anaphase II The chromatids separate into independent daughter chromosomes, one former chromatid moving toward each pole.

(h) Telophase II The chromosomes finish moving to opposite poles. Nuclear envelopes re-form, and the chromosomes become extended again (not shown here).

(i) Four haploid cells Cytokinesis results in four haploid cells, each containing one member of each pair of homologous chromosomes (shown here in the condensed state).

two sister chromatids) from each homologous pair moves to each pole of the dividing cell. Thus, at the end of anaphase I, the cluster of chromosomes at each pole contains one member of every pair of homologous chromosomes.

After Telophase I and Cytokinesis, There Are Two Haploid Daughter Cells

In *telophase I*, the spindle microtubules disappear and the nuclear envelope may reappear (**Fig. 8-12d**). In most cases, cytokinesis takes place and divides the cell into two daughter cells. Each daughter cell has only one of each pair of homologous chromosomes, and is therefore haploid. Each chromosome, however, still consists of two sister chromatids.

Meiosis II Separates Sister Chromatids into Four Haploid Daughter Cells

Meiosis II usually begins immediately after meiosis I ends. There is no intervening interphase and the chromosomes are not replicated again. Typically, the chromosomes remain condensed. However, if the chromosomes do relax at the end of meiosis I, they condense again before meiosis II.

Meiosis II is virtually identical to mitosis, although it occurs in haploid cells. During *prophase II*, the spindle microtubules re-form (**Fig. 8-12e**). The duplicated chromosomes attach to spindle microtubules as they would in mitosis. Each sister chromatid now has a kinetochore by which it attaches to spindle microtubules that extend to one or the other pole of the cell. During *metaphase II*, the duplicated chromosomes line up at the cell's equator (**Fig. 8-12f**). During *anaphase II*, the sister chromatids separate and move to opposite poles (**Fig. 8-12g**). As *telophase II* and cytokinesis conclude meiosis II, nuclear envelopes form, the chromosomes relax into their extended, noncondensed state, and the cytoplasm divides (**Figs. 8-12h,i**). Usually, both daughter cells of meiosis I undergo meiosis II, producing a total of four haploid cells from the original parental diploid cell.

Table 8-1 summarizes mitotic and meiotic cell division, showing both the similarities and differences between the two types of cell division.

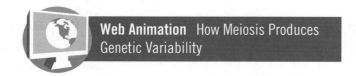

Web Animation How Meiosis Produces Genetic Variability

8.7 How Do Meiotic Cell Division and Sexual Reproduction Produce Genetic Variability?

Life has been shaped by evolution, and a crucial prerequisite for evolutionary change is genetic variability among organisms. (We will explore this theme at length in Unit 3.) Mutations are the ultimate source of new genes, but meiosis and sexual reproduction play key roles in ensuring that individuals differ in the combinations of genes that they carry.

Shuffling of Homologues Creates Novel Combinations of Chromosomes

One important mechanism by which meiosis produces genetic diversity is the random assortment of homologues to daughter cells at meiosis I. Remember that the two members of each homologous pair of chromosomes have very *similar*, but usually not *identical*, DNA sequences. At metaphase I, paired homologues line up at the cell's equator. For each pair of homologues, one chromosome faces one pole and the other chromosome faces the opposite pole. However, which chromosome faces which pole is random. To see the effects of

Table 8-1	A Comparison of Mitotic and Meiotic Cell Divisions in Animal Cells	
Feature	**Mitotic Cell Division**	**Meiotic Cell Division**
Cells in which it occurs	Body cells	Gamete-producing cells
Final chromosome number	Diploid—$2n$; two copies of each type of chromosome (homologous pairs)	Haploid—$1n$; one member of each homologous pair
Number of daughter cells	Two, identical to the parent cell and to each other	Four, containing recombined chromosomes due to crossing over
Number of cell divisions per DNA replication	One	Two
Function in animals	Development, growth, repair, and maintenance of tissues; asexual reproduction	Gamete production for sexual reproduction

this random arrangement of homologues, let's consider meiosis in mosquitoes, which have three pairs of homologous chromosomes ($n = 3$, $2n = 6$). At metaphase I, the chromosomes can align in four configurations (**Fig. 8-13a**), yielding eight possible combinations of chromosomes ($2^3 = 8$) when they separate during anaphase I (**Fig. 8-13b**). Each of these chromosome clusters will then undergo meiosis II to produce two gametes (see Fig. 8-12g–i). A single mosquito, with three pairs of homologous chromosomes, therefore produces gametes that each contain one of eight possible sets of chromosomes. A single human, with 23 pairs of homologous chromosomes, can theoretically produce gametes that contain one of more than 8 million (2^{23}) different combinations of maternal and paternal chromosomes.

Crossing Over Creates Chromosomes with Novel Combinations of Genetic Material

During prophase I of meiosis, homologous chromosomes pair up, bound together by proteins that hold the homologues together so that they match up exactly along their entire length. Enzymes cut open the DNA of the paired homologues and graft the cut ends of the DNA together, often joining part of a chromatid of the maternal homologue to part of a chromatid of the paternal homologue, and vice versa. This mutual exchange of DNA is known as **crossing over** (**Fig. 8-14**) and is visible under a light microscope as chiasmata where the chromatids cross one another.

The result of crossing over is genetic **recombination**—the formation of new combinations of genes on a chromosome. In fact, these new combinations may

(a) The four possible chromosome arrangements at metaphase of meiosis I

(b) The eight possible sets of chromosomes after meiosis I

▲ **Figure 8-13** Random separation of homologues during meiosis produces genetic variability

◀ **Figure 8-14 Crossing over** Chromatids of different members of a homologous pair of chromosomes exchange DNA at chiasmata.

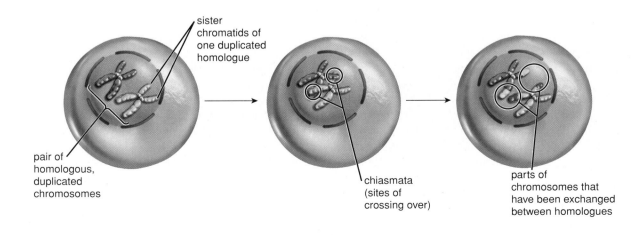

sister chromatids of one duplicated homologue

pair of homologous, duplicated chromosomes

chiasmata (sites of crossing over)

parts of chromosomes that have been exchanged between homologues

scientific inquiry

Carbon Copies—Cloning in Nature and the Lab

Cloning is the creation of one or more individual organisms **(clones)** that are genetically identical to a preexisting individual. How are clones produced? Why is cloning such a hot—and controversial—topic in the news?

Cloning in Nature: The Role of Mitotic Cell Division

As you now know, there are two types of cell division: mitotic division and meiotic division. Sexual reproduction relies on meiotic cell division, the production of gametes, and fertilization, and usually produces genetically unique offspring. In contrast, asexual reproduction relies on mitotic cell division. Because mitotic cell division creates daughter cells that are genetically identical to the parent cell, offspring produced by asexual reproduction are genetically identical to their parents—they are clones.

Cloning Plants: A Familiar Application in Agriculture

Humans have been in the cloning business a lot longer than you might think. For example, consider navel oranges, which don't have seeds. Without seeds, how do they reproduce? Navel orange trees are propagated by cutting a piece of stem from an adult navel tree and grafting it onto the top of the root of a seedling orange tree, usually of a different type. Therefore, the cells of the aboveground, fruit-bearing parts of the resulting tree are clones of the original navel orange stem.

All navel oranges apparently originated from a single mutant bud of an orange tree discovered in Brazil in the early 1800s, and propagated asexually ever since. Two navel orange trees were brought from Brazil to Riverside, California, in the 1870s. (One of them is still there, producing fruit!) All American navels are clones of these trees.

Cloning Adult Animals

Animal cloning isn't a recent development either. In the 1950s, John Gurdon and his colleagues inserted nuclei from early frog embryos into eggs, and some of the resulting cells developed into complete frogs. By the 1990s, several labs had been able to clone mammals using embryonic nuclei, but it wasn't until 1996 that Dr. Ian Wilmut of the Roslin Institute in Edinburgh, Scotland, cloned the first adult mammal, the famous Dolly (**Fig. E8-2**).

Why is it important to clone an adult animal? Only in adults can we see the traits that we wish to propagate, such as the speed and agility of Scamper, the champion barrel-racing horse. Cloning an adult would produce "offspring" that are genetically identical to the adult. Therefore, insofar as the valuable traits of the adult are genetically determined, all of its clones would also express those same valuable traits. Cloning of embryos would usually not be useful, because the embryonic cells would have been produced by sexual reproduction in the first place, and usually no one could tell if the embryo had any especially desirable traits.

For some medical applications, too, cloning adults is essential. Suppose that a pharmaceutical company genetically engineered a cow that secreted a valuable molecule, such as an antibiotic, in its milk (see Chapter 12). Genetic engineering is extremely expensive and somewhat hit-or-miss, so the company may successfully produce only one profitable cow. This cow could then be cloned, creating a whole herd of antibiotic-producing cows.

Cloning: An Imperfect Technology

Unfortunately, cloning mammals is inefficient and beset with difficulties. An egg is subjected to severe trauma when its nucleus is sucked out or destroyed, and a new nucleus is inserted. Often, the egg may simply die. Molecules in the cytoplasm that are needed to control development may be lost or moved to the wrong places, so that even if the egg survives and divides, it may not develop properly. If the eggs develop into viable embryos, the embryos must then be implanted into the uterus of a surrogate mother. Many clones die or are aborted during gestation, often with serious or fatal consequences for the surrogate mother. Given the high failure rate—

it took 277 tries to produce Dolly—cloning mammals is an expensive and risky proposition. Even if the clone survives gestation and birth, it may have defects, commonly a deformed head, lungs, or heart. For example, one of Snuppy's "brothers" aborted, while a second died of pneumonia 22 days after birth, despite the best of care. Dolly developed arthritis when she was 5½; and was euthanized with a serious lung disease when she was 6½, both relatively young ages for a sheep, although no one knows if these health problems occurred because she was a clone.

The Future of Cloning

A new technology, called *chromatin transfer*, appears to reduce the likelihood of defective clones. Many researchers think that DNA in "old" cells is in a different chemical state than the DNA of a newly fertilized egg. Although inserting an old nucleus into an enucleated egg helps to rejuvenate the old DNA, it doesn't always do the trick. In chromatin transfer, the membranes of donor cells are made leaky. The permeable cells are then incubated with a "mitotic extract" derived from rapidly dividing, and therefore "young," cells. This remodels the DNA of the older cell and causes it to condense, much like DNA does during prophase of mitotic cell division. The rejuvenated cell is then fused with an enucleated egg, as with more traditional cloning procedures. The biotech firm Viagen cloned Scamper using chromatin transfer.

Modern technology has now successfully cloned cows, cats, dogs, sheep, horses, and a variety of other animals. As the process becomes more routine, it also raises ethical questions. Although hardly anyone objects to navel oranges and few would refuse antibiotics or other medicinal products derived from cloned livestock, some people think that cloning pets is a frivolous luxury, especially when you consider that every 9 seconds an unwanted dog or cat is euthanized in the United States. And what about human cloning? We will explore this aspect of cloning more fully in "Send in the Clones, *Revisited*" at the end of the chapter.

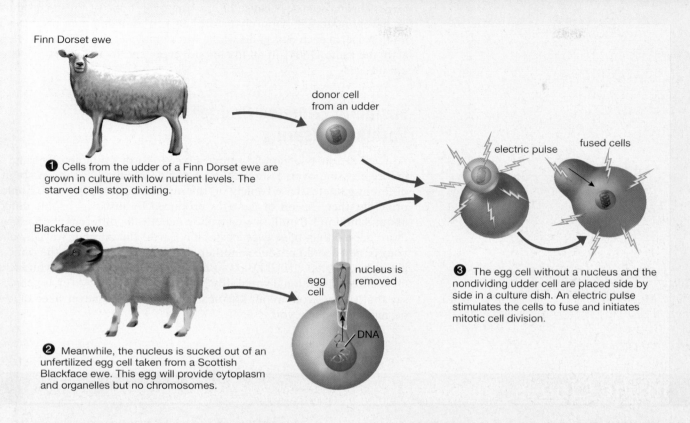

Finn Dorset ewe

donor cell from an udder

❶ Cells from the udder of a Finn Dorset ewe are grown in culture with low nutrient levels. The starved cells stop dividing.

Blackface ewe

electric pulse   fused cells

egg cell   nucleus is removed

DNA

❸ The egg cell without a nucleus and the nondividing udder cell are placed side by side in a culture dish. An electric pulse stimulates the cells to fuse and initiates mitotic cell division.

❷ Meanwhile, the nucleus is sucked out of an unfertilized egg cell taken from a Scottish Blackface ewe. This egg will provide cytoplasm and organelles but no chromosomes.

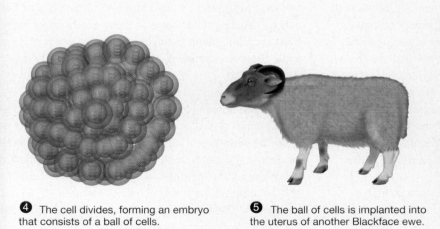

❹ The cell divides, forming an embryo that consists of a ball of cells.

❺ The ball of cells is implanted into the uterus of another Blackface ewe.

❻ The Blackface ewe gives birth to Dolly, a female Finn Dorset lamb, a genetic twin of the Finn Dorset ewe.

▲ **Figure E8-2 The making of Dolly**

have never existed before in all of Earth's history, because homologous chromosomes cross over in new and different places at each meiotic division. In humans, although one in 8 million gametes should have identical sets of maternal and paternal chromosomes, because of crossing over, none of those chromosomes will be purely maternal or purely paternal. Even though a man produces about 100 million sperm each day, in his whole life he may never produce two that carry exactly the same DNA! In all likelihood, every sperm and every egg is genetically unique.

Fusion of Gametes Creates Genetically Variable Offspring

At fertilization, two haploid gametes fuse to form a diploid offspring. Even if we ignore crossing over, a single human could potentially produce about 8 million different gametes, based solely on the random separation of the homologues. In principle, then, fusion of gametes produced by just two people could produce about 8 million × 8 million = 64 trillion genetically different children, which is far more people than have ever existed on Earth! Put another way, the chances that your parents could produce another child who is genetically the same as you are about $1/8,000,000 \times 1/8,000,000$, or about 1 in 64 trillion. When you include the almost endless additional variability produced by crossing over, we can confidently say that (unless you are an identical twin), there has never been, and never will be, anyone just like you.

Send in the Clones Revisited

Why are clones genetically identical to their genetic donors, while children are so different from their parents and from one another? As you've learned, barring the occasional mutation, mitotic cell division produces daughter cells that are genetically identical to the parent cell. Meiotic cell division, in contrast, produces haploid cells that have a random sampling of one member of each pair of homologous chromosomes. Crossing over further shuffles the genes. Thus, when a sperm fertilizes an egg, the offspring receive half their genes from the mother and half from the father. Two children of the same parents will never be genetically the same as either parent or as one another (except for identical twins). That, of course, is one reason why many famous racehorses, including Seabiscuit, War Admiral, and Secretariat, never sired any offspring who could run as fast as they could.

However, during cloning a researcher takes the nucleus from a donor cell and implants the nucleus into an egg in which the nucleus has been removed or destroyed. The resulting cell divides by mitotic cell division to form the cloned organism, so all of the cells of the clone have the same genes as the donor did.

Everything about an organism—structure, metabolism, personality, learning ability, and other aspects of behavior—is influenced by both genes and environment. The uterus in which a fetus develops and the conditions in which the organism lives (nutrition, family conditions, friends, world events, and so on) are important in ways that we have barely begun to understand. Nevertheless, clones are likely to resemble their genetic donors, both in appearance and behavior, far more than any child resembles a parent, a brother, or a sister.

Is cloning good or bad? Or does it depend on the organism, the reason for cloning, and the uses to which cloning is put? People will probably be debating the answers to these questions for many years. One thing is certain: Send in the clones? In the words of Broadway songwriter Stephen Sondheim, "Don't bother, they're here!"

Consider This

What about human cloning? There are two potentially separate issues. First, should human "therapeutic cloning" for medical purposes be permitted? Cells produced from this type of cloning would be genetically identical to those of the human nuclear donor, and under the right conditions, would include stem cells that might be able to differentiate into any type of mature cell, including brain or heart cells. These cells could then be used to repair damaged organs, such as after a heart attack or a stroke, in the person who donated the nucleus. Some people are fervent supporters of therapeutic cloning for medical treatments, while others are equally fervently opposed, on the grounds that it is not acceptable to create a human embryo for the sole purpose of serving as a cell donor and then destroying that embryo.

The fame, potential fortune, and medical benefits that can arise from therapeutic human cloning sometimes prove too much for scientists to handle. Woo-Suk Hwang, who cloned Snuppy and was named one of the "people who matter" by *Time* magazine in 2004, also claimed to have isolated stem cells from human therapeutic clones, which would have been yet another remarkable achievement. However, this claim has been proven false, and Hwang has been forced to resign his position at the Seoul National University. (Independent investigators determined that Snuppy is indeed a clone of Tai.) Despite his forced resignation, Hwang appears poised for a possible comeback. In August 2006, Hwang and many of his former collaborators started up an independent research lab, focusing on animal cloning. Hwang would like to resume human cloning, but it remains to be seen if the South Korean government will restore his license for human cloning.

The second issue is whether humans should be the subject of "reproductive cloning," in which the purpose is to produce a living cloned child. Assuming that the technology exists to clone people, is it a good idea? Snuppy is the only survivor of 1,095 embryos implanted into 123 surrogate mothers. A cloned Brahma bull, Second Chance, was in intensive care for 2 weeks with respiratory and cardiovascular disorders. With today's limited technology, it's hard to argue with Rudolf Jaenisch of the Whitehead Institute for Biomedical Research, who stated that human reproductive cloning is "just criminal." But what about the potential of cloning in 2010, or 2050, if the technology is "perfected"? What do you think about the merits and pitfalls of therapeutic and reproductive cloning?

Chapter Review

Summary of Key Concepts

For additional study help and activities, go to www.mybiology.com.

8.1 Why Do Cells Divide?

Growth of multicellular eukaryotic organisms and replacement of cells that die during an organism's life occur through cell division and differentiation of the daughter cells. Asexual reproduction of prokaryotic unicellular organisms, eukaryotic unicellular organisms, and some multicellular eukaryotic organisms occurs through cell division. A specialized type of cell division, called meiotic cell division, is required for sexual reproduction. In animals, meiotic cell division produces gametes (sperm or eggs) with half the DNA of an ordinary body cell. Fusion of gametes creates a fertilized egg that has a genetic makeup different from that of either parent.

8.2 What Occurs During the Prokaryotic Cell Cycle?

The prokaryotic cell cycle consists of growth, DNA replication, and division by binary fission. The two daughter cells produced by binary fission are genetically identical to the parent cell.

8.3 How Is the DNA in Eukaryotic Cells Organized?

Each eukaryotic chromosome consists of a single DNA double helix combined with a variety of proteins. During most of the life of a cell, the chromosomes are extended and accessible for transcribing their genes. During cell division, the chromosomes condense into short, thick structures. Eukaryotic cells typically contain pairs of homologous chromosomes. Homologues are virtually identical in appearance because they carry the same genes with similar nucleotide sequences. Cells with pairs of homologous chromosomes are diploid. Cells with only a single member of each chromosome pair are haploid.

8.4 What Occurs During the Eukaryotic Cell Cycle?

The eukaryotic cell cycle consists of interphase and cell division. During interphase, the cell grows and duplicates its DNA; it may also differentiate during interphase. Some cells remain permanently in interphase and do not divide again. Most cells that do divide, do

so by mitotic cell division, which produces two daughter cells that are genetically identical to their parent cell. Specialized reproductive cells undergo meiotic cell division, producing four daughter cells that contain half the DNA of the parent cell.

Multicellular eukaryotic organisms begin life as a fertilized egg. Mitotic cell divisions and differentiation of daughter cells produce the organism's body. At some point in the life of the organism, meiotic cell division produces gametes that fuse with other gametes to produce a new fertilized egg, starting the life cycle anew.

Web Animation **Cell Division in Humans**

8.5 How Does Mitotic Cell Division Produce Genetically Identical Daughter Cells?

The chromosomes are duplicated during interphase, prior to mitosis. A replicated chromosome consists of two identical sister chromatids that remain attached to one another at the centromere during the early stages of mitosis. Mitosis consists of four phases (see Fig. 8-9):

- **Prophase:** The chromosomes condense and their kinetochores attach to the spindle microtubules that form at this time.

- **Metaphase:** The chromosomes move along their attached spindle microtubules to the equator of the cell.

- **Anaphase:** The two chromatids of each duplicated chromosome separate and move along the spindle microtubules to opposite poles of the cell.

- **Telophase:** The chromosomes relax into their extended state, and nuclear envelopes reform around each new daughter nucleus.

Usually, cytokinesis occurs during telophase, producing two new daughter cells, each containing one of the nuclei produced during mitosis.

BioFlix Mitosis

Web Animation Mitosis

8.6 How Does Meiotic Cell Division Produce Haploid Cells?

During interphase before meiotic cell division, chromosomes are duplicated. The cell then undergoes two divisions—meiosis I and meiosis II—to produce four haploid daughter cells (see Fig. 8-12):

- **Meiosis I:** During prophase I, homologous duplicated chromosomes, each consisting of two chromatids, pair up and exchange parts by crossing over. During metaphase I, homologues move together as a pair to the cell's equator. The two members of each pair face opposite poles of the cell. Homologous chromosomes separate during anaphase I, and two nuclei form during telophase I. Each daughter nucleus receives only one member of each pair of homologues, and is therefore haploid. The sister chromatids remain attached to each other throughout meiosis I.

- **Meiosis II:** Both daughter nuclei go through meiosis II, which resembles mitosis in a haploid cell. The duplicated chromosomes

move to the cell's equator during metaphase II. The two chromatids of each chromosome separate and move to opposite poles of the cell during anaphase II. During telophase II, four haploid nuclei are produced, two from each of the nuclei formed by meiosis I. Cytokinesis completes the production of four haploid cells, each containing one of the nuclei formed during telophase II.

BioFlix Meiosis

Web Animation Meiosis

8.7 How Do Meiotic Cell Division and Sexual Reproduction Produce Genetic Variability?

The random shuffling of homologous maternal and paternal chromosomes creates new chromosome combinations. Crossing over creates chromosomes with DNA sequences that may never have occurred on a single chromosome before. Because of crossing over, a parent probably never produces any two gametes that are completely identical. The fusion of two genetically unique gametes creates genetically unique offspring.

Web Animation How Meiosis Produces Genetic Variability

Key Terms

anaphase *p. 122*
asexual reproduction *p. 113*
autosome *p. 116*
binary fission *p. 114*
cell cycle *p. 112*
cell division *p. 111*
cell plate *p. 123*
centriole *p. 121*
centromere *p. 116*
chiasma (plural, **chiasmata**) *p. 124*

chromatid *p. 116*
chromosome *p. 112*
clone *p. 128*
cloning *p. 128*
crossing over *p. 127*
cytokinesis *p. 118*
daughter cell *p. 111*
deoxyribonucleic acid (DNA) *p. 112*
diploid *p. 116*
double helix *p. 112*

gamete *p. 113*
gene *p. 112*
haploid *p. 117*
homologous chromosome *p. 116*
homologue *p. 116*
interphase *p. 117*
kinetochore *p. 122*
meiosis *p. 118*
meiotic cell division *p. 118*
metaphase *p. 122*

mitosis *p. 118*
mitotic cell division *p. 118*
nucleotide *p. 112*
prophase *p. 120*
recombination *p. 127*
sex chromosome *p. 116*
sexual reproduction *p. 113*
spindle microtubule *p. 121*
stem cell *p. 112*
telophase *p. 122*

Thinking Through the Concepts

Suggested answers to end-of-chapter and figure-based questions can be found at the end of the text.

Fill-in-the-Blank

1. Prokaryotic cells divide by a process called _____.

2. Growth and development of multicellular organisms occur through _____ cell division and _____ of the resulting daughter cells.

3. Most plants and animals have pairs of chromosomes that have similar appearance and genetic composition. These pairs are called _____. Pairs of chromosomes that are the same in both males

and females are called _____. Pairs that are different in males and females are called _____.

4. The four phases of mitosis are _____, _____, _____, and _____. Division of the cytoplasm into two cells, called _____, usually occurs during which phase? _____

5. Meiotic cell division produces _____ haploid daughter cells from each diploid parental cell. The first haploid nuclei are produced at the end of _____. In animals, the haploid daughter cells produced by meiotic cell division are called _____.

6. During _____ of meiosis I, homologous chromosomes intertwine, forming structures called _____. These structures are the sites of what event? _____.

7. Three processes that promote genetic variability of offspring during sexual reproduction are _____, _____, and _____.

Review Questions

1. Define mitosis and cytokinesis. What changes in cell structure would result if cytokinesis does not occur after mitosis?

2. Diagram the stages of mitosis. How does mitosis ensure that each daughter nucleus receives a full set of chromosomes?

3. Define the following terms: homologous chromosome (homologue), centromere, kinetochore, chromatid, diploid, haploid.

4. Describe and compare cytokinesis in animal cells and in plant cells.

5. Diagram the events of meiosis. At which stage do homologous chromosomes separate?

6. In what ways are mitosis and meiosis similar? In what ways are they different?

7. Describe how meiosis produces genetic variability. If an animal had a haploid number of 2 (no sex chromosomes), how many genetically different types of gametes could it produce (assume no crossing over)? If it had a haploid number of 5?

Applying the Concepts

1. Most nerve cells in the adult human central nervous system, as well as heart muscle cells, remain in interphase and do not divide. In contrast, cells lining the inside of the small intestine divide frequently. Discuss this difference in terms of why damage to the nervous system and heart muscle cells (such as damage caused by a stroke or heart attack) is so dangerous. What do you think might happen to tissues such as the intestinal lining if some disorder or drug blocked mitosis in all cells of the body?

2. IS THIS SCIENCE? Many therapies for cancer, including chemotherapy and radiation, have side effects that include hair loss and damage to the lining of the stomach and intestines, producing nausea. Cells in hair follicles and the intestinal lining divide frequently. What can you infer about the mechanisms of these treatments? There are claims in magazines and on the Internet that a variety of herbs can cure cancer with few or no side effects. Do you think that these claims are reasonable? Why or why not? What experiments could you perform to test these claims?

3. Some animal species can reproduce either asexually or sexually, depending on the state of the environment. Asexual reproduction tends to occur in stable, favorable environments; sexual reproduction is more common in unstable or unfavorable environments. Discuss the advantages and disadvantages of asexual and sexual reproduction from an evolutionary perspective.

For additional resources, go to www.mybiology.com.

chapter 9

Patterns of Inheritance

Olympic volleyball silver medalist Flo Hyman was struck down by Marfan syndrome at the height of her career.

Case Study   Sudden Death on the Court— Marfan Syndrome

Flo Hyman, over 6 feet tall, graceful and athletic, was one of the best female volleyball players of all time. Star of the 1984 silver medal American Olympic volleyball team, Hyman later joined a professional Japanese team. In 1986, taken out of a game for a short breather, she died at the age of 31, sitting quietly on the bench. How could this happen to someone so young and fit?

Based on his physical appearance, President Abraham Lincoln is also believed to have had Marfan syndrome.

Flo Hyman had a genetic disorder called Marfan syndrome, which affects about 1 in 5,000 to 10,000 people, probably including Abraham Lincoln (photo at left), the classical composer and pianist Sergei Rachmaninoff, and the Egyptian pharaoh Akhenaten. People with Marfan syndrome are typically tall and slender, with unusually long limbs and large hands and feet. For some people with Marfan syndrome, these characteristics lead to fame and fortune. Unfortunately, Marfan syndrome can also be deadly.

An autopsy showed that Hyman died from a rupture of her aorta, the massive artery that carries blood from the heart to most of the body. Why did Hyman's aorta break? What does a weak aorta have in common with tallness and large hands? Marfan syndrome is caused by a mutation in the gene that encodes a protein called fibrillin, an essential component of connective tissue. Many parts of your body contain connective tissue, including the tendons that attach muscles to bones, ligaments (such as the fibers that hold the lens in place in your eye), and the tough walls of arteries. As its name implies, fibrillin forms long fibers that give elasticity and strength to connective tissue. Defective fibrillin molecules produce weak ligaments, tendons, and artery walls, sometimes with tragic consequences. Normal fibrillin binds to and traps certain types of growth factors that would otherwise promote cell division (see Chapter 8) and hence the growth of various parts of the body. When bound by fibrillin, these growth factors cannot stimulate cell division, so the elongation of bone and connective tissue is controlled. Recent research suggests that the defective fibrillin produced by people with Marfan syndrome cannot trap these growth factors, and, as a result, their arms, legs, hands, and feet tend to become unusually long.

As we described in Chapter 8, diploid organisms, including people, generally have two copies of each gene, one on each homologous chromosome. A single defective copy of the fibrillin gene is enough to cause Marfan syndrome. Further, the children of a person with Marfan syndrome each have a 50:50 chance of inheriting the disease. Do all inherited disorders have this same pattern of inheritance? If not, what do these facts reveal about the inheritance of Marfan syndrome? To find out, we must go back in time to a monastery in Moravia and visit the garden of Gregor Mendel. ■

9.1 What Is the Physical Basis of Inheritance?

We have known for hundreds of years that some of the characteristics of parents are inherited by their offspring. As you learned in Chapters 2 and 8, genetic information is contained in molecules of DNA. Segments of DNA from a few hundred

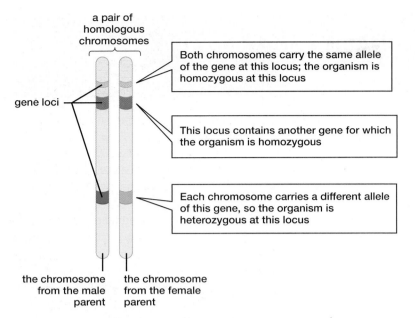

a pair of
homologous
chromosomes

Both chromosomes carry the same allele
of the gene at this locus; the organism is
homozygous at this locus

gene loci

This locus contains another gene for which
the organism is homozygous

Each chromosome carries a different allele
of this gene, so the organism is
heterozygous at this locus

the chromosome
from the male
parent

the chromosome
from the female
parent

▲ **Figure 9-1 The relationships among genes, alleles, and chromosomes** Homologues carry the same gene loci, but may have the same or different alleles at corresponding loci.

to many thousands of nucleotides long are the units of inheritance—the **genes**—that store genetic information in a cell. In Chapters 10 and 11, you will learn how the sequence of nucleotides in DNA encodes the information needed to produce proteins, cells, and entire organisms. For now, we will focus on the fact that chromosomes consist of DNA (together with a variety of proteins), which means that genes are parts of chromosomes. Chromosomes are passed from cell to cell and organism to organism during reproduction. **Inheritance,** then, occurs when genes are transmitted from parent to offspring.

We begin our exploration with a brief overview of genes and chromosomes. In this chapter, we confine our discussion to diploid organisms, including most plants and animals, that reproduce sexually by the fusion of haploid gametes.

Genes Are Segments of DNA at Specific Locations on Chromosomes

A gene's physical location on a chromosome is called its **locus** (plural, loci). Each member of a pair of homologous chromosomes carries the same genes, located at the same loci (Fig. 9-1). However, the nucleotide sequences of a given gene may differ in different members of a species, producing, for example, different human blood types, or Labrador retriever dogs that are black, yellow, or brown. In fact, a single organism often has different nucleotide sequences of a gene on each of its two homologues. These different versions of a gene at a given locus are called **alleles** (see Fig. 9-1).

Mutations Are the Source of Alleles

The only way to get different alleles of a gene is for the nucleotide sequence of the gene to change—in other words, for a mutation to occur. We will explore the molecular basis of mutations more thoroughly in Chapter 10. For now, think of genes as very long sentences, written in an alphabet of nucleotides instead of letters. If you were asked to type a sentence that was thousands of letters long, you'd probably make some mistakes. Although the mechanisms are obviously different, cells also make mistakes when they replicate the DNA of their chromosomes. Toxic chemicals, radiation, or ultraviolet light can also change the nucleotide sequence of a gene. These "nucleotide misspellings" are mutations. If a mutation occurs in the cells that become sperm or eggs, it can be passed from parent to offspring.

An allele, then, is simply a mutated gene. Most of the alleles in an organism's DNA first appeared in the gametes of the organism's ancestors, perhaps thousands or even millions of years ago, and have been inherited, generation after generation, ever since. A few alleles, which we will call "new mutations" may have occurred in the reproductive cells of an organism's parents. A new mutation, therefore, will first appear in the DNA of the body cells of this particular organism and may alter its metabolism, structures or behavior, even though none of its relatives—even its parents, brothers, or sisters—are affected. In the "Health Watch: Muscular Dystrophy" feature box on p. 150, we will see that a relatively common genetic disorder, Duchenne muscular dystrophy, frequently arises as a new mutation.

An Organism's Two Alleles May Be the Same or Different

A diploid organism has pairs of homologous chromosomes. Each member of a pair contains the same gene loci. Therefore, the organism has two copies of each gene. If both homologous chromosomes have the *same* allele at a locus, the organism is said to be **homozygous** at that locus (*homo* means "same"). For example, the chromosomes illustrated in Figure 9-1 are homozygous at two loci. If two homologous chromosomes have *different* alleles at a locus, the organism is **heterozygous** at that locus (*hetero* means "different"). The chromosomes in

Figure 9-1 are heterozygous at one locus. Organisms that are heterozygous for a specific trait are sometimes called *hybrids*.

Whether an individual is homozygous or heterozygous at a locus affects the genetic makeup of the gametes produced by that individual. Because homologous chromosomes separate during meiosis (see Chapter 8), each gamete receives one member of each pair of homologous chromosomes. As a result, each gamete has only one allele for each gene. If an individual is homozygous at a particular locus, all of the gametes it produces contain the same allele at that locus. In contrast, an individual that is heterozygous at a locus produces two kinds of gametes with respect to that allele: half contain one allele, and half contain the other allele.

9.2 How Were the Principles of Inheritance Discovered?

The common patterns of inheritance and many essential facts about genes, alleles, and the distribution of alleles in gametes and zygotes during sexual reproduction were deduced by an Austrian monk, Gregor Mendel (**Fig. 9-2**), in the mid-1800s. This was long before DNA, chromosomes, or meiosis had been discovered. Because his experiments are succinct, elegant examples of science in action, let's follow Mendel's paths of discovery.

Doing It Right: The Secrets of Mendel's Success

There are three key steps to any successful experiment in biology: (1) choosing the right organism with which to work, (2) designing and performing the experiment correctly, and (3) analyzing the data properly. Mendel was the first geneticist to employ all three steps in the study of inheritance.

Mendel chose the edible pea as the subject for his experiments in inheritance (**Fig. 9-3**). Stamens, the male reproductive structures of a flower, produce pollen. Each pollen grain contains sperm. Pollination allows sperm to fertilize the egg, located within the ovary at the base of the flower's female reproductive structure, called the carpel. In pea flowers, the petals enclose all of the reproductive structures, preventing another flower's pollen from entering (see Fig. 9-3), so the egg cells in a pea flower must be fertilized by sperm from the pollen of the same flower. This is called **self-fertilization.**

Mendel, however, often wanted to mate two different pea plants to see what types of offspring they would produce. To do this, he opened one pea flower and removed its stamens, preventing self-fertilization. Then, he dusted the sticky tip of the carpel with pollen from a flower of another plant. When sperm from one organism fertilize eggs from a different organism, the process is called **cross-fertilization.**

Mendel's experimental design was brilliant. In contrast to earlier scientists, who usually tried to understand inheritance by comparing entire organisms to their offspring, Mendel studied individual characteristics, called *traits*. Further, he chose traits with unmistakably different forms, such as white versus purple flowers. He also began by studying only one trait at a time. Finally, Mendel followed the inheritance of these traits for several generations, counting the numbers of offspring with each type of trait. By analyzing these numbers, the basic patterns of inheritance became clear. Today, quantifying experimental results and applying statistical analysis are essential tools in virtually every field of biology, but in Mendel's time, numerical analysis was an innovation.

9.3 How Are Single Traits Inherited?

Even in Mendel's day, many varieties of pea plants were true-breeding for different forms of a single trait. The term *true-breeding* refers to organisms with a trait, such as purple flowers, that is always inherited by all of their offspring that result

▲ **Figure 9-2 Gregor Mendel** A portrait of Mendel, painted in about 1888, after he had completed his pioneering genetics experiments.

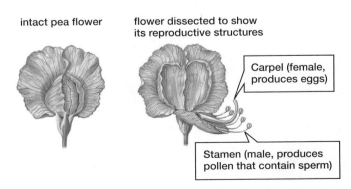

intact pea flower   flower dissected to show its reproductive structures

Carpel (female, produces eggs)

Stamen (male, produces pollen that contain sperm)

▲ **Figure 9-3 Flowers of the edible pea** In the intact pea flower (left), the lower petals form a container enclosing the reproductive structures. Pollen normally cannot enter the flower from outside, so peas usually self-fertilize. If the flower is opened (right), it can be cross-pollinated by hand.

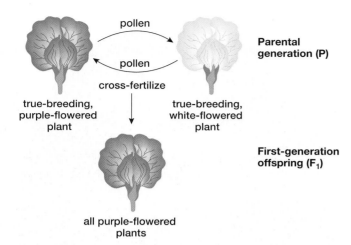

▲ Figure 9-4 **Cross of pea plants that are true-breeding for white or purple flowers** All of the offspring bear purple flowers.

Web Animation Crosses Involving Single Traits

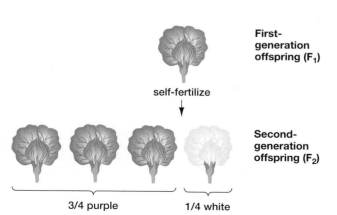

3/4 purple   1/4 white

▲ Figure 9-5 **Self-fertilization of F₁ plants with purple flowers** Three-quarters of the offspring bear purple flowers and one-quarter bear white flowers.

from self-fertilization. In his earliest experiments, Mendel cross-fertilized plants that possessed different true-breeding traits, saved the seeds that the plants produced, and grew the seeds to observe the traits of the resulting offspring.

In one of these experiments, Mendel cross-fertilized true-breeding, white-flowered plants with true-breeding, purple-flowered plants. The plants he crossed were the parental generation, denoted by the letter P. When he grew the resulting seeds, he found that all of the first-generation offspring (the "first filial," or F_1, generation) produced purple flowers (**Fig. 9-4**). What had happened to the white color? The flowers of the F_1 offspring were just as purple as the flowers of their purple parents. The white color seemed to have disappeared in these offspring.

Mendel then allowed the F_1 flowers to self-fertilize, collected the seeds, and planted them the next spring. In the second generation (F_2), Mendel counted 705 plants with purple flowers and 224 plants with white flowers. These numbers are approximately three-fourths purple and one-fourth white, or a ratio of 3 purple to 1 white (**Fig. 9-5**). This result showed that the gene that produced white flowers had not disappeared in the F_1 generation, but had only been "hidden."

Mendel allowed the F_2 plants to self-fertilize and produce a third (F_3) generation. He found that all of the white-flowered F_2 plants were true-breeding; that is, they produced only white-flowered offspring. For as many generations as he had time and patience to raise, white-flowered parents always gave rise only to white-flowered offspring.

Allowing the purple-flowered F_2 plants to self-fertilize showed that they were of two types: About one-third were true-breeding for purple, and the other two-thirds produced both purple- and white-flowered offspring, again in the ratio of about 3 purple to 1 white. Therefore, the F_2 generation included one-quarter true-breeding purple plants, one-half hybrid purple, and one-quarter true-breeding white.

The Inheritance of Dominant and Recessive Alleles on Homologous Chromosomes Can Explain the Results of Mendel's Crosses

Mendel's results, supplemented by modern knowledge of the behavior of genes and chromosomes, allow us to develop a five-part hypothesis to explain the inheritance of single traits:

1. Each trait is determined by pairs of distinct physical units that we now call *genes*. Each organism has two alleles for each gene, such as the gene that determines flower color. One allele of the gene is present on each homologous chromosome. True-breeding, white-flowered peas have different alleles of the "flower-color" gene than true-breeding, purple-flowered peas.

2. When two different alleles are present in an organism, one—the **dominant** allele—may mask the expression of the other—the **recessive** allele. The recessive allele, however, is still present. In the edible pea, the allele for purple flowers is dominant, and the allele for white flowers is recessive.

3. The pairs of alleles on homologous chromosomes separate from each other during gamete formation, so each gamete receives only one allele from each pair. This conclusion is known as Mendel's **law of segregation:** The two alleles of a gene segregate (separate) from one another during meiosis. When a sperm fertilizes an egg, the resulting offspring receives one allele from the father (in his sperm) and one from the mother (in her egg).

4. Which allele ends up in any given gamete is determined by chance. Because homologous chromosomes separate at random during meiosis, the distribution of alleles to the gametes is also random.

5. True-breeding (homozygous) organisms have two copies of the same allele for a given gene. Therefore, all of the gametes from a homozygous individual have the same allele for that gene (**Fig. 9-6a**). Hybrid (heterozygous)

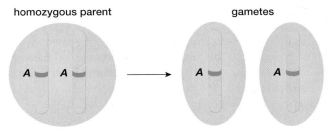

(a) Gametes produced by a homozygous parent

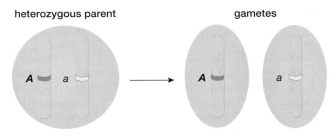

(b) Gametes produced by a heterozygous parent

▲ **Figure 9-6 The distribution of alleles in gametes (a)** All of the gametes produced by homozygous organisms contain the same allele. **(b)** Half of the gametes produced by heterozygous organisms contain one allele, and half of the gametes contain the other allele.

organisms have two different alleles for a given gene. Half of the organism's gametes will contain one allele for that gene and half will contain the other allele (**Fig. 9-6b**).

Let's see how this hypothesis explains the results of Mendel's experiments. We assign the uppercase letter P to the dominant allele for purple flower color and the lowercase letter p to the recessive allele for white. (By Mendel's convention, the dominant allele is represented by an uppercase letter.) A homozygous purple-flowered plant has two alleles for purple flowers (PP), whereas a white-flowered plant has two alleles for white flowers (pp). All of the sperm and eggs produced by a PP plant carry the P allele; all of the sperm and eggs of a pp plant carry the p allele (**Fig. 9-7a**). Mendel's F_1 hybrid offspring were produced when P sperm fertilized p eggs or when p sperm fertilized P eggs. In both cases, the F_1 offspring were Pp. Because P is dominant over p, all of the offspring were purple (**Fig. 9-7b**). Mendel's F_2 offspring resulted from self-fertilization of the heterozygous F_1 plants. Each gamete produced by a heterozygous Pp plant has an equal chance of receiving either the P allele or the p allele. That is, the heterozygous plant produces equal numbers of P and p sperm and equal numbers of P and p eggs. When a Pp plant self-fertilizes, each type of sperm has an equal chance of fertilizing each type of egg (**Fig. 9-7c**). Therefore, Mendel's F_2 generation contained three types of off-spring (PP, Pp, and pp), and the three types were in the approximate proportions of one-quarter PP (homozygous purple), one-half Pp (heterozygous purple), and one-quarter pp (homozygous white).

▶ **Figure 9-7 Segregation of alleles and fusion of gametes predict the distribution of alleles and traits in Mendel's experiment with flower color in peas (a)** Parental generation. All of the gametes of the homozygous parents contain the same allele: only P alleles in the gametes of PP parents and only p alleles in the gametes of pp parents. **(b)** F_1 generation. Fusion of gametes containing the P allele with gametes containing the p allele produces only Pp offspring. **(c)** F_2 generation. Half of the gametes of heterozygous Pp parents contain the P allele and half contain the p allele. Fusion of these gametes produces PP, Pp, and pp offspring.

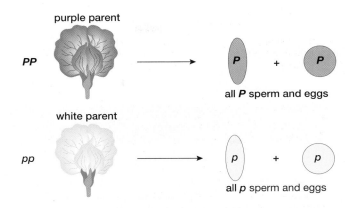

(a) **Gametes produced by homozygous parents**

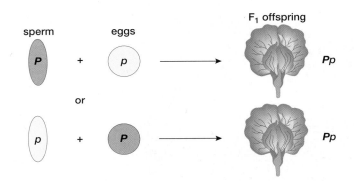

(b) **Fusion of gametes produces F_1 offspring**

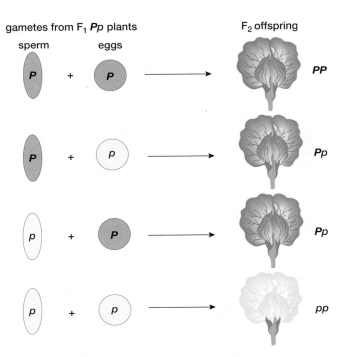

(c) **Fusion of gametes from the F_1 generation produces F_2 offspring**

Two plants that look alike may actually carry different combinations of alleles. The combination of alleles carried by an organism (for example, *PP* or *Pp*) is its **genotype.** The organism's traits, including its outward appearance, behavior, digestive enzymes, blood type, or any other observable or measurable feature, make up its **phenotype.** Even though *PP* and *Pp* plants have different genotypes, they have the same phenotype for purple flower color. Therefore, the F_2 generation of Mendel's peas contained three genotypes (one-quarter *PP*, one-half *Pp*, and one-quarter *pp*), but only two phenotypes (three-quarters purple and one-quarter white).

Simple "Genetic Bookkeeping" Can Predict the Genotypes and Phenotypes of Offspring

The **Punnett square method,** named after R. C. Punnett, a famous geneticist of the early 1900s, is a convenient way to predict the genotypes and phenotypes of offspring. **Figure 9-8** shows how to use a Punnett square to determine the proportions of offspring that arise from the self-fertilization of an organism that is heterozygous for a single trait. As you use this "genetic bookkeeping" technique, keep in mind that, in a real experiment, the offspring will occur only in *approximately* the predicted proportions because sperm and eggs with different genotypes encounter one another randomly. Let's consider an example. Each time a baby is conceived, it has a 50:50 chance of becoming a boy or a girl. However, many families with two children do not have one girl and one boy. The 50:50 ratio of girls to boys occurs only if we average the genders of the children in many families.

Mendel's Hypothesis Can Predict the Outcome of New Types of Single-Trait Crosses

You have probably recognized that Mendel used the scientific method, making an observation and using it to formulate a hypothesis. But does Mendel's hypothesis accurately predict the results of further experiments? Based on his hypothesis that

Sudden Death on the Court— Marfan Syndrome

Continued

When a person with Marfan syndrome has children with a person without the syndrome, those children have a 50% chance of inheriting the condition. Do you think that Marfan is inherited as a dominant or a recessive allele? Why? Check your reasoning in "Sudden Death on the Court—Marfan Syndrome, *Revisited*" at the end of the chapter.

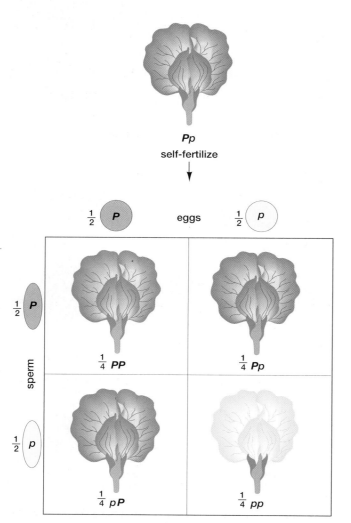

◄ **Figure 9-8 The Punnett square method** The Punnett square method allows you to predict both genotypes and phenotypes of specific crosses. Here we use it for a cross between plants that are heterozygous for a single trait, flower color.

1. Assign letters to the different alleles. Use uppercase letters for dominant and lowercase for recessive.
2. Determine all of the types of genetically different gametes that can be produced by the male and female parents.
3. Draw the Punnett square, with the columns labeled with the egg genotypes and the rows labeled with the sperm genotypes. (We have included the fractions of these genotypes with each label.)
4. Fill in the genotype of the offspring in each box by combining the genotype of the sperm in its row with the genotype of the egg in its column. (We have placed the fractions in each box.)
5. Count the number of offspring with each genotype. (Note that *Pp* is the same as *pP*.)
6. Convert the number of offspring of each genotype to a fraction of the total number of offspring. In this example, out of four fertilizations, only one is predicted to produce the *pp* genotype, so one-quarter of the total number of offspring produced by this cross is predicted to be white. To determine phenotypic fractions, add the fractions of genotypes that would produce a given phenotype. For example, purple flowers are produced by ¼ *PP* + ¼ *Pp* + ¼ *pP*, for a total of three-quarters of the offspring.

QUESTION: *If you crossed a heterozygous* Pp *plant with a homozygous recessive* pp *plant, what would be the expected ratio of offspring? How does this differ from the offspring of a* PP × pp *cross? Try working this out before you read further in the text.*

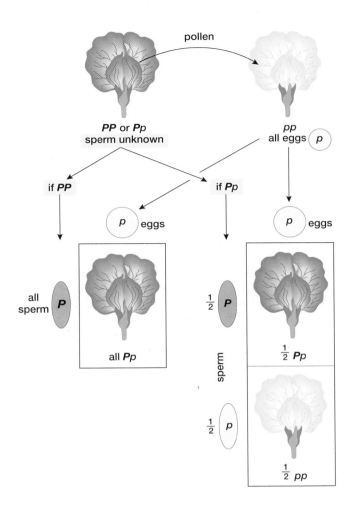

◄ **Figure 9-9 Punnett square of a test cross** An organism with a dominant phenotype may be either homozygous or heterozygous. Crossing such an organism with a homozygous recessive organism can determine whether the dominant organism was homozygous (left) or heterozygous (right).

Web Animation Crosses Involving Multiple Traits

the heterozygous F₁ plants had one allele for purple flowers and one for white (*Pp*), Mendel predicted the outcome of cross-fertilizing these *Pp* plants with homozygous recessive white plants (*pp*): There should be equal numbers of *Pp* (purple) and *pp* (white) offspring. This is indeed what he found. Setting up such a "test cross" in a Punnett square can help you to see why (**Fig. 9-9**).

9.4 How Are Multiple Traits Inherited?

Mendel next turned to the more complex question of how multiple traits are inherited. He crossed pea plants that differed in two traits, for example, seed color (yellow or green) and seed shape (smooth or wrinkled) (**Fig. 9-10**). From earlier crosses of plants with these single traits, he already knew that the "smooth" allele of the gene that controls seed shape (*S*) is dominant to the "wrinkled" allele (*s*), and that the "yellow" allele of the seed color gene (*Y*) is dominant to the "green" allele (*y*).

Mendel crossed a true-breeding plant with smooth, yellow seeds (*SSYY*) to a true-breeding plant with wrinkled, green seeds (*ssyy*). All F₁ offspring had the genotype *SsYy* and had smooth, yellow seeds. Mendel allowed these F₁ plants to self-fertilize, and observed that the F₂ generation consisted of 315 plants with smooth yellow seeds, 101 with wrinkled yellow seeds, 108 with smooth green seeds, and 32 with wrinkled green seeds, a ratio of about 9:3:3:1. Mendel crossed a variety of plants that were heterozygous for two traits, and always found that the F₂ generation had phenotypic ratios of about 9:3:3:1.

Trait	Dominant form	Recessive form
Seed shape	smooth	wrinkled
Seed color	yellow	green
Pod shape	inflated	constricted
Pod color	green	yellow
Flower color	purple	white
Flower location	at leaf junctions	at tips of branches
Plant size	tall (about 6 feet)	dwarf (about 8 to 16 inches)

▲ **Figure 9-10 Traits of pea plants studied by Gregor Mendel**

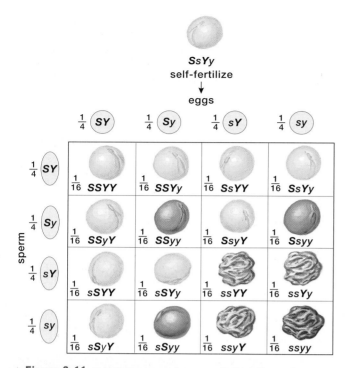

▲ Figure 9-11 Predicting genotypes and phenotypes for a cross between parents that are heterozygous for two traits
In pea seeds, yellow color (*Y*) is dominant to green (*y*), and smooth shape (*S*) dominant to wrinkled (*s*). In this cross, both parents are heterozygous for each trait (or a single individual heterozygous for both traits self-fertilizes). There are now 16 boxes in the Punnett square. In addition to predicting all of the genotypic combinations and the 9:3:3:1 overall phenotypic ratio, the Punnett square predicts three-quarters yellow seeds, one-quarter green seeds, three-quarters smooth seeds, and one-quarter wrinkled seeds, just as we would expect from crosses made of each trait separately.   **QUESTION:** Use Punnett squares to answer this question: Can the genotype of a plant with smooth yellow seeds (both dominant traits) be revealed by crossing it with a plant bearing wrinkled green seeds?

Mendel Concluded That Multiple Traits Are Inherited Independently

Mendel realized that these results could be explained if the genes for seed color and seed shape were inherited independently. If so, then for *each trait*, three-quarters of the offspring should show the dominant phenotype and one-quarter should show the recessive phenotype. This result is just what Mendel observed. He found 423 plants with smooth seeds (of either color) and 133 with wrinkled seeds (a ratio of about 3:1), and 416 plants with yellow seeds (of either shape) and 140 with green seeds (also about 3:1). **Figure 9-11** shows how a Punnett square can be used to determine the outcome of a cross between organisms that are heterozygous for two traits.

The independent inheritance of two or more distinct traits is called the **law of independent assortment.** Multiple traits are inherited independently because the alleles of one gene are distributed to gametes independently of the alleles of other genes. The physical basis for this independence is found in the events of meiosis (see Chapter 8). When paired homologous chromosomes line up during metaphase I, homologues face the poles of the cell at random, and the orientation of any one homologous pair does not influence the orientation of other pairs. Therefore, when the homologues separate during anaphase I, which allele of a gene on homologous pair 1 moves "north" does not affect which allele of a gene on homologous pair 2 moves "north"; that is, the alleles of genes on different chromosomes are distributed, or "assorted," independently (**Fig. 9-12**). Note that only genes that are located on different chromosomes assort independently.

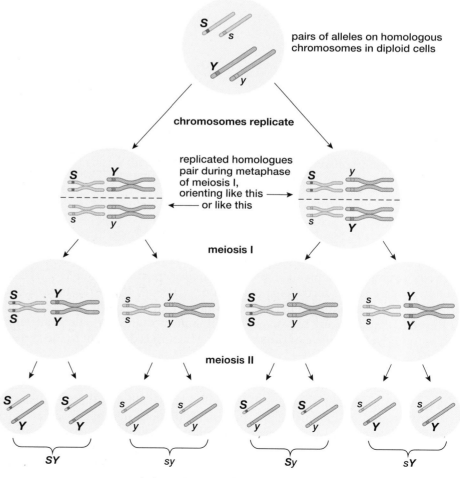

► Figure 9-12 Independent assortment of alleles
Chromosome movements during meiosis produce independent assortment of alleles, as shown here for two genes. Each possible combination is equally likely, producing gametes in the proportions one-quarter *SY*, one-quarter *sy*, one-quarter *sY*, and one-quarter *Sy*.   **QUESTION:** *If the genes for seed color and seed shape were on the* same *chromosome* rather than on *different chromosomes, would they assort independently? Why or why not?*

In an Unprepared World, Genius May Go Unrecognized

Gregor Mendel presented his hypotheses about inheritance in 1865, and they were published the following year. His paper did not mark the beginning of genetics. In fact, it didn't make any impression at all during his lifetime. Mendel's experiments, which eventually spawned one of the most important scientific theories in all of biology, simply vanished from the scene. Apparently, very few biologists read his paper, and those who did failed to recognize its significance or discounted it because it contradicted prevailing ideas of inheritance.

It was not until 1900 that three biologists—Carl Correns, Hugo de Vries, and Erich Tschermak—working independently of one another and knowing nothing of Mendel's work, rediscovered the principles of inheritance. No doubt to their intense disappointment, when they searched the scientific literature before publishing their results, they found that Mendel had beaten them to it more than 30 years earlier. To their credit, they graciously acknowledged the important work of the Austrian monk, who had died in 1884.

9.5 How Are Genes Located on the Same Chromosome Inherited?

As you learned earlier in this chapter, each chromosome contains many genes (see Fig. 9-1). Chromosomes, not genes, are separated during meiosis. Therefore, genes on the same chromosome do not usually assort independently.

Genes on the Same Chromosome Tend to Be Inherited Together

Genetic **linkage** is the inheritance of genes as a group because they are on the same chromosome. One of the first pairs of linked genes to be discovered was found in the sweet pea, a different species from Mendel's garden pea. In sweet peas, the gene for flower color and the gene for pollen grain shape are located on the same chromosome. Thus, alleles of these genes generally assort together into gametes during meiosis and so they are inherited together.

Consider a sweet pea plant with red flowers and long pollen that has the pair of homologous chromosomes shown in **Figure 9-13**. Note that the purple allele of the flower color gene and the long allele of the pollen shape gene are both located on one homologous chromosome. The red allele of the flower color gene and the round allele of the pollen shape gene are both located on the other homologue. Therefore, the gametes produced by this sweet pea plant are likely to have *either* purple and long alleles *or* red and round alleles. Therefore, a cross of two sweet pea plants, both of which are heterozygous for pollen color and pollen shape, would be expected to produce only purple/long and red/round flowers, and no purple/round or red/long flowers. This pattern of inheritance violates the law of independent assortment, which states that the alleles for flower color and pollen shape should assort independently. Instead, genes on the same chromosome tend to stay together during meiosis and to be inherited together.

Crossing Over Can Create New Combinations of Linked Alleles

However, genes on the same chromosome do not *always* stay together. For example, in the sweet pea cross just described, the F_2 generation usually does include a few plants in which the genes for flower color and pollen shape have been inherited as if they were *not* linked. That is, some of the offspring will have purple flowers and round pollen, and some will have red flowers and long pollen. How can this be?

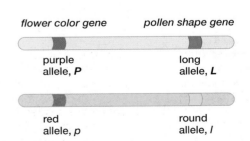

▲ **Figure 9-13 Homologous chromosomes of the sweet pea, showing the genes for flower color and pollen shape**

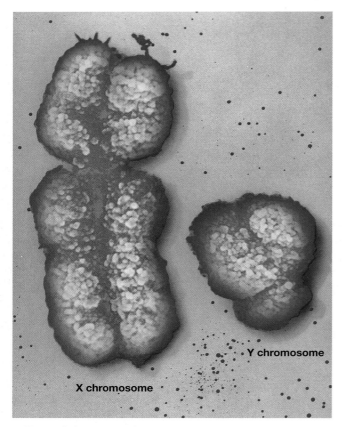

▲ **Figure 9-14 Photomicrograph of human sex chromosomes** Notice the small size of the Y chromosome, which carries relatively few genes.

As you learned in Chapter 8, during prophase I of meiosis, homologous chromosomes sometimes exchange parts, a process called **crossing over** (see Fig. 8-14). During meiosis, each pair of homologous chromosomes usually exchanges at least one segment of DNA. This exchange of corresponding segments of DNA during crossing over produces new allele combinations on both homologous chromosomes. Then, when recombined homologous chromosomes separate at anaphase I, the chromosomes that each haploid daughter cell receives have combinations of alleles that are different from those of the parent cell. Not surprisingly, how often genes are separated by crossing over depends on how far apart they are on the chromosome. Genes located right next to each other are almost never separated. In contrast, genes on opposite ends of a chromosome are almost always separated by crossing over.

Let's look at the sweet pea chromosomes once again. If crossing over occurs between the color locus and the pollen shape locus, then there will be new combinations of alleles on the recombined chromosomes. If the purple color allele (*P*) and long shape allele (*L*) were originally on one homologue, and the red color allele (*p*) and round shape allele (*l*) were on the other homologue, crossing over will produce a few chromosomes with the purple/round allele combination (*Pl*), and a few chromosomes with the red/long allele combination (*pL*). Although all of the F₁ generation will still have the *PpLl* genotype, a few will receive that genotype by having *Pl* and *pL* chromosomes instead of *PL* and *pl* chromosomes. Hence, a few plants in the F₂ generation will receive the *Pl* chromosome from both parents, and have the *PPll* genotype and purple/round phenotype. A few others will receive the *pL* chromosome from both parents, and have the *ppLL* genotype and red/long phenotype.

9.6 How Is Sex Determined?

In animals of many species, an individual's sex is determined by a special pair of chromosomes, the **sex chromosomes.** In mammals, for example, males and females have the same total number of chromosomes, but females have two identical sex chromosomes, called X chromosomes, whereas males have one X chromosome and one Y chromosome (**Fig. 9-14**). The rest of the chromosomes, called **autosomes,** are found in both sexes.

How is the sex of a fertilized egg determined? Even though the Y is much smaller than the X, the X and Y chromosomes act as homologues, pairing up during prophase of meiosis I and separating during anaphase I. During gamete formation, the sex chromosomes segregate. Each egg receives one X chromosome (because females are XX) and each sperm receives either an X or a Y chromosome (because males are XY). Thus, an egg that is fertilized by a Y-bearing sperm yields a male offspring, and an egg fertilized by an X-bearing sperm yields a female offspring (**Fig. 9-15**).

9.7 How Are Sex-Linked Genes Inherited?

Genes that are found on one sex chromosome but not on the other are called **sex-linked** genes. In many animal species, the Y chromosome carries only a few genes. In humans, the Y chromosome contains a few dozen genes, many of which play a role in male reproduction. In contrast, the X chromosome contains about 1,000 genes, few of which have any specific role in female reproduction. Most of the genes on the human X chromosome have no counterpart on the Y chromosome, including genes for color vision, blood clotting, and certain structural proteins in muscles. Because they have two X chromosomes, females can be either

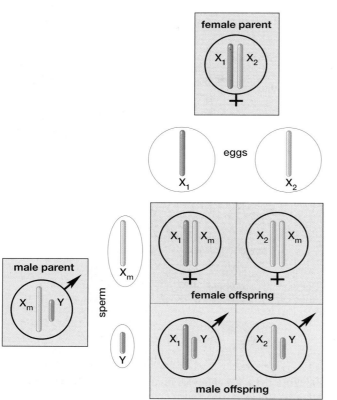

◀ **Figure 9-15 Sex determination in mammals** Male offspring receive their Y chromosome from the father. Female offspring receive the father's X chromosome (labeled Xₘ). Both male and female offspring receive an X chromosome (either X₁ or X₂) from the mother.
QUESTION: Will a man transmit X chromosome genes to his sons, his daughters, or both?

homozygous or heterozygous for genes on the X chromosome, and normal dominant versus recessive relationships among alleles apply. In contrast, males have only a single X chromosome, and thus express *all* of the alleles they have on their X chromosome, whether those alleles are dominant or recessive. Therefore, if a woman inherits one copy of a defective, recessive allele on one X chromosome, she will be phenotypically normal because her other X chromosome bears a functional, dominant allele. If a man inherits one defective, recessive allele on his X chromosome, he will show the defective phenotype. Not surprisingly, sex-linked traits appear far more frequently in males.

In humans, the most familiar genetic defects caused by recessive alleles of X-chromosome genes are hemophilia and red-green color blindness (**Fig. 9-16**). Red-green color blindness is caused by recessive alleles of either of two genes located on the X chromosome. The normal alleles of these genes encode proteins that allow one set of cones (color-vision cells) in the retina of the eye to be most sensitive to red light and another set to be most sensitive to green light. There are several defective recessive alleles of these genes. In the most extreme cases, the defective alleles encode proteins that make both sets of cones equally sensitive to both red and green light. Therefore, the affected person cannot distinguish red from green.

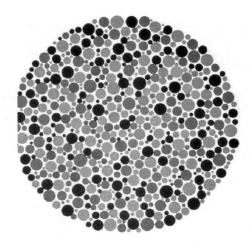

▲ **Figure 9-16 Color blindness, a sex-linked recessive trait**
This figure, called an *Ishihara chart* after its inventor, distinguishes color-vision defects. People with red-deficient vision see a 6, and those with green-deficient vision see a 9. People with normal color vision see 96.

9.8 Do the Mendelian Rules of Inheritance Apply to All Traits?

So far, we have looked at only the simplest kinds of traits: those that are completely controlled by a single gene, those for which there are only two possible alleles of each gene, and those in which one allele is completely dominant to the other, recessive, allele. Most traits, however, are influenced in more varied and subtle ways.

Incomplete Dominance Produces Intermediate Phenotypes

When one allele is completely dominant over a second allele, heterozygotes with one dominant allele have the same phenotype as homozygotes with two dominant alleles (see Figs. 9-7 and 9-8). However, relationships between alleles are not always this simple. When the heterozygous phenotype is intermediate between the two homozygous phenotypes, the pattern of inheritance is called **incomplete dominance.** In humans, hair texture is influenced by a gene with two incompletely dominant alleles, which we will call C_1 and C_2. A person with two copies of the C_1 allele has curly hair; two copies of the C_2 allele produce straight hair. Heterozygotes, with the C_1C_2 genotype, have wavy hair. Two wavy-haired people could have children with any of the three hair types, with probabilities of one-quarter curly (C_1C_1), one-half wavy (C_1C_2) and one-quarter straight (C_2C_2) (**Fig. 9-17**).

A Single Gene May Have Multiple Alleles

A single individual can have only two alleles of any gene, one on each homologous chromosome. However, if we could sample all members of a species, we would typically find dozens of alleles for every gene. For example, there are hundreds of different alleles that cause Marfan syndrome.

Human blood types are a familiar example of multiple alleles of a single gene. The blood types A, B, AB, and O arise as a result of three different alleles (for simplicity, we will designate them *A, B,* and *o*) of a single gene located on chromosome 9. This gene codes for an enzyme that adds sugar molecules to the ends of recognition proteins that protrude from the surfaces of red blood cells. Alleles *A* and *B* code for enzymes that add different sugars to the recognition proteins. (We will call the resulting molecules "recognition proteins *A* and *B*.") Allele *o* codes for a nonfunctional enzyme that doesn't add any sugar. A person may have one of six genotypes: *AA, BB, AB, Ao, Bo,* or *oo* (**Table 9-1**).

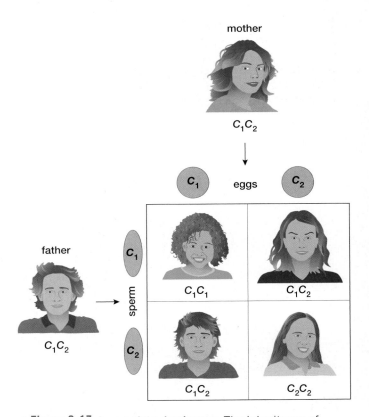

▲ **Figure 9-17 Incomplete dominance** The inheritance of hair texture in humans provides an example of incomplete dominance. In such cases, we use uppercase letters for both alleles, here C_1 and C_2. Homozygotes may have curly hair (C_1C_1) or straight hair (C_2C_2). Heterozygotes (C_1C_2) have wavy hair. The children of a man and a woman, both with wavy hair, may have curly, straight, or wavy hair, in the approximate ratio of one-quarter curly:one-half wavy:one-quarter straight.

Table 9-1   Human Blood Group Characteristics

Blood Type	Genotype	Red Blood Cells	Has Plasma Antibodies to:	Can Receive Blood from:	Can Donate Blood to:	Frequency in United States
A	AA or Ao	A glycoprotein	B glycoprotein	A or O (no blood with B glycoprotein)	A or AB	40%
B	BB or Bo	B glycoprotein	A glycoprotein	B or O (no blood with A glycoprotein)	B or AB	10%
AB	AB	Both A and B glycoproteins	Neither A nor B glycoprotein	AB, A, B, O (universal recipient)	AB	4%
O	oo	Neither A nor B glycoprotein	Both A and B glycoproteins	O (no blood with A or B glycoprotein)	O, AB, A, B (universal donor)	46%

Alleles *A* and *B* are dominant to *o*. Therefore, people with genotypes *AA* or *Ao* have only type A recognition proteins and have type A blood. Those with genotypes *BB* or *Bo* synthesize only type B recognition proteins and have type B blood. Homozygous recessive *oo* individuals lack both types of proteins and have type O blood. In people with type AB blood, both enzymes are present, so the plasma membranes of their red blood cells have both A *and* B recognition proteins. When heterozygotes express phenotypes of *both* of the homozygotes (in this case, both A and B proteins), the pattern of inheritance is called **codominance,** and the alleles are said to be *codominant* to one another.

People make antibodies to the type of recognition protein(s) that they lack. These antibodies are proteins in blood plasma that bind to the sugars on foreign recognition proteins. The antibodies cause red blood cells that bear foreign recognition proteins to clump together and to rupture. The resulting clumps and fragments can clog small blood vessels and damage vital organs such as the brain, heart, lungs, or kidneys. This means that blood type must be determined and matched carefully before a blood transfusion is made.

Type O blood, lacking any sugars, is not attacked by antibodies in A, B, or AB blood, so it can be transfused safely to all other blood types. (The antibodies present in transfused blood become too diluted to cause problems.) People with type O blood are called "universal donors." But type O blood carries antibodies to both A and B recognition proteins, so type O individuals can receive transfusions of only type O blood.

A Single Trait May Be Influenced by Several Genes

If you look around your class, you are likely to see people of varied heights, skin colors, and body builds, to consider just a few obvious traits. These traits are governed not by single genes, but by interactions among two or more genes, a phenomenon called **polygenic inheritance.**

In some cases, polygenic inheritance appears to be additive: Several incompletely dominant genes with similar effects contribute to the trait. This may be the case with skin color, in which multiple genes influence the concentration and

◀Figure 9-18 **Polygenic inheritance of skin color** The combination of polygenic inheritance and environmental effects (especially exposure to sunlight) produces a very fine gradation of human skin colors.

packaging of a pigment, called melanin, in the skin. In other cases, multiple genes with quite different effects all contribute to a trait. For example, human eye color is controlled by at least three genes. One gene has two alleles, for brown and blue color; one gene has two alleles, for green and blue; and one gene determines the presence or absence of a central brown area in the iris. Even this complicated set of genes fails to explain why people may have lighter or darker brown eyes, or how anyone could have gray eyes. Obviously, there are more genes awaiting discovery!

In polygenic inheritance, the more genes that contribute to a single trait, the greater the number of phenotypes and the finer the distinctions among them. When more than three genes contribute to a trait, it is usually difficult to classify the phenotypes reliably. For example, at least three, and possibly dozens, of genes, affect human skin pigmentation. Exposure to the sun further alters skin color, with the result that humans show virtually continuous variation from very dark to very light skin (**Fig. 9-18**).

The Environment Influences the Expression of Genes

An organism's phenotype is not just the sum of its genes. The environment in which an organism lives also plays a large role in determining phenotype. A striking example of environmental effects on gene action is provided by the Himalayan rabbit, which, like the Siamese cat, has pale body fur, but black ears, nose, tail, and feet. The Himalayan rabbit actually has the genotype for black fur everywhere on its body. However, the enzyme that produces the black pigment is inactive at temperatures above about 93°F (34°C). Because most of a rabbit's body surface is warmer than 93°F, most of the rabbit's fur is pale. Its ears, nose, tail, and feet, however, are usually cooler, and therefore have black fur (**Fig. 9-19**).

All traits that are influenced by genes are also influenced by the environment. Even traits that seem to be under rigid genetic control, such as the colors of flowers, vary somewhat depending on environmental conditions such as light intensity, nutrients, and temperature. More complex traits have correspondingly complex interactions between genetic and environmental influences. For example, the polygenic trait of human skin color is modified by the environmental effects of sun exposure. Height, another polygenic trait, can be reduced by poor nutrition. By the same token, even the most sophisticated human behaviors depend on physical structures (such as the brain) whose construction is influenced by genes. In fact, genetic differences have been shown to influence human traits as diverse as intelligence, sexuality, and susceptibility to cancer.

9.9 How Are Human Genetic Disorders Investigated?

Medical geneticists are especially interested in genes that influence our susceptibility to disease. Because experimentally crossbreeding humans with specific traits is out of the question, human geneticists study the inheritance of disease-causing alleles by searching medical, historical, and family records to study past crosses. Records extending across several generations can be arranged in the

<blockquote>

Sudden Death on the Court—Marfan Syndrome

Continued

A single gene may influence several different traits, a phenomenon called **pleiotropy**. Marfan syndrome is an example of pleiotropy. The defective fibrillin gene causes increased height, long limbs, large hands and feet, weak walls in the aorta, and often dislocated lenses in one or both eyes—multiple traits controlled by the actions of a single gene.

</blockquote>

▲ Figure 9-19 **Environmental influence on phenotype** Expression of the gene for black fur in the Himalayan rabbit depends on an interaction between genotype and environment. The gene is expressed only on the cooler parts of the rabbit's body.

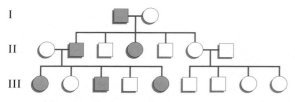

(a) A pedigree for a dominant trait

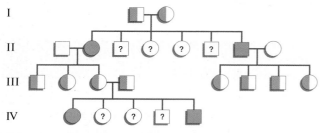

(b) A pedigree for a recessive trait

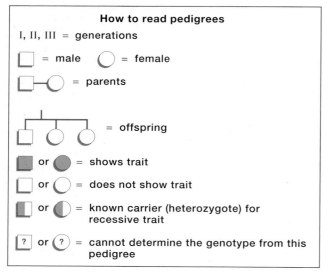

How to read pedigrees

I, II, III = generations

☐ = male ○ = female

☐─○ = parents

= offspring

■ or ◑ = shows trait

☐ or ○ = does not show trait

◨ or ◖ = known carrier (heterozygote) for recessive trait

☐? or ⊘ = cannot determine the genotype from this pedigree

▲ **Figure 9-20 Family pedigrees** *(a)* A pedigree for a dominant trait. Note that any offspring showing a dominant trait *must* have at least one parent with the trait (see Figs. 9-8 and 9-9). *(b)* A pedigree for a recessive trait. Any individual showing a recessive trait must be homozygous recessive. If that person's parents did not show the trait, then *both* of the parents must be heterozygotes (carriers). Note that the genotype cannot be determined for some offspring, who may be either heterozygotes or homozygous dominants. **QUESTION:** If it were possible to do breeding experiments with humans, how could you determine if the offspring denoted by question marks in part (b) are heterozygotes or homozygous dominants?

form of family **pedigrees,** diagrams that show the genetic relationships among a set of related individuals (**Fig. 9-20**).

Careful analysis of pedigrees reveals whether a particular trait is inherited in a dominant, recessive, or sex-linked pattern. Since the mid-1960s, analysis of human pedigrees, combined with molecular genetic technology, has produced great strides in understanding human genetic diseases. For instance, geneticists now know the genes responsible for dozens of inherited diseases, including sickle-cell anemia, Marfan syndrome, and muscular dystrophy. Research in molecular genetics promises to increase our ability to predict genetic diseases and perhaps even to cure them, a topic we will explore further in Chapter 12.

9.10 How Are Single-Gene Disorders Inherited?

Many common human traits, including freckles, long eyelashes, cleft chin, and widow's peak hairline, are inherited in a simple Mendelian fashion: Each trait is controlled by a single gene with a dominant and a recessive allele. Here, we focus on a few examples of medically important disorders with simple Mendelian inheritance.

Some Human Genetic Disorders Are Caused by Recessive Alleles

Many genes encode the information needed by cells to synthesize enzymes or structural proteins. A defective allele of such a gene will cause the synthesis of damaged or inactive proteins. In many cases, the effects of a defective allele are masked when one normal allele is present and generates enough functional protein. Thus, a person who is heterozygous (with one normal and one defective allele) will remain healthy and have a phenotype that is indistinguishable from that of a person with two copies of the normal allele. In other words, if the defective allele is recessive, only people who inherit two copies of the defective allele will have the disorder. Muscular dystrophy, a fatal degeneration of the muscles of young boys, is a recessive genetic disorder that we describe in "Health Watch: Muscular Dystrophy" on p. 150.

People who are heterozygous for a harmful recessive allele have a normal phenotype, but they can pass the defective allele to their offspring. Geneticists estimate that each of us carries 5 to 15 harmful recessive alleles that would cause a serious genetic defect in a homozygote. When a couple has a child, each of the defective alleles in each parent has a 50:50 chance of being passed on, so the probability is very high that the child will inherit some harmful recessive alleles. Fortunately, it is unlikely that both parents carry defective alleles of the *same* gene, so the child is unlikely to be homozygous for a harmful recessive allele.

Recessive disorders in offspring are more common if the parents are closely related (especially if they are first cousins or closer). Related couples have inherited some of their genes from a recent common ancestor, so they are much more likely to carry defective recessive alleles of the same genes. Each child borne by a couple who share the same harmful recessive allele has a 1 in 4 chance of being homozygous for, and therefore affected by, the genetic disorder.

A Defective Allele for Hemoglobin Synthesis Causes Sickle-Cell Anemia

Sickle-cell anemia results from a mutation in the hemoglobin gene. The hemoglobin protein, found in red blood cells, transports oxygen. In sickle-cell anemia, a change of a single nucleotide results in an incorrect amino acid at a crucial position in the hemoglobin molecule. When oxygen concentrations are low (such as in exercising muscles), sickle-cell hemoglobin molecules stick together. The clumps force the red blood cell out of its normal disk shape (Fig. 9-21a) into a longer, sickle shape (Fig. 9-21b). Sickled red blood cells clump, clogging capillaries. Tissues downstream of the clog do not receive enough oxygen. Paralyzing strokes can occur if blood vessels in the brain become blocked. The sickled cells are also more fragile than normal red blood cells, so they die before their time, causing anemia.

People homozygous for the sickle-cell allele synthesize only defective hemoglobin. Therefore, many of their red blood cells become sickled. Although heterozygotes have about half normal and half abnormal hemoglobin, usually they have few sickled cells and are not seriously affected—in fact, many world-class athletes are heterozygous for the sickle-cell allele. Because only homozygous recessive people usually show symptoms, sickle-cell anemia is considered a recessive disorder.

About 20% to 40% of sub-Saharan Africans and 8% of African Americans are heterozygous for sickle-cell anemia, but the allele is extremely rare in Caucasians. Why? People who are heterozygous for the sickle-cell allele are less susceptible to malaria, a potentially fatal disease caused by a parasite transmitted to people by mosquito bites. Malaria is a major killer in the warmer regions of the world, including much of Africa. Worldwide, there are hundreds of millions of cases of malaria each year, causing about a million deaths, primarily among young children. Geneticists have calculated that the sickle-cell allele arose in Africa as at least three independent mutations that occurred between 75,000 and 150,000 years ago. Since that time, people who were homozygous for the normal hemoglobin allele often contracted malaria. People who were homozygous for the sickle-cell allele often developed severe anemia and blood clots. However, natural selection favored heterozygotes, who had resistance to malaria. Therefore, the sickle-cell allele is common in Africa. In colder parts of the world, where malaria is uncommon, the sickle-cell allele has only harmful effects, and so it would be selected against.

Some Human Genetic Disorders Are Caused by Dominant Alleles

Many other genetic diseases are caused by dominant alleles, in which a single defective allele is enough to cause the disorder. For dominant diseases to be inherited, at least one parent must have the disease and that parent must remain healthy enough, long enough, to mature and have children. In some dominant diseases, such as Huntington's disease (see below), symptoms do not appear until later in life, often after the affected person has reproduced. Occasionally, a defective dominant allele may result from a new mutation in an unaffected person's eggs or sperm.

How can a defective allele be dominant to the normal allele? Some dominant alleles may encode proteins that carry out new, toxic reactions. Other dominant alleles may encode proteins that are overactive, performing their functions at inappropriate times and places. Finally, some dominant alleles may encode defective proteins that interfere with the actions of normal proteins.

Huntington's disease, for example, is a dominant inherited disorder that causes a slow, progressive deterioration of parts of the brain, resulting in a loss of coordination, flailing movements, personality disturbances, and eventual death. The symptoms of Huntington's disease typically do not appear until 30 to 50 years of age. Therefore, many people pass the allele to their children before they suffer the first symptoms. In 1983, molecular genetic technology and painstaking pedigree analysis were used to localize the Huntington gene to a small part of chromosome 4. Geneticists finally isolated the Huntington gene in 1993 and a few years later identified the gene's product, a protein they named huntingtin. The function of normal huntingtin remains unknown. Mutant huntingtin seems both to interfere with the action of normal huntingtin and to form large aggregates in nerve cells that ultimately kill the cells.

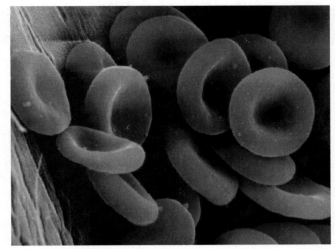

(a) Normal red blood cells

(b) Sickled red blood cells

▲ Figure 9-21 Sickle-cell anemia (a) Normal red blood cells are disk-shaped with indented centers. (b) The red blood cells of a person with sickle-cell anemia take on a slender, curved shape resembling a sickle.

health watch

Muscular Dystrophy

When Olympic weightlifter Hossein Reza Zadeh of Iran won the gold medal in 2004 with a "clean and jerk" lift of over 580 pounds (263 kilograms), he exerted tremendous forces on his own body (Fig. E9-1). How could his muscles withstand these stresses? For that matter, why don't the muscles in your body rip apart when you climb a flight of stairs?

▲ Figure E9-1 Hossein Reza Zadeh wins the gold with a lift of 580 pounds

Muscle cells are firmly tied together by a gigantic protein called **dystrophin.** The almost 3,700 amino acids of dystrophin form a supple yet strong rod that connects the cytoskeleton inside a muscle cell to proteins in its plasma membrane, which in turn attach to proteins that form a fibrous support surrounding each muscle. Thus, when a muscle contracts, the forces are distributed fairly evenly throughout each cell in the muscle and to the extracellular support proteins, so the muscle cells remain intact.

Unfortunately, about 1 in 3,500 boys synthesizes seriously defective dystrophin proteins. As these boys use their muscles, the lack of functioning dystrophin means that ordinary muscle contraction tears the muscle cells. The cells die and are replaced by fat and connective tissue (Fig. E9-2). By the age of 7 or 8, the boys can no longer walk. Death usually occurs in the early 20s because of heart and respiratory difficulties.

These boys suffer from **muscular dystrophy,** which literally means "degeneration of the muscles." The most severe form is called Duchenne muscular dystrophy; a less severe, but eventually still fatal, form is called Becker

muscular dystrophy, after the physicians who first described these disorders. Muscular dystrophy is caused by a defective allele of the dystrophin gene (usually called the *DMD* gene because of its involvement in *Duchenne Muscular Dystrophy*).

Although 1 in 3,500 boys has muscular dystrophy, girls almost never do. Why not? You should be able to guess, based on what you've learned in this chapter—the dystrophin gene is on the X chromosome, and muscular dystrophy alleles are recessive. Therefore, a boy will suffer muscular dystrophy if he has a defective dystrophin allele on his single X chromosome, but a girl, with two X chromosomes, would need two defective copies to suffer the disorder. This virtually never happens, because a girl would have to receive a defective dystrophin allele from both her mother, on one of her X chromosomes, and from her father, on his X chromosome. With their early disability and death, boys with muscular dystrophy almost never reproduce.

This scenario makes genetic sense, but it may seem to defy the concept of evolution by natural selection. How can a lethal allele be so common? Shouldn't natural selection have

9.11 How Do Errors in Chromosome Number Affect Humans?

In Chapter 8 we examined the intricate mechanisms of meiosis by which each sperm or egg normally receives only one chromosome from each homologous pair. Not surprisingly, this elaborate dance of the chromosomes occasionally misses a step, resulting in gametes that have too many or too few chromosomes. Such errors in meiosis, called **nondisjunction,** can affect the number of either sex chromosomes or autosomes. Most embryos that arise from gametes with abnormal chromosome numbers spontaneously abort, accounting for 20% to 50% of all miscarriages, but some embryos with abnormal chromosome numbers survive to birth or beyond. Abnormal numbers of chromosomes can be diagnosed before birth by examining the chromosomes of fetal cells (see "Health Watch: Prenatal Genetic Screening" on p. 205).

Abnormal Numbers of Sex Chromosomes Cause Some Disorders

Because the X and Y chromosomes pair up during meiosis, sperm normally carry either an X or a Y chromosome. Nondisjunction of sex chromosomes in a male,

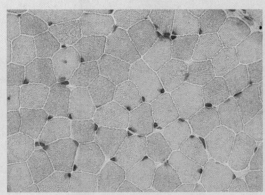

(a) Normal muscle

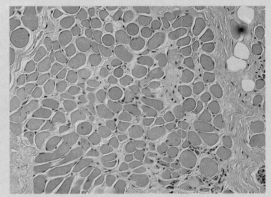

(b) Degenerating muscle in muscular dystrophy

▲ Figure E9-2 Defective dystrophin proteins cause muscle degeneration *(a)* A normal muscle is packed with specialized muscle cells, with very little space between the cells. *(b)* In muscular dystrophy, the individual cells are now widely scattered in the midst of fat, white blood cells, and connective tissue.

almost completely eliminated the defective dystrophin alleles, even if they are expressed only in boys? Actually, natural selection *does* rapidly eliminate the defective alleles. However, the dystrophin gene is enormous— about 2.2 *million* nucleotides long, compared to about 27 *thousand* nucleotides for the average human gene. In fact, the dystrophin gene occupies about 2% of the entire X chromosome.

Why does this matter? Remember, alleles arise as mutations in DNA. Returning to our analogy of genes as long sentences written in an alphabet of nucleotides, the dystrophin "sentence" is 2.2 million letters long! It is a tribute to the astounding accuracy of DNA copying during cell division that we don't all suffer from muscular dystrophy, and you shouldn't be surprised to learn that the mutation rate for the dystrophin gene is a hundred times

greater than average—about 1 mutation in every 10,000 gametes. Therefore, about one-third of the boys with muscular dystrophy receive new mutations that occurred in the reproductive cells of one of their parents, and two-thirds inherit preexisting alleles present on one of their mother's X chromosomes. The new mutations counterbalance natural selection, resulting in the steady incidence of about 1 in 3,500 boys.

Is muscular dystrophy hopeless, then? Right now, there are no cures, although several treatments are available that slow the progress of the disorder in some cases, prolong life, and make the affected boys more comfortable. An exciting possibility based on stem cells, however, has been discovered in dogs with a genetic disorder very similar to Duchenne muscular dystrophy. As we described in Chapter 8, stem cells can differentiate into many different types of mature cells. Giulio Cossu and colleagues at the San Raffaele Scientific Institute in Milan, Italy, found that injecting these dogs with stem cells taken from the blood vessels of healthy dogs allows them to synthesize normal dystrophin and retain muscle function long after the dogs would normally be unable to walk. The Muscular Dystrophy Association is currently funding research with similar human stem cells, for use in possible clinical trials.

however, produces sperm that have two sex chromosomes (XX, YY, or XY) or that have no sex chromosomes at all (these are designated O). Similarly, nondisjunction in a female can produce XX or O eggs instead of normal eggs with a single X. When these defective sperm or eggs fuse with normal gametes, the resulting offspring have abnormal numbers of sex chromosomes (Table 9-2). The most common abnormalities are XO, XXX, XXY, and XYY. (Genes on the X chromosome are essential to survival, and any embryo without at least one X chromosome spontaneously aborts very early in development.)

Turner Syndrome (XO)

About one in every 3,000 phenotypically female babies has only one X chromosome, a condition known as Turner syndrome. At puberty, hormone deficiencies prevent most XO females from menstruating or developing secondary sexual characteristics, such as enlarged breasts. Hormone treatment can promote physical development. However, because most women with Turner syndrome lack functional ovaries, hormone treatment cannot reverse their infertility. Other symptoms of Turner syndrome include short stature, folds of skin around the neck, and increased risk of cardiovascular disease, kidney defects, and hearing loss. Because women with Turner syndrome have only one X chromosome, they have more X-linked recessive disorders, such as hemophilia and color blindness, than do XX women.

Table 9-2	Effects of Nondisjunction of the Sex Chromosomes During Meiosis		
Nondisjunction in Father			
Sex Chromosomes of Defective Sperm	Sex Chromosomes of Normal Egg	Sex Chromosomes of Offspring	Phenotype
0 (none)	X	XO	Female—Turner syndrome
XX	X	XXX	Female—Trisomy X
YY	X	XYY	Male—Jacob syndrome
XY	X	XXY	Male—Klinefelter syndrome
Nondisjunction in Mother			
Sex Chromosomes of Normal Sperm	Sex Chromosomes of Defective Egg	Sex Chromosomes of Offspring	Phenotype
X	0 (none)	XO	Female—Turner syndrome
Y	0 (none)	YO	Dies as embryo
X	XX	XXX	Female—Trisomy X
Y	XX	XXY	Male—Klinefelter syndrome

Trisomy X (XXX)

About 1 in every 1,500 women has three X chromosomes, a condition known as trisomy X or triple X. Most such women have no detectable defects. There is, however, a higher incidence of below-normal intelligence among women with trisomy X. Unlike women with Turner syndrome, most trisomy X women are fertile and, interestingly enough, almost always bear normal XX and XY children. An unknown mechanism operating during meiosis seems to prevent the extra X chromosome from being included in their eggs.

Klinefelter Syndrome (XXY)

About 1 male in every 1,000 is born with two X chromosomes and one Y chromosome. Most of these men go through life never realizing that they have an extra X chromosome. However, at puberty, some of these men show mixed secondary sexual characteristics, including partial breast development, broadening of the hips, and small testes. These symptoms are known as Klinefelter syndrome. XXY men are usually infertile, because of a low sperm count, but are not impotent. They are usually diagnosed when an affected man and his partner seek medical help for their inability to conceive a baby.

Jacob Syndrome (XYY)

Another common type of sex chromosome abnormality is XYY, present in about 1 male in every 1,000. You might expect that having an extra Y chromosome, which has few genes, would not make very much difference, and this seems to be true in most cases. However, XYY males usually have high levels of testosterone, often have severe acne, and are tall (about two-thirds of XYY males are more than 6 feet tall, compared with the average male height of 5 feet 9 inches). XYY males also tend to have slightly lower than average IQ scores.

Abnormal Numbers of Autosomes Cause Some Disorders

Nondisjunction of autosomes produces eggs or sperm that are missing an autosome or that have two copies of an autosome. Fusion of one of these abnormal gametes with a normal one (that bears one copy of each autosome) leads to an embryo with either one or three copies of the affected autosome. Embryos that have only one copy of any autosome abort so early in development that the woman never knows she was pregnant. Embryos with three copies of an autosome (trisomy) also usually abort spontaneously. However, a small fraction of embryos with three copies of chromosomes 13, 18, or 21 can develop sufficiently to be born. A baby with trisomy 21 can live into adulthood.

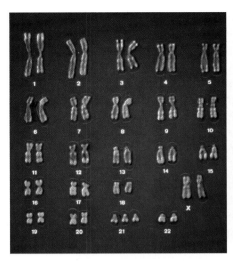

(a) Karyotype showing three copies of chromosome 21

(b) Girls with Down syndrome

◀ **Figure 9-22** **Trisomy 21, or Down syndrome** *(a)* Chromosomes of a Down syndrome child reveal three copies of chromosome 21. *(b)* These girls have the small mouth and distinctively shaped eyes typical of Down syndrome.

Trisomy 21 (Down Syndrome)

In about 1 of every 900 births, the child inherits an extra copy of chromosome 21, a condition called trisomy 21 or Down syndrome. Children with Down syndrome have several distinctive physical characteristics, including weak muscle tone, a small mouth held partially open because it cannot accommodate the tongue, and distinctively shaped eyelids (**Fig. 9-22**). More serious defects include low resistance to infectious diseases, heart malformations, and varying degrees of mental retardation, often severe.

The frequency of nondisjunction in gametes increases with age. There is some increase in defective sperm with increasing age of the father, and nondisjunction in sperm accounts for about 25% of Down syndrome cases. The mother's age, however, is a more significant factor in the probability of Down syndrome (**Fig. 9-23**). Since the 1970s, it has become more common for couples to delay having children, increasing the probability of trisomy 21. Nevertheless, older women as a group account for far fewer babies than younger women, so the majority of Down syndrome babies are born to young women.

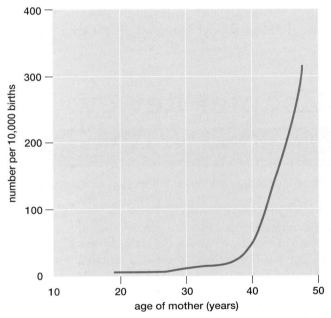

▲ **Figure 9-23** **Down syndrome frequency increases with maternal age** The shape of the red line shows that a low percentage of 20-year-old mothers gives birth to a child with Down syndrome, but the percentage begins to grow before age 30 and increases dramatically after age 35.

Sudden Death on the Court—Marfan Syndrome   Revisited

As you now know, the patterns of inheritance differ for various characteristics, depending on whether they are controlled by alleles that are dominant, recessive, incompletely dominant, and so on. Medical examinations revealed that Flo Hyman's father and sister have Marfan syndrome, but her mother and brother do not. Does this prove that Hyman inherited the defective allele from her father? As you learned in this chapter, diploid organisms, including people, generally have two alleles of each gene, one on each homologous chromosome. One defective fibrillin allele is enough to cause Marfan syndrome. Further, the children of a person with Marfan syndrome have a 50% chance of inheriting the disease. What can we conclude from these data?

First, if even one defective fibrillin allele produces Marfan syndrome, then Hyman's mother must carry two normal alleles, because she does not have Marfan syndrome. Second, because new mutations

(continued)

are rare and Hyman's father has Marfan syndrome, it is very likely that Hyman inherited a defective fibrillin allele from her father. The fact that her sister also has Marfan syndrome makes this virtually certain. Third, we can conclude that Marfan syndrome is inherited as a dominant condition because even one defective allele is enough to cause Marfan syndrome. Finally, if Hyman had borne children, any children who inherited her defective allele would develop Marfan syndrome. Therefore, on average, half of her children would have had Marfan syndrome. (Try working this out with a Punnett square.)

At the beginning of this chapter, we suggested that Hyman, Lincoln, Rachmaninoff, and Akhenaten all had Marfan syndrome. Evaluating the evidence critically, you may wonder how anyone could know if Lincoln, Rachmaninoff, or Akhenaten had the disorder. Well, you're right—nobody really knows. The "diagnosis" is based on photographs and descriptions. For example, Akhenaten was the first pharaoh to be depicted with long head, arms, and legs, all typical of Marfan syndrome. As for Lincoln, one visitor to the White House is said to have remarked, "Mr. President, what long legs you have!" To which Lincoln reportedly replied, "Just long enough to reach the ground, Madam." Although other genetic conditions can contribute to tall stature, long limbs, and large hands, the fact that these men were otherwise healthy, successful, and fertile points to Marfan syndrome.

Consider This

At this time, Marfan syndrome cannot be detected in an embryo by a simple biochemical test. Duchenne muscular dystrophy, however, can be detected, in both heterozygotes and homozygotes, in adults, children, and embryos. Suppose that a couple's first child is a son with Duchenne muscular dystrophy, and that they strongly desire another child. Genetic testing reveals that the mother carries the defective dystrophin allele, meaning that there would be a 50:50 chance that any future sons would also inherit the disease. There are several ways of preventing the birth of another son with muscular dystrophy. For example, they could use amniocentesis and prenatal diagnosis of an embryo followed by therapeutic abortion of an affected male (see Chapter 12). They could employ in vitro fertilization combined with screening of the embryos before implantation into the mother's womb (see Chapter 25). Should they use these technologies? Should they refrain from having further children? Should they have additional children without testing? Should they adopt? What would you do?

Chapter Review

Summary of Key Concepts

For additional study help and activities, go to www.mybiology.com.

9.1 What Is the Physical Basis of Inheritance?
The units of inheritance are genes, which are segments of DNA found at specific locations (loci) on chromosomes. Genes may exist in two or more slightly different, alternative forms, called alleles. When both homologous chromosomes carry the same allele at a given locus, the organism is homozygous for that particular gene. When the two homologous chromosomes have different alleles at a given locus, the organism is heterozygous for that gene.

9.2 How Were the Principles of Inheritance Discovered?
Gregor Mendel deduced many principles of inheritance in the mid-1800s, before the discovery of DNA, genes, chromosomes, or meiosis. He did this by choosing an appropriate experimental subject, designing his experiments carefully, following offspring for several generations, and analyzing his data statistically.

9.3 How Are Single Traits Inherited?
Traits are inherited in patterns that depend on the alleles that parents pass on to their offspring. Each parent provides its offspring with one allele of every gene, so the offspring inherits a pair of alleles for every gene. The combination of alleles in an offspring determines whether it displays a particular phenotype for a trait. Dominant alleles mask the expression of recessive alleles. Because recessive alleles can be masked, organisms with different genotypes may have the same phenotype. That is, organisms that are heterozygous for a dominant gene have the same phenotype as do homozygous dominant organisms. Because each allele segregates randomly during meiosis, we can predict the relative proportions of offspring with a particular trait. The Punnett square is a simple method of predicting the offspring from such genetic experiments.

Web Animation   Crosses Involving Single Traits

9.4 How Are Multiple Traits Inherited?
If the genes for two traits are located on different chromosome pairs, they assort independently of one another into the egg or sperm. Thus, crossing two organisms that are heterozygous at two loci on different chromosomes can produce offspring with nine different genotypes. If the alleles are typical dominant and recessive alleles, these progeny will display only four different phenotypes.

Web Animation   Crosses Involving Multiple Traits

9.5 How Are Genes Located on the Same Chromosome Inherited?
Genes located on the same chromosome are linked and tend to be inherited together. Unless two linked alleles are separated by crossing over during meiosis, they pass together to the offspring.

9.6 How Is Sex Determined?

In many animals, sex is determined by sex chromosomes, often designated X and Y. Female mammals have two X chromosomes, whereas males have one X and one Y chromosome. The Y chromosome has many fewer genes than the X chromosome.

9.7 How Are Sex-Linked Genes Inherited?

Because males have only one copy of X chromosome genes, recessive alleles on the X chromosome are more likely to be phenotypically expressed in males.

9.8 Do the Mendelian Rules of Inheritance Apply to All Traits?

Not all inheritance follows the simple dominant-recessive pattern:

- In incomplete dominance, heterozygotes have a phenotype that is intermediate between the two homozygous phenotypes.
- Codominance occurs when heterozygotes express the phenotypes of both of the homozygotes, such as both the A and B recognition proteins of the human AB blood type.
- Many traits are determined by polygenic inheritance, in which several different genes at different loci contribute to the phenotype.
- Many genes have multiple effects on the phenotype (pleiotropy).
- The environment influences the phenotypic expression of all traits.

9.9 How Are Human Genetic Disorders Investigated?

Studying the genetics of humans is similar to studying the genetics of other animals, except that experimental crosses are not feasible. Analysis of family pedigrees and, more recently, molecular genetic techniques help to determine the mode of inheritance of human traits.

9.10 How Are Single-Gene Disorders Inherited?

Many genetic disorders are inherited as recessive traits, with only homozygous recessive persons showing symptoms of the disease. Heterozygotes carry the recessive allele but do not express the trait. Many other diseases are inherited as simple dominant traits. In such cases, only one copy of the dominant allele is needed to cause the disease symptoms.

9.11 How Do Errors in Chromosome Number Affect Humans?

Errors in chromosome distribution during meiosis, called nondisjunction, result in gametes with abnormal numbers of sex chromosomes or autosomes. Many people with abnormal numbers of sex chromosomes have distinguishing physical characteristics. Abnormal numbers of autosomes typically lead to spontaneous abortion early in pregnancy. In rare instances, babies trisomic for chromosomes 13, 18, or 21 are born. Down syndrome (trisomy 21) is the most common trisomy; Down babies often survive to adulthood, but have mental and physical deficiencies.

Key Terms

allele *p. 136*
autosome *p. 144*
codominance *p. 146*
cross-fertilization *p. 137*
crossing over *p. 144*
dominant *p. 138*
dystrophin *p. 150*
gene *p. 136*

genotype *p. 140*
heterozygous *p. 136*
homozygous *p. 136*
incomplete dominance *p. 145*
inheritance *p. 136*
law of independent assortment *p. 142*

law of segregation *p. 138*
linkage *p. 143*
locus (plural, loci) *p. 136*
muscular dystrophy *p. 150*
nondisjunction *p. 150*
pedigree *p. 148*
phenotype *p. 140*

pleiotropy *p. 147*
polygenic inheritance *p. 146*
Punnett square method *p. 140*
recessive *p. 138*
self-fertilization *p. 137*
sex chromosome *p. 144*
sex-linked *p. 144*

Thinking Through the Concepts

Suggested answers to end-of-chapter and figure-based questions can be found at the end of the text.

Fill-in-the-Blank

1. The physical position of a gene on a chromosome is called its _____. Alternative forms of a gene are called _____. These alternative forms of genes arise as _____, which are changes in the nucleotide sequence of a gene.

2. An organism is described as *Rr* and red. *Rr* is the organism's _____, while red color is its _____. This organism would be (homozygous / heterozygous) for this color gene.

3. The inheritance of multiple traits depends on the locations of the genes that control the traits. If the genes are on different chromosomes,

then the traits are inherited (as a group / independently). If the genes are located close together on a single chromosome, then the traits tend to be inherited (as a group / independently). Genes on the same chromosome are said to be _____.

4. Many organisms, including mammals, have both autosomes and sex chromosomes. In mammals, males have _____ sex chromosomes and females have _____ sex chromosomes. The sex of offspring depends on which chromosome is present in the (sperm / egg).

5. Genes that are present on one sex chromosome but not the other are called _____.

6. If the phenotype of heterozygotes is intermediate between the phenotypes of the two homozygotes, this pattern of inheritance is called _____. If heterozygotes express phenotypes of both homozygotes (not intermediate, but showing both traits), this is called _____. In _____, many genes, usually with similar effects on phenotype, control the inheritance of a trait.

7. Abnormal numbers of chromosomes arise through a process called _____. Four syndromes in humans caused by abnormal numbers of sex chromosomes are _____, _____, _____, and _____. The most common syndrome caused by abnormal numbers of autosomes is _____, which is caused by an extra copy of chromosome number _____.

Review Questions

1. Define the following terms: gene, allele, dominant, recessive, true-breeding, homozygous, heterozygous, cross-fertilization, self-fertilization.

2. Explain why genes located on the same chromosome are said to be linked. Why do alleles of linked genes sometimes separate during meiosis?

3. Would you expect genes that are close together on a chromosome, or genes that are far apart on a chromosome, to be more tightly linked? Why?

4. Define polygenic inheritance. Why is it possible for a couple to have offspring that are notably different in eye or skin color than either parent?

5. What is sex linkage? In mammals, which sex would be most likely to show recessive sex-linked traits?

6. What is the difference between a phenotype and a genotype? Does knowledge of an organism's phenotype always allow you to determine the genotype? What type of experiment would you perform to determine the genotype of a phenotypically dominant individual?

7. Define nondisjunction, and describe the common syndromes caused by nondisjunction of sex chromosomes and autosomes.

Applying the Concepts

1. Sometimes the term "gene" is used rather casually. Compare the terms "allele" and "gene."

2. Mendel's numbers seem almost too perfect to be real; some believe he may have fudged a bit on his data. Perhaps he continued to collect data until the numbers matched his predicted ratios, then stopped. Recently, there has been much publicity over violations of scientific ethics, including researchers plagiarizing others' work, using other scientists' methods to develop lucrative patents, or just plain fabricating data. How important an issue is this for society? What are the boundaries of ethical scientific behavior? How should the scientific community or society "police" scientists? What punishments would be appropriate for violations of scientific ethics?

3. Although American society has been described as a "melting pot," people often engage in "assortative mating," in which they marry others of similar height, socioeconomic status, race, and IQ. Discuss the consequences to society of assortative mating among humans. Would society be better off if people mated more randomly? Discuss why or why not.

4. IS THIS SCIENCE? Many disorders, including heart disease, Alzheimer's disease, and breast cancer, have both genetic and environmental components. For example, most dietary guidelines recommend reducing the amount of salt in the diet, because excessive salt consumption can cause high blood pressure (hypertension), which in turn can cause heart disease. However, high dietary salt causes hypertension principally in genetically "salt-sensitive" people. Unfortunately, most people don't know if they are salt sensitive or not. Do you think that dietary recommendations against salt are justified if only a relatively small proportion of the population is affected? What about laws regulating the amount of salt in "junk food"? How do you think that scientific data should be used in cases such as this?

Genetics Problems

1. In certain cattle, hair color can be red (homozygous R_1R_1), white (homozygous R_2R_2), or roan (a mixture of red and white hairs, heterozygous R_1R_2).

    a. When a red bull is mated to a white cow, what genotypes and phenotypes of offspring could be obtained?

    b. If one of the offspring in part (a) were mated to a white cow, what genotypes and phenotypes of offspring could be produced? In what proportion?

2. The palomino horse is golden in color. Unfortunately for horse fanciers, palominos do not breed true. In a series of matings between palominos, the following offspring were obtained:

    Palomino: 65

    Cream-colored: 32

    Chestnut (reddish brown): 34

    What is the probable mode of inheritance of palomino coloration?

3. In the edible pea, tall (T) is dominant to short (t), and green pods (G) are dominant to yellow pods (g). List the gametes and offspring phenotypes that would be produced in the following crosses:

    a. $TtGg \times TtGg$

    b. $TtGg \times TTGG$

    c. $TtGg \times Ttgg$

4. In tomatoes, round fruit (R) is dominant to long fruit (r), and smooth skin (S) is dominant to fuzzy skin (s). A true-breeding round, smooth tomato ($RRSS$) is crossed with a true-breeding long, fuzzy tomato ($rrss$). All of the F_1 offspring are round and smooth ($RrSs$). When these F_1 plants are crossed with each other, the following F_2 generation is obtained:

    Round, smooth: 43   Long, fuzzy: 13

    Are the genes for skin texture and fruit shape likely to be on the same chromosome or on different chromosomes? Explain your answer.

5. In the tomatoes of Genetics Problem 4, an F_1 offspring ($RrSs$) is mated with a homozygous recessive plant ($rrss$). The following offspring are obtained:

    Round, smooth: 583   Round, fuzzy: 21

    Long, fuzzy: 602   Long, smooth: 16

    What is the most likely explanation for this distribution of phenotypes?

6. In humans, hair color is controlled by two interacting genes. The same pigment, melanin, is present in both brown-haired and blond-haired people, but brown hair has much more of it. The allele for brown hair (B) is dominant to the allele for blond (b). Whether any melanin can be synthesized depends on another gene. The dominant allele (M) allows melanin synthesis; the recessive allele (m) prevents melanin synthesis. Homozygous recessives (mm) are albino. What will be the expected proportions of phenotypes in the children of the following parents?

    a. $BBMM \times BbMm$

    b. $BbMm \times BbMm$

    c. $BbMm \times bbmm$

7. In humans, one of the genes determining color vision is located on the X chromosome. The dominant allele (C) produces normal color vision; red-green color blindness is caused by a recessive allele (c). If a man with normal color vision marries a color-blind woman, what is the probability of their having a color-blind son? A color-blind daughter?

8. In the couple described in Genetics Problem 7, the woman gives birth to a color-blind but otherwise normal daughter. The husband sues for a divorce on the grounds of adultery. Will his case stand up in court? Explain your answer.

For additional resources, go to www.mybiology.com.

chapter 10

Q At a Glance

DNA: The Molecule of Heredity

Ordinary bull or incredible hulk? A tiny change in DNA makes all the difference.

Case Study Muscles, Mutations, and Myostatin

No, the bull shown at the top of the chapter-opening photo hasn't been pumping iron—he's a Belgian Blue, and they always have bulging muscles. What makes a Belgian Blue look like a bodybuilder, compared to an ordinary bull, such as the Hereford below it?

When a mammal develops, its cells divide many times, enlarge, and become specialized for a specific function. The size, shape, and cell types in any organ are precisely regulated during development, so you don't wind up with a head the size of a basketball, or have hair growing on your liver. Muscle development is no exception. When you were very young, cells destined to form your muscles multiplied, fused together to form long, relatively thick cells with multiple nuclei, and synthesized the specialized proteins that cause muscles to contract and thereby move your skeleton. A protein called *myostatin*, found in all mammals, puts the brakes on this process. The word "myostatin" literally means "to make muscles stay the same," and that is exactly what myostatin does. As muscles develop, myostatin slows down—and eventually stops—the multiplication of these pre-muscle cells. A bodybuilder can bulk up by lifting weights, which *enlarges* the muscle cells, but doesn't usually add many *more* cells.

Belgian Blues have more muscle cells than ordinary cattle do. Why? You may have already guessed: They don't produce normal myostatin. And why not? Proteins such as myostatin are synthesized from the genetic directions contained in deoxyribonucleic acid, or DNA. The DNA of a Belgian Blue is very slightly different from the DNA of normal cattle—it has a change, or mutation, in the DNA of its myostatin gene. As a result, it produces defective myostatin. Belgian Blue pre-muscle cells multiply more than normal, producing remarkably buff cattle. Other animals, including some breeds of dogs (photo at left), may also have myostatin mutations, with similar effects.

How does DNA contain the instructions for traits such as muscle size, flower color, or gender? How are these instructions usually passed unchanged from generation to generation? And why do the instructions sometimes change? The answers to these questions lie in the structure and function of DNA. ∎

Like Belgian Blue cattle, "bully" whippets have defective myostatin, causing them to have enormous muscles.

10.1 What Is the Structure of DNA?

For well over a century, biologists have known that inherited characteristics are transmitted from parents to offspring in discrete units called genes, and that genes are parts of chromosomes. Then, in the mid-twentieth century, the efforts of many researchers revealed that genes are made of **deoxyribonucleic acid,** or **DNA,** and that a gene is a particular segment of the DNA of a specific chromosome. Further research revealed that most genes encode the information needed to spell out the amino acid sequence of a protein.

Discovering that genes are made of DNA was a great breakthrough, but it was only the beginning of learning how inheritance works. For example, how is information stored in DNA, how do cells use that information to synthesize proteins, and how is DNA transmitted from parent to offspring? To find out, biologists needed to decipher the structure of the DNA molecule.

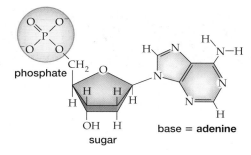

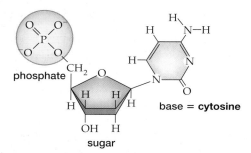

▲ **Figure 10-1 DNA nucleotides** All DNA nucleotides contain identical phosphate and deoxyribose sugar groups, and one of four nitrogen-containing bases: adenine, guanine, thymine, or cytosine.

Web Animation DNA Structure

DNA Is Composed of Four Different Subunits

DNA molecules are composed of four small subunits called **nucleotides.** Each nucleotide in DNA consists of three parts: (1) a phosphate group, (2) a sugar called deoxyribose, and (3) one of four possible nitrogen-containing **bases: adenine (A), guanine (G), thymine (T),** or **cytosine (C)** (Fig. 10-1). Note that the phosphate and sugar groups of all nucleotides are the same; the nucleotides differ from one another in which base they contain.

A DNA Molecule Contains Two Nucleotide Strands

Many researchers, working over scores of years, sought to understand the molecular basis of inheritance. Some discovered the nucleotides. Others showed that DNA is the molecule of inheritance in organisms as diverse as bacteria and people. Still others deciphered the basic structure of DNA: long chains of nucleotides, twisted like a corkscrew. Finally, in the early 1950s, James Watson and Francis Crick developed a model for the three-dimensional structure of DNA (see "Scientific Inquiry: The Discovery of the Double Helix" on p. 162).

Combining knowledge of how complex molecules bond together with an intuition that "important biological objects come in pairs," Watson and Crick proposed that a DNA molecule consists of two separate chains of nucleotides called **strands.** Within each strand, the phosphate group of one nucleotide bonds to the sugar of the next nucleotide in the strand. This bonding pattern produces a "backbone" of alternating, covalently bonded sugars and phosphates. The nucleotide bases protrude from this **sugar-phosphate backbone** (Fig. 10-2a).

All the nucleotides in a DNA strand are oriented in the same direction, so the two ends of the strand differ. One end has a "free," or unbonded, sugar, and the other end has a free phosphate. (Think of a long line of cars stopped on a crowded one-way street at night: The cars' headlights always point forward, and their tail lights always point backward. If the cars are jammed tightly together, a pedestrian standing in front of the line of cars will see only the headlights on the first car; a pedestrian at the back of the line will see only the tail lights of the last car.)

Hydrogen Bonds Hold the Two DNA Strands Together in a Double Helix

The two DNA strands are held together by hydrogen bonds that form between the bases that protrude from the sugar-phosphate backbones of each strand. These bonds give DNA a ladder-like structure, with the sugar-phosphate backbones forming the sides and the bases forming the rungs (see Fig. 10-2a). However, the DNA strands are not straight. Instead, they are twisted together to form a **double helix,** much like a ladder twisted into a circular staircase shape (Fig. 10-2b). The two strands in a DNA double helix are oriented in opposite directions. (Again, imagine an evening traffic jam, this time on a two-lane road: A traffic helicopter pilot overhead sees only the headlights on cars in one lane and only tail lights in the other lane.)

Take a closer look at the way the bases pair up to form each rung of the double helix ladder (see Fig. 10-2a). Adenine forms hydrogen bonds only with thymine, and guanine forms hydrogen bonds only with cytosine. These A–T and G–C pairs are called **complementary base pairs.** All of the bases of the two strands of a DNA double helix are complementary to each other. For example, if one strand reads A-T-T-C-C-A-G-G-C-T, then the other strand must read T-A-A-G-G-T-C-C-G-A.

10.2 How Does DNA Encode Information?

Look again at the DNA structure shown in Figure 10-2. The relative simplicity of the structure and the small number of different bases raise the question of how

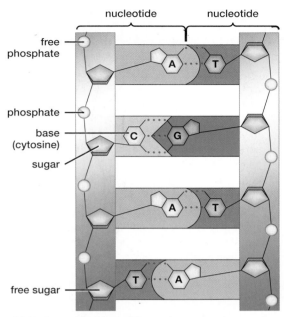

nucleotide   nucleotide

free phosphate

phosphate

base (cytosine)

sugar

free sugar

(a) Hydrogen bonds hold complementary base pairs together in DNA

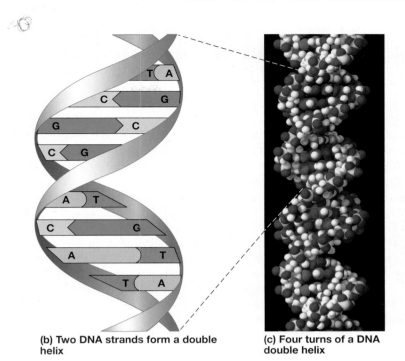

(b) Two DNA strands form a double helix

(c) Four turns of a DNA double helix

▲ **Figure 10-2 The Watson-Crick model of DNA structure** *(a)* Hydrogen bonding between complementary base pairs holds the two strands of DNA together. Three hydrogen bonds (red dotted lines) hold guanine to cytosine, and two hydrogen bonds hold adenine to thymine. Note that each strand has a free phosphate (yellow ball) on one end and a free sugar (blue pentagon) on the opposite end. Further, the two strands run in opposite directions. *(b)* Strands of DNA wind together in a double helix, like a twisted ladder, with the sugar-phosphate backbone forming the sides and the complementary base pairs forming the rungs. *(c)* A space-filling model of the DNA double helix.   **QUESTION:** Which do you think would require more energy to break apart: an A–T base pair or a C–G base pair? Why?

DNA can encode the vast amount of information required to build and operate an organism. How, for example, can the color of a bird's feathers, the size and shape of its beak, its ability to make a nest, and its song all be determined by a molecule with just four simple parts?

The answer is that it's not the *number* of different subunits but their *sequence* that's important. Within a DNA strand, the four types of bases can be arranged in any order, and each sequence of bases represents a unique set of genetic instructions. An analogy might help: You don't need lots of unique letters to make up a language. English has 26 letters, but Hawaiian has only 12, and computers use only two "letters" (0 and 1, or "on" and "off"). Nevertheless, all three can spell out thousands of different words. A stretch of DNA just 10 nucleotides long can have more than a million possible sequences of the four bases. Because organisms have millions (in bacteria) to billions (in plants or animals) of nucleotides, their DNA molecules can encode a staggering amount of information.

Of course, to make sense, both the letters of a language and the nucleotides of DNA must be in the correct order. Just as "friend" and "fiend" mean different things, and "fliend" doesn't mean anything, different sequences of bases in DNA may encode very different pieces of information, or no information at all.

10.3 How Is DNA Copied?

Recall from Chapter 8 that all of the trillions of cells of your body are the offspring of other cells, going all the way back to when you were a fertilized egg. What's more, nearly every cell of your body contains identical genetic information—the same genetic information that was present in that fertilized egg. Even after you reach adulthood, millions of your cells continue to divide. For example, skin cells constantly flake off or are worn away by rubbing on your clothes, and must be replaced. Thus, skin cells divide about once a day. Each daughter cell receives a nearly perfect copy of the parent cell's genetic information. Consequently, before a cell divides, it must have two exact copies of its DNA. A process known as **DNA replication** produces these two identical DNA double helices.

Muscles, Mutations, and Myostatin

Continued

All "normal" mammals have a DNA sequence that encodes a functional myostatin protein, which limits their muscle growth. Belgian Blue cattle have a mutation that changes a "friendly" gene to a nonsensical "fliendly" one that no longer codes for a functional protein, so they have excessive muscle development.

scientific inquiry

The Discovery of the Double Helix

In the early 1950s, many biologists realized that the key to understanding inheritance lay in the structure of DNA. They also knew that whoever deduced the correct structure of DNA would receive recognition, very possibly the Nobel Prize. Linus Pauling of Caltech was the person most likely to solve the puzzle of DNA structure. He probably knew more about the chemistry of large organic molecules than any person alive. Pauling, however, had two main handicaps. First, for years he had concentrated on protein research, and therefore he had little data about DNA. Second, he was active in the peace movement. At that time, some government officials, including Senator Joseph McCarthy (remembered today for his vigorous attacks on people he accused of being Communists), considered such activity to be potentially subversive and dangerous to national security. This latter handicap may have proved decisive.

The second most likely competitors were the British scientists Rosalind Franklin and Maurice Wilkins, who were experts in using X-rays to determine the structures of large molecules. In fact, Franklin and Wilkins' X-ray studies made them the only scientists who had very good data about the general shape of the DNA molecule. Unfortunately for them, their methodical approach was slow.

The door was open for the eventual discoverers of the double helix, James Watson and Francis Crick, who had neither Pauling's tremendous understanding of chemical bonds nor Franklin and Wilkins' expertise in X-ray analysis. Watson and Crick did no experiments in the usual sense of the word. Instead, they spent their time thinking about DNA, trying to construct a molecular model that made sense and fit the data. Because they were working in England and because Wilkins shared Franklin's data with them (perhaps against her wishes),

Watson and Crick were familiar with all of the X-ray information relating to DNA.

This X-ray information was just what Pauling lacked. Because of Pauling's presumed subversive tendencies, the U.S. State Department refused to issue him a passport to leave the United States, so he could neither attend meetings at which Wilkins presented the X-ray data nor visit England to talk with Franklin and Wilkins directly. Watson and Crick knew that Pauling was working on DNA structure and were driven by the fear that he would beat them to it. In his book *The Double Helix*, Watson recounts his belief that, if Pauling could have seen the X-ray pictures, "in a week at most, Linus would have the structure."

Now, you might think, "But wait just a minute! That's not fair. If the goal of science is to advance knowledge, then everyone should have access to all of the data. If Pauling was the best, he should have discovered the double helix first." Perhaps so. But science is an activity of scientists, who are, after all, people. Although virtually all scientists want to see the progress of science and benefits for humanity, each individual also wants to be the one responsible for that progress and to receive the credit and the glory. Linus Pauling remained in the dark about the X-ray pictures of DNA and the clues they gave about its structure.

When Watson and Crick discovered the double helix structure of DNA (**Fig. E10-1**), Watson described it in a letter to Max Delbruck, a friend and adviser at Caltech. He asked Delbruck not to

reveal the contents of the letter to Pauling until their structure was formally published. Delbruck, perhaps more of a model scientist, firmly believed that scientific discoveries belong in the public domain and promptly told Pauling all about it. Showing himself to be not only a great scientist but also a great person, Pauling graciously congratulated Watson and Crick on their brilliant solution to the DNA structure. The race was over.

▲ **Figure E10-1　The discovery of DNA** James Watson and Francis Crick with a model of DNA structure. **QUESTION:** When developing their model, Watson and Crick knew the identity of the four bases in DNA, and that the diameter of a DNA molecule is the same from one end to the other (that is, it doesn't get alternately fatter and thinner). If "biological objects come in pairs," how did that knowledge help them to decide which bases might form pairs in the center of the double helix? Hint: Look at the structure of nucleotides in Figure 10-1.

DNA Replication Produces Two DNA Double Helices, Each with One Original Strand and One New Strand

Web Animation　DNA Replication

How does a cell accurately replicate its DNA? In their paper describing DNA structure, Watson and Crick included one of the greatest understatements in the history of science: "It has not escaped our notice that the specific [base] pairing we have postulated immediately suggests a possible copying mechanism for the genetic material." In fact, base pairing is the foundation of DNA replication. Remember, the rules for base pairing are that an adenine on one strand must pair with a thymine on the other strand, and a cytosine must pair with a guanine. If one strand reads ATG, for example, then the other strand must read TAC. Therefore, the base sequence of one strand contains all the information needed to replicate the other strand.

Conceptually, DNA replication is quite simple (**Fig. 10-3**). The ingredients are (1) the parental DNA strands, (2) free nucleotides that were previously synthesized in the cytoplasm and then imported into the nucleus, and (3) a variety of enzymes that unwind the parental DNA double helix and synthesize new DNA strands.

First, enzymes called **DNA helicases** ("enzymes that break apart the double helix") pull apart the parental DNA double helix so that the bases of the two parental DNA strands no longer form base pairs with one another. Then, enzymes called **DNA polymerases** ("enzymes that make a DNA polymer") move along each separated parental DNA strand, matching each base on the strand with free nucleotides that have a complementary base. For example, DNA polymerase pairs an exposed adenine base with a thymine base. DNA polymerase also connects these free nucleotides together to form new DNA strands: One new daughter strand is complementary to each of the parental DNA strands. We will go into more detail on how DNA is replicated in the next section of this chapter, "What Are the Mechanisms of DNA Replication?"

When replication is complete, one parental DNA strand and its complementary daughter DNA strand wind together into one double helix, while the other parental strand and its daughter strand entwine in a second double helix. DNA replication thus keeps, or *conserves*, one parental DNA strand and produces one new daughter strand in each new double helix. Hence, the process is called **semiconservative replication** (**Fig. 10-4**).

If no mistakes have been made, the base sequences of both new DNA double helices are identical to the base sequence of the original, parental DNA double helix and, of course, to each other.

10.4 What Are the Mechanisms of DNA Replication?

DNA replication involves three major actions (**Fig. 10-5**). First, as we have seen, the DNA double helix must be opened up, so that the base sequence can be "read." Then, new DNA strands with base sequences complementary to the two original strands must be synthesized. In eukaryotic cells, these new DNA strands are synthesized in fairly short pieces. Therefore, the third step in DNA replication is to stitch the pieces together to form a continuous strand of DNA. Each step is carried out by a distinct set of enzymes.

DNA Helicase Separates the Parental DNA Strands

As mentioned earlier, the enzyme DNA helicase breaks the hydrogen bonds between complementary base pairs that hold the two parental DNA strands together. Breaking the bonds separates and unwinds the double helix, forming a *replication bubble* (**Fig. 10-5**, steps ❶ and ❷). Within the replication bubble, the nucleotide bases of the parental DNA strands are no longer paired with one

▶ **Figure 10-4 DNA replication is semiconservative** When DNA is replicated, each new double helix consists of one "conserved" parental DNA strand (blue) and one newly synthesized daughter DNA strand (red); hence, the process is called "semiconservative."

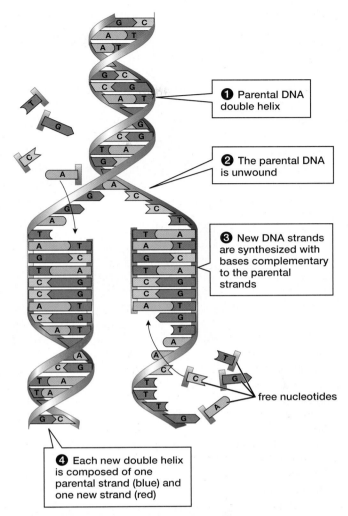

❶ Parental DNA double helix

❷ The parental DNA is unwound

❸ New DNA strands are synthesized with bases complementary to the parental strands

free nucleotides

❹ Each new double helix is composed of one parental strand (blue) and one new strand (red)

▲ **Figure 10-3 Basic features of DNA replication**

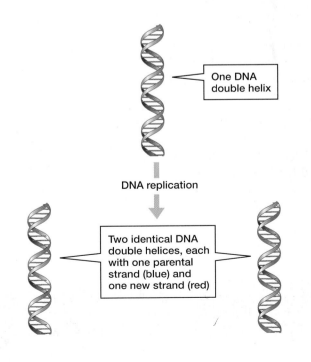

One DNA double helix

DNA replication

Two identical DNA double helices, each with one parental strand (blue) and one new strand (red)

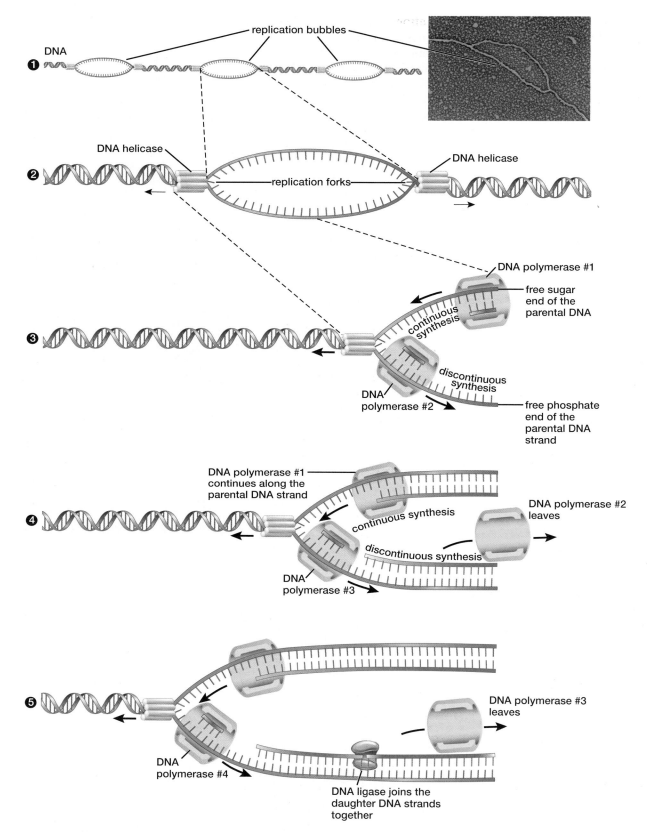

▲ **Figure 10-5 The mechanism of DNA replication** ❶ DNA helicase enzymes separate the parental strands of a chromosome to form replication bubbles. ❷ Each replication bubble consists of two replication forks, with "unwound" DNA strands between the forks. ❸ DNA polymerase enzymes synthesize new pieces of DNA. ❹ DNA helicase moves along the parental DNA double helix, unwinding it and enlarging the replication bubble. DNA polymerases in the replication bubble synthesize daughter DNA strands. ❺ DNA ligase joins small DNA segments into a single daughter strand.   **QUESTION:** During DNA synthesis, why doesn't DNA polymerase move toward the replication fork on both strands?

another. Each replication bubble contains a replication "fork" at each end, where the two parental DNA strands are just beginning to unwind.

To help visualize this process, imagine that you are driving down a two-lane, undivided road; each lane represents a single strand of a DNA double helix. The two DNA strands of the double helix point in opposite directions (see Fig. 10-2), just as cars in each lane of the road travel in opposite directions. A replication bubble is analogous to the two lanes' separating, with a wide median between them. A little farther down the road, the median disappears, and the road once again becomes undivided. The median is the replication bubble, and the places where the median begins and where it ends are the forks.

Eukaryotic chromosomes are very long; human chromosomes, for example, are about 23 million to 246 million nucleotide pairs long. If these chromosomes were replicated as single, continuous pieces, starting at one end and proceeding to the other end, they would take about 5 to 57 days to copy. To replicate an entire chromosome in a reasonable length of time (less than a day for cells in the skin and intestinal lining), a large number of DNA helicase enzymes must open up replication bubbles simultaneously.

DNA Polymerase Synthesizes New DNA Strands

Replication bubbles are essential because they allow a second enzyme, DNA polymerase, to gain access to the bases of each DNA strand (Fig. 10-5, step ❸). At each replication fork, DNA polymerase synthesizes two new DNA strands that are complementary to the two parental strands. During this process, DNA polymerase recognizes an unpaired base in the parental strand and matches it up with a free nucleotide that has the correct complementary base. For example, DNA polymerase pairs up an exposed cytosine base in the parental strand with a guanine base in a free nucleotide. Then, DNA polymerase links the phosphate of the incoming free nucleotide to the sugar of the previously added nucleotide in the growing daughter strand. In this way, DNA polymerase synthesizes the sugar-phosphate backbone of the daughter strand.

DNA polymerase always moves toward the "free phosphate" end of a parental DNA strand (see Fig. 10-2; note that the backbone of each strand has a free phosphate on one end and a free sugar on the other end). Because the two strands of the parental DNA double helix are oriented in opposite directions, the new complementary DNA strands will also be synthesized in opposite directions (see Fig. 10-5, step ❸). Returning to our road analogy, a DNA polymerase enzyme stays on its "own" side of the DNA road, driving in the direction of the free phosphate end of the strand.

DNA helicase and DNA polymerase work together (Fig. 10-5, steps ❸ and ❹). As a DNA helicase enzyme moves along the parental DNA, it unwinds the double helix and separates the strands. Because the two DNA strands run in opposite directions, as a DNA helicase enzyme moves toward the free phosphate end of one strand, it is simultaneously moving toward the free sugar end of the other strand.

Now visualize two DNA polymerases "landing" on the two separated strands of DNA (see Fig. 10-5, step ❸). One DNA polymerase (call it polymerase #1) can follow behind the helicase toward the free phosphate end, synthesizing a continuous, complete new DNA strand. On the other strand, however, DNA polymerase #2 moves *away from* the helicase, and therefore can synthesize only *part* of a new DNA strand. As the helicase continues to unwind more of the double helix, additional DNA polymerases (#3, #4, and so on) must land on this strand and in turn synthesize more pieces of DNA (see Fig. 10-5, steps ❹ and ❺). Thus, DNA synthesis on this strand is discontinuous.

DNA Ligase Joins Together Segments of DNA

Multiple DNA polymerases synthesize pieces of DNA of varying lengths, as many as a million pieces for a single human chromosome. How are all of these pieces sewn together? This is the job of the third major enzyme, **DNA ligase** ("an enzyme that ties DNA together"; see Fig. 10-5, step ❺). Many DNA ligase enzymes bond together the sugar-phosphate backbones of these fragments of DNA until each daughter strand consists of one long, continuous DNA polymer.

Proofreading Produces Almost Error-Free Replication of DNA

Ideally, DNA polymerase synthesizes daughter DNA strands by incorporating only nucleotides with bases that are complementary to the bases of the parental strands. However, nothing in life is perfect, and DNA polymerase incorporates incorrect nucleotides about once in every 10,000 base pairs, partly because replication is so fast (up to 50 nucleotides per second in some eukaryotic cells). Nevertheless, completed DNA strands contain only about one mistake in every *billion* base pairs. This phenomenal accuracy is ensured by a variety of DNA repair enzymes that "proofread" each daughter strand during and after its synthesis. For example, some forms of DNA polymerase recognize a base-pairing mistake as it is made. This type of DNA polymerase will pause, remove the incorrect nucleotide, attach a new nucleotide with the correct complementary base, and then continue synthesizing more DNA.

Mistakes Do Happen

In addition to mistakes made during DNA replication, the DNA in each cell in your body loses about 10,000 bases every day, simply due to spontaneous chemical breakdown at normal human body temperatures. Cosmic radiation, sunlight, and a variety of chemicals also damage DNA. For example, smoking causes lung cancer because it contains several chemicals that damage DNA, including the DNA that encodes some of the proteins that control cell division (see "Health Watch: Cancer—Mitotic Cell Division Run Amok" in Chapter 8, p. 119). Cellular repair enzymes fix most of the damaged DNA, but some mistakes inevitably remain. These mistakes, which change the sequence of nucleotides in DNA, are **mutations** (see also Chapters 8 and 11, where we describe some of the consequences of mutations).

Some mutations occur more frequently than others. For example, some types of DNA polymerase are more likely to make a mistake replicating an A–T nucleotide pair than a C–G pair. Ultraviolet light typically causes mutations in genes with adjacent thymines or cytosines. Nevertheless, when we consider all sources of mutations collectively, mutations are essentially random changes in the nucleotide sequence of DNA.

How is an entire organism affected by a mutation? Think of a mutation as a typographical error in DNA. In some cases, the "DNA typo" is still understandable, like typing "mutaition" instead of "mutation." In other cases, the DNA typo makes things completely wrong, like typing "no" instead of "go." Similarly, most real-life mutations are either neutral (the organism's phenotype is not altered very much, if at all) or harmful (the organism is less likely to survive and reproduce). In rare instances, however, mutations may be beneficial. For example, some people have mutations that help to protect them against AIDS. Mutations in many species of bacteria provide resistance to antibiotics (a definite advantage to the bacteria, although a major disadvantage to people infected by them). Mutations that are beneficial, at least in certain environments, may be favored by natural selection, and are the basis for the evolution of life on Earth (see Unit 3). We will explore the effects of mutations at the cellular and organismic levels more thoroughly in Chapter 11.

Muscles, Mutations, and Myostatin

Continued

In addition to Belgian Blues, several other breeds of cattle, including Maine Anjou, Piedmontese, Limousine, Charolais, and Blonde d'Aquitaine, have excessive muscle development caused by mutated myostatin genes. From what you've learned about the "language" of DNA and the mechanisms of DNA replication, would you expect that all of these cattle would have the same mutation? Check your reasoning in the "Muscles, Mutations, and Myostatin, *Revisited*" section at the end of the chapter.

Muscles, Mutations, and Myostatin Revisited

Belgian Blue cattle have a mutation in their myostatin gene, causing their cells to stop synthesizing the myostatin protein about halfway through the process. Several other breeds of "double-muscled" cattle have the same mutation, but some Maine Anjou, Piedmontese, Limousine, and Charolais cattle have totally different mutations. What they all have in common is that their myostatin proteins are nonfunctional. This is an important feature of the "language" of DNA: The nucleotide words must be spelled just right, or at least really close, for the resulting proteins to function. Any one of an enormous number of possible mistakes will render the proteins useless.

Humans have myostatin, too, and, not surprisingly, mutations can occur in the human myostatin gene. As you learned in Chapter 9, a child inherits two copies of most genes, one from each parent. In 1999, a child was born in Germany who inherited a mutated myostatin gene from both parents. Although the mutation is different from the one in Belgian Blue cattle, it also results in short, inactive,

myostatin proteins. Even at 7 months, this boy had well-developed calf, thigh, and buttock muscles. At 4 years, he could hold a 7-pound dumbbell in each hand, with his arms fully extended horizontally out to his sides (try it—it's not that easy for many adults).

You know that mutations may be neutral, harmful, or beneficial. Into which category do myostatin mutations fall? Belgian Blue cattle are born so muscular, and consequently so large, that they usually must be delivered by cesarean section. A few become so muscular that their muscles get in the way, and they can hardly walk. So far, the German boy appears to be healthy. What will happen as he grows up? Will he become a super-athlete, or suffer debilitating health effects as he ages, or both? Only time will tell.

Consider This

If a person becomes a super-athlete as a result of a known mutation, such in as the myostatin gene, is it "fair" to allow him or her to compete against people who don't have this mutated allele?

Chapter Review

Summary of Key Concepts

For additional study help and activities, go to www.mybiology.com.

10.1 What Is the Structure of DNA?

DNA consists of nucleotides that are linked together into long strands. Each nucleotide consists of a phosphate group, the sugar deoxyribose, and a nitrogen-containing base. There are four types of bases in DNA: adenine, guanine, thymine, and cytosine. The sugar of one nucleotide is linked to the phosphate of the next nucleotide, forming a sugar-phosphate backbone for each strand. The bases stick out from this backbone. Two nucleotide strands wind together to form a DNA double helix, which resembles a twisted ladder. The sugar-phosphate backbones form the sides of the ladder. The bases of each strand pair up in the middle of the helix, held together by hydrogen bonds, forming the rungs of the ladder. Only complementary base pairs can bond together in the helix: Adenine bonds with thymine, and guanine bonds with cytosine.

Web Animation DNA Structure

10.2 How Does DNA Encode Information?

Genetic information is encoded as the sequence of nucleotide bases in a DNA molecule, much as the meaning of a word is determined by its sequence of letters. Because DNA molecules are usually millions to billions of nucleotides long, DNA can encode huge amounts of information in its base sequence.

10.3 How Is DNA Copied?

Before a cell can reproduce, it must replicate its DNA so that each daughter cell will receive all of the genetic information contained in the parent cell. During DNA replication, DNA helicase enzymes

unwind the two parental DNA strands. Then DNA polymerase enzymes bind to each parental DNA strand. Free nucleotides form hydrogen bonds with complementary bases on the parental strands, and DNA polymerase links the free nucleotides together to form new DNA strands. Therefore, the sequence of bases in each newly formed strand is complementary to the sequence of a parental strand.

Replication is semiconservative because, when DNA replication is complete, the two new DNA double helices consist of one conserved parental DNA strand and one newly synthesized, complementary daughter strand. The new DNA double helices are therefore duplicates of the parental DNA double helix.

Web Animation DNA Replication

10.4 What Are the Mechanisms of DNA Replication?

The enzyme DNA helicase unwinds the double helix, forming a replication bubble. One DNA polymerase enzyme then binds to each of the unwound strands of DNA. Because the two parental DNA strands are oriented in opposite directions, the DNA polymerase on one parental strand can synthesize a long, continuous daughter strand (following the DNA helicase as it unwinds the DNA double helix), but the DNA polymerase on the other parental strand can synthesize only a short daughter strand (because it moves away from the DNA helicase). As more DNA polymerase molecules bind to this parental strand, they each synthesize a short daughter strand. Finally, DNA ligase connects the sugar-phosphate backbones of these short daughter strands to form a complete, intact strand (see Fig. 10-5).

DNA polymerase and other repair enzymes "proofread" the DNA, correcting most of the mistakes made during replication. Nevertheless, mistakes in DNA replication and damage from environmental agents such as chemicals in cigarette smoke or ultraviolet light in sunlight can change the base sequence of DNA. These sequence changes are called mutations. Most mutations are neutral or harmful, but a few may prove beneficial in specific environments, and are the basis of evolution by natural selection.

Key Terms

adenine (A) *p. 160*
base *p. 160*
complementary base pair *p. 160*
cytosine (C) *p. 160*
deoxyribonucleic acid (DNA) *p. 159*
DNA helicase *p. 163*
DNA ligase *p. 166*
DNA polymerase *p. 163*
DNA replication *p. 161*
double helix *p. 160*
guanine (G) *p. 160*
mutation *p. 166*
nucleotide *p. 160*
semiconservative replication *p. 163*
strand *p. 160*
sugar-phosphate backbone *p. 160*
thymine (T) *p. 160*

Thinking Through the Concepts

Suggested answers to end-of-chapter and figure-based questions can be found at the end of the text.

Fill-in-the-Blank

1. DNA consists of subunits called _____. Each subunit consists of three parts: _____, _____, and _____.

2. The subunits of DNA are assembled by linking the _____ of one nucleotide to the _____ of the next. The resulting polymer is usually called a(n) _____. As it is found in chromosomes, two DNA polymers are wound together into a structure called a(n) _____.

3. The "base-pairing rule" in DNA is that adenine pairs with _____, whereas guanine pairs with _____. Bases that can form pairs in DNA are called _____.

4. When DNA is replicated, two new DNA double helices are formed, each consisting of one parental strand and one new, daughter strand. For this reason, DNA replication is called _____.

5. The DNA double helix is unwound by an enzyme called _____. Daughter DNA strands are synthesized by the enzyme _____.

6. Sometimes, mistakes are made during DNA replication. If uncorrected, these mistakes are called _____.

Review Questions

1. Draw the general structure of a nucleotide. Which parts are identical in all nucleotides, and which can vary?

2. Name the four types of nitrogen-containing bases found in DNA.

3. Which bases are complementary to one another? How are they held together in the double helix of DNA?

4. Describe the structure of DNA. Where are the bases, sugars, and phosphates in the structure?

5. Describe the process of DNA replication.

Applying the Concepts

1. IS THIS SCIENCE? Several companies market shampoos that contain DNA, with claims that these shampoos might strengthen the hair and nourish the scalp, promote hair growth, or add bounce to hair. Assuming that the shampoos can accomplish any or all of those things, do you think that DNA is likely to be the active ingredient? Why or why not?

2. As you learned in "Scientific Inquiry: The Discovery of the Double Helix," scientists in different laboratories often compete with one another to make new discoveries. Do you think this competition helps promote scientific discoveries? Sometimes, researchers in different laboratories collaborate with one another. What advantages does collaboration offer over competition? What factors might provide barriers to collaboration and lead to competition?

3. Genetic information is encoded in the sequence of nucleotides in DNA. Let's suppose that the nucleotide sequence on one strand of a double helix encodes the information needed to synthesize a hemoglobin molecule. Do you think that the sequence of nucleotides on the other strand of the double helix also encodes useful information? Why or why not? Why do you think DNA is double stranded?

For additional resources, go to www.mybiology.com.

chapter 11

Gene Expression and Regulation

Although she was diagnosed early in life with cystic fibrosis, Alice Martineau hoped that ". . . people will realise when they hear the music, I am a singer-songwriter who just happens to be ill."

🔍 At a Glance

Case Study   Cystic Fibrosis

If all you knew was her music, you'd think Alice Martineau had it made—a young, pretty singer-songwriter, popular in the London music scene, signed to a recording contract with Sony Music Entertainment. But genetics dealt Alice Martineau, who died in March 2003 at the age of 30, a double whammy—two copies of a defective recessive allele for a crucially important protein: CFTR. Martineau, like 30,000 Americans, 3,000 Canadians, and 20,000 Europeans, had cystic fibrosis, the most common serious recessive genetic disorder in North America and Europe. Before modern medical care, most people with cystic fibrosis died by age 4 or 5; even now, the average life span is less than 40 years.

CFTR is a channel protein that is selectively permeable to chloride and is found in many parts of the body, including the sweat glands, lungs, and intestines. Let's look at its role in perspiration first. Sweat is mostly water, and it cools the body by evaporating from the skin. However, when it is first secreted by the sweat glands deep in the skin, sweat also contains a lot of salt (sodium chloride), about as much

Lisa Bentley, sometimes called the Iron Queen, wins another triathlon.

as in blood. If you perspire slowly enough, most of the salt is reclaimed as the sweat moves through tubes leading from the secreting cells to the surface of the skin. Although the exact mechanism is not yet fully understood, the CFTR protein is crucial for sodium chloride to flow from the sweat, through the cells that make up the walls of the tubes, and back into the blood. These cells are quite impermeable to water, so salt is reclaimed while the water remains in the tubes, and the resulting sweat isn't very salty. Mutations in the *CFTR* gene, which encodes the CFTR protein, cause cystic fibrosis. (Note that we are following the standard usage in genetics, by italicizing the names of genes, such as *CFTR,* and using regular type for the names of proteins, such as CFTR.) The defective CFTR proteins block reabsorption of chloride, and perhaps sodium too, with the result that salt stays in the sweat.

Canadian triathlete Lisa Bentley (photo at left) has a fairly mild case of cystic fibrosis, but during a 9-hour triathlon, she produces copious amounts of very salty sweat. It is a constant challenge for Bentley to keep her body supplied with salt during the race. Nevertheless, she has won numerous triathlons, including the Australian Ironman Triathlon five straight years, from 2002 through 2006.

Except during prolonged strenuous exercise, salty sweat isn't very harmful. Unfortunately, the cells lining the airways in the lungs have the same CFTR proteins that sweat glands have. Normally, the airways are covered with a thin film of watery mucus, which traps bacteria and debris. "Natural antibiotic" proteins in the fluid kill many bacteria. Most bacteria—both live and dead—are swept out of the lungs by cilia on the cells lining the airways. No one is certain just how the CFTR protein regulates the amount or composition of the fluid. However, defective CFTR proteins cause the airways to be covered with dehydrated mucus that may also have reduced antibiotic properties. The result is that the mucus becomes so thick that the cilia can't move it out of the lungs. Therefore, the airways are partially clogged and bacteria multiply, causing chronic lung infections. Not surprisingly, people with cystic fibrosis cough frequently, trying to clear their airways. Alice Martineau credited coughing with strengthening her vocal cords and helping to produce her deep, strong voice.

In this chapter, we examine the processes by which the instructions in genes are translated into proteins. When a gene mutates, how does that affect the structure and function of the encoded protein? Why might different mutations in the same gene have different consequences? ∎

11.1 How Is the Information in DNA Used in a Cell?

Information, by itself, doesn't *do* anything. For example, a blueprint may describe the structure of a house in great detail, but unless that information is translated into action, no house will ever be built. Likewise, although the base sequence of DNA, the "molecular blueprint" of every cell, contains an incredible amount of information, DNA cannot carry out any action on its own. So how does DNA determine whether you are male or female, have brown or blue eyes, or have normal lung function or cystic fibrosis?

Most Genes Contain Information for the Synthesis of a Single Protein

Most chromosomes contain hundreds or thousands of genes, each of which occupies a particular position in the DNA of the chromosome. In most cases, a **gene** is a stretch of DNA encoding the instructions for the synthesis, or manufacture, of a single protein, a fact that leads to the general rule "one gene, one protein." Proteins, in turn, are the molecular workhorses of the cell, forming many of its cellular structures and the enzymes that catalyze its chemical reactions.

Technically, the "one gene, one protein" relationship should really be expressed as "one gene, one polypeptide." A polypeptide is a chain of amino acids. Although many proteins consist of a single polypeptide, others are composed of more than one polypeptide subunit. For example, DNA polymerase, a key enzyme in DNA replication (see pp. 163–165), is composed of more than a dozen polypeptides, each one encoded by a different gene. Nonetheless, many biochemists informally use the terms polypeptide and protein interchangeably, and this text uses protein to refer to all polypeptides.

Proteins are Synthesized Through the Processes of Transcription and Translation

DNA does not directly guide protein synthesis. Instead, an intermediary, **ribonucleic acid (RNA),** carries information from the nucleus to the cytoplasm. RNA is similar to DNA but differs structurally in three respects: RNA is single stranded, whereas DNA is double-stranded; RNA has the sugar ribose instead of deoxyribose in its backbone; and RNA contains the base uracil instead of the base thymine (Table 11-1).

Protein synthesis occurs in two steps, called transcription and translation (Fig. 11-1).

1. In **transcription** (Fig. 11-1a), the information contained in the DNA of a specific gene is copied into one of three types of RNA: messenger RNA (mRNA), transfer RNA (tRNA), or ribosomal RNA (rRNA). In eukaryotic cells, transcription occurs in the nucleus.

2. The base sequence of mRNA encodes the amino acid sequence of a protein. During **translation** (Fig. 11-1b), ribosomes convert the base sequence in mRNA to the amino acid sequence of a protein. In eukaryotic cells, ribosomes are found in the cytoplasm, so translation occurs there as well.

It is easy to confuse the terms transcription and translation, but by comparing their common English meanings with their biological meanings, it may be easier to distinguish their differences. In English, to *transcribe* means to make a written copy of something, almost always in the same language. In a courtroom, for example, verbal testimony is transcribed into a written copy, and both the witnesses' testimony and the transcriptions are in the same language. In biology,

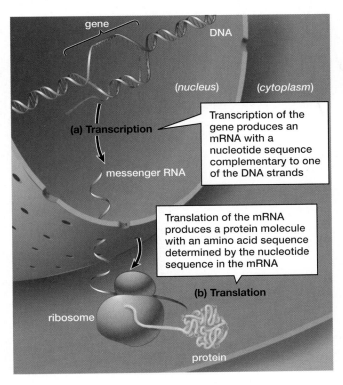

▲ **Figure 11-1** **Genetic information flows from DNA to RNA to protein**

Transcription of the gene produces an mRNA with a nucleotide sequence complementary to one of the DNA strands

Translation of the mRNA produces a protein molecule with an amino acid sequence determined by the nucleotide sequence in the mRNA

Table 11-1	A Comparison of DNA and RNA		
	DNA	**RNA**	
Strands	2	1	
Sugar	Deoxyribose	Ribose	
Types of Bases	Adenine (A), thymine (T) cytosine (C), guanine (G)	Adenine (A), uracil (U) cytosine (C), guanine (G)	
Base Pairs	DNA:DNA A–T T–A C–G G–C	RNA:DNA A–T U–A C–G G–C	RNA:RNA A–U U–A C–G G–C
Function	Contains genes; the sequence of bases in most genes determines the amino acid sequence of a protein	**Messenger RNA (mRNA):** carries the code for a protein-coding gene from DNA to ribosomes **Ribosomal RNA (rRNA):** Combines with proteins to form ribosomes, the structures that link amino acids to form a protein **Transfer RNA (tRNA):** carries amino acids to the ribosomes	

transcription is the process of copying information from DNA to RNA using the common "language" of nucleotides. In contrast, the common meaning of *translation* is to convert words from one language to a different language. Similarly, in biology, translation means to convert information from the "nucleotide language" of RNA to the "amino acid language" of proteins.

11.2 What Are the Functions of RNA?

There are three types of RNA: **messenger RNA (mRNA), transfer RNA (tRNA),** and **ribosomal RNA (rRNA).** Each plays a different role in converting the nucleotide sequence of DNA into the amino acid sequence of a protein (Fig. 11-2).

Messenger RNA Carries the Code for a Protein from the Nucleus to the Cytoplasm

All RNA is produced by transcription from DNA, but only mRNA carries the code for the amino acid sequence of a protein (Fig. 11-2a). In eukaryotic cells, mRNA molecules are synthesized in the nucleus and enter the cytoplasm through the pores in the nuclear envelope. In the cytoplasm, mRNA binds to ribosomes, which synthesize a protein specified by the mRNA base sequence. The DNA remains safely stored in the nucleus, like a valuable document in a library, while mRNA, like a photocopy, carries the information to the cytoplasm to be used in protein synthesis.

The base sequence of mRNA carries the information for the amino acid sequence of a protein

(a) Messenger RNA (mRNA)

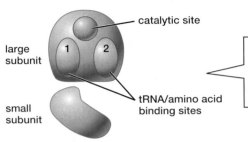

catalytic site

large subunit

tRNA/amino acid binding sites

small subunit

rRNA combines with proteins to form ribosomes; the small subunit binds mRNA; the large subunit binds tRNA and catalyzes peptide bond formation between amino acids during protein synthesis

(b) Ribosome: contains ribosomal RNA (rRNA)

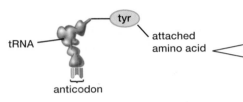

tyr

tRNA

attached amino acid

anticodon

Each tRNA carries a specific amino acid (in this example, tyrosine [tyr]) to a ribosome during protein synthesis; the anticodon of tRNA pairs with a codon of mRNA, ensuring that the correct amino acid is incorporated into the protein

(c) Transfer RNA (tRNA)

Ribosomal RNA and Proteins Form Ribosomes

Ribosomes carry out translation, the process of synthesizing the proteins encoded in mRNA base sequences. Ribosomal RNA and many different proteins combine to form ribosomes. Each ribosome consists of two subunits—one small and one large (**Fig. 11-2b**). The small subunit has binding sites for mRNA, a "start" tRNA (see the next section for a description of tRNA), and several other proteins that cooperate in reading mRNA and starting protein synthesis. The large subunit has two binding sites for tRNA molecules and one catalytic site where peptide bonds join amino acids together into a protein. The two subunits remain separate unless they are actively synthesizing proteins. During protein synthesis, the two subunits come together, clasping an mRNA molecule between them.

Transfer RNA Molecules Carry Amino Acids to the Ribosomes

Each cell synthesizes many different types of transfer RNA, one (or sometimes several) for each amino acid. Twenty different enzymes in the cytoplasm, one for each amino acid, recognize the tRNA molecules and attach the correct amino acid to each tRNA (**Fig. 11-2c**). These "loaded" tRNA molecules deliver their amino acids to the ribosome, where they are incorporated into the growing protein chain.

11.3 What Is the Genetic Code?

We will investigate both transcription and translation in more detail in sections 11.4 and 11.5. First, however, let's see how geneticists broke the language barrier; namely, how does a cell translate the language of base sequences in DNA and messenger RNA into the language of amino acid sequences in proteins? This translation relies on a "dictionary" called the genetic code.

The **genetic code** translates the sequence of bases in nucleic acids into the sequence of amino acids in proteins. But which combinations of bases code for which amino acids? Both DNA and RNA contain four different bases: A, G, C, and T (in DNA) or U (in RNA) (see Table 11-1). However, proteins are made of 20 different amino acids. Therefore, one base cannot code for one amino acid because there are simply not enough different types of bases. A system in which a sequence of two bases codes for an amino acid won't work either, because there would be 16 possible combinations, which still isn't enough to code for all 20 amino acids. A three-base, or *triplet*, sequence, however, gives 64 possible combinations, which is more than enough. Under the assumption that nature operates as economically as possible, biologists hypothesized that the genetic code must be triplet; that is, three bases should specify a single amino acid. In 1961, Francis Crick and three coworkers demonstrated that this hypothesis is correct.

For any language to be understood, its users must know what the words mean, where words start and stop, and where sentences begin and end. To decipher the "words" of the genetic code, in 1961 Marshall Nirenberg and Heinrich Matthaei ground up bacteria and isolated the components needed to synthesize proteins. To this mixture, they added artificial mRNA, which allowed them to control which "words" were to be translated. They could then see which amino acids were incorporated into the resulting proteins. For example, an mRNA strand composed entirely of uracil (UUUUUUU . . .) directed the mixture to synthesize a protein composed solely of the amino acid phenylalanine. Therefore, the triplet UUU must specify phenylalanine. Because the genetic code was deciphered by using these artificial mRNAs, it is usually written in terms of the base triplets in mRNA (rather than in DNA) that code for each amino acid (**Table 11-2**). These mRNA triplets are called **codons.**

What about punctuation? How does a cell recognize where codons start and stop, and where the code for an entire protein starts and stops? All proteins originally begin with the same amino acid, methionine (although it may be removed after the protein is synthesized). Methionine is specified by the codon AUG, which is known as the **start codon.** Three codons—UAG, UAA, and UGA—are **stop codons,** which denote the end of the protein. When the ribosome encounters

Table 11-2 The Genetic Code (Codons of mRNA)

First Base		Second Base								Third Base
		U		**C**		**A**		**G**		
U	UUU	Phenylalanine (Phe)	UCU	Serine (Ser)	UAU	Tyrosine (Tyr)	UGU	Cysteine (Cys)	U
	UUC	Phenylalanine	UCC	Serine	UAC	Tyrosine	UGC	Cysteine	C
	UUA	Leucine (Leu)	UCA	Serine	UAA	Stop	UGA	Stop	A
	UUG	Leucine	UCG	Serine	UAG	Stop	UGG	Tryptophan (Trp)	G
C	CUU	Leucine	CCU	Proline (Pro)	CAU	Histidine (His)	CGU	Arginine (Arg)	U
	CUC	Leucine	CCC	Proline	CAC	Histidine	CGC	Arginine	C
	CUA	Leucine	CCA	Proline	CAA	Glutamine (Gln)	CGA	Arginine	A
	CUG	Leucine	CCG	Proline	CAG	Glutamine	CGG	Arginine	G
A	AUU	Isoleucine (Ile)	ACU	Threonine (Thr)	AAU	Asparagine (Asp)	AGU	Serine (Ser)	U
	AUC	Isoleucine	ACC	Threonine	AAC	Asparagine	AGC	Serine	C
	AUA	Isoleucine	ACA	Threonine	AAA	Lysine (Lys)	AGA	Arginine (Arg)	A
	AUG	Methionine (Met) Start	ACG	Threonine	AAG	Lysine	AGG	Arginine	G
G	GUU	Valine (Val)	GCU	Alanine (Ala)	GAU	Aspartic acid (Asp)	GGU	Glycine (Gly)	U
	GUC	Valine	GCC	Alanine	GAC	Aspartic acid	GGC	Glycine	C
	GUA	Valine	GCA	Alanine	GAA	Glutamic acid (Glu)	GGA	Glycine	A
	GUG	Valine	GCG	Alanine	GAG	Glutamic acid	GGG	Glycine	G

a stop codon, it releases both the newly synthesized protein and the mRNA. Because all codons consist of three bases, and the beginning and end of a protein are specified, then punctuation ("spaces") between codons is unnecessary. Why? Consider what would happen if English used only three-letter words: A sentence such as THEDOGSAWTHECAT would be perfectly understandable, even without spaces between the words.

Because the genetic code has three stop codons, 61 nucleotide triplets remain to specify only 20 amino acids. Therefore, most amino acids are specified by several different codons. For example, six different codons—UUA, UUG, CUU, CUC, CUA, and CUG—all specify leucine (see Table 11-2). However, each individual codon specifies one, and only one, amino acid; for instance, GUU always specifies valine, never leucine, glycine, or any other amino acid.

We have now introduced the pathway of information flow from DNA to RNA to protein. We have also described the genetic code by which cells convert the language of nucleotides in DNA and RNA into the language of amino acids in proteins. Now let's examine the two major processes in this pathway—transcription and translation—in more detail.

11.4 How Is the Information in a Gene Transcribed into RNA?

Transcription copies the genetic information of DNA into RNA, so in eukaryotic cells, transcription occurs in the nucleus. We can view transcription as a process consisting of (1) initiation, (2) elongation, and (3) termination. These three steps correspond to the three major parts of most genes: (1) a promoter region at the beginning of the gene, where transcription is started, or initiated; (2) the "body" of the gene, where the RNA strand is elongated; and (3) a termination signal at the end of the gene, where RNA synthesis stops.

Transcription Begins When RNA Polymerase Binds to the Promoter of a Gene

The enzyme **RNA polymerase** synthesizes RNA. To initiate transcription, RNA polymerase must first locate the **promoter** region, an untranscribed sequence of DNA bases that marks the beginning of the gene. As we will see shortly, other proteins binding near the promoter may increase or decrease the ability of RNA polymerase to attach to the promoter. When RNA polymerase binds to the promoter, the DNA double helix at the beginning of the gene partially unwinds and transcription begins (**Fig. 11-3**, step ❶).

Elongation Generates a Growing Strand of RNA

RNA polymerase travels down one of the DNA strands, called the **template strand,** unwinding the DNA double helix as it goes. RNA polymerase synthesizes a single strand of RNA with bases that are complementary to those in the template strand of the DNA (**Fig. 11-3**, step ❷). Base pairing between RNA and DNA is the same as between two strands of DNA, except that uracil in RNA replaces thymine and pairs with adenine in DNA (see Table 11-1). After about 10 nucleotides have been added to the growing RNA chain, the first nucleotides in the RNA molecule separate from the DNA template strand, allowing the two DNA strands to rewind into a double helix once again. As transcription continues to elongate the RNA molecule, one end of the RNA drifts away from the

▶ **Figure 11-3 Transcription is the synthesis of RNA from instructions in DNA** A gene is a segment of a chromosome's DNA. One of the DNA strands will serve as the template for the synthesis of an RNA molecule with bases complementary to the bases in this DNA strand.   **QUESTION:** Suppose you could design an RNA polymerase that would transcribe both strands of a DNA double helix at the same time. Do you think this would be useful for a cell?

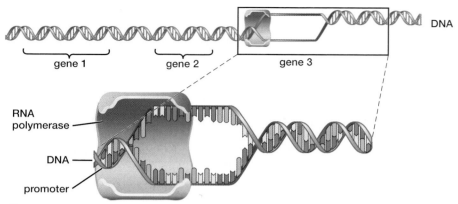

DNA

gene 1   gene 2   gene 3

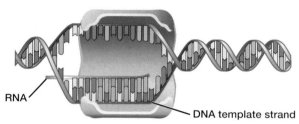

RNA polymerase

DNA

promoter

❶ **Initiation:** RNA polymerase binds to the promoter region of DNA near the beginning of a gene, separating the double helix near the promoter.

RNA

DNA template strand

❷ **Elongation:** RNA polymerase travels along the DNA template strand (blue), unwinding the DNA double helix and synthesizing RNA by catalyzing the addition of ribose nucleotides into an RNA molecule (red). The nucleotides in the RNA are complementary to the template strand of the DNA.

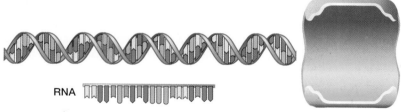

termination signal

❸ **Termination:** At the end of the gene, RNA polymerase encounters a DNA sequence called a termination signal. RNA polymerase detaches from the DNA and releases the RNA molecule.

RNA

❹ **Conclusion of transcription:** After termination, the DNA completely rewinds into a double helix. The RNA molecule is free to move from the nucleus to the cytoplasm for translation, and RNA polymerase may move to another gene and begin transcription once again.

DNA while RNA polymerase keeps the other end temporarily attached to the DNA template strand (**Fig. 11-4**).

Transcription Stops When RNA Polymerase Reaches the Termination Signal

RNA polymerase continues along the template strand until it reaches a sequence of DNA bases known as the termination signal (**Fig. 11-3**, step ❸). At this point,

RNA polymerase releases the completed RNA molecule and detaches from the DNA (**Fig. 11-3**, step **4**). The RNA polymerase is then free to bind to another promoter and synthesize another RNA molecule.

Transcription Is Selective

Some genes are transcribed in all cells because they encode essential proteins, such as those that comprise the electron transport chain of the mitochondria (see pp. 103–104). Many genes, however, are transcribed in only specific types of cells. For example, every cell in your body contains the gene for the protein hormone insulin, but that gene is transcribed only in certain cells in your pancreas.

How do cells regulate which genes are transcribed? In many cases, proteins that bind to "control regions" of DNA, found near the promoter of a specific gene, block or enhance the binding of RNA polymerase. Different cells contain different proteins that regulate transcription, thereby restricting transcription to the genes needed by a particular type of cell at a particular time. For example, a specific set of proteins, found only in the pancreas, activates transcription of the insulin gene. Cells outside the pancreas do not contain these proteins, so they do not produce insulin.

We describe additional mechanisms by which cells regulate which genes are transcribed later in this chapter, in sections 11.6 and 11.7.

11.5 How Is the Information in Messenger RNA Translated into Protein?

Once transcription has produced a messenger RNA molecule with a specific base sequence, the mRNA is used during translation to direct the synthesis of a protein with the amino acid sequence encoded by the mRNA. Decoding the base sequence of mRNA is the job of tRNA and ribosomes in the cytoplasm. Remember that triplets of bases in mRNA, called codons, specify which amino acid should be incorporated into the protein, and that tRNA transports amino acids to the ribosome. The ability of tRNA to deliver the correct amino acid depends on base pairing between each codon of mRNA and a set of three complementary bases in tRNA, appropriately called the **anticodon** (see Fig. 11-2). For example, suppose that the codon UAC occurs in mRNA. This codon will form base pairs with the anticodon AUG of a tRNA that has tyrosine attached to it, so the ribosome would add tyrosine to the protein.

Like transcription, translation has three steps: (1) initiation of protein synthesis, (2) elongation of the protein chain, and (3) termination of translation.

Initiation: Translation Begins When tRNA and mRNA Bind to a Ribosome

The first AUG codon in mRNA specifies the start of translation. Because AUG also codes for methionine (see Table 11-2), all newly synthesized proteins begin with this amino acid. An *initiation complex*, composed of a small ribosomal subunit, a methionine tRNA, and several proteins, binds to the first AUG codon in an mRNA molecule (**Fig. 11-5**, steps **1** and **2**). The AUG codon in mRNA forms base pairs with the UAC anticodon of the methionine tRNA. A large ribosomal subunit then attaches to the small subunit, sandwiching the mRNA between the two subunits, completing the assembly of the ribosome (**Fig. 11-5**, step **3**).

Elongation: Amino Acids Are Added One at a Time to the Growing Protein Chain

The large ribosomal subunit has two tRNA binding sites and a catalytic site. When the first binding site is occupied by a tRNA and its attached amino acid, a second

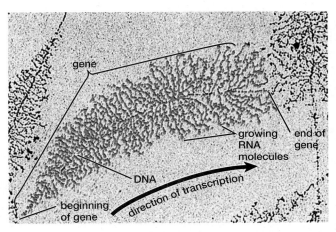

gene

growing RNA molecules

end of gene

DNA

beginning of gene

direction of transcription

▲ **Figure 11-4 RNA transcription in action** This electron micrograph shows RNA transcription. In the treelike structure in the middle of the micrograph, the central "trunk" (blue) is DNA and the "branches" (red) are RNA molecules. The beginning of the gene is on the left. Many RNA polymerase molecules (not visible in the micrograph) are traveling along the DNA, synthesizing RNA as they go. The short RNA molecules on the left have just begun to be synthesized, whereas the long RNA molecules on the right have almost been completed.

QUESTION: Why do you think so many mRNA molecules are being synthesized from the same gene? (Hint: Do organisms always need to have the same amounts of all types of proteins?)

BioFlix   Protein Synthesis

Web Animation Translation

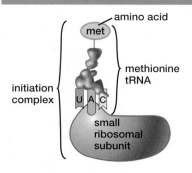

1 A tRNA with an attached methionine amino acid binds to a small ribosomal subunit, forming an initiation complex.

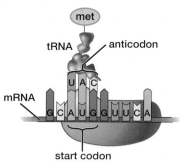

2 The initiation complex binds to an mRNA molecule. The methionine (met) tRNA anticodon (UAC) base-pairs with the start codon (AUG) of the mRNA.

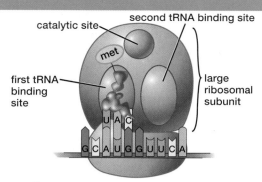

3 The large ribosomal subunit binds to the small subunit. The methionine tRNA binds to the first tRNA site on the large subunit.

Elongation:

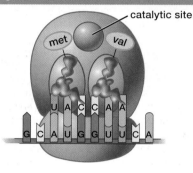

4 The second codon of mRNA (GUU) base-pairs with the anticodon (CAA) of a second tRNA carrying the amino acid valine (val). This tRNA binds to the second tRNA site on the large subunit.

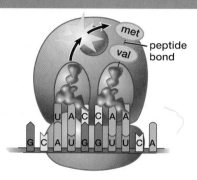

5 The catalytic site on the large subunit catalyzes the formation of a peptide bond linking the amino acids methionine and valine. The two amino acids are now attached to the tRNA in the second binding site.

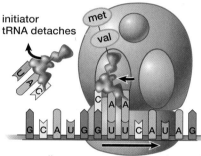

ribosome moves one codon to the right

6 The "empty" tRNA is released and the ribosome moves down the mRNA, one codon to the right. The tRNA that is attached to the two amino acids is now in the first tRNA binding site and the second tRNA binding site is empty.

Termination:

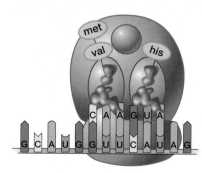

7 The third codon of mRNA (CAU) base-pairs with the anticodon (GUA) of a tRNA carrying the amino acid histidine (his). This tRNA enters the second tRNA binding site on the large subunit.

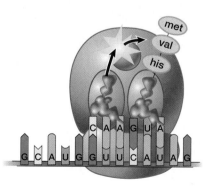

8 The catalytic site forms a peptide bond between valine and histidine, leaving the peptide attached to the tRNA in the second binding site. The tRNA in the first site leaves, and the ribosome moves one codon over on the mRNA.

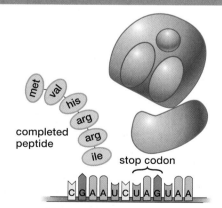

9 This process repeats until a stop codon is reached; the mRNA and the completed peptide are released from the ribosome, and the subunits separate.

▲ **Figure 11-5 Translation is the process of protein synthesis** Translation decodes the base sequence of an mRNA into the amino acid sequence of a protein. **QUESTION:** Examine panel **9**. If all of the guanine molecules (G) visible in this mRNA sequence were changed to uracil (U), how would the translated protein differ from the one shown?

178

tRNA, with an anticodon complementary to the next codon of the mRNA, moves into the second binding site. The catalytic site forms a peptide bond between the two amino acids (**Fig. 11-5**, steps **4** and **5**). The two-amino-acid chain remains attached to the second tRNA. The empty methionine tRNA leaves the ribosome, which then moves to the next codon, shifting the growing protein chain from the second to the first binding site (**Fig. 11-5**, step **6**). A new tRNA moves into the second binding site, carrying its amino acid to the ribosome (**Fig. 11-5**, step **7**). This amino acid is joined to the chain (**Fig. 11-5**, step **8**), the empty tRNA leaves the first binding site, and the ribosome shifts to the fourth codon. The whole process repeats over and over again as the ribosome moves along the mRNA one codon at a time.

Termination: A Stop Codon Signals the End of Translation

A stop codon in the mRNA molecule signals the ribosome to terminate protein synthesis. Stop codons do not bind to a tRNA. Instead, proteins called *release factors* bind to the ribosome when it encounters a stop codon, forcing the ribosome to release the finished protein chain and the mRNA (**Fig. 11-5**, step **9**). The ribosome disassembles into large and small subunits, which can then be used to translate another mRNA.

SUMMING UP: Transcription and Translation

Let's summarize how a eukaryotic cell decodes the genetic information stored in its DNA to synthesize a protein (**Fig. 11-6**).

a. With a few exceptions, such as the genes for tRNA and rRNA, each gene codes for a single protein.

b. Transcription of a protein-coding gene produces a messenger RNA (mRNA) molecule that is complementary to the template strand of the DNA of the gene. Starting from the first AUG, each codon in the mRNA is a sequence of three bases that specifies either an amino acid or a stop.

c. Enzymes in the cytoplasm attach the appropriate amino acid (based on the tRNA's anticodon) to each tRNA.

d. The mRNA moves out of the nucleus to a ribosome in the cytoplasm. Transfer RNAs carry their attached amino acids to the ribosome. There, the bases in tRNA anticodons bind to the complementary bases in mRNA codons, so the amino acids attached to the tRNAs line up in the sequence specified by the codons. The ribosome joins the amino acids together with peptide bonds to form a protein. When a stop codon is reached, the finished protein is released from the ribosome.

11.6 How Do Mutations Affect Gene Function?

Mistakes during DNA replication, as well as ultraviolet rays in sunlight, chemicals in cigarette smoke, and a host of other environmental factors, may change the sequence of bases in DNA. These changes are called **mutations,** and may result in a variety of different alterations in the DNA base sequence. Sometimes, during DNA replication, an incorrect pair of nucleotides is incorporated into the growing DNA double helix. These **nucleotide substitutions** are also called **point mutations,** because individual nucleotides in the DNA sequence are changed. An **insertion mutation** occurs when one or more new nucleotide pairs are inserted into a gene. A **deletion mutation** occurs when one or more nucleotide pairs are removed from a gene.

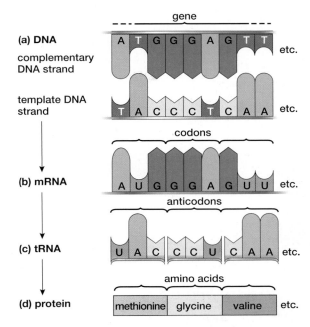

▲ **Figure 11-6 Complementary base pairing is critical to decoding genetic information** *(a)* DNA contains two strands. RNA polymerase uses the template strand to synthesize an RNA molecule. *(b)* Bases in the template strand are transcribed into a complementary mRNA. Codons are sequences of three bases that specify an amino acid or a stop during protein synthesis. *(c)* Unless it is a stop codon, each mRNA codon forms base pairs with the anticodon of a tRNA molecule that carries a specific amino acid. *(d)* The amino acids are linked to form a protein.

Cystic Fibrosis

Continued

The chain of events, from DNA bases to mRNA codons to tRNA anticodons to amino acids, decodes the information in DNA and results in the synthesis of a protein with a specific amino acid sequence, determined by the base sequence within a gene. Because a protein will usually function properly only if it has the right amino acid sequence, the accuracy of each step in this process is crucial. Thus, the CFTR protein can help to keep the lungs clear and healthy only if it is synthesized in its entirety and has the correct amino acid sequence. Otherwise, cystic fibrosis will result.

Of course, accurate transcription and translation will produce a functional protein only when the original repository of information—the base sequence of DNA—is intact. Section 11.6 describes what happens to protein structure and function if the base sequence of a gene is changed.

Mutations May Have a Variety of Effects on Protein Structure and Function

The effects of a mutation on protein structure and function depend on the type of mutation. For example, deletions and insertions of one or two nucleotides usually have catastrophic effects on a gene, because all of the codons that follow the deletion or insertion are altered. Think of our sample English sentence, THEDOGSAWTHECAT. Deleting or inserting a letter (deleting the first E, for example) makes all of the following three-letter words nonsensical: THD OGS AWT HEC AT. The protein synthesized from an mRNA containing such a mutation will almost always be nonfunctional.

Point mutations (nucleotide substitutions) within a protein-coding gene, however, will not always render the code useless. Nucleotide substitutions can produce at least four different outcomes. As an example, let's consider mutations that occur in the gene encoding beta-globin, one of the subunits of hemoglobin, the oxygen-carrying protein in red blood cells (Table 11-3). The other type of subunit in hemoglobin is called alpha-globin; a normal hemoglobin molecule consists of two alpha and two beta subunits. In all but the last of the following possible outcomes, we consider the results of mutations that occur in the sixth codon of beta-globin (CTC in DNA, transcribed to GAG in mRNA). This codon specifies glutamic acid—a charged, hydrophilic, water-soluble amino acid.

1. *The protein may be unchanged.* Remember that most amino acids can be encoded by several different codons. If a mutation changes the beta-globin DNA base sequence from CTC to CTT, this sequence still codes for glutamic acid. Therefore, the protein synthesized from the mutated gene remains unchanged (see Mutation 1 in Table 11-3).

2. *The new protein may be functionally equivalent to the original one.* Many proteins have regions whose exact amino acid sequence is relatively unimportant. In beta-globin, the amino acids on the outside of the protein must be hydrophilic to keep the protein dissolved in the cytoplasm of red blood cells. Exactly *which* hydrophilic amino acids are on the outside usually doesn't matter very much. For example, a family in the Japanese town of Machida has a mutation from CTC to GTC, replacing glutamic acid (hydrophilic) with glutamine (also hydrophilic; see Mutation 2 in Table 11-3). Hemoglobin containing this mutant beta-globin protein functions well. Mutations in which the amino acids of a protein are unchanged or are functionally equivalent are called *neutral mutations.*

3. *Protein function may be changed by an altered amino acid sequence.* A mutation from CTC to CAC replaces glutamic acid (hydrophilic) with valine (hydrophobic; see Mutation 3 in Table 11-3). This substitution is the genetic defect that causes sickle-cell anemia (see pp. 148–149). The valines on the

Table 11-3	Effects of Mutations in the Hemoglobin Gene					
	DNA (template strand)	mRNA	Amino Acid	Properties of Amino Acid	Functional Effect on Protein	Disease
Original codon 6	CTC	GAG	Glutamic acid	Hydrophilic	Normal protein function	None
Mutation 1	CTT	GAA	Glutamic acid	Hydrophilic	Neutral, normal protein function	None
Mutation 2	GTC	CAG	Glutamine	Hydrophilic	Neutral, normal protein function	None
Mutation 3	CAC	GUG	Valine	Hydrophobic	Loses water solubility; compromises protein function	Sickle-cell anemia
Original codon 17	TTC	AAG	Lysine	Hydrophilic	Normal protein function	None
Mutation 4	ATC	UAG	Stop codon	Ends translation after amino acid 16	Synthesizes only part of the protein; eliminates protein function	Beta-thalassemia

outside of the hemoglobin molecules cause them to clump together, distorting the shape of the red blood cells.

4. *Protein function may be destroyed by a premature stop codon.* A particularly catastrophic mutation occasionally occurs in the 17th codon of the beta-globin gene (TTC in DNA, AAG in mRNA). A mutation from TTC to ATC (UAG in mRNA) results in a stop codon, halting translation before the beta-globin protein is completed (see Mutation 4 in Table 11-3). People who inherit this mutant gene from both their mother and their father do not synthesize any functional beta-globin protein; instead, they manufacture hemoglobin consisting entirely of alpha-globin subunits. This "pure alpha" hemoglobin does not bind oxygen very well. This condition, called beta-thalassemia, can be fatal unless treated with regular blood transfusions.

Mutations Are the Raw Material for Evolution

Mutations are the ultimate source of all genetic differences among individuals, providing raw material for evolution. Without mutations, all individuals would share the same DNA sequence. Although most mutations are neutral or harmful to the individuals who carry them, a mutation may occasionally improve an individual's ability to survive and reproduce. If such a beneficial mutation occurs in a cell that gives rise to gametes (sperm or eggs), it may be passed on to future generations. If organisms possessing the mutant gene produce more offspring than individuals that lack the mutation, the mutant gene and the characteristics it bestows will become more common over time. This process, known as *natural selection*, is a major cause of evolutionary change, which is described in Unit 3.

11.7 Are All Genes Expressed?

All of the 21,000 genes in the human genome are present in each body cell, but individual cells *express* (transcribe and translate) only a small fraction of these genes. The particular set of genes that is expressed depends on the type of cell and the needs of the organism. This regulation of gene expression is crucial for proper functioning of individual cells and entire organisms.

Gene Expression Differs from Cell to Cell and over Time

In organisms with more than one cell, the set of genes that is expressed largely depends on the function of a cell. For example, in humans and other mammals, the cells of hair follicles synthesize keratin, the protein from which hair is made. Muscle cells, in contrast, synthesize large amounts of the proteins actin and myosin, which are necessary for muscle contraction, but they do not synthesize keratin.

Gene expression also changes over time, depending on the body's needs from moment to moment. For example, immediately after a baby's birth, the milk-producing cells in a woman's breasts begin to express the gene that encodes casein, the major protein in milk. This change in gene expression allows the mother to produce large amounts of protein-rich milk to feed her baby. Circumstances may even dictate that some genes are never expressed. A human male, for example, does not express the casein gene. However, he will pass a copy of this gene to his daughters, who will express it if they bear children.

Environmental Cues Influence Gene Expression

Changes in an organism's environment also help determine which genes are transcribed. In birds living in temperate climates, for example, the longer spring days stimulate the sex organs (testes or ovaries) to enlarge and produce sex hormones. The sex hormones in turn cause the birds to produce eggs and sperm, to sing, to mate, and to build nests. The proliferation of cells in the sex organs, the

Cystic Fibrosis

Continued

Why hasn't natural selection essentially eliminated mutated *CFTR* alleles? Perhaps because the mutated alleles provide some protection against cholera and typhoid. The normal CFTR protein is activated by cholera toxin, causing excessive chloride secretion from intestinal cells. Water follows by osmosis, and so cholera victims have debilitating, and often fatal, diarrhea. Defective CFTR proteins cannot be activated by cholera toxin. The CFTR protein is also the site by which typhoid bacteria enter cells, but typhoid bacteria cannot enter via mutated CFTR proteins. Such protection, of course, cannot make up for the devastating effects of cystic fibrosis. However, heterozygotes, with one normal and one mutated *CFTR* allele, have nearly normal CFTR function in lungs, sweat glands, and intestines, but might be less severely affected by cholera and typhoid. This "heterozygote advantage" may explain the high frequency of mutated *CFTR* alleles.

production of hormones by these cells, and the effects of those hormones on other cells throughout the body all result, directly or indirectly, from changes in gene expression caused by a change in the environment.

11.8 How Is Gene Expression Regulated?

A cell may regulate gene expression in many different ways. It may alter the rate of transcription of mRNA, how long a given mRNA molecule lasts before it is broken down, how fast the mRNA is translated into protein, or how long the protein lasts before it is degraded. Here, we will describe a few ways in which transcription is regulated.

Regulatory Proteins That Bind to Promoters Alter the Transcription of Genes

As we described earlier in this chapter, transcription may be regulated by proteins, often combined with other molecules, that enhance or inhibit the ability of RNA polymerase to bind to the promoter region of a gene. Many steroid hormones act in this way (see p. 444 in Chapter 23 for a complete description of steroid hormone action). In birds, for example, the sex hormone estrogen regulates expression of the gene for albumin (the protein in egg whites). The albumin gene is not transcribed in winter, when birds are not breeding and estrogen levels are low. During the breeding season, however, estrogen enters cells in the female reproductive system and binds to a receptor protein. The estrogen–receptor combination then binds to the DNA in a region near the promoter of the albumin gene. This attachment makes it easier for RNA polymerase to bind to the promoter. As a result, the cells begin to transcribe large amounts of albumin mRNA, which is translated into the albumin protein needed to make eggs. Sex hormones act in a similar manner in humans. In "Health Watch: Mutations and Gender," we describe what happens when the receptor for the male sex hormone, testosterone, is defective.

Some Regions of Chromosomes Are Condensed and Not Normally Transcribed

Certain parts of eukaryotic chromosomes are in a highly condensed, compact state in which most of the DNA is inaccessible to RNA polymerase. Some of these tightly condensed regions contain genes that are not currently being transcribed. When the product of a gene is needed, the portion of the chromosome containing that gene becomes decondensed—loosened so that the nucleotide sequence is accessible to RNA polymerase and transcription can occur.

Entire Chromosomes May Be Inactivated and Not Transcribed

In some cases, almost an entire chromosome may be condensed, making it largely inaccessible to RNA polymerase. An example occurs in the sex chromosomes of female mammals. Male mammals usually have an X and a Y chromosome (XY), and females usually have two X chromosomes (XX). As a consequence, females have the capacity to synthesize mRNA from genes on their two X chromosomes, while males, with only one X chromosome, may produce only half as much. In 1961, the geneticist Mary Lyon hypothesized that perhaps one of the two X chromosomes in women was inactivated in some way, so that its genes were not expressed. We now know that about 85% of the genes on an inactivated X chromosome are coated with a special type of RNA molecule, called Xist, that condenses the chromosome and prevents gene transcription. The condensed X chromosome, called a Barr body after its discover, Murray Barr, appears as a discrete spot in the nuclei of the cells of female mammals (Fig. 11-7).

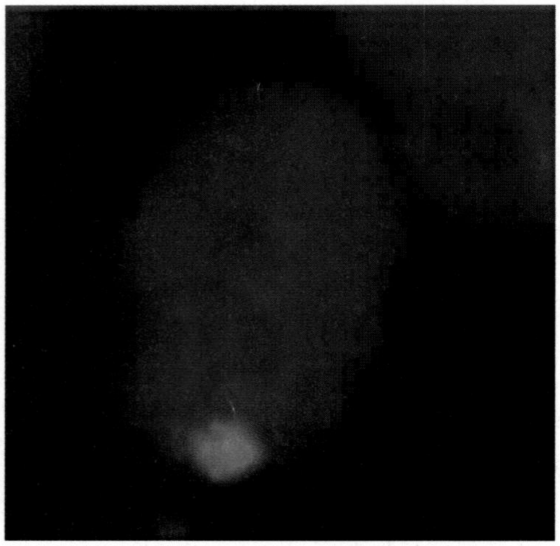

▲ Figure 11-7 A Barr body The red spot at the bottom of the nucleus is an inactivated X chromosome called a Barr body. In this fluorescence micrograph, the Barr body is stained with a dye that binds to the Xist RNA coating the inactivated X chromosome.

Mutations and Gender

Sometime in her early to mid-teens, a girl usually goes through puberty: Her breasts swell, her hips widen, and she begins to menstruate. In rare instances, however, a girl may develop all of the outward signs of womanhood but not menstruate. Eventually, when it becomes clear that she isn't merely developing a bit late, she reports this symptom to her physician, who may take a blood sample to do a chromosome test. In some cases, the chromosome test gives what might seem to be an impossible result: The girl's sex chromosomes are XY, a combination that would normally give rise to a boy. The reason she has not begun to menstruate is that she lacks ovaries and a uterus, but instead has testes that have remained inside her abdominal cavity. She has about the same concentrations of *androgens* (male sex hormones, such as testosterone) circulating in her blood as would be found in a boy her age. The problem is that her cells cannot respond to them—a rare condition called *androgen insensitivity*.

Androgen insensitivity is caused by a mutation in a gene located on the X chromosome. The affected gene codes for a protein known as an androgen receptor. In normal males, androgen receptor proteins are present in the cytoplasm of many body cells. Male hormones such as

testosterone bind to the receptor molecules. The hormone–receptor combination enters the cell nucleus and binds to DNA, stimulating the transcription of genes that help to produce many male features, including the formation of a penis, the descent of the testes into sacs outside the body cavity, and sexual characteristics that develop at puberty, such as a beard and increased muscle mass.

There are more than 200 mutant alleles of the androgen receptor gene. The most serious are usually insertions, deletions, or point mutations creating a premature stop codon—and, as you now know, these types of mutations are likely to have catastrophic effects on protein structure and function (see Table 11-3). Therefore, a person with XY chromosomes who inherits one of these mutant androgen receptor alleles will be unable to make a functional receptor protein. Even though genetically a male with both X and Y chromosomes, her cells will be unable to respond to the testosterone that the testes produce, and male characteristics will not develop. Thus, a mutation that changes the nucleotide sequence of a single gene, causing a single type of defective protein to be produced, can cause a person who is genetically male (XY) to look like and perceive herself to be a woman (**Fig. E11-1**).

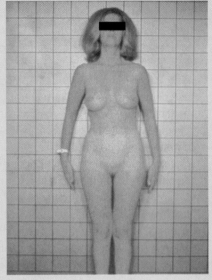

▲ **Figure E11-1 Androgen insensitivity leads to female features** This individual looks female but is genetically male because she has an X and a Y chromosome. As a result, she has testes (located in her abdomen) that produce testosterone, and she lacks ovaries and a uterus. A mutation in her androgen receptor gene prevents her cells from responding to the testosterone produced by her testes and causes her female appearance.

Usually, fairly large clusters of cells have the same X chromosome inactivated, because every cell in a given cluster is descended from the same cell that was produced early in development. As a result, the bodies of female mammals (including women) are composed of patches of cells in which one of the X chromosomes is fully active, and patches of cells in which the other X chromosome is active. The results of this phenomenon are easily observed in calico cats (**Fig. 11-8**). The X chromosome of a cat contains a gene encoding an enzyme that produces fur pigment. This gene comes in two versions, one producing orange fur and the other producing black fur. If one X chromosome in a female cat has the orange version of the fur color gene and the other X chromosome has the black version, the cat will have patches of orange and black fur. These patches represent areas of skin that developed from cells in the early embryo in which different X chromosomes were inactivated. Calico coloring is almost exclusively found in female cats. Because male cats usually have only one X chromosome, which is active in all of their cells, normal male cats can have black fur or orange fur, but not both.

▶ **Figure 11-8 Inactivation of the X chromosome regulates gene expression** This female calico cat carries a gene for orange fur on one X chromosome and a gene for black fur on her other X chromosome. Inactivation of different X chromosomes produces the black and orange patches. The white color is due to an entirely different gene that prevents pigment formation altogether.

Cystic Fibrosis Revisited

Researchers have identified more than 1,000 different mutations in the *CFTR* gene. In the most common mutation, the gene is missing three nucleotides—exactly one codon—causing the protein to lack the amino acid phenylalanine at a crucial site. This change causes the CFTR protein to be degraded before it ever leaves the endoplasmic reticulum, where it is synthesized. Other mutations convert any of several codons to a stop codon, causing protein synthesis to stop long before the protein is complete. Still other mutations substitute one amino acid for another, rendering the protein less functional. Some of these point mutations make the protein almost totally ineffective, while others only slow down chloride transport to varying extents.

People who are heterozygous, with one normal *CFTR* allele and one copy of any of these mutations, produce enough normal CFTR protein for adequate chloride transport. Therefore, they are phenotypically normal—that is, they produce watery secretions in their lungs and do not develop cystic fibrosis. A person with two defective alleles will have no functional CFTR proteins and will develop the disease.

Genetic diseases such as cystic fibrosis can't be cured in the way that an infection can be cured by killing the offending bacteria or viruses. Typically, genetic diseases are treated by replacing the lost function, such as giving insulin to diabetics who can't synthesize their own, or by relieving the symptoms. In cystic fibrosis, the most common treatments relieve some of the symptoms. These treatments include medicines that open the airways (similar to those used by people with asthma), frequent administration of antibiotics, and physical therapy to drain the lungs. In the next chapter, we will see that biotechnology offers some hope of replacing the lost function, at least in the lungs, by delivering functioning *CFTR* genes to the cells lining the airways.

What ultimately happens to a person with cystic fibrosis? Well, some of the mutations damage the CFTR protein less than others. For example, triathlete Lisa Bentley has relatively mild cystic fibrosis. Bentley carefully controls her diet, especially her salt intake. Vigorous exercise helps to clear out her lungs. She also scrupulously avoids situations where she might be exposed to contagious diseases, even the common cold. Thus, her cystic fibrosis hasn't kept her from becoming one of the finest athletes in the world. Alice Martineau, on the other hand, had one of the most severe mutations. By the age of 30, Martineau needed a triple transplant—heart, lung, and liver. She died before a suitable donor could be found.

Consider This

The vast majority of research in biology and medicine is funded by taxes distributed to government agencies, such as the National Institutes of Health (NIH) in the United States. The NIH funds research on genetic disorders such as cystic fibrosis, muscular dystrophy, and Marfan syndrome; infectious diseases such as AIDS and tuberculosis; and a host of other conditions, including heart disease and cancer. The NIH also funds research on diseases, such as malaria, that are uncommon in the United States but that claim hundreds of thousands of lives each year in impoverished tropical countries. Even though the NIH spends billions of dollars each year on biomedical research, faster progress could certainly be made on most of these diseases if more money were spent on them. How do you think that scarce NIH funds would be spent? Should they be spent in proportion to disease severity and incidence in the United States, which might mean that almost all NIH funds would be spent on heart disease and cancer? Or should the funds be spent according to the prospective life span of the victims, so that diseases of the young, such as muscular dystrophy, cystic fibrosis, or tuberculosis, receive much more funding? Does the United States have a responsibility to help people in other countries suffering from diseases that are rare in America?

Chapter Review

Summary of Key Concepts

For additional study help and activities, go to www.mybiology.com.

11.1 How Is the Information in DNA Used in a Cell?

Genes are segments of DNA that usually code for a single protein. Transcription copies the information in DNA into ribonucleic acid, or RNA. Translation converts the "nucleotide language" of RNA into the "amino acid language" of proteins.

11.2 What Are the Functions of RNA?

There are three types of RNA: messenger RNA (mRNA), ribosomal RNA (rRNA), and transfer RNA (tRNA). mRNA carries the genetic information of a gene from the nucleus to the cytoplasm, where organelles called ribosomes use the information to synthesize a protein. Ribosomes contain a combination of rRNA and proteins organized into large and small subunits. There are many different tRNAs. Each tRNA binds a specific amino acid and carries it to a ribosome for incorporation into a protein.

11.3 What Is the Genetic Code?

The nucleotide sequence in messenger RNA specifies the amino acid sequence of a protein, according to a set of "translation rules" called the genetic code. The genetic code consists of codons—

sequences of three bases in mRNA—that specify either an amino acid in a protein chain, the beginning of protein synthesis (start codon), or the end of protein synthesis (stop codon).

11.4 How Is the Information in a Gene Transcribed into RNA?

Using a strand of DNA as a template, transcription produces a strand of RNA. Within an individual cell, only certain genes are transcribed. During transcription, RNA polymerase binds to the promoter region of the DNA of a gene and synthesizes a single strand of RNA. This RNA is complementary to the template strand in the gene's DNA double helix. Transcription is selective, because each cell transcribes only a subset of its genes into RNA.

BioFlix Protein Synthesis
Web Animation Transcription

11.5 How Is the Information in Messenger RNA Translated into Protein?

The start codon of mRNA binds to the small subunit of a ribosome, along with the corresponding (methionine) tRNA. A large subunit binds to the small subunit, forming the complete protein-synthesizing machine. Transfer RNAs deliver the appropriate amino acids to the ribosome for incorporation into the growing protein. Base pairing between the anticodons of the tRNAs and the codons of the mRNA puts the amino acids in the order coded in the mRNA. Two tRNAs, each carrying an amino acid, bind simultaneously to the ribosome's large subunit, which then catalyzes the formation of peptide bonds between the amino acids. As each new amino acid is attached, one tRNA detaches, and the ribosome moves down the mRNA one codon and binds to another tRNA that carries the amino acid specified by that codon. Addition of amino acids to the growing protein continues until a stop codon is reached, signaling the ribosome to disassemble and to release both the mRNA and the newly formed protein.

BioFlix Protein Synthesis
Web Animation Translation

11.6 How Do Mutations Affect Gene Function?

A mutation is a change in the nucleotide sequence of a gene. Mutations can be caused by mistakes in base pairing during replication, by chemical agents, and by environmental factors such as radiation. Common types of mutations include changes in a nucleotide base pair (point mutations) and insertions or deletions of nucleotide base pairs. Mutations are generally neutral or harmful, but in rare cases a beneficial mutation will encode features that will be favored by natural selection.

11.7 Are All Genes Expressed?

To be expressed, a gene is transcribed and translated, and the resulting protein performs some action within the cell. Which genes are expressed in a cell at any given time depends on the function of the cell, the developmental stage of the organism, and the environment.

11.8 How Is Gene Expression Regulated?

Cells regulate gene expression by altering the rate of transcription of mRNA, how long a given mRNA molecule lasts before it is broken down, how fast the mRNA is translated into protein, how long the protein lasts, and how fast a protein enzyme catalyzes a reaction. The rate of transcription may be regulated by stimulating or repressing transcription of an individual gene, by condensing or exposing large parts of chromosomes, or by condensing entire chromosomes.

Key Terms

anticodon *p. 177*
codon *p. 174*
deletion mutation *p. 179*
gene *p. 171*
genetic code *p. 174*
insertion mutation *p. 179*

messenger RNA (mRNA) *p. 172*
mutation *p. 179*
nucleotide substitution *p. 179*
point mutation *p. 179*
promoter *p. 175*
ribonucleic acid (RNA) *p. 171*

ribosomal RNA (rRNA) *p. 172*
ribosome *p. 173*
RNA polymerase *p. 175*
start codon *p. 174*
stop codon *p. 174*
template strand *p. 175*

transcription *p. 171*
transfer RNA (tRNA) *p. 172*
translation *p. 171*

Thinking Through the Concepts

Suggested answers to end-of-chapter and figure-based questions can be found at the end of the text.

Fill-in-the-Blank

1. Synthesis of RNA from the instructions in DNA is called _____. Synthesis of a protein from the instructions in messenger RNA is called _____. Which organelle in the cell is the site of protein synthesis? _____

2. The three types of RNA are _____, _____, and _____.

3. The genetic code uses _____ (how many?) bases to code for a single amino acid. This short sequence of bases in messenger RNA is called a(n) _____. The complementary sequence of bases in transfer RNA is called a(n) _____.

4. The enzyme _____ synthesizes RNA from the instructions in DNA. DNA has two strands, but for any given gene, usually only one strand, called the _____ strand, is transcribed. To begin transcribing a gene, this enzyme binds to a specific sequence of DNA bases located at the beginning of the gene. This DNA sequence is called the _____. Transcription ends when the enzyme encounters a DNA sequence at the end of the gene called the _____.

5. Protein synthesis begins when messenger RNA binds to a ribosome. Translation begins with the _____ codon of messenger RNA and continues until a _____ codon is reached. Individual amino acids are brought to the ribosome by _____ RNA. These amino acids are linked into protein by _____ bonds.

6. There are several different types of mutations in DNA. If one nucleotide is substituted for another, this is called a _____ mutation. _____ mutations occur if nucleotides are added in the middle of a gene. _____ mutations occur if nucleotides are removed from the middle of a gene.

Review Questions

1. How does RNA differ from DNA?

2. What are the three types of RNA? What is the function of each?

3. Define the following terms: genetic code, codon, anticodon. What is the relationship among the bases in DNA, the codons of mRNA, and the anticodons of tRNA?

4. Diagram and describe protein synthesis.

5. Explain how both transcription and translation require complementary base pairing.

6. How is gene expression regulated?

7. Define mutation, and give one example of how a gene might mutate. Would you expect most mutations to be beneficial or harmful? Explain your answer.

Applying the Concepts

1. IS THIS SCIENCE? About 40 years ago, some researchers reported that they could transfer learning from one animal (*Planaria*, a type of flatworm) to another by feeding trained animals to untrained animals. Further, they claimed that RNA was the active molecule of learning. Given your knowledge of the roles of RNA and protein in cells, do you think that a *specific* memory (for example, remembering that females usually have two X chromosomes) could be encoded by a *specific* molecule of RNA and that this RNA could transfer that memory to another person? In other words, someday, could you learn biology by popping an RNA pill? If you could, how would the pill work? If you think the RNA pill wouldn't work, can you propose a reasonable hypothesis for the *Planaria* results? How would you test your hypothesis?

2. As you learned in this chapter, many factors influence gene expression, including hormones. The use of anabolic steroids and growth hormones among athletes has created controversy in recent years. Hormones certainly affect gene expression, but, in the broadest sense, so do vitamins and foods. What do you think are appropriate guidelines for the use of hormones? Should athletes take steroids and growth hormones? Should children at risk of being unusually short be given growth hormones? Should parents be allowed to request growth hormones for their children of normal height in the hope of producing a future basketball player?

3. Androgen insensitivity, caused by a mutation on the X chromosome, is inherited as a simple recessive trait, because one copy of the normal androgen receptor allele produces sufficient amounts of androgen receptors. Given this information and your knowledge of the chromosomal basis of inheritance, can androgen insensitivity be inherited, or must it arise as a new mutation each time it occurs? If it can be inherited, would inheritance be through the mother or through the father? Why?

For additional resources, go to www.mybiology.com.

Biotechnology

DNA profiling proved that Earl Ruffin, shown here with three of his grandchildren, was innocent of the rape and assault for which he spent 21 years in prison.

Case Study   Guilty or Innocent?

At about 2 a.m., December 5, 1981, in Norfolk, Virginia, a mother of three was awakened by a stranger in her bedroom. Before attacking her, he warned, "You scream, you are dead." At one point, she caught a glimpse of his face, lit by the street lamp outside her bedroom window. After the assault, he ordered her to take a shower. Despite her desperate fear, she carefully avoided washing away evidence that might be used to find her assailant.

A few weeks later, she encountered a man in an elevator who she believed to be her rapist. The man, Julius Earl Ruffin, 28, was arrested and tried for rape and assault. The victim testified that she recognized Ruffin's face. Further, his blood type, called B-secretor—which only about only 8% of men possess— matched the semen sample taken from the victim on the night of the rape (secretors "secrete" their blood antigens into body fluids, including semen). Despite major inconsistencies between Ruffin's appearance and the victim's initial descriptions of her assailant, and despite testimony by Ruffin's girl- friend and brother that they had been with Ruffin on the night of the assault, Ruffin was convicted and sentenced to life in prison on each of multiple counts. Ruffin's sentence put an innocent man behind bars and left the real rapist to roam the streets and strike again.

Although he was terribly wronged, fate eventually smiled on Ruffin. Remember, the victim had not washed away her rapist's semen when she showered. Although this contributed to Ruffin's initial conviction, it also proved to be essential for his eventual exoneration. A second stroke of good fortune for Ruffin was that Mary Jane Burton was the forensic scientist assigned to his case. Instead of following normal procedures and returning all of the samples to police investigators, who usually destroyed evidence after a conviction, she kept biological mate- rial in her case files. Finally, Ruffin was fortunate that the Innocence Project, founded in 1992 by Barry Scheck and Peter Neufeld of the Benjamin Cardozo School of Law at Yeshiva University (photo at left), applied the power of biotechnology to cases like his.

You've probably already guessed how Ruffin's innocence was proved—by DNA evidence. In this chapter, we investigate biotechnology, which pervades so much of modern life. How do crime scene investigators decide that two DNA samples match? How can biotechnology diagnose inherited diseases? Should biotechnology be used to change the genetic makeup of crops, livestock, or even people? ∎

Peter Neufeld and Barry Scheck, cofounders of the Innocence Project.

12.1 What Is Biotechnology?

In its broadest sense, **biotechnology** is any alteration of organisms, cells, or bio- logical molecules to achieve specific practical goals. By this definition, biotech- nology is not new. For example, people have been using yeast cells to produce bread, beer, and wine for the past 10,000 years. Selectively breeding plants and animals that have desirable traits—crops that bear large fruits, docile cattle that grow rapidly and give lots of milk, cooperative dogs that protect and herd sheep—has an equally long history. For example, 8,000- to 10,000-year-old frag- ments of squash, found in a dry cave in Mexico, have larger seeds and thicker rinds than wild squash, suggesting selective breeding. Similarly, prehistoric art and animal remains indicate that dogs, sheep, goats, pigs, and camels were domesticated and selectively bred beginning at least 10,000 years ago. Selective breeding causes domestic plants and animals to differ genetically from their wild relatives; for instance, the short legs, floppy ears, erect tails, and characteristic barks of beagles are genetically determined and differ enormously from the comparable characteristics of wolves, from which all dogs were derived.

Even today, selective breeding remains an important part of biotechnology. Modern biotechnology, however, frequently uses **genetic engineering,** a term that refers to more direct methods for modifying genetic material. Genetically engineered cells or organisms may have genes deleted, added, or changed. Genetic engineering can be used to learn more about how cells and genes work, to develop better treatments for diseases, to produce valuable biological molecules, and to improve plants and animals for agriculture.

A key tool in genetic engineering is **recombinant DNA,** which is DNA that has been altered to contain genes or parts of genes from different organisms. Large amounts of recombinant DNA can be grown in bacteria, viruses, or yeast, and then transferred into other species. Plants and animals that express DNA that has been modified or derived from other species are called **transgenic,** or **genetically modified, organisms (GMOs).**

Since its development in the 1970s, recombinant DNA technology has grown explosively. Today, researchers in almost every area of biology routinely use recombinant DNA technology in their experiments. In the pharmaceutical industry, genetic engineering has become the preferred way to manufacture many products, including several human hormones, such as insulin, and some vaccines, such as one against hepatitis B.

Modern biotechnology also includes many methods of manipulating DNA, whether or not the DNA is subsequently put into a cell or an organism. For example, determining the nucleotide sequence of specific pieces of DNA is crucial to forensic science and the diagnosis of inherited disorders.

This chapter provides an overview of modern biotechnology. Our principal emphasis is on applications of biotechnology and their impacts on society, but we also briefly describe some of the important methods used in those applications. We organize our discussion around five major themes: (1) recombinant DNA mechanisms found in nature; (2) biotechnology in criminal forensics, principally DNA matching; (3) biotechnology in agriculture, specifically the production of transgenic plants and animals; (4) the Human Genome Project; and (5) biotechnology in medicine, focusing on the diagnosis and treatment of inherited disorders. We conclude the chapter with a discussion of ethical issues surrounding the use of biotechnology.

12.2 How Does DNA Recombine in Nature?

Web Animation Genetic Recombination in Bacteria

Although most people think that a species' genetic makeup does not change, except for the occasional mutation, genetic reality is far more fluid. Many natural processes can transfer DNA from one organism to another, sometimes even to organisms of different species. Recombinant DNA technologies used in the laboratory are often based on these naturally occurring processes.

Sexual Reproduction Recombines DNA

Sexual reproduction literally recombines DNA from two different organisms. As we saw in Chapter 8, homologous chromosomes exchange DNA by crossing over during meiosis I. Thus, each chromosome in a gamete usually contains a mixture of alleles from the two parental chromosomes. In this sense, every egg and every sperm contain recombinant DNA, derived from the organism's two parents. When a sperm fertilizes an egg, the resulting offspring also contains recombinant DNA.

Transformation May Combine DNA from Different Bacterial Species

To some people, DNA recombination within a species during sexual reproduction is "natural" and therefore good, but recombination in the laboratory between different species is "unnatural" and therefore bad. Recombination of DNA between species, however, is not confined to the lab—it also occurs in nature.

Bacteria can undergo several types of recombination that allow gene transfer between species (Fig. 12-1). In a process called **transformation,** for example, bacteria pick up pieces of DNA from the environment. The DNA may be part of the chromosome of another bacterium (Fig. 12-1b), even from another species, or may be in the form of tiny circular DNA molecules called **plasmids** (Fig. 12-1c). A single bacterium may contain dozens or even hundreds of copies of a plasmid. When the bacterium dies, it releases its plasmids into the environment, where they may be taken up by bacteria of the same or different species. In addition, living bacteria can often pass their plasmids di-

▶ Figure 12-1 **Transformation in bacteria** *(a)* In addition to their large circular chromosome, bacteria commonly possess small rings of DNA called plasmids, which often carry additional useful genes. Bacterial transformation occurs when living bacteria take up *(b)* fragments of chromosomes or *(c)* plasmids.

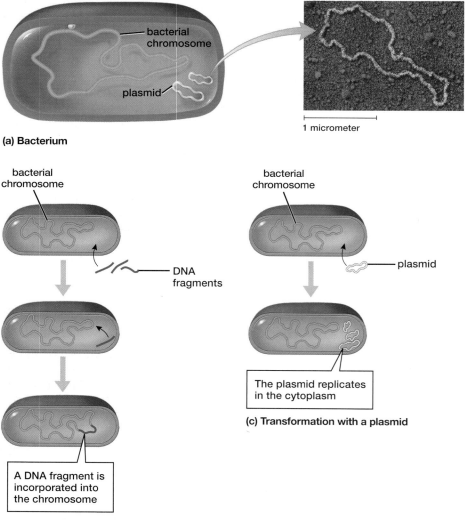

(a) Bacterium

bacterial chromosome

plasmid

1 micrometer

bacterial chromosome

bacterial chromosome

DNA fragments

plasmid

The plasmid replicates in the cytoplasm

(c) Transformation with a plasmid

A DNA fragment is incorporated into the chromosome

(b) Transformation with a DNA fragment

rectly to other living bacteria. Sometimes, plasmids may pass from a bacterium to a yeast cell, thereby moving genes from a prokaryotic cell to a eukaryotic cell!

Viruses May Transfer DNA Between Species

Viruses, which are often little more than genetic material encased in a protein coat, transfer their genetic material to cells that they infect, called *host cells*. Once inside the host cell, the viral genes replicate and use the infected cell's enzymes and ribosomes to synthesize viral proteins. The replicated genes and new viral proteins assemble inside the cell, forming new viruses that are released to infect other cells (**Fig. 12-2**).

Some viruses can transfer genes from one organism to another. In these instances, the virus inserts its DNA into a host cell's chromosome. The viral DNA may remain there for days, months, or even years. Every time the cell divides, it replicates the viral DNA along with its own DNA. When new viruses are finally produced, some of the host's genes may be incorporated into the viral DNA. If such recombinant viruses infect other cells and insert their DNA into the new host cells' chromosomes, pieces of the previous host cell's DNA will also be inserted. Most of the time, viruses move host DNA between different individuals of a single, or at least very closely related, species. However, some viruses infect multiple, very different, host species, and may transfer

◀ Figure 12-2 Viruses may transfer genes between cells

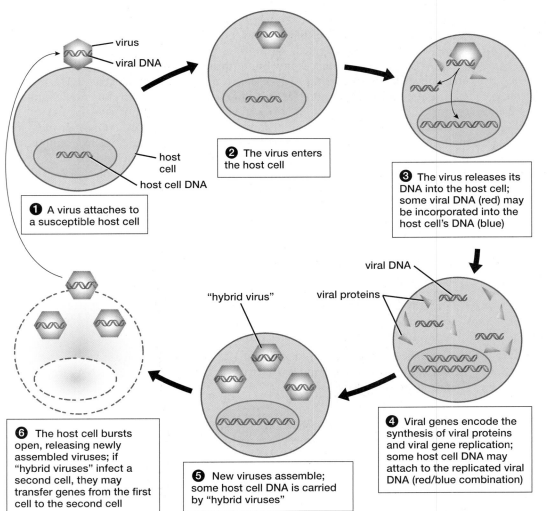

virus
viral DNA

❶ A virus attaches to a susceptible host cell

host cell
host cell DNA

❷ The virus enters the host cell

❸ The virus releases its DNA into the host cell; some viral DNA (red) may be incorporated into the host cell's DNA (blue)

viral DNA
viral proteins

❹ Viral genes encode the synthesis of viral proteins and viral gene replication; some host cell DNA may attach to the replicated viral DNA (red/blue combination)

"hybrid virus"

❺ New viruses assemble; some host cell DNA is carried by "hybrid viruses"

❻ The host cell bursts open, releasing newly assembled viruses; if "hybrid viruses" infect a second cell, they may transfer genes from the first cell to the second cell

genes from one to another. In some instances, such as the influenza and bird-flu viruses, gene transfer between viruses that infect multiple species can produce extremely lethal recombined viruses.

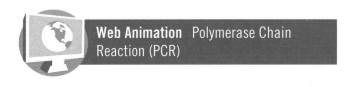

Web Animation Polymerase Chain Reaction (PCR)

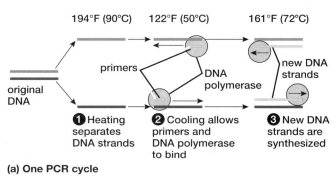

194°F (90°C)   122°F (50°C)   161°F (72°C)

original DNA

primers

DNA polymerase

new DNA strands

❶ Heating separates DNA strands

❷ Cooling allows primers and DNA polymerase to bind

❸ New DNA strands are synthesized

(a) One PCR cycle

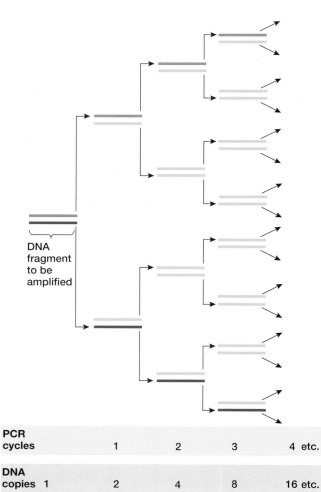

DNA fragment to be amplified

PCR cycles	1	2	3	4 etc.
DNA copies 1	2	4	8	16 etc.

(b) Each PCR cycle doubles the number of copies of the DNA

12.3 How Is Biotechnology Used in Forensic Science?

The applications of biotechnology vary, depending on the goals of those who use it. Forensic scientists need to identify victims and criminals; biotechnology firms need to identify specific genes and insert them into organisms such as bacteria, cattle, or crop plants; and biomedical firms and physicians need to detect defective alleles and, ideally, devise ways to fix them or to insert normally functioning alleles into patients. We begin by describing a few common methods of manipulating DNA, using their application to forensics as a specific example.

The Polymerase Chain Reaction Amplifies DNA

Developed by Kary Mullis of the Cetus Corporation in 1986, the **polymerase chain reaction (PCR)** produces virtually unlimited amounts of selected pieces of DNA. PCR is so crucial to molecular biology that it earned Mullis a share in the Nobel Prize for Chemistry in 1993. Let's see how PCR amplifies a specific piece of DNA.

When we described DNA replication in Chapter 10, we omitted some of its real-life complexity. One of the things we did not discuss is crucial to PCR: by itself, DNA polymerase doesn't know where to start copying a strand of DNA. When a DNA double helix is unwound during DNA replication, enzymes put a little piece of complementary RNA, called a *primer*, on each strand. DNA polymerase recognizes this primed region of DNA as the place to start replicating the rest of the DNA strand.

PCR generally uses artificial primers made of DNA. These DNA primers are manufactured in a DNA synthesizer, a machine that can be programmed to make short pieces of DNA with any desired sequence of nucleotides. PCR needs two DNA primers, one complementary to the beginning of one strand of the DNA segment to be copied and one complementary to the beginning of the other strand. These primers "tell" DNA polymerase where to start copying.

In a small test tube, DNA is mixed with primers, free nucleotides, and a special DNA polymerase, isolated from microbes that live in hot springs (see "Scientific Inquiry: Hot Springs and Hot Science"). PCR consists of the following steps, repeated as many times as necessary to generate enough copies of the DNA (**Fig. 12-3**):

1. The test tube is heated to about 194°F (90°C) (**Fig. 12-3a**, step ❶). High temperatures break the hydrogen bonds between complementary bases, separating the DNA into single strands.

2. The temperature is lowered to about 122°F (50°C), which allows the two primers to form complementary base pairs with the original DNA strands (**Fig. 12-3a**, step ❷).

◀ **Figure 12-3 PCR copies a specific DNA sequence** *(a)* The polymerase chain reaction consists of a cycle of heating, cooling, and warming that is repeated 20 to 30 times. *(b)* Each cycle doubles the amount of target DNA. After just 20 cycles, a million copies of the target DNA have been synthesized. **QUESTION:** Why do you think that the reaction is warmed up to 161°F (72°C) for DNA synthesis [part (a) of the figure]? Hint: Reread "Scientific Inquiry: Hot Springs and Hot Science." Given the normal living conditions of *Thermus aquaticus,* do you think that its DNA polymerase would work most rapidly at 122°F (50°C) or 161°F (72°C)?

scientific inquiry

Hot Springs and Hot Science

At a hot spring, such as those found in Yellowstone National Park, water literally boils out of the ground, gradually cooling as it flows to the nearest stream (**Fig. E12-1**). You might think that such springs, scalding hot and often containing poisonous metals and sulfur compounds, must be lifeless. However, closer examination often reveals a diversity of microorganisms, each adapted to a different temperature zone within the spring. Back in 1966, in a Yellowstone hot spring, Thomas Brock of the University of Wisconsin discovered *Thermus aquaticus*, a bacterium that lives in water as hot as 176°F (80°C).

When Kary Mullis first developed the polymerase chain reaction, he encountered a major technical difficulty. The DNA solution must be heated almost to the boiling point to separate the double helix into single strands,

then cooled so DNA polymerase can synthesize new DNA, and this process must be repeated over and over again. "Ordinary" DNA polymerase, like most proteins, is ruined by high temperatures. Therefore, new DNA polymerase had to be added after every heat cycle, which was expensive and labor-intensive. Enter *Thermus aquaticus*. Like other organisms, it replicates its DNA when it reproduces. But because it lives in hot springs, it has evolved a particularly heat-resistant DNA polymerase. When DNA polymerase from *T. aquaticus* is used in PCR, it needs to be added to the DNA solution only once, at the start of the reaction. Sequencing the human genome, proving Earl Ruffin innocent, and doing reseach in modern molecular genetic research would all be far more difficult and expensive without DNA polymerase from hot spring bacteria.

▲ **Figure E12-1 Thomas Brock surveys Mushroom Spring** The colors in hot springs arise from minerals dissolved in the water and from various types of microbes that live at different temperatures.

3. The temperature is raised to 158° to 161°F (70 to 72°C). DNA polymerase, directed by the primers, uses the free nucleotides to make copies of the DNA segment bounded by the primers (**Fig. 12-3a**, step ❸).

4. This cycle is repeated as many times as desired.

As Figure 12-3b shows, the amount of DNA produced by PCR doubles at each step of the cycle. Therefore, 20 PCR cycles would make about a million copies, and a little over 30 cycles would make a billion copies. Each cycle takes only a few minutes, so PCR can produce billions of copies of a gene or DNA segment in a single afternoon, starting, if necessary, from a single molecule of DNA. The DNA is then available for forensics, sequencing, cloning, making transgenic organisms, or many other purposes.

Differences in Short Tandem Repeats Can Identify Individuals

In many criminal investigations, PCR is used to amplify the DNA so that there is enough to compare the DNA left at a crime scene with the suspect's DNA. How is this comparison done? After years of painstaking work, forensics experts have found that small segments of DNA, called **short tandem repeats (STRs),** can be used to identify people with astonishing accuracy. Think of STRs as very short, stuttering genes (**Fig. 12-4**). Each STR contains a *short* sequence of nucleotides (2 to 5 nucleotides long) that is *repeated* (about 5 to 15 times) in *tandem* (all the repetitions are right after one another). Just like any gene, different people may have different alleles of the STRs. In the case of an STR, each allele is simply a different number of repeats of the same few nucleotides. In 1999, British and American law enforcement agencies agreed to use the same set of 10 to 13 STRs that vary greatly among individuals. A perfect match of 10 STRs in a suspect's DNA and the DNA

Guilty or Innocent?

Continued

In 2002, when investigators located the semen samples from the Earl Ruffin case, they needed to find out if the samples collected from the rape victim in 1981 came from Ruffin. To do this, lab technicians used two techniques that have become commonplace in virtually all DNA labs. First, they amplified the DNA so that they had enough material to analyze. Then, they determined whether the DNA from the semen samples matched Ruffin's DNA.

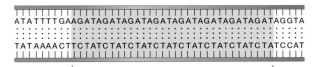

Eight side-by-side (tandem) repeats of the same four-nucleotide sequence

▲ **Figure 12-4 Short tandem repeats** This STR consists of the sequence AGAT, repeated from 7 to 13 times in different individuals.

found at a crime scene means that there is less than one chance in a *trillion* that the two DNA samples came from different people. What's more, the DNA around STRs doesn't degrade very rapidly, so even old DNA samples, such as those in the Ruffin case, usually have intact STRs.

Forensics labs use PCR primers that amplify only the DNA immediately surrounding the STRs. Because STR alleles vary in how many times they repeat, they vary in length: An STR with more repeats has more nucleotides and is longer. Therefore, a forensic lab needs to identify each of the 10 to 13 crucial STRs in a DNA sample and to determine their lengths.

Gel Electrophoresis Separates DNA Segments

Modern forensics labs use sophisticated and expensive machines to determine the number of STR repeats. Most of these machines, however, are based on two methods that are used in molecular biology labs around the world: first, separating the DNA by size, and second, labeling specific DNA segments.

The mixture of DNA pieces is separated by a technique called **gel electrophoresis** (**Fig. 12-5**). First, the mixture of DNA fragments is loaded into shal-

▶ Figure 12-5 Gel electrophoresis is used to separate and identify segments of DNA

❶ DNA samples are pipetted into wells (shallow slots) in the gel. Electrical current is sent through the gel (negative at the end with the wells, positive at the opposite end).

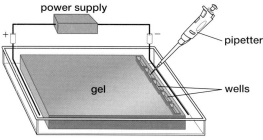

❷ Electrical current moves the DNA segments through the gel. Smaller pieces of DNA move farther toward the positive electrode.

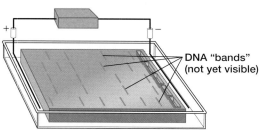

❸ The gel is placed on special nylon "paper." Electrical current drives the DNA out of the gel onto the nylon.

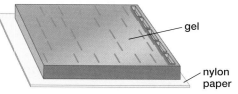

❹ The nylon paper with the DNA bound to it is bathed in a solution of labeled DNA probes (red) that are complementary to specific DNA segments in the original DNA sample.

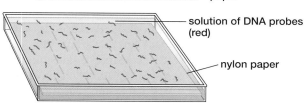

❺ Complementary DNA segments are labeled by the probes (red bands).

low grooves, or wells, in a slab of agarose, a carbohydrate purified from certain types of seaweed (**Fig. 12-5**, step ❶). The agarose forms a *gel*, which is simply a meshwork of fibers with tiny holes of various sizes between the fibers. The gel is put into a chamber with electrodes connected to each end. One electrode is made positive and the other negative. Therefore, a current flows between the electrodes *through the gel*. How does this separate pieces of DNA? Remember, the phosphate groups in the backbones of DNA are negatively charged. When electrical current flows through the gel, the negatively charged DNA fragments move toward the positively charged electrode. Because smaller pieces of DNA slip through the holes in the gel more easily than larger pieces do, they move more rapidly toward the positively charged electrode. Eventually the DNA fragments are separated by size, forming distinct bands on the gel (**Fig. 12-5**, step ❷).

DNA Probes Are Used to Label Specific Nucleotide Sequences

Unfortunately, the DNA bands are invisible, so how can anyone identify which band contains a specific STR? Well, how does *nature* identify sequences of DNA? Right—by base-pairing! Usually, the two strands of the DNA double helix are separated during gel electrophoresis. This allows pieces of synthetic DNA, called **DNA probes**, to base-pair with specific DNA fragments in the sample. DNA probes are short pieces of single-stranded DNA that are complementary to the nucleotide sequence of a given STR (or any other DNA of interest). The DNA probes are labeled, either by radioactivity or by attaching colored molecules to them. Therefore, a given DNA probe will label certain DNA sequences and not others (**Fig. 12-6**).

When the gel is finished running, the technician transfers the single-stranded DNA segments out of the gel and onto a piece of paper made of nylon (**Fig. 12-5**, step ❸). Then, the paper is bathed in a solution containing a specific DNA probe (**Fig. 12-5**, step ❹), which will base-pair with, and therefore bind to, only a specific STR, making this STR visible (**Fig. 12-5**, step ❺).

Every Person Has a Unique DNA Profile

Until the early 1990s, forensic scientists ran DNA samples from a crime scene and from the suspects side by side on a gel, to see which suspect's DNA matched that found at the scene. In modern STR analysis, however, the suspect and crime scene DNA samples can be run on different gels, in different states or countries, and even years apart. Why? As we mentioned earlier, people have different numbers of tandem repeats in their STRs, so that every person on Earth (except for identical twins or triplets) has a unique set of STRs. When DNA samples are run on STR gels, they produce a pattern, called a **DNA profile** (**Fig. 12-7**).

In many states, anyone convicted of certain crimes (for example, assault or burglary) must give a blood sample. Forensic technicians then determine the criminal's DNA profile. This DNA profile is coded (by the number of repeats for all the STR genes) and stored in computer files, usually at the FBI. (On *CSI: Crime Scene Investigation* and other TV crime shows, when you hear the actors refer to "CODIS," that acronym stands for "Combined DNA Index System," a DNA profile database kept on FBI computers.) Because all forensic labs use the same STRs, computers can easily determine if DNA left at a crime scene matches one of the millions of profiles stored in the CODIS database. If there aren't any matches, the crime scene DNA profile will remain on file. Sometimes, years later, a newly convicted criminal's DNA profile will match an archived crime scene profile, and a cold case will be solved.

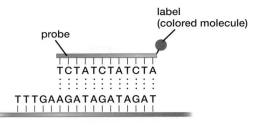

STR #1: The probe base-pairs and binds to the DNA

TCTATCTATCTA

ACTGAATGAATGAATGAATG

STR #2: The probe cannot base-pair with the DNA, so it does not bind

▲ **Figure 12-6 DNA probes identify specific DNA sequences** A short, single-stranded piece of DNA is labeled with a colored molecule (red ball). This labeled DNA will base-pair with a target strand of DNA with a complementary base sequence (top), but not with a noncomplementary strand (bottom).

Guilty or Innocent?

Continued

In the Earl Ruffin case, the DNA profiles of Ruffin and the rapist were very different, conclusively proving that Ruffin did not commit the crime. But could the profile of DNA from the semen help to find the real rapist? See the "Guilty or Innocent? *Revisited*" section at the end of the chapter to see how DNA profiling can solve cold cases.

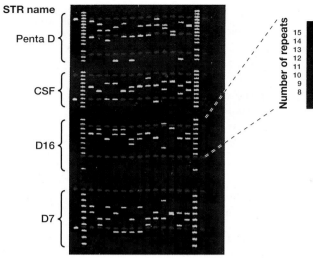

STR name

Penta D

CSF

D16

D7

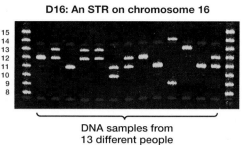

D16: An STR on chromosome 16

Number of repeats

15
14
13
12
11
10
9
8

DNA samples from
13 different people

▲ **Figure 12-7 DNA profiling** The lengths of short tandem repeats of DNA form characteristic patterns on a gel. This gel displays four different STRs (Penta D, CSF, and so on). The evenly spaced yellow bands on the far left and far right sides of the gel are standards that show the number of repeats of the individual STRs. DNA samples from 13 people were run between these standards, resulting in one or two bands in each vertical lane. In the enlargement of the D16 STR, reading from left to right, the first person's DNA has 12 repeats, the second person's has 13 and 12, the third has 11, and so on. Although different people have the same number of repeats in *some* STRs, none has the same number of repeats in *all* the STRs. *(Photo courtesy of Dr. Margaret Kline, National Institute of Standards and Technology.)* **QUESTION:** For any single person, a given STR always has either one or two bands. Why? Further, single bands are always about twice as bright as each band of a pair. For example, in the D16 STR, the single bands of the first and third DNA samples are twice as bright as the pairs of bands of the second, fourth, and fifth samples. Why?

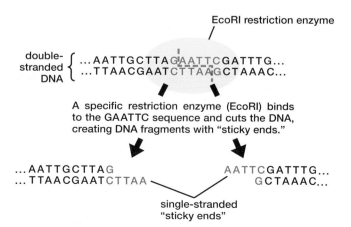

EcoRI restriction enzyme

double-
stranded
DNA { ...AATTGCTTAGAATTCGATTTG...
...TTAACGAATCTTAAGCTAAAC...

A specific restriction enzyme (EcoRI) binds to the GAATTC sequence and cuts the DNA, creating DNA fragments with "sticky ends."

...AATTGCTTAG   AATTCGATTTG...
...TTAACGAATCTTAA   GCTAAAC...

single-stranded
"sticky ends"

▲ **Figure 12-8 Restriction enzymes cut DNA at specific nucleotide sequences**

12.4 How Is Biotechnology Used in Agriculture?

The main goals of agriculture are to grow as much food as possible, as cheaply as possible, with minimal loss from pests such as insects and weeds. Many commercial farmers and seed suppliers have turned to biotechnology to achieve these goals.

Many Crops Are Genetically Modified

According to the U.S. Department of Agriculture, in 2006, about 61% of the corn, 83% of the cotton, and 89% of the soybeans grown in the United States were transgenic; that is, they contained genes from other species. Crops are most commonly modified to improve their resistance to herbicides and insects.

Many herbicides kill plants by inhibiting an enzyme that is used by plants, but not animals, to synthesize essential amino acids. Without these amino acids, the plants cannot synthesize proteins, and they die. Many herbicide-resistant transgenic crops have been given a bacterial gene encoding an enzyme that functions even in the presence of these herbicides, so the plants continue to synthesize normal amounts of amino acids and proteins. Herbicide-resistant crops allow farmers to use herbicides to kill weeds without harming their crops. Less competition from weeds means more water, nutrients, and light for the crops, and hence larger harvests.

The insect resistance of many crops has been enhanced by giving them a gene, called *Bt*, from the bacterium *Bacillus thuringiensis*. The protein encoded by the *Bt* gene damages the digestive tract of insects, but not mammals. Transgenic *Bt* crops therefore suffer far less damage from insects, and farmers can apply much less pesticide to their fields.

How would a seed company make a transgenic plant? Let's examine the process, using insect-resistant *Bt* plants as an example.

The Desired Gene Is Cloned

Cloning a gene usually involves two tasks: (1) obtaining the gene and (2) inserting it into a plasmid so that huge numbers of copies of the gene can be made.

There are two common ways of obtaining a gene. One is to isolate the gene from the organism that makes it. For a long time, this was the only practical method. More recently, biotechnologists have been able to synthesize genes—or modified versions of them—in the lab, using PCR or DNA synthesizers.

Once the *Bt* gene has been obtained, why insert it into a plasmid? Recall that plasmids, which are small circles of DNA, can be taken up by bacteria (see Fig. 12-1), and are replicated when the bacteria multiply. Therefore, once the *Bt* gene has been inserted into a plasmid, producing huge numbers of copies of the gene is as simple as raising lots of bacteria.

Restriction Enzymes Cut DNA at Specific Nucleotide Sequences

Genes are inserted into plasmids through the action of **restriction enzymes** isolated from bacteria (**Fig. 12-8**). Each restriction enzyme cuts DNA at a specific nucleotide sequence. Many restriction enzymes make a staggered cut, snipping the

DNA in a different location on each of the two strands, so that single-stranded sections hang off the ends of the DNA. These single-stranded regions are commonly called "sticky ends," because they can base-pair with, and thus stick to, other single-stranded pieces of DNA with complementary bases.

Cutting Two Pieces of DNA with the Same Restriction Enzyme Allows the Pieces to Be Joined Together

The *Bt* gene is inserted into a plasmid by cutting the DNA on either side of the *Bt* gene and splitting open the circle of the plasmid with the same restriction enzyme (**Fig. 12-9**, step ❶). As a result, the ends of the *Bt* gene and the opened-up plasmid both have complementary bases in their "sticky ends" and can base-pair with each other. When the cut *Bt* genes and plasmids are mixed, some of the *Bt* genes will be temporarily inserted between the ends of the plasmids. Adding DNA ligase (see pp. 164–166) permanently bonds the *Bt* genes into the plasmid (**Fig. 12-9**, step ❷).

Plasmid-Transformed Bacteria Insert the *Bt* Gene into a Plant

Certain bacteria can enter the cells of specific types of plants. These bacteria are transformed with the recombinant plasmids (**Fig. 12-9**, step ❸). When the transformed bacteria enter a plant cell, the plasmids insert their DNA, including the *Bt* gene, into the plant cell's chromosomes (**Fig. 12-9**, step ❹). Thereafter, anytime that plant cell divides, all of its daughter cells inherit the *Bt* gene. Hormones stimulate the transgenic plant cells to divide and differentiate into entire plants. These plants are bred to one another, or to other plants of the same species, to create commercially valuable plants that resist insect attack (**Fig. 12-10**).

Genetically Modified Plants May Produce Medicines

Similar techniques can be used to insert medically useful genes into plants, producing medicines down on the "pharm." For example, it might be possible to engineer plants to produce human antibodies that would combat various diseases. As you will learn in Chapter 22, when bacteria or viruses invade your body, it takes several days for your immune system to respond and produce enough antibodies to overcome the infection. Meanwhile, you feel terrible, and might even die. A direct injection of large quantities of the right antibodies might cure the disease much more rapidly. Plant-derived antibodies against bacteria that cause tooth decay and non-Hodgkins lymphoma (a cancer of the lymphatic system) are in clinical trials. Ideally, such "plantibodies" could be produced very cheaply, making such therapies available to rich and poor alike.

Genetically Modified Animals May Be Useful in Agriculture, Industry, and Medicine

Making transgenic animals usually involves injecting the desired DNA into a fertilized egg. The egg is allowed to divide a few times in culture before being implanted into a surrogate mother. If the offspring are healthy and express the foreign gene, they are then bred together to produce homozygous transgenic organisms. So far, it has proven difficult to produce commercially valuable transgenic livestock, but several companies are working on it.

One example is Nexia Biotechnologies, which has engineered goats to carry the genes for spider silk, and to secrete silk protein into their milk. The resulting BioSteel® can be spun into silk that is five times stronger than steel and twice as strong as Kevlar, a fiber used in bulletproof vests. Companies such as Aqua Bounty Technologies have developed salmon and trout with modified or added growth-hormone genes. These fish grow much faster than wild-type fish, and do not display any obvious ill effects. However, whether fish farms should be allowed to grow these fish remains controversial, principally because of concerns about whether they would outcompete native fish if they escaped into the wild. To try to address these concerns, Aqua Bounty produces sterile fish that cannot breed.

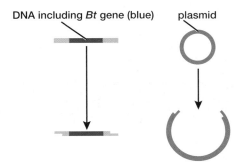

DNA including *Bt* gene (blue)   plasmid

❶ The DNA containing the *Bt* gene and the plasmid are cut with the same restriction enzyme.

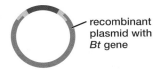

recombinant plasmid with *Bt* gene

❷ *Bt* genes and plasmids, both with the same complementary sticky ends, are mixed together; DNA ligase bonds the *Bt* genes into the plasmids.

bacterium
bacterial chromosome
recombinant plasmids

❸ Bacteria are transformed with the recombinant plasmids.

plant chromosome   plant cell

Bt gene

❹ Transgenic bacteria enter the plant cells, and *Bt* genes are inserted into the chromosomes of the plant cells.

▲ Figure 12-9 Using plasmids to insert DNA into a plant cell

▲ Figure 12-10 *Bt* **plants resist insect attack** Transgenic cotton plants expressing the *Bt* gene (right) resist attack by bollworms, which eat cotton seeds. The transgenic plants therefore produce far more cotton than the nontransgenic plants (left).

Web Animation Human Genome Sequencing

Because medicines are generally much more valuable than meat, many researchers are developing animals that will produce medicines. For example, there are transgenic livestock whose milk contains alpha-1-antitrypsin (a protein that may prove valuable in treating cystic fibrosis); erythropoietin (a hormone that stimulates red blood cell production), and clotting factors (for treating hemophilia).

12.5 How Is Biotechnology Used to Learn About the Human Genome?

To begin to understand how our genes influence our lives, the Human Genome Project was launched in 1990, to determine the nucleotide sequence of all of the DNA in our entire set of genes, called the human **genome.** By 2003, this joint project of molecular biologists in several countries had sequenced the human genome with an accuracy of about 99.99%. To many people's surprise, the human genome contains only about 21,000 genes, comprising approximately 2% of the DNA. Some of the other 98% consists of promoters and regions that regulate how often individual genes are transcribed, but scientists still do not understand the function of much of our DNA.

What good is it to sequence the human genome? First, many genes were discovered whose functions are completely unknown. The genetic code allows biologists to predict the amino acid sequences of the proteins these previously unknown genes encode. Comparing these proteins to already well-understood proteins will enable us to find out what many of these genes do.

Second, the Human Genome Project has had an enormous impact on medical research. In 1990, fewer than 100 genes known to be associated with human diseases had been discovered. By 2006, this number had jumped to more than 1,800, mostly as a result of the Human Genome Project.

Third, there is no single "human genome." Your DNA is mostly identical to everyone else's, but each of us also carries our own unique set of alleles. Some of those alleles may cause diseases such as Marfan syndrome, sickle-cell anemia, or cystic fibrosis, or may make it more likely that people will develop conditions such as alcoholism, schizophrenia, heart disease, or Alzheimer's disease. One major impact of the Human Genome Project will be to help diagnose genetic disorders or predispositions and, we hope, to devise treatments or even cures for them.

Fourth, the Human Genome Project, along with companion projects that have sequenced the genomes of organisms as diverse as bacteria, mice, and chimpanzees, help us to appreciate our place in the evolution of life on Earth. For example, the DNA of humans and chimps differs by only about 1.2%. Comparing the similarities and differences may help biologists to understand what genetic differences help to make us human, and why we are susceptible to certain diseases that chimps are not.

12.6 How Is Biotechnology Used for Medical Diagnosis and Treatment?

For more than a decade, biotechnology has been routinely used to diagnose some inherited disorders. Potential parents can learn if they are carriers of a genetic disorder, and often an embryo can be diagnosed early in a pregnancy (see "Health Watch: Prenatal Genetic Screening" on p. 205). More recently, medical researchers have begun to use biotechnology to attempt to cure, or at least treat, genetic diseases.

DNA Technology Can Be Used to Diagnose Inherited Disorders

People inherit a genetic disease because they inherit one or more dysfunctional alleles. Defective alleles have different nucleotide sequences than functional alleles. Two methods are currently used to find out if a person carries a normal allele or a malfunctioning allele.

Restriction Enzymes May Cut Different Alleles at Different Locations

Remember that any given restriction enzyme cuts DNA only at a specific nucleotide sequence. Therefore, a defective allele may be cut by a particular restriction enzyme, while normal alleles are left intact, or vice versa. This difference can be used to diagnose sickle-cell anemia.

Sickle-cell anemia is caused by a point mutation in which adenine replaces thymine near the beginning of the globin gene (see Table 11-3 on p. 180). A restriction enzyme called MstII cuts DNA just "ahead" of the globin gene locus (cut #1 in **Fig. 12-11a,b**). It also makes a cut in about the middle of both the normal and sickle-cell alleles (cut #3). Crucially, MstII also makes a cut near the beginning of the normal globin allele (cut #2), but, because of the point mutation, cannot make this cut in the sickle-cell allele (Fig. 12-11b). How can this difference be used to diagnose sickle-cell anemia? A DNA probe is synthesized that is complementary to the part of the globin allele spanning the unique cut site. When sickle-cell DNA is cut with MstII and run on a gel, this probe labels a single band near the top of the gel, consisting of very large pieces of DNA (**Fig. 12-11c**). When normal DNA is cut with MstII, the probe labels two bands, one with small pieces of DNA (near the bottom of the gel) and one with pieces that are not quite as large as the sickle-cell pieces.

The genotypes of parents, children, and fetuses can be determined by this simple test. Someone who is homozygous for the normal globin allele will have two bands. Someone who is homozygous for the sickle-cell allele will have one band. A heterozygote will have three bands (see Fig. 12-11c).

Different Alleles Bind to Different DNA Probes

The Chapter 11 case study introduced you to cystic fibrosis, a disease caused by a defect in a transport protein, called CFTR, that normally helps to move chloride ions across cells. There are more than 1,000 different *CFTR* alleles, all at the same locus, each encoding a slightly different, defective CFTR protein.

How can you diagnose a disorder with a thousand different alleles? Most of these are extremely rare; about 90% of the cases of cystic fibrosis are produced by only 32 defective alleles. Although 32 alleles is a lot, each of these defective alleles has a different nucleotide sequence. Therefore, one strand of each allele will form perfect base-pairs only with its own complementary strand. Several companies now produce cystic fibrosis *arrays,* which are pieces of specialized filter paper to which single-stranded DNA probes are bound. Each probe is complementary to a different *CFTR* allele (**Fig. 12-12a**). A person's DNA is tested for cystic fibrosis by cutting it into small pieces, separating them into single strands, and labeling the strands with a colored molecule (**Fig. 12-12b**). The array is then bathed in the resulting solution of labeled DNA fragments. Only a perfect complementary strand of the person's DNA will bind to any given probe on the array, thereby showing which *CFTR* alleles the person possesses (**Fig. 12-12c**).

▶ **Figure 12-11 Diagnosing sickle-cell anemia with restriction enzymes** The globin gene locus is shown in red; adjacent DNA on the chromosome is shown in yellow. *(a)* The restriction enzyme MstII cuts the normal globin allele somewhat "ahead" of the globin locus (cut site #1), near the beginning of the allele (cut site #2), and about in the middle of the allele (cut site #3). A DNA probe (blue) is synthesized that is complementary to the DNA on both sides of cut site #2. This probe can therefore label one large and one small piece of DNA from the normal allele. *(b)* MstII cuts the sickle-cell allele only at cut sites #1 and #3. The DNA probe will label only one, very large piece of DNA from the sickle-cell allele. *(c)* DNA is extracted from thousands of cells (to provide enough DNA to see on a gel) and cut with MstII. The cut DNA is separated on the gel and labeled with the DNA probe. The large single pieces of DNA from sickle-cell alleles run close to the beginning (top) of the gel, while the smaller pieces from normal alleles run farther into the gel (lower down). The pattern of DNA bands shows the genotype of the person from whom the DNA sample was obtained. **QUESTION:** Why do heterozygotes have three bands? Would the bands be the same intensity (brightness) as the bands from homozygotes?

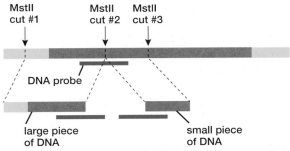

(a) MstII cuts the normal globin allele into two pieces that can be labeled by a probe

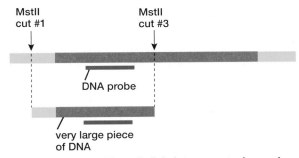

(b) MstII cuts the sickle-cell allele into one very large piece that can be labeled by the same probe

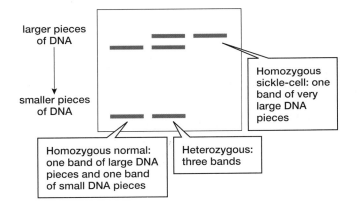

(a) Linear array of probes for cystic fibrosis

DNA probe for normal *CFTR* allele

DNA probes for 10 different mutant *CFTR* alleles

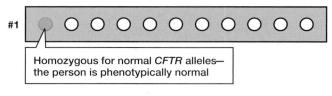

colored molecule

piece of patient's DNA

ATCATCTTTGGTG

(b) *CFTR* allele labeled with a colored molecule

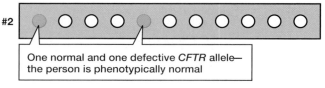

#1

Homozygous for normal *CFTR* alleles— the person is phenotypically normal

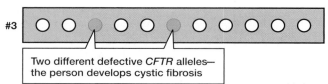

#2

One normal and one defective *CFTR* allele— the person is phenotypically normal

#3

Two different defective *CFTR* alleles— the person develops cystic fibrosis

(c) Linear arrays with labeled DNA samples from three different people

▲ **Figure 12-12 A cystic fibrosis diagnostic array (a)** A "linear array" for cystic fibrosis consists of special paper to which DNA probes complementary to the normal *CFTR* allele (far left spot) and several of the most common defective *CFTR* alleles (the other 10 spots) are attached. **(b)** A patient's *CFTR* alleles are isolated and labeled with colored molecules. **(c)** The array is bathed in a solution of the patient's labeled DNA. Different spots on the array are labeled, depending on which *CFTR* alleles the patient possesses.

Although not yet practical for routine medical use, an expanded version of this type of DNA analysis may one day offer customized medical care. Different people have different alleles of hundreds of genes; these differences may cause them to be more or less susceptible to many diseases, or to respond more or less well to various treatments. Someday, physicians might be able to use an array containing hundreds, even thousands, of probes of disease-related alleles, to determine which susceptibility alleles each patient carries and tailor medical care accordingly. If this sounds too much like science fiction, consider that arrays containing probes for thousands of human genes have already been manufactured (**Fig. 12-13**). Several companies produce small arrays tailored to investigate gene activity in specific diseases, such as breast cancer. These have not yet revolutionized medical care, but some hospitals use them in an attempt to provide patients with treatments that are most likely to succeed with their specific types of breast cancer.

DNA Technology Can Be Used to Treat Disease

DNA technology can be used to treat disease as well as diagnose it. Thanks to recombinant DNA technology, several medically important proteins are now routinely made in bacteria. The first human protein made by recombinant DNA technology was insulin. Prior to 1982, when recombinant human insulin was first licensed for use, the insulin needed by diabetics was extracted from the pancreases of cattle or pigs slaughtered for meat. Although the insulin from these animals is very similar to human insulin, the slight differences cause an allergic reaction in about 5% of diabetics. Recombinant human insulin does not cause allergic reactions.

Some of the types of human proteins produced by recombinant DNA technology are described in **Table 12-1**. These proteins may prevent or cure a variety of diseases, but they cannot cure inherited disorders. They merely treat the symptoms. They are also easy to use, because they only need to be injected into the bloodstream, and then circulate throughout the body, causing appropriate responses in the appropriate cells. What about diseases, such as cystic fibrosis, in which the functional protein must be delivered *inside* cells? And what about actually *curing* inherited disorders? Biotechnology offers great promise here too, although progress has been painfully slow so far. Let's look at two specific examples of how biotechnology may treat, or even cure, inherited diseases.

Using Biotechnology to Treat Cystic Fibrosis

Cystic fibrosis causes devastating effects in the lungs, where the lack of chloride transport causes the normally thin, watery fluid lining the airways to become thick and clogged with mucus. Several research groups are developing methods to deliver the normal *CFTR* allele to the cells of the lungs, and to get them to synthesize functional proteins and insert these proteins into their plasma membranes.

The researchers first disable a suitable virus, so that the treatment itself doesn't cause a disease. Cold viruses are often used, because they normally infect cells of the

Table 12-1	Examples of Medical Products Produced by Recombinant DNA Methods

Type of Product	Purpose	Product	Typical Method of Genetic Engineering
Human hormones	Used in treatment of diabetes, growth deficiency	Humulin™ (human insulin)	Human gene inserted into bacteria
Human cytokines (regulators of immune system function)	Used in bone marrow transplants and to treat cancers and viral infections, including hepatitis and genital warts	Leukine™ (granulocyte-macrophage colony stimulating factor)	Human gene inserted into yeast
Antibodies (immune system proteins)	Used to fight infections, cancers, diabetes, organ rejection, and multiple sclerosis	Herceptin™ (antibodies to a protein expressed in some breast cancer cells)	Recombinant antibody genes inserted into cultured hamster cell line
Viral proteins	Used to generate vaccines against viral diseases and for diagnosing viral infections	Energiz-B™ (hepatitis B vaccine)	Viral gene inserted into yeast

respiratory tract. The DNA of the normal *CFTR* allele is then inserted into the DNA of the virus. The recombinant viruses are sprayed into the patient's nose. The viruses enter cells of the lungs and release the normal *CFTR* allele into the cells. The cells then manufacture CFTR proteins under the instructions of the normal allele, insert them into their plasma membranes, and transport chloride into the fluid lining the lungs.

The clinical trials under way for such treatments have been reasonably successful, but for only a few weeks. The patients' immune systems probably attack the viruses and eliminate them—and the helpful genes they carry. Because lung cells are continually replaced over time, a single dose "wears off" as the modified cells die. Researchers are trying both to increase the expression of the normal *CFTR* alleles in the viruses and to extend the effective lifetime of a single treatment.

Using Biotechnology to Cure Severe Combined Immune Deficiency

Like the cells of the lung, the vast majority of cells in the human body eventually die and are replaced by new cells. In many cases, the new cells come from special populations of cells called **stem cells.** As we noted in Chapter 8, when stem cells divide, they produce daughter cells that can differentiate into several different types of mature cells, perhaps even any cell type in the entire body. We will discuss this potential of stem cells in greater detail in Chapter 25. For now, we look at a more limited role for stem cells: producing just one or two types of cells.

All the cells of the immune system (mostly white blood cells) originate in the bone marrow. As mature cells die, they are replaced by new cells that arise from division of stem cells in the bone marrow. Severe combined immune deficiency (SCID) is a rare inherited disorder in which a child fails to develop an immune system. About 1 in 80,000 children is born with some form of SCID. Infections that would be trivial in a normal child become life-threatening in a child with SCID. In some cases, if the child has an unaffected relative with a similar genetic makeup, a bone marrow transplant from the healthy relative can give the child normal stem cells, so that he or she can develop a functioning immune system. Most SCID victims, however, die before their first birthday.

Although there are several forms of SCID, most are recessive, single-gene defects. In some cases, children are homozygous recessive for a defective allele that normally codes for an enzyme called adenosine deaminase. In 1990, the first test of human gene therapy was performed on such a SCID patient, 4-year-old Ashanti DeSilva. Some of her white blood cells were removed, genetically altered with a virus containing a functional version of her defective allele, and then returned to her bloodstream. Now an adult, Ashanti is healthy, with a reasonably functional immune system. However, as the altered white blood cells die, they must be replaced with new ones; therefore, Ashanti needs repeated treatments. She is also given regular injections of a form of adenosine deaminase. Nevertheless, Ashanti receives only a 4-year-old's dose of adenosine deaminase, so the gene therapy, although not perfect, is certainly making a major difference.

In 2005, Italian researchers completely cured Ashanti's type of SCID in six children. Instead of inserting a normal copy of the adenosine deaminase gene into mature white blood cells, the Italian team inserted the gene into stem cells. Because the "cured" stem cells should continue to multiply and churn out new white blood cells, these children will probably have functioning immune systems for the rest of their lives. (In 1990, when Ashanti received her pioneering treatments, stem cell research was just in its infancy. At that time, it was not possible to isolate her stem cells and correct their adenosine deaminase genes.)

12.7 What Are Some of the Major Ethical Issues of Modern Biotechnology?

Modern biotechnology offers the promise—some would say the threat—of greatly changing our lives and the lives of many other organisms on Earth. As Spider-Man noted, "With great power comes great responsibility." Can humanity

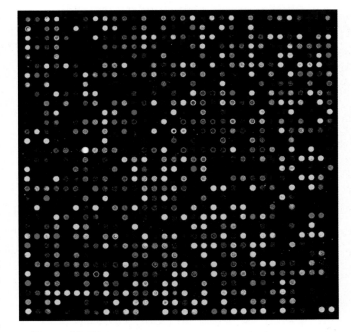

▲ **Figure 12-13 A human DNA microarray** Each spot contains a DNA probe for a specific human gene. In most research applications, messenger RNA is isolated from a subject (for example, from a human cancer), and labeled with a fluorescent dye. The mRNA is then poured onto the array, and each base pairs with its complementary template DNA probe. Genes that are particularly active in the cancer will "light up" the corresponding DNA probe.

handle the responsibility of biotechnology? Here we will explore two controversies: the use of genetically modified organisms (GMOs) in agriculture and the prospect of genetically modifying human beings.

Should Genetically Modified Organisms Be Permitted in Agriculture?

Transgenic crops offer clear advantages to farmers. Herbicide-resistant crops allow farmers to rid their fields of weeds, which may reduce harvests by 10% or more, through the use of powerful, nonselective herbicides, at virtually any stage of crop growth. Insect-resistant crops decrease the need to apply synthetic pesticides, saving the cost of the pesticides, tractor fuel, and labor. Transgenic crops also have the potential to be more nutritious than "standard" crops (see "Health Watch: Golden Rice"). Nevertheless, many people strenuously object to transgenic crops or livestock, fearing that GMOs may be hazardous to human health or dangerous to the environment.

Are Foods from GMOs Dangerous to Eat?

In most cases, there is no reason to think that GMO foods are dangerous to eat. For example, the Bt protein is not toxic to mammals, and should not prove a danger to human health. If growth-enhanced livestock are ever developed, they will simply have more meat, composed of exactly the same proteins that exist in nontransgenic animals.

On the other hand, some people might be allergic to genetically modified plants. In the 1990s, a gene from Brazil nuts was inserted into soybeans in an attempt to improve the balance of amino acids in soybean proteins. It was soon discovered that people allergic to Brazil nuts would probably also be allergic to the transgenic soybeans. Needless to say, these transgenic plants never made it to the farm. The U.S. Food and Drug Administration now monitors all new transgenic crop plants for allergenic potential. Interestingly, several groups of researchers are using genetic engineering to try to make nonallergenic peanuts (for people who are allergic to normal peanuts) by modifying or removing the genes that encode the most allergenic peanut proteins.

In 2003, the U.S. Society of Toxicology studied the risks of genetically modified plants and concluded that the current transgenic plants pose no significant dangers to human health. The society also recognized that past safety does not guarantee future safety, and recommended continued testing and evaluation of all new genetically modified plants.

Are GMOs Hazardous to the Environment?

The environmental effects of GMOs are more problematic. One clear benefit of *Bt* crops is that farmers can apply less pesticide to their fields. A study of Chinese farmers in 2002 and 2003 found that planting *Bt* rice reduced pesticide use by 80%, saving the farmers money, reducing the risk of unintentionally killing beneficial insects or other animals, and reducing the incidence of pesticide poisoning of the farmers themselves. A 10-year study in Arizona showed that *Bt* cotton allowed farmers to use less pesticide while obtaining the same yields of cotton.

On the other hand, *Bt* or herbicide-resistance genes might spread outside a farmer's fields. Because these genes are incorporated into the genome of the transgenic crop, the genes will be in its pollen, too. A farmer cannot control where pollen from a transgenic crop will go. In 2006, researchers at the U.S. Environmental Protection Agency identified "escaped" herbicide-resistant grasses more than 2 miles away from a test plot in Oregon. Based on genetic analyses, some of the herbicide resistance genes escaped in pollen (most grasses are wind pollinated) and some escaped in seeds (most grasses have very lightweight seeds).

Does this matter? Many crops, including corn, sunflowers, and, in Eastern Europe and the Middle East, wheat, barley and oats, have wild relatives living nearby. Suppose these wild plants interbred with transgenic crops and became resistant to herbicides or pests. Would they become significant weed problems?

health watch

Golden Rice

Rice is the principal food for about two-thirds of the people on Earth (Fig. E12-2). Rice provides carbohydrates and some protein, but is a poor source of many vitamins, including vitamin A. Unless people eat enough fruits and vegetables, they often lack sufficient vitamin A, and may suffer from poor vision, immune system defects, and damage to the respiratory, digestive, and urinary tracts. According to the World Health Organization, more than 100 million children suffer from vitamin A deficiency; as a result each year 250,000 to 500,000 children become blind, principally in Asia, Africa, and Latin America. Vitamin A deficiency typically strikes the poor, because rice may be all they can afford to eat. In 1999 biotechnology provided a possible remedy: rice genetically engineered to contain beta-carotene, a pigment that makes daffodils bright yellow, and that the human body easily converts into vitamin A.

▲ **Figure E12-2 A field of dreams?** For hundreds of millions of people, rice provides the major source of calories, but not enough vitamins and minerals. Can biotechnology improve the quality of rice, and hence the quality of life for these people?

Creating a rice strain with high levels of beta-carotene wasn't simple. However, funding from the Rockefeller Institute, the European Community Biotech Program, and the Swiss Federal Office for Education and Science enabled molecular biologists Ingo Potrykus and Peter Beyer to tackle the task. They inserted three genes into the rice genome, two from daffodils and one from a bacterium. As a result, "Golden Rice" grains synthesize beta-carotene (Fig. E12-3, upper right).

The trouble was, the original Golden Rice didn't make enough beta-carotene, so people would have had to eat enormous amounts to get enough vitamin A. The Golden Rice community didn't give up. Researchers at the biotech company Syngenta set about increasing the beta-carotene levels. It turns out that daffodils aren't the best source for genes that direct beta-carotene synthesis. Golden Rice 2, with genes from corn, produces over 20 times more beta-carotene than the original Golden Rice (compare the rice in the upper right and left-hand sections of Fig. E12-3). About 3 cups of cooked Golden Rice 2 should provide enough beta-carotene to equal the full recommended daily amount of vitamin A.

In addition, Syngenta and multiple patent holders have given the technology—free—to research centers in the Philippines, India, China, and Vietnam, with the hope that they will modify native rice varieties for local use. Further, any individual farmer who produces less than about $10,000 worth of Golden Rice doesn't have to pay any fees to Syngenta or the other patent holders. Finally, Syngenta has donated Golden Rice 2 to the Humanitarian Rice Board for experiments and planting in Southeast Asia.

Is Golden Rice 2 the best way, or the only way, to solve the problems of malnutrition in poor

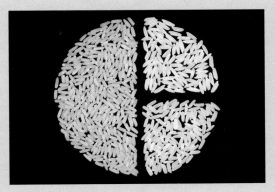

▲ **Figure E12-3 Golden Rice** Conventional milled rice is white or very pale tan (lower right). The original Golden Rice (upper right) was pale golden-yellow because of its increased beta-carotene content. Second-generation Golden Rice 2 (left) is much deeper yellow, because it contains about 20 times more beta-carotene than original Golden Rice.

people? Perhaps not. For one thing, many poor people's diets are deficient in many nutrients, not just vitamin A. To help solve that problem, the Bill and Melinda Gates Foundation is funding research by Peter Beyer to increase the levels of vitamin E, iron, and zinc in rice. Further, not all poor people eat mostly rice. In parts of Africa, sweet potatoes are the main source of starches. Eating orange, instead of white, sweet potatoes, has dramatically increased vitamin A intake for many of these people. Finally, in many parts of the world, governments and humanitarian organizations have started vitamin A supplementation programs. In some parts of Africa and Asia, as many as 80% of the children receive large doses of vitamin A a few times when they are very young. Some day, the combination of these efforts may result in a world in which no children suffer blindness from the lack of a simple nutrient in their diets.

Would they displace other plants in the wild, because they would be less likely to be eaten by insects? Even if transgenic crops have no close relatives in the wild, bacteria and viruses can carry genes from one plant to another, even between unrelated species. Could viruses spread unwanted genes into wild plant populations? No one really knows the answers to these questions.

In 2002, a committee of the U.S. National Academy of Sciences pointed out that crops modified by both traditional methods and recombinant DNA technologies may cause major changes in the environment. The committee also found that the United States does not have an adequate system for monitoring changes

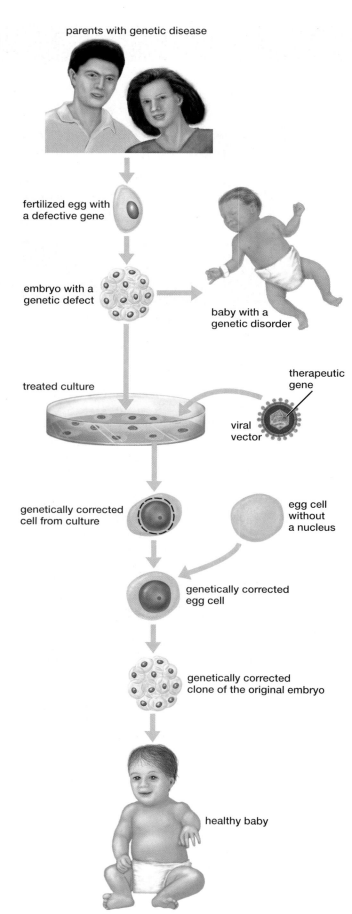

parents with genetic disease

fertilized egg with
a defective gene

embryo with a
genetic defect

baby with a
genetic disorder

treated culture

therapeutic
gene

viral
vector

genetically corrected
cell from culture

egg cell
without
a nucleus

genetically corrected
egg cell

genetically corrected
clone of the original embryo

healthy baby

in ecosystems that might be caused by transgenic crops. It recommended more thorough screening of transgenic plants before they are used commercially, and sustained ecological monitoring after commercialization.

What about transgenic animals? Unlike pollen, most domesticated animals, such as cattle or sheep, are relatively immobile. Further, most have few wild relatives, so the dangers to natural ecosystems appear minimal. However, some transgenic animals, especially fish, have the potential to pose more significant threats. If they escaped, they might disperse rapidly and be nearly impossible to recapture. If they were more aggressive, grew faster, or matured faster than wild fish, they might rapidly replace native populations. This is why Aqua Bounty plans to market only sterile transgenic salmon and trout, so that any escapees would die without reproducing and thus have minimal impact on natural ecosystems.

Should the Genomes of People Be Changed by Biotechnology?

Many of the ethical implications of human applications of biotechnology are fundamentally the same as those associated with other medical procedures. For example, long before biotechnology enabled prenatal testing for sickle-cell anemia, trisomy 21 (Down syndrome) could be diagnosed in embryos by simply counting the chromosomes in cells taken from the amniotic fluid (see Health Watch: "Prenatal Genetic Screening"). Whether parents should use such information as a basis for abortion or to prepare to care for the affected child generates enormous controversy. Other ethical concerns have arisen purely as a result of advances in biotechnology. For instance, should people be allowed to select, or even change, the genomes of their offspring?

On July 4, 1994, a girl in Colorado was born with Fanconi anemia, a genetic disorder that is fatal without a bone marrow transplant. Her parents wanted another child—a very special child. They wanted one without Fanconi anemia, of course, but they also wanted a child who could serve as a donor for their daughter. They went to Yury Verlinsky of the Reproductive Genetics Institute for help. Verlinsky used the parents' sperm and eggs to create dozens of embryos in culture. The embryos were then tested both for the genetic defect and for tissue compatibility with the couple's daughter. Verlinsky chose an embryo with the desired genotype and implanted it into the mother's uterus. Nine months later, a son was born. Blood from his umbilical cord provided cells to transplant into his sister's bone marrow. Today, the girl's bone marrow failure has been cured, although she will always have anemia and many accompanying symptoms. Was this an appropriate use of genetic screening? Should dozens of embryos be created, knowing that the vast majority will be discarded? Is this ethical if it is the only way to save the life of another child?

Today's technology allows physicians only to select among existing embryos, not to change their genomes. But technologies do exist to alter the genomes of, for example, bone marrow stem cells to cure SCID. What if biotechnology could change the genes of fertilized eggs (**Fig. 12-14**)? For example, suppose it were possible to insert alleles encoding functional CFTR proteins into human eggs, thereby preventing cystic fibrosis. Would this be an ethical change to the human genome? What about modifying the genome to make bigger and stronger football players? If and when the technology is developed to cure diseases, it will be difficult to prevent it from being used for nonmedical purposes. Who will determine what is an appropriate use and what is a trivial vanity?

◀ **Figure 12-14 Human cloning technology might allow permanent correction of genetic defects** In this process, human embryos would be derived from eggs fertilized in culture dishes using sperm and eggs from a man and woman, one or both of whom have a genetic disorder. When an embryo containing a defective allele grows into a small cluster of cells, a single cell would be removed from the embryo and a normal allele inserted into its DNA. The repaired nucleus would be injected into another egg (taken from the same woman) whose nucleus had been removed. The repaired egg cell would then be implanted in the woman's uterus for normal development.

health watch

Prenatal Genetic Screening

Prenatal (before-birth) diagnosis of many genetic disorders, including cystic fibrosis, sickle-cell anemia, and Down syndrome, requires samples of fetal cells (cells from the fetus) or chemicals produced by the developing fetus. Two main techniques are used to obtain these samples—*amniocentesis* and *chorionic villus sampling*. New techniques for isolating fetal cells from the mother's blood are currently under development.

Amniocentesis

A human fetus is surrounded by a watery liquid, which is enclosed in a waterproof membrane called the amnion. As the fetus grows, it sheds some cells into the surrounding fluid. When a fetus is 16 weeks or older, this amniotic fluid can be collected by a procedure called amniocentesis. A physician determines the position of the fetus by ultrasound scanning, inserts a sterilized needle through the abdominal wall, the uterus, and the amnion, and withdraws 10 to 20 milliliters of fluid (**Fig. E12-4**). Biochemical analysis may be performed on the fluid immediately, but there are very few cells in the fluid sample. For any analyses requiring cells, such as testing for Down syndrome, the cells must first be allowed to multiply in a laboratory culture for a week or two.

Chorionic Villus Sampling

The chorion, a membrane produced by the fetus, has many small projections called villi. In chorionic villus sampling (CVS), a physician inserts a small tube through the mother's vagina and up into the uterus, and suctions off a few villi (see Fig. E12-4). CVS has two major advantages over amniocentesis. First, it can be done much earlier in pregnancy—as early as the eighth week. This is especially important if the woman is contemplating a therapeutic abortion if the fetus has a major defect. Second, the sample contains enough fetal cells that analyses can be performed immediately rather than a week or two later. However, chorionic cells are more likely to have abnormal numbers of chromosomes (even when the fetus is normal). CVS also appears to have slightly greater risks than amniocentesis. Finally, CVS cannot detect certain disorders, such as

spina bifida. For these reasons, CVS is much less commonly performed than amniocentesis.

Fetal Cells in Maternal Blood

A tiny number of fetal cells cross the placenta and enter the mother's bloodstream as early as the sixth week of pregnancy. Separating fetal cells (as few as one per milliliter of blood) from the huge numbers of maternal cells is challenging, but can be done. Several companies now offer paternity testing of maternal blood samples, but practical, unambiguous genetic screening for inherited disorders still seems to be several years in the future.

Analyzing the Samples

Several types of analyses can be performed on the fetal cells or amniotic fluid. Biochemical analysis may determine the concentration of chemicals in the amniotic fluid. Many metabolic disorders can be detected by abnormally low concentrations of key enzymes, or by abnormally high concentrations of the reactants that those enzymes are supposed to break down. Other tests require whole cells. For example, analysis of the

chromosomes of the fetal cells can show if the correct number of chromosomes is present, or if any chromosomes show structural abnormalities.

Recombinant DNA techniques can be used to analyze the DNA of fetal cells to detect many defective alleles, such as those for cystic fibrosis or sickle-cell anemia. Prior to the development of PCR, fetal cells typically had to be grown in culture for as long as 2 weeks before cells had multiplied sufficiently. Now, one can extract DNA from just a few cells and use PCR to amplify the gene of interest. After a few hours, enough DNA is available for techniques such as restriction enzyme analysis, which is used to detect the allele that causes sickle-cell anemia (see Fig. 12-11). If the infant is homozygous for the sickle-cell allele, some therapeutic measures can be taken. In particular, regular doses of penicillin greatly reduce bacterial infections that otherwise kill about 15% of homozygous children. Further, knowing that a child has the disorder ensures correct diagnosis and rapid treatment during a "sickling crisis," when malformed red blood cells clump and block blood flow.

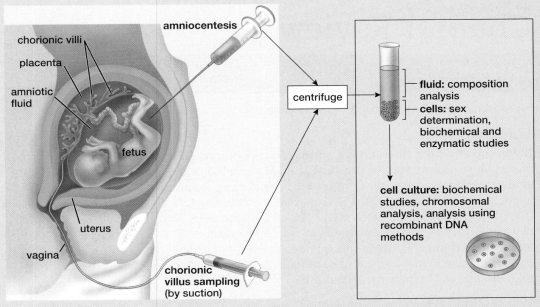

▲ **Figure E12-4 Prenatal cell sampling techniques** Two methods of obtaining fetal cell samples—amniocentesis and chorionic villus sampling—and some of the tests performed on the fetal cells.

Guilty or Innocent? Revisited

In order for Earl Ruffin's innocence to be proven, the semen collected from the rape victim needed to be located and its DNA compared to Ruffin's to see if the DNA matched. For over a decade, Ruffin and the Innocence Project tried to find out if the rape kit still existed. Finally, in 2002, a Virginia state attorney found it, preserved, along with hundreds of other samples, by forensic scientist Mary Jane Burton. Analysis of STRs showed that Ruffin was not the rapist. The DNA profile indicated that another man, by that time in prison for a different rape, was the real perpetrator. On February 12, 2003, after 21 years in prison, Earl Ruffin was freed.

What of the other people involved in the Ruffin case? Many people find it almost impossible to admit that they were wrong. To her enormous credit, the woman who was raped that December night is not one of them. She wrote to Ruffin, "I thank God for DNA testing. I do not know how to express my sorrow and devastation." She has asked Virginia legislators to support a bill that would pay Ruffin monetary compensation for his time spent in prison.

Mary Jane Burton didn't live to see the fruits of her labors, because she died in 1999. However, two other innocent men, Arthur Whitfield and Marvin Anderson, are now free because of Burton's meticulous work. Although she was not identified, the forensic lab in Patricia Cornwell's best-selling novel, *Postmortem*, was based partly on Burton's, with whom Cornwell briefly worked.

Police and district attorneys, of course, also use DNA technology as an investigative tool. In 1990, three elderly women in Goldsboro, North Carolina, were raped; two were murdered. DNA evidence indicated that all three crimes were committed by the same assailant, known only as the "Night Stalker." In 2001, the Goldsboro police created a DNA profile of the Night Stalker from the evidence they had carefully stored for over a decade. They sent the profile to the North Carolina DNA database and discovered a match. Faced with indisputable DNA evidence, the Night Stalker confessed. He is now in prison.

Who are the "heroes" in these stories? There are the obvious ones, of course—Mary Jane Burton, the professors and students of the Innocence Project, and the members of the Goldsboro Police Department. But what about Thomas Brock, who discovered *Thermus aquaticus* in Yellowstone hot springs (see "Scientific Inquiry: Hot Springs and Hot Science" on p. 193)? How about Kary Mullis, who discovered PCR, or the hundreds of biologists, chemists, and mathematicians who developed procedures for gel electrophoresis, labeling DNA, and statistical analysis of sample matching?

Consider This

Scientists often say that science is worthwhile for its own sake, and that it is difficult or impossible to predict which discoveries will lead to the greatest benefits for humanity. Nonscientists, when asked to pay the costs of scientific projects, are sometimes skeptical of such claims. How do you think public support of science should be allocated? Forty years ago, would *you* have voted to give Thomas Brock public funds to see what types of organisms lived in hot springs?

Chapter Review

Summary of Key Concepts

For additional study help and activities, go to www.mybiology.com.

12.1 What Is Biotechnology?

Biotechnology is any alteration of organisms, cells, or biological molecules to achieve specific practical goals. Modern biotechnology alters genetic material by means of genetic engineering. In many applications, recombinant DNA is produced by combining DNA from different organisms. DNA is transferred between organisms by vectors, such as bacterial plasmids, to create genetically modified, or transgenic, organisms. Some major goals of genetic engineering are to increase our understanding of gene function, to treat disease, and to improve agriculture.

12.2 How Does DNA Recombine in Nature?

DNA recombination occurs naturally through processes such as bacterial transformation, viral infection, and crossing over during sexual reproduction.

Web Animation   Genetic Recombination in Bacteria

12.3 How Is Biotechnology Used in Forensic Science?

Specific regions of DNA can be amplified by the polymerase chain reaction (PCR). The most common regions used in forensics are short tandem repeats (STRs). The STRs are separated by gel electrophoresis and made visible with DNA probes. The pattern of STRs, called a DNA profile, is unique to each individual and can be used to match DNA found at a crime scene with DNA from suspects.

Web Animation   Polymerase Chain Reaction (PCR)

12.4 How Is Biotechnology Used in Agriculture?

Many crop plants have been modified by the addition of genes that promote herbicide resistance or pest resistance. Plants may also be modified to produce human antibodies or other proteins. Transgenic animals may be produced as well, with properties such as faster growth, increased production of valuable products such as milk, or the ability to produce human antibodies or other proteins.

12.5 How Is Biotechnology Used to Learn About the Human Genome?

Techniques of biotechnology were used to discover the complete nucleotide sequence of the human genome. This knowledge will be used to learn the identities and functions of new genes, to discover medically important genes, to explore genetic variability among individuals, and to better understand the evolutionary relationships between humans and other organisms.

Web Animation   Human Genome Sequencing

12.6 How Is Biotechnology Used for Medical Diagnosis and Treatment?

Biotechnology may be used to diagnose genetic disorders such as sickle-cell anemia or cystic fibrosis. In the diagnosis of sickle-cell anemia, restriction enzymes cut normal and defective globin alleles in different locations. The resulting DNA fragments of different lengths may then be separated and identified by gel electrophoresis. In the diagnosis of cystic fibrosis, DNA probes complementary to various cystic fibrosis alleles are placed on a DNA array. Base-pairing of a patient's DNA to specific probes on the array identifies which alleles are present in the patient.

Inherited diseases are caused by defective alleles of crucial genes. Genetic engineering may be used to insert functional alleles of these genes into normal cells, stem cells, or even into eggs to correct the genetic disorder.

12.7 What Are Some of the Major Ethical Issues of Modern Biotechnology?

The use of genetically modified organisms in agriculture is controversial for two major reasons: consumer safety concerns and their potentially harmful effects on the environment. In general, GMOs contain proteins that are harmless to mammals, are readily digested, or are already found in other foods. The transfer of potentially allergenic proteins to normally nonallergenic foods can be avoided by thorough testing. Environmental effects of GMOs are more difficult to predict. It is possible that foreign genes, such as those for pest resistance or herbicide resistance, might be transferred to wild plants, with resulting damage to agriculture and/or disruption of ecosystems. If they escape, highly mobile transgenic animals might displace their wild relatives.

Genetically selecting or modifying human embryos is highly controversial. As technologies improve, society may be faced with decisions about the extent to which parents should be allowed to correct or enhance the genomes of their children.

Key Terms

biotechnology *p. 188*
DNA probe *p. 195*
DNA profile *p. 195*
gel electrophoresis *p. 194*
genetic engineering *p. 189*

genetically modified organism
(GMO) *p. 189*
genome *p. 198*
plasmid *p. 190*

polymerase chain reaction
(PCR) *p. 192*
recombinant DNA *p. 189*
restriction enzyme *p. 196*
short tandem repeat (STR) *p. 193*

stem cell *p. 201*
transformation *p. 190*
transgenic *p. 189*
virus *p. 191*

Thinking Through the Concepts

Suggested answers to end-of-chapter and figure-based questions can be found at the end of the text.

Fill-in-the-Blank

1. _____ are organisms that contain DNA that was derived from other species or that was modified, usually using recombinant DNA technology.

2. _____ is the process whereby bacteria pick up DNA, from the same or other species, from their environment. This DNA may be part of a chromosome, or it may be tiny circles of DNA called _____.

3. The _____ is a technique for multiplying DNA in the laboratory.

4. Matching DNA samples in forensics uses a specific set of small "genes" called _____. The alleles of these genes in different people vary in the number of _____ of the allele. The pattern of these alleles that a given person possesses is called his or her _____.

5. Pieces of DNA can be separated according to size by a process known as _____. The identity of a specific sample of DNA is usually determined by binding a synthetic piece of DNA called a(n) _____, which binds to the sample DNA by _____.

Review Questions

1. Describe three natural forms of genetic recombination, and discuss the similarities and differences between recombinant DNA technology and these natural forms of genetic recombination.

2. What is a plasmid? How are plasmids involved in bacterial transformation?

3. What is a restriction enzyme? How can restriction enzymes be used to splice a piece of human DNA within a plasmid?

4. Describe the polymerase chain reaction.

5. What is a short tandem repeat? How are short tandem repeats used in forensic science?

6. How does gel electrophoresis separate pieces of DNA?

7. How are DNA probes used to identify specific nucleotide sequences of DNA? How are they used in the diagnosis of genetic disorders?

8. Describe several uses of genetic engineering in agriculture.

9. Describe several uses of genetic engineering in human medicine.

10. Describe amniocentesis and chorionic villus sampling, including the advantages and disadvantages of each. What are their medical uses?

Applying the Concepts

1. **IS THIS SCIENCE?** In a 2004 Web survey conducted by the Canadian Museum of Nature, 84% of the people polled said they would eat a genetically modified banana that contained a vaccine against an infectious disease, while only 47% said they would eat a GM banana with extra vitamin C produced by the action of a rat gene. Do you think that this difference in acceptance of GMOs is scientifically valid? Why or why not?

2. As you may know, many insects have evolved resistance to common pesticides. Do you think that insects might evolve resistance to *Bt* crops? If this is a risk, do you think that *Bt* crops should be planted anyway? Why or why not?

3. Discuss the ethical issues that surround the release of genetically modified organisms (plants, animals, or bacteria) into the environment. What could go wrong? What precautions might prevent the problems you listed from occurring? What benefits do you think would justify the risks?

4. If you were contemplating having a child, would you want both yourself and your partner tested for the cystic fibrosis allele? If both of you were found to carry this allele, how would you deal with this knowledge?

For additional resources, go to www.mybiology.com.

Evolution

The ghostly grandeur of ancient bones evokes images of a lost world. Fossil remnants of extinct creatures, such as this *Triceratops* dinosaur skeleton, provide clues for the biologists who attempt to reconstruct the history of life.

chapter 13

Principles of Evolution

This massive, earthbound ostrich has useless wings, a legacy of its evolutionary heritage.

🔍 At a Glance

Case Study   What Good Are Wisdom Teeth?

Have you had your wisdom teeth removed yet? If not, it's probably only a matter of time. Almost all of us will visit an oral surgeon to have our wisdom teeth extracted. There's just not enough room in our jaws for these rearmost molars, and removing them is the best way to prevent the pain, infections, and gum disease that can accompany the emergence of wisdom teeth. Removal is harmless because we don't really need wisdom teeth—they're pretty much useless.

If you've already suffered through a wisdom tooth extraction, you may have found yourself wondering why we even have these useless molars. Biologists hypothesize that we have them because our apelike ancestors had them and we inherited them, even though we don't need them. The presence in a living species of structures that have no current function, but that *are* useful in other living species, demonstrates shared ancestry among these species.

Some excellent evidence of the connection between useless traits and evolutionary ancestry is provided by flightless birds. Consider the ostrich, a bird that can grow to 8 feet tall and weigh 300 pounds. These massive creatures cannot fly. Nonetheless, they have wings, just as sparrows and ducks do. Why do ostriches have wings that serve no function? Because the common ancestor of sparrows, ducks, and ostriches had wings, as do all of its descendants, even those that cannot fly. The bodies of today's organisms may contain now-useless hand-me-downs from their ancestors. ∎

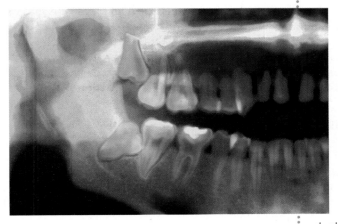

Squeezed into a jaw that is too short to contain them, wisdom teeth often become impacted—unable to erupt through the surface of the gum. The leftmost upper and lower teeth in this x-ray image are impacted wisdom teeth.

13.1 How Did Evolutionary Thought Evolve?

When you began studying biology, you may not have seen a connection between your wisdom teeth and an ostrich's wings. But the connection is there, provided by the concept that unites all of biology: **evolution,** or change over time in the characteristics of populations.

Modern biology is based on our understanding that life has evolved, but early scientists did not recognize this fundamental principle. The main ideas of evolutionary biology became widely accepted only after the publication of Charles Darwin's work in the late nineteenth century. Nonetheless, the intellectual foundation on which these ideas rest developed gradually over the centuries before Darwin's time. (You may wish to refer to the timeline in **Figure 13-1** as you read the following historical account.)

▶ **Figure 13-1 A timeline of the roots of evolutionary thought** Each bar represents the life span of a key figure in the development of modern evolutionary biology

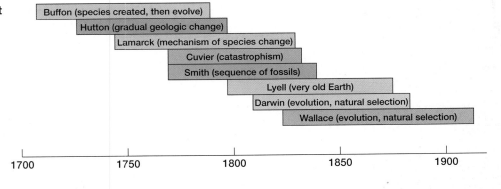

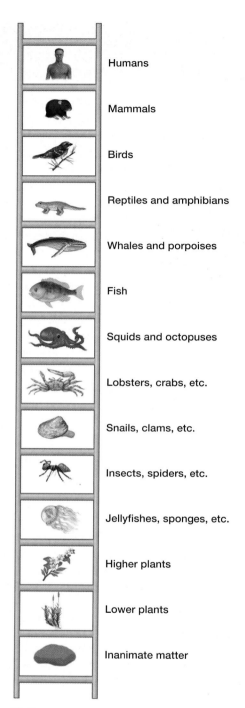

Humans

Mammals

Birds

Reptiles and amphibians

Whales and porpoises

Fish

Squids and octopuses

Lobsters, crabs, etc.

Snails, clams, etc.

Insects, spiders, etc.

Jellyfishes, sponges, etc.

Higher plants

Lower plants

Inanimate matter

▲ Figure 13-2 **Aristotle's ladder of Nature** In Aristotle's view, fixed, unchanging species can be arranged in order of increasing closeness to perfection, with inferior types at the bottom and superior types above.

Early Biological Thought Did Not Include the Concept of Evolution

Pre-Darwinian science, heavily influenced by theology, held that all organisms were created simultaneously by God and that each distinct life-form remained fixed and unchanging from the moment of its creation. This explanation of how life's diversity arose was elegantly expressed by the ancient Greek philosophers, especially Plato and Aristotle. Plato (427–347 B.C.) proposed that each object on Earth was merely a temporary reflection of its divinely inspired "ideal form." Plato's student Aristotle (384–322 B.C.) categorized all organisms into a linear hierarchy that he called the "ladder of Nature" (**Fig. 13-2**).

These ideas formed the basis of the view that the form of each type of organism is permanently fixed. This view reigned unchallenged for nearly 2,000 years. By the eighteenth century, however, several lines of newly emerging evidence began to erode the dominance of this static view of creation.

Exploration of New Lands Revealed a Staggering Diversity of Life

The Europeans who explored and colonized Africa, Asia, and the Americas were often accompanied by naturalists who observed and collected the plants and animals of these previously unknown (to Europeans) lands. By the 1700s, the accumulated observations and collections of the naturalists had begun to reveal the true scope of life's variety. The number of species, or different types of organisms, was much greater than anyone had suspected.

Stimulated by the new evidence of life's incredible diversity, some eighteenth-century naturalists began to take note of some fascinating patterns. They noticed, for example, that each area had its own distinctive set of species. In addition, the naturalists saw that some of the species in a given location closely resembled one another, yet differed in some characteristics. To some scientists of the day, the differences between the species of different geographical areas and the existence of clusters of similar species within areas seemed inconsistent with the idea that species were fixed and unchanging.

A Few Scientists Speculated That Life Had Evolved

A few eighteenth-century scientists went so far as to speculate that species had, in fact, changed over time. For example, the French naturalist Georges Louis LeClerc (1707–1788), known by the title Comte de Buffon, suggested that perhaps the original creation provided a relatively small number of founding species and that some modern species had been "conceived by Nature and produced by Time"—that is, they had changed over time through natural processes.

Fossil Discoveries Showed That Life Has Changed over Time

As Buffon and his contemporaries pondered the implications of new biological discoveries, developments in geology cast further doubt on the idea of permanently fixed species. Especially important was the discovery, during excavations for roads, mines, and canals, of rock fragments that resembled parts of living organisms. People had known of such objects since the fifteenth century, but most thought they were ordinary rocks that wind, water, or people had worked into lifelike forms. As more and more organism-shaped rocks were discovered, however, it became obvious that they were **fossils,** the preserved remains or traces of organisms that had died long ago (**Fig. 13-3**). Many fossils are bones, wood, shells, or their impressions in mud that have been *petrified*—converted to stone. Fossils also include other kinds of preserved traces, such as tracks, burrows, pollen grains, eggs, and feces.

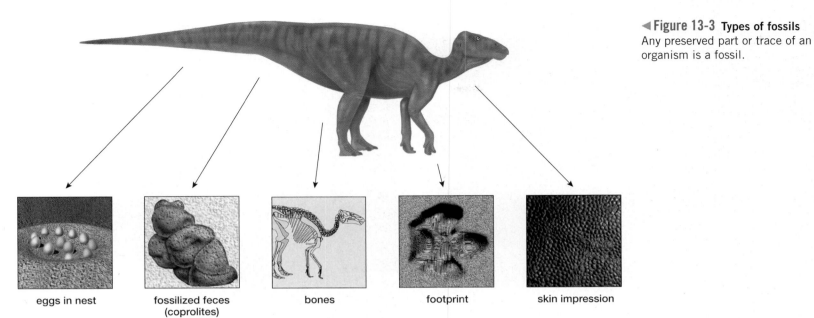

◀ **Figure 13-3** **Types of fossils**
Any preserved part or trace of an
organism is a fossil.

eggs in nest   fossilized feces   bones   footprint   skin impression
(coprolites)

By the beginning of the nineteenth century, some pioneering investigators realized that the manner in which fossils were distributed in rock was also significant. Many rocks occur in layers, with newer layers positioned over older layers. The British surveyor William Smith (1769–1839), who studied rock layers and the fossils embedded in them, recognized that certain fossils were always found in the same layers of rock. Further, the organization of fossils and rock layers was consistent: Fossil type A could always be found in a rock layer resting beneath a younger layer containing fossil type B, which in turn rested beneath a still-younger layer containing fossil type C, and so on (**Fig. 13-4**).

Scientists of the period also discovered that fossil remains showed a remarkable progression. Most fossils found in the oldest layers were very different from modern organisms, and the resemblance to modern organisms gradually increased in progressively younger rocks. Many of the fossils were from plant or animal species that had gone *extinct;* that is, no members of the species still lived on Earth.

Putting all of these facts together, some scientists came to an inescapable conclusion: Different types of organisms had lived at different times in the past.

Some Scientists Devised Nonevolutionary Explanations for Fossils

Despite the growing fossil evidence, many scientists of the period did not accept the proposition that species changed and new ones had arisen over time. To account for extinct species while preserving the notion of a single creation by God, Georges Cuvier (1769–1832) advanced the idea of *catastrophism*. Cuvier, a French paleontologist, hypothesized that a vast supply of species was created initially. Successive catastrophes (such as the Great Flood described in the Bible) produced layers of rock and destroyed many species, fossilizing some of their remains in the process. The organisms of the modern world, he speculated, are the species that survived the catastrophes.

Geology Provided Evidence That Earth Is Exceedingly Old

Cuvier's hypothesis of a world shaped by successive catastrophes was challenged by the work of the geologist Charles Lyell (1797–1875). Lyell, building on the earlier thinking of James Hutton (1726–1797), considered the forces of wind, water,

▶ Figure 13-4 **Fossils of extinct organisms** Fossils provide strong support for the idea that today's organisms were not created all at once but arose over time by the process of evolution. If all species had been created simultaneously, we would not expect **(a)** trilobites to be found in older rock layers than **(b)** seed ferns, which in turn would not be expected in older layers than **(c)** dinosaurs, such as *Allosaurus*. Trilobites became extinct about 230 million years ago, seed ferns about 150 million years ago, and dinosaurs 65 million years ago.

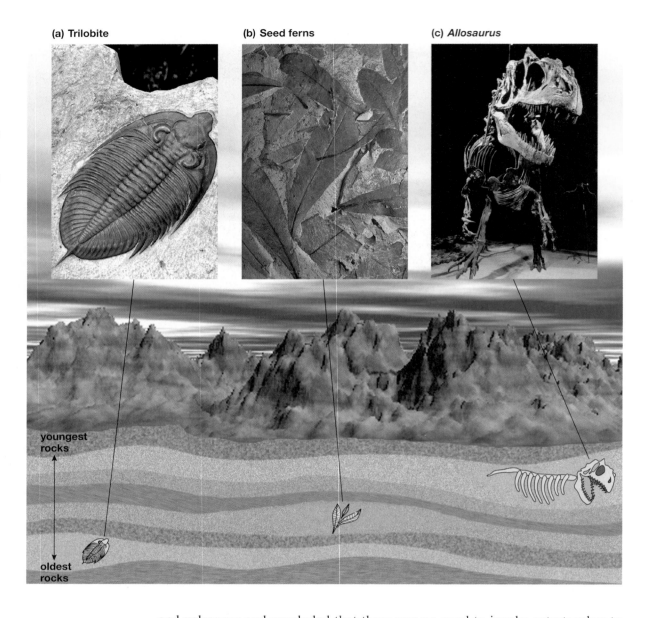

(a) Trilobite   **(b) Seed ferns**   **(c) *Allosaurus***

youngest rocks

oldest rocks

and volcanoes and concluded that there was no need to invoke catastrophes to explain the findings of geology. Don't flooding rivers lay down layers of sediment? Don't lava flows produce layers of basalt? Shouldn't we conclude, then, that layers of rock are evidence of ordinary natural processes, occurring repeatedly over long periods of time? This concept, that Earth's present landscape was produced by past action of the same gradual geological processes that we observe today, is called *uniformitarianism*. Acceptance of uniformitarianism by scientists of the time had a profound impact, because the idea implies that Earth is very old.

Before the 1830 publication of Lyell's evidence in support of uniformitarianism, few scientists suspected that Earth could be more than a few thousand years old. Counting generations in the Old Testament, for example, yields a maximum age of 4,000 to 6,000 years. An Earth this young poses problems for the idea that life has evolved. For example, ancient writers such as Aristotle described wolves, deer, lions, and other organisms that were identical to those present in Europe more than 2,000 years later. If organisms had changed so little over that time, how could whole new species possibly have arisen if Earth was created only a couple of thousand years before Aristotle's time?

But if, as Lyell suggested, rock layers thousands of feet thick were produced by slow, natural processes, then Earth must be old indeed, many millions of years old. Lyell, in fact, concluded that Earth was eternal. (Modern geologists estimate that Earth is about 4.5 billion years old; see "Scientific Inquiry: How Do We Know How Old a Fossil Is?" on p. 260)

Lyell (and his intellectual predecessor Hutton) showed that there was enough time for evolution to occur. But what was the mechanism? What process could cause evolution?

Some Pre-Darwin Biologists Proposed Mechanisms for Evolution

One of the first scientists to propose a mechanism for evolution was the French biologist Jean Baptiste Lamarck (1744–1829). Lamarck was impressed by the sequences of organisms in the rock layers. He observed that older fossils tend to be simple, whereas younger fossils tend to be more complex and more like existing organisms. In 1801, Lamarck hypothesized that organisms evolved through the inheritance of acquired characteristics, a process in which the bodies of living organisms are modified through the use or disuse of parts, and these modifications are inherited by offspring. Why would bodies be modified? Lamarck proposed that all organisms possess an innate drive for perfection. For example, if ancestral giraffes tried to improve their lot by stretching upward to feed on leaves that grow high up in trees, their necks became slightly longer as a result. Their offspring would inherit these longer necks and then stretch even farther to reach still higher leaves. Eventually, this process would produce modern giraffes with very long necks indeed.

Today, we understand how inheritance works and can see that Lamarck's proposed evolutionary process could not work as he described it. Acquired characteristics are not inherited. The fact that a prospective father pumps iron doesn't mean that his child will look like a champion body-builder. Remember, though, that in Lamarck's time the principles of inheritance had not yet been discovered. (Gregor Mendel was born a few years before Lamarck's death, and his work with inheritance in pea plants was not universally recognized until 1900—see Chapter 9.) In any case, Lamarck's insight that inheritance plays an important role in evolution had an important influence on the later biologists who discovered the key mechanism of evolution.

Darwin and Wallace Proposed a Mechanism of Evolution

By the mid-1800s, a growing number of biologists had concluded that present-day species had evolved from earlier ones. But how? In 1858, Charles Darwin and Alfred Russel Wallace, working separately, provided convincing evidence that evolution was driven by a simple yet powerful process.

Although their social and educational backgrounds were very different, Darwin and Wallace were quite similar in some respects. Both had traveled extensively in the tropics and had studied the plants and animals living there. Both found that some species differed in only a few features (Fig. 13-5). Both were familiar with the fossils that had been discovered, which showed a trend toward increasing complexity through time. Finally, both were aware of the studies of Hutton and Lyell, who had proposed that Earth is extremely ancient. These facts suggested to both Darwin and Wallace that species change over time. Both men sought a mechanism that might cause such evolutionary change.

▼ Figure 13-5 **Darwin's finches, residents of the Galapagos Islands** Darwin studied a group of closely related species of finches on the Galapagos Islands. Each species specializes in eating a different type of food and has a beak of characteristic size and shape, because natural selection has favored the individuals best suited to exploit each local food source efficiently.

(a) Large ground finch, beak suited to large seeds

(b) Small ground finch, beak suited to small seeds

(c) Warbler finch, beak suited to insects

(d) Vegetarian tree finch, beak suited to leaves

scientific inquiry

Charles Darwin—Nature Was His Laboratory

Like many students, Charles Darwin excelled only in subjects that intrigued him. Although his father was a physician, Darwin was uninterested in medicine and unable to stand the sight of surgery. He eventually obtained a degree in theology from Cambridge University, although theology too was of minor interest to him. What he really liked to do was to tramp over the hills, observing plants and animals, collecting new specimens, scrutinizing their structures, and categorizing them.

In 1831, when Darwin was 22 years old (Fig. E13-1), he secured a position as "gentleman companion" to Captain Robert Fitzroy of the HMS *Beagle*. The *Beagle* soon embarked on a 5-year surveying expedition along the coastline of South America and then around the world.

Darwin's voyage on the *Beagle* sowed the seeds for his theory of evolution. In addition to his duties as companion to the captain, Darwin served as the expedition's official naturalist, whose task was to observe and collect geological and biological specimens. The *Beagle* sailed to South America and made many stops along its coast. There Darwin observed the plants and

▲ **Figure E13-1** **A painting of Charles Darwin as a young man**

animals of the tropics and was stunned by the greater diversity of species compared with that of Europe.

Although he had boarded the *Beagle* convinced of the permanence of species,

Darwin's experiences soon led him to doubt it. He discovered a snake with rudimentary hind limbs, calling it "the passage by which Nature joins the lizards to the snakes" (Fig. E13-2). Another snake he encountered vibrated its tail like a rattlesnake but had no rattles and therefore made no noise. Similarly, Darwin noticed that penguins used their wings to paddle through the water rather than fly through the air. If a creator had individually created each animal in its present form, to suit its present environment, what could be the purpose behind these makeshift arrangements?

Perhaps the most significant stopover of the voyage was the month spent on the Galapagos Islands off the northwestern coast of South America. There, Darwin found huge tortoises. Different islands were home to distinctively different types of tortoises. Darwin also found several types of finches and, as with the tortoises, different islands had slightly different finches. Could the differences in these organisms have arisen after they became isolated from one another on separate islands? The diversity of tortoises and finches haunted him for years afterward.

Of the two, Darwin was the first to propose a mechanism for evolution, which he described in a paper he wrote in 1842. But he did not submit the paper for publication, perhaps because he was fearful of the controversy that publication would cause. Some historians wonder if Darwin would ever have publicized his ideas had he not received, sixteen years later, a draft of a paper by Wallace, which outlined ideas remarkably similar to Darwin's own. Darwin realized that he could delay no longer.

In separate but similar papers that were presented to the Linnaean Society in London in 1858, Darwin and Wallace each described the same mechanism for evolution. Initially, their papers had little impact. The secretary of the society, in fact, wrote in his annual report that nothing very interesting happened that year. Fortunately, the next year, Darwin published his monumental book, *On the Origin of Species by Means of Natural Selection,* which attracted a great deal of attention to the new ideas about how species evolve. (To learn more about Darwin's life, see "Scientific Inquiry: Charles Darwin—Nature Was His Laboratory.")

13.2 How Does Natural Selection Work?

Darwin and Wallace proposed that life's huge variety of excellent designs arose by a process of descent with modification, in which individuals in each generation differ slightly from the members of the preceding generation. Over long stretches of time, these small differences accumulate to produce major transformations.

▲ **Figure E13-2 The vestigial remnants of hind legs in a snake** Some snakes have small "spurs" (large photo, at arrow) where their distant ancestors had rear legs. In some species, these vestigial structures even retain claws (inset).

In 1836, Darwin returned to England after 5 years on the *Beagle* and became established as one of the foremost naturalists of his time. But the problem of how isolated populations come to differ from each other gnawed constantly at his mind. Part of the solution came to him from an unlikely source: the writings of an English economist and clergyman, Thomas Malthus. In his *Essay on the Principle of Population,* Malthus wrote, "It may safely be pronounced, therefore, that [human] population, when unchecked, goes on doubling itself every 25 years, or increases in a geometrical ratio."

Darwin realized that a similar principle holds true for plant and animal populations. In fact, most organisms can reproduce much more rapidly than can humans (consider rabbits, dandelions, and houseflies) and consequently could produce overwhelming populations in short order. Nonetheless, the world is not chest-deep in rabbits, dandelions, or flies: Natural populations do not grow "unchecked" but tend to remain approximately constant in size. Clearly, vast numbers of individuals must die in each generation, and most must not reproduce.

From his experience as a naturalist, Darwin realized that members of a species typically differ from one another. Further, which individuals die without reproducing in each generation is not arbitrary, but depends to some extent on the structures and abilities of the organisms. This observation was the source of the theory of evolution by natural selection. As Darwin's colleague Alfred Wallace put it, "Those which, year by year, survived this terrible destruction must be, on the whole, those which have some little superiority enabling them to escape each special form of death to which the great majority succumbed." Here you see the origin of the expression "survival of the fittest." That "little superiority" that confers greater success might be better resistance to cold, more efficient digestion, or any of hundreds of other advantages, some very subtle.

Everything now fell into place. Darwin wrote, "It at once struck me that under these circumstances favorable variations would tend to be preserved, and unfavorable ones to be destroyed." If a favorable variation were inheritable, then the entire species would eventually consist of individuals possessing the favorable trait. With the continual appearance of new variations (due, as we now know, to mutations), which in turn are subject to further natural selection, "the result . . . would be the formation of new species. Here, then, I had at last got a theory by which to work."

When Darwin finally published *On the Origin of Species* in 1859, his evidence had become truly overwhelming. Although its full implications would not be realized for decades, Darwin's theory of evolution by natural selection has become a unifying concept for virtually all of biology.

Darwin and Wallace's Theory Rests on Four Postulates

The chain of logic that led Darwin and Wallace to their proposed process of evolution turns out to be surprisingly simple and straightforward. It is based on four postulates about **populations,** or all the individuals of one species in a particular area.

Postulate 1: Individual members of a population differ from one another in many respects.

Postulate 2: At least some of the differences among members of a population are due to characteristics that may be passed from parent to offspring.

Postulate 3: In each generation, some individuals in a population survive and reproduce successfully but others do not.

Postulate 4: The fate of individuals is not determined entirely by chance or luck. Instead, an individual's likelihood of survival and reproduction depends on its characteristics. Individuals with advantageous traits survive longest and leave the most offspring, a process known as **natural selection.**

Darwin and Wallace understood that if all four postulates were true, populations would inevitably change over time. If members of a population have different traits, and if the individuals that are best suited to their environment leave more offspring, and if those individuals pass their favorable traits to the next

▲ Figure 13-6 **Variation in a population of snails** Although these snails are all members of the same population, no two are exactly alike. **QUESTION:** Is sexual reproduction required to generate the variability in structures and behaviors that is necessary for natural selection?

generation, then the favorable traits will be more common in subsequent generations. The characteristics of the population will change slightly with each generation. This process is evolution by natural selection.

Are the four postulates true? Darwin thought so, and devoted much of *On the Origin of Species* to describing supporting evidence. Let's briefly examine each postulate, in some cases with the advantage of knowledge that was not available to Darwin and Wallace.

Postulate 1: Individuals in a Population Vary

The accuracy of postulate 1 is apparent to anyone who has glanced around a crowded room. People differ in size, eye color, skin color, and many other physical features. Similar variability is present in populations of other organisms, although it may be less obvious to the casual observer (**Fig. 13-6**). We now know that the variations in natural populations arise purely by chance, as a result of random mutations in DNA (see p. 166). Thus, the differences among individuals extend to the molecular level. The reason that DNA tests can match blood from a crime scene to a suspect is that each person's exact DNA sequence is unique.

Postulate 2: Traits Are Passed from Parent to Offspring

The principles of genetics had not yet been discovered when Darwin published *On the Origin of Species*. Therefore, although observation of people, pets, and farm animals seemed to show that offspring generally resemble their parents, Darwin and Wallace did not have scientific evidence in support of postulate 2. Mendel's later work, however, demonstrated conclusively that particular traits can be passed to offspring. Since Mendel's time, genetics researchers have produced an incredibly detailed picture of how inheritance works.

Postulate 3: Some Individuals Fail to Survive and Reproduce

Darwin and Wallace's formulation of postulate 3 was heavily influenced by Thomas Malthus's *Essay on the Principle of Population* (1798), which described the perils of unchecked growth of human populations. Darwin was keenly aware that organisms can produce far more offspring than are required merely to replace the parents. He calculated, for example, that a single pair of elephants would multiply to a population of 19 million in 750 years if each descendant had six offspring.

But we aren't overrun with elephants. The number of elephants, like the number of individuals in most natural populations, tends to remain relatively constant. Therefore, more organisms must be born than survive long enough to reproduce. In each generation, many individuals must die young. Even among those that survive, many must fail to reproduce, produce few offspring, or produce less-vigorous offspring that, in turn, fail to survive and reproduce. As you might expect, whenever biologists have measured reproduction in a population, they have found that some individuals have more offspring than others.

Postulate 4: Survival and Reproduction Are Not Determined by Chance

If unequal reproduction is the norm in populations, what determines which individuals leave the most offspring? A large amount of scientific evidence has shown that reproductive success depends on an individual's characteristics. For example, scientists have found that larger male elephant seals in a California population had more offspring than smaller males (because females preferred to mate with large males). In a Colorado population, snapdragon plants with white flowers had more offspring than plants with yellow flowers (because pollinators found white flowers more attractive). In a hospital patient's body, antibiotic-resistant tuberculosis bacteria reproduced more rapidly than antibiotic-sensitive ones (because the patient was being treated with antibiotics). These results, and hundreds of other similar ones, show that in the competition to survive and reproduce, winners are determined not by chance but by the traits they possess.

Natural Selection Modifies Populations over Time

Observation and experiment suggest that the four postulates of Darwin and Wallace are sound. Logic suggests that the resulting consequence ought to be change over time in the characteristics of populations. In *On the Origin of Species*, Darwin proposed the following example: "Let us take the case of a wolf, which preys on various animals, securing [them] by . . . fleetness. . . . The swiftest and slimmest wolves would have the best chance of surviving, and so be preserved or selected. . . . Now if any slight innate change of habit or structure benefited an individual wolf, it would have the best chance of surviving and of leaving offspring. Some of its young would probably inherit the same habits or structure, and by the repetition of this process, a new variety might be formed." The same logic applies to the wolf's prey; the fastest or most alert would be most likely to avoid predation and would pass on these traits to its offspring.

Notice that natural selection acts on individuals within a population. Natural selection's influence on the fates of individuals eventually has consequences for the population as a whole. Over generations, the population changes as the percentage of individuals inheriting favorable traits increases. An individual cannot evolve, but a population can.

Although it is easier to understand how natural selection would cause changes within a species, under the right circumstances, the process might produce entirely *new* species. We will discuss the circumstances that might give rise to new species in Chapter 14.

13.3 How Do We Know That Evolution Has Occurred?

Today, evolution is an accepted scientific theory. You may recall from Chapter 1 that a scientific theory is a general explanation of important natural phenomena, developed through extensive, reproducible observations. An overwhelming body of evidence supports the conclusion that evolution has occurred. The key lines of evidence come from fossils, comparative anatomy (the study of how body structures differ among species), embryology, biochemistry, and genetics.

Fossils Provide Evidence of Evolutionary Change over Time

If it is true that many fossils are the remains of species ancestral to modern species, we might expect to find progressive series of fossils that start with an ancient organism, progress through several intermediate stages, and culminate in a modern species. Such series have indeed been found. For example, fossils of the ancestors of modern whales illustrate stages in the evolution of an aquatic species from land-dwelling ancestors (**Fig. 13-7**). Series of fossil giraffes, elephants, horses, and mollusks also show the evolution of body structures over time. These fossil series suggest that new species evolved from, and replaced, previous species. Certain sequences of fossil snails have such slight gradations in body structures between successive rock layers that paleontologists cannot easily decide where one species leaves off and the next one begins.

What Good Are Wisdom Teeth?

Continued

The common ancestry of all birds explains why even flightless birds have wings, but does not explain why some bird species have become flightless. Flightlessness evolved by natural selection acting on flying birds. The ability to fly provides obvious benefits, such as the ability to escape earthbound predators, so at first glance it might seem unlikely that natural selection could ever favor the loss of such a valuable trait. But flying is costly; flight uses a great deal of energy and requires growth of large flight muscles. So, if a population of birds found itself in an environment in which the benefits of flying were diminished, natural selection might favor individuals that did not pay the high cost of flight. For example, in such conditions non-flyers could devote the energy saved by not flying to increased reproduction. What kinds of environments might reduce the benefits of flight and thereby favor the evolution of flightlessness? Environments with few or no predators could meet this criterion; a reduced need to escape predation would remove one of the main advantages of flying. Large predators are often absent from oceanic islands, and bird populations on islands have proved especially likely to evolve flightlessness.

▶ **Figure 13-7 The evolution of the whale** During the past 50 million years, whales have evolved from four-legged land dwellers, to semi-aquatic paddlers, to fully aquatic swimmers with shrunken hind legs, to today's sleek ocean dwellers. **QUESTION:** The fossil history of some kinds of modern organisms, such as sharks and crocodiles, shows that their structure and appearance have changed very little over hundreds of millions of years. Is this lack of change evidence that such organisms have not evolved over that time?

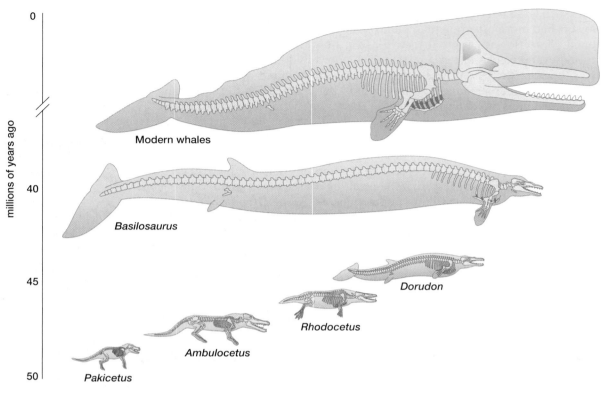

millions of years ago

0

Modern whales

40

Basilosaurus

45

Dorudon

Rhodocetus

Ambulocetus

50

Pakicetus

Web Animation Analogous and Homologous Structures

Comparative Anatomy Gives Evidence of Descent with Modification

Fossils provide snapshots of the past that allow biologists to trace evolutionary changes, but careful examination of today's organisms can also uncover evidence of evolution. Comparing the bodies of organisms of different species can reveal similarities that can be explained only by shared ancestry and differences that could result only from evolutionary change during descent from a common ancestor. In this way, the study of comparative anatomy has supplied strong evidence that different species are linked by a common evolutionary heritage.

Homologous Structures Provide Evidence of Common Ancestry

The same body structure may be modified by evolution to serve different functions in different species. The forelimbs of birds and mammals, for example, are variously used for flying, swimming, running over several types of terrain, and grasping objects, such as branches and tools. Despite this enormous diversity of function, the internal anatomy of all bird and mammal forelimbs is remarkably similar (Fig. 13-8). It seems inconceivable that the same bone arrangements would be used to serve such diverse functions if each animal had been created separately. Such similarity is exactly what we would expect, however, if bird and mammal forelimbs were derived from a common ancestor. Through natural selection, each has been modified and now performs a particular function. Such internally similar structures are called **homologous structures,** meaning that they have the same evolutionary origin despite any differences in current function or appearance.

Functionless Structures Are Inherited from Ancestors

Evolution by natural selection also helps explain the curious circumstance of **vestigial structures** that serve no apparent purpose. Examples include such things as molar teeth in vampire bats (which live on a diet of blood and therefore don't chew their food) and pelvic bones in whales and certain snakes (Fig. 13-9). Both of these vestigial structures are clearly homologous to structures that are found in—and used by—other vertebrates (animals with a backbone). Their continued

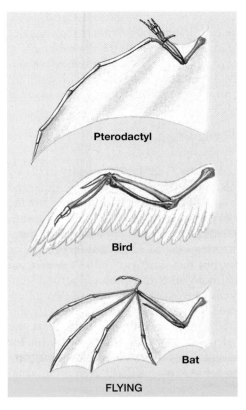
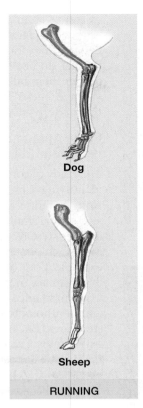
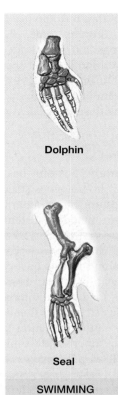
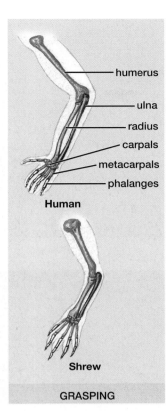

Pterodactyl

Dolphin

Dog

humerus

ulna

radius

carpals

metacarpals

phalanges

Human

Bird

Bat

FLYING

Seal

SWIMMING

Sheep

RUNNING

Shrew

GRASPING

▲ Figure 13-8 Homologous structures Despite wide differences in function, the forelimbs of all of these animals contain the same set of bones, inherited from a common ancestor. The different colors of the bones highlight the correspondences among the various species.

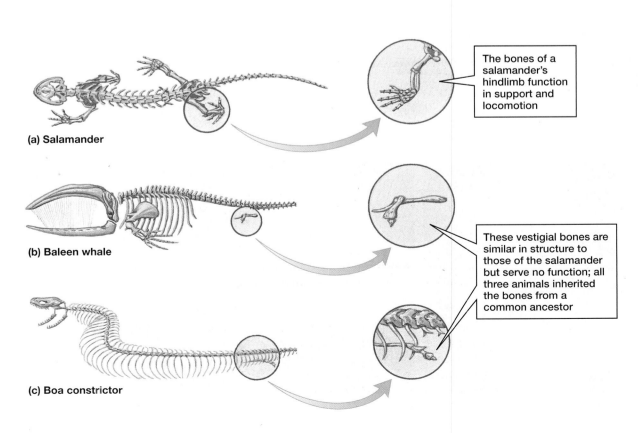

(a) Salamander

The bones of a salamander's hindlimb function in support and locomotion

(b) Baleen whale

These vestigial bones are similar in structure to those of the salamander but serve no function; all three animals inherited the bones from a common ancestor

(c) Boa constrictor

◀ Figure 13-9 Vestigial structures Many organisms have vestigial structures that serve no apparent function. The (a) salamander, (b) baleen whale, and (c) boa constrictor all inherited hind limb bones from a common ancestor. These bones remain functional in the salamander but are vestigial in the whale and snake.

(a) Damselfly

(b) Swallow

▲ **Figure 13-10 Analogous structures** Convergent evolution can produce outwardly similar structures that differ anatomically, such as the wings of **(a)** insects and **(b)** birds. **QUESTION:** Are a peacock's tail and a dog's tail homologous structures or analogous structures?

▶ **Figure 13-11 Embryological stages reveal evolutionary relationships** Early embryonic stages of a **(a)** lemur, **(b)** pig, and **(c)** human, showing strikingly similar anatomical features.

existence in animals that have no use for them is best explained as a sort of "evolutionary baggage." For example, the ancestral mammals from which whales evolved had four legs and a well-developed set of pelvic bones (see Fig. 13-7). Whales do not have hind legs, yet they have small pelvic and leg bones embedded in their sides. During whale evolution, losing the hind legs provided an advantage, better streamlining the body for movement through water. The result is the modern whale with small, useless pelvic bones.

Some Anatomical Similarities Result from Evolution in Similar Environments

The study of comparative anatomy has demonstrated the shared ancestry of life by identifying a host of homologous structures that different species have inherited from common ancestors, but comparative anatomists have also identified many anatomical similarities that do not stem from common ancestry. Instead, these similarities stem from **convergent evolution,** in which natural selection causes non-homologous structures that serve similar functions to resemble one another. For example, both birds and insects have wings, but this similarity did not arise from evolutionary modification of a structure that both birds and insects inherited from a common ancestor. Instead, the similarity arose from modification of two different, non-homologous structures that eventually gave rise to superficially similar structures. Because natural selection favored flight in both birds and insects, the two groups evolved superficially similar structures—wings—that are useful for flight. Such outwardly similar but non-homologous structures are called **analogous structures** (Fig. 13-10). Analogous structures are typically very different in internal anatomy, because the parts are not derived from common ancestral structures.

Embryological Similarity Suggests Common Ancestry

In the early 1800s, German embryologist Karl von Baer noted that all vertebrate embryos (developing organisms in the period from fertilization to birth or hatching) look quite similar to one another early in their development (Fig. 13-11). In their early embryonic stages, fish, turtles, chickens, mice, and humans all develop tails and gill slits (also called gill grooves). But among this group of animals, only fish retain gills as adults, and only fish, turtles, and mice retain substantial tails.

Why do vertebrates that are so different have similar developmental stages? The only plausible explanation is that ancestral vertebrates possessed genes that directed the development of gills and tails. All of their descendants still have those genes. In fish, these genes are active throughout development, resulting in adults with fully developed tails and gills. In humans and chickens, these genes are active only during early developmental stages, and the structures are lost or inconspicuous in adults.

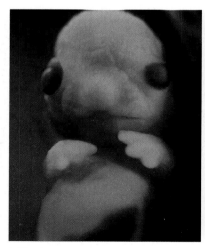

(a) Lemur

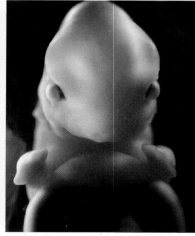

(b) Pig

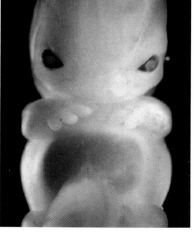

(c) Human

Modern Biochemical and Genetic Analyses Reveal Relatedness Among Diverse Organisms

Biologists have been aware of anatomical and embryological similarities among organisms for centuries, but it took the emergence of modern technology to reveal similarity at the molecular level. Biochemical similarities among organisms provide perhaps the most striking evidence of their evolutionary relatedness. Just as relatedness is revealed by homologous anatomical structures, it is revealed by homologous molecules.

Today's scientists have access to a powerful tool for revealing molecular homologies: DNA sequencing. It is now possible to quickly determine the sequence of nucleotides in a DNA molecule and to compare the DNA of different organisms. For example, consider the gene that encodes the protein cytochrome *c* (see Chapters 10 and 11 for information on DNA and how it encodes proteins). Cytochrome *c* is present in all plants and animals (and many single-celled organisms) and performs the same function in all of them. The sequence of nucleotides in the gene for cytochrome *c* is similar in these diverse species (**Fig. 13-12**). The widespread presence of the same complex protein, encoded by the same gene and performing the same function, is evidence that the common ancestor of plants and animals had cytochrome *c* in its cells. At the same time, though, the sequence of the cytochrome *c* gene differs slightly in different species, showing that variations arose during the independent evolution of Earth's multitude of plant and animal species.

Some biochemical similarities are so fundamental that they extend to all living cells. For example:

- All cells have DNA as the carrier of genetic information.
- All cells use RNA, ribosomes, and approximately the same genetic code to translate that genetic information into proteins.
- All cells use roughly the same set of 20 amino acids to build proteins.
- All cells use ATP as a cellular energy carrier.

What Good Are Wisdom Teeth?
Continued

Just as anatomical homology can lead to vestigial structures such as human wisdom teeth and the wings of flightless birds, genetic homology can lead to vestigial DNA sequences. For example, many mammal species produce an enzyme, L-gulono-gamma-lactone oxidase, that catalyzes the last step in the production of vitamin C. The species that produce the enzyme are able to do so because they all inherited the gene that encodes it from a common ancestor. Humans, however, do not produce L-gulono-gamma-lactone oxidase, so we can't produce vitamin C ourselves and must consume it in our diets. But even though we don't produce the enzyme, our cells do contain a stretch of DNA with a sequence very similar to that of the enzyme-producing gene present in rats and most other mammals. The human version, though, does not encode the enzyme (or any protein). We inherited this stretch of DNA from an ancestor that we share with other mammal species, but in us the sequence has undergone a change that rendered it nonfunctional. It remains as a vestigial trait, evidence of our shared ancestry.

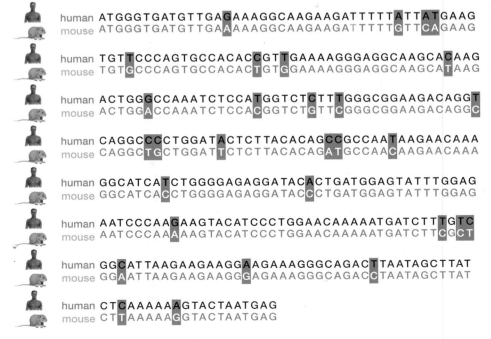

◀ **Figure 13-12 Molecular similarity shows evolutionary relationships** The DNA sequences of the genes that code for cytochrome *c* in a human and a mouse. Of the 315 nucleotides in the gene, only 30 (shaded blue) differ between the two species.

(a) Gray wolf

(b) Diverse dogs

▲ **Figure 13-13 Dog diversity illustrates artificial selection** A comparison of **(a)** the ancestral dog (the gray wolf, *Canis lupus*) and **(b)** various breeds of dog. Artificial selection by humans has caused great divergence in the size and shape of dogs in only a few thousand years.

The most plausible explanation for such widespread sharing of complex and specific biochemical traits is that the traits are homologies. That is, they arose only once, in the common ancestor of all living things, from which all of today's organisms inherited them.

13.4 What Is the Evidence That Populations Evolve by Natural Selection?

We have seen that evidence of evolution comes from many sources. But what is the evidence that evolution occurs by the process of natural selection?

Controlled Breeding Modifies Organisms

One line of evidence supporting evolution by natural selection is **artificial selection**, the breeding of domestic plants and animals to produce specific desirable features. The various dog breeds provide a striking example of artificial selection (**Fig. 13-13**). Dogs descended from wolves, and even today, the two will readily crossbreed. With few exceptions, however, modern dogs do not resemble wolves. Some breeds are so different from one another that they would be considered separate species if they were found in the wild. Humans produced these radically different dogs in a few thousand years by doing nothing more than repeatedly selecting individuals with desirable traits for breeding. Therefore, it is quite plausible that natural selection could, by an analogous process acting over hundreds of millions of years, produce the spectrum of living organisms. Darwin was so impressed by the connection between artificial selection and natural selection that he devoted a chapter of *Origin of Species* to the topic.

Evolution by Natural Selection Occurs Today

The logic of natural selection gives us no reason to believe that evolutionary change is limited to the past. After all, inherited variation and competition for access to resources are certainly not limited to the past. If Darwin and Wallace were correct that those conditions lead inevitably to evolution by natural selection, then scientific observers and experimenters ought to be able to detect evolutionary change as it occurs. And they have. We consider next some examples that give us a glimpse of natural selection at work.

Brighter Coloration Can Evolve When Fewer Predators Are Present

On the island of Trinidad, guppies live in streams that are also inhabited by several species of larger, predatory fish that frequently dine on guppies (**Fig. 13-14**). In upstream portions of these streams, however, the water is too shallow for the predators, and guppies are free of danger from predators. When scientists compared male guppies in an upstream area with ones in a downstream area, they found that the upstream guppies were much more brightly colored than the downstream guppies. The scientists knew that the source of the upstream population was guppies that had found their way up into the shallower waters many generations earlier.

The explanation for the difference in coloration between the two populations stems from the sexual preferences of female guppies. The females prefer to mate with the most brightly colored males, so the brightest males have a large advantage when it comes to reproduction. In predator-free areas, male guppies with the bright colors that females prefer have more offspring than duller males. Bright color, however, makes guppies more conspicuous to predators, and therefore more likely to be eaten. Thus, where predators are common, they act as agents of natural selection by eliminating the bright-colored males before they can reproduce. In these areas, the duller males have the advantage and produce more offspring. The color difference between the upstream and downstream guppy populations is a direct result of natural selection.

Natural Selection Can Lead to Pesticide Resistance

Natural selection is also evident in numerous instances of insect pests evolving resistance to the pesticides with which we try to control them. For example, a few decades ago, Florida homeowners were dismayed to realize that roaches were ignoring a formerly effective poison bait called Combat®. Researchers discovered that the bait had acted as an agent of natural selection. Roaches that liked it were consistently killed; those that survived inherited a rare mutation that caused them to dislike glucose, a type of sugar found in the corn syrup used as bait in Combat. By the time researchers identified the problem in the early 1990s, the formerly rare mutation had become common in Florida's urban roach population. (For additional examples of how humans influence evolution, see "Earth Watch: People Promote High-Speed Evolution" on p. 226.)

Unfortunately, the evolution of pesticide resistance in insects is a common example of natural selection in action. Such resistance has been documented in more than 500 species of crop-damaging insects, and virtually every pesticide has fostered the evolution of resistance in at least one insect species. We pay a heavy price for this evolutionary phenomenon. The additional pesticides that farmers apply in their attempts to control resistant insects cost almost $2 billion each year in the United States alone and add millions of tons of poisons to Earth's soil and water.

Experiments Can Demonstrate Natural Selection

In addition to observing natural selection in the wild, scientists have also devised numerous experiments that confirm the action of natural selection. For example, one group of evolutionary biologists released small groups of *Anolis sagrei* lizards onto 14 small Bahamian islands that were previously uninhabited by lizards (**Fig. 13-15**). The original lizards came from a population on Staniel Cay, an island with tall vegetation, including plenty of trees. In contrast, the islands to which the small colonial groups were introduced had few or no trees and were covered mainly with small shrubs and other low-growing plants.

The biologists returned to those islands 14 years after releasing the colonists and found that the original small groups of lizards had given rise to thriving populations of hundreds of individuals. On all 14 of the experimental islands, lizards had legs that were shorter and thinner than lizards from the original source population on Staniel Cay. In just over a decade, it appeared, the lizard populations had changed in response to new environments.

Why had the new lizard populations evolved shorter, thinner legs? Long legs allow greater speed, but shorter legs allow for more agility and maneuverability on narrow surfaces. So, natural selection favors legs that are as long and thick as possible while still allowing sufficient maneuverability. When the lizards were moved from an environment with thick-branched trees to an environment with only thin-branched bushes, the individuals with formerly favorable long legs were at a disadvantage. In the new environment, more agile, shorter-legged individuals were better able to escape predators and survive to produce a greater number of offspring. Thus, members of subsequent generations had shorter legs on average.

▲ **Figure 13-14 Guppies evolve to become more colorful in predator-free environments** Male guppies (top) are more brightly colored than females (bottom). Some male guppies are more colorful than others. In some environments, brighter males are selected; in other environments, duller males are selected.

▲ **Figure 13-15 Anole leg size evolves in response to a changed environment**

People Promote High-Speed Evolution

You probably don't think of yourself as a major engine of evolution. Nonetheless, as you go about the routines of your daily life, you are contributing to what is perhaps today's most significant cause of rapid evolutionary change. Human activity has changed Earth's environments tremendously, and when environments change, populations adapt or perish. The biological logic of natural selection, spelled out so clearly by Darwin, tells us that environmental change leads inevitably to evolutionary change. Thus, by changing the environment, humans have become a major agent of natural selection.

Unfortunately, many of the evolutionary changes we have caused have turned out to be bad news for us. Our liberal use of pesticides has selected for resistant pests that frustrate efforts to protect our food supply. By overmedicating ourselves with antibiotics and other drugs, we have selected for resistant "supergerms" and diseases that are ever more difficult to treat. Heavy fishing in the world's oceans has favored smaller fish that can slip through nets more easily, thereby selecting for slow-growing individuals that remain small even as mature adults. As a result, fish of many commercially important species are now so small that our ability to extract food from the sea is compromised.

Our use of pesticides, antibiotics, and fishing technology has caused evolutionary changes that threaten our health and welfare, but the scope of these changes may be dwarfed by those that will arise from human-caused modification of Earth's climate. Human activities, especially activities that use energy derived from fossil fuels, modify the climate by contributing to global warming. In coming years, species' evolution will be influenced by environmental changes associated with a warming climate, such as reduced ice and snow, longer growing seasons, and shifts in the life cycles of other species that provide food or shelter.

There is growing evidence that global warming is already causing evolutionary change. Warming-related evolution has been found in populations as diverse as mosquitoes, birds, and squirrels. For example, researchers have discovered that, in northern populations of a mosquito species, the insects' genetically programmed response to changing day length has shifted during the past four decades. (Mosquitoes use day length as a cue to tell them what time of year it is.) As a result, shorter days are now required to stimulate mosquito larvae to enter their overwintering pupal stage, and the transition takes place much later in the autumn. The delay allows the larvae to take advantage of the longer feeding and growing season produced by global warming.

Climate change is also affecting the evolution of bird migration, for example by causing European blackcap birds to evolve into genetically distinct populations with different migration patterns. Historically, these birds have bred in Austria and Germany and migrated south into Spain and Morocco for the winter. Since the early 1960s, however, increasing numbers of blackcaps have instead spent the nonbreeding season in southern England, where winters have become milder and food more abundant. Individuals in this population have a selective advantage over birds that spend the winter farther away from the breeding grounds; they reach their nesting area earlier in the spring, gain better territories, and produce more offspring. Further, laboratory studies have demonstrated that blackcaps that overwinter in England prefer to mate with each other, and that their offspring inherit the tendency to migrate to England. Thus, the proportion of blackcaps wintering in England is increasing, and this group of birds is evolving to become increasingly distinct from other blackcaps.

In Canada, red squirrels in a colony closely monitored by scientists now produce litters 18 days earlier, on average, than they did 10 years ago (Fig. E13-3). The change is tied to a warming climate, because spring now arrives earlier and spruce trees produce earlier crops of seeds, the squirrels' only food. Squirrels that breed earlier gain a competitive edge by better exploiting the warmer weather and more abundant food. And, because the time at which a squirrel gives birth

▲ Figure E13-3 Red squirrels have evolved in response to global warming

is influenced by the animal's genetic makeup, early-breeding squirrels pass to their offspring the genes that confer this advantage. As a result, the genetic makeup of the squirrel population is changing, and early-breeding squirrels are becoming more common than later-breeding ones.

The available evidence suggests that global climate change will have an enormous evolutionary impact, potentially affecting the evolution of almost every species. How will these evolutionary changes affect us and the ecosystems on which we depend? This question is not readily answerable, because the path of evolution is not predictable. We can hope, however, that careful monitoring of evolving species and increased understanding of evolutionary processes will help us take appropriate steps to safeguard our health and well-being as Earth warms.

Selection Acts on Random Variation to Favor the Phenotypes That Work Best in Particular Environments

Two important points underlie the evolutionary changes just described:

- **The variations on which natural selection works are produced by chance mutations.** The bright coloration in Trinidadian guppies, distaste for glucose in Florida cockroaches, and shorter legs in Bahamian lizards were not *produced* by the female mating preferences, poisoned corn syrup, or thinner branches. The mutations that produced each of these beneficial traits arose spontaneously.

- **Natural selection favors organisms that are best adapted to a particular environment.** Natural selection is not a process for producing ever-greater degrees of perfection. Natural selection does not select for the "best" in any absolute sense, but only for what is best in the context of a particular environment, which varies from place to place and which may change over time. A trait that is advantageous under one set of conditions may become disadvantageous if conditions change. For example, in the presence of poisoned corn syrup, a distaste for glucose yields an advantage to a cockroach, but under natural conditions, avoiding glucose would cause the insect to bypass good sources of food.

What Good Are Wisdom Teeth? Revisited

Wisdom teeth are but one of dozens of human anatomical structures that appear to serve no function. Darwin himself noted many of these "useless, or nearly useless" traits in the very first chapter of *Origin* and declared them to be prime evidence that humans had evolved from earlier species.

Another vestigial structure is the appendix, a narrow finger-like projection attached to the large intestine. The appendix is homologous with the tip of the *cecum*, which is a portion of the large intestine that is used for cellulose digestion in many plant-eating mammals. Although the appendix produces some white blood cells, it does not perform its ancestral digestive function. Thus, digestive function is not harmed in any of the roughly 300,000 Americans per year whose diseased appendixes are surgically removed.

Body hair is another functionless human trait. It seems to be an evolutionary relic of the fur that kept our distant ancestors warm (and that still warms our closest evolutionary relatives, the great apes). Not only do we retain useless body hair, we also still have erector pili, the muscle fibers that allow other mammals to puff up their fur for better insulation. In humans, these vestigial structures just give us goose bumps.

Though humans don't have and don't need a tail, we nonetheless have a tailbone. The tailbone consists of a few tiny vertebrae fused into a small structure at the base of the backbone, where a tail would be if we had one. People born without a tailbone or who have theirs surgically removed suffer no ill effects.

Consider This

Advocates of creationism argue that there are no vestigial organs because if a structure can do *anything*, it cannot be considered functionless, even if its removal has no effect. Thus, according to this view, wisdom teeth are not evidence of evolution, because they *can* be used to chew if not removed. Do you find this argument persuasive?

Chapter Review

Summary of Key Concepts

For additional study help and activities, go to www.mybiology.com.

13.1 How Did Evolutionary Thought Evolve?
Historically, the most common explanation for the origin of species was the divine creation of each species in its present form, and species were believed to remain unchanged after their creation. This view was challenged by evidence from fossils, geology, and biological exploration of the tropics. Since the middle of the nineteenth century, scientists have realized that species originate and evolve by the operation of natural processes that change the genetic makeup of populations.

13.2 How Does Natural Selection Work?

Charles Darwin and Alfred Russel Wallace independently proposed the theory of evolution by natural selection. Their theory expresses the logical consequences of four postulates about populations. If populations are variable and the variable traits can be inherited, and if there is differential (unequal) reproduction based on the traits of individuals, the characteristics of successful individuals will be "naturally selected" and become more common over time.

13.3 How Do We Know That Evolution Has Occurred?

Many lines of evidence indicate that evolution has occurred, including the following:

- Sequences of fossils have been discovered that show a graded series of changes in form. Both of these observations would be expected if modern species evolved from older species.

- Species thought to be related through evolution from a common ancestor possess many similar anatomical structures. An example is the forelimbs of amphibians, reptiles, birds, and mammals.

- Stages in early embryological development are quite similar among very different types of vertebrates.

- Similarities in such biochemical traits as the use of DNA as the carrier of genetic information support the notion of

descent of related species through evolution from common ancestors.

Web Animation   Analogous and Homologous Structures

13.4 What Is the Evidence That Populations Evolve by Natural Selection?

Similarly, many lines of evidence indicate that natural selection is the chief mechanism driving changes in the characteristics of species over time, including the following:

- Inheritable traits have been changed rapidly in populations of domestic animals and plants by selectively breeding organisms with desired features (artificial selection). The immense variations in species produced in a few thousand years of artificial selection by humans makes it almost inevitable that much larger changes would be wrought by hundreds of millions of years of natural selection.

- Evolution can be observed today. Both natural and human activities drastically change the environment over short periods of time. Inherited characteristics of species have been observed to change significantly in response to such environmental changes.

Key Terms

analogous structure *p. 222*
artificial selection *p. 224*
convergent evolution *p. 222*

evolution *p. 211*
fossil *p. 212*

homologous structure *p. 220*
natural selection *p. 217*

population *p. 217*
vestigial structure *p. 220*

Thinking Through the Concepts

Suggested answers to end-of-chapter and figure-based questions can be found at the end of the text.

Fill-in-the-Blank

1. The flipper of a seal is homologous with the _____ of a bird, and both of these are homologous with the _____ of a human. The wing of a bird and the wing of a butterfly are described as _____ structures that arose as a result of _____ evolution. Remnants of structures in animals that have no use for them, such as the small hind leg bones of whales, are described as _____ structures.

2. The finding that all organisms share the same genetic code provides evidence that all descended from a _____. Further evidence is provided by the fact that all cells use roughly the same set of _____ to build proteins, and all cells use the molecule _____ as an energy carrier.

3. Georges Cuvier espoused a concept called _____ to explain layers of rock with embedded fossils. Charles Lyell, building

on the work of James Hutton, proposed an alternative explanation called _____, which states that layers of rock and many other geologic features can be explained by gradual processes that occurred in the past just as they do in the present. This concept provided important support for evolution because it required that Earth be extremely _____.

4. The process by which inherited characteristics of populations change over time is called _____. Variability among individuals is the result of chance changes called _____ that occur in the hereditary molecule _____.

5. The process by which individuals with traits that provide an advantage in their natural habitats are more successful at reproducing is called _____. People who breed animals or plants can produce large changes in their characteristics in a relatively short time, a process called _____.

6. On the basis of observed similarity within families, Darwin postulated that _____. The work of _____ provided the first experimental evidence for this postulate.

Review Questions

1. Selection acts on individuals, but only populations evolve. Explain why this statement is true.

2. Distinguish between catastrophism and uniformitarianism. How did these hypotheses contribute to the development of evolutionary theory?

3. Describe Lamarck's hypothesis of inheritance of acquired characteristics. Why is it invalid?

4. What is natural selection? Describe how natural selection might have caused differential reproduction among the ancestors of a fast-swimming predatory fish, such as the barracuda.

5. Describe how evolution occurs through the interactions among the relatively constant size of natural populations, the reproductive potential of a species, variation among individuals of a species, inheritance, and natural selection.

6. What is convergent evolution? Give an example.

7. How do biochemistry and molecular genetics contribute to the evidence that evolution occurred?

Applying the Concepts

1. IS THIS SCIENCE? In discussions of untapped human potential, it is commonly said that the average person uses only 10% of his or her brain. Is this conclusion likely to be correct? Explain your answer in terms of natural selection.

2. IS THIS SCIENCE? Why is evolution by natural selection a scientific theory, whereas special creation (which states that all species were simultaneously created by God) is not?

3. Does evolution through natural selection produce "better" organisms in an absolute sense? Are we climbing the "ladder of Nature"? Defend your answer.

For additional resources, go to www.mybiology.com.

chapter 14

How Populations Evolve

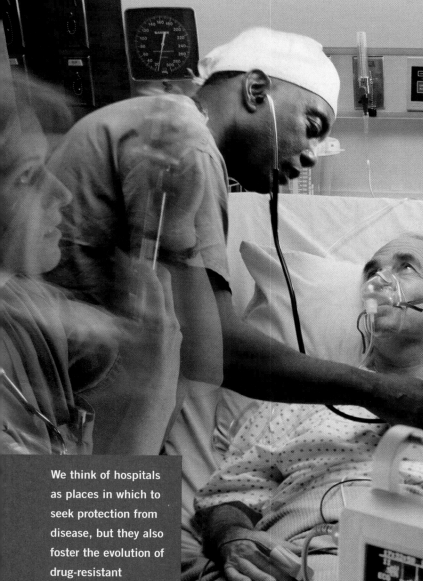

We think of hospitals as places in which to seek protection from disease, but they also foster the evolution of drug-resistant supergerms.

Case Study Evolution of a Menace

When you come down with a bad cold, do you drag yourself to the health center to ask for some antibiotics? If you do, you have probably noticed that physicians are increasingly unlikely to prescribe antibiotics, especially for illnesses such as colds and flu. (These maladies are caused by viruses and are thus not readily cured by antibiotics, which target bacteria.) This increased caution stems from fears that overuse of antibiotics has contributed to the rise of drug-resistant bacteria.

The bacteria that cause disease are becoming less susceptible to antibiotic drugs. For example, more than half of the skin infections treated in U.S. emergency rooms are caused by bacteria that do not respond to formerly effective antibiotics. Drug resistance has also appeared in the bacteria that cause tuberculosis, a disease that kills almost 2 million people each year. Although tuberculosis can be deadly, it is generally treatable. In an increasing number of cases, however, the disease does not respond to any of the drugs commonly used to treat it. Such multidrug-resistant tuberculosis is extremely difficult, sometimes impossible, to cure.

Multidrug-resistant tuberculosis is a frightening and increasingly widespread threat to public health in many parts of the world, including the United States. Drug resistance is also becoming common in many other dangerous bacteria, including those that cause food poisoning, blood poisoning, dysentery, pneumonia, gonorrhea, meningitis, and urinary tract infections. We are experiencing a global onslaught of resistant "supergerms," and are facing the specter of diseases that cannot be cured, even by our best medicines.

Many physicians and scientists believe that the most effective way to combat the rise of resistant diseases is to reduce the use of antibiotics. Why might such a strategy be effective? Because the upsurge of antibiotic resistance is a consequence of evolutionary change in populations of bacteria, and the agent of this change is natural selection applied by antibiotic drugs. To understand how this crisis arose and to devise a strategy to resolve it, we must have a clear understanding of the mechanisms by which populations evolve. ■

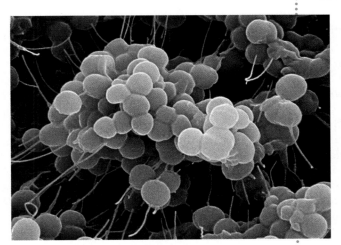

Staphylococcus aureus, a common source of human infections, is among the many bacterial species that have evolved resistance to antibiotics.

14.1 How Are Populations, Genes, and Evolution Related?

If you live in an area with a seasonal climate and you own a dog or cat, you have probably noticed that your pet's fur gets thicker and heavier as winter approaches. Has the animal evolved? No. The changes that we see in an individual organism over the course of its lifetime are not evolutionary changes. Instead, evolutionary changes occur from generation to generation, causing descendants to be different from their ancestors.

Furthermore, we cannot detect evolutionary change across generations by looking at a single set of parents and offspring. For example, if you observed that a 6-foot-tall man had an adult son who stood 5 feet tall, could you conclude that humans were evolving to become shorter? Obviously not. Rather, if you wanted to learn about evolutionary change in human height, you would begin by measuring many humans of many generations to see if the average height is changing over time. Clearly, evolution is a property not of individuals but of populations. (A **population** is a group that includes all the members of a species living in a given area.)

The recognition that evolution is a population-level phenomenon was one of Darwin's key insights. But populations are composed of individuals, and the actions and fates of individuals determine which characteristics will be passed to descendant populations. In this fashion, inheritance provides the link between the lives of individual organisms and the evolution of populations. We therefore begin our discussion of the processes of evolution by reviewing some principles of genetics as they apply to individuals. We then extend those principles to the genetics of populations.

Genes and the Environment Interact to Determine Traits

Each cell of every organism contains genetic information encoded in the DNA of its chromosomes. Recall from Chapter 10 that a *gene* is a segment of DNA located at a particular place on a chromosome. The sequence of nucleotides in a gene encodes the sequence of amino acids in a protein, usually an enzyme that catalyzes a particular reaction in the cell. At a given gene's location, different members of a species may have slightly different nucleotide sequences, called *alleles*. Different alleles generate different forms of the same enzyme. In this way, various alleles of the genes that influence eye color in humans, for example, help produce eyes that are brown, or blue, or green, and so on.

In any population of organisms, there are usually two or more alleles of each gene. An individual of a diploid species whose alleles of a particular gene are both the same is *homozygous* for that gene, and an individual with different alleles for that gene is *heterozygous*. The specific alleles borne on an organism's chromosomes (its *genotype*) interact with the environment to influence the development of its physical and behavioral traits (its *phenotype*).

Let's illustrate these principles with an example. A black hamster's coat is colored black because a chemical reaction in its hair follicles produces a black pigment. When we say that a hamster has the allele for a black coat, we mean that a particular stretch of DNA on one of its chromosomes contains a sequence of nucleotides that codes for the enzyme that catalyzes this reaction. A hamster with the allele for a brown coat has a different sequence of nucleotides at the corresponding chromosomal position. That different sequence codes for an enzyme that cannot produce black pigment. If a hamster is homozygous for the black allele or is heterozygous (one black allele and one brown allele), its fur contains the pigment and is black. But if a hamster is homozygous for the brown allele, its hair follicles produce no black pigment and its coat is brown (Fig. 14-1). Because the hamster's coat is black even when only one copy of the black allele is present, the black allele is considered *dominant* and the brown allele *recessive*.

The Gene Pool Is the Sum of the Genes in a Population

In studying evolution, looking at the process from the point of view of a gene has proven to be an enormously effective tool. In particular, evolutionary biologists have made excellent use of the tools of a branch of genetics called *population genetics*, which deals with the frequency, distribution, and inheritance of alleles in populations. To take advantage of this powerful aid to understanding evolution, you will need to learn a few of the basic concepts of population genetics.

Population genetics defines the **gene pool** as the sum of all the genes in a population. In other words, the gene pool consists of all the alleles of all the genes in all the individuals of a population. Each particular gene can also

▼ **Figure 14-1 Alleles, genotype, and phenotype in individuals**
An individual's particular combination of alleles is its genotype. The word *genotype* can refer to the alleles of a single gene (as shown here), to a set of genes, or to all of an organism's genes. An individual's phenotype is determined by its genotype and environment. Phenotype can refer to a single trait, a set of traits, or all of an organism's traits.

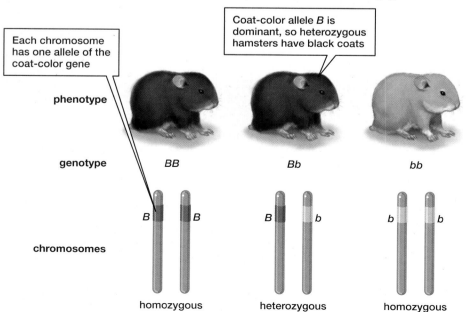

Each chromosome has one allele of the coat-color gene

Coat-color allele *B* is dominant, so heterozygous hamsters have black coats

phenotype

genotype   *BB*   *Bb*   *bb*

chromosomes   *B*   *B*   *B*   *b*   *b*   *b*

           homozygous   heterozygous   homozygous

be considered to have its own gene pool, which consists of all the alleles of that specific gene in a population (Fig. 14-2). If we added up all the copies of each allele of that gene in all the individuals in a population, we could determine the relative proportion of each allele, a number called the **allele frequency.** For example, the population of 25 hamsters portrayed in Fig. 14-2 contains 50 alleles of the gene that controls coat color (because hamsters are diploid and each hamster thus has two copies of each gene). Twenty of those 50 alleles are of the type that codes for black coats, so the frequency of that allele in the population is 40%, because 20/50 = 0.40, or 40%.

Evolution Is the Change of Allele Frequencies Within a Population

A casual observer might define evolution on the basis of changes in the outward appearance or behaviors of the members of a population. A population geneticist, however, looks at a population and sees a gene pool that just happens to be divided into the packages that we call individual organisms. So any outward changes that we observe in the individuals that make up the population can also be viewed as the visible expression of underlying changes to the gene pool. A population geneticist therefore defines evolution as the changes in allele frequencies that occur in a gene pool over time. Evolution is a change in the genetic makeup of a population over generations.

The Equilibrium Population Is a Hypothetical Population in Which Evolution Does Not Occur

It is easier to understand what causes populations to evolve if the characteristics of a population that would *not* evolve are considered first. In 1908, English mathematician Godfrey H. Hardy and German physician Wilhelm Weinberg independently developed a simple mathematical model now known as the **Hardy-Weinberg principle.** This model showed that, under certain conditions, allele frequencies and genotype frequencies in a population will remain constant no matter how many generations pass. In other words, this population will not evolve. Population geneticists use the term **equilibrium population** for this idealized, nonevolving population in which allele frequencies do not change, as long as the following conditions are met:

- There must be no mutation.

- There must be no **gene flow** between populations. That is, there must be no movement of alleles into or out of the population (as would be caused, for example, by the movement of organisms into or out of the population).

- The population must be very large.

- All mating must be random, with no tendency for certain genotypes to mate with specific other genotypes.

- There must be no natural selection. That is, all genotypes must reproduce with equal success.

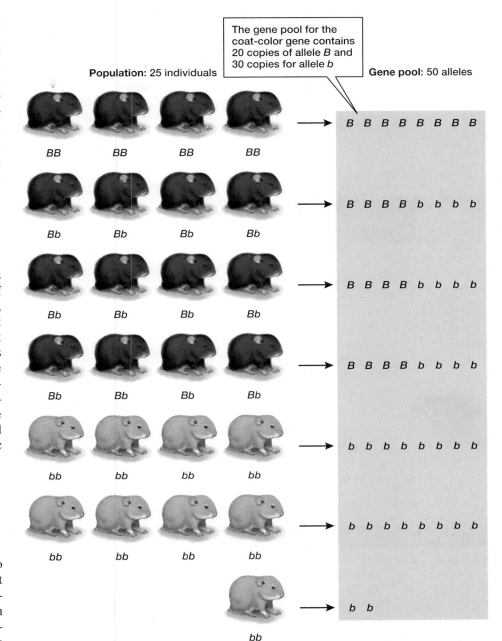

Population: 25 individuals

The gene pool for the coat-color gene contains 20 copies of allele *B* and 30 copies for allele *b*

Gene pool: 50 alleles

BB BB BB BB → B B B B B B B B

Bb Bb Bb Bb → B B B B b b b b

Bb Bb Bb Bb → B B B B b b b b

Bb Bb Bb Bb → B B B B b b b b

bb bb bb bb → b b b b b b b b

bb bb bb bb → b b b b b b b b

bb → b b

▲ Figure 14-2 **A gene pool** In diploid organisms, each individual in a population contributes two alleles of each gene to the gene pool.

Under these conditions, allele frequencies within a population will remain the same indefinitely. If one or more of these conditions is violated, then allele frequencies will change. The population will evolve.

As you might expect, few if any natural populations are truly in equilibrium. What, then, is the importance of the Hardy-Weinberg principle? The conditions specified by the Hardy-Weinberg principle are useful starting points for studying the mechanisms of evolution. In the following sections, we examine some of the conditions, show that natural populations often fail to meet them, and illustrate the consequences of such failures. In this way, we can better understand both the inevitability of evolution and the processes that drive evolutionary change.

14.2 What Causes Evolution?

Population genetics theory predicts that the Hardy-Weinberg equilibrium can be disturbed by deviations from any of its five conditions. Most evolutionary biologists agree, however, that the most important sources of evolutionary change are mutation, small population size, and natural selection.

Mutations Are the Original Source of Genetic Variability

A population remains in genetic equilibrium only if there are no **mutations** (changes in DNA sequence). Most mutations occur during cell division, when a cell must make a copy of its DNA. Sometimes, there are errors in the copying process and the copied DNA does not match the original. Most such errors are quickly corrected by cellular systems that identify and repair DNA copying mistakes, but some changes in nucleotide sequence slip past the repair systems. An unrepaired mutation in a cell that gives rise to gametes may be passed to offspring and enter the gene pool of a population.

Inherited Mutations Are Rare but Important

How significant is mutation in changing the gene pool of a population? For any given gene, only a tiny proportion of a population inherits a mutation from the previous generation. For example, a mutant version of a typical human gene will appear in only about 1 out of every 100,000 gametes produced and, because new individuals are formed by the fusion of two gametes, in only about 1 of every 50,000 newborns. Therefore, mutation by itself generally causes only very small changes in the frequency of any particular allele.

Despite the rarity of inherited mutations of any particular gene, the cumulative effect of mutations is essential to evolution. Most organisms have a large number of different genes, so even if the rate of mutation is low for any one gene, the sheer number of possibilities means that each new generation of a population is likely to include some mutations. For example, humans have about 21,000 different genes, so each person carries about 42,000 alleles. Thus, even if each allele has, on average, only a 1 in 100,000 chance of mutation, most newborn individuals will probably carry one or two mutations overall. These mutations are the source of new alleles—new variations on which other evolutionary processes can work. As such, they are the foundation of evolutionary change. Without mutations there would be no evolution.

Mutations Are Not Goal Directed

A mutation does not arise as a result of, or in anticipation of, environmental necessities. A mutation simply happens and may in turn produce a change in a structure or function of an organism. Whether that change is helpful or harmful or neutral, now or in the future, depends on environmental conditions over which the organism has little or no control (**Fig. 14-3**). The mutation provides a potential for evolutionary

Web Animation Agents of Change

Evolution of a Menace

Continued

If mutations are rare and random, why do mutant alleles that confer antibiotic resistance arise so commonly in bacterial populations? The seemingly inevitable emergence of resistance results in large measure from the huge size of bacteria populations and from their very short generation times. A single drop of human saliva contains about 150 million bacteria, and an entire human body contains hundreds of trillions. With so many bacteria present, even an extremely rare resistance mutation that occurs in only a tiny percentage of the population will be present in some individuals. Also, because many mutations occur during cell division (when DNA replicates and copying errors can occur), rapid reproduction creates many opportunities for mutations to arise. Bacteria reproduce very rapidly, as often as every 15 minutes in some species. This rapid reproduction, taking place in huge populations, results in a high likelihood that mutations leading to antibiotic resistance will be present in bacterial populations.

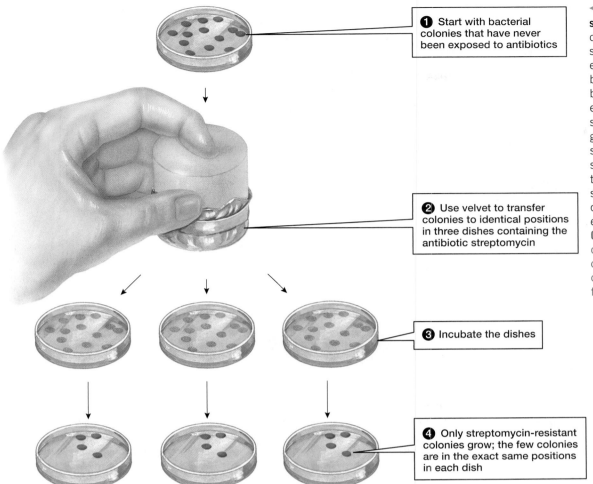

1 Start with bacterial colonies that have never been exposed to antibiotics

2 Use velvet to transfer colonies to identical positions in three dishes containing the antibiotic streptomycin

3 Incubate the dishes

4 Only streptomycin-resistant colonies grow; the few colonies are in the exact same positions in each dish

◄ **Figure 14-3 Mutations occur spontaneously** This experiment demonstrates that mutations occur spontaneously and not in response to environmental pressures. When bacterial colonies that have never been exposed to antibiotics are exposed to the antibiotic streptomycin, only a few colonies grow. The observation that these surviving colonies grow in the exact same positions in all dishes shows that the mutations for resistance to streptomycin were present in the original dish before exposure to the environmental pressure, streptomycin. **QUESTION:** If it were true that mutations do occur in response to the presence of an antibiotic, how would the result of this experiment have differed from the actual result?

change. Other processes, especially natural selection, may act to spread the mutation through the population or to eliminate it from the population.

Allele Frequencies May Drift in Small Populations

Allele frequencies in populations can be changed by chance events other than mutations. For example, if bad luck prevents some members of a population from reproducing, their alleles will ultimately be removed from the gene pool, altering its makeup. What kinds of bad-luck events can randomly prevent some individuals from reproducing? Seeds can fall into a pond and never sprout; flowers can be destroyed by a hailstorm; or organisms can be killed by a fire or by a volcanic eruption. Any event that arbitrarily cuts lives short or otherwise allows only a random subset of a population to reproduce can cause random changes in allele frequencies (**Fig. 14-4**). The process by which chance events change allele frequencies is called **genetic drift.**

To see how genetic drift works, imagine a population of 20 hamsters in which the frequency of the black coat-color allele B is 50% and the frequency of the brown coat-color allele b is 50% (Fig. 14-4, top). If all of the hamsters in the population were to interbreed to yield another population of 20 animals, the frequencies of the two alleles would not change in the next generation. But if we instead allow only two, randomly chosen hamsters (the ones circled in Fig. 14-4, top) to breed and become the parents of the next generation of 20 animals, allele frequencies might be quite different in generation 2 (Fig. 14-4, center; the frequency of B has decreased and the frequency of b has increased). And if breeding in the second generation

▶ **Figure 14-4 Genetic drift** If chance events prevent some members of a population from reproducing, allele frequencies can change randomly. **QUESTION:** Explain why the distribution of genotypes in Generation 2 is as shown.

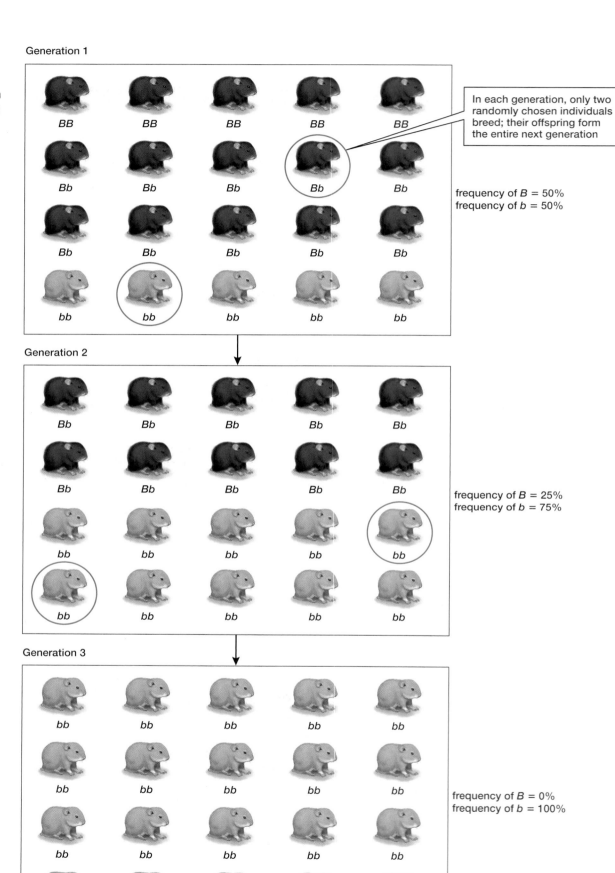

Generation 1

BB   BB   BB   BB   BB

Bb   Bb   Bb   Bb   Bb

Bb   Bb   Bb   Bb   Bb

bb   bb   bb   bb   bb

In each generation, only two randomly chosen individuals breed; their offspring form the entire next generation

frequency of B = 50%
frequency of b = 50%

Generation 2

Bb   Bb   Bb   Bb   Bb

Bb   Bb   Bb   Bb   Bb

bb   bb   bb   bb   bb

bb   bb   bb   bb   bb

frequency of B = 25%
frequency of b = 75%

Generation 3

bb   bb   bb   bb   bb

bb   bb   bb   bb   bb

bb   bb   bb   bb   bb

bb   bb   bb   bb   bb

frequency of B = 0%
frequency of b = 100%

were again restricted to two randomly chosen hamsters (circled in Fig. 14-4, center), allele frequencies might change again in the third generation (Fig. 14-4, bottom). Allele frequencies will continue to change in random fashion for as along as reproduction is restricted to a random subset of the population. Note that the changes caused by genetic drift can include the disappearance of an allele from the population, as illustrated by the disappearance of the *B* allele in generation 3 of the example shown in Fig. 14-4.

Population Size Matters

Genetic drift occurs to some extent in all populations, but it occurs more rapidly and has a bigger effect in small populations than in large ones. If a population is sufficiently large, chance events are unlikely to significantly alter its genetic composition, because random removal of a few individuals' alleles won't have a big impact on allele frequencies in the population as a whole. In a small population, however, a particular allele may be carried by only a few organisms. Chance events could eliminate most or all examples of such an allele from the population.

To see how population size affects genetic drift, imagine two populations of amoebas in which each amoeba is either red or blue, and color is controlled by two alleles (*A* and *a*) of a gene. Half of the amoebas in each of our two populations are red and half are blue. One population, however, has only four individuals in it, whereas the other has 10,000.

Now let's picture reproduction in our imaginary populations. Let's select, at random, half of the individuals in each population and allow them to reproduce by dividing. To do so, each reproducing amoeba splits in half to yield two amoebas, each of which is the same color as the parent. In the large population, 5,000 amoebas reproduce, yielding a new generation of 10,000. What are the chances that all 10,000 members of the new generation will be red? Just about nil. In fact, it would be extremely unlikely for even 3,000 amoebas to be red or for 7,000 to be red. The most likely outcome is that about half will be red and half blue, just as in the original population. In this large population, then, we would not expect a major change in allele frequencies from generation to generation.

One way to test this prediction is to write a computer program that simulates how the allele frequencies of the alleles could change over generations. **Figure 14-5a** shows the results from four runs of such a simulation. Notice that the frequency of allele *A*, encoding red color, remains close to 50%, consistent with the expectation that half of the amoebas would be red.

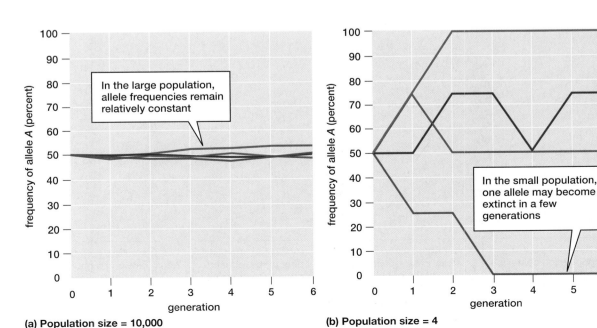

(a) **Population size = 10,000**

(b) **Population size = 4**

◀ **Figure 14-5 The effect of population size on genetic drift** Each colored line represents one computer simulation of the change over time in the frequency of allele *A* in a **(a)** large and a **(b)** small population. In each population, half the alleles were *A* (50%), and randomly chosen individuals reproduced. **EXERCISE:** Sketch a graph that shows the result you would predict if the simulation were run four times with a population size of 20.

In the small population, the situation is different. Only two amoebas reproduce, and there is a 25% chance that both reproducers will be red. (This outcome is as likely as flipping two coins and having both come up heads.) If only red amoebas reproduce, then the next generation will consist entirely of red amoebas—a relatively likely outcome. It is thus possible, within a single generation, for the allele for blue color to disappear from the population.

Figure 14-5b shows the fate of allele *A* in four runs of a simulation of our small population. In one of the four runs (red line), allele *A* reaches a frequency of 100% in the second generation, meaning that all of the amoebas in the third and following generations are red. In another run, the frequency of *A* drifts to zero in the fourth generation (blue line), and the population subsequently is all blue. Thus, one of the two amoeba phenotypes disappeared in half of the simulations.

A Population Bottleneck Can Cause Genetic Drift

Two causes of genetic drift, the population bottleneck and the founder effect, further illustrate the effect that small population size can have on the allele frequencies of a species. In a **population bottleneck,** a population is drastically reduced, as a result of a natural catastrophe or overhunting, for example. Then, only a few individuals are available to contribute genes to the next generation. Population bottlenecks can rapidly change allele frequencies and can reduce genetic variability by eliminating alleles (**Fig. 14-6**). Even if the population later increases, the genetic effects of the bottleneck may remain for hundreds or thousands of generations.

A special case of a population bottleneck is the **founder effect,** which occurs when isolated colonies are founded by a small number of organisms. A flock of birds, for instance, that becomes lost during migration or is blown off course by a storm may settle on an isolated island. The small founder group may, by chance, have allele frequencies that are very different from the frequencies of the parent population. If they do, the gene pool of the future population in the new location will be quite unlike that of the larger population from which it sprang. For example, a set of genetic defects known as Ellis–van Creveld syndrome is far more common

Web Animation The Bottleneck Effect

▶ **Figure 14-6 Population bottlenecks reduce variation** A population bottleneck may drastically reduce genotypic and phenotypic variation because the few organisms that survive may all carry similar sets of alleles. **QUESTION:** If a population grows large again after a bottleneck, genetic diversity will ultimately increase. Why?

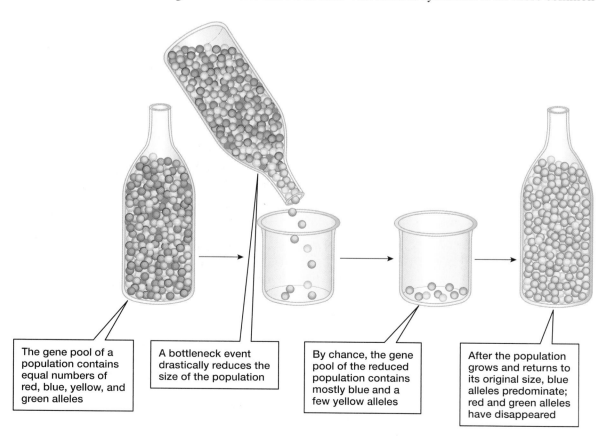

The gene pool of a population contains equal numbers of red, blue, yellow, and green alleles

A bottleneck event drastically reduces the size of the population

By chance, the gene pool of the reduced population contains mostly blue and a few yellow alleles

After the population grows and returns to its original size, blue alleles predominate; red and green alleles have disappeared

among the Amish inhabitants of Lancaster County, Pennsylvania, than among the general population (Fig. 14-7). Today's Lancaster County Amish are descended from only 200 or so eighteenth-century immigrants, and one couple among these immigrants is known to have carried the Ellis–van Creveld allele. In such a small founder population, this single occurrence meant that the allele was carried by a comparatively high proportion of the Amish founder population (1 or 2 carriers out of 200, versus perhaps 1 in 1,000 in the general population). This high initial allele frequency, combined with subsequent genetic drift, has led to extraordinarily high levels of Ellis–van Creveld syndrome among this Amish group.

All Genotypes Are Not Equally Beneficial

In a hypothetical equilibrium population, individuals of all genotypes survive and reproduce equally well; no genotype has any advantage over the others. This condition, however, is probably met only rarely, if ever, in real populations. Even though some alleles are neutral, in the sense that organisms possessing any of several alleles are equally likely to survive and reproduce, clearly not all alleles are neutral in all environments. Any time an allele provides, in Alfred Russel Wallace's words, "some little superiority," natural selection favors the individuals who possess it. That is, those individuals have higher reproductive success. This phenomenon is illustrated by an example concerning an antibiotic drug.

Antibiotic Resistance Evolves by Natural Selection

The antibiotic penicillin first came into widespread use during World War II, when it was used to combat infections in wounded soldiers. Suppose that an infantryman, brought to a field hospital after suffering a gunshot wound in his arm, develops a bacterial infection in that arm. A medic resolves to treat the wounded soldier with an intravenous drip of penicillin. As the antibiotic courses through the soldier's blood vessels, millions of bacteria die before they can reproduce. A few bacteria, however, carry a rare allele that codes for an enzyme that destroys any penicillin that comes into contact with the bacterial cell. (This allele is a variant of a gene that normally codes for an enzyme that breaks down the bacterium's waste products.) The bacteria carrying this rare allele are able to survive and reproduce, and their offspring inherit the penicillin-destroying allele. After a few generations, the frequency of the penicillin-destroying allele has soared to nearly 100%, and the frequency of the normal, waste-processing allele has declined to near zero. As a result of natural selection imposed by the antibiotic's killing power, the population of bacteria within the soldier's body has evolved. The gene pool of the population has changed, and natural selection, in the form of bacterial destruction by penicillin, has caused the change.

Penicillin Resistance Illustrates Key Points About Evolution

The example of penicillin resistance highlights some important features of natural selection and evolution.

Natural selection does not cause genetic changes in individuals. The allele for penicillin resistance arose spontaneously, long before penicillin was dripped into the soldier's vein. Penicillin did not cause resistance to appear; its presence merely favored the survival of bacteria with penicillin-destroying alleles over that of bacteria with waste-processing alleles.

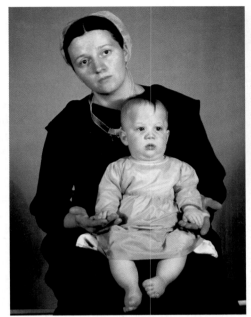

(a) A child with Ellis–van Creveld syndrome

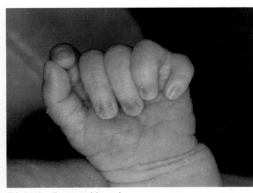

(b) A six-fingered hand

▲ Figure 14-7 A human example of the founder effect
(a) An Amish woman with her child, who suffers from a set of genetic defects known as Ellis–van Creveld syndrome. Symptoms of the syndrome include short arms and legs, *(b)* extra fingers, and, in some cases, heart defects. The founder effect accounts for the prevalence of Ellis–van Creveld syndrome among the Amish residents of Lancaster County, Pennsylvania.

(a) A contest for dominance

(b) Drinking at a water hole

▲ **Figure 14-8 A compromise between opposing environmental pressures (a)** A male giraffe with a long neck is at a definite advantage when establishing dominance during combat. **(b)** But a giraffe's long neck forces it to assume an extremely awkward and vulnerable position when drinking. Thus, drinking and male–male contests place opposing evolutionary pressures on neck length.

Web Animation Natural Selection for Antibiotic Resistance

Natural selection acts on individuals, but it is populations that are changed by evolution. The agent of natural selection, penicillin, acted on individual bacteria. As a result, some individuals reproduced and some did not. However, it was the population as a whole that evolved as its allele frequencies changed.

Evolution is change in allele frequencies of a population, owing to unequal success at reproduction among organisms bearing different alleles. In evolutionary terminology, the **fitness** of an organism is measured by its reproductive success. In our example, the penicillin-resistant bacteria had greater fitness than the normal bacteria did, because the resistant bacteria produced greater numbers of viable (able to survive) offspring.

Evolution is not progressive; it does not make organisms "better." The traits favored by natural selection change as the environment changes. The resistant bacteria were favored only because of the presence of penicillin in the soldier's body. At a later time, when the environment of the soldier's body no longer contained penicillin, the resistant bacteria may have been at a disadvantage relative to other bacteria that could process waste more effectively. Similarly, the long necks of male giraffes are helpful when the animals battle to establish dominance, but are a hindrance to drinking (Fig. 14-8). The length of male giraffe necks represents an evolutionary compromise between the advantage of being able to win contests with other males and the disadvantage of vulnerability while drinking water. (The necks of female giraffes are long—though not as long as male necks—because successful males pass the alleles for long necks to daughters as well as to sons.)

14.3 How Does Natural Selection Work?

Natural selection is not the only evolutionary force. As we have seen, mutation provides variability in heritable traits, and the chance effects of genetic drift may change allele frequencies. Further, evolutionary biologists are now beginning to appreciate the power of random catastrophes in shaping the history of life on Earth; massively destructive events may exterminate thriving and failing species alike. Nevertheless, it is natural selection that shapes the evolution of populations as they adapt to their changing environment. For this reason, we next examine natural selection in more detail.

Natural Selection Stems from Unequal Reproduction

To most people, **natural selection** is synonymous with the phrase "survival of the fittest." Natural selection evokes images of wolves chasing caribou or of lions snarling angrily in competition over a zebra carcass. Natural selection, however, is not about survival alone. It is also about reproduction. It is certainly true that if an organism is to reproduce, it must survive long enough to do so. In some cases, it is also true that a longer-lived organism has more chances to reproduce. But no organism lives forever, and the only way that its genes can continue into the future is through successful reproduction. When an organism dies without reproducing, its genes die with it. An organism that reproduces lives on, in a sense, through the genes that it has passed to its offspring. Therefore, although evolutionary biologists often discuss survival, partly because survival is usually easier to observe than reproduction, the main issue of natural selection is *differences in reproduction*: Individuals bearing certain alleles leave more offspring (who inherit those alleles) than do other individuals with different alleles.

Natural Selection Acts on Phenotypes

Although we have defined evolution as changes in the genetic composition of a population, it is important to recognize that natural selection cannot act directly on the genotypes of individual organisms. Rather, natural selection acts on phenotypes, the structures and behaviors displayed by the members of a population. This selection of phenotypes, however, inevitably affects the genotypes present in a population, because phenotypes and genotypes are closely tied. For example, we know that a pea plant's height is strongly influenced by the plant's alleles of certain genes. If a population of pea plants were to encounter environmental conditions that favored taller plants, then taller plants would leave more off-spring. These offspring would carry the alleles that contributed to their parents' height. Thus, if natural selection favors a particular phenotype, it will necessarily also favor the underlying genotype.

Some Phenotypes Reproduce More Successfully Than Others

As we have seen, natural selection simply means that some phenotypes reproduce more successfully than do others. This simple process is such a powerful agent of change because only the fittest phenotypes pass traits to subsequent generations. But what makes a phenotype fit? Successful phenotypes are those that have the best adaptations to their particular environment. **Adaptations** are characteristics that help an individual survive and reproduce.

An Environment Has Nonliving and Living Components

Individual organisms must cope with an environment that includes not only physical factors but also the other organisms with which the individual interacts. The nonliving (abiotic) component of the environment includes factors such as climate, availability of water, and minerals in the soil. The abiotic environment plays a large role in determining the traits that help an organism to survive and reproduce. However, adaptations also arise because of interactions with other organisms—the living (biotic) component of the environment. As Darwin wrote, "The structure of every organic being is related . . . to that of all other organic beings, with which it comes into competition for food or residence, or from which it has to escape, or on which it preys." A simple example illustrates this concept.

Consider a buffalo grass plant growing in a small patch of soil in the eastern Wyoming plains. Its roots must be able to take up enough water and minerals for growth and reproduction, and to that extent, it must be adapted to its abiotic environment. But even in the dry prairies of Wyoming, this requirement is relatively trivial, provided that the plant is alone and protected in its square yard of soil. In reality, however, many other plants—other buffalo grass plants as well as other types of grasses, sagebrush bushes, and annual wildflowers—also sprout in that same patch of soil. If our buffalo grass is to survive, it must compete with the other plants for resources. Its long, deep roots and efficient methods of mineral uptake have evolved not so much because the plains are dry as because the buffalo grass must share the dry prairies with other plants. Further, buffalo grass must also coexist with animals that wish to eat it, such as the cattle that graze the prairie (and the bison that grazed it in the past). As a result, buffalo grass is extremely tough. Silica compounds reinforce its leaves, an adaptation that discourages grazing. Over time, tougher, hard-to-eat plants survived better and reproduced more than did less-tough plants—another adaptation to the biotic environment.

Competition Acts as an Agent of Selection

As the buffalo grass example shows, one of the major agents of natural selection in the biotic environment is **competition** with other organisms for scarce resources. Competition for resources is most intense among members of the same

▲ **Figure 14-9 Competition between males favors the evolution, through sexual selection, of structures for ritual combat** Two male bighorn sheep spar during the fall mating season. In many species, the losers of such contests are unlikely to mate, while winners enjoy tremendous reproductive success. **QUESTION:** If we studied a population of bighorn sheep and were able to identify the father and mother of each lamb born, would you predict that the difference in number of offspring between the most reproductively successful adult and the least successful adult would be greater for males or for females?

▲ **Figure 14-10 The peacock's showy tail has evolved through sexual selection** The ancestors of today's peahens were apparently picky when deciding on a male with which to mate, favoring males with longer and more colorful tails.

species. As Darwin wrote in *On the Origin of Species*, "The struggle almost invariably will be most severe between the individuals of the same species, for they frequent the same districts, require the same food, and are exposed to the same dangers." In other words, no competing organism has such similar requirements for survival as does another member of the same species. Different species may also compete for the same resources, although generally to a lesser extent than do individuals within a species.

Both Predator and Prey Act as Agents of Selection

When two species interact extensively, each exerts strong selection on the other. When one evolves a new feature or modifies an old one, the other typically evolves new adaptations in response. This constant, mutual feedback between two species is called **coevolution.** Perhaps the most familiar form of coevolution is found in predator–prey relationships.

Predation includes any situation in which one organism eats another. In some instances, coevolution between predators (those who do the eating) and prey (those who are eaten) is a sort of "biological arms race," with each side evolving new adaptations in response to "escalations" by the other. Darwin used the example of wolves and deer: Wolf predation selects against slow or careless deer, thus leaving faster, more-alert deer to reproduce and continue the species. In their turn, alert, swift deer select against slow, clumsy wolves, because such predators cannot acquire enough food.

Sexual Selection Favors Traits That Help an Organism Mate

In many animal species, males have conspicuous features such as bright colors, long feathers or fins, or elaborate antlers. Males may also exhibit bizarre courtship behaviors or sing loud, complex songs. Although these extravagant features typically play a role in mating, they also seem to be at odds with efficient survival and reproduction. Exaggerated ornaments and displays may help males gain access to females, but they also make the males more conspicuous and thus vulnerable to predators. Darwin was intrigued by this apparent contradiction. He coined the term **sexual selection** to describe the special kind of selection that acts on traits that help an animal acquire a mate.

Darwin recognized that sexual selection could be driven either by sexual contests among males or by female preference for particular male phenotypes. Male–male competition for access to females can favor the evolution of features that provide an advantage in fights or ritual displays of aggression (**Fig. 14-9**). In many species, the winners of such male–male contests experience increased mating success, either because winners gain control of groups of females, or because winners gain control of resources, such as territories, that are attractive to breeding females. Female mate choice provides a second source of sexual selection. In animal species in which females actively choose their mates from among males, females often seem to prefer males with the most elaborate ornaments or most extravagant displays (**Fig. 14-10**). Why?

A popular hypothesis is that male structures, colors, and displays that do not enhance survival might instead provide a female with an outward sign of a male's condition. Only a vigorous, energetic male can survive when burdened with conspicuous coloration or a large tail that might make him more vulnerable to predators. Conversely, males that are sick or under parasitic attack are dull and frumpy compared with healthy males. A female that chooses the brightest, most ornamented male is also choosing the healthiest, most vigorous male. By doing so, she gains fitness if, for example, the most vigorous male provides superior parental care to offspring or if he carries alleles for disease resistance that will be inherited by offspring and help ensure their survival. Females thus gain a reproductive advantage by choosing the most highly ornamented males, and the traits (including the exaggerated male ornament) of these flashy males will be passed to subsequent generations.

earth watch

Cloning Endangered Species

Environmentalists tend to be skeptical of claims that biotechnology can help solve environmental problems, but one group of scientists is determined to use bioengineering to rescue endangered species. Researchers at the biotech company Advanced Cell Technologies (ACT) have embarked on an ambitious plan to clone species that are in danger of extinction. They had their first success a few years ago with the birth of a gaur, a wild ox native to India and Southeast Asia (**Fig. E14-1**). Gaurs are very rare and in danger of extinction because their habitat is disappearing rapidly.

The gaur was cloned using a preserved skin cell from an individual that had died 8 years earlier. The genetic material from this cell was inserted into a cow egg from which the nucleus had been removed (see "Scientific Inquiry: Carbon Copies—Cloning in Nature and the Lab" on pp. 128–129). The resulting cloned embryo was implanted in the uterus of a surrogate mother cow. (Cows were chosen to be the egg donor and surrogate mother because of easy availability and to avoid posing undue risk to any gaur.) The embryo developed properly and the cow gave birth to the gaur calf. Unfortunately, the calf died 2 days later of dysentery, an infection that the researchers say was unrelated to the cloning procedure.

Buoyed by their first success, the cloning team subsequently cloned a banteng, a highly endangered relative of the gaur, also native to Southeast Asia. Two banteng embryos were cloned from a cell taken from the frozen tissue of an animal that had died in 1980. The embryos were carried to term by cow surrogate mothers. One of the clones survived and now lives with the banteng herd at the San Diego Zoo. Zookeepers hope that, when the cloned animal matures, it will breed and add much-needed genetic diversity to the captive herd.

The scientists at ACT plan to continue with the cloning project. They recently received permission from the Spanish government to clone the bucardo, a mountain goat species of Europe that is already extinct (the last captive individual recently died). The scientists have spoken of being especially inspired by the idea of resurrecting an extinct species (they will use preserved bucardo tissues that are in storage in Spain). The researchers have also focused on the endangered and charismatic giant panda, as well as endangered big cats such as tigers, cheetahs, and leopards.

The idea of cloning endangered species is highly controversial among conservationists. Many feel that producing new animals one at a time can do little to truly help endangered species and that, in any case, the ability to artificially reproduce members of a species does little good if the species' habitat has been destroyed. Advocates reply that cloning may be a way to keep a species alive until efforts to restore its habitat are successful and to preserve genetic diversity by saving the genotypes of the most-threatened populations of a species. Many who wish to continue efforts to clone endangered species argue that it doesn't make sense to abandon any approach to conserving species, because we will need more than one strategy to solve such a difficult problem.

▲ **Figure E14-1** **The gaur, an endangered Asian ox, was recently cloned**

14.4 What Is a Species?

Although Darwin brilliantly explained how evolution shapes complex, amazingly well-designed organisms, his ideas did not fully explain life's diversity. In particular, the process of natural selection cannot by itself explain how living things came to be divided into groups, with each group distinctly different from all other groups. When we look at big cats, we don't see a continuous array of different tiger phenotypes that gradually grades into a lion phenotype. We see lions and tigers as separate, distinct types with no overlap. Each distinct type is known as a species.

(Populations of some species are very small. To find out about an unorthodox approach to preserving such endangered species, see "Earth Watch: Cloning Endangered Species").

Biologists Need a Clear Definition of Species

Before we can study the origin of species, we must first clarify our definition of the term. Throughout most of human history, "species" was a poorly defined concept. In pre-Darwinian Europe, the word "species" simply referred to one of the "kinds" produced by the biblical creation. In this view, humans could not possibly know the criteria of the creator, but could only attempt to distinguish among species on the basis of visible differences in structure. In fact, "species" is Latin for "appearance."

On a coarse scale, it is easy to use quick visual comparisons to distinguish species. For example, warblers are clearly different from eagles, which are obviously different from ducks. But it is far more difficult to distinguish among different species of warblers, eagles, or ducks. How do scientists make these finer distinctions?

Species Are Groups of Interbreeding Populations

Today, biologists define a **species** as a group of populations that evolves independently. Each species follows a separate evolutionary path because alleles do not move between the gene pools of different species. This definition, however, does not clearly state the standard by which such evolutionary independence is judged. The most widely used standard defines species as "groups of actually or potentially interbreeding natural populations, which are reproductively isolated from other such groups." This definition, known as the *biological species concept*, is based on the observation that **reproductive isolation** (inability to successfully breed outside the group) ensures evolutionary independence.

The biological species concept has at least two major limitations. First, because the definition is based on patterns of sexual reproduction, it does not help us determine species boundaries among asexually reproducing organisms. Second, it is not always practical or even possible to directly observe whether members of two different groups interbreed. Thus, a biologist who wishes to determine if a group of organisms is a separate species must often make the determination without knowing for sure if group members breed with organisms outside the group.

Despite the limitations of the biological species concept, most biologists accept it for identifying species of sexually reproducing organisms. Nonetheless, scientists who study bacteria and other organisms that reproduce asexually must use alternative definitions of species.

Appearance Can Be Misleading

Biologists have found that some organisms with very similar appearances belong to different species. For example, the cordilleran flycatcher and the Pacific-slope flycatcher are so similar that even experienced bird-watchers cannot tell them apart (Fig. 14-11). Until recently, these birds were considered to be a single species. However, research revealed that the two kinds of birds do not interbreed and are in fact two different species.

Conversely, differences in appearance do not always mean that two populations belong to different species. For example, bird field guides published in the 1970s

Evolution of a Menace

Continued

In bacteria, there are sometimes exceptions to the general principle that the gene pools of different species are isolated from one another. These exceptions play a role in the spread of antibiotic resistance. Some bacterial genes are carried on small, circular pieces of DNA called plasmids. Within species, plasmids (and the genes they carry) are commonly transferred from one individual to another by a process called conjugation. On occasion, conjugation may occur between members of different species. If some of the genes on a plasmid transferred between species are useful to the receiving individual, and if that individual prospers and has many descendants, genes can jump across the barrier between species. Unfortunately, genes that confer antibiotic resistance have jumped between species in just this fashion, hastening the spread of resistance to additional disease-causing species.

▶ **Figure 14-11 Members of different species may be similar in appearance** *(a)* The cordilleran flycatcher and *(b)* Pacific-slope flycatcher are different species.

(a) Cordilleran flycatcher

(b) Pacific-slope flycatcher

list the myrtle warbler and Audubon's warbler as distinct species (Fig. 14-12). These birds differ in geographical range and in the color of their throat feathers. More recently, scientists decided that these birds are local varieties of the same species. The reason: Where their ranges overlap, these warblers interbreed, and the offspring are just as vigorous and fertile as the parents.

14.5 How Is Reproductive Isolation Between Species Maintained?

What prevents different species from interbreeding? The traits that prevent interbreeding and maintain reproductive isolation are called **isolating mechanisms.** Isolating mechanisms provide a clear benefit to individuals. Any individual that mates with a member of another species will probably produce no offspring (or unfit or sterile offspring), thereby wasting its reproductive effort and failing to contribute to future generations. Thus, natural selection favors traits that prevent mating across species boundaries. Mechanisms that prevent mating between species are called **premating isolating mechanisms.**

When premating isolating mechanisms fail or have not yet evolved, members of different species may mate. If, however, all resulting hybrid offspring die during development, then the two species are still reproductively isolated from one another. Even if hybrid offspring are able to survive, if these hybrids are less fit than their parents or are themselves infertile, the two species may still remain separate, with little or no gene flow between them. Mechanisms that prevent the formation of vigorous, fertile hybrids between species are called **postmating isolating mechanisms.**

Premating Isolating Mechanisms Prevent Mating Between Species

Reproductive isolation can be maintained by a variety of mechanisms, but those that prevent mating attempts are especially effective. We next describe the most important types of such premating isolating mechanisms.

Members of Different Species May Be Prevented from Meeting

Members of different species cannot mate if they never get near one another. *Geographical isolation* prevents interbreeding between populations that do not come into contact because they live in different, physically separated places. However, we cannot determine if geographically separated populations are actually distinct species. Should the physical barrier separating the two populations disappear (a new channel might connect two previously isolated lakes, for example), the reunited populations might interbreed freely and not be separate species after all. If they cannot interbreed, then other mechanisms, such as those we consider next, must have developed during their isolation. Geographical isolation, therefore, is usually considered to be a mechanism that allows new species to form rather than a mechanism that maintains reproductive isolation between species.

Different Species May Occupy Different Habitats

Two populations that use different resources may spend time in different habitats within the same general area and thus exhibit *ecological isolation*. White-crowned and white-throated sparrows, for example, have extensively overlapping ranges. The white-throated sparrow, however, frequents dense thickets, whereas the white-crowned sparrow inhabits fields and meadows, seldom penetrating far into dense growth. The two species may coexist within a few hundred yards of one another and yet seldom meet during the breeding season. A more dramatic example is provided by the more than 750 species of fig wasp (Fig. 14-13). Each species of fig wasp breeds in (and pollinates) the fruits of a particular species of fig, and each fig species hosts one and only one species of pollinating wasp.

(a) Myrtle warbler

(b) Audubon's warbler

▲ Figure 14-12 **Members of a species may differ in appearance** *(a)* The myrtle warbler and *(b)* Audubon's warbler are members of the same species.

▲ Figure 14-13 **Ecological isolation** This female fig wasp is carrying fertilized eggs from a mating that took place within a fig. She will find another fig of the same species, enter it through a pore, lay eggs, and die. Her offspring will hatch, develop, and mate within the fig. Because each species of fig wasp reproduces only in its own particular fig species, each wasp species is reproductively isolated.

Although ecological isolation may slow down interbreeding, it seems unlikely that it could prevent gene flow entirely. Other mechanisms normally also contribute to reproductive isolation.

Different Species May Breed at Different Times

Even if two species occupy similar habitats, they cannot mate if they have different breeding seasons, a phenomenon called *temporal (time-related) isolation*. For example, the spring field cricket and the fall field cricket both occur in many areas of North America but, as their names suggest, the former species breeds in spring and the latter in autumn. As a result, the two species do not interbreed.

In plants, the reproductive structures of different species may mature at different times. For example, Bishop pines and Monterey pines grow together near Monterey on the California coast, but the two species release their sperm-containing pollen (and have eggs ready to receive the pollen) at different times: The Monterey pine releases pollen in early spring, the Bishop pine in summer. For this reason, the two species do not interbreed under natural conditions.

Different Species May Have Different Courtship Signals

Among animals, elaborate courtship colors and behaviors not only serve as recognition and evaluation signals between males and females of the same species, but also prevent mating with members of other species. Signals and behaviors that differ from species to species create *behavioral isolation*. The distinctive plumage and vocalizations of a male bird, for example, may attract females of his own species, but females of other species may treat these displays with indifference. For example, the extravagant plumes and arresting pose of a courting male greater bird of paradise are conspicuous indicators of his species and there is little chance that females of another species will be mistakenly attracted (**Fig. 14-14**). Among frogs, males are often impressively indiscriminate, jumping on every female in sight, regardless of the species, when the spirit moves them. Females, however, approach only male frogs that utter the call appropriate to their species. If they do find themselves in an unwanted embrace, they utter the "release call," which causes the male to let go. As a result, few hybrids are produced.

Species' Differing Sexual Organs May Foil Mating Attempts

In rare instances, a male and a female of different species may attempt to mate. Such an attempt, however, is likely to fail. Among animal species with internal fertilization (in which the sperm is deposited inside the female's reproductive tract), the male's and female's sexual organs simply may not fit together. Among plants, differences in flower size or structure may prevent pollen transfer between species because the differing flowers may attract different pollinators. Isolating mechanisms of this type are called *mechanical incompatibilities*.

Postmating Isolating Mechanisms Limit Hybrid Offspring

Premating isolation sometimes fails. When it does, members of different species may mate, and the sperm of one species may reach the egg of another species. Such matings, however, often fail to produce vigorous, fertile hybrid offspring, owing to postmating isolating mechanisms.

One Species' Sperm May Fail to Fertilize Another Species' Eggs

Even if a male inseminates a female of a different species, his sperm may not be able to fertilize her eggs, an isolating mechanism called *gametic incompatibility*. For example, in animals with internal fertilization, fluids in the female reproductive tract may weaken or kill sperm of other species. Among plants, similar chemical incompatibility may prevent the germination of pollen from one species that lands on the stigma (pollen-catching structure) of the flower of another species.

▲ **Figure 14-14 Behavioral isolation** The mate-attraction display of a male greater bird of paradise includes distinctive posture, movements, plumage, and vocalizations that do not resemble those of other bird of paradise species.

Table 14-1	Mechanisms of Reproductive Isolation

Premating isolating mechanisms: factors that prevent organisms of two populations from mating

- **Geographical isolation:** The populations cannot interbreed because a physical barrier separates them.
- **Ecological isolation:** The populations do not interbreed even if they are within the same area because they occupy different habitats.
- **Temporal isolation:** The populations cannot interbreed because they breed at different times.
- **Behavioral isolation:** The populations do not interbreed because they have different courtship and mating rituals.
- **Mechanical incompatibility:** The populations cannot interbreed because their reproductive structures are incompatible.

Postmating isolating mechanisms: factors that prevent organisms of two populations from producing vigorous, fertile offspring after mating

- **Gametic incompatibility:** Sperm from one population cannot fertilize the eggs of another population.
- **Hybrid inviability:** Hybrid offspring fail to survive to maturity.
- **Hybrid infertility:** Hybrid offspring are sterile or have low fertility.

Hybrid Offspring May Survive Poorly

If cross-species fertilization does occur, the resulting hybrid may be unable to survive, a situation called *hybrid inviability*. The genetic instructions directing development of the two species may be so different that hybrids abort early in development. For example, captive leopard frogs can be induced to mate with wood frogs and the matings generally yield fertilized eggs. The resulting embryos, however, inevitably fail to survive more than a few days.

In other animal species, a hybrid might survive but display behaviors that are mixtures of the two parental types. In attempting to do some things the way species *A* does them and other things the way species *B* does them, the hybrid may be hopelessly uncoordinated and therefore unable to reproduce. Hybrid offspring of different species of lovebirds, for example, have great difficulty learning to carry nest materials during flight and probably could not reproduce in the wild.

Hybrid Offspring May Be Infertile

Most animal hybrids, such as the mule (a cross between a horse and a donkey) and the liger (a zoo-based cross between a male lion and a female tiger), are sterile (**Fig. 14-15**). *Hybrid infertility* prevents hybrids from passing on their genetic material to offspring. A common reason for hybrid infertility is the failure of chromosomes to pair properly during meiosis, so that eggs and sperm fail to develop.

Table 14-1 summarizes the different types of isolating mechanisms.

14.6 How Do New Species Form?

Despite his exhaustive exploration of the process of natural selection, Charles Darwin never proposed a complete mechanism of **speciation,** the process by which new species form. One scientist who did play a large role in describing the process of speciation was Ernst Mayr of Harvard University, an ornithologist (expert on birds) and a pivotal figure in the history of evolutionary biology. Mayr developed the biological species concept discussed earlier. He was also among the first to recognize that speciation depends on two factors acting on a pair of populations: isolation and genetic divergence.

- **Isolation of populations.** If individuals move freely between two populations, interbreeding and the resulting gene flow will cause changes in one population to soon become widespread in the other as well. Thus, two populations cannot grow increasingly different unless something happens to block interbreeding between them. Speciation depends on isolation.

- **Genetic divergence of populations.** It is not sufficient for two populations simply to be isolated. They will become separate species only if, during the

▲ **Figure 14-15 Hybrid infertility** This liger, the hybrid offspring of a lion and a tiger, is sterile. The gene pools of its parent species remain separate.

Web Animation Speciation

period of isolation, they evolve sufficiently large genetic differences. The differences must be large enough that, if the isolated populations are reunited, they can no longer interbreed and produce vigorous, fertile offspring. That is, speciation is complete only if divergence results in evolution of an isolating mechanism. Such differences can arise by chance (genetic drift), especially if at least one of the isolated populations is small. Large genetic differences can also arise through natural selection, if the isolated populations experience different environmental conditions.

Geographical Separation of a Population Can Lead to Speciation

New species can arise when an impassible barrier physically separates different parts of a population.

Organisms May Colonize Isolated Habitats

A small population can become isolated if it moves to a new location. For example, some members of a population of land-dwelling organisms might colonize an oceanic island. The colonists might be birds, flying insects, fungal spores, or wind-borne seeds blown by a storm. More earthbound organisms could reach the island on a drifting "raft" of vegetation torn from the mainland coast. Whatever the means, we know that such colonization must occur regularly, given the presence of living things on even the remotest islands.

Isolation by colonization need not be limited to islands. For example, different coral reefs may be separated by miles of open ocean, so any reef-dwelling sponges, fishes, or algae that were carried by ocean currents to a distant reef would be effectively isolated from their original populations. Any bounded habitat, such as a lake, a mountaintop, or a parasite's host can isolate arriving colonists.

Geological and Climate Changes May Divide Populations

Isolation can also result from landscape changes that divide a population. For example, rising sea levels might transform a coastal hilltop into an island, isolating the residents. New rock from a volcanic eruption can divide a previously continuous sea or lake, splitting populations. A river that changes course can also divide populations, as can a newly formed mountain range. Climate shifts, such as those that happened in past ice ages, can change the distribution of vegetation and strand portions of populations in isolated patches of suitable habitat. You can probably imagine many other scenarios that could lead to the geographical subdivision of a population.

Over the history of Earth, many populations have been divided by continental drift. Earth's continents float on molten rock and slowly move about the surface of the planet. On a number of occasions during Earth's long history, continental landmasses have broken into pieces that subsequently moved apart (see Fig. 15-11 on p. 270). Each of these breakups must have split a multitude of populations. The bird group known as the ratites provides evidence of such a split. Ratites are large, flightless birds, including the ostrich of Africa, the rhea of South America, and the emu of Australia. The ancestor of all the ratite species lived on the ancient supercontinent of Gondwana. When Gondwana broke apart, different portions of the ancestral ratite population were isolated on separate drifting continents.

Natural Selection and Genetic Drift May Cause Isolated Populations to Diverge

If two populations become geographically isolated for any reason, there will be no gene flow between them. If the pressures of natural selection differ in the separate locations, then the populations may accumulate genetic differences. Alternatively, genetic differences may arise if one or more of the separated populations is small enough for genetic drift to occur, which may be especially likely in the aftermath of a founder event (in which a few individuals become

(a) Ahinahina

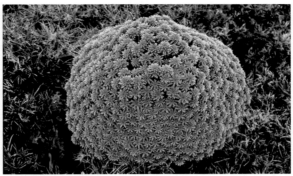

(b) Waialeale dubautia

(c) Kupaoa

(d) Na'ena'e 'ula

◀**Figure 14-16 Adaptive radiation**
About 30 species of silversword plants inhabit the Hawaiian Islands. These species are found nowhere else, and all of them descended from a single ancestral population within a few million years. This adaptive radiation has led to a collection of closely related species of diverse form and appearance, with an array of adaptations for exploiting the many different habitats in Hawaii, from warm, moist rain forests to cool, barren volcanic mountaintops.

isolated from the main body of the species). In either case, genetic differences between the separated populations may eventually become large enough to make interbreeding impossible. At that point, the two populations will have become separate species. Most evolutionary biologists believe that geographical isolation followed by speciation has been the most common source of new species, especially among animals.

Under Some Conditions, Many New Species May Arise

In some cases, many new species arise in a relatively short time. This process, called **adaptive radiation,** can occur when populations of one species invade a variety of new habitats and evolve in response to the differing environmental pressures in those habitats.

Adaptive radiation has occurred many times and in many groups of organisms, typically when species encounter a wide variety of unoccupied habitats. For example, episodes of adaptive radiation took place when some wayward finches colonized the Galapagos Islands, when a population of cichlid fish reached isolated Lake Malawi in Africa, and when an ancestral tarweed plant species arrived at the Hawaiian Islands (**Fig. 14-16**). These events gave rise to adaptive radiations of 13 species of Darwin's finches in the Galapagos, more than 300 species of cichlids in Lake Malawi, and 30 species of silversword plants in Hawaii. In these examples, the invading species faced no competitors except other members of their own species, and all the available habitats and food sources were rapidly exploited by new species that evolved from the original invaders.

14.7 What Causes Extinction?

Every living organism must eventually die, and the same is true of species. Just like individuals, species are "born" (through the process of speciation), persist for some period of time, and then perish. The ultimate fate of any species is **extinction,** the death of the last of its members. In fact, at least 99.9% of all the

Endangered Species—from Gene Pools to Gene Puddles

Many of Earth's species are in danger. According to the World Conservation Union, more than 15,000 species of plants and animals alone are currently threatened with extinction. For most of these endangered species, the main threat is habitat destruction. When a species' habitat shrinks, its population size almost invariably follows suit.

Many people, organizations, and governments are concerned about the plight of endangered species and are working to protect them and their habitats. The hope is that these efforts will not only protect endangered species, but will also restore their numbers so that they are no longer in danger of extinction. Unfortunately, however, a population that has already become small enough to warrant endangered status is likely to undergo evolutionary changes that increase its chances of going extinct. The principles of evolutionary genetics that we've explored in this chapter can help us understand these changes.

One problem is that, in small populations, mating choices are limited and a high proportion of matings may be between close relatives. This inbreeding increases the odds that offspring will be homozygous for harmful recessive alleles. These less-fit individuals may die before reproducing, further reducing the size of the population.

The greatest threat to small populations, however, stems from their inevitable loss of genetic diversity (**Fig. E14-2**). From our discussion of population bottlenecks, it is apparent that, when populations shrink to very small sizes, many of the alleles that were present in the original population will not be represented in the gene pool of the remnant population. Furthermore, we have seen that genetic drift in small populations will cause many of the surviving alleles to subsequently disappear permanently from the population (see Fig. 14-5b). Because genetic drift is a random process, many of the lost alleles will be advantageous ones that were previously favored by natural selection. Inevitably, the number of different alleles in the population grows ever smaller. As ecologist Thomas Foose aptly put it, "gene pools are being converted into gene puddles." Even if the size of an endangered population eventually begins to grow, the damage has already been done; lost genetic diversity is regained only very slowly.

Why does it matter if a population's genetic diversity is low? There are two main risks. First, the fitness of the population as a whole is reduced by the loss of advantageous alleles that underlie adaptive traits. A less-fit population is unlikely to thrive. Second, a genetically impoverished population lacks the variation that will allow it to adapt when environmental conditions change. When the environment changes, as it inevitably will, a genetically uniform species is less likely to contain individuals well suited to survive and reproduce under the new conditions. A species unable to adapt to changing conditions is at very high risk of extinction.

What can be done to preserve the genetic diversity of endangered species? The best solution, of course, is to preserve plenty of diverse types of habitat so that species never become endangered. The human population, however, has grown so large and has appropriated so large a share of Earth's resources that this solution is impossible in many places. For many species, the only solution is to ensure that areas of preserved habitat are large enough to hold populations of sufficient size to contain most of a threatened species' total genetic diversity. If, however, circumstances dictate that preserved areas are small, it is important that the small areas be linked by corridors of the appropriate habitat, so that gene flow among populations in the small preserves can increase the spread of new and beneficial alleles.

▲ **Figure E14-2 Endangered by habitat destruction and loss of genetic diversity** Only a few hundred Sumatran rhinoceroses remain.

species that have ever existed are now extinct. The natural course of evolution, as revealed by fossils, is continual turnover of species as new ones arise and old ones become extinct.

The immediate cause of extinction is probably always environmental change, in either the living or the nonliving parts of the environment. Two major environmental factors that may drive a species to extinction are competition among species and habitat destruction (see "Earth Watch: Endangered Species—from Gene Pools to Gene Puddles").

Interactions with Other Species May Drive a Species to Extinction

As described earlier, interactions such as competition and predation serve as agents of natural selection. In some cases, these same interactions can lead to extinction rather than to adaptation.

Organisms compete for limited resources in all environments. If a species' competitors evolve superior adaptations and the species doesn't evolve fast enough to keep up, it may become extinct. A particularly striking example of extinction through competition occurred in South America, beginning about 2.5 million years ago. At that time, the isthmus of Panama rose above sea level and formed a land bridge between North America and South America. After the previously separated continents were connected, the mammal species that had evolved in isolation on each continent were able to mix. Many species did indeed expand their ranges, as North American mammals moved southward and South American mammals moved northward. As they moved, each species encountered resident species that occupied the same kinds of habitats and exploited the same kinds of resources. The ultimate result of the ensuing competition was that the North American species diversified and underwent an adaptive radiation that displaced the vast majority of the South American species, many of which went extinct. Clearly, evolution had bestowed on the North American species a set of adaptations that enabled their descendants to exploit resources more efficiently and effectively than their South American counterparts could.

Habitat Change and Destruction Are the Leading Causes of Extinction

Habitat change, both contemporary and prehistoric, is the single greatest cause of extinctions. Present-day habitat destruction due to human activities is proceeding at a rapid pace. Many biologists believe that we are presently in the midst of the fastest-paced and most widespread episode of species extinction in the history of life on Earth. Loss of tropical forests is especially devastating to species diversity. As many as half the species presently on Earth may be lost during the next 50 years as the tropical forests that contain them are cut for timber and to clear land for cattle and crops. We will discuss extinctions due to prehistoric habitat change in Chapter 15.

Evolution of a Menace Revisited

The evolution of antibiotic resistance in populations of bacteria is a direct consequence of natural selection applied by antibiotic drugs. When a population of disease-causing bacteria begins to grow in a human body, physicians try to halt population growth by introducing an antibiotic drug to the bacteria's environment. Although many bacteria are killed, some surviving bacteria have genomes with a mutant allele that confers resistance. Bacteria carrying the "resistance allele" produce a disproportionately large share of offspring, which inherit the allele. Soon, resistant bacteria predominate within the population.

By introducing massive quantities of antibiotics into the bacteria's environment, humans have accelerated the pace of the evolution of antibiotic resistance. Each year, U.S. physicians write more than 100 million prescriptions for antibiotics; the Centers for Disease Control and Prevention estimates that about half of these prescriptions are unnecessary.

Although medical use and misuse of antibiotics is the most important source of natural selection for antibiotic resistance, antibiotics also pervade the environment outside our bodies. Our food supply, especially meat, contains a portion of the 20 million pounds of antibiotics that are fed to farm animals each year. In addition, Earth's soils and water are laced with antibiotics that enter the environment through human and animal wastes, and from the antibacterial soaps and cleansers that are now routinely used in many households and workplaces. As a result of this massive alteration of the environment, resistant bacteria are now found not only in hospitals and the bodies of sick people but also in our food, water, and soil. Susceptible bacteria are under constant attack, and resistant strains have little competition. In our fight against disease, we have rashly overlooked some basic principles of evolutionary biology and are now paying a heavy price.

Consider This

Because natural selection acts only on existing variation among phenotypes, antibiotic resistance could not evolve if bacteria in natural populations did not already carry alleles that help them resist attack by antibiotic chemicals. Why are such alleles present (albeit at low levels) in bacterial populations? (*Hint:* Almost all medically useful antibiotics were originally derived from fungi or bacteria.) Conversely, if resistance alleles are beneficial, why are they rare in natural populations of bacteria?

Chapter Review

Summary of Key Concepts

For additional study help and activities, go to www.mybiology.com.

14.1 How Are Populations, Genes, and Evolution Related?

Evolution is change in frequencies of alleles in a population's gene pool. Allele frequencies in a population will remain constant over generations only if the following conditions are met: (1) There is no mutation; (2) there is no gene flow; (3) the population is very large; (4) all mating is random; and (5) all genotypes reproduce equally well (that is, there is no natural selection). These conditions are rarely, if ever, met in nature. Understanding what happens when they are not met helps reveal the mechanisms of evolution.

14.2 What Causes Evolution?

The most important sources of evolutionary change are mutation, small population size, and natural selection.

- Mutations are random, undirected changes in DNA composition. Although most mutations are neutral or harmful to the organism, some prove advantageous in certain environments. Mutations are rare and do not change allele frequencies very much, but they provide the raw material for evolution.

- In any population, chance events kill or prevent reproduction by some of the individuals. If the population is small, chance events may eliminate a disproportionate number of individuals who bear a particular allele, thereby greatly changing the allele frequency in the population. This process is called genetic drift.

- The survival and reproduction of organisms are influenced by their phenotypes. Because phenotype depends at least partly on genotype, natural selection tends to favor the reproduction of certain alleles at the expense of others.

    Web Animation Agents of Change
    Web Animation The Bottleneck Effect

14.3 How Does Natural Selection Work?

Natural selection is driven by differences in reproductive success among different genotypes. Natural selection proceeds from the interactions of organisms with both the biotic and abiotic parts of their environments. When two or more species exert mutual environmental pressures on each other for long periods of time, both of them evolve in response. Such coevolution can result from any type of relationship between organisms, including competition and predation. Phenotypes that help organisms mate can evolve by sexual selection.

    Web Animation Natural Selection for Antibiotic Resistance

14.4 What Is a Species?

According to the biological species concept, a species consists of all the populations of organisms that are potentially capable of interbreeding under natural conditions and that are reproductively isolated from other populations.

14.5 How Is Reproductive Isolation Between Species Maintained?

Reproductive isolation between species may be maintained by one or more of several mechanisms, collectively known as premating isolating mechanisms and postmating isolating mechanisms. Premating isolating mechanisms include geographical isolation, ecological isolation, temporal isolation, behavioral isolation, and mechanical incompatibility. Postmating isolating mechanisms include gametic incompatibility, hybrid inviability, and hybrid infertility.

14.6 How Do New Species Form?

Speciation, the formation of new species, takes place when gene flow between two populations is reduced or eliminated and the populations diverge genetically. Most commonly, speciation follows geographical isolation and subsequent genetic divergence of the separated populations through genetic drift or natural selection.

    Web Animation Speciation

14.7 What Causes Extinction?

Factors that cause extinction include competition among species and habitat destruction.

Key Terms

adaptation *p. 241*
adaptive radiation *p. 249*
allele frequency *p. 233*
coevolution *p. 242*
competition *p. 241*
equilibrium population *p. 233*
extinction *p. 249*

fitness *p. 240*
founder effect *p. 238*
gene flow *p. 233*
gene pool *p. 232*
genetic drift *p. 235*
Hardy-Weinberg principle *p. 233*
isolating mechanism *p. 245*

mutation *p. 234*
natural selection *p. 240*
population *p. 232*
population bottleneck *p. 238*
postmating isolating
    mechanism *p. 245*
predation *p. 242*

premating isolating mechanism
    p. 245
reproductive isolation *p. 244*
sexual selection *p. 242*
speciation *p. 247*
species *p. 244*

Thinking Through the Concepts

Suggested answers to end-of-chapter and figure-based questions can be found at the end of the text.

Fill-in-the-Blank

1. The _____ provides a simple mathematical model for a nonevolving population, also called a(n) _____ population, in which _____ frequencies do not change over time. Are such populations likely to be found in nature? _____

2. Different versions of the same gene are called _____. These versions arise as a result of changes in the sequence of _____ that form the gene. These changes are caused by _____. An individual with two identical copies of a given gene is described as being _____ for that gene, while an individual with two different versions of that gene is described as _____.

3. An organism's _____ refers to the specific alleles found within its chromosomes, while the traits that these alleles produce are called its _____. Which of these does natural selection act on? _____

4. A random form of evolution is called _____. This form of evolution is most significant in populations that are _____. Two important causes of this form of evolution are the _____ and _____. Which of these would apply to a population started by a breeding pair that was stranded on an island? _____

5. Competition is most intense between members of _____. Predators and their prey act as agents of _____ on one another, resulting in a form of evolution called _____. This results in the evolution of characteristics called _____ that help both predators and their prey survive and reproduce.

6. The evolutionary fitness of an organism is measured by its success at _____. The fitness of an organism can change if its _____ changes.

7. Fill in the following with the appropriate isolating mechanism: occurs when members of two populations have different courtship behaviors: _____; occurs when hybrid offspring fail to survive to reproduce:_____; occurs when members of two populations have different breeding seasons: _____; occurs when sperm from one species fails to fertilize the eggs of another species: _____; occurs when the sexual organs of two species are incompatible: _____.

Review Questions

1. What is a gene pool? How would you determine the allele frequencies in a gene pool?

2. Define *equilibrium population*. Outline the conditions that must be met for a population to stay in genetic equilibrium.

3. How does population size affect the likelihood of changes in allele frequencies by chance alone? Can significant changes in allele frequencies (that is, evolution) occur as a result of genetic drift?

4. If you measured the allele frequencies of a gene in a population over time and found large differences from those predicted by the Hardy-Weinberg principle for an equilibrium population, would that prove that natural selection is occurring in the population you are studying? Explain your answer.

5. People like to say that "you can't prove a negative." Study the experiment in Fig. 14-3 again, and comment on what it demonstrates.

6. What is sexual selection? How is sexual selection similar to and different from other forms of natural selection?

7. Many of the oak tree species in central and eastern North America hybridize (interbreed). Are they "true species"?

8. What are the two major types of reproductive isolating mechanisms? Give examples of each type and describe how they work.

Applying the Concepts

1. **IS THIS SCIENCE?** A question commonly asked by new biology students is: "If humans descended from apes, why are there still apes?" On the basis of your understanding of the process of evolutionary change, provide a scientifically informed answer to this question.

2. **IS THIS SCIENCE?** Southern Wisconsin is home to several populations of gray squirrels (*Sciurus carolinensis*) with black fur. Some observers claim that the black squirrels are a separate species. Design a study to test the hypothesis that the gray squirrels and black squirrels are separate species.

3. In many countries, conservationists are trying to design national park systems so that "islands" of natural habitat (the big parks) are connected by thin "corridors" of undisturbed habitat. The idea is that this arrangement will allow animals and plants to migrate between parks. Why would such migration be important?

For additional resources, go to www.mybiology.com.

chapter 15

The History of Life on Earth

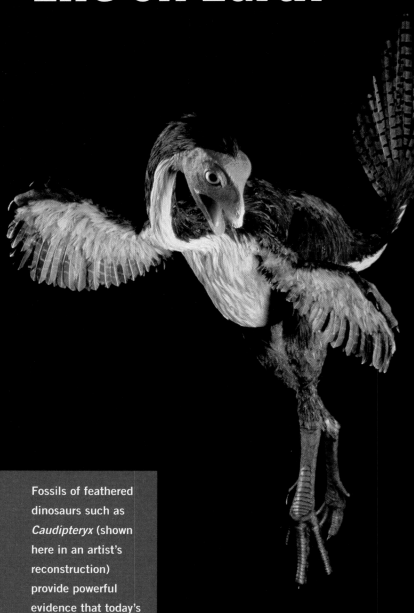

Fossils of feathered dinosaurs such as *Caudipteryx* (shown here in an artist's reconstruction) provide powerful evidence that today's birds descended from dinosaur ancestors.

Case Study   Dinosaurs Singing in the Backyard?

Dinosaurs no longer walk the earth, but their descendants are lurking outside your window. The sparrows singing in your backyard, the pigeons perched atop city buildings, and the crows soaring overhead all descended from dinosaurs. Although a small, delicate, flying bird may not seem to have much in common with a large, heavy-limbed, earthbound dinosaur, birds are the closest living relatives of the extinct dinosaurs. Still, given the dissimilarity of the two groups, how can we know that dinosaurs are the ancestors of birds? The key evidence comes from fossils, which provide a record of life's history.

Paleontologists (scientists who study fossils) have long known that birds share ancestry with reptiles. This shared ancestry was revealed by fossils of *Archaeopteryx*, the earliest known bird. *Archaeopteryx* had feathers very much like those of modern birds, but also had some reptilian traits. Unlike modern birds, *Archaeopteryx* had teeth, clawed fingers, and a long, bony tail. These traits suggest that *Archaeopteryx* represents an early moment in the history of birds, when birds still retained many structures that were later lost as birds evolved to become the efficient flying machines that we see today.

Beautifully preserved fossils of *Archaeopteryx*, the first of which was discovered in 1861, provide a window to the past and reveal the link between birds and other reptiles. However, evidence of the link between birds and the particular group of reptiles known as dinosaurs was a long time coming. Although the hypothesis that birds descended from dinosaurs is not new, convincing fossil evidence in support of the hypothesis was not found until recently.

Can you guess the main prediction of the birds-from-dinosaurs hypothesis of bird evolution? What fossil evidence would support the prediction? Think about these questions as we review a broad outline of the history of life on Earth. ■

This 150-million-year-old fossil impression is the oldest evidence of feathers.

15.1 How Did Life Begin?

Pre-Darwinian thought held that all species were simultaneously created by God a few thousand years ago. Further, until the nineteenth century, most people thought that new members of species sprang up all the time through **spontaneous generation** from both nonliving matter and other, unrelated forms of life. In 1609, a French botanist wrote, "There is a tree . . . frequently observed in Scotland. From this tree leaves are falling; upon one side they strike the water and slowly turn into fishes, upon the other they strike the land and turn into birds." Medieval writings abound with similar observations. Microorganisms were thought to arise spontaneously from broth, maggots from meat, and mice from mixtures of sweaty shirts and wheat.

Experiments Refuted Spontaneous Generation

In 1668, the Italian physician Francesco Redi disproved the maggots-from-meat hypothesis simply by keeping flies (whose eggs hatch into maggots) away from uncontaminated meat (see "Scientific Inquiry: Controlled Experiments, Then and Now" on pp. 6–7). In the mid-1800s, Louis Pasteur in France and John Tyndall in England

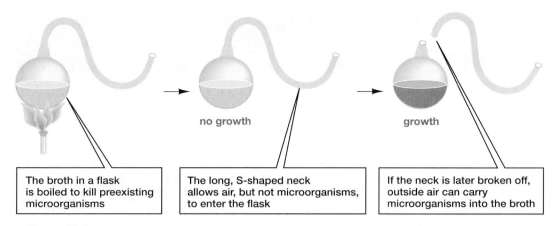

| The broth in a flask is boiled to kill preexisting microorganisms | The long, S-shaped neck allows air, but not microorganisms, to enter the flask | If the neck is later broken off, outside air can carry microorganisms into the broth |

▲ **Figure 15-1 Spontaneous generation refuted**
Louis Pasteur's experiment disproving the spontaneous generation of microorganisms in broth.

disproved the broth-to-microorganism idea (**Fig. 15-1**). Although their work effectively demolished the notion of spontaneous generation, it did not address the question of how life on Earth originated in the first place. Or, as the biochemist Stanley Miller put it, "Pasteur never proved it didn't happen once, he only showed that it doesn't happen all the time."

The First Living Things Arose from Nonliving Ones

For almost half a century, the subject lay dormant. Eventually, biologists returned to the question of the origin of life. In the 1920s and 1930s, Alexander Oparin in Russia and John B. S. Haldane in England noted that today's oxygen-rich atmosphere would not have permitted the spontaneous formation of the complex organic molecules necessary for life. Oxygen reacts readily with other molecules, disrupting chemical bonds. Thus, an oxygen-rich environment tends to keep molecules simple.

Oparin and Haldane speculated that the atmosphere of the young Earth must have contained very little oxygen and that, under such atmospheric conditions, complex organic molecules could have arisen through ordinary chemical reactions. Some kinds of molecules could persist in the lifeless environment of early Earth better than others, and would therefore become more common over time. This chemical version of the "survival of the fittest" is called *prebiotic* (meaning "before life") evolution. In the scenario envisioned by Oparin and Haldane, prebiotic chemical evolution gave rise to progressively more complex molecules and eventually to living organisms.

Organic Molecules Can Form Spontaneously Under Prebiotic Conditions

Inspired by the ideas of Oparin and Haldane, Stanley Miller and Harold Urey set out in 1953 to simulate prebiotic evolution in the laboratory. They knew that, on the basis of the chemical composition of the rocks that formed early in Earth's history, geochemists had concluded that the early atmosphere probably contained virtually no oxygen gas, but did contain other substances, including methane (CH_4), ammonia (NH_3), hydrogen (H_2), and water vapor (H_2O). Miller and Urey simulated the oxygen-free atmosphere of early Earth by mixing these components in a flask. An electrical discharge mimicked the intense energy of early Earth's lightning storms. In this experimental microcosm, the researchers found that simple organic molecules appeared after just a few days (**Fig. 15-2**). The experiment showed that small molecules likely present in the early atmosphere can combine to form larger organic molecules if electrical energy is present. (Recall from Chapter 5 that reactions that synthesize biological molecules from smaller ones are endergonic—they consume energy.) Similar experiments by Miller and others have produced amino acids, short proteins, nucleotides, adenosine triphosphate (ATP), and other molecules characteristic of living things.

In recent years, new evidence has convinced most geochemists that the actual composition of Earth's early atmosphere probably differed from the mixture of gases used in the pioneering Miller-Urey experiment. This improved understanding of the early atmosphere, however, has not undermined the basic finding of the Miller-Urey experiment. Additional experiments with more realistic (but still oxygen-free) simulated atmospheres have also yielded organic molecules. In addition, these experiments have shown that electricity is not the only suitable energy source. Other energy sources that were available on early Earth, such as heat or ultraviolet (UV) light, have also been shown to drive the formation of organic molecules in experimental simulations of prebiotic conditions. Thus, even though we may never know exactly what the earliest atmosphere was like, we can be confident that organic molecules formed spontaneously on early Earth.

Additional organic molecules probably arrived from space when meteorites and comets crashed into Earth's surface. Analysis of present-day meteorites recovered from impact craters on Earth has revealed that some meteorites contain relatively high concentrations of amino acids and other simple organic molecules. Laboratory experiments suggest that these molecules could have formed in interstellar space before plummeting to Earth. When small molecules known to be present in space were placed under space-like conditions of very low temperature and pressure and bombarded with UV light, larger organic molecules were produced.

Organic Molecules Can Accumulate Under Prebiotic Conditions

Prebiotic synthesis was neither very efficient nor very fast. Nonetheless, in a few hundred million years, large quantities of organic molecules accumulated in the early Earth's oceans. Today, most organic molecules have a short life because they are either digested by living organisms or they react with atmospheric oxygen. Early Earth, however, lacked both life and free oxygen, so molecules would not have been exposed to these threats.

Still, the prebiotic molecules must have been threatened by the sun's high-energy UV radiation, because early Earth lacked an ozone layer. The ozone layer is a region high in today's atmosphere that is enriched with ozone (O_3) molecules, which form when incoming solar energy splits some O_2 molecules in the outer atmosphere into individual oxygen atoms (O), which then react with O_2 to form O_3 (ozone). The resulting high-altitude layer of ozone molecules absorbs some of the sun's UV light before it reaches Earth's surface. Early Earth, however, had no ozone layer, because there was little or no oxygen gas in the atmosphere and therefore no ozone formation.

Before the ozone layer formed, UV bombardment must have been fierce. UV radiation, as we have seen, can provide energy for the formation of organic molecules but it can also break them apart. Some places, however, such as those beneath rock ledges or at the bottoms of even fairly shallow seas, would have been protected from UV radiation. In these locations, organic molecules may have accumulated.

Clay May Have Catalyzed the Formation of Larger Organic Molecules

In the next stage of prebiotic evolution, simple molecules combined to form larger molecules. The chemical reactions that formed the larger molecules required that the reacting molecules be packed closely together. Scientists have proposed several processes by which the required high concentrations might have been achieved on early Earth. One possibility is that small molecules accumulated on the surfaces of clay particles, which may have a small electrical charge that attracts dissolved molecules with the opposite charge. Clustered on a clay particle, small molecules would have been sufficiently close together to allow chemical reactions between them. Researchers have demonstrated the plausibility of this scenario with experiments in which adding clay to solutions of dissolved small biological molecules catalyzed the formation of larger, more complex molecules. Such molecules might have formed on clay at the bottom of early Earth's oceans or lakes and gone on to become the building blocks of the first living organisms.

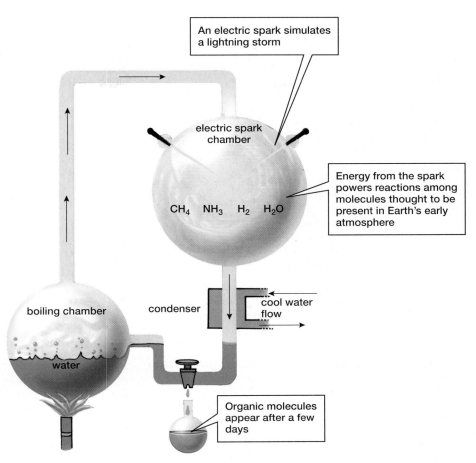

An electric spark simulates a lightning storm

electric spark chamber

Energy from the spark powers reactions among molecules thought to be present in Earth's early atmosphere

CH_4 NH_3 H_2 H_2O

boiling chamber

condenser

cool water flow

water

Organic molecules appear after a few days

▲ **Figure 15-2 The experimental apparatus of Stanley Miller and Harold Urey** Life's very earliest stages left no fossils, so evolutionary historians have pursued a strategy of re-creating in the laboratory the conditions that may have prevailed on early Earth. The mixture of gases in the spark chamber simulates Earth's early atmosphere. **QUESTION:** How would the experiment's result change if oxygen (O_2) were included in the spark chamber?

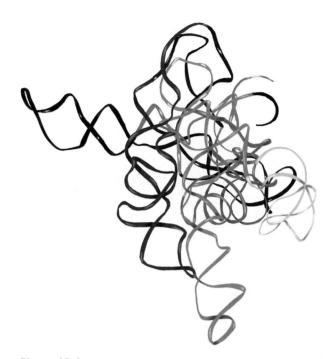

▲ **Figure 15-3 A ribozyme** This RNA molecule, isolated from the single-celled organism *Tetrahymena*, acts like an enzyme, catalyzing metabolic reactions. Different parts of the molecule, represented here by different colors, play particular roles in the catalytic process.

RNA May Have Been the First Self-Reproducing Molecule

Although all living organisms use DNA to encode and store genetic information, it is unlikely that DNA was the earliest informational molecule. DNA can reproduce itself only with the help of large, complex protein enzymes, but the instructions for building these enzymes are encoded in DNA itself. For this reason, the origin of DNA's role as life's information storage molecule poses a "chicken and egg" puzzle: DNA requires proteins, but those proteins require DNA. It is thus difficult to construct a plausible scenario for the origin of self-replicating DNA from prebiotic molecules. It is therefore likely that the current DNA-based system of information storage evolved from an earlier system.

RNA Can Act as a Catalyst

A prime candidate for the first self-replicating informational molecule is RNA. In the 1980s, Thomas Cech and Sidney Altman, working with the single-celled organism *Tetrahymena*, discovered a cellular reaction that was catalyzed not by a protein, but by a small RNA molecule. Because this special RNA molecule performed a function previously thought to be performed only by protein enzymes, Cech and Altman decided to give their catalytic RNA molecule the name **ribozyme** (Fig. 15-3).

In the years since their discovery, researchers have found dozens of naturally occurring ribozymes that catalyze a variety of different reactions, including cutting other RNA molecules and splicing together different RNA fragments. Ribozymes have also been found in the protein-manufacturing machinery of cells, where they help catalyze the attachment of amino acid molecules to growing proteins. In addition, researchers have been able to synthesize different ribozymes in the laboratory, including some that can catalyze the replication of small RNA molecules.

Earth May Once Have Been an RNA World

The discovery that RNA molecules can act as catalysts for diverse reactions, including RNA replication, provides support for the hypothesis that life arose in an "RNA world." According to this view, the current era of DNA-based life was preceded by one in which RNA served as both the information-carrying genetic molecule and the enzyme catalyst for its own replication. This RNA world may have emerged after hundreds of millions of years of prebiotic chemical synthesis, during which RNA nucleotides would have been among the molecules synthesized. After reaching a sufficiently high concentration, perhaps on clay particles, the nucleotides probably bonded together to form short RNA chains.

Let's suppose that, purely by chance, one of these RNA chains was a ribozyme that could catalyze the production of copies of itself. This first self-reproducing ribozyme probably wasn't very good at its job and produced copies with lots of errors. These mistakes were the first mutations. Like modern mutations, most undoubtedly ruined the catalytic abilities of the "daughter molecules," but a few may have been improvements. Such improvements set the stage for the evolution of RNA molecules, as variant ribozymes with increased speed and accuracy of replication reproduced, copying themselves and displacing less efficient molecules. Molecular evolution in the RNA world proceeded until, by some still unknown chain of events, RNA gradually receded into its present role as an intermediary between DNA and protein enzymes.

Membrane-Like Vesicles May Have Enclosed Ribozymes

Self-replicating molecules alone do not constitute life; these molecules must be contained within some kind of enclosing membrane. The precursors of the earliest biological membranes may have been simple structures that formed spontaneously from purely physical, mechanical processes. For example, chemists have shown that if water containing proteins and lipids is agitated to simulate waves beating against ancient shores, the proteins and lipids combine to form hollow structures called *vesicles*. These hollow balls resemble living cells in several respects. They have a well-defined outer boundary that separates their internal contents from the external solution. If the composition of the vesicle is right, a "membrane" forms that is remarkably similar in appearance to a real cell membrane. Under certain conditions, vesicles can absorb material from the external solution, grow, and even divide.

If a vesicle happened to surround the right ribozymes, it would form something resembling a living cell. We could call it a **protocell,** structurally similar to a cell but not a living thing. In the protocell, ribozymes and any other enclosed molecules would have been protected from free-roaming ribozymes in the primordial soup. Nucleotides and other small molecules might have diffused across the membrane and been used to synthesize new ribozymes and other complex molecules. After sufficient growth, the vesicle may have divided, with a few copies of the ribozymes becoming incorporated into each daughter vesicle. If this process occurred, the path to the evolution of the first cells would be nearly at its end.

Was there a particular moment when a nonliving protocell gave rise to something alive? Probably not. Like most evolutionary transitions, the change from protocell to living cell was a continuous process, with no sharp boundary between one state and the next.

But Did All This Happen?

The scenario just discussed, although plausible and consistent with many research findings, is by no means certain. One of the most striking aspects of origin-of-life research is a great diversity of assumptions, experiments, and contradictory hypotheses. Researchers disagree as to whether life arose in quiet pools, in the sea, in moist films on the surfaces of crystals, or in hot deep-sea vents. A few researchers even argue that life arrived on Earth from space. Can we draw any conclusions from the research conducted so far? No, but we can make a few observations.

First, the experiments of Miller and others show that amino acids, nucleotides, and other organic molecules, along with simple membrane-like structures, would have formed in abundance on the early Earth. Second, chemical evolution had long periods of time and huge areas of the Earth available to it. Given sufficient time and a sufficiently large pool of reactant molecules, even extremely rare events can occur many times. Given the vast expanses of time and space available, each small step on the path from primordial soup to living cell had ample opportunity to take place.

Most biologists accept that the origin of life was probably an inevitable consequence of the working of natural laws. We should emphasize, however, that this proposition cannot be definitively tested. The origin of life left no record, and researchers exploring this mystery can proceed only by developing a hypothetical scenario and then conducting laboratory investigations to determine if the scenario's steps are chemically and biologically possible and plausible.

15.2 What Were the Earliest Organisms Like?

When Earth first formed about 4.5 billion years ago, it was quite hot (**Fig. 15-4**). A multitude of meteorites smashed into the forming planet, and the kinetic energy of these extraterrestrial rocks was converted into heat on impact. Still more heat was released by the decay of radioactive atoms. The rock composing Earth melted, and heavier elements such as iron and nickel sank to the center of the planet, where they remain molten even today. It must have taken hundreds of millions of years for Earth to cool enough to allow water to exist as a liquid. Nonetheless, it appears that life arose in fairly short order once liquid water was available.

The oldest fossil organisms found so far are in rocks that are about 3.5 billion years old. (Their age was determined using radiometric dating techniques; see "Scientific Inquiry: How Do We Know How Old a Fossil Is?" on p. 260.) Chemical

scientific inquiry

How Do We Know How Old a Fossil Is?

Early geologists could date rock layers and their accompanying fossils only in a *relative* way: fossils found in deeper layers of rock were generally older than those found in shallower layers. With the discovery of radioactivity, it became possible to determine *absolute* dates. The nuclei of radioactive elements spontaneously break down, or decay, into other elements. For example, carbon-14 (usually written ^{14}C) decays by emitting an electron to become nitrogen-14 (^{14}N). Each radioactive element decays at a rate that is independent of temperature, pressure, or the chemical compound of which the element is a part. The time it takes for half of a radioactive element's nuclei to decay at this characteristic rate is called its *half-life*. The half-life of ^{14}C for example, is 5,730 years.

How are radioactive elements used in determining the age of rocks? If we know the rate of decay and measure the proportion of decayed nuclei to undecayed nuclei, we can estimate how much time has passed since these radioactive elements became trapped in the rock. This process is called *radiometric dating*.

A particularly straightforward radiometric dating technique measures the decay of potassium-40 (^{40}K), which has a half-life of about 1.25 billion years, into argon-40 (^{40}Ar). Potassium-40 is commonly found in volcanic rocks such as granite and basalt, and the argon-40 it decays into is a gas. Let's suppose that a volcano erupts with a massive lava flow, covering the countryside. All of the ^{40}Ar, being a gas, will bubble out of the molten lava, so when the lava first cools and solidifies into rock, it will not contain any ^{40}Ar (**Fig. E15-1**). Over time, however, any ^{40}K present in the hardened lava will decay into ^{40}Ar, with half of the ^{40}K decaying every 1.25 billion years. This ^{40}Ar gas will be trapped in the rock. A geologist could take a sample of the rock and measure the proportion of ^{40}K to ^{40}Ar to determine the rock's age. For example, if the analysis finds equal amounts of the two elements, the geologist will conclude that the lava hardened 1.25 billion years ago (see Fig. E 15-1). With appropriate care, such age estimates are quite reliable. If a fossil is found

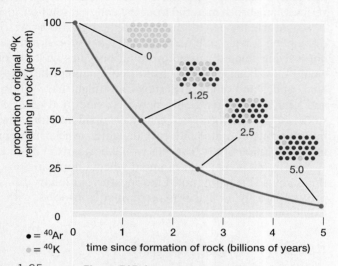

$\bullet = ^{40}Ar$
$\circ = ^{40}K$

▲ **Figure E15-1** **The relationship between time and the decay of radioactive ^{40}K to ^{40}Ar EXERCISE:** Uranium-235, with a half-life of 713 million years, decays to lead-207. If you analyze a rock and find that it contains uranium-235 and lead-207 in a ratio of 3:1, how old is the rock?

beneath a lava flow dated at, say, 500 million years, then we know that the fossil is at least that old.

traces in older rocks have led some paleontologists to believe that life is even older, perhaps as old as 3.9 billion years.

The period in which life began is known as the Precambrian era. This interval was designated by geologists and paleontologists, who have devised a hierarchical naming system of eras, periods, and epochs to delineate the immense span of geological time (**Table 15-1**).

▶ **Figure 15-4** **Early Earth** Life began on a planet characterized by abundant volcanic activity, frequent electrical storms, repeated meteorite strikes, and an atmosphere that lacked oxygen gas.

Table 15-1	**The History of Life on Earth**				

Era	Period	Epoch	Millions of Years Ago	Major Events	
Cenozoic	Quaternary	Recent	0.01–present	Evolution of genus *Homo*	
		Pleistocene	1.8–0.01		
	Tertiary	Pliocene	5–1.8		
		Miocene	23–5	Widespread flourishing of birds, mammals, insects, and flowering plants	
		Oligocene	38–23		
		Eocene	54–38		
		Paleocene	65–54		
Mesozoic	Cretaceous		146–65	Mass extinction of marine and terrestrial life, including dinosaurs	
				Flowering plants appear and become dominant	
	Jurassic		208–146	Dominance of dinosaurs and conifers	
				First birds	
	Triassic		245–208	First mammals and dinosaurs	
				Forests of gymnosperms and tree ferns	
Paleozoic	Permian		286–245	Massive marine extinctions, including trilobites	
				Flourishing of reptiles and decline of amphibians	
	Carboniferous		360–286	Forests of tree ferns and club mosses	
				Dominance of amphibians and insects	
				First reptiles and conifers	
	Devonian		410–360	Fishes and trilobites flourish	
				First amphibians, insects, seeds, and pollen	
	Silurian		440–410	Many fishes, trilobites, mollusks	
				First vascular plants	
	Ordovician		505–440	Dominance of arthropods and mollusks in the sea	
				Invasion of the land by plants and arthropods	
				First fungi	
	Cambrian		544–505	Marine algae flourish	
				Origin of most marine invertebrate phyla	
				First fishes	
Precambrian			About 1,000	First animals (soft-bodied marine invertebrates)	
			1,200	First multicellular organisms	
			2,000	First eukaryotes	
			2,200	Accumulation of free oxygen in the atmosphere	
			3,500	Origin of photosynthesis (in cyanobacteria)	
			3,900–3,500	First living cells (prokaryotes)	
			4,000–3,900	Appearance of the first rocks on Earth	
			4,600	Origin of the solar system and Earth	

The First Organisms Were Anaerobic Prokaryotes

The first cells to arise in Earth's oceans were **prokaryotes,** cells whose genetic material was not contained within a nucleus. These cells probably obtained nutrients and energy by absorbing organic molecules from their environment. There was no oxygen gas in the atmosphere, so the cells must have metabolized the organic molecules anaerobically. You will recall from Chapter 7 that anaerobic metabolism yields only small amounts of energy.

Thus, the earliest cells were primitive anaerobic bacteria. As these bacteria multiplied, they must have eventually used up the organic molecules produced by prebiotic chemical reactions. Simpler molecules, such as carbon dioxide and water, would still have been very abundant, as was energy in the form of sunlight. What was lacking, then, was not materials or energy itself, but energetic molecules—molecules in which energy is stored in chemical bonds.

Some Organisms Evolved the Ability to Capture the Sun's Energy

Eventually, some cells evolved the ability to use the energy of sunlight to drive the synthesis of complex, high-energy molecules from simpler molecules. In other words, photosynthesis appeared. Photosynthesis requires a source of hydrogen, and the very earliest photosynthetic bacteria probably used hydrogen sulfide gas dissolved in water for this purpose (much as today's purple photosynthetic bacteria do). Eventually, however, Earth's supply of hydrogen sulfide (which is produced mainly by volcanoes) must have run low. The shortage of hydrogen sulfide set the stage for the evolution of photosynthetic bacteria that were able to use the planet's most abundant source of hydrogen: water (H_2O).

Photosynthesis Increased the Amount of Oxygen in the Atmosphere

Water-based photosynthesis converts water and carbon dioxide to energetic molecules of sugar, releasing oxygen as a by-product. The emergence of this new method for capturing energy introduced significant amounts of free oxygen to the atmosphere for the first time. At first, the newly liberated oxygen was quickly consumed by reactions with other molecules in the atmosphere and in Earth's crust, or surface layer. One especially common reactive atom in the crust was iron, and much of the new oxygen combined with iron atoms to form huge deposits of iron oxide (also known as rust). As a result, iron oxide is abundant in rocks formed during this period.

After all the accessible iron had turned to rust, the concentration of oxygen gas in the atmosphere began to increase. Chemical analysis of rocks suggests that significant amounts of oxygen first appeared in the atmosphere about 2.3 billion years ago, produced by bacteria that were probably very similar to modern cyanobacteria. (Because Earth's supply of oxygen molecules is continually recycled, you will undoubtedly breathe in some oxygen molecules today that were expelled 2 billion years ago by one of these early cyanobacteria!)

Aerobic Metabolism Arose in Response to the Oxygen Crisis

Oxygen is potentially very dangerous to living things, because it can react with organic molecules, breaking them down. Many of today's anaerobic bacteria perish when exposed to what is for them a deadly poison—oxygen. The accumulation of oxygen in the atmosphere of early Earth probably exterminated many organisms and fostered the evolution of cellular mechanisms for detoxifying oxygen. This crisis for evolving life also provided the environmental pressure for the next great advance in the Age of Microbes: the ability to use oxygen in metabolism. This ability not only provides a defense against the chemical action of oxygen, but actually channels oxygen's destructive power through aerobic respiration to generate useful energy for the cell (see pp. 101–104 for more information on aerobic respiration). Because the amount of energy available to a cell is vastly increased when oxygen is used to metabolize food molecules, aerobic cells had a significant selective advantage.

Some Organisms Acquired Membrane-Enclosed Organelles

Hordes of bacteria would offer a rich food supply to any organism that could eat them. There are no fossils of the first predatory cells to roam the seas, but paleobiologists speculate that once a suitable prey population (such as these bacteria) appeared, predation would have evolved quickly. According to the most widely accepted hypothesis, these early predators were prokaryotes that had evolved to become larger than typical bacteria. In addition, they had lost the rigid cell wall that surrounds most bacterial cells, so that their flexible plasma membrane was in contact with the surrounding environment. Thus, the predatory cells were able to envelop smaller bacteria in an infolded pouch of membrane and in this fashion engulf whole bacteria as prey.

These early predators were probably capable of neither photosynthesis nor aerobic metabolism. Although they could capture large food particles, namely bacteria, they metabolized them inefficiently. By about 1.7 billion years ago, however, one predator probably gave rise to the first eukaryotic cell. Eukaryotic cells differ from prokaryotic cells in that they have an elaborate system of internal membranes, many of which enclose organelles such as a nucleus that contains the cell's genetic material. Organisms composed of one or more eukaryotic cells are known as **eukaryotes.**

The Internal Membranes of Eukaryotes May Have Arisen Through Infolding of the Plasma Membrane

The internal membranes of eukaryotic cells may have originally arisen through inward folding of the cell membrane of a single-celled predator. If, as is true of most of today's bacteria, the DNA of the eukaryotes' ancestor was attached to the inside of its cell membrane, an infolding of the membrane near the point of DNA attachment may have pinched off and become the precursor of the cell nucleus.

In addition to the nucleus, other key eukaryotic structures include the organelles used for energy metabolism: mitochondria and (in photosynthetic organisms only) chloroplasts. How did these organelles evolve?

Mitochondria and Chloroplasts May Have Arisen from Engulfed Bacteria

The **endosymbiont hypothesis** proposes that early eukaryotic cells acquired the precursors of mitochondria and chloroplasts by engulfing certain types of bacteria. These cells and the bacteria trapped inside them (*endo* means "within") gradually entered into a *symbiotic* relationship, a close association between different types of organisms over an extended time. How might this have happened?

Let's suppose that an anaerobic predatory cell captured an aerobic bacterium for food, as it often did, but for some reason failed to digest this particular prey. The aerobic bacterium remained alive and well. In fact, it was better off than ever, because the cytoplasm of its predator-host was chock-full of half-digested food molecules, the remnants of anaerobic metabolism. The aerobe absorbed these molecules and used oxygen to metabolize them, thereby gaining enormous amounts of energy. So abundant were the aerobe's food resources, and so bountiful its energy production, that the aerobe must have leaked energy, probably as ATP or similar molecules, back into its host's cytoplasm. The anaerobic predatory cell with its symbiotic bacteria could now metabolize food aerobically, gaining a great

 Web Animation Endosymbiosis

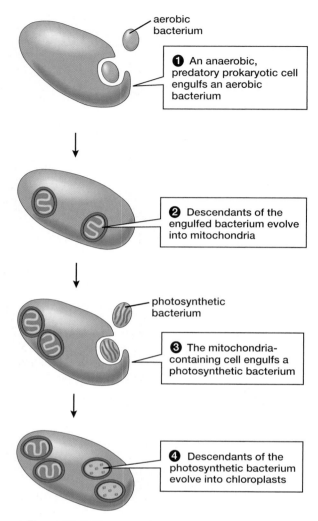

① An anaerobic, predatory prokaryotic cell engulfs an aerobic bacterium

aerobic bacterium

② Descendants of the engulfed bacterium evolve into mitochondria

photosynthetic bacterium

③ The mitochondria-containing cell engulfs a photosynthetic bacterium

④ Descendants of the photosynthetic bacterium evolve into chloroplasts

▲ **Figure 15-5 The probable origin of mitochondria and chloroplasts in eukaryotic cells QUESTION:** Scientists have identified a living bacterium believed to be descended from the endosymbiont that gave rise to mitochondria. Would you expect the DNA sequence of this modern bacterium to be most similar to the sequence of DNA from a plant chloroplast, an animal cell nucleus, or a plant mitochondrion?

advantage over other anaerobic cells and leaving a greater number of offspring. Eventually, the endosymbiotic bacterium lost its ability to live independently of its host, and the mitochondrion was born (Fig. 15-5, steps ① and ②).

One of these successful new cellular partnerships managed a second feat: It captured a photosynthetic bacterium and similarly failed to digest its prey. The bacterium flourished in its new host and gradually evolved into the first chloroplast (Fig. 15-5, steps ③ and ④). Other eukaryotic organelles may have also originated through endosymbiosis. Many biologists believe that cilia, flagella, centrioles, and microtubules may all have evolved from a symbiosis between a spirilla-like bacterium (a form of bacterium with an elongated corkscrew shape) and an early eukaryotic cell.

Evidence for the Endosymbiont Hypothesis Is Strong

Several types of evidence support the endosymbiont hypothesis. A particularly compelling line of evidence is the many distinctive biochemical features shared by eukaryotic organelles and living bacteria. In addition, mitochondria, chloroplasts, and centrioles each contain their own minute supply of DNA, which many researchers interpret as remnants of the DNA originally contained within the engulfed bacteria.

Another kind of support comes from *living intermediates*, organisms alive today that are similar to hypothetical ancestors and thus help show that a proposed evolutionary pathway is plausible. For example, the amoeba *Pelomyxa palustris* lacks mitochondria but hosts a permanent population of aerobic bacteria that carry out much the same role. Similarly, a variety of corals, some clams, a few snails, and at least one species of *Paramecium* harbor a permanent collection of photosynthetic algae in their cells (Fig. 15-6). These examples of modern cells that host bacterial endosymbionts suggest that similar symbiotic associations could have occurred almost 2 billion years ago and led to the first eukaryotic cells.

15.3 What Were the Earliest Multicellular Organisms Like?

Once predation had evolved, increased size became an advantage. In the marine environments to which life was restricted, a larger cell could easily engulf a smaller cell and would also be difficult for other predatory cells to ingest. Larger organisms can also generally move faster than smaller ones, making successful predation and escape more likely. But enormous single cells have problems. Oxygen and nutrients going into the cell and waste products going out must diffuse through the plasma membrane. The larger a cell becomes, the less surface membrane is available per unit volume of cytoplasm.

There are only two ways that an organism larger than a millimeter or so in diameter can survive. First, it can have a low metabolic rate so that it doesn't need much oxygen or produce much carbon dioxide. This strategy seems to work for certain very large single-celled algae. Alternatively, an organism can be multicellular; that is, it can consist of many small cells packaged into a larger, unified body.

Some Algae Became Multicellular

The oldest fossils of multicellular organisms are about 1.2 billion years old and include impressions of the first multicellular algae, which arose from single-celled eukaryotic cells containing chloroplasts. Multicellularity would have provided at least two advantages for these seaweeds. First, large, many-celled algae would have been difficult for single-celled predators to engulf. Second, specialization of cells would have provided the potential for staying in one place in the brightly lit waters of the shoreline, as rootlike structures burrowed in sand or clutched onto rocks, while leaflike structures floated above, bathed in the sunlight necessary for photosynthesis. The green, brown, and red algae lining our shores today—some, such as the brown kelp, more than 200 feet long—are the descendants of these early multicellular algae.

Animal Diversity Arose in the Precambrian Era

In addition to fossil algae, billion-year-old rocks have yielded fossil traces of animal tracks and burrows. This evidence of early animal life notwithstanding, fossils of animal bodies first appear in Precambrian rocks laid down between 610 million and 544 million years ago. Some of these ancient invertebrate animals (animals lacking a backbone) are quite different in appearance from any animals that appear in later fossil layers and may represent types of animals that left no descendants. Other fossils in these rock layers, however, appear to be ancestors of today's animals. Ancestral sponges and jellyfish appear in the oldest layers, followed later by ancestors of worms, mollusks, and arthropods.

The full range of modern invertebrate animals, however, does not appear in the fossil record until the Cambrian period, marking the beginning of the Paleozoic era, about 544 million years ago. (The phrase "fossil record" is a shorthand reference to the entire collection of all fossil evidence that has been found to date.) These Cambrian fossils reveal an adaptive radiation (see p. 249) that had already yielded a diverse array of complex body plans. Almost all of the major groups of animals on Earth today were already present in the early Cambrian. The apparently sudden appearance of so many different kinds of animals suggests that the earlier evolutionary history that produced such an impressive range of different animal forms is not preserved in the fossil record.

The early diversification of animals was probably driven in part by the emergence of predatory lifestyles. Coevolution of predator and prey led to the evolution of new features in many kinds of animals. By the Silurian period (440 million to 410 million years ago), mud-skimming, armored trilobites were preyed on by ammonites and the chambered nautilus, which still survives in almost unchanged form in deep Pacific waters (Fig. 15-7).

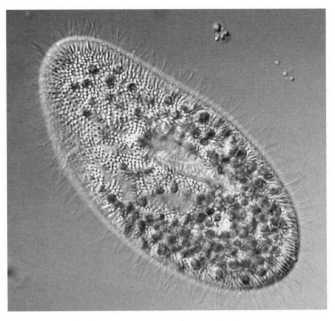

▲ **Figure 15-6 Symbiosis within a modern cell** The ancestors of the chloroplasts in today's plant cells may have resembled *Chlorella*, the green, photosynthetic, single-celled algae living symbiotically within the cytoplasm of the *Paramecium* pictured here.

(a) Silurian scene

(b) Trilobite

(c) Ammonite

(d) *Nautilus*

◄ **Figure 15-7 Diversity of ocean life during the Silurian period** **(a)** Life characteristic of the oceans during the Silurian period, 440 million to 410 million years ago. Among the most common fossils from that time are **(b)** the trilobites and their predators, the nautiloids, and **(c)** the ammonites. **(d)** This living *Nautilus* is very similar in structure to the Silurian nautiloids, showing that a successful body plan may exist virtually unchanged for hundreds of millions of years.

Many animals of the Paleozoic era were more mobile than their evolutionary predecessors. Predators gain an advantage from an ability to travel over wide areas in search of suitable prey; and the ability to make a speedy escape is an advantage for prey. The evolution of efficient movement was often associated with the evolution of greater sensory capabilities and more complex nervous systems. Senses for detecting touch, chemicals, and light became highly developed, along with nervous systems capable of handling the sensory information and directing appropriate behaviors.

In many species, mobility was enhanced in part by the origin of hard external body coverings known as an **exoskeletons.** Exoskeletons improved mobility by providing hard surfaces to which muscles attached; these attachments made it possible for animals to use their muscles to move appendages used to swim about or move over the sea floor. Exoskeletons also provided support for animals' bodies and protection from predators.

About 530 million years ago, one group of animals—the fishes—developed a new form of body support and muscle attachment: an internal skeleton. These early fishes were inconspicuous members of the ocean community, but by 400 million years ago, fishes were a diverse and prominent group. By and large, the fishes proved to be faster than the invertebrates, with more acute senses and larger brains. Eventually, they became the dominant predators of the open seas.

Web Animation Evolutionary Timescales

15.4 How Did Life Invade the Land?

A compelling subplot in the long tale of life's history is the story of life's invasion of land after more than 3 billion years of a strictly watery existence. In moving to solid ground, organisms had many obstacles to overcome. Life in the sea provides buoyant support against gravity, but on land an organism must bear its weight against the crushing force of gravity. The sea provides ready access to life-sustaining water, but a terrestrial organism must find adequate water. Sea-dwelling plants and animals can reproduce by means of mobile sperm or eggs, or both, which swim to each other through the water. The gametes of land-dwellers, however, must be protected from drying out.

Despite the obstacles to life on land, the vast empty spaces of the Paleozoic landmass represented a tremendous evolutionary opportunity. The potential rewards of terrestrial life were especially great for plants. Water strongly absorbs light, so even in the clearest water, photosynthesis is limited to the upper few hundred meters of depth, and usually much less. Out of the water, the dazzling brightness of the sun permits rapid photosynthesis. Furthermore, terrestrial soils are rich storehouses of nutrients, whereas seawater tends to be low in certain nutrients, particularly nitrogen and phosphorus. Finally, the Paleozoic sea swarmed with plant-eating animals, but the land was devoid of animal life. The plants that first colonized the land would have had ample sunlight, untouched nutrient sources, and no predators.

Some Plants Became Adapted to Life on Dry Land

In moist soils at the water's edge, a few small green algae began to grow, taking advantage of the sunlight and nutrients. They didn't have large bodies to support against the force of gravity, and, living right in the film of water on the soil, they could easily obtain water. About 475 million years ago, some of these algae gave rise to the first multicellular land plants. Initially simple, low-growing forms, land plants rapidly evolved solutions to two of the main difficulties of plant life on land: obtaining and conserving water and staying upright despite gravity and winds. Water-resistant coatings on aboveground parts reduced water loss by evaporation, and rootlike structures delved into the soil, mining water and minerals. Specialized cells formed tubes called vascular tissues to conduct water from roots to leaves. Extra-thick walls surrounding certain cells enabled stems to stand erect.

Primitive Land Plants Retained Swimming Sperm and Required Water to Reproduce

Reproduction out of water presented challenges. As do animals, plants produce sperm and eggs, which must be able to meet to produce the next generation. The first land plants had swimming sperm, presumably much like those of today's mosses and ferns. Consequently, the earliest plants were restricted to swamps and marshes, where the sperm and eggs could be released into the water, or to areas with abundant rainfall, where the ground would occasionally be covered with water. Later, plants with swimming sperm prospered during periods in which the climate was warm and moist. For example, the Carboniferous period (360 million to 286 million years ago) was characterized by vast forests of giant tree ferns and club mosses (Fig. 15-8). The coal we mine today is derived from the fossilized remains of those forests.

Seed Plants Encased Sperm in Pollen Grains

Meanwhile, some plants inhabiting drier regions had evolved a means of reproduction that no longer depended on water. The eggs of these plants were retained on the parent plant, and the sperm were encased in drought-resistant pollen grains that traveled on the wind from plant to plant. When the pollen grains landed near an egg, they released sperm cells directly into living tissue, eliminating the need for a surface film of water. The fertilized egg remained on the parent plant, where it developed inside a seed, which provided protection and nutrients for the developing embryo within.

The earliest seed-bearing plants appeared in the late Devonian period (375 million years ago) and produced their seeds along branches, without any specialized structures to hold them. By the middle of the Carboniferous period, however, a new kind of seed-bearing plant had arisen. These plants, called **conifers,** protected their developing seeds inside cones. Conifers, which as wind-pollinated plants did not depend on water for reproduction, flourished and spread during the Permian period (286 to 245 million years ago), when mountains rose, swamps drained, and the climate became much drier. The conifers' good fortune, however, was not shared by the tree ferns and giant club mosses, which, with their swimming sperm, largely went extinct.

Flowering Plants Enticed Animals to Carry Pollen

About 140 million years ago, during the Cretaceous period, the flowering plants appeared, having evolved from a group of conifer-like plants. Many flowering plants are pollinated by insects and other animals, and this mode of pollination seems to have conferred an evolutionary advantage. Flower pollination by animals can be far more efficient than pollination by wind. Wind-pollinated plants must produce an enormous amount of pollen because the vast majority of pollen grains fail to reach their target. Flowering plants also evolved other advantages, including more rapid reproduction and, in some cases, faster growth. Today, flowering plants dominate the land, except in cold northern regions, where conifers still prevail.

Some Animals Became Adapted to Life on Dry Land

Soon after land plants evolved, providing potential food sources for other organisms, animals emerged from the sea. The first animals to move onto land were **arthropods** (the group that today includes insects, spiders, scorpions, centipedes, and crabs). Why arthropods? The answer seems to be that they already possessed certain structures that, purely by chance, were suited to life on land. Foremost among these structures was an exoskeleton, such as the shell of a lobster or crab.

▲ Figure 15-8 **The swamp forest of the Carboniferous period** The treelike plants in this artist's reconstruction are tree ferns and giant club mosses, most species of which are now extinct.

▲ **Figure 15-9 A fish that walks on land** Some modern fishes, such as this mudskipper, walk on land. As did the ancient lobefin fishes that gave rise to amphibians, mudskippers use their strong pectoral fins to move across dry areas in their swampy habitats. **QUESTION:** Does the mudskipper's ability to walk on land constitute evidence that lobefin fishes were the ancestors of amphibians?

Dinosaurs Singing in the Backyard?

Continued

Just as fossils of *Archaeopteryx*, the ancient bird with some reptile-like traits, provide evidence that reptiles are the ancestors of birds, fossils show that lobefinned fish are the ancestors of amphibians. For example, paleontologists recently discovered fossils of a lobefin dubbed *Tiktaalik*. Like other lobefins, *Tiktaalik* had scales, gills, and fins, but it also had amphibian-like features, such as a movable neck and arm-like joints (shoulders, wrists, and elbows) in its fin bones.

Exoskeletons are both waterproof and strong enough to support a small animal against the force of gravity.

For millions of years, arthropods had the land and its plants to themselves, and for tens of millions of years more, they were the dominant land animals. Dragonflies with a wingspan of 28 inches (70 centimeters) flew among the Carboniferous tree ferns, while millipedes 6.5 feet (2 meters) long munched their way across the swampy forest floor. Eventually, however, the arthropods' splendid isolation came to an end.

Amphibians Evolved from Lobefin Fishes

About 400 million years ago, a group of Silurian fishes called the lobefins appeared, probably in freshwater. **Lobefins** had two important features that would later enable their descendants to colonize land: (1) stout, fleshy fins with which they crawled about on the bottoms of shallow, quiet waters, and (2) an outpouching of the digestive tract that could be filled with air, like a primitive lung. One group of lobefins colonized very shallow ponds and streams, which shrank during droughts and often became oxygen poor. By taking air into their lungs, these lobefins could obtain oxygen anyway. Some began to use their fins to crawl from pond to pond in search of prey or water, as some modern fish do today (**Fig. 15-9**).

The benefits of feeding on land and moving from pool to pool favored the evolution of a group of animals that could stay out of water for longer periods and that could move about more effectively on land. With improvements in lungs and legs, **amphibians** evolved from lobefins, first appearing in the fossil record about 350 million years ago. To an amphibian, the Carboniferous swamp forests were a kind of paradise: no predators to speak of, abundant prey, and a warm, moist climate. As had the insects and millipedes, some amphibians evolved gigantic size, including salamanders more than 10 feet (3 meters) long.

Despite their success, the early amphibians were not fully adapted to life on land. Their lungs were simple sacs without very much surface area, so they had to obtain some of their oxygen through their skin. Therefore, their skin had to be kept moist, a requirement that restricted them to swampy habitats where they wouldn't dry out. Further, amphibian sperm and eggs could not survive in dry surroundings and had to be deposited in watery environments. So, although amphibians could move about on land, they could not stray too far from the water's edge. Along with the tree ferns and club mosses, amphibians declined when the climate turned dry at the beginning of the Permian period about 286 million years ago.

Reptiles Evolved from Amphibians

As the conifers were evolving on the fringes of the swamp forests, a group of amphibians was also evolving adaptations to drier conditions. These amphibians ultimately gave rise to the **reptiles,** which had four major adaptations to life on land. First, reptiles evolved shelled, waterproof eggs that enclosed a supply of water for the developing embryo. Thus, eggs could be laid on land without the reptiles' having to venture back to the dangerous swamps full of fish and amphibian predators. Second, reptiles evolved internal fertilization, in which a male deposits sperm inside a female's body, thereby avoiding the need to release swimming sperm into water. Third, ancestral reptiles evolved scaly, water-resistant skin that helped prevent the loss of body water to the dry air. Finally, reptiles evolved improved lungs that were able to provide the entire oxygen supply for an active animal. As the climate dried during the Permian period, reptiles became the dominant land vertebrates, relegating amphibians to the swampy backwaters where most remain today.

A few tens of millions of years later, the climate returned to more moist and stable conditions. This period saw the evolution of some very large reptiles, in particular the dinosaurs. The variety of dinosaur forms was enormous—from predators (**Fig. 15-10**) to plant eaters, from those that dominated the land to others that took to the air, to still others that returned to the sea. Dinosaurs were among the most successful animals ever, if we consider persistence as a measure of success. They flourished for more than 100 million years, until about 65 million years ago, when the last dinosaurs went extinct. No one is certain why they died out, but the aftereffects of a gigantic meteorite's impact with Earth seem to have been the final blow (as discussed in section 15.5).

Even during the age of dinosaurs, many reptiles remained quite small. One major difficulty faced by small reptiles is maintaining a high body temperature. Being active on land is helped by a warm body that maximizes the efficiency of the nervous system and muscles. But a warm body loses heat to the environment unless the air is also warm. Heat loss is a big problem for small animals, which have a larger surface area per unit of weight than do larger animals. Many species of small reptiles have retained slow metabolisms and have coped with the heat-loss problem by developing lifestyles in which they remain active only when the air is sufficiently warm. Two groups of small reptiles, however, independently followed a different evolutionary pathway: They developed insulation. In both groups, insulating structures arose from evolutionary modification of reptilian scales. One group evolved feathers, and another evolved hair.

Reptiles Gave Rise to Both Birds and Mammals

In ancestral birds, insulating feathers helped retain body heat. Consequently, these animals could be active in cool habitats and during the night, when their scaly relatives became sluggish. Later, some ancestral birds evolved longer, stronger feathers on their forelimbs, perhaps under selection for better ability to glide from trees or to jump after insect prey. Ultimately, feathers evolved into structures capable of supporting powered flight. Fully developed, flight-capable feathers are present in 150-million-year-old fossils, so the earlier insulating structures that eventually developed into flight feathers must have been present well before that time.

Unlike birds, which retained the reptilian habit of laying eggs, **mammals** evolved live birth and the ability to feed their young with secretions of the mammary (milk-producing) glands. Ancestral mammals also developed hair, which provided insulation. Because the uterus, mammary glands, and hair do not fossilize, we may never know when these structures first appeared, or what their intermediate forms looked like. Recently, however, a team of paleontologists found bits of fossil hair preserved in coprolites, which are fossilized animal feces. These coprolites, found in the Gobi Desert of China, were deposited by an anonymous predator 55 million years ago, so mammals have presumably had hair for at least that many years.

The earliest fossil mammal unearthed thus far is almost 200 million years old. Early mammals thus coexisted with the dinosaurs. They were mostly small creatures. The largest known mammal from the dinosaur era was about the size of a modern raccoon, but most early mammal species were far smaller than that. When the dinosaurs went extinct, however, mammals colonized the habitats left empty by the extinctions. Mammal species prospered, diversifying into the array of modern forms that we see today.

15.5 What Role Has Extinction Played in the History of Life?

If there is a lesson in the great tale of life's history, it is that nothing lasts forever. The story of life can be read as a long series of evolutionary dynasties, with each new dominant group rising, ruling the land or the seas for a time, and, inevitably,

▲ **Figure 15-10 A reconstruction of a Cretaceous forest** By the Cretaceous period, flowering plants dominated terrestrial vegetation. Dinosaurs, such as the predatory pack of 6-foot-long *Velociraptors* shown here, were the preeminent land animals. Although small by dinosaur standards, *Velociraptors* were formidable predators with great running speed, sharp teeth, and deadly, sickle-like claws on their hind feet.

Dinosaurs Singing in the Backyard?

Continued

Archaeopteryx fossils convincingly demonstrate that birds descended from reptile ancestors, but it has been challenging to determine which particular group of reptiles gave rise to birds. Today, most paleontologists agree that birds descended from dinosaurs. If this hypothesis is correct, it implies that feathers could well have been present in dinosaurs before birds evolved. Thus, finding fossils of dinosaurs with feathers would provide strong support for the hypothesis, and the search for such fossils was a longtime focus of scientists interested in the origin of birds. The predicted fossils of feathered dinosaurs were finally uncovered in 1998. (See the "Dinosaurs Singing in the Backyard? *Revisited*" section at the end of the chapter for more information on these fossils.)

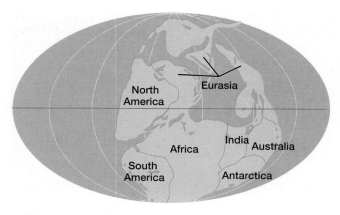

(a) 340 million years ago

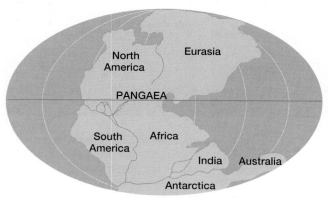

(b) 225 million years ago

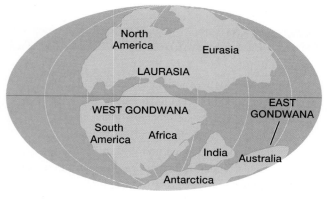

(c) 135 million years ago

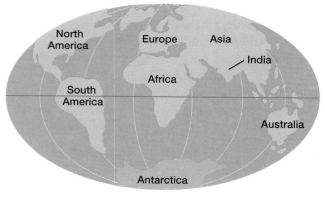

(d) Present

falling into decline and extinction. Dinosaurs are the most famous of these fallen dynasties, but the list of extinct groups known only from fossils is impressively long. Despite the inevitability of extinction, however, the overall trend has been for species to arise at a faster rate than they disappear, so the number of species on Earth has tended to increase over time.

Evolutionary History Has Been Marked by Periodic Mass Extinctions

Over much of life's history, dynastic succession has proceeded in a steady, relentless manner. This slow and steady turnover of species, however, has been interrupted by episodes of **mass extinction.** These mass extinctions are characterized by the relatively sudden disappearance of a wide variety of species over a large part of Earth. The largest episode of all, which occurred 245 million years ago at the end of the Permian period, wiped out more than 90% of the world's species, and life came perilously close to disappearing altogether.

Climate Change Contributed to Mass Extinctions

Mass extinctions have had a profound impact on the course of life's history, repeatedly redrawing the picture of life's diversity. What could have caused such dramatic changes in the fortunes of so many species? Many evolutionary biologists believe that changes in climate must have played an important role. When the climate changes, as it has done many times over the course of Earth's history, organisms that are adapted for survival in one climate may be unable to survive in a drastically different climate. In particular, at times when warm climates gave way to drier, colder climates with more variable temperatures, species may have gone extinct after failing to adapt to the harsh new conditions.

One cause of climate change is the shifting positions of continents. These movements are sometimes called *continental drift.* Continental drift is caused by **plate tectonics,** in which the earth's surface, including the continents and the seafloor, is divided into plates that rest atop a viscous but fluid layer and move slowly about. As the plates wander, their positions may change in latitude (**Fig. 15-11**). For example, 340 million years ago, much of North America was located at or near the equator, an area characterized by consistently warm and wet tropical weather. But as time has passed, plate tectonics has carried the continent up into temperate and arctic regions. As a result, the once tropical climate was replaced by a regime of seasonal changes, cooler temperatures, and less rainfall. Plate tectonics continues today; the Atlantic Ocean, for example, widens by a few centimeters each year.

Catastrophic Events May Have Caused the Largest Mass Extinctions

Geological data indicate that most mass extinction events coincided with periods of climatic change. To many scientists, however, the rapidity of mass extinctions suggests that the slow process of climate change could not, by itself, be responsible for such large-scale disappearances of species. Perhaps more sudden events also play a role. For example, catastrophic geological events, such as massive volcanic eruptions, could have had devastating effects. Geologists have found evidence of

◀ **Figure 15-11 Continental drift from plate tectonics** The continents are passengers on plates moving on Earth's surface as a result of plate tectonics. **(a)** About 340 million years ago, much of what is now North America was positioned at the equator. **(b)** All the plates eventually fused together into one gigantic landmass, which geologists call Pangaea. **(c)** Gradually Pangaea broke up into Laurasia and Gondwanaland, which itself eventually broke up into West and East Gondwana. **(d)** Further plate motion eventually resulted in the current positions of the modern-day continents.

past volcanic eruptions so huge that they make the 1980 Mount St. Helens explosion look like a firecracker by comparison. Even such gigantic eruptions, however, would directly affect only a relatively small portion of Earth's surface.

The search for the causes of mass extinctions took a fascinating turn in the early 1980s when Luis and Walter Alvarez proposed that the extinction event of 65 million years ago, which wiped out the dinosaurs and many other species, was caused by the impact of a huge meteorite. The Alvarezes' idea was met with great skepticism when it was first introduced, but geological research since that time has generated a great deal of evidence that a massive impact did indeed occur 65 million years ago. In fact, researchers have identified the Chicxulub crater, a 100-mile-wide crater buried beneath the Yucatan Peninsula of Mexico, as the impact site of a giant meteorite—10 miles in diameter—that collided with Earth just at the time that dinosaurs disappeared.

Could this immense meteorite strike have caused the mass extinction that coincided with it? No one knows for sure, but scientists suggest that such a massive impact would have thrown so much debris into the atmosphere that the entire planet would have been darkened for a period of years. With little light reaching the planet, temperatures would have dropped precipitously and the photosynthetic capture of energy (upon which all life ultimately depends) would have declined drastically. The worldwide "impact winter" would have spelled doom for the dinosaurs and a host of other species.

15.6 How Did Humans Evolve?

Humans have been present for only a tiny portion of Earth's long history (Fig. 15-12). Nonetheless, scientists are intensely interested in the origin and evolution of humans, especially in the evolution of the gigantic human brain. The outline of human evolution that we present in this section is a synthesis of current thought on the subject, but we note that it is speculative, because the fossil evidence of human evolution is comparatively scarce. Paleontologists disagree about the interpretation of the fossil evidence, and many ideas may have to be revised as new fossils are found.

Humans Inherited Some Early Primate Adaptations for Life in Trees

Humans are members of a mammal group known as **primates,** which also includes lemurs, monkeys, and apes. The oldest primate fossils are 55 million years old, but because primate fossils are relatively rare compared with those of many other animals, the first primates probably arose considerably earlier but left no fossil record. Early primates probably fed on fruits and leaves, and were adapted for life in the trees. Many modern primates retain the tree-dwelling lifestyle of their ancestors (Fig. 15-13). The common heritage of humans and other primates is reflected in a set of physical characteristics that was present in the earliest primates and that persists in many modern primates, including humans.

Binocular Vision Provided Early Primates with Accurate Depth Perception

One of the earliest primate adaptations seems to have been large, forward-facing eyes (see Fig. 15-13). Jumping from branch to branch is risky business unless an animal can accurately judge where the next branch is located. Accurate depth perception was made possible by binocular vision, provided by forward-facing eyes with overlapping fields of view. Another key adaptation was color vision. We cannot, of course, tell if a fossil animal had color vision, but because modern primates have excellent color vision, it seems reasonable to assume that earlier primates did as well. Many primates feed on fruit, and color vision helps to detect ripe fruit among a bounty of green leaves.

Dinosaurs Singing in the Backyard?

Continued

The mass extinction of 65 million years ago may have doomed the dinosaurs, but it ultimately proved to be beneficial for the dinosaurs' close relatives, the birds. The relatively sudden extinction of so many species meant that the resources used by those extinguished species became available to the surviving species. Birds were among the survivors, and their ability to exploit the newly available resources led to an expansion of bird populations and a rapid proliferation of bird species. This post-extinction bird "explosion" is responsible for today's diversity of bird species.

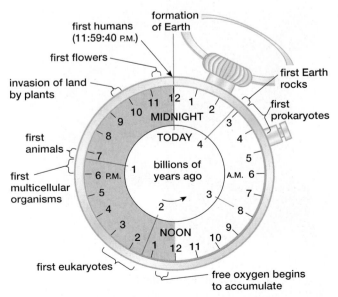

▲ Figure 15-12 Earth's history projected on a 24-hour clock

(b) Lemur

(a) Tarsier

(c) Macaque

▲ **Figure 15-13** **Representative primates** The *(a)* tarsier, *(b)* lemur, and *(c)* lion-tail macaque monkey all have a relatively flat face, with forward-looking eyes providing binocular vision. All also have color vision and grasping hands. These features, retained from the earliest primates, are shared by humans.

Early Primates Had Grasping Hands

Early primates had long, grasping fingers that could wrap around and hold onto tree limbs. This adaptation to tree dwelling was the basis for later evolution of human hands that could perform both a precision grip (used by modern humans for delicate maneuvers such as manipulating small objects, writing, and sewing) and a power grip (used for powerful actions, such as swinging a club or thrusting with a spear).

A Large Brain Facilitated Hand–Eye Coordination and Complex Social Interactions

Primates have brains that are larger, relative to their body size, than the brains of almost all other animals. No one really knows for certain which environmental factors favored the evolution of large brains. It seems reasonable, however, that controlling and coordinating rapid locomotion through trees, the dexterous movements of the hands in manipulating objects, and binocular, color vision would be facilitated by increased brain power. Most primates also have fairly complex social systems, which require relatively high intelligence. If sociality promoted increased survival and reproduction, then there would have been environmental pressures for the evolution of larger brains.

The Oldest Hominid Fossils Are from Africa

On the basis of comparisons of DNA from modern chimps, gorillas, and humans, researchers estimate that the **hominid** line (humans and their fossil relatives) diverged from the ape lineage sometime between 5 million and 8 million years ago. The fossil record, however, suggests that the split must have occurred at the early end of that range. Paleontologists working in the African country of Chad in 2002 discovered fossils of a hominid, *Sahelanthropus tchadensis*, that lived more than 6 million years ago (**Fig. 15-14**). *Sahelanthropus* is clearly a hominid, as it shares several anatomical features with later members of the group. But because this oldest known member of our family also exhibits other features that are more characteristic of apes, it may represent a point on our family tree that is close to the split between apes and hominids.

In addition to *Sahelanthropus*, two other hominid species—*Ardipithecus ramidus* and *Orrorin tugenensis*—are known from fossils appearing in rocks that are between 4 million and 6 million years old. Our knowledge of these hominids is limited, however, because only a few specimens have been found thus far, and most of these recent discoveries typically include only small portions of skeletons. A more extensive record of early hominid evolution does not begin until about 4 million years ago. That date marks the beginning of the fossil record of the genus *Australopithecus* (**Fig. 15-15**), a group of African hominid species with brains larger than those of their prehominid forebears but still much smaller than those of modern humans.

▲ **Figure 15-14** **The earliest hominid** This nearly complete skull of *Sahelanthropus tchadensis*, which is more than 6 million years old, is the oldest hominid fossil yet found.

The Earliest Hominids Could Stand and Walk Upright

The earliest australopithecines (as the various species of *Australopithecus* are collectively known) had legs that were shorter, relative to their height, than those of modern humans, but their knee joints allowed them to straighten their legs fully, permitting efficient bipedal (upright, two-legged) locomotion. Footprints almost 4 million years old, discovered in Tanzania by anthropologist Mary Leakey, show that even the earliest australopithecines could, and at least sometimes did, walk

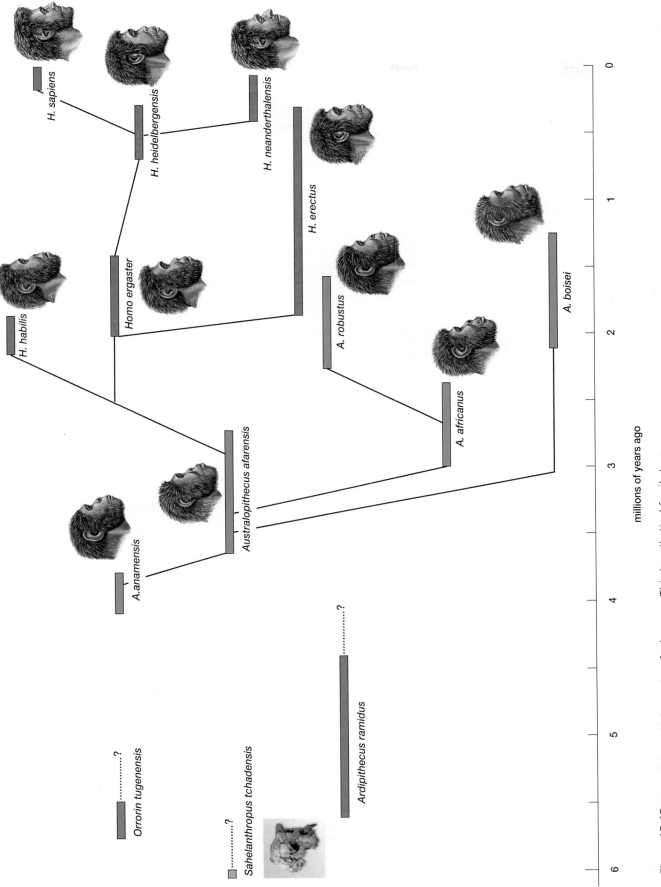

▲ **Figure 15-15 A possible evolutionary tree for humans** This hypothetical family tree shows facial reconstructions of representative specimens. Although many paleontologists consider this to be the most likely human family tree, there are several alternative interpretations of the known hominid fossils. Fossils of the earliest hominids are scarce and fragmentary, so the relationship of these species to later hominids remains unknown.

upright. Upright posture may have evolved even earlier. The discoverers of *Sahelanthropus* and *Orrorin* argue that the leg and foot bones of these earliest hominids have characteristics that indicate bipedal locomotion, but this conclusion will remain speculative until more complete skeletons of these species are found.

The reasons for the evolution of bipedal locomotion among the early hominids remain poorly understood. Perhaps hominids that could stand upright gained an advantage in gathering or carrying food in their forest habitat. Whatever its cause, the early evolution of upright posture was extremely important in the evolutionary history of hominids, because it freed their hands from use in walking. Later hominids were thus able to carry weapons, manipulate tools, and eventually achieve the cultural revolutions produced by modern *Homo sapiens*.

Several Species of *Australopithecus* Emerged in Africa

The oldest australopithecine species, represented by fossilized teeth, skull fragments, and arm bones, was unearthed near an ancient lake bed in Kenya from sediments that were dated as being between 3.9 million and 4.1 million years old. It was named *Australopithecus anamensis* by its discoverers (*anam* means "lake" in the local Turkana language). The second most ancient australopithecine, called *Australopithecus afarensis*, was discovered in the Afar region of Ethiopia. Fossil remains of this species as old as 3.9 million years have been unearthed. The *A. afarensis* line apparently gave rise to other *Australopithecus* species, all of which had gone extinct by 1.2 million years ago. Before disappearing, however, one of these species gave rise to a new branch of the hominid family tree, the genus *Homo* (see Fig. 15-15).

The Genus *Homo* Diverged from the Australopithecines 2.5 Million Years Ago

Hominids that are sufficiently similar to modern humans to be placed in the genus *Homo* first appear in African fossils that are about 2.5 million years old. Among the earliest African *Homo* fossils are *H. habilis* (see Fig. 15-15), a species whose body and brain were larger than those of the australopithecines but that retained the apelike long arms and short legs of their australopithecine ancestors. In contrast, the skeletal anatomy of *H. ergaster*, a species whose fossils first appear 2 million years ago, has limb proportions more like those of modern humans. This species is believed by many paleoanthropologists (scientists who study human origins) to be on the evolutionary branch that led ultimately to our own species, *H. sapiens*. In this view, *H. ergaster* was the common ancestor of two distinct branches of hominids. The first branch led to *H. erectus*, which was the first hominid species to leave Africa. The second branch from *H. ergaster* ultimately led to *H. heidelbergensis*, some of which migrated to Europe and gave rise to the Neanderthals, *H. neanderthalensis*. Meanwhile, back in Africa, another branch split off from the *H. heidelbergensis* lineage. This branch became *H. sapiens*, modern humans.

Neanderthals Had Large Brains and Excellent Tools

Neanderthals first appeared in the European fossil record about 150,000 years ago. By about 70,000 years ago, they had spread throughout Europe and western Asia. By 30,000 years ago, however, the species was extinct.

Contrary to the popular image of a hulking, stoop-shouldered "caveman," Neanderthals were quite similar to modern humans in many ways. Although more heavily muscled, Neanderthals walked fully erect, were dexterous enough to manufacture finely crafted stone tools, and had brains that, on average, were slightly larger than those of modern humans. Many European Neanderthal fossils show heavy brow ridges and a broad, flat skull, but others, particularly from areas around the eastern shores of the Mediterranean Sea, are somewhat more physically similar to *H. sapiens*.

Despite the physical and technological similarities between Neanderthals and *H. sapiens*, there is no solid archaeological evidence that Neanderthals ever developed an advanced culture that included such characteristically human endeavors as art, music, and rituals. Some anthropologists argue that, because their skeletal anatomy shows that they were physically capable of making the sounds required for speech, Neanderthals might have acquired language. This interpretation of Neanderthal anatomy, however, is not unanimously accepted. In general, the available evidence of the Neanderthal way of life is limited and open to different interpretations, and anthropologists are engaged in a sometimes heated debate about how advanced Neanderthal culture became.

Though some anthropologists argue that Neanderthals were simply a variety of *H. sapiens*, most agree that Neanderthals were a separate species. Dramatic evidence in support of this hypothesis has come from researchers who have isolated DNA from Neanderthal and *H. sapiens* skeletons that are more than 20,000 years old. These extractions of ancient DNA have allowed researchers to compare the nucleotide sequences of Neanderthal genes with the sequences of the same genes in both fossil and modern humans. The comparisons have shown that Neanderthal sequences are very different from those of both modern and fossil humans, but that modern and fossil humans share similar sequences. These findings indicate that the evolutionary branch leading to Neanderthals diverged from the ancestral human line hundreds of thousands of years before the emergence of modern *H. sapiens*. This early divergence supports the conclusion that Neanderthals were a distinct species, *H. neanderthalensis*.

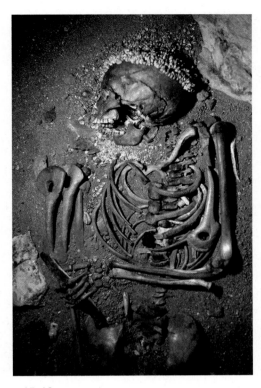

▲ **Figure 15-16 Paleolithic burial** This 24,000-year-old grave shows evidence that Cro-Magnon people ritualistically buried their dead. The body was covered with a dye known as red ocher, then buried with a headdress made of snail shells and a flint tool in its hand.

Modern Humans Emerged Less Than 200,000 Years Ago

The fossil record shows that anatomically modern humans appeared in Africa at least 160,000 years ago and possibly as long as 195,000 years ago. The location of these fossils suggests that *Homo sapiens* originated in Africa, but most of our knowledge about our own early history comes from European and Middle Eastern fossils of *H. sapiens*, collectively known as Cro-Magnons (after the district in France in which their remains were first discovered). Cro-Magnons appeared about 90,000 years ago. They had domed heads, smooth brows, and prominent chins (just like us). Their tools were precision instruments similar to the stone tools used until recently in many parts of the world.

Behaviorally, Cro-Magnons seem to have been similar to, but more sophisticated than, Neanderthals. Artifacts from 30,000-year-old Cro-Magnon archaeological sites include elegant bone flutes, graceful carved ivory sculptures, and evidence of elaborate burial ceremonies (**Fig. 15-16**). Perhaps the most remarkable accomplishment of Cro-Magnons is the magnificent art left in caves in places such as Altamira in Spain and Lascaux and Chauvet in France (**Fig. 15-17**). The oldest cave paintings found so far are more than 30,000 years old, and even the oldest ones make use of sophisticated artistic techniques. No one knows exactly why these paintings were made, but they attest to minds as capable as our own.

Cro-Magnons and Neanderthals Lived Side by Side

Cro-Magnons coexisted with Neanderthals in Europe and the Middle East for perhaps as many as 50,000 years before the Neanderthals disappeared. Some researchers believe that Cro-Magnons interbred extensively with Neanderthals, so Neanderthals were essentially absorbed into the human genetic mainstream. Other scientists disagree, citing mounting evidence such as the fossil DNA described earlier, and suggest that later-arriving Cro-Magnons simply overran and displaced the less-well-adapted Neanderthals.

▲ **Figure 15-17 The sophistication of Cro-Magnon people** Cave paintings by Cro-Magnons have been remarkably preserved by the relatively constant underground conditions of a cave in Lascaux, France.

Neither hypothesis does a good job of explaining how the two kinds of hominids managed to occupy the same geographical areas for such a long time. The persistence in one area of two similar but distinct groups for tens of thousands of years seems inconsistent with both interbreeding and direct competition. Perhaps the competition between *H. neanderthalensis* and *H. sapiens* was indirect, so that the two species were able to coexist for a time in the same habitat, until the superior ability of *H. sapiens* to exploit the available resources slowly drove Neanderthals to extinction.

The Evolutionary Origin of Large Brains May Be Related to Meat Consumption

The main physical features that distinguish us from our closest relatives, the apes, are our upright posture and large, highly developed brains. As described earlier, upright posture arose very early in hominid evolution, and hominids walked upright for several million years before large-brained *Homo* species arose. What circumstances might have caused the evolution of increased brain size? Many explanations have been proposed, but little direct evidence is available; hypotheses about the evolutionary origins of large brains are necessarily speculative.

One proposed explanation for the origin of large brains suggests that they evolved in response to increasingly complex social interactions. In particular, fossil evidence suggests that, beginning about 2 million years ago, hominid social life began to include a new type of activity—the cooperative hunting of large game. The resulting access to significant amounts of meat must have fostered a need to develop methods for distributing this valuable, limited resource among group members. Some anthropologists hypothesize that the individuals best able to manage this social interaction would have been more successful at gaining a large share of meat and using their share advantageously. Perhaps this social management was best accomplished by individuals with larger, more powerful brains, and natural selection therefore favored such individuals. Observations of chimpanzee societies have shown that the distribution of group-hunted meat often involves intricate social interactions in which meat is used to form alliances, repay favors, gain access to sexual partners, placate rivals, and so on. Perhaps the mental skill required to plan, assess, and remember such interactions was the driving force behind the evolution of our large, clever brains.

Dinosaurs Singing in the Backyard? Revisited

The first feathered dinosaur fossils to be discovered were found in rock deposits near the little village of Sihetun in northeastern China. In 1998, fossil hunters there extracted the exquisitely preserved remains of some previously undiscovered types of dinosaurs. Plainly visible along the margin of some of these clearly dinosaurian fossil skeletons were impressions of what appeared to be feathers. For the first time, scientists had solid evidence that some dinosaurs had feathers.

In the years since the first discoveries in China, many additional feathered dinosaurs have been unearthed, including some with especially well-preserved fossil feathers. The fossils include more than a dozen different species of feathered dinosaurs. To most paleontologists, these fossils collectively demonstrate that feathers did not originate with birds, but were present in a group of dinosaurs, known as theropods, that gave rise to birds. Some paleontologists argue that all

theropods probably had feathers, including *Velociraptor* (famously portrayed in *Jurassic Park*) and the fearsome predator *Tyrannosaurus*.

Eventually all dinosaurs, feathered and unfeathered, went extinct. Their evolutionary cousins the birds, however, lived on and evolved into the diverse group that brightens our backyards today.

Consider This

An alternative hypothesis to explain the existence of feathered dinosaurs is convergent evolution. Is it possible that dinosaurs and birds are unrelated, and that each group separately evolved feathers? What is the likelihood of convergent evolution of feathers? What evidence might, if found, cause you to reject the conclusion that birds are descended from dinosaurs and accept the hypothesis of convergent evolution?

Chapter Review

Summary of Key Concepts

For additional study help and activities, go to www.mybiology.com.

15.1 How Did Life Begin?

Before life arose, lightning, ultraviolet light, and heat formed organic molecules from water and the components of primordial Earth's atmosphere. These molecules probably included nucleic acids, amino acids, short proteins, and lipids. By chance, some molecules of RNA may have had enzymatic properties, catalyzing the assembly of copies of themselves from nucleotides in Earth's waters. These may have been the forerunners of life. Protein-lipid vesicles enclosing these ribozymes may have formed the first protocells.

15.2 What Were the Earliest Organisms Like?

The oldest fossils, about 3.5 billion years old, are of prokaryotic cells that fed by absorbing organic molecules that had been synthesized in the environment. Because there was no free oxygen in the atmosphere, their energy metabolism must have been anaerobic. As the cells multiplied, they depleted the organic molecules that had been formed by prebiotic synthesis. Some cells developed the ability to synthesize their own food molecules by using simple inorganic molecules and the energy of sunlight. These earliest photosynthetic cells were probably ancestors of today's cyanobacteria.

Photosynthesis releases oxygen as a by-product, and by about 2.2 billion years ago significant amounts of free oxygen were accumulating in the atmosphere. Aerobic metabolism, which generates more cellular energy than does anaerobic metabolism, probably arose about this time.

Eukaryotic cells had evolved by about 1.7 billion years ago. The first eukaryotic cells probably arose as symbiotic associations between predatory prokaryotic cells and other bacteria. Mitochondria may have evolved from aerobic bacteria engulfed by predatory cells. Similarly, chloroplasts may have evolved from photosynthetic cyanobacteria.

Web Animation Endosymbiosis

15.3 What Were the Earliest Multicellular Organisms Like?

Multicellular organisms evolved from eukaryotic cells and first appeared, in the seas, about 1.2 billion years ago. Multicellularity offers several advantages, including greater size. In plants, increased size offered some protection from predation. Specialization of cells allowed plants to anchor themselves in the nutrient-rich, well-lit waters of the shore. For animals, multicellularity allowed more efficient predation and more effective escape from predators. These in turn provided environmental pressures for faster locomotion, improved senses, and greater intelligence.

15.4 How Did Life Invade the Land?

The first land organisms were probably algae. The first multicellular land plants appeared about 475 million years ago. Life on land required special adaptations for support of the body, reproduction, and the acquisition, distribution, and retention of water, but the land also offered abundant sunlight and protection from aquatic herbivores. Soon after land plants evolved, arthropods invaded the land. Absence of predators and abundant land plants for food probably facilitated the invasion of the land by animals.

The earliest land vertebrates evolved from lobefin fishes, which had leglike fins and a primitive lung. A group of lobefins evolved into the amphibians about 350 million years ago. Reptiles evolved from amphibians, with several further adaptations for life on land: internal fertilization, waterproof eggs that could be laid on land, water-resistant skin, and better lungs. Birds and mammals evolved independently from separate groups of reptiles. A major advance in the evolution of both birds and mammals was insulation over the body surface in the form of feathers and hair.

Web Animation Evolutionary Timescales

15.5 What Role Has Extinction Played in the History of Life?

The history of life has been characterized by constant turnover of species as species go extinct and are replaced by new ones. Mass extinctions, in which large numbers of species disappear within a relatively short time, have occurred periodically. Mass extinctions were probably caused by some combination of climate change and catastrophic events, such as volcanic eruptions and meteorite impacts.

15.6 How Did Humans Evolve?

One group of mammals evolved into the tree-dwelling primates. Some primates descended from the trees, and these were the ancestors of apes and humans. The australopithecines arose in Africa about 4 million years ago. These hominids walked erect, had larger brains than did their forebears, and fashioned primitive tools. One group of australopithecines gave rise to a line of hominids in the genus *Homo*, which in turn gave rise to modern humans.

Key Terms

amphibian *p. 268*
arthropod *p. 267*
conifer *p. 267*
endosymbiont hypothesis *p. 263*
eukaryotes *p. 263*

exoskeleton *p. 266*
hominid *p. 272*
lobefin *p. 268*
mammal *p. 269*

mass extinction *p. 270*
plate tectonics *p. 270*
primate *p. 271*
prokaryote *p. 262*

protocell *p. 259*
reptile *p. 268*
ribozyme *p. 258*
spontaneous generation *p. 255*

Thinking Through the Concepts

Suggested answers to end-of-chapter and figure-based questions can be found at the end of the text.

Fill-in-the-Blank

1. Because there was no oxygen in the earliest atmosphere, the first cells must have derived energy by _____ metabolism of organic molecules. Oxygen was introduced into the atmosphere when some microbes developed the ability to _____ and released oxygen as a by-product. Oxygen was _____ to many of the earliest cells, but some evolved the ability to use oxygen in _____ respiration, which provided far more _____.

2. The molecule _____ became a candidate for the first self-replicating, information-carrying molecule when Tom Cech and Sidney Altman discovered that some of these molecules can act as _____, which they called _____.

3. Complex cells that contain a nucleus and other organelles are called _____ cells. A compelling explanation for the origin of these complex cells is the _____ hypothesis. This hypothesis suggests that mitochondria arose when a predatory anaerobic cell engulfed a cell that obtained energy through _____ metabolism. One observation that supports this hypothesis is that mitochondria have their own _____.

4. The sperm of early land plants had to reach the egg by _____, limiting them to _____ environments. An important adaptation to dry land was the evolution of _____, which enclosed sperm in a drought-resistant coat.

5. Early plants that protected their seeds within cones are called _____. These relied on _____ to carry their pollen. Later, some plants evolved _____, which attracted animals, particularly _____, that carried their pollen. Animal pollination is much more _____ than wind pollination.

6. The first animals to live on land were _____ because their external skeletons, also called _____, supported the animals' weight, while protecting their bodies from _____.

7. Amphibians gave rise to _____, which had four important adaptations to life on dry land: shelled, waterproof _____; _____ fertilization; scaly, water-resistant_____; and more efficient _____.

Review Questions

1. What is the evidence that life might have originated from nonliving matter on early Earth?

2. If the first cells with aerobic metabolism were so much more efficient at producing energy, why didn't they drive to extinction cells with only anaerobic metabolism?

3. Explain the endosymbiont hypothesis for the origin of chloroplasts and mitochondria.

4. Name two advantages of multicellularity for early algae.

5. What advantages and disadvantages would terrestrial existence have had for the first plants to invade the land? For the first land animals?

6. Outline the major adaptations that emerged during the evolution of land-dwelling vertebrates, from amphibians to reptiles to birds and mammals. Explain how these adaptations increased the fitness of the various groups for life on land.

7. Outline the evolution of humans from early primates. Include in your discussion such features as binocular vision, grasping hands, bipedal locomotion, tool making, and brain expansion.

Applying the Concepts

1. IS THIS SCIENCE? The "panspermia" hypothesis proposes that life did not originate on Earth, but instead arose elsewhere in the galaxy. According to the hypothesis, cells arrived on Earth via meteorites or pieces of comets that fell to Earth from space. Is panspermia a plausible alternative to other hypotheses for the origin of life on Earth? What kinds of evidence would provide support for the panspermia hypothesis?

2. Do you think that studying our ancestors can shed light on the behavior of modern humans? Why or why not?

3. Starting with a protocell, what would you consider to be the most significant evolutionary events that lead to the profusion of life that exists today? Explain your answer.

For additional resources, go to www.mybiology.com.

The Diversity of Life

The saola, unknown to science until 1992, is one of a number of previously undiscovered species recently found in the mountains of Vietnam.

Case Study   Hidden Treasures

This brush-finch is a member of a species that was unknown to science until 2006.

The steep, rain-drenched slopes of Vietnam's Annamite Mountains are remote and forbidding, cloaked in tropical mists that lend an air of mystery and concealment to the forested mountains. As it turns out, this remote refuge conceals a most astonishing biological surprise: the saola, a hoofed, horned mammal that was unknown to science until 1992. The discovery of a new species of large mammal at this late date was a complete shock. After centuries of human exploration and exploitation in every corner of the world's forests, deserts, and savannas, scientists were certain that no large-sized mammal species could have escaped detection. As long ago as 1812, French naturalist Georges Cuvier wrote that "there is little hope of discovering new species of large quadrupeds." And yet, the saola—3 feet high at the shoulder, weighing up to 200 pounds, and sporting 20-inch black horns—remained hidden in Annamite Mountain forests, outside the realm of scientific knowledge.

It is surprising that the saola stayed hidden from science for so long because people tend to notice large animals. As the discovery of the saola showed, however, even relatively conspicuous organisms may remain unknown if they live in a sufficiently remote area. The Annamite Mountains, isolated by inhospitable terrain and the wars fought in Vietnam during the last century, concealed the saola and other undiscovered plant, mammal, and reptile species. Similarly, the remote Foja Mountains of New Guinea hid a new bird species and dozens of new species of butterflies, frogs, and flowering plants, which were not discovered until a 2005 scientific expedition to the region. A 2006 survey of unexplored underwater habitats in Indonesia revealed more than 50 new species of fish, coral, and shrimp. Such discoveries show that we have incomplete knowledge of the diversity of life, even for intensively studied, conspicuous groups of organisms.

Continuing scientific exploration yields a steady trickle of new birds and mammals, and a somewhat larger stream of new fish, reptiles, amphibians, and flowering plants. Is the situation similar for Earth's smaller and less photogenic organisms? We will return to this question following a survey of life's known diversity. ■

Web Animation Taxonomic Classification

16.1 How Are Organisms Named and Classified?

Organisms are placed into categories on the basis of their evolutionary relationships. There are eight major categories: domain, kingdom, phylum, class, order, family, genus, and species. These categories form a nested hierarchy in which each level includes all of the other levels below it. Each domain contains a number of kingdoms; each kingdom contains a number of phyla; each phylum includes a number of classes; each class includes a number of orders; and so on. As we move down the hierarchy, smaller and smaller groups are included. Each category is increasingly narrow and specifies groups whose common ancestor is increasingly recent. Table 16-1 describes some examples of classifications of specific organisms.

Each Species Has a Unique, Two-Part Name

The scientific name of an organism is formed from the two smallest categories—its genus and its species. Each genus includes a group of very closely related

Table 16-1   **Classification of Selected Organisms, Reflecting Their Degree of Relatedness***

	Human	Chimpanzee	Wolf	Fruit Fly	Sequoia Tree	Sunflower
Domain	**Eukarya**	**Eukarya**	**Eukarya**	**Eukarya**	**Eukarya**	**Eukarya**
Kingdom	**Animalia**	**Animalia**	**Animalia**	**Animalia**	**Plantae**	**Plantae**
Phylum	**Chordata**	**Chordata**	**Chordata**	Arthropoda	Coniferophyta	Anthophyta
Class	**Mammalia**	**Mammalia**	**Mammalia**	Insecta	Coniferosida	Dicotyledoneae
Order	**Primates**	**Primates**	Carnivora	Diptera	Coniferales	Asterales
Family	Hominidae	Pongidae	Canidae	Drosophilidae	Taxodiaceae	Asteraceae
Genus	*Homo*	*Pan*	*Canis*	*Drosophila*	*Sequoiadendron*	*Helianthus*
Species	*sapiens*	*troglodytes*	*lupus*	*melanogaster*	*giganteum*	*annuus*

*Boldface categories are those that are shared by more than one of the organisms classified. Genus and species names are always italicized or underlined.

species, and each species within a genus includes populations of organisms that can potentially interbreed under natural conditions. Thus, the genus *Sialia* (bluebirds) includes the eastern bluebird (*Sialia sialis*), the western bluebird (*Sialia mexicana*), and the mountain bluebird (*Sialia currucoides*)—very similar birds that do not interbreed (**Fig. 16-1**).

Each two-part scientific name is unique, so referring to an organism by its scientific name rules out any chance of ambiguity or confusion. For example, the bird *Gavia immer* is commonly known in North America as the common loon, in Great Britain as the northern diver, and by still other names in non-English-speaking countries. But the Latin scientific name *Gavia immer* is recognized by biologists worldwide, overcoming language barriers and allowing precise communication.

Note that, by convention, scientific names are always <u>underlined</u> or *italicized*. The first letter of the genus name is always capitalized, and the first letter of the species name is always lowercase. The species name is never used alone but is always paired with its genus name.

Classification Originated as a Hierarchy of Categories

Aristotle (384–322 B.C.) was among the first to attempt to formulate a logical, standardized language for naming living things. Based on characteristics such as structural complexity, behavior, and degree of development at birth, he classified about 500 organisms into 11 categories (see Fig. 13-2 on p. 212). Aristotle's categories

◀ **Figure 16-1 Three species of bluebird**
Despite their obvious similarity, these three species of bluebird remain distinct because they do not interbreed. The three species shown are **(a)** the eastern bluebird (*Sialia sialis*), **(b)** the western bluebird (*Sialia mexicana*), and **(c)** the mountain bluebird (*Sialia currucoides*).

(a) Eastern bluebird

(b) Western bluebird

(c) Mountain bluebird

formed a hierarchical structure, with each category more inclusive than the one beneath it, a concept that is still used today.

Building on this foundation more than 2,000 years later, Swedish naturalist Carl von Linné (1707–1778)—who called himself Carolus Linnaeus, a Latinized version of his given name—laid the groundwork for the modern classification system. He placed each organism into a series of hierarchically arranged categories on the basis of its resemblance to other life-forms, and introduced the scientific name composed of genus and species.

Nearly 100 years later, Charles Darwin (1809–1882) published *On the Origin of Species*, which demonstrated that all organisms are connected by common ancestry (see p. 220). Biologists then began to recognize that the categories ought to reflect the pattern of evolutionary relatedness among organisms. The more categories two organisms share, the closer their evolutionary relationship.

Biologists Identify Features That Reveal Evolutionary Relationships

Biologists seek to reconstruct the tree of life, but they must do so without much direct knowledge of evolutionary history. Because they can't see into the past, they must infer the past as best they can, on the basis of similarities among living organisms. Not just any similarity will do, however. Some observed similarities, such as the streamlined shapes of dolphins and sharks, stem from convergent evolution (see p. 222) in organisms that are not closely related, and such similarities are not useful for classification. Instead, biologists value the similarities that exist because two kinds of organisms both inherited a characteristic from a common ancestor. Therefore, the scientists who devise classifications must distinguish informative similarities caused by common ancestry from uninformative similarities that result from convergent evolution. In the search for informative similarities, biologists look at many kinds of characteristics.

Anatomical Similarities Play a Key Role in Classification

Historically, the most important and useful distinguishing characteristics have been anatomical. Biologists look carefully at similarities in both external body structure (see Fig. 16-1) and internal structures, such as skeletons and muscles. For example, homologous structures such as the finger bones of dolphins, bats, seals, and humans provide evidence of a common ancestor (see Fig. 13-8 on p. 221). To detect relationships among more closely related species, biologists may use microscopes to discern finer details—for example, the number and shape of the "teeth" on the tongue-like radula of a snail; the shape and position of the bristles on a marine worm; or the external structure of pollen grains of a flowering plant (Fig. 16-2).

▶ Figure 16-2 **Microscopic structures may be used to classify organisms** *(a)* The "teeth" on a snail's tonguelike radula (a structure used in feeding), *(b)* the bristles on a marine worm, and *(c)* the shape and surface features of pollen grains are characteristics potentially useful in classification. Such finely detailed structures can reveal similarities between species that are not apparent in larger and more obvious structures.

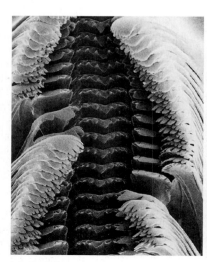
(a) **Radula**

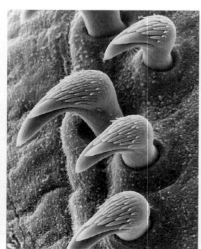
(b) **Bristles**

(c) **Pollen grains**

Molecular Similarities Are Also Useful for Classification

The anatomical characteristics shared by related organisms are expressions of underlying genetic similarities, so it stands to reason that evolutionary relationships among species must also be reflected in genetic similarities. Of course, direct genetic comparisons were not possible for most of the history of biology. Since the 1980s, however, advances in the techniques of molecular genetics have revolutionized studies of evolutionary relationships.

As a result of these technical advances, today's biologists can use the nucleotide sequences of DNA (that is, organisms' genotypes) to investigate relatedness among different types of organisms. Closely related species have similar DNA sequences. For example, the DNA sequences of chimpanzees and humans are extremely similar, showing that these two species are very closely related.

16.2 What Are the Domains of Life?

Web Animation Tree of Life

Before 1970, all forms of life were classified into two kingdoms: Animalia and Plantae. All bacteria, fungi, and photosynthetic eukaryotes were considered to be plants, and all other organisms were classified as animals. As scientists learned more about fungi and microorganisms, however, it became apparent that the two-kingdom system oversimplified evolutionary history. To help rectify this problem, in 1969, Robert H. Whittaker proposed a five-kingdom classification that was eventually adopted by most biologists.

The Five-Kingdom System Improved Classification

Whittaker's five-kingdom system placed all prokaryotic organisms into a single kingdom and divided the eukaryotes into four kingdoms. The designation of a separate kingdom (called Monera) for the prokaryotes reflected growing recognition that the evolutionary pathway of these tiny, single-celled organisms had diverged from that of the eukaryotes early in the history of life. Among the eukaryotes, the five-kingdom system recognized three kingdoms of multicellular organisms (Plantae, Fungi, and Animalia) and placed all of the remaining, mostly single-celled eukaryotes into a single kingdom (Protista).

Because it more accurately reflected current understanding of evolutionary history, the five-kingdom system was an improvement over the old two-kingdom system. As understanding continued to grow, however, our view of life's most fundamental categories needed yet another revision. The pioneering work of microbiologist Carl Woese has shown that biologists had overlooked a fundamental event in the early history of life, one that demands a new and more accurate classification of living things.

A Three-Domain System More Accurately Reflects Life's History

Woese and other biologists interested in the evolutionary history of microorganisms have studied the biochemistry of prokaryotic organisms. These researchers, focusing on nucleotide sequences of the RNA that is found in the organisms' ribosomes, discovered that the supposed kingdom Monera actually included two very different kinds of organisms. Woese has dubbed these two groups the Bacteria and the Archaea (Fig. 16-3).

Despite superficial similarities in their appearance under the microscope, the Bacteria and the Archaea are radically different. These two groups are no more closely related to one another than either one is to any eukaryote. The tree of life split into three parts very early in the history of life, long before the appearance of plants, animals, and fungi. As a result of this new understanding, the five-kingdom system has been replaced by a classification that divides life into three domains: **Bacteria, Archaea,** and **Eukarya** (Fig. 16-4).

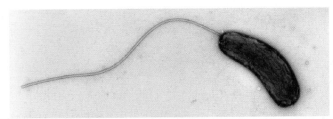

(a) A bacterium

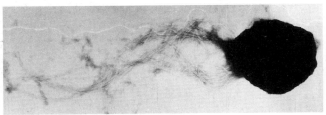

(b) An archaean

▲ **Figure 16-3 Two domains of prokaryotic organisms** Although superficially similar in appearance, *(a) Vibrio cholerae* and *(b) Methanococcus jannaschi* are less closely related than a mushroom and an elephant. *Vibrio* is in the domain Bacteria, and *Methanococcus* is in the Archaea.

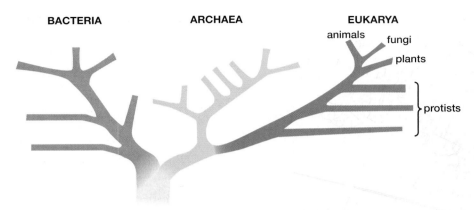

▲ **Figure 16-4 The tree of life** The three domains of life represent the earliest branches in evolutionary history.

Kingdom-Level Classification Remains Unsettled

The move to a three-domain classification system has required systematists to reexamine the kingdoms within each domain, and the process of establishing kingdoms is still under way. If we accept that the striking differences between plants, animals, and fungi demand that each of these evolutionary lineages retains its kingdom status, then the logic of classification requires that we also assign kingdom status to groups that branch off of the tree of life earlier than these three groups of multicellular eukaryotes. Following this logic, biologists recognize about 15 kingdoms among the Bacteria and three or so kingdoms among the Archaea. Biologists also recognize additional kingdoms within the Eukarya, reflecting a number of very early evolutionary splits within the diverse array of single-celled eukaryotes formerly lumped together in the kingdom Protista. Biologists, however, have yet to reach a consensus about the precise definitions of new prokaryotic and eukaryotic kingdoms, though new information about the evolutionary history of single-celled organisms is emerging rapidly. Thus, kingdom-level classification is in a state of transition as biologists strive to incorporate the latest information.

This text's descriptions of the diversity of life sidestep the unsettled state of life's kingdoms. The prokaryotic domains Archaea and Bacteria are discussed without reference to kingdom-level relationships. Among the eukaryotes, fungi, plants, and animals are treated as distinct evolutionary units, and the generic term "protist" designates the diverse collection of eukaryotes that are not members of these three kingdoms.

16.3 Bacteria and Archaea

Earth's first organisms were prokaryotes, single-celled microbes that lacked organelles such as the nucleus, chloroplasts, and mitochondria. (See pp. 58–60 for a comparison of prokaryotic and eukaryotic cells.) For the first 1.5 billion years or more of life's history, all life was prokaryotic. Even today, prokaryotes are extraordinarily abundant. A drop of seawater contains hundreds of thousands of prokaryotic organisms, and a spoonful of soil contains billions. The average human body is home to trillions of prokaryotes, which live on the skin, in the mouth, and in the stomach and intestines. In terms of abundance, prokaryotes are Earth's predominant form of life.

Bacteria and Archaea Are Fundamentally Different

Two of life's three domains, Bacteria and Archaea, consist entirely of prokaryotes. Bacteria and archaea are superficially similar in appearance under the microscope, but have striking differences in their structural and biochemical features that reveal the ancient evolutionary separation between them. For example, the cell walls of bacterial cells contain molecules of the polymer *peptidoglycan*, which helps strengthen the cell wall. Peptidoglycan is unique to bacteria, and the cell walls of archaea do not contain it. Bacteria and archaea also differ in the structure and composition of plasma membranes, ribosomes, and RNA polymerases, as well as in the mechanics of basic processes such as transcription and translation.

Classification of Prokaryotes Within Each Domain Is Difficult

The sharp biochemical differences between archaea and bacteria make distinguishing the two domains a straightforward matter, but classification within each domain poses challenges. Prokaryotes are tiny and structurally simple, and simply do

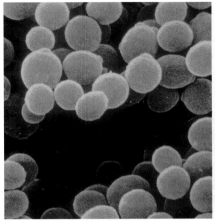

(a) Spherical

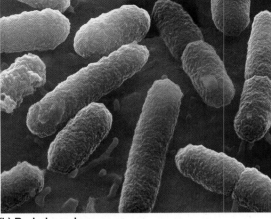

(b) Rod-shaped

◄ **Figure 16-5 Three common prokaryote shapes**

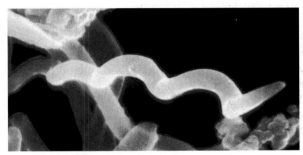

(c) Corkscrew-shaped

not exhibit the huge array of anatomical and developmental differences that can be used to infer the evolutionary history of plants, animals, and other eukaryotes. Consequently, prokaryotes have been classified on the basis of such features as shape, means of locomotion, pigments, nutrient requirements, the appearance of colonies (groups of individuals that descended from a single cell), and staining properties.

In recent years, however, our understanding of the evolutionary history of the prokaryotic domains has been greatly expanded by comparisons of DNA and RNA nucleotide sequences. Prokaryote classification, however, is a rapidly changing field, and consensus on kingdom-level classification has thus far proved elusive. With new DNA sequence data being generated at a furious pace, and with new and distinctive types of bacteria and archaea being discovered and described on a regular basis, the revision of prokaryote classification schemes is likely to continue for some time to come.

Prokaryote Adaptations Provide Mobility and Protection

Both bacteria and archaea are usually very small, ranging from about 0.2 to 10 micrometers in diameter. (In comparison, the diameters of eukaryotic cells range from about 10 to 100 micrometers.) About 250,000 average-sized bacteria or archaea could congregate on the period at the end of this sentence, although a few species of bacteria are larger. The largest known bacterium (*Thiomargarita namibiensis*) is as much as 700 micrometers in diameter, making it visible to the naked eye.

The cell walls that surround prokaryotic cells give characteristic shapes to different types of bacteria and archaea. The most common shapes are spherical, rod shaped, and corkscrew shaped (**Fig. 16-5**).

Some Prokaryotes Are Mobile

Many bacteria and archaea adhere to a surface or drift passively in liquid surroundings, but some can move about. Many of these mobile prokaryotes have **flagella** (singular, flagellum). Prokaryote flagella may appear singly at one end of a cell, in pairs

Hidden Treasures

Continued

Biologists agree that the prokaryote species discovered so far represent only a small fraction of the total number of species. Prokaryotes are so small that it is hard to find and see single individuals, even using powerful microscopes. And if a biologist does manage to view one, she will probably still be unable to tell if it's a new species, because different prokaryote species can be virtually identical in shape and size. To conclusively document a new species, a biologist must first culture it in the lab by taking a small sample of material containing unknown prokaryotes, supplying it with suitable conditions and nutrients, and waiting for the organisms to reproduce and form colonies. The colonies can then be examined to determine if they contain a new species. Unfortunately, most prokaryotes have proved to be difficult or impossible to culture; it is very difficult to determine the conditions that will prompt unknown prokaryotes to thrive and reproduce. Therefore, although a number of new species have been discovered by microbiologists who devised new methods of prokaryote culture, the search for new species has become, in large measure, a search for new DNA sequences. In this approach, researchers procure a sample likely to contain prokaryotes, and determine the sequence of all DNA present in the sample. By comparing the sequences found to those of known prokaryote species, researchers can make an educated guess as to whether new species are present. This approach is far faster than the culture method, but yields less conclusive species identifications and little information about the biology of new species.

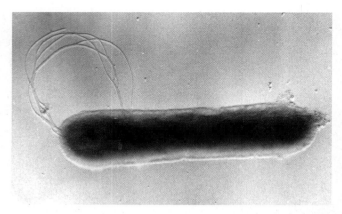

▲ **Figure 16-6 The prokaryote flagellum** An archaean of the genus *Aquifex* uses its flagella to move toward favorable environments.

(one at each end of the cell), as a tuft at one end of the cell (**Fig. 16-6**), or scattered over the entire cell surface. Flagella can rotate rapidly, propelling the organism through its liquid environment. The use of flagella to move allows prokaryotes to disperse into new habitats, migrate toward nutrients, and leave unfavorable environments.

Many Prokaryotes Form Films on Surfaces

The cell walls of some prokaryote species are surrounded by sticky layers of protective slime, composed of polysaccharide or protein, which protect the cells and help them adhere to surfaces. In many cases, slime-secreting prokaryotes of one or more species aggregate in colonies to form communities known as *biofilms*. One familiar biofilm is dental plaque, which is formed by the bacteria that inhabit the mouth. The protection afforded by biofilms helps defend the embedded bacteria against a variety of attacks, including those launched by antibiotics and disinfectants. As a result, biofilms formed by bacteria harmful to humans can be very difficult to eradicate. Many infections of the human body take the form of biofilms, including those responsible for tooth decay, gum disease, and ear infections.

Protective Endospores Allow Some Bacteria to Withstand Adverse Conditions

When environmental conditions become inhospitable, many rod-shaped bacteria form protective structures called **endospores.** An endospore, which forms inside a bacterium, contains genetic material and a few enzymes encased within a thick protective coat (**Fig. 16-7**). Metabolic activity ceases until the spore encounters favorable conditions, at which time metabolism resumes and the spore develops into an active bacterium.

Endospores are resistant to even extreme environmental conditions. Some can withstand boiling for an hour or more. Endospores are also able to survive for extraordinarily long periods. In the most extreme example of such longevity, scientists recently discovered bacterial spores that had been sealed inside rock for 250 million years. After being carefully extracted from their rocky "tomb," the spores were incubated in test tubes. Amazingly, live bacteria developed from the ancient spores, which were older than the oldest dinosaur fossils.

Endospores are one of the main reasons that the bacterial disease anthrax has become an agent of biological terrorism. The bacterium that causes anthrax forms endospores, which provide the means by which terrorists can disperse the bacteria. The spores can be stored indefinitely and can survive harsh conditions that would kill other kinds of bacteria. Anthrax spores can even survive detonation of an explosive device that would disperse them into the atmosphere, where they would remain viable until inhaled by a potential victim.

Prokaryotes Are Specialized for Specific Habitats

Prokaryotes occupy virtually every habitat, including those where extreme conditions keep out other forms of life. Many archaea live in extremely hot environments, including springs where the water actually boils or in deep-ocean vents, where superheated water is spewed through cracks in Earth's crust at temperatures of up to 230°F (110°C). Prokaryotes can also survive at the extremely high pressures found deep beneath Earth's surface and in very cold environments, such as in Antarctic sea ice.

Even extreme chemical conditions fail to impede invasion by prokaryotes. Thriving colonies of bacteria and archaea live in the Dead Sea, where a salt concentration seven times that of the oceans precludes all other life, and in waters that are as acidic as vinegar or as alkaline as household ammonia. Of course, rich bacterial communities also reside in a full range of more moderate habitats, including in and on the healthy human body.

No single species of prokaryote, however, is as versatile as these examples may suggest. In fact, most prokaryotes are specialists. One species of archaea that

▲ **Figure 16-7 Spores protect some bacteria** Resistant endospores have formed inside bacteria of the genus *Clostridium*, which causes the potentially fatal food poisoning called botulism. **QUESTION:** What might explain the observation that most endospore-forming bacteria are species that live in soil?

inhabits deep-sea vents, for example, grows optimally at 223°F (106°C) and stops growing altogether at temperatures below 194°F (90°C). Clearly, this species could not survive in a less extreme habitat. Bacteria that live on the human body are also specialized; different species colonize the skin, the mouth, the respiratory tract, the large intestine, and the urogenital tract.

Prokaryotes Use Diverse Metabolic Pathways

Prokaryotes are able to colonize diverse habitats partly because they have evolved diverse methods of acquiring energy and nutrients from the environment. For example, unlike eukaryotes, many prokaryotes are **anaerobes;** their metabolisms do not require oxygen. Their ability to inhabit oxygen-free environments allows prokaryotes to exploit habitats that are off-limits to eukaryotes. Some anaerobes, such as many of the archaea found in hot springs and the bacterium that causes tetanus, are actually poisoned by oxygen. Others are opportunists, engaging in anaerobic respiration when oxygen is lacking and switching to aerobic respiration (a more efficient process) when oxygen becomes available. Many prokaryotes, of course, are strictly aerobic, and require oxygen at all times.

Whether aerobic or anaerobic, different prokaryote species can extract energy from an amazing array of substances. Prokaryotes subsist not only on the sugars, carbohydrates, fats, and proteins that we generally think of as foods, but also on compounds that are inedible or even poisonous to humans, including petroleum, methane (the main component of natural gas), and solvents such as benzene and toluene. Some prokaryotes can even metabolize inorganic molecules, including hydrogen, sulfur, ammonia, iron, and nitrite. The process of metabolizing inorganic molecules sometimes yields by-products that are useful to other organisms. For example, certain bacteria release sulfates or nitrates, crucial plant nutrients, into the soil.

Some species of bacteria use photosynthesis to capture energy directly from sunlight. Like green plants, photosynthetic bacteria possess chlorophyll. Most species produce oxygen as a by-product of photosynthesis, but some, known as the sulfur bacteria, use hydrogen sulfide instead of water in photosynthesis, releasing sulfur instead of oxygen. No photosynthetic archaea are known.

Prokaryotes Reproduce by Binary Fission

Most prokaryotes reproduce asexually by a simple form of cell division called binary fission, which produces genetically identical copies of the original cell (see Fig. 8-3b on p. 114). Under ideal conditions, a prokaryotic cell can divide about once every 20 minutes, potentially giving rise to sextillions (10^{21}) of offspring in a single day. This rapid reproduction allows prokaryotes to exploit temporary habitats, such as a mud puddle or warm pudding. Rapid reproduction also allows bacterial populations to evolve quickly. Recall that many mutations, the source of genetic variability, are the result of mistakes in DNA replication during cell division (see p. 166). Thus, the rapid, repeated cell division of prokaryotes provides ample opportunity for new mutations to arise and also allows mutations that enhance survival to spread quickly.

Prokaryotes Affect Humans and Other Organisms

Although they are largely invisible to us, prokaryotes play a crucial role in life on Earth. Plants and animals (including humans) are utterly dependent on prokaryotes. Prokaryotes help plants and animals obtain vital nutrients and help break down and recycle wastes and dead organisms. We could not survive without prokaryotes, but their impact on us is not always beneficial. Some of humanity's most deadly diseases stem from microbes.

Prokaryotes Play Important Roles in Animal Nutrition

Many eukaryotic organisms depend on close associations with prokaryotes. For example, most animals that eat leaves—including cattle, rabbits, koalas, and deer—cannot themselves digest cellulose, the principal component of plant cell walls. Instead, these animals depend on certain bacteria that have the unusual ability to break down cellulose. Some of these bacteria live in the animals' digestive tracts, where they help liberate nutrients from plant tissue that the animals are unable to break down themselves. Without the bacteria, leaf-eating animals could not survive.

Prokaryotes also have important impacts on human nutrition. Many foods, including cheese, yogurt, and sauerkraut, are produced by the action of bacteria. Bacteria also inhabit your intestines. These bacteria feed on undigested food and synthesize such nutrients as vitamin K and vitamin B_{12}, which the human body absorbs.

Prokaryotes Capture the Nitrogen Needed by Plants

Humans could not live without plants, and plants are entirely dependent on bacteria. In particular, plants are unable to capture nitrogen, required to synthesize proteins and nucleic acids, from that element's most abundant reservoir, the atmosphere. Plants need nitrogen to grow. To acquire it, they depend on **nitrogen-fixing bacteria,** which live both in soil and in specialized *nodules*, small, rounded lumps on the roots of certain plants (legumes, which include alfalfa, soybeans, lupines, and clover; see Fig. 17-18 on p. 330). The nitrogen-fixing bacteria capture nitrogen gas (N_2) from air trapped in the soil and combine it with hydrogen to produce ammonium (NH_4^+), a nitrogen-containing nutrient that plants can use directly.

Prokaryotes Are Nature's Recyclers

Prokaryotes play a crucial role in recycling waste. Most prokaryotes obtain energy by breaking down complex organic (carbon-containing) molecules. Such prokaryotes find a plentiful source of organic molecules in the waste products and dead bodies of plants and animals. By consuming and thereby decomposing these wastes, prokaryotes ensure that wastes do not accumulate in the environment. In addition, decomposition by prokaryotes releases the nutrients contained in wastes. Once released, the nutrients become available for reuse by living organisms.

Prokaryotes perform their recycling service wherever organic matter is found. They are important decomposers in lakes and rivers, in the oceans, and in the soil and groundwater of forests, grasslands, deserts, and other terrestrial environments. The recycling of nutrients by prokaryotes and other decomposers provides the basis for continued life on Earth.

Prokaryotes Can Clean Up Pollution

Many of the pollutants that are produced as by-products of human activity are organic compounds. As such, these pollutants can potentially serve as food for archaea and bacteria. Many of them are, in fact, consumed; the range of compounds consumed by prokaryotes is staggering. Nearly anything that people can synthesize—including detergents, many toxic pesticides, and harmful industrial chemicals such as benzene and toluene—can be broken down by some type of prokaryote.

Even oil can be broken down by prokaryotes. Soon after the tanker *Exxon Valdez* dumped 11 million gallons of crude oil into Prince William Sound, Alaska, in 1989, researchers sprayed oil-soaked beaches with a fertilizer that encouraged the growth of natural populations of oil-eating bacteria. Within 15 days, the oil deposits on these beaches were noticeably reduced in comparison with unsprayed areas.

The practice of manipulating conditions to stimulate breakdown of pollutants by living organisms is known as *bioremediation*. Improved methods of bioremediation could dramatically increase our ability to clean up toxic waste sites and polluted groundwater. A great deal of current research is therefore aimed at identifying prokaryote species that are especially effective for bioremediation and at discovering practical methods for manipulating these organisms to improve their effectiveness.

Some Anaerobic Bacteria Produce Dangerous Poisons

Some bacteria produce toxins that attack the nervous system. Examples of such pathogens include *Clostridium tetani*, which causes tetanus, and *C. botulinum*, which causes botulism (a sometimes lethal food poisoning). Both of these bacterial species are anaerobes that survive as spores until introduced into a favorable, oxygen-free environment. For example, a deep puncture wound may allow tetanus bacteria to penetrate a human body and reach a place where they will be protected from contact with oxygen. As they multiply, the bacteria release their paralyzing poison into the bloodstream. For botulism bacteria, a sealed container of canned food that has been improperly sterilized may provide a haven. Thriving on the nutrients in the can, these anaerobes produce a toxin so potent that a single gram could kill 15 million people.

Humans Battle Bacterial Diseases Old and New

The feeding habits of certain bacteria threaten our health and well-being. These **pathogenic** (disease-producing) bacteria synthesize toxic substances that cause disease symptoms. (So far, no pathogenic archaea have been identified.)

Bacterial diseases have had a significant impact on human history. Perhaps the most infamous example is bubonic plague, or "Black Death," which killed 100 million people during the mid-fourteenth century. In many parts of the world, one-third or more of the population died. Plague is caused by a highly infectious bacterium that is spread by fleas that feed on infected rats and then move to human hosts. Although bubonic plague has not reemerged as a large-scale epidemic, about 2,000 to 3,000 people worldwide are still diagnosed with the disease each year.

Perhaps the most frustrating pathogens are those that come back to haunt us long after we believed that we had them under control. Tuberculosis, a bacterial disease once almost vanquished in developed countries, is again on the rise in the United States and elsewhere. Two sexually transmitted bacterial diseases, gonorrhea and syphilis, have reached epidemic proportions around the globe. Cholera, a water-transmitted bacterial disease that flourishes when raw sewage contaminates drinking water or fishing areas, is under control in developed countries but remains a major killer in poorer parts of the world.

Common Bacterial Species Can Be Harmful

Some pathogenic bacteria are so widespread and common that we cannot expect to ever be totally free of their damaging effects. For example, different species of the abundant streptococcus bacterium produce several diseases. One streptococcus causes strep throat. Another causes pneumonia by stimulating an allergic reaction that clogs the lungs with fluid. Yet another streptococcus has gained fame as the "flesh-eating bacterium." About 800 Americans each year develop necrotizing fasciitis (as the "flesh-eating" infection is more properly known), and about 15% of these victims die. The streptococci enter through broken skin and produce toxins that either destroy flesh directly or stimulate an overwhelming and misdirected attack by the immune system against the body's own cells. A limb can be destroyed in hours, and in some cases, only amputation can halt the rapid tissue destruction. In other cases, these rare strep infections sweep through the body, causing death within a matter of days.

One of the most common bacterial inhabitants of the human digestive system, *Escherichia coli*, is also capable of doing harm. Different populations of *E. coli* may differ genetically, and some genetic differences can transform this normally benign species into a pathogen. One particularly notorious strain, known as O157:H7, infects about 70,000 Americans each year, about 60 of whom die from its effects. Most O157:H7 infections result from consumption of contaminated beef. About a third of the cattle in the United States carry O157:H7 in their intestinal tracts, and the bacteria can be transmitted to humans when a slaughterhouse inadvertently grinds some gut contents into hamburger. Once in a human digestive system, O157:H7 bacteria attach firmly to the wall of the intestine and begin to release a toxin that causes intestinal bleeding and that spreads to and damages other organs as well. The best defense against O157:H7 is to cook all meat thoroughly.

16.4 Protists

The third domain, Eukarya, includes all eukaryotic organisms. The most conspicuous Eukarya are members of the kingdoms Plantae, Fungi, and Animalia, which we discuss later in this chapter. The remaining eukaryotes constitute a diverse collection of organisms collectively known as **protists.** The term "protist" does not describe a true evolutionary unit united by shared features, but is a term of convenience that means "any eukaryote that is not a plant, animal, or fungus."

Most Protists Are Single-Celled

Most protists are single-celled and invisible to us as we go about our daily lives. If we could somehow shrink to their microscopic scale, we might be more impressed with their spectacular and beautiful forms, their varied and active lifestyles, their astonishingly diverse modes of reproduction, and the structural and physiological innovations that are possible within the limits of a single cell. In reality, however, their small size makes them challenging to observe. A microscope and a good supply of patience are required to appreciate the beauty and intricacy of protists.

Although most protists are single-celled, some are visible to the naked eye, and a few are genuinely large. Some larger protists are aggregations or colonies of single-celled individuals, whereas others are multicellular organisms.

Protists Affect Humans and Other Organisms

Protists have important impacts, both positive and negative, on human lives. The primary positive impact actually benefits all living things and stems from the ecological role of photosynthetic marine protists. Just as plants do on land, photosynthetic protists in the oceans capture solar energy and make it available to other organisms in the ecosystem. Thus, the marine ecosystems on which humans depend for food in turn depend on protists. Further, as they use photosynthesis to capture energy, the protists release oxygen gas that helps replenish atmospheric oxygen.

On the negative side of the ledger, many human diseases are caused by parasitic protists. The diseases caused by protists include some of humanity's most prevalent ailments and some of its deadliest afflictions. Protists also cause a number of plant diseases, some of which attack crops that are important to humans. In addition to causing diseases, some marine protists release toxins that can accumulate to harmful levels in coastal areas.

The following sections include information about a small sample of the protists responsible for these helpful and harmful effects.

Stramenopiles Include Photosynthetic and Nonphotosynthetic Protists

The *stramenopiles* form a group whose shared ancestry was discovered through genetic comparison. The group is designated as a kingdom by some biologists. All members of the group have fine, hairlike projections on their flagella (though in many stramenopiles, flagella are present only at certain stages of the life cycle). Despite their shared evolutionary history, however, stramenopiles display a wide range of different forms. Some are photosynthetic and some are not; most are single-celled, but some are multicellular. Two major stramenopile groups are the diatoms (**Fig. 16-8**) and the brown algae.

▲ **Figure 16-8 Some representative diatoms** This photomicrograph illustrates the intricate beauty and variety of the glassy walls of diatoms.

Diatoms Encase Themselves Within Glassy Walls

The *diatoms*, photosynthetic stramenopiles found in both freshwater and saltwater, produce protective shells of silica (glass), some of exceptional beauty (see Fig. 16-8). These shells consist of top and bottom halves that fit together like a pillbox or petri dish. Accumulations of diatoms' glassy walls over millions of years have produced fossil deposits of "diatomaceous earth" that may be hundreds of meters thick. This slightly abrasive substance is widely used in products such as toothpaste and metal polish.

Diatoms form part of the **phytoplankton,** the single-celled photosynthesizers that float passively in the upper layers of Earth's lakes and oceans. Phytoplankton play an immensely important ecological role. Marine phytoplankton account for nearly 70% of all the photosynthetic activity on Earth, absorbing carbon dioxide, recharging the atmosphere with oxygen, and supporting the complex web of aquatic life.

Brown Algae Live in Cool Coastal Waters

Though most photosynthetic protists, such as diatoms, are single-celled, some form multicellular organisims that are commonly known as *seaweeds*. Although some seaweeds seem to resemble plants, they are not closely related to plants and lack many of the distinctive features of the plant kingdom. For example, none of the seaweeds have roots or shoots, and none form embryos during reproduction.

The stramenopiles include one group of seaweeds, the *brown algae,* that is named for the brownish-yellow pigments that, in combination with green chlorophyll, increase the seaweed's light-gathering ability. Almost all brown algae are marine. The group includes the dominant seaweed species that dwell along rocky shores in the temperate (cooler) oceans of the world, including the eastern and western coasts of the United States. Brown algae live in habitats ranging from near the shore, where they cling to rocks that are exposed at low tide, to far offshore. Several species use gas-filled floats to support their bodies (**Fig. 16-9a**). Some of the giant kelp found along the Pacific coast reach heights of 325 feet (100 meters) and may grow more than 6 inches (15 centimeters) in a single day. With their dense growth and towering height, kelp form undersea forests that provide food, shelter, and breeding areas for an enormous variety of marine animals (**Fig. 16-9b**).

Alveolates Include Parasites, Predators, and Phytoplankton

The *alveolates* are single-celled organisms that have distinctive, small cavities beneath the surface of their cells. Like the stramenopiles, the alveolates form a distinct group that may eventually be given kingdom status. Also like the stramenopiles, the evolutionary link among the alveolates was long obscured by the variety of structures and ways of life among the group members, but was revealed by genetic comparisons. Some alveolates are photosynthetic, some are parasitic, and some are predatory. The major alveolate groups are the dinoflagellates, apicomplexans, and ciliates.

Dinoflagellates Swim by Means of Two Whiplike Flagella

Although most *dinoflagellates* are photosynthetic, there are also some nonphotosynthetic species. Dinoflagellates are named for the motion created by their two whiplike flagella. One flagellum encircles the cell, and the second pro-jects behind it. Some dinoflagellates are enclosed only by a cell membrane; others have cellulose walls that resemble armor plates (**Fig. 16-10**). Although some species live in freshwater, dinoflagellates are especially abundant in the ocean, where they are an important component of the phytoplankton and a food

(a) *Fucus*

(b) **Kelp forest**

▲ **Figure 16-9   Brown algae** *(a) Fucus*, a genus found near shores, is shown here exposed at low tide. Notice the gas-filled floats, which provide buoyancy in water. *(b)* The giant kelp *Macrocystis* forms underwater forests off southern California.

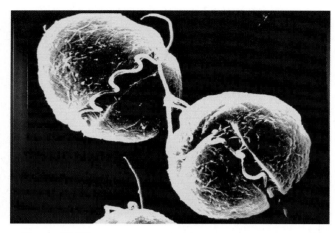

▲ **Figure 16-10** **Dinoflagellates** Two dinoflagellates covered with protective cellulose armor. Visible on each is a flagellum in a groove that encircles the body.

▲ **Figure 16-11** **A red tide** The explosive reproductive rate of certain dinoflagellates under the right environmental conditions can produce dinoflagellate concentrations so great that their microscopic bodies dye the seawater red or brown.

source for larger organisms. Many dinoflagellates are bioluminescent, producing a brilliant blue-green light when disturbed. Specialized dinoflagellates live within the tissues of corals, some clams, and even other protists, where they provide their hosts with nutrients from photosynthesis and remove carbon dioxide.

Warm water that is rich in nutrients may bring on a dinoflagellate population explosion. Dinoflagellates can become so numerous that the water is dyed red by the color of their bodies, causing a "red tide" (**Fig. 16-11**). During red tides, fish die by the thousands, suffocated by clogged gills or by the oxygen depletion that results from the decay of billions of dinoflagellates. But dinoflagellate population explosions can benefit oysters, mussels, and clams, which have a feast, filtering millions of the protists from the water and consuming them. In the process, however, their bodies accumulate concentrations of a nerve poison produced by the dinoflagellates. Humans who eat these mollusks may be stricken with potentially lethal paralytic shellfish poisoning.

Apicomplexans Are Parasitic and Have No Means of Locomotion

All *apicomplexans* are parasitic, living inside the bodies and sometimes inside the individual cells of their hosts. They form infectious spores, resistant structures transmitted from one host to another through food, water, or the bite of an infected insect. As adults, apicomplexans have no means of locomotion. Many have complex life cycles, a common feature of parasites. A well-known example is the malarial parasite *Plasmodium*. Parts of its life cycle are spent in the stomach, and later the salivary glands, of the female *Anopheles* mosquito. When the mosquito bites a human, it passes the *Plasmodium* to the unfortunate victim. The apicomplexan develops in the victim's liver, then enters the blood, where it reproduces rapidly in red blood cells. The release of large quantities of spores through the rupture of the blood cells causes the recurrent fever of malaria. Uninfected mosquitoes may acquire the parasite by feeding on the blood of a malaria victim; they can then spread the parasite when they bite another person.

Ciliates Are the Most Complex of the Alveolates

Ciliates, which inhabit freshwater and saltwater, represent the peak of unicellular complexity. They possess many specialized organelles, including cilia, the short hairlike outgrowths after which they are named. The cilia may cover the cell or may be localized. In the well-known freshwater genus *Paramecium*, rows of cilia cover the organism's entire body surface. Their coordinated beating propels the cell through the water at a rate of 1 millimeter per second—a protistan speed record. Although only a single cell, *Paramecium* responds to its environment as if it had a nervous system. Confronted with a noxious chemical or a physical barrier, the cell immediately backs up by reversing the beating of its cilia and then proceeds in a new direction. Some ciliates, such as *Didinium*, are accomplished predators (**Fig. 16-12**).

Cercozoans Have Thin Pseudopods and Elaborate Shells

Protists in a number of different groups possess flexible plasma membranes that they can extend in any direction to form finger-like projections called *pseudopods*, which they use in locomotion and for engulfing food. The pseudopods of *cercozoans* are thin and threadlike. In most species in this group,

◀**Figure 16-12** **A microscopic predator** In this scanning electron micrograph, the predatory ciliate *Didinium* attacks a *Paramecium*. Note that the cilia of *Didinium* are confined to two bands, whereas *Paramecium* has cilia over its entire body. Ultimately, the predator will engulf and consume its prey. This microscopic drama could take place on a pinpoint, with room to spare.

the pseudopods extend through hard shells. The largest group within the cercozoans is the foramineferans.

Fossil Foraminiferan Shells Form Chalk

The *foraminiferans* are primarily marine protists that produce beautiful shells.
Their shells are constructed mostly of calcium carbonate (chalk; **Fig. 16-13**). These
elaborate shells are pierced by myriad openings through which pseudopods extend. The chalky shells of dead foraminiferans, sinking to the ocean bottom and
accumulating over millions of years, have resulted in immense deposits of limestone, such as those that form the famous White Cliffs of Dover, England.

Amoebozoans Inhabit Aquatic and Terrestrial Environments

Amoebozoans move by extending finger-shaped pseudopods, which may also be
used for feeding. They generally do not have shells. The major groups of amoebozoans are the amoebas and the slime molds.

Amoebas Have Thick Pseudopods and No Shells

Amoebas are common in freshwater lakes and ponds (**Fig. 16-14**). Many amoebas
are predators that stalk and engulf prey, but some species are parasites. One parasitic form causes amoebic dysentery, a disease that is prevalent in warm climates. The dysentery-causing amoeba multiplies in the intestinal wall, triggering
severe diarrhea.

Slime Molds Are Decomposers That Inhabit the Forest Floor

The physical form of *slime molds* seems to blur the boundary between a colony
of different individuals and a single, multicellular individual. The life cycle of the
slime mold consists of two phases: a mobile feeding stage and a stationary reproductive stage called a *fruiting body*. There are two types of slime mold: acellular
and cellular. We focus here on the acellular slime molds.

The acellular slime molds consist of a mass of cytoplasm, called a plasmodium,
that may spread thinly over an area of several square yards. Although the mass
contains thousands of diploid nuclei, the nuclei are not confined in separate cells
surrounded by plasma membranes, which is why these protists are described as
"acellular" (without cells). The plasmodium oozes through decaying leaves and
rotting logs, engulfing food such as bacteria and particles of organic material. The
mass may be bright yellow or orange; a large plasmodium can be rather startling
(**Fig. 16-15a**). Dry conditions or starvation stimulate the plasmodium to form a
fruiting body, in which haploid spores are produced (**Fig. 16-15b**). The spores are

▲ **Figure 16-13 Foraminiferans** The chalky shells of
foraminiferans show numerous interior chambers.

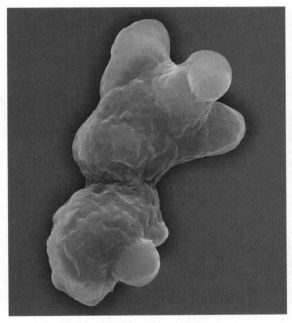

▲ **Figure 16-14 Amoebas** Amoebas, like the one shown
dividing here, use their pseudopods to move and capture prey.

◄ **Figure 16-15 The acellular slime
mold *Physarum* (a)** *Physarum*
oozes over a stone on the damp
forest floor. **(b)** When food becomes
scarce, the mass differentiates into
fruiting bodies in which spores are
formed.

(a) Plasmodium

(b) Fruiting bodies

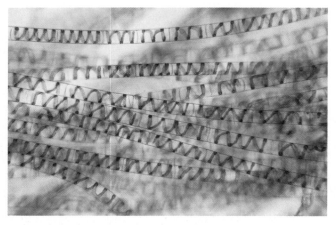

▲ Figure 16-16 **A green alga** *Spirogyra* is a filamentous green alga composed of strands only one cell thick. The alga's unusual spiral-shaped chloroplasts are clearly visible within the cells.

Web Animation Evolution of Plant Structure

dispersed and germinate under favorable conditions, eventually giving rise to a new plasmodium.

Green Algae Live Mostly in Ponds and Lakes

Green algae, a large and diverse group of photosynthetic protists, include both multicellular and unicellular species. Most species live in freshwater ponds and lakes, but some live in the seas. Some green algae, such as *Spirogyra*, form thin filaments from long chains of cells (**Fig. 16-16**). Other species of green algae form colonies containing clusters of cells that are somewhat interdependent and constitute a structure intermediate between unicellular and multicellular forms. These colonies range from a few cells to a few thousand cells, as in species of *Volvox*. Most green algae are microscopic or barely visible to the naked eye, but some marine species are much larger. For example, the green alga *Ulva*, or sea lettuce, is similar in size to lettuce leaves.

The green algae are of special interest because, unlike other groups that contain multicellular, photosynthetic protists, green algae are closely related to plants. Plants share a common ancestor with some types of green algae, and many researchers believe that the very earliest plants were similar to today's multicellular green algae.

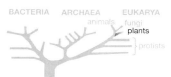

16.5 Plants

What distinguishes members of the plant kingdom from other organisms? Perhaps the most noticeable feature of plants is their green color. The color comes from the presence of the pigment chlorophyll in many plant tissues. Chlorophyll plays a crucial role in photosynthesis, the process by which plants use energy from sunlight to convert water and carbon dioxide to sugar. Chlorophyll and photosynthesis, however, are not unique to plants; they are also present in many types of protists and prokaryotes. Other features of plants are more distinctive:

- Plant reproduction features *alternation of generations*. Separate diploid and haploid generations alternate with one another. (Recall that a diploid organism has two sets of chromosomes; a haploid organism has one set.) When plants in the diploid phase of their life cycle reproduce, the offspring are haploid. When these haploid plants themselves reproduce, the offspring are diploid, and the cycle begins again. Alternation of generations is described in greater detail on pp. 338–339.

- Plants have dependent embryos. Zygotes develop into multicellular embryos that are initially retained within and receive nutrients from the tissues of the parent plant. That is, a plant embryo is attached to and dependent on its parent during the early stages of its growth and development.

- Plants have roots or rootlike structures that anchor the plant and/or absorb water and nutrients from the soil, and stems or stemlike structures that support the aboveground portion of the plant body.

- Plants have a waxy cuticle that covers the surfaces of leaves and stems and that limits the evaporation of water.

Two major groups of land plants arose from ancient algal ancestors. One group, the **nonvascular plants** (also known as bryophytes), requires a moist environment to reproduce and thus straddles the boundary between aquatic and terrestrial life, much like the amphibians of the animal kingdom. The other group, the **vascular plants** (also known as tracheophytes), has been able to colonize drier habitats.

Nonvascular Plants Lack Conducting Structures

Nonvascular plants retain some characteristics of their algal ancestors. For example, they lack true roots, leaves, and stems. They do possess rootlike anchoring structures called rhizoids that bring water and nutrients into the plant body, but lack the well-developed structures for conducting water and nutrients that characterize true roots, leaves, and stems. They must instead rely on slow diffusion or poorly developed conducting tissues to distribute water and other nutrients. As a result, their body size is limited. Size is also limited by the absence of any stiffening agent in their bodies. Without such material, they cannot grow upward very far. Most nonvascular plants are less than 1 inch (2.5 centimeters) tall.

Nonvascular Plants Include the Hornworts, Liverworts, and Mosses

The nonvascular plants include three phyla: hornworts, liverworts, and mosses. Hornworts and liverworts are named for their shapes. Hornworts generally have a spiky shape that appears hornlike to some observers (Fig. 16-17a). Certain liverwort species have a lobed form reminiscent of the shape of a liver (Fig. 16-17b). Hornworts and liverworts are most abundant in areas where moisture is plentiful, such as moist forests and near the banks of streams and ponds.

Mosses are the most diverse and abundant of the nonvascular plants (Fig. 16-17c). Like hornworts and liverworts, mosses are most likely to be found in moist habitats. Some mosses, however, have a waterproof covering (a waxy cuticle) that retains moisture, reducing water loss. These mosses can survive in deserts, on bare rock, and in far northern and southern latitudes where humidity is low and liquid water is scarce for much of the year.

The Reproductive Structures of Nonvascular Plants Are Protected

Among the nonvascular plants' adaptations to terrestrial existence are their enclosed reproductive structures, which prevent the gametes from drying out. There are two types of structures—one in which eggs develop and one in which sperm are formed. In some nonvascular plant species, both egg-producing and sperm-producing structures are located on the same plant; in other species, each individual plant is either male or female. In all nonvascular plants, the sperm must swim to the egg (which emits a chemical attractant) through a film of water. Nonvascular plants that live in drier areas must time their reproduction to coincide with rains.

Vascular Plants Have Conducting Vessels That Also Provide Support

Vascular plants are distinguished by specialized groups of conducting cells called vessels, which characterize true roots, stems, and leaves. The vessels are impregnated with a stiffening substance, lignin, and serve both supportive and conducting functions. Vessels allow vascular plants to grow taller than nonvascular plants, both because of the extra support provided by lignin and because the conducting cells allow water and nutrients absorbed by the roots to move to the upper portions of the plant. The vascular plants can be divided into two groups: the seedless vascular plants and the seed plants.

(a) Hornwort

(b) Liverwort

(c) Moss

▶ Figure 16-17 Nonvascular plants The plants shown here are less than a half inch (about 1 centimeter) in height. *(a)* Hornworts are named for their elongated, hornlike structures. *(b)* Liverworts grow in moist, shaded areas. This is the female plant, bearing umbrella-like structures, which hold the eggs. Sperm must swim up the stalks through a film of water to fertilize the eggs. *(c)* Moss plants, showing the stalks that carry spore-bearing capsules. **QUESTION:** Why are all nonvascular plants short?

(a) Club moss

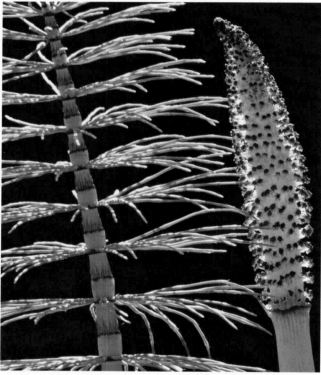

(b) Horsetail

(c) Fern

Seedless Vascular Plants Include the Club Mosses, Horsetails, and Ferns

Like the nonvascular plants, *seedless vascular plants* have swimming sperm and require water for reproduction. As their name implies, they do not produce seeds but rather propagate by spores. The present-day seedless vascular plants—the club mosses, horsetails, and ferns—are much smaller than their ancestors, which dominated the landscape hundreds of millions of years ago. The bodies of ancient seedless vascular plants—transformed by heat, pressure, and time—are burned today as coal. The seedless vascular plants were once Earth's largest and most conspicuous plants, but today the more versatile seed plants have become dominant.

The club mosses are now limited to representatives a few inches in height. Their leaves are small and scale-like (**Fig. 16-18a**). Modern horsetails form a single genus, *Equisetum*, that contains only 15 species, most less than 3 feet tall. Horsetail leaves are reduced to tiny scales on the branches (**Fig. 16-18b**). The ferns, with 12,000 species, are the most diverse of the seedless vascular plants, and are the only seedless vascular plants that have broad leaves (**Fig. 16-18c**).

Seed Plants Dominate the Land, Aided by Two Important Adaptations: Pollen and Seeds

The *seed plants* are distinguished from nonvascular plants and seedless vascular plants by their production of pollen and seeds. **Pollen** grains are tiny structures that carry sperm-producing cells. Pollen grains are dispersed by wind or by animal pollinators such as bees. In this way, sperm move through the air to fertilize egg cells. This airborne transport means that the distribution of seed plants is not limited by the need for water through which sperm can swim to the egg. Seed plants are fully adapted to life on dry land.

Analogous to the eggs of birds and reptiles, **seeds** consist of an embryonic plant, a supply of food for the embryo, and a protective outer coat. The *seed coat* maintains the embryo in a state of suspended animation or dormancy until conditions are proper for growth. The stored food helps sustain the emerging plant until it develops roots and leaves and can make its own food by photosynthesis. Some seeds possess elaborate adaptations that allow them to be dispersed by wind, water, and animals.

Seed plants are grouped into two general types: gymnosperms, which lack flowers, and angiosperms, the flowering plants.

Gymnosperms Are Nonflowering Seed Plants

Gymnosperms evolved earlier than the flowering plants. One group, the **conifers** (**Fig. 16-19a**), still dominates large areas of our planet, whereas other gymnosperms, such as the ginkgos and cycads (**Figs. 16-19b,c**), have declined to a small remnant of their former range and abundance.

Conifers, whose 500 species include pines, firs, spruce, hemlocks, and cypresses, are most abundant in the cold latitudes of the far north and at high elevations where conditions are dry. Not only is rainfall limited in these areas, but water in the soil remains frozen and unavailable during the long winters. Conifers are adapted to dry, cold conditions in three ways. First, conifers retain green leaves throughout the year, enabling these plants to continue photosynthesizing and growing slowly during times when most other plants become dormant. For this reason, conifers are often called evergreens. Second, conifer leaves are actually thin needles covered with a thick, waterproof surface that minimizes evaporation. Finally, conifers produce an "antifreeze" in their sap that enables them to

◄ **Figure 16-18 Seedless vascular plants** Seedless vascular plants are found in moist woodland habitats. **(a)** The club mosses (sometimes called ground pines) grow in temperate forests. This specimen is releasing spores. **(b)** The giant horsetail (genus *Equisetum*) extends long, narrow branches in a series of rosettes. Its leaves are insignificant scales. At right is a spore-forming structure. **(c)** The leaves of this deer fern are emerging from coiled fiddleheads.

(a) Conifer

(b) Gingko

(c) Cycad

▲ **Figure 16-19 Gymnosperms** *(a)* The needle-shaped leaves of conifers such as this pine are protected by a waxy surface layer. *(b)* This ginkgo, or maidenhair tree, is female and bears fleshy seeds the size of large cherries. *(c)* Common in the age of dinosaurs, cycads are now limited to about 160 species. Like ginkgos, cycads have separate sexes.

continue transporting nutrients in below-freezing temperatures. This substance gives them their fragrant piney scent.

Angiosperms Are Flowering Seed Plants

Modern flowering plants, or **angiosperms,** have dominated Earth for more than 100 million years. The group is incredibly diverse, with more than 230,000 species (**Fig. 16-20**). Angiosperms range in size from the diminutive duckweed (**Fig. 16-20a**) to the towering eucalyptus tree (**Fig. 16-20b**). From desert cactus to tropical orchids to grasses to parasitic mistletoe, angiosperms rule over the plant kingdom.

Three major adaptations have contributed to the enormous success of angiosperms: flowers, fruits, and broad leaves.

Flowers Attract Pollinators Both male and female gametes form in **flowers,** which may have evolved when gymnosperm ancestors formed an association with animals (most likely insects) that carried their pollen from plant to plant. According to this scenario, the relationship between these ancient gymnosperms and their animal pollinators was so beneficial that natural selection favored the evolution of showy flowers that advertised the presence of pollen to insects and other animals.

Fruits Encourage Seed Dispersal The ovary surrounding the seed of an angiosperm matures into a **fruit.** Just as flowers encourage animals to transport pollen, so, too, many fruits entice animals to disperse seeds. If an animal eats a fruit, many of the enclosed seeds may pass through the animal's digestive tract unharmed, perhaps to fall at a suitable location for germination. Not all fruits, however, depend on edibility for dispersal. Some fruits (called burs) disperse by clinging to animal fur. Other fruits, such as those of maples, form wings that carry the seed through the air.

Broad Leaves Capture More Sunlight The third feature that gives angiosperms an advantage in warmer, wetter climates is broad leaves. When water is plentiful, broad leaves provide an advantage by collecting more sunlight for photosynthesis. In regions with seasonal variation in growing conditions, many trees and shrubs drop their leaves during periods when water is in short supply (winter in temperate climates, the dry season in tropical climates), because being leafless reduces evaporative water loss.

Hidden Treasures

Continued

Groups consisting mainly of large, conspicuous, land-dwelling organisms have historically attracted the attention of scientists and naturalists, so almost all species of trees have already been discovered and new ones are found only rarely. When new tree species are discovered, they are typically found in isolated, inaccessible locations. For example, the remote mountains that concealed the saola are also home to the golden Vietnamese cypress, a conifer tree that was unknown to science until 2002. The species lives in dense forests on steep limestone ridges in a rainy, roadless region. It eluded discovery by scientists for so long because its geographic range is limited to an area that is very difficult for botanists to visit. Sometimes, however, a tree species remains undetected because it is not especially distinctive and is simply overlooked. Thus, it was not until 2007 that researchers discovered the Catacol whitebeam, a small tree species that is found only on a small island off the coast of Scotland.

BACTERIA ARCHAEA EUKARYA
animals fungi
plants
protists

16.6 Fungi

When you think of a *fungus*, you probably picture a mushroom. Most fungi, however, do not produce mushrooms. And even in fungus species that do produce mushrooms, the mushrooms are just temporary reproductive structures that extend from a main body that is typically concealed beneath the soil or inside a piece of decaying wood.

(a) Duckweed

(c) Grass

(d) Butterfly weed

(b) Eucalyptus

▲ **Figure 16-20 Flowering plants** *(a)* The smallest angiosperm is the duckweed, found floating on ponds. These specimens are about 1/8 inch (3 millimeters) in diameter. *(b)* The largest angiosperms are eucalyptus trees, which can reach 325 feet (100 meters) in height. *(c)* Inconspicuous flowers and wind pollination are features of grasses. *(d)* More conspicuous flowers, such as those shown on this butterfly weed and eucalyptus tree [inset in part (b)], entice insects and other animals that carry pollen between individual plants. **QUESTION:** What are the advantages of pollination by animals?

Fungi Have Distinctive Adaptations

Fungi can be distinguished from other organisms by several distinctive features:

- A fungus's body consists of slender threads. The body is a **mycelium** (plural, mycelia) (**Fig. 16-21a**), which is an interwoven mass of one-cell-thick, threadlike filaments called **hyphae** (singular, hypha; **Fig. 16-21b**). Periodically, the hyphae grow together and differentiate into reproductive structures that project above the surface beneath which the mycelium grows. These structures, including mushrooms, puffballs, and molds, represent only a fraction of the complete fungal body, but are typically the only part of the fungus that we can easily see.

- Fungal cells are surrounded by cell walls. The cell walls of fungi differ from those of plants and protists, in that fungal cell walls are strengthened by *chitin*, the same substance found in the exoskeletons of arthropods.

- Fungi obtain nutrients by secreting enzymes that digest complex molecules outside their bodies. The molecules are thus broken into smaller subunits that can be absorbed. Some fungi digest the bodies of dead organisms. Others are parasitic, feeding on living organisms and causing disease.

- Fungi propagate by spores. Unlike plants and animals, fungi do not form embryos. Instead, fungi reproduce by means of **spores**—tiny, lightweight reproductive packages that are extraordinarily mobile, even though most lack a means for self-propulsion. Spores are distributed far and wide as hitchhikers on the outside of animal bodies, as passengers inside the digestive systems of animals that have eaten them, or as airborne drifters, cast aloft by chance or shot into the atmosphere by elaborate reproductive structures (**Fig. 16-22**). If a spore lands on a moist, nutrient-rich surface, it may germinate into a new fungal hypha.

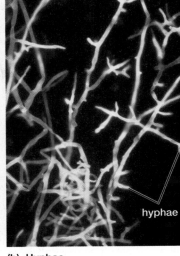

(a) Mycelium   **(b) Hyphae**

▲ **Figure 16-21 The filamentous body of a fungus** *(a)* A fungal mycelium spreads over decaying vegetation. *(b)* The mycelium is composed of microscopic hyphae, only one cell thick. **QUESTION:** How is the structure of a fungus adapted to help it obtain nutrients?

Most Fungi Can Reproduce Both Sexually and Asexually

In general, fungi are capable of both asexual and sexual reproduction. For the most part, fungi reproduce asexually under stable conditions, with sexual reproduction occurring mainly under conditions of environmental change or stress. Both asexual and sexual reproduction ordinarily involve the production of spores within special fruiting bodies that project above the mycelium.

Fungi Affect Humans and Other Organisms

Although many fungi are at least somewhat more apparent to humans than are prokaryotes and protists, their relatively low profile can nonetheless cause us to overlook their importance to human affairs. Our relationship with the fungi has consequences, both helpful and harmful, that are of great importance to our well-being.

Fungi Play a Key Role in Ecosystems

Fungi play a major role in the destruction of dead plant tissue. Alone among organisms, fungi can digest both lignin and cellulose, the molecules that make up wood. When a tree or other woody plant dies, only fungi are capable of decomposing its remains.

Fungi are Earth's undertakers, consuming not only dead wood but the dead of all kingdoms. The fungi that are *saprophytes* (feeding on dead organisms) return the dead tissues' component substances to the ecosystems from which they came. The extracellular digestive activities of saprophytic fungi liberate nutrients such as carbon, nitrogen, and phosphorus compounds and minerals that can be used by plants. If fungi and bacteria were suddenly to disappear, the consequences would

▲ **Figure 16-22 Some fungi can eject spores** A ripe earthstar mushroom, struck by a drop of water, releases a cloud of spores that will be dispersed by air currents.

▲ **Figure 16-23 A delicious fungus** The morel, an edible delicacy. (Consult an expert before sampling any wild fungus—some are deadly!)

be disastrous. Nutrients would remain locked in the bodies of dead plants and animals, the recycling of nutrients would grind to a halt, soil fertility would rapidly decline, and waste and organic debris would accumulate. In short, ecosystems would collapse.

Many Antibiotics Are Derived from Fungi

Fungi also have positive impacts on human health. The modern era of life-saving antibiotic medicines was ushered in by the discovery of penicillin, which is produced by a fungus. Penicillin is still used, along with other fungi-derived antibiotics such as oleandomycin and cephalosporin, to combat bacterial diseases. Other important drugs are also derived from fungi, including cyclosporin, which is used to suppress the immune response during organ transplants so that the body is less likely to reject the transplanted organs.

Fungi Make Important Contributions to Gastronomy

Fungi make important contributions to human nutrition. The most obvious components of this contribution are the fungi that we consume directly: cultivated mushrooms, wild mushrooms such as morels (Fig. 16-23), and other fungi such as the rare and prized truffle. The role of fungi in cuisine, however, also has less visible manifestations. For example, some of the world's most famous cheeses, including Roquefort, Camembert, Stilton, and Gorgonzola, gain their distinctive flavors from fungi that grow on them as they ripen. Perhaps the most important and pervasive fungal contributors to our food supply, however, are the single-celled fungi known as yeasts.

Among the many foods and beverages that depend on yeasts for their production are bread, wine, and beer. All derive their special qualities from fermentation by yeasts. In fermentation, yeasts extract energy from sugar and, as by-products of the metabolic process, emit carbon dioxide and ethyl alcohol. When yeasts are induced to consume the fruit sugars in grape juice, the sugars are converted to alcohol, and wine is the result. If yeasts instead consume the sugar in sprouted grain, beer is the result. (Beer brewers capture fermentation's carbon dioxide by-product along with the alcohol, giving the beer its characteristic bubbly carbonation.) In bread, the alcohol evaporates during baking but the carbon dioxide is trapped in the dough, where it forms bubbles that give bread its airy texture.

Fungi Attack Plants That Are Important to People

Fungi cause the majority of plant diseases, and some of the plants that they infect are important to humans. For example, fungal pathogens have a devastating effect on the world's food supply. Especially damaging are the plant pests descriptively called rusts and smuts, which cause billions of dollars' worth of damage to grain crops annually (Fig. 16-24).

Fungi Cause Human Diseases

The kingdom Fungi includes parasitic species that attack humans directly. Some of the most familiar fungal diseases are those caused by fungi that attack the skin, resulting in athlete's foot, jock itch, vaginal infections, and ringworm. These diseases, though unpleasant, are not life threatening and can usually be treated with antifungal ointments. Fungi can also infect the lungs if victims inhale spores of fungal species, such as those that cause valley fever and histoplasmosis. Like other fungal infections, these diseases can, if promptly diagnosed, be controlled with antifungal drugs. If untreated, however, they can develop into serious, systemic infections. Singer Bob Dylan, for instance, became gravely ill with histoplasmosis when a fungus infected the pericardial membrane surrounding his heart.

Fungi Can Produce Toxins

In addition to their role as agents of infectious disease, some fungi produce toxins that are dangerous to humans. Of particular concern are toxins produced by

▲ **Figure 16-24 Corn smut** This fungal pathogen destroys millions of dollars' worth of corn each year. However, even a pest such as corn smut has its admirers. In Mexico this fungus is known as *huitlacoche* and is considered to be a great delicacy.

fungi that grow on grains and other foodstuffs that have been stored in overly moist conditions. For example, molds of the genus *Aspergillus* produce highly toxic, carcinogenic compounds known as aflatoxins. Some foods, such as peanuts, seem especially susceptible to attack by *Aspergillus*. Since aflatoxins were discovered in the 1960s, food growers and processors have developed methods for reducing the growth of *Aspergillus* in stored crops, so aflatoxins have been largely eliminated from the nation's peanut butter supply.

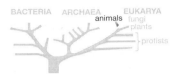

16.7 Animals

It is difficult to devise a concise definition of the term "animal." No single feature fully characterizes animals, so the group is defined by a list of characteristics. None of these characteristics is unique to animals, but together they distinguish animals from members of other kingdoms:

- Animals are multicellular.
- Animals obtain their energy by consuming the bodies of other organisms.
- Animals typically reproduce sexually. Although animal species exhibit a tremendous diversity of reproductive styles, most are capable of sexual reproduction.
- Animal cells lack a cell wall.
- Animals are motile (able to move about) during some stage of their lives. Even the relatively stationary sponges have a free-swimming larval stage (a juvenile form).
- Most animals are able to respond rapidly to external stimuli as a result of the activity of nerve cells, muscle tissue, or both.

Most Animals Lack Backbones

It's easy to overlook the differences among the multitude of small, boneless animals in the world. Even Carolus Linnaeus, the originator of modern biological classification, recognized only two phyla of animals without backbones (insects and worms). Today, however, biologists recognize about 27 phyla of boneless animals.

For convenience, biologists often place animals in one of two major categories: **vertebrates,** those with a backbone (or vertebral column), and **invertebrates,** those lacking a backbone. The vertebrates—fish, amphibians, reptiles, birds, and mammals—are perhaps the most conspicuous animals from a human point of view, but less than 3% of all known animal species are vertebrates. The vast majority of animals are invertebrates.

The earliest animals probably originated from colonies of protists whose members had become specialized to perform distinct roles within the colony. In our survey of the kingdom Animalia, we will begin with the sponges, whose body plan most closely resembles the probable ancestral protist colonies.

Sponges Have a Simple Body Plan

Sponges (phylum Porifera) are found in most marine and aquatic environments. Most of Earth's 5,000 or so sponge species live in saltwater, where they inhabit ocean waters warm and cold, deep and shallow. In addition, some sponges live in freshwater habitats such as lakes and rivers. Adult sponges live attached to rocks or other underwater surfaces. They do not generally move, though researchers have demonstrated that some species, at least when captive in aquaria, are able to move about (very slowly—a few millimeters per day). Sponges come in a variety of shapes and sizes. Some species have a well-defined shape, but others grow free-form over underwater rocks (**Fig. 16-25**). The largest sponges can grow to more than 3 feet (1 meter) in height.

(a) Encrusting sponge

(b) Tubular sponge

▲ **Figure 16-25 Sponges** Some sponges, such as **(a)** this fire sponge, grow in free-form pattern over undersea rocks. **(b)** Tiny appendages attach this tubular sponge to rocks. **QUESTION:** Sponges are often described as the most "primitive" of animals. How can such a primitive organism have become so diverse and abundant?

Sponges have a simple body plan that reflects their probable evolutionary descent from an ancestral colony of single-celled protists. The body is perforated by numerous tiny pores, through which water enters, and by fewer, large openings, through which it is expelled. Within the sponge, water travels through canals. As water passes through the sponge, oxygen is extracted, microorganisms are filtered out and taken into individual cells where they are digested, and wastes are released. An internal skeleton composed of calcium carbonate (chalk), silica (glass), or protein spines provides support for the sponge's body.

Because sponges remain in one spot and have no protective shell, they are vulnerable to predators such as fish, turtles, and sea slugs. Many sponges, however, have evolved chemical defenses against predation. The bodies of these sponges contain chemicals that are toxic or distasteful to potential predators. Fortunately for people, a number of these chemicals have proved to be valuable medicines. For example, the drug spongistatin, a compound first isolated from a sponge, is an emerging treatment for the fungal infections that frequently sicken AIDS patients. Other medicines derived from sponges include some promising new anticancer drugs. The discovery of these and other medicines has raised hopes that, as more species are screened by researchers, chemicals unique to sponges will become the basis of new drugs.

Cnidarians Are Well-Armed Predators

Cnidarians (phylum Cnidaria)—sea jellies, sea anemones, corals, and hydrozoans—come in a bewildering and beautiful variety of forms (Fig. 16-26). In cnidarians, body parts are arranged in a circle around the mouth and digestive cavity. This arrangement of parts is well suited to these animals, which are either fixed in place or carried randomly by water currents, because it enables them to respond to prey or threats from any direction.

▶ **Figure 16-26 Cnidarians** *(a)* A red-spotted sea anemone spreads its tentacles to capture prey. *(b)* A sea jelly drifts in open water. *(c)* A close-up of coral reveals bright yellow individuals in various stages of tentacle extension. At the lower right, areas where the coral has died expose the calcium carbonate skeleton that supports the animals and forms the reef. A crab with long banded legs sits on the coral, holding tiny white sea anemones in its claws. Their stinging tentacles help protect the crab.

(a) Anemone

(b) Sea jelly

(c) Coral

Cnidarian tentacles are armed with cells containing structures that, when stimulated by contact, explosively inject poisonous or sticky filaments into prey. The venom of some cnidarians can cause painful stings in humans unfortunate enough to come into contact with them, and the stings of a few sea jelly species can even be life threatening. However, the function of stinging cells is not to sting human swimmers, but rather to capture prey. Although all cnidarians are preda- tory, none hunt actively. Instead, they wait for their victims to blunder, by chance, into the grasp of their enveloping tentacles. Stung and firmly grasped, the prey is forced through an expansible mouth into a digestive sac.

Like sponges, cnidarians are confined to watery habitats, and most species are marine. One group of cnidarians, the corals, is of particular ecological importance (see Fig. 16-26c), forming undersea habitats that are the basis of an ecosystem of stunning diversity and unparalleled beauty (see pp. 617–618). Corals form large colonies, and each member of the colony secretes a hard skeleton of calcium car- bonate. The skeletons persist long after the organisms die, serving as a base to which other individuals may attach themselves. The cycle continues until, after thousands of years, massive coral reefs are formed.

Coral reefs are found in both cold and warm oceans. Cold-water reefs form in deep waters and, though widely distributed, have only recently attracted the at- tention of researchers and are not yet well studied. The more familiar warm- water coral reefs are restricted to clear, shallow waters in the tropics.

Annelids Are Composed of Identical Segments

Charles Darwin, one of the greatest of all biologists, devoted substantial time to the study of earthworms. Darwin was impressed by earthworms' role in improv- ing soil fertility. More than a million earthworms may live in an acre of land, where they tunnel through soil, consuming and excreting soil particles and or- ganic matter. These actions help ensure that air and water can move easily through the soil and that organic matter is continually mixed into it, creating con- ditions that are favorable for plant growth. In Darwin's view, the activity of the earthworms has had such a significant beneficial impact on agriculture that "it may be doubted whether there are many other animals which have played so im- portant a part in the history of the world."

Earthworms are examples of *annelids* (phylum Annelida), the segmented worms. A prominent feature of annelids is the division of the body into a series of repeating segments. Externally, these segments are delineated by ringlike de- pressions on the surface. Internally, most of the segments contain identical copies of nerves, excretory structures, and muscles. Segmentation is advantageous for locomotion, because the body compartments are each controlled by separate muscles and collectively are capable of more complex movement than could be achieved with only one set of muscles to control the whole body.

The annelid phylum includes three main subgroups: the oligochaetes, the poly- chaetes, and the leeches. The oligochaetes include the familiar earthworm and its relatives. Polychaetes live primarily in the ocean. Some polychaetes have paired fleshy paddles on most of their segments, used in locomotion. Others live in tubes from which they project feathery gills that both exchange gases and sift the water for microscopic food (**Fig. 16-27a,b**). Leeches live in freshwater or moist terrestrial habitats and are either carnivorous or parasitic (**Fig. 16-27c**). Carnivorous leeches prey on smaller invertebrates; parasitic leeches suck the blood of larger animals.

Most Mollusks Have Shells

If you have ever enjoyed a bowl of clam chowder, a plate of oysters on the half shell, sautéed scallops, or escargot, you are indebted to *mollusks* (phylum Mol- lusca). The mollusks includes species with a range of lifestyles, from passive forms that pass their adult lives in one spot, filtering microorganisms from water, to active, voracious predators of the ocean depths. Mollusks also include the

(a) Tubeworm gills

(b) Deep-sea polychaete

(c) Leech

▲ **Figure 16-27 Annelids** *(a)* A polychaete annelid projects brightly spiraling gills from a tube, made by the worm and attached to rock. When the gills retract, the tube is covered by a "trap door" visible as a reddish collar encircling the top of the tube. *(b)* The polychaete known as the Pompeii worm lives near deep-sea vents where the water temperature may reach 175°F (80°C). *(c)* This leech, a freshwater annelid, shows numerous segments. The whitish disk is a sucker encircling its mouth, allowing it to attach to its prey.

▶ **Figure 16-28 Gastropod mollusks** *(a)* A Florida tree snail displays a brightly striped shell and eyes at the tip of stalks that are retracted instantly if touched. *(b)* Spanish shawl sea slugs prepare to mate.

(a) Snail

(b) Sea slugs

largest invertebrate animals (giant squids) and the most intelligent ones (octopuses). Mollusks are very diverse; in terms of number of known species, the 50,000 mollusks are second (albeit a distant second) only to the arthropods. With the exception of some snails and slugs, mollusks are water dwellers.

Most mollusks protect their bodies with hard shells of calcium carbonate. Others, however, lack shells and escape predation by moving swiftly or, if caught, by tasting terrible. With the exception of some snails and slugs, mollusks are water dwellers. Among the many subgroups of mollusks, we will discuss three in more detail: snails and their relatives, clams and their relatives, and octopuses and their relatives.

Gastropods Are One-Footed Crawlers

Snails and slugs—collectively known as *gastropods*—crawl on a muscular foot, and many are protected by shells that vary widely in form and color (**Fig. 16-28a**). Not all gastropods are shelled, however. Sea slugs, for example, lack shells, but their brilliant colors warn potential predators that they are poisonous or foul tasting (**Fig. 16-28b**).

Gastropods feed with a radula, a flexible ribbon of tissue studded with spines that is used to scrape algae from rocks or to grasp larger plants or prey. Most snails use gills, typically enclosed in a cavity beneath the shell, for respiration. Gases can also diffuse readily through the skin of most gastropods; sea slugs rely on this mode of gas exchange. The few gastropod species that live in terrestrial habitats (including the destructive garden snails and slugs) use a simple lung for breathing.

Bivalves Are Filter Feeders

Included among the *bivalves* are scallops, oysters, mussels, and clams (**Fig. 16-29**). Bivalves lend exotic variety to the human diet and are important members of the

▶ **Figure 16-29 Bivalve mollusks** *(a)* This swimming scallop parts its hinged shells. The upper shell is covered with an orange encrusting sponge. Tiny blue eyespots that detect light rim the tissue along the shell openings. *(b)* Mussels attach to rocks in dense aggregations exposed at low tide. White barnacles are attached to the mussel shells and surrounding rock.

(a) Scallop

(b) Mussels

nearshore marine community. Bivalves possess two shells connected by a flexible hinge. A strong muscle clamps the shells closed in response to danger; this muscle is what you are served when you order scallops in a restaurant.

Clams use a muscular foot for burrowing in sand or mud. In mussels, which live attached to rocks, the foot is smaller and helps secrete threads that anchor the animal to the rocks. Scallops lack a foot and move by a sort of jet propulsion achieved by flapping their shells together. Bivalves are filter feeders, using their gills as both respiratory and feeding structures. Water circulates over the gills, which are covered with a thin layer of mucus that traps microscopic food particles. Food is conveyed to the mouth by the beating of cilia on the gills.

Cephalopods Are Marine Predators

The *cephalopods* include octopuses, nautiluses, cuttlefish, and squids (Fig. 16-30). The largest invertebrate, the giant squid, which can be up to 60 feet (18 meters) long, belongs to this group. All cephalopods are predatory carnivores, and all are marine. In these mollusks, the foot has evolved into tentacles with well-developed sensory abilities and suction disks for detecting and grasping prey. Prey grasped by the tentacles may be immobilized by a paralyzing venom in the saliva before being torn apart by beaklike jaws.

Cephalopods have highly developed brains and sensory systems. The cephalopod eye rivals our own in complexity and exceeds it in efficiency of design. The cephalopod brain, especially that of the octopus, is (for an invertebrate brain) exceptionally large and complex. It is enclosed in a skull-like case of cartilage and endows the octopus with highly developed capabilities to learn and remember. In the laboratory, octopuses can rapidly learn to associate certain symbols with food and to open a screw-cap jar to obtain it.

Arthropods Are the Dominant Animals on Earth

In terms of both number of individuals and number of species, no other animal phylum comes close to the *arthropods* (phylum Arthropoda), which include insects, arachnids, and crustaceans. About 1 million arthropod species have been discovered, and scientists estimate that millions more remain undescribed.

All arthropods have an exoskeleton, an external skeleton that encloses the arthropod body like a suit of armor. This external skeleton protects against predators and is responsible for arthropods' greatly increased agility relative to their wormlike ancestors. The exoskeleton is thin and flexible in places, allowing movement of the paired, jointed appendages. By providing stiff but flexible appendages and rigid attachment sites for muscles, the exoskeleton makes possible the flight of the bumblebee and the intricate, delicate manipulations of the spider as it weaves its web.

Arthropods are segmented, but their segments tend to be few and specialized for different functions such as sensing the environment, feeding, and movement. For example, in insects, sensory and feeding structures are concentrated on the front segment, known as the *head*, and digestive structures are largely confined to the *abdomen*, which is the rear segment. Between the head and the abdomen is the *thorax*, the segment to which structures used in locomotion, such as wings and walking legs, are attached.

Insects Are the Only Flying Invertebrates

The number of described *insect* species is about 850,000, roughly three times the total number of known species in all other classes of animals combined (Fig. 16-31). Insects have three pairs of legs, usually supplemented by two pairs of wings. Insects' capacity for flight distinguishes them from all other invertebrates and has contributed to their enormous success. As anyone who has pursued a fly can testify, flight helps in escaping from predators. It also allows the insect to find widely dispersed food.

During their development, insects undergo **metamorphosis,** a radical change from a juvenile body form to an adult body form. In insects with complete metamorphosis,

(a) Octopus

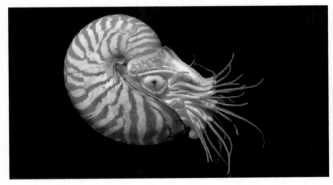
(b) Squid

(c) Nautilus

▲ Figure 16-30 Cephalopod mollusks *(a)* An octopus can crawl rapidly by using its eight suckered tentacles. It can alter its color and skin texture to blend with its surroundings. In emergencies this mollusk can jet backward by vigorously contracting its mantle. *(b)* The squid moves by jet propulsion, ejecting a jet of water that propels the animal backward through the water. *(c)* The chambered nautilus secretes a shell with internal, gas-filled chambers that provide buoyancy. Note the well-developed eye and the tentacles used to capture prey.

▶ **Figure 16-31 Insects** *(a)* The rose aphid sucks sugar-rich juice from plants. *(b)* A mating pair of Hercules beetles. Only the male has the large "horns." *(c)* A June beetle displays its two pairs of wings as it comes in for a landing. The outer wings protect the abdomen and the inner wings, which are relatively thin and fragile. *(d)* Caterpillars are larval forms of moths or butterflies. This caterpillar larva of the Australian fruit-sucking moth displays large eyespot patterns that may frighten potential predators, who mistake them for the eyes of a large animal.

(a) Aphid

(b) Beetles mating

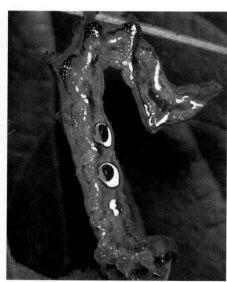

(c) Beetle flying

(d) Moth larva

the immature stage, called a **larva** (plural, larvae), is worm shaped (for example, the maggot of a housefly or the caterpillar of a moth or butterfly; see **Fig. 16-31d**). The larva hatches from an egg, grows by eating voraciously, and repeatedly **molts** (sheds its exoskeleton). The larva then forms a nonfeeding stage called a **pupa.** Encased in an outer covering, the pupa undergoes a radical change in body form, emerging in its adult, winged form. The adults mate and lay eggs, continuing the cycle. Metamorphosis may include a change in diet as well as in shape, thereby eliminating competition for food between adults and juveniles, and, in some cases, allowing the insect to exploit different foods when they are most available. For instance, a caterpillar that feeds on new green shoots in springtime metamorphoses into a butterfly that drinks nectar from the summer's blooming flowers. Some insects, such as grasshoppers and crickets, undergo a more gradual metamorphosis (called incomplete metamorphosis), hatching as young that bear some resemblance to the adult, then gradually acquiring more adult features as they grow and molt.

Most Arachnids Are Predatory Meat Eaters

The *arachnids* include spiders, mites, ticks, and scorpions (**Fig. 16-32**). All arachnids have eight walking legs, and most are carnivorous. Many subsist on a liquid diet of blood or predigested prey. For example, spiders, the most numerous arachnids, first immobilize their prey with a paralyzing venom. They then inject digestive enzymes into the helpless victim (typically an insect) and suck in the resulting soup. Arachnid eyes are particularly sensitive to movement, and in some species they probably can form images. Most spiders have eight eyes placed in such a way as to give them a panoramic view of predators and prey.

Among the distinctive features of spiders is their production of protein threads known as silk. Spiders manufacture silk in special glands in their abdomens and

(b) Scorpion

(a) Spider

(c) Ticks

◄ **Figure 16-32 Arachnids** *(a)* The tarantula is among the largest of spiders but is relatively harmless. *(b)* Scorpions, found in warm climates, including deserts of the southwestern United States, paralyze their prey with venom from a stinger at the tip of the abdomen. A few species can harm humans. *(c)* Ticks before (*left*) and after (*right*) feeding on blood. The uninflated exoskeleton is flexible and folded, allowing the animal to enlarge while feeding.

use it to perform a variety of functions, such as building webs that capture prey, wrapping up and immobilizing captured prey, constructing protective shelters for themselves, making cocoons to surround their eggs, and making "draglines" that connect a spider to a web or other surface and support its weight if it drops from its perch. Spider silk is an amazingly light, strong, and elastic fiber. It can be stronger than a steel wire of the same size, yet is as elastic as rubber.

Most Crustaceans Are Aquatic

The *crustaceans*, including crabs, crayfish, lobster, shrimp, and barnacles, are the only arthropods whose members live primarily in the water (**Fig. 16-33**). Crustaceans range in size from microscopic maxillopods that live in the spaces between grains of sand to the largest of all arthropods, the Japanese crab, with legs spanning nearly 12 feet (4 meters). Crustaceans have two pairs of sensory antennae, but the rest of their appendages are highly variable in form and number, depending on the habitat and lifestyle of the species.

The Chordates Include Both Invertebrates and Vertebrates

Humans are members of the *chordate* phylum (phylum Chordata), which we share not only with birds and apes but also with the tunicates (sea squirts) and small fishlike creatures called lancelets. What characteristics do we share with these creatures that seem so different from us? All chordates are united by four features that all possess at some stage of their lives:

- The dorsal, hollow **nerve cord** of chordates is hollow and lies above the digestive tract, running lengthwise along the dorsal (upper) portion of the body. In contrast, the nerve cords of other animals are solid and lie in a ventral, below-the-digestive-tract position. During embryonic development in chordates, the nerve cord develops a thickening at one end that becomes a brain.

- The **notochord** is a stiff but flexible rod, located between the digestive tract and the nerve cord, that extends along the length of the body. It provides support for the body and an attachment site for muscles. In many chordates, the notochord is present only during early stages of development and disappears as a skeleton develops.

▶ **Figure 16-33 Crustaceans** *(a)* The microscopic water flea is common in freshwater ponds. Notice the eggs developing within the body. *(b)* The sowbug, found in dark, moist places such as under rocks, leaves, and decaying logs, is one of the few crustaceans whose ancestors successfully invaded the land. *(c)* The hermit crab protects its soft abdomen by inhabiting an abandoned snail shell. *(d)* The goose-neck barnacle uses a tough, flexible stalk to anchor itself to rocks, boats, or even animals such as whales. Other types of barnacles attach with shells that resemble miniature volcanoes (see Fig. 16-29b). Early naturalists thought barnacles were mollusks until the jointed legs, seen extending into the water, were observed.

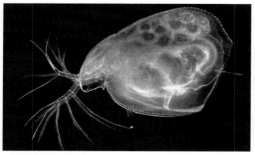
(a) Water flea

(b) Sowbug

(c) Hermit crab

(d) Barnacles

- **Pharyngeal gill slits** are located in the pharynx (the cavity behind the mouth). They may form functional openings for gills (organs for gas exchange in water) or may appear only as grooves during an early stage of development.

- The posterior portion of a chordate body extends past the anus to form a **post-anal tail.** Other animals lack this kind of tail, because their digestive tracts extend the full length of the body.

This list of distinctive chordate structures may seem puzzling because, although humans are chordates, at first glance we seem to lack every feature except the nerve cord. But evolutionary relationships are sometimes seen most clearly during early stages of development, and it is during our embryonic life that we develop, and lose, our notochord, our gill slits, and our tails.

The Invertebrate Chordates Live in the Seas

The invertebrate chordates lack the backbone that is the defining feature of vertebrates. These chordates comprise two groups, the lancelets and the tunicates. The small (2 inches, or about 5 centimeters, long) fishlike lancelet spends most of its time half-buried in the sandy sea bottom, filtering tiny food particles from the water. All four chordate features are present in the adult lancelet.

The tunicates form a larger group of marine invertebrate chordates that includes the sea squirts. It is difficult to imagine a less likely relative of humans than the immobile, filter-feeding, vaselike sea squirt (**Fig. 16-34**). Its ability to move is limited to forceful contractions of its saclike body, which can send a jet of seawater into the face of anyone who plucks it from its undersea home; hence, the name sea squirt. Although adult sea squirts are immobile, their larvae swim actively and possess the four chordate features.

Vertebrates Have a Backbone

In vertebrates, the embryonic notochord is normally replaced during development by a backbone, or **vertebral column.** The vertebral column is composed of bone or cartilage, a tissue that resembles bone but is less brittle and more flexible. This column supports the body, offers attachment sites for muscles, and pro-

▲ **Figure 16-34 An invertebrate chordate** This adult sea squirt (a type of tunicate) has lost its tail and notochord during metamorphosis and has assumed a sedentary life.

tects the delicate nerve cord and brain. It is also part of a living internal skeleton that can grow and repair itself.

The sequence of evolutionary events that led from invertebrate chordates to the first, fishlike vertebrates remains shrouded in mystery, because fossils of intermediate forms have never been discovered. Today, vertebrates are represented by fish, amphibians, reptiles and birds, and mammals.

Ray-finned Fishes   Just as our size bias makes us overlook the most diverse invertebrate groups, our habitat bias makes us overlook the most diverse vertebrates. The most diverse and abundant vertebrates are not the birds or the predominantly terrestrial mammals. The vertebrate diversity crown belongs instead to the lords of the oceans and freshwater, the ray-finned fishes. This enormously successful group has spread to nearly every watery habitat, both freshwater and saltwater.

The ray-finned fishes include not only a large number of species but also a huge variety of different forms and lifestyles (Fig. 16-35). These range from snakelike eels to flattened flounders; from sluggish bottom-feeders that probe the sea floor to speedy, streamlined predators that range in open water; from brightly colored reef-dwellers to translucent, luminescent deep-sea dwellers; from the 3,000-pound mola to the tiny stout infantfish, which weighs in at about 0.00003 ounce (1 milligram).

Ray-finned fishes are an extremely important source of food for humans; we collectively consume a prodigious number of fish. Unfortunately, however, our appetite for ray-finned fish, combined with increasingly effective high-tech methods for finding and catching them, has had a devastating impact on fish populations. Biologists report that populations of almost all economically important ray-finned fish species have declined drastically. Large, predatory fish such as tuna and cod have been especially hard hit; today's populations of these species contain less than 10% of the number present when commercial fishing began. If overfishing continues, fish stocks are likely to collapse. The solution to this problem—catch fewer fish—is simple in concept but very difficult to implement in practice due to economic and political factors.

Amphibians   The amphibians straddle the boundary between aquatic and terrestrial existence (Fig. 16-36). The limbs of amphibians show varying degrees of adaptation to movement on land, from the belly-dragging crawl of salamanders to the long leaps of frogs and toads. A three-chambered heart (in contrast to the two-chambered heart of fishes) circulates blood more efficiently, and lungs replace gills in most adult forms. Amphibian lungs, however, are poorly developed and must be supplemented by the skin, which serves as an additional respiratory organ. This respiratory function requires that the skin remain moist, a constraint that greatly restricts the range of amphibian habitats on land.

Amphibians are also tied to moist habitats by their breeding behavior, which requires water. Their fertilization is external and takes place in water, where the sperm can swim to the eggs. The eggs must remain moist, as they are protected only by a jelly-like coating that leaves them vulnerable to water loss by evaporation. Different amphibian species keep their eggs moist in different ways, but many species simply lay their eggs in water. In some amphibian species, fertilized

▼ **Figure 16-35 Ray-finned fishes** Ray-finned fishes have colonized nearly every aquatic habitat. **(a)** A female deep-sea angler fish and two males (the much smaller fish on the right side of the photo). The female attracts prey with a living, luminescent lure that projects just above her mouth. She is ghostly white because in the 2,000-meter depth where anglers live, no light penetrates, so colors are superfluous. The much smaller males remain as permanent parasites attached to the female, always available to fertilize her eggs. **(b)** This tropical green moray eel lives in rocky crevices. A small fish (a banded cleaner goby) on its lower jaw eats parasites that cling to the moray's skin. **(c)** The tropical seahorse may anchor itself with its tail, which is adapted for grasping, while the animal feeds on small crustaceans.

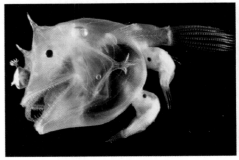

(a) Angler fish

(b) Moray eel

(c) Sea horse

(a) Tadpole

(b) Frog

(c) Salamander

▲ **Figure 16-36 Amphibians** In many amphibian species, individuals make the transition from **(a)** the completely aquatic larval tadpole to **(b)** the adult animal, here a bullfrog, leading a semiterrestrial life. **(c)** The red salamander is restricted to moist habitats in the eastern United States. Salamanders hatch in a form that closely resembles the adult. **QUESTION:** What advantages might amphibians gain from their two-part "double life"?

eggs develop into aquatic larvae such as the tadpoles of most frogs and toads. These aquatic larvae undergo a dramatic transformation into semiterrestrial adults, a metamorphosis that gives the amphibians their name, which means "double life." Their double life and their thin, permeable skin have made amphibians particularly vulnerable to pollutants and to environmental degradation, as described in "Earth Watch: Frogs in Peril" on p. 312.

Reptiles and Birds   The reptiles include lizards and snakes, turtles, alligators, and crocodiles (**Fig. 16-37**). This evolutionary group also includes the birds (**Fig. 16-38**). Some reptiles, particularly desert dwellers such as tortoises and lizards, are completely independent of their aquatic origins. This independence was achieved through a series of adaptations, of which three are outstanding: (1) Reptiles evolved a tough, scaly skin that resists water loss and protects the body. (2) Reptiles evolved internal fertilization, in which the male deposits sperm within the female's body. (3) Reptiles evolved a shelled egg, which can be buried in sand or dirt, far from water that might contain hungry predators. The shell prevents the egg from drying out on land. An internal membrane encloses the embryo in the watery environment that all developing animals require.

One very distinctive group of reptiles is the birds. Although birds have traditionally been classified as a group separate from reptiles, biologists have shown that birds are really a subset of an evolutionary group that includes both birds and the groups that have been traditionally designated as reptiles. Birds are distinguished from other modern reptiles by feathers, which are essentially a highly specialized version of reptilian body scales. Modern birds retain scales on their legs—evidence of the ancestry they share with the rest of the reptiles.

Bird anatomy and physiology are dominated by adaptations that help them to fly. In particular, birds are exceptionally light for their size. Hollow bones reduce the weight of the bird skeleton to a fraction of that of other vertebrates, and many bones present in other reptiles have been lost in the course of evolution or fused with other bones. The shelled egg that contributed to the reptiles' success on land frees the mother bird from carrying her developing offspring internally. Feathers form lightweight extensions to the wings and the tail for the lift and control required for flight, and they also provide lightweight protection and insulation for the body.

▼ **Figure 16-37 Reptiles** **(a)** The mountain king snake has a color pattern very similar to that of the venomous coral snake, which potential predators avoid. This mimicry helps the harmless king snake elude predation. **(b)** The outward appearance of the American alligator, found in swampy areas of the South, is almost identical to reconstructions of 150-million-year-old fossil alligators. **(c)** The tortoises of the Galapagos Islands, Ecuador, may live to be more than 100 years old.

(a) Snake

(b) Alligator

(c) Tortoise

(a) Hummingbird

(b) Frigate bird

(c) Ostrich

Mammals One branch of the reptile evolutionary tree gave rise to a group that evolved hair and diverged to form the mammals. In most mammals, fur protects and insulates the warm body. Legs that evolved for running rather than crawling make many mammals fast and agile. In contrast to birds, whose bodies reflect the requirements of flight, mammals have evolved a remarkable diversity of form.

Mammals are named for the milk-producing mammary glands used by all female members of this group to suckle their young. In addition to these unique glands, the mammalian body has sweat, scent, and sebaceous (oil-producing) glands, none of which are found in other vertebrates. The mammals include three main groups: monotremes, marsupials, and placental mammals.

Monotremes are found only in Australia and New Guinea, and include only three species: the platypus (**Fig. 16-39a**) and two species of spiny anteaters, also

▲ **Figure 16-38 Birds** *(a)* The delicate hummingbird beats its wings about 60 times per second and weighs about 0.15 ounce (4 grams). *(b)* This young frigate bird, a fish-eater from the Galapagos Islands, has nearly outgrown its nest. *(c)* The ostrich is the largest of all birds, weighing more than 300 pounds (136 kilograms). Its eggs weigh more than 3 pounds (1,500 grams). **QUESTION:** Although the earliest birds could fly, many bird species—such as the ostrich—cannot. Why do you suppose flightlessness has evolved repeatedly among birds?

(a) Platypus

(b) Wallaby

◀ **Figure 16-39 Mammals**
(a) Monotremes, such as this platypus, lay leathery eggs resembling those of reptiles. Platypuses live in burrows that they dig on the banks of rivers, lakes, or streams. *(b)* Marsupials, such as the wallaby, give birth to extremely immature young who immediately grasp a nipple and develop within the mother's protective pouch (inset). *(c)* A humpback whale, a placental mammal, gives its offspring a boost. *(d)* A bat, the only mammal capable of true flight, navigates at night by using a kind of sonar. Large ears aid in detecting echoes as its high-pitched cries bounce off nearby objects.

(c) Whale

(d) Bat

Frogs in Peril

Frogs and toads have lived in Earth's ponds and swamps for nearly 150 million years, somehow surviving the Cretaceous catastrophe that extinguished the dinosaurs and so many other species about 65 million years ago. Their evolutionary longevity, however, doesn't protect them from the environmental changes wrought by human activities. During the past two decades, herpetologists (biologists who study reptiles and amphibians) from around the world have documented an alarming decline in amphibian populations. Thousands of species of frogs, toads, and salamanders are dramatically decreasing in number, and many have apparently gone extinct.

This is a worldwide phenomenon; population crashes have been reported from every part of the globe. Yosemite toads and yellow-legged frogs are disappearing from the mountains of California. Tiger salamanders have been nearly wiped out in the Colorado Rockies. Leopard frogs are becoming rare in the United States. In Costa Rica, the golden toad was common in the early 1980s but has not been seen since 1989. The gastric brooding frog of Australia fascinated biologists by swallowing its eggs, brooding them in its stomach, and later regurgitating fully formed offspring. This species was abundant and seemed safe in a national park. Suddenly, in 1980, the gastric brooding frog disappeared and hasn't been seen since.

The causes of the worldwide decline in amphibian diversity are not fully understood, but researchers have discovered that frogs and toads in many places are succumbing to infection by a pathogenic fungus. The fungus has been found in the skin of dead and dying frogs in widespread locations, including Australia, Central America, and the western United States. In those places, discovery of the fungus has coincided with massive frog and toad die-offs, and most herpetologists agree that the fungus is causing the deaths.

It seems unlikely, however, that the fungus alone is responsible for the worldwide decline of amphibians. For one thing, die-offs have occurred in many places where the fungus has not been found. In addition, many herpetologists believe that the fungal epidemic would not have arisen if the frogs and toads had not first been weakened by other stresses. So, if the fungus is not doing all of the damage on its own, what are the other possible causes of amphibian decline? All of the most likely causes stem from human modification of the biosphere—the portion of Earth that sustains life.

Habitat destruction, especially the draining of wetlands that are especially hospitable to amphibian life, is one major cause of the decline. Amphibians are also vulnerable to toxic substances in the environment. For example, researchers found that frogs exposed to trace amounts of atrazine, a widely used herbicide that is found in virtually all freshwater in the United States, suffered severe damage to their reproductive tissues. The unique biology of amphibians makes them especially susceptible to poisons in the environment. Amphibian bodies at all stages of life are protected only by a thin, permeable skin that pollutants can easily penetrate (**Fig. E16-1**). To make matters worse, the double life of many amphibians exposes their permeable skin to a wide range of aquatic and terrestrial habitats and to a correspondingly wide range of environmental toxins.

Amphibian eggs can also be damaged by ultraviolet (UV) light, according to research by Andrew Blaustein, an ecologist at Oregon State University. Blaustein demonstrated that the eggs of some species of frogs in the Pacific Northwest are sensitive to damage from UV light and that the most sensitive species are experiencing the most drastic declines. Unfortunately, many parts of Earth are subject to increasingly intense UV radiation levels, because atmospheric pollutants have caused a thinning of the protective ozone layer.

Many scientists believe that the troubles of amphibians signal an overall deterioration of Earth's ability to support life. According to this

▲ **Figure E16-1 Amphibians in danger** The corroboree toad, shown here with its eggs, is rapidly declining in its native Australia. Tadpoles are developing within the eggs. The thin water-permeable and gas-permeable skin of the adult and the jelly-like coating around the eggs provide little, if any, protection against air and water pollutants.

line of reasoning, the highly sensitive amphibians are providing an early warning of environmental degradation that will eventually affect more resilient organisms as well. Equally worrisome is the observation that amphibians are not just sensitive indicators of the health of the biosphere but are also crucial components of many ecosystems. They may keep insect populations in check, in turn serving as food for larger carnivores. Their decline will further disrupt the balance of these delicate communities.

Margaret Stewart, an ecologist at the State University of New York, Albany, aptly summarized the problem: "There's a famous saying among ecologists and environmentalists: 'Everything is related to everything else.'. . . You can't wipe out one large component of the system and not see dramatic changes in other parts of the system."

know as echidnas. Unlike other mammals, monotremes lay eggs rather than giving birth to live young. Newly hatched monotremes are tiny and helpless and feed on milk secreted by the mother. Monotremes, however, lack nipples. Milk from the mammary glands oozes through ducts on the mother's abdomen and soaks the fur around the ducts. The young then suck the milk from the fur.

In all mammals except the monotremes, embryos develop in the uterus, a muscular organ in the female reproductive tract. The lining of the uterus combines with membranes derived from the embryo to form the *placenta*, a structure that allows gases, nutrients, and wastes to be exchanged between the circulatory systems of the mother and embryo. In *marsupials*, embryos develop in the uterus for only a short period. Marsupial young are born at a very immature stage of development. Immediately after birth, they crawl to a nipple, firmly grasp it, and, nourished by milk, complete their development (**Fig. 16-39b**). In most, but not all, marsupial species, this postbirth development takes place in a protective pouch.

Most mammal species are *placental* mammals, so named because their placentas are far more complex than those of marsupials. Compared to marsupials, placental mammals retain their young in the uterus for a much longer period, so that offspring complete their embryonic development before being born. The placental mammals have evolved a remarkable diversity of form. The bat, mole, impala, whale, seal, monkey, and cheetah exemplify the radiation of mammals into nearly all habitats, with bodies finely adapted to their varied lifestyles (**Figs. 16-39c,d**). The largest groups of placental mammals, in terms of number of species, are the bats and the rodents. About 20% of mammal species are bats, and almost 40% are rodents (a group that includes rats, mice, squirrels, hamsters, guinea pigs, porcupines, beavers, woodchucks, chipmunks, and voles).

Hidden Treasures Revisited

Although finding a previously unknown bird or monkey might require a trek to an inaccessible tropical forest, that is not the case for the vast majority of undiscovered species. When it comes to the less conspicuous inhabitants of Earth, undiscovered species abound. Some are indeed hidden by inaccessibility, because they live beneath the soil surface or in the ocean depths. But many are hidden only by their small size and lack of a broad scientific effort to find them. As a consequence, we are scientifically aware of only a fraction of the planet's species of fungi, bacteria, archaea, protists, and invertebrate animals.

The gap between our knowledge of species and their actual number was recently illustrated by scientists who sequenced the DNA in a sample of water from the Sargasso Sea. The sample contained 1.2 million different, mostly previously unknown genes, suggesting the presence of around 1,800 microscopic species. If 1,800 mostly unknown species are present in one test tube of seawater, the current total of about 5,000 known species of bacteria worldwide must represent only a tiny fraction of true bacterial diversity.

When a new species is found, whether in a remote mountain forest or in a test tube of water, it must be named. Typically, the scientist who discovers and describes a new species is entitled to choose its Latin name. Scientists usually choose a name that describes a trait of the species or perhaps the location where it was found. Sometimes, however, the naming process becomes a financial transaction. For example, in return for a contribution that supports efforts to discover and conserve endangered species, the conservation organization BIOPAT will allow you to name a newly discovered species. You pick a name, which is then given an appropriate Latin ending and published in a scientific journal.

In perhaps the most extraordinary example of purchased species naming rights, a newly discovered monkey was named after an online casino. In return for a contribution of $650,000, the new species was named *Callicebus aureipalatii*; the species name is Latin for "golden palace." The payment for naming the Golden Palace monkey will be used for management of the Madidi National Park in Bolivia, where the new species was discovered

Consider This

The All Species Foundation is a nonprofit organization that promotes the goal of finding and naming all of Earth's undiscovered species within the next 25 years. According to the foundation, this task "deserves to be one of the great scientific goals of the new century." The foundation estimates the cost of the job at between $700 and $2,000 per species, with perhaps millions of undiscovered species remaining to found. Do you think the search for undiscovered species should continue? What value or benefit to humans does the search for new species provide?

Chapter Review

Summary of Key Concepts

For additional study help and activities, go to www.mybiology.com.

16.1 How Are Organisms Named and Classified?

Organisms are classified and placed into hierarchical categories that reflect their evolutionary relationships. The eight major categories, in order of decreasing inclusiveness, are domain, kingdom, phylum, class, order, family, genus, and species. The scientific name of an organism, such as *Homo sapiens*, is its genus name followed by its species name. Anatomical and genetic similarities are used to classify organisms. Genetic similarities among organisms are a measure of evolutionary relatedness.

Web Animation   Taxonomic Classification

16.2 What Are the Domains of Life?

The three domains of life, each representing one of three main branches of the tree of life, are Bacteria, Archaea, and Eukarya. Each domain contains a number of kingdoms, but the details of kingdom-level classification are in a period of transition and remain unsettled. Within the domain Eukarya, however, the kingdoms Fungi, Plantae, and Animalia are universally accepted as valid groups.

Web Animation   Tree of Life

16.3 Bacteria and Archaea

Members of the domains Bacteria and Archaea are unicellular and prokaryotic. Archaea and bacteria are not closely related and differ in fundamental features, including cell wall composition and membrane structure. A cell wall determines the characteristic shapes of prokaryotes: round, rodlike, or spiral. Certain types of bacteria can form spores that disperse widely and withstand inhospitable environmental conditions. Prokaryotes obtain energy in a variety of ways. Some rely on photosynthesis. Others break down inorganic molecules to obtain energy. Many are anaerobic, able to obtain energy from fermentation when oxygen is not available.

Some bacteria are pathogenic, causing disorders such as pneumonia, tetanus, botulism, and the venereal diseases gonorrhea and syphilis. Most bacteria, however, are harmless to humans and play important roles in natural ecosystems. Bacteria and archaea have colonized nearly every habitat on Earth, including hot, acidic, very salty, and anaerobic environments. Some live in the digestive tracts of ruminant animals, where they break down cellulose. Nitrogen-fixing bacteria enrich the soil and aid in plant growth; many others live off the dead bodies and wastes of other organisms, liberating nutrients for reuse.

16.4 Protists

Most protists are single, highly complex eukaryotic cells, but some form colonies and some, such as seaweeds, are multicellular. Photosynthetic protists form much of the phytoplankton, which plays a key ecological role. Protists exhibit diverse modes of nutrition, reproduction, and locomotion. Protist groups include the stra-menopiles (diatoms and brown algae), the alveolates (dinoflagellates, apicomplexans, and ciliates), cercozoans, amoebozoans, and green algae (the closest relatives of plants). Some protists, especially types of apicomplexans, cause human diseases.

16.5 Plants

Two groups of plants, nonvascular plants and vascular plants, arose from ancient algal ancestors. Nonvascular plants, including the liverworts and mosses, are small land plants that lack conducting vessels. Although some have adapted to dry areas, most live in moist habitats. Reproduction in nonvascular plants requires water through which the sperm swims to the egg.

In vascular plants a system of vessels, stiffened by lignin, conducts water and nutrients absorbed by the roots into the upper portions of the plant and supports the body as well. Owing to this support system, seedless vascular plants, including the club mosses, horsetails, and ferns, can grow larger than nonvascular plants.

Vascular plants with seeds have two other major adaptive features: pollen and seeds. Seed plants are often classified into two categories: gymnosperms and angiosperms. Gymnosperms include ginkgos, cycads, and the highly successful conifers. Angiosperms, the flowering plants, dominate much of the land today. In addition to pollen and seeds, angiosperms produce flowers and fruits. The flower allows angiosperms to utilize animals as pollinators. In contrast to wind, animals can in some cases carry pollen farther and with greater accuracy and less waste. Fruits may attract animal consumers, which incidentally disperse the seeds in their feces.

Web Animation   Evolution of Plant Structure

16.6 Fungi

Fungal bodies generally consist of filamentous hyphae that form large, intertwined mycelia. A cell wall of chitin surrounds fungal cells. Fungi obtain energy by secreting digestive enzymes outside their bodies and absorbing the liberated nutrients. Fungal reproduction depends on spores, and most species reproduce both sexually and asexually.

The majority of plant diseases are caused by parasitic fungi. Some parasitic fungi can help control insect crop pests. Others can cause human diseases, including ringworm, athlete's foot, and common vaginal infections. Some fungi produce toxins that can harm humans. Nonetheless, fungi add variety to the human food supply, and fermentation by fungi helps make wine, beer, and bread.

Fungi are extremely important decomposers in ecosystems. Their filamentous bodies penetrate rich soil and decaying organic material, liberating nutrients through extracellular digestion.

16.7 Animals

Animals are multicellular, sexually reproducing, motile organisms. Most can perceive and react rapidly to environmental stimuli and are motile at some stage in their lives. Their cells lack cell walls.

The bodies of sponges are typically free-form in shape and cannot move. Sponges have relatively few types of cells. Digestion occurs exclusively within the individual cells.

Cnidarians (jellyfish, sea anemones, corals, and hydrozoans) are radially symmetrical and use armed tentacles to capture prey.

The annelids, the segmented worms, include earthworms, marine tubeworms, and leeches.

Arthropods, which include the insects, arachnids, and crustaceans, are the most diverse and abundant organisms on Earth. They have invaded nearly every available terrestrial and aquatic habitat. Jointed appendages and well-developed nervous systems make possible complex, finely coordinated behavior. The exoskeleton (which conserves water and provides support) helps enable the insects and arachnids to inhabit dry land. The diversification of insects has been enhanced by their ability to fly. Crustaceans, which include the largest arthropods, are restricted to moist, usually aquatic habitats.

The mollusks (gastropods, bivalves, and cephalopods) lack a skeleton. Some forms protect the soft, moist, muscular body with a single shell (many gastropods and a few cephalopods) or a pair of hinged shells (the bivalves). The lack of a waterproof external covering limits mollusks to aquatic and moist terrestrial habitats. The body plan of gastropods and bivalves limits the complexity of their behavior, but the cephalopod's tentacles are capable of precisely controlled movements. The octopus has the most complex brain and the best-developed learning capacity of any invertebrate.

The chordate phylum includes two invertebrate groups, the lancelets and tunicates, as well as the familiar vertebrates. All chordates possess a notochord, a nerve cord, pharyngeal gill slits, and a post-anal tail at some stage in their development. Vertebrates have a backbone that is part of a living internal skeleton.

During the evolutionary transitions from fishes to amphibians to reptiles, some vertebrates evolved adaptations that helped them colonize dry land. All amphibians have legs, and most have simple lungs for breathing in air rather than in water. Most are confined to relatively damp terrestrial habitats by their need to keep their skin moist, use of external fertilization, and the requirement that their eggs and larvae develop in water. Reptiles are well adapted to the driest terrestrial habitats, thanks to well-developed lungs, dry skin covered with relatively waterproof scales, internal fertilization, and shelled eggs with their own water supply. Birds and mammals are also fully terrestrial and have additional adaptations. The bird body is adapted for flight, with feathers and hollow bones. Mammals have insulating hair and give birth to live young that are nourished with milk from the mothers' mammary glands. The mammalian nervous system is the most complex in the animal kingdom, providing mammals with enhanced learning ability that helps them adapt to changing environments.

Key Terms

anaerobe p. 287
angiosperm p. 297
Archaea p. 283
Bacteria p. 283
conifer p. 296
endospore p. 286
Eukarya p. 283
flagellum (plural, flagella) p. 285
flower p. 297

fruit p. 297
gymnosperm p. 296
hypha (plural, hyphae) p. 299
invertebrate p. 301
larva (plural, larvae) p. 306
metamorphosis p. 305
molt p. 306
mycelium (plural, mycelia) p. 299

nerve cord p. 307
nitrogen-fixing bacteria p. 288
nonvascular plant p. 294
notochord p. 307
pathogenic p. 289
pharyngeal gill slit p. 308
phytoplankton p. 291
pollen p. 296
post-anal tail p. 308

protist p. 290
pupa p. 306
seed p. 296
spore p. 299
vascular plant p. 294
vertebral column p. 308
vertebrate p. 301

Thinking Through the Concepts

Suggested answers to end-of-chapter and figure-based questions can be found at the end of the text.

Fill-in-the-Blank

1. The five-kingdom classification has been replaced with a system that divides life into three major categories called _____. The two categories that consist of prokaryotic cells are _____ and _____. The category that consists of organisms composed of eukaryotic cells is _____. All pathogenic prokaryotes identified to date belong to the _____. These pathogens cause disease primarily by secreting _____.

2. Fill in the blanks with the correct protist. Photosynthetic stramenopiles with glassy shells: _____; alveolates with two whiplike flagella, with some causing "red tides": _____; alveolates that are the most complex of all single cells: _____; alveolates whose shells can form limestone deposits: _____; extend pseudopods, most lack shells; this group includes the slime molds: _____.

3. Scientists hypothesize that the ancestors of plants were _____. There are two major types of plants; those that lack conducting vessels are called _____, and those with vessels are _____. All plants exhibit a complex life cycle described as _____.

4. Seedless vascular plants must reproduce when conditions are wet because their sperm must _____. Two adaptations that allow seed plants to thrive on dry land are _____ and _____. The seed plants fall into two major categories: the nonflowering _____ and the flowering _____. Flowers were favored by natural selection because they _____.

5. The portions of a fungus that are visible to the naked eye are often structures specialized for _____. These structures release tiny _____, which are dispersed to produce new fungi. Most fungi consist of a(n) _____ that is composed of microscopic threads called _____. Single-celled fungi that cause breads to rise and wines to ferment are _____.

6. Animals that lack a backbone are described as _____; those that possess a backbone are _____. The vast majority of all animals fall into which of these two groups? _____. The simplest animals are _____, whose bodies resemble a colony of _____. Sea anemones and corals are _____. Earthworms and leeches are _____.

7. Three major groups within the phylum Mollusca are the two-shelled clams and scallops, called _____; the foot-crawling snails and slugs, called _____; and the tentacled squid and octopuses, called _____. Members of the largest animal phylum are called _____. Three important groups within this phylum are the six-legged, often flying _____; the eight-legged spiders and mites called _____; and the mostly aquatic _____.

8. Mammals are named for their milk-producing _____. Three major groups of mammals are the egg-laying _____; the _____, which carry their developing young in a pouch; and the _____ mammals, whose young develop more fully in the uterus. Almost 40% of all mammals are _____.

Review Questions

1. What features would you study to determine whether a dolphin is more closely related to a fish or to a bear?

2. Only a small fraction of the total number of species on Earth has been scientifically described. Why?

3. Describe some of the ways in which bacteria obtain energy and nutrients.

4. What are nitrogen-fixing bacteria, and what role do they play in ecosystems?

5. What is the major ecological role played by unicellular algae?

6. What portion of the fungal body is represented by mushrooms, puffballs, and similar structures? Why are these structures elevated above the ground?

7. List the adaptations that helped plants become successful terrestrial organisms.

8. The number of species of flowering plants is greater than the number of species in the rest of the plant kingdom. What feature(s) are responsible for the enormous success of angiosperms? Explain why.

9. Distinguish between vertebrates and invertebrates.

10. List four distinguishing features of chordates.

11. Describe the ways in which amphibians are adapted to life on land. In what ways are amphibians still restricted to a watery or moist environment?

12. List the adaptations that distinguish reptiles from amphibians and helped reptiles adapt to life in dry terrestrial environments.

Applying the Concepts

1. IS THIS SCIENCE? Biologist E. O. Wilson has estimated that 27,000 species go extinct each year. Some skeptics, however, point out that scientists have documented the disappearance of only 1,100 species during the past 500 years. Does this observation demonstrate that Wilson's estimate is inaccurate? Explain your answer.

2. There are many areas of disagreement about the classification of organisms. For example, there is no consensus about whether the red wolf is a species distinct from the gray wolf or about how many kingdoms are within the domain Bacteria. What difference does it make whether biologists consider the red wolf a species, or into which kingdom a bacterial species falls? As Shakespeare put it, "What's in a name?"

3. The internal structure of the cells of many protists is much more complex than that of cells of multicellular organisms. Does this mean that the protist is engaged in more complex activities than the multicellular organism? If not, why is the protistan cell much more complicated?

For additional resources, go to www.mybiology.com.

unit four

Plant Anatomy and Physiology

Flowers provide beauty and delight to human admirers. The real function of flowers, however, is to attract animal pollinators.

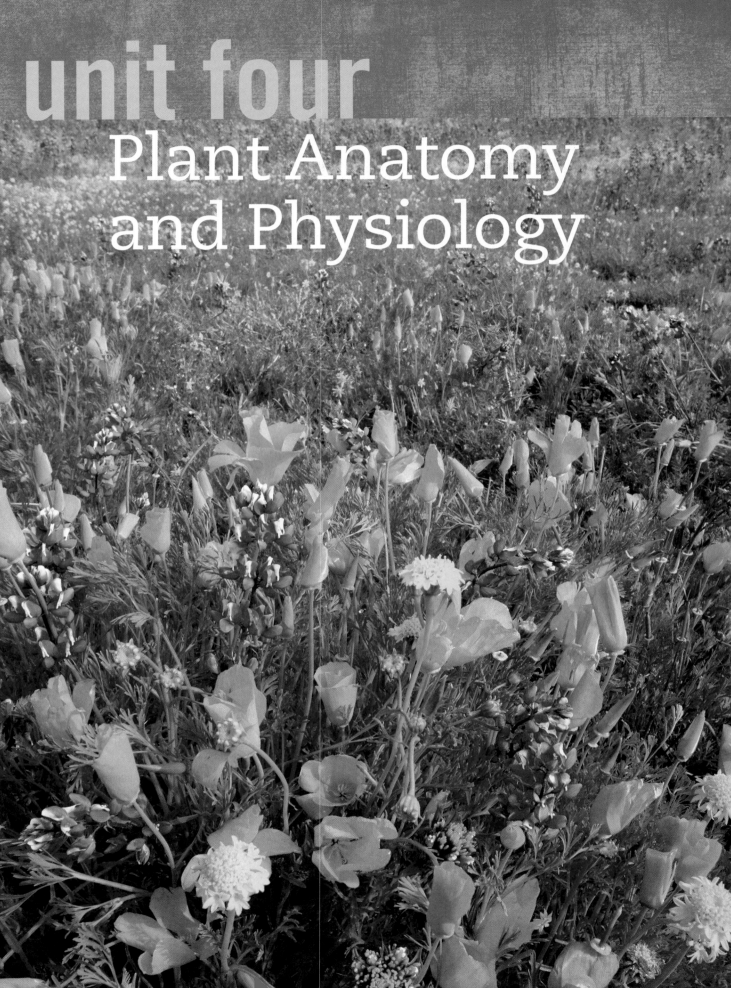

chapter 17

Plant Form and Function

Did the red hue of an autumn leaf evolve for the pleasure of human observers or for protection?

Q At a Glance

Case Study   Sunscreen for Maple Trees?

In autumn, the back roads of New England fill with "leaf peepers" who have come to gaze at hillsides ablaze with brilliant colors. The spectacular fall foliage of northern regions amazes us with its beauty, but why do the leaves of these deciduous trees change color before dropping off? At one level, we can say that the color changes because the green chlorophyll in the leaves breaks down, revealing other pigments whose color was previously masked. But that answer leaves open the question of why the other pigments are there in the first place.

Particularly puzzling are the pigments known as anthocyanins, which provide the bright red color of many autumn leaves. Anthocyanins are produced just before the leaves drop. Why would dying leaves invest precious energy in making a new pigment?

One hypothesis is that anthocyanins act as a kind of sunscreen, protecting leaves from harmful sunlight. When a leaf is exposed to more light than it can use in photosynthesis, the energy in the "extra" light can damage the leaf and reduce photosynthesis. In autumn, when metabolism declines and the photosynthetic machinery begins to break down, leaves become especially susceptible to the harmful effects of bright sunlight. Anthocyanins might absorb some light and limit the damage.

In one test of this hypothesis, researchers used the red-osier dogwood, a species in which some leaves turn bright red in the fall but other leaves remain green until they drop. The researchers exposed both kinds of autumn leaves to intense light and recorded their levels of photosynthesis. As predicted by the hypothesis, photosynthesis decreased more drastically, and took longer to recover, in green leaves. The researchers concluded that anthocyanins protect leaves from the harmful effects of bright sunlight. But why protect photosynthesis in a leaf that's about to die anyway? Consider that question as we explore the plant body. ∎

Are the red areas of these pepperwood leaves less likely to suffer damage than the green areas are?

Web Animation Plant Anatomy

17.1 How Are Plant Bodies Organized?

The evolutionary history of the plant kingdom has produced many distinctly different types of plants. In this chapter, which discusses how plant bodies grow and survive, we focus mainly on the flowering plants, or angiosperms, which have become the most diverse and widespread of plant groups.

Flowering Plants Have a Root System and a Shoot System

Flowering plant bodies consist of two major parts, the root system and the shoot system (**Fig. 17-1**). **Roots** are branched portions of the plant body that are usually embedded in soil. They anchor the plant in the ground, absorb and transport water and nutrients from the soil, and store surplus sugars manufactured during photosynthesis.

The rest of the plant is the shoot system, usually located aboveground. The shoot system consists of leaves, buds, and (in season) flowers and fruits, all borne on **stems,** which are typically branched. The functions of shoots include photosynthesis, reproduction, and transport of materials between different parts of the plant body.

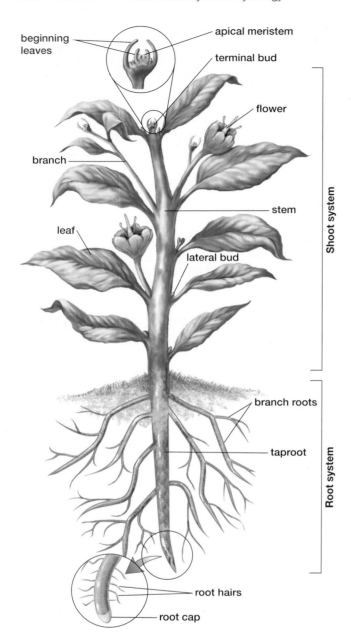

▲ **Figure 17-1 The structure of a typical flowering plant**

Flowering Plants Can Be Divided into Two Groups

Flowering plants fall into two main groups: the **monocots,** which include grasses, lilies, palms, and orchids, and the **dicots,** which include broadleaf trees (including those that drop their leaves in winter), bushes, and many garden flowers. Monocots and dicots differ in the structure of their flowers, leaves, vascular tissue, root pattern, and seeds. These differences are illustrated in **Figure 17-2.** For now, don't worry about unfamiliar terms in the figure. Look it over and refer to it later as we examine the parts of flowering plants in more detail.

17.2 How Do Plants Grow?

Animals and plants develop in dramatically different ways. One difference is in the timing of growth. In animals, growth is generally confined to a fixed period early in life. For example, you began growing at conception and grew until you reached your adult height, at which point you stopped growing (upward, at least!). In contrast, flowering plants keep growing throughout their lives, never reaching a stable final body form.

The sites of growth also differ between animals and plants. In animals, growth takes place in all of the body's tissues. As you grew from a baby to an adult, all parts of your body became larger. Most plants, however, grow longer only at the tips of their branches and roots. As a result, structures that develop at a particular spot on a plant's body remain in exactly the same place relative to the surroundings, even as the plant continues to grow. Thus, a tree branch does not move farther from the ground each year, even though the tree grows taller. What causes plants to grow this way?

During Growth, Meristem Cells Give Rise to Differentiated Cells

Plant bodies are composites of two fundamentally different kinds of cells: meristem cells and differentiated cells. **Meristem cells** are capable of mitotic cell division. Some of the offspring, or daughter cells, of dividing meristem cells lose the ability to divide and become **differentiated cells** that make up the nongrowing portions of the plant body. The continued divisions of meristem cells keep a plant growing throughout its life, and their differentiated daughter cells form more stable or permanent parts of the plant, such as mature leaves or the trunks of trees.

Plants grow through the division and differentiation of two major types of meristem cells: apical meristems and lateral meristems. **Apical meristems** ("tip meristems") are located at the tips of roots and shoots, including main stems and branches (see Fig. 17-1 and also Fig. 17-8 later in this chapter). **Lateral meristems** ("side meristems"), also called **cambia** (singular, cambium), form cylinders that run lengthwise inside roots and stems (see Fig. 17-13 later in this chapter).

Different Processes Are Responsible for Growth in Length and Diameter

Plant growth takes two forms: primary growth and secondary growth. **Primary growth** occurs by division and differentiation of apical meristem cells. This type of growth takes place in the growing tips of roots and shoots. Primary growth is responsible for both the increases in length and the development of specialized plant structures such as leaves and flowers. The elongation of roots and shoots through primary growth allows them to enter new space from which to collect light, nutrients, and water.

Secondary growth, which is responsible for increases in diameter, occurs by the division and differentiation of lateral meristem cells. Secondary growth causes the stems and roots of dicots (and some monocots) to become thicker and woodier as they age. Although in this chapter we discuss secondary growth of stems only, keep in mind that secondary growth also occurs in roots.

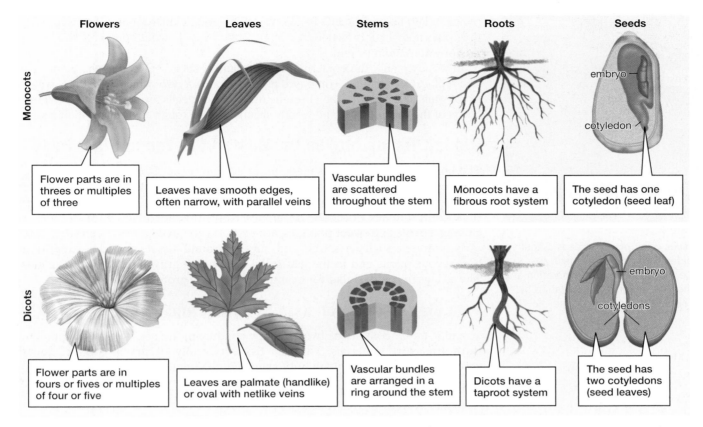

Flowers | **Leaves** | **Stems** | **Roots** | **Seeds**

Monocots

Flower parts are in threes or multiples of three

Leaves have smooth edges, often narrow, with parallel veins

Vascular bundles are scattered throughout the stem

Monocots have a fibrous root system

The seed has one cotyledon (seed leaf)

Dicots

Flower parts are in fours or fives or multiples of four or five

Leaves are palmate (handlike) or oval with netlike veins

Vascular bundles are arranged in a ring around the stem

Dicots have a taproot system

The seed has two cotyledons (seed leaves)

▲ **Figure 17-2 Monocots and dicots** The major differences between the two major classes of flowering plants, the monocots and the dicots.

17.3 What Are the Tissues and Cell Types of Plants?

The structures of plants, including roots, stems, and leaves, consist of three types of tissue: dermal, ground, and vascular. Within each type there are subtypes composed of specialized cells (Fig. 17-3). **Dermal tissue** covers the outer surfaces of the plant body. **Ground tissue** makes up most of the body of young plants and includes all nondermal and nonvascular tissues. Its varied functions include photosynthesis, support, and storage. **Vascular tissue** transports water, nutrients, sugars, and hormones throughout the plant.

Some flowering plants, described as *herbaceous*, are soft bodied and have flexible stems. Herbaceous plants include lettuce, beans, and wheat. Such plants usually live only 1 year. Other flowering plants, such as trees and bushes, are described as woody. Most are perennial (living many years) and develop hard, thickened, woody stems as a result of secondary growth. As you will see, different types of tissue are present in herbaceous and woody plants.

Dermal Tissue Covers the Plant Body

An important part of the dermal tissue that covers a plant's outer surface is the **epidermis,** the outermost cell layer that covers the leaves, stems, and roots of all young plants (Fig. 17-4). The epidermis of the aboveground parts of a plant is generally composed of tightly packed, thin-walled cells and is covered with a waterproof, waxy layer known as the **cuticle.** The cuticle reduces the evaporation of water from

▶ **Figure 17-3 The structure of the root and shoot** Both the root and shoot of a flowering plant consist of dermal, ground, and vascular tissues. (The structure of the leaves, vascular tissue system, and roots indicate that our sample plant is a dicot.)
EXERCISE: On this drawing, circle all of the locations at which primary growth will occur, and draw arrows pointing to some of the locations at which secondary growth will occur.

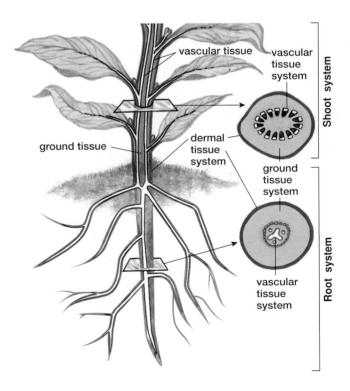

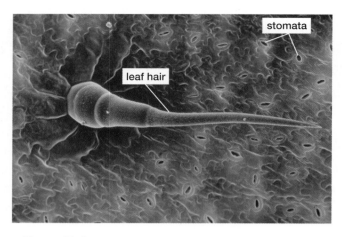

▲ **Figure 17-4 Dermal tissues cover plant surfaces** Shoot epidermis, such as the epidermis of this zinnia leaf, is covered with a waxy cuticle that reduces the evaporation of water. The leaf hair seen here also reduces evaporation by slowing the movement of air across the leaf surface.

the plant and helps protect it from invasion by disease microorganisms. In contrast, the epidermis of roots has no cuticle, because a cuticle would prevent the absorption of water and nutrients.

In woody plants, the epidermis of stems and roots is replaced as the plants age. The replacement tissue is composed primarily of *cork cells*, which have thick, waterproof walls and die once they become mature. Cork cells form the protective outer layers of the bark of trees and woody shrubs and the tough covering of their roots.

Ground Tissue Makes Up Most of the Young Plant Body

The bulk of a young plant is made up of ground tissue. Ground tissue cells typically carry out most of the metabolic activities of the plant. Depending on their location within the plant body, ground tissue cells may photosynthesize, store sugars and starches, or secrete hormones. Roots that are adapted for storage, including carrots and sweet potatoes, are packed with ground tissue cells that store carbohydrates, such as starch and sugar. Ground tissue provides support in herbaceous plants and in the leaf stalks and young growing stems of all plants. The strings in celery stalks, for example, are mostly ground tissue.

Vascular Tissue Consists of Xylem and Phloem

Vascular tissue consists of two complex conducting tissues: xylem and phloem. Both transport materials. Vascular tissue typically occurs in strands, called *vascular bundles*, that contain both xylem and phloem.

Xylem Conducts Water and Dissolved Nutrients from the Roots to the Rest of the Plant

Xylem conducts water and nutrients from roots to shoots in tubes that are made from cells called *tracheids* and *vessel elements* (**Fig. 17-5**). Both tracheids and vessel elements are tube-shaped cells that die after they develop. Their cytoplasm and plasma membranes then disintegrate, leaving behind hollow tubes made of cell walls.

Tracheids have thick cell walls and slanted ends that resemble the tips of hypodermic needles. In a shoot or root, tracheids are stacked on top of one another with their slanted ends overlapping. Porous sections in the overlapping portions allow water and nutrients to pass from one tracheid to the next.

Vessel elements also meet end to end, but are larger in diameter than tracheids. Perforations connect adjoining vessel elements. In some cases, the ends of vessel el-

▶ **Figure 17-5 Xylem** *(a)* Xylem is a mixture of ground tissue and two types of conducting cells, tracheids and vessel elements. Both tracheids and vessel elements have porous side walls, allowing water and dissolved minerals to move sideways between adjacent conducting tubes. *(b)* A micrograph of xylem.

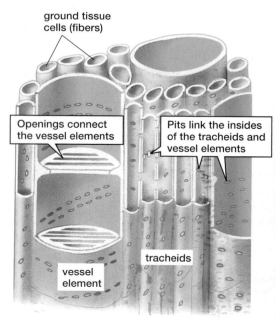

(a) Xylem structure

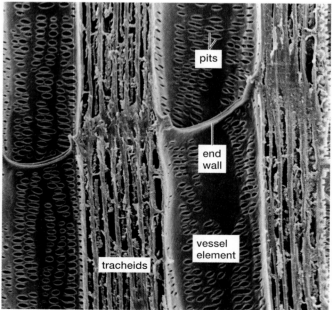

(b) Longitudinal section of xylem

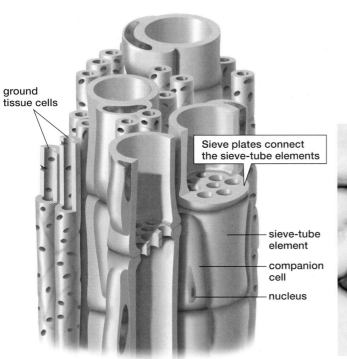

ground
tissue cells

Sieve plates connect
the sieve-tube elements

sieve-tube
element

companion
cell

nucleus

(a) Phloem structure

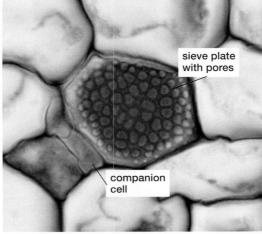

sieve plate
with pores

companion
cell

(b) Cross section of phloem

◄**Figure 17-6 Phloem (a)** Phloem
is a mixture of ground tissue, sieve-
tube elements, and companion
cells. Sieve-tube elements, stacked
end to end, form the conducting
tubes of phloem. **(b)** A micrograph
of one end of a sieve-tube element,
showing the sieve plate.

ements completely disintegrate, leaving an open tube (see Fig. 17-5). Thus, vessel elements form wide-diameter, relatively unobstructed pipelines from root to leaf.

Phloem Conducts Substances Throughout the Plant

Plants must move substances that they synthesize, such as sugars, amino acids, and
hormones, from the parts of the plant at which the substances are produced to the
parts at which they are needed. The transported substances, dissolved in water, are
conducted by **phloem.** In phloem, *sieve tubes*, each consisting of a strand of cells
called *sieve-tube elements*, are scattered among supportive ground tissue (**Fig. 17-6**).
Where sieve-tube elements meet end to end, holes form in the cell walls, linking the
cells to form a continuous conducting tube. As sieve-tube elements mature, most of
their internal contents disintegrate, leaving behind only a thin layer of cytoplasm
that lines the plasma membrane. Sieve-tube elements thus lack most organelles and
must be nourished by smaller, adjacent *companion cells*.

17.4 How Do Roots Grow and What Do They Do?

As a seed sprouts, the primary root—the first root to develop—grows down into
the soil. Many dicots, such as carrots and dandelions, develop a taproot. A
taproot consists of the primary root, which usually becomes longer and stouter
with time, and many smaller roots that grow out from the sides (**Fig. 17-7a**). In
contrast, in monocots such as grasses and palms, the primary root soon dies off,
replaced by many new roots that emerge from the base of the stem. These new
roots are called **fibrous roots** (**Fig. 17-7b**).

Roots Elongate by Primary Growth

In young roots, divisions of the apical meristem give rise to four distinct regions: an
outer envelope of epidermis, a vascular cylinder at the core of the root, a cortex
between the two, and a protective root cap. The **root cap** lies at the very tip of the root
and prevents the apical meristem from being scraped off as the root pushes down

(a) A taproot system   **(b) A fibrous root system**

▲ **Figure 17-7 Taproots and fibrous roots (a)** Dicots typically
have a long central taproot with many smaller roots
branching from it. **(b)** Monocots typically have many fibrous
roots of equal size.

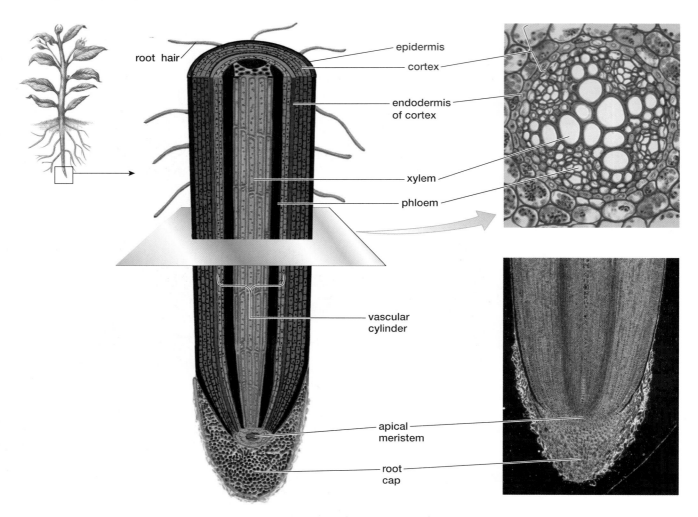

▲ **Figure 17-8 Primary growth in roots** Primary growth in roots results from cell divisions in the apical meristem, located near the tip of the root. Four regions are formed: the root cap, at the very tip of the root, and the epidermis, cortex, and vascular cylinder. **QUESTION:** Of the cells labeled in the cross-section photo at top right, which form ground tissue, which form vascular tissue, and which form epidermal tissue?

▲ **Figure 17-9 Root hairs** Root hairs, shown here in a sprouting radish, greatly increase a root's surface area for the absorption of water and nutrients from the soil.

between particles of soil (Fig. 17-8). Root-cap cells have thick cell walls and secrete a slimy lubricant that helps ease the way between soil particles. Nevertheless, root-cap cells wear away and must be continuously replaced by new cells from the meristem.

The Epidermis of the Root Is Very Permeable to Water

The root's outermost covering of cells is the epidermis, which contacts the soil and any air or water trapped among the soil particles. The walls of the epidermal cells are highly permeable to water, allowing water to penetrate into the interior of the root. Many epidermal cells have **root hairs,** which are long, thin extensions that grow into the surrounding soil (Fig. 17-9). By increasing the surface area, root hairs increase the root's ability to absorb water and nutrients. Root hairs may add dozens of square yards of surface area to the roots of even small plants.

The Cortex Controls the Absorption of Water and Nutrients

Cortex is a type of ground tissue that makes up most of the inside of a young root. The cortex consists of an outer mass of large, loosely packed cells just beneath the epidermis and an inner layer of smaller, close-fitting cells called the **endodermis** that encircles the vascular cylinder (see Fig. 17-8). The cells of the endodermis have highly specialized cell walls. Where they contact one another, a waxy, waterproof substance prevents water from seeping through the spaces between cells. But where the cells contact the cortex or vascular cylinder, the cell walls are permeable. Water and dissolved nutrients thus cannot pass between endodermal cells, and can move from the cortex into the vascular cylinder only by moving across the membranes of the endodermal cells. These membranes are therefore able to regulate the types and amounts of materials that the roots absorb.

The Vascular Cylinder Contains Xylem and Phloem and Meristem for Branch Roots

The **vascular cylinder** of a root contains the conducting tissues of xylem and phloem, which transport water and dissolved materials within the plant. The outer-most layer of the vascular cylinder retains the ability to divide and form the apical meristems of new roots that grow as branches from existing roots (**Fig. 17-10**). When a new branch root begins to grow, it must break out through the cortex and epidermis of the primary root. It does so by crushing the cells that lie in its path and se-creting enzymes that digest them away. The vascular tissues of the branch root connect with the vascular tissues of the primary root.

17.5 How Do Stems Grow and What Do They Do?

Like roots, stems develop from a small group of actively dividing cells. These cells lie at the tip of the young shoot within the **terminal bud** and make up the apical meristem. The daughter cells of the apical meristem differentiate into the special-ized cell types of stem, buds, leaves, and flowers.

Most young stems are composed of four tissues: epidermis (dermal tissue), cortex (ground tissue), pith (ground tissue), and vascular tissue (**Fig. 17-11**). As Figure 17-2 illustrates, monocots and dicots differ somewhat in the arrangement of vascular tissues. We will discuss only dicot stems here.

The Epidermis of the Stem Reduces Water Loss While Allowing Carbon Dioxide to Enter

In the stem (and leaves), the epidermis is exposed to air, making it a potential pathway for water loss. Epidermal cells of the stem secrete a waxy cuticle that reduces evaporation of water. The cuticle, how-ever, also reduces the diffusion of carbon dioxide and oxygen into and out of the plant. Hence, the epi-dermis is perforated with adjustable pores called **stomata** (singular, stoma) that permit and regulate the movement of these gases. Stomata are discussed in more detail later in this chapter.

The Cortex and Pith Support the Stem, Store Food, and Photosynthesize

The two types of ground tissue in stems are cor-tex (located between the epidermis and vascular tissues) and pith (inside the vascular tissues at the center of the stem). They are similar in most respects. Cortex and pith perform three major functions: support, storage, and, in some cases, photosynthesis:

- **Support:** In very young stems, water fills the central vacuoles of cortex and pith cells, causing pressure that stiffens the cells, much as air inflates a tire. Somewhat older stems gain support from other ground cells with thickened cell walls, which don't depend on water pressure for strength.

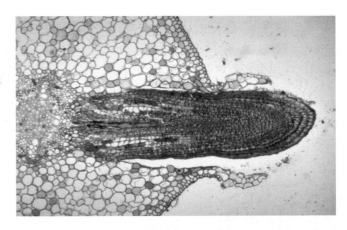

▲ **Figure 17-10 Branch roots** A branch root emerges from a primary root. The central axis of this branch root is already differentiating into vascular tissue.

Web Animation Primary and Secondary Growth

▼ **Figure 17-11 A young dicot stem** In cross-section, vascular tissue forms a ring of vascular bundles in dicots, as in the sunflower stem shown in the photomicrograph.

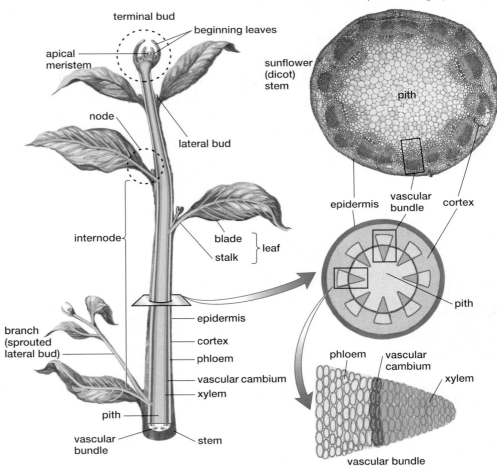

- **Storage:** Cells in both cortex and pith convert the sugar produced by photosynthesis into starch and store the starch as a food reserve.

- **Photosynthesis:** In many stems, the outer layers of cortex cells contain chloroplasts and carry out photosynthesis. In some desert plants, such as cacti, the leaves are reduced or absent, and the cortex of the stem is the only green photosynthetic part of the plant.

Vascular Tissues in Stems Transport Water, Dissolved Nutrients, and Hormones

The xylem and phloem of stems, like those of roots, transport water, nutrients, sugars, and hormones. These vascular tissues are continuous in root, stem, and leaf, interconnecting all parts of the plant. The xylem and phloem in young stems arise from the apical meristem and are called *primary xylem* and *primary phloem*.

Branches Form from Lateral Buds That Contain Meristem Cells

As the shoot grows, small clusters of meristem cells are left behind on the surface of the stem. Some of these meristem cells form structures that develop into leaves, and others form **lateral buds** that grow into branches (**Fig. 17-12**). Lateral buds are located just above the attachment points of the leaves. When stimulated by appropriate hormones, the meristem cells of a lateral bud are activated and the bud sprouts, growing into a branch. As the branch grows, it duplicates the development of the stem: It has an apical meristem at its tip and produces its own leaves and lateral buds as it grows.

▲ **Figure 17-12 Lateral buds** Lateral buds are located on the surface of a stem, just above the points where leaves are attached. A lateral bud's apical meristem generates an outward-growing branch, replicating the structure of the main stem.

Secondary Growth Produces Thicker, Stronger Stems

In dicot shrubs and trees, woody stems may survive for up to hundreds of years, becoming thicker and stronger each year. This secondary growth in stem thickness results from cell division in two lateral meristems: the vascular cambium and cork cambium.

Vascular Cambium Produces Secondary Xylem and Phloem

The **vascular cambium** is a cylinder of meristem cells located between the primary xylem and primary phloem. Daughter cells of the vascular cambium produced toward the inside of the stem differentiate into *secondary xylem*, and those produced toward the outside of the stem differentiate into *secondary phloem* (**Fig. 17-13**). Because the center of the stem is already filled with pith and primary xylem, newly formed secondary xylem pushes the vascular cambium and all outer tissues farther out, increasing the diameter of the stem.

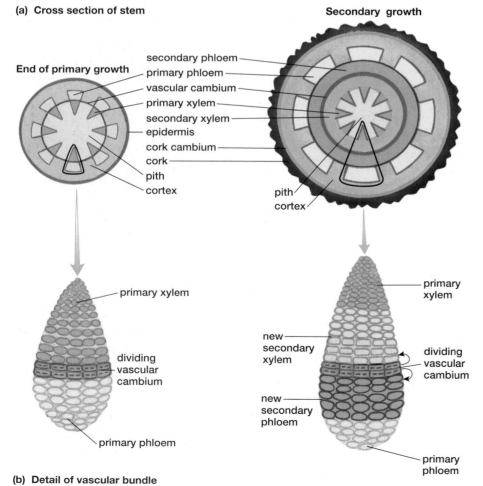

(a) Cross section of stem

Secondary growth

End of primary growth

- secondary phloem
- primary phloem
- vascular cambium
- primary xylem
- secondary xylem
- epidermis
- cork cambium
- cork
- pith
- cortex

pith
cortex

- primary xylem
- dividing vascular cambium
- primary phloem

- primary xylem
- new secondary xylem
- dividing vascular cambium
- new secondary phloem
- primary phloem

(b) Detail of vascular bundle

◄ **Figure 17-13 Secondary growth** *(a)* Cross-section of a dicot stem at the end of primary growth (left) and during early secondary growth (right). *(b)* Cells of the vascular bundle. When vascular cambium cells (green) divide during secondary growth, daughter cells formed on the inside of the vascular cambium differentiate into secondary xylem (blue). Cells formed on the outside of the cambium differentiate into secondary phloem (orange). **QUESTION:** If a tree is "girdled" by removing a strip of bark completely around its trunk, it will probably die. Why?

The secondary xylem, with its thick cell walls, forms the wood that makes up most of the trunk of a tree. Young xylem, called sapwood and located just inside the vascular cambium, transports water and minerals. Older xylem, called heartwood, contributes to the strength of the trunk but no longer carries water and solutes.

Phloem cells are much weaker than xylem cells. As they die over time, the phloem cells are crushed between the hard xylem on the inside of the trunk and the tough cork on the outside (described below). Only a thin strip of recently formed phloem remains alive and functioning.

Tree Rings Reflect Seasonal Changes

Tree rings form from seasonal changes in vascular cambium growth. In trees adapted to temperate climates, such as oaks and pines, cell division in the vascular cambium ceases during the cold of winter. In spring, the cambium cells divide, forming new xylem and phloem. The young cells grow by absorbing water and swelling while their cell walls are still soft. As the cells mature, their cell walls harden, preventing further growth. Because water is readily available in spring, xylem cells formed then swell considerably and are large when mature. As summer progresses, however, water becomes scarcer, so new xylem cells absorb less water and are therefore smaller when they mature.

As a result, cross-sections of tree trunks show a pattern of alternating pale rings (large cells formed in spring) and dark rings (small cells formed in summer). These alternating bands form the annual growth rings of temperate trees. You can determine a cut tree's approximate age by counting the dark regions.

Secondary Growth Causes the Epidermis to Be Replaced by Woody Cork

Recall that epidermal cells are mature, differentiated cells that can no longer divide. Therefore, as new secondary xylem and phloem enlarge the stem, the epidermis cannot expand. Instead, the epidermis splits and dies. Then some cells in the cortex form a new lateral meristem, the **cork cambium** (see Fig. 17-13). The cells of the cork cambium divide, forming daughter cells toward the outside of the stem. These daughter cells, called *cork cells*, or simply cork, develop tough, waterproof walls that protect the trunk from drying out and from physical damage. Cork cells die as they mature and may form a protective layer 2 feet thick in some tree species. As the trunk expands from year to year, the outermost layers of cork split apart or peel off, accommodating the growth.

The term "bark" refers to all tissues outside the vascular cambium: phloem, cork cambium, and cork. The complete removal of a strip of bark all the way around a tree, called girdling, will kill the tree because it severs the phloem. With the phloem gone, sugars synthesized in the leaves cannot reach the roots. Deprived of energy, the roots no longer take up water and nutrients, and the tree dies.

Some Specialized Stems Produce New Plants or Store Water or Food

Many plants have stems that are modified to perform functions very different from the original one of raising leaves up to the light. Strawberries, for example, grow stems that snake out over the soil, sprouting new strawberry plants where lateral buds touch the soil (Fig. 17-14a). Some plants, such as the baobab tree (Fig. 17-14b), store water in aboveground stems. Many other plants store carbohydrates in underground stems. For example, a potato is actually a storage stem. Each eye of a potato is a lateral bud, ready to send up a branch that will use the energy stored as starch in the potato to power its growth.

▶ Figure 17-14 **Some adaptations of stems** *(a)* The beach strawberry can reproduce via runners, horizontal stems that spread out over the surface of the sand. If a lateral bud of a runner touches the soil, it will sprout roots and develop into a complete plant. *(b)* The baobab tree develops an enormous water-storing trunk. The baobab grows in dry regions, so when it rains, it is to the tree's advantage to store all the water it can get.

Sunscreen for Maple Trees?

Continued

Plants of some species produce anthocyanins in their stems as well as in their leaves. Red, anthocyanin-containing stems are present mainly in plants that use stems for photosynthesis, as might be expected if anthocyanin serves to promote photosynthesis by protecting photosynthesizing tissue from the harmful effects of sunlight. In support of this explanation of red stems, researchers have found that stems get redder (produce more anthocyanin) when they grow in bright sunlight than when they grow in the shade, and that, when plants grow in bright sunlight, red stems photosynthesize more effectively than green stems.

(a) Strawberry plants

(b) A baobab tree

17.6 What Is the Structure of Leaves and What Do They Do?

Plants use photosynthesis to capture the energy of sunlight, and most plant photosynthesis takes place in **leaves.** To carry out photosynthesis, leaves require water, carbon dioxide, and sunlight. Sunlight is best gathered by leaves with a large surface area, and a porous leaf would best permit carbon dioxide (CO_2) to diffuse into the leaf from the air. But a large, porous leaf would allow too much water, which is transported from the soil to the leaf through the xylem, to evaporate, so a leaf must be reasonably waterproof as well. The leaves of flowering plants represent an elegant compromise among these conflicting demands.

Leaves Have Two Major Parts

A typical leaf of a flowering plant consists of a broad, flat blade, connected to the stem by a stalk. The stalk positions the blade in space, usually orienting the leaf for maximum exposure to the sun. Inside the stalk are xylem and phloem that link the vascular tissues in the stem to those in the leaf blade. Within the blade, the vascular tissues branch, forming the veins of the leaf.

The leaf epidermis consists of a layer of nonphotosynthetic, transparent cells that secrete a cuticle on their outer surfaces that reduces evaporation (**Fig. 17-15**). The epidermis and its cuticle are pierced by adjustable pores, the stomata, which regulate the diffusion of CO_2 and water into and out of the leaf. Each stoma is surrounded by two sausage-shaped **guard cells,** which adjust the size of the opening.

Beneath the epidermis lie the loosely packed cells of the **mesophyll.** These cells contain chloroplasts and perform most of the photosynthesis of the leaf. The openness of the leaf interior allows CO_2 to diffuse easily to all of the mesophyll cells. The branching vascular tissues are embedded within the mesophyll tissue, with fine veins reaching very close to each photosynthetic cell. Thus, each mesophyll cell receives energy from sunlight transmitted through the clear epidermis, carbon dioxide from air that enters through the stomata, and water from the xylem. The sugars produced in the mesophyll are carried away to the rest of the plant by the phloem.

Specialized Leaves May Provide Support, Store Food, or Even Capture Insects

In some plants, modified leaves serve functions unrelated to photosynthesis or water conservation. The common edible pea, for example, grasps fences, mailbox posts, or other plants with clinging tendrils. Unlike the tendrils of grapes, which are derived from stems, pea tendrils are slender, supple leaflets. Some plants, such as

Sunscreen for Maple Trees?

Continued

In the leaves of many plant species, the red pigment anthocyanin is found in both epidermis cells and in mesophyll cells. Anthocyanin's presence in the epidermis cells on a leaf's outer surface is consistent with the pigment's hypothesized function as a protective sunscreen. But a sunscreen function would not be well served by anthocyanin molecules in the mesophyll cells in a leaf's interior, so the pigment's presence there suggests that it may also have other functions. "Sunscreen for Maple Trees? *Revisited*" at the end of this chapter discusses one such function.

▶ **Figure 17-15 A typical dicot leaf** *(a)* The cells of a leaf's epidermis lack chloroplasts and are transparent, allowing sunlight to penetrate to the chloroplast-containing mesophyll cells beneath. The stomata that pierce the epidermis and the loose, open arrangement of the mesophyll cells ensure that CO_2 can diffuse into the leaf from the air and reach all of the photosynthetic cells. *(b)* A micrograph of a section of a leaf.

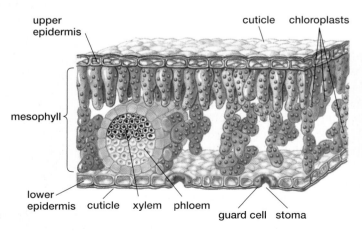

(a) Leaf structure

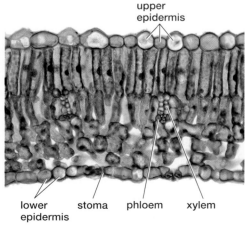

(b) Leaf cross section

onions, daffodils, and tulips, use thick, fleshy leaves as food-storage organs; the bulbs of these plants are composed of modified leaves. A few plants have turned the table on the animals and have become predators. For example, sundews (Fig. 17-16) have leaves that are modified into snares for trapping unwary insects.

17.7 How Do Plants Acquire Nutrients?

Nutrients are elements essential to normal life. Plants require relatively large quantities of the following nutrients: carbon (obtained from CO_2), hydrogen (from water), oxygen (from air and water), phosphorus (from phosphate ions in soil), nitrogen (from nitrate and ammonium ions in soil), and magnesium, calcium, and potassium (from ions in soil). Plants also require very small quantities of such nutrients as iron, chlorine, copper, manganese, zinc, boron, and molybdenum.

Carbon dioxide and oxygen usually enter a plant by diffusion from the air into leaves, stems, and roots. Water and the remaining nutrients, collectively called **minerals,** are extracted from the soil by roots.

Roots Take Up Minerals Dissolved in Water

Soil consists of bits of pulverized rock, air, water, and organic matter. Although both the rock particles and the organic matter contain essential minerals, only minerals dissolved in water are accessible to roots. The concentration of minerals in the soil water is very low, usually much lower than the concentration in plant cells and fluids. For example, the concentration of potassium ions (K^+) in soil water is at least 10 times lower than that in root cells, so diffusion cannot move potassium into a root. In general, minerals are moved into a root against their concentration gradients by active transport, using energy from sugar transported from the leaves to the roots. (Recall from Chapter 3 that the movement of molecules from areas of low concentration to areas of high concentration requires energy.)

Fungi and Bacteria Help Plants Acquire Nutrients

Many minerals are not abundant enough in soil water to support plant growth, even though plenty of minerals may be bound up in the surrounding rock particles. One nutrient—nitrogen—is almost always in short supply both in rock particles and in soil water. Most plants have evolved beneficial relationships with other organisms that help the plants acquire scarce nutrients such as nitrogen, phosphorus, and minerals. Examples of such relationships include root–fungus associations, called mycorrhizae, and root–bacteria associations formed in nodules of legumes.

Fungal Mycorrhizae Help Plants Acquire Minerals

Most land plants form symbiotic relationships with fungi to form root–fungus complexes called **mycorrhizae,** which help the plant extract and absorb nutrients. Fungal strands intertwine between the root cells and extend out into the soil (Fig. 17-17). In some way that is not yet understood, the fungus renders nutrients, such as phosphates and certain minerals, accessible for uptake by the roots, perhaps by converting rock-bound minerals into simple soluble compounds that root plasma membranes can transport. The fungus, in return, receives sugars, amino acids, and vitamins from the plant. In some cases, this mutually beneficial relationship allows both the fungus and the plant to grow in places where neither could survive alone, including deserts and rocky soils that are low in nutrients.

Bacteria-Filled Nodules on Their Roots Help Legumes Acquire Nitrogen

Amino acids, nucleic acids, and chlorophyll all contain nitrogen, so plants most acquire large amounts of nitrogen to synthesize these crucial molecules. Although nitrogen gas (N_2) makes up about 79% of the atmosphere and diffuses from the atmosphere into the air spaces in soil, plants cannot use it. Plants can take up nitrogen only in the form of ammonium ions (NH_4^+) or nitrate ions

▲ Figure 17-16 **A specialized leaf** A lacewing is trapped in the sticky, enzyme-laden hairs of a sundew leaf. Nutrients from the insect's body help sundews to thrive in nitrogen-poor soils.

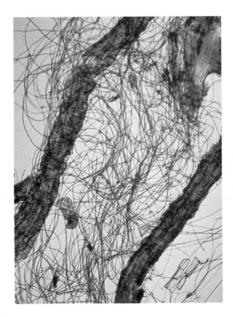

▲ Figure 17-17 **Mycorrhizae, a widespread root–fungus symbiosis** A tangled meshwork of fungal strands surrounds and penetrates into the roots of most plants.

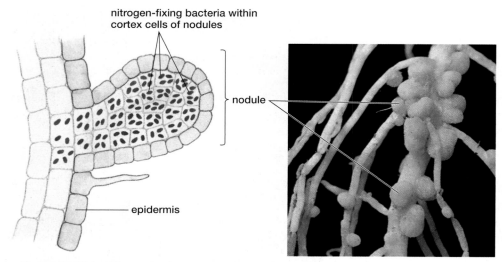

nitrogen-fixing bacteria within
cortex cells of nodules

nodule

epidermis

▲ Figure 17-18 Nitrogen fixation in legumes A diagram and photograph of root nodules, which contain nitrogen-fixing bacteria.

BioFlix   **Water Transport in Plants**

Web Animation Plant Transport Mechanisms

(NO$_3^-$), but they don't have the enzymes needed to carry out **nitrogen fixation,** the conversion of N$_2$ gas into ammonium or nitrate. Various **nitrogen-fixing bacteria,** however, do have these enzymes. Some of these bacteria are free living in the soil, but they do not routinely manufacture a lot of extra ammonium and liberate it into the soil.

Some plants, particularly legumes (peas, clover, and soybeans), enter into a mutually beneficial relationship with certain species of nitrogen-fixing bacteria. By secreting chemicals into the soil, legumes attract nitrogen-fixing bacteria to their roots. Once there, the bacteria enter the root hairs. The bacteria then digest channels through the cytoplasm of the epidermal cells and into underlying cortex cells. As both bacteria and their host cortex cells multiply, a swelling, or nodule, forms and houses the root–bacteria complex (Fig. 17-18).

Within the nodule, a cooperative relationship develops. The plant transports sugars from its leaves down to the root cortex for storage, just as it normally would. The bacteria in the cortex use some of the sugar's energy to power their metabolic processes, including nitrogen fixation. The bacteria obtain so much energy that they produce more ammonium than they need. The surplus ammonium diffuses into the cytoplasm of their host cells, providing the plant with a steady supply of usable nitrogen. Surplus ammonium also diffuses into the surrounding soil, where the ammonium becomes available to other types of plants. Farmers plant legumes not only for their commercial value but also to enrich the soil with ammonium for future crops.

17.8 How Do Plants Acquire Water and Transport Water and Nutrients?

Nearly 99% of the water absorbed by the roots of plants evaporates through the stomata of leaves and, to a lesser extent, stems in a process called **transpiration.** Transpiration drives the movement of water through the plant body by pulling water up through the xylem of the roots and stem and into the leaves. (Transpiration also has important effects on ecosystems; see "Earth Watch: Plants Help Regulate the Distribution of Water" on p. 332.)

Transpiration Provides the Force for Water Movement in Xylem

After entering the root xylem, water and dissolved minerals still must be moved to the uppermost reaches of the plant. (In redwood trees, the distance may be more than 300 feet.) How do plants overcome the force of gravity and make water flow upward? Transpiration provides the necessary force.

As a leaf transpires, the concentration of water in its mesophyll tissue drops. This drop causes water to move by osmosis from the leaf's xylem into the mesophyll cells. The water molecules leaving the xylem are attached by hydrogen bonds to other water molecules in the same xylem tube. Therefore, when one water molecule leaves, it pulls adjacent water molecules up the xylem. As these water molecules move upward, other water molecules farther down the tube are pulled up to replace them. This process continues all the way to the roots, where water is pulled into the xylem (Fig. 17-19).

This transpiration-driven upward movement of water is made possible by the strength with which the water in the xylem holds together. Just as individually weak cotton threads woven together make the strong fabric of your jeans, the network of individually weak hydrogen bonds in water collectively produces a very high cohesion, or tendency to resist being separated. The column of water within the xylem is at least as strong—and as unbreakable—as a steel wire of the same diameter. Supplementing the cohesion between water molecules is the attraction of water molecules to the cell walls of the thin xylem tubes. This force helps the water column resist the downward pull of gravity.

Water Enters Roots Mainly by Pressure Differences Created by Transpiration

The upward movement of water ultimately causes soil water to move into the roots. Water from the soil moves first into the epidermis and outer cortex layers of the root, traveling between the cells along a pathway created by their porous cell walls. From there, the water moves across the band of endodermal cells, entering them by osmosis and continuing onward into the vascular cylinder at the center of the root.

The main force powering the flow of water from the soil to the root's vascular cylinder is low water pressure in the cylinder. Pressure is reduced in the vascular cylinder because the water there is drawn continually away, pulled upward through the xylem to replace water lost through transpiration from the leaves. Thus, the force generated by evaporation from the leaves is transmitted down the xylem to the roots. This force is strong enough that roots can absorb water even from quite dry soils.

Adjustable Stomata Control the Rate of Transpiration

Although transpiration provides the force that transports water and minerals to the leaves at the top of the plant, it is also the plant's largest source of water loss—a loss that may threaten the very survival of the plant, especially in hot, dry weather. Most water is transpired through the stomata of the leaves and stem, so you might think that a plant could prevent water loss simply by closing its stomata. Don't forget, however, that photosynthesis requires carbon dioxide from the air, which diffuses into the leaf primarily through open stomata. A plant that simply closed its stomata would be unable to photosynthesize. Instead, a plant must balance carbon dioxide uptake and water loss by opening and closing its stomata. In general, stomata open during the day, when sunlight allows photosynthesis, and close at night, conserving water. They will also close in the sunlight if the plant is losing too much water.

The Pressure-Flow Theory Explains Sugar Movement in Phloem

Sugars manufactured by photosynthesis in leaves must be moved to other parts of the plant. The sugars, dissolved in water to form sap that also contains amino acids and hormones, travel through phloem. After reaching their destination, the sugars may nourish nonphotosynthetic structures such as roots or flowers, or may be stored in roots or stems for later use.

What drives the movement of the sap that carries sugars through the phloem? The most widely accepted explanation for fluid transport the phloem of flowering plants is the pressure-flow theory, which states that sap is propelled

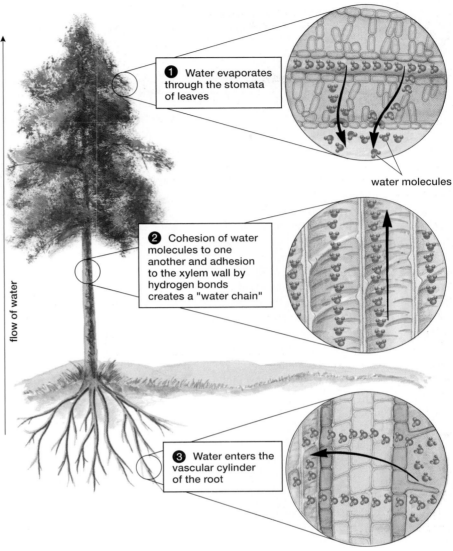

① Water evaporates through the stomata of leaves

water molecules

② Cohesion of water molecules to one another and adhesion to the xylem wall by hydrogen bonds creates a "water chain"

③ Water enters the vascular cylinder of the root

flow of water

▲ **Figure 17-19 Transpiration powers the movement of water**

earth watch

Plants Help Regulate the Distribution of Water

Water is probably the most important environmental factor influencing the distribution of plants: Cacti inhabit deserts because they can withstand drought; orchids and mahogany trees need the frequent drenching rains of the rain forest. However, people often overlook the flip side of the relationship between plants and water: Plants actually help control the amount and distribution of rainfall, soil water, and even river flow. How? Through transpiration.

Consider the Amazon rain forest (Fig. E17-1). There, an acre of soil supports hundreds of towering trees, each bearing millions of leaves. The surface area of the leaves dwarfs the surface area of the soil, so up to 75% of all water evaporating from the acre of forest is due to leaf transpiration. This transpiration raises the humidity of the air and causes rain to fall. In fact, about half of the

water transpired from the leaves falls again as rain. As a result, as much as one-half or more of the total rainfall in the rain forest consists of water recycled by transpiration. Thus, the high humidity and frequent showers that the rain forest needs to survive are partly created by the forest itself.

If the rain forest in an area is cut down, less water evaporates in that area, so less rain falls, and new rain forest tree seedlings cannot grow. An entirely different plant community becomes established on the disturbed land. Even nearby areas of undisturbed forest can be harmed by the overall decline in rainfall and humidity.

As the example of the Amazon rain forest suggests, plants have an enormous influence on nonliving aspects of the biosphere, including humidity, rainfall, soil water, and river flow. Human activities such as deforestation that disturb or

▲ Figure E17-1 The Amazon rain forest The rain forest community helps mold its own environment.

destroy plant communities can have severe, often unpredictable impacts on water distribution in ecosystems. These impacts can in turn affect climate, weather, and water quality and availability.

▼ Figure 17-20 Pressure differences power the movement of sugars A photosynthesizing leaf is a sugar source; a developing fruit is a sugar sink. Water pressure drives sap from sources to sinks. (The numbers in the figure correspond to explanations in the text).

through phloem sieve tubes by differences in water pressure. These water pressure differences are created indirectly by differences in the production and use of sugar in different parts of the plant. Any portion of the plant that produces more sugar than it uses is called a sugar *source*; a mature leaf is a good example. Conversely, any structure that uses up more sugar than it produces is a sugar *sink*. Developing fruits are good examples of sugar sinks. Phloem sieve tubes carry sap away from sugar sources (which have excess sugar) and toward sugar sinks (where sugar is required).

The pressure-flow theory is illustrated in Figure 17-20: ❶ The sugar produced by a source cell is moved by active transport into a phloem sieve tube. As a result, the sugar concentration of the sap increases in that portion of the sieve tube. ❷ Water (from xylem) follows the sugar into the sieve tube by osmosis. The entering water increases the pressure inside the sieve tube, because the rigid walls of the tube prevent it from expanding. ❸ The increased water pressure forces the sugar-rich sap through the phloem sieve tubes into regions of lower pressure. ❹ Cells of a sugar sink actively transport sugar out of the phloem. Water follows by osmosis, creating a region of lower water pressure in this portion of the sieve tube. The sap thus moves from a source region where water pressure is high to a sink region where water pressure is lower, carrying the sugar with it. Think of (and thank) the pressure-flow process the next time you are enjoying a strawberry, or whatever fruit is your favorite sugar sink.

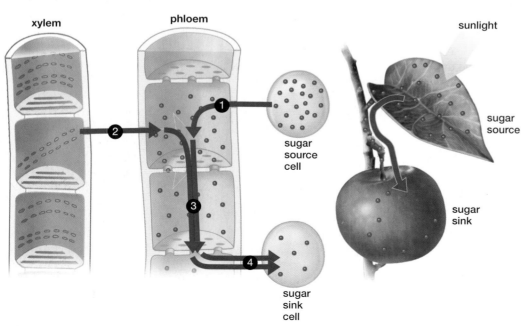

The protective effect of the bright red anthocyanin pigments in some autumn leaves may not be limited to the pigments' ability to absorb harmful wavelengths of sunlight. Anthocyanins may also absorb free radicals—highly reactive molecules that can damage cells. Scientists have long known that, in test tubes, anthocyanin molecules act as antioxidants, reacting with free radicals and rendering them harmless. Now, evidence is growing that the antioxidant properties of anthocyanins work in living leaves as well.

To test the hypothesis that anthocyanin protects leaves from free radicals, one group of researchers developed a technique to produce free radicals in chloroplasts and to measure how long the radicals persisted. The technique was applied to a tropical shrub with two-toned, red and green leaves in which the red parts of the leaves contained anthocyanins and the green parts did not (see photo on p. 319). In the green portions of leaves, free radicals increased for 10 minutes and then began a slow decline. In contrast, free radicals in the red parts of leaves subsided to low levels in less than 5 minutes.

The bright red color of autumn leaves apparently acts as both sunscreen and antioxidant, preserving the leaves' ability to photosynthesize. But why does a plant need photosynthesis in a dying leaf? One possibility is increased ability to recover vital nutrients from the leaf. Perennial plants salvage scarce nutrients from dying leaves and pump them into woody tissues for storage over the winter. This process requires energy, which the plant derives from photosynthesis. Perhaps, by protecting leaves during their final days, anthocyanin allows the plant to continue photosynthesis long enough to gain the energy it needs to salvage nutrients for use in the coming spring.

Consider This

Plants of many species do not produce anthocyanin in autumn leaves. Thus, the leaves of birches, aspens, ash, and other tree species do not turn red in the fall, but instead turn yellow. This observation raises a question: If anthocyanins provide valuable protection, why don't all plants have them? One hypothesis is that the high concentrations of yellow carotenoid pigments present in yellow autumn leaves provide protection similar to that provided by anthocyanin. Another hypothesis is that trees of some species do not need the protection that anthocyanin provides. Can you design an experiment to test these hypotheses? Can you think of other hypotheses to explain why some autumn leaves turn red, but others do not?

Chapter Review

Summary of Key Concepts

For additional study help and activities, go to www.mybiology.com.

17.1 How Are Plant Bodies Organized?

The body of a plant consists of root and shoot. The underground roots anchor the plant in the soil, absorb and transport water and minerals from the soil, and store surplus photosynthetic products. The aboveground shoot consists of stem, leaves, buds, flowers, and fruit. Shoot functions include photosynthesis, transport of materials, and reproduction. Flowering plants include the monocots and dicots.

Web Animation   Plant Anatomy

17.2 How Do Plants Grow?

Plant bodies are composed of meristem cells and differentiated cells. Meristem cells retain the capacity for mitotic cell division. Differentiated cells arise from divisions of meristem cells, become specialized for particular functions, and usually do not divide. Meristem cells are located in apical meristems at the tips of roots and shoots and in lateral meristems in stems and roots. Primary growth (growth in length and differentiation of parts) results from the division and differentiation of apical meristem cells. Secondary growth (growth in diameter) results from the division and differentiation of lateral meristem cells.

17.3 What Are the Tissues and Cell Types of Plants?

Plant bodies consist of dermal, ground, and vascular tissues. Dermal tissue forms the outer covering of the plant body. The dermal tissue of leaves and of primary roots and stems is a single cell layer of epidermis. Dermal tissue after secondary growth is a covering of cork.

Ground tissue consists of cell types that are involved in photosynthesis, support, or storage. Ground tissue makes up most of a young plant during primary growth. During secondary growth of stems and roots, ground tissue becomes an increasingly small part of the plant body.

Vascular tissue consists of xylem, which transports water and nutrients from the roots to the shoots, and phloem, which transports water, sugars, amino acids, and hormones throughout the plant body.

17.4 How Do Roots Grow and What Do They Do?

Primary growth in roots produces an outer epidermis, an inner vascular cylinder of conducting tissues, and cortex between the two. The apical meristem near the tip of the root is protected by the root cap. Cells of the root epidermis absorb water and nutrients from the soil. Root hairs are extensions of epidermal cells that increase the surface area for absorption. Most cortex cells store surplus sugars (usually in the form of starch) produced by photosynthesis. The innermost layer of cortex cells is the endodermis, which controls the movement of water and nutrients from the soil into the vascular cylinder. The vascular cylinder contains the conducting tissues: xylem and phloem.

17.5 How Do Stems Grow and What Do They Do?

Primary growth in dicot stems results in a structure consisting of an outer, waterproof epidermis; supporting and photosynthetic cells of cortex beneath the epidermis; vascular tissues of xylem and phloem; and supporting and storage cells of pith at the center. Leaves and lateral buds grow at intervals along the surface of the stem. Under appropriate conditions, lateral buds may sprout into a branch.

Secondary growth in stems results from cell divisions in the vascular cambium and cork cambium. Vascular cambium produces secondary xylem and secondary phloem, increasing the stem's diameter. Cork cambium produces waterproof cork cells that cover the outside of the stem.

Web Animation   Primary and Secondary Growth

17.6 What Is the Structure of Leaves and What Do They Do?

Leaves are the main photosynthetic organs of plants. The blade of a leaf consists of a waterproof outer epidermis surrounding mesophyll cells, which have chloroplasts and which carry out photosynthesis, and the xylem and phloem, which carry water, minerals, and photosynthetic products to and from the leaf. The epidermis is perforated by adjustable stomata that regulate the exchange of gases and water.

17.7 How Do Plants Acquire Nutrients?

Most minerals are taken up from the soil water by active transport into the root hairs. These minerals diffuse into the root and then into the xylem.

Most plants have fungal mycorrhizae associated with their roots that help absorb soil nutrients. Plants can absorb nitrogen only as ammonium or nitrate, which are scarce in most soils. Legumes have evolved a cooperative relationship with nitrogen-fixing bacteria that invade legume roots. The plant provides the bacteria with sugars, and the bacteria use some of the energy in those sugars to convert atmospheric nitrogen to ammonium, some of which the plant absorbs.

17.8 How Do Plants Acquire Water and Transport Water and Nutrients?

Water and dissolved nutrients from the soil are transported in xylem. The water within xylem tubes holds together almost as if it were a solid chain, because water molecules are attracted to one another by hydrogen bonds. As water molecules evaporate from the leaves during transpiration, the hydrogen bonds pull other water molecules up the xylem to replace them. This movement continues down the xylem to the root, where water loss from the vascular cylinder promotes water movement across the endodermis from the soil water by osmosis. In the root, water moves by osmosis across the plasma membranes of the endodermal cells into the vascular cylinder. The water pressure gradient caused by loss of water through transpiration is the primary force drawing water into the root.

Sugars manufactured by photosynthesis are dissolved in water and the resulting sap is transported in phloem. The sap in phloem near sugar sources is more concentrated than is sap in phloem near sugar sinks. Movement of water by osmosis into phloem with concentrated sap results in higher water pressure that forces the sap toward parts of the plant where sap concentration, and therefore water pressure, is lower.

BioFlix   Water Transport in Plants
Web Animation   Plant Transport Mechanisms

Key Terms

apical meristem *p. 320*	fibrous roots *p. 323*	mycorrhizae *p. 329*	stem *p. 319*
cambium (plural, cambia) *p. 320*	ground tissue *p. 321*	nitrogen fixation *p. 330*	stomata *p. 325*
cork cambium *p. 327*	guard cell *p. 328*	nitrogen-fixing bacterium *p. 330*	taproot *p. 323*
cortex *p. 324*	lateral bud *p. 326*	phloem *p. 323*	terminal bud *p. 325*
cuticle *p. 321*	lateral meristem *p. 320*	primary growth *p. 320*	transpiration *p. 330*
dermal tissue *p. 321*	leaf *p. 328*	root *p. 319*	vascular cambium *p. 326*
dicot *p. 320*	meristem cell *p. 320*	root cap *p. 323*	vascular cylinder *p. 325*
differentiated cell *p. 320*	mesophyll *p. 328*	root hair *p. 324*	vascular tissue *p. 321*
endodermis *p. 324*	mineral *p. 329*	secondary growth *p. 320*	xylem *p. 322*
epidermis *p. 321*	monocot *p. 320*		

Thinking Through the Concepts

Suggested answers to end-of-chapter and figure-based questions can be found at the end of the text.

Fill-in-the-Blank

1. A mutually beneficial relationship between soil fungi and plant roots is called _____. In this relationship, the _____ provide mineral nutrients to the _____. Another beneficial relationship occurs between roots and nitrogen-fixing _____. These microorganisms inhabit structures on the root called _____, and produce surplus _____, which is absorbed by the roots.

2. Water travels upward through plant roots, stems, and leaves within hollow tubes called _____, which are found in the _____ bundles. Water molecules are interconnected by forces called _____, which allow a chain of molecules to be pulled up through the plant. The pull is exerted by evaporation from leaves, a process called _____.

3. Plants require two gases, _____ and _____, which are obtained from air by the process of _____. Plants also require a variety of nutrients collectively called _____. Most of these nutrients are moved into root cells by the process of _____.

4. Monocot plants have a _____ root system, the veins in their leaves are _____, their flower parts come in multiples of _____, and they have one _____, a trait that gives them their name.

5. Dicot plant roots form a _____ system, the veins in their leaves are _____, their flower parts come in multiples of _____, and they have _____ cotyledons, a trait that gives them their name.

6. In the Amazon rain forest, about (what fraction) _____ of the rain that falls has recently transpired through plant leaves? Transpiration occurs through openings in the leaves called _____. As a result of transpiration, rain forest vegetation has a major impact on both rainfall and _____.

Review Questions

1. Describe the locations and functions of the three main types of tissues systems in plants.

2. Distinguish between primary growth and secondary growth, and describe the cell types involved in each.

3. Distinguish between meristem cells and differentiated cells. Which meristems cause primary growth? Which ones cause secondary growth? Where is each type located?

4. Diagram the internal structure of a root after primary growth, labeling and describing the function of epidermis, cortex, endodermis, xylem, and phloem.

5. How do xylem and phloem differ?

6. What are the main functions of roots, stems, and leaves?

7. What types of cells form root hairs? What is the function of root hairs?

8. Diagram the internal structure of leaves. What structures regulate water loss and gas exchange by a leaf?

Applying the Concepts

1. IS THIS SCIENCE? A Web site reports that "It is a known fact that music plays an important role in plant growth. But plants are choosy about the kind of music they hear. Plants thrive if soothing instrumental music is played in the background, but they shrivel and die if exposed to heavy metal or rock music." State the testable hypotheses presented in this claim. On the basis of your knowledge of plant growth, are the hypotheses likely to be correct? For each hypothesis, make a list of predictions and describe a controlled experiment to test one of them.

2. The desert tends to have two categories of plants: small grasses or herbs, and shrubs or small trees. The grasses and herbs typically have fibrous roots. The shrubs and trees have taproots. What advantages can you think of for each type of root? How does each type of root allow for survival in a desert environment?

3. Grasses (monocots) form their primary meristem near the ground surface rather than at the tips of branches the way dicots do. How does this feature allow you to grow a lawn and mow it every week in the summer? What would happen if you had a dicot lawn and tried to mow it?

For additional resources, go to www.mybiology.com.

The Plant Life Cycle

The indescribably delicious taste of a fresh, ripe tomato is a sensation rarely experienced by people who buy their produce in a supermarket. New research on plant hormones may help change that sad situation.

Case Study · Fountain of Youth for Fruit

People who enjoy the beauty of plant reproductive structures stand to benefit from new experimental varieties that last for weeks in a vase.

A great new product is coming soon to a supermarket near you—a tomato that never ripens. Why would anyone want a nonripening tomato? To help get better-tasting tomatoes into your kitchen, believe it or not. Today, tomatoes are generally picked green and hard to avoid damage during shipping. It's difficult to control how these green tomatoes ripen, so a tomato's condition on arrival at your supermarket is unpredictable. It might be too green and hard or too soft and squishy, and it certainly won't taste as good as it would have if it had ripened normally on the vine.

Our liberation from troublesome supermarket tomatoes may come from tomato plants with a mutation known as *never-ripe*. These plants bear fruit that ripens extremely slowly. Although consumers obviously have no use for a permanently unripe tomato, researchers hope that, once they fully understand how mutations such as *never-ripe* affect ripening, it will be possible to exert fine control over the ripening process. Tomatoes could remain on the vine until they were red and tasty, and then ripening could be halted so that no further softening occurred during harvest and shipping. At home, you could leave your perfectly ripe tomatoes on the windowsill for weeks without fear of spoilage. Similarly, flowers cut from plants with this kind of mutation might last for weeks in a vase.

Although such sophisticated control of ripening is not yet possible, it may one day soon become a routine procedure, because scientists have made significant progress in understanding how *never-ripe* and similar mutations work. We will describe some of their findings later in this chapter, after some exploration of the plant life cycle. ■

18.1 What Is the Life Cycle of Plants?

Living things reproduce, and plants are no exception. Plant reproduction affects us on a daily basis, when we consume structures produced by reproducing plants, such as seeds and fruits. In addition to its positive impact on human nutrition (and human taste buds), plant reproduction has some distinctive features, as you will discover in this chapter's description of the plant life cycle.

Asexual Reproduction Requires Only One Parent

Many plants can reproduce either sexually, in which male and female gametes combine to give rise to a new plant, or asexually. In asexual reproduction, a new plant grows from part of an existing plant. If you have houseplants, you may have observed asexual reproduction when you took a cutting from a stem of a favorite plant, placed it in a pot of soil, and waited for a new plant to grow from the cutting. Asexual reproduction is also common in nature. For example, a grass plant may produce underground stems that grow horizontally away from the plant. Then, at a point some distance from the orignal plant, these stems grow roots that extend down into the soil and a shoot that extends upward and sprouts leaves. Eventually, the connecting underground stem may die, leaving two separate plants.

In asexual reproduction, the offspring plant is produced by mitotic division of the parent plant's cells, so the offspring is genetically identical to the parent. Asexual reproduction is more rapid and efficient than sexual reproduction, and can allow plants to spread quickly when circumstances permit. Nonetheless, most plants reproduce sexually at least some of the time.

Two Types of Plant Bodies Alternate in the Sexual Life Cycle

Sexual reproduction in plants involves two distinct types of plant bodies. As plant reproduction proceeds, the two body types alternate; this kind of life cycle is known as **alternation of generations** (Fig. 18-1). In alternation of generations, each body type produces reproductive cells that give rise to the other body type.

The alternating body types are called the **sporophyte** and the **gametophyte.** The sporophyte body is composed of diploid cells and produces **spores** (cells that can develop into adult organisms); the gametophyte body consists of haploid cells and produces gametes. All plants have both a sporophyte phase and a gametophyte phase, but the relative size and conspicuousness of the two phases differ among different kinds of plants. For example, in mosses the gametophyte phase is larger and more conspicuous than the sporophyte phase, but in ferns the reverse is true. In seed plants, the gametophyte phase is microscopic and lives surrounded by the tissues of the sporophyte.

▼ **Figure 18-1 The life cycle of a fern—a nonflowering plant**
Ferns typify the alternation-of-generations life cycle found in all plants. One generation is the haploid gametophyte (blue arrows). The other generation is the diploid sporophyte (yellow arrows). The photo at top left shows a portion of the underside of a sporophyte leaf, the site of reproductive structures that release spores to the environment.

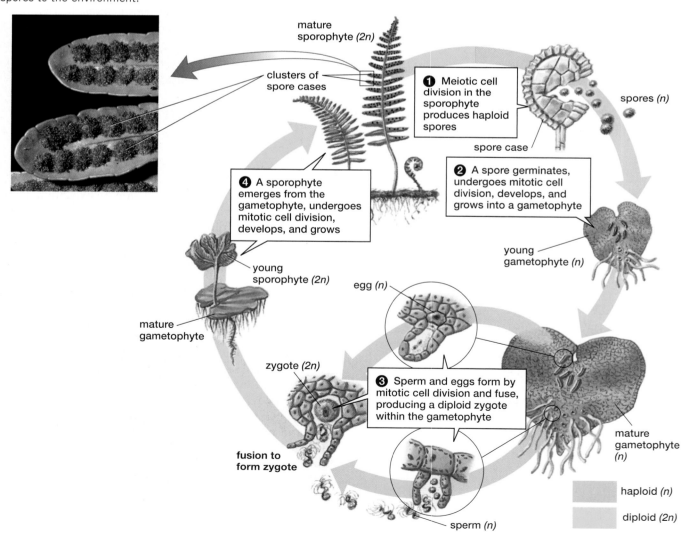

mature sporophyte (2n)

clusters of spore cases

❶ Meiotic cell division in the sporophyte produces haploid spores

spores (n)

spore case

❹ A sporophyte emerges from the gametophyte, undergoes mitotic cell division, develops, and grows

❷ A spore germinates, undergoes mitotic cell division, develops, and grows into a gametophyte

young sporophyte (2n)

young gametophyte (n)

mature gametophyte

egg (n)

zygote (2n)

❸ Sperm and eggs form by mitotic cell division and fuse, producing a diploid zygote within the gametophyte

mature gametophyte (n)

fusion to form zygote

sperm (n)

haploid (n)

diploid (2n)

In Ferns, the Sporophyte and Gametophyte Are Separate, Independent Plants

To better picture how alternation of generations works, consider the life cycle of a fern. In ferns, the leafy plant body that grows up from the forest floor is the plant's diploid sporophyte phase. To reproduce, the fern sporophyte bears reproductive cells that undergo meiosis to produce haploid spores (**Fig. 18-1**, step ❶).

The almost microscopic fern spore is carried away from the sporophyte leaf, drifting on currents of air or water. If it lands on soil, it may begin to grow and develop, dividing repeatedly by mitosis to form a small, multicellular, haploid plant—the fern gametophyte (**Fig. 18-1**, step ❷). Fern gametophytes are heart shaped and usually measure less than an inch across; most casual observers would not recognize them as ferns.

The gametophyte fern plant produces gametes—both sperm and eggs (**Fig. 18-1**, step ❸). Because all of the gametophyte's cells are haploid, these gametes are produced without further meiosis. To reach an egg, sperm either swim through a thin film of water or are splashed by raindrops from one plant to the next. Sperm and egg fuse to form a zygote that develops into a new diploid sporophyte plant, beginning anew the cycle of alternation of generations (**Fig. 18-1**, step ❹).

Alternation of Generations Is Less Obvious in Flowering Plants

In flowering plants, gametophytes are microscopic and do not live independently of the sporophyte. Instead, the gametophytes develop within **flowers**, reproductive structures that are part of the sporophyte body. The gametophytes can be female or male; female gametophytes produce eggs and male gametophytes produce sperm.

The female and male gametophytes grow from two types of spores that are formed by meiosis within the flowers of the sporophyte generation (**Fig. 18-2**, steps ❶ and ❷). One type of spore, the **megaspore**, arises from a precursor cell called the *megaspore mother cell*. The megaspore undergoes a few mitotic divisions and develops into the female gametophyte, a small cluster of cells called an **embryo sac** (**Fig. 18-2**; steps ❸ and ❹). The embryo sac remains permanently within the flower and includes an egg cell.

The other type of spore, the **microspore**, arises from a precursor cell called the *microspore mother cell*. The microspore develops into the male gametophyte: a tough, watertight **pollen grain** that enables flowering plants to reproduce in dry environments (see Fig. 18-2, steps ❸ and ❹). The pollen grain, which contains the sperm, drifts on the wind or is carried by an animal from one

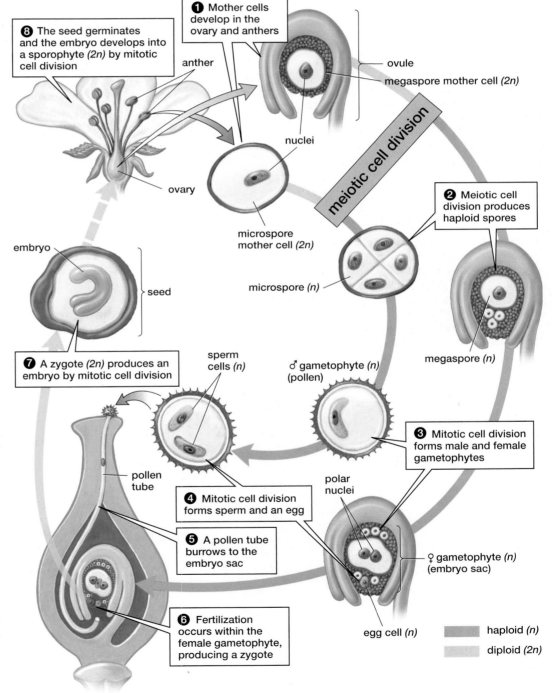

▼ **Figure 18-2 The life cycle of a flowering plant** Although this cycle shows the same basic stages as the life cycle of a fern (see Fig. 18-1), the haploid stages (blue arrows) are much smaller and cannot live independently of the diploid sporophyte (yellow arrows).

❽ The seed germinates and the embryo develops into a sporophyte *(2n)* by mitotic cell division

anther

ovule
megaspore mother cell *(2n)*

❶ Mother cells develop in the ovary and anthers

nuclei

ovary

meiotic cell division

microspore mother cell *(2n)*

❷ Meiotic cell division produces haploid spores

microspore *(n)*

megaspore *(n)*

embryo

seed

♂ gametophyte *(n)* (pollen)

❸ Mitotic cell division forms male and female gametophytes

❼ A zygote *(2n)* produces an embryo by mitotic cell division

sperm cells *(n)*

polar nuclei

pollen tube

❹ Mitotic cell division forms sperm and an egg

❺ A pollen tube burrows to the embryo sac

♀ gametophyte *(n)* (embryo sac)

❻ Fertilization occurs within the female gametophyte, producing a zygote

egg cell *(n)*

haploid *(n)*

diploid *(2n)*

flower to another. On the recipient flower, the pollen grain elongates, burrowing through the flower's tissues to the female gametophyte within (**Fig. 18-2**, step **5**). This miniature male gametophyte then liberates its sperm inside the female gametophyte, where fertilization takes place (**Fig. 18-2**, step **6**). The resulting zygote develops into an embryonic plant that becomes enclosed in a drought-resistant seed, which also includes a food reserve and a protective outer coating (**Fig. 18-2**, step **7**). The embryo in the seed develops into a new sporophyte (**Fig. 18-2**, step **8**).

We describe flowering plant gametophytes and reproduction in greater detail in section 18.3.

Web Animation Reproduction in Flowering Plants

18.2 What Is the Structure of Flowers?

The flower is a sexual display that enhances a plant's reproductive success. By enticing animals to transfer pollen from one plant to another, flowers enable stationary plants to "court" distant members of their own species. This critical advantage has allowed the flowering plants to become the dominant plants on land.

Flowers Evolved from Leaves

Evolution produces new structures by modifying old ones, and flower parts are actually highly modified leaves, shaped by natural selection into structures that foster the transfer of pollen from one plant to another. **Complete flowers,** such as those of petunias, roses, and lilies, consist of a central axis on which four sets of modified leaves are attached: sepals, petals, stamens, and carpels (**Fig. 18-3**). The **sepals** are located at the base of the flower. In dicots, the sepals are typically green and leaflike, but in monocots, most sepals resemble the petals (see Fig. 18-3b). In either case, sepals enclose and protect the flower bud as the other three structures develop. Just above the sepals are the **petals,** which are usually brightly colored and fragrant, advertising the location of the flower.

▼ **Figure 18-3 A complete flower** *(a)* A complete flower has four parts: sepals, petals, stamens (the male reproductive structures), and at least one carpel (the female reproductive structure). This drawing shows a complete dicot flower. *(b)* The amaryllis is a complete monocot flower, with three sepals (virtually identical to the petals), three petals, six stamens (composed of filaments and anthers), and three fused carpels (composed of stigma and style).

Flowers Incorporate Male and Female Reproductive Structures

The male reproductive structures, the **stamens,** are attached just above the petals. Most stamens consist of a long and slender *filament* that supports an **anther,** the structure that produces pollen. The female reproductive structures, the **carpels,**

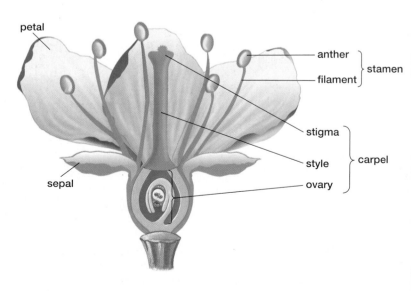

(a) A dicot flower in cross-section

(b) An amaryllis (monocot) flower

occupy the uppermost position in the flower. A typical carpel is somewhat vase shaped, with a sticky **stigma** for catching pollen mounted atop an elongated **style.** The style connects the stigma with the bulbous **ovary.** Inside the ovary are one or more **ovules,** in which the female gametophytes develop. When mature, each ovule will become a seed, and the ovary will develop into a protective, adhesive, or edible enclosure called the **fruit.**

As you may know from your own gardening experience, not all flowers are complete. **Incomplete flowers** lack one or more of the four floral parts. For example, in plants that have separate male and female flowers, the male flowers lack carpels, and the female flowers lack stamens (**Fig. 18-4**). (Male and female flowers may be borne on the same plant, as in pecan trees or cucumbers, or on different plants, as in willow trees or asparagus). Incomplete flowers may also lack sepals or petals. For example, the flowers of grasses have no sepals or petals.

18.3 What Are the Gametophytes of Flowering Plants?

The gametophytes of flowering plants are small and inconspicuous. All but invisible to the casual observer, the gametophyte generation consists of pollen grains (male) and embryo sacs (female) that develop within flowers.

The Pollen Grain Is the Male Gametophyte

Each anther produces thousands of haploid microspores, each of which divides by mitosis to produce a pollen grain. As the pollen grain matures, it produces two haploid sperm cells. A tough, protective surface coat develops around the pollen grain (**Fig. 18-5**; see also Fig. 18-2, step ❸).

When the pollen grains have matured, they are released from the anther. In wind-pollinated flowers, the pollen grains spill out and are carried widely by wind currents. In animal-pollinated flowers, the pollen grains adhere weakly to the anther until a pollinator comes along and brushes or picks them off.

The Embryo Sac Is the Female Gametophyte

Within the ovules in the ovary, large haploid megaspores undergo an unusual set of mitotic divisions. Three nuclear divisions produce a total of eight haploid nuclei, but plasma membranes then divide the cytoplasm into seven, not eight, cells: three small cells at each end, each containing one nucleus, and one remaining large cell in the middle containing two nuclei, called **polar nuclei** (see Fig. 18-2, step ❹). This seven-celled organism is the embryo sac—the haploid female gametophyte. The **egg** is the central small cell at the bottom of the embryo sac, located near an opening in the ovule (see Fig. 18-2, step ❹).

18.4 How Does Pollination Lead to Fertilization?

Sexual reproduction in plants requires **fertilization,** the fusion of sperm and egg. But fertilization can occur only after **pollination,** when a pollen grain lands on a stigma.

Pollinators May Be Attracted by Feeding Opportunities

Plants of some species are pollinated by the wind; grains of pollen drift through the air and some, by chance, land on flowers of the same species. Most wind-pollinated flowers, such as those of grasses and oaks, are inconspicuous and unscented. Many are scarcely more than naked stamens that liberate pollen to the wind (**Fig. 18-6**).

▲ **Figure 18-4 Some plants have separate male and female flowers** Plants of the squash family, such as these zucchinis, bear separate female (left) and male (right) flowers. **QUESTION:** In species with separate male and female flowers on the same plant, why would natural selection favor individuals whose male and female flowers bloom at different times?

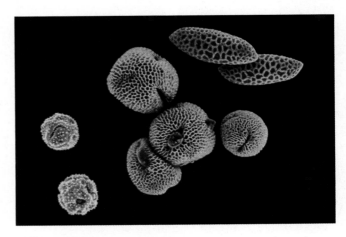

▲ **Figure 18-5 Pollen grains** The tough outer coverings of many pollen grains are elaborately sculpted in species-specific shapes and patterns. The pollen grains in this color-enhanced scanning electron micrograph are from a dandelion (yellow), a geranium (orange), and a tiger lily (fuchsia).

▲ **Figure 18-6 A wind-pollinated flower** The flowers of grasses and many deciduous trees are wind pollinated, with anthers (yellow structures) exposed to the wind.

In contrast to the generally plain flowers of wind-pollinated plants, flowers that are pollinated by animals are often colorful, conspicuous, and fragrant. These showy flowers typically contain food, usually nectar, which attracts foraging animals such as beetles, bees, moths, butterflies, and hummingbirds. As the animals move from flower to flower to consume the food, they unwittingly distribute pollen.

Some Flowers Have Adaptations That Help Ensure Pollination by Insects

We can thank bees for most sweet-smelling flowers, because sweet "flowery" odors attract these pollinators. Bees also have good color vision and can recognize nectar-producing flowers by color. Typically, flowers that attract bees are ones we see as white, blue, or yellow. Bees, however, do not see exactly the same range of colors that humans do, and many bee-attracting flowers have markings that we do not see because they reflect UV (ultraviolet) light, which is invisible to our eyes but can be seen by bees (Fig. 18-7).

Many flowers that are pollinated by moths or butterflies have nectar-containing tubes that accommodate the long tongues of these insects. Flowers pollinated by night-flying moths are open at night; most are white, and some give off strong, musky odors that help the moths locate the flowers in the dark.

Many beetles prefer to feed on animal material, and beetle-pollinated flowers typically smell like dung or rotting flesh, which attracts these scavenging insects. Their names, including "carrion flower," "skunk cabbage," and "corpse flower" (Fig. 18-8), describe their odors.

The corpse flower and its relative, the skunk cabbage common in U.S. swamps, share another adaptation: Their flowers heat up! By metabolizing stored food, mostly fat, eastern skunk cabbage flowers can reach temperatures up to 75°F (24°C) higher than the air around them. The heat probably attracts pollinators and certainly helps broadcast foul-smelling scents. These flowers deceive their pollinators; they smell like nutrient-rich rotting meat, but offer no food at all.

Birds Can Be Pollinators, Too

Hummingbirds are among the few important vertebrate pollinators. Hummingbirds do not use their sense of smell to find food, and most hummingbird-pollinated flowers do not synthesize fragrant chemicals. They do, however, often produce more nectar than other flowers do. The abundant nectar attracts hummingbirds, which use a great deal of energy and, hence, will favor flowers that provide lots of nectar. Hummingbird-pollinated flowers protect their large nectar supplies from insects, which would become sated on the abundant sugar and fail to transfer pollen to another flower. Adaptations to hummingbird pollination include flower shapes that provide no place for insects to land and rest while dining, and a deep, tubular structure that matches the long bills and tongues of hummers. Most hummingbird-pollinated flowers are red or orange.

Pollination May Be Followed by Fertilization

When a pollen grain, carried on the wind or by an animal pollinator, lands on the stigma of a flower of the same species, a remarkable chain of events begins (Fig. 18-9; see also Fig. 18-2, steps ❺ and ❻). A cell of the pollen grain elongates to form a tube, which grows down the style toward an ovule in the ovary. If all goes well, the pollen tube reaches the pore in the ovule and breaks into the embryo sac, where the tube's tip ruptures, releasing the two sperm (Fig. 18-9, step ❷).

human vision   "bee vision"

◀ **Figure 18-7 Ultraviolet patterns guide bees to nectar** The spectra of color vision for humans and bees overlap considerably but differ at their extremes. Humans are sensitive to red wavelengths, which bees do not perceive; bees can see UV light, which is invisible to the human eye. Many flowers photographed under *(a)* ordinary daylight and *(b)* under UV light show striking differences in color patterns. Bees can see the UV patterns that presumably direct them to the nectar- and pollen-containing centers of the flowers.

Inside the embryo sac, one sperm fertilizes the egg, forming a diploid zygote. The second sperm enters the large central cell, and its nucleus fuses with both polar nuclei, forming a *triploid* nucleus (having three sets of chromosomes) in the embryo sac's large central cell. This process, in which the egg fuses with one sperm and the two polar nuclei fuse with a second sperm, is called **double fertilization** and is unique to flowering plants (**Fig. 18-9**, step ❸). (The other five cells of the embryo sac degenerate soon after fertilization.)

18.5 How Do Seeds and Fruits Develop?

After double fertilization, the diploid zygote and the cell with a triploid nucleus continue to develop, forming an embryo and a food reserve, respectively, which will become incorporated into a seed. The plant will encase the seed in a fruit which will help the seed disperse to a new location distant from the parent plant. (Some fruits may also one day become medicines; see "Health Watch: Edible Vaccines" on p. 345.)

▲ **Figure 18-8 The corpse flower is overwhelming in size and scent** Scavenging insects such as flies and beetles are attracted to the odor of rotting meat produced by a flowering *Amorphophallus titanum*. The species is native to Sumatra, but is sometimes cultivated indoors in the United States. Thousands of visitors flock to see (and smell) blooming specimens. The "bloom," which is often over 6 feet tall, actually contains hundreds of separate male and female flowers.

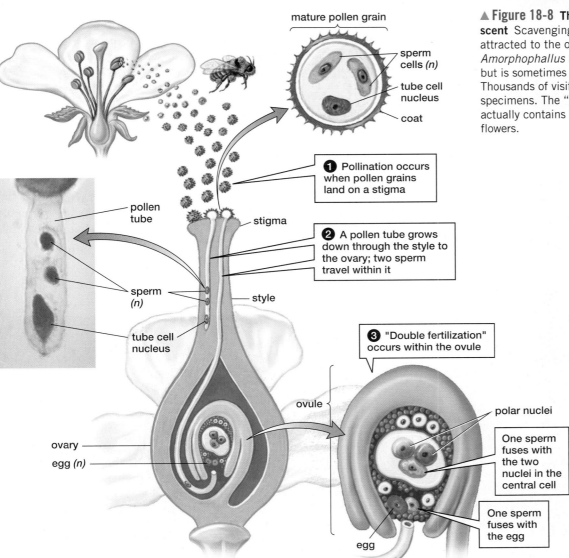

mature pollen grain

sperm cells *(n)*

tube cell nucleus

coat

❶ Pollination occurs when pollen grains land on a stigma

pollen tube

stigma

❷ A pollen tube grows down through the style to the ovary; two sperm travel within it

sperm *(n)*

style

tube cell nucleus

❸ "Double fertilization" occurs within the ovule

ovule

polar nuclei

One sperm fuses with the two nuclei in the central cell

ovary

egg *(n)*

One sperm fuses with the egg

egg

pollen tube

◀ **Figure 18-9 Pollination and fertilization of a flower QUESTION:** How does double fertilization prevent plants from wasting energy?

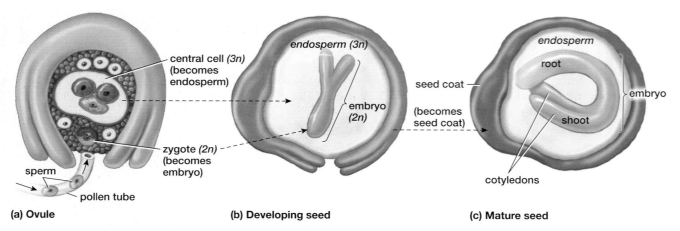

(a) Ovule **(b) Developing seed** **(c) Mature seed**

▲ **Figure 18-10** **Seed development** *(a)* Seed formation begins after one sperm fuses with the egg, forming a diploid zygote, and the second sperm fuses with the central cell. *(b)* The endosperm develops from the triploid central cell, which undergoes many mitotic cell divisions as it absorbs nutrients from the parent plant. Then the embryo develops, absorbing nutrients from the endosperm. *(c)* The two cotyledons of dicots absorb endosperm as the seed develops.

Web Animation Fruits and Seeds: Structure and Development

▼ **Figure 18-11** **Development of fruit and seed from flower parts** The fruit and seed coat are derived from the parent plant. The ovary wall ripens into the fruit flesh, and the wall of the ovule forms the seed coat. In many plants, such as the pepper shown here, each ovary contains many ovules, so each fruit contains many seeds. The zygote within each seed develops into an embryo.

The Seed Develops from the Ovule and Embryo Sac

Drawing on the resources of the parent plant, the embryo sac and the surrounding cells of the ovule develop into a **seed.** The outer layers of the ovule develop into the **seed coat,** the outer covering of the seed. Meanwhile, within the seed, the zygote develops into an embryo (**Figs. 18-10a,b**). In addition, the cell with a triploid nucleus undergoes repeated mitotic divisions. The resulting triploid daughter cells absorb nutrients from the parent plant and form the **endosperm,** a food-storage tissue that surrounds the developing embryo.

Both dicot and monocot embryos consist of three parts: the shoot, the root, and one (monocots) or two (dicots) **cotyledons,** or seed leaves (**Fig. 18-10c**). The cotyledons absorb food molecules from the endosperm and transfer them to other parts of the embryo.

The Fruit Develops from the Ovary Wall and Helps Disperse Seeds

The seed is surrounded by the ovary, which develops into a fruit (**Fig. 18-11**). The function of most fruits is to disperse seeds. A plant increases its chances of reproductive success if its seeds are dispersed far enough that the young plants don't compete with the parents for light and nutrients.

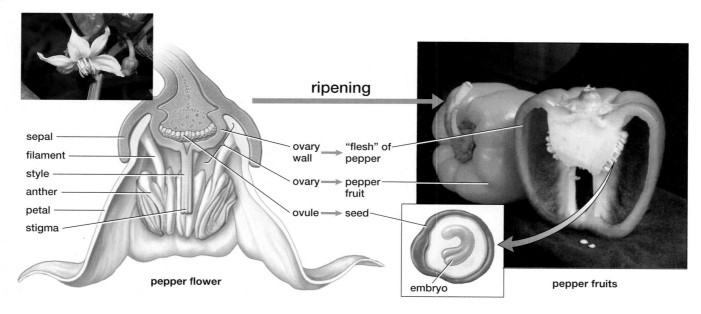

ripening

sepal	ovary wall → "flesh" of pepper
filament	
style	ovary → pepper fruit
anther	
petal	ovule → seed
stigma	

pepper flower

embryo

pepper fruits

health watch

Edible Vaccines

One of medicine's greatest life-saving successes has been the fight against infectious diseases. A key tool in this battle has been vaccines—injections that make people immune to a disease by stimulating the body's immune system. Vaccines have eradicated smallpox, nearly eliminated polio, and drastically reduced the incidence of many other infectious diseases. But the benefits of vaccination have not reached everyone. For example, about 20% of the world's children are not vaccinated against the widespread diseases that are most likely to harm them, such as measles and diphtheria. The World Health Organization estimates that vaccinating these children would prevent 2 million deaths each year.

Most of the unvaccinated children live in the world's poorest countries. Their chances of being vaccinated would be increased if they could be provided with vaccines that were inexpensive, were easy to administer, and could be stored without refrigeration. Few existing vaccines meet those criteria. But what if vaccines could be delivered in fruits? What if a child could be vaccinated by eating a banana (**Fig. E18-1**)? The techniques of biotechnology are bringing this idea ever closer to reality.

Much of plant biotechnology is based on geneticists' ability to insert foreign genes into plants. Many modern vaccines work by introducing to a person's body harmless proteins from a disease-causing bacterium or virus. The body's immune system builds defenses against these proteins, and these defenses jump into action if the disease-causing organism later invades the body. To create edible vaccines,

researchers are attempting to insert genes for the appropriate bacterial or viral proteins into plants that produce edible fruits.

For an edible vaccine to work, the modified plant must actually produce the vaccine proteins, and the proteins must stimulate an immune response when eaten. Recent research results have demonstrated that both of these conditions can be met. Researchers have successfully inserted genes from bacteria that cause food poisoning, hepatitis B, and cholera into tomato and potato plants, and they have demonstrated that the plants produce the bacterial proteins. Additional research has shown that mice that eat the modified plants mount an immune response that protects them from later infections. Apparently, the tough cell walls of plant cells protect the vaccine proteins from being broken down by the digestive juices of the stomach, and they are absorbed as they pass along the intestine.

Will the edible vaccines work in humans as they do in mice? A few successful human trials have been completed. In the largest of these, volunteers who ate raw potatoes containing proteins from the Norwalk virus (which causes diarrhea) mounted an immune response to the virus. It is not known, however, if the subjects are protected against infection. Nonetheless, it is now clear that vaccine proteins can be introduced into edible plant parts, and that these edible vaccines do induce the expected immune response.

The potential of edible vaccines is enormous. Instead of struggling to store and administer perishable vaccines that must be injected, health care providers in poor, remote villages

▲ **Figure E18-1 Edible vaccination?** Will eating a banana someday replace injections for some vaccines? Perhaps, but only if significant problems can be overcome.

could instead grow some special tomatoes or bananas. Before that can happen, though, some significant problems must be overcome. One of the biggest is figuring out how to ensure that the dosage of vaccine in a fruit is consistent, predictable, and sufficient. And large-scale clinical trials are required to ensure the vaccines' safety and effectiveness. But biotechnologists have already accomplished "proof of concept," and they are making steady progress in overcoming the remaining hurdles.

Although you may think of familiar foods such as bananas, apples, and oranges when you think of fruit, fruits assume a huge variety of shapes and sizes. The outer layers of fruits may be fleshy, hard, winged, or even spiked like a medieval mace. Each type of fruit disperses seeds in a different way.

Lightweight Fruits Can Be Dispersed by Wind

Dandelions and maples produce lightweight fruits with surfaces that catch the wind (**Fig. 18-12**). (Each individual hairy tuft on a dandelion ball is a separate fruit.) Each fruit typically contains a single small seed, because having only one

▶ **Figure 18-12 Wind-dispersed fruits** *(a)* Dandelion fruits have light, feathery tufts that catch the breezes. *(b)* Maple fruits are miniature glider-helicopters, silently whirling away from the tree as they fall.

(a) Dandelion fruits   **(b) Maple fruits**

▲ **Figure 18-13 Water-dispersed fruit** After a long journey at sea, this coconut was washed high onto a beach by a storm. The massive food reserves of coconuts allow seedlings to grow on barren, sandy beaches.

Fountain of Youth for Fruit

Continued

When we domesticate plants, we protect them from the natural processes that shaped their evolution. In the protected environments of orchards, farm fields, and biotechnology labs, we can allow our captive plants to acquire traits that we find useful, but that would spell doom for a wild plant. Consider the *never-ripe* mutation. What would happen if this mutation occurred in a wild plant? The fruits of the mutant plant would not ripen, and so would not perform their function of dispersing seeds. The plant would have few or no offspring, and *never-ripe* mutants would be rare or nonexistent in the next generation. So even if we ultimately find *never-ripe* tomatoes on supermarket shelves, we won't find *never-ripe* fruits in the wild.

▲ **Figure 18-14 The cocklebur fruit uses hooked spines to hitch a ride on furry animals**

seed reduces weight and lets the fruit remain aloft longer. These featherweight fruits help the seed travel away from the parent plant, from a few yards for maples to miles for a milkweed or dandelion on a windy day.

Floating Fruits Can Be Dispersed by Water

Many fruits can float on water for a time and may be dispersed by streams or rivers. The coconut fruit, however, is the ultimate floater. Round, buoyant, and watertight, the coconut drops off its parent palm, rolls to the sea, and floats for weeks or months until it washes ashore on some distant island (**Fig. 18-13**). There it sprouts, perhaps establishing a new coconut colony on a formerly barren island.

Clingy or Tasty Fruits Can Be Dispersed by Animals

Many fruits use animals as agents of seed dispersal. Two quite distinct strategies have evolved for dispersal by animals: grabbing an animal as it passes by, or enticing it to eat the fruit, which contains indigestible seeds.

Anyone who takes a long-haired dog on a walk through an abandoned field knows about fruits that hitchhike on animal fur. Burdocks, burr clover, foxtails, and sticktights all develop fruits with prongs, hooks, spines, or adhesive hairs (**Fig. 18-14**). The parent plants hold these fruits very loosely, so even slight contact with fur pulls the fruit free of the plant and leaves it stuck on the animal. Some of these adhering fruits may fall off the next time the animal brushes against a tree or rock or grooms its fur.

Unlike hitchhiker fruits, edible fruits benefit the animals that disperse the enclosed seeds. An animal that consumes an edible fruit is nourished by the sugar-rich flesh surrounding the seeds. In consuming the fruit, the animal also disperses seeds. Some edible fruits, such as peaches and plums, contain large, hard seeds that animals usually do not eat, but instead discard. Other fruits, including blackberries, raspberries, strawberries, and tomatoes, have small seeds that are swallowed along with the fruit flesh. The seeds then pass through the animal's digestive tract without being harmed. In addition to being transported away from its parent, a seed that is swallowed and excreted benefits in another way: It ends up with its own supply of fertilizer.

18.6 How Do Seeds Germinate and Grow?

The stage at which a seed or spore begins to grow and develop—when it sprouts—is known as **germination.** Many seeds do not germinate immediately after they mature, but instead enter a period of **dormancy,** during which they will not germinate. Dormancy may last for months or years, as the seed awaits conditions favorable for growth.

Seed Dormancy Helps Ensure Germination at an Appropriate Time

Seed dormancy avoids two problems. First, if a seed were to germinate while still enclosed in a fruit and hanging from a tree or vine, it might exhaust its food reserves before it ever touched the ground. Second, environmental conditions that are suitable for seedling growth (such as adequate moisture and favorable temperatures) may not coincide with seed maturation. Seeds that mature in the late summer in temperate climates, for example, face the harsh winter to come. Spending the winter as a dormant seed is clearly preferable to death by freezing as a tender young sprout. In the warm, moist tropics, seed dormancy is much less common than in temperate regions, because tropical environmental conditions are suitable for germination throughout the year.

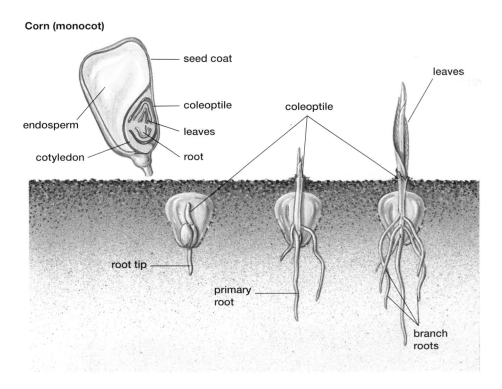

Corn (monocot)

- seed coat
- endosperm
- coleoptile
- leaves
- cotyledon
- root
- coleoptile
- leaves
- root tip
- primary root
- branch roots

◀ Figure 18-15 Seed germination The delicate shoot tip of a monocot is protected within a tough coleoptile, which pushes up through the soil when the seed sprouts.

The Root Emerges First

During germination, the embryo breaks dormancy and emerges from the seed. The embryo absorbs water, which makes it swell and burst its seed coat. The root is usually the first structure to emerge from the seed coat, growing rapidly and absorbing water and minerals from the soil. Much of the water is transported to cells in the shoot. As its cells elongate, the shoot lengthens, pushing up through the soil.

The Shoot Tip Must Be Protected

The growing shoot faces a serious difficulty: It must push through the soil without scraping away the apical meristem and tender leaflets at its tip. Unlike roots, which must always contend with tip abrasion and whose apical meristem is protected by a root cap (see Fig. 17-8 on p. 324), shoots spend most of their time in the air and do not develop permanent protective caps. Instead, germinating shoots have other ways of coping with the abrasion of sprouting.

In monocots, a tough sheath called the **coleoptile** encloses the shoot tip like a glove (Fig. 18-15). The coleoptile "glove" pushes aside the soil particles as it grows. Once out in the air, the coleoptile tip degenerates, allowing the tender shoot "finger" to emerge.

Dicots do not have coleoptiles. Instead, the dicot shoot forms a hook. The bend of the hook, encased in epidermal cells with tough cell walls, leads the way through the soil, clearing the path for the downward-pointing apical meristem with its delicate new leaves.

Cotyledons Nourish the Sprouting Seed

Food stored in the seed provides the energy for sprouting. In dicots, the cotyledons absorb the endosperm while the seed is developing and are swollen and full of food by the time the seed germinates. In many dicots, the elongating shoot carries the cotyledons out of the soil into the air, where they become green and photosynthetic and transfer both previously stored food and newly synthesized sugars to the shoot (Fig. 18-16). In other dicots, the cotyledons stay below the ground, shriveling up as the

▲ Figure 18-16 Cotyledons nourish the developing plant In the squash family, the cotyledons expand into photosynthetic leaves (the pair of smooth oval leaves). The first true leaf (crinkled single leaf) develops a little later. Eventually, the cotyledons shrivel up.

scientific inquiry

How Were Plant Hormones Discovered?

Everyone who keeps houseplants on a windowsill knows that the plants bend toward the window as they grow, in response to the sunlight streaming in. More than a hundred years ago, Charles Darwin and his son Francis studied this phenomenon of growth toward light, or phototropism, and in so doing started a chain of scientific events that culminated in the discovery of the first known plant hormone.

First, the Darwins Determined the Direction of Information Transfer

The Darwins illuminated grass coleoptiles (protective sheaths surrounding monocot seedlings) from various angles. They noted that a

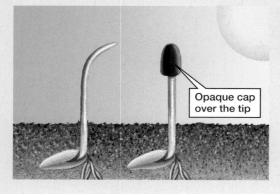

Opaque cap over the tip

region of the coleoptile a few millimeters below the tip bent toward the light, causing the tip to point toward the light source. But when they covered the tip with an opaque cap, the coleoptile didn't bend (see the figure at bottom left).

A clear cap allowed the stem to bend, as did covering the bending region below the tip with an opaque sleeve.

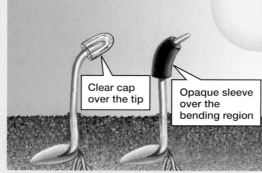

Clear cap over the tip

Opaque sleeve over the bending region

The Darwins concluded that (1) the tip of the coleoptile detects the direction of light, and (2) the bending occurs farther down the coleoptile; therefore, (3) the tip transmits information about the light direction down to the bending region.

How does the coleoptile bend? Although the Darwins didn't know this, the coleoptile bends as

a result of unequal elongation of cells on opposite sides of the shaft. The cells on the darker side elongate faster than those facing the light, bending the shaft toward the light. Information transmitted from the tip to the bending region causes unequal cell elongation.

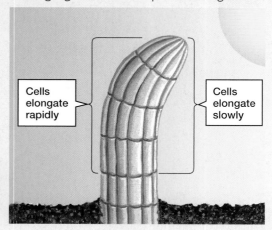

Cells elongate rapidly

Cells elongate slowly

Next, Peter Boysen-Jensen Demonstrated That the Information Is Transferred by Chemicals

About 30 years after the Darwins' experiments, Peter Boysen-Jensen cut the tips off coleoptiles

embryo absorbs their stored food. Monocots retain most of their food reserve in the endosperm until germination, when it is digested and absorbed by the cotyledon as the embryo grows. The cotyledon remains below ground.

Development Is Regulated Throughout the Life Cycle

Once out in the air, the shoot rapidly spreads its leaves to the sun. Simultaneously, the root system extends into the soil. The apical meristem cells of the shoot and root divide, giving rise to the mature structures discussed in Chapter 17. Eventually, this plant, too, will mature, flower, and set seed, renewing the cycle of life. Regulation of this cycle—ensuring that shoots grow upward while roots grow downward, that plants produce flowers at the proper time of year, and that fruits ripen—is accomplished by plant hormones.

18.7 What Are Plant Hormones, and How Do They Act?

Plant cells are miniature factories that produce a diverse array of molecules. Some of these molecules are **plant hormones,** chemicals produced by cells in one part of a plant body and transported to other parts of the plant, where they exert specific effects. Each hormone can cause a variety of responses in plant cells, depending on

Web Animation Plant Hormones

and found that the remaining stump neither elongated nor bent toward the light. If he replaced the tip and placed the patched-together coleoptile in the dark, it elongated straight up. In the light, it showed normal phototropism. When he inserted a thin layer of porous gelatin that prevented direct contact but permitted diffusion of substances between the severed tip and the stump, he still observed elongation and bending. In contrast, an impenetrable barrier eliminated these responses.

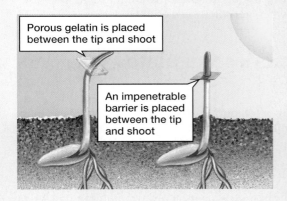

Boysen-Jensen concluded that a chemical is produced in the tip and moves down the shaft, causing cell elongation. In the dark, the chemical that causes the cells to elongate diffuses straight down from the tip and causes the coleoptile to elongate straight up. Presumably, light causes the chemical to become more concentrated on the "shady" far side of the shaft, so cells on the shady side elongate faster than do cells on the "sunny" near side, causing the shaft to bend toward the light.

Finally, the Chemical Auxin Was Identified

The next step was to isolate and identify the chemical. In the 1920s Frits Went devised a way to collect the elongation-promoting chemical. He cut off the tips of oat coleoptiles and placed them on a block of agar (a porous, gelatinous material) for a few hours. Went hypothesized that the chemical would migrate out of the coleoptiles into the agar.

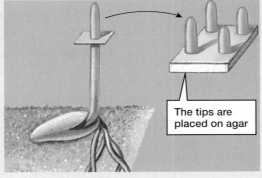

He then cut up the agar, now presumably loaded with the chemical, and placed small pieces on the tops of coleoptile stumps growing in darkness. When he put a piece of agar squarely on top of a stump, the stump elongated straight up. All the stump cells received equal amounts of the chemical and elongated at the same rate. If he placed a piece on one side of a cut stump, the stump would invariably bend away from the side with the agar.

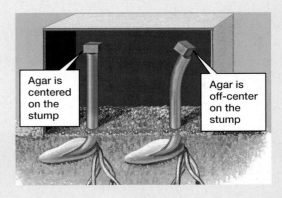

It was apparent that cells on the side under the agar received more of the chemical and were more stimulated to elongate. Went called the chemical *auxin,* from a Greek word meaning "to increase." Kenneth Thimann later purified auxin and determined its molecular structure.

factors such as the type of target cell, the developmental stage of the plant, the concentration of the hormone, and the presence of other hormones.

Plant physiologists have identified five major classes of plant hormones: auxins, gibberellins, cytokinins, ethylene, and abscisic acid (Table 18-1).

Auxins influence the elongation of cells. In coleoptiles and other parts of the shoot, high concentrations of auxins cause cells to elongate (see "Scientific Inquiry: How Were Plant Hormones Discovered?"). In roots, *low* concentrations stimulate elongation, whereas slightly higher concentrations inhibit elongation. Auxins also affect many other aspects of plant development, such as the formation of conducting tissues (xylem and phloem) and the development of fruits. Auxins also stimulate root branching. When applied to stems, auxins cause them to grow roots, an effect that is helpful to people who wish to grow new plants from cuttings.

Gibberellins, like auxins, promote elongation of cells in stems. In some plants, gibberellins stimulate flowering, fruit development, seed germination, and bud sprouting.

Cytokinins promote cell division in many plant tissues. Consequently, they stimulate the sprouting of buds and the development of fruit, endosperm, and the embryo. Cytokinins also stimulate plant metabolism, thus delaying the aging of plant parts, especially leaves.

Ethylene is an unusual plant hormone in that it is a gas at normal environmental temperatures. Ethylene is best known for its ability to cause fruits to ripen. It also regulates the timing of leaf, flower, and fruit drop.

Table 18-1	Hormone Actions in Plants
Hormone	**Functions**
Auxins	Promote elongation of cells in coleoptiles and shoots, phototropism, gravitropism in shoots and roots, root growth and branching, development of vascular tissue and fruit, and ethylene production in fruit; retard senescence (aging) in leaves and fruit
Gibberellins	Promote germination of seeds, sprouting of buds, and elongation of stems; stimulate flowering and development of fruit
Cytokinins	Promote sprouting of lateral buds; prevent leaf senescence; promote cell division; stimulate the development of fruit, endosperm, and embryo
Ethylene	Promotes ripening of fruit and abscission (falling off) of fruits, flowers, and leaves
Abscisic acid	Maintains dormancy in seeds and buds

Abscisic acid helps plants withstand unfavorable environmental conditions. For example, it inhibits the activity of gibberellins, thus helping maintain dormancy in buds and seeds at times when germination would be dangerous.

18.8 How Do Hormones Regulate the Plant Life Cycle?

In the rest of this chapter, we follow a temperate-zone plant through a year in its life, illustrating how hormones regulate growth and development at each stage of its life cycle.

Abscisic Acid Maintains Seed Dormancy

We begin with a seed maturing within a juicy fruit on a warm autumn day. In its warm, moist surroundings, the seed experiences ideal conditions for germination, yet it remains dormant until the following spring. In many seeds, this dormancy is enforced by abscisic acid. Abscisic acid slows down the metabolism of the embryo within the seed, preventing its growth. The seeds of some desert plants contain especially high concentrations of abscisic acid. In these seeds, only a hard rain can wash out the abscisic acid, freeing the embryo from the hormone's inhibitory effects and allowing the seed to germinate. In the seeds of high-latitude plants, which typically require a prolonged period of winter cold to break dormancy, the breakdown of abscisic acid is triggered by chilling.

Gibberellin Stimulates Germination

Eventually, a seed's period of dormancy comes to a close and it germinates, beginning the development of a new plant. Germination is induced by hormones, especially gibberellin. Cells in a seed begin to manufacture gibberellin in response to the same environmental conditions that cause abscisic acid to break down. As the seed germinates, gibberellin initiates the synthesis of enzymes that digest the food reserves of the endosperm and cotyledons, making sugars, lipids, and amino acids available to the growing embryo.

Auxin Controls the Orientation of the Sprouting Seedling

When the growing embryo breaks out of the seed coat, it immediately faces a crucial problem: Which way is up? The roots must burrow downward, while the shoot must grow upward to find the light. Auxin controls the responses of both roots and shoots to light and gravity.

Auxin Stimulates Shoot Elongation Away from Gravity and Toward Light

As a shoot first emerges from a seed buried underground, auxin is synthesized in the shoot tip. The auxin moves down the shaft of the stem and stimulates cell

Fountain of Youth for Fruit

Continued

Fruits from tomato plants with the *never-ripe* mutation fail to ripen, presumably because some aspect of the normal ripening process is disrupted. The disruption occurs because some protein that is ordinarily present in tomato plants is not produced if the gene that encodes it is mutated. On the basis of what you have learned so far in this chapter, can you make an educated guess about the likely function of the missing protein? (The answer will be revealed in "Fountain of Youth for Fruit, *Revisited*" at the end of this chapter.)

elongation. If the stem is not exactly vertical, organelles in the cells of the stem detect the direction of gravity and cause auxin to accumulate on the stem's lower side (**Fig. 18-17**, top). Therefore, the cells on the lower side elongate rapidly, forcing the stem to bend upward. Once the shoot tip is vertical, auxin becomes evenly distributed around the stem. The stem then grows straight up, emerging from the soil into the light. This is an example of **gravitropism**—directional growth with respect to gravity.

Auxin also influences a growing shoot's **phototropism** (tendency to grow toward light). Ordinarily, light's effect on the distribution of auxin is the same as the effect of gravity, because growth toward the brightest light (the sun) is in roughly the same direction as growth away from the pull of gravity. For example, if a young shoot still buried underground is close enough to the surface that some light penetrates down to it, both light and gravity cause auxin to be transported to the lower side of the shoot and promote upward bending. Thus, under normal conditions gravitropism and phototropism add to each other's effects.

Auxin May Control the Direction of Root Growth

Gravitropism in roots is less well understood. According to one hypothesis, the direction of root growth is, as with shoot tip growth, controlled by auxin. Auxin is transported from the shoot down to the root and, if the root is not vertical, the root cap senses the direction of gravity and causes the auxin to accumulate on the lower side of the root (see **Fig. 18-17**, bottom). In roots, unlike

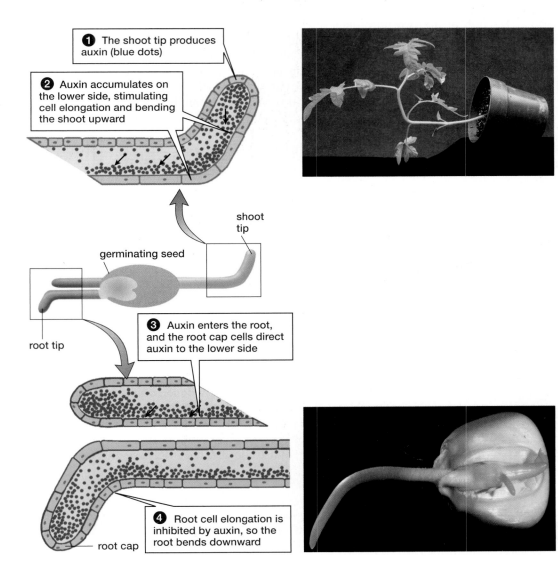

❶ The shoot tip produces auxin (blue dots)

❷ Auxin accumulates on the lower side, stimulating cell elongation and bending the shoot upward

shoot tip

germinating seed

root tip

❸ Auxin enters the root, and the root cap cells direct auxin to the lower side

❹ Root cell elongation is inhibited by auxin, so the root bends downward

root cap

◀ **Figure 18-17 Gravitropism in shoots and roots** The tomato plant in the top photo was placed on its side in the dark for less than a day. In this short time, faster elongation of cells on the lower side of the stem brought the plant to a nearly vertical orientation. In the bottom photo, the embryonic root of a germinating corn seed grows downward after the seed was placed on it side.

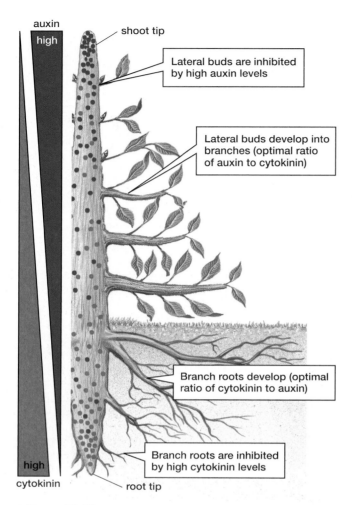

▲ **Figure 18-18 Role of auxin and cytokinin in lateral bud sprouting** The interplay of auxin (blue dots) and cytokinin (red dots) controls sprouting of lateral buds and the development of branch roots. Auxin is produced by shoot tips and moves downward; cytokinin is produced by root tips and moves upward.

in shoots, moderate concentrations of auxin *inhibit* cell elongation. Therefore, cell elongation in the lower side of the root is inhibited, but remains unaffected in the upper side of the root. As a result, the root bends downward. When the root tip points directly downward, the auxin distribution becomes equal on all sides, and the root continues to grow straight downward.

Stem and Root Branching Is Influenced by Auxin and Cytokinin

As a young plant's roots and shoots grow longer, they also begin to branch. Auxin and cytokinin together control this sprouting of lateral buds and growth of branch roots. Auxin is produced by the shoot tip and is transported down the stem, gradually decreasing in concentration as it moves downward. Cytokinin is produced by the roots and is transported up the stem, decreasing in concentration as it moves upward. Therefore, the relative concentrations of these two hormones vary along the length of the plant, and tissues at different positions experience differing hormonal influences.

In stems, auxin by itself inhibits lateral bud sprouting. However, a combination of auxin and cytokinin in particular proportions promotes sprouting. In a growing shoot, the lateral buds closest to the shoot tip receive a great deal of auxin but, because they are so far from the roots, they receive very little cytokinin. Therefore, they remain dormant. Lateral buds lower on the stem receive less auxin but receive more cytokinin. Those stimulated by the optimal combination of the two hormones sprout. In many types of plants, the interaction between auxin and cytokinin produces an orderly progression of bud sprouting from the bottom to the top of the shoot (**Fig. 18-18**).

In roots, the response to auxin and cytokinin differs from that of the stem. Growth of branch roots is stimulated by auxin and inhibited by cytokinin. Thus, branching is inhibited at the root tips where cytokinin is produced and stimulated at root locations closer to the aboveground shoot tip where auxin is produced.

Hormones Coordinate the Development of Seeds and Fruit

Flowering is followed by pollination, fertilization, and the development of seeds and fruit. When a flower is pollinated, the pollen releases auxin or gibberellin that stimulates the ovary to begin developing into a fruit. If fertilization follows, the developing seeds release additional auxin or gibberellin (or both) into the surrounding ovary tissues. In response, cells of the ovary multiply and grow larger, storing starches and other food materials and forming a mature fruit.

When the seeds mature, the fruit ripens: If it is of a type that disperses seeds via animal consumption, the fruit becomes softer, sweeter (as starches are converted to sugar), and more brightly colored, making it more noticeable and attractive to animals (**Fig. 18-19**). The produce section of your supermarket is full of brightly colored fruits—adapted to attract animals that might disperse seeds. Similarly, if you walk through a field in the autumn, you will see a few blue fruits on the dying plants, and many orange or red ones. But almost no fruits are green when ripe.

Ripening in many fruits is stimulated by ethylene, which also causes the breakdown of green chlorophyll, revealing the attractive pigments that signal a ripe fruit. Ethylene is synthesized by fruit cells in response to a surge of auxin that is released by the seeds when they are mature (a mechanism by which seed development and fruit development are coordinated). Because ethylene is a gas,

◀ **Figure 18-19 Ripe fruit is attractive to animal seed dispersers** The prickly pear cactus fruit is green, hard, and bitter before it ripens, so animals are discouraged from eating it. After the seeds mature, the fruit becomes soft, red, and tasty, attracting animals such as this desert tortoise.

a ripe fruit continually leaks ethylene into the air. When you store fruit in a closed container, ethylene released from one fruit will hasten ripening in the rest.

The discovery of the role of ethylene in ripening revolutionized modern fruit and vegetable marketing. For example, bananas grown in Central America can be picked green and tough and shipped to North American markets. When the green bananas reach their destination, grocers expose them to ethylene and market the resulting perfectly ripe fruit. Not all fruits, however, can be ripened after separation from the plant. Grapes and strawberries, for example, are not sensitive to ethylene and must ripen on the plant.

Senescence and Dormancy Prepare the Plant for Winter

As its fruits ripen, our plant's seeds are ready to begin anew the progression of hormone-controlled growth and development that we have described in this section. But for plants that live longer than a single year, there are some additional developments. Uneaten fruits drop to the ground. Leaves may be shed as well. In many cases, leaves would be a liability in winter, unable to photosynthesize but still allowing water to evaporate. So fruits and leaves drop from the plant, after undergoing a rapid aging process called **senescence.**

As winter approaches, cytokinin production in the roots slows down, and fruits and leaves produce less auxin. Perhaps driven by these hormonal changes, much of the organic material in leaves is broken down to simple molecules that are transported to the roots for winter storage. A layer of thin-walled cells called an **abscission layer** forms at the base of the leaf stalk (**Fig. 18-20**). Ethylene, released by both aging leaves and ripening fruit, stimulates the abscission layer to produce an enzyme that digests the cell wall that holds the leaf or fruit to the stem. When the abscission layers weaken, leaves and fruits fall from the branches.

Other changes also prepare the plant for winter. Buds, which developed into new leaves and branches during spring and summer, now become tightly wrapped up and dormant, waiting out the winter. Dormancy in buds, as in seeds, is enforced by abscisic acid. Metabolism slows to a crawl, and the plant enters its long winter sleep, waiting for the signals of warmth and longer days in spring before awakening once again.

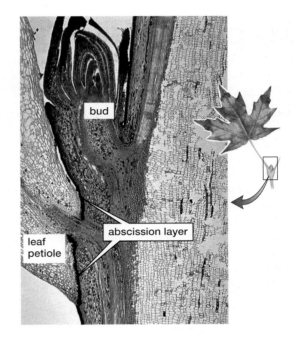

▲ **Figure 18-20 The abscission layer** This cross-section shows the abscission layer forming at the base of a maple leaf. A new lateral bud is visible above the dying leaf.

Fountain of Youth for Fruit Revisited

As you may have deduced, the peculiar properties of *never-ripe* tomatoes are related to the mutant plants' response to ethylene, which normally stimulates tomato fruits to ripen. Although the plants produce normal amounts of ethylene, their fruits fail to respond to the gas's hormonal signal. This failure to respond is due to a lack of *receptors* for ethylene. (Receptors are special protein molecules that initiate the appropriate response when a hormone molecule binds to them.) Tomato plants with the *never-ripe* mutation do not manufacture the ethylene-receptor protein, so their fruits never receive the signal that initiates ripening.

The *never-ripe* mutation suggests that scientists might one day manage fruit ripening by switching the genes for ethylene receptors on and off as needed. Plant breeders also hope to make use of ethylene-resistance genes for purposes beyond improved control of fruit ripening. For example, ethylene's role in promoting the onset of flower senescence means that the flowers of ethylene-resistant plants stay

fresh for much longer before wilting. Researchers have successfully transferred the tomato *never-ripe* gene into petunia plants, and the resulting flowers last two to three times longer than those of normal petunia plants.

Unfortunately, however, a different experiment on *never-ripe* petunias and tomatoes revealed that their insensitivity to ethylene also reduced the number of roots that grow from the cuttings that breeders use to propagate them. So some problems must be overcome before the *never-ripe* gene can give us flowers that last longer in a vase and fruits that last longer in the refrigerator.

Consider This

If we do achieve the ability to control fruit ripening by switching genes for ethylene receptors on and off, what problems is a plant likely to have during periods when the gene is switched off?

Chapter Review

Summary of Key Concepts

For additional study help and activities, go to www.mybiology.com.

18.1 What Is the Life Cycle of Plants?

The life cycle of plants, called alternation of generations, includes both a multicellular diploid form (the sporophyte generation) and a multicellular haploid form (the gametophyte generation).

In flowering plants, gametophytes are small. The male gametophyte is the pollen grain, a drought-resistant structure that can be carried from plant to plant by wind or animals. The female gametophyte is also reduced and is retained within the body of the sporophyte. In this way, flowering plants can reproduce independently of liquid water.

18.2 What Is the Structure of Flowers?

Complete flowers consist of four parts: sepals, petals, stamens (male reproductive structures), and carpels (female reproductive structures). The sepals form the outer covering of the flower bud. Most petals (and in some cases, the sepals) are brightly colored and attract pollinators to the flower. The stamen consists of a filament that bears at its tip an anther, in which pollen grains (the male gametophytes) develop. The carpel consists of the ovary, in which one or more embryo sacs (the female gametophytes) develop, and a style. The style bears at its end a sticky stigma, to which pollen adheres during pollination.

Incomplete flowers lack one or more of the four floral parts.

Web Animation Reproduction in Flowering Plants

18.3 What Are the Gametophytes of Flowering Plants?

Pollen grains develop in the anthers. Each haploid microspore divides mitotically to form pollen grains. A pollen grain contains two sperm cells. The embryo sac develops within the ovules of the ovary. A haploid megaspore undergoes three sets of mitotic divisions to produce the eight nuclei of the embryo sac. These eight nuclei come to reside in only seven cells. One of these cells, with a single nucleus, is the egg cell.

18.4 How Does Pollination Lead to Fertilization?

Pollination is the transfer of pollen from anther to stigma. When a pollen grain lands on a stigma, it grows a tube that extends through the style to the embryo sac. The two sperm cells travel down the style within the tube, eventually entering the embryo sac. One sperm fuses with the egg to form a diploid zygote, which gives rise to the embryo. The other sperm fuses with the large cell containing the polar nuclei to produce a triploid cell. This cell gives rise to the endosperm, a food-storage tissue within the seed.

18.5 How Do Seeds and Fruits Develop?

The embryo develops a root, shoot, and cotyledons. Cotyledons digest and absorb food from the endosperm and transfer it to the growing embryo. The seed is enclosed within a fruit, which develops from the ovary wall. The fruit helps disperse the seeds away from the parent plant, by either wind, water, traveling in an animal's digestive tract, or by clinging to a passerby.

Web Animation Fruits and Seeds: Structure and Development

18.6 How Do Seeds Germinate and Grow?

Seeds typically remain dormant for some time after fruit ripens. Seed germination requires warmth and moisture. On germination, the root generally emerges first, followed by the shoot. Energy for germination comes from food stored in the endosperm. That food is transferred to the embryo by the cotyledons.

18.7 What Are Plant Hormones, and How Do They Act?

Plant hormones are chemicals that are produced by cells in one part of a plant body and transported to other parts of the plant, where they exert specific effects. The five major classes of plant hormones are auxins, gibberellins, cytokinins, ethylene, and abscisic acid. The major functions of these hormones are summarized in Table 18-1.

Web Animation Plant Hormones

18.8 How Do Hormones Regulate the Plant Life Cycle?

Dormancy in seeds is enforced by abscisic acid. Falling levels of abscisic acid and rising levels of gibberellin trigger germination. As the seedling develops, it grows directionally with respect to light (phototropism) and gravity (gravitropism). Auxin is responsible for phototropism and gravitropism in shoots and for gravitropism in roots. In shoots, auxin stimulates the elongation of cells. In roots, similar concentrations of auxin inhibit elongation.

Branching in stems results from the interplay of two hormones, auxin (produced in shoot tips and transported downward) and cytokinin (synthesized in roots and transported upward). High concentrations of auxin inhibit the growth of lateral buds. An optimum concentration of both auxin and cytokinin stimulates the growth of lateral buds.

Developing seeds produce auxin, which diffuses into the surrounding ovary tissues and causes the production of a fruit. As the seed matures, a surge of auxin stimulates fruit cells to release another hormone, ethylene, which causes the fruit to ripen. During ripening, starches are converted to sugars, the fruit softens and develops bright colors, and, commonly, an abscission layer forms at the base of the fruit stalk.

Several changes prepare perennial plants of temperate zones for winter. Leaves and fruits undergo a rapid aging process called senescence, including the formation of an abscission layer. Senescence occurs when levels of auxin and cytokinin fall, and, perhaps, when ethylene concentrations rise. Other parts of the plant, including buds, become dormant. Dormancy in buds is due to high concentrations of abscisic acid.

Key Terms

abscisic acid *p. 350*
abscission layer *p. 353*
alternation of generations *p. 338*
anther *p. 340*
auxin *p. 349*
carpel *p. 340*
coleoptile *p. 347*
complete flower *p. 340*
cotyledon *p. 344*
cytokinin *p. 349*
dormancy *p. 346*

double fertilization *p. 343*
egg *p. 341*
embryo sac *p. 339*
endosperm *p. 344*
ethylene *p. 349*
fertilization *p. 341*
flower *p. 339*
fruit *p. 341*
gametophyte *p. 338*
germination *p. 346*
gibberellin *p. 349*

gravitropism *p. 351*
incomplete flower *p. 341*
megaspore *p. 339*
microspore *p. 339*
ovary *p. 341*
ovule *p. 341*
petal *p. 340*
phototropism *p. 351*
plant hormone *p. 348*
polar nuclei *p. 341*
pollen grain *p. 339*

pollination *p. 341*
seed *p. 344*
seed coat *p. 344*
senescence *p. 353*
sepal *p. 340*
spore *p. 338*
sporophyte *p. 338*
stamen *p. 340*
stigma *p. 341*
style *p. 341*

Thinking Through the Concepts

Suggested answers to end-of-chapter and figure-based questions can be found at the end of the text.

Fill-in-the-Blank

1. A pollen grain is formed in the _____ and is the _____ of a flowering plant. Pollination occurs when pollen lands on the _____ of a plant of the same species. The pollen grain then grows a tube through the _____ and eventually penetrates the ovary. In the ovary, the pollen tube enters a pore in the _____ and breaks into the embryo sac, which is the _____ of the plant. Here, it releases two _____ cells.

2. Flowering plants are said to exhibit "double fertilization" because one sperm cell fuses with the _____ to produce the new embryo, while the other sperm fuses with two _____, producing a large cell that divides to form _____. This substance provides _____ for the developing embryo.

3. In plants, the gametophyte produces eggs and sperm by the type of cell division called _____. These gametes have a(n) _____ number of chromosomes. The _____ generation produces spores by _____. Spores have the _____ number of chromosomes. The plant cycle is described by the term _____.

4. As the seed develops, the outer layers of the ovule form the _____. The embryo develops from the _____. The three parts of the embryo are the _____, _____, and _____. The _____ transfer endosperm nourishment to the embryo.

5. The emergence of the embryo from the seed is called _____. During this process, the _____ emerges first, and the _____ protects its growing tip. The emerging shoot of monocots is protected by the _____, whereas in dicots, the shoot forms a _____ that protects its leaves.

6. A newly formed seed typically will not germinate because _____ must first be washed from the seed by a hard rain or broken down by cold temperatures. After this occurs, germination is stimulated by the hormone _____. The root of the sprouting seedling grows downward and the shoot grows upward because of the distribution of the hormone _____. This same hormone stimulates branching in _____. When the plant produces fruit, ripening is often stimulated by _____.

Review Questions

1. Diagram the plant life cycle, comparing ferns with flowering plants. Which stages are haploid, and which are diploid? At which stage are gametes formed?

2. What are the advantages of the reduced gametophytes of flowering plants compared with the larger gametophytes of ferns?

3. Diagram a complete flower. Where are the male and female gametophytes formed? What are these gametophytes called?

4. How does the female gametophyte with its egg cell develop in a flowering plant? How does this structure allow double fertilization to occur?

5. What does it mean when we say that pollen is the male gametophyte? How is pollen formed?

6. What is the endosperm? From which cell of the embryo sac is it derived?

7. What did the Darwins, Boysen-Jensen, and Went each contribute to our understanding of phototropism? Do their experiments truly prove that auxin is the hormone that controls phototropism? What other experiments would you like to see done?

8. How can one hormone, an auxin, cause shoots to grow up and roots to grow down?

9. Which hormones cause fruit development? From where do these hormones come? Which hormone causes fruit ripening?

10. Which hormone(s) is (are) involved in leaf and fruit drop? In bud dormancy?

1. **IS THIS SCIENCE?** A common piece of advice is to place an apple in a bag of stored potatoes to keep the potatoes from sprouting. People interested in storing apples, however, are advised to keep them away from potatoes, whose presence will hasten softening of the apples. Use your knowledge of plant hormones to generate hypotheses that would explain why this advice is useful. Design experiments to test each hypothesis.

2. Charles Darwin once described a flower that produced nectar at the bottom of a tube 25 centimeters (10.5 inches) deep. He predicted that there must be a moth or other animal with a 25-centimeter-long tongue to match, and he was right. Such specialization almost certainly means that this particular flower could be pollinated only by that specific moth. What are the advantages and disadvantages of such specialization?

3. Agent Orange, a combination of two synthetic auxins, was used in Vietnam to defoliate the rain forest during the Vietnam War. Being similar to natural growth hormones, how can synthetic auxins harm or kill plants?

For additional resources, go to www.mybiology.com.

The animal body is an exquisite expression of the elegance with which evolution has linked form to function

Animal Anatomy and Physiology

chapter 19

Homeostasis and the Organization of the Animal Body

Under water, a
human body faces
conditions that
challenge its ability
to maintain the
internal conditions
necessary for
survival.

Case Study Surviving a Dive

On a July day in 2005, Patrick Musimu breathed in a lungful of air and plunged beneath the surface of the Red Sea. Heavily weighted to speed his descent, he plummeted rapidly, sinking to a depth of 687 feet. After reaching his target depth, he released the weights and began to ascend. As he burst through the water's surface, he was greeted by wild cheers from the crew aboard his boat. Musimu had just set a new record for the deepest dive without an air tank.

Musimu's record-breaking dive subjected his body to extraordinary stress. At the dive's deepest point, his body was experiencing more than 300 pounds of pressure per square inch. Such pressure deforms and stresses tissues and organs. A diver's lungs collapse, and air is forced out of the lungs and into the blood, where it dissolves. When the diver ascends and pressure decreases, the dissolved gases emerge from the blood—rapidly. The rapid exit of these gases from the blood can be dangerous or even deadly, especially if bubbles form in blood vessels and block the flow of blood.

A further insult to a deep diver's body is lack of oxygen. The oxygen inhaled at the surface before submerging can sustain the average person's body for only a few minutes; even the world's best breath-holders must resurface after less than 7 minutes of lying motionless under water. A deep diver whose body exhausts its oxygen supply while hundreds of feet from the surface is in serious peril.

Although Patrick Musimu emerged in good health from his confrontation with the dangers of deep diving, not all divers are so fortunate. In 2002, Audrey Mestre was ascending from a then-record dive to 561 feet when her body went limp. By the time rescue divers brought her to the surface, she was dead.

For humans, diving deep is extremely risky. But for members of some other species, even the dive that killed Audrey Mestre would be a routine excursion. Weddell's seals, for example, can dive to 2,200 feet and stay submerged for more than an hour. Elephant seals can go as deep as 5,000 feet and stay under for up to 2 hours. How do these deep-diving animals accomplish such feats, which are far beyond the abilities of even the best-conditioned human athlete? Consider this question as you read about the animal body's organization and the processes by which it maintains the internal conditions necessary for life. ■

Weddell's seals routinely dive to depths that would kill a human.

19.1 How Is the Animal Body Organized?

An animal body is composed of cells. The body of a small animal such as a roundworm might contain several thousand cells; a human body contains trillions of cells. The cells in a body are arranged into numerous different body parts, each of which has a distinctive size, shape, and combination of specialized cell types appropriate to its function. This match between structure and function is the result of natural selection, which has favored the evolution of body parts that are well adapted to perform tasks that increase an animal's ability to survive and reproduce. Natural selection has also helped ensure that a body's parts work together with an extraordinary degree of precision and integration.

Body structure and organization can be described at different levels of organization. **Tissues** are the basic building blocks of bodies; a tissue is composed of dozens to billions of structurally similar cells that act in concert to perform a particular function (**Fig. 19-1**). Tissues combine to form **organs,** in which at least two

▶ **Figure 19-1 Cells, tissues, organs, and organ systems** The animal body is composed of cells, which make up tissues, which combine to form organs that work in harmony as organ systems.

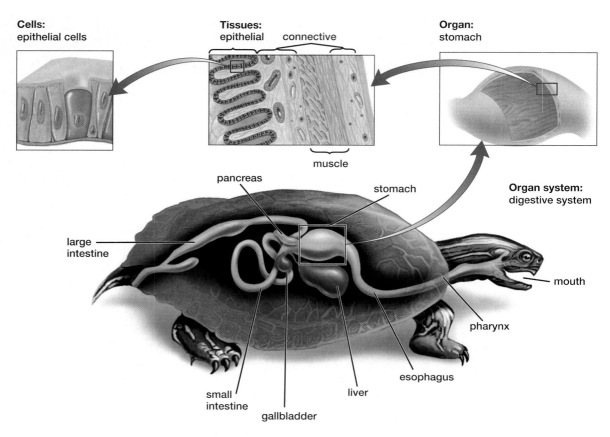

Cells: epithelial cells

Tissues: epithelial   connective

muscle

Organ: stomach

Organ system: digestive system

pancreas

stomach

large intestine

mouth

pharynx

esophagus

liver

small intestine

gallbladder

tissue types work together to perform a specific function. Examples of organs include the stomach, small intestine, kidneys, and urinary bladder. Organs in turn are organized into **organ systems** that consist of two or more individual organs that work together to perform a common function. For example, the digestive system includes the stomach, small intestine, large intestine, and other organs.

19.2 How Do Tissues Differ?

In the other chapters of this unit, you will read about the structure and function of the organ systems found in animal bodies. In these descriptions, you will discover that organ systems include a wide variety of different tissues. All of these tissues, however, can be grouped into just four types. Here we present a brief overview of the four types of animal tissue: epithelial tissue, connective tissue, muscle tissue, and nerve tissue.

Epithelial Tissue Forms Sheets

Epithelial tissue (also called epithelium) forms sheets that cover the body and line body cavities such as the mouth, the stomach, and the bladder. The cells of epithelial tissues are packed closely together, creating barriers (such as the skin) that resist the movement of substances across them or barriers (such as the lining of the small intestine) that allow the movement of specific substances across them.

Epithelial Tissue Covers the Body and Lines Its Cavities

There are many types of epithelial tissues, and the structure of each type is related to its function. For example, the epithelium that lines the lungs, where gas exchange takes place, consists of thin, flattened cells arranged in a single layer, which gas molecules can easily cross (**Fig. 19-2a**). Another form of epithelium consists of elongated cells, often with cilia, that can secrete mucus (**Fig. 19-2b**). This

(a) Thin epithelial tissue

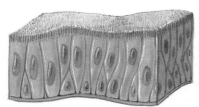

(b) Ciliated epithelial tissue

▲ **Figure 19-2 Epithelial tissue** *(a)* Thin, flattened cells in a single layer form the epithelial tissue that lines lungs and blood vessels, where exchange of materials is important. *(b)* Elongated epithelial cells bearing cilia and capable of secreting mucus line the passage to the lungs and tubes of the reproductive organs.

type of epithelium lines the passage leading to the lungs; the mucus traps dust particles, and the cilia transport the particles away from the lungs. This type of epithelium also lines the tubes of the reproductive organs, where the cilia transport gametes to their destinations.

An important property of many epithelial tissues is that they are continuously lost and replaced by mitotic cell division. For example, consider the abuse suffered by the epithelium that lines your mouth. Scalded by coffee and scraped by corn chips, it would be destroyed within a few days if it were not continuously replacing itself. Similarly, the stomach lining, ground down by food and attacked by acids and protein-digesting enzymes, is completely replaced every 2 to 3 days. Your skin's outer surface is renewed about twice a month.

Some Epithelial Tissues Form Glands

During development, the cells of some epithelial tissues change shape and function to form **glands,** clusters of cells that are specialized to secrete (release) substances. Glands are classified into two broad categories: exocrine glands and endocrine glands.

Exocrine glands remain connected to the epithelium by a passageway, or duct. Examples of exocrine glands are *sweat glands* and *sebaceous* (oil-secreting) *glands*, both of which are found in the skin. Exocrine glands called *salivary glands* release saliva into the mouth, and exocrine glands are also abundant in the stomach lining, where some secrete protective mucus and others secrete substances, such as acid or enzymes, that help digest food.

Endocrine glands are not connected to epithelium by a duct, but instead become separated from the epithelium that produced them. Most products of endocrine glands are hormones, which are secreted into the extracellular fluid that surrounds the glands and then diffuse into nearby capillaries.

Connective Tissues Have Diverse Structures and Functions

Connective tissues serve mainly to support and bind other tissues. Most connective tissues include large quantities of extracellular substances, typically secreted by the connective tissue cells themselves. The properties of this extracellular material differ among the different kinds of connective tissue and play a large role in determining the characteristics of each connective tissue type.

Some connective tissues that support and nourish epithelial tissues consist mainly of loosely woven fibers. For example, a connective tissue called the *dermis* lies beneath the epithelial tissue of the skin and contains fibers of an extracellular protein called *collagen*, as well as capillaries and fluid-filled spaces that nourish the epithelium (see Fig. 19-9 on p. 363). Other fibrous connective tissues known as *tendons* and *ligaments* attach muscles to bones and bones to bones, respectively. The strength necessary for these functions comes from densely packed, parallel strands of collagen.

Cartilage is a flexible and resilient connective tissue that consists of widely spaced cells surrounded by a thick, nonliving matrix. The matrix is made of collagen secreted by the cartilage cells (**Fig. 19-3**). Cartilage covers the ends of bones at joints, provides the supporting framework for the passages through which we breathe, supports the ear and nose, and forms shock-absorbing pads between the vertebrae.

Bone resembles cartilage but is hardened by deposits of calcium phosphate. Bone material is deposited in concentric layers around small central canals, each of which contains a blood capillary (**Fig. 19-4**).

Long-term energy storage in the animal body is provided by fat cells, collectively called *adipose tissue* (**Fig. 19-5a**). Adipose tissue is especially important in the physiology of animals adapted to cold environments, because it not only stores energy but also serves as insulation (**Fig. 19-5b**).

Although they are liquids, *blood* and *lymph* are considered connective tissues because they are composed largely of extracellular fluids. Blood consists of red

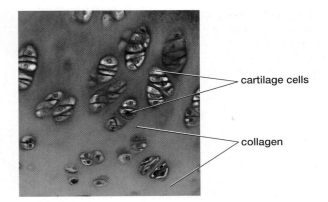

▲ **Figure 19-3 Cartilage** This specialized connective tissue supports the body and protects soft tissues. It is more flexible than bone. The plasma membrane of the cartilage cells is stained dark purple. The material stained pale purple is the matrix of collagen secreted by the cartilage cells.

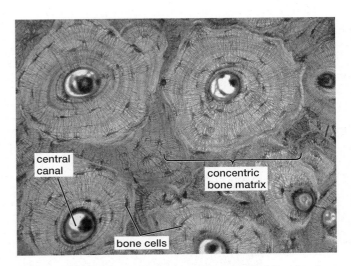

▲ **Figure 19-4 Bone consists of cells embedded in a hard matrix** Concentric circles of bone, deposited around a central canal that contains a blood vessel, are clearly visible in this micrograph. Bone cells appear as dark spots trapped in small chambers within the hard matrix that the cells themselves deposit.

▶ **Figure 19-5 Adipose tissue**
(a) Adipose tissue, such as that shown here from a mouse, is made up almost exclusively of fat cells; very little extracellular material is present. Each fat cell contains a droplet of oil that takes up most of the volume of the cell. *(b)* A hooded seal at 4 days of age has doubled her birth weight by drinking mother's milk that consists of 61% fat. At 100 pounds, she is almost too fat to move. The fat will feed and insulate the pup as the ice floes break up and she dives into icy water to learn to hunt and feed on her own. **QUESTION:** Why are young mammals so often fatter than adults?

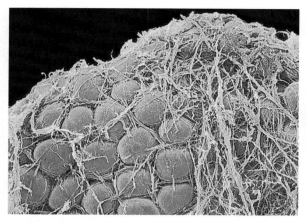

(a) Fat cells

(b) A seal pup

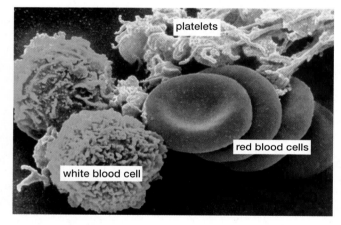

▲ **Figure 19-6 Blood** Blood contains three types of cellular components, shown here in a color-enhanced scanning electron micrograph, which are suspended in plasma.

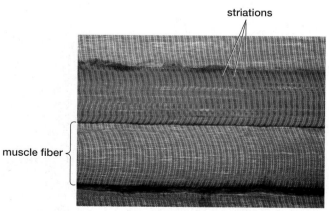

▲ **Figure 19-7 Muscle tissue consists of contractile cells called muscle fibers** A regular arrangement of fibrous proteins in this skeletal muscle tissue appears as stripes, or striations, when viewed under a microscope.

blood cells, white blood cells, and cell fragments called platelets, all suspended in extracellular fluid called *plasma* (**Fig. 19-6**). Blood carries dissolved oxygen and other nutrients to cells and carries away carbon dioxide and other cellular waste products. Blood also carries hormones throughout the body. Lymph consists largely of fluid that has leaked out of blood vessels. It drains into lymph vessels that return it to the blood.

Muscle Tissue Has the Ability to Contract

The long, thin cells of **muscle tissue** contract (shorten) when stimulated, then relax passively (**Fig. 19-7**). The three types of muscle tissue are skeletal, cardiac, and smooth. *Skeletal muscle* is normally under voluntary control, and has a striped appearance due to the orderly arrangement of fibrous proteins in its cells. (The stripes are sometimes called *striations*.) As its name implies, the main function of skeletal muscle is to move the skeleton, as when you walk or turn the pages of this text. *Cardiac muscle* is located only in the heart. Like skeletal muscle, cardiac muscle is striated, but unlike skeletal muscle, it is spontaneously active and not under voluntary control. Cardiac muscle cells are interconnected such that electrical signals spread rapidly through the heart, stimulating the cardiac muscle cells to contract in a coordinated fashion. *Smooth muscle*, so named because it lacks the striations seen in cardiac and skeletal muscles, is embedded in the walls of the digestive tract, the uterus, the bladder, and large blood vessels. It produces slow, sustained contractions that are usually involuntary.

Nerve Tissue Transmits Electrical Signals

You owe your ability to sense and respond to the world to **nerve tissue,** which transmits electrical signals and makes up the brain, the spinal cord, and the nerves that travel from them to all parts of the body. Nerve tissue is composed of two types of cells: neurons and glial cells. **Neurons** are specialized to generate electrical signals and to conduct these signals to other neurons, muscles, or glands. *Glial cells* surround, support, and protect neurons and regulate the composition of the extracellular fluid, allowing neurons to function optimally.

A neuron has four major parts, each with a specialized function (**Fig. 19-8**). The *dendrites* of a neuron receive signals from other neurons or from the external environment. The *cell body* directs the maintenance and repair of the cell. The *axon* conducts the electrical signal to its target cell, and the *synaptic terminals* transmit the signal to the target cell.

19.3 How Are Tissues Combined into Organs?

An organ is formed from at least two tissue types that function together. Some organs include all four tissue types. In this section, we examine the components and functions of a representative animal organ, the skin.

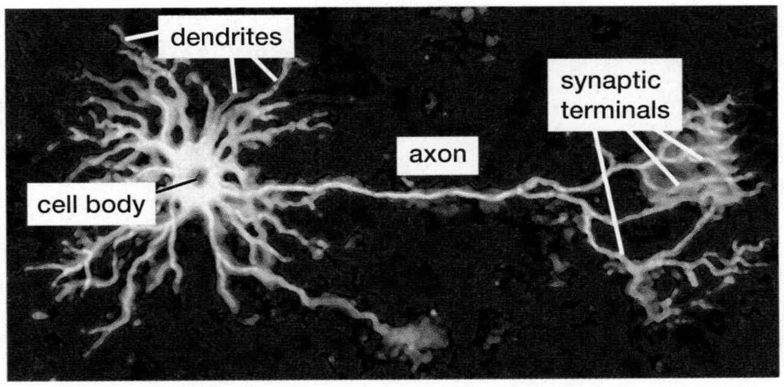

Skin Is an Organ That Contains All Four Tissue Types

In animal skin, an outer epithelium sits on top of connective tissue that contains a blood supply, nerves, in some cases muscle, and glandular structures derived from the epithelium.

The *epidermis*, or outer layer of the skin, is a specialized epithelial tissue (Fig. 19-9). It is covered by a protective layer of dead cells produced by underlying living epidermal cells. These dead cells are packed with the protein keratin, which helps keep the skin both airtight and relatively waterproof. Immediately beneath the epidermis lies a layer of connective tissue, the *dermis*. Blood vessels—arterioles, venules, and capillaries—spread through the dermis and carry the blood that nourishes both the dermal and epidermal tissue. Lymph vessels within the dermis collect and carry off extracellular fluid. A network of collagen fibers provides flexible support. Various nerve endings responsive to temperature, touch, pressure, vibration, and pain are scattered throughout the dermis and epidermis.

The dermis is also packed with glands derived from epithelial tissue. Glands called *hair follicles* produce hair from protein-containing secretions. Sweat glands produce watery secretions that cool the skin and excrete substances such as salts and urea. Sebaceous glands secrete an oily substance that lubricates the epithelium.

In addition to the epithelial, connective, and nerve tissues, some areas of the skin also contain muscle tissue. Tiny muscles attached to hair follicles can cause the hairs of the skin to "stand on end" in response to signals from neurons. Although this reaction is useless in humans, in most other mammals erecting individual hairs effectively increases the thickness of the fur, thereby helping to retain heat in cold weather.

Organ Systems Consist of Two or More Interacting Organs

Organ systems include at least two organs that function together. For example, the skin is part of the integumentary system, which also includes the hair and the nails and serves as a barrier between the environment and the inside of the body. Another example is the digestive system, in which the mouth, esophagus, stomach, intestines, and other organs that supply digestive enzymes, such as the liver and pancreas, all function together to convert food into nutrient molecules (see Fig. 19-1). The major organ systems of the vertebrate body are listed in **Table 19-1.**

▲ **Figure 19-8 A nerve cell** A nerve cell has four major parts, each specialized for a specific function. Nerve cells vary tremendously in shape, depending on their functions in the nervous system.

Surviving a Dive

Continued

As described at the beginning of this chapter, one of the dangers of deep diving comes from gas bubbles that can form when a body ascends from depth. Large bubbles in vulnerable tissues and organs can be deadly, but even small bubbles can cause unpleasant symptoms. For example, many divers have suffered from a malady commonly known as "skin bends," in which small gas bubbles accumulate in skin; the resulting tissue damage leads to a blotchy appearance, swelling, itching, and sometimes an overpowering sensation that tiny insects are crawling over the skin.

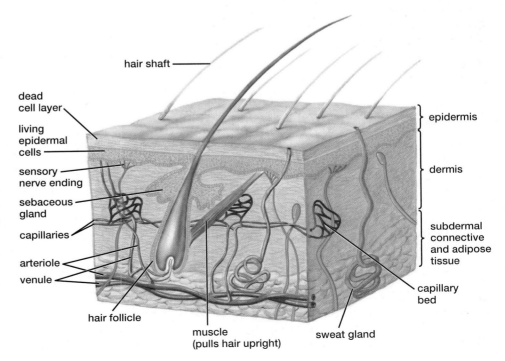

▶ **Figure 19-9 Skin** Mammalian skin, a representative organ, in cross-section. **QUESTION:** Why is skin considered to be an organ, but blood is considered to be a tissue?

Table 19-1 Major Vertebrate Organ Systems

Organ System	Major Structures	Physiological Role	Organ System	Major Structures	Physiological Role
Circulatory system	Heart, blood vessels, blood	Transports nutrients, gases, hormones, metabolic wastes; also assists in temperature control	Endocrine system	A variety of hormone-secreting glands, including the hypothalamus, pituitary, thyroid, pancreas, and adrenals	Controls physiological processes, typically in conjunction with the nervous system
Lymphatic/ immune system	Lymph, lymph nodes and vessels, white blood cells	Carries fat and excess fluids to blood; destroys invading microbes	Nervous system	Brain, spinal cord, peripheral nerves	Controls physiological processes in conjunction with the endocrine system; senses the environment; directs behavior
Digestive system	Mouth, esophagus, stomach, small and large intestines, glands producing digestive secretions	Supplies the body with nutrients that provide energy and materials for growth and maintenance	Muscular system	Skeletal muscle / Smooth muscle / Cardiac muscle	Moves the skeleton / Controls movement of substances through hollow organs (digestive tract, large blood vessels) / Initiates and implements heart contractions
Excretory system	Kidneys, ureters, bladder, urethra	Maintains homeostatic conditions within the bloodstream; filters out cellular wastes, certain toxins, and excess water and nutrients	Skeletal system	Bones, cartilage, tendons, ligaments	Provides support for the body, attachment sites for muscles, and protection for internal organs
Respiratory system	Nose, trachea, lungs (in mammals, birds, reptiles, amphibians), gills (in fish and some amphibians)	Provides an area for gas exchange between the blood and the environment; allows oxygen acquisition and carbon dioxide elimination	Reproductive system	Male (mammal): testes, seminal vesicles, penis / Female (mammal): ovaries, oviducts, uterus, vagina, mammary glands	Male: produces sperm, inseminates female / Female (mammal): produces egg cells, sustains developing embryo

19.4 How Do Animals Maintain Internal Constancy?

To function properly, an organ system must be situated in stable surroundings of just the right moisture level, temperature, and chemical composition. But an animal's external environment is rarely characterized by such conditions. Animals inhabit all kinds of environments, from scorching deserts to icy polar seas, from windy mountaintops to underground rivers. To survive, an animal body must be able to maintain a constant, suitable internal environment, regardless of its external environment.

Survival of a body ultimately depends on survival of the cells that make up the body. In animal bodies, these cells are bathed in slightly salty liquid. This wet, salty environment is required for the cells' survival and must be maintained even if the animal's external environment consists of dry air or fresh, salt-free water. What's more, it's not enough that the internal environment simply be wet and salty. Cells also require a very specific, complex mixture of dissolved substances and a specific temperature range. Because the body's cells cannot survive if the conditions of the internal environment deviate from a small range of acceptable states, the cells devote a large portion of their energy to actions that serve to keep the environment within that range.

This constancy of the internal environment was first recognized by French physiologist Claude Bernard in the mid-nineteenth century, and the term **homeostasis** was later coined to describe it. Although the word "homeostasis" (derived from Greek words meaning "to stay the same") may suggest a static, unchanging state, the internal environment in fact seethes with activity as the body continuously adjusts to internal and external changes. The end result of all this activity, however, is that conditions are kept within the narrow range that cells require to function. Conditions are maintained within the appropriate range by a host of mechanisms that are collectively known as feedback systems.

Negative Feedback Reverses the Effects of Changes

The most important mechanism governing homeostasis is **negative feedback,** in which the response to a change counteracts the change. In other words, an input causes an output response that affects or "feeds back" to the initial input and decreases its effects. Because the initial change triggers a response that reverses its effects, the overall result is to return the system to its original condition. This kind of feedback is called "negative" because it reverses or negates the initial change.

A negative feedback mechanism requires a *control center* that manages the overall process and establishes a set point, a *sensor* that detects deviations from the set point, and an *effector* that causes changes. The manner in which a control center, sensor, and effector work together is illustrated by a familiar example of negative feedback: your home thermostat.

In a thermostat, an input—the temperature dropping below a set point established by the thermostat setting—is detected by a thermometer (the sensor in this system). The thermometer responds with an output, signaling a control device (the control center) that switches on a heater. The heater (the effector) restores the temperature to the set point, and the heater is switched off. The response to the initial change counteracted that change. Continuously repeated on–off cycles keep your home's temperature near the set point. (**Fig. 19-10a**).

How do people maintain their internal temperature within a narrow range despite extreme fluctuations in the temperature around them? Negative feedback keeps a person's body temperature close to 98.6°F (37°C). The control center of the temperature control system is in the hypothalamus, a region of

Web Animation Homeostasis

Surviving a Dive

Continued

In a human body that is submerged in deep water, the level of dissolved oxygen in the blood deviates from its set point. The deviation generates negative feedback signals. However, the mechanisms for restoring oxygen to the correct level cannot respond to the feedback, in part because the body's normal reservoirs of oxygen, the lungs, cannot be replenished by breathing. A body cannot long survive this disruption of homeostasis, but some bodies, like those of trained deep divers such as Patrick Musimu, can survive it for longer than others.

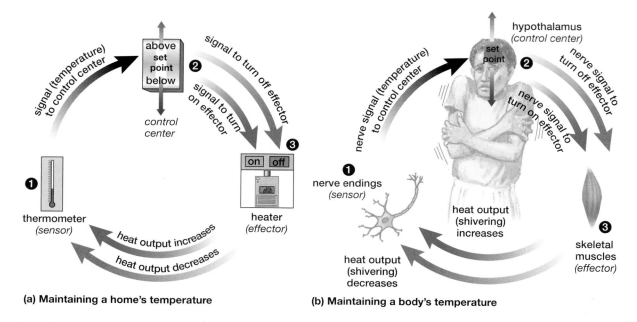

(a) Maintaining a home's temperature

(b) Maintaining a body's temperature

▲ **Figure 19-10 Negative feedback** Both *(a)* our homes and *(b)* our bodies use negative feedback to maintain appropriate temperatures. ❶ A sensor measures the temperature and sends a signal to ❷ a control center, where the temperature is compared to a set point. If the temperature is below the set point, the control center sends an "on" signal (green arrow) to ❸ an effector, which in response acts to increase temperature. If the temperature is above the set point, the control center sends an "off" signal (red arrow) to the effector. **QUESTION:** What would happen if a cold, shivering mammal ingested a poison that destroyed all of the nerve endings that detect heat? More generally, what happens if the sensor in a negative feedback system loses its ability to receive signals?

the brain (**Fig. 19-10b**). Nerve endings in the abdomen, spinal cord, skin, and large veins act as temperature sensors and transmit this information to the hypothalamus. When your body temperature drops, the hypothalamus activates various effector mechanisms that raise your body temperature. These mechanisms include shivering (which generates heat through muscular activity), restricting blood supply to the skin (which reduces heat loss), and elevating metabolic rate (which generates heat). When normal body temperature is restored, sensors signal the hypothalamus to switch off these temperature control mechanisms.

Negative feedback mechanisms abound in physiological systems. In the chapters that follow, you will find many examples in which homeostasis is maintained by negative feedback, including blood oxygen content, water balance, and blood sugar levels.

Positive Feedback Amplifies the Effects of Changes

In contrast to a negative feedback system, a change in a **positive feedback** system produces a response that intensifies the original change. The end result is that change tends to proceed in the same direction rather than reversing to return to a set point. Positive feedback can cause chain reactions. For example, in nuclear fission, each particle that is split from an atom triggers the splitting of another atom, the pieces of which trigger the splitting of other atoms, and so on. When controlled, the chain reaction supplies nuclear power. When deliberately set out of control, it produces an atomic explosion. A familiar biological example of positive feedback is population growth; each offspring gives rise to still more offspring. Ecologist Paul Ehrlich coined the apt expression "population bomb" to describe unchecked population growth.

Positive feedback mechanisms are rare in animal physiology. One example of positive feedback, however, controls events during childbirth. The early contractions of labor begin to force the baby's head against the cervix, located at the base of the uterus. This pressure causes the cervix to dilate (open). Stretch receptors in the cervix respond to this expansion by signaling the hypothalamus, which responds by triggering the release of a hormone (oxytocin) that stimulates more and stronger uterine contractions. Stronger contractions create further pressure on the cervix, which in turn prompts the release of more hormone. The feedback cycle is finally terminated by the expulsion of the baby and its placenta.

The Body's Organ Systems Act in Concert

To maintain the internal environment in the narrow range required for life, the organ systems of the body work together in a coordinated manner. Numerous feedback mechanisms are constantly at work, responding to inputs that continuously change as an animal's activities and external environment change. If these control mechanisms worked independently, each without regard to the activities of the other organ systems, it would not be possible to maintain homeostasis in the body as a whole.

Fortunately, evolution has ensured that the different organ systems work together. For example, the systems that take substances into the body (for example, the digestive system) act in concert with the systems responsible for transporting substances within the body (for example, the circulatory system) and systems that remove substances from the body (such as the excretory system). This kind of coordinated action is possible because the body has ways of sending signals from one part of the body to another. Each part of the body is connected to all the others by an elaborate network of vessels and nerves that carries molecules and messages to the appropriate locations. Molecules are transported to the sites where they are needed and away from locations where their presence would be harmful. Messages travel from sensors to effectors and back again, allowing feedback mechanisms to maintain homeostasis.

Surviving a Dive Revisited

A key element of homeostasis in animal bodies is maintenance of appropriate levels of dissolved gases, especially oxygen. Maintaining homeostasis with respect to dissolved oxygen is a difficult challenge for a human body that is underwater for more than a few moments. Human bodies are, after all, adapted for life on land. Nonetheless, a person can, by undergoing a training program, extend the range of circumstances in which underwater survival is possible. Such training can increase a body's ability to store oxygen, decrease the rate at which oxygen is consumed, and increase tolerance to pain caused by high water pressure. Even the most intense training, however, causes comparatively small physiological changes and is not a reliable guarantor of survival at depth, as was tragically demonstrated by the death of the deep diver Audrey Mestre.

Unlike humans, air-breathing mammals that habitually dive deep do not rely for survival on small physiological changes induced by training. Instead, they rely on evolved modifications of their bodies. What adaptations make it possible for diving mammals to maintain homeostasis during deep, long dives? One important feature is streamlining, which reduces the rate at which a diving animal uses oxygen. Streamlined shapes allow these animals to experience only a small amount of drag as they swim through the water. In addition, diving mammals such as seals and dolphins spend most of their underwater time gliding, rarely moving their tails or limbs. With their streamlined bodies and efficient movement, diving mammals conserve energy, reducing their demand for oxygen.

In addition to minimizing their use of oxygen, diving mammals also carry more oxygen with them when they dive than do humans. For example, a Weddell seal's body contains almost three times as much oxygen, per pound of body weight, as a human body. This extraordinary capacity for storing oxygen is due to extremely high concentrations of the protein myoglobin. Like the hemoglobin in our blood, myoglobin in the seal's muscles binds oxygen molecules and later releases them to supply cells. Abundant myoglobin acts as a reservoir of oxygen, allowing diving mammals to remain submerged through long dives in deep water.

Consider This

Researchers have discovered that when a person holds his or her breath, heart rate decreases and blood flow to the limbs is reduced. These effects are increased if a person's face is immersed in cold water. Why might these responses have evolved?

Chapter Review

Summary of Key Concepts

For additional study help and activities, go to www.mybiology.com.

19.1 How Is the Animal Body Organized?
The animal body is composed of organ systems consisting of two or more organs. Organs, in turn, are made up of tissues. A tissue is a group of cells that forms a structural and functional unit that is specialized for a specific task.

19.2 How Do Tissues Differ?
Animal tissues include epithelial, connective, muscle, and nerve tissue. Epithelial tissue covers internal and external body surfaces and also gives rise to glands. Connective tissue contains considerable extracellular material. Muscle tissue is specialized for movement. Nerve tissue is specialized for the generation and conduction of electrical signals.

19.3 How Are Tissues Combined into Organs?
Skin is an organ that includes epidermis (epithelial tissue), dermis (connective tissue), nerve endings (nerve tissue), and tiny muscles (muscle tissue). Animal organ systems include the circulatory, lymphatic/immune, digestive, excretory, respiratory, endocrine, nervous, muscular, skeletal, and reproductive systems, as summarized in Table 19-1. The organ systems contain many feedback mechanisms that work in concert to maintain homeostasis in the body as a whole.

19.4 How Do Animals Maintain Internal Constancy?
Conditions in an animal's internal environment tend to remain within a narrow range that permits the continuation of life. This homeostasis is maintained mostly through negative feedback, in which a change triggers a response that counteracts the change and restores conditions to a set point. Positive feedback, in which a change initiates events that intensify the change, is rare in physiological processes.

Web Animation Homeostasis

Key Terms

bone *p. 361*
cartilage *p. 361*
connective tissue *p. 361*
endocrine gland *p. 361*

epithelial tissue *p. 360*
exocrine gland *p. 361*
gland *p. 361*
homeostasis *p. 365*

muscle tissue *p. 362*
negative feedback *p. 365*
nerve tissue *p. 362*
neuron *p. 362*

organ *p. 359*
organ system *p. 360*
positive feedback *p. 366*
tissue *p. 359*

Thinking Through the Concepts

Suggested answers to end-of-chapter and figure-based questions can be found at the end of the text.

Fill-in-the-Blank

1. The process by which the body maintains internal conditions within the narrow range required by cells to function is called _____. The most important mechanism governing this process is _____.

2. The three levels of organization of the animal body, from smallest to most inclusive, are _____, _____, and _____.

3. Fill in the appropriate tissue type—supports and binds other tissues: _____; forms glands: _____; includes the blood: _____; includes the dermis of the skin: _____; covers the body and lines its cavities: _____; can shorten when stimulated: _____; includes glial cells: _____; includes adipose tissue: _____.

4. Glands that have no ducts connecting them to the epithelium are called _____ glands. Three examples of this type of gland are _____, _____, and _____. Glands with ducts are called _____ glands, and most of these secrete _____.

5. Fill in the appropriate type(s) of muscle—contracts rhythmically and spontaneously: _____; is controlled voluntarily: _____; contains orderly arrangements of fibrous proteins: _____, _____; is not under voluntary control: _____, _____; is found in walls of the digestive tract: _____; moves the skeleton: _____.

Review Questions

1. Define homeostasis, and explain how negative feedback helps maintain it. Describe one example of homeostasis in the human body.

2. Explain positive feedback, and provide one physiological example.

3. Explain why body temperature in humans cannot be maintained at *exactly* 98.6°F (37°C) at all times.

4. Describe the structure and functions of epithelial tissue.

5. What property distinguishes connective tissue from all other tissue types? List five types of connective tissue, and briefly describe the function of each type.

6. Describe the skin as an organ. Include the various tissues that compose it and the role of each.

Applying the Concepts

1. **IS THIS SCIENCE?** Promotional materials for a sports drink say that "water is not enough" for people who are exercising and that it's important to instead consume a liquid that contains flavoring, carbohydrates, potassium, calcium, and sodium. Do you think that consuming a sports drink after exercise can help maintain homeostasis? Explain your answer.

2. Third-degree burns are usually painless. Skin regenerates only from the edges of these wounds. Second-degree burns regenerate from cells located at the burn edges, in hair follicles, and in sweat glands. First-degree burns are painful but heal rapidly from undamaged epidermal cells. From this information, draw the depth of first-, second-, and third-degree burns on Figure 19-9.

3. Imagine you are a health care professional teaching a prenatal class for fathers. Create a design for a machine with sensors, electrical currents, motors, and so on, that would illustrate the feedback relationships involved in labor and childbirth that a layperson could understand.

For additional resources, go to www.mybiology.com.

chapter 20
Circulation and Respiration

Many student smokers plan to quit, but only one in three will succeed.

🔍 At a Glance

Case Study   Lives Up in Smoke

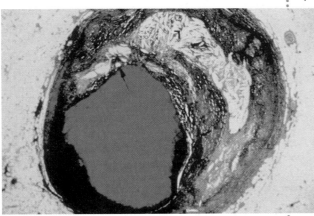

Increased risk of heart disease is among the consequences of smoking. The coronary artery (an artery supplying the heart) shown here is blocked by a blood clot that prevents blood from reaching a portion of the heart muscle. The blockage will cause a heart attack.

On virtually every college campus, a visitor to an academic building will find clusters of students lingering just outside the doors, lighting up between classes. Within the cloud of smoke that surrounds them, coughs punctuate their conversation. Like most adult smokers, these students probably picked up the habit in high school. Every day in the United States, more than 4,000 young people under age 18 light up their first cigarette. According to the Centers for Disease Control and Prevention, about 22% of U.S. high school students use tobacco, and about 10% are frequent smokers. "I started as a social smoker in high school," says a junior at the University of Illinois. "The next thing you know, I'm addicted." She's not alone; more than 30% of the students at the university have used tobacco during the past month. For many of these users, smoking in college is the beginning of a lifelong struggle with tobacco addiction.

Smoking is dangerous, and young smokers run a grave risk of harming their health. They are far more likely than nonsmokers of the same age to suffer frequent and severe respiratory infections, and smoking may retard the growth of their lungs. In general, smoking increases the incidence of diseases of the respiratory system. Less obviously, smoking also increases the likelihood of diseases of the circulatory system (the heart and blood vessels). The diseases that kill the greatest number of Americans are all linked to smoking.

Many student smokers are well aware of the dangers of smoking and say they will stop "when the time comes." Are these new smokers likely to quit? We will return to this question after reviewing the structure and function of the organ systems most likely to be damaged by smoking. ■

Web Animation Circulatory System Features

20.1 What Are the Major Features and Functions of Circulatory Systems?

Billions of years ago, the first cells were nurtured by the sea in which they had evolved. The seawater carried dissolved nutrients, which diffused into the cells, and washed away the wastes that had diffused out. Today, microorganisms and some simple multicellular animals still rely almost exclusively on diffusion for the exchange of nutrients and wastes with the environment. Sponges, for example, rely on seawater that circulates through pores in their bodies, bringing the environment close to each cell.

In larger, more complex animals, many cells are distant from the outside world. This distance suggests a question: How can nutrients reach each cell and how can cells avoid being poisoned by their own wastes? The answer lies in the circulatory system, a sort of internal sea that transports to each cell a fluid rich in food and oxygen and carries away wastes produced by the cells.

Circulatory systems have three main components. A fluid, **blood,** serves as a medium of transport. A system of channels, **blood vessels,** conducts blood throughout the body. A pump, the **heart,** keeps the blood circulating.

Animal Circulatory Systems May Be Open or Closed

Circulatory systems differ among animal species, but can be grouped into two basic types: open and closed. In an *open circulatory system*, blood flows though blood vessels but at some point leaves the vessels and moves freely through

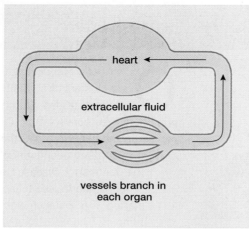

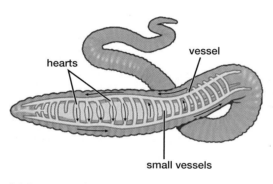

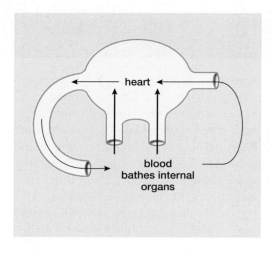

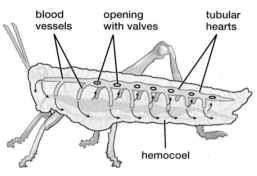

(a) Open circulatory system

(b) Closed circulatory system

▲ **Figure 20-1 Open and closed circulatory systems** *(a)* In the open circulatory system of insects and other arthropods, hearts pump blood through vessels into the hemocoel, where blood directly bathes the other organs. *(b)* In a closed circulatory system, blood remains confined to blood vessels. In the earthworm, five contractile vessels serve as hearts and pump blood through major vessels, from which smaller vessels branch.

tissues. Animals with an open circulatory system have one or more hearts, a network of blood vessels, and a large open space, called a *hemocoel*, within the body (**Fig. 20-1a**). When the heart contracts, it pumps blood through vessels that release the blood into the hemocoel. Within the hemocoel, tissues and internal organs are bathed in blood. As the heart relaxes and expands, it draws blood from the hemocoel back into the vessels and to the heart. Animals with open circulatory systems include crustaceans, spiders, insects, snails, and clams.

Some invertebrates, including the earthworm, and all vertebrates, including humans, have *closed circulatory systems* (**Fig. 20-1b**). In a closed circulatory system, blood remains confined to blood vessels and flows through them in a continuous circuit under pressure generated by the pumping of a heart or hearts. Closed circulatory systems allow more rapid blood flow, more efficient transport of wastes and nutrients, and higher blood pressure than is possible in open systems.

In the closed circulatory systems of vertebrates, vessels that carry blood away from the heart are called **arteries.** The blood in arteries ultimately flows to **capillaries,** which are microscopic vessels that penetrate tissues. The walls of capillaries are extremely thin, so dissolved substances can be exchanged between blood in capillaries and the fluid that surrounds the capillaries. This fluid, known as the **extracellular fluid,** bathes nearly all of the body's cells, and consists primarily of water and dissolved nutrients, hormones, gases, wastes, and small proteins from the blood. After passing though capillaries, blood moves back toward the heart through vessels called **veins.**

The Vertebrate Circulatory System Transports Many Substances

The vertebrate circulatory system helps sustain all of the other organ systems in the body by performing a variety of transport functions:

- Transporting oxygen from the lungs or gills to the rest of the body.

- Transporting carbon dioxide from the tissues to the lungs or gills.

- Transporting nutrients from the digestive system to the tissues.

- Transporting waste products and toxic substances to the liver (where many of them are detoxified) and to the kidney for excretion.

- Transporting hormones from the glands and organs that produce them to the tissues on which they act.

- Helping to regulate body temperature by adjusting blood flow.

- Helping to protect the body from bacteria and viruses by circulating the cells and molecules of the immune system.

In the next few sections, we look at the different parts of the vertebrate circulatory system, focusing on humans as a representative vertebrate.

20.2 How Does the Vertebrate Heart Work?

The circulatory system requires a dependable pump. Blood must be moved through the body continuously throughout an animal's life. In vertebrates, blood movement is driven by muscle contractions in the heart.

The Vertebrate Heart Consists of Muscular Chambers

In vertebrate hearts, muscular chambers called **atria** (singular, atrium) collect blood, which moves from the atria into the **ventricles,** chambers whose contractions circulate blood through the body. The number of atria and ventricles differs among different classes of vertebrates.

Fish hearts have two chambers: one atrium and one ventricle. Blood travels through the fish body in a single loop (**Fig. 20-2a**). Amphibians and most reptiles have three-chambered hearts and two circulatory loops, one for the lungs and one for the rest of the body (**Fig. 20-2b**). Because blood from both circuits mixes in the heart, the two circuits are not completely separate. In alligators, crocodiles, birds, and mammals, the heart has four chambers, and the two circulatory loops are completely separate (**Fig. 20-2c**).

Four-chambered hearts, including the human heart, can be thought of as two separate pumps. In each pump, an atrium receives and briefly stores the blood and then passes it to a ventricle that propels it through the body (**Fig. 20-3**).

One pump, consisting of the right atrium and right ventricle, deals with oxygen-depleted blood, pumping it to the lungs for replenishment. Oxygen-depleted blood flows from the body to the right atrium through two large veins, the superior vena

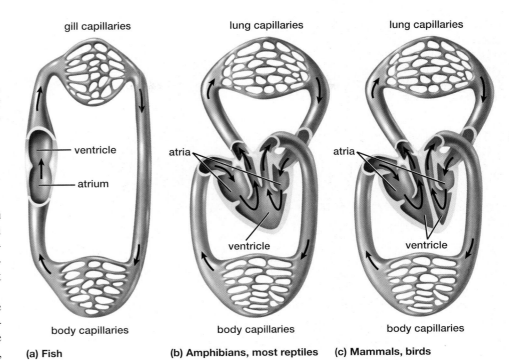

(a) Fish   **(b) Amphibians, most reptiles**   **(c) Mammals, birds**

▲ **Figure 20-2 Vertebrate hearts (a)** Fishes have two-chambered hearts. **(b)** Amphibians and most reptiles have a heart with two atria, from which blood empties into a single ventricle. **(c)** The hearts of birds and mammals are actually two separate pumps that prevent mixing of oxygenated and deoxygenated blood. Note that in this and in subsequent illustrations, oxygenated blood is depicted as bright red, and deoxygenated blood is colored blue.

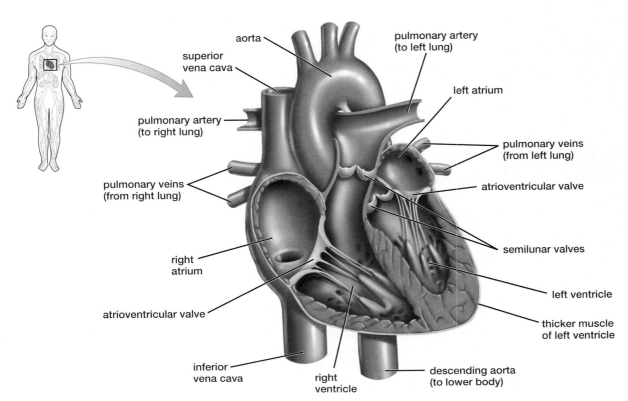

◀ **Figure 20-3 The human heart and its valves and vessels** The heart appears here as if it were in a body facing you, so that right and left appear reversed. Note the thickened walls of the left ventricle, which must pump blood much farther through the body than does the right ventricle, which propels blood to the lungs. One-way valves, called semilunar valves, are located between the aorta and the left ventricle and between the pulmonary artery and the right ventricle. Atrioventricular valves separate the atria and ventricles.

▶ Figure 20-4 The cardiac cycle
QUESTION: How would this cycle be altered if the right atrioventricular valve were damaged badly enough to prevent it from functioning?

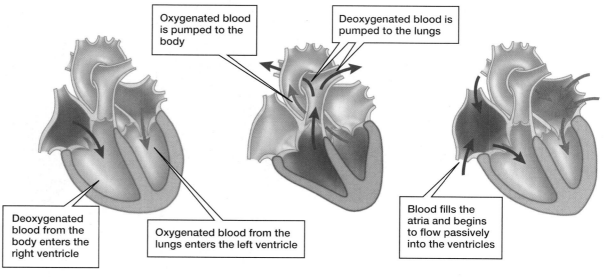

Oxygenated blood is pumped to the body

Deoxygenated blood is pumped to the lungs

Deoxygenated blood from the body enters the right ventricle

Oxygenated blood from the lungs enters the left ventricle

Blood fills the atria and begins to flow passively into the ventricles

❶ Atria contract, forcing blood into the ventricles

❷ Then the ventricles contract, forcing blood through the arteries to the lungs and the rest of the body

❸ The cycle ends as the heart relaxes

BioFlix Muscle Contraction

Web Animation Structure of the Human Heart

Web Animation Function of the Human Heart

Web Animation Measuring Blood Pressure

cava and the inferior vena cava. When full, the right atrium contracts, forcing the blood into the right ventricle. Then, the right ventricle contracts, sending the oxygen-depleted blood to the lungs via pulmonary arteries.

The other pump, consisting of the left atrium and ventricle, moves oxygen-rich blood. Oxygen-rich blood from the lungs enters the left atrium through pulmonary veins and is then squeezed into the left ventricle. Strong contractions of the left ventricle, the heart's most muscular chamber, send the oxygenated blood coursing out through a major artery, the *aorta*, to the rest of the body.

The Atria and Ventricles Contract in a Coordinated Cycle

Your heart beats about 100,000 times each day, its chambers alternately contracting and relaxing. The two atria contract at the same time, emptying their contents into the ventricles. A fraction of a second later, the two ventricles contract simultaneously, forcing blood into arteries that exit the heart. Both atria and ventricles then relax briefly before the cycle repeats (Fig. 20-4). At a normal resting heart rate, this cycle lasts just under 1 second. Blood pressure changes as the cycle proceeds. *Systolic* pressure is measured during ventricular contraction and *diastolic* pressure is measured between contractions (Fig. 20-5).

Valves Prevent Blood from Moving in the Wrong Direction

Maintaining proper, one-way blood flow presents challenges. When the ventricles contract, blood must move only out through the arteries and not back into the atria. Then, once blood has left the ventricles and entered the arteries, it must be prevented from flowing back as the heart relaxes. These needs are met by four one-way *valves* (see Figs. 20-3 and 20-4). Pressure in one direction opens the valves, but reverse pressure forces them tightly closed. **Atrioventricular valves** separate the atria from the ventricles. **Semilunar valves** allow blood to enter the pulmonary artery and the aorta when the ventricles contract, but prevent it from returning as the ventricles relax. Closing valves make the "lub-dub" sound of a beating heart heard through a stethoscope.

Heart valves may be attacked and damaged by bacterial infections, allowing blood to flow inappropriately between the chambers. In people whose valves have been damaged, valve replacement, using an artificial valve or a valve transplanted from a pig or a cow, can be lifesaving.

▶Figure 20-5 **Measuring blood pressure** Blood pressure is measured with an inflatable blood pressure cuff and a stethoscope. The cuff is inflated until its pressure closes off the arm's main artery, blood ceases to flow, and no pulse can be detected below the cuff. Then the pressure is gradually reduced. When the pulse is first audible in the artery, the pressure pulses created by the contracting left ventricle are just overcoming the pressure in the cuff and blood is flowing. This is the upper reading: the systolic pressure. Cuff pressure is then further reduced until no pulse is audible, indicating that blood is flowing continuously through the artery and that the pressure between ventricular contractions is just overcoming the cuff pressure. This is the lower reading: the diastolic pressure. The numbers are in millimeters of mercury (abbreviated mmHg), a standard measure of pressure also used in barometers. A healthy person should have blood pressure below 120/80 (systolic below 120 mmHg and diastolic pressure below 80 mmHg). **EXERCISE:** Sketch a graph that shows how blood pressure inside an artery changes during the cardiac cycle. On the graph, label the points that correspond to the systolic and diastolic blood pressure measured by the blood pressure cuff.

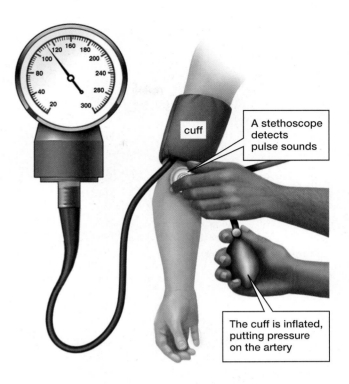

cuff

A stethoscope detects pulse sounds

The cuff is inflated, putting pressure on the artery

Electrical Impulses Coordinate the Sequence of Contractions

A second challenge is to coordinate the contractions of all four chambers. The atria must contract first and empty their contents into the ventricles, so that the atria can refill while the ventricles contract. For this sequence to be maintained, there must be a delay between contraction of the atria and contraction of the ventricles. The necessary coordination of contractions is initiated by a *pacemaker*, a cluster of specialized heart muscle cells that produce spontaneous electrical signals at a regular rate. The electrical signals pass rapidly among the interconnected muscle cells of the heart and cause them to contract.

The heart's primary pacemaker is the **sinoatrial (SA) node,** located in the wall of the right atrium (**Fig. 20-6**). From the SA node, an electrical impulse

◀Figure 20-6 **The heart's pacemaker and its connections**

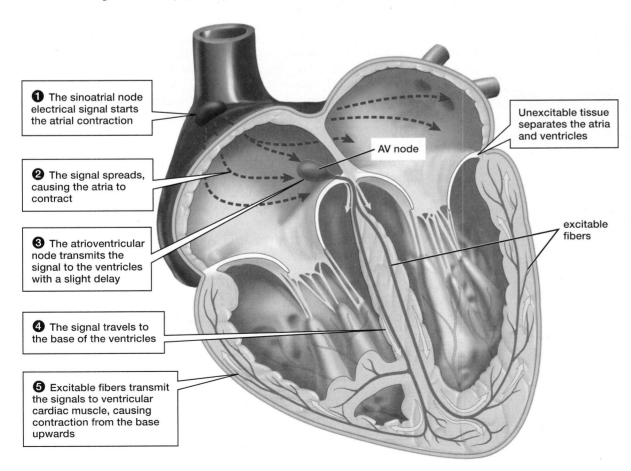

❶ The sinoatrial node electrical signal starts the atrial contraction

❷ The signal spreads, causing the atria to contract

❸ The atrioventricular node transmits the signal to the ventricles with a slight delay

❹ The signal travels to the base of the ventricles

❺ Excitable fibers transmit the signals to ventricular cardiac muscle, causing contraction from the base upwards

AV node

Unexcitable tissue separates the atria and ventricles

excitable fibers

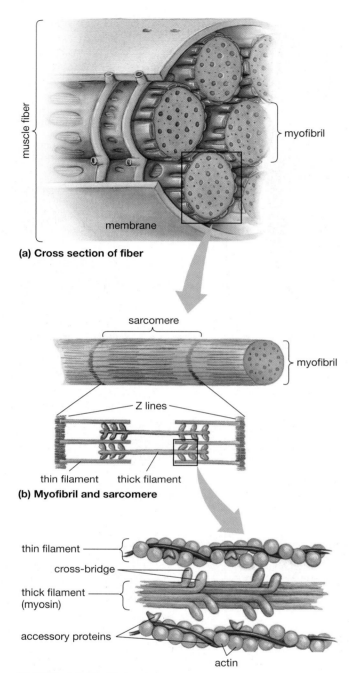

(a) Cross section of fiber

muscle fiber

myofibril

membrane

sarcomere

myofibril

Z lines

thin filament

thick filament

(b) Myofibril and sarcomere

thin filament

cross-bridge

thick filament (myosin)

accessory proteins

actin

(c) Thick and thin filaments

▲ **Figure 20-7 A muscle fiber** **(a)** Each muscle fiber is surrounded by a membrane and contains numerous myofibrils. **(b)** Each myofibril consists of sarcomeres, attached end to end by proteins called Z lines. **(c)** Within each sarcomere, thick and thin filaments alternate and can be connected by cross-bridges.

creates a wave of contraction that sweeps through the muscles of the right and left atria, which contract together. Then the signal reaches a barrier of unexcitable tissue between the atria and the ventricles. There the excitation is channeled through the **atrioventricular (AV) node,** a small mass of specialized muscle cells located in the floor of the right atrium (see Fig. 20-6). The impulse is delayed at the AV node, postponing the ventricular contraction for about 0.1 second after the atria contract. This delay gives the atria time to complete the transfer of blood into the ventricles before ventricular contraction begins. From the AV node, the signal to contract spreads along excitable fibers to the base of the two ventricles. The impulse then travels rapidly from these fibers up through the muscle cells of the ventricles, causing the ventricles to contract in unison.

Disorders can interfere with the events that produce smooth, regular heartbeats. If the pacemaker fails, uncoordinated, irregular contractions called *fibrillation* occur. Fibrillation of the ventricles can be fatal, because blood is not pumped out of the heart to the brain and other organs but is merely sloshed around. People who experience fibrillation may be treated with a defibrillating machine, which applies a jolt of electricity to the heart, synchronizing the contractions of the ventricular muscles and sometimes allowing the pacemaker to resume its normal coordinating function.

The Heart's Contractions Result from Movement of Filaments in Muscle Cells

Like the muscles that enable body movement, the muscle tissue that makes up the heart consists of cells known as *muscle fibers.* Each muscle fiber contains many *myofibrils*, cylindrical structures that extend from one end of the fiber to the other (**Fig. 20-7a**). The myofibrils are composed of subunits called *sarcomeres*, which are aligned end to end along the length of the myofibril (**Fig. 20-7b**).

Within each sarcomere is a beautifully precise arrangement of filaments of the proteins *actin* and *myosin.* Actin molecules form the *thin filaments*, and myosin forms the *thick filaments.* The thick filaments are suspended between the thin filaments and are capable of linking temporarily to the thin filaments using small myosin projections called cross-bridges (**Fig. 20-7c**). The thin filaments are attached to fibrous protein bands called *Z lines,* which separate adjacent sarcomeres. The regular arrangement of thick and thin filaments in all the myofibrils gives heart (and skeletal) muscle fibers their striped appearance.

Muscle contraction is controlled by a process that depends on the molecular structure of the thin filament. The actin protein that makes up most of the thin filament is formed from a double chain of subunits and resembles a twisted double strand of pearls. Each of the subunits has a site that can form a bond with myosin. In a relaxed muscle cell, however, these binding sites are blocked by molecules of accessory proteins (see Fig. 20-7c).

When a muscle contracts, the accessory proteins of the thin filament move aside, exposing the binding sites on the actin. As soon as the sites are exposed, myosin (in the thick filaments) binds the actin (in the thin filaments), forming cross-bridges. Using energy from the splitting of ATP, the cross-bridges repeatedly bend, release, and reattach farther along, much like a sailor pulling in an anchor line hand over hand (**Fig. 20-8a**). The thin filaments are pulled past the thick filaments, shortening the sarcomere and contracting the muscle (**Fig. 20-8b**).

20.3 What Is Blood?

Blood, which has been called the "river of life," transports dissolved nutrients, gases, hormones, and wastes through the body. It has two major components: a fluid called **plasma** and cellular components (red blood cells, white blood cells, and platelets) that are suspended in the plasma (**Fig. 20-9**). The cellular components develop in the bone marrow before moving into the blood. On average, they account for 40% to

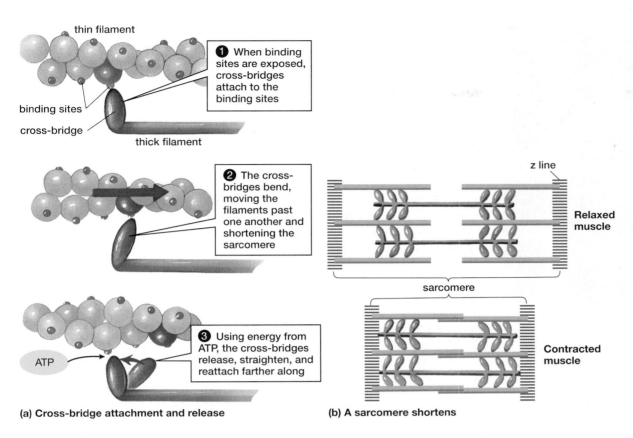

① When binding sites are exposed, cross-bridges attach to the binding sites

thin filament

binding sites

cross-bridge

thick filament

② The cross-bridges bend, moving the filaments past one another and shortening the sarcomere

③ Using energy from ATP, the cross-bridges release, straighten, and reattach farther along

ATP

(a) Cross-bridge attachment and release

z line

Relaxed muscle

sarcomere

Contracted muscle

(b) A sarcomere shortens

◀ **Figure 20-8 Muscle contraction** *(a)* The cycle of cross-bridge attachment, bending, release, and reattachment repeats over and over again. *(b)* The thick and thin filaments slide past one another, shortening the muscle cell.

45% of blood volume, and the remainder is plasma. The average human has 5 to 6 quarts of blood, which account for about 8% of a person's total body weight.

Plasma Is Primarily Water and Dissolved Substances

Water makes up about 90% of the straw-colored plasma. Dissolved in the plasma are proteins, hormones, nutrients (glucose, vitamins, amino acids, lipids), gases (carbon dioxide, oxygen), salts (sodium, calcium, potassium, magnesium), and wastes, such as urea. Plasma proteins are the most abundant of the dissolved substances.

Red Blood Cells Carry Oxygen from the Lungs to the Rest of the Body

The most abundant cells in the blood are the oxygen-carrying **red blood cells,** or erythrocytes. In fact, red blood cells make up about 99% of all blood cells. The shape of a red blood cell resembles a ball of clay that has been squeezed between your thumb and forefinger (**Fig. 20-9a**). This shape provides a larger surface area

▼ **Figure 20-9 Blood cells** *(a)* This false-color scanning electron micrograph shows the shape of red blood cells. *(b)* This composite photo of stained white blood cells shows the five different types. *(c)* Here a single megakaryocyte is budding off dozens of platelets.

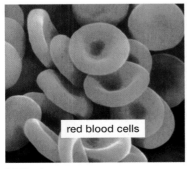

red blood cells

(a) Erythrocytes

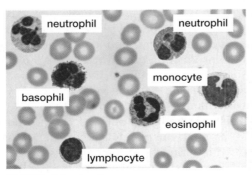

neutrophil

neutrophil

monocyte

basophil

eosinophil

lymphocyte

(b) White blood cells

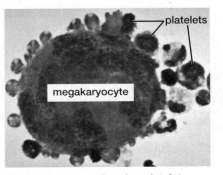

platelets

megakaryocyte

(c) Megakaryocyte forming platelets

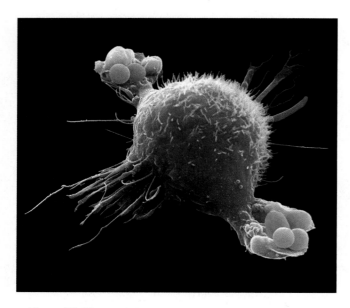

▲ **Figure 20-10   A white blood cell attacks bacteria** The bacteria (small green spheres) are *Escherichia coli*, intestinal bacteria that can cause disease if they enter the bloodstream.

than would a spherical cell of the same volume, and thereby increases the cell's ability to absorb and release oxygen through its plasma membrane.

Red blood cells get their red color from **hemoglobin,** an iron-containing protein that accounts for about one-third the weight of each red blood cell and carries about 97% of the blood's oxygen. One hemoglobin molecule can bind and carry up to four molecules of oxygen, permitting blood to hold far more oxygen than would be possible if all the gas were simply dissolved in plasma. Hemoglobin picks up oxygen in the blood vessels of the lungs, where oxygen concentration is high, and releases it in the blood vessels of other tissues of the body, where the oxygen concentration is lower. As hemoglobin releases oxygen, the hemoglobin molecule undergoes a slight change in shape, which alters its color. As a result, the bright cherry-red color of oxygen-rich blood changes to a dark maroon-red that appears bluish through the skin.

Red blood cells are short-lived; their average life span is about 4 months. Every second, more than 2 million red blood cells die and are replaced by new ones from the bone marrow. Dead or damaged red blood cells are removed from circulation, primarily in the liver and spleen, and broken down to release their iron. The iron is recycled in the hemoglobin of new red blood cells.

White Blood Cells Help Defend the Body Against Disease

White blood cells, or leukocytes, make up less than 1% of blood cells but play a key role in the body's resistance to disease. The five types of white blood cells—*neutrophils, eosinophils, basophils, lymphocytes,* and *monocytes* (see Fig. 20-9b)—together play a key role in the body's defense against invasion by disease-causing microbes. For example, lymphocytes give rise to cells that produce the immune response against disease (as described in Chapter 22). Neutrophils and monocytes travel through blood vessels to wounds where bacteria have gained entry, and then ooze out through narrow openings in the capillary walls to attack and engulf foreign particles, including bacteria. Some monocytes that have emerged from capillaries change into amoeba-like, particle-engulfing cells called *macrophages* (**Fig. 20-10**). The macrophages and neutrophils that attack bacteria typically die in the process, and their dead bodies accumulate and contribute to the white substance, called *pus,* seen at infection sites.

Platelets Are Cell Fragments That Aid in Blood Clotting

Platelets, which are crucial to blood clotting, are pieces of large cells called *megakaryocytes.* Megakaryocytes remain in the bone marrow, where they pinch off membrane-enclosed chunks of their cytoplasm that we call platelets (see Fig. 20-9c). The platelets then enter the blood and play a central role in blood clotting.

Blood clotting is a complex process that keeps us from bleeding to death from normal wear and tear on the body. Clotting starts when platelets contact an irregular surface, such as a damaged blood vessel. The ruptured surface of an injured blood vessel causes platelets to adhere and partially block the opening. The adhering platelets and the injured tissues initiate a complex sequence of reactions among circulating plasma proteins. These reactions result in the production of a fibrous network of molecules of the protein fibrin. This protein web immobilizes the fluid portion of the blood, causing it to solidify in much the same way that gelatin does as it cools. The web traps red blood cells, further increasing the density of the clot (**Fig. 20-11**). Platelets adhere to the fibrous mass and send out sticky projections that attach to one another. Within half an hour, the platelets contract, pulling the mesh tighter and forcing liquid out. This action creates a denser, stronger clot (on the skin it is called a *scab*) and also constricts the wound, pulling the damaged surfaces closer together in a way that promotes healing.

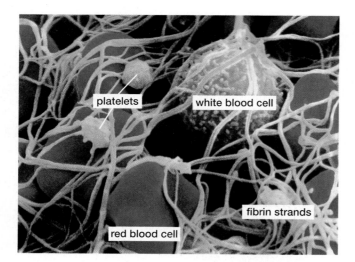

platelets

white blood cell

fibrin strands

red blood cell

▲ **Figure 20-11   Blood clotting** Threadlike fibrin proteins produce a tangled sticky mass that traps red blood cells and eventually forms a clot.

20.4 What Are the Types and Functions of Blood Vessels?

The river of life flows in well-defined channels called blood vessels. Some of the major blood vessels of the human circulatory system are diagrammed in **Figure 20-12**. As blood leaves the heart, it travels from arteries to arterioles to capillaries to venules to veins, which return it finally to the heart. These vessels are shown in **Figure 20-13**. The consequences of impaired vessel function are discussed in "Health Watch: Matters of the Heart" on pp. 382–383.

Arteries and Arterioles Carry Blood Away from the Heart

Arteries carry blood away from the heart. These vessels have thick walls embedded with smooth muscle and elastic connective tissue (see Fig. 20-13). With each surge of blood from the ventricles, the arteries expand slightly, like thick-walled balloons. As their elastic walls recoil between heartbeats, the arteries help pump the blood and maintain a steady flow through the smaller vessels. Arteries branch into vessels of smaller diameter called **arterioles.**

Capillaries Are Microscopic Vessels Through Which Nutrients and Wastes Are Exchanged

The entire circulatory system is an elaborate device for allowing each cell of the body to exchange nutrients and wastes by diffusion. This diffusion occurs in capillaries, the smallest of all blood vessels. Here wastes, nutrients, gases, and hormones are exchanged between the blood and the extracellular fluid.

Capillaries are finely adapted to their role of exchange; their walls are only one cell thick. Thus, substances that can cross a capillary cell's plasma membrane can easily move into and out of capillaries. For example, fatty acids, fat-soluble hormones, oxygen, and carbon dioxide diffuse readily across capillary cell plasma membranes, whereas charged particles (such as sodium, potassium, calcium, and chloride ions) and small molecules such as water, glucose, amino acids, and urea diffuse through tiny gaps between adjacent cells in the capillary wall. White blood cells can also ooze through these openings. However, large plasma proteins, red blood cells, and platelets are unable to leave the capillaries because they are too large.

Capillaries are so narrow that red blood cells must pass through them in single file (**Fig. 20-14**). Consequently, all blood is sure to pass very close to the capillary walls, where exchange occurs. In addition, capillaries are so numerous that no body cell is more than 100 micrometers (about as thick as four pages of this book) from a capillary. A typical person has about 50,000 miles (80,600 kilometers) of capillaries! The speed of blood flow drops very quickly as it moves through this narrow, almost endless capillary network, allowing more time for diffusion.

The flow of blood in capillaries is regulated by tiny rings of smooth muscle called *precapillary sphincters*, which surround the junctions between arterioles and capillaries (see Fig. 20-13). The sphincters open and close in response to local chemical changes that signal the needs of nearby tissues.

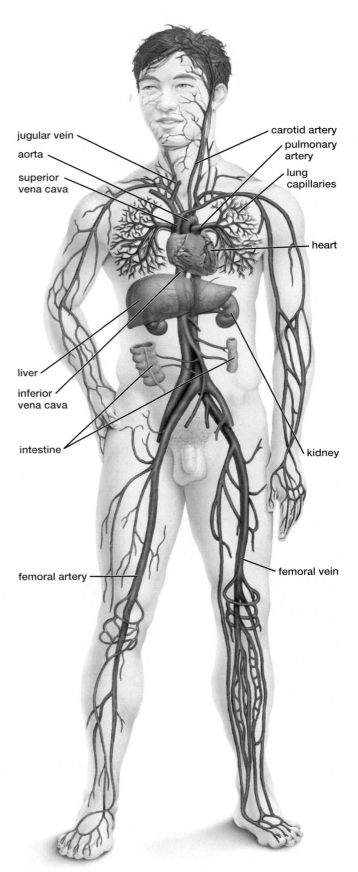

▶ **Figure 20-12 The human circulatory system** Veins (shown in blue) carry blood to the heart, and arteries (shown in red) conduct blood away from the heart. In general, veins carry deoxygenated blood and arteries carry oxygenated blood, but the pulmonary veins (which carry oxygenated blood from the lungs to heart) and pulmonary arteries (which carry deoxygenated blood from the heart to the lungs) are exceptions.

▶ **Figure 20-13 Structures and interconnections of blood vessels** Oxygenated blood moves from arteries to arterioles to capillaries. Capillaries empty deoxygenated blood into venules, which empty into veins. Precapillary sphincters regulate the movement of blood from arterioles into capillaries.

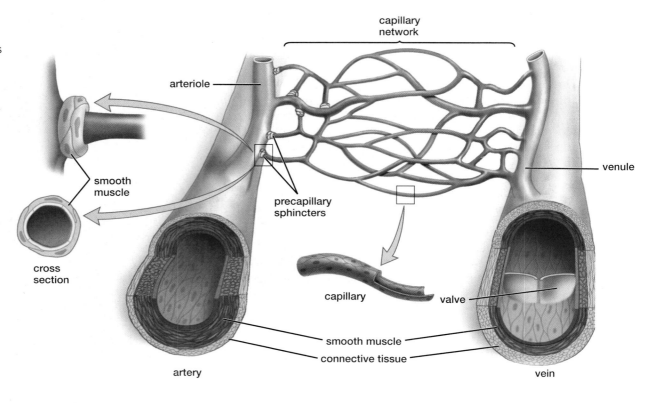

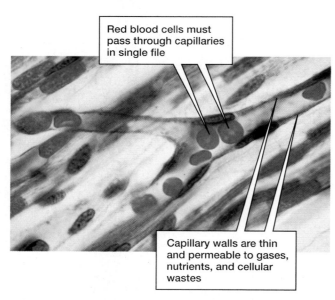

Red blood cells must pass through capillaries in single file

Capillary walls are thin and permeable to gases, nutrients, and cellular wastes

▲ **Figure 20-14 Red blood cells flow through a capillary**
QUESTION: Why does oxygen move out of capillaries in body tissues while carbon dioxide moves in (instead of the other way around)?

Venules and Veins Carry Blood Back to the Heart

After picking up carbon dioxide and other cellular wastes from cells, capillary blood drains into larger vessels called **venules,** which empty into still larger veins (see Fig. 20-13). Veins provide a low-resistance pathway by which blood can return to the heart. The walls of veins are thinner, less muscular, and more expandable than those of arteries, although both contain a layer of smooth muscle. To prevent blood from flowing away from the heart, veins are equipped with valves that allow blood to flow in only one direction, toward the heart (Fig. 20-15).

Because blood pressure in the veins is too low to push blood back to the heart, contractions of skeletal muscle during exercise and breathing help return blood to the heart by squeezing the veins and forcing blood through them. When you sit or stand for long periods, the lack of muscular activity allows blood to accumulate in the veins of the lower legs. This is why you may find your feet swollen after a long airplane flight. Long periods of inactivity can also contribute to varicose veins, in which the valves become stretched and weakened, and the veins become permanently swollen and often visible through the skin.

20.5 How Does the Lymphatic System Work with the Circulatory System?

The **lymphatic system** consists of a network of lymph vessels that empty into the circulatory system, numerous small lymph nodes, and two additional organs: the thymus and the spleen (Fig. 20-16). Although not strictly part of the circulatory system, the lymphatic system is closely associated with it. The lymphatic system

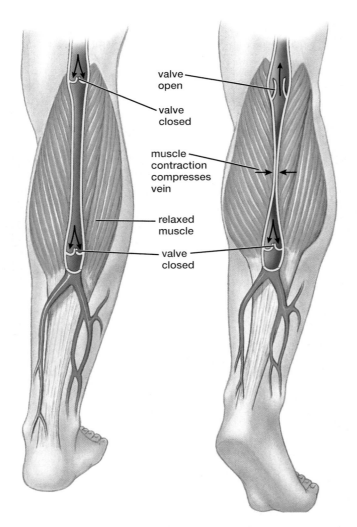

▲ **Figure 20-15** **Valves direct the flow of blood in veins** Veins and venules have one-way valves that maintain blood flow toward the heart. When the vein is compressed by nearby muscles, the valves allow blood to flow toward the heart but then clamp shut to prevent backflow.

removes excess fluid and dissolved substances that leak from the capillaries, transports fats from the small intestine to the bloodstream, and defends the body by exposing bacteria and viruses to white blood cells.

Lymphatic Vessels Resemble the Capillaries and Veins of the Circulatory System

The smallest lymph vessels are lymph capillaries. Like blood capillaries, lymph capillaries form a complex network of microscopically narrow, thin-walled vessels into which substances can move readily. Unlike the walls of blood capillaries, lymph capillary walls have specialized openings between the cells that act as one-way valves. These openings allow relatively large particles to be carried into lymph capillaries along with fluid. Also unlike blood capillaries, which form a

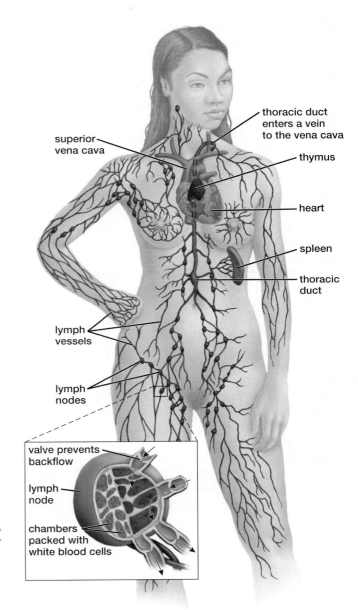

▲ **Figure 20-16** **The human lymphatic system** Lymph vessels, lymph nodes, and the thymus and spleen. Lymph returns to the circulatory system by way of the thoracic duct, which empties into a vein that in turn empties into the vena cava, a large vein. *(Inset)* A cross section of a lymph node. The node is filled with channels lined with white blood cells that attack foreign matter in the lymph.

health watch

Matters of the Heart

Cardiovascular disease, which impairs the heart and blood vessels, is the leading cause of death in the United States, killing nearly 1 million Americans annually. And no wonder. Your heart must contract vigorously more than 2.5 billion times during your lifetime without once stopping to rest, forcing blood through vessels whose total length would encircle the globe twice. Because these vessels may become constricted, weakened, or clogged, the cardiovascular system is a prime candidate for malfunction.

Atherosclerosis Obstructs Blood Vessels

Atherosclerosis causes the walls of the large arteries to thicken and lose their elasticity. This change is caused by deposits called *plaques*, which are composed of cholesterol and other fatty substances as well as calcium and fibrin. Plaques are deposited within the wall of the artery between the smooth muscle and the tissue that lines the artery (**Fig. E20-1**). A plaque may rupture through the lining into the interior of the vessel, stimulating platelets to adhere to the vessel wall and initiate blood clots. These clots further obstruct the artery and may completely block it (see the photo at the beginning of this chapter). Or a clot may break loose and clog a narrower artery "downstream." Arterial clots are responsible for the most serious consequences of atherosclerosis: heart attacks and strokes.

A *heart attack* occurs when one of the coronary arteries is blocked. (Coronary arteries supply the heart muscle itself.) Deprived of

nutrients and oxygen, a heart muscle whose blood supply is curtailed by a blocked artery dies rapidly and painfully. Heart attacks are the major cause of death from atherosclerosis. About 1.1 million Americans suffer heart attacks each year, and about half a million people die from them. But atherosclerosis also causes plaques and clots to form in arteries throughout the body. If a clot or plaque obstructs an artery that supplies the brain, it can cause a stroke, in which brain function is lost in the area deprived of blood and its vital oxygen and nutrients.

Treatment of Atherosclerosis

The exact cause of atherosclerosis is unclear, but it is promoted by high blood pressure, cigarette smoking, genetic predisposition, obesity, diabetes, lack of exercise, and high blood levels of a certain type of cholesterol bound to a carrier molecule called low-density lipoprotein (LDL). If LDL-bound cholesterol levels are too high, cholesterol may be deposited in arterial walls. In contrast, cholesterol bound to high-density lipoprotein (HDL) is metabolized or excreted and hence is often called "good" cholesterol.

Treatment of atherosclerosis includes the use of drugs or changes in diet and lifestyle to lower blood pressure and blood cholesterol levels. If a person has survived a heart attack or suffers from *angina* (chest pain caused by insufficient blood flow to the heart), he or she may be a candidate for *angioplasty*, a procedure to widen an obstructed coronary artery (**Fig. E20-2**). In

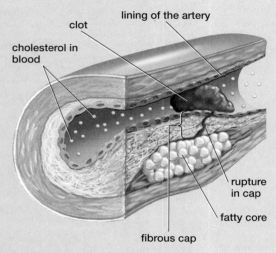

▲ **Figure E20-1 Plaques clog arteries** When the fibrous cap ruptures, a blood clot forms, obstructing the artery.

angioplasty, a thin, flexible tube is threaded through an artery in the upper leg or arm and guided into the clogged artery. The tube is typically tipped with a small balloon that, when inflated, compresses the plaque. The angioplasty tube may be equipped with small rotating blades that shear off the plaque as it is compressed. Alternatively, the tube may carry a high-speed, diamond-coated drill that grinds the plaque into microscopic pieces that are carried away in the blood. After the plaque is removed, a wire mesh tube, or *stent*, is often inserted into the artery to help keep it open.

continuous interconnected network, lymph capillaries dead-end in the body's tissues.

Materials collected by the lymph capillaries flow into larger lymph vessels. Large lymph vessels have somewhat muscular walls, but, as in blood-transporting veins, most of the force for lymph flow comes from the contraction of nearby muscles, such as those used in breathing and walking. As in blood-transporting veins, the direction of flow is regulated by one-way valves.

The Lymphatic System Returns Fluids to the Blood

As described earlier, dissolved substances are exchanged between the blood capillaries and body cells through the extracellular fluid that bathes nearly all of the body's cells. In an average person, the amount of fluid that leaves the blood capillaries each day exceeds the amount that is reabsorbed by them by about

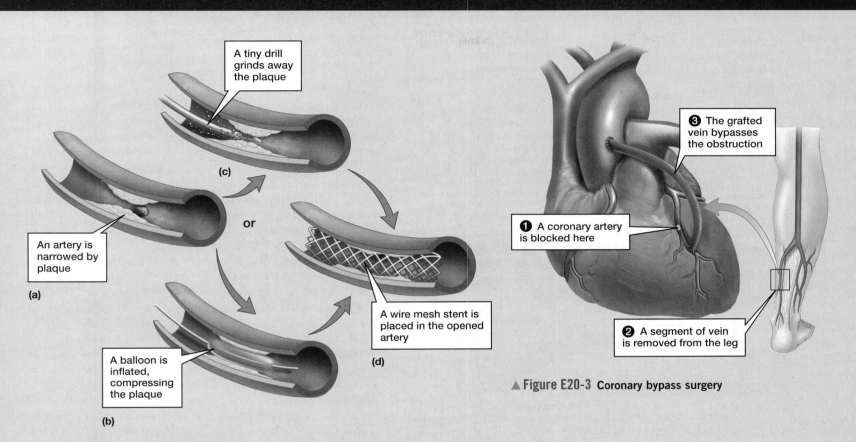

▲ **Figure E20-2** **Angioplasty unclogs arteries** **(a)** A narrowed artery may be opened by **(b)** inflating a tiny balloon or **(c)** drilling out the plaque. Following angioplasty, **(d)** a metal mesh stent is often inserted to maintain the opening.

▲ **Figure E20-3** **Coronary bypass surgery**

Coronary bypass surgery consists of bypassing one or more obstructed coronary arteries with a piece of vein (usually obtained from the patient's leg; **Fig. E20-3**) or with an artery (often from the patient's forearm). But many patients who might benefit from bypass surgery do not have vessels suitable for grafting, In an effort to help these patients, researchers are working to develop artificial vessels. For example, when tubes molded from collagen protein are placed in a nutrient broth with living blood vessel cells, the cells invade and cover the tubes to form at least temporarily functional vessels. In another promising method, a plastic mold serves as a scaffold for vessels grown from the patient's own cells. These techniques are promising, but they need more development and testing. It will be years before artificial vessels are available for use in coronary bypass surgery.

3 quarts. The lymphatic system returns this excess fluid to the blood. As extracellular fluid accumulates, its pressure forces the fluid through the one-way openings in the lymph capillary walls (**Fig. 20-17**). The lymphatic system transports this fluid, now called **lymph,** back to the circulatory system.

The Lymphatic System Transports Fats from the Small Intestine to the Blood

After a fatty meal, the cells of the small intestine absorb globules of digested fat. These globules are released from the intestinal cells into the extracellular fluid, but they are too large to diffuse into blood capillaries. They can, however, easily move through the openings between lymph capillary cells. Once in the lymph, they are dumped into veins that carry them to the superior vena cava, a large vein that enters the heart.

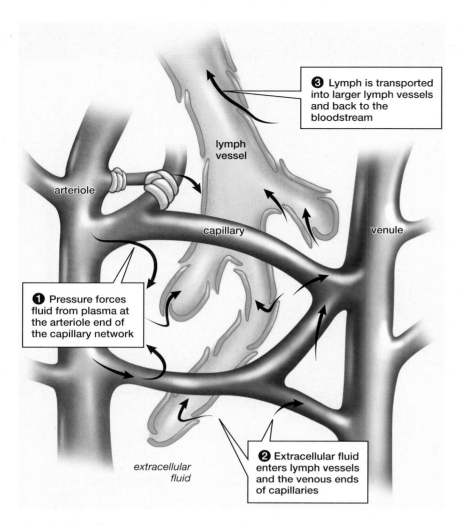

③ Lymph is transported into larger lymph vessels and back to the bloodstream

lymph vessel

arteriole

capillary

venule

① Pressure forces fluid from plasma at the arteriole end of the capillary network

extracellular fluid

② Extracellular fluid enters lymph vessels and the venous ends of capillaries

◀ **Figure 20-17 Lymph capillary structure** Lymph capillaries end blindly in the body tissues, where pressure from the accumulation of extracellular fluid forces the fluid into the lymph capillaries.

The Lymphatic System Helps Defend the Body Against Disease

In addition to its other roles, the lymphatic system helps defend the body against foreign invaders, such as bacteria and viruses. The system includes patches of connective tissue, located in the linings of the respiratory, digestive, and urinary tracts, that contain large numbers of white blood cells. The largest of these patches are the *tonsils*, located in the cavity behind the mouth. Inserted at intervals within the large lymph vessels are kidney bean-shaped structures about 1 inch (2.5 centimeters) long called **lymph nodes** (see Fig. 20-16). In the nodes, lymph is forced through channels lined with masses of white blood cells that recognize and destroy foreign particles. The white blood cells are killed in the process of destroying invaders, and the painful swelling of lymph nodes that accompanies certain diseases is largely a result of the accumulation of dead white blood cells and dead virus-infested cells they have engulfed.

The thymus and the spleen are often considered part of the lymphatic system (see Fig. 20-16). The **thymus,** an organ in which certain types of white blood cells mature, is located beneath the breastbone, slightly above the heart. The thymus is particularly active in infants and young children but decreases in size and importance in early adulthood. The **spleen** is located in the left side of the abdominal cavity, between the stomach and diaphragm. Just as the lymph nodes filter lymph, the spleen filters blood, exposing it to white blood cells that destroy foreign particles and aged red blood cells.

Lives Up in Smoke

Continued

Many people know that smoking increases the risk of lung cancer, but fewer are aware that smoking is also a risk factor for other forms of cancer. For example, cancers of the lymphatic system are more likely to occur in smokers than in nonsmokers. *Lymphoma*, as cancers of the lymphatic system are collectively known, affects the infection-fighting white blood cells that are abundant in the lymphatic system. In people with lymphoma, cancerous white blood cells accumulate and form tumors in the structures of the lymphatic system, most commonly in lymph nodes.

20.6 How Are Oxygen and Carbon Dioxide Exchanged in Animal Bodies?

One of the key functions of circulatory systems is to transport oxygen to the tissues and carbon dioxide away from the tissues. To accomplish this function, circulatory systems must be able to acquire oxygen gas from the external environment and deliver carbon dioxide gas there. This process of gas exchange is called **respiration,** and the organ systems that exchange gases are called *respiratory systems*. As you might expect, respiratory systems are usually closely connected to circulatory systems.

Animal respiratory systems are diverse, and include both systems that function in terrestrial environments and systems that operate in aquatic environments. In all cases, respiratory systems rely on diffusion of gases across a respiratory surface.

Respiratory surfaces are large and moist. They must have a large surface area in contact with the environment to allow adequate gas exchange, and they must remain moist, because gases must be dissolved in water to diffuse into or out of cells.

Aquatic Animals May Have Gills

Some very small animals that live in moist environments don't have specialized respiratory systems. Instead, gases are exchanged across the body surface. Most larger animals, however, have specialized respiratory structures. For example, many animals that live in water have respiratory organs called **gills.** Gills usually have elaborately branched or folded shapes, which maximize surface area (**Fig. 20-18**).

In fish, delicate gill tissue is typically protected beneath a bony flap. A fish creates a continuous flow of water over the gills by pumping water into its mouth and ejecting it through an opening just behind the gills. The gills themselves consist of a series of filaments (Fig. 20-19). Each filament is attached to a bony *gill arch* and is served by an incoming blood vessel that carries oxygen-depleted blood and an outgoing vessel that contains oxygen-rich blood. Filaments are covered with thin folds of tissue called *lamellae* (singular, lamella). Capillaries lie just beneath the outer membranes of the lamellae, so that blood passes close to the lamellae's surface, where gases are exchanged.

Terrestrial Animals Have Internal Respiratory Structures

Land-dwelling animals live surrounded by air, which has a far higher oxygen concentration than does water. Extracting oxygen from dry air, however, presents special challenges. Respiratory surfaces must remain moist, but moist tissues on the outside of the body lose water continuously by evaporation and tend to dry out. As a result, land animals have evolved internal structures in which respiratory organs are protected from drying. In land-dwelling vertebrates, the respiratory structures are lungs. **Lungs** are chambers containing moist gas-exchange membranes that are protected within the body, where water loss is minimized and the body wall provides support.

20.7 How Does the Human Respiratory System Work?

The respiratory system in humans and other lung-breathing vertebrates has two parts: the conducting portion and the gas-exchange portion. The conducting portion consists of a series of passageways that carry air into and out of the gas-exchange portion, where gases move to and from the blood.

The Conducting Portion Carries Air to the Lungs

The conducting portion of the respiratory system brings air to the lungs. Air enters through the nose or the mouth, passes through the nasal cavity or oral cavity into the **pharynx,** and then travels through the **larynx** (Fig. 20-20a). The pharynx is

▲ Figure 20-18 **External gills of a mollusk**

Web Animation Human Respiratory Anatomy

▼ Figure 20-19 **Gills exchange gases with water** The gill reaches its greatest complexity in fish. *(a)* Fish pump water in through their mouths and out over their gills. *(b)* Water flows past a dense array of paired filaments attached to bony gill arches. *(c)* Lamellae protrude from each gill filament. Water flows over the lamellae in the opposite direction from the blood flowing through them in capillary beds.

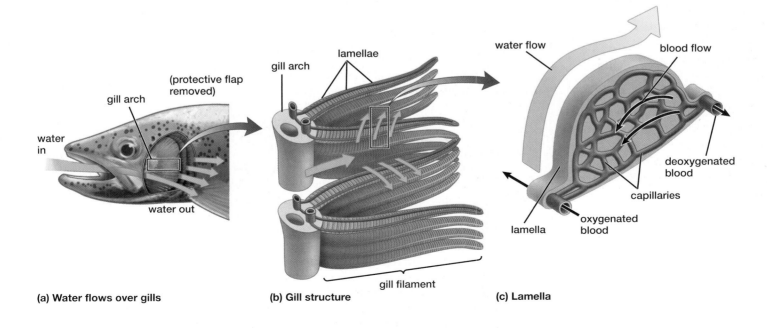

(a) Water flows over gills

(b) Gill structure

(c) Lamella

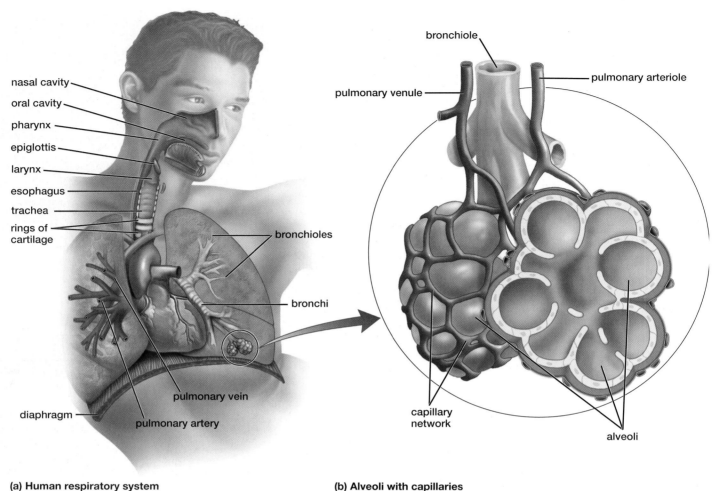

(a) Human respiratory system

(b) Alveoli with capillaries

▲ **Figure 20-20 The human respiratory system *(a)*** The passages and vessels of the respiratory system and its connection with the circulatory system. ***(b)*** Close-up of alveoli (interiors shown in cut-away section) and their surrounding capillaries. **QUESTION:** The nasal passages of most mammals direct air over a large surface area of membrane that covers complex ridges of bone. Why?

shared by the respiratory and digestive systems, so both food and air pass through it. Food is prevented from entering the rest of the respiratory system by the *epiglottis*, a flap of tissue supported by cartilage that guards the opening from the pharynx to the larynx.

During normal breathing, the epiglottis is tilted upward, as shown in Figure 20-20a, allowing air to flow freely into the larynx. During swallowing, the epiglottis tilts downward and covers the larynx, directing substances into the esophagus instead. If a person attempts to inhale and swallow at the same time, the epiglottis may not tilt downward and food may become lodged in the larynx, blocking air from entering the lungs.

Inhaled air continues past the larynx into the **trachea,** a flexible tube whose walls are reinforced with semicircular bands of stiff cartilage. Within the chest, the trachea splits into two large branches called **bronchi** (singular, bronchus), one leading to each lung. Inside the lung, each bronchus branches repeatedly into ever-smaller tubes called **bronchioles.** Bronchioles lead finally to the microscopic **alveoli** (singular, alveolus), tiny air sacs where gas exchange occurs (**Fig. 20-20b**). The microscopic (0.2 millimeter in diameter) alveoli give magnified lung tissue the appearance of sponge cake.

During its passage through the conducting portion, air is warmed and moistened. Much of the dust and bacteria it carries is trapped in mucus secreted by cells that line the respiratory passages. The mucus, with its trapped debris, is continuously swept upward toward the pharynx by cilia that line the bronchioles, bronchi, and trachea. Upon reaching the pharynx, the mucus is coughed up or swallowed.

Gas Exchange Occurs in the Alveoli

The lung has an enormous moist surface for gas exchange. Each lung is packed with 1.5 million to 2.5 million alveoli, which together provide a huge surface area for diffusion—about 800 square feet (75 square meters) in an adult human, or roughly 80 times the total skin surface area. The alveoli, clustered at the end of each bronchiole like a bunch of grapes, are entirely enmeshed in capillaries (see Fig. 20-20b). Both the alveolar wall and the adjacent capillary walls are only one cell thick, so the air in the lungs is extremely close to the blood in the capillaries. A thin layer of watery fluid lines each alveolus. Gases dissolve in this fluid and diffuse through the alveolar and capillary membranes (**Fig. 20-21**).

Blood is pumped to the lungs after returning from the body's tissues, so the blood surrounding the alveoli is low in oxygen (which has been used up by cellular respiration in the body cells; see pp. 101–105) and high in carbon dioxide (which has been released by the respiring cells). Thus, oxygen diffuses from the air in the alveoli, where its concentration is high, into the blood, where its concentration is low. Conversely, carbon dioxide diffuses out of the blood, where its concentration is high, into the air in the alveoli, where its concentration is low. The blood, now rich in oxygen and purged of carbon dioxide, returns to the heart, which pumps it to the body tissues (**Fig. 20-22**).

Carbon Dioxide and Oxygen Are Transported in Different Ways

As described earlier in this chapter, almost all of the oxygen transported in blood is bound to hemoglobin molecules. Hemoglobin, however, carries only a relatively small proportion (about 20%) of the carbon dioxide in the blood. (After releasing its oxygen in oxygen-starved tissues, hemoglobin picks up some carbon dioxide for the return trip to the lungs.) Some of the blood's remaining carbon dioxide (about 10% of the total) is dissolved in plasma, but the largest portion (about 70%) is converted into bicarbonate ions (HCO_3^-) for transport.

Bicarbonate is formed by a reaction that occurs when water and carbon dioxide come together in the presence of the enzyme carbonic anhydrase:

$$CO_2 + H_2O \xrightarrow{\text{carbonic anhydrase}} H^+ + HCO_3^-$$

Carbonic anhydrase is present inside red blood cells, so that is the main location for bicarbonate formation. Bicarbonate ions then diffuse from the cells into the surrounding plasma. When bicarbonate-rich blood flows past the alveoli (where carbon dioxide concentration is low), the reaction proceeds in reverse, consuming bicarbonate and releasing carbon dioxide that diffuses into the alveoli.

The Lungs Are Protected Within an Airtight Cavity

The chest cavity that surrounds the lungs is airtight—bounded by neck muscles and connective tissue on top and by the dome-shaped, muscular **diaphragm** on the bottom. Within the wall of the chest, the rib cage surrounds and protects the lungs. Lining the rib cage and surrounding the lungs is a double layer of membranes. These membranes contribute to the airtight seal between the lungs and the chest wall.

Air Is Inhaled Actively and Exhaled Passively

We breathe in two stages: **inhalation,** when air is actively drawn into the lungs, and **exhalation,** when it is passively expelled from the lungs. Inhalation is accomplished by enlarging the chest cavity. To do so, the diaphragm muscles contract, drawing the diaphragm downward. The rib muscles also contract, lifting the ribs

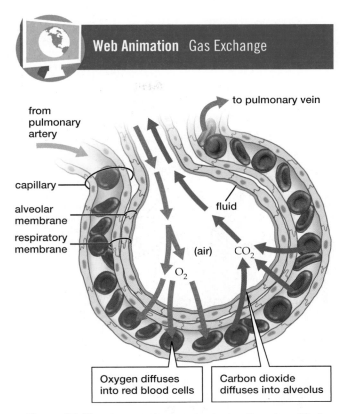

Web Animation Gas Exchange

from pulmonary artery

to pulmonary vein

capillary

alveolar membrane

respiratory membrane

fluid

(air) CO_2

O_2

Oxygen diffuses into red blood cells

Carbon dioxide diffuses into alveolus

▲ Figure 20-21 **Gas exchange between alveoli and capillaries** The alveoli and capillary walls are only one cell thick, very close to one another, and the cells are coated in a thin layer of fluid. This arrangement allows gases to dissolve and diffuse easily between the lungs and the circulatory system.

Lives Up in Smoke

Continued

Smoking interferes with the body's mechanism for protecting the lungs from airborne debris. As smoke is inhaled, toxic substances such as nicotine and sulfur dioxide paralyze the cilia that line the respiratory tract. These ciliary sweepers normally remove inhaled particles, but smoking inhibits them just when they are most needed. With the cilia out of action, the billions of microscopic carbon particles that make up the visible portion of cigarette smoke stick to the walls of the respiratory tract and enter the lungs. In response to the resulting irritation, the respiratory tract produces more mucus in an effort to trap the foreign particles. But without the cilia to sweep it along, the mucus builds up and can obstruct the airways. The familiar "smoker's cough" is an attempt to expel it. (Smoking has many other negative effects; see "Health Watch: Smoking—A Life and Breath Decision" on p. 389.)

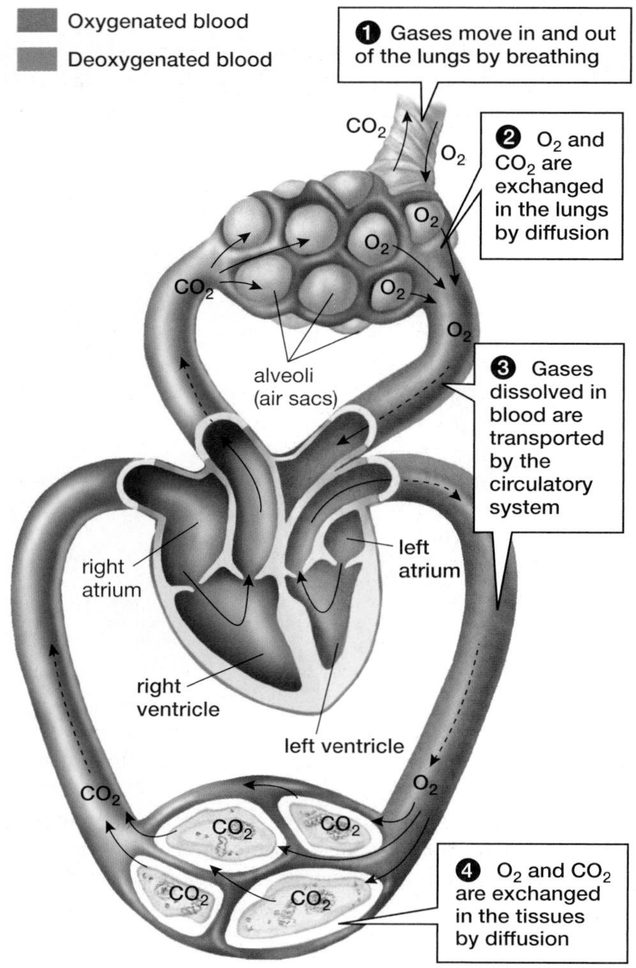

■ Oxygenated blood
■ Deoxygenated blood

1 Gases move in and out of the lungs by breathing

2 O_2 and CO_2 are exchanged in the lungs by diffusion

CO_2
O_2

O_2

CO_2
O_2

O_2

O_2

alveoli (air sacs)

3 Gases dissolved in blood are transported by the circulatory system

right atrium

left atrium

right ventricle

left ventricle

CO_2

O_2

CO_2
CO_2

CO_2
CO_2

4 O_2 and CO_2 are exchanged in the tissues by diffusion

▲ **Figure 20-22 An overview of gas exchange**

up and outward (**Fig. 20-23a**). When the chest cavity expands, the lungs expand with it as surface tension holds them tightly against the moist inner wall of the chest. As the lungs expand, their increased volume decreases the air pressure inside them to below atmospheric pressure. Because gases always move from an area of higher pressure to an area of lower pressure, the outside air, which is at atmospheric pressure, moves into the lungs.

Exhalation occurs automatically when the muscles that cause inhalation are relaxed (**Fig. 20-23b**). The relaxed diaphragm bends upward, and the ribs move down and inward, decreasing the size of the chest cavity and forcing air out of the lungs. More air can be forced out by contracting the abdominal muscles. After exhalation, the lungs still contain some air. This air prevents the thin alveoli from collapsing and fills the space within the conducting portion of the respiratory system.

Breathing Rate Is Controlled by the Respiratory Center of the Brain

We breathe automatically, without conscious thought. The muscles used in breathing, however, are not self-activating like the heart muscle. Instead, each contraction is stimulated by impulses from nerve cells. These impulses originate in the **respiratory center,** which is located in the medulla, the part of the brain that lies just above the spinal cord. Nerve cells in the respiratory center generate cyclic bursts of impulses that cause alternating contraction and relaxation of the respiratory muscles.

The respiratory center receives input from several sources and adjusts breathing rate and volume to meet the body's changing needs. The respiratory rate is regulated to maintain a constant level of carbon dioxide in the blood, as monitored by carbon dioxide receptors in the medulla. For example, when you run up stairs, increased activity in your muscle cells produces an elevated level of carbon dioxide in your blood, signaling a need for more oxygen to support cellular respiration in your muscles. The receptors in the medulla detect the elevated carbon dioxide levels and stimulate an increase in the rate and depth of breathing. These receptors are so sensitive that an increase in carbon dioxide of only 0.3% can double the breathing rate.

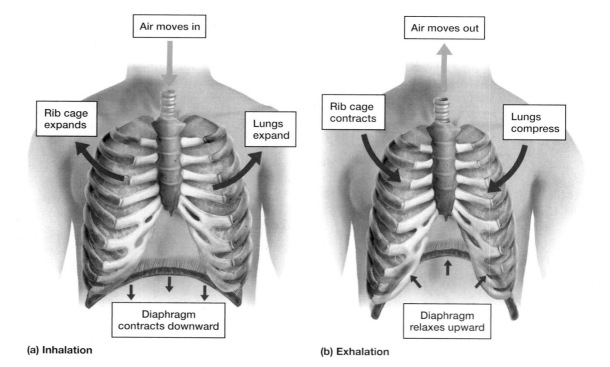

► **Figure 20-23 The mechanics of breathing** *(a)* During inhalation, the diaphragm moves downward and the ribs move up and outward. The size of the chest cavity increases, causing air to rush in. *(b)* During exhalation, relaxation of the muscles that moved the diaphragm and ribs allows the diaphragm to bend upward and the rib cage to move inward, forcing air out of the lungs.

Air moves in

Rib cage expands

Lungs expand

Diaphragm contracts downward

(a) Inhalation

Air moves out

Rib cage contracts

Lungs compress

Diaphragm relaxes upward

(b) Exhalation

health watch

Smoking—A Life and Breath Decision

About 440,000 people in the United States die of smoking-related diseases each year; these diseases include lung cancer, emphysema, chronic bronchitis, heart disease, stroke, and other forms of cancer.

Tobacco smoke has a dramatic impact on the human respiratory tract. In smokers, microscopic smoke particles accumulate in the alveoli over the years until the lungs of a heavy smoker are literally blackened (compare the normal lung in **Fig. E20-4a** to the diseased lung in **Fig. E20-4b**). Adhering to the particles are about 200 different toxic substances, of which more than a dozen are known or probable carcinogens (cancer-causing substances). The longer the delicate tissues of the lungs are exposed to the carcinogens on the trapped particles, the greater the chance that cancer will develop (**Fig. E20-4c**).

Some smokers will develop *chronic bronchitis*, a persistent lung infection characterized by coughing, swelling of the lining of the respiratory tract, an increase in mucous production, and a decrease in the number and activity of cilia. The result: a decrease in air flow to the alveoli. *Emphysema* (see Fig. E20-4b)

develops when toxic substances in cigarette smoke cause the body to produce substances that lead to brittle and ruptured alveoli. The loss of the alveoli, where gas exchange occurs, leads to oxygen deprivation of all body tissues. In an individual with emphysema, breathing becomes increasingly labored and loss of respiratory effectiveness may ultimately be fatal.

Carbon monoxide is present at high levels in cigarette smoke and binds tenaciously to red blood cells in place of oxygen. This binding reduces the blood's oxygen-carrying capacity and thereby increases the work the heart must do. Chronic bronchitis and emphysema compound this problem, making smokers twice as likely as nonsmokers to suffer a heart attack. Smoking also causes atherosclerosis (see "Health Watch: Matters of the Heart" on pp. 382–383). As a result, smokers are 70% more likely than nonsmokers to die of heart disease.

The carbon monoxide in cigarette smoke may also contribute to the reproductive problems of women who smoke during pregnancy, including infertility, miscarriage, lower birth weight of their babies, and, for their children, more learning and behavioral problems.

"Passive smoking," or breathing secondhand smoke, poses health hazards for both children and adults. Children whose parents smoke are more likely to contract bronchitis, pneumonia, ear infections, coughs, and colds. Their lung capacity is often decreased, and they are more likely to develop asthma and allergies as well. For children with asthma, the number and severity of asthma attacks are increased by secondhand smoke. Among adults, nonsmoking spouses of smokers face a 30% higher risk of both heart attack and lung cancer than do spouses of nonsmokers. A recent study links even relatively infrequent exposure to secondhand smoke with atherosclerosis. Government agencies report that secondhand smoke is responsible for an estimated 3,000 lung-cancer deaths and at least 35,000 deaths from heart disease in nonsmokers in the United States each year. For smokers who quit, however, healing begins immediately and the chances of heart attack, lung cancer, and numerous other smoking-related illnesses gradually diminish.

(a) Normal lung

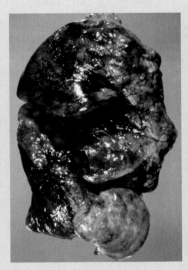

(b) A smoker's lung

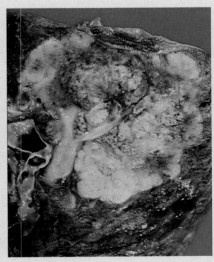

(c) A lung cancer

◀ **Figure E20-4 Smoking damages the lungs** *(a)* A normal lung. *(b)* Lung of a smoker who died of emphysema is both blackened and collapsed. *(c)* A lung cancer is visible as a pale mass; the lung tissue surrounding it is blackened by trapped smoke particles.

Lives Up in Smoke Revisited

Mark Twain once said "Quitting smoking is easy, I've done it a thousand times." Nicotine is a powerfully addictive drug, as likely to lead to addiction as cocaine or heroin. Researchers have found that, like cocaine and heroin, nicotine activates the brain's reward center. The brain adjusts by becoming less sensitive, requiring larger quantities of nicotine to experience the same rewarding effect, and causing the reward center to feel understimulated when nicotine is withdrawn. Withdrawal symptoms can include nicotine craving, depression, anxiety, irritability, difficulty concentrating, headaches, and disturbed sleep. So it's no wonder that for some people, the only way to quit smoking is to not start smoking at all. Although at least 70% of smokers would like to quit, only about 2.5% of smokers are successful each year. By age 60, about 33% will have succeeded, most after having made two or three attempts.

The power of addiction has deadly consequences. About one-third of smokers who fail to quit will die from a smoking-related cause (for details, see "Health Watch: Matters of the Heart" on pp. 382–383 and "Health Watch: Smoking—A Life and Breath Decision" on p. 389).

The danger is well known to smokers, who may nonetheless lack sufficient motivation to overcome addiction. A junior at the University of Illinois expresses a common attitude among young smokers: "I know it's bad for me and it could kill me. I'm twenty-one years old, I'm not dumb to it. My grandpa died of cancer from smoking all his life." A nonsmoking student explains: "People are going to do it no matter what. You can show them pictures of [cancerous] lungs or rotting teeth but it's not going to affect people all the same way. . . . We have the right to choose how we live our lives."

Consider This

Nearly 90% of lung cancers are caused by smoking. Direct medical costs for treating these cancers are approximately $5 billion annually in the United States. This expense (and the expense of treating other smoking-caused diseases) contribute to the rapidly rising cost of health insurance and to the resulting increase in the number of uninsured people. In your view, should those who exercise their right to smoke bear some responsibility for offsetting these costs to society?

Chapter Review

Summary of Key Concepts

For additional study help and activities, go to www.mybiology.com.

20.1 What Are the Major Features and Functions of Circulatory Systems?

Circulatory systems transport blood rich in dissolved nutrients and oxygen close to each cell, where nutrients can be released and wastes absorbed by diffusion. All circulatory systems have three major parts: (1) blood—a fluid; (2) vessels—a system of channels to conduct the blood; and (3) a heart—a pump to circulate the blood. Vertebrate circulatory systems transport gases, hormones, and wastes; distribute nutrients; help regulate body temperature; and defend the body against disease.

Web Animation   Circulatory System Features

20.2 How Does the Vertebrate Heart Work?

In the mammalian four-chambered heart, blood is pumped separately to the lungs and to the rest of the body, maintaining complete separation of oxygenated and deoxygenated blood. Deoxygenated blood from the body collects in the right atrium and passes to the right ventricle, which pumps it to the lungs. Oxygenated blood from the lungs enters the left atrium, passes to the left ventricle, and is pumped to the rest of the body.

The heartbeat consists of two stages: (1) atrial contraction, followed by (2) ventricular contraction. Valves within the heart maintain the direction of blood flow. The contractions of the heart

are initiated and coordinated by the sinoatrial (SA) node, the heart's pacemaker.

BioFlix Muscle Contraction

Web Animation   Structure of the Human Heart

Web Animation   Function of the Human Heart

Web Animation   Measuring Blood Pressure

20.3 What Is Blood?

Blood is composed of both fluid and cellular materials. The fluid plasma consists of water that contains proteins, hormones, nutrients, gases, and wastes. Red blood cells, or erythrocytes, are packed with a large iron-containing protein called hemoglobin, which carries oxygen. White blood cells, or leukocytes, fight infection. Platelets are cell fragments that are important for blood clotting.

20.4 What Are the Types and Functions of Blood Vessels?

Blood leaving the heart travels (in sequence) through arteries, arterioles, capillaries, venules, veins, and then back to the heart. Each vessel is specialized for its role. Elastic, muscular arteries help pump the blood. The thin-walled capillaries are the sites of exchange of materials between the body cells and the blood. Veins provide a path of low resistance back to the heart, with one-way valves that maintain the direction of blood flow.

20.5 How Does the Lymphatic System Work with the Circulatory System?

The human lymphatic system consists of lymphatic vessels, lymph nodes, and the thymus and spleen. The lymphatic system removes excess extracellular fluid that leaks through blood capillary walls. It transports fats to the bloodstream from the small intestine and fights infection by filtering the lymph through lymph nodes, where white blood cells ingest foreign invaders, such as viruses and bacteria. The thymus, which is most active in young children, stores some types of white blood cells as they mature. The spleen filters blood past white blood cells, which remove bacteria and damaged blood cells.

20.6 How Are Oxygen and Carbon Dioxide Exchanged in Animal Bodies?

Respiration makes possible the exchange of oxygen and carbon dioxide between the body and the environment by diffusion of these gases across a moist surface. In moist environments, animals with small bodies may rely exclusively on diffusion through the body surface. Larger, more active animals have evolved specialized respiratory systems. Animals in aquatic environments have evolved gills. Terrestrial vertebrates have evolved lungs that provide internal protection to moist respiratory surfaces.

20.7 How Does the Human Respiratory System Work?

The human respiratory system consists of a conducting portion and a gas-exchange portion. Air passes first through the conducting portion, consisting of the nose and mouth, pharynx, larynx, trachea, bronchi, and bronchioles, and then into the gas-exchange portion, composed of microscopic sacs called alveoli. Blood within a dense capillary network surrounding the alveoli releases carbon dioxide and absorbs oxygen from the air.

Most of the oxygen in the blood is bound to hemoglobin within red blood cells. Hemoglobin binds oxygen in alveolar capillaries and releases it at the lower oxygen concentrations of the body tissues. Carbon dioxide diffuses into the blood from the body tissues; some is bound to hemoglobin but most is transported as bicarbonate ions in the plasma.

To inhale, we actively draw air into the lungs by contracting the diaphragm and the rib muscles, which expand the chest cavity. The relaxation of these muscles causes the chest cavity to decrease in size and the air to be exhaled. Respiration is controlled by nerve impulses that originate in the medulla's respiratory center. The respiration rate is modified by receptors in the medulla that monitor carbon dioxide levels in the blood.

Web Animation   Human Respiratory Anatomy

Web Animation   Gas Exchange

Key Terms

- -

alveolus (plural, alveoli) p. 386
arteriole p. 379
artery p. 372
atrioventricular (AV) node p. 376
atrioventricular valve p. 374
atrium (plural, atria) p. 373
blood p. 371
blood clotting p. 378
blood vessel p. 371
bronchiole p. 386

bronchus (plural, bronchi) p. 386
capillary p. 372
diaphragm p. 387
exhalation p. 387
extracellular fluid p. 372
gill p. 384
heart p. 371
hemoglobin p. 378
inhalation p. 387
larynx p. 385

lung p. 385
lymph p. 383
lymph node p. 384
lymphatic system p. 380
pharynx p. 385
plasma p. 376
platelet p. 378
red blood cell p. 377
respiration p. 384
respiratory center p. 388

semilunar valve p. 374
sinoatrial (SA) node p. 375
spleen p. 384
thymus p. 384
trachea p. 386
vein p. 372
ventricle p. 373
venule p. 380
white blood cell p. 378

Thinking Through the Concepts

- -

Suggested answers to end-of-chapter and figure-based questions can be found at the end of the text.

Fill-in-the-Blank

1. The vessels where wastes and nutrients are exchanged between the blood and body cells are the _____. These vessels form a network that connects the _____ (small vessels that carry oxygenated blood) with the _____ (small vessels that carry deoxygenated blood).

2. Systolic blood pressure is produced by contraction of the _____. The pressure that occurs during the pause between heartbeats is called _____ pressure. The chamber of the heart with the most muscular wall is the _____. The chamber that pumps blood to the lungs is the _____.

3. Lymph closely resembles _____. Lymph is forced through vessels by _____. One function of the lymphatic system is to restore fluid to the _____. The lymphatic system includes numerous small chambers called _____ , which contain masses of white blood cells. It also includes an organ near the heart called the _____, and a blood-filtering organ called the _____.

4. Which part of the conducting portion of the human respiratory system is shared by the digestive tract? _____. What structure normally keeps food from entering the larynx? _____. After passing through the larynx, air travels in sequence through the _____, _____, _____, and, finally, to the gas-exchange portion of the respiratory system, called the _____.

5. In the lungs, oxygen diffuses into blood vessels called_____ . In the blood, most of the oxygen binds to a protein called _____. Carbon dioxide is primarily transported in blood in the form of _____. All body cells require oxygen and release carbon dioxide because they are generating energy using by a process called _____.

6. List three diseases that are far more common in smokers than in nonsmokers: _____, _____, _____. List three ailments caused by passive smoking that are more common in the children of smokers: _____, _____, _____.

Review Questions

1. Trace the flow of blood through the vertebrate circulatory system, starting and ending with the right atrium.

2. What are five functions of the vertebrate circulatory system?

3. Describe three important functions of the lymphatic system.

4. Distinguish among plasma, extracellular fluid, and lymph.

5. Describe veins, capillaries, and arteries, noting their similarities and differences.

6. Explain in detail what causes the vertebrate heart to beat.

7. Describe the formation of an atherosclerotic plaque. What are the risks associated with atherosclerosis?

8. Trace the route taken by air in the vertebrate respiratory system, listing the structures through which it flows and the point at which gas exchange occurs.

9. What events occur during human inhalation? Exhalation? Which of these is always an active process?

10. Trace the pathway of an oxygen molecule in the human body, starting with the alveoli and ending with a body cell.

11. Describe the effects of smoking on the human respiratory system.

12. Explain how the structure and arrangement of alveoli make them well suited for their role in gas exchange.

Applying the Concepts

1. **IS THIS SCIENCE?** In an effort to improve stamina and endurance, some long-distance runners and bicyclists withdraw some of their own blood, store it in a freezer for a few weeks, and then inject it back into their bodies just before a race. On the basis of your understanding of circulatory and respiratory systems, do you think that this technique is likely to effectively enhance performance? Why or why not?

2. Nicotine is a drug in tobacco that supplies the effects that smokers crave. Discuss the advantages and disadvantages of low-nicotine cigarettes.

3. Mary, a strong-willed 3-year-old, threatens to hold her breath until she dies if she doesn't get her way. Can she carry out her threat? Explain.

For additional resources, go to www.mybiology.com.

chapter 21

Nutrition, Digestion, and Excretion

Now free from anorexia, former supermodel Carré Otis still walks the runways, but no longer in the top fashion shows. Instead, she travels a path to a healthier life.

Case Study Dying to Be Thin

ormer supermodel Carré Otis explains, "The sacrifices I made were life threatening. I had entered a world that seemed to support a 'whatever it takes' mentality to maintain abnormal thinness." For many models, performers, and others in the public eye, meeting expectations for thinness is a continuing battle that can lead to eating disorders. At 5 feet 10 inches, Otis once weighed only 100 pounds, giving her a body mass index (BMI) of 14.3. (The World Health Organization considers a BMI below 16 as "starvation.") Now maintaining a healthy weight—at the expense of her modeling career—Otis has become a spokesperson for the National Eating Disorders Association, hoping that she can help others avoid the damage her body suffered.

Ana Carolina Reston at the peak of her modeling career—glamorous, but dying.

Otis suffered from two eating disorders, anorexia and bulimia. People with anorexia typically eat very little, and often exercise almost nonstop in an effort to lose still more weight. About half of all anorexics also develop bulimia—binge eating of relatively large amounts of food, followed by self-induced vomiting or overdosing with laxatives to purge the food from their bodies. Anorexics lose muscle mass and often damage their digestive, cardiac, endocrine, and reproductive systems.

Sometimes, the damage proves fatal. In October 2006, Ana Carolina Reston (photo at left), one of Brazil's leading models, was hospitalized for a kidney malfunction. After three weeks in intensive care, she died from multiple organ failure and septicemia (massive infection throughout the bloodstream). At 5 feet 8 inches and 88 pounds, her BMI was only 13.4. Two other extremely thin models, the Uruguayan sisters Luisel and Eliana Ramos, died in 2006 and 2007, respectively, from heart failure probably brought on by anorexia.

Anorexia and bulimia typically strike teenage girls and women in their 20s, who often feel pressured to conform to unrealistic ideals of body size and shape. Because many people with anorexia or bulimia never consult a physician, no one really knows how common these disorders might be. In the United States, health authorities estimate that between 200,000 and a few million people, mostly young women, suffer from anorexia, bulimia, or other eating disorders.

In this chapter, you will learn about the processes of nutrition, digestion, and excretion. As you do, think about how anorexia and bulimia affect the structures and functions of the digestive and urinary tracts. Besides not enough Calories, what specific nutrients might be missing from a "starvation" diet? Why would anorexia and bulimia damage many organs throughout the body? Why are anorexics at risk for heart attacks? ■

21.1 How Do Animals Regulate the Composition of Their Bodies?

In Chapter 19, we introduced the concept of homeostasis—keeping an organism's body within the narrow range of conditions that allows it to survive and reproduce. Nutrition, digestion, and excretion play crucial roles in homeostasis.

A **nutrient** is any substance that an animal needs but cannot synthesize in its own body, and hence must acquire from its environment as it eats or drinks. Nutrients provide animals with both the materials with which to construct their bodies and the energy to fuel their life processes. An animal may obtain nutrients directly in usable form (for example, water, sodium, or glucose); combined into large, complex molecules such as fats or proteins; or as parts of the bodies of plants or other animals that they eat. **Digestion** is the process whereby an animal physically grinds up and chemically breaks down its food, producing small, simple molecules that can be absorbed into the circulatory system. **Nutrition** is a more comprehensive term that includes taking food into the body, converting it into usable forms, absorbing the resulting molecules from the digestive tract into the circulatory system, and using the nutrients in the animal's own metabolism.

When an animal eats or drinks, it never obtains precisely the right mixture of water, minerals, carbohydrates, fats, and proteins that it needs to build and sustain itself. Some components of its food may be indigestible—hair, bone, and cellulose, for example, cannot be digested by most animals. Other substances in food may be harmful, including toxins produced by many plants. An animal's own metabolism produces carbon dioxide and some highly toxic molecules, such as ammonia, that must be eliminated. Finally, an animal may simply consume too much of otherwise useful substances, such as water, sodium, or potassium. **Excretion** is the disposal of these indigestible, toxic, or surplus materials. There is a great diversity of excretory structures and functions in the animal kingdom. As a general rule, however, indigestible food is expelled from the digestive tract as feces (see section 21.5). Carbon dioxide and, in some animals, ammonia and some other toxic molecules are excreted by the respiratory tract (lungs or gills) or the skin. Surplus minerals and most toxic substances, whether eaten or produced by the animal's own metabolism, are excreted by the urinary tract (see sections 21.6, 21.7, and 21.8).

21.2 What Nutrients Do Animals Need?

Animal nutrients fall into six major categories: lipids, carbohydrates, proteins, minerals, vitamins, and water.

The Primary Sources of Energy Are Lipids and Carbohydrates

Cells require a continuous supply of energy to stay alive and perform their functions. In animals, energy is provided mostly by three kinds of nutrients: lipids, carbohydrates and, to a lesser extent, proteins. These molecules, or parts of them, are used in glycolysis and cellular respiration, and the energy derived from them is used to produce ATP (see pp. 100–105).

Energy in food can be measured in **calories,** defined as the energy required to raise the temperature of 1 gram of water by 1 degree Celsius. However, this unit is so small—a single Big Mac with cheese contains 700,000 calories—that it is customary to use **Calories** (with a capital *C*) instead; a Calorie contains 1,000 calories (lowercase *c*). The unit that you see in the "Nutritional Information" tables on cereal boxes and in fast-food restaurants is the Calorie; a Big Mac, for example, contains 700 Calories. The average human body at rest burns about 1,550 Calories per day (usually somewhat more if you're young and/or male; less if you're older and/or female), and people burn more Calories when exercising than when resting (Table 21-1). In a really fit athlete, vigorous exercise can raise energy consumption from a resting rate of about 1 Calorie per minute to nearly 20 Calories per minute.

Lipids Include Fats, Phospholipids, and Cholesterol

Lipids are a diverse group of molecules that includes triglycerides (fats and oils), phospholipids, and cholesterol (see pp. 30–33). Fats and oils are used primarily as a source of energy. Phospholipids are important components of all cellular membranes. Cholesterol is used to manufacture cellular membranes, several hormones including estrogen and testosterone, and bile (which aids in fat digestion).

Table 21-1	Approximate Energy Consumed by a 150-Pound Person Performing Different Activities				
		Time to "Work Off"			
Activity	Calories/Hour	500 Calories (Cheeseburger)	300 Calories (Ice Cream Cone)	70 Calories (Apple)	40 Calories (1 Cup Broccoli)
Running (6 mph)	700	43 min	26 min	6 min	3 min
Cross-country skiing (moderate)	560	54 min	32 min	7.5 min	4 min
Roller skating	490	1 hr 1 min	37 min	8.6 min	5 min
Bicycling (11 mph)	420	1 hr 11 min	43 min	10 min	6 min
Walking (3 mph)	250	2 hr	1 hr 12 min	17 min	10 min
Frisbee® playing	210	2 hr 23 min	1 hr 26 min	20 min	11 min
Studying	100	5 hr	3 hr	42 min	24 min

Dying to Be Thin

Continued

In her efforts to lose weight, Ana Reston ate almost nothing but tomatoes and apples for several months. These foods contain virtually no fat, so she would have lacked essential fatty acids, damaging her cell membranes.

▲ Figure 21-1 Fat provides lightweight energy storage
Although their diet consists mostly of sugar (in the nectar of flowers), ruby-throated hummingbirds convert sugars to fat for energy storage prior to migrating in the fall.

Animals of some species can synthesize all of the types of lipids they need. Others must acquire specific lipid building blocks, called **essential fatty acids,** from their food. For example, humans are unable to synthesize linoleic acid, which is required for the synthesis of certain phospholipids. Therefore, we must obtain this essential fatty acid from our diet, mainly from vegetable oils such as safflower or sunflower oil. In most developed nations, obesity is an increasingly serious health problem, so many people rightly try to limit the amount of fat in their diets. However, a truly fat-free diet would be lethal.

Fats Store Energy in Concentrated Form

Humans and most other animals store energy primarily as fat. When an animal eats more Calories than it uses, most of the excess carbohydrates, fats, or proteins are converted to fat for storage. Fats have two major advantages as energy-storage molecules. First, they contain more than twice as much energy per unit weight as either carbohydrates or protein (about 9 Calories per gram for fats compared with about 4 Calories per gram for carbohydrates and proteins). Second, lipids are hydrophobic; that is, they do not dissolve in water. Fat deposits, therefore, do not cause water to accumulate in the body. For both these reasons, fats store more calories with less weight than do other molecules.

Minimizing weight allows an animal to move faster and farther (important for escaping predators, hunting prey, and migrating) and to use less energy for movement (important when food supplies are limited). For example, ruby-throated hummingbirds (Fig. 21-1) migrate across the Gulf of Mexico in the fall. Obviously, the open ocean doesn't provide anything for a hummingbird to eat. Therefore, a ruby-throat that weighs 2 to 3 grams in early summer puts on about 2 grams of fat before migrating. If it stored carbohydrate or protein instead, it would have to gain almost 6 grams to provide the same amount of energy. It probably could barely fly, and certainly couldn't make it across the Gulf before collapsing from exhaustion.

Because people evolved under the same food constraints as other animals did, we have a strong tendency to eat when food is available, even if we aren't really hungry. In addition, foods that are high in fat and sugar usually taste the best. Many people now have access to almost unlimited, high-calorie food, and have jobs that do not require much exercise. Under these circumstances, we often need considerable willpower to avoid becoming overweight.

Carbohydrates Are a Source of Quick Energy

Carbohydrates include simple sugars, as well as longer chains of sugars called polysaccharides (see p. 29). The simple sugar glucose is the primary source of energy for most cells, but the typical diet contains little glucose. During digestion, glucose is derived from the breakdown of more complex carbohydrates, such as sucrose and starch.

Animals, including humans, store the carbohydrate **glycogen**—a large, highly branched chain of glucose molecules—in the liver and muscles. Glycogen provides much of the energy used during exercise, but people typically store less than a pound of it, the equivalent of less than 2,000 Calories. Therefore, marathon runners can go about 20 miles before their glycogen is used up. The expression "hitting the wall" describes the extreme fatigue that long-distance runners experience when their supply of glycogen is gone.

Proteins Provide Amino Acids for Building New Proteins

Protein in the diet serves mainly as a source of amino acids to make the body's own proteins. Dietary protein is broken down in the digestive tract to yield amino acids. In your cells, the amino acids are linked in specific sequences to form the many different proteins specific to your body. Any excess protein in the diet is broken down to extract energy for immediate use or for storage as fat.

Humans can synthesize only 12 of the 20 amino acids commonly used in proteins. The other eight, called **essential amino acids,** must be supplied in the diet. Two additional amino acids are usually synthesized in fairly small amounts, and so are essential for growing children but usually not for adults. Although animal proteins almost always contain sufficient amounts of all of the essential amino acids, many plant proteins are deficient in some, so vegetarians must take steps to avoid protein deficiency. Generally, they need to make sure their diet includes a variety of plants (for example, legumes, grains, and corn) whose proteins collectively provide all of the essential amino acids. Protein deficiency can cause a variety of debilitating conditions, including kwashiorkor, which is seen in some impoverished countries (**Fig. 21-2**).

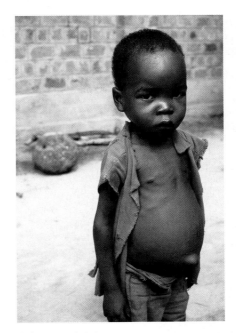

▲ **Figure 21-2 Kwashiorkor** Symptoms of kwashiorkor, caused by protein deficiency, include a swollen abdomen and emaciated arms and legs.

Minerals Are Elements Required by the Body

The term "mineral" has very different meanings in geology and in nutrition. In geology, a mineral is a homogeneous, usually crystalline, element or compound, such as the beautiful crystals often displayed in natural history museums. In nutrition, a **mineral** is specifically a chemical element (not a compound) required for proper bodily function (**Table 21-2**). Because animals cannot manufacture elements, minerals must be obtained through the diet, either from food or dissolved in drinking water. Essential minerals include calcium, magnesium, and phosphorus, which are major constituents of bones and teeth. Others, such as sodium and potassium, are essential for muscle contraction and the conduction of nerve impulses. Iron is used in the production of hemoglobin, and iodine is found in hormones produced by the thyroid gland. In addition, trace amounts of several other minerals, including zinc, copper, and selenium, are required, typically as parts of enzymes.

Vitamins Play Many Roles in Metabolism

Vitamins are a diverse group of organic compounds that animals require in very small amounts. The body cannot synthesize most vitamins (or cannot synthesize them in adequate amounts), so they are normally obtained from food. Convenience foods—doughnuts, soda, and French fries, for example—usually contain lots of Calories but not many vitamins. The vitamins considered essential in human nutrition are listed in **Table 21-3**. These vitamins are often grouped into two categories: water soluble and fat soluble.

Water-Soluble Vitamins

Water-soluble vitamins include vitamin C and the eight compounds that make up the B-vitamin complex. These substances dissolve in the water of

Table 21-2 **Minerals, Sources, and Functions in Humans**

Mineral	Dietary Sources	Major Functions in Body	Deficiency Symptoms
Calcium	Milk, cheese, green vegetables, legumes	Bone and tooth formation Blood clotting Nerve impulse transmission	Stunted growth Rickets, osteoporosis Convulsions
Phosphorus	Milk, cheese, meat, poultry, grains	Bone and tooth formation Acid-base balance	Weakness Demineralization of bone Loss of calcium
Potassium	Meats, milk, fruits	Acid-base balance Body water balance Nerve function	Muscular weakness Paralysis
Chlorine	Table salt	Formation of gastric juice Acid-base balance	Muscle cramps Apathy Reduced appetite
Sodium	Table salt	Acid-base balance Body water balance Nerve function	Muscle cramps Apathy Reduced appetite
Magnesium	Whole grains, green leafy vegetables	Activation of enzymes in protein synthesis	Growth failure Behavioral disturbances Weakness, spasms
Iron	Eggs, meats, legumes, whole grains, green vegetables	Constituent of hemoglobin and enzymes involved in energy metabolism	Iron-deficiency anemia (weakness, reduced resistance to infection)
Fluorine	Fluoridated water, tea, seafood	Strengthening teeth and probably bone	High frequency of tooth decay
Zinc	Widely distributed in foods	Constituent of enzymes involved in digestion	Reduced growth rate Small sex glands
Iodine	Seafish and shellfish, dairy products, many vegetables, iodized salt	Constituent of thyroid hormones	Goiter (enlarged thyroid)
Chromium	Fruits, vegetables, whole grains	Metabolism of sugar and fats	Reduced glucose tolerance Elevated insulin in blood

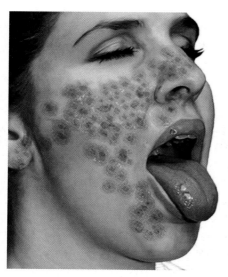

▲ **Figure 21-3 Pellagra** Scaly, reddish brown skin lesions and a red and swollen tongue are caused by a deficiency of niacin, a B vitamin.

the blood plasma and are excreted by the kidneys, so they are not stored in the body in any appreciable amounts. Water-soluble vitamins generally work together with enzymes to promote essential chemical reactions in the body's cells.

Because each vitamin participates in several metabolic processes, a deficiency of a single vitamin can have wide-ranging effects (see Table 21-3). For example, deficiency of niacin, a B vitamin, causes pellagra, associated with cracked, scaly skin as well as digestive and nervous system disorders (Fig. 21-3). In 1996, the U.S. Food and Drug Administration (FDA) ordered folic acid, another B vitamin, to be added to grain foods such as bread, pasta, and rice. The addition of folic acid has reduced the incidence of neural tube defects (serious birth defects of the brain and spinal cord linked to folic acid deficiency in pregnant women) by about 20%. Researchers believe that supplementing food with folic acid has also contributed to a decline in both stroke and heart disease.

Fat-Soluble Vitamins

The fat-soluble vitamins are A, D, E, and K. Vitamin A is used to produce the light-capturing molecule in the retina of the eye, and vitamin A deficiency can cause poor night vision or, in severe cases, blindness. Vitamin D is important for bone formation. Several recent studies have found a high incidence of vitamin D deficiency in the United States, including in urban adolescents, postmenopausal

Table 21-3	**Vitamins, Sources, and Functions in Humans**		
Vitamin	**Dietary Sources**	**Functions in Body**	**Deficiency Symptoms**
Water soluble			
B-complex			
Vitamin B$_1$ (thiamin)	Milk, meat, bread	Coenzyme in metabolic reactions	Beriberi (muscle weakness, peripheral nerve changes, edema, heart failure)
Vitamin B$_2$ (riboflavin)	Widely distributed in foods	Constituent of coenzymes in energy metabolism	Reddened lips, cracks at corner of mouth, lesions of eye
Niacin	Liver, lean meats, grains, legumes	Constituent of two coenzymes in energy metabolism	Pellagra (skin and gastrointestinal lesions; nervous mental disorders)
Vitamin B$_6$ (pyridoxine)	Meats, vegetables, whole-grain cereals	Coenzyme in amino acid metabolism	Irritability, convulsions, muscular twitching, dermatitis, kidney stones
Pantothenic acid	Milk, meat	Constituent of coenzyme A, with a role in energy metabolism	Fatigue, sleep disturbances, impaired coordination
Folic acid	Legumes, green vegetables, whole wheat	Coenzyme involved in nucleic and amino acid metabolism	Anemia, gastrointestinal disturbances, diarrhea, retarded growth, birth defects
Vitamin B$_{12}$	Meats, eggs, dairy products	Coenzyme in nucleic acid metabolism	Pernicious anemia, neurological disorders
Biotin	Legumes, vegetables, meats	Coenzymes required for fat synthesis, amino acid metabolism, and glycogen formation	Fatigue, depression, nausea, dermatitis, muscular pains
Others			
Choline	Egg yolk, liver, grains, legumes	Constituent of phospholipids, precursor of the neurotransmitter acetylcholine	None reported in humans
Vitamin C (ascorbic acid)	Citrus fruits, tomatoes, green peppers	Maintenance of cartilage, bone, and dentin (hard tissue of teeth); collagen synthesis	Scurvy (degeneration of skin, teeth, gums, blood vessels; epithelial hemorrhages)
Fat soluble			
Vitamin A (retinol)	Beta-carotene in green, yellow, and red vegetables; retinol added to dairy products	Constituent of visual pigment; maintenance of epithelial tissues	Night blindness, permanent blindness
Vitamin D	Cod-liver oil, eggs, dairy products	Promotes bone growth and mineralization; increases calcium absorption	Rickets (bone deformities) in children; skeletal deterioration
Vitamin E (tocopherol)	Seeds, green leafy vegetables, margarines, shortenings	Antioxidant, prevents cellular damage	Possibly anemia
Vitamin K	Green leafy vegetables; product of intestinal bacteria	Important in blood clotting	Bleeding, internal hemorrhages

women (who may suffer more bone fractures as a result), and African Americans. People with dark skin are at particular risk because vitamin D is synthesized in skin exposed to sunlight, and dark pigmentation reduces the penetration of sunlight to the cells beneath that synthesize vitamin D. Pediatricians are seeing an alarming increase in rickets (**Fig. 21-4**), particularly in African American children, as a consequence of vitamin D deficiency. Vitamin E is an *antioxidant*, which may help to protect the body against damaging substances that are formed as cells use oxygen to produce high-energy molecules such as ATP. Vitamin K helps regulate blood clotting.

Fat-soluble vitamins can be stored in body fat and may accumulate in the body over time. For this reason, some fat-soluble vitamins (vitamin A, for example) may be toxic if excessive amounts are eaten.

Dying to Be Thin

Continued

People with anorexia usually don't eat a nutritionally balanced diet. Vitamin deficiencies are very common. If their diets lack the essential amino acids, they will suffer from deficiencies of protein metabolism. If anorexics fail to ingest enough of the right minerals, they can develop severe imbalances of sodium, potassium, calcium, and magnesium in their blood and extracellular fluid, which can result in nervous and cardiac disorders.

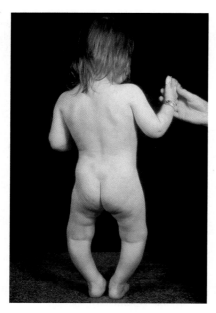

▲ **Figure 21-4 Rickets** The bone deformities (particularly bow legs) that are characteristic of rickets result from vitamin D deficiency.

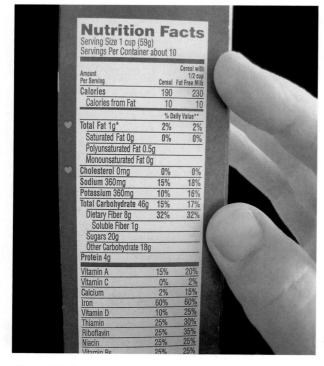

▲ **Figure 21-5 Food labeling** The U.S. government requires quite complete nutritional labeling of foods, such as in this sample. **QUESTION:** Does this food seem to be a good nutritional choice? Explain.

Web Animation Mechanical and Chemical Breakdown of Food

The Human Body Is About Two-Thirds Water

Water is a crucial part of both the structure and physiology of all animals. For example, water is the principal component of saliva, blood, lymph, extracellular fluid, and the cytoplasm within each cell. Most metabolic reactions occur in a watery solution, and water directly participates in the hydrolysis reactions that break down proteins, carbohydrates, and fats into simpler molecules (see p. 27). As we will see later in this chapter, the kidneys excrete wastes dissolved in the water of urine.

The average adult human requires about 10 cups (2.5 liters) of water per day, but this need can increase dramatically with exercise, high temperatures, or low humidity, as water evaporates from sweat and from our lungs when we breathe. Although people can often survive for weeks without food, death occurs in a few days without water, because we lose so much every day. Water intake occurs mostly through eating and drinking. There is enough water in the typical diet for about half of the usual daily requirement, with the rest obtained by drinking fluids.

Nutritional Guidelines Help People Obtain a Balanced Diet

Most people in the United States are fortunate to live amidst an abundance of food. However, the amazing diversity of foods in a typical U.S. supermarket and the easy availability of fast food can lead to poor nutritional choices. To help people make informed choices, the U.S. government has recently placed nutritional guidelines called "My Pyramid" on an interactive Web site. Another source of information is the nutritional labeling required on commercially packaged foods. These labels provide complete information about Calorie, fiber, fat, sugar, and vitamin content (**Fig. 21-5**). Also, most fast-food chains provide nutritional information about their products.

Are You Too Heavy?

A simple way to determine whether your weight is likely to pose a health risk is to calculate your body mass index (BMI). The BMI takes into account your weight and height to arrive at an estimate of body fat. This simple calculation assumes that you have an average amount of muscle, so it does not apply to bodybuilders or marathon runners. The formula is as follows: weight (in kilograms)/height^2 (in meters), but you can calculate your BMI by multiplying your weight (in pounds) by 703, then dividing by your height^2 (in inches). Or, simply type "BMI" on your favorite Internet search engine, and you will find many sites that calculate it for you. A BMI between 18.5 and 25 is considered healthy. People with anorexia usually have a BMI of 17.5 or lower. Unless you are a bodybuilder and have far more muscle than average, a BMI between 25 and 30 indicates that you are probably overweight and a BMI over 30 indicates that you are probably obese. "Health Watch: Obesity and the Brain–Gut Connection" on p. 408 explores some potential future treatments for obesity.

21.3 What Are the Major Processes of Digestion?

Animals eat the bodies of other organisms, but these organisms may resist becoming food. Plants, for example, support each cell with a wall of indigestible cellulose. Animals may be covered with indigestible fur, scales, or feathers. In addition, the complex lipids, proteins, and carbohydrates in food cannot be used directly. These nutrients must be broken down before they can be used by the cells of the animal that has consumed them; after being broken down, they are recombined into new molecules. Animals have various types of digestive tracts, each finely tuned for a unique diet and lifestyle. Amid this diversity, however, all digestive systems must accomplish certain tasks:

- **Ingestion.** The food must be brought into the digestive tract through an opening, usually called a mouth.

- **Mechanical breakdown.** In most animals, food must be physically broken down into smaller pieces. The particles produced by mechanical breakdown provide a large surface area for attack by digestive enzymes.

- **Chemical breakdown.** The particles of food must be exposed to digestive enzymes that break down large molecules into smaller subunits.

- **Absorption.** The small molecules must be transported out of the digestive tract and into the body cells. In a few animals, the small molecules may be absorbed directly into the body cells.

- **Elimination.** Indigestible materials must be expelled from the body.

21.4 What Types of Digestive Systems Are Found in Non-Human Animals?

The animal kingdom displays a remarkable diversity of digestive systems, ranging from digestion inside single cells in sponges, to relatively simple sacs in jellyfish and anemones, through an array of tubular digestive systems with two openings in animals as different as earthworms, insects, and humans.

In Sponges, Digestion Occurs Within Individual Cells

Sponges are the only animals that rely exclusively on individual cells to digest their food (Fig. 21-6). As you might suspect, this limits their food to microscopic organisms or particles. Sponges circulate seawater through pores in their bodies.

▼ **Figure 21-6 Intracellular digestion in a sponge** *(a)* Tube sponges in the Virgin Islands. *(b)* Single-celled microorganisms are filtered from the water, and *(c)* are trapped on the outside of the fringed collar of a collar cell, engulfed by phagocytosis, and digested inside the cytoplasm of the collar cell.

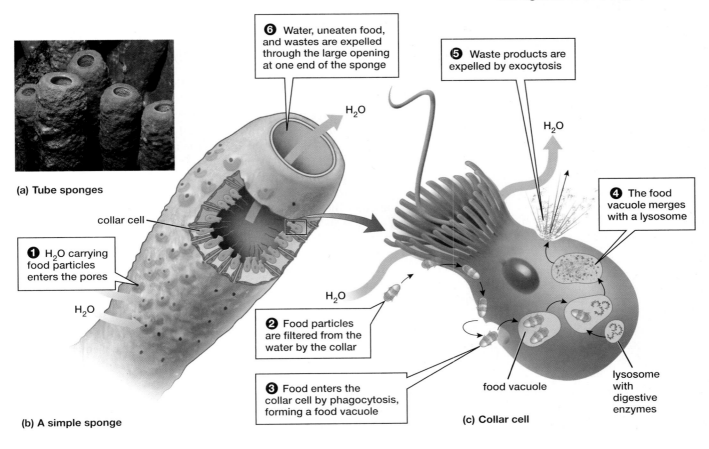

❻ Water, uneaten food, and wastes are expelled through the large opening at one end of the sponge

❺ Waste products are expelled by exocytosis

❹ The food vacuole merges with a lysosome

(a) Tube sponges

collar cell

❶ H₂O carrying food particles enters the pores

❷ Food particles are filtered from the water by the collar

❸ Food enters the collar cell by phagocytosis, forming a food vacuole

food vacuole

lysosome with digestive enzymes

(b) A simple sponge

(c) Collar cell

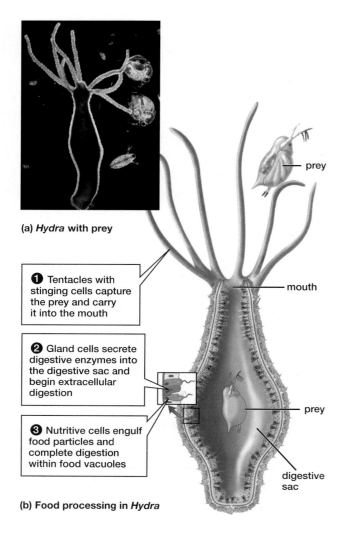

(a) Hydra with prey

❶ Tentacles with stinging cells capture the prey and carry it into the mouth

❷ Gland cells secrete digestive enzymes into the digestive sac and begin extracellular digestion

❸ Nutritive cells engulf food particles and complete digestion within food vacuoles

mouth

prey

digestive sac

(b) Food processing in Hydra

◄ **Figure 21-7 Digestion in a sac** *(a)* A *Hydra* (a cnidarian) has just captured and ingested a waterflea (*Daphnia*, a microscopic crustacean). *(b)* Within the digestive cavity of *Hydra*, enzymes digest the prey into smaller particles and nutrients.

Fringes of plasma membrane on specialized *collar cells* filter microscopic organisms from the water and ingest the prey by phagocytosis (see p. 50). Because the food is so small, mechanical breakdown, which is essential in most digestive systems, is unnecessary. Rather, phagocytosis encloses the food in a small sac called a food vacuole, which basically serves as a temporary, miniature stomach. The vacuole fuses with other sacs called lysosomes, which contain digestive enzymes, and the food is then digested within the vacuole into smaller molecules that can be absorbed into the cell cytoplasm. Undigested remnants of food remain in the vacuole, which eventually expels its contents back into the seawater.

Jellyfish and Their Relatives Have Digestive Systems Consisting of a Sac with a Single Opening

In most other animals, digestion takes place in a chamber within the body where enzymes break down chunks of food. One of the simplest of these chambers is found in cnidarians, such as sea anemones, *Hydra,* and jellyfish. These animals possess a digestive sac with a single opening for ingesting food and ejecting wastes (**Fig. 21-7**). This opening is generally referred to as the mouth, but it also serves as the anus. The animal's stinging tentacles capture its prey, which is then moved into the digestive sac where enzymes break it down. Cells lining the cavity absorb the nutrients and engulf small food particles. The undigested remains are eventually expelled through the mouth.

Most Animals Have Digestive Systems Consisting of a Tube with Several Specialized Compartments

Most animal species, including worms, mollusks, arthropods, and vertebrates, have a digestive system that is principally a one-way tube through the body. It begins with a mouth and ends with an anus. Such digestive systems usually consist of a series of specialized regions that process food in an orderly sequence, first grinding it up, then breaking it down with enzymes, absorbing the nutrient molecules into the circulatory system, and finally excreting the undigested wastes. This orderly processing in a tube allows the animal to eat more frequent meals than does a saclike digestive system.

The earthworm, which continuously ingests soil as it burrows, is a good example (**Fig. 21-8**). A tubular digestive system is essential to its way of life. Soil and bits of vegetation enter the mouth and pass through the pharynx and the esophagus to the crop, a thin-walled storage organ. The crop collects the food and gradually passes it to the gizzard. There, bits of sand and muscular contractions grind the food into smaller particles. The food then travels to the intestine, where enzymes digest it into simple molecules that can be absorbed by the cells lining the intestine and, ultimately, into the circulatory system. Indigestible material passes out through the anus.

Different species of animals show a remarkable diversity of digestive tracts. There are specializations for capturing and ingesting food, such as the slender bills and long tongues of hummingbirds, the sharp canines and meat-slicing molars of

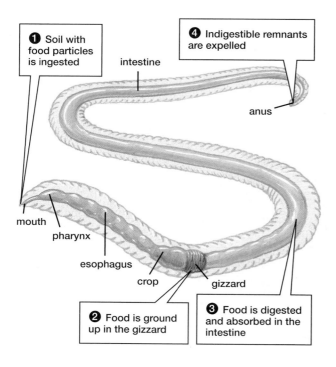

❶ Soil with food particles is ingested

intestine

❹ Indigestible remnants are expelled

anus

mouth

pharynx

esophagus

crop

gizzard

❷ Food is ground up in the gizzard

❸ Food is digested and absorbed in the intestine

◄ **Figure 21-8 Tubular digestive systems** The earthworm has a one-way digestive system that passes food through a series of compartments, each specialized to play a specific role in breaking down food and absorbing it.

tigers, and the pincers of lobsters and ants. The digestive tracts of carnivores tend to be short and simple, because meat is fairly easy to digest. Herbivores, however, often have chambers housing bacteria that digest the abundant cellulose found in plants, because the herbivores themselves cannot produce cellulose-digesting enzymes. Some animals, such as spiders, inject enzymes into their prey, so that the prey is actually digested outside of the spider's body and the spider slurps up the resulting liquid diet.

21.5 How Do Humans Digest Food?

Like most other animals, humans have a tubular digestive tract with several compartments in which food is broken down—first physically and then chemically (Fig. 21-9). Nearly everything of nutritional value is extracted and absorbed into

◀ **Figure 21-9 The human digestive system** The digestive system includes both the digestive tube, consisting of the oral cavity, pharynx, esophagus, stomach, small intestine, and large intestine; and organs such as the salivary glands, liver, gallbladder, and pancreas, which produce and store digestive secretions. **QUESTION:** Why is the stomach both muscular and expandable?

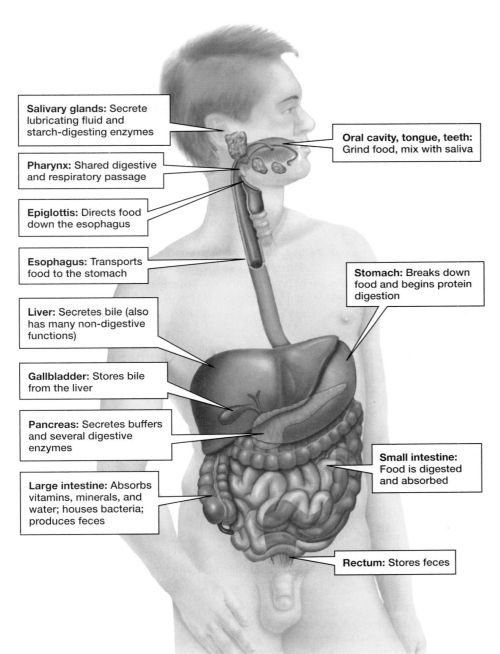

Salivary glands: Secrete lubricating fluid and starch-digesting enzymes

Pharynx: Shared digestive and respiratory passage

Epiglottis: Directs food down the esophagus

Esophagus: Transports food to the stomach

Liver: Secretes bile (also has many non-digestive functions)

Gallbladder: Stores bile from the liver

Pancreas: Secretes buffers and several digestive enzymes

Large intestine: Absorbs vitamins, minerals, and water; houses bacteria; produces feces

Oral cavity, tongue, teeth: Grind food, mix with saliva

Stomach: Breaks down food and begins protein digestion

Small intestine: Food is digested and absorbed

Rectum: Stores feces

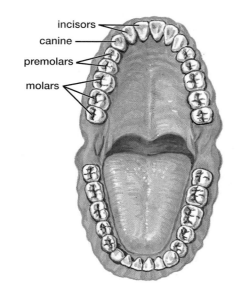

incisors
canine
premolars
molars

◄ **Figure 21-10** **Teeth begin the mechanical breakdown of food** Human teeth allow us to process a wide range of foods. The human mouth contains teeth specialized for a variety of functions: flat incisors for biting, pointed canines for tearing, and premolars and molars for crushing and grinding.

the circulatory system. Digesting and absorbing food requires coordinated action from the various structures of the digestive system.

Breakdown of Food Begins in the Mouth

You take a bite, you salivate, and you begin chewing. This begins both the mechanical and chemical breakdown of food. In humans and other mammals, the mechanical work is done mostly by teeth. In adult humans, 32 teeth of varying shapes and sizes tear, cut, and grind food into small pieces (**Fig. 21-10**).

While the food is being pulverized by the teeth, the first phase of chemical digestion begins as three pairs of salivary glands pour saliva into the mouth. Saliva has many functions. It contains the digestive enzyme amylase, which begins the breakdown of starches into sugar. Saliva also contains a bacteria-killing enzyme and antibodies that help to guard against infection. It lubricates food to facilitate swallowing and dissolves some food molecules, such as acids and sugars, carrying them to taste buds on the tongue. The taste buds bear sensory receptors that help to identify the type and the quality of the food.

The Pharynx Connects the Mouth to the Rest of the Digestive System

With the help of the muscular tongue, the food is manipulated into a mass and pressed backward into the **pharynx,** a muscular cavity connecting the mouth with the esophagus (**Fig. 21-11a**). The pharynx also connects the nose and mouth with the larynx, which leads to the trachea, the tube that conducts air to the lungs. This arrangement occasionally causes problems, as anyone who has ever choked on a piece of food can attest. Normally, however, the swallowing reflex (triggered by food entering the pharynx) elevates the larynx, so that a flap of tissue called the epiglottis blocks off the opening to the larynx and guides food into the esophagus (**Fig. 21-11b**).

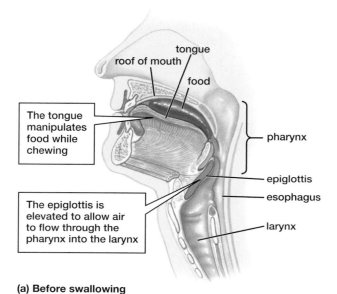

tongue
roof of mouth
food

The tongue manipulates food while chewing

The epiglottis is elevated to allow air to flow through the pharynx into the larynx

pharynx

epiglottis
esophagus

larynx

(a) Before swallowing

The Esophagus Conducts Food to the Stomach

Swallowing forces food into the **esophagus,** a muscular tube that propels the food from the mouth to the stomach. Muscles surrounding the esophagus produce a wave of contraction that begins just above the swallowed food and progresses down the esophagus, forcing the food toward the stomach. This muscular action, called **peristalsis,** occurs throughout the digestive tract, and is so effective that a person can actually swallow while upside-down. Mucus secreted by cells that line the esophagus helps to protect the lining from abrasion and lubricates the food during its passage.

The Stomach Stores and Breaks Down Food

The human **stomach** is an expandable muscular sac capable of holding as much as a gallon of food and liquids. The stomach has three primary functions. First, it stores food and releases it gradually into the small intestine at a rate suitable for

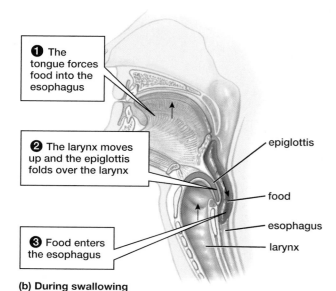

❶ The tongue forces food into the esophagus

❷ The larynx moves up and the epiglottis folds over the larynx

❸ Food enters the esophagus

epiglottis

food

esophagus
larynx

(b) During swallowing

◄ **Figure 21-11** **The challenge of swallowing** *(a)* Swallowing is complicated by the fact that both the esophagus (part of the digestive system) and the larynx (part of the respiratory system) open into the pharynx. *(b)* During swallowing, the larynx moves upward beneath the epiglottis. The epiglottis folds down over the larynx, sealing off the opening to the respiratory system and directing food down the esophagus instead.

proper digestion and absorption. Second, the stomach assists in the mechanical breakdown of food. Its muscular walls produce a variety of churning movements that break up large pieces of food. The third function of the stomach is to break food down chemically.

Glands in the stomach lining secrete hydrochloric acid, a protein called pepsinogen, and mucus. The hydrochloric acid gives the fluid in the stomach a pH of 1 to 3 (about the same as lemon juice). This kills many microbes that are inevitably swallowed along with food. Pepsinogen is the inactive precursor of a protein-digesting enzyme called pepsin. The stomach's acidity converts pepsinogen to pepsin, which then begins digesting the proteins in food. Why not just secrete pepsin in the first place? The glands secrete pepsinogen because pepsin would digest the very cells that manufacture it before it ever got into the stomach. Finally, mucus coats the stomach lining and serves as a barrier to self-digestion. The protection, however, is not perfect, so the cells lining the stomach must be replaced every few days.

Food in the stomach is gradually converted to a thick, acidic liquid called **chyme,** which consists of partially digested food and digestive secretions. Peristaltic waves (about three per minute) propel the chyme toward the small intestine. A ring of muscle at the lower end of the stomach allows only about a teaspoon of chyme to enter the small intestine with each contraction. It takes 2 to 6 hours, depending on the size of the meal, to empty the stomach completely.

Although digestion begins in the stomach, almost no absorption of nutrients occurs there. Only a few substances, including water, alcohol, and some other drugs, can enter the bloodstream through the stomach wall.

Most Digestion Occurs in the Small Intestine

The **small intestine** is about 1 inch in diameter and 10 feet long in a living human adult. The small intestine digests food into small molecules and absorbs these molecules into the bloodstream. The first role of the small intestine—digestion—is accomplished with the aid of secretions from three sources: the liver, the pancreas, and the cells of the small intestine itself.

The Liver and Gallbladder Provide Bile

The **liver** has many functions, including storing glycogen and detoxifying many poisonous substances. Its role in digestion is to produce **bile,** a complex mixture of bile salts, other salts, water, and cholesterol. Bile is stored in the **gallbladder** (see Fig. 21-9) and released into the small intestine through a tube called the bile duct. Although they help in the digestion of lipids, bile salts are not enzymes. Rather, much like dish detergent, bile salts have a hydrophobic part that interacts with fats and a hydrophilic part that dissolves in water. As a result, bile salts disperse chunks of fat into microscopic particles, exposing a large surface area to attack by lipid-digesting enzymes produced by the pancreas.

The Pancreas Secretes Digestive Substances

The **pancreas** consists of two major types of cells. One type produces hormones that help to regulate blood sugar (as we shall see in Chapter 23), and the other produces a digestive secretion called **pancreatic juice.** About a quart of pancreatic juice is released into the small intestine each day. This secretion contains water, sodium bicarbonate, and several digestive enzymes that break down carbohydrates, lipids, and proteins. Sodium bicarbonate (the active ingredient in baking soda) neutralizes the acidic chyme in the small intestine, producing a slightly basic pH. In contrast to the stomach's digestive enzymes, which require an acidic pH, pancreatic digestive enzymes require a slightly basic pH to function properly.

The Intestinal Wall Completes Digestion and Absorbs Nutrients

The wall of the small intestine contains cells that complete the digestive process and absorb the small molecules that result. Digestive enzymes are actually

---

Dying to Be Thin

Continued

When someone with bulimia vomits, the contents of the stomach erupt back through the esophagus, pharynx, and mouth. These structures do not have the thick mucous layer that protects the stomach, so the stomach acid burns away the cells of their linings. Further, the acid dissolves the enamel of the teeth, so that prolonged bulimia can lead to significant tooth loss.

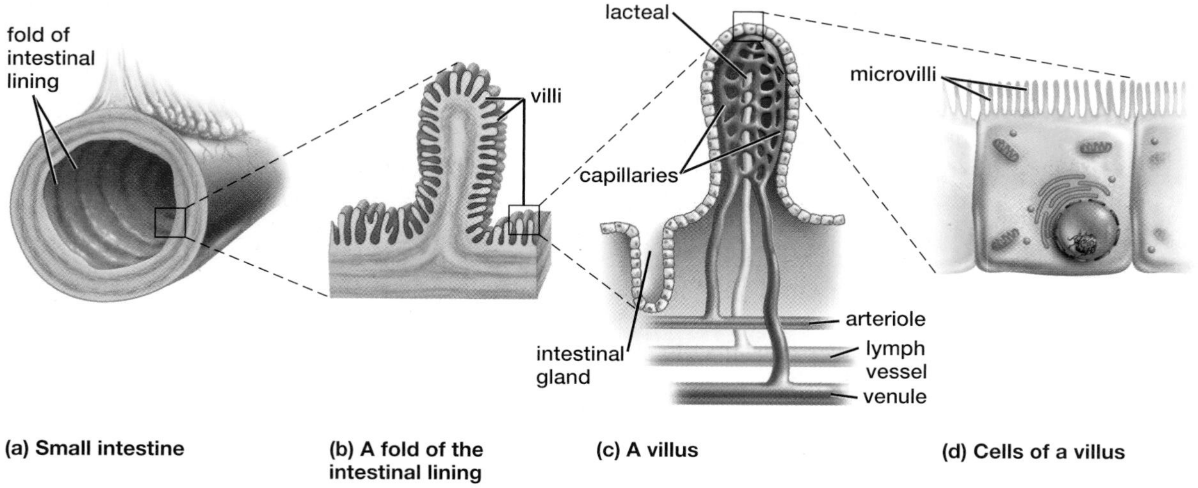

(a) Small intestine

(b) A fold of the intestinal lining

(c) A villus

(d) Cells of a villus

▲ **Figure 21-12 The small intestine** *(a)* The folds of the small intestine maximize the surface area available to absorb nutrients. *(b)* Large folds in the intestinal lining are themselves carpeted with tiny, finger-like projections called villi, *(c)* which enclose a network of capillaries and a lymph vessel, or lacteal. *(d)* If we use a microscope to zoom in on one villus, we see that the epithelial cells on its surface are sheathed in plasma membranes that have yet another level of microscopic projections, microvilli. **QUESTION:** What might the anatomy of the small intestine be like if its folds, villi, and microvilli had not evolved?

embedded in the plasma membranes of the cells that line the small intestine, so the final phase of digestion occurs as the nutrient is being absorbed into the cell. As in the stomach, the small intestine is protected from digesting itself by mucus, which is secreted by specialized cells in its lining.

Most Absorption Occurs in the Small Intestine

The small intestine is the major site of nutrient absorption into the blood. The small intestine has numerous folds and projections, giving it an internal surface area about 600 times greater than a smooth tube of the same length (**Fig. 21-12a,b**). Finger-like projections called **villi** (singular, villus) cover the entire surface of the intestinal wall (**Fig. 21-12c**). Villi, which range from 0.5 to 1.5 millimeters in length, give the intestinal lining a velvety appearance to the naked eye. Each individual cell of a villus bears a fringe of microscopic projections called **microvilli** (**Fig. 21-12d**). Collectively, these projections of the small intestine wall give it a surface area of more than 2,700 square feet (about 250 square meters)—almost the size of a tennis court.

Contractions of the circular muscles of the intestine slosh the chyme back and forth, bringing nutrients into contact with the absorptive surface of the small intestine. Within each villus is a network of blood capillaries and a single lymph capillary, called a *lacteal* (see Fig. 21-12c). Most nutrients, including water, simple sugars, amino acids, vitamins, and minerals, pass through the cells lining the small intestine and enter the capillaries. The breakdown products of fat (glycerol and fatty acids), however, take a different route. After diffusing into the cells lining the small intestine, they are resynthesized into fats, coated with protein, and then released as particles into the extracellular fluid within the villi. These particles are far too large to enter the blood capillaries, but they can pass through the lacteal wall. Suspended in lymph, the particles move from the lacteals through other lymph vessels, and are eventually delivered to the bloodstream when the lymph vessels empty into the veins (see pp. 382–384).

The Large Intestine Absorbs Water, Minerals, and Vitamins, and Forms Feces

The **large intestine** in a living human adult is about 5 feet long and 3 inches in diameter. The large intestine has two parts. For most of its length it is called the **colon;** its final 6-inch compartment is called the **rectum.** The leftovers of digestion flow from the small intestine into the colon: a mixture of water, minerals, indigestible fibers (mostly cellulose, from the cell walls of vegetables and fruits), and small amounts of undigested fats and proteins. The colon contains a flourishing bacterial population that lives on these nutrients. These bacteria synthesize vitamin B_{12}, thiamin, riboflavin, and, most importantly, vitamin K, which would otherwise be deficient in a typical diet. Cells lining the large intestine absorb these vitamins as well as water and minerals.

After absorption is complete, the result is the semisolid **feces,** consisting mostly of indigestible wastes and bacteria (about one-third of the dry weight of feces). The feces are transported by peristaltic movements until they reach the rectum, where expansion of this chamber stimulates the urge to defecate.

Digestion Is Controlled by the Nervous System and Hormones

When you eat, your body coordinates the events that convert food into nutrients circulating in your blood. The secretions and muscular activity of the digestive tract are regulated by both nerves and hormones.

Sensory Signals Initiate Digestion

The sight, smell, taste, and sometimes just the thought of food generate signals from the brain that act on many parts of the digestive tract. For example, nerve impulses stimulate the salivary glands and cause the stomach to begin secreting acid and mucus. As food moves through the digestive tract, its bulk stimulates local nervous reflexes that cause peristalsis.

Hormones Help Regulate Digestive Activity

There are at least a couple of dozen hormones that control appetite, satiation, and digestion. Here, we discuss only four major hormones, all secreted by the digestive system. These enter the bloodstream, circulate through the body, and act on specific receptors within the digestive tract. Like most hormones, they are regulated by negative feedback. For example, nutrients in chyme such as amino acids and peptides from protein digestion stimulate cells in the stomach lining to release the hormone *gastrin* into the bloodstream. Gastrin travels back to the stomach cells and stimulates further acid secretion, which promotes protein digestion. When the stomach contents become sufficiently acidic, gastrin secretion is inhibited, which in turn inhibits further acid secretion.

In response to chyme, the cells of the small intestine release *secretin* and *cholecystokinin*. These hormones stimulate the release of digestive fluids into the small intestine: bicarbonate and digestive enzymes from the pancreas and bile from the liver and gallbladder. *Gastric inhibitory peptide*, produced by cells of the small intestine in response to fatty acids and sugars in chyme, stimulates the pancreas to release insulin, which stimulates many body cells to absorb sugar. Gastric inhibitory peptide (as its name suggests) also inhibits both acid production and peristalsis in the stomach. As a result, it slows the rate at which chyme is pumped into the small intestine, providing additional time for digestion and absorption to occur.

A host of other hormones also regulate appetite, the sensation of fullness, and the rate at which food is metabolized. Some of these offer hope for controlling weight gain and obesity, as we discuss in the "Health Watch: Obesity and the Brain–Gut Connection" on p. 408.

21.6 What Are the Functions of Urinary Systems?

We may load our digestive tracts with pepperoni pizza, hot fudge sundaes, and coffee, but our cells must remain bathed in a precisely regulated solution of salts and nutrients. The digestive system is relatively unselective, so that any molecule that *can* be absorbed into the circulatory system through the small intestine *is* absorbed, whether the molecule is useful or toxic, needed immediately or already available in excess. How does the animal body, then, maintain homeostasis, fine-tuning and

health watch

Obesity and the Brain–Gut Connection

In 2006, medical experts estimated that about 65% of American adults were overweight; about half of these were obese (defined as a body mass index of more than 30—that's a 5-foot 6-inch woman weighing about 185 pounds, or a 5-foot 10-inch man weighing about 210 pounds). Overweight people have traditionally been told to eat less and exercise more. While that's good advice, it clearly isn't stemming the rise in obesity in the United States and many other countries around the world. And obesity isn't just a matter of personal health or fitting into airline seats. In the United States, every year, obesity is estimated to cost more than $100 billion in medical expenses and reduced productivity on the job, and to contribute to 300,000 deaths.

For at least the past half-century, diet books have appeared on bookshelves with depressing frequency, each claiming to be the answer to the prayers of the overweight. Diet pills have been around for decades, too, but most either have little long-term effect or are dangerous. Many overweight people diet strenuously for a few weeks, lose a few pounds, and find themselves a few months later just as heavy as before. Can anything really be done?

Maybe soon, something can be. Researchers have discovered a bewildering array of hormones that are produced by the digestive tract and act on the brain, influencing appetite, satiation, and metabolism. In 1994, Jeffrey Friedman of Rockefeller University discovered a hormone, synthesized by fat cells, that travels to the brain and suppresses appetite. Perhaps optimistically, Friedman called the hormone *leptin*, after a Greek word meaning thin. Many people hoped that a simple injection of leptin would suppress appetite in overweight people, so they could comfortably eat less and lose weight. No such

luck. It turns out that obese people make plenty of leptin, more than lean people do, but their brains don't respond to it.

Since then, researchers have found additional eating-related hormones, some of which look a lot more promising than leptin. Two of these hormones, with opposing effects on appetite and satiation, are ghrelin and peptide YY (often called PYY) (**Fig. E21-1**). When the stomach has been empty for a few hours, it churns out ghrelin, which travels to the brain and stimulates hunger. After a meal, ghrelin levels drop dramatically. Food in the intestines causes PYY to be produced, which travels to the brain to reduce hunger. Unfortunately, in dieters and overweight people, both ghrelin and PYY act in ways that make dieting hard and eating easy. Ghrelin levels increase when people diet and lose weight, which makes them feel hungrier and hungrier. Obese people also generally produce less PYY after a meal, so they don't feel full even though they've eaten enough food.

These two hormones are attractive targets for weight control. In 2006, researchers at the

Scripps Research Institute successfully vaccinated rats against ghrelin. The vaccines caused the rats to produce antibodies that bind ghrelin, so it can't get to the brain and stimulate hunger. The vaccinated rats ate normally, but gained less weight than control rats, indicating that their metabolism might have increased. Also in 2006, Nastech Pharmaceutical Company started clinical trials of a PYY nasal spray in people. When sprayed into the nose, PYY apparently penetrates to the brain and reduces appetite.

Will these new treatments help people to lose weight and keep it off? Or will other pathways in the body compensate for too much PYY or not enough ghrelin? Appetite, metabolism, and satiation are controlled by complex, interacting mechanisms, and it's too early to tell if treatments targeting ghrelin and PYY will work. Nevertheless, someday biomedical science may solve enough of the puzzle of weight gain and weight loss to produce effective treatments. For hundreds of millions of people around the world, that day can't come soon enough.

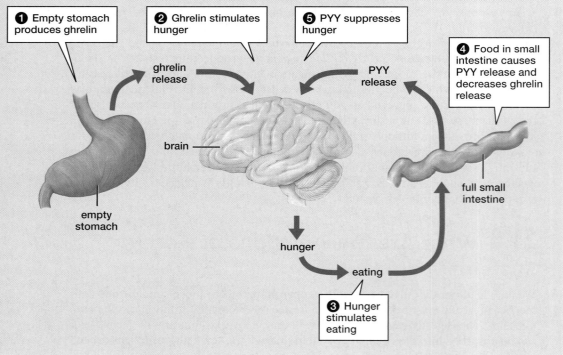

▶ **Figure E21-1 Hunger and satiation are regulated by hormones** The digestive tract produces ghrelin and PYY, which interact to regulate appetite, fullness, and probably metabolism.

❶ Empty stomach produces ghrelin

❷ Ghrelin stimulates hunger

❺ PYY suppresses hunger

❹ Food in small intestine causes PYY release and decreases ghrelin release

ghrelin release

PYY release

brain

empty stomach

full small intestine

hunger

❸ Hunger stimulates eating

eating

precisely regulating its internal environment? The skin, digestive tract, and respiratory system all play a role. However, eliminating harmful substances and excess nutrients while retaining useful substances is primarily the domain of the urinary system.

Whether we consider flatworms, fishes, or people, all urinary systems function similarly. First, the blood or other body fluids are filtered, with water and small dissolved molecules moving from the blood into the urinary system. Next, nutrients are selectively reabsorbed back into the blood. Often, highly toxic substances that must be removed very quickly are actively secreted from the blood into the urinary system. Finally, wastes and excess nutrients, dissolved in variable amounts of water, are excreted from the body.

21.7 What Types of Urinary Systems Are Found in Non-Human Animals?

There are nearly as many types of urinary systems (in invertebrates, these are usually called excretory systems) as there are phyla of animals. In a few, such as sponges, individual cells merely dump their wastes into the surrounding water. Most animals, however, have complex urinary systems, often under nervous and hormonal control, that precisely regulate which substances are excreted and which are kept in the body's fluids. These mechanisms ensure that the internal chemical composition of the animal remains fairly constant, despite enormous changes in its diet or living conditions. Here, we discuss only two of the many types of excretory systems in invertebrates.

Flame Cells Filter Fluids in Flatworms

The freshwater flatworm lives in streams. Because it constantly absorbs water by osmosis, it must excrete the water or else it will explode. Therefore, the major function of its excretory system is to regulate water balance. The flatworm's excretory system consists of a network of tubes that branch throughout the body (Fig. 21-13). At intervals, the branches end blindly in single-celled bulbs called flame cells. Water and dissolved substances are filtered from the body into the bulbs, where a cluster of beating cilia (that reminded their discoverers of the flickering flame of a candle) produces a current that forces the fluid through the tubes. Within the tubes, more wastes are added and some nutrients are reabsorbed. The resulting solution is expelled through pores in the body surface. Flatworms also have a large surface area through which many cellular wastes leave by diffusion.

Nephridia Filter Fluids in Earthworms

Earthworms, mollusks, and several other invertebrates have simple filtering structures called *nephridia* (singular, nephridium) that resemble the filtering structures that we will examine in vertebrate kidneys. The earthworm body is composed of repeating segments, and nearly every segment contains its own pair of nephridia. Fluid fills the body cavity that surrounds the earthworm's internal organs. This fluid collects both wastes and nutrients from the blood and tissues. The fluid is moved by cilia into a narrow, tubelike portion of the nephridium (Fig. 21-14). Here, salts and other dissolved nutrients are absorbed back into the blood, leaving water and wastes behind. The resulting urine is stored in an enlarged bladder-like portion of the nephridium and is excreted through an excretory pore in the body wall.

21.8 How Does the Human Urinary System Work?

Urinary systems of all vertebrates, including humans, face major challenges. Urine formation begins by filtering the blood. Size, however, is the only criterion for filtration—anything small enough to fit through the pores of the filter will leave the

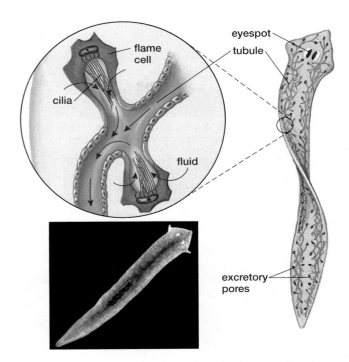

▲ **Figure 21-13 The simple excretory system of a flatworm** In the flatworm (inset), hollow flame cells direct excess water and dissolved wastes into a network of tubes. The beating cilia of the flame cells help move the fluid to excretory pores.

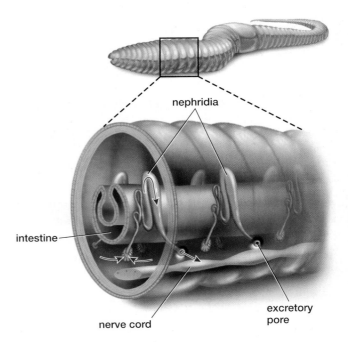

▲ **Figure 21-14 The excretory system of the earthworm** This system consists of structures called nephridia, one pair per segment. Fluid is drawn into one end of the nephridia, and urine is released at the other end through an excretory pore.

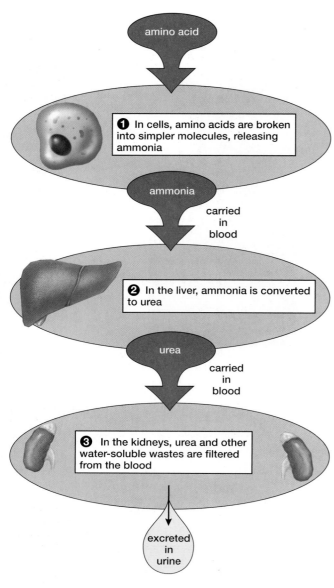

◀ Figure 21-15 A flow diagram showing the formation and excretion of urea QUESTION: In some animals, ammonia is not converted into urea but, instead, circulates in the blood until it is excreted. In what types of environments would you expect to find such animals?

blood and enter the urinary system. Therefore, not only wastes, but also water and nutrients that the body cannot afford to lose, are filtered out. In humans and many other vertebrates, the kidneys have evolved complex internal structures and metabolic abilities that eliminate wastes while retaining most of the water and nutrients.

The Human Urinary System Produces, Transports, and Excretes Urine

The **kidneys** are organs in which the fluid portion of the blood is collected and filtered. From this fluid, water and important nutrients are then reabsorbed into the blood. The remaining fluid, called **urine**—consisting of toxic substances, cellular waste products, excess vitamins, salts, some hormones, and water—stays behind to be excreted from the body. The rest of the urinary system channels and stores urine until it is eliminated from the body.

The Urinary System Is Crucial for Homeostasis

The urinary system of humans and other vertebrates helps maintain homeostasis in the body in several ways. These include:

- Regulating blood levels of minerals and other ions such as sodium, potassium, chloride, and calcium.

- Regulating the water content of the blood.

- Maintaining the proper pH of the blood.

- Retaining important nutrients such as glucose and amino acids in the blood.

- Eliminating cellular waste products such as urea. When amino acids are used in cells as a source of energy or for the synthesis of new molecules, ammonia (NH_3) is produced as a by-product. Ammonia is very toxic. In mammals, the liver converts ammonia to urea, a far less toxic substance (Fig. 21-15). In the kidneys, urea is filtered from the blood and ultimately excreted in the urine.

Web Animation Human Urinary System Anatomy

The Urinary System Consists of the Kidneys, Ureter, Urinary Bladder, and Urethra

Human kidneys are paired organs located on either side of the spinal column, slightly above the waist (Fig. 21-16). Each is approximately 5 inches long, 3 inches wide, and 1 inch thick. Blood enters each kidney through a renal artery. After the blood has been filtered, it exits through a renal vein. The kidneys produce urine that leaves each kidney through a narrow, muscular tube called the **ureter.** Using peristaltic contractions, the ureters transport urine to the **urinary bladder.** This hollow, muscular chamber collects and stores the urine. During urination, contraction of the bladder forces the urine out of the body through the **urethra,** a single narrow tube about 1.5 inches long in women and about 8 inches long in men.

Urine Is Formed in the Nephrons of the Kidneys

Each kidney contains a solid outer layer where urine forms and an inner chamber that collects urine and funnels it into the ureter (Fig. 21-17). The outer layer of each kidney contains about a million tiny individual tubes, called **nephrons,** which filter the blood, process the filtered fluid, and form urine. Each nephron

Dying to Be Thin

Continued

The malnutrition, vomiting, and laxative use that accompany anorexia and bulimia often wreak havoc with the body's absorption of minerals such as sodium, potassium, and calcium. Despite heroic efforts, the kidneys often cannot maintain the proper concentrations of these minerals in the bloodstream. When mineral homeostasis fails, nervous disorders, heart attacks, or multiple organ failure may result.

▶ **Figure 21-16 The human urinary system** Diagrammatic view of the human urinary system and its blood supply. **QUESTION:** Why is there such an extensive blood supply to the kidneys and the individual nephrons?

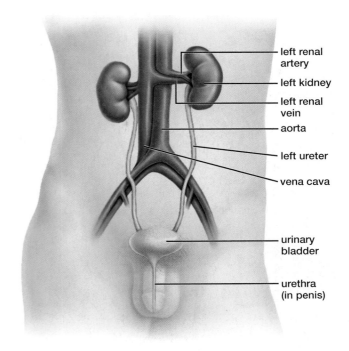

left renal artery
left kidney
left renal vein
aorta
left ureter
vena cava
urinary bladder
urethra (in penis)

has three major parts (**Fig. 21-18**): (1) the **glomerulus,** a dense knot of capillaries from which fluid is filtered from the blood and collected into (2) a surrounding cuplike structure, called **Bowman's capsule,** which funnels the fluid into (3) a long, twisted **tubule.** The Bowman's capsule channels fluid into the **proximal tubule.** The fluid then moves through the **loop of Henle** and the **distal tubule.** Different portions of the tubule selectively modify the fluid as it travels through them. In the tubule, nutrients are selectively reabsorbed from the fluid back into the blood, while wastes and some of the water are left behind to form urine. Additional wastes are also secreted into the tubule from the blood. Finally, the distal tubules of multiple nephrons drain into a **collecting duct,** which conducts urine into the renal pelvis, a hollow, funnel-like structure in the center of the kidney that connects with the ureter.

Blood Is Filtered by the Glomerulus

Urine formation starts with the process of **filtration** (**Fig. 21-19**, step ❶). Blood enters each nephron by an arteriole that branches from the renal artery. Within the cup-shaped portion of the nephron—Bowman's capsule—the arteriole branches into numerous capillaries that form the mass of the glomerulus (see Fig. 21-18). The walls of these capillaries are extremely permeable to water and small dissolved molecules, but blood cells and most proteins are too large to be filtered out, so they remain in the blood. Blood pressure within the capillaries drives water and dissolved substances from the blood out through the capillary walls. The resulting watery fluid, called the **filtrate,** is collected in Bowman's capsule, beginning its journey through the rest of the nephron.

Web Animation Urine Formation

The Filtrate Is Converted to Urine in the Tubules of the Nephron

The filtrate collected in Bowman's capsule contains a mixture of wastes, essential nutrients, and a lot of water. In fact, in the average adult human, 8 quarts (about 7.5 liters) of fluid are filtered into the nephrons *every hour*. When you consider that a human has only a little over 5 quarts (about 5 liters) of blood, and only a little over 3 quarts (about 3 liters) of that is water, you can appreciate how important it is for the kidneys to reclaim almost all of the water that is filtered. Normally, a person produces only about 1.5 quarts (1.5 liters) of urine each day, so more than 99% of the water is returned to the blood.

Overall, therefore, the nephrons must restore the nutrients and most of the water to the blood while retaining wastes for elimination. This task is accomplished by two processes: tubular reabsorption and tubular secretion.

Tubular Reabsorption Moves Water and Nutrients from the Nephron to the Blood From Bowman's capsule, the filtrate passes through the proximal tubule, which is surrounded by capillaries. Most of the water and nutrients in the filtrate move from the proximal tubule into the capillaries in a process called **tubular reabsorption** (**Fig. 21-19**, step ❷). The cells of the proximal tubule actively transport salts and other nutrients, such as amino acids and glucose, out of the tubule and into the surrounding extracellular fluid. The nutrient molecules then diffuse from the extracellular fluid into the adjacent capillaries. Water follows the nutrients

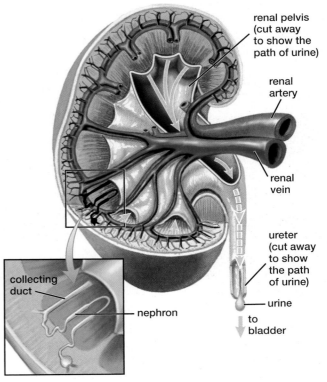

renal pelvis (cut away to show the path of urine)
renal artery
renal vein
ureter (cut away to show the path of urine)
urine
to bladder
collecting duct
nephron

enlargement of a single nephron and collecting duct

▶ **Figure 21-17 Cross section of a kidney** The cross section shows the blood supply and internal structure of a kidney. A single nephron (inset), considerably enlarged, is drawn to show the orientation of nephrons in the kidney and their connection to a collecting duct. Collecting ducts empty urine into the renal pelvis, which drains into the ureter. Yellow arrows show the pathway of urine flow.

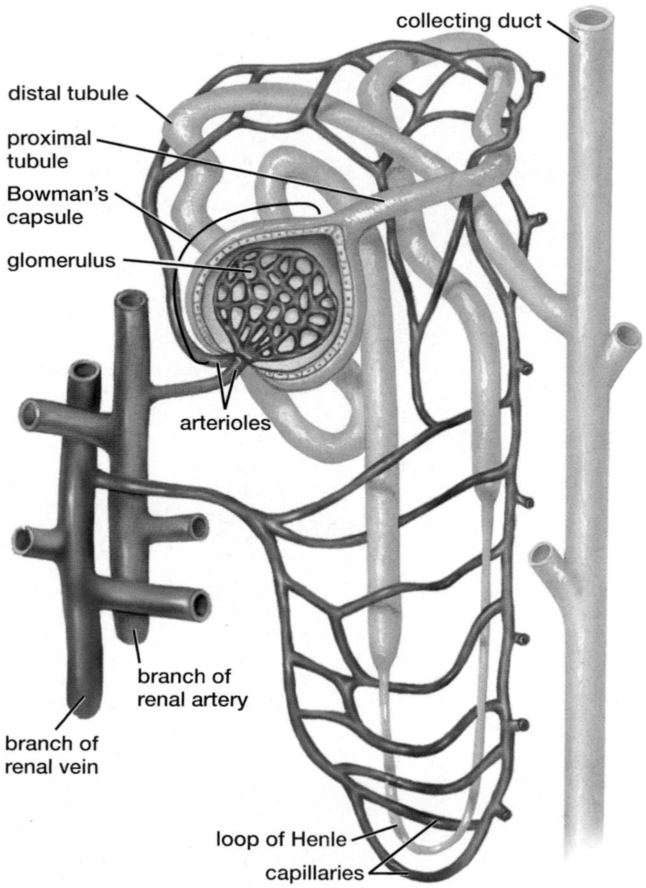

collecting duct

distal tubule

proximal tubule

Bowman's capsule

glomerulus

arterioles

branch of renal artery

branch of renal vein

loop of Henle

capillaries

▲ **Figure 21-18 An individual nephron and its blood supply**

out of the tubule and into the capillaries by osmosis. Wastes such as urea remain in the tubule and become more concentrated as water leaves.

Tubular Secretion Moves Wastes from the Blood into the Nephron   In **tubular secretion**, wastes such as hydrogen ions, potassium, ammonia, and many drugs are moved from the capillaries into the nephron (**Fig. 21-19**, step ❸). Typically, the cells of the distal tubule actively transport wastes from the surrounding extracellular fluid into the tubule. Lowering the concentration of wastes in the extracellular fluid produces a concentration gradient so that the wastes passively diffuse out of the capillaries.

Why bother with tubular secretion, when these wastes are mostly small molecules that can enter the nephron by filtration anyway? Because many of these wastes, such as acid and ammonia, are extremely toxic, even the rapid filtration of the blood into Bowman's capsule cannot remove them from the blood fast enough. Tubular secretion speeds up the process of ridding the body of these dangerous substances.

Urine Becomes Concentrated in the Collecting Ducts

Finally, in the collecting ducts, **concentration** of urine may occur through the removal of water (**Fig. 21-19**, step ❹). When mammals (including humans) and birds need to conserve water, they can produce urine that has a higher concentration of dissolved materials than their blood has.

Urine can become concentrated because there is a concentration gradient of salts and urea in the extracellular fluid that surrounds the nephrons and the collecting ducts. (The concentration gradient is created and maintained by the loops of Henle.) As filtrate travels through the collecting ducts to the renal pelvis, it passes through areas of increasingly concentrated extracellular fluid. As the difference in concentration between the filtrate and the surrounding fluid increases, water leaves the filtrate by osmosis and is carried off by the surrounding capillaries. The filtrate in the collecting duct, now called urine, can become as concentrated as the surrounding fluid, which may be four times as concentrated as blood. The urine remains concentrated because the rest of the excretory system is fairly impermeable to water, salts, and urea.

▶ **Figure 21-19 Urine formation in the nephron and collecting duct**

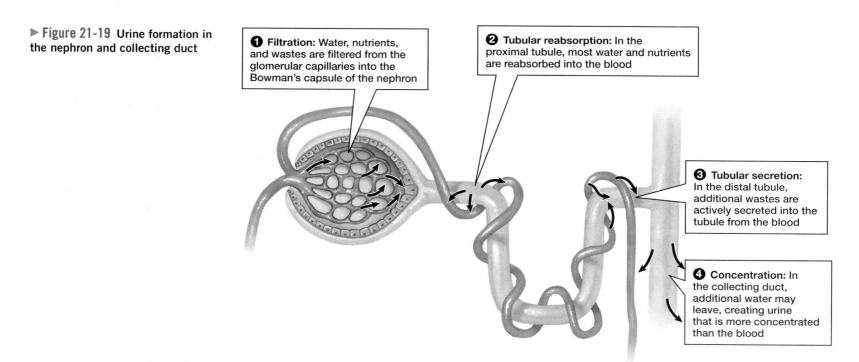

❶ **Filtration:** Water, nutrients, and wastes are filtered from the glomerular capillaries into the Bowman's capsule of the nephron

❷ **Tubular reabsorption:** In the proximal tubule, most water and nutrients are reabsorbed into the blood

❸ **Tubular secretion:** In the distal tubule, additional wastes are actively secreted into the tubule from the blood

❹ **Concentration:** In the collecting duct, additional water may leave, creating urine that is more concentrated than the blood

Negative Feedback Regulates the Water Content of the Blood

Maintaining the proper volume of water in the body is a key function of the urinary system, as we have seen, but the appropriate amount of water to retain changes continually as conditions change. The kidneys, therefore, must reabsorb more water when a person is perspiring heavily and less water when the person has just drunk a lot of water.

The amount of water reabsorbed into the blood is controlled by negative feedback (see pp. 365–366). One of these feedback mechanisms is based on the amount of *antidiuretic hormone (ADH)* circulating in the blood. This hormone (secreted by the pituitary gland) allows more water to be reabsorbed from the urine. It does so by increasing the permeability of the distal tubule and the collecting duct to water. The release of ADH is regulated by receptor cells in the brain that monitor the concentration of the blood and by receptors in the heart that monitor blood volume.

Let's look at an example. A lost traveler staggers through the hot desert, perspiring heavily and losing water with every breath. As he becomes dehydrated, his blood volume falls, and the osmotic concentration of his blood rises, triggering release of ADH by the pituitary gland (**Fig. 21-20**). The ADH increases the permeability of the distal tubule and the collecting duct, thereby increasing the reabsorption of water. Water is returned to the blood, leaving urine that is more concentrated than the blood.

Eventually, our traveler finds an oasis and overindulges in the cool, clear water of a spring. His blood volume rises and its osmotic concentration falls, triggering a decrease in his ADH output. Reduced ADH makes his distal tubules and collecting ducts less permeable to water, so less water is reabsorbed from them. He will now produce urine that is more dilute than the blood. In extreme cases, urine flow may exceed 1 quart (about 1 liter) per hour. As the proper water level in his blood is restored, the increased osmotic concentration of the blood and decreased blood volume will again stimulate some ADH release, thus maintaining homeostasis by keeping the blood water content within narrow limits.

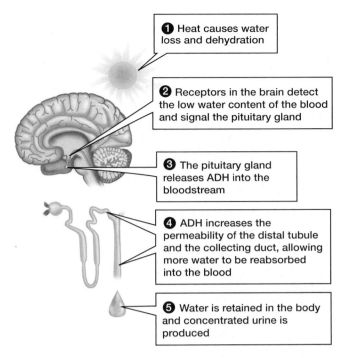

❶ Heat causes water loss and dehydration

❷ Receptors in the brain detect the low water content of the blood and signal the pituitary gland

❸ The pituitary gland releases ADH into the bloodstream

❹ ADH increases the permeability of the distal tubule and the collecting duct, allowing more water to be reabsorbed into the blood

❺ Water is retained in the body and concentrated urine is produced

▲ Figure 21-20 **Dehydration stimulates ADH release and water retention QUESTION:** Alcohol inhibits ADH release. How does alcohol consumption affect the body's water balance?

Dying to be Thin Revisited

In the most extreme cases, eating disorders, as you might predict, can cause death from starvation. However, long before they literally starve to death, people with eating disorders seriously damage their bodies, and often die of other causes. According to Carré Otis, "It was common for the young girls I worked with to have a heart attack," probably brought about by massive imbalances in the concentrations of sodium and potassium in the blood, caused by malnutrition. At age 30, Otis herself required surgery to repair her damaged heart.

How do people develop eating disorders? Anorexia and bulimia usually arise from a faulty self-image or from the perceived demands of family, friends, or career. At her first foreign modeling assignment in 2004, Ana Reston was told that she was fat, even though she weighed only about 110 pounds. Friends and associates trace her descent to multiple organ failure from that time.

Genes apparently play a role, too. Women whose mother or sister suffers from anorexia are 12 times more likely to develop anorexia, and 4 times more likely to develop bulimia, than women without such a family history. Not surprisingly, if one member of a pair of identical twins develops an eating disorder, the other is highly likely to as well. However,

although genetic contributions to personality traits such as perfectionism, anxiety, or low self-esteem are suspected, no one has yet identified any specific genes that might predispose anyone to eating disorders.

Eating disorders are difficult to treat. Victims are usually given nutritional therapy to help them recover from malnutrition. Psychotherapy is often necessary, and antidepressant drugs are helpful in some cases. Because many victims hide or deny their problem, and because treatment is expensive, the majority of sufferers are inadequately treated. Successes do occur, however. The TV show *American Idol* may have saved the life of one of its stars, Katharine McPhee, who suffered from bulimia. Worried that repeated vomiting might eat away at her vocal cords and hurt her chances to win the *Idol* singing competition, McPhee spent 3 months, 6 days a week, at an eating disorder clinic. It worked: She is no longer bulimic, was the *Idol* runner-up, and released her first CD in January 2007.

Consider This

In 2006, the Madrid Fashion Week in Spain banned any models with a BMI below 18 (a 5-foot 8-inch model weighing about 120 pounds).

Among women who couldn't meet the new BMI requirements are Spain's most famous model, Esther Canadas, who has a BMI of about 14. In 2007, a fashion show in Milan mandated a minimum BMI of 18.5. Most modeling agencies, of course, strenuously object to the BMI rule.

Thousands of young women become anorexic, many of them because they think that society finds extreme thinness to be desirable. Is there any way to change this message? Are there any appropriate measures that a free society can or should take to reverse or limit this message?

Chapter Review

Summary of Key Concepts

For additional study help and activities, go to www.mybiology.com.

21.1 How Do Animals Regulate the Composition of Their Bodies?

Nutrition, digestion, and excretion are crucial to homeostasis, maintaining internal conditions that sustain life. Nutrients are substances that an animal requires in its diet, providing both energy and materials to construct its body. Digestion is the process of converting food into molecules that can be absorbed by the circulatory system. Through excretion by the digestive and respiratory tracts, the skin, and especially the urinary tract, animals retain needed nutrients and dispose of surplus nutrients, wastes, and toxic molecules in their food.

21.2 What Nutrients Do Animals Need?

Each animal species has specific nutritional requirements. These requirements include molecules that can be broken down to liberate energy, such as lipids, carbohydrates, and protein; chemical building blocks used to construct complex molecules; minerals; vitamins needed for the chemical reactions of metabolism; and water.

21.3 What Are the Major Processes of Digestion?

Digestive systems must accomplish five tasks: ingestion, mechanical and chemical breakdown of food, absorption, and elimination of wastes. Digestive systems convert the complex molecules of the bodies of ingested organisms into simpler molecules that can be used.

Web Animation   Mechanical and Chemical Breakdown of Food

21.4 What Types of Digestive Systems Are Found in Non-Human Animals?

In sponges, individual cells ingest and digest food. The simplest multicellular digestive system is a saclike cavity such as that of cnidarians, in which there is a single opening that serves as both mouth and anus. Most animals have a tubular digestive system with specialized parts where food is processed in an orderly sequence.

21.5 How Do Humans Digest Food?

In humans, digestion begins in the mouth, where food is mechanically broken down by chewing, and chemical digestion is initiated by saliva. Food is then conducted to the stomach by peristaltic waves of the esophagus. In the acidic environment of the stomach, food is churned into smaller particles, and protein digestion begins. Gradually, the liquefied food, now called chyme, is released to the small intestine. There, it is neutralized by sodium bicarbonate from

the pancreas. Secretions from the liver, pancreas, and the cells of the intestine complete the breakdown of proteins, fats, and carbohydrates. In the small intestine, the simple molecular products of digestion are absorbed into the bloodstream for distribution to the body cells. The large intestine absorbs water, minerals, and vitamins, and converts indigestible material to feces.

Digestion is regulated by the nervous system and hormones. The smell and taste of food and the action of chewing trigger the secretion of saliva in the mouth and the production of gastrin by the stomach. Gastrin stimulates stomach acid production. As chyme enters the small intestine, additional hormones are produced by intestinal cells, including secretin, which causes sodium bicarbonate production to neutralize the acidic chyme. Other hormones stimulate bile release and cause the pancreas to secrete digestive enzymes into the small intestine.

21.6 What Are the Functions of Urinary Systems?

The urinary system plays a crucial role in homeostasis, regulating the water and mineral content of the blood, as well as blood pH. The organs of the urinary system help retain nutrients and eliminate cellular wastes and toxic substances.

21.7 What Types of Urinary Systems Are Found in Non-Human Animals?

In the invertebrate flatworm, wastes and excess water are filtered from the body fluids into tubules. Cilia on flame cells move the fluids through the tubules and expel them through pores in the body surface. In the earthworm, nephridia (which resemble individual nephrons) filter fluid that bathes the organs and blood vessels, storing excess water and wastes, which are excreted through pores.

21.8 How Does the Human Urinary System Work?

The urinary system of vertebrates (including humans) helps maintain homeostasis in the body in several ways, including (1) regulating the blood levels of important minerals and other ions; (2) regulating the water content of the blood by negative feedback involving antidiuretic hormone (ADH), produced in the pituitary gland of the brain; (3) maintaining proper pH of the blood; (4) retaining important nutrients; and (5) eliminating cellular waste products such as urea.

The urinary system of humans and other vertebrates consists of kidneys, ureters, bladder, and urethra. Kidneys produce urine, which is conducted by the ureters to the bladder, a storage organ. Urine passes out of the body through the urethra.

Each kidney contains more than a million individual nephrons in its outer layer. Urine formed in the nephrons enters collecting ducts that empty into the renal pelvis, from which it is funneled into the ureter. Each nephron is served by an arteriole that branches from the renal artery. The arteriole further branches into a mass of capillaries called the glomerulus. There, water and dissolved substances are filtered from the blood by pressure. The filtrate is collected in the cup-shaped Bowman's capsule and conducted along the tubular portion of the nephron. During tubular reabsorption, nutrients are actively pumped out of the filtrate through the walls of the tubule. Nutrients then enter capillaries that surround the tubule, and water follows by osmosis. Some wastes remain in the filtrate; others are pumped into the tubule by tubular secretion. The tubule forms the loop of Henle, which creates a salt concentration gradient surrounding it. After completing its passage through the tubule, the filtrate enters the collecting duct, which passes through the concentration gradient. Final passage of the filtrate through this gradient via the collecting duct concentrates the urine.

Web Animation Human Urinary System Anatomy

Web Animation Urine Formation

Key Terms

bile *p. 405*

Bowman's capsule *p. 411*

calorie *p. 395*

Calorie *p. 395*

chyme *p. 405*

collecting duct *p. 411*

colon *p. 406*

concentration *p. 412*

digestion *p. 395*

distal tubule *p. 411*

esophagus *p. 404*

essential amino acid *p. 397*

essential fatty acid *p. 396*

excretion *p. 395*

feces *p. 407*

filtrate *p. 411*

filtration *p. 411*

gallbladder *p. 405*

glomerulus *p. 411*

glycogen *p. 397*

kidney *p. 410*

large intestine *p. 406*

liver *p. 405*

loop of Henle *p. 411*

microvillus (plural, **microvilli**) *p. 406*

mineral *p. 397*

nephron *p. 410*

nutrient *p. 394*

nutrition *p. 395*

pancreas *p. 405*

pancreatic juice *p. 405*

peristalsis *p. 404*

pharynx *p. 404*

proximal tubule *p. 411*

rectum *p. 406*

small intestine *p. 405*

stomach *p. 404*

tubular reabsorption *p. 411*

tubular secretion *p. 412*

tubule *p. 411*

ureter *p. 410*

urethra *p. 410*

urinary bladder *p. 410*

urine *p. 410*

villus (plural, **villi**) *p. 406*

vitamin *p. 397*

Thinking Through the Concepts

Suggested answers to end-of-chapter and figure-based questions can be found at the end of the text.

Fill-in-the-Blank

1. A substance that an animals needs to build or operate its body, but that it cannot synthesize itself, is called a(n) _____. _____ is the process of physically and chemically breaking down food into molecules that can be absorbed into the circulatory system. Indigestible material, waste products of cellular metabolism, toxic substances, and substances eaten in excess of the body's needs are eliminated through _____, which occurs in the _____, _____, _____, and _____ (major organs or organ systems of the body).

2. The primary sources of energy for animals are _____ and _____. Organic molecules needed in very small amounts that an animal cannot synthesize itself (in sufficient quantities) are called _____. _____ are important components of bone, teeth, and the dissolved materials in the blood and extracellular fluid.

3. Amino acids and fatty acids that the body requires but cannot synthesize are called _____.

4. Digestion includes five major processes: _____, _____, _____, _____, and _____.

5. Most animals have a tubular digestive tract. The major cavities of the human digestive tract are the _____, _____, _____, stomach, _____, and _____.

6. Glands in the stomach wall produce three major secretions: _____, _____, and _____.

7. Enzymes from the _____ empty into the small intestine; this gland also produces an acid-neutralizing buffer called _____. The small intestine also receives _____ from the liver and gallbladder; this secretion, although not an enzyme, is important in the digestion of _____ (type of nutrient).

8. In humans, urine is produced in tiny tubules of the kidney, called _____. Blood is first filtered into the beginning of the tubules, the _____. The filtrate is then processed through tubular absorption and secretion. Finally, it is concentrated in the _____. Urine is stored in the _____ and leaves the body through the _____.

9. If you begin to dehydrate, a gland called the _____ releases the hormone _____, which causes your kidneys to reabsorb water and produce concentrated urine.

Review Questions

1. List six general types of nutrients, and describe the role of each in nutrition.

2. List and describe the function of the three principal secretions of the stomach.

3. List the substances secreted into the small intestine, and describe the origin and function of each.

4. Describe the structural and functional adaptations of the human small intestine that ensure good digestion and absorption.

5. Control of the human digestive tract involves messages that coordinate activity in one chamber with those taking place in subsequent chambers. List the coordinating events you discovered in this chapter in the appropriate order, beginning with tasting, chewing, and swallowing a piece of meat and ending with residue that enters the large intestine. What initiates each process?

6. What are the major functions of the urinary system in any animal?

7. Describe the processes of filtration, tubular reabsorption, and tubular secretion.

Applying the Concepts

1. IS THIS SCIENCE? Humor writer Dave Barry described Calories as "tiny units of measurement that cause food to taste good." Is this pure humor, or is there some scientific truth to this statement? Use evolutionary concepts in your answer. How is this statement related to the current obesity problem in affluent countries such as the United States?

2. One of the common remedies for constipation (difficulty eliminating feces) is a laxative solution that contains magnesium salts. In the large intestine, magnesium salts are absorbed very slowly by the intestinal wall, remaining in the intestinal tract for long periods of time. Thus, the salts affect water movement in the large intestine. On the basis of this information, explain the laxative action of magnesium salts.

3. Some employers require their employees to submit to urine tests before they can be employed and at random intervals during their employment. Refusal to take the test or failure to "pass" the test could be grounds for termination. What is the purpose of a urine test? How would you feel if you had to undergo a urine test to obtain or keep a job? Explain your answers.

For additional resources, go to mybiology.com.

chapter 22

Defenses Against Disease

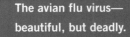

The avian flu virus—
beautiful, but deadly.

Case Study | Avian Flu

Almost everyone has heard about avian flu, commonly called "bird flu." As its name implies, the avian flu virus primarily infects birds, usually chickens and ducks. Until recently, most avian flu viruses weren't very deadly even to birds, and usually didn't infect people at all. Then in 1997, a particularly deadly strain appeared in Hong Kong, infecting 18 people, 6 of whom died. Health authorities concluded that the victims caught the virus by handling infected chickens. To prevent further human disease, the government ordered every chicken in Hong Kong—about 1.5 million birds—to be killed. Nevertheless, deadly avian flu continues to spread. By October 2006, infected migratory birds, including ducks and swans, had been found across Europe, including Russia, Greece, Italy, France, Denmark, Sweden, and Germany. Avian flu has struck people not only in southeast Asia, but as far away as Egypt, Iraq, and Turkey. As of August 2007, about 60% of the 320 confirmed human cases of avian flu have been fatal (although the mortality rate may actually be somewhat lower, because no one knows how many people with less severe symptoms never saw a physician and were never correctly diagnosed).

So far, avian flu isn't very efficient at passing from birds to people, and it seems almost totally incapable of being transmitted from person to person (or else Hong Kong, with more than 16,000 people per square mile, would have had many more cases). But viruses mutate, and flu viruses mutate faster than most. Mutations allowed relatively benign avian flu viruses to become deadly to chickens, and to infect people. Additional mutations might very well allow avian flu to become efficient at moving from person to person. If so, millions of people might die. This possibility has health authorities scrambling for solutions. Several TV shows, such as the crime series *Numb3rs*, have used the threat of avian flu as a plot device.

What's so deadly about avian flu, compared to "regular" human flu viruses? After all, flu viruses have been infecting people for thousands of years. Most reasonably healthy people who catch the flu become miserably ill, but recover even without any medical treatment. In this chapter, we explore the marvelous defense mechanisms of the human immune system that fight off dozens of would-be infections every day, including most cases of the flu. We will see that the immune system recognizes viruses as "foreign," and mobilizes an army of cells and molecules to eliminate them from the body. As you learn about the immune response, consider a few questions: What is it about a virus that labels it as foreign? How do you become immune to a disease? Why doesn't immunity or vaccination against human flu viruses provide protection against avian flu? ■

People and poultry, packed close together, provide the perfect opportunity for avian flu to jump from birds to people.

22.1 How Does the Body Defend Against Invasion?

Our bodies are under constant attack from microbes, including viruses, bacteria, fungi, and protists. Since the early 1980s, many viruses have emerged as threats to people, including HIV (human immunodeficiency virus), Ebola virus, hantavirus, West Nile, SARS (severe acute respiratory syndrome), and avian flu. We have witnessed outbreaks of relatively rare but deadly strains of the common intestinal bacterium *Escherichia coli* (commonly known as *E. coli*) and the respiratory bacterium *Streptococcus pyogenes* ("flesh-eating bacteria"). Meanwhile, hundreds of "cold virus" strains and new strains of flu continue to threaten us.

Given the prevalence and diversity of disease-causing organisms, you might wonder, "Why don't we get sick more often?" Fortunately for us, the human body has multiple lines of defense against microbial attack (**Fig. 22-1**). First, nonspecific external barriers to invasion keep microbes out of the body. Second, internal defenses, also nonspecific, combat invading microbes. Third, the immune system directs its assault—the immune response—against specific microbes.

The Skin and Mucous Membranes Are Nonspecific External Barriers to Invasion

The best defense against disease is to prevent microbes from entering the body in the first place. In animal bodies, this first line of defense is performed by the two surfaces that are exposed to the environment: the skin and the mucous membranes of the digestive, respiratory, and urogenital tracts.

The Skin Is an Inhospitable Environment for Microbial Growth

The outer surface of the skin consists of dry, dead cells filled with tough proteins similar to those in hair and nails. Most microbes that land on the skin cannot obtain the water and nutrients they need to survive. Those few that gain a foothold on skin are likely to be shed before they can do harm, because skin cells are constantly sloughed off and replaced from below. The skin is also protected by secretions from sweat glands and oil glands. These secretions contain natural antibiotics, such as lactic acid, that inhibit the growth of bacteria and fungi. These multiple defenses make the unbroken skin an extremely effective barrier against microbial invasion.

Mucus, Ciliary Action, and Secretions Defend Mucous Membranes Against Microbes

The warm, moist mucous membranes that line our digestive, respiratory, and urogenital tracts are much more hospitable to microbes than is the dry, oily skin, but these membranes nonetheless possess effective defenses. First, as their name implies, mucous membranes secrete mucus, which traps microbes that enter the body through the nose or mouth (**Fig. 22-2**). Further, the mucus contains antibacterial enzymes that kill bacteria by weakening their cell walls. Then, cilia on the membranes sweep up the mucus, microbes and all, until it is either coughed or sneezed out of the body or swallowed.

If microbes are swallowed, they enter the stomach, where they encounter both extreme acidity and protein-digesting enzymes, both of which can kill microbes. Farther along in the digestive tract, the intestine contains bacteria that are normally both harmless to people and secrete substances that destroy invading bacteria or fungi.

Despite these defenses, many disease organisms do enter the body through the mucous membranes or through cuts in the skin. What happens then?

Nonspecific Internal Defenses Combat Invaders

Invading microbes that penetrate the skin or mucous membranes encounter an array of internal defenses. Some of these defenses are nonspecific; that is, they attack a wide variety of microbes rather than targeting specific invaders.

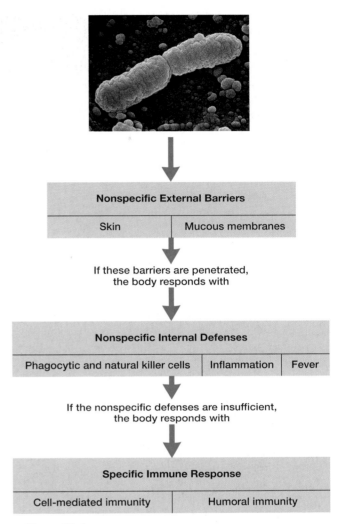

Nonspecific External Barriers	
Skin	Mucous membranes

If these barriers are penetrated, the body responds with

Nonspecific Internal Defenses		
Phagocytic and natural killer cells	Inflammation	Fever

If the nonspecific defenses are insufficient, the body responds with

Specific Immune Response	
Cell-mediated immunity	Humoral immunity

▲ **Figure 22-1** Levels of defense against infection

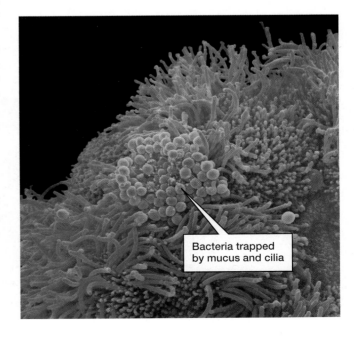

Bacteria trapped by mucus and cilia

▶ **Figure 22-2** **The protective function of mucus** Mucus traps microbes and debris in the respiratory tract. In this colored micrograph, a cluster of bacteria is caught in mucus atop the cilia. The cilia lining the walls of the respiratory tract then sweep both mucus and foreign matter out of the body.

Nonspecific internal defenses are the body's second line of defense. They fall into three main categories:

- The body has a standing army of **phagocytic cells,** which directly destroy microbes by engulfing and digesting them, and **natural killer cells,** which destroy body cells that have been infected by viruses. These two types of white blood cells mop up most of the steady trickle of microbes that make it past the body's external barriers.

- An injury, with its combination of tissue damage and relatively massive invasion of microbes, provokes an **inflammatory response.** The inflammatory response simultaneously recruits phagocytic cells and natural killer cells to the site of injury and walls off the injured area, isolating the infected tissue from the rest of the body.

- If a population of microbes succeeds in establishing a major infection, the body may produce a **fever,** which slows down microbial reproduction and enhances the body's own fighting abilities.

Phagocytic Cells and Natural Killer Cells Destroy Invading Microbes

The body contains several types of phagocytic white blood cells that can engulf and digest microbes. One important type is the **macrophage** (literally, "big eater"). These are white blood cells that circulate in the blood, exit the bloodstream through small openings in capillary walls (**Fig. 22-3a**), and patrol the extracellular fluid. Macrophages ingest microbes by phagocytosis, consuming some of the bacteria and foreign substances that penetrate the mucous membranes or skin (**Fig. 22-3b**). They also eat debris from dead cells, thereby cleaning up a wound. As we discuss later, macrophages also play a crucial role in the immune response.

Natural killer cells strike primarily at the body's own cells that either have become cancerous or have been invaded by viruses. Viruses enter a cell and force the cell's own metabolic processes to manufacture more viruses (see Fig. 12-2 on p. 191). Therefore, viral infections can be stopped by killing the infected cells before the viruses have had enough time to multiply. Virus-infected cells usually have some viral proteins on their surfaces. Cancerous cells often have unusual proteins on their surfaces, too. Natural killer cells recognize and kill cells bearing these "foreign" proteins. Unlike macrophages, which eat their victims, natural killer cells secrete enzymes that attack the infected or cancerous cell. They also secrete proteins that open up holes in the target cell's membrane. Chewed on by enzymes and shot full of holes, the infected or cancerous cell soon dies.

The Inflammatory Response Defends Against Local Infections

The inflammatory response occurs when cells are damaged. Inflammation, which literally means "to set on fire," causes the injured tissues to become warm, red, swollen, and painful. The inflammatory response attracts phagocytic cells to the affected area, promotes blood clotting, and causes pain sensations that stimulate behaviors to protect the wounded area.

The inflammatory response begins when damaged cells release chemicals that cause nearby cells (called *mast cells*) to release *histamine* and other chemicals into the wounded area (**Fig. 22-4**). Histamine both increases blood flow, making the area warm, and makes capillary walls leaky. With extra blood flowing through leaky capillaries, fluid seeps from the capillaries into the tissues around the wound, so the area becomes swollen.

Web Animation Inflammation

▼ **Figure 22-3 The attack of the macrophages**

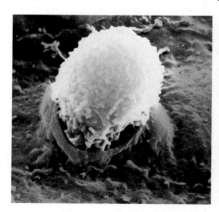

(a) A macrophage leaves a capillary and enters a wound

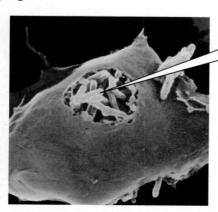

Bacteria visible through a hole in the macrophage membrane

(b) A macrophage stuffed with bacteria that it has ingested

Meanwhile, other chemicals released by injured cells initiate blood clotting (see p. 378). The clotting plugs up damaged blood vessels, preventing microbes from entering the bloodstream. At the same time, it seals off the wound from the outside world, limiting the entry of more microbes. Still other chemicals, some released by wounded cells and others produced by the microbes themselves, attract macrophages and other phagocytic cells to the wound. These cells squeeze out through the leaky capillary walls and ingest bacteria, dirt, and cellular debris. Finally, pain is caused by swelling and the chemicals released by the injured tissue. When you have a cold, the swollen, painful throat and the sneezing, runny nose, and coughing are a direct result of an inflammatory response.

After entering the wound, the phagocytic cells engulf microbes, dirt, and damaged cells (see Figs. 22-3b and 22-4). Unfortunately, if a phagocyte eats too many microbes, it dies. If tissue damage is too severe or the wound is too dirty, the phagocytes may be unable to complete the cleanup job. In that case, the fluid around the wound turns into pus, a thick mixture of microbes, tissue debris, and living and dead white blood cells.

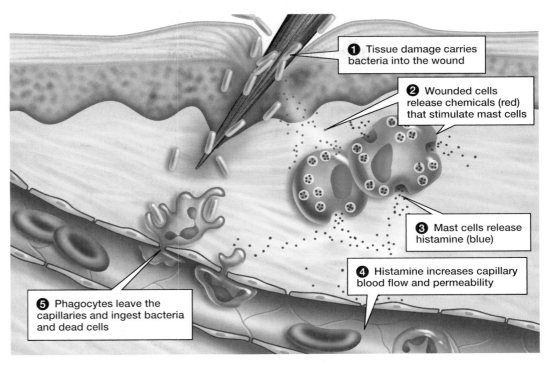

1. Tissue damage carries bacteria into the wound

2. Wounded cells release chemicals (red) that stimulate mast cells

3. Mast cells release histamine (blue)

4. Histamine increases capillary blood flow and permeability

5. Phagocytes leave the capillaries and ingest bacteria and dead cells

▲ Figure 22-4 **The inflammatory response** QUESTION: Why does the inflammatory response cause swelling?

Fever Combats Large-Scale Infections

If invading microbes are able to breach the defenses just discussed and mount a large-scale infection, they may trigger a fever. Although high fevers are dangerous, and even moderate ones are very unpleasant, fever is an important part of the body's defense against infection.

The onset of fever is controlled by the hypothalamus, a part of the brain containing temperature-sensing nerve cells that act as the body's thermostat. For most people, the thermostat is set at about 98.6°F (37°C). Some white blood cells, as they respond to an infection, release chemicals that travel in the bloodstream to the hypothalamus and raise the thermostat's set point. The body responds with shivering, increased metabolism, constriction of blood vessels in the skin, and behaviors such as burrowing under blankets—activities that increase body temperature. Other chemicals reduce the concentration of iron in the blood.

Fever both enhances the body's normal defenses and harms invading microbes. The higher temperature increases the activity of phagocytic white blood cells, while simultaneously slowing the reproduction of bacteria. Many bacteria also need more iron at higher temperatures, so iron deficiency further curtails their multiplication. Fever also helps fight viral infections by increasing the infected cells' production of a protein called *interferon*, which travels to uninfected cells and increases their resistance to viral attack.

22.2 What Are the Key Characteristics of the Immune System?

The external barriers of the skin and mucous membranes and the internal defenses of phagocytic cells, natural killer cells, the inflammatory response, and fever are *nonspecific*. Their role is to block or overcome *any* microbes that might invade the body. Unfortunately, however, these nonspecific defenses are not impregnable. When they fail to do the job, the body mounts a highly specific and coordinated

Avian Flu

Continued

Sneezing and coughing are uncomfortable, but at least they rid your body of some of the viruses, right? Perhaps, but they are also efficient mechanisms for spreading viruses. From an evolutionary perspective, the most successful viruses are the ones that spread rapidly to new hosts, whether they kill their current victim or not. As we will explore in "Avian Flu, *Revisited*" at the end of the chapter, the reason that avian flu viruses haven't killed thousands of people is because they haven't yet evolved an effective means of moving from one person to another.

immune response directed against the *particular* organism that has successfully colonized the body.

The essential features of the immune response were recognized more than 2,000 years ago by Greek historian Thucydides. He observed that occasionally someone would contract a disease, recover, and never catch that particular disease again—the person had become immune. With rare exceptions, however, immunity to one disease confers no protection against other diseases. Thus, an immune response attacks one specific type of microbe, overcomes it, and provides future protection against that microbe but no others.

The Immune System Consists of Cells and Molecules Dispersed Throughout the Body

Unlike most other systems (such as the digestive system, for example), the immune system is not composed of physically attached structures. Instead, as befits its mission of patrolling the entire body for microbial invaders and then destroying them, the immune system is distributed throughout the body, with concentrations of cells in certain locations. Thus, the **immune system** consists of the organs where immune cells are produced and reside (including the bone marrow, thymus, lymph nodes, and spleen), the immune cells themselves (Table 22-1), and the proteins they produce, including antibodies and cytokines. The immune response results from the interactions among these cells and molecules, with the result that microbes and foreign molecules are eliminated from the body. (Don't worry if you don't understand the roles of all the cells in Table 22-1 right now. However, you will find it helpful to refer to the table as you read the remainder of the chapter.)

Immune Cells Originate in the Bone Marrow

The cells of the human immune system include macrophages, which were introduced earlier in this chapter, and about 2 trillion specialized white blood cells called **lymphocytes.** There are two types of lymphocytes, **B cells** and **T cells.** Like all white blood cells, B cells and T cells arise from precursor cells in the bone marrow. Some of these lymphocyte precursors are released into the bloodstream and travel to the thymus, where they complete their differentiation into T cells (the designation "T" stands for "thymus"). In contrast, B cells differentiate in the bone marrow ("B" stands for "bone").

Although they play quite different roles, immune responses produced by both B cells and T cells include the same three fundamental steps. First, the cells of the

Table 22-1	The Body's Cellular Armory
Mast cells	Connective tissue cells that release histamine; important in the inflammatory response
Natural killer cells	White blood cells that destroy infected or cancerous cells
Macrophages	White blood cells that both engulf invading microbes and help alert other immune cells to the invasion by presenting antigens on their surfaces
B cells	White blood cells that produce antibodies
Plasma cells	Offspring of B cells that secrete antibodies into the bloodstream
Memory B cells	Offspring of B cells that provide future immunity against invasion by the same antigen
T cells	White blood cells that regulate the immune response or kill certain types of cells
Helper T cells	Offspring of T cells that stimulate immune responses by both B cells and cytotoxic T cells
Cytotoxic T cells	Offspring of T cells that destroy specific targeted cells, usually foreign eukaryotic cells, infected body cells, or cancerous body cells
Memory T cells	Offspring of T cells that provide future immunity against invasion by the same antigen

immune system recognize the invader; second, they launch an attack; and third, they retain a memory of the invader that allows them to ward off future infections.

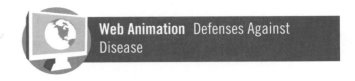

Web Animation Defenses Against Disease

22.3 How Does the Immune System Recognize Invaders?

To understand how the immune system recognizes invaders and initiates a response, we must answer three related questions: How do immune cells recognize foreign cells and molecules? How can immune cells produce specific responses to so many different types of cells and molecules? How do immune cells avoid mistaking the body's own cells and molecules for invaders?

Immune Cells Recognize Invaders' Complex Molecules

From the perspective of the immune system, hepatitis viruses and humans differ from one another because each contains specific, complex molecules that the others don't. The immune system can also distinguish intravenous (IV) solutions from rattlesnake venom because the venom contains specific, complex molecules that IV solutions lack. (Most common IV solutions are principally water, salts, and glucose.) Finally, *your* immune cells distinguish *your* body's cells and molecules from those of all the other organisms on Earth because some of your complex molecules are unique to you (unless you have an identical twin), and some of the complex molecules of all other organisms are unique to them. These large, complex molecules, usually proteins or polysaccharides, are called **antigens,** because they can be "*anti*body *gen*erating" molecules; that is, they can provoke an immune response that includes the production of antibodies (see the next section). Antigens may be individual molecules dissolved in your body fluids (for example, in the venom injected by a rattlesnake) or they may be located on the surfaces of cells (for example, on invading microbes).

Antibodies and T-Cell Receptors Recognize and Bind to Antigens

Cells of the immune system generate two types of proteins that recognize, bind, and then help to destroy specific antigens. **Antibody** proteins are produced by B cells and **T-cell receptor** proteins are produced by T cells. Antibodies may remain attached to the surfaces of the B cells that produced them or they may be released into the blood plasma. In contrast, T-cell receptors always remain attached to the surfaces of the T cells that produced them; they are never secreted into the plasma.

Antibodies Both Recognize and Help to Destroy Invaders

Antibodies are Y-shaped molecules composed of two pairs of peptide chains: one pair of identical large (heavy) chains and one pair of identical small (light) chains. Both heavy and light chains consist of a **constant region,** which is similar in all antibodies of the same type, and a **variable region,** which differs among individual antibodies (**Fig. 22-5**). The combination of light and heavy chains results in antibodies with two functional parts: the two "arms" of the Y and the "stem" of the Y. The variable regions (at the tips of the arms) form binding sites for specific antigens. These binding sites are a lot like the active sites of enzymes (see pp. 79–80). Each binding site has a particular size, shape and electrical charge, so only certain molecules can fit in and bind. The binding sites are so specific that each antibody can bind only a few, very similar, types of antigen molecules.

Antibodies may act as *receptors*, binding to specific antigens and eliciting a response to them, or as *effectors*, helping to destroy the cells or molecules that bear the antigens. As a receptor, the stem of an antibody anchors it to the plasma membrane of the B cell that produced it, while its two arms stick out from the B cell,

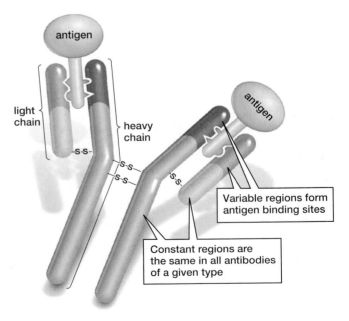

▲ **Figure 22-5 Antibody structure** Antibodies are Y-shaped proteins composed of two pairs of peptide chains (light chains and heavy chains). The variable regions of the light and heavy chains on each "arm" of the Y form binding sites for foreign molecules, which are called antigens. **QUESTION:** Why do antibody molecules have both constant and variable regions?

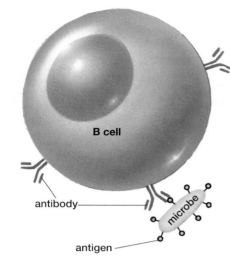

(a) Antibody receptor function

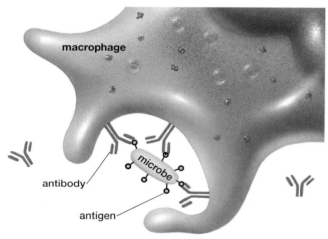

(b) Antibody effector function

▲ **Figure 22-6 Antibodies can serve as either receptors or effectors in the immune response** *(a)* As receptors, antibodies attached to the surface of a cell recognize and bind to foreign antigens, triggering responses in immune cells, such as the B cell depicted here. *(b)* As effectors, circulating antibodies bind to foreign antigens on a microbe, thus helping the immune system to identify and destroy them. These antibodies may promote phagocytosis of the microbe by a macrophage, as shown here (also see Figs. 22-10 and 22-11).

sampling the blood and lymph for antigen molecules (**Fig. 22-6a**). When an arm of the antibody encounters an antigen with a compatible chemical structure, it binds to it. This binding triggers a response in the B cell. As effectors, some antibodies are secreted from B cells and circulate in the bloodstream. There they neutralize poisonous antigens (such as snake vemon), destroy microbes that bear antigens, or attract macrophages that engulf the antigen-bearing microbes (**Fig. 22-6b**). These functions of antibodies are described in more detail in section 22.4.

T-Cell Receptors Recognize Invaders and Help to Trigger the Immune Response

T-cell receptors, found only on the surfaces of T cells, have both similarities to and differences from antibodies. Like antibodies, they consist of peptide chains that form highly specific binding sites for antigens. Unlike antibodies, T-cell receptors are not released into the bloodstream. A T-cell receptor triggers a response in the T cell only when it encounters an antigen on the surface of a cancerous or infected cell, or on the surface of a macrophage that has ingested an invading microbe. Also unlike antibodies, T-cell receptors do not directly contribute to the destruction of invading microbes or toxic molecules. Instead, they alter the activity of the T cell to which they are attached, as we describe later.

The Immune System Can Recognize Millions of Different Molecules

During your lifetime, your body will be challenged by a multitude of different invaders. Your classmates will sneeze cold and flu viruses into the air you breathe. Your food may contain bacteria or molds. A mosquito carrying West Nile virus may bite you. There is no escape from these assaults. Fortunately, your immune system recognizes and responds to virtually all of the millions of antigens that you might encounter, because your immune cells produce millions of different antibodies and T-cell receptors.

The ways in which the immune system recognizes so many antigens are complex and fascinating. Antibodies and T-cell receptors are proteins, and proteins are encoded by genes. But there are only about 21,000 genes in the human genome. This means that a relatively small number of genes must code for millions of antibodies and T-cell receptors.

Antibody Genes Are Assembled from Several Segments of DNA

There are no genes for entire antibody molecules. Instead, B cells have genes that code for *individual parts* of antibodies—variable regions (V), constant regions (C), and joining (J) and diversity (D) regions that connect the two (**Fig. 22-7**). For example, humans have about 150 genes for the variable region of the light chain, and 5 genes for the joining region. There are about 200 genes for the variable region of the heavy chain, and 50 and 6 genes, respectively, for the diversity and joining regions (**Fig. 22-7a**).

As each B cell develops, it randomly cuts out and discards all but one gene of each type, and assembles two unique "antibody genes" from the genes it keeps—a "heavy chain gene" consisting of one variable, one diversity, one joining, and one constant region, and a "light chain gene" consisting of one variable, one joining, and one constant region (**Fig. 22-7b**). As its name implies, the constant region in each chain is the same for any antibody of a particular type; for example, all antibodies located on the surface of a B cell have the same type of constant region for their heavy chain. The random selection of antibody-component genes yields about 3 million unique combinations. Further diversity arises because only a random chunk of each joining region is actually used in any given antibody. Although no one knows the real number, some immunologists estimate that perhaps 15 to 20 *billion* unique antibodies are possible, each with a different antigen-binding site. The result: Each B cell probably produces an antibody that is different from the one produced by every other B cell (except its own daughter cells).

It may be helpful to think of antibody generation in terms of dealing cards for 5-card poker. Although a single deck contains only 52 cards, they can be dealt out in 2,598,960 unique 5-card hands. Similarly, each mature B cell is "dealt" an "antibody hand" consisting of a few parts randomly selected from pools of a few hundred parts, resulting in billions of possible unique antibodies.

T-cell receptors are made from different genes, but the process is similar. More parts are available for T-cell receptor genes, so there may be as many as a quadrillion (10^{15}) different possible T-cell receptors!

Antibodies Are Not Tailor-Made for Antigens

At any given time, a human body contains an army of perhaps 100 million different antibodies and even more T-cell receptors. With that much diversity, an antigen will almost always be confronted by an antibody that can bind it. But it is important to recognize that the immune system does not *design* antibodies (or T-cell receptors) to fit invading antigens, as a tailor might design custom clothes for a client. Instead, the immune system randomly synthesizes millions of different antibodies. The array of antibodies is simply there, waiting, much like clothes in a department store. Given enough racks of clothing from which to choose, most of us will find something suitable that fits. Similarly, virtually every antigen can be bound by at least a few antibodies because of the immense numbers of different antibodies present in the body.

The Immune System Distinguishes "Self" from "Non-Self"

The surfaces of the body's own cells bear large proteins and polysaccharides, just as those of microbes do. So why doesn't your immune system bind these antigens and destroy your own cells? The key seems to be the continuous presence of the body's own antigens during the time when the immune cells differentiate. Some newly formed immune cells do indeed produce antibodies or T-cell receptors that can bind the body's own proteins and polysaccharides, treating them as antigens. However, if these immature immune cells contact molecules that bind to their antibodies or T-cell receptors, the immune cells undergo a special kind of "programmed cell death," in which they essentially commit cellular suicide. In this way, potentially dangerous immune cells, which would attack the body's own cells, are eliminated. Because immature cells are unlikely to encounter foreign antigens during this brief sensitive period of their development, only those bearing antibodies to "self" usually encounter antigens. Thus, the immune system distinguishes "self" from "non-self" by retaining only those immune cells that do not respond to the body's own molecules.

Some cell-surface proteins, collectively called the *major histocompatibility complex* (MHC), are unique to each person (except identical twins, who have the same genes and hence the same MHC proteins). Because they are different from those of everyone else, your MHC proteins act as antigens in other people's bodies. That is why transplants are rejected. The recipient's immune system recognizes MHC proteins on the donor's cells as foreign and destroys the transplanted tissue. To avoid rejection, physicians must find a transplant donor whose MHC

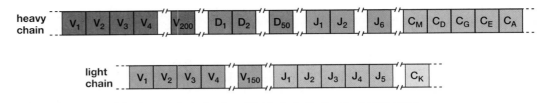

(a) Genes for parts of the heavy chain (top) and light chain (bottom) of antibodies

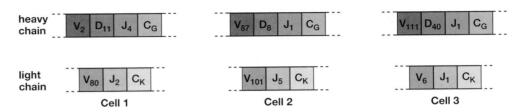

(b) Complete antibody genes in three different B cells

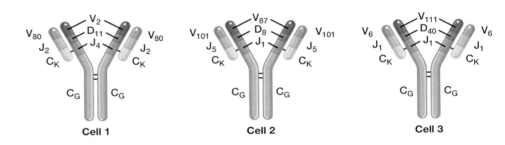

(c) Antibodies synthesized by these three B cells

▲ **Figure 22-7 Recombination of segments of DNA form antibody genes** **(a)** The chromosomes in cells that give rise to B cells have many copies of genes that code for parts of antibodies, illustrated here as variable (V), diversity (D), joining (J), and constant (C) region genes. Each colored band represents an individual gene for an antibody part. Only a few of the many possible variable, diversity, and joining genes are shown here. For example, for the diversity gene of the heavy chain, we show only the D_1, D_2, and D_{50} genes out of a total of 50 D genes; the diagonal lines on the chromosomes denote the many other genes that are not shown. **(b)** Genes for antibody parts are spliced together to form genes for complete antibodies. For the heavy chain, there is one gene each for the variable, diversity, joining, and constant regions; and for the light chain, one each for the variable, joining, and constant regions. In this illustration, the complete antibody genes assembled in cells 1, 2, and 3 are composed of different genes for the antibody parts, as signified by the different numbers assigned for each gene type. **(c)** As a result, the variable regions for the antibodies produced by cells 1, 2, and 3 differ from one another.

proteins are as similar as possible to those of the recipient. They also use drugs to suppress the transplant recipient's immune system.

22.4 How Does the Immune System Launch an Attack?

If the body is invaded and its nonspecific defenses are breached, the immune system mounts two types of attack. **Humoral immunity** is provided by B cells and the circulating antibodies they secrete into the bloodstream. Invaders or the toxic molecules they produce are attacked by these antibodies while they are outside the body's own cells. **Cell-mediated immunity** is produced by a type of T cell called the *cytotoxic T cell,* which attacks body cells that have become infected. Both humoral and cell-mediated immunity are stimulated by helper T cells. Humoral and cell-mediated immunity are not completely independent, but we consider them separately for ease of understanding.

An Immune Response Takes Time to Develop

You may not be able to "find" the immune system, the way you can find your stomach or brain, but it is actually quite large—if you could gather all of the immune cells in your body together into a single ball, it would be about the size of your head. Now, although you have millions of different antibodies and T-cell receptors constantly present in your body, you actually have only one or a few cells bearing each type of antibody or T-cell receptor. If there were thousands or millions of each type of cell, you couldn't fit your immune system in your body! The benefit to having millions of different types of immune cells, even though there are only a few cells of each type, is that almost any invader will provoke an immune response. The drawback is that the immune response takes some time to become effective, while cells recognizing the invader multiply and differentiate. In fact, it usually takes 1 or 2 weeks to mount a really good immune response to the first exposure to an invading microbe (**Fig. 22-8**). In the meantime, you become ill, and may even die, if the development of the immune response loses the race to the multiplying microbe and the damage it causes to your body.

Humoral Immunity Is Produced by Antibodies Dissolved in Blood

Each B cell bears a specific type of antibody on its surface. During an infection, the antibodies borne by a few B cells can bind to antigens on the invader. Antigen–antibody binding causes these B cells—but no others—to divide rapidly. This process is called **clonal selection** because the resulting population of cells is composed of "clones" (genetically identical to the parent B cells) that have been "selected" to multiply by the presence of particular invading antigens (**Fig. 22-9**).

The daughter cells differentiate into two cell types: plasma cells and memory B cells. **Plasma cells** become enlarged and packed with rough endoplasmic reticulum (see Fig. 22-9b). As you learned in Chapter 4, the ribosomes embedded in rough ER manufacture proteins. As a result, plasma cells can churn out huge quantities of antibodies, each bearing the same antigen-binding site as the parent B cell did. These antibodies are released into the bloodstream (hence the name "humoral" immunity; to the ancient Greeks, blood was one of the four "humors," or body fluids). **Memory B cells** do not release antibodies but play an important role in future immunity to the particular invader that stimulated their production (as we will soon see).

Clonal selection, division of activated B cells, differentiation of the daughter cells into plasma and memory cells, and massive antibody secretion by plasma cells all take time. This is why it may take a couple of weeks to completely recover from an infection (see Fig. 22-8).

Web Animation   Humoral Immunity

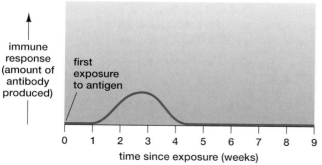

immune response (amount of antibody produced)

first exposure to antigen

time since exposure (weeks)

▲ **Figure 22-8 An effective immune response takes time to develop**

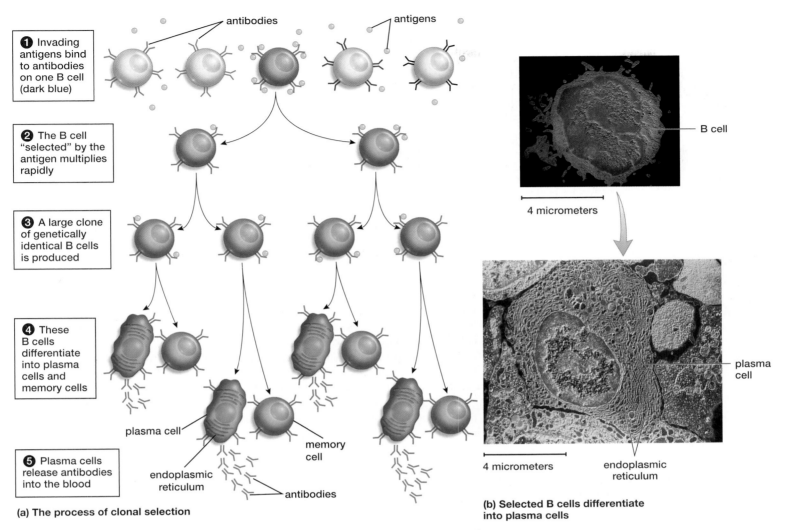

1 Invading antigens bind to antibodies on one B cell (dark blue)

antibodies   antigens

2 The B cell "selected" by the antigen multiplies rapidly

3 A large clone of genetically identical B cells is produced

4 These B cells differentiate into plasma cells and memory cells

5 Plasma cells release antibodies into the blood

plasma cell

endoplasmic reticulum

antibodies

memory cell

(a) The process of clonal selection

B cell

4 micrometers

plasma cell

4 micrometers   endoplasmic reticulum

(b) Selected B cells differentiate into plasma cells

▲ **Figure 22-9   Clonal selection among B cells by invading antigens   (a)** The immune system contains millions of B cells, each with a unique antibody (top row). An invading antigen binds to the antibody on only one B cell, stimulating it to divide. Its daughter cells differentiate into plasma cells and memory B cells. **(b)** Ordinary B cells are small, with little endoplasmic reticulum (top). As they differentiate into plasma cells, they become much larger (bottom), with the extra cell volume almost entirely filled with rough endoplasmic reticulum that synthesizes antibodies.

Humoral Antibodies Have Multiple Modes of Action

Humoral immunity is produced by antibodies in the blood that combat invading molecules or microbes in three principal ways. First, the circulating antibodies may bind to a foreign molecule, virus, or cell, and render it harmless, a process called neutralization. For example, if the active site of a toxic enzyme in snake venom is covered with antibodies, it cannot harm your body (**Fig. 22-10**). Many viruses, such as those that cause rabies, gain entry into your body's cells when a protein on the virus binds to a specific receptor protein on the surface of a cell. If antibodies cover up this viral protein, the virus cannot enter the cell.

Second, antibodies may coat the surface of an invading molecule, virus, or cell, and make it easier for phagocytic cells to destroy it (**Fig. 22-11**). Remember, the variable regions on the "arms" of an antibody bind to antigens on invaders, so the constant regions that make up the "stems" stick out into the blood or extracellular fluid. Macrophages recognize the antibody stems, engulf the antibody-coated invaders, and digest them.

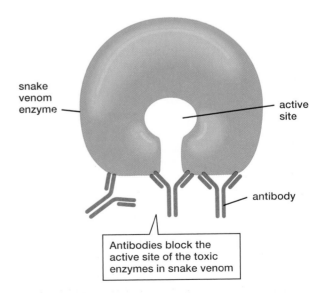

snake venom enzyme

active site

antibody

Antibodies block the active site of the toxic enzymes in snake venom

▲ **Figure 22-10   Antibodies neutralize toxic molecules**

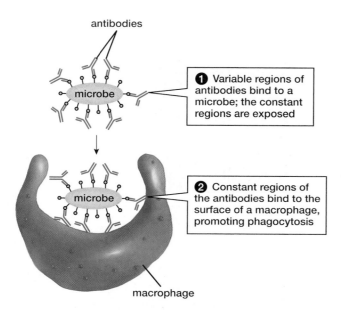

1. Variable regions of antibodies bind to a microbe; the constant regions are exposed

2. Constant regions of the antibodies bind to the surface of a macrophage, promoting phagocytosis

antibodies

microbe

microbe

macrophage

▲ **Figure 22-11 Antibodies promote phagocytosis of invading microbes**

Avian Flu

Continued

People overcome flu, whether avian flu or the typical human varieties, because their cell-mediated immunity destroys infected cells before the flu viruses have completed their life cycle. In essence, the body sacrifices some of its cells to save all the rest.

Third, when antibodies bind to antigens on the surface of a microbe, they attract other proteins, called *complement proteins*. In a complex set of interactions involving over a dozen molecules, some complement proteins punch holes in the plasma membrane of the microbe, killing it. Other complement proteins promote phagocytosis of the invaders.

Humoral Immunity Fights Invaders That Are Outside Cells

Antibodies are large proteins that usually do not cross plasma membranes. Therefore, humoral immunity can only defend against invaders that are in the blood or extracellular fluid. Bacteria (most of which never enter the body's cells), toxic molecules in venoms or released by bacteria, and some fungi and protists are therefore susceptible to the humoral immune response. Invaders that penetrate into the body's cells, as viruses do, are vulnerable to antibodies only when they are outside a body cell, and are safe from antibody attack when they are inside a cell. Fighting viral infections, therefore, requires the help of the cell-mediated immune response.

Cell-Mediated Immunity Is Produced by Cytotoxic T Cells

Cell-mediated immunity, produced by **cytotoxic T cells,** is the body's primary defense against cells that are cancerous or have been infected by viruses. Although the process is quite complex, in essence it works like this: When a cell is infected by a virus, some pieces of viral proteins are brought to the surface of the infected cell and "displayed," protruding outward from the cell membrane. Cytotoxic T cells randomly drift about, occasionally bumping into the displayed viral proteins. If a cytotoxic T cell has a T-cell receptor that binds to the viral protein, the cytotoxic T cell undergoes clonal selection. If its differentiated daughter cells encounter a body cell displaying the same antigen, they squirt pore-forming proteins onto the surface of the infected cell, punching holes in its plasma membrane and killing it. Why kill off your own body's cells? Remember that viruses multiply only *inside cells*. If the infected cell is killed before the virus has had enough time to finish multiplying, then no new viruses are produced, and the rest of the body's cells are spared new infection.

Both Humoral and Cell-Mediated Immunity Are Enhanced by Helper T Cells

Helper T cells bear receptors that bind to antigens displayed either on the surfaces of infected cells or macrophages that have engulfed and digested invading microbes. As with B cells and cytotoxic T cells, only helper T cells bearing the "right" T-cell receptor can bind to any particular antigen. When its receptor binds an antigen, a helper T cell multiplies rapidly. Its daughter cells differentiate and release chemicals, called **cytokines,** that stimulate cell division and differentiation in both B cells and cytotoxic T cells. Although our earlier discussion implied that B cells and cytotoxic T cells can mount an immune response by themselves, they are actually quite ineffective without "help" from helper T cells. In fact, both B cells and cytotoxic T cells make a significant contribution to defense against disease only if they simultaneously bind antigen *and* receive stimulation by cytokines from helper T cells. As you may know, HIV, which causes AIDS, kills helper T cells. Without helper T cells, the immune system cannot fight off diseases that would otherwise be trivial.

Figure 22-12 summarizes the major characteristics of the humoral and cell-mediated immune responses, and the role of helper T cells in stimulating both types of response.

22.5 How Does the Immune System Remember Its Past Victories?

After you have recovered from a disease, you often remain immune to the particular strain of microbe that made you ill; this immunity may last for years, perhaps a lifetime. Plasma cells and cytotoxic T cells directly fight disease organisms, but they

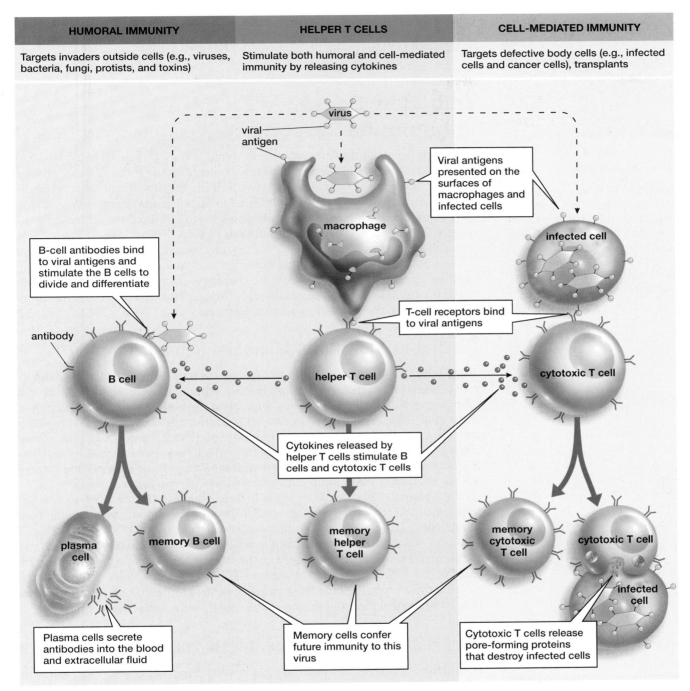

HUMORAL IMMUNITY	HELPER T CELLS	CELL-MEDIATED IMMUNITY
Targets invaders outside cells (e.g., viruses, bacteria, fungi, protists, and toxins)	Stimulate both humoral and cell-mediated immunity by releasing cytokines	Targets defective body cells (e.g., infected cells and cancer cells), transplants

virus

viral antigen

Viral antigens presented on the surfaces of macrophages and infected cells

macrophage

infected cell

B-cell antibodies bind to viral antigens and stimulate the B cells to divide and differentiate

antibody

T-cell receptors bind to viral antigens

B cell

helper T cell

cytotoxic T cell

Cytokines released by helper T cells stimulate B cells and cytotoxic T cells

plasma cell

memory B cell

memory helper T cell

memory cytotoxic T cell

cytotoxic T cell

infected cell

Plasma cells secrete antibodies into the blood and extracellular fluid

Memory cells confer future immunity to this virus

Cytotoxic T cells release pore-forming proteins that destroy infected cells

▲ Figure 22-12 A summary of humoral and cell-mediated immunity, and the role of helper T cells in stimulating both responses

usually live only a few days. However, during the early stages of a disease—while B cells, cytotoxic T cells, and helper T cells are dividing rapidly—some of their daughter cells differentiate into **memory cells** that may survive for many years (see Figs. 22-9 and 22-12). Then, if the body is reinvaded by this same type of microbe, the memory cells will recognize the invader and mount an immune response. There are *far more memory cells* than there were original B, cytotoxic T, or helper T cells that responded to the first infection by this type of microbe. Further, *each memory cell responds faster* than its original parent cells could. Therefore, at a second infection, memory cells usually produce an immune response that is so fast and so large that the body fends off the attack before the microbe can produce any noticeable disease symptoms—you have become immune (Fig. 22-13).

Why, then, do you repeatedly suffer from colds and flu? The problem is that there are hundreds of different cold viruses, and flu viruses mutate rapidly. Thus,

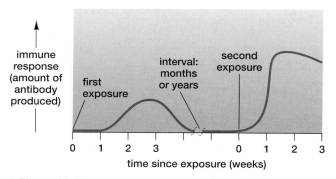

▲ **Figure 22-13 Acquired immunity** The immune response to the first exposure to an invading microbe (left) is relatively slow, as tiny numbers of immune cells must undergo clonal selection, multiply, differentiate, and finally kill the invader. The response to a second invasion by the same microbe (right) activates numerous memory cells left behind during the first immune response, and is consequently much faster and much larger. Usually, no disease symptoms result from the second exposure.

your next cold will be caused by an entirely different virus than your most recent cold, and this year's flu will probably be caused by a mutated version of the virus that you fought off last year.

22.6 How Does Medical Care Assist the Immune Response?

We may have made the immune system sound invincible, but, as you know, it is not. An infection is really a race between the invading microbes and the immune response. If the initial infection is massive or if the microbes produce particularly toxic products, full activation of the immune response may come too late. Further, many microbes have evolved defenses that help them evade the immune system. In the end, many infections kill their victims.

For most of human evolutionary history, the battle against disease was fought by the immune system alone. Now, however, the immune system has a powerful assistant: medical treatment. Scientific and medical advances have greatly reduced the toll taken by infectious diseases. We describe here two of the most important medical tools: antibiotics and vaccinations.

Antibiotics Slow Down Microbial Reproduction

Antibiotics help combat infection by slowing down the multiplication of many invaders, including bacteria, fungi, and protists. Although antibiotics usually do not destroy every single disease-causing microbe in the body, they may kill enough of them to give the immune system time to finish the job. One problem with antibiotics, however, is that they are potent agents of natural selection. The occasional mutant microbe that is resistant to an antibiotic will pass on the genes for resistance to its offspring. The result: Resistant mutants proliferate while susceptible microbes die off. Eventually, many antibiotics become ineffective in treating diseases.

Until recently, little could be done to help people suffering from viral infections except to treat the symptoms and hope that the immune system would triumph. Now, drugs are available that target different stages of the viral cycle of infection, which includes attachment to a host cell, replication of viral parts using the host cell's machinery, assembly of the virus within the host cell, and the release of the virus into the extracellular fluid to infect new cells. These antiviral drugs are not prescribed for most viruses, but they are used to treat HIV, severe herpes virus infections, and in some cases, the flu virus, including avian flu.

Vaccinations Stimulate the Development of Memory Cells

A **vaccine** stimulates an immune response by exposing a person to antigens produced by a disease organism. Vaccines often consist of weakened or killed disease microbes (which therefore cannot cause disease) or antigens from the disease organism that are synthesized using genetic engineering techniques. When the body is exposed to these antigens, it produces an army of memory cells that confer immunity against living, dangerous microbes of the same type. Learn more about vaccines in "Scientific Inquiry: How Vaccines Were Discovered."

Today, many diseases, including polio, diphtheria, typhoid fever, and measles, are controlled through vaccination. Smallpox, which formerly killed about 30% of its victims, has been completely eradicated since 1980 as a result of a massive vaccination program sponsored by the World Health Organization. There are two legal repositories of smallpox viruses, one in Russia and one in the United States. These repositories are intended to safeguard against the remote possibility that "bioterrorists" possess stocks of smallpox virus, which they could release to infect an unvaccinated and therefore vulnerable population. In this case, live smallpox virus might be needed to test the effectiveness of vaccines in cell cultures before

scientific inquiry

How Vaccines Were Discovered

Smallpox has been feared since ancient times. This highly contagious disease formerly killed about 30% of its victims and left most of the survivors disfigured with pitted skin. The pits are caused by blisters that fill with pus before forming scabs. Writings from India dating back to 1100 B.C. describe protecting against smallpox by exposing healthy people to the pus from blisters of smallpox victims. Although some died, many recipients of this earliest form of vaccination developed only mild symptoms and were able to resist later exposure to the disease. By the eighteenth century, people in England had noticed that victims of cowpox (a related but far less serious disease) did not get smallpox. In 1796, the English surgeon and experimental biologist Edward Jenner obtained fluid from cowpox blisters on the hands of a milkmaid and inoculated an 8-year-old boy with this bacteria-laden material (**Fig. E22-1**). A few months later, in a very risky experiment, Jenner inoculated the boy with material from a smallpox lesion—fortunately, the boy remained healthy. After repeating these results, Jenner published his findings in 1798. This early form of vaccination was rapidly adopted in Europe and eventually worldwide, dramatically reducing deaths from smallpox.

Somewhat surprisingly, nearly a century passed before vaccination was applied to other infectious diseases. The French microbiologist

Louis Pasteur, among the first to recognize the role of microbes in causing disease, experimented with chicken cholera in the late 1800s. He grew cholera bacteria in culture medium and found that they caused the fatal disease when injected into chickens. The story goes that when Pasteur injected chickens with bacteria from an old culture that had been left over the summer holidays, the chickens became ill but didn't die. Assuming something had gone wrong, he grew a fresh culture. As luck would have it, he inoculated some of the chickens that had survived the earlier injection with the spoiled culture. To his surprise, these chickens remained healthy, while "naïve" chickens died. Pasteur hypothesized that weakened bacteria in the spoiled culture protected against later infection by healthy bacteria. He coined the term "vaccine" (from the Latin "vacca," meaning "cow"), in recognition of Jenner's pioneering work with cowpox. Pasteur later applied the technique to anthrax in sheep and then to rabies in humans, saving a young boy who had been bitten by a rabid dog.

▲ **Figure E22-1 The first vaccination against smallpox** This painting commemorates Edward Jenner's pioneering (and dangerous) experiment of exposing a boy to smallpox after he had been vaccinated with cowpox.

Pasteur's findings with chicken cholera were both a stroke of luck and a flash of genius. As Pasteur himself remarked, "Chance favors the prepared mind." This insight is just as relevant to scientific inquiry now as it was over a century ago.

using the vaccines in humans. In May 2007, the World Health Assembly expressed the hope that enough research can be completed in the next few years so that the smallpox stocks can be destroyed in 2010.

22.7 What Happens When the Immune System Malfunctions?

The immune system is always on guard, protecting the body against threats to health. In some cases, however, the immune system attacks inappropriately, undermining health instead of promoting it. In addition, the immune system, like other body systems, is itself subject to diseases and malfunctions that decrease its effectiveness.

Allergies Are Misdirected Immune Responses

More than 35 million Americans suffer from **allergies,** immune responses to substances that are not harmful and to which many other people do not respond. Common allergies include those to pollen, dust, mold spores, animal dander, bee

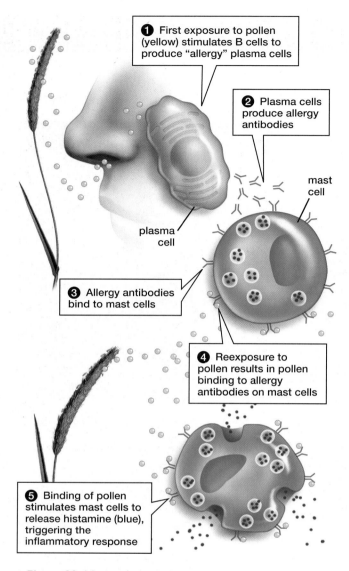

❶ First exposure to pollen (yellow) stimulates B cells to produce "allergy" plasma cells

❷ Plasma cells produce allergy antibodies

mast cell

plasma cell

❸ Allergy antibodies bind to mast cells

❹ Reexposure to pollen results in pollen binding to allergy antibodies on mast cells

❺ Binding of pollen stimulates mast cells to release histamine (blue), triggering the inflammatory response

▲ Figure 22-14 An allergic reaction

stings, and some foods. In people with allergies, a foreign substance, such as pollen, enters the bloodstream and is recognized as an antigen by a specific type of B cell that produces "allergy antibodies" (Fig. 22-14). These B cells proliferate, producing plasma cells that secrete these allergy antibodies into the plasma.

The stems of allergy antibodies attach to the plasma membranes of histamine-containing mast cells in the respiratory and digestive tracts (see Table 22-1). If pollen antigens later bind to these attached antibodies, they trigger the release of histamine, which (as you learned earlier) causes leaky capillaries and other symptoms of inflammation. In the respiratory tract, histamine also increases mucous secretion. Thus, airborne substances such as pollen grains, which typically enter the nose and throat, often trigger allergic reactions that include the runny nose, sneezing, and congestion of "hay fever" (see "Health Watch: The Perils of Pollen" for more information about pollen allergies). Food allergies usually cause cramps and diarrhea. In severe cases, such as peanut allergies, the inflammatory response in the airways is so strong that the airways completely close, causing death by suffocation. More than 100 people in the United States die each year from peanut allergies.

An Autoimmune Disease Is an Immune Response Against the Body's Own Cells

Our immune systems rarely mistake our own cells for invaders. Occasionally, however, something goes awry, and "anti-self" antibodies are produced. The result is an **autoimmune disease,** in which the immune system attacks a component of one's own body. Some types of anemia, for example, are caused by antibodies that destroy a person's red blood cells. Many cases of type 1 diabetes begin when the immune system attacks the insulin-secreting cells of the pancreas. Rheumatoid arthritis results when the immune system attacks cartilage in the joints. Unfortunately, at present there are no known cures for autoimmune diseases. For some diseases, replacement therapy can alleviate the symptoms—for instance, by administering insulin to diabetics or blood transfusions to anemia victims. Alternatively, the autoimmune response can be reduced with drugs that suppress the immune response. Immune suppression, however, also reduces responses to the everyday assaults of disease microbes, so this therapy has major drawbacks.

Immune Deficiency Diseases Occur When the Body Cannot Mount an Effective Immune Response Against Invaders

There are two very different disorders in which the immune system cannot combat routine infections, one inherited and the other acquired. About 1 in 80,000 children is born with **severe combined immune deficiency (SCID),** a family of genetic defects in which few or no immune cells are formed. Much more frequent is **acquired immune deficiency syndrome (AIDS),** in which a viral infection destroys a formerly functional immune system.

Severe Combined Immune Deficiency

A child with SCID may survive the first few months of postnatal life, protected by antibodies acquired from the mother during pregnancy or in her milk. Once these antibodies are lost, however, common infections can prove fatal, because the child, lacking an immune system, cannot generate an effective immune response. One form of therapy is to transplant bone marrow (from which immune cells arise) from a healthy donor into the child. In some children, marrow transplants have resulted in enough immune cell production to confer normal immune responses. As we described in Chapter 12, genetic engineering has been used to create a functioning immune system in a handful of children with SCID. The most recent technology, first used in clinical trials in 2005, might provide a permanent cure for some types of SCID.

health watch

The Perils of Pollen

Are you one of the millions who fight "hay fever" every summer? Hay fever occurs when your immune system reacts to certain pollens as if they were disease-causing organisms, mounting an inflammatory response against them. So your eyes water, your nose runs, and you sneeze a lot. From personal and scientific perspectives, there are three interesting questions about pollen allergies. First, how could such an apparently maladaptive response ever evolve? Second, are all types of pollen equally offensive? Third, and of most immediate concern to the millions of people who suffer from pollen allergies, what can we do about it?

Immunologists and evolutionary biologists think that allergies evolved as a defense against parasites. Many parasites, such as amoebas and various types of worms, enter the body through the respiratory, digestive, or urogenital tracts. Sneezing, coughing, vomiting, and diarrhea are all probably mechanisms to eject parasites from the body. Allergic responses to pollen or peanuts are simply a mistake—a violent attempt to rid the body of things that actually wouldn't do any harm.

While a given individual may react to various types of pollen, certain types of pollen are more problematic than others. As you learned in Chapter 18, plants with showy flowers are usually pollinated by insects or other animals. These plants tend to have flowers with relatively heavy, sticky pollen, so that even if you are allergic to, say, rose pollen, you probably aren't ever seriously exposed to it unless you directly smell the flower. Wind-pollinated plants, however—such as grasses and ragweed (**Fig. E22-2**)—tend to have inconspicuous flowers that produce huge amounts of very lightweight pollen that can travel

for miles. Thus, one is much more likely to encounter wind-borne pollen and, subsequently, experience the common, miserable sensations associated with an allergic response as a result of the exposure.

But even with wind-borne pollen, an individual will tend to develop allergies to only a relatively small number of types of pollen; for example, you may be allergic to grasses but not to ragweed. Why? As you know, antibodies bind only to specific antigens. Thus, if your body doesn't produce the specific allergy antibodies that can bind to proteins or polysaccharides on ragweed pollen, you won't be allergic to it.

So, what can the unfortunate millions do about pollen allergies? The typical antihistamines, as their name implies, block the actions of the histamine released by mast cells during an allergic response. Some allergy medications mimic the actions of norepinephrine, and help to keep airways open even during the allergic response. Both of these tend to be only partially effective, and sometimes have side effects such as "rebound" airway swelling if they are used too often. A more permanent solution is allergy vaccination, in which the immune system is "trained" to ignore pollen or other allergens. This treatment consists of a series of injections of tiny amounts of the allergen, such as pollen, mold, or dust mites. For many, allergic symptoms eventually subside, and the beneficial effects usually last at least 5 years.

Researchers have found that three major changes occur during the course of allergy vaccinations. First, chemicals that promote inflammation are reduced. Second, the body produces soluble antibodies to the allergen that

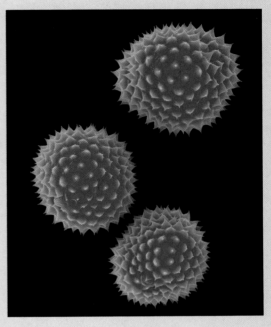

▲ **Figure E22-2** Ragweed pollen—a common cause of pollen allergies

are different from the allergy antibodies that bind to mast cells and promote inflammation. These nonallergy antibodies bind to the allergen and cover up the sites by which it would bind to the allergy antibodies on mast cells, so they do not release histamine. Finally, the allergy antibodies to the allergen are eventually reduced. Although the current treatment involves getting a large number of injections, new versions of allergy vaccines that are now in clinical trials may provide protection with just three or four shots. Allergy vaccinations can help people with severe allergies resume a more active and comfortable life.

Acquired Immune Deficiency Syndrome

AIDS is much more common than SCID. The World Health Organization estimates that, in 2006, nearly 3 million people died of AIDS and almost 4.5 million more became infected, bringing the total infected population to almost 40 million. About 95% of the infected people live in the developing world, with the highest rates in sub-Saharan Africa. In the United States, the Centers for Disease Control and Prevention estimates that about 1.2 million people are infected with AIDS, with 40,000 new cases each year.

AIDS is caused by **human immunodeficiency viruses** 1 and 2 (HIV-1 and HIV-2; **Fig. 22-15**). These viruses infect and destroy helper T cells. As you now know, helper T cells are essential for both cell-mediated and humoral immune

Web Animation Effects of HIV on the Immune System

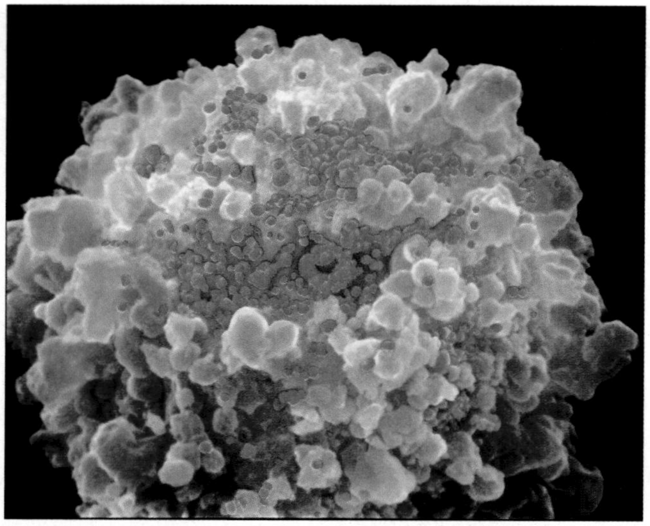

▲ **Figure 22-15 HIV: The virus that causes AIDS** The red specks in this colorized scanning electron micrograph are HIV that have just emerged from the large, green helper T cell.

responses. AIDS does not kill people directly, but AIDS victims become increasingly susceptible to other diseases as their helper T cell populations decline.

In the early stages of HIV infection, the patient may have a fever, rash, muscle aches, headaches, and enlarged lymph nodes as the immune system fights the infection. After several months, viral replication slows. In most cases, enough helper T cells remain that the patient is able to resist disease and may feel quite well. This stage can persist for years. Eventually, however, the number of helper T cells begins to drop. When the helper T cell count drops to about one-fourth of the normal number, the patient is considered to have AIDS. In an untreated patient, HIV levels then skyrocket, helper T-cell numbers continue to fall, and the patient falls victim to other infections. The life expectancy for untreated AIDS victims is about 1 to 2 years.

Because HIV cannot survive for very long outside the body, it can be transmitted only by direct contact of broken skin or mucous membranes with body fluids containing the virus—including blood, semen, vaginal secretions, and breast milk. HIV infection can be spread by sexual activity, by sharing needles among intravenous drug users, or by blood transfusions (this is rare in developed countries that screen all donated blood for anti-HIV antibodies). A woman infected with HIV can transmit the virus to her child during pregnancy, childbirth, or through breast-feeding. Nearly all people infected with HIV eventually develop AIDS.

For persons infected with AIDS, there are two categories of therapy. First, infections that result from the impaired immune system must be treated as they would be in any patient. Second, drugs are available that slow the multiplication of HIV and, therefore, slow the progress of AIDS. Combinations of drugs targeting different stages of viral replication have been particularly effective, and a complete AIDS treatment regimen has now been combined into a once-a-day pill. Unfortunately, HIV can mutate into forms resistant to the drugs, and in some patients, the drugs have severe side effects.

Clearly, the best solution would be to develop an AIDS vaccine, but this has proven frustratingly difficult. First, HIV destroys helper T cells, which are crucial components of the immune response to a vaccine. Second, HIV "hides" from the immune response inside the helper T cells that it has infected but not yet destroyed. Third, HIV has an incredibly high mutation rate—perhaps a thousand times faster than that of flu viruses—so a single AIDS patient may harbor multiple strains of HIV because of the mutations that occurred within his or her body after the initial infection. A vaccine might protect against some but not all of these strains. Despite billions of dollars invested in research and in animal and human trials, no HIV vaccine has yet proven both safe and effective.

22.8 How Does the Immune System Combat Cancer?

Cancer is one of the most dreaded words in the English language, and with good reason. Cancer causes one out of every four deaths, and is second only to heart disease as the leading cause of mortality in the United States. A sobering 40% of people in the United States will eventually contract cancer. Despite decades of intensive research, the quest for a cancer cure remains unfulfilled. Why can't we cure or prevent many cancers?

Unlike most other diseases, cancer is not usually an invasion of the body by a foreign organism. Instead, cancer is essentially a failure of the mechanisms that control the reproduction of the body's own cells (see "Health Watch: Cancer—Mitotic Cell Division Run Amok" on p. 119). Any population of cells that has escaped these regulatory mechanisms, and therefore multiplies at an abnormal rate, is called a *tumor*. The cells in a *benign tumor* usually remain in one area, but the cells in a *malignant tumor* reproduce uncontrollably and spread to other areas of the body. As a malignant tumor grows, it uses increasing amounts of the body's energy and nutrient supplies, and literally squeezes out vital organs nearby. **Cancer** is a disease caused by uncontrolled growth of malignant tumor cells.

Most Cancerous Cells Are Recognized as Foreign

Cancer cells form in our bodies every day. We cannot avoid some carcinogens, such as gamma rays from the sun, radioactivity from the rocks beneath our feet, and naturally produced carcinogens in our food. Fortunately, natural killer cells and cytotoxic T cells destroy nearly all cancer cells before they have a chance to proliferate and spread.

How are cancer cells weeded out? Cancer cells are, of course, "self" cells (the body's own cells), and the immune system does not respond to "self." However, the processes that cause cancer often cause new and slightly different proteins to appear on the surfaces of cancer cells. Natural killer and cytotoxic T cells encounter these new proteins, recognize them as "non-self" antigens, and destroy the cancer cells (**Fig. 22-16**).

Unfortunately, some cancer cells evade detection because they do not bear antigens that allow the immune system to recognize them as foreign. Other cancers, such as leukemia, suppress the immune system; and some simply grow so fast that the immune response can't keep up. If the tumor grows and spreads, the individual's health depends on medical treatment.

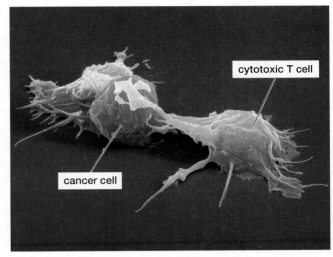

▲ **Figure 22-16 Cell-mediated immunity defends against cancer** Cytotoxic T cells attack a cancer cell in this colorized scanning electron micrograph. **QUESTION:** Cancer consists of the body's own cells, so why does the immune system attack cancer cells?

Treatments for Cancer Depend on Distinguishing and Selectively Killing Cancerous Cells

The three main forms of medical treatment for cancer are surgery, radiation, and chemotherapy. The surgical removal of the tumor is the first step in the treatment of many cancers. Unfortunately, the surgeon may not be able to see and remove every bit of the tumor. Surgery also cannot treat cancer that has spread throughout the body. Alternatively, cancerous cells can be bombarded with radiation in an effort to kill them. Unlike surgery, radiation may destroy even microscopic clusters of cancer cells. Like surgery, however, radiation cannot be used to treat widespread cancers because irradiating the whole body would damage a great deal of healthy tissue as well.

Chemotherapy, or drug treatment, is commonly used to supplement surgery and radiation, or to combat cancers that cannot be treated in any other way. Chemotherapy drugs attack the machinery of cell division, and so they are somewhat selective for cancer cells, which divide much more frequently than normal cells do. Nonetheless, chemotherapy inevitably also kills some healthy, rapidly dividing cells. Damage to dividing cells in patients' hair follicles and intestinal lining by chemotherapy produces its well-known side effects of hair loss, nausea, and vomiting.

A tremendous amount of research has been devoted to the search for cancer treatments that are more effective and have fewer side effects. Developing vaccines against cancer is a high priority. Other approaches include developing therapies that would stimulate the immune system to attack tumors, or to develop antibody molecules that recognize tumor cells. These antibody molecules could be used to deliver drugs and other treatments directly to the tumor cells without affecting healthy cells. Research and clinical trials continue, and cancer patients may one day reap the benefits of innovative new treatments.

Avian Flu Revisited

Flu is caused by a rather simple virus. Flu viruses consist of only 11 genes (made of RNA, not DNA) and a few proteins, surrounded by a coat studded with two crucial proteins—one to help the virus get into a host cell and one to help it get out. The "entry protein," called hemagglutinin, attaches to cells of the respiratory tract, stimulates the cells to engulf the virus by endocytosis, and frees the viral RNA inside the

host cell. The viral RNA then hijacks the cell's metabolic machinery to make more flu viruses. A protein called neuraminidase is the "exit protein," allowing the newly formed viruses to escape from the host cell and move on to infect more cells.

For a virus to be truly dangerous to human populations, it must have two properties: infectivity and transmissibility in humans. Infectivity

(continued)

means that the virus can enter human cells, multiply there, and cause disease symptoms. Transmissibility means that the virus can spread easily from one human to another. As of 2007, avian flu had made the jump to infecting humans, but its person-to-person transmissibility is very low.

What makes one virus infectious to birds and another infectious to humans? As you may have guessed, the "original" *avian* flu virus has a different entry protein than the *human* flu virus has. The hemagglutinin of the original avian flu virus can bind to molecules on the surfaces of cells in the respiratory tract of birds, but not to those of people. Unfortunately, however—as evidenced by the avian flu's ability to make the jump to humans—the avian flu hemagglutinin has mutated so that it can now bind to human cells.

Why do the current strains of avian flu have very low person-to-person transmissibility? Human flu viruses bind to, and infect, cells in the nasal passages, trachea, and bronchi. When new viruses are manufactured in these cells, they are sneezed or coughed out, easily infecting nearby people. Avian flu viruses infect the same structures in birds, but can only bind to and infect cells deep in the lungs in humans. Therefore, newly manufactured avian flu viruses almost never escape a human victim's body. No one really knows how many mutations it would take to allow a deadly avian flu virus to infect the upper respiratory tract of humans and thus become easily transmissible, but public health officials fear that it may not be more than one or two.

In most cases, the human immune system destroys human flu viruses; why, then, doesn't the immune system destroy avian flu viruses? The entry and exit proteins on the surface of the virus do act as antigens—they are recognized by antibodies and T-cell receptors as foreign, and provoke an immune response. However, as Fig. 22-13 shows, a really good immune response to the first exposure to an antigen takes a week or two, while the response to a second exposure is much faster. For many diseases, including human flu, there's an intermediate scenario—partial immunity against rapidly mutating viruses. For example, if you caught the human flu last winter, you will have memory cells for that particular flu strain lying in wait

this winter, ready to recognize the entry and exit proteins and mount a fast, enormous immune response. Meanwhile, however, the human flu virus will probably have mutated and bear slightly different entry and exit proteins. The memory cells from last year do respond, but not at maximum speed. You still get the flu, but the illness usually isn't fatal. Partial immunity, therefore, is very useful.

However, your immune system has never "seen" avian flu, and so it can't provide partial immunity. The race is on, then, between your immune system's response to its first exposure to avian flu and the speed of replication of the virus. To make matters worse, the avian flu virus seems deadlier than human flu viruses, attacking not only the respiratory tract, but also the digestive tract, and even the brain.

Therefore, another race is on, right now. Before a massive epidemic starts, governments and pharmaceutical companies are developing vaccines against avian flu. One, by GlaxoSmithKline, could be ready by the time you read this. Of course, it will be the "wrong" vaccine because it cannot possibly target mutations that haven't yet happened, which would cause the avian entry protein to bind to cells in a human's upper respiratory tract and allow rapid person-to-person transmission. The vaccine, however, would provide partial immunity, and thus might save millions of lives. This is a race we can't afford to lose.

Consider This

Many bacteriologists and immunologists are currently involved in "dual-use" research that both increases our understanding of disease mechanisms and how to combat them, but also might be used by bioterrorists to make more effective weapons. For example, suppose someone found out which mutation(s) would make avian flu more readily transmissible from person to person. This knowledge could be used to make better vaccines, but could also be used to deliberately create a deadly strain of virus. How do you think that governments should regulate such research? How would you evaluate the statement by Peter Palese of the Mount Sinai School of Medicine that "Nature is still the most effective terrorist"?

Chapter Review

Summary of Key Concepts

For additional study help and activities, go to www.mybiology.com.

22.1 How Does the Body Defend Against Invasion?
The human body has three lines of defense against invasion by microbes: (1) the nonspecific barriers of skin and mucous membranes; (2) nonspecific internal defenses, including phagocytosis, killing by natural killer cells, inflammation, and fever; and (3) the immune response.

If microbes do enter the body, white blood cells travel to the site of entry and engulf the invading cells. Natural killer cells secrete proteins that kill infected or cancerous cells. Injuries stimulate the inflammatory response, which attracts phagocytic white blood cells, increases blood flow, and makes capillaries leaky. Later, blood clots

wall off the injury site. Fever is caused by raising the set point of the body's thermostat. The high temperature from a fever inhibits bacterial growth and accelerates the immune response.

Web Animation Inflammation

22.2 What Are the Key Characteristics of the Immune System?
B cells, T cells, and other white blood cells interact to carry out the immune response. Immune responses have three steps: (1) recognition, (2) attack, and (3) memory.

22.3 How Does the Immune System Recognize Invaders?

Antibodies (on B cells) and T-cell receptors (on T cells) recognize antigens and trigger the immune response. Antibodies are Y-shaped proteins composed of a constant region and a variable region. Antibodies both detect and actively destroy antigens. Each B cell synthesizes only one type of antibody, unique to that particular cell and its progeny. Each antibody has specific sites that bind only one or a few types of antigen. The immune system produces millions of different antibodies and T-cell receptors, each capable of binding only one or a few very similar types of antigens. Normally, immune cells that synthesize antibodies or T-cell receptors that can bind to molecules of your own body die early in development. Therefore, immune responses usually occur only to foreign cells and molecules, and not to the cells and molecules of your own body.

Web Animation Defenses Against Disease

22.4 How Does the Immune System Launch an Attack?

Antigens from an invader bind to and activate only those B and T cells with the complementary antibodies or T-cell receptors. In humoral immunity, the presence of antigens stimulates B cells with complementary antibodies to divide rapidly to produce plasma cells that synthesize massive quantities of the antibody. Antibodies circulating in the blood destroy antigens and antigen-bearing microbes by inactivating them or coating them. Coating microbes or toxic molecules with antibodies promotes phagocytosis by white blood cells. In cell-mediated immunity, T cells with the proper receptors bind antigens and divide rapidly. Cytotoxic T cells bind to antigens on microbes, infected cells, or cancer cells and then kill the cells. Helper T cells stimulate both the B-cell and cytotoxic T-cell responses.

Web Animation Humoral Immunity

22.5 How Does the Immune System Remember Its Past Victories?

Some progeny cells of both B and T cells are long-lived memory cells. If the same antigen reappears in the bloodstream, these memory cells are immediately activated, divide rapidly, and cause an immune response that is much faster and stronger than the original response. Usually no symptoms of the disease are felt.

22.6 How Does Medical Care Assist the Immune Response?

Antibiotics kill cellular microbes or slow down their reproduction, thus allowing the immune system more time to respond and exterminate the invaders. Vaccinations are injections of antigens from disease organisms, typically the weakened or dead microbes themselves. These antigens evoke an immune response, providing memory and a rapid response should an infection with live microbes of the same type occur later.

22.7 What Happens When the Immune System Malfunctions?

Allergies are immune responses to normally harmless foreign substances, such as pollen, dust, or certain foods. Allergens bind to "allergy antibodies" on a special class of B cells, causing them to release allergy antibodies into the plasma. Allergy antibodies bind to mast cells. A second exposure to the allergen causes the mast cells to release histamine, which causes a local inflammatory response. Autoimmune diseases arise when the immune system destroys some of the body's own cells. Immune deficiency diseases occur when the immune system cannot respond strongly enough to ward off normally minor diseases.

Infection with human immunodeficiency viruses can cause AIDS. These viruses invade and destroy helper T cells. Without helper T cells to stimulate the immune responses of B cells and cytotoxic T cells, a person with AIDS is extremely susceptible to infections, which are usually eventually fatal.

Web Animation Effects of HIV on the Immune System

22.8 How Does the Immune System Combat Cancer?

Cancer is a population of the body's cells that divides without control. Cancerous cells may be recognized as "different" by the immune system and destroyed by natural killer cells and cytotoxic T cells. Some cancerous cells may evolve the capacity to evade the immune system, some attack immune cells, and others multiply too fast for the immune system to keep up. In these cases, cancer develops and requires aggressive medical treatment.

Key Terms

Thinking Through the Concepts

Fill-in-the-Blank

1. External defenses against microbial invasion include the _____, and the mucous membranes that line the _____, _____, and _____ tracts.

2. Nonspecific internal defenses against disease include _____, which engulf and digest microbes; _____, which destroy cells that have been infected by viruses; the _____, provoked by injury; and _____, an elevation of body temperature that slows microbial reproduction and enhances the body's defenses.

3. The specific immune response is stimulated when the body is invaded by complex proteins or polysaccharides collectively called _____. These molecules bind to one of two types of protein receptors of the immune system: _____ or _____.

4. An antibody consists of four protein chains, two _____ chains and two _____ chains. Each is composed of a _____ region and a _____ region. The _____ regions form the binding site for antigen.

5. _____ immunity is provided by B cells and their daughter cells, called _____, which secrete antibodies into the blood plasma. _____ immunity is provided by T cells. _____ T cells kill body cells that have been infected by viruses. _____ T cells produce cytokines that stimulate immune responses in both B cells and T cells. Protection against future invasions by microbes bearing the same antigens is provided by _____ cells of both B and T types.

6. In medical practice, a(n) _____ provides antigens that stimulate an immune response, to protect against infection without actually causing disease.

7. A(n) _____ occurs when the immune system produces a response to a harmless substance such as pollen. In a(n) _____, the immune system cannot mount an effective response, even to dangerous infections; these may be inborn or acquired. When the immune system attacks a person's own body, this is called a(n) _____.

Review Questions

1. List the human body's three lines of defense against invading microbes. Which are nonspecific (that is, act against all types of invaders), and which are specific (act only against a particular type of invader)?

2. How do natural killer cells and cytotoxic T cells destroy their targets?

3. Describe humoral immunity and cell-mediated immunity. Include in your answer the types of immune cells involved in each, the location of antibodies and receptors that bind foreign antigens, and the ways in which invading cells are destroyed.

4. How does the immune system construct so many different antibodies?

5. How does the immune system distinguish "self" from "non-self"?

6. Diagram the structure of an antibody. What parts bind to antigens? Why does each antibody bind only to a specific antigen?

7. What are memory cells? How do they contribute to long-lasting immunity to specific diseases?

8. What is a vaccine? How does it confer immunity to a disease?

9. Compare and contrast the defensive response to a wound (such as that caused by a splinter) with an allergic reaction. Compare the cells involved, the substances produced, and the symptoms experienced.

10. Distinguish between autoimmune diseases and immune deficiency diseases, and give one example of each.

11. Describe the causes and eventual outcome of AIDS. How do AIDS treatments work? How is the HIV virus spread?

12. Why is cancer sometimes fatal?

Applying the Concepts

1. **IS THIS SCIENCE?** Refer to the essay in this chapter titled "Scientific Inquiry: How Vaccines Were Discovered." Ancient people (and even people of Jenner's time in the late 1700s) knew nothing of disease-causing microorganisms and nothing about the immune system, yet they practiced a primitive form of vaccination. Compare their likely methods of scientific inquiry with those of today. How were they similar and how were they different?

2. There are smallpox virus stocks in two laboratories—one in the United States and one in Russia. A debate is raging about whether these stocks should be destroyed. In brief, one side argues that having smallpox around is too dangerous. The other side argues that we may be able to learn things from smallpox that may help us conquer future diseases. If bioterrorists possess and disseminate smallpox, we will need the stocks to research better vaccines and treatments. Do you believe the smallpox stocks should be destroyed? Why or why not?

3. Organ transplant patients typically receive the drug cyclosporine, which inhibits the production of a regulatory molecule (interleukin-2) that stimulates helper T cells to proliferate. How does cyclosporine prevent the rejection of transplanted organs? Some patients who received successful transplants many years ago and have been taking cyclosporine are now developing various kinds of cancers. Propose a hypothesis to explain this phenomenon.

For additional resources, go to www.mybiology.com.

chapter 23

Chemical Control of the Animal Body: The Endocrine System

Are massive muscles worth the health risks of anabolic steroids?

Q At a Glance

Case Study   Anabolic Steroids—Fool's Gold?

It was summer of 2003 when officials at the United States Anti-Doping Agency received an anonymous tip. The informant, now known to be Trevor Graham—the coach of several world-class sprinters—claimed that an undetectable steroid was being used by track and field athletes. As proof, he mailed the agency a used syringe, still containing traces of the substance. Officials asked Dr. Don Catlin and his team of scientists at the UCLA Olympic Analytic Laboratory to identify it.

Thanks to extensive knowledge of chemistry and biology, millions of dollars of high-tech equipment, and dogged persistence, Catlin's world-renowned lab can test for more than 200 substances banned by athletic associations. Of these, the most notorious are anabolic steroids. *Anabolic steroid* is a term used to describe any of dozens of performance-enhancing drugs whose chemical composition resembles that of the male sex hormone testosterone. Bodybuilders often take them, and several Olympic athletes have been caught illegally exploiting them. Most anabolic steroids leave a chemical signature in the body that can be detected in urine months after a person stops using them. A new steroid that would be missed by standard lab tests would give its users an enormous advantage.

Catlin and his colleagues set about analyzing a tiny amount of the drug that was rinsed from the syringe Graham had mailed to them. Based on its chemical structure, Catlin dubbed it tetrahydrogestrinone, or THG. The molecule was new to science, proving that creating new, undetectable versions of synthetic testosterone is a big enough business to engage the efforts of skilled chemists, and suggesting that more versions of performance-enhancing drugs may be in the works, if indeed they are not already in widespread use.

In the "game" of drug use and detection, Don Catlin finds himself competing against both ambitious athletes and clever chemists.

The U.S. Food and Drug Administration (FDA) quickly banned the new substance. Because officials keep urine samples collected at important athletic competitions in cold storage for years, they retrieved samples for retesting. The results were disturbing. Although THG was probably manufactured by only one California lab, BALCO, the substance was found in urine samples from several champion track and field athletes. Hammer thrower Melissa Price, shot put champion Kevin Toth, and runners Regina Jacobs and Michelle Collins tested positive for THG and received 2- to 4-year bans from competition. The British Olympic Association permanently banned sprinter Dwain Chambers from its Olympic teams.

Why do athletes take anabolic steroids? Are their effects solely beneficial, from the athlete's point of view? As you read this chapter, notice how many different effects the same hormone can have throughout the body. Why might the multiple effects of anabolic steroids produce health risks? ∎

23.1 How Do Animal Cells Communicate?

The individual cells of an animal's body are in continuous communication with one another. In most cases, cells communicate by releasing chemical messenger molecules that affect other cells, either nearby or far away. A messenger molecule alters the physiology of a cell by binding to **receptors,** which are specialized proteins located either on the surface of, or inside, the cell. When a messenger molecule binds to a receptor, the recipient cell responds in a way that is determined by the messenger, the receptor, and the type of cell. These responses can be as varied as muscle contraction, electrical activity in nerve cells, secretion of milk in lactating women, and active transport of salt by cells in the kidney. Every cell has several, probably dozens, of receptors, each capable of binding to a different messenger and stimulating a different response in the cell.

Although chemical communication between cells is essential throughout the bodies of all animals, almost all animals have two large, specialized organ systems in which chemical communication is especially important: the endocrine and nervous systems. In Chapter 24, we will discuss the structure and functions of the nervous system. Here, we focus on the **endocrine system,** which is an integrated group of secretory structures, called glands, that regulate almost all aspects of animal physiology, including growth and development, metabolism, and reproduction.

23.2 How Do Animal Hormones Work?

Hormones are chemical messengers that are secreted by the specialized cells of **endocrine glands.** An endocrine gland may be a well-defined cluster of cells whose principal, or only, function is hormone secretion, as is the case with the thyroid and pituitary glands. Alternatively, an endocrine gland may consist of clusters of cells, or even numerous scattered individual cells, that are embedded in organs that have multiple functions, such as in the pancreas, stomach, ovary, or testes. In all cases, the secretory cells of an endocrine gland are embedded within a network of capillaries. The cells secrete their hormones into the extracellular fluid surrounding the capillaries, where the hormones then diffuse into the capillaries and are carried throughout the body by the bloodstream (**Fig. 23-1**).

This arrangement, of course, has the potential to create problems. Because almost all cells have a blood supply, a hormone in the bloodstream will reach nearly all of the body's cells, but it would not be useful for every cell to respond to a given hormone. Oxytocin, for example, stimulates the contraction of uterine muscles during childbirth, but it does not cause most of the other muscles of the body to contract. The uterine muscles, therefore, are **target cells** for oxytocin, whereas a woman's biceps are not. How can a hormone that reaches all cells affect some of them but not others? Right—by binding to a select set of receptors. If a cell doesn't have receptors for a given hormone, it can't respond. Because each cell in the body has a limited set of receptors, each cell can therefore respond to some hormones but not to others (see Fig. 23-1).

The changes induced by hormones may be prolonged and irreversible, as in the onset of puberty or the transformation of a caterpillar into a butterfly. More typically, the changes are temporary and reversible, and help to regulate the physiological systems of the animal body on a day-to-day, or sometimes even a second-by-second, time course.

Many Different Molecules Serve as Hormones

A wide variety of molecules can function as hormones, but vertebrate hormones fall into three general classes: **steroid hormones,** which are synthesized from cholesterol; **peptide hormones,** which are chains of amino acids ranging in length from 3 to about 180 amino acids; and **amino acid-derived hormones,** which consist of modified amino acids. As you can see in **Table 23-1,** most vertebrate hormones are peptides.

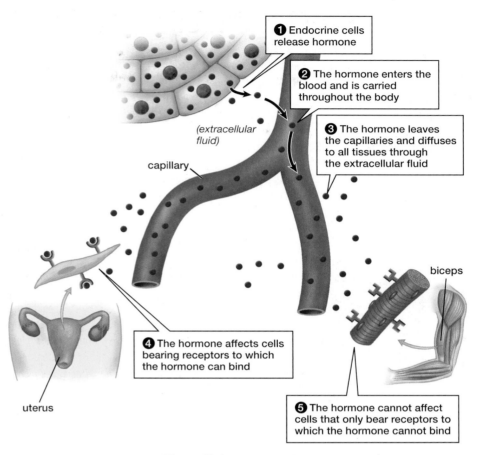

1 Endocrine cells release hormone

2 The hormone enters the blood and is carried throughout the body

3 The hormone leaves the capillaries and diffuses to all tissues through the extracellular fluid

(extracellular fluid)

capillary

biceps

4 The hormone affects cells bearing receptors to which the hormone can bind

uterus

5 The hormone cannot affect cells that only bear receptors to which the hormone cannot bind

▲ **Figure 23-1 Hormone release, distribution, and reception** The cells of endocrine glands secrete hormones into the extracellular fluid. From there, the hormones diffuse into the capillaries and circulate in the bloodstream. When they reach capillary beds in other parts of the body, they diffuse out into the extracellular fluid again, potentially contacting every cell in the body. A hormone influences only those cells with receptors to which the hormone can bind. For example, smooth muscle cells in the uterus, but not skeletal muscle cells in the biceps, have receptors for oxytocin.

Web Animation Modes of Action of Hormones

Table 23-1   Mammalian Endocrine Glands and Hormones

Endocrine Gland	Hormone	Type of Chemical	Principal Functions
Hypothalamus (via posterior pituitary)	Antidiuretic hormone (ADH)	Peptide	Promotes reabsorption of water from the kidneys; constricts arterioles
	Oxytocin	Peptide	In females, stimulates the contraction of uterine muscles during childbirth, milk ejection, and maternal behaviors; in males, causes sperm ejection
Hypothalamus (to anterior pituitary)	Releasing and inhibiting hormones	Peptide	At least nine hormones; releasing hormones stimulate the release of hormones from the anterior pituitary; inhibiting hormones reduce the release of hormones from the anterior pituitary
Anterior pituitary	Follicle-stimulating hormone (FSH)	Peptide	In females, stimulates the growth of follicles, the secretion of estrogen, and perhaps ovulation; in males, stimulates the development of sperm
	Luteinizing hormone (LH)	Peptide	In females, stimulates ovulation, the growth of the corpus luteum, and the secretion of estrogen and progesterone; in males, stimulates the secretion of testosterone
	Thyroid-stimulating hormone (TSH)	Peptide	Stimulates the thyroid to release thyroxine
	Adrenocorticotropic hormone (ACTH)	Peptide	Stimulates the adrenal cortex to release hormones, especially glucocorticoids
	Growth hormone	Peptide	Stimulates growth, protein synthesis, and fat metabolism; inhibits sugar metabolism
	Prolactin	Peptide	Stimulates milk synthesis in and secretion from mammary glands
Thyroid	Thyroxine	Amino acid derivative	Increases the metabolic rate of most body cells; increases body temperature; regulates growth and development
	Calcitonin	Peptide	Inhibits the release of calcium from bones; decreases blood calcium concentration
Parathyroid	Parathyroid hormone	Peptide	Stimulates the release of calcium from bones; increases blood calcium concentration
Pancreas	Insulin	Peptide	Decreases blood glucose levels by increasing the uptake of glucose into cells and converting glucose to glycogen, especially in the liver; regulates fat metabolism
	Glucagon	Peptide	Converts glycogen to glucose, raising blood glucose levels
Ovaries[a]	Estrogen	Steroid	Causes the development of female secondary sexual characteristics and the maturation of eggs; promotes the growth of the uterine lining
	Progesterone	Steroid	Stimulates development of the uterine lining and the formation of the placenta
Testes[a]	Testosterone	Steroid	Stimulates development of genitalia and male secondary sexual characteristics; stimulates the development of sperm
Adrenal medulla	Epinephrine and norepinephrine	Amino acid derivative	Increase levels of sugar and fatty acids in the blood; increase metabolic rate; increase rate and force of the contractions of the heart; constrict some blood vessels
Adrenal cortex	Glucocorticoids (cortisol)	Steroid	Increase blood sugar; regulate sugar, lipid, and fat metabolism; have anti-inflammatory effects
	Mineralocorticoids (aldosterone)	Steroid	Increase reabsorption of salt in the kidney
	Testosterone	Steroid	Causes masculinization of body features, growth

Other Sources of Hormones

Pineal gland	Melatonin	Amino acid derivative	Regulates seasonal reproductive cycles and sleep–wake cycles; may regulate the onset of puberty
Thymus	Thymosin	Peptide	Stimulates maturation of cells of the immune system
Kidney	Renin	Peptide	Acts on blood proteins to produce a hormone (angiotensin) that regulates blood pressure
	Erythropoietin	Peptide	Stimulates red blood cell synthesis in bone marrow
Heart	Atrial natriuretic peptide (ANP)	Peptide	Increases salt and water excretion by the kidneys; lowers blood pressure
Digestive tract[b]	Secretin, gastrin, cholecystokinin, and others	Peptide	Control the secretion of mucus, enzymes, and salts in the digestive tract; regulate peristalsis
Fat cells	Leptin	Peptide	Regulates appetite; stimulates immune function; promotes blood vessel growth; required for onset of puberty

[a]See Chapter 25.

[b]See Chapter 21.

Hormones Act by Binding to Receptors on or in Target Cells

Hormone receptors may be located either on the plasma membranes of target cells or inside the cells, often within the nucleus. We will describe many specific hormonal actions as we proceed through this chapter. Here, we describe the general principles that govern hormone action on target cells.

Some Hormones Bind to Receptors on the Surfaces of Target Cells

Peptide and amino acid-derived hormones are usually soluble in water but not in lipids. Therefore, they cannot readily penetrate plasma membranes (which are composed largely of phospholipids). Instead, these hormones bind to receptors on the surface of the target cell's plasma membrane (**Fig. 23-2**, step **❶**). Hormone–receptor binding stimulates production of another molecule, called a **second messenger,** inside the cell (**Fig. 23-2**, step **❷**). The second messenger transfers the signal from the first messenger—the hormone—to other molecules within the cell (**Fig. 23-2**, step **❸**). A common second messenger is cyclic adenosine monophosphate (cyclic AMP), a nucleotide that regulates many cellular activities (see p. 36). The second messenger then initiates a chain of biochemical reactions (**Fig. 23-2**, step **❹**). The end result of the reactions varies depending on the hormone, the second messenger, and the target cell, but in many cases, the target cells synthesize or secrete substances.

Hormones that bind to receptors on the target cell's surface often exert rapid but short-lived effects. For example, when you're excited or stressed, your adrenal glands secrete epinephrine (also called adrenaline), which binds to receptors on your heart and liver, stimulating synthesis of cyclic AMP. The cyclic AMP triggers changes in metabolism that make your heart beat faster and harder and cause your liver to dump glucose into your bloodstream—both useful in escaping

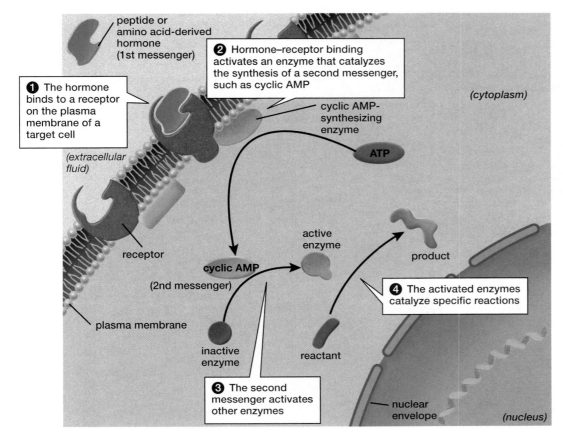

◄Figure 23-2 Actions of peptide and amino acid-derived hormones on target cells Peptide and amino acid-derived hormones usually stimulate target cells by binding to receptors on the plasma membrane, which causes the cell to synthesize a second messenger molecule that sets off a cascade of intracellular metabolic reactions. For example, in liver cells, epinephrine (an amino acid-derived hormone) stimulates synthesis of cyclic AMP (a second messenger), which activates an enzyme that converts glycogen to glucose (a metabolic reaction).

from predators or taking an exam. Within minutes after the stressful situation ends, so does the secretion of epinephrine and the altered metabolism of the heart and liver cells.

Some Hormones Bind to Receptors Inside Target Cells

Steroid hormones are lipid soluble, so they pass freely through the plasma membrane (**Fig. 23-3**, step **1**). Once inside, these hormones bind to receptors inside target cells (**Fig. 23-3**, step **2**). Usually, the receptors are either in the nucleus or move into the nucleus after binding to the hormone. The hormone–receptor complex then binds to the DNA of the promoter region of specific genes (**Fig. 23-3**, step **3**) and stimulates the transcription of messenger RNA (mRNA) (**Fig. 23-3**, step **4**). The mRNA then travels to the cytoplasm and directs protein synthesis (**Fig. 23-3**, step **5**). In hens, for example, the steroid hormone estrogen stimulates transcription of the albumin gene, causing the synthesis of albumin (egg white), which is packaged in the egg as a food supply for the developing chick. Hormones that bind to intracellular receptors may take several minutes or even days to exert their full effects.

Hormone Release Is Regulated by Feedback Mechanisms

The release of most hormones is controlled by negative feedback. As you may recall from Chapter 19, **negative feedback** is a response to a change that tends to counteract the change, and to restore the system to its original condition. For example, suppose you have jogged a few miles on a hot, sunny day and have lost a quart of water through perspiration. In response to the loss of water from your bloodstream, your pituitary gland releases antidiuretic hormone (ADH), which causes your kidneys to reabsorb water and to produce very concentrated urine (see p. 413). If, however, you arrive home and drink two quarts of Gatorade™, you will more than replace the water you lost in sweat. Continued retention of this excess water could raise your blood pressure and possibly damage your heart. Negative feedback ensures that ADH secretion is turned off when the water content of your blood returns to normal, so that your kidneys can eliminate the excess water. Look for other examples of negative feedback throughout this chapter.

Anabolic Steroids—Fool's Gold?

Continued

Anabolic steroids bind to the same intracellular receptors as testosterone. However, many anabolic steroids are far more powerful than testosterone in activating these receptors. Further, if you inject testosterone into your body, it will be completely metabolized within a few hours, whereas some anabolic steroids last for weeks. Therefore, the strength and duration of hormone–receptor action in target cells is much, much greater for anabolic steroids than for natural testosterone.

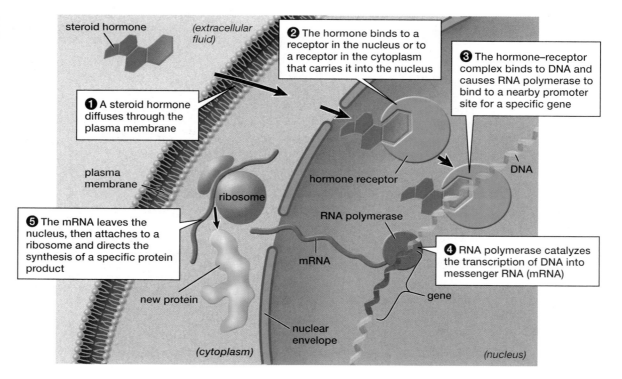

▶**Figure 23-3 Steroid hormone action on target cells** Steroid hormones often stimulate target cells by binding to intracellular receptors. This process creates a hormone–receptor complex that activates gene transcription and ultimately results in the synthesis of new, or increased amounts of, specific proteins.

steroid hormone

(extracellular fluid)

2 The hormone binds to a receptor in the nucleus or to a receptor in the cytoplasm that carries it into the nucleus

3 The hormone–receptor complex binds to DNA and causes RNA polymerase to bind to a nearby promoter site for a specific gene

1 A steroid hormone diffuses through the plasma membrane

plasma membrane

ribosome

hormone receptor

DNA

RNA polymerase

5 The mRNA leaves the nucleus, then attaches to a ribosome and directs the synthesis of a specific protein product

mRNA

4 RNA polymerase catalyzes the transcription of DNA into messenger RNA (mRNA)

new protein

gene

nuclear envelope

(cytoplasm)

(nucleus)

In a few cases, hormone release is controlled by **positive feedback,** in which the response to a change enhances the change. For example, when a baby is ready to be born, his or her lungs release a protein that starts a cascade of events that results in the contraction of the mother's uterine muscles. Uterine contraction pushes the baby's head against the cervix (the ring of connective tissue between the uterus and the vagina), which causes the cervix to stretch. Stretching the cervix sends nervous signals to the mother's brain, which in turn causes the release of the hormone oxytocin. Oxytocin stimulates continued contractions of the uterine muscles, pushing the baby harder against the cervix, which stretches further, causing still more oxytocin to be released. Note that positive feedback cannot continue indefinitely without causing significant damage or even death. In the case of childbirth, the positive feedback between oxytocin release and uterine contractions is eventually terminated by negative feedback: After the infant is born, the cervix is no longer stretched, so oxytocin release stops.

23.3 What Are the Structures and Functions of the Mammalian Endocrine System?

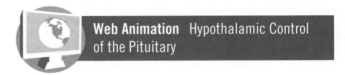

Web Animation Hypothalamic Control of the Pituitary

In the following sections, we focus on the functions of the major endocrine glands and organs: the hypothalamus–pituitary complex, the thyroid gland, the pancreas, the sex organs, and the adrenal glands. These and other hormone-secreting organs are illustrated in **Figure 23-4** and are described in Table 23-1.

◀ **Figure 23-4 The major sources of hormones in mammals**

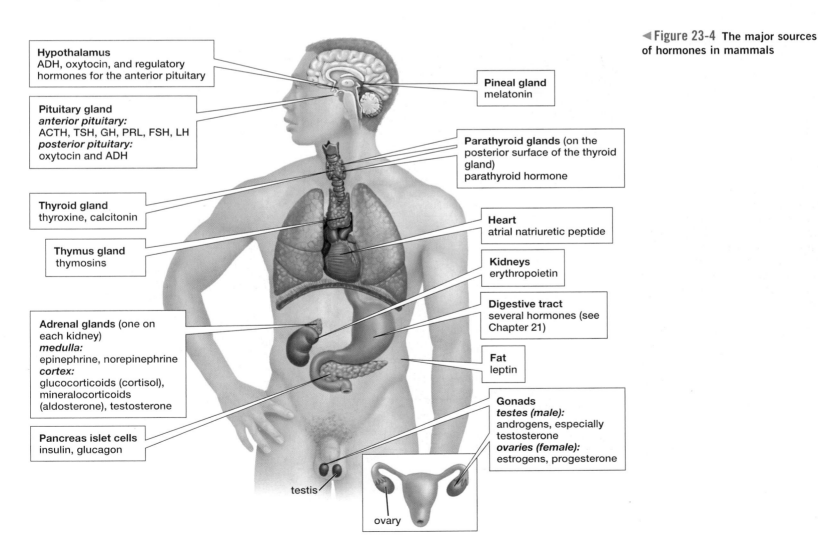

Hypothalamus
ADH, oxytocin, and regulatory hormones for the anterior pituitary

Pituitary gland
anterior pituitary:
ACTH, TSH, GH, PRL, FSH, LH
posterior pituitary:
oxytocin and ADH

Thyroid gland
thyroxine, calcitonin

Thymus gland
thymosins

Adrenal glands (one on each kidney)
medulla:
epinephrine, norepinephrine
cortex:
glucocorticoids (cortisol), mineralocorticoids (aldosterone), testosterone

Pancreas islet cells
insulin, glucagon

Pineal gland
melatonin

Parathyroid glands (on the posterior surface of the thyroid gland)
parathyroid hormone

Heart
atrial natriuretic peptide

Kidneys
erythropoietin

Digestive tract
several hormones (see Chapter 21)

Fat
leptin

Gonads
testes (male):
androgens, especially testosterone
ovaries (female):
estrogens, progesterone

testis

ovary

Hormones of the Hypothalamus and Pituitary Gland Regulate Many Functions Throughout the Body

The hypothalamus and pituitary gland coordinate the action of many key hormonal systems. The **hypothalamus** is a part of the brain that contains clusters of specialized nerve cells called **neurosecretory cells.** Neurosecretory cells synthesize peptide hormones, store them, and release them when stimulated. The **pituitary gland** is a pea-sized gland that dangles from the hypothalamus by a stalk (**Fig. 23-5**). The pituitary consists of two distinct parts: the anterior pituitary and the posterior pituitary. The anterior pituitary is a true endocrine gland, composed of several types of hormone-secreting cells enmeshed in a network of capillaries. The posterior pituitary, however, develops as an outgrowth of the hypothalamus. It consists mainly of a capillary bed and the endings of neurosecretory cells whose cell bodies are in the hypothalamus. The hypothalamus controls the release of hormones from both parts of the pituitary.

Hormones from the Hypothalamus Control Hormone Release in the Anterior Pituitary

Neurosecretory cells of the hypothalamus produce at least nine peptide hormones that regulate the release of hormones from the anterior pituitary (**Fig. 23-5**, green step ❶). These hypothalamic hormones are called releasing hormones or inhibiting hormones, depending on whether they stimulate or inhibit the release

▼ **Figure 23-5** **The hypothalamus–pituitary system.** The left side of the diagram (green circled numbers) shows the relationship between the hypothalamus and the anterior pituitary, and the right side (blue circled numbers) shows the relationship between the hypothalamus and the posterior pituitary. (Left) One group of neurosecretory cells in the hypothalamus controls hormone release in the anterior lobe of the pituitary by secreting releasing hormones (green circles) or inhibiting hormones. These hormones enter a capillary network that carries them to the anterior pituitary. There, releasing hormones stimulate endocrine cells bearing the appropriate receptors to secrete their hormones (red squares). Inhibiting hormones (not shown) reduce the secretion of hormones by the anterior pituitary. (Right) The posterior lobe of the pituitary is an extension of the hypothalamus. Neurosecretory cells in the hypothalamus have endings on a capillary bed in the posterior lobe, where they release either oxytocin or antidiuretic hormone (ADH), shown as blue triangles.

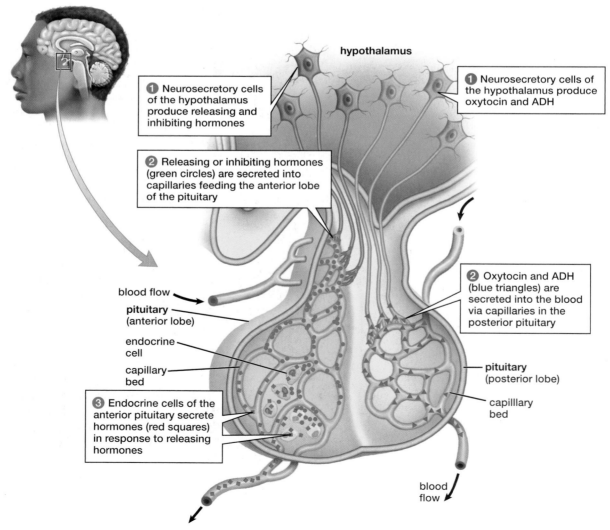

hypothalamus

❶ Neurosecretory cells of the hypothalamus produce releasing and inhibiting hormones

❶ Neurosecretory cells of the hypothalamus produce oxytocin and ADH

❷ Releasing or inhibiting hormones (green circles) are secreted into capillaries feeding the anterior lobe of the pituitary

❷ Oxytocin and ADH (blue triangles) are secreted into the blood via capillaries in the posterior pituitary

blood flow

pituitary (anterior lobe)

endocrine cell

capillary bed

❸ Endocrine cells of the anterior pituitary secrete hormones (red squares) in response to releasing hormones

pituitary (posterior lobe)

capilllary bed

blood flow

of hormones by endocrine cells of the anterior pituitary. Releasing and inhibiting hormones are secreted into a capillary bed that directly feeds into a second capillary bed in the anterior pituitary (**Fig. 23-5,** green step ❷). There, they diffuse out of the capillaries and bind to receptors on the surfaces of the pituitary endocrine cells. Some of these hypothalamic hormones, such as growth hormone-releasing hormone, stimulate the release of pituitary hormones (in this case the release of growth hormone; **Fig. 23-5,** green step ❸). Others, such as the inhibiting hormone somatostatin, inhibit the release of pituitary hormones (again, in this case growth hormone).

The Anterior Pituitary Produces and Releases Several Hormones

The anterior pituitary produces several peptide hormones. Four of these regulate hormone production in other endocrine glands. *Follicle-stimulating hormone* (FSH) and *luteinizing hormone* (LH) stimulate the production of sperm and testosterone in the testes of males, and of eggs, estrogen, and progesterone in the ovaries of females. (We will discuss the roles of FSH and LH in greater detail in Chapter 25.) *Thyroid-stimulating hormone* (TSH) stimulates the thyroid gland to release its hormones, and ACTH, or *adrenocorticotropic hormone,* causes the release of hormones from the adrenal cortex. We discuss the effects of thyroid and adrenal cortical hormones later in this chapter.

Unlike the hormones described in the preceding paragraph, the remaining hormones of the anterior pituitary do not act on other endocrine glands. *Prolactin,* in conjunction with other hormones, stimulates the development of the milk-producing mammary glands in the breasts during pregnancy. *Growth hormone* regulates the body's growth by acting on nearly all of the body's cells—increasing protein synthesis, fat utilization, and carbohydrate storage. During childhood, it stimulates bone growth, which influences the ultimate size of the adult.

Much of the normal variation in human height is due to differences in the secretion of growth hormone from the anterior pituitary. Too little growth hormone causes some cases of *dwarfism;* too much can cause *gigantism* (**Fig. 23-6**). A major advance in the treatment of pituitary dwarfism occurred when molecular biologists successfully inserted the gene for human growth hormone into bacteria, causing them to synthesize large quantities of the hormone. Now, many children with underactive pituitary glands can achieve normal height, by receiving injections of human growth hormone produced by these bacteria.

The Posterior Pituitary Releases Hormones Synthesized in the Hypothalamus

The hypothalamus contains two types of neurosecretory cells that grow thin strands, called axons, into the posterior pituitary (**Fig. 23-5,** blue step ❶). The axons end in a capillary bed in the posterior pituitary into which they release hormones that are then carried by the bloodstream to the rest of the body (**Fig. 23-5,** blue step ❷). These neurosecretory cells synthesize and release either *antidiuretic hormone* (ADH) or *oxytocin.*

Antidiuretic hormone helps prevent dehydration. As you learned in Chapter 21, ADH acts by increasing the water permeability of the collecting ducts of nephrons in the kidney. This increased permeability causes water to be reabsorbed from the urine and returned to the bloodstream. Interestingly, alcohol inhibits the release of ADH and, as a consequence, increases urination. Therefore, people who drink strong alcoholic beverages may lose more liquid than they consume and, thus, may become dehydrated. In fact, dehydration is thought to contribute to the headache and generally miserable sensations known as a hangover.

As we described earlier, oxytocin stimulates contraction of uterine muscles during childbirth, helping to push the fetus from the uterus. Oxytocin also triggers the "milk letdown reflex" in nursing mothers by causing muscle tissue within

▲ **Figure 23-6 When the anterior pituitary malfunctions** An improperly functioning anterior pituitary produces either too much or too little growth hormone. Too little causes dwarfism, and too much causes gigantism. **QUESTION:** Why is it easier to treat dwarfism than gigantism?

▶ **Figure 23-7 Hormones and breast-feeding**
Feedback between an infant and its mother
regulates the control of milk letdown by oxytocin
during breast-feeding. The cycle begins with the
infant's suckling and continues until the infant is
full and stops suckling. With the nipple no longer
being stimulated, oxytocin release stops, the
muscles relax, and milk flow ceases. **QUESTION:**
Explain how the negative feedback that regulates
milk letdown actually involves two individuals.

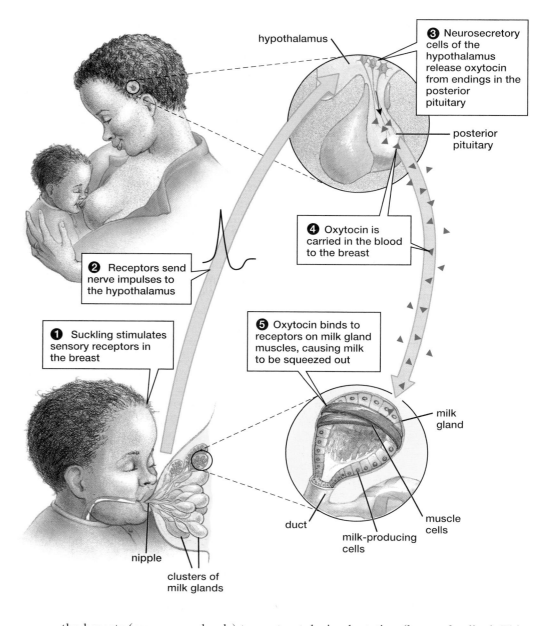

hypothalamus

❸ Neurosecretory
cells of the
hypothalamus
release oxytocin
from endings in the
posterior
pituitary

posterior
pituitary

❹ Oxytocin is
carried in the blood
to the breast

❷ Receptors send
nerve impulses to
the hypothalamus

❶ Suckling stimulates
sensory receptors in
the breast

❺ Oxytocin binds to
receptors on milk gland
muscles, causing milk
to be squeezed out

milk
gland

muscle
cells

duct

milk-producing
cells

nipple

clusters of
milk glands

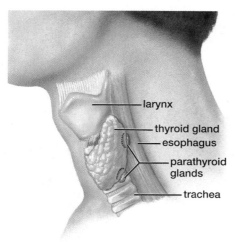

larynx

thyroid gland

esophagus

parathyroid
glands

trachea

▲ **Figure 23-8 The thyroid and parathyroid glands** The thyroid
gland wraps around the front of the larynx in the neck. The tiny
parathyroid glands sit in the back of the thyroid.

the breasts (mammary glands) to contract during lactation (breast-feeding). This
reflex ejects milk from the saclike milk glands into the nipples (**Fig. 23-7**). In both
males and females, oxytocin may also play a role in emotions, including both romantic and maternal love (see the case study, "How Do I Love Thee?", in
Chapter 24).

The Thyroid and Parathyroid Glands Influence Metabolism and Calcium Levels

Lying at the front of the neck, nestled just below the larynx (**Fig. 23-8**), the **thyroid
gland** produces two major hormones: *thyroxine* and *calcitonin*. The *parathyroid
gland* consists of two pairs of small disks of endocrine cells, one pair on each side
of the back of the thyroid gland; these cells release *parathyroid hormone*.

Parathyroid Hormone and Calcitonin Regulate Calcium Metabolism

The proper concentration of calcium is essential to nerve and muscle function.
Parathyroid hormone and calcitonin work together to maintain nearly constant
calcium levels in the blood and body fluids. The skeleton serves as a "bank" into

which calcium can be deposited or withdrawn as necessary. If blood calcium levels drop, parathyroid hormone causes the bones to release calcium. It also causes the kidneys to reabsorb more calcium during urine production and to return the calcium to the blood. The increased blood calcium then inhibits further release of parathyroid hormone in a negative feedback loop. If blood calcium gets too high, the thyroid releases calcitonin, which inhibits the release of calcium from bone. In most vertebrates, calcitonin is important in regulating calcium concentrations in the blood, and may even promote bone growth. In humans, however, the actions of calcitonin appear to be minor compared to those of parathyroid hormone.

Thyroxine Influences Energy Metabolism

Thyroxine, often referred to as thyroid hormone, is an iodine-containing amino acid derivative that raises the metabolic rate of most body cells. It also stimulates the synthesis of enzymes that break down glucose and provide energy. These cellular effects cause the body to increase oxygen consumption and heart rate. Thyroxine's influence on metabolic rate helps regulate body temperature and responses to stress. Exposure to cold, for example, greatly increases thyroid hormone production, which in turn increases metabolic rate, warming the body.

In juvenile animals (including humans), thyroxine helps regulate growth by stimulating both metabolic rate and nervous system development. Undersecretion of thyroxine early in life can cause *cretinism,* a condition characterized by mental retardation and dwarfism. Fortunately, early diagnosis and thyroxine supplementation can prevent this condition. Conversely, too much thyroxine in developing vertebrates speeds up development, which was discovered almost a century ago, when J. F. Gudernatsch showed that thyroxine supplementation causes early metamorphosis of tadpoles into tiny frogs.

Thyroxine Release Is Controlled by the Hypothalamus

Levels of thyroxine in the bloodstream are finely tuned by negative feedback. Thyroxine release is stimulated by thyroid-stimulating hormone (TSH) from the anterior pituitary, which in turn is stimulated by a releasing hormone from the hypothalamus. The amount of TSH released from the pituitary is regulated by thyroxine levels in the blood: High levels of thyroxine inhibit the secretion of TSH, thus inhibiting further release of thyroxine from the thyroid.

A diet deficient in iodine can reduce the production of thyroxine, triggering a dramatic increase in the number of thyroxine-producing cells, thereby enlarging the thyroid gland. The hypertrophied gland may bulge from the neck, producing a condition called goiter (**Fig. 23-9**). The widespread use of iodized salt has now all but eliminated this condition in developed countries.

The Pancreas Has Both Digestive and Endocrine Functions

The **pancreas** produces both digestive secretions and hormones. As you may recall from Chapter 21, the pancreas produces bicarbonate and a number of enzymes that are released into the small intestine, promoting the digestion of all types of food. The endocrine portion of the pancreas consists of clusters of cells called **islet cells.** Each islet cell produces one of two hormones: *insulin* or *glucagon.*

Insulin and Glucagon Control Glucose Levels in the Blood

Insulin and glucagon work in opposition to regulate carbohydrate and fat metabolism: Insulin reduces the blood glucose level, and glucagon increases it (**Fig. 23-10**). Together the two hormones help keep the blood glucose level nearly constant. If blood glucose rises (for example, after you have eaten; **Fig. 23-10,**

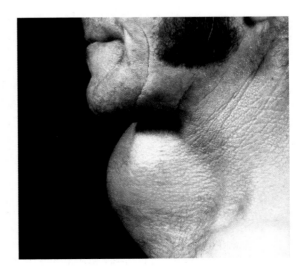

▲ **Figure 23-9 Goiter** An iodine-deficient diet often causes enlargement of the thyroid gland.

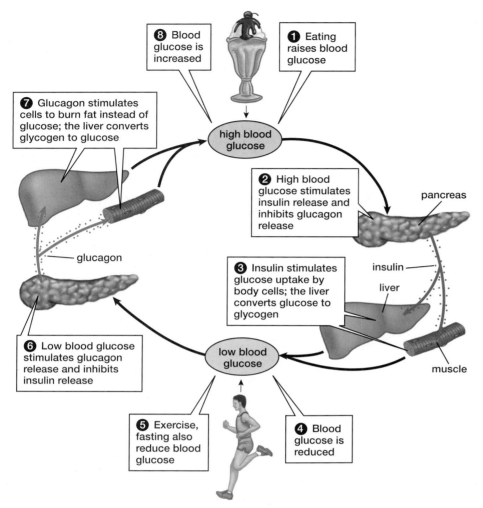

Figure 23-10 The pancreas controls blood glucose levels
The pancreatic islet cells contain two populations of
hormone-producing cells: One produces insulin (green
arrows); the other produces glucagon (blue arrows). These
two hormones cooperate in a two-part negative-feedback
loop to control blood glucose concentrations (black arrows).

step ❶), insulin is released by the pancreas (**Fig. 23-10,**
step ❷). Insulin causes body cells to take up glucose
(**Fig. 23-10, step ❸**) and either metabolize it for energy
or convert it to fat or glycogen (a starch-like molecule
stored primarily in the liver and muscles). When blood
glucose levels drop (for example, after you have
skipped breakfast or have run a 10-kilometer race;
Fig. 23-10, steps ❹ and ❺), insulin secretion is inhibited
and glucagon secretion is stimulated (**Fig. 23-10, step ❻**).
Glucagon activates an enzyme in the liver that
breaks down glycogen, releasing glucose into the blood
(**Fig. 23-10, step ❼**). Glucagon also promotes lipid
breakdown, which releases fatty acids that can be metabolized for energy. These actions increase blood glucose levels (**Fig. 23-10, step ❽**), which inhibits glucagon
secretion. If blood glucose rises too far, insulin is secreted once again.

Diabetes Results from Malfunctions of the Insulin Control System

Defects in insulin synthesis or release by the pancreas,
or in the ability of target cells to respond to insulin in
the blood, result in *diabetes mellitus,* a condition in
which blood glucose levels are high and fluctuate
wildly with sugar intake. The lack of functional insulin
responses in diabetics causes the body to rely heavily
on fats as an energy source, leading to high levels of
lipids—including cholesterol—circulating in the blood.
Severe diabetes causes fat deposits to accumulate in
the blood vessels, predisposing individuals to high
blood pressure and heart disease; for this reason, diabetes is a major cause of heart attacks in the United
States. Fat deposits in small blood vessels in the kidneys and the retina of the
eye may cause kidney failure and blindness, respectively. Insulin replacement
therapy greatly improves both glucose and lipid metabolism and thus profoundly improves the health of diabetics. However, insulin replacement therapy requires daily blood testing and insulin injections and cannot fully mimic
natural control of energy metabolism. Potentially important advances in the
treatment of diabetes that may both avoid the daily inconveniences and improve health are described in "Health Watch: Closer to a Cure for Diabetes."

The Sex Organs Produce Both Gametes and Sex Hormones

The sex organs (**testes** in males and **ovaries** in females) synthesize gametes
(sperm or eggs) and sex hormones. The testes secrete several male sex hormones,
all steroids, of which the most important is *testosterone.* The ovaries secrete two
types of steroid hormones: *estrogen* and *progesterone.* The role of the sex hormones in sperm and egg production, the menstrual cycle, and pregnancy is discussed in Chapter 25.

Sex Hormone Levels Increase During Puberty

Sex hormones play a key role in *puberty,* the phase of life during which the reproductive systems of both sexes become mature and functional. Puberty begins
when (for reasons not fully understood) the hypothalamus starts to secrete increasing amounts of hormones that stimulate the anterior pituitary to secrete
more luteinizing hormone (LH) and follicle-stimulating hormone (FSH) into the

Closer to a Cure for Diabetes

The most severe kind of diabetes, type 1, occurs when a person's immune system attacks and kills the insulin-producing islet cells of the pancreas. This form of diabetes often strikes early in life, triggering a lifelong daily regimen of multiple blood tests and insulin injections. For the millions of people who suffer from this form of diabetes, recent research offers real hope.

In 1999, researchers led by James Shapiro of the University of Alberta, Canada, isolated islet cells from brain-dead donors and injected them into veins leading to the livers of seven patients with type 1 diabetes. Some of the islet cells colonized the livers and started secreting insulin. Although some initially required insulin therapy after the transplant, all seven patients eventually became completely independent of insulin injections. Unfortunately, however, because the islet cells are from donors, the recipients are required to take immunosuppressant drugs for the duration of their lives in order to prevent tissue rejection.

Because immune suppression is a major threat to the recipient's health, and because islet cell donors are in very short supply compared to the millions of type 1 diabetics, alternate forms of treatment continue to be sought. In 2006, scientists at Novocell, a biotech firm in San Diego, California, reported the first two steps in bypassing immune rejection and obtaining unlimited numbers of islet cells. First, they encapsulated islet cells from deceased donors into polyethyleneglycol (PEG), a large, inert polymer that has a long history of safe use in humans as an ingredient in skin creams and laxatives. They injected the cells just beneath the skin of human volunteers and found that fairly small molecules, such as sugar and insulin, could diffuse into and out of the PEG capsules, but immune cells and antibodies could not. Because the immune system couldn't attack the transplanted islet cells, the cells were not rejected.

Although a milestone, this still didn't solve the problem of a shortage of transplantable islet cells. However, the Novocell researchers also found a way to cause embryonic stem (ES) cells to differentiate into insulin-secreting cells. Recall from Chapter 8 that stem cells have the potential to continuously generate more cells, so the right stem-cell line could some day produce a virtually unlimited supply of transplantable, islet-like cells (**Fig. E23-1**). Unfortunately, these ES-derived cells did not completely mimic functional islet cells. Normally, islet cells form part of a negative feedback loop—too much glucose in the blood causes the islet cells to secrete lots of insulin, which stimulates other cells of the body to take up glucose. This in turn reduces the glucose concentration in the blood, so the islet cells slow down their insulin secretion. In contrast, the differentiated stem

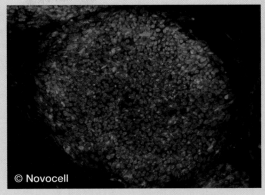
© Novocell

▲ **Figure E23-1   A potential cure for type 1 diabetes** Embryonic stem cells such as these can be stimulated with just the right mix of growth factors to cause them to differentiate into insulin-producing cells that resemble fetal pancreatic islet cells. Photo © Novocell.

cells behaved like islet cells in fetuses—they produced insulin, but did not release it in response to glucose. The researchers are hopeful that a few more manipulations can produce fully functional islet-like cells that release insulin when stimulated by glucose.

Researchers and physicians rightly caution that these treatments are many years away from routine clinical practice. Nevertheless, encapsulated ES-derived "islet cells" might just prove to be the magic bullet against type 1 diabetes.

---

bloodstream. LH and FSH stimulate target cells in the testes or ovaries to produce higher levels of sex hormones.

The resulting increase in circulating sex hormones ultimately affects tissues throughout the body that bear the appropriate receptors. In females, estrogen released by the ovaries stimulates the growth of breasts and the maturation of the female reproductive system, including production of mature egg cells. Progesterone prepares the reproductive tract to receive and nourish the fertilized egg, and is also secreted by the ovaries during pregnancy.

In males, testosterone, secreted by the testes, promotes sperm cell production and stimulates the development of male secondary sexual attributes, including the growth of body and facial hair; the development of a larger larynx ("voice box"), which lowers the voice; and increased muscle growth.

Although there is a surge of sex hormone production at puberty, sex hormones are present from the fetal stage onward. They influence brain development in both sexes, and they continue to have an effect on both brain and behavior throughout life.

Anabolic Steroids—Fool's Gold?

Continued

For athletes, the point of anabolic steroids is to mimic the muscle-building actions of testosterone, only more so. However, the far-reaching effects of testosterone, in both men and women—on cholesterol metabolism, bone growth, mood, and behavior—illustrate a difficulty faced by medical practitioners dating back to Hippocrates: It is extremely difficult to find a drug that does only one thing. In the "Anabolic Steroids—Fool's Gold? *Revisited*" section at the end of the chapter, we explore a few of the common, sometimes deadly, side effects of anabolic steroids.

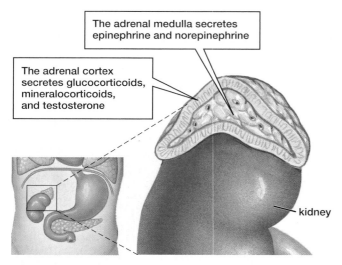

The adrenal medulla secretes epinephrine and norepinephrine

The adrenal cortex secretes glucocorticoids, mineralocorticoids, and testosterone

kidney

▲ **Figure 23-11 The adrenal glands** On top of each kidney sits an adrenal gland, composed of an outer cortex and an inner medulla, each composed of very dissimilar cells.

▲ **Figure 23-12 Cushing's syndrome in an elderly horse** Oversecretion of glucocorticoids occurs in many mammals. In horses, the typical symptoms include lethargy, swayback, long, curly hair that doesn't shed in spring, hoof problems, and excessive drinking and urination (as much as 20 gallons a day—four times the normal rate).

The Adrenal Glands Secrete Hormones That Regulate Metabolism and Responses to Stress

The **adrenal glands** consist of two very different parts: the adrenal cortex and the adrenal medulla (**Fig. 23-11**).

The Hormones of the Adrenal Cortex Regulate Glucose and Salt Metabolism

The outer layer of the adrenal gland forms the *adrenal cortex*. The cortex secretes two major types of steroid hormones, called glucocorticoids and mineralocorticoids, along with small amounts of testosterone. As their name implies, *glucocorticoids* help control glucose metabolism. Their release is stimulated by *adrenocorticotropic hormone* (ACTH) from the anterior pituitary. ACTH release is stimulated by hormones produced by the hypothalamus in response to stimuli such as trauma, infection, or exposure to temperature extremes. In some respects, the glucocorticoids act similarly to glucagon, raising blood glucose levels by stimulating glucose production and promoting the use of fats instead of glucose for energy production. Glucocorticoids also reduce inflammation, which is why a synthetic version, called hydrocortisone, can be purchased in your local pharmacy to reduce swelling and itching from rashes or insect bites.

A few thousand people each year develop tumors of either the pituitary gland (causing excess ACTH secretion) or the adrenal cortex. In either case, excessive glucocorticoid secretion may cause Cushing's syndrome, characterized by multiple symptoms such as weight gain, fragile skin, weakening of the bones, excessive hair growth, and increased thirst and urination. Cushing's may strike many mammals, including dogs and horses (**Fig. 23-12**).

Mineralocorticoid hormones regulate the mineral (salt) content of the blood. The most important mineralocorticoid is *aldosterone*, which helps to control sodium concentrations. Sodium ions are the most abundant positive ions in blood and extracellular fluid. The sodium ion gradient across plasma membranes (high outside, low inside) is a factor in many cellular events, including the production of electrical signals by nerve cells. If blood sodium falls, the adrenal cortex releases aldosterone, which causes the kidneys and sweat glands to retain sodium. When salt and other sources of dietary sodium, combined with aldosterone-induced sodium conservation, return blood sodium concentrations back to within normal limits, further aldosterone secretion is shut off—yet another example of negative feedback.

The Hormones of the Adrenal Medulla Prepare the Body for Stress and Exercise

The *adrenal medulla* is located in the center of each adrenal gland. It produces two hormones—*epinephrine* and a smaller quantity of *norepinephrine* (also called *adrenaline* and *noradrenaline*, respectively)—in response to stress or exercise. These hormones, which are amino acid derivatives, prepare the body for emergency action. They increase heart and respiratory rates, increase blood pressure, cause blood glucose levels to rise, and direct blood flow away from the digestive tract and toward the brain and muscles. They also cause the air passages of the lungs to expand, allowing larger volumes of air to enter and leave the lungs. For this reason, epinephrine is often administered to asthmatics, whose airways become constricted during an asthma attack.

Embryologically, the adrenal medulla is actually part of the nervous system. It is activated by the sympathetic nervous system, which prepares the body for "fight or flight." We will cover this concept in greater depth in Chapter 24.

Hormones Are Also Produced by the Pineal Gland, Thymus, Kidneys, Digestive Tract, Heart, and Fat Cells

The **pineal gland** is located between the two hemispheres of the brain, just above and behind the hypothalamus (see Fig. 23-4). Named for its resemblance to a pine cone, the pineal gland is smaller than a pea. In 1646, philosopher René Descartes described it as "the seat of the rational soul." Since then, scientists have learned more about this organ, but many of its functions are still poorly understood.

The pineal gland produces the hormone *melatonin,* an amino acid derivative. Melatonin is secreted in a daily rhythm, which in mammals is regulated by light falling on the eyes. In some vertebrates, such as the frog, the pineal gland itself contains photoreceptive cells, and the skull above it is thin, so the pineal can detect sunlight and thus day length. By responding to day lengths characteristic of different seasons, the pineal gland appears to regulate the seasonal reproductive cycles of many mammals. Despite years of research, there is considerable uncertainty about the functions of the pineal gland and melatonin in humans. In children, the pineal appears to suppress the onset of puberty by secreting high amounts of melatonin, although no convincing mechanism has been shown. A great deal of evidence suggests that the pineal plays a role in sleep–wake cycles, although how important a role is still unclear. However, in keeping with a role in sleep–wake cycles, melatonin can reduce jet lag when taken just before the "desired" sleep time during long-distance travel.

The **thymus** is located in the chest cavity behind the breastbone. The thymus produces the hormone *thymosin,* which stimulates the reproduction and development of the specialized white blood cells known as T cells, which play crucial roles in the immune response (see pp. 422–423). T cells are formed in the bone marrow and migrate to the thymus, where, under the influence of thymosin and other hormones, they multiply and differentiate into helper T cells and cytotoxic T cells. The thymus is extremely large in infants, but, under the influence of sex hormones, steadily decreases in size after puberty. As a result, aged people produce fewer new T cells each day than adolescents do, and become more susceptible to new diseases.

Many organs that are not primarily endocrine glands also produce hormones. For example, the kidney secretes *erythropoietin,* a peptide hormone that is released when the blood is not transporting enough oxygen. Erythropoietin stimulates the bone marrow to produce more oxygen-carrying red blood cells. Endurance athletes, who need all the oxygen they can get, sometimes attempt to cheat by taking erythropoietin or a synthetic version, called darbepoietin. In the 2002 Winter Olympics, cross-country skiers from Russia and Spain were stripped of their medals when it was discovered that they had taken darbepoietin (see the Chapter 7 case study, "When Athletes Boost Their Blood Counts: Do Cheaters Prosper?").

The stomach and small intestine produce several peptide hormones. As we described in Chapter 21, these hormones help to control digestion and appetite, and may be involved in regulation of body weight (see "Health Watch: Obesity and the Brain–Gut Connection," on page 408).

Even the heart releases a hormone. *Atrial natriuretic peptide* inhibits the release of ADH and aldosterone and increases the excretion of sodium. This heart-derived hormone is released in response to increased blood volume. Excess blood volume overfills the atria, stretching their walls and stimulating the release of atrial natriuretic peptide. By reducing reabsorption of water and salt by the kidneys, atrial natriuretic peptide helps to lower blood volume.

Finally, who would have thought that fat could be an endocrine organ? In 1995, Jeffrey Friedman and colleagues at Rockefeller University discovered just that: The peptide hormone *leptin* is released by fat cells. Mice genetically engineered to lack the gene for leptin became obese (**Fig. 23-13**), and leptin injections caused them to lose weight. The researchers hypothesized that by releasing leptin, fat tissue "tells" the body how much fat it has stored and therefore how much to eat: If the mice already have lots of stored fat, then high leptin levels would cause them

▲ **Figure 23-13 Leptin helps regulate body fat** The obese mouse on the left has been genetically engineered to lack the gene for the hormone leptin.

to eat less. Unfortunately, trials of leptin as a human weight-loss aid have not been encouraging. Many obese people have high levels of leptin but seem to be relatively insensitive to it. However, researchers have discovered additional functions for leptin, which appears to stimulate the growth of new capillaries and to speed wound healing. It also stimulates the immune system and is required for the onset of puberty and the development of secondary sexual characteristics.

Anabolic Steroids—Fool's Gold? Revisited

Like natural testosterone, anabolic steroids help to increase muscle mass. However, naturally occurring hormones are usually present in minuscule amounts. Taking relatively large doses, especially of synthetic hormones that are often much more potent than the natural hormones they mimic, can mean trouble. Even if they're never caught, steroid abusers risk losing their health.

In males, above-normal amounts of testosterone-like anabolic steroids create a negative feedback effect that reduces the natural production of testosterone. Tricked by these testosterone mimics, the anterior pituitary releases smaller amounts of hormones that are required for testes development and sperm production, so the testes often shrink and sperm count drops. Some anabolic steroids can be converted into estrogen, causing partial breast development.

In females, anabolic steroids promote male-like bodily changes, including deepening of the voice, increased facial hair, and even pattern baldness. Testosterone-like hormones also interfere with egg development and ovulation, often causing irregularities in the menstrual cycle. In both sexes, anabolic steroids cause acne and may depress the immune system. Mood swings and sudden aggressiveness are sufficiently common to have produced the slang term "roid rage." Anabolic steroids have been linked to increases in blood pressure and decreases in the "good" form of cholesterol (HDL)—both are risk factors for heart attacks and strokes. Anabolic steroids can cause bone growth to shut down prematurely, so young steroid abusers may never reach their full potential height.

Illegal "designer steroids" also aren't rigorously tested for health and safety. THG, for example, potently activates progesterone receptors at concentrations seven times lower than natural progesterone does. Because progesterone activates sperm, suppresses the immune sys-

tem, and can increase sleepiness, appetite, and fat storage, athletes who take THG may unknowingly expose themselves to myriad harmful effects beyond those attributable to THG's testosterone-like actions.

Anabolic steroids and other performance-enhancing drugs have tarnished the achievements of athletes in many sports. Steroid abusers have been stripped of Olympic medals, seen their world records erased from the record books, and been banned from competition, sometimes for life. To Don Catlin, desecration of the Olympics is especially offensive. "The notion of the Olympic Games to me is the cleanest, purest kind of event ever," he says. "People in every country in the world can compete and the best man or woman crosses the finish line first. What could be worse than to think they are tainted?"

Consider This

Some athletes say that they would willingly risk long-term damage to their bodies in order to win Olympic gold. Others shrug off the dangers, thinking that "It won't happen to me." Some commentators have suggested that there will always be some successful, undetected, drug-abusing athletes. Therefore, they suggest that professional sports should let everyone use any drug they want, in order to level the playing field—if the athletes want to ruin their health, that's their problem. But what about high school and college athletes? Based on a large poll, the Centers for Disease Control and Prevention concluded that about 6% of U.S. high school athletes use anabolic steroids at some time in their careers. Some of these students are as young as 14 or 15 years old, with bodies that are still developing rapidly. Should drug testing in high school sports be routine, not so much to ensure fair competition but to protect the athletes' health? Why or why not?

Chapter Review

Summary of Key Concepts

For additional study help and activities, go to www.mybiology.com.

23.1 How Do Animal Cells Communicate?

Generally, the cells of an animal body communicate with one another by releasing chemical messenger molecules that bind to receptors on other cells, either nearby or far away. The binding of messengers to receptors stimulates responses in the recipient cells.

23.2 How Do Animal Hormones Work?

Hormones are produced by endocrine glands, which are clusters of cells embedded within a network of capillaries. A hormone is secreted into the extracellular fluid and diffuses into the capillaries. The hormone is then transported in the bloodstream to other parts of the body. Although the hormone may potentially contact all of the cells of the body, it evokes responses only in target cells that have appropriate receptors. Hormones are synthesized either from amino acids (amino acid derivatives and peptide hormones) or from lipids (steroid hormones).

Most hormones act on their target cells in one of two ways: (1) Peptide hormones and amino acid derivatives mostly bind to receptors on the surfaces of target cells and activate intracellular second messengers, such as cyclic AMP. The second messengers then alter the metabolism of the cell. (2) Steroid hormones diffuse through the plasma membranes of all cells, but bind to receptors that are present only in the nucleus or cytoplasm of target cells. In the nucleus, the hormone–receptor complex promotes the transcription of specific genes.

Hormone release is commonly regulated through negative feedback, a process in which a hormone causes changes that inhibit further secretion of that hormone. In rare instances, such as childbirth, hormone release may be temporarily controlled by positive feedback, but in all known cases, hormone release is eventually reduced or stopped by negative feedback.

Web Animation Modes of Action of Hormones

23.3 What Are the Structures and Functions of the Mammalian Endocrine System?

The major endocrine glands of the human body are the hypothalamus–pituitary complex, the thyroid gland, the pancreas, the sex organs, and the adrenal glands. Other sources of hormones include the pineal gland, thymus, kidneys, digestive tract, heart, and fat cells. The functions of the hormones released by the major endocrine glands of the human body are summarized in Table 23-1.

Web Animation Hypothalamic Control of the Pituitary

Key Terms

adrenal gland *p. 452*
amino acid-derived hormone *p. 441*
endocrine gland *p. 441*
endocrine system *p. 441*
hormone *p. 441*

hypothalamus *p. 446*
islet cell *p. 449*
negative feedback *p. 444*
neurosecretory cell *p. 446*
ovary *p. 450*
pancreas *p. 449*

peptide hormone *p. 441*
pineal gland *p. 453*
pituitary gland *p. 446*
positive feedback *p. 445*
receptor *p. 440*
second messenger *p. 443*

steroid hormone *p. 441*
target cell *p. 441*
testis (plural, testes) *p. 450*
thymus *p. 453*
thyroid gland *p. 448*

Thinking Through the Concepts

Suggested answers to end-of-chapter and figure-based questions can be found at the end of the text.

Fill-in-the-Blank

1. Hormones are molecules released by cells that are parts of the _____. These cells are embedded in capillary beds, so the hormones enter the bloodstream and move throughout the body. Only specific cells of the body, called _____, can respond to any given hormone, because only these cells bear proteins, called _____, that can bind the hormone.

2. Most hormones fall into three chemical classes: _____, _____, and _____. _____ and _____ hormones are mostly water soluble and bind to receptors on the surfaces of cells. These typically stimulate the synthesis of intracellular molecules called _____, which activate enzymes and change the cell's metabolism. _____ hormones are lipid soluble and bind to receptors in the cytoplasm or nucleus. The hormone–receptor complex typically binds to DNA and causes _____.

3. _____ is a response to a change in conditions that counteracts the change and tends to restore the original condition. _____ is a response to a change in conditions that tends to enhance the change. The release of most hormones is regulated by _____.

4. A part of the brain called the _____ controls the activity of the pituitary gland. Specialized nerve cells in this brain area, called _____, release the hormones _____ or _____ from the endings of their axons in the posterior lobe of the pituitary. Other specialized nerve cells release _____ into capillary beds that drain into the anterior pituitary, where they stimulate or reduce the secretion of hormones by endocrine cells in the anterior pituitary.

5. The major hormones produced by the anterior pituitary gland are _____, _____, _____, _____, _____, and _____.

6. The pancreas releases two hormones that help to regulate sugar metabolism: _____ and _____. _____ is released when blood glucose levels become too high; it causes many cells of the body to take up glucose. When the pancreas produces too little of this hormone, or body cells cannot respond to it, a disorder called _____ results. _____ is released when blood glucose levels become too low; it causes the liver to break down the starch-like storage molecule, _____, and release glucose into the blood.

7. In male mammals, the sex organs, called the _____, release the male sex hormone _____. The sex organs of female mammals, called the _____, release two major female hormones: _____ and _____. The chemical type of all three of these hormones is _____.

8. The adrenal gland consists of two parts, the outer _____ and the inner _____. The outer layer releases three major types of steroid hormones, _____, _____, and _____. The inner layer releases the amino acid-derived hormones _____ and _____.

Review Questions

1. What chemical classes of hormones usually attach to membrane receptors? What cellular events follow?

2. When a hormone attaches to a receptor in the nucleus of a target cell, what cellular events follow? What chemical class of hormone usually attaches to an intracellular receptor?

3. Diagram the process of negative feedback, and give an example of negative feedback in the control of hormone action.

4. What are the major endocrine glands in the human body, and where are they located?

5. Describe the structure and functions of the hypothalamus–pituitary complex.

6. Describe how the hormones of the pancreas act together to regulate the concentration of glucose in the blood.

7. Compare the adrenal cortex and adrenal medulla by answering the following questions: Where are they located within the adrenal gland? Which hormones do they produce? Which organs do their hormones target?

Applying the Concepts

1. IS THIS SCIENCE? A Web site offering a "dietary supplement" consisting of a combination of "human growth hormone-releasing factor" and a component of cell membranes makes the following claims: "Regenerate your skin, muscle, hair, and bones to their former youthful levels; protect yourself from toxins and diseases; restore your potency, libido, and sexuality." Research the following, using the Internet: (a) Are dietary supplements regulated by the U.S. Food and Drug Administration? (b) For dietary supplements, are vendors required to base all claims such as those listed here on solid scientific evidence? (c) Must dietary supplement labels include warnings about possible side effects?

2. Some parents who are interested in college sports scholarships for their children are asking physicians to prescribe growth hormone treatments. What biological and ethical problems do you foresee for parents, children, physicians, coaches, and college scholarship boards if human growth hormone were routinely administered to children who were not deficient in it?

3. Certain industrial and pharmaceutical products are "estrogen mimics" that can produce hormonal effects on wildlife and perhaps people. Many of these products are also extremely useful as plasticizers, detergents, and flame retardants. Do you think that chemicals that mimic the actions of estrogen should be regulated or even banned? Why or why not?

4. Suggest a hypothesis about the endocrine system to explain why many birds lay their eggs in the spring. Why would egg farmers keep lights on at night in their chicken coops?

For additional resources, go to www.mybiology.com.

chapter 24

The Nervous System and the Senses

Love: "A fire sparkling in lovers' eyes . . . a madness most discreet?" Or just the right mixture of chemicals deep within lovers' brains?

Case Study How Do I Love Thee?

*"But soft! What light through yonder window breaks?
It is the east, and Juliet is the sun."*
—*Romeo and Juliet*, Act II, scene II

In Shakespeare's *Romeo and Juliet*, two teenagers fall in love the first time they meet. A few hours later, as Romeo watches Juliet gazing out of her window, he sees her as the sun that lights up his life. *Romeo and Juliet* is one of the finest expressions of the power of romantic love, for which the lovers defy their families, risk their fortunes and their futures, and, ultimately, give up their lives.

Romance is, of course, not the only manifestation of love. A mother's love for her child is just as strong. People have also defined their lives and willingly died for love of God and love of country. Just what *is* love? Are all these types of love distinct or are they related? What happens in the brain when two lovers meet or when a mother cradles her infant?

No one knows for sure—not in people, anyway. Perhaps surprisingly, neuroscientists know a lot about love—or at least monogamy, pair bonding, and sex—in a small rodent called the prairie vole. If Juliet had been a prairie vole, her first encounter with Romeo would have released a flood of oxytocin—the same hormone that causes uterine contractions during childbirth. The oxytocin would have bound to receptors in a tiny part of her brain called the nucleus accumbens, causing nerve cells there to release a chemical called dopamine. Because of the dopamine, she would have felt wonderful and, what's more, she would have linked that euphoric feeling to Romeo. In a "Romeo vole," some of the molecules and brain regions would have differed,

Prairie voles provide insights into the neurochemical basis of emotional love.

but the end result would have been similar: a torrent of dopamine, giving him the ultimate high, and he would have been convinced that he could attain that feeling again only with Juliet. So, the two voles would mate and bond for life. They would build a nest, live together, and raise their young.

How do humans and other animals perceive their world, whether the sun's warmth or the faces and scents of their lovers? How do they evaluate what they perceive, and become calm or excited, fearful or smitten? Finally, how do they respond with appropriate behaviors such as resting, eating, or mating? Although most perceptions and behaviors are not completely understood, the answers to these questions are to be found in the senses and the nervous system. ■

24.1 How Do Nerve Cells Work?

BioFlix How Neurons Work

Our study of the nervous system begins with the individual nerve cell, or **neuron** (**Fig. 24-1**). Although many neurons are similar to the one shown in Figure 24-1, others are very different in structure. Some can be 60 feet long (such as in a blue whale), and some are only a tenth of an inch long, or even less. Some are fairly simple in shape, and others are extremely bushy. Despite this huge variety of structures, most neurons perform the same four functions:

1. Receive information from the internal or external environment or from other neurons.

2. Process this information, often along with information from other sources, and produce an electrical signal.

3. Conduct the electrical signal, sometimes for a considerable distance, to a junction where the neuron meets another cell.

4. Communicate with other cells, including other neurons, or the cells in muscles or glands.

The Functions of a Neuron Are Localized in Separate Parts of the Cell

A "typical" vertebrate neuron has four distinct parts that carry out the four functions just listed: dendrites, the cell body, the axon, and the synaptic terminals (see Fig. 24-1).

Dendrites Respond to Stimuli

Dendrites, branched tendrils that extend outward from the cell body, perform the "receive information" function (**Fig. 24-1**, step ❷ and step ❽). Their branched shape provides a large surface area for receiving signals. Depending on the location and function of the neuron, dendrites may respond to chemicals released by other neurons; to stimuli from the internal environment, such as body temperature, blood pH, or the position of a joint; or to stimuli from the external environment, such as touch, odor, light, or heat.

The Cell Body Processes Signals from the Dendrites

Electrical signals travel down the dendrites to the neuron's **cell body,** which performs the "process information" function (**Fig. 24-1**, step ❸). The cell body adds up, or integrates, signals from the dendrites. If the sum of these signals is sufficiently large and positive, the cell body produces an **action potential**, the electrical output signal of the neuron (**Fig. 24-1**, step ❹). The cell body also contains the organelles that are found in most cells, such as the nucleus, endoplasmic reticulum, and Golgi apparatus, and performs typical cellular activities such as synthesizing proteins and coordinating the cell's metabolism.

Axons Conduct Action Potentials Long Distances

In a typical neuron, a long, thin fiber called an **axon** extends outward from the cell body. The axon conducts action potentials (**Fig. 24-1**, step ❺) from the cell body to the synaptic terminals at the axon's end (see below), where it contacts other cells (**Fig. 24-1**, step ❶ and step ❼). Single axons may stretch from your spinal cord to your toes, a distance of about 3 feet (1 meter), making neurons the longest cells in your body. Some axons are wrapped with insulation called *myelin*, which allows for faster conduction of action potentials (**Fig. 24-1**, step ❻). Myelin is formed from non-neuronal cells that wrap around the axon. Multiple axons are often bundled together into **nerves**, much like bundles of wires in an electrical cable.

At Synapses, Signals Are Transmitted from One Cell to Another

The site where a neuron communicates with another cell is called a **synapse.** A typical synapse consists of (1) the **synaptic terminal,** which is a swelling at the end of the axon of the "sending" neuron; (2) a dendrite or cell body of a "receiving" neuron, gland, or muscle cell; and (3) a small gap that separates the two (**Fig. 24-1**, step ❼; see also Fig. 24-3 later in this chapter). Most synaptic terminals contain a chemical called a **neurotransmitter** that is released in response to an action potential reaching the terminal. At the synapse, the output of the first cell becomes the input to the second cell.

24.2 How Do Nerve Cells Produce and Transmit Signals?

Although there are many exceptions, as a general rule, information is carried *within* a neuron by electrical signals, whereas information is transmitted *between* neurons by chemicals that are released from one neuron and received by a second neuron.

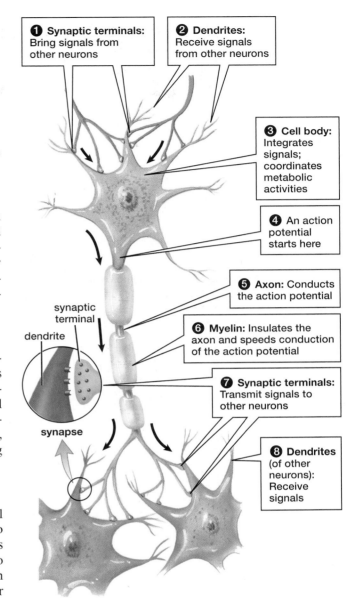

❶ **Synaptic terminals:** Bring signals from other neurons

❷ **Dendrites:** Receive signals from other neurons

❸ **Cell body:** Integrates signals; coordinates metabolic activities

❹ An action potential starts here

❺ **Axon:** Conducts the action potential

❻ **Myelin:** Insulates the axon and speeds conduction of the action potential

❼ **Synaptic terminals:** Transmit signals to other neurons

❽ **Dendrites** (of other neurons): Receive signals

synaptic terminal

dendrite

synapse

▲ **Figure 24-1 A neuron, showing its specialized parts and their functions**

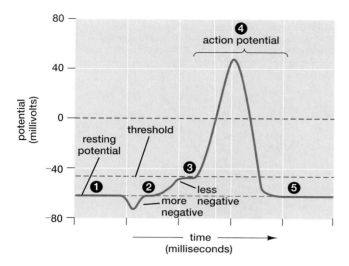

▲ **Figure 24-2 Electrical events in a neuron ❶** A neuron maintains a voltage across its plasma membrane called the resting potential, about –60 mV with respect to the outside. **❷** Stimulation from the environment or other cells may make the neuron more negative (downward deflection) or less negative (upward deflection). **❸** If the potential becomes about 10 to 20 mV less negative, the neuron reaches threshold, and **❹** produces a brief positive potential called an action potential. **❺** After a millisecond or two, the voltage across the neuron's plasma membrane returns to the resting potential.

How Do I Love Thee?

Continued

Different animal species, and even different individuals within a species, have very different numbers and types of receptors for neurotransmitters. Monogamous, pair-bonding prairie voles, for example, have a very high density of receptors for dopamine, the "reward" neuro-transmitter, in certain parts of their brains. Promiscuous montane voles, in contrast, have much lower concentra-tions of dopamine receptors in these same brain regions. In humans, scientists have found that people with social phobias, who avoid many social situations for fear of embarrassment, have fewer dopamine recep-tors in reward areas of the brain than people who are good at social interactions.

Electrical Signals Convey Information Within a Single Neuron

In the 1930s, biologists developed ways to record electrical events inside individual neurons. They found that unstimulated, inactive neurons maintain a constant electri-cal voltage, or *potential*, across their plasma membranes, similar to the voltage across the poles of a battery. This voltage, called the **resting potential**, is always negative inside the cell and ranges from –40 to –90 millivolts (mV; thousandths of a volt).

When a neuron is stimulated, the negative potential inside the neuron changes. It can become more negative or less negative. If the potential becomes sufficiently less negative, it reaches a level called the **threshold** and triggers an action potential (**Fig. 24-2**). During an action potential, the voltage inside the neuron rapidly rises to about +50 mV. Action potentials last a few milliseconds (thousandths of a second) before the cell's negative resting potential is restored. The plasma membranes of axons are specialized to conduct action potentials from the cell body to the synap-tic terminals. Unlike electrical voltages in metal wires, which decrease with dis-tance, action potentials are conducted from cell body to axon terminal—as far as 60 feet in a blue whale—with no change in voltage.

Neurotransmitter Chemicals Allow Neurons to Communicate with One Another at Synapses

A synapse includes the synaptic terminal of one neuron, called the **presynaptic neuron;** an area of specialized plasma membrane on the surface of a second neu-ron, called the **postsynaptic neuron;** and a tiny gap between the two (**Fig. 24-3**).

The synaptic terminal at the end of an axon contains dozens of membrane-enclosed sacs, called vesicles, each full of neurotransmitter molecules. When an ac-tion potential traveling down an axon (**Fig. 24-3**, step **❶**) reaches the synaptic terminal (**Fig. 24-3**, step **❷**), the inside of the terminal becomes positively charged. This causes vesicles to release neurotransmitter into the gap between the cells (**Fig. 24-3**, step **❸**). The outer surface of the plasma membrane of the postsynaptic neuron, just across the gap, is packed with receptor proteins. The neurotransmitter molecules diffuse across the gap and bind to these receptors (**Fig. 24-3**, step **❹**).

Excitatory and Inhibitory Potentials Are Produced at Synapses

The binding of neurotransmitter molecules to receptors on the postsynaptic neu-ron opens pores, or channels, in the postsynaptic plasma membrane, causing a small, brief change in voltage: the **postsynaptic potential (PSP).** Depending on the type of channels that open and the type of ions that flow, PSPs can be either *excitatory* (EPSPs) or *inhibitory* (IPSPs) (**Fig. 24-3**, step **❺**). EPSPs make the neu-ron less negative inside and thus more likely to produce an action potential. IPSPs make the neuron more negative inside and less likely to produce an action potential (see Fig. 24-2).

Postsynaptic potentials travel to the cell body, where they determine whether an action potential will be produced. How? The dendrites and cell body of a sin-gle neuron may receive EPSPs and IPSPs from the synaptic terminals of thou-sands of presynaptic neurons. The PSPs that reach the postsynaptic cell body at approximately the same time are "added up," or integrated. If the excitatory and inhibitory potentials, when added together, raise the voltage inside the neuron above threshold, the postsynaptic cell produces an action potential.

Neurotransmitter Action Is Usually Brief

Consider what would happen if a presynaptic neuron started stimulating a postsy-naptic cell and never stopped. You might, for example, contract your biceps mus-cle, flex your arm, and have it stay flexed forever! Not surprisingly, the nervous system has several ways of terminating neurotransmitter action (**Fig. 24-3**, step **❻**). Some neurotransmitters—notably, acetylcholine, the transmitter that stimulates

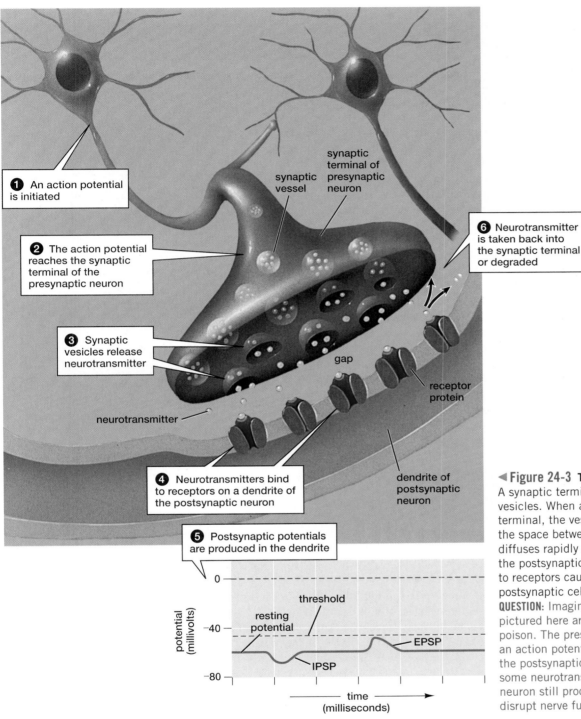

① An action potential is initiated

② The action potential reaches the synaptic terminal of the presynaptic neuron

③ Synaptic vesicles release neurotransmitter

synaptic vessel

synaptic terminal of presynaptic neuron

⑥ Neurotransmitter is taken back into the synaptic terminal or degraded

gap

receptor protein

neurotransmitter

④ Neurotransmitters bind to receptors on a dendrite of the postsynaptic neuron

dendrite of postsynaptic neuron

⑤ Postsynaptic potentials are produced in the dendrite

0

threshold

resting potential

potential (millivolts)

−40

EPSP

IPSP

−80

time (milliseconds)

◀ **Figure 24-3 The structure and function of the synapse**
A synaptic terminal contains many neurotransmitter-filled vesicles. When an action potential enters the synaptic terminal, the vesicles release their neurotransmitter into the space between the neurons. The neurotransmitter diffuses rapidly across the gap and binds to receptors on the postsynaptic cell. In many cases, a transmitter binding to receptors causes a change in the resting potential of the postsynaptic cell, called a postsynaptic potential (PSP). **QUESTION:** Imagine an experiment in which the neurons pictured here are bathed in a solution containing a nerve poison. The presynaptic neuron is stimulated and produces an action potential, but this does not result in a PSP in the postsynaptic neuron. When the experimenter adds some neurotransmitter to the synapse, the postsynaptic neuron still produces no PSP. How does the poison act to disrupt nerve function?

skeletal muscle cells—are rapidly broken down by enzymes in the synapse. Many other neurotransmitters are transported back into the presynaptic neuron.

The Nervous System Uses Many Neurotransmitters

During the past few decades, researchers have become increasingly aware that the neurons of the brain synthesize and respond to a vast array of chemicals, including many hormones. At least 50 neurotransmitters have been identified, and the list is growing. In Table 24-1, we list a few well-known neurotransmitters and some of their functions. In "Health Watch: Neurotransmitters, Drugs, and Disease" on p. 462, we explore the role of neurotransmitters in addiction and neurological diseases.

health watch

Neurotransmitters, Drugs, and Diseases

Chances are, you know someone who is addicted. How can substances such as nicotine, alcohol, and cocaine so profoundly influence people's lives? A big part of the answer lies in the effects of these drugs on neurotransmitters and in how the nervous system adapts to those insidious effects.

Addictive drugs activate the brain's "reward circuitry," creating feelings of intense pleasure. Cocaine is a good example. Synapses in the brain that use the neurotransmitters dopamine, serotonin, or norepinephrine contribute to our energy level and our overall sense of well-being. Normally, the presynaptic neuron, after releasing one of these neurotransmitters, immediately starts pumping it back in, thus limiting its effects.

Researchers have found that cocaine works by blocking the pump. The result? When a person takes cocaine, these neurotransmitters remain in their synapses much longer and reach higher concentrations than normal, so their effects are enhanced. The user feels euphoric and energetic. However, eventually the brain attempts to restore the normal state. During repeated cocaine use, the postsynaptic neurons decrease their number of receptors for these neurotransmitters. When fewer receptors are present, the high levels of neurotransmitter caused by cocaine are now *required*, just for the user to feel normal. If cocaine is withdrawn, the postsynaptic neurons are inadequately stimulated, and the user experiences an emotional "crash" that can be relieved only by more cocaine. Over time, more and more cocaine is needed just to feel okay, let alone to get high—the user has become an addict (**Fig. E24-1**).

Alcohol stimulates receptors that produce inhibitory postsynaptic potentials, and blocks receptors that produce excitatory postsynaptic potentials. When a person drinks frequently, the brain compensates by decreasing the inhibitory receptors and increasing the excitatory receptors. Without alcohol, an alcoholic feels jittery and nervous—in short, overstimulated. In extreme cases, withdrawal from alcohol can cause convulsions. Nicotine and other components of cigarette smoke also interfere with normal synaptic transmission, producing a variety of addictive effects. To overcome addictions, drug users must undergo the misery caused by a nervous system that is deprived of a drug to which it has adapted. Although receptors eventually return to normal levels, for unknown reasons, drug cravings often recur periodically.

You may also know someone with Parkinson's or Alzheimer's disease. Both diseases are caused by the death of specific neurons in the brain and the loss of their neurotransmitters, which normally communicate with other neurons. In Parkinson's disease, dopamine-releasing neurons in the midbrain die, interfering with the complex control system that underlies smooth movements. Parkinson's patients experience tremors and have difficulty initiating movement. In Alzheimer's disease, neurons in the temporal lobes that produce the neurotransmitter acetylcholine die in large numbers. Memory loss is a prominent symptom of Alzheimer's.

Health experts estimate that as many as 5% of Americans suffer from depression. In many cases, depression is caused by too little serotonin in the

▲ **Figure E24-1 Drug addiction** An addict experiences extreme physical and emotional distress without his or her drug because the nervous system has adapted to it.

brain. Prozac® and several other antidepressants selectively block the re-uptake of serotonin into presynaptic neurons, enhancing the neurotransmitter's effects. The drug Ecstasy causes a temporary, massive increase in serotonin in synapses. Users report feelings of pleasure, increased energy, heightened sensory awareness, and improved rapport with other people. Increasing evidence from both animal and human research suggests that Ecstasy users may incur long-term damage to serotonin-producing neurons, and they may suffer from deficits in learning and memory.

Table 24-1 Some Important Neurotransmitters

Neurotransmitter	Location in the Nervous System	Some Functions
Acetylcholine	Motor neuron-to-muscle synapse; autonomic nervous system, many areas of the brain	Activates skeletal muscles; activates target organs of the parasympathetic nervous system
Dopamine	Midbrain	Important in emotional rewards and the control of movement
Norepinephrine (noradrenaline)	Sympathetic nervous system	Activates target organs of the sympathetic nervous system
Serotonin	Midbrain, pons, and medulla	Influences mood, sleep
Glutamate	Many areas of the brain and spinal cord	Major excitatory neurotransmitter in the CNS
Glycine	Spinal cord	Major inhibitory neurotransmitter in the spinal cord
GABA (gamma amino butyric acid)	Throughout brain	Major inhibitory neurotransmitter in the brain
Endorphins	Many areas of the brain and spinal cord	Influence mood, reduce pain sensations
Nitric oxide	Many areas of the brain	Important in forming memories

24.3 How Do Nervous Systems Process Information?

The individual neuron uses a language of action potentials, yet somehow this basic language allows even simple animals to perform a variety of complex behaviors. One key to the versatility of the nervous system is the presence of networks of neurons that range from dozens to billions of cells. As in computers, simple elements can perform amazing feats when connected properly.

Information Processing in the Nervous System Requires Four Basic Operations

At a minimum, a nervous system must perform four operations:

1. Determine the nature of a stimulus.
2. Determine and signal the intensity of a stimulus.
3. Integrate information from many sources.
4. Initiate and direct appropriate responses.

Let's examine each of these operations.

The Nature of a Stimulus Is Determined by Connections Between the Senses and the Brain

If action potentials are the units of information of neurons, and if all action potentials are basically the same, then how can the brain of a pond snail or a person determine *what* a stimulus is—light, sound, or odor—or *how strong* a stimulus is?

All nervous systems interpret what a stimulus is by monitoring which neurons are firing action potentials. For example, your brain interprets action potentials that occur in the axons of your optic nerves (originating in the eyes and traveling to the visual areas of the brain) as the sensation of light. Supposedly, a German physiologist once sat in a dark room and poked himself in the eye, slightly damaging his retina and producing action potentials that traveled to his brain. (As they say in TV ads showing cars speeding on racetracks, do *not* try this yourself!) The result? He "saw stars" because his brain interpreted *any* action potential in his optic nerve as light. Thus, you can easily distinguish the sound of music from the taste of coffee, or the bitterness of coffee from the sweetness of sugar, because these different stimuli normally result in action potentials in different axons that end up in different areas of your brain.

The Intensity of a Stimulus Is Coded by the Frequency of Action Potentials

All action potentials are roughly the same size and duration, so information about the intensity of a stimulus (for example, the loudness of a sound) cannot be encoded in a single action potential. Instead, intensity is coded in two other ways (**Fig. 24-4**). First, intensity can be signaled by the frequency of action potentials in a single neuron. The more intense the stimulus, the faster the neuron produces action potentials, or *fires*. Second, most nervous systems have many neurons that can respond to the same input. Stronger stimuli tend to excite more of these neurons, whereas weaker stimuli excite fewer, so intensity can also be signaled by the number of similar neurons that fire at the same time. Thus, a gentle touch may cause a single touch receptor in your skin to fire action potentials very slowly (**Fig. 24-4a**), whereas a hard poke may cause multiple touch receptors to fire, some very rapidly (**Fig. 24-4b**).

The Nervous System Processes Information from Many Sources

Your brain is continuously bombarded by sensory stimuli that originate both inside and outside the body. The brain must evaluate these inputs, determine which ones are important, and decide how to respond. Nervous systems, like individual

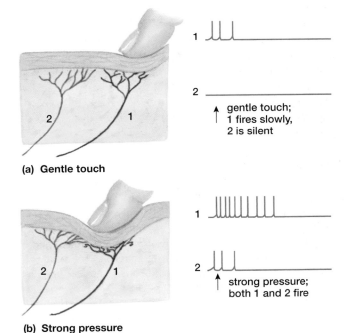

(a) **Gentle touch**

1 ‖‖

2 ———

↑ gentle touch;
1 fires slowly,
2 is silent

(b) **Strong pressure**

1 ‖‖‖‖‖‖‖‖

2 ‖‖‖

↑ strong pressure;
both 1 and 2 fire

▲ **Figure 24-4 Signaling the intensity of a stimulus** The intensity of a stimulus is signaled by the rate at which individual sensory neurons produce action potentials and by the number of sensory neurons activated. In this example, increasing pressure on the skin first causes faster firing and then causes an adjacent receptor to be activated. **QUESTION:** On the basis of this information, in what way would you predict that skin areas that are especially sensitive to touch differ from less sensitive areas?

neurons, integrate information from many sources: A large number of neurons may all funnel their signals to fewer neurons. For example, many sensory neurons may converge onto a smaller number of brain cells. Some of these brain cells act as "decision-making cells," adding up the postsynaptic potentials that result from the synaptic activity of the sensory neurons, and changing their activity accordingly.

The Nervous System Produces Outputs to Muscles and Glands

Action potentials from the decision-making neurons may travel to other parts of the brain, to the spinal cord, or to the sympathetic and parasympathetic nervous systems (described later in this chapter). Ultimately, the output of the nervous system will stimulate activity in the muscles or glands that actually produce observable behaviors.

24.4 How Are Nervous Systems Organized?

Most behaviors are controlled by neuron-to-muscle pathways composed of four elements:

1. **Sensory neurons,** which respond to a stimulus, either internal or external to the body.
2. **Interneurons,** which receive signals from sensory neurons, hormones, neurons that store memories, and many other sources. Based on this input, interneurons often activate motor neurons.
3. **Motor neurons,** which receive instructions from sensory neurons or interneurons and activate muscles or glands.
4. **Effectors,** usually muscles or glands that perform the response directed by the nervous system.

Simple behaviors, such as reflexes (see pp. 468–469), may be controlled by activity in as few as two or three neurons ultimately stimulating a single muscle. Complex behaviors are organized by interconnected neural pathways, in which several types of sensory input (along with memories, hormones, and other factors) converge on a set of interneurons. By integrating the postsynaptic potentials from multiple sources, the interneurons "decide" what to do and stimulate motor neurons to direct the appropriate activity in muscles and glands.

Complex Nervous Systems Are Centralized

In the animal kingdom, there are really only two nervous system designs (Fig. 24-5): a diffuse nervous system, such as that of cnidarians (*Hydra*, jellyfish, and their relatives), and a centralized nervous system, found to varying degrees in more complex organisms.

Not surprisingly, nervous system design is highly correlated with the animal's lifestyle. Radially symmetrical cnidarians have no "front end," so there has been no evolutionary pressure to concentrate the senses in one place. A *Hydra* sits anchored to an underwater stone or twig, and prey or predators are equally likely to come from

▼ **Figure 24-5 Nervous system organization** *(a)* The diffuse nervous system of *Hydra* contains a few concentrations of neurons, particularly at the bases of tentacles, but no brain. Neural signals are conducted in virtually all directions throughout the body. *(b)* The flatworm has a nervous system that is less diffuse, with a cluster of ganglia in the head. *(c)* The octopus has a large, complex brain and learning capabilities rivaling those of some mammals.

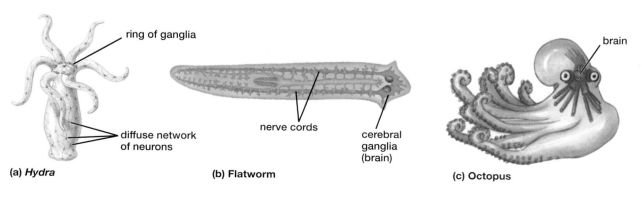

ring of ganglia

diffuse network
of neurons

nerve cords

cerebral
ganglia
(brain)

brain

(a) Hydra   **(b) Flatworm**   **(c) Octopus**

any direction. Cnidarian nervous systems are composed of a network of neurons, often called a *nerve net*, woven through the animal's tissues (**Fig. 24-5a**). Here and there we find a cluster of neurons, called a ganglion, but nothing resembling a real brain.

Almost all other animals are bilaterally symmetrical, with definite head and tail ends. Because the head is usually the first part of the body to encounter food, danger, and potential mates, it is advantageous to have sense organs concentrated there. Sizable ganglia evolved that integrate the information gathered by the senses and initiate appropriate action. Over evolutionary time, the major sense organs of animals with increasingly complex nervous systems became localized in the head, and the ganglia became centralized into a brain. This trend, called *cephalization*, is clearly seen in the invertebrates (**Fig. 24-5b, 24-5c**). Cephalization reaches its peak in the vertebrates, in which nearly all the cell bodies of the nervous system are localized in the brain and spinal cord.

24.5 What Is the Structure of the Human Nervous System?

The nervous systems of all mammals, including humans, can be divided into two parts—central and peripheral—each of which has further subdivisions (**Fig. 24-6**). The **central nervous system (CNS)** consists of the **brain** and **spinal cord.** The **peripheral nervous system (PNS)** consists of neurons that lie outside the CNS and the axons that connect these neurons with the CNS.

The Peripheral Nervous System Links the Central Nervous System to the Rest of the Body

The nerves of the peripheral nervous system link the brain and spinal cord to muscles, sensory organs, and the organs of the digestive, respiratory, excretory, and circulatory systems. The peripheral nerves contain axons of sensory neurons that bring

◄ **Figure 24-6 The organization and functions of the mammalian nervous system**

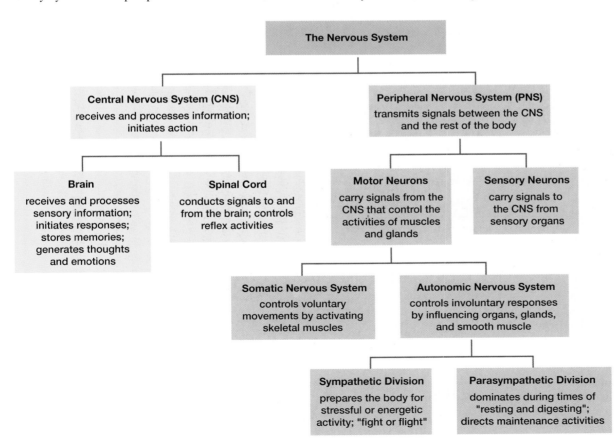

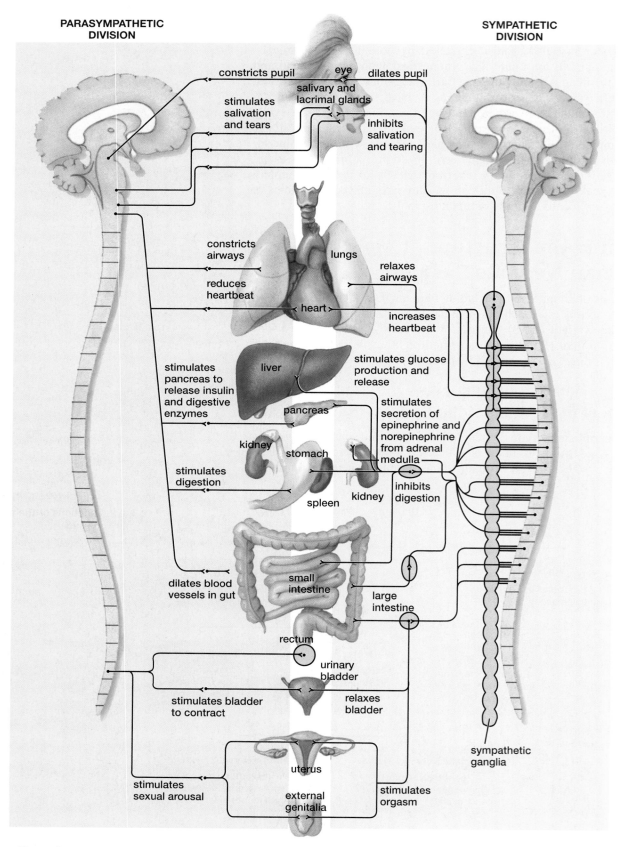

PARASYMPATHETIC DIVISION

SYMPATHETIC DIVISION

constricts pupil

eye

dilates pupil

stimulates salivation and tears

salivary and lacrimal glands

inhibits salivation and tearing

constricts airways

lungs

relaxes airways

reduces heartbeat

heart

increases heartbeat

stimulates pancreas to release insulin and digestive enzymes

liver

stimulates glucose production and release

pancreas

stimulates secretion of epinephrine and norepinephrine from adrenal medulla

kidney

stomach

stimulates digestion

spleen

kidney

inhibits digestion

dilates blood vessels in gut

small intestine

large intestine

rectum

urinary bladder

stimulates bladder to contract

relaxes bladder

uterus

stimulates sexual arousal

external genitalia

stimulates orgasm

sympathetic ganglia

▲ **Figure 24-7 The autonomic nervous system** The autonomic nervous system has two divisions, sympathetic and parasympathetic, which supply nerves to many of the same organs but generally produce opposite effects. Activation of the autonomic nervous system is commanded by signals from the hypothalamus, medulla, and pons.

sensory information *to* the central nervous system from all parts of the body. Peripheral nerves also contain the axons of motor neurons that carry signals *from* the central nervous system to the organs and muscles. The motor portion of the peripheral nervous system, through which the brain controls the body, can be subdivided into two parts: the somatic nervous system and the autonomic nervous system.

The Somatic Nervous System Controls Voluntary Movement

Motor neurons of the **somatic nervous system** form synapses on skeletal muscles and control voluntary movement. As you take notes, lift a coffee cup, or adjust your iPod, your somatic nervous system is in charge. The cell bodies of somatic motor neurons are located in the spinal cord, and their axons go directly to the muscles they control.

The Autonomic Nervous System Controls Involuntary Actions

Motor neurons of the **autonomic nervous system** send signals to the heart, smooth muscle, and glands, and produce mostly involuntary responses. The autonomic nervous system is controlled by the hypothalamus, medulla, and pons, parts of the brain that are described later in this chapter. It consists of two divisions: the *sympathetic division* and the *parasympathetic division*. The two divisions of the autonomic nervous system affect most of the same organs, but usually produce opposite effects (Fig. 24-7).

The neurons of the sympathetic division release the neurotransmitter norepinephrine (noradrenaline) onto their target organs, preparing the body for stressful or energetic activity, such as fighting, escaping, or taking an exam. During such "fight or flight" activities, the sympathetic nervous system reduces activity of the digestive tract, redirecting some of its blood supply to the muscles of the arms and legs. The heart rate accelerates. The pupils of the eyes open wider, admitting more light, and the air passages in the lungs expand, letting in more air.

The neurons of the parasympathetic division release acetylcholine onto their target organs. The parasympathetic division dominates during maintenance activities that can be carried on at leisure, often called "rest and digest." Under its control, the digestive tract becomes active, the heart rate slows, and air passages in the lungs constrict because less oxygen is needed.

The Central Nervous System Consists of the Spinal Cord and Brain

The spinal cord and brain make up the central nervous system (CNS). The CNS receives and processes sensory information, generates thoughts, and directs responses. It consists primarily of interneurons—perhaps as many as 100 billion of them! The brain and spinal cord are protected by a bony armor, consisting of the *skull*, which surrounds the brain, and the backbone, or *vertebral column*, a chain of bones (the vertebrae) that protects the spinal cord. Beneath the bones lies a triple layer of connective tissue called the *meninges*. Between the layers of the meninges, the cerebrospinal fluid, a clear, lymphlike liquid, cushions the brain and spinal cord and nourishes the cells of the CNS.

The Spinal Cord Controls Many Reflexes and Conducts Information to and from the Brain

About as thick as your little finger, the spinal cord (Fig. 24-8; see also Fig. 24-7) extends from the base of the brain to the lower back. Nerves carrying axons of

▼ **Figure 24-8 The spinal cord** In cross section, the spinal cord has an outer region of myelin-covered axons (white matter) that travel to and from the brain and an inner, butterfly-shaped region of dendrites and the cell bodies of interneurons and motor neurons (gray matter). The cell bodies of sensory neurons lie outside the cord in the dorsal root ganglion. The peripheral (spinal) nerves arise from the merger of the dorsal root (containing axons of sensory neurons) and the ventral root (containing axons of motor neurons).

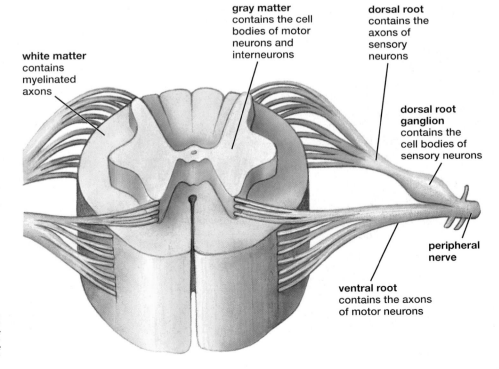

gray matter contains the cell bodies of motor neurons and interneurons

dorsal root contains the axons of sensory neurons

white matter contains myelinated axons

dorsal root ganglion contains the cell bodies of sensory neurons

peripheral nerve

ventral root contains the axons of motor neurons

sensory neurons emerge from the dorsal (back) part of the spinal cord, and nerves carrying axons of motor neurons emerge from the ventral (front) part. The nerves from the two sides of the spinal cord merge to form the peripheral nerves that innervate most of the body.

In the center of the spinal cord is a butterfly-shaped area of *gray matter* (see Fig. 24-8). Gray matter consists of the cell bodies of several types of neurons that control voluntary muscles and the autonomic nervous system, plus neurons that communicate with the brain and other parts of the spinal cord. The gray matter is surrounded by *white matter*, formed of myelin-coated axons of neurons that extend up or down the spinal cord. These axons carry sensory signals from internal organs, muscles, and the skin up to the brain and carry signals from the brain that direct the motor portions of the peripheral nervous system.

Reflex Pathways Include Neurons in the Spinal Cord

The simplest type of behavior in animals is the **reflex,** a largely involuntary movement of a body part in response to a stimulus. Reflexes occur without involving conscious portions of the brain; in vertebrates, many are controlled entirely by the spinal cord and peripheral neurons. Examples of human reflexes include the familiar knee-jerk and pain-withdrawal reflexes, both of which are produced by neurons in the spinal cord. Although reflexes like these do not require the brain, other pathways inform the brain of pricked fingers and may trigger other, more complex behaviors (cursing, for example!).

Let's examine the simple pain-withdrawal reflex, which involves neurons of both the central and the peripheral nervous systems (**Fig. 24-9**). If you lean your hand on a tack, the tissue damage activates pain sensory neurons (**Fig. 24-9**, step ❶; see also Section 24-7). The axons of these pain sensory neurons travel up a peripheral nerve and enter the spinal cord through a dorsal root (**Fig. 24-9**, step ❷). The pain sensory neuron stimulates an interneuron, which stimulates a motor neuron, both located within the gray matter of the spinal cord (**Fig. 24-9**, step ❸; in this diagram, you can see that interneurons are literally "between other neurons," in this case between a sensory neuron and a motor neuron). The motor neuron sends an axon out the ventral root,

Web Animation Reflex Arc

▶ **Figure 24-9 The pain-withdrawal reflex** This simple reflex circuit begins with a sensory neuron that has pain-sensitive endings in the skin and a long axon leading to the spinal cord. This sensory neuron stimulates an interneuron in the spinal cord, which in turn stimulates a motor neuron in the cord, causing withdrawal of the injured body part. Some interneurons also send a signal to the brain, informing it of the painful event. **QUESTION:** Why does a paralyzed victim of a spinal cord injury, when pricked with a pin on a paralyzed part of his or her body, often exhibit a normal pain-withdrawal reflex but feel no pain?

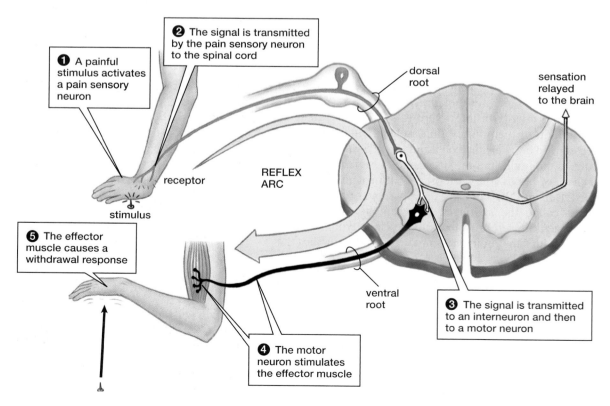

❶ A painful stimulus activates a pain sensory neuron

❷ The signal is transmitted by the pain sensory neuron to the spinal cord

dorsal root

sensation relayed to the brain

receptor

REFLEX ARC

stimulus

❺ The effector muscle causes a withdrawal response

ventral root

❸ The signal is transmitted to an interneuron and then to a motor neuron

❹ The motor neuron stimulates the effector muscle

through a peripheral nerve, and stimulates a skeletal muscle (**Fig. 24-9**, step **④**) that contracts, withdrawing your hand away from the tack (**Fig. 24-9**, step **⑤**).

Many spinal cord interneurons also have axons that extend up to the brain (see Fig. 24-9). Action potentials carried along these axons alert the brain to the painful event. The brain, in turn, sends action potentials down axons in the white matter to interneurons and motor neurons in the gray matter. These signals can modify spinal reflexes. With enough training or motivation, you can suppress the pain-withdrawal reflex. To rescue a child from a burning crib, for example, you could reach into the flames.

Some Complex Actions Are Coordinated by the Spinal Cord

In addition to simple reflexes, the wiring for some fairly complex activities also resides within the spinal cord. All the neurons and interconnections needed for the basic movements of walking and running, for example, are contained in the spinal cord. The advantage of this semi-independent arrangement between brain and spinal cord is probably an increase in speed and coordination, because messages do not have to travel all the way up to the brain and back down again merely to swing one leg forward while walking. The brain's role in these "semi-automatic" behaviors is to initiate, guide, and modify the activity of spinal motor neurons, based on conscious decisions (Where are you going? How fast should you walk?). To maintain balance, the brain also uses sensory input from the muscles to command motor neurons to adjust the way the muscles move.

The Brain Consists of Many Parts Specialized for Specific Functions

All vertebrate brains have the same general structure, consisting of three parts: the hindbrain, midbrain, and forebrain (**Fig. 24-10a**). In nonmammalian vertebrates, the three divisions remain prominent (**Fig. 24-10b, c**). However, in mammals—particularly in humans—the brain regions are significantly modified. Some, such as the midbrain, are reduced in size; others, especially the cerebrum in the forebrain, are greatly enlarged (**Fig. 24-10d, e**). The major structures of the human brain are shown in **Fig. 24-11**.

The Hindbrain Controls Breathing, Sleep, and Movement

In humans, the **hindbrain** contains the medulla, the pons, and the cerebellum (**Fig. 24-11a**). In both structure and function, the *medulla* is very much like an enlarged extension of the spinal cord, with neuron cell bodies at its center, surrounded by a layer of myelin-covered axons. It controls several automatic functions, such as breathing, heart rate, blood pressure, and swallowing. Certain neurons in the *pons*, located above the medulla, appear to influence transitions between sleep and wakefulness and between stages of sleep. Other neurons influence the rate and pattern of breathing. The *cerebellum* is crucial in coordinating movements of the body. It receives information from command centers in the areas of the forebrain that control movement and from position sensors in muscles and joints. By comparing what the command centers order with information from the position sensors, the cerebellum guides smooth, accurate motions and body positioning. The cerebellum is also involved in learning and memory storage for behaviors. As you take notes in class or play the piano, your cerebellum is instructing your forebrain about the order and timing of muscle movements in your hand.

▶ **Figure 24-10 A comparison of vertebrate brains (a)** The brains of modern vertebrate embryos, thought to be similar to the brains of the distant ancestors of today's vertebrates, consist of three distinct regions: the forebrain (tan), midbrain (blue), and hindbrain (pink). **(b)** The brain of an adult shark maintains this basic organization. **(c)** In the goose, the midbrain is reduced, and the cerebrum and cerebellum are larger. **(d, e)** In mammals, especially humans, the cerebrum is very large compared to the other brain regions.

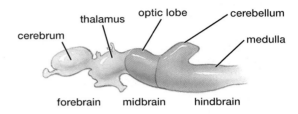

(a) Embryonic vertebrate brain

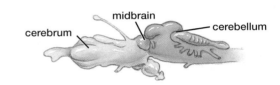

(b) Shark brain

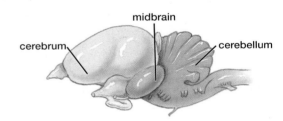

(c) Goose brain

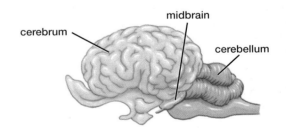

(d) Horse brain

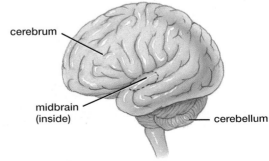

(e) Human brain

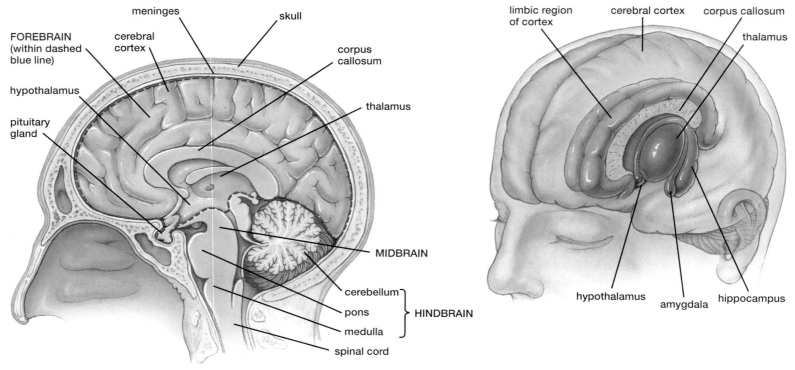

(a) A section through the human brain

(b) The limbic system

▲ **Figure 24-11 The human brain** *(a)* A section through the midline of the human brain reveals some of its major structures. *(b)* In this view of the limbic system and the thalamus, the left cerebral cortex is transparent.

How Do I Love Thee?

Continued

Dopamine-containing neurons in the midbrain send axons to several parts of the forebrain, such as the frontal cortex and the limbic system, which includes the amygdala and hippocampus. When people fall in love, dopamine release in these areas contributes to emotional attachment, lack of fear and critical judgment of the loved one, and fond memories of times spent together.

Web Animation Brain Structure

The Midbrain Filters Input

The **midbrain** is quite small in humans (**Fig. 24-11a**). It contains an auditory relay center and a center that controls reflex movements of the eyes. Another important relay system, the *reticular formation*, passes through it. The neurons of the reticular formation extend all the way from the central core of the medulla up into lower regions of the forebrain. The reticular formation receives input from virtually every sense, from every part of the body, and from many areas of the brain as well. Your reticular formation filters sensory inputs before they reach the conscious regions of the brain, so you can read and concentrate in the presence of such distracting stimuli as music and the smell of coffee. The fact that a mother awakens upon hearing the faint cry of her infant but sleeps through loud traffic noise outside the window testifies to the effectiveness of the reticular formation in screening inputs to the brain.

The Forebrain Gives Us Our Emotions, Thoughts, and Perceptions

The **forebrain,** also called the *cerebrum*, includes the thalamus, the limbic system, and the cerebral cortex (see Fig. 24-11).

The *thalamus* is a complex relay station that channels sensory information from all parts of the body to the limbic system and cerebral cortex. Signals from the cerebellum that allow the cerebral cortex to direct coordinated movements are also channeled through the thalamus.

The *limbic system* is a diverse group of structures located in an arc between the thalamus and cerebral cortex. These structures work together to produce our most basic emotions and drives, including fear, rage, tranquillity, hunger, thirst, pleasure, and sexual responses. Portions of the limbic system are also important in the formation of memories. The limbic system includes the hypothalamus, the amygdala, and the hippocampus as well as nearby regions of the cerebral cortex (**Fig. 24-11b**).

The *hypothalamus* contains many clusters of neurons. Some of these release hormones into the blood, and others control the release of a variety of hormones from the pituitary gland. Other regions of the hypothalamus direct the activities of the autonomic nervous system. The hypothalamus, through its hormone production and neural connections, acts as a major coordinating center, controlling body temperature, food intake, water balance, the menstrual cycle, and the sleep–wake cycle.

Stimulating different parts of the *amygdala* in experimental animals produces behaviors characteristic of pleasure, fear, rage, and sexual arousal. Conscious people whose amygdalas have been electrically stimulated report feelings of rage or fear. Damage to the amygdala early in life eliminates the ability both to feel fear and to recognize fearful facial expressions in other people.

Electrical stimulation of the *hippocampus* also elicits emotional responses, including rage and sexual arousal. The hippocampus is also important in the formation of long-term memory and thus plays a role in learning.

In humans, the largest part of the brain by far is the *cerebral cortex*, the outer layer of the forebrain (see Fig. 24-11a). The cerebral cortex and underlying parts of the forebrain are divided into two halves, called *cerebral hemispheres*. These halves communicate with each other by means of a large band of axons, the *corpus callosum* (see Fig. 24-11b). The cerebral cortex is the most sophisticated information-processing center known, with billions of neurons packed into a sheet just a few millimeters thick. The cortex is folded into *convolutions*, which greatly increase its area. In the cortex, cell bodies of neurons predominate, giving this outer layer—in preserved specimens—a gray color; hence the term "gray matter." These neurons receive sensory information, process it, create memories for future use, direct voluntary movements, and allow us to plan and think, using mechanisms that neurobiologists do not fully understand.

The cerebral cortex is divided into four anatomical regions: the *frontal*, *parietal*, *occipital*, and *temporal lobes* (**Fig. 24-12**). Functionally, the cortex contains *primary sensory areas*, regions where signals originating in sensory organs such as the ears and eyes are received and converted into subjective impressions—for instance, sound and light. Nearby *association areas* interpret the sounds as speech or music, for example, and the visual stimuli as recognizable objects or the words on this page. Association areas also link the stimuli with memories stored in the cortex and generate commands to produce speech. Although they are superficially similar, the association areas of the brain do not always have the same function in the right and left hemispheres (see p. 472).

Primary sensory areas in the parietal lobe interpret sensations of touch that originate in all parts of the body; these body parts are "mapped" in an orderly sequence (see Fig. 24-12). In an adjacent region of the frontal lobe, *primary motor areas* command movements in corresponding areas of the body by stimulating the motor neurons in the spinal cord that control the muscles, allowing you to walk to class, throw a Frisbee®, or type a term paper. Like the primary sensory area, the primary motor area also has adjacent association areas, including the motor association area (also called the premotor area), which seems to be responsible for directing the motor area to produce movements. Behind the bones of the forehead lie the association areas of the frontal lobes, which are important in complex functions such as decision making, predicting the consequences of actions, controlling aggression, and planning for the future.

Damage to the cortex from trauma, stroke, or a tumor results in specific deficits, such as problems with speech, difficulty reading, or the inability to sense or move specific parts of the body. Most brain cells of adults cannot be replaced, so if a brain region is destroyed, these deficits may be permanent. Fortunately, however, training can sometimes allow undamaged regions of the cortex to take control over and restore some of the lost functions.

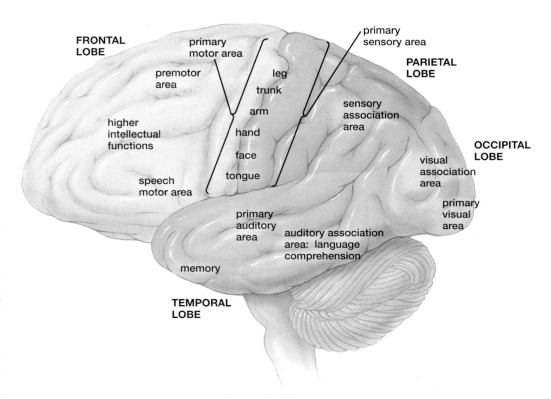

▲ **Figure 24-12 The cerebral cortex** Regions of the human left cerebral cortex. A map of the right cerebral cortex would be similar, except that speech and language are less well developed there.

24.6 How Does the Brain Produce the Mind?

Many people find it difficult to accept the idea that a few pounds of soft tissue in the skull produce the human mind, with its thoughts, emotions, memories, and self-awareness. This "mind–brain problem" has intrigued generations of philosophers and, more recently, neurobiologists. Beginning with observations of individuals with head injuries and progressing to sophisticated surgical, physiological, biochemical, and imaging experiments, the outlines of how the brain creates the mind are beginning to emerge.

The "Left Brain" and "Right Brain" Are Specialized for Different Functions

Although the two hemispheres of the forebrain are similar in appearance, it has been known since the early 1900s that this symmetry does not extend to brain function. Historically, information about the differences in hemisphere function came from studies of accident or stroke victims with localized damage to one hemisphere. A great deal has also been learned from studies of patients who have had their corpus callosums (which connects the two hemispheres) severed, a surgical procedure that is performed in certain cases of uncontrollable epilepsy to prevent the spread of seizures through the brain.

Beginning in the 1950s, Roger Sperry of the California Institute of Technology studied people whose hemispheres had been separated by cutting the corpus callosum. In his research, Sperry made use of the knowledge that axons within each optic tract (which are not severed by the surgery) follow a pathway that causes the left half of each visual field to be perceived by the right cerebral hemisphere and the right half to be seen by the left hemisphere (Fig. 24-13). Sperry used an ingenious device that projected different images onto the left and right visual fields and thus sent different signals to each hemisphere.

When Sperry projected an image of a nude figure onto just the left visual field, the subjects would blush and smile but would claim to have seen nothing. The same figure projected onto the right visual field was readily described verbally. These and later experiments revealed that—in right-handed people—the left hemisphere is almost always dominant in speech, reading, writing, language comprehension, mathematical ability, and logical problem solving. The right side of the brain is superior to the left in musical skills, artistic ability, recognizing faces, spatial visualization, and the ability to recognize and express emotions.

Recent experiments, however, indicate that the left–right dichotomy is not as rigid as was once believed. An individual who has suffered a stroke that disrupted the blood supply to the left hemisphere typically shows symptoms such as loss of speaking ability. However, these deficits can often be partially overcome through training, even though the left hemisphere itself has not recovered. This observation suggests that the right hemisphere has some latent language capabilities.

Learning and Memory Involve Biochemical and Structural Changes in Specific Parts of the Brain

Although the cellular mechanisms of learning and memory remain poorly understood, biologists have nonetheless made significant progress in understanding some aspects of these phenomena.

▼ **Figure 24-13 Specialization of the cerebral hemispheres** Each half of the retina of each eye perceives the opposite visual field. The axons from the half-retinas that see the left visual field send information to the right hemisphere, and vice versa. Therefore, with a quick glance at the word "heart" (before your eyes could move), the right hemisphere would perceive "he" and the left hemisphere would perceive "art." In addition to perceiving different parts of the visual field, the two hemispheres typically control the opposite sides of the body and are specialized for a variety of functions.

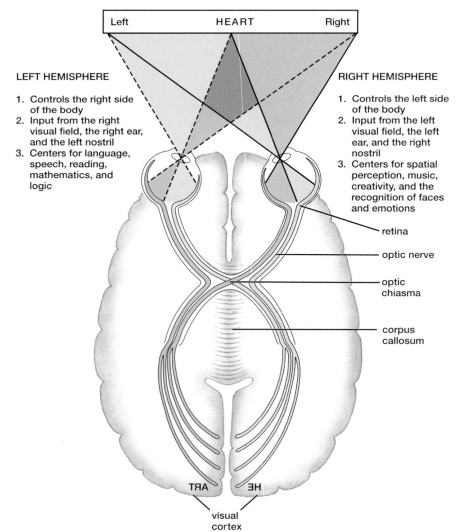

LEFT HEMISPHERE

1. Controls the right side of the body
2. Input from the right visual field, the right ear, and the left nostril
3. Centers for language, speech, reading, mathematics, and logic

RIGHT HEMISPHERE

1. Controls the left side of the body
2. Input from the left visual field, the left ear, and the right nostril
3. Centers for spatial perception, music, creativity, and the recognition of faces and emotions

retina

optic nerve

optic chiasma

corpus callosum

visual cortex

Memory May Be Brief or Long Lasting

Learning has two phases: *working memory* and *long-term memory*. For example, if you look up a number in the phone book, you will probably remember the number long enough to dial it but will then promptly forget it. This is working memory. But if you call the number frequently, eventually you will remember the number more or less permanently—an example of long-term memory.

Some working memory requires the repeated activity of a particular neural circuit in the brain. As long as the circuit is active, the memory stays. In other cases, working memory depends on short-term biochemical changes within neurons that temporarily strengthen synapses. In contrast, long-term memory requires the formation of new, long-lasting synapses or the long-term strengthening of existing but weak synapses.

The Temporal Lobes Are Important for Memory

There is strong evidence that the hippocampi, deep within the brain's temporal lobes, are involved in learning. For example, researchers have detected intense electrical activity in the hippocampi during learning. Even more striking are the results of hippocampal damage. A person in whom both hippocampi (one in each hemisphere) are destroyed retains much of his or her former memories but is unable to learn new information after the loss.

For example, one patient whose hippocampi were surgically removed in an attempt to control seizures remains unable to recall his address or find his way home after many years at the same residence. He can entertain himself indefinitely by reading the same magazine over and over, and people whom he sees daily require reintroduction at each encounter. This and other examples suggest that the hippocampus is responsible for transferring information from working memory to long-term memory.

The temporal lobes of the cerebral hemispheres seem to be important in recalling long-term memories. In a famous series of experiments in the 1940s, neurosurgeon Wilder Penfield electrically stimulated the temporal lobes of conscious patients undergoing brain surgery. The patients did not merely recall memories but felt that they were experiencing the past events right there in the operating room!

Discrete Areas of the Brain Perform Specific Functions

Until about 100 years ago, the mind was considered to be a subject more appropriate for philosophers than for scientists, in part because tools for studying the brain did not yet exist. New tools and new discoveries, however, are rapidly changing our view of how the mind works.

During recent decades, we have begun to understand the neural bases of some psychological phenomena. Many forms of mental illness, such as schizophrenia, manic depression, and autism, were once thought to be caused by childhood trauma or inept parenting. They are now recognized as the results of biochemical imbalances or structural abnormalities in the brain. Many traits, such as tendencies toward shyness or alcoholism, are now known to be caused both by genetic predispositions and environmental factors.

A striking illustration of how the physical structure of the brain is related to personality was unwittingly provided by Phineas Gage in 1848 (**Fig. 24-14**). A railroad construction foreman, Gage was setting an explosive charge when it triggered prematurely. The blast blew a 13-pound steel rod through his skull, damaging both of his frontal lobes.

Although Gage survived for many years after his accident, his personality changed radically. Before the accident, Gage was conscientious, industrious, and well liked. After his recovery, he became impetuous, profane, and incapable of working toward a goal. Subsequent research has implicated the frontal lobes in emotional expression, control of aggression, recognition of appropriate behavior, and the ability to work for delayed rewards.

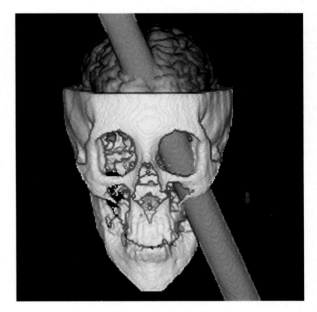

▲ **Figure 24-14 A revealing accident** Studies of the skull of Phineas Gage have enabled scientists to create this computer-generated reconstruction of the path taken by the steel rod that was blown through his head by an explosion.

scientific inquiry

Neuroimaging—Observing the Brain in Action

For most of human history, the brain has been a mystery. Now, however, imaging techniques provide exciting insights into brain function. These techniques include *PET* (positron emission tomography) and *fMRI* (functional magnetic resonance imaging). PET and fMRI are both based on the fact that regions of the brain that are most active have higher energy demands; they use more glucose and attract a greater flow of oxygenated blood than do less active areas. In PET scans, scientists inject the subject with a radioactive substance, such as a radioactive form of glucose, and then monitor levels of radioactivity that reflect differences in metabolic rate. These are translated by computer into colors on images of the brain. By monitoring radioactivity while a specific task is performed, scientists can identify which parts of the brain are most active during that task. In contrast, fMRI detects differences in levels of oxygenated and deoxygenated blood among different brain regions. The presence of oxygen bound to the hemoglobin in blood causes it to respond differently to the powerful magnetic field and pulses of radio waves generated by an fMRI machine. Active brain regions can be distinguished with fMRI without using radioactivity and over much shorter time spans than those required by PET.

Using fMRI or PET, researchers can observe changes as the brain performs a specific reasoning task or responds to an odor or a visual or auditory stimulus. PET and fMRI have confirmed that different aspects of the processing of language occur in distinct areas of the cerebral cortex. Using fMRI, scientists analyzed the frontal lobe areas used in generating words in individuals who spoke two languages. In subjects who had grown up speaking two languages, the same region of the frontal lobe was used in speaking both languages. In contrast, subjects who had learned a second language later in life used different but adjacent frontal lobe areas for the two languages.

Functional MRI scans have also been used to determine the parts of the brain that are most active during various emotional states. For example, when a person is frightened, the amygdala lights up (**Fig. E24-2a**). When people in love view photos of their lovers, other areas in the brain are activated (**Fig. E24-2b**). Interestingly, most of these same areas are also activated by drugs of abuse, such as cocaine.

By the way, never let anyone tell you that you use only a small fraction of your brain! Although PET and fMRI images sometimes make it appear that only a small area of the brain is active, this is because activity of other regions is subtracted out during the imaging process, to show where brain activity has *changed* as a result of stimulation. The images shown here merely highlight areas where activity is more intense than in a resting state.

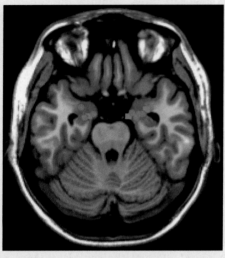

(a) Activation of the amygdala (red) by a frightening stimulus

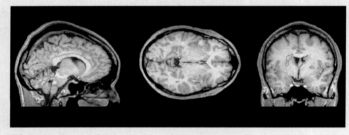

(b) Activation of the forebrain when a person views a photo of a loved one

▲ **Figure E24-2 Localization of emotions in the human brain** *(a)* A frightening experience activates the amygdala, a part of the forebrain that apparently produces emotions such as fear and rage. *(b)* Looking at photos of lovers activates multiple parts of the brain, including areas in the cerebral cortex (left) and structures deep within the forebrain (right).

Studies of other people with brain injuries have revealed that many parts of the brain are highly specialized. One patient with very localized damage to the left frontal lobe was unable to name fruits and vegetables (although he could name everything else). Victims of damage to other areas of the brain are unable to recognize faces, suggesting that the brain has regions specialized to recognize specific categories of things.

In the past, the insights that could be gained from patients with brain damage were limited, because the exact extent of the damage remained unknown until revealed by autopsy. Today, however, new techniques for visualizing brain structure and activity permit insight into the functioning of normal, as well as diseased, brains, as we explore in "Scientific Inquiry: Neuroimaging—Observing the Brain in Action."

24.7 How Does the Nervous System Sense the Environment?

The word "receptor" is used in several contexts in biology. In this portion of the chapter, we discuss **sensory receptors,** which consist of a specialized cell (typically a neuron) that produces an electrical response to a particular stimulus. All sensory receptors produce electrical signals, but each receptor type is specialized to produce its signal only in response to a particular type of stimulus (**Table 24-2**). The stimuli that activate sensory receptors may arise from the world around us (for example, light or sound) or from within our own bodies (for example, the position of a joint or the oxygen concentration in the blood). Therefore, sensory receptors are our only links to both our external and internal environments. As Aristotle observed in the fourth century B.C., "Nothing is understood by the intellect which is not first perceived by the senses."

Stimulating a sensory receptor causes an electrical signal called a **receptor potential.** Unlike action potentials, which are always the same size, receptor potentials vary in size with the strength of the stimulus—the stronger the stimulus, the larger the receptor potential. Although the numbers of cells, their connections, and the mechanisms by which they respond to stimuli vary considerably, the final output of all sensory perception is a burst of action potentials heading to the brain. Larger receptor potentials produce a higher frequency of action potentials. Therefore, *which axons* fire action potentials informs the brain about *what type of stimulus* has occurred, and the *frequency of action potentials* informs the brain about the *intensity of the stimulus* (see Fig. 24-4).

Mechanoreceptors Respond to Physical Displacement or Movement

Mechanoreceptors are found throughout the body. They include receptors in the skin that respond to touch, vibration, or pressure; stretch receptors in many internal organs; position sensors in the joints; and receptors in the inner ear that allow us to detect sound (see the following section).

The skin of humans and most other vertebrates is exquisitely sensitive to touch. Embedded in and directly beneath the skin are several types of mechanoreceptor neurons, each with a dendrite that produces a receptor potential when its membrane is stretched or dented (**Fig. 24-15**). The dendrites of some touch receptors are free nerve endings that can produce sensations of touch, itching, or tickling. The endings of other receptors are enclosed in layers of connective tissue—for example, the Pacinian corpuscle, which responds to rapid changes in pressure such as vibrations or a sharp poke. The density of mechanoreceptors in the skin

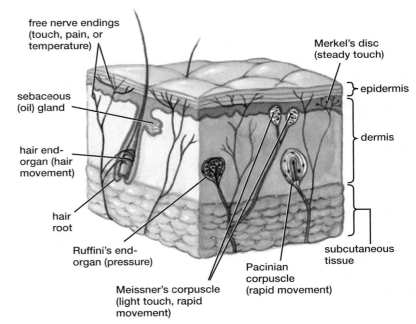

▼ **Figure 24-15 Receptors in the human skin** The diversity of receptors in the skin allows us to perceive touch, pressure, vibration, tickling, itching, pain, heat, and cold.

free nerve endings (touch, pain, or temperature)

Merkel's disc (steady touch)

epidermis

sebaceous (oil) gland

dermis

hair end-organ (hair movement)

hair root

Ruffini's end-organ (pressure)

Meissner's corpuscle (light touch, rapid movement)

Pacinian corpuscle (rapid movement)

subcutaneous tissue

Table 24-2	**Some Vertebrate Receptor Types**		
Type of Receptor	**Sensory Cell Type**	**Stimulus**	**Location**
Thermoreceptor	Free nerve ending	Heat, cold	Skin
Mechanoreceptor	Hair cell	Vibration, motion, gravity	Inner ear
	Nerve endings in skin	Vibration, pressure, touch	Skin
	Nerve endings in muscles or joints	Stretch	Muscles, tendons
Photoreceptor	Rod, cone	Light	Retina of eye
Chemoreceptor	Olfactory receptor	Odor (airborne molecules)	Nasal cavity
	Taste receptor	Taste (waterborne molecules)	Tongue
Pain receptor	Free nerve ending	Chemicals released by tissue injury	Widespread in the body

varies tremendously over the surface of the body. Each square centimeter of fingertip has dozens of touch receptors, but on the back there may be less than one per square centimeter.

Mechanoreceptors in many hollow organs, such as the stomach and bladder, signal fullness by responding to stretch. Mechanoreceptors in the joints allow us to know whether our elbows are bent—and by how much—without having to look. Imagine how annoying it would be if you needed to watch your fork on its way to your mouth to avoid stabbing yourself in the face!

Mechanoreceptors are probably also the way in which many animals detect Earth's magnetic field. Perceiving magnetic fields allows birds to migrate on cloudy days when they can't see the sun and are flying over unfamiliar terrain, or even over the ocean. How can animals detect magnetic fields? A variety of animals have tiny magnetic particles in their brains or other body parts. One hypothesis is that Earth's magnetic field moves these particles within specialized cells, distorting the cells' plasma membranes, similar to having the membrane poked by a mechanical stimulus.

The Perception of Sound Is a Specialized Type of Mechanoreception

Sound is produced by any vibrating object—a drum, vocal cords, or the speaker of your CD player. These vibrations, or sound waves, are transmitted through the air and intercepted by our ears, which convert them to signals that our brains interpret as the direction, pitch, and loudness of sound. The mammalian ear consists of a variety of structures (for example, the outer ear and eardrum) that transmit vibrations to specialized mechanoreceptor cells deep in the inner ear.

The Ear Converts Sound Waves into Electrical Signals

The ear of humans and most other vertebrates consists of three parts: the outer, middle, and inner ear (Fig. 24-16a). The *outer ear* consists of the *pinna* and the *auditory canal*. The pinna is the flap of skin-covered cartilage attached to the sur-

▼ **Figure 24-16 The human ear** *(a)* Overall anatomy of the ear. *(b)* The cochlea in cross section. *(c)* The hairs of hair cells span the gap between the tectorial and basilar membranes. Sound vibrations move the membranes relative to one another, bending the hairs and producing a receptor potential in the hair cells.

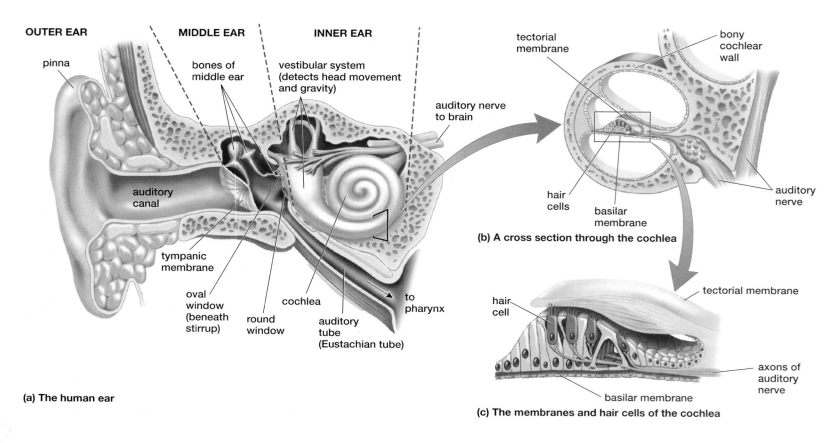

(a) The human ear

(b) A cross section through the cochlea

(c) The membranes and hair cells of the cochlea

face of the head. The pinna collects sound waves and modifies them in various ways. Humans and other fairly large animals determine sound direction by differences in *when* sound arrives at the two ears and in *how loud* it is in each ear. The shape and mobility of the pinna further contribute to sound localization. Bats have probably the most precise sound localization in the animal kingdom. Insect-eating bats emit extremely high-pitched shrieks that reflect off moths and other insect prey. Large ears, a highly developed auditory cortex, and an enormous cerebellum (for precise control of flying) combine to allow the bats to intercept their flying prey in pitch darkness.

The air-filled auditory canal conducts the sound waves to the *middle ear*, consisting of the *tympanic membrane*, or *eardrum*; three tiny bones called the *hammer, anvil*, and *stirrup*; and the auditory tube (also called the *Eustachian tube*). The auditory tube connects the middle ear to the pharynx and equalizes the air pressure between the middle ear and the atmosphere.

Sound waves traveling down the auditory canal vibrate the tympanic membrane, which in turn vibrates the hammer, the anvil, and the stirrup. These bones transmit vibrations to the *inner ear*, which contains the spiral-shaped *cochlea*. The stirrup bone transmits vibrations to the fluid within the cochlea by vibrating a membrane on the cochlea called the *oval window*. The *round window* is a second membrane that allows fluid within the cochlea to shift back and forth as the stirrup bone vibrates the oval window.

Sound Is Converted into Electrical Signals in the Cochlea

The cochlea, in cross section, consists of three fluid-filled canals (**Fig. 24-16b**). The central canal houses the receptors and the supporting structures that activate them in response to sound vibrations. The floor of this central canal is the *basilar membrane*, on top of which sit mechanoreceptors called *hair cells*. Hair cells have small cell bodies topped by hairlike projections that resemble stiff cilia. Some of these hairs are embedded in a gelatinous structure called the *tectorial membrane* (**Fig. 24-16b, c**).

The oval window passes vibrations from the bones of the middle ear to the fluid in the cochlea, which in turn vibrates the basilar membrane relative to the tectorial membrane. This movement bends the hairs of the hair cells, producing receptor potentials. The hair cells then release neurotransmitters onto neurons whose axons form the auditory nerve. Action potentials triggered in these axons travel to auditory processing centers within the brain.

The structures of the inner ear allow us to perceive *loudness* (the magnitude of sound vibrations) and *pitch* (the frequency of sound vibrations). Soft sounds cause small vibrations, which bend the hairs only slightly and result in a small receptor potential and a low rate of action potentials in axons of the auditory nerve. Loud sounds cause large vibrations, which cause greater bending of the hairs, a larger receptor potential, and a high rate of action potentials in the axons of the auditory nerve. Loud sounds sustained for a long time can damage the hair cells (**Fig. 24-17a**), resulting in hearing loss, a fate suffered by many rock musicians and their fans. In fact, many sounds in our everyday environment have the potential to damage hearing, especially if exposure to them is prolonged (**Fig. 24-17b**).

The perception of pitch is a little more complex. The basilar membrane resembles a harp in shape and stiffness: narrow and stiff at the end near the oval window but wider and more flexible near the tip of the cochlea. In a harp, the short, tight strings produce high notes and the long, looser strings produce low notes. In the basilar membrane, the progressive change in shape and stiffness causes each portion to vibrate most strongly to a particular frequency of sound: high notes near the oval window and low notes near the tip of the cochlea. The brain interprets signals from hair cells near the oval window as high-pitched sound, whereas signals from hair cells located progressively closer to the tip of the cochlea are interpreted as increasingly lower in pitch. Young people with undamaged cochleas can detect sounds from about 30 vibrations per second (very low pitched) to about 20,000 vibrations per second (very high pitched).

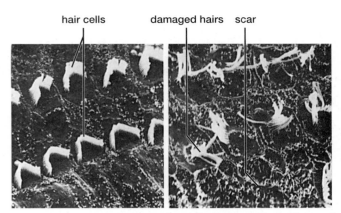

(a) Normal hair cells (left) and hair cells damaged by loud sound (right)

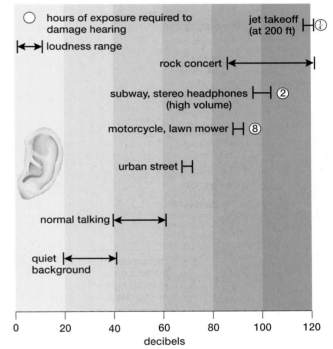

(b) The loudness of everyday sounds

▲ **Figure 24-17 Loud sounds can damage hair cells**
(a) Scanning electron micrographs (SEMs) show the effect of intense sound on the hair cells of the inner ear. (Left) The hair cells in a normal guinea pig, with hairs emerging from each receptor in a V-shaped pattern. (Right) After 24-hour exposure to a sound level approached by loud rock music (2,000 vibrations per second at 120 decibels), many of the hairs are damaged or missing, leaving scars. Hair cells in humans do not regenerate, so such hearing loss is permanent. (SEMs by Robert S. Preston, courtesy of Professor J. E. Hawkins, Kresge Hearing Research Institute, University of Michigan Medical School.) *(b)* Sound levels of everyday noises and their potential to damage hearing. Sound intensity is measured in decibels on a logarithmic scale, so a 20-decibel sound is 10 times as loud as a 10-decibel sound, a 30-decibel sound is 100 times as loud, and so on. You feel pain at sound intensities above 120 decibels. (Source: Deafness Research Foundation, National Institute on Deafness and Other Communication Disorders.)

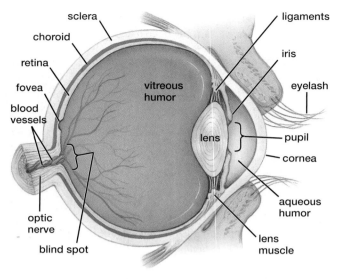

▲ **Figure 24-18** The human eye

Photoreception Perceives the World of Light

The visual systems of different animal species vary in their ability to provide sharp, accurate representations of the appearance of the real world; in fact, several types of eyes have evolved independently. All forms of vision, however, use *photoreceptors*. These sensory cells contain receptor molecules called *photopigments* (because they are colored), which change shape when they absorb light. This shape change sets off a series of chemical reactions inside the photoreceptor cell that ultimately produces a receptor potential.

The Mammalian Eye Collects and Focuses Light, Converting It into Electrical Signals

Mammalian eyes consist of two major parts (**Fig. 24-18**): (1) a variety of structures that serve one of four functions—holding the eye in a fairly fixed shape, allowing light to enter the eye, controlling the amount of light that enters, and focusing the light rays; and (2) the retina, which contains the photoreceptors that actually respond to the incoming light.

Light first encounters the *cornea*, a transparent covering over the front of the eyeball that collects light waves and begins to focus them. Behind the cornea, light passes through a chamber filled with a watery fluid called *aqueous humor*, which provides nourishment for both the lens and cornea. The amount of light entering the eye is adjusted by the *iris*, a pigmented muscular tissue. The iris regulates the size of the *pupil*, the opening in its center. Light passing through the pupil strikes the *lens*, a structure that resembles a flattened sphere composed of transparent protein fibers. The lens is suspended behind the pupil by muscles that regulate its shape and allow fine focusing of the image. Behind the lens is a much larger chamber filled with *vitreous humor*, a clear jellylike substance that helps to maintain the shape of the eyeball.

After passing through the vitreous humor, light finally reaches the retina, a thin tissue that lines the back of the eyeball. Here, the light energy is converted into action potentials that are conducted to the brain. Behind the retina is the *choroid,* a darkly pigmented tissue. The choroid's rich blood supply helps nourish the cells of the retina. Its dark pigment absorbs stray light whose reflection inside the eyeball would interfere with clear vision. Surrounding the outer portion of the eyeball is the *sclera*, a tough connective tissue layer that is visible as the white of the eye and is continuous with the cornea.

The Lens Focuses Light from Distant and Nearby Objects on the Retina

Although focusing begins at the cornea, whose rounded contour bends light rays, the lens is responsible for final, sharp focusing. The shape of the lens is adjusted by its encircling muscle. When viewed from the side, the lens is either flattened, to focus on distant objects, or rounded, to focus on nearby objects (**Fig. 24-19a**).

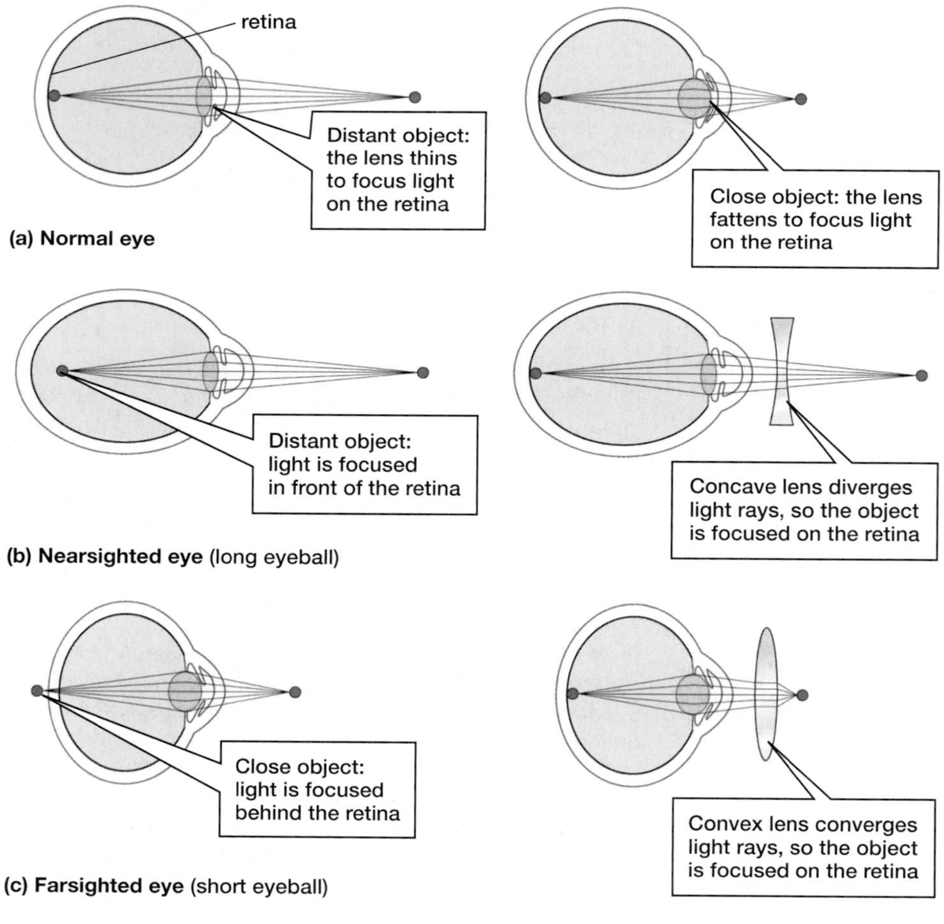

(a) Normal eye

Distant object: the lens thins to focus light on the retina

Close object: the lens fattens to focus light on the retina

(b) Nearsighted eye (long eyeball)

Distant object: light is focused in front of the retina

Concave lens diverges light rays, so the object is focused on the retina

(c) Farsighted eye (short eyeball)

Close object: light is focused behind the retina

Convex lens converges light rays, so the object is focused on the retina

◀ **Figure 24-19 Focusing in the human eye** *(a)* The lens changes shape to focus on objects at different distances. *(b)* Nearsightedness is corrected by eyeglasses with concave lenses. *(c)* Farsightedness is corrected by eyeglasses with convex lenses. **QUESTION:** Many nearsighted and farsighted people choose to correct their vision problems with laser surgery on their corneas rather than with corrective lenses. For nearsightedness and farsightedness, how should corneas be reshaped to correct the problem?

▶ **Figure 24-20** **The human retina** The human retina has rods and cones (photoreceptors), signal-processing neurons, and ganglion cells. Each rod and cone bears a long extension packed with discs of membrane in which the light-sensitive molecules are embedded.

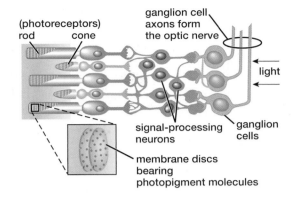

If your eyeball is too long or your cornea is too rounded, you will be nearsighted—the light from distant objects will focus in front of the retina. The eyes of farsighted people, with eyeballs that are too short or corneas that are too flat, focus light from nearby objects behind the retina. These conditions can be corrected by either contact lenses or glasses of the appropriate shape (**Fig. 24-19b, c**). Both nearsightedness and farsightedness can also be corrected with laser surgery that reshapes the cornea, producing a new corneal curvature that acts much like a corrective lens.

The Retina Detects Light and Produces Action Potentials in the Optic Nerve

The photoreceptors, called *rods* and *cones* after their shapes, gather light at the rear of the retina (**Fig. 24-20**). Between the receptors and incoming light are several layers of neurons that process the signals from the photoreceptors. These neurons enhance our ability to detect edges, movement, dim light, and changes in light intensity. The retinal layer nearest the vitreous humor consists of *ganglion cells*, whose axons make up the *optic nerve*. Electrical signals from the photoreceptors and intervening neurons are converted to action potentials in the ganglion cells. To reach the brain, ganglion cell axons pass back through the retina at a location called the *blind spot* (**Fig. 24-21**; see also Fig. 24-18). Because the blind spot lacks receptors, objects focused there seem to disappear. You can locate your blind spot by using the star and circle shown in **Figure 24-22**.

Rods and Cones Differ in Distribution and Light Sensitivity

Photoreception in both rods and cones begins with the absorption of light by photopigment molecules that are embedded in membranes of the photoreceptors (see Fig. 24-20). Light hitting a photopigment molecule changes its shape, setting off chemical reactions that ultimately produce a receptor potential in the photoreceptor.

Although cones are located throughout the retina, they are concentrated in the *fovea*, the area of the retina where the lens focuses images most sharply (see Figs. 24-18 and 24-21). The fovea looks like a dent near the center of the retina because the layers of signal-processing neurons are spread apart, although they still retain their interconnections. This arrangement allows light to reach the cones of the fovea without having to pass through so many other layers of cells.

Human eyes have three varieties of cones, each containing a slightly different photopigment. Each type of photopigment is most strongly stimulated by a particular wavelength of light, corresponding roughly to red, green, or blue. The brain distinguishes color according to the relative intensity of stimulation of different cones. For example, the sensation of yellow is caused by fairly equal stimulation of red and green cones. About 4% to 8% of males have difficulty distinguishing red from green because they possess a defective gene for the red or green photopigment on the X chromosome (see pp. 144–145). These people are sometimes referred to as being "color-blind," but are more accurately described as "color-deficient." True color blindness, in which an individual perceives the world in black and white and shades of gray, is extremely rare.

Rods are most abundant outside the fovea, around the edge of the retina. Rods are far more sensitive to light than are cones and so are largely responsible for our vision in dim light. Unlike cones, rods do not distinguish colors. In moonlight, which is too dim to activate the cones, the world appears in shades of gray.

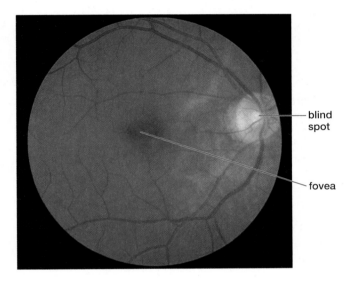

▲ **Figure 24-21** **A living human retina** A portion of the human retina, photographed through the cornea and lens of a living person. The blind spot and fovea are visible. Blood vessels supply oxygen and nutrients. Notice that they are dense over the blind spot (where they don't interfere with vision) and scarcer near the fovea.

 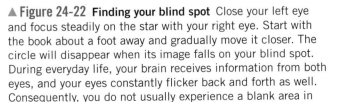

▲ **Figure 24-22** **Finding your blind spot** Close your left eye and focus steadily on the star with your right eye. Start with the book about a foot away and gradually move it closer. The circle will disappear when its image falls on your blind spot. During everyday life, your brain receives information from both eyes, and your eyes constantly flicker back and forth as well. Consequently, you do not usually experience a blank area in your visual field.

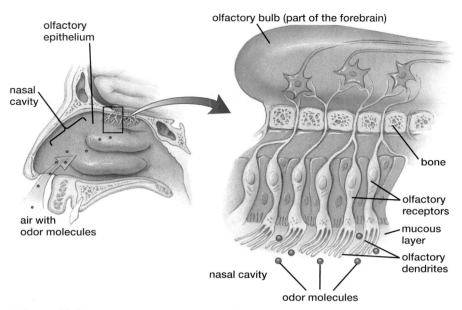

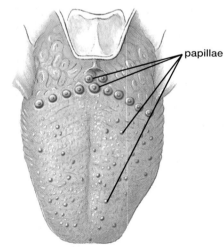

(a) The human tongue

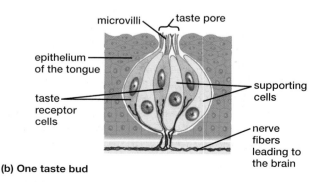

(b) One taste bud

▲ **Figure 24-23 Human olfactory receptors** The receptors for olfaction in humans are neurons bearing hairlike projections that protrude into the nasal cavity. The projections are embedded in a mucous layer in which odor molecules dissolve before contacting the receptors.

Not all animals have both rods and cones. Animals that are active almost entirely during the day (certain lizards, for example) may have all-cone retinas, whereas many nocturnal animals (such as rats and ferrets) and those dwelling in dimly lit habitats (such as deep-sea fishes) have mostly or entirely rods.

Chemoreceptors Detect Chemicals in the Internal or External Environment

Chemoreceptors allow animals to locate food, avoid poisonous materials, find mates, and maintain homeostasis. Internal chemoreceptors in certain large blood vessels and in the hypothalamus of the brain monitor levels of crucial molecules such as sugar, water, oxygen, and carbon dioxide in the blood; they also stimulate activities that maintain these levels within narrow limits. As we will see, certain chemoreceptors respond to molecules released by damaged cells, producing the sensation of pain. For perceiving chemicals in the external environment, terrestrial vertebrates, including people, have two distinct senses: the sense of *smell* or *olfaction* responds to chemicals in the air, while the sense of *taste* detects chemicals dissolved in water or saliva.

Olfactory Receptors Detect Airborne Chemicals

In humans and most other vertebrates, receptors for olfaction are neurons located in a patch of tissue, known as the olfactory epithelium, in the upper portion of each nasal cavity (Fig. 24-23). This tissue is smaller in humans than in many other mammals. Dogs, for example, have a sense of smell that is hundreds of times more sensitive than ours. Olfactory receptors have hairlike dendrites that protrude into the nasal cavity and are embedded in a layer of mucus. Odor molecules in the air diffuse into the mucous layer and bind with receptor protein molecules on the dendrites. Humans produce about 500 different types of receptor proteins, but each olfactory receptor neuron expresses only one type. Each receptor protein is specialized to bind a particular type of molecule and stimulate the olfactory receptor to send a message to the brain. Many odors are complex mixtures of molecules that stimulate several receptor proteins, so our perception of odors arises as a result of the brain interpreting signals that arise from many different chemoreceptors.

Taste Receptors Detect Chemicals That Contact the Tongue

The human tongue bears about 10,000 *taste buds*, structures embedded in small bumps, called *papillae*, that cover the surface of the tongue (Fig. 24-24a). Each taste bud is a cluster of 60 to 80 taste receptors surrounded by supporting cells in a small pit. The cells in the pit communicate with the mouth through a taste pore (Fig. 24-24b). Microvilli (thin membrane projections) of taste receptors protrude through the pore. Dissolved chemicals enter the pore and bind to receptor proteins on the microvilli, producing a receptor potential.

Four major types of taste receptors have long been known: sweet, sour, salty, and bitter. A fifth type, *umami* (a Japanese word loosely translated as "delicious"), was identified in 2000. The umami receptor responds to glutamate, which is found in the seasoning monosodium glutamate (MSG), often described as tasting "meaty" and used to enhance the flavor of foods. Although it was once believed

◀ **Figure 24-24 Human taste receptors** **(a)** The human tongue is covered with papillae, bumps in which taste buds are embedded. Small papillae are located on the front two-thirds of the tongue; larger ones with more taste buds are far in the back. **(b)** Each taste bud consists of supporting cells surrounding taste receptors, whose microvilli have receptor proteins that bind tasty molecules, producing a receptor potential.

▶ **Figure 24-25 Pain perception** An injury, such as a tack stab, damages both cells and blood vessels. The damaged cells and vessels release substances that activate pain receptor neurons.

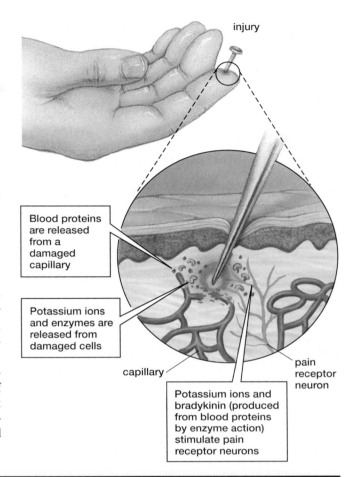

that each type of taste bud is concentrated on a specific region of the tongue, recent research has shown that the different types are quite evenly distributed.

We perceive a great variety of tastes through two mechanisms. First, a particular substance may stimulate two or more taste receptor types to different degrees, making the substance taste "sweet and sour," for example. Second, and more important, a substance being tasted usually releases molecules into the air inside the mouth. These odor molecules diffuse to the olfactory receptors, which contribute an odor component to the basic flavor. To prove that what we call taste is really mostly smell, try holding your nose (and closing your eyes) while you eat different flavors of gourmet jelly beans. The flavors—from grape, cherry, and pear to buttered popcorn—will be indistinguishably sweet.

Pain Is a Specialized Chemical Sense

Pain perception is a special kind of chemoreception (**Fig. 24-25**). When cells and capillaries are damaged by a cut or a burn, their contents flow into the extracellular fluid. The cell contents include potassium ions, which produce receptor potentials in *pain receptors*. Damaged cells also release enzymes that convert certain blood proteins into a chemical called bradykinin, which also activates pain receptors.

Each part of the body has a separate set of pain receptor neurons that provide input to particular brain cells. Hence, the brain can identify the location of the pain. Drugs that provide pain relief, such as morphine or Demerol, block synapses in the pain pathways of the brain or spinal cord. The brain can also reduce its perception of pain by releasing its own narcotic-like chemicals called endorphins.

How Do I Love Thee?   Revisited

"... heaven is here,
Where Juliet lives."

—*Romeo and Juliet*, Act III, scene III

What happens in the human brain when we fall in love? Although people aren't just big prairie voles, you may be surprised—and perhaps disappointed—to find out that there are some striking similarities in both brain function and hormones during emotional encounters. For example, areas of the human brain that contain oxytocin and dopamine respond to pictures of the faces of lovers and one's own children, but not to the faces of equally attractive and familiar people to whom the observer feels no emotional bond. Some of these areas are the same ones that are activated in prairie voles and that appear to be important in motivation and reward. In humans, as in voles, oxytocin probably plays an important role in attraction and commitment. Oxytocin reduces stress and inhibits the amygdala, a part of the brain involved in feeling fear. Oxytocin levels increase in women and men during sexual encounters. In a recent study, an oxytocin nasal spray was found to promote trust, even between complete strangers.

What about the different kinds of love? Brain scans reveal similarities as well as differences between romantic and parental love. Some of the same brain areas are activated by photos of lovers and children; other areas are activated by one or the other, but not both. Still other brain areas, particularly those involved in critical decision-making and

social judgments, are turned off, at least while seeing the lover or child. The result is that lovers and children almost always seem to be better than they really are. This is probably especially important for parental love of newborns. Although often very cute, newborn infants can hardly reciprocate love the way adults can.

Will neurobiological explanations ruin the magic of love? Anthropologist Helen Fisher doesn't think so. "You can know every ingredient in a piece of chocolate cake, and . . . it's [still] wonderful. In the same way, you can know all the ingredients of romantic love and still feel that passion."

In some ways, love strongly resembles addiction. For example, both activate "reward circuits" in the brain that release dopamine. Dr. Fisher notes that people involved in romantic love, like addicts, experience tolerance, withdrawal, and relapse. With both drugs and newfound love, small doses may spark interest, but soon small doses aren't nearly enough—you need more and more to satisfy your craving. If you don't get your love or drug "fix," you go into withdrawal—crying, depression, and desperately seeking a new "dose." Finally, there's relapse. Just as reformed drug addicts may develop sudden, urgent cravings if they encounter the drug again, people who have fallen "out of love" may still feel a surge of emotion if they see their former lovers again. But does love really resemble addiction to drugs? Or is it the other way around? It is extremely unlikely that humans have evolved reward centers and dopamine neurons so that

(continued)

we can become cocaine addicts. Rather, these circuits probably evolved to promote sex, pair bonding, and infant care, and drugs like cocaine are addictive because they chemically hijack the same reward circuits that make people in love feel "high."

Chapter Review

Summary of Key Concepts

For additional study help and activities, go to www.mybiology.com.

24.1 How Do Nerve Cells Work?

Nervous systems are composed of billions of neurons. A neuron has four major specialized functions, which are reflected in its structure: (1) Dendrites receive information from the environment or from other neurons. (2) The cell body adds together electrical signals from the dendrites and "decides" whether to produce an action potential. The cell body also coordinates the neuron's metabolic activities. (3) The axon conducts the action potential to its synaptic terminal. (4) Synaptic terminals transmit the signal to other neurons, to glands, or to muscles.

BioFlix How Neurons Work

24.2 How Do Nerve Cells Produce and Transmit Signals?

The inside of an unstimulated neuron has a negative voltage called the resting potential. Signals received from other neurons are small, rapidly fading changes in voltage called postsynaptic potentials. Inhibitory and excitatory postsynaptic potentials (IPSPs and EPSPs) make the neuron less likely or more likely, respectively, to produce an action potential. If postsynaptic potentials, added together within the cell body, bring the neuron to threshold, an action potential is triggered. The action potential is a positive charge that travels, undiminished in magnitude, along the axon to the synaptic terminal.

At a synapse, neurotransmitters from the presynaptic neuron, released in response to an action potential, bind to receptors in the postsynaptic neuron's plasma membrane, opening ion channels there. Ions flow, producing either an EPSP or an IPSP, depending on the type of ion channels opened.

24.3 How Do Nervous Systems Process Information?

Information processing in the nervous system requires four operations. The nervous system must (1) determine the type of stimulus, (2) signal the intensity of the stimulus, (3) integrate information from many sources, and (4) initiate and direct appropriate responses.

24.4 How Are Nervous Systems Organized?

Neural pathways usually have four elements: (1) sensory neurons, (2) interneurons, (3) motor neurons, and (4) effectors. Overall, nervous systems consist of numerous interconnected neural pathways, which may be either diffuse or centralized.

24.5 What Is the Structure of the Human Nervous System?

The human nervous system consists of the central nervous system and the peripheral nervous system. The peripheral nervous system is further subdivided into sensory and motor portions. The motor portions consist of the somatic nervous system (which controls voluntary movement) and the autonomic nervous system (which directs involuntary responses).

Within the central nervous system, the spinal cord contains (1) neurons controlling voluntary muscles and the autonomic nervous system, and neurons communicating with the brain and other parts of the spinal cord; (2) axons leading to and from the brain; and (3) neural pathways for reflexes and certain simple behaviors.

The brain consists of three regions: the hindbrain, midbrain, and forebrain, each further subdivided into distinct parts. The hindbrain in humans consists of the medulla and pons, which control involuntary functions (such as breathing), and the cerebellum, which coordinates complex motor activities (such as typing). In humans, the small midbrain contains the reticular formation, a filter and relay for sensory stimuli. The forebrain includes the thalamus, a sensory relay station that shuttles information to and from conscious centers in the forebrain. The diverse structures of the limbic system of the forebrain are involved in emotion, learning, and the control of instinctive behaviors such as sex, feeding, and aggression. The cerebral cortex of the forebrain is the center for information processing, memory, and initiation of voluntary actions. It includes primary sensory and motor areas and association areas that analyze sensory information and plan movements.

Web Animation Reflex Arc

Web Animation Brain Structure

24.6 How Does the Brain Produce the Mind?

Each cerebral hemisphere is specialized. The left hemisphere dominates speech, reading, writing, language comprehension, mathematical ability, and logical problem solving. The right hemisphere specializes in recognizing faces and spatial relationships, generating artistic and musical abilities, and recognizing and expressing emotions.

Working memory is electrical or chemical and lasts for only a short time. Long-term memory is acquired through structural changes that increase the effectiveness of synapses. The hippocampi, located in the temporal lobes, are important sites for learning, which involves the transfer of information into long-term memory.

24.7 How Does the Nervous System Sense the Environment?

Sensory receptors are specialized cells that produce receptor potentials in response to sensory stimuli. The receptor potentials either directly in the receptor itself, or indirectly in neurons that the receptors synapse on, trigger action potentials that travel to the brain.

A variety of mechanoreceptors detect stimuli such as touch, vibration, pressure, stretch, or sound. In most cases, a mechanoreceptor produces a receptor potential in response to deformation or stretching of its plasma membrane.

In the vertebrate ear, air vibrates the tympanic membrane, which transmits vibrations to the bones of the middle ear and then to the oval window of the fluid-filled cochlea. Within the cochlea, vibrations bend the hairs of hair cells, which are vibration-sensitive mechanoreceptors. This bending produces receptor potentials in the hair cells that cause action potentials in the axons of the auditory nerve, which leads to the brain.

In the vertebrate eye, light enters the cornea and passes through the pupil to the lens, which focuses an image on the fovea of the retina. Two types of photoreceptor, rods and cones, are located in the retina. They produce receptor potentials in response to light. These signals are processed through several layers of neurons in the retina and are translated into action potentials in the optic nerve, which leads to the brain. Rods are more abundant and more light sensitive than are cones and provide vision in dim light. Cones, which are concentrated in the fovea, provide color vision.

Humans detect chemicals in the environment by olfaction and taste. Each type of olfactory or taste receptor responds to only one or a few specific types of molecules, allowing discrimination among tastes and odors. Olfactory neurons of vertebrates are located in the upper nasal cavity. Taste receptors are located in clusters called taste buds on the tongue. Pain is a special chemical sense in which pain receptors respond to chemicals released by damaged cells.

Key Terms

action potential *p. 459*
autonomic nervous system *p. 467*
axon *p. 459*
brain *p. 465*
cell body *p. 459*
central nervous system (CNS) *p. 465*
dendrite *p. 459*
effector *p. 464*

forebrain *p. 470*
hindbrain *p. 469*
interneuron *p. 464*
midbrain *p. 470*
motor neuron *p. 464*
nerve *p. 459*
neuron *p. 458*
neurotransmitter *p. 459*

peripheral nervous system (PNS) *p. 465*
postsynaptic neuron *p. 460*
postsynaptic potential (PSP) *p. 460*
presynaptic neuron *p. 460*
receptor potential *p. 475*
reflex *p. 468*

resting potential *p. 460*
sensory neuron *p. 464*
sensory receptor *p. 475*
somatic nervous system *p. 467*
spinal cord *p. 465*
synapse *p. 459*
synaptic terminal *p. 459*
threshold *p. 460*

Thinking Through the Concepts

Suggested answers to end-of-chapter and figure-based questions can be found at the end of the text.

Fill-in-the-Blank

1. The parts of an individual nerve call, also called a(n) _____, are specialized to perform different functions. The "input" end of a nerve cell, called a(n) _____, receives information from the environment or from other nerve cells. The _____ contains the nucleus and other typical organelles of a eukaryotic cell. Electrical signals are sent down the _____, a long, thin strand that leads to the _____, where the nerve cell sends out its signal to other cells.

2. When they are not being stimulated, neurons have an electrical charge across their membranes called the resting potential. This potential is _____ (what charge?) inside. When a neuron receives a sufficiently large stimulus and reaches a potential called the _____, it produces an action potential. This causes the neuron to become _____ (what charge?) inside.

3. When an action potential reaches a synaptic terminal, it causes the release of a chemical called a(n)_____. This chemical binds to _____ on the postsynaptic cell, causing a change in potential. If the postsynaptic cell becomes less negative, this change in potential is called a(n) _____. If the postsynaptic cell becomes more negative, it is called a(n) _____ .

4. All action potentials are the same size and duration, so the strength or intensity of a stimulus is not coded by changes in individual action potentials. Instead, intensity is encoded by the _____ of action potentials in a single neuron, or by the _____ of similar neurons that respond to the stimulus.

5. The _____ is part of the peripheral nervous system that activates skeletal muscles. Smooth muscles and glands are activated by the _____, which has two divisions: the _____, active during "fight or flight," and the _____, active during "rest and digest."

6. The human hindbrain consists of three parts: the _____, the _____, and the _____. One of these, the _____, is important for coordinating complex movements such as typing.

7. The forebrain includes the thalamus, limbic system, and cerebral cortex. The _____ serves as a relay center connecting most of the senses with the cerebral cortex. The _____ is a group of structures that together control emotions such as fear, rage, and hunger.

8. The cerebral cortex consists of four lobes: the _____, _____, _____, and _____. The visual centers are located in the _____ lobe. The primary motor areas are in the _____ lobe.

9. Sensory receptors are finely tuned to respond to specific stimuli. _____ respond to physical deformation, such as touch or pressure. _____ respond to light. The receptors in the taste buds and olfactory epithelium are examples of _____.

10. The mammalian ear consists of three parts: the outer, middle, and inner ear. The flap of skin on the outer ear, often folded or trumpet shaped, is the _____. This structure collects sound waves and funnels them down the auditory canal to a flexible membrane called the _____. This membrane connects to three small bones, the _____, _____, and _____. The final bone vibrates the _____, the beginning of the cochlea. Within the cochlea, the vibrations finally move the cilia of specialized mechanoreceptors called

_____.

11. The retina of the human eye contains two types of photoreceptors, called _____ and _____. _____ are larger, are more sensitive to light, and provide black-and-white vision. _____ are mostly located in the central region of the retina, called the _____. These photoreceptors are smaller, are less sensitive to light, but provide color vision.

Review Questions

1. List four major parts of a neuron, and explain the specialized function of each part.

2. Diagram a synapse. How are signals transmitted from one neuron to another at a synapse?

3. How does the brain perceive the intensity of a stimulus? The type of stimulus?

4. What are the four elements of a neuron-to-muscle pathway? Describe how these elements function in the human pain-withdrawal reflex.

5. Draw a cross section of the spinal cord. What types of neurons are located in the spinal cord? Explain why severing the cord paralyzes the body below the level where it is severed.

6. Describe the functions of the following parts of the human brain: medulla, cerebellum, reticular formation, thalamus, limbic system, and cerebral cortex.

7. What structure connects the two cerebral hemispheres? Describe the distinct intellectual functions of the hemispheres in most right-handed people.

8. Distinguish between long-term memory and working memory.

9. What are the names of the specific receptors used for taste, vision, hearing, smell, and touch?

10. Why are we apparently able to distinguish hundreds of different flavors if we have only five types of taste receptors?

11. How are we able to distinguish so many different odors?

12. Describe the structure and function of the various parts of the human ear by tracing the route of a sound wave from the air outside the ear to the cells that cause action potentials in the auditory nerve.

13. How does the structure of the inner ear allow for the perception of pitch? Of sound intensity?

14. Diagram the overall structure of the human eye. Label the cornea, iris, lens, sclera, retina, and choroid. Describe the function of each structure.

15. How does the lens change shape to allow focusing of distant objects? What defect makes focusing on distant objects impossible, and what is this condition called? What type of lens can be used to correct it, and how does it do so?

16. List the similarities and differences between rods and cones.

17. Describe how pain is signaled by tissue damage.

Applying the Concepts

1. IS THIS SCIENCE? More than 1,900 years ago, the Roman poet Juvenal wrote that "You should pray for a healthy mind in a healthy body." As a rule, only the last part, "a healthy mind in a healthy body," is remembered now, and it is often interpreted to mean that you need a healthy body to produce a healthy mind. Is this correct? Research the Internet for studies linking or refuting the notion that general health, proper nutrition, or exercise help to promote an active mind.

2. In Parkinson's disease, which afflicts at least one million Americans, cells degenerate in a small part of the brain that is important in the control of movement. Some physicians have reported improvement after injecting cells taken from the same

brain region of aborted fetuses into the appropriate parts of the brain of a Parkinson's patient. This is an enormously expensive procedure. Do you think this type of surgery is the answer for Parkinson's victims? Discuss this from ethical, financial, and practical perspectives.

3. Explain the statement "Your sensory perceptions are purely a creation of your brain." Discuss the implications for communicating with other humans, with other animals, and with intelligent life from another universe.

For additional resources, go to www.mybiology.com.

chapter 25

Animal Reproduction and Development

John, who struggles with the lasting effects of fetal alcohol syndrome, helps his adoptive mother Teresa Kellerman to educate people about the dangers of drinking during pregnancy.

🔍 At a Glance

Case Study   The Faces of Fetal Alcohol Syndrome

"The guilt is tremendous. . . . I did it again and again. . . . I don't know how to tell them. It was something I could have prevented." Debbie, the young mother pictured here with her youngest child, Sabrina (photo at left), has had seven children. One of her older daughters, Cory, has been diagnosed with fetal alcohol syndrome (FAS), the most serious type of alcohol damage. At age three she was hyperactive and talked like a one-year-old. Doctors believe that Sabrina is almost certainly a victim as well. Her face bears the characteristic features of FAS, including small eyes, a short nose, and a small head. At 7 months, she was weak, began having seizures, and could not eat solid food because she was unable to close her upper lip around the spoon. As this pregnant mother repeatedly got drunk, so did her developing children.

Debbie regrets drinking while she was pregnant with her daughter, Sabrina, whose nervous system was damaged by alcohol.

John's mother drank while she was pregnant with him, and was drunk when she delivered him. John was fortunate, however, to have been adopted by a truly remarkable woman, Teresa Kellerman, who has become a leading parent advocate for children and their families who have been damaged by alcohol. The damage to children born to mothers who drink is irreparable. "Without intervention they end up homeless, jobless, addicted, arrested, pregnant or getting someone pregnant, living on the streets or dead, so many of them," says Kellerman, who founded the Fetal Alcohol Syndrome Community Resource Center in Tucson, Arizona.

Debbie entered rehabilitation and has resolved to stay sober and be a good parent to Sabrina and Cory. But even with the best parenting, neither John, Cory, nor Sabrina, nor the thousands of other children born each year with full-blown FAS, are ever likely to live without supervision. Tens of thousands of others with milder damage may become functional but will never reach their normal potential.

As you study reproduction and development, think about how alcohol reaches a developing child when a pregnant woman drinks. How is the embryo nourished by its mother? To what extent is the embryo protected from environmental insults, whether alcohol and other drugs, toxins in cigarette smoke, or industrial pollutants? ■

25.1 How Do Animals Reproduce?

Animals reproduce either sexually or asexually. In **asexual reproduction,** a single animal produces offspring, usually through repeated mitotic cell divisions in some part of its body. The offspring are therefore genetically identical to the parent. In **sexual reproduction,** the **gonads** of sexually mature animals produce haploid gametes through meiotic cell division. Two gametes—usually from separate parents—fuse and produce a diploid cell, which then undergoes repeated mitotic cell divisions to produce a diploid offspring. Because the offspring receives genes from both of its parents, it is not genetically identical to either.

In Asexual Reproduction, A Single Organism Reproduces by Itself

Asexual reproduction is efficient in effort (no need to hunt for mates, court members of the opposite sex, or battle rivals), materials (no wasted gametes), and genes (the offspring usually have all the genes of their single parent). This ef-

▶ Figure 25-1 **Budding** The offspring of some cnidarians, such as these anemones, grow as buds that appear as miniature adults sprouting from the body of the parent. **QUESTION:** Which type of cell division gives rise to the cells of the bud's body?

ficiency has probably been the driving force behind the independent evolution of a variety of methods of asexual reproduction.

Budding Produces a Miniature Version of the Adult

One common method of asexual reproduction in animals is **budding.** Many sponges and cnidarians, for example, reproduce in this manner (**Fig. 25-1**). A miniature version of the animal—a bud—grows directly on the body of the adult. When it has grown large enough, the bud breaks off and becomes independent.

Fission Followed by Regeneration Can Produce a New Individual

Many animals are capable of regeneration, the ability to regrow lost body parts. Regeneration is part of reproduction in species that reproduce by **fission.** Several annelid and flatworm species can reproduce by dividing into two pieces, each of which regenerates an entire body (**Fig. 25-2**). Some corals and jellyfish can also split themselves in half lengthwise and regenerate two new individuals.

Eggs May Develop Without Fertilization

The females of some animal species can reproduce by a process in which egg cells develop into adults without being fertilized. In fact, some fish, including relatives of the mollies and platys that are popular in tropical fish tanks, and some lizards, such as the whiptail of the southwestern United States and Mexico, have populations consisting entirely of asexually reproducing females, with no males at all! Still other animals, such as some aphids, can reproduce either sexually or asexually, depending on the season of the year or the availability of food (**Fig. 25-3**).

Sexual Reproduction Requires the Union of Sperm and Egg

No one is certain why sexual reproduction evolved and flourished. However, sex has an important outcome: The genetic recombination that results from sexual reproduction creates novel genotypes—and therefore novel phenotypes—that are an important source of variation on which natural selection may act. Although this is difficult to prove, many biologists suspect that genetic recombination is the selective advantage favoring sexual reproduction.

In most animal species, an individual is either male or female, defined by the gamete that each produces. The female gonad, called the **ovary,** produces **eggs,** which are large, nonmotile cells containing food reserves that provide nourishment for the embryo. The male gonad, called the **testis** (plural, testes), produces small, motile **sperm,** which have almost no cytoplasm and hence no food reserves. The union of sperm and egg is called **fertilization.**

In some animal species, including earthworms and many snails, single individuals produce both sperm and eggs. Such individuals are called *hermaphrodites.* Most hermaphrodites still engage in sex, with the sperm of one individual fertilizing the eggs of another. Some hermaphrodites, however, can fertilize their own eggs. These

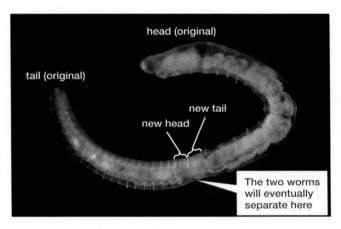

▲ Figure 25-2 **Fission followed by regeneration** This marine annelid (segmented worm) can reproduce by splitting its body and regenerating each half.

▶ Figure 25-3 **A female aphid gives live birth** In spring and early summer when food is abundant, aphid females reproduce asexually. In fall, they reproduce sexually, as the females mate with males. Aphids have evolved the ability to exploit the advantages of both asexual reproduction (rapid population growth during times of abundant food, no energy spent in seeking a mate, no wasted gametes) and sexual reproduction (genetic recombination, which produces diverse offspring).

▶ **Figure 25-4 Environmental cues may synchronize spawning** *(a)* In the Great Barrier Reef of Australia, thousands of corals spawn simultaneously, creating this "blizzard" effect. Spawning in these corals is linked to the phase of the moon. *(b)* Close-up of a package of sperm and eggs erupting from a spawning coral.

packages of gametes

single package of gametes

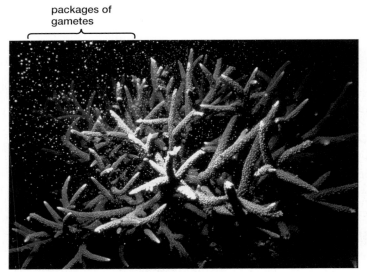

(a) Synchronized spawning in corals

(b) A coral releases gametes

▲ **Figure 25-5 Courtship rituals synchronize spawning** During spawning, male and female fighting fish perform a courtship dance that culminates with both partners releasing gametes simultaneously. The male will collect the fertilized eggs in his mouth and spit them into the nest of bubbles visible above the courting fish. **QUESTION:** In addition to ensuring synchronized release of gametes, what other advantage is gained by performing courtship rituals?

animals, including tapeworms and some types of snails, may find themselves isolated from other members of their species—circumstances under which the ability to fertilize oneself is advantageous.

Gametes Are Brought Together in Various Ways

For species with separate sexes and for hermaphrodites that cannot self-fertilize, reproduction requires that sperm and eggs from different animals be brought together for fertilization. How the union of sperm and egg is accomplished depends on the mobility of the animals and on whether they breed in water or on land.

External Fertilization Occurs Outside the Parents' Bodies

In *external fertilization*, sperm and eggs unite outside the bodies of the parents. In almost all species that use external fertilization, both sperm and eggs are released into water, and sperm swim to reach the eggs. The reproductive behavior of males and females must be synchronized, so that sperm and eggs are in the same place at the same time.

Many animals rely at least in part on environmental cues to synchronize breeding. For example, seasonal changes in day length often stimulate the physiological changes that lead to readiness for breeding. As a result, breeding is restricted to certain times of year. More precise synchrony, however, is required to coordinate the actual release of sperm and egg. For example, many corals of Australia's Great Barrier Reef synchronize spawning by the phase of the moon. On the fourth or fifth night after the full moons of November and December, all of the corals of a particular species on an entire reef release a blizzard of sperm and eggs into the water (**Fig. 25-4**).

Even when gamete release is synchronized, many sedentary aquatic animals such as corals and mussels waste enormous quantities of sperm and eggs because the gametes may be released relatively far apart. In more mobile species, mating behaviors help ensure synchrony of spawning in both time and place. Most fish, for example, have courtship rituals in which the male and female come very close together and release their gametes simultaneously (**Fig. 25-5**). Frogs and toads carry this ritual one step further by assuming a characteristic mating pose (**Fig. 25-6**). At

◀ **Figure 25-6 Golden toads mating** By clutching the female, the smaller male stimulates her to release eggs. Golden toads are now believed to be extinct.

the edges of ponds and lakes, the male mounts the female and stimulates her to release her eggs, which the male fertilizes by releasing a cloud of sperm above the eggs.

Internal Fertilization Occurs Within the Female's Body

In *internal fertilization*, sperm and eggs unite within the body of the female. Internal fertilization is a key adaptation to life on land, because sperm quickly die outside of a liquid medium. Even in aquatic environments, internal fertilization increases the likelihood of successful fertilization, because the sperm and eggs are confined together in a small space rather than relying on chance encounters within a large volume of water.

Internal fertilization usually occurs by **copulation,** a behavior in which the male deposits sperm directly into the reproductive tract of the female (**Fig. 25-7**). To be fertilized, a mature egg must be released into the female reproductive tract during the limited time when sperm are present. Most mammals, for example, copulate only at certain seasons of the year or when the female signals readiness to mate. The season or mating signal typically coincides with ovulation (the release of an egg from an ovary). In some animals, such as rabbits, courtship and mating stimulate ovulation. An alternative method, employed by many female snails and insects, is to store sperm for days, weeks, or even months, thus ensuring a supply of sperm whenever eggs are produced.

25.2 How Does the Human Reproductive System Work?

Humans and all other mammals have separate sexes, copulate, and fertilize their eggs internally. Unlike most other mammalian species, however, human reproduction is not restricted by season. Men produce sperm more or less continuously, and women ovulate (release an egg cell) about once a month.

The Male Reproductive System Includes the Testes and Accessory Structures

The male reproductive system includes the gonads (the testes), which produce sperm, along with several glands and ducts that secrete substances that activate and nourish sperm, store them, and conduct them to the female reproductive tract (**Table 25-1** and **Fig. 25-8**).

(a) Beetles mate on a dandelion flower

(b) King penguins mate in the snow

▲ Figure 25-7 Internal fertilization allows reproduction on land

Web Animation Human Reproductive System

Table 25-1	The Human Male Reproductive Tract
Structure	**Function**
Testis (male gonad)	Produces sperm and testosterone
Epididymis and vas deferens (ducts)	Store sperm; conduct sperm from the testes to the urethra
Urethra (duct)	Conducts semen from the vas deferens and urine from the urinary bladder to the tip of the penis
Penis	Deposits sperm in the female reproductive tract
Seminal vesicles (glands)	Secrete fluid into semen
Prostate gland	Secretes fluid into semen
Bulbourethral glands	Secrete fluid into semen

▶ Figure 25-8 **The human male reproductive tract** The male testes hang beneath the abdominal cavity in the scrotum. Sperm pass from the testis to the epididymis, through the vas deferens and urethra to the tip of the penis. Along the way, fluids are added from the seminal vesicles, the bulbourethral glands, and the prostate gland.

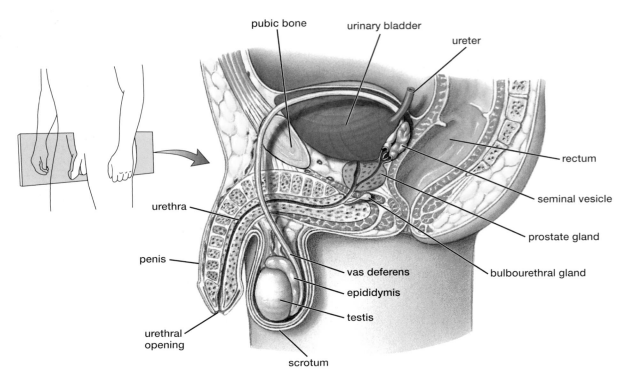

(a) A section through the testis

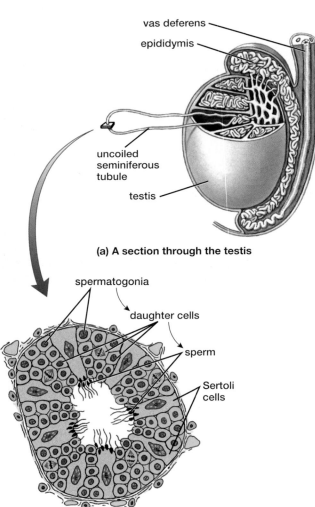

(b) Cross-section of a seminiferous tubule

Sperm Are Produced in the Testes

The testes are located in the **scrotum,** a pouch that hangs outside the abdominal cavity. This location keeps the testes about 7°F (4°C) cooler than the core of the body, providing the optimal temperature for sperm development. Coiled, hollow **seminiferous tubules,** in which the sperm are produced, nearly fill each testis (**Fig. 25-9a**). In the spaces between the tubules are cells that produce the male sex hormone, testosterone.

Just inside the wall of each seminiferous tubule lie *spermatogonia* (singular, spermatogonium), diploid cells from which sperm arise, and much larger *Sertoli cells*, which nourish developing sperm and regulate their growth (**Fig. 25-9b**). When a spermatogonium undergoes mitotic cell division, it forms two types of daughter cells (**Fig. 25-10**). One daughter cell remains a spermatogonium, ensuring a steady supply of spermatogonia throughout a male's life. The other daughter cell becomes committed to **spermatogenesis,** a series of developmental events that produces haploid sperm.

Spermatogenesis begins as this "committed" daughter cell grows and differentiates into a *primary spermatocyte*. Each primary spermatocyte then undergoes meiotic cell division (see pp. 123–126). At the end of meiosis I, each primary spermatocyte gives rise to two haploid *secondary spermatocytes*. Meiosis II produces two *spermatids* from each secondary spermatocyte. Thus, each primary spermatocyte generates four spermatids. The spermatids undergo further differentiation to become sperm, which enter the central cavity of the seminiferous tubule.

A human sperm is unlike any other human cell. Most of the cytoplasm disappears, leaving a haploid nucleus that nearly fills the *head* (**Fig. 25-11**). Atop the nucleus lies a specialized lysosome called the *acrosome*. The acrosome contains enzymes that dissolve protective layers around the egg, enabling the sperm to enter and fertilize it. Behind the head is the *midpiece*, which is packed with mito-

◀ Figure 25-9 **Sperm are produced in the testes** *(a)* A section of the testis, showing the seminiferous tubules, epididymis, and vas deferens. *(b)* The walls of the seminiferous tubules are lined with Sertoli cells and spermatogonia. As spermatogonia divide, the daughter cells move inward. Mature sperm are freed into the central cavity of the tubules for transport to the penis.

chondria. The mitochondria provide the energy needed to move the *tail*, which protrudes out the back. Whiplike movements of the tail, which is really a long flagellum, propel the sperm through the female reproductive tract.

Sperm Production Begins at Puberty

Spermatogenesis begins at *puberty*, a period of rapid growth and transition to sexual maturity. At puberty, cells in the testes begin to produce testosterone, which initiates spermatogenesis. Testosterone also stimulates the development of *secondary sexual characteristics* (such as the growth of facial hair in males), maintains sexual drive, and is required for successful intercourse (copulation).

Accessory Structures Contribute to Semen and Conduct the Sperm Outside the Body

The seminiferous tubules merge to form the **epididymis,** a long, coiled tube (see Fig. 25-9a). The epididymis leads into a duct called the **vas deferens,** which leaves the scrotum and enters the abdominal cavity. Most of the hundreds of millions of sperm produced each day are stored in the epididymis and vas deferens. The vas deferens joins the **urethra,** which connects the bladder to the tip of the **penis.** The urethra conducts urine out of the body during urination, and conducts sperm out of the body during ejaculation, a reflex that occurs as a result of sexual stimulation.

The fluid ejaculated from the penis, called *semen*, is about 5% sperm, mixed with secretions from three glands that empty into the vas deferens or urethra: the *seminal vesicles*, the *prostate gland*, and the *bulbourethral glands* (see Fig. 25-8 and Table 25-1). About 60% of the semen is fluid produced by the seminal vesicles. The alkaline pH of the fluid protects the sperm from the acidity of urine in the man's urethra and from acidic secretions in the woman's vagina. The fluid contains fructose, which provides energy for sperm swimming, and prostaglandins, which are hormone-like molecules that stimulate uterine contractions, which help to transport sperm up the female reproductive tract. The prostate gland produces a nutrient-rich secretion that comprises about 30% of the semen volume and includes enzymes that increase the fluidity of the semen after it is released into the vagina, allowing the sperm to swim more freely. Bulbourethral glands secrete a small amount of alkaline mucus into the urethra; this mucus neutralizes any remaining traces of acidic urine.

The Female Reproductive System Includes the Ovaries and Accessory Structures

The female reproductive tract is almost entirely contained within the abdominal cavity (Table 25-2 and Fig. 25-12). It consists of a pair of gonads—the ovaries—and structures that accept sperm, conduct the sperm to the egg, and nourish the developing **embryo,** an organism in its early stages of development.

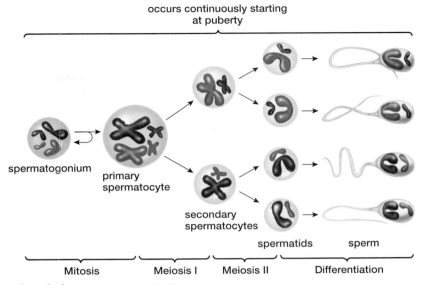

occurs continuously starting at puberty

spermatogonium · primary spermatocyte · secondary spermatocytes · spermatids · sperm

| Mitosis | Meiosis I | Meiosis II | Differentiation |

▲ **Figure 25-10 Spermatogenesis** A spermatogonium divides by mitotic cell division; one of its two daughter cells remains a spermatogonium, while the other enlarges and becomes a primary spermatocyte, committed to meiotic cell division followed by differentiation, producing haploid sperm. Although only 4 chromosomes are shown for clarity, in humans, the diploid number is 46 and the haploid number is 23.

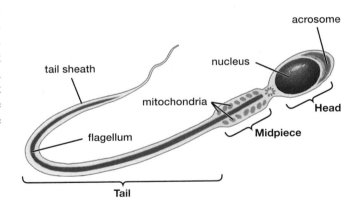

acrosome · nucleus · tail sheath · mitochondria · Head · flagellum · Midpiece · Tail

▲ **Figure 25-11 A human sperm cell** A mature sperm is a cell equipped only with the elements needed for locomotion and fertilization. **QUESTION:** What function does each part of the sperm perform?

Table 25-2 The Human Female Reproductive Tract

Structure	Function
Ovary (female gonad)	Produces eggs, estrogen, and progesterone
Fimbria (opening of uterine tube)	Bears cilia that sweep the egg into the oviduct
Uterine tube	Conducts the egg to the uterus; site of fertilization
Uterus	Muscular chamber where the fetus develops
Cervix	Closes off the lower end of the uterus
Vagina	Receptacle for semen; birth canal

▶ **Figure 25-12 The human female reproductive tract** Eggs are produced in the ovaries and enter the uterine tube. Sperm and egg usually meet in the uterine tube, where fertilization and very early development occur. The early embryo embeds in the lining of the uterus, where development continues. The vagina receives sperm and serves as the birth canal.

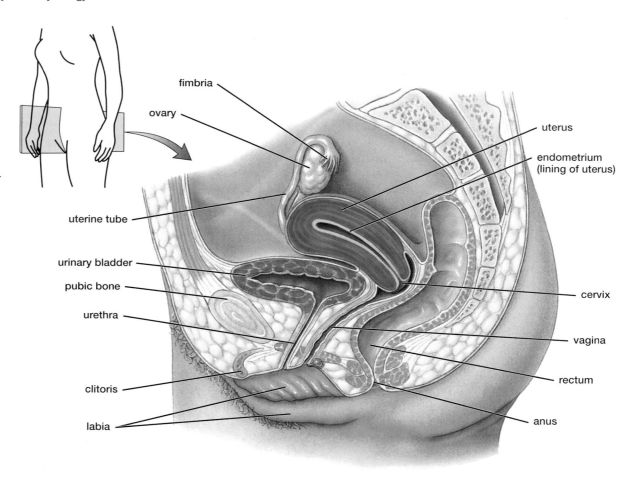

Egg Production in the Ovaries Begins Before Birth

The formation of egg cells, or the process of **oogenesis** (Fig. 25-13), begins in the developing ovaries of a female **fetus** (an embryo that is sufficiently developed to be recognizably human). The process begins with the formation of precursor egg cells called *oogonia* (singular, oogonium). By the end of the third month of fetal development, all of the oogonia have enlarged and differentiated to become *primary oocytes*. As fetal development continues, meiotic cell division begins in all the primary oocytes, but is halted at prophase of meiosis I.

At birth, a lifetime supply of primary oocytes is already in place, and no new ones are generated later in life. The ovaries start out with about 2 million primary oocytes. Many primary oocytes die each day, but at puberty (usually 11 to 14 years of age) about 400,000 still remain. This is plenty, because only a few oocytes resume meiotic cell division during each month of a woman's reproductive span (from puberty to *menopause* at about age 50).

Egg Development Resumes After Puberty

Each primary oocyte is surrounded by a layer of much smaller cells that, together with the oocyte, make up a **follicle** (Fig. 25-14, step ❶). Roughly every month after puberty, the hormonal changes of the **menstrual cycle** stimulate the development

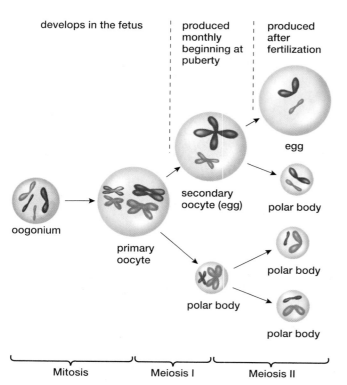

◀ **Figure 25-13 Oogenesis** An oogonium enlarges to form a primary oocyte. At meiosis I, almost all of the cytoplasm is included in one daughter cell, the secondary oocyte. The other daughter cell is a small polar body that contains chromosomes but little cytoplasm. At meiosis II, almost all of the cytoplasm of the secondary oocyte is included in the egg, and a second small polar body discards the remaining "extra" chromosomes. The first polar body may also undergo the second meiotic division. In humans, meiosis II does not occur unless a sperm penetrates into the egg.

of about a dozen follicles (**Fig. 25-14**, steps **2**, **3**). The small follicle cells multiply, providing nourishment for the developing oocyte. In response to hormones secreted by the anterior pituitary, they also release estrogen into the bloodstream. Normally, only one follicle completely matures during each menstrual cycle. We discuss the menstrual cycle in more detail on p. 494.

A Mature Oocyte Leaves the Ovary

As a follicle develops, its primary oocyte completes meiosis I, dividing into a single *secondary oocyte* and a *polar body*, a small cell that is little more than a discarded set of chromosomes (see Fig. 25-13). Meiosis II will not occur until fertilization. As the follicle matures, it grows, eventually erupting through the surface of the ovary and releasing the secondary oocyte, a process called **ovulation** (**Fig. 25-14**, step **4**). The secondary oocyte then travels through the tube positioned above the ovary, called the **uterine tube,** or *oviduct*. For convenience, we will refer to the ovulated secondary oocyte as the *egg*.

Some of the follicle cells accompany the egg, but most remain in the ovary (**Fig. 25-14**, step **5**), where they enlarge, forming the **corpus luteum** (**Fig. 25-14**, step **6**). The corpus luteum secretes the hormones estrogen and progesterone, which stimulate growth of the uterine lining. If the egg is not fertilized, the corpus luteum degenerates a few days later (**Fig. 25-14**, step **7**).

Fertilization Usually Occurs in the Uterine Tube

Each ovary is adjacent to, but not attached to, a uterine tube (see Fig. 25-12). The open end of the uterine tube is fringed with ciliated "fingers" called *fimbriae*, which nearly surround the ovary. The cilia create a current that sweeps the egg into the uterine tube. Once inside the tube, it may encounter a sperm and be fertilized. The **zygote,** as the fertilized egg is called, is swept down the uterine tube by muscular contractions and ciliary action, and released into the pear-shaped **uterus.** There it will develop for 9 months.

The Uterus Has Two Functions: Nourishment of the Embryo and Childbirth

The wall of the uterus has two layers. The inner lining, or *endometrium*, is richly supplied with blood vessels. The endometrium will form the mother's contribution to the **placenta,** the structure that transfers oxygen, carbon dioxide, nutrients, and wastes between fetus and mother. The outer muscular wall of the uterus, the *myometrium*, contracts strongly during delivery, squeezing the infant out into the world. The outer end of the uterus is nearly closed off by the **cervix,** a ring of connective tissue. The cervix holds a developing baby in the uterus, expanding only at the onset of labor to permit passage of the child. Beyond the cervix is the **vagina,** which opens to the outside. The vagina serves both as the receptacle for the penis and sperm during intercourse and as the birth canal.

Uterine Tissues Are Shed If the Egg Is Unfertilized

If an egg is fertilized, it encounters a rich environment for growth, because hormones secreted by developing follicles and

▼ **Figure 25-14 Follicle development** For convenience, this diagram shows all of the stages of follicle development throughout a complete menstrual cycle, going clockwise from lower right. In a real ovary, all the stages would never be present at the same time, and the follicle would not move around the ovary as it develops.

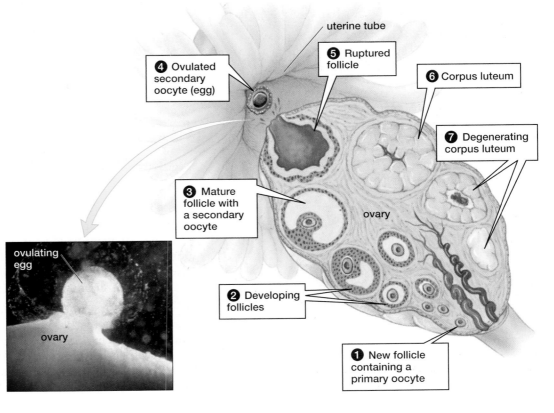

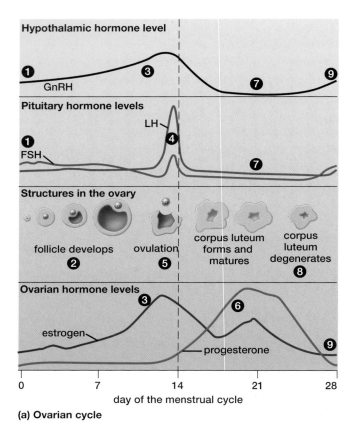

(a) Ovarian cycle

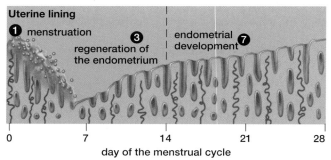

(b) Uterine cycle

▲ **Figure 25-15 Hormonal control of the menstrual cycle** The menstrual cycle consists of two related cycles: **(a)** the ovarian cycle and **(b)** the uterine cycle. These cycles are generated by interactions among the hormones of the hypothalamus, the anterior pituitary, and the ovaries. The circled numbers refer to the interactions discussed in the text.

Web Animation Hormonal Control of the Menstrual Cycle

the corpus luteum stimulate the lining of the uterus to grow an extensive network of blood vessels and nutrient-producing glands. If the egg is not fertilized, however, the corpus luteum disintegrates, hormone levels fall, and the enlarged endometrium disintegrates as well. The myometrium of the uterus contracts (sometimes causing menstrual cramps) and expels the endometrial tissue. The resulting flow of tissue and blood is called **menstruation.**

Hormonal Interactions Control the Menstrual Cycle

The menstrual cycle (**Fig. 25-15**) actually consists of two cycles: (1) the ovarian cycle (**Fig. 25-15a**), in which interactions of hormones produced by the hypothalamus (in the brain), the anterior pituitary gland, and the ovaries drive the development of follicles, maturation of oocytes, and conversion of the follicle cells after ovulation into the corpus luteum; and (2) the uterine cycle (**Fig. 25-15b**), in which estrogen and progesterone produced by the ovaries drive the development of the endometrium of the uterus. The numbers in the descriptions that follow correspond to the numbers in the illustration of the ovarian cycle (Fig. 25-15a). In the "typical" 28-day menstrual cycle, the beginning of menstruation is designated as day 1, because this is an easily observed phenomenon, even though the hormonal events that drive the cycle actually begin a day or two earlier.

❶ The hypothalamus spontaneously releases gonadotropin releasing hormone (GnRH; top panel). GnRH stimulates the anterior pituitary to release follicle-stimulating hormone (FSH; blue line, second panel) and luteinizing hormone (LH; red line, second panel). Typically, at this time, the endometrium of the uterus is still being sloughed off (step ❶ in **Fig. 25-15b**).

❷ FSH stimulates the development of several follicles within each ovary (third panel). Under the combined influences of FSH and LH from the anterior pituitary and estrogen from the follicles, the follicles grow. The primary oocyte within each follicle enlarges, storing food and other substances that will be used by the fertilized egg during early development. Usually, only one follicle completes development each month.

❸ As the follicle matures, it secretes increasing amounts of estrogen (purple line, fourth panel). Estrogen has three effects. First, it promotes the continued development of the follicle and of the primary oocyte within it (third panel). Second, it stimulates the growth of the uterine lining (step ❸ in **Fig. 25-15b**). Third, estrogen stimulates more release of GnRH (top panel).

❹ Increased GnRH stimulates a surge of LH at about day 14 of the cycle (second panel). The increase in LH has three important consequences. First, it triggers the resumption of meiosis I in the oocyte, producing the secondary oocyte and the first polar body.

❺ Second, the LH surge causes the follicle to mature and release the oocyte (ovulation; third panel). Third, it transforms the remnants of the follicle into the corpus luteum.

❻ The corpus luteum secretes both estrogen (purple line, fourth panel) and progesterone (green line, fourth panel).

❼ The combination of estrogen and progesterone inhibits GnRH production by the hypothalamus, reducing the release of FSH and LH, and thereby preventing the development of more follicles. Simultaneously, estrogen and progesterone stimulate the uterine lining to develop a network of blood vessels and nutrient-producing glands to prepare to nourish the developing embryo (step ❼ in **Fig. 25-15b**).

❽ If the egg is not fertilized, the corpus luteum starts to disintegrate about 12 days after ovulation (third panel). This occurs because the corpus luteum cannot survive without LH (or a similar hormone released by the developing embryo, described below). Because progesterone secreted by the corpus luteum shuts down LH production, the corpus luteum actually causes its own destruction.

9 With the corpus luteum gone, estrogen and progesterone levels plummet. Deprived of stimulation by estrogen and progesterone, the endometrium of the uterus dies, and its blood and tissue are shed. This shedding forms the menstrual flow that begins about the 28th day of the cycle. The reduced level of circulating progesterone no longer inhibits the hypothalamus, so the spontaneous release of GnRH resumes. The release of GnRH in turn stimulates the release of FSH and LH, initiating the development of a new set of follicles and restarting the cycle (back to step **1** in both the ovarian and uterine cycles).

The Embryo Sustains Pregnancy

During early pregnancy, the embryo itself prevents the menstrual cycle. Shortly after the ball of cells formed by the dividing fertilized egg implants in the endometrium, it starts secreting an LH-like hormone called *chorionic gonadotropin* (CG). This hormone travels in the bloodstream to the ovary, where it keeps the corpus luteum alive and secreting estrogen and progesterone. These hormones continue to stimulate the development of the endometrium, nourishing the embryo. The early embryo releases so much CG that the hormone is excreted in the mother's urine; most pregnancy tests check for the presence of CG in urine.

Copulation Allows Internal Fertilization

As terrestrial mammals, humans use internal fertilization to deposit sperm into the moist environment of the female's reproductive tract.

During Copulation, Sperm Are Ejected into the Female's Vagina

The male role in copulation begins with erection of the penis. Before erection, the penis is relaxed (flaccid), because the arterioles that supply it are constricted, allowing little blood flow (**Fig. 25-16a**). Under psychological and physical stimulation, the arterioles dilate and blood flows into spaces in the tissue within the penis. As these tissues swell, they squeeze off the veins that drain the penis (**Fig. 25-16b**). Blood pressure increases, causing an erection. After the penis is inserted into the vagina, movements further stimulate touch receptors on the penis, triggering male orgasm, a feeling of intense pleasure and release that is accompanied by ejaculation. Ejaculation forces semen out of the penis and into the vagina when muscles encircling the epididymis, vas deferens, and urethra contract. On average, about three-quarters of a teaspoon of semen (about 3 to 4 milliliters), containing 300 million to 400 million sperm, are ejaculated.

Copulation induces similar changes in females. Sexual excitement causes increased blood flow to the vagina and to the external parts of the reproductive tract. These external parts consist of the **labia,** paired folds of skin, and the **clitoris,** a small rounded projection (see Fig. 25-12). Stimulation of the clitoris (for example, by the penis) may cause the clitoris to become engorged with blood and give rise to orgasm, a series of rhythmic contractions of the vagina and uterus accompanied by sensations of pleasure and release. Female orgasm is not necessary for fertilization.

Intimate contact during copulation creates a situation in which disease organisms can readily be transmitted. Ever since the "sexual revolution," which began in the 1960s, many people have had multiple sexual partners, and the incidence of *sexually transmitted diseases* (STDs) has greatly increased (see "Health Watch: Sexually Transmitted Diseases" on pp. 496–497).

During Fertilization, the Sperm and Egg Nuclei Unite

Neither sperm nor egg lives very long on its own. Sperm, under ideal conditions, may live for 2 days, and an unfertilized egg remains viable for only a day or so. Therefore, fertilization can succeed only if copulation occurs within about a day before or after ovulation.

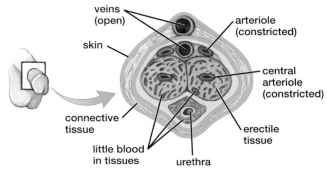

(a) Relaxed

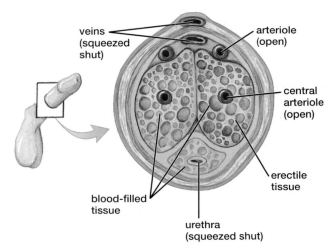

(b) Erect

▲ **Figure 25-16 Changes in blood flow within the penis cause erection (a)** Most of the time, the smooth muscles that encircle the arterioles leading into the penis are contracted, limiting blood flow. **(b)** During sexual excitement, these muscles relax, and blood flows into spaces within the penis. The swelling penis squeezes the veins through which blood leaves the penis, thereby increasing the pressure exerted by fluids within the penis and causing it to become elongated and firm.

Sexually Transmitted Diseases

Sexually transmitted diseases (STDs) are caused by viruses, bacteria, protists, or arthropods that infect the sexual organs and reproductive tract. As their name implies, they are transmitted primarily through sexual contact.

Bacterial Infections

Gonorrhea is a common STD often called "the clap." The U.S. Centers for Disease Control and Prevention (CDC) estimates that there are about 700,000 new cases of gonorrhea each year in the United States. Gonorrhea bacteria penetrate membranes lining the urethra, anus, cervix, uterus, uterine tubes, and throat. Males may experience painful urination and discharge of pus from the penis; female symptoms are often mild and include a vaginal discharge or painful urination. Gonorrhea can be treated with antibiotics. However, because many infected individuals experience few or no symptoms, they may not seek treatment and may, therefore, continue to infect their sexual partners. Gonorrhea can lead to infertility by blocking the uterine tubes with scar tissue. The bacteria also attack the eyes of newborns of infected mothers and were once a major cause of blindness. Today, most newborns are immediately given antibiotic eyedrops to kill the bacteria.

Syphilis is relatively rare; in the United States, fewer than 35,000 cases are reported to the CDC each year. Syphilis bacteria penetrate the mucous membranes of the genitals, lips, anus, or breasts. Because they don't survive prolonged exposure to air, syphilis bacteria are spread almost entirely by intimate contact. Syphilis begins with a sore at the site of infection. If untreated, syphilis bacteria spread through the body, damaging many organs including the skin, kidneys, heart, and brain, in some cases with fatal results. As with gonorrhea, syphilis is easily cured with antibiotics, but many people in the early stages of syphilis infection have mild symptoms and do not seek treatment. Syphilis can be transmitted to the fetus during pregnancy. Some infected infants are stillborn or die shortly after birth; others suffer damage to the skin, teeth, bones, liver, and central nervous system.

Chlamydia is the most common bacterial STD, with about 2.8 million new cases in the United States annually. Chlamydia causes inflammation of the urethra in males and of the urethra and cervix in females. In many cases, there are no obvious symptoms, so the infection goes untreated and spreads. The chlamydia bacterium can infect and block the uterine tubes, resulting in sterility. Chlamydial infection can cause eye inflammation in infants born to infected mothers and is a major cause of blindness in developing countries.

Viral Infections

Acquired immune deficiency syndrome, or **AIDS,** is caused by the human immuno-deficiency virus (HIV) and is discussed in Chapter 22. It is spread primarily by sexual activity, contaminated blood and needles, and from mother to newborn. Because syphilis usually causes sores through which the HIV virus can enter the genitals, syphilis increases the chance of HIV infection by two- to fivefold. HIV attacks the immune system, leaving the victim vulnerable to a variety of infections. There is no cure, but drug combinations can prolong life considerably.

Genital herpes is extremely common; the CDC estimates that 20% of American adults have had genital herpes infections. Genital herpes causes painful blisters on the genitals and surrounding skin and is transmitted by intercourse, primarily when blisters are present. Herpes remains in the body, emerging unpredictably, possibly in response to stress. Antiviral drugs can reduce the severity of outbreaks. A pregnant woman with an active

---

When it leaves the ovary, the egg is surrounded by follicle cells. These cells, now called the *corona radiata*, form a barrier between the sperm and the egg (**Fig. 25-17a**). A second barrier, the jelly-like *zona pellucida*, lies between the corona radiata and the egg. Recent research suggests that the human egg releases a chemical attractant that lures the sperm toward it.

In the uterine tube, hundreds of sperm (out of the hundreds of millions that were ejaculated) reach the egg and encircle the corona radiata (**Fig. 25-17b**). Each sperm releases enzymes from its acrosome. These enzymes weaken both the corona radiata and the zona pellucida, allowing the sperm to wriggle through to the egg. When the first sperm contacts the surface of the egg, the plasma membranes of egg and sperm fuse, and the sperm's head is drawn into the egg's cytoplasm.

As the sperm enters, it triggers two important changes in the egg. First, vesicles near the surface of the egg release chemicals into the zona pellucida, reinforcing it and preventing additional sperm from entering the egg. Second, the egg undergoes meiosis II, producing a haploid nucleus. The haploid nuclei of sperm and egg then fuse, forming a diploid nucleus that contains all the genes of a new human being.

Defects in the male or female reproductive systems can prevent fertilization. For example, a blocked uterine tube can prevent sperm from reaching the egg. Men who produce too few sperm are unlikely to father a child through sexual

case of genital herpes can transmit the virus to the developing fetus, in very rare cases causing mental or physical disability or stillbirth. Herpes can also be transmitted during childbirth.

Human papillomavirus (HPV) (Fig. E25-1a) infects an estimated 50% of sexually active individuals, including 80% of women, at some time in their lives. Most show no symptoms and recover from the infection without knowing they had it. The virus may cause warts on the labia, vagina, cervix, or anus in females and on the penis, scrotum, groin, or thighs in males. The warts usually disappear, or they can be removed. HPV is of concern because it can cause cervical cancer, which kills nearly 4,000 women each year in the United States. In 2006, the U.S. Food and Drug Administration approved a vaccine against the forms of HPV that cause most cases of cervical cancer. If given to young women before they become sexually active, the vaccine could greatly reduce cervical cancer rates in the future.

Protist and Arthropod Infections

Trichomoniasis is caused by a protist that colonizes the mucous membranes lining the urinary tract and genitals of both males and females. About 7.5 million Americans contract trichomoniasis each year. Symptoms may include itching or burning sensations, and sometimes a discharge from the penis or vagina. The protist is usually spread by intercourse but

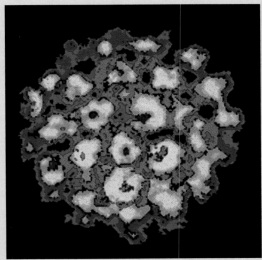

(a) The human papillomavirus

(b) Pubic lice

▲ **Figure E25-1 Agents of sexually transmitted diseases** *(a)* The DNA of some strains of the human papilloma virus becomes inserted into chromosomes of cells of the cervix. Proteins synthesized from the instructions in the viral DNA promote uncontrolled cell division, causing cancer. *(b)* Pubic lice are called "crabs" because of their shape and the tiny claws on the ends of their legs, which they use to cling to pubic hair.

can also be acquired from contaminated clothing and toilet articles. Trichomoniasis can be cured with the drug metronidazole. Prolonged infections can cause sterility.

Crab lice, also called pubic lice (Fig. E25-1b), are tiny insects that live and lay their eggs in pubic hair. Their mouthparts are

adapted for penetrating skin and sucking blood and body fluids, a process that causes severe itching. "Crabs" are not only irritating, they can also spread infectious diseases. Crab lice are usually transmitted through sexual contact. They can be killed by insecticides such as permethrin or pyrethrin.

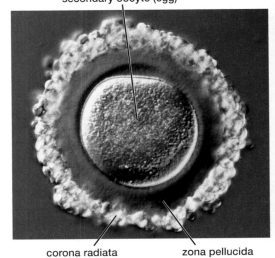

secondary oocyte (egg)

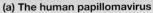

corona radiata   zona pellucida

(a) An ovulated secondary oocyte

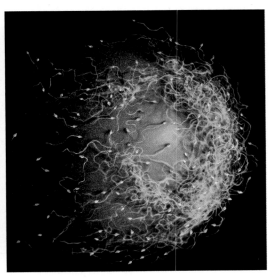

(b) Sperm surrounding an oocyte

◀ **Figure 25-17 The secondary oocyte and fertilization** *(a)* A human secondary oocyte shortly after ovulation. Sperm must digest their way through the small follicular cells of the corona radiata and the clear zona pellucida to reach the oocyte itself. *(b)* Sperm on the surface of an oocyte, attacking its defensive barriers. **QUESTION:** Why is it important that many sperm reach the egg?

scientific inquiry

High-Tech Reproduction

A surprisingly large number of couples have difficulty conceiving a child, but our ability to manipulate biological processes has led to major advances in assisted reproduction. For women who do not ovulate regularly, injection of fertility drugs or hormones stimulates ovulation. However, fertility drugs usually cause multiple ovulation, with the result that the rate of multiple births in the United States has quadrupled since 1971. This is a significant drawback, because multiple births are much riskier than single births, for both the mother and her children.

About 40,000 babies born in the United States each year are conceived in a shallow glass dish by *in vitro fertilization* (IVF; *in vitro* literally means "in glass"). IVF is a complex and delicate procedure. First, the woman is given fertility drugs to stimulate multiple ovulation. Surgeons then thread a hollow needle up to the ovaries and suck out the ripe follicles, from which oocytes are extracted. The oocytes are incubated in a dish with freshly collected sperm (about 75,000 sperm per oocyte). In 2 to 3 days, some of the oocytes will have been fertilized, begun dividing, and reached the eight-cell stage. Two to four of these embryos are sucked into a tube and expelled very gently into the uterus. (Extra embryos can be frozen for later use in case the first attempt at implantation is unsuccessful.) Transplanting multiple embryos increases the success rate for implantation, but of course also increases the probability of multiple births. The popularity of IVF, which can cost about $10,000 to $15,000 per attempt, is a testimony to the strong biological drive to have children. With an average success rate of about 30%, a typical conception via IVF costs about $35,000.

Intracytoplasmic sperm injection (ICSI) allows men whose sperm are incapable of swimming to and fertilizing an egg to father their own children. In ICSI, immature sperm, or even sperm precursor cells, are extracted from the testes. (Apparently, sperm are capable of fertilization very early in their developmental process, and further maturation mainly provides them with the ability to swim to the egg and penetrate its protective layers.) Then a tiny, sharp pipette is used to pierce the plasma membrane of an egg cell and inject a single sperm directly into the egg's cytoplasm (**Fig. E25-2**).

Assisted reproduction provides opportunities—and controversies—that could only have been imagined just a few years ago. For example, a widow can be impregnated by her dead husband's frozen sperm. Recently, twins were born to a woman whose ovaries could not produce eggs. Her children came from donor eggs that had been frozen for 2 years. Alternatively, some couples produce viable eggs and sperm, but the woman cannot carry a child to term (for instance, if she has had a hysterectomy). If they are lucky or wealthy, another woman, called a surrogate mother, may be willing to have their embryo, usually produced by IVF, implanted in her uterus for development.

Using assisted reproductive technology, parents can also influence the gender of their child. This may be medically important if the parents are carriers of sex-linked disorders, but some parents are just seeking to balance their families. Because the Y chromosome is so small, sperm carrying a Y chromosome have 2.8% less DNA than X-bearing sperm, and this difference can be used to sort the sperm. So far, sperm

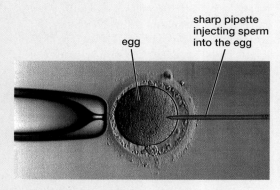

▲ **Figure E25-2** **Injecting a sperm into an egg cell** An egg, held in place by a glass pipette, is injected with a sperm cell that is placed directly into the egg's cytoplasm.

sorting has been most successful at increasing the percentage of X sperm and thereby increasing the probability that a couple will have a girl. More recently, it has become possible to determine the gender of an IVF embryo before implantation. Although not yet routine, this method would allow a couple to select the gender of their child with absolute certainty.

As these examples show, biomedical technology has the potential to alter how humans reproduce (wealthy humans, at least). In 2006, an article in the popular science magazine *New Scientist* half-jokingly suggested that in the not-too-distant future, sex might be for fun, but for reproduction, people would use IVF combined with preimplantation diagnosis and/or genetic engineering. Assuming that such a future becomes possible, do you think that this is a good idea?

intercourse. Today, some couples seek high-technology help in the form of in vitro fertilization (see "Scientific Inquiry: High-Tech Reproduction").

25.3 How Does Animal Development Proceed?

Development, the transformation from a fertilized egg to a multicellular organism, is a beautiful and awe-inspiring process that has been carefully described for a number of animal species. In textbooks, the changes in the developing embryo are generally depicted in stages, but the stages are just convenient "snapshots." The actual development is a smooth, continuous process. A general description of vertebrate development follows.

The Fertilized Egg Develops into a Hollow Ball of Cells

The formation of an embryo begins with **cleavage,** a series of mitotic cell divisions of the fertilized egg. An egg is a very large cell. Cleavage reduces cell size while increasing cell number. Unlike most mitotic cell divisions—in which cells divide, grow, duplicate genetic material, then divide again—cell divisions during cleavage skip the growth phase. Consequently, as cleavage progresses, the cytoplasm is split up into ever-smaller cells whose sizes approach those of cells in the adult organism. Finally, a solid ball of small cells, the *morula,* is formed. The morula as a whole is about the same size as the fertilized egg. As cleavage continues, a cavity opens within the morula. The cells of the morula then become the outer covering of a hollow (typically spherical) structure, the **blastula** (Fig. 25-18a).

Gastrulation Forms Three Tissue Layers

In the next stage of development, many cells of the blastula migrate to new locations, a process called **gastrulation.** Gastrulation begins when a slit called the blastopore forms on one side of the blastula. Blastula cells then migrate through the blastopore in a continuous sheet, much as if you punched in an underinflated basketball. The resulting indentation enlarges to form a cavity that will eventually become the digestive tract (Fig. 25-18b).

During gastrulation, three embryonic tissue layers are formed. The cells that move through the blastopore to line the future digestive cavity become **endoderm** (meaning "inner skin"). The cells remaining on the outside will form the epidermis of the skin and the nervous system and are called **ectoderm** ("outer skin"). Some cells migrate between the endoderm and ectoderm, forming a third layer, the **mesoderm** ("middle skin"). Mesoderm gives rise to muscles, the circulatory system, and the skeleton, including the *notochord*, a supporting rod found at some stage in all chordates (vertebrates and their relatives). The three-layered embryo is called a **gastrula** (Fig. 25-18c).

Adult Structures Develop During Organogenesis

Gradually, the ectoderm, mesoderm, and endoderm arrange themselves into the organs that will be present in the adult (Table 25-3). This process is called **organogenesis.**

Organogenesis proceeds through two major processes. First, a series of "master switch" genes turn on and off, each controlling the transcription and translation of the many genes involved in producing, say, an arm or a backbone. Some of these master genes are first turned on by proteins and RNA deposited in the egg during oogenesis. The activity of other master genes is regulated by chemical signals from nearby structures. For example, the vertebrate retina is actually an outgrowth of the brain. When it reaches the ectoderm where the eye will form, the developing retina releases chemicals that "reprogram" the cells of the ectoderm to produce the lens of the eye.

ectoderm   mesoderm   endoderm

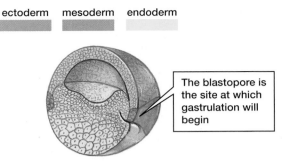

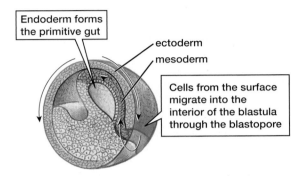

The blastopore is the site at which gastrulation will begin

(a) The blastula just before gastrulation

Endoderm forms the primitive gut

ectoderm

mesoderm

Cells from the surface migrate into the interior of the blastula through the blastopore

(b) Cells migrate at the start of gastrulation

primitive gut

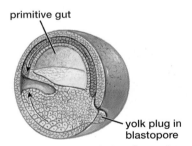

yolk plug in blastopore

(c) A three-layered gastrula has formed

▲ **Figure 25-18 Gastrulation in the frog** *(a)* In the blastula before gastrulation begins, the three embryonic tissue layers have not yet formed. Colors indicate the future fate of the cells after differentiation in the gastrula. *(b)* When gastrulation begins, cells migrate from the surface into the blastula to form the endoderm and mesoderm layers of the gastrula. The cells remaining on the surface will form ectoderm. *(c)* The resulting three-layered embryo is now termed a gastrula.

Table 25-3	Derivation of Adult Tissues and Organs from Embryonic Cell Layers
Embryonic Layer	**Adult Tissue**
Ectoderm	Epidermis of the skin; lining of the mouth and nose; hair; glands of the skin (sweat, sebaceous, and mammary glands); nervous system; lens of the eye; inner ear
Mesoderm	Dermis of the skin; muscle, skeleton; circulatory system; gonads; kidneys; outer layers of the digestive and respiratory tracts
Endoderm	Lining of the digestive and respiratory tracts; liver; pancreas

A second major developmental process is to prune away superfluous cells, much as Michelangelo chiseled away "extra" marble, leaving behind his superb statue of David. In development, this sculpting requires the death of excess cells. For example, a 5-week-old human embryo has a tail and webbed hands and feet. Two weeks later, the webbing cells have died, leaving separate fingers, and the tail cells are dying, shortening the tail.

At least two processes seem to cause cell death during development. Some cells die unless they receive a signal that promotes survival. Neurons that control muscles, for example, are programmed to die unless they form a synapse with a skeletal muscle, which releases a chemical that prevents the death of the neuron. For other cells, the situation is just the reverse; they live unless they receive a signal that induces death.

Development in Reptiles, Birds, and Mammals Depends on Extraembryonic Membranes

Fish live and reproduce in water, usually by spawning. Although many amphibians spend their adult lives on land, they too lay their eggs in water. In both cases, the embryo obtains nutrients from the yolk of the egg and oxygen from the water, and releases its wastes, including carbon dioxide, into the water. Fully terrestrial vertebrate life was not possible until the evolution of the **amniotic egg.** This innovation arose first in reptiles and persists today in that group and its descendants, the birds and mammals. It allows these groups to complete their development into the adult form in their own "private pond." The amniotic egg is characterized by four membranes, called **extraembryonic membranes:** the *chorion*, the *amnion*, the *allantois*, and the *yolk sac*. In reptiles and birds, the **chorion** lines the shell and exchanges oxygen and carbon dioxide between the embryo and the air. The **amnion** encloses the embryo in a watery environment. The **allantois** surrounds and isolates wastes. The **yolk sac** contains the yolk. In placental mammals (all mammals except marsupials, such as kangaroos, and monotremes, such as platypuses), the embryo develops within the mother's body until birth. Nevertheless, all four extraembryonic membranes still persist, remnants of the reptilian genetic program for development, and in fact are essential for mammalian development. In mammals, the amnion still cradles the embryo in its own private pond. The chorion forms the embryo's contribution to the placenta. The blood vessels of the umbilical cord develop from the remnants of the allantois. Although the yolk sac in mammals contains no yolk, it produces red blood cells for the early embryo.

Web Animation Human Development

25.4 How Do Humans Develop?

Figure 24-19 summarizes the stages of human embryonic and fetal development. You may want to refer to this figure as we describe human development in the next few sections.

Differentiation and Growth Are Rapid in the First Two Months

Usually, a human egg is fertilized in the uterine tube and undergoes a few cleavage divisions on its way to the uterus, a journey that takes about 4 days (**Fig. 25-20**). By about 1 week after fertilization, the zygote has developed into a hollow ball of cells, known as the **blastocyst** (the mammalian version of a blastula). One section of the blastocyst wall is thickened to form a group of cells called the *inner cell mass*. The inner cell mass will become the embryo, and each of its cells has the potential to develop into any type of tissue. Researchers hope that such **stem cells** may in future be used to replace damaged

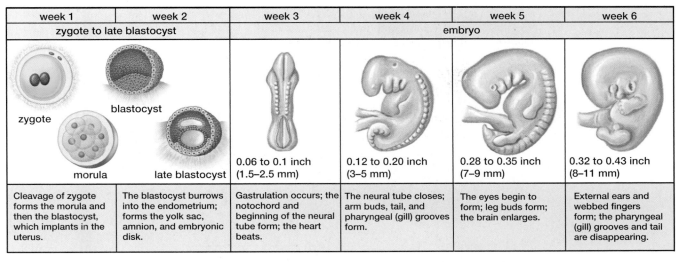

week 1	week 2	week 3	week 4	week 5	week 6
zygote to late blastocyst		embryo			
		0.06 to 0.1 inch (1.5–2.5 mm)	0.12 to 0.20 inch (3–5 mm)	0.28 to 0.35 inch (7–9 mm)	0.32 to 0.43 inch (8–11 mm)
Cleavage of zygote forms the morula and then the blastocyst, which implants in the uterus.	The blastocyst burrows into the endometrium; forms the yolk sac, amnion, and embryonic disk.	Gastrulation occurs; the notochord and beginning of the neural tube form; the heart beats.	The neural tube closes; arm buds, tail, and pharyngeal (gill) grooves form.	The eyes begin to form; leg buds form; the brain enlarges.	External ears and webbed fingers form; the pharyngeal (gill) grooves and tail are disappearing.

week 7	week 8	week 10	week 12	week 16
embryo		fetus		
0.67 to 0.79 inch (1.7–2.0 cm)	0.90 to 1.10 inches (2.3–2.8 cm)	1.25–1.75 inches (3.2–4.4 cm)	2–3 inches (5–7.6 cm)	4–5 inches (10.2–12.7 cm)
Webbed toes form; bones begin to stiffen; the back straightens; the eyelids begin to form.	All the major organs and male genitals begin to form; the arms can bend; fingers are distinct. Facial features and outer ears take shape.	After 8 weeks, the embryo is called a fetus. Red blood cells form; toes separate; eyelids have developed; major brain parts are present; the hands can form fists.	The neck is well defined; all organs are present; male and female genitals are present; arms and legs move; teeth begin to form; a heartbeat can be detected electronically.	Sucking and swallowing movements occur; the liver and pancreas begin functioning. The body has grown relative to the head; major organs continue developing. The mother may feel movement; weight is about 5 oz.

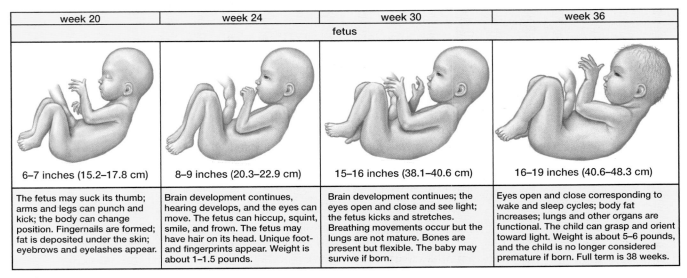

week 20	week 24	week 30	week 36
fetus			
6–7 inches (15.2–17.8 cm)	8–9 inches (20.3–22.9 cm)	15–16 inches (38.1–40.6 cm)	16–19 inches (40.6–48.3 cm)
The fetus may suck its thumb; arms and legs can punch and kick; the body can change position. Fingernails are formed; fat is deposited under the skin; eyebrows and eyelashes appear.	Brain development continues, hearing develops, and the eyes can move. The fetus can hiccup, squint, smile, and frown. The fetus may have hair on its head. Unique foot- and fingerprints appear. Weight is about 1–1.5 pounds.	Brain development continues; the eyes open and close and see light; the fetus kicks and stretches. Breathing movements occur but the lungs are not mature. Bones are present but flexible. The baby may survive if born.	Eyes open and close corresponding to wake and sleep cycles; body fat increases; lungs and other organs are functional. The child can grasp and orient toward light. Weight is about 5–6 pounds, and the child is no longer considered premature if born. Full term is 38 weeks.

▲ Figure 25-19 A calendar of development from blastocyst to birth

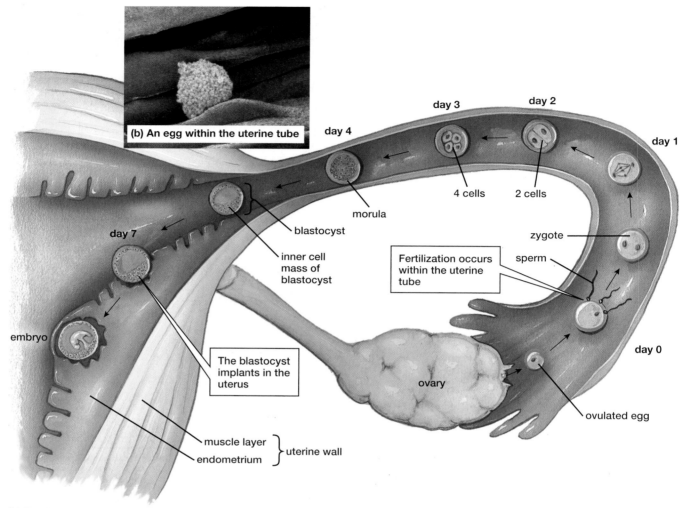

(b) An egg within the uterine tube

day 3

day 2

day 1

day 4

4 cells

2 cells

morula

zygote

blastocyst

sperm

day 7

inner cell mass of blastocyst

Fertilization occurs within the uterine tube

embryo

The blastocyst implants in the uterus

day 0

ovary

ovulated egg

muscle layer

endometrium

} uterine wall

(a) The first week of development

▲ **Figure 25-20 The journey of the egg** *(a)* The egg is fertilized in the uterine tube and slowly travels down to the uterus. Along the way, the zygote divides a few times, until a hollow blastocyst is formed. When the blastocyst reaches the uterine endometrium, it burrows in. *(b)* The egg, surrounded by cells from the follicle (the corona radiata), is cradled in the uterine tube on its way to the uterus.

adult tissues, as described in "Health Watch: The Promise of Stem Cells." The outer wall of the blastocyst (the future chorion) adheres to the uterus and burrows into the endometrium, a process called **implantation** (**Fig. 25-21a**). The complex intermingling of the chorion and the endometrium forms the placenta, which will remove wastes and provide nourishment for the embryo for the rest of the pregnancy. In addition, the chorion secretes the hormone chorionic gonadotropin (CG), which prevents the death of the corpus luteum. The continued secretion of progesterone and estrogen by the corpus luteum sustains the pregnancy for the first couple of months, until the placenta takes over the secretion of these hormones.

After Implantation, Two Cavities Form and Gastrulation Occurs

A few days after implantation, the inner cell mass splits to form two fluid-filled sacs (**Fig. 25-21b**). One sac, bounded by a membrane called the amnion, forms the amniotic cavity. The amnion eventually grows around the embryo, containing the watery environment in which all animal embryos develop. The second sac is called the *yolk sac*, although in humans it contains no yolk. The two sacs are separated by a double layer of cells called the *embryonic disk*, which will eventually develop into the embryo. The side of the embryonic disk facing the amniotic cavity consists of cells that will become ectoderm, and the side facing the yolk sac consists of future endoderm cells. Gastrulation, which forms the ectoderm, mesoderm, and endoderm, begins at the end of the second week after fertilization.

The Promise of Stem Cells

The nucleus of almost every cell in your body contains your entire genetic blueprint. During development, whether a cell becomes muscle, bone, or brain is determined by complex factors in the cell's surroundings that determine which genes are active. Although most of the cells of an early embryo can differentiate into any of the approximately 200 cell types in the body, most adult cells cannot—they have become committed to a specific developmental pathway and cannot be forced out of that path. In contrast to these "committed" adult cells, a stem cell has not yet differentiated and so can give rise to more than one cell type. The medical potential of stem cells is vast—victims of heart attack, stroke, spinal cord injury, and degenerative diseases from arthritis to Parkinson's disease would benefit if their damaged tissues could be regenerated.

There are two sources of stem cells. *Embryonic stem cells* (ESCs) are derived from the inner cell mass of the blastocyst (see Fig. 25-21). In 1998, James Thompson and fellow researchers at the University of Wisconsin first isolated human ESCs, induced them to grow in culture dishes, and then to differentiate into a variety of human tissues (**Fig. E25-3**). The appeal of ESCs is that they can produce any cell type in the body.

Recent research has shown that most adult tissues, including muscle, skin, liver, brain, heart, and blood, contain small numbers of stem cells, called *adult stem cells* (ASCs). In fact, stem cells from bone marrow, which normally produce both red and white blood cells, have been used for decades in transplants to treat diseases such as leukemia. Although it was once thought that ASCs

could only differentiate into a few cell types, researchers have recently been able to coax them into forming more varieties than originally believed possible. Bone marrow stem cells, for example, can differentiate into three types of brain cells, and both cardiac and skeletal muscle cells.

How might stem cells be used in medicine? In one recent experiment, researchers induced heart attacks in mice, and then injected adult stem cells into the damaged heart. The ASCs multiplied and partially repaired the heart muscle. As we described in "Health Watch: Closer to a Cure for Diabetes" in Chapter 23 (p. 451), researchers have been able to coax ESCs into differentiating into insulin-secreting cells, and they hope that this might be the first step toward a permanent cure for type 1 diabetes.

However, there are significant obstacles to surmount before stem cells can be used for effective, routine therapies. One obstacle is the time it would take for a patient's own stem cells to multiply, differentiate, and repair damaged tissue. For example, in principle, heart muscle stem cells could be stimulated to reproduce and replace muscle killed by a heart attack. Alternatively, doctors might harvest stem cells from an area where they are abundant (say, bone marrow), then treat them with specific differentiating factors and inject them into the damaged body part. However, both of these approaches take a long time to produce enough ASCs to be medically useful. Someone with a serious heart attack can't wait weeks or months.

A second obstacle is that ASCs cannot be differentiated into every medically useful cell type, at

least not yet. ESCs *can* differentiate into all cell types. However, because embryos must be destroyed to obtain ESCs (at least with today's technology), many people object to this procedure, even when the embryos are "extras" that are routinely created for in vitro fertilization and usually would be discarded.

A third obstacle is that the immune system will reject ASCs or ESCs that are not genetically identical, or at least a very close match, to the recipient. This has led to proposals for *therapeutic cloning*, which would involve inserting a cell nucleus from an adult donor who needed tissue repair into an egg whose nucleus had been removed, to create embryonic stem cells that would not be rejected. Because one cannot rule out the possibility that this process, if sufficiently developed, could be used to produce a human clone of the donor, it has been highly controversial.

To avoid both immune rejection and destruction of embryos, some parents have turned to a readily available source of "self" stem cells for their children: the placenta. As you know, much of the placenta is formed from the chorion of the embryo. Further, the placenta has many stem cells. Therefore, parents may have a sample of their newborn's placenta frozen at extremely low temperatures, so that stem cells with the child's exact genetic makeup will be available to repair damaged tissue later in life. In the future, researchers hope that genetic engineering techniques could be used to modify cell surface proteins to create stem cells without rejection problems.

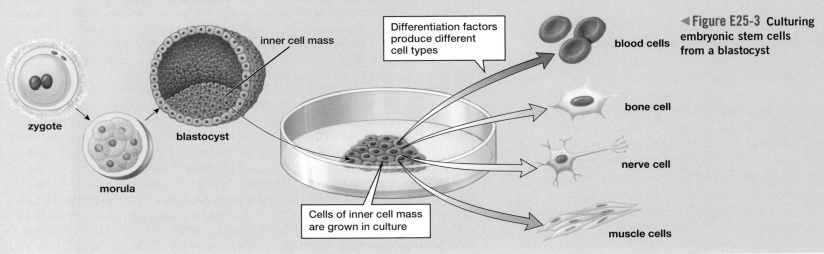

inner cell mass

Differentiation factors produce different cell types

blood cells

◀ **Figure E25-3** Culturing embryonic stem cells from a blastocyst

zygote

blastocyst

morula

bone cell

nerve cell

Cells of inner cell mass are grown in culture

muscle cells

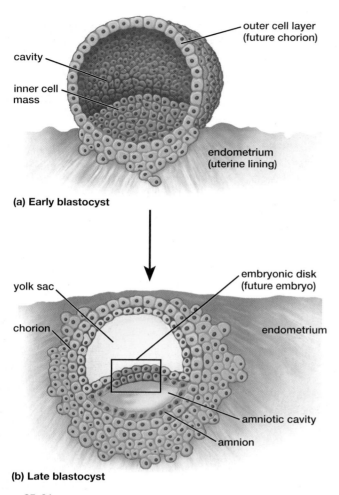

(a) Early blastocyst

outer cell layer (future chorion)

cavity

inner cell mass

endometrium (uterine lining)

yolk sac

embryonic disk (future embryo)

chorion

endometrium

amniotic cavity

amnion

(b) Late blastocyst

▲ **Figure 25-21 Implantation of the blastocyst** *(a)* As the blastocyst burrows into the uterine lining, the outer layer forms the chorion, the embryo's contribution to the future placenta. *(b)* A few days later, the blastocyst has completely submerged beneath the uterine lining. The inner cell mass begins to develop into the yolk sac, the amnion, and the embryonic disk.

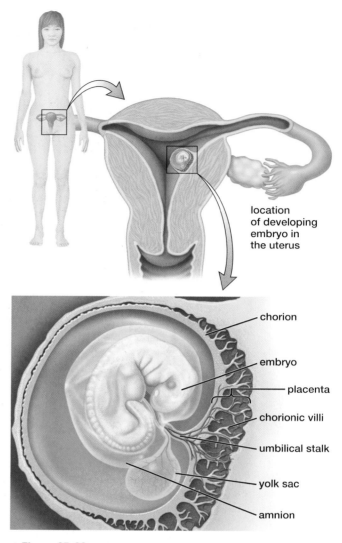

location of developing embryo in the uterus

chorion

embryo

placenta

chorionic villi

umbilical stalk

yolk sac

amnion

▲ **Figure 25-22 Human development during the fourth week** The embryo bulges into the uterus, and the placenta is restricted to one side. The umbilical stalk (future umbilical cord) exchanges wastes and nutrients.

The Faces of Fetal Alcohol Syndrome

Continued

In the developing brain, alcohol disrupts the ability of cells to recognize their neighbors, migrate to their correct locations, and develop the correct connections to other brain cells. The incorrect cell locations and connections can never be repaired. For example, the corpus callosum, the band of axons interconnecting the two cerebral hemispheres (see pp. 470–471), is often smaller in children with FAS, so the left and right sides of the brain cannot communicate effectively with one another. Many brain cells die if correct connections are never made, resulting in a reduction in the size of many brain structures, including the cerebellum, which is important in controlling movement.

Body Structures Begin to Form During Weeks Three to Six

During the third week of development, the embryo, enclosed in its amniotic sac, curls to form the future head. The chorion extends tiny fingers called *chorionic villi* (singular, villus) into the endometrium of the uterus. Embryonic blood vessels invade the villi, carrying blood pumped by the embryonic heart, which has just begun to beat.

As the embryo grows during the fourth week, the endoderm begins to form the digestive tract. A rudimentary tail is visible. Within the embryo, ectoderm is forming structures that will become the brain and spinal cord.

By the beginning of the fourth week, the placenta has become restricted to one side of the embryo (**Fig. 25-22**). The embryo is completely surrounded by the amnion, which is penetrated by the umbilical stalk (the future *umbilical cord*), which links the embryo to the placenta and exchanges wastes and nutrients.

During the fifth and sixth weeks, the embryo develops a notochord, a prominent tail, and pharyngeal (gill) grooves (**Fig. 25-23a**). These three structures are reminders that we share ancestry with other vertebrates that retain them in adulthood. In humans, however, they disappear as development continues. By the seventh week, the embryo has rudimentary eyes, and the fingers and toes on its tiny hands and feet are beginning to become separate (**Fig. 25-23b**). Especially notable at this stage is the rapid growth of the head, which is nearly as large as the rest of the body.

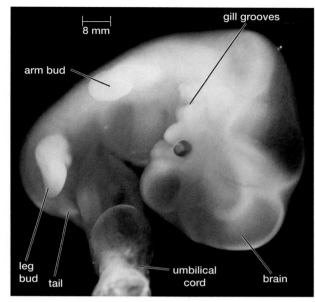

(a) Fifth week

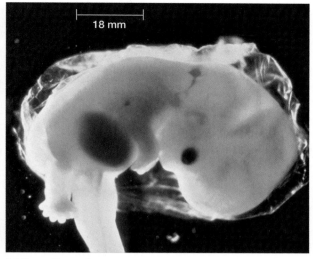

(b) Seventh week

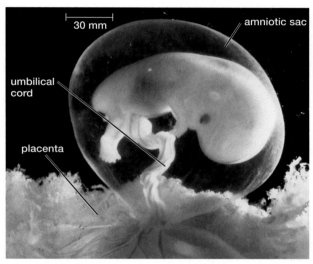

(c) Eighth week

After Two Months, the Embryo Is Recognizably Human

As the second month draws to an end, nearly all of the major organs have formed. The gonads appear and develop into testes or ovaries, depending on the presence or absence of the Y chromosome. Sex hormones are produced, either testosterone from the testes or estrogen from the ovaries. These hormones will affect the future development of several embryonic organs, including the reproductive organs and certain regions of the brain. After the second month of development, the embryo is called a fetus, signifying that the developing embryo has taken on a generally human appearance (Fig. 25-23c).

Rapidly dividing and differentiating cells are extremely susceptible to damage from a variety of causes, including nutritional deficiencies, drugs, medications, industrial pollutants, and radiation. Therefore, during the explosive growth and development of the early embryo, it is particularly vulnerable to many different insults, including alcohol and many drugs.

Growth and Development Continue During the Last Seven Months of Pregnancy

The fetus continues to grow and develop for another 7 months. The brain continues to develop rapidly, and the head remains disproportionately large. As the brain and spinal cord grow, they begin to generate behavior. As early as the third month of pregnancy, the fetus begins to move and respond to stimuli. Some instinctive behaviors appear, including sucking, which will have obvious importance soon after birth. Structures that the fetus will need when it emerges from the uterus, such as the lungs, stomach, and intestine, enlarge and become functional, although they will not be used until after birth.

The Placenta Exchanges Materials Between Mother and Fetus

During the first few weeks of pregnancy, embryonic cells obtain nutrients directly from the nearby cells of the endometrium. Eventually, however, the embryo is nourished through the placenta, which develops from interlocking tissues of the embryo and the mother (Fig. 25-24). The embryo's contribution to the placenta comes from the chorion, which penetrates the mother's endometrium with its chorionic villi. Blood vessels of the umbilical cord connect the embryo's circulatory

▲ Figure 25-23 Human development during the fifth through eighth weeks (a) At the end of the fifth week, the human embryo is about half head. Fingers and toes have begun to develop. A tail and pharyngeal (gill) grooves are clearly visible, evidence of our evolutionary relationship to other vertebrates. (b) By the seventh week, the human form has been more clearly defined by the selective death of cells that form the tail and connect the fingers and toes. The tail has nearly disappeared, and fingers and toes are separated from one another. (c) At the end of the eighth week, the embryo is clearly human in appearance and is now called a fetus. Most of the major organs of the adult body have begun to develop.

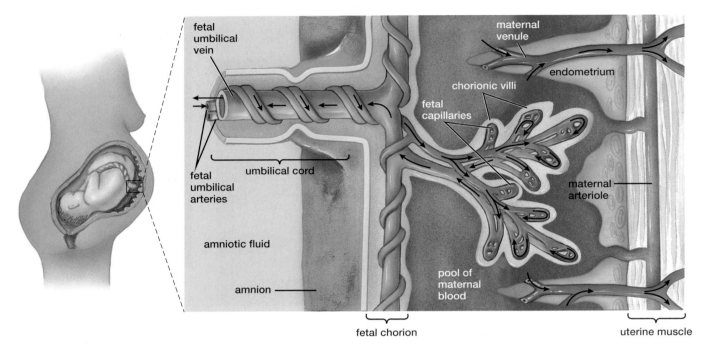

▲ **Figure 25-24 The placenta** The placenta forms from both the chorion of the embryo and the endometrium of the mother. Capillaries of the endometrium break down, releasing blood to form pools within the placenta. Meanwhile, the chorion develops projections (the chorionic villi) that extend into these pools of maternal blood. Blood vessels from the umbilical cord branch extensively within the villi. The maternal and fetal blood supplies are separate. Umbilical arteries carry deoxygenated blood from the fetus to the placenta, and umbilical veins carry oxygenated blood to the fetus. **QUESTION:** A few types of mammals, such as kangaroos, lack a true placenta. What would you predict about the nature of development in such mammals?

The Faces of Fetal Alcohol Syndrome

Continued

Substances that are soluble in water can easily reach all the cells of the body via the blood and extracellular fluid. Substances that are soluble in lipids can easily cross the phospholipid bilayer of the plasma membrane (see p. 46). Alcohol, being soluble both in water and lipids, easily passes through capillary walls—whether in the brain of an adult drinker, the placenta of a pregnant woman, or the body of an embryo. As one FAS support group notes, "What you drink, baby drinks too."

system with a dense network of capillaries in the chorionic villi, which are bathed in pools of maternal blood. This arrangement permits many small molecules to diffuse between fetal blood and maternal blood. Oxygen diffuses from maternal blood to fetal blood, and carbon dioxide from fetal blood to maternal blood. Nutrients, some aided by active transport, travel from mother to fetus. Fetal urea diffuses into the mother's blood, to be filtered out by the mother's kidneys.

The placenta does not allow fetal and maternal blood to mix. The membranes of the fetal capillaries and the maternal chorionic villi allow exchange by diffusion, but they also block the passage of most large proteins and cells. Many harmful chemicals, however—including caffeine, nicotine, and most drugs (both medicinal and "recreational")—can penetrate the placenta.

Some disease-causing organisms, including the viruses that cause AIDS, German measles, and syphilis, can cross the placenta and cause devastating problems for the fetus. Any drug, even aspirin, has the potential to harm the fetus, and any woman who thinks she may be pregnant should seek medical advice about any drugs she takes.

Pregnancy Culminates in Labor and Delivery

During the last months of pregnancy, the fetus usually becomes positioned head downward in the uterus with the crown of the skull resting against (and being supported by) the cervix. Near the end of the ninth month, give or take a couple of weeks, the process of birth normally begins (**Fig. 25-25**). Birth results from a complex interplay between uterine stretching caused by the growing fetus, and fetal and maternal hormones that stimulate **labor,** the contractions of the uterus that result in birth.

Unlike skeletal muscles, uterine muscles can contract spontaneously, and stretching enhances these contractions. As the baby grows, it stretches the uterine muscles, which occasionally contract weeks before delivery. No one knows exactly what triggers labor in humans, but it is probably stimulated by one or more chemicals produced by the fetus as it becomes capable of living outside the womb. Whatever the initial stimulus, the uterine muscles become more likely to contract. As the uterus contracts, it pushes the head of the fetus against the cervix, stretching it. This has two effects. First, expanding the cervix is essential for childbirth, so that the fetus's head can fit through. Second, stretching the cervix sends nervous signals to the mother's brain, causing the release of the hormone oxytocin. Oxytocin stimulates contrac-

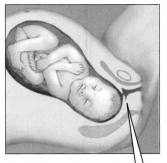
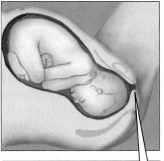
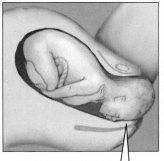
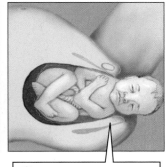

❶ The baby orients head downward, facing the mother's side; the cervix begins to thin and expand in diameter (dilate)

❷ The cervix dilates completely to 10 centimeters (almost 4 inches wide), and the baby's head enters the vagina, or birth canal; the baby rotates to face the mother's back

❸ The baby's head emerges

❹ The baby rotates to the side once again as the shoulders emerge

▲ Figure 25-25 Human childbirth

tions of the uterine muscles, pushing the baby harder against the cervix, which stretches further, causing still more oxytocin to be released. This positive feedback cycle continues until the baby emerges from the vagina.

After a brief rest, uterine contractions resume, causing the uterus to shrink remarkably. During these contractions, the placenta is sheared from the uterus and expelled through the vagina as the *afterbirth*. The muscles that surround fetal blood vessels in the umbilical cord contract, shutting off blood flow to the placenta. (Tying off the cord is standard practice but is not usually necessary; if it were, other mammals would not survive birth.) A new human being has been born.

Pregnancy Hormones Stimulate Milk Secretion

As the fetus grows, nourished by nutrients diffusing through the placenta, changes in the mother's breasts prepare her to continue nourishing her child after it is born. During pregnancy, large quantities of estrogen and progesterone (acting together with several other hormones) stimulate the **mammary glands,** milk-producing glands in the breasts, to grow, branch, and develop the capacity to secrete milk (**Fig. 25-26**). Each mammary gland has a milk duct that leads to the nipple. The actual secretion of milk, called *lactation*, is promoted by the pituitary hormone prolactin.

25.5 How Can People Limit Fertility?

Natural selection favors individuals who reproduce successfully and whose offspring survive long enough to reproduce themselves. During most of human evolution, child mortality was high, and natural selection favored those who produced enough children to offset this high mortality rate. Today, people no longer need to have many children to ensure that a few will survive to adulthood. We nevertheless retain the reproductive drives with which evolution has endowed us. As a result, each week sees nearly 1.5 million new people added to our increasingly crowded planet. Controlling birthrates has become an environmental necessity. On the individual level, birth control allows people to plan their families and provide the best opportunities for themselves and their children.

Historically, limiting fertility has not been easy. In the past, some cultures advocated inventive, if bizarre, techniques such as swallowing froth from the mouth of a camel or placing crocodile dung in the vagina. Since the 1970s, however, effective techniques have been developed for **contraception** (the prevention of pregnancy). These include methods that prevent production of sperm or eggs, prevent fertilization, or prevent implantation of a blastocyst in the uterus. The choice of a birth control method should be made in consultation with a health professional, who can provide more complete information.

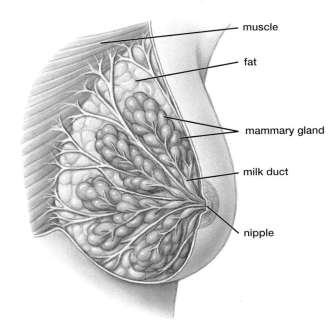

muscle

fat

mammary gland

milk duct

nipple

▲ Figure 25-26 The structure of the human breast During pregnancy, both fatty tissue and the milk-secreting glands and ducts increase in size.

Sterilization Provides Permanent Contraception

In the long run, the most foolproof method of contraception is sterilization, in which the pathways through which sperm or egg must travel are interrupted. Sterilization is generally permanent. Sometimes, however, in a delicate and expensive operation, a surgeon can reconnect pathways that have been severed.

In men, the vas deferens leading from each testis may be severed in an operation called a *vasectomy*. Sperm are still produced, but they cannot reach the penis during ejaculation. The surgery is performed under a local anesthetic, and vasectomy has no known effects on health or sexual performance. Sperm are still produced, but they cannot leave the epididymis, where they die. Phagocytic white blood cells and the cells that make up the lining of the epididymis remove the debris.

The somewhat more complex operation of *tubal ligation* renders a woman infertile by cutting her uterine tubes. Eggs are still produced, but sperm cannot travel to the egg, nor can the egg reach the uterus. The breakdown products of the eggs are absorbed by the body.

Temporary Birth Control Methods Are Reversible

Temporary methods of birth control prevent pregnancy in the immediate future, while leaving open the option of later pregnancies. Temporary birth control methods (Table 25-4) use one or more of three mechanisms: (1) preventing ovulation, (2) preventing sperm and egg from meeting, and (3) preventing implantation in the uterus. Although temporary methods of male contraception are not yet widely available in the United States, clinical trials are under way in other countries.

Synthetic Hormones Prevent Ovulation

As you learned earlier in this chapter, ovulation is triggered by a midcycle surge of LH. One way to prevent ovulation is to suppress LH release by providing a continuous supply of estrogen and progesterone. This is the basis for a variety of contraception techniques, such as the birth control pill. Several other delivery systems for estrogen and progesterone (generally in synthetic form) are now available, as described in Table 25-4.

Fertilization Will Not Occur If the Sperm Cannot Contact the Egg

There are several ways to prevent the encounter of sperm and egg, including what are called barrier methods. With barrier methods, one of a variety of devices block the opening of the cervix, preventing the entry of sperm. Alternatively, the male condom prevents sperm from being deposited in the vagina. A female condom is now available that completely lines the vagina. Condoms are particularly effective against the spread of sexually transmitted diseases. Barrier devices are more effective when combined with a *spermicide* (sperm-killing substance), in case the barriers are breached, allowing a few sperm to enter the vagina or uterus.

Less reliable ways of preventing sperm from reaching the egg include the use of spermicides alone and the *rhythm method* (abstaining from intercourse during the ovulatory period of the menstrual cycle). In practice, the rhythm method has a high failure rate because the menstrual cycle usually varies somewhat from month to month. The rhythm method can be made somewhat more reliable by carefully monitoring changes in body temperature, which rises a few tenths of a degree right after ovulation.

Highly unreliable techniques include *withdrawal* (removing the penis from the vagina before ejaculation) and *douching* (attempting to wash sperm out of the vagina after intercourse).

Some Birth Control Methods Work Through Multiple Mechanisms

Some birth control methods may simultaneously prevent ovulation, hinder survival or motility of sperm, or prevent implantation of the blastocyst in the uterus.

Table 25-4	Temporary Birth Control Techniques		
Method	**Technique and Mechanisms**	**Failure Rates[a]**	**Protection from STD**
Hormonal methods that prevent ovulation			
Birth control pill	Pill containing either estrogen and synthetic progesterone (combination pill) or progesterone only (minipill); taken daily	0.1% to 3%	None
Contraceptive patch	Skin patch containing synthetic estrogen and progesterone; replaced weekly	<1%[b]	None
Contraceptive injection	Injection of synthetic progesterone that blocks ovulation for 3 months; repeated at 3-month intervals	0.3%	None
Vaginal ring	Flexible plastic ring impregnated with synthetic estrogen and progesterone; inserted into the vagina around the cervix, and remains in place for 3 weeks; replaced every 4 weeks	0.3% to 8%	None
Methods that prevent sperm and egg from meeting			
Abstinence	Deciding not to be sexually active	0%	Excellent
Condom (male)	Thin, disposable latex sheath placed over penis just before intercourse; prevents sperm from entering vagina; is more effective with spermicide	3% to 15%	Good
Condom (female)	Lubricated polyurethane pouch inserted into vagina; prevents sperm from entering the cervix; is more effective with spermicide	5% to 21%	Poor
Sponge	Domed disposable sponge impregnated with spermicide; is inserted into the vagina and works for 24 hours	9% to 20%	Poor
Diaphragm/cervical cap	Reusable, flexible domed rubber-like barrier; spermicide is placed within the dome, and the diaphragm (larger) or cap (smaller) is fitted over the cervix just before intercourse	6% to 14%	Poor
Rhythm	Measuring body temperature and cervical mucous changes to estimate the time of ovulation and avoiding intercourse during the fertile period	2% to 20% (rarely performed correctly)	None
Spermicide	Sperm-killing foam is placed in vagina before intercourse, forming a chemical barrier to sperm	6% to 26%	Little or none
Methods with multiple actions			
IUD (intrauterine device)	Small plastic device treated with hormones or copper and inserted through the cervix into the uterus; reduces sperm motility and survival; blocks implantation	0.6% to 2.0%	None
"Morning after" pill	Concentrated dose of the hormones in birth control pills, taken within 72 hours after intercourse; delays or prevents ovulation	25%	None

[a]Percentage of women becoming pregnant per year. The low and high numbers, respectively, indicate the differences between consistent and correct use and use in a more typical way that is not always consistent or correct.

[b]Preliminary findings report that the patch is as effective as the pill, and more likely to be used properly; however, women weighing more than 200 pounds may find it less effective.

The *intrauterine device (IUD)*, a small copper or plastic loop, squiggle, or shield inserted through the cervix and into the uterus, interferes with sperm motility or survival, thereby reducing the likelihood of sperm reaching the egg. IUDs also alter the uterine lining and lessen the chance of implantation of a blastocyst, should fertilization occur. The "morning after" pill, which contains high doses of hormones similar to those in birth control pills, appears to act primarily by delaying or preventing ovulation. Some studies suggest that the morning after pill may also interfere with the development of the corpus luteum and possibly prevent implantation. Despite its name, the morning after pill is most effective if taken immediately after intercourse; its effectiveness declines with time, but it can prevent pregnancy in some cases as long as 3 to 5 days later.

Abortion Removes the Embryo from the Uterus

Abortion is usually performed by dilating the cervix and removing the embryo and placenta by suction. Most abortions are performed during the first 3 months of pregnancy. Alternatively, abortion can be induced during the first 7 weeks of pregnancy by the drug RU-486 (mifepristone). RU-486 blocks the action of progesterone, which is required to maintain the uterine lining during pregnancy.

Abortion is a controversial procedure. Science can describe the events underlying pregnancy and fetal development, but it cannot resolve ethical dilemmas, such as when a fetus becomes a "person" or about the relative merits of fetal rights versus maternal rights.

The Faces of Fetal Alcohol Syndrome Revisited

When John Kellerman was born, the delivery room reeked of the alcohol polluting his amniotic fluid: John was born drunk. Even today, his adoptive mother, Teresa Kellerman, explains that, without medication, his behavior resembles that of a drunken person: silly, volatile, and lacking impulse control.

The writings of young people with FAS poignantly convey the day-to-day difficulties that these innocent people face over a lifetime. In his poem "Help," John writes:

When my brain is not working right,
I need to let someone I trust
Help me
To stay safe and get calm again.
When my brain is not working right,
I feel like I am on a FASD Train
Going downhill,
And the engineer is asleep
And I can't wake him up,
And I can't put on the brakes.

CJ, an FAS victim from Canada, explains:

"I am small, I have a different face . . . and Lots of Learning problems.
Moms do not do this on purpose
Do not be mean or mad or blame them
I am 17 years old and have had FAS all my life
I will have it forever
It will never go away
Don't be mean or mad or blame me. . . .

Consider This

In about 20 states, women who use illegal drugs or abuse alcohol during pregnancy may be subject to legal action to protect fetal rights. How do you think society should approach the dilemma of pregnant women who often unwittingly damage their unborn children through alcohol or other drugs? Is this child abuse? Based on what you know about development, how would you deal with a friend who continued to smoke, drink, or take other drugs during pregnancy?

Chapter Review

Summary of Key Concepts

For additional study help and activities, go to www.mybiology.com.

25.1 How Do Animals Reproduce?

Animals reproduce either sexually or asexually. In sexual reproduction, haploid gametes, usually from two separate parents, unite and produce offspring that are genetically different from either parent. In asexual reproduction, a single animal produces offspring that are usually genetically identical to itself.

Among animals that engage in sexual reproduction, the female produces large, nonmotile eggs, and the male produces small, motile sperm. Fertilization occurs, depending on the species, outside the bodies of the animals (external fertilization) or inside the body of the female (internal fertilization). For external fertilization, the animals must be in water so that the sperm can swim to meet the egg. In most species, internal fertilization requires copulation.

25.2 How Does the Human Reproductive System Work?

The male reproductive system consists of paired testes, which produce sperm and testosterone, and structures that conduct the sperm to the female's reproductive tract and secrete fluids that provide the proper pH for sperm survival, activate sperm swimming, and provide energy. Spermatogenesis and testosterone production are nearly continuous, beginning at puberty and lasting until death.

The female reproductive system consists of paired ovaries, which produce eggs as well as the hormones estrogen and progesterone, and structures that conduct sperm to the egg and receive and nourish the embryo during prenatal development. In human females, oogenesis, hormone production, and development of the lining of the uterus vary in a monthly menstrual cycle. The cycle is controlled by hormones from the hypothalamus (GnRH), anterior pituitary (FSH and LH), and ovaries (estrogen and progesterone).

During copulation, the male ejaculates semen into the female's vagina. The sperm move through the vagina and uterus into the uterine tube, where fertilization usually takes place. The unfertilized egg is surrounded by two barriers, the corona radiata and the zona pellucida. Enzymes released from the acrosomes in the head of sperm digest these layers, permitting sperm to reach the egg. Only one sperm enters the egg and fertilizes it.

Web Animation Human Reproductive System

Web Animation Hormonal Control of the Menstrual Cycle

25.3 How Does Animal Development Proceed?

Animals develop through several stages. *Cleavage*: The fertilized egg undergoes mitotic cell divisions with little intervening growth, so the egg cytoplasm is partitioned into smaller cells. Cleavage divisions form the morula, a solid ball of cells. *Blastula formation:* A cavity then opens up within the morula, forming the blastula, a hollow ball of cells. *Gastrulation*: A dimple forms in the blastula, and cells migrate from the surface into the interior of the ball, eventually forming a three-layered gastrula. The three cell layers of ectoderm, mesoderm, and endoderm give rise to all adult tissues. *Organogenesis:* The cell layers of the gastrula form organs characteristic of the animal species.

25.4 How Do Humans Develop?

Human embryonic development follows the same principles as the development of other mammals. A fertilized egg (zygote) develops into a hollow blastocyst and implants in the endometrium of the uterus. The blastocyst's outer wall will become the chorion and will form the embryonic contribution to the placenta, and the inner cell mass develops into the embryo. During gastrulation, cells migrate and differentiate into ectoderm, mesoderm, and endoderm. By the end of the second month, the major organs have formed, the embryo appears human, and is called a fetus. In the next 7 months, until birth, the fetus continues to grow. The lungs, stomach, intestine, and kidneys enlarge and become functional. The stages of human development are summarized in Fig. 25-19.

During pregnancy, mammary glands in the mother's breasts enlarge under the influence of estrogen, progesterone, and other hormones. After about 9 months, uterine contractions are triggered by a complex interplay of uterine stretch, chemical signals from the fetus, and oxytocin release. As a result, the uterus expels the baby and then the placenta.

Web Animation Human Development

25.5 How Can People Limit Fertility?

Permanent contraception can be achieved by sterilization: severing the vas deferens in males (vasectomy) or the uterine tubes in females (tubal ligation). Temporary birth control techniques include synthetic hormones that prevent ovulation, such as birth control pills. Barrier methods that prevent sperm and egg from meeting include the diaphragm, the cervical cap, and the condom, accompanied by spermicide. The rhythm method requires abstinence at the time of ovulation. Intrauterine devices may block sperm movement and also prevent implantation of the early embryo. The "morning after" pill may prevent ovulation or development of the corpus luteum or it may prevent implantation. Abortion causes the expulsion of the embryo from the uterus.

Key Terms

allantois *p. 500*

acquired immune deficiency syndrome (AIDS) *p. 496*

amnion *p. 500*

amniotic egg *p. 500*

asexual reproduction *p. 486*

blastocyst *p. 500*

blastula *p. 499*

budding *p. 487*

cervix *p. 493*

chlamydia *p. 496*

chorion *p. 500*

cleavage *p. 499*

clitoris *p. 495*

contraception *p. 507*

copulation *p. 489*

corpus luteum *p. 493*

crab lice *p. 497*

development *p. 498*

ectoderm *p. 499*

egg *p. 487*

embryo *p. 491*

endoderm *p. 499*

epididymis *p. 491*

extraembryonic membranes *p. 500*

fertilization *p. 487*

fetus *p. 492*

fission *p. 487*

follicle *p. 492*

gastrula *p. 499*

gastrulation *p. 499*

genital herpes *p. 496*

gonad *p. 486*

gonorrhea *p. 496*

human papillomavirus (HPV) *p. 497*

implantation *p. 502*

labia *p. 495*

labor *p. 506*

mammary glands *p. 507*

menstrual cycle *p. 492*

menstruation *p. 494*

mesoderm *p. 499*

oogenesis *p. 492*

organogenesis *p. 499*

ovary *p. 487*

ovulation *p. 493*

penis *p. 491*

placenta *p. 493*

scrotum *p. 490*

seminiferous tubule *p. 490*

sexual reproduction *p. 486*

sexually transmitted disease (STD) *p. 496*

sperm *p. 487*

spermatogenesis *p. 490*

stem cell *p. 500*

syphilis *p. 496*

testis (plural, testes) *p. 487*

trichomoniasis *p. 497*

urethra *p. 491*

uterine tube *p. 493*

uterus *p. 493*

vagina *p. 493*

vas deferens *p. 491*

yolk sac *p. 500*

zygote *p. 493*

Thinking Through the Concepts

Suggested answers to end-of-chapter and figure-based questions can be found at the end of the text.

Fill-in-the-Blank

1. Reproduction by a single animal, without the need for sperm fertilizing an egg, is called _____ reproduction. A(n) _____ is a new individual that grows on the body of the adult and eventually breaks off to become independent. During _____, an adult animal splits in two, with each of the two halves regenerating the rest of a complete organism.

2. In mammals, the male gonad is called the _____. It produces both sperm and the sex hormone _____. Within the male gonad, spermatogenesis occurs within the hollow, coiled _____.

3. A sperm consists of three regions: the head, midpiece, and tail. The head contains very little cytoplasm, and consists mostly of the _____, in which the chromosomes are found, and a specialized lysosome, the _____. Organelles in the midpiece, the _____, provide energy for movement of the tail.

4. Sperm are stored in the _____ and _____ until ejaculation, when the sperm, mixed with fluids from three glands—the _____, _____, and _____—flow through the _____ to the tip of the penis.

5. In mammals, the female gonad is the _____. It produces eggs and two hormones, _____ and _____. Although usually referred to as an "egg," female mammals actually ovulate a cell called a _____, which has completed only meiosis I. Meiosis I also produces a much smaller cell, the _____, that serves mainly as a way to discard chromosomes. The "egg" is swept up by a ciliated structure, called the fimbriae, that forms the entrance to the _____. Fertilization usually occurs in this structure. The fertilized egg then implants in the _____, where it will develop until birth.

6. Oocytes develop in a multicellular structure, the follicle. After ovulation, most of the follicle cells remain in the ovary, forming a temporary endocrine "gland," the _____. This "gland" disintegrates a few days after ovulation unless stimulated by a hormone, _____, which is secreted by the developing embryo.

7. A fertilized egg is a very large cell. One of the first events of animal development is the division of this cell, a process called _____, to produce a solid ball of much smaller cells, the _____. Soon, a cavity opens in this ball of cells, producing a hollow ball called the _____. Movements of cells in this hollow ball during gastrulation produce the three embryonic tissue layers, the _____, _____, and _____.

8. The evolution of the amniotic egg allowed development to occur in fully terrestrial animals that do not lay eggs in water. The amniotic egg contains four extraembryonic membranes. In both birds and mammals, the _____ encloses the embryo in a watery environment. In mammals, the _____ forms the embryo's part of the placenta. The allantois forms the blood vessels of the _____, which connects the embryo to the placenta. Finally, the _____ contains stored food in bird eggs, but in mammals, this is "empty" and principally forms the lining of the digestive tract.

Review Questions

1. List the advantages and disadvantages of asexual reproduction, sexual reproduction, external fertilization, and internal fertilization, including an example of an animal that uses each type.

2. Describe the structures that surround the egg.

3. What is the role of the corpus luteum in the menstrual cycle? In early pregnancy? What determines its survival after ovulation?

4. List the structures, in order, through which sperm pass on their way from the seminiferous tubules of the testis to the uterine tube of the female.

5. Describe the interactions among hormones secreted by the pituitary gland and ovaries that produce the menstrual cycle.

6. Name two structures derived from each of the three embryonic tissue layers—endoderm, ectoderm, and mesoderm.

7. Explain how the structure of the placenta prevents mixing of fetal and maternal blood while allowing the exchange of substances between the mother and the fetus.

8. How do changes in the breast prepare a mother to nurse her newborn? How do hormones influence these changes and stimulate milk production?

Applying the Concepts

1. **IS THIS SCIENCE?** An organization called Clonaid claims to have produced many healthy human clones. Investigate and evaluate these claims. Argue for and against the use of human cloning for reproductive purposes.

2. Many drugs can cross the placenta. Some drugs that do not cause significant side effects in adults may cause devastating birth defects if taken during pregnancy. For many drugs, the possible effects on developing fetuses remain uncertain, because it is so difficult to devise well-designed, ethical clinical trials to detect such effects. What should a pregnant woman, or a woman planning a pregnancy, do?

3. Fertility drugs greatly increase the incidence of multiple births. Multiple fetuses are much more likely to be born prematurely, with developmental problems. The medical cost of caring for multiple premature infants is staggering. When fertility drugs produce multiple embryos, the physician can selectively eliminate some of these embryos early in development, so the remaining few have a better chance to develop fully and normally. Given these facts, discuss the ethical implications of taking fertility drugs.

For additional resources, go to www.mybiology.com.

chapter 26

Animal Behavior

What makes a man sexy? Could it be his symmetry?

Q At a Glance

Case Study   Sex and Symmetry

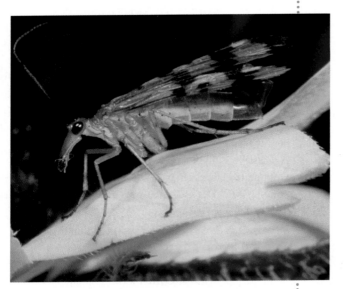

The most symmetrical male scorpionflies mate more often than other males.

What makes a man sexy? According to a growing body of research, it's his symmetry. Female sexual preference for symmetrical males was first documented in insects. For example, biologist Randy Thornhill found that symmetry accurately predicts the mating success of male Japanese scorpionflies (photo at left). In Thornhill's experiments and observations, the most successful males were those whose left and right wings were equal or nearly equal in length. Males with one wing longer than the other were less likely to copulate; the greater the difference between the two wings, the lower the likelihood of success.

Thornhill's work with scorpionflies led him to wonder if the effects of male symmetry also extend to humans. To test the hypothesis that female humans find symmetrical males more attractive, Thornhill and colleagues began by measuring symmetry in some young adult males. Each man's degree of symmetry was assessed by measurements of his ear length and the width of his foot, ankle, hand, wrist, elbow, and ear. From these measurements, the researchers derived an index that summarized the degree to which the size of these features differed between the right and left sides of the body.

The researchers next gathered a panel of female observers who were unaware of the nature of the study and showed them photos of the faces of the measured males. As predicted by the researchers' hypothesis, the panel judged the most symmetrical men to be most attractive. A survey of the male subjects revealed that the more symmetrical men also tended to begin having sex earlier in life and to have had a larger number of sexual partners. Apparently, a man's sexual activity and attractiveness to women are correlated with his body symmetry.

Why might females prefer symmetrical males? Consider this question as you read about animal behavior. ■

26.1 How Do Innate and Learned Behaviors Differ?

Behavior is any observable activity of a living animal. For example, a moth flies toward a bright light, a honeybee flies toward a cup of sugar-water, and a housefly flies toward a piece of rotting meat. Bluebirds sing, wolves howl, and frogs croak. Mountain goats butt heads in ritual combat; chimpanzees groom one another; ants attack a termite that approaches an anthill. Humans smoke cigarettes, play tennis, and plant gardens. Even the most casual observer sees many examples of animal behavior each day, and a careful observer encounters a virtually limitless number of fascinating behaviors.

Innate Behaviors Can Be Performed Without Prior Experience

Innate behaviors are performed in reasonably complete form the first time an animal of the right age and motivational state encounters a particular stimulus. (The proper motivational state for feeding, for example, would be hunger.) Scientists can demonstrate that a behavior is innate by depriving an animal of the opportunity to learn it. For example, red squirrels, which in the wild bury nuts in

(a) A cuckoo chick ejects an egg

(b) A foster parent feeds a cuckoo

◀ **Figure 26-1 Innate behavior**
(a) The cuckoo chick, just hours after it hatches and before its eyes have opened, evicts the eggs of its foster parents from the nest. **(b)** The parents, responding to the stimulus of a cuckoo chick's wide-gaping mouth, feed the chick, unaware that it is not related to them. **QUESTION:** The cuckoo chick benefits from its innate behavior, but the foster parent harms itself with its innate response to the cuckoo chick's begging. Why hasn't natural selection eliminated this disadvantageous innate behavior?

the fall for retrieval during the winter, can be raised from birth in a bare cage on a liquid diet, providing them with no experience of nuts, digging, or burying. Nonetheless, such a squirrel will, when presented with nuts for the first time, carry one to the corner of its cage, and then make covering and patting motions with its forefeet. Nut burying is therefore an innate behavior.

Innate behaviors can also be recognized by their occurrence immediately after birth, before any opportunity for learning presents itself. The cuckoo, for example, lays its eggs in the nest of another bird species, to be raised by the unwitting adoptive parent. Soon after a cuckoo egg hatches, the cuckoo chick performs the innate behavior of shoving the nest owner's eggs (or baby birds) out of the nest, eliminating its competitors for food (**Fig. 26-1**).

Learned Behaviors Require Experience

Natural selection may favor innate behaviors in many circumstances. For instance, it is clearly to the advantage of a red-winged blackbird chick to perform begging calls as soon as possible after hatching, because the calls stimulate the parent to feed the chick. But in other circumstances, rigidly fixed behavior patterns may be less useful. For example, a male red-winged blackbird presented with a stuffed female blackbird will often attempt to copulate with the stuffed bird, a behavior that obviously will produce no offspring. In many situations, a degree of behavioral flexibility is advantageous.

The capacity to make changes in behavior on the basis of experience is called **learning.** This deceptively simple definition encompasses a vast array of phenomena. A toad learns to avoid distasteful insects (**Fig. 26-2**); a baby shrew learns which adult is its mother; a human learns to speak a language; a sparrow learns to use the stars for navigation. Each of the many examples of animal learning represents the outcome of a unique evolutionary history, so learning is as diverse as animals themselves.

There Is No Sharp Distinction Between Innate and Learned Behaviors

Although the terms "innate" and "learned" can help us describe and understand behaviors, these words also have the potential to lull us into an oversimplified view of animal behavior. In practice, no behavior is totally innate or totally learned. Instead, all behaviors are mixtures of the two.

Web Animation Observing Behavior: Homing in Digger Wasps

① A naive toad is presented with a bee.

② While trying to eat the bee, the toad is stung painfully on the tongue.

③ Presented with a harmless robber fly, which resembles a bee, the toad cringes.

④ The toad is presented with a dragonfly.

⑤ The toad immediately eats the dragonfly, demonstrating that the learned aversion is specific to bees and insects resembling bees.

▲ **Figure 26-2 Learning in a toad**

Seemingly Innate Behavior Can Be Modified by Experience

Behaviors that seem to be performed correctly on the first attempt without prior experience can later be modified by experience. For example, a newly hatched herring gull chick pecks at a red spot on its parent's beak (**Fig. 26-3**), an innate behavior that causes the parent to regurgitate food for the chick to eat. Biologist Niko Tinbergen studied this pecking behavior and found that the pecking response of very young chicks was triggered by the long, thin shape and red color of the parent's bill. In fact, when Tinbergen offered newly hatched chicks a thin, red rod with white stripes painted on it, they pecked at it more often than at a real beak. Within a few days, however, the chicks learned enough about the appearance of their parents that they began pecking more frequently at models more closely resembling the parents. After one week, the young gulls recognized their parents' appearance sufficiently well to prefer models of their own species to models of a closely related species. Eventually, the young birds learned to beg only from their own parents.

Habituation, a simple form of learning in which an animal's response declines if a harmless stimulus is repeated, can also fine-tune an organism's innate responses to environmental stimuli. For example, young birds crouch down when a hawk flies over but ignore harmless birds such as geese. Early observers hypothesized that only the very specific shape of predatory birds provoked crouching. Niko Tinbergen and Konrad Lorenz, two of the founding fathers of the scientific study of animal behavior, tested this hypothesis by using an ingenious model. When moved in one direction, the model resembled a goose, and chicks ignored it (**Fig. 26-4**). When its movement was reversed, however, the model resembled a hawk and elicited crouching behavior. Further research revealed that newborn chicks instinctively crouch when *any* object moves over their heads. Over time, their response habituates to things that soar by harmlessly and frequently, such

▲ **Figure 26-3 Innate behaviors can be modified by experience** A herring gull chick pecks at the red spot on its mother's bill, causing her to regurgitate food.

as leaves, songbirds, and geese. Predators are much less common, and the novel shape of a hawk continues to elicit instinctive crouching. Thus, learning modifies the innate response, making it more advantageous.

Learning May Be Governed by Innate Constraints

Learning always occurs within boundaries that help increase the chances that only the appropriate behavior is acquired. For example, even though young robins hear the singing of sparrows, warblers, finches, and other bird species that share nesting areas with robins, the young birds do not imitate the songs of these other species. Instead, young robins learn only the songs of adult robins. The robin's ability to learn songs is limited to those of its own species, and the songs of other species are excluded from the learning process.

The innate constraints on learning are perhaps most strikingly illustrated by **imprinting,** a special form of learning in which an animal's nervous system is rigidly programmed to learn a certain thing only at a certain period of development. This causes a strong association to be formed during a particular stage, called a *sensitive period*, in the animal's life. During this stage, the animal is primed to learn specific information, which is then incorporated into behaviors that are not easily altered by further experience.

Imprinting is best known in birds such as geese, ducks, and chickens. These birds learn to follow the animal or object that they most frequently encounter during an early sensitive period. In nature, a mother bird is likely to be nearby during the sensitive period, so her offspring imprint on her. In captivity, however, these birds may imprint on a toy train or other moving object (**Fig. 26-5**). If given a choice, however, they select an adult of their own species.

All Behavior Arises out of Interaction Between Genes and the Environment

Many early ethologists (people who study animal behavior) saw innate behaviors as rigidly controlled by genetic factors and viewed learned behaviors as determined exclusively by an animal's environment. Today, however, ethologists realize that, just as no behavior is wholly innate or wholly learned, no behavior can be caused strictly by genes or strictly by the environment. Instead, all behavior develops out of an interaction between genes and the environment. The relative contributions of heredity and learning vary among animal species and among behaviors within an individual.

The precise nature of the link among genes, environments, and behaviors is not well understood in most cases. The chain of events between the transcription of genes and the performance of a behavior may be so complex that we will never decipher it fully. Nonetheless, a great deal of evidence demonstrates the existence of both genetic and environmental components in the development of behaviors. For example, consider bird migration. Even though it is well known that migratory birds must learn by experience how to navigate with celestial cues, this learning is not the only factor involved.

Bird Migration Behavior Has an Inherited Component

At the close of the summer, many birds leave their breeding habitats and head for their winter territory, which may be hundreds or even thousands of miles away. Many of these migrating birds are traveling for the first time, because they were hatched only a few months earlier. Amazingly, these naive birds depart at the proper time, head in the proper direction, and locate the proper wintering location, even though they often do not simply follow more experienced birds (which typically depart a few weeks in advance of the first-year birds). Somehow, these young birds execute a very difficult task the first time they try it. Thus, it seems that birds must be born with the ability to migrate; it must be "in their genes." Indeed, birds hatched and raised in isolation indoors still orient in the proper migratory direction when autumn comes, apparently without the need for any learning or experience.

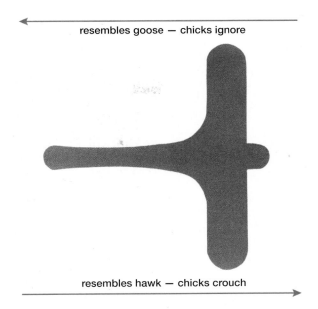

resembles goose — chicks ignore

resembles hawk — chicks crouch

▲ **Figure 26-4 Habituation modifies innate responses** The model used by Konrad Lorenz and his student Niko Tinbergen to investigate the response of chicks to the shape of objects flying overhead. The chicks' response depended on which direction the model moved. Moving toward the right, the model resembles a predatory hawk, but when it moves left, it resembles a harmless goose.

▲ **Figure 26-5 Konrad Lorenz and imprinting** Konrad Lorenz, who won a Nobel prize for his pioneering studies of animal behavior, is followed by goslings that imprinted on him shortly after they hatched. They follow him as they would their mother.

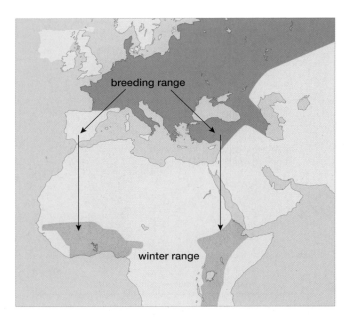

▲ **Figure 26-6 Genes influence migratory behavior** Blackcap warblers from western Europe begin their fall migration by flying in a southwesterly direction, but those from eastern Europe fly to the southeast when they begin migrating. **QUESTION:** If young blackcap warblers from a wild population in western Europe were transported to eastern Europe and reared to adulthood in a normal environment, in which direction would you expect them to orient?

The conclusion that birds must have a genetically controlled ability to migrate in the right direction has been further supported by hybridization experiments with blackcap warblers. This species breeds in Europe and migrates to Africa, but populations from different areas travel by different routes. Blackcaps from western Europe travel in a southwesterly direction to reach Africa, whereas birds from eastern Europe travel to the southeast (Fig. 26-6). If birds from the two populations are crossbred in captivity, however, the hybrid offspring exhibit migratory orientation due south, which is intermediate between the orientations of the two parents. This result suggests that parental genes—of which offspring inherit a mixture—influence migratory direction.

26.2 How Do Animals Communicate?

Animals frequently broadcast information. The sounds uttered, movements made, and chemicals emitted by animals can reveal their location, level of aggression, readiness to mate, and so on. If this information evokes a response from other individuals, and if that response tends to benefit the sender and the receiver, then a communication channel can form. **Communication** is defined as the production of a signal by one organism that causes another organism to change its behavior in a way beneficial to one or both.

Although animals of different species may communicate (picture a cat, its tail erect and bushy, hissing at a dog), most animals communicate only with members of their own species. Potential mates must communicate, as must parents and offspring. Communication is also often used to help resolve the conflicts that arise when members of a species compete directly with one another for food, space, and mates.

The ways in which animals communicate are astonishingly diverse and use all of the senses. In the following sections, we look at communication by visual displays, sound, chemicals, and touch.

Visual Communication Is Most Effective over Short Distances

Animals with well-developed eyes, from insects to mammals, use visual signals to communicate. Visual signals can be *active*, in which a specific movement (such as baring fangs or lowering the head) conveys a message (Fig. 26-7). Alternatively, visual signals may be *passive*, in which case the size, shape, or color of the animal conveys important information, commonly about its sex and reproductive state. For example, when female mandrills become sexually receptive, they develop a large, brightly colored swelling on their buttocks (Fig. 26-8). Active and passive signals can be combined, as illustrated by the lizard in Figure 26-9.

Like all forms of communication, visual signals have both advantages and disadvantages. On the plus side, they are instantaneous, and active signals can be rapidly changed to convey a variety of messages in a short period. Visual communication is quiet and unlikely to alert distant predators, although the signaler does make itself conspicuous to those nearby. On the negative side, visual signals are generally ineffective in dense vegetation or in darkness, although female fireflies do communicate with potential mates at night by using species-specific patterns of flashes. Finally, visual signals are limited to close-range communication.

▲ **Figure 26-7 Active visual signals** The wolf signals aggression by lowering its head, ruffling the fur on its neck and along its back, facing its opponent with a direct stare, and exposing its fangs. These signals can vary in intensity, communicating different levels of aggression.

Communication by Sound Is Effective over Longer Distances

The use of sound overcomes many of the shortcomings of visual displays. Like visual displays, sound signals reach receivers almost instantaneously. But unlike visual signals, sound can be transmitted through darkness, dense forests, and murky water. Sound signals can also be effective over longer distances than visual signals. For example, the low, rumbling calls of African elephants can be heard by elephants several miles away, and the songs of humpback whales are audible for hundreds of miles. Likewise, the howls of a wolf pack carry for miles on a still night. Even the small kangaroo rat produces a sound (by striking the desert floor with its hind feet) that is audible 150 feet (45 meters) away. The advantages of long-distance transmission, however, are offset by an important disadvantage: Predators and other unwanted receivers can also detect a sound signal from a distance, and can use the signal to find the location of the signaler.

Sound signals are similar to visual displays in that they can be varied to convey rapidly changing messages. (Think of speech and the emotional nuances conveyed by the human voice during a conversation.) Changes in motivation can be signaled by a change in the loudness or pitch of the sound. An individual can convey different messages by variations in the pattern, volume, and pitch of the sound produced. In a study of vervet monkeys in Kenya in the 1960s, ethologist Thomas Struhsaker found that they produced different calls in response to threats from each of their major predators: snakes, leopards, and eagles. In 1980 other researchers reported that the response of other vervet monkeys to each of these calls is appropriate to the particular predator. For example, the "bark" that warns of a leopard or other four-legged carnivore causes monkeys on the ground to take to trees and those in trees to climb higher. The "rraup" call, signaling an eagle or other hunting bird, causes monkeys on the ground to look upward and take cover, whereas monkeys already in trees drop to the shelter of lower, denser branches. The "chutter" call that indicates the presence of a snake causes the monkeys to stand up and search the ground for the predator.

The use of sound is by no means limited to birds and mammals. Male crickets produce species-specific songs that attract female crickets of the same species. The annoying whine of the female mosquito as she prepares to bite alerts nearby males that she may soon have the blood meal necessary for laying eggs. Male water striders vibrate their legs, sending species-specific patterns of vibrations through the water, attracting mates and repelling other males (**Fig. 26-10**). From these rather simple signals to the complexities of human language, sound is one of the most important forms of communication.

Chemical Messages Persist Longer But Are Hard to Vary

Chemical substances that are produced by individuals and that influence the behavior of other members of the species are called **pheromones.** Pheromones can carry messages over long distances and, unlike sound, take very little energy to produce. Pheromones may not even be detected by other species, whereas predators might be attracted to visual or sound signals. Like a signpost, a pheromone persists over time and can convey a message after the animal has departed. Wolf packs, hunting over areas of nearly 400 square miles (about 1,000 square kilometers), warn other packs of their presence by marking the boundaries of their travels with urine that contains pheromones. As anyone who has walked a male dog can attest, the domesticated dog reveals its wolf ancestry by staking out its neighborhood with urine that carries the chemical message "I live in this area."

Chemical communication requires animals to synthesize a different substance for each message. As a result, chemical signaling systems communicate fewer different messages than do sight- or sound-based systems. In addition, pheromone signals cannot easily convey rapidly changing messages. Nonetheless, chemicals effectively convey a few simple but critical messages.

▲ **Figure 26-8 A passive visual signal** The female mandrill's colorfully swollen buttocks serve as a passive visual signal that she is fertile and ready to mate.

▲ **Figure 26-9 Active and passive visual signals combined** An *Anolis* lizard raises his head (an active visual signal) to reveal a brilliantly colored throat pouch (a passive visual signal).

▲ **Figure 26-10 Communication by vibration** By vibrating its legs, a water strider sends signals that radiate out over the surface of the water.

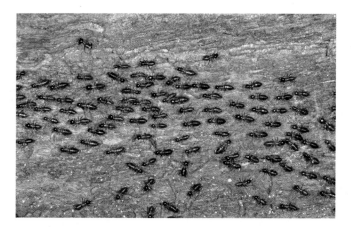

▲ **Figure 26-11 Communication by chemical messages** A trail of pheromones, secreted by termites from their own colony, orients foraging termites toward a source of food.

Many pheromones cause an immediate change in the behavior of the animal that detects them. For example, foraging termites that discover food lay a trail of pheromones from the food to the nest, and other termites follow the trail (**Fig. 26-11**).

Communication by Touch Helps Establish Social Bonds

Communication by physical contact often serves to establish and maintain social bonds among group members. This function is especially apparent in humans and other primates, which have many gestures—including kissing, nuzzling, patting, petting, and grooming—with important social functions (**Fig. 26-12a**). Touch may even be essential to human well-being. For example, research has shown that when the limbs of premature human infants were stroked and moved for 45 minutes daily, the infants were more active, responsive, and emotionally stable and gained weight more rapidly than did premature infants who received standard hospital treatment.

Communication by touch is not limited to primates, however. In many other mammal species, close physical contact helps cement the bond between parent and offspring. Species in which sexual activity is preceded or accompanied by physical contact can be found throughout the animal kingdom (**Fig. 26-12b**).

26.3 How Do Animals Compete for Resources?

The contest to survive and reproduce stems from the scarcity of resources relative to the reproductive potential of populations. The resulting competition underlies many of the most frequent types of interactions between animals.

Aggressive Behavior Helps Secure Resources

One of the most obvious manifestations of competition for resources such as food, space, or mates is **aggression,** or antagonistic behavior, between members of the same species. Although the expression "survival of the fittest" evokes images of the strongest animal emerging triumphantly from among the dead bodies of its competitors, in reality most aggressive encounters between members of the same species end without physical damage to the participants. Natural selection has favored the evolution of symbolic displays or rituals for resolving conflicts.

▶ **Figure 26-12 Communication by touch** *(a)* An adult olive baboon grooms a juvenile. Grooming both reinforces social relationships and removes debris and parasites from the fur. *(b)* Touch is also important in sexual communication. These land snails (*Helix*) engage in courtship behavior that will culminate in mating.

(a) Baboons

(b) Land snails

(a) A male baboon

(b) Sarcastic fringeheads

◀Figure 26-13 **Aggressive displays** *(a)* Threat display of the male baboon. Despite the potentially lethal fangs so prominently displayed, aggressive encounters between baboons rarely cause injury. *(b)* The aggressive display of many male fish, such as these sarcastic fringeheads, includes elevating the fins and flaring the gill covers, thus making the body appear larger.

During fighting, even the victorious animal can be injured, and might not survive to pass on its genes. Aggressive *displays*, in contrast, allow the competitors to assess each other and acknowledge a winner on the basis of size, strength, and motivation rather than wounds inflicted.

During aggressive displays, animals may exhibit weapons, such as claws and fangs (**Fig. 26-13a**), and often make themselves appear larger (**Fig. 26-13b**). Competitors often stand upright and erect their fur, feathers, ears, or fins (see Fig. 26-7). The displays are typically accompanied by intimidating sounds (growls, croaks, roars, chirps) whose loudness can help decide the winner. Fighting tends to be a last resort when displays fail to resolve a dispute.

In addition to aggressive visual and vocal displays, many animal species engage in ritualized combat. Deadly weapons may clash harmlessly (**Fig. 26-14**) or may not be used at all. In many cases, these encounters involve shoving rather than slashing. The ritual thus allows contestants to assess the strength and the motivation of their rivals, and the loser slinks away in a submissive posture that minimizes the size of its body.

▲ Figure 26-14 **Displays of strength** Ritualized combat of fiddler crabs. Oversized claws, which could severely injure another animal, grasp harmlessly. Eventually one crab, sensing greater vigor in his opponent, retreats unharmed.

Dominance Hierarchies Help Manage Aggressive Interactions

Aggressive interactions use a lot of energy, can cause injury, and can disrupt other important tasks, such as finding food, watching for predators, or raising young. Thus, there are advantages to resolving conflicts with minimal aggression. In a **dominance hierarchy,** each animal establishes a rank that determines its access to resources. Although aggressive encounters occur frequently while the dominance hierarchy is being established, once each animal learns its place in the hierarchy, disputes are infrequent, and the dominant individuals obtain most access to the resources needed for reproduction, including food, space, and mates. For example, domestic chickens, after squabbling, sort themselves into a reasonably stable "pecking order." Thereafter, all birds in the group defer to the dominant bird, all but the dominant bird give way to the second most dominant, and so on. In wolf packs, one member of each sex is the dominant, or "alpha," individual to whom all others of that sex are subordinate. Among male bighorn sheep, dominance is reflected in horn size (**Fig. 26-15**).

▲ Figure 26-15 **A dominance hierarchy** The dominance hierarchy of the male bighorn sheep is signaled by the size of the horns; these rams increase in status from right to left. The backward-curving horns, clearly not designed to inflict injury, are used in ritualized combat.

▲ **Figure 26-16** **A feeding territory** Acorn woodpeckers live in communal groups that excavate acorn-sized holes in dead trees, stuffing the holes with green acorns for dining during the lean winter months. The group defends the trees vigorously against other groups of acorn woodpeckers and against acorn-eating birds of other species, such as jays.

▲ **Figure 26-17** **Defense of a territory by song** A male seaside sparrow announces ownership of his territory.

Animals May Defend Territories That Contain Resources

In many animal species, competition for resources takes the form of **territoriality,** the defense of an area where important resources are located. The defended area may include places to mate, raise young, feed, or store food. Territorial animals generally restrict most or all of their activities to the defended area and advertise their presence there. Territories may be defended by males, females, a mated pair, or entire social groups (as in the defense of a nest by social insects). However, territorial behavior is most commonly seen in adult males, and territories are usually defended against members of the same species, who compete most directly for the resources being protected.

Territories are as diverse as the animals defending them. For example, a territory can be a tree where a woodpecker stores acorns (**Fig. 26-16**), a small depression in the sand used as a nesting site by a cichlid fish, a hole in the sand that is home to a crab, or an area of forest providing food for a squirrel.

Territoriality Reduces Aggression

Acquiring and defending a territory requires considerable time and energy, yet territoriality is seen in animals as diverse as worms, arthropods, fish, birds, and mammals. The fact that organisms as distantly related as worms and humans independently evolved similar behavior suggests that territoriality provides some important advantages. Although the particular benefits depend on the species and the type of territory it defends, one widespread benefit is reduced aggression. As with dominance hierarchies, once a territory is established through aggressive interactions, relative peace prevails if boundaries are recognized and respected. The saying "good fences make good neighbors" also applies to nonhuman territories.

Competition for Mates May Be Based on Territories

For males of many species, successful territorial defense has a direct impact on reproductive success. In these species, males defend territories, and females are attracted to high-quality territories, which might have features such as large size, abundant food, and secure nesting areas. Males who successfully defend the best territories have the greatest chance of mating and passing on their genes. For example, male stickleback fish that defend large territories are more successful in attracting mates than are males that defend small territories. Females that select males with the best territories increase their own reproductive success and pass their genetic traits (typically including their mate-selection preferences) to their offspring.

Animals Advertise Their Occupancy

Territories are advertised through sight, sound, and smell. If a territory is small enough, its owner's mere presence, reinforced by aggressive displays toward intruders, can provide sufficient defense. A mammal that has a territory but cannot always be present may use pheromones to scent-mark the boundaries of its territory. Male rabbits use pheromones secreted by glands in their chin and by anal glands to mark their territories. Hamsters rub the areas around their dens with secretions from special glands in their flanks.

Vocal displays are a common form of territorial advertisement. Male sea lions defend a strip of beach by swimming up and down in front of it, calling continuously. Male crickets produce a specific pattern of chirps to warn other males away from their burrows. Birdsong is a striking example of territorial defense. The husky trill of the male seaside sparrow is part of an aggressive display, warning other males to steer clear of his territory (**Fig. 26-17**). In fact, male sparrows that are unable to sing are unable to defend territories. The importance of singing to seaside sparrows' territorial defense was elegantly demonstrated by ornithologist M. Victoria McDonald, who captured territorial males and performed an operation that left them temporarily unable to sing but still able to utter the other, shorter and quieter signals in their vocal repertoires. The songless males were unable to defend territories or attract mates, but regained their lost territories when they recovered their singing ability.

26.4 How Do Animals Find Mates?

In many sexually reproducing animal species, mating involves copulation or other close contact between males and females. Before animals can successfully mate, however, they must identify one another as members of the same species, as members of the opposite sex, and as being sexually receptive. In many species, finding an appropriate potential partner is only the first step. Often the male must demonstrate his quality before the female will accept him as a mate. The need to fulfill all of these requirements has resulted in the evolution of a diverse and fascinating array of courtship behaviors.

Signals Encode Sex, Species, and Individual Quality

Individuals that waste energy and gametes by mating with members of the wrong sex or wrong species are at a disadvantage in the contest to reproduce. Thus, natural selection favors behaviors by which animals communicate their sex and species to potential mates.

Many Mating Signals Are Acoustic

Animals often use sounds to advertise their sex and species. Consider the raucous nighttime chorus of male tree frogs, each singing a species-specific song. Male grasshoppers and crickets also advertise their sex and species by their calls, as does the female mosquito with her high-pitched whine.

Signals that advertise sex and species may also be used by potential mates in comparisons among rival suitors. For example, the male bellbird uses its deafening song to defend large territories and attract females from great distances. A female flies from one territory to another, alighting near each male in his tree. The male, beak gaping, leans directly over the flinching female and utters an earsplitting note. The female apparently endures this noise to compare the songs of the various males, perhaps choosing the loudest as a mate.

Visual Mating Signals Are Also Common

Many species use visual displays for courting. The firefly, for example, flashes a message that identifies its sex and species. Male fence lizards bob their heads in a species-specific rhythm, and females distinguish and prefer the rhythm of their own species. The elaborate construction projects of the male gardener bowerbird and the scarlet throat of the male frigate bird serve as flashy advertisements of sex, species, and male quality (Fig. 26-18). Sending these extravagant signals must be risky, because they make it much easier for predators to locate the sender. For males, the added risk is an evolutionary necessity, because females won't mate with males that lack the appropriate signal. Females, in contrast, typically do not need to attract males or assume the risk associated with a conspicuous signal, so in many species females are drab in comparison to males (Fig. 26-19).

The intertwined functions of sex recognition and species recognition, advertisement of individual quality, and synchronization of reproductive behavior commonly require a complex series of signals, both active and passive, by both

(a) A bowerbird bower

(b) A male frigatebird

▲ **Figure 26-18 Sexual displays** *(a)* During courtship, a male gardener bowerbird builds a bower out of twigs and decorates it with colorful items that he gathers. *(b)* A male frigate bird inflates his scarlet throat pouch to attract passing females. **QUESTION:** The male bowerbird provides no protection, feeding, or other care to his mate or offspring. Why, then, do females carefully compare the bowers of different males before choosing a mate?

▲ **Figure 29-19 Sex differences in guppies** As in many animal species, the male guppy (left) is brighter and more colorful than the female.

Sex and Symmetry

Continued

A male's success at using a visual signal to attract a mate may depend in part on the signal's symmetry. For example, a male house finch has a patch of bright red feathers on the crown of his head. Males whose crown patches are brightly colored are more likely to attract a mate than are males with dull patches, but males with patches that are both colorful *and* highly symmetrical have the highest mating success of all.

sexes. Such signals are beautifully illustrated by the complex underwater "ballet" executed by the male and female three-spined stickleback fish (**Fig. 26-20**).

Chemical Signals Can Bring Mates Together

Pheromones can also play an important role in reproductive behavior. A sexually receptive female silk moth, for example, sits quietly and releases a chemical message that can be detected by males up to 3 miles (5 kilometers) away. The exquis-

▼ **Figure 26-20 Courtship of the three-spined stickleback**

❶ A male, inconspicuously colored, leaves the school of males and females to establish a breeding territory.

❷ As his belly takes on the red color of the breeding male, he displays aggressively at other red-bellied males, exposing his red underside.

❸ Having established a territory, the male begins nest construction by digging a shallow pit that he will fill with bits of algae cemented together by a sticky secretion from his kidneys.

❹ After he tunnels through the nest to make a hole, his back begins to take on the blue courting color that makes him attractive to females.

❺ An egg-carrying female displays her enlarged belly to him by assuming a head-up posture. Her swollen belly and his courting colors are passive visual displays.

❻ Using a zigzag dance, he leads her to the nest.

❼ After she enters, he stimulates her to release eggs by prodding at the base of her tail.

❽ He enters the nest as she leaves and deposits sperm, which fertilize the eggs.

itely sensitive and selective receptors on the antennae of the male silk moth respond to just a few molecules of the substance, allowing him to travel upwind along a concentration gradient to find the female (**Fig. 26-21a**).

Water is an excellent medium for dispersing chemical signals, and fish commonly use a combination of pheromones and elaborate courtship movements to ensure the synchronous release of gametes. Mammals, with their highly developed sense of smell, often rely on pheromones released by the female during her fertile periods to attract males (**Fig. 26-21b**).

26.5 Why Do Animals Play?

Many animals play. Pygmy hippopotamuses push one another, shake and toss their heads, splash in the water, and pirouette on their hind legs. Otters delight in elaborate acrobatics. Bottlenose dolphins balance fish on their snouts, throw objects, and carry them in their mouths while swimming. Baby vampire bats chase, wrestle, and slap each other with their wings. Pigface, a giant African softshell turtle who lived at the National Zoo in Washington, D.C., for more than 50 years, would spend hours each day batting a ball around his enclosure. Even octopuses have been seen playing a game: pushing objects away from themselves and into a current, then waiting for the objects to drift back, only to push them back into the current to start the cycle over again.

Animals Play Alone or with Other Animals

Play can be solitary, as when a single animal manipulates an object, such as a cat with a ball of yarn, or the dolphin with its fish, or a macaque monkey making and playing with a snowball. Or, play can be social. Often, young of the same species play together, but parents may join them (**Fig. 26-22a**). Social play typically includes chasing, fleeing, wrestling, kicking, and gentle biting (**Fig. 26-22b,c**).

Play seems to lack any clear immediate function, and is abandoned in favor of feeding, courtship, and escaping from danger. Young animals play more frequently than do adults. Play typically borrows movements from other behaviors (attacking, fleeing, stalking, and so on) and uses considerable energy. Also, play is potentially dangerous. Many young humans and other animals are injured, and some are killed, during play. In addition, play can distract an animal from the presence of danger while making it conspicuous to predators. So, why do animals play?

▼ **Figure 26-21 Pheromone detectors** *(a)* Male moths find females not by sight but by following airborne pheromones released by females. These odors are sensed by receptors on the male's huge antennae, whose enormous surface area maximizes the chances of detecting the female scent. *(b)* When dogs meet, they typically sniff each other near the base of the tail. Scent glands there broadcast information about the bearer's sex and interest in mating. **QUESTION:** Female dogs use a pheromone to signal readiness to mate, but female mandrills (see Fig. 26-8) signal mating readiness with a visual signal. What differences would you predict between the two species' methods of searching for food?

(a) Antennae detect pheromones

(b) Noses detect pheromones

(a) Chimpanzees

(b) Polar bears

(c) Fox cubs

▲ Figure 26-22 Young animals at play

Play Aids Behavioral Development

It is likely that play has survival value and that natural selection has favored those individuals who engage in playful activities. One of the best explanations for the survival value of play is the "practice hypothesis." It suggests that play allows young animals to gain experience in behaviors that they will use as adults. By performing these acts repeatedly in play, the animal practices skills that will later be important in hunting, fleeing, or social interactions.

Recent research supports and extends that proposal. Play is most intense early in life when the brain develops and crucial neural connections form. John Byers, a zoologist at the University of Idaho, has observed that species with large brains tend to be more playful than species with small brains. Because larger brains are generally linked to greater learning ability, this relationship supports the idea that adult skills are learned during juvenile play. Watch children roughhousing or playing tag, and you will see how play fosters strength and coordination and develops skills that might have helped our hunting ancestors survive. Quiet play with other children, with dolls, blocks, and other toys, helps children prepare to interact socially, nurture their own children, and deal with the physical world.

Although Shakespeare tells us "play needs no excuse," there is good evidence that the tendency to play has evolved as an adaptive behavior in animals capable of learning. Play is quite literally serious fun.

26.6 What Kinds of Societies Do Animals Form?

Sociality is a widespread feature of animal life. Most animals interact at least a little with other members of their species. Many spend the bulk of their lives in the company of others, and a few species have developed complex, highly structured societies. Social interaction can be cooperative or competitive, but is typically a mixture of the two.

Group Living Has Advantages and Disadvantages

Living in a group has both costs and benefits, and a species will not evolve social behavior unless the benefits of doing so outweigh the costs.

On the negative side, social animals may encounter:

- Increased competition within the group for limited resources.
- Increased risk of infection from contagious diseases.
- Increased risk that offspring will be killed by other members of the group.
- Increased risk of being spotted by predators.

Benefits to social animals include:

- Increased abilities to detect, repel, and confuse predators.
- Increased hunting efficiency or increased ability to spot localized food resources.
- Advantages resulting from the potential for division of labor within the group.
- Increased likelihood of finding mates.

Sociality Varies Among Species

The degree to which animals of the same species cooperate varies from one species to the next. Some types of animals, such as the mountain lion, are basically solitary; interactions between adults consist of brief aggressive encounters and mating. Other types of animals cooperate on the basis of changing needs. For example, the coyote is solitary when food is abundant, but hunts in packs in the winter when food becomes scarce.

Loose social groups, such as pods of dolphins, schools of fish, flocks of birds, and herds of musk oxen (Fig. 26-23), can provide benefits. For example, the characteristic spacing of fish in schools or the V-pattern of geese in flight provides a hydrodynamic or aerodynamic advantage for each individual in the group, reducing the energy required for swimming or flying. Some biologists hypothesize that herds of antelope or schools of fish confuse predators; their myriad bodies make it difficult for the predator to focus on and pursue a single individual.

A Few Species Form Complex Societies

At the other end of the social spectrum are a few highly integrated cooperative societies, primarily among insects and mammals. As you read the following section, you may notice that some cooperative societies are based on behavior that seems to sacrifice the individual for the good of the group. There are many examples: Young, mature Florida scrub jays may remain at their parents' nest and help them raise subsequent broods instead of breeding; worker ants often die in defense of their nest; Belding ground squirrels may sacrifice their own lives to warn the rest of their group of an approaching predator. These behaviors are examples of **altruism**—behavior that decreases the reproductive success of one individual to benefit another.

Forming Groups with Relatives Fosters the Evolution of Altruism

How could altruistic behavior evolve? When individuals perform self-sacrificing deeds, why aren't the alleles that contribute to this behavior eliminated from the gene pool? One possibility is that other members of the group are close relatives of the altruistic individual. Because close relatives share alleles, the altruistic individual may promote the survival of its own alleles through behaviors that maximize the survival of its close relatives. This concept is called **kin selection.** Kin selection helps explain the self-sacrificing behaviors that contribute to the success of cooperative societies. Cooperative behavior is illustrated in the following sections, which describe two examples of complex societies, one in an insect species and one in a mammal species.

Honeybees Live Together in Rigidly Structured Societies

Perhaps the most puzzling of all animal societies are those of the bees, ants, and termites. Scientists have long struggled to explain the evolution of a social structure in which most individuals never breed, but instead labor intensively to feed and protect the offspring of a different individual. Whatever its evolutionary explanation, the intricate organization of a social insect colony makes a compelling story. In these communities, the individual is a mere cog in an intricate, smoothly running machine and could not survive by itself.

Individual social insects are born into one of several castes within the society. These castes are groups of similar individuals that perform a specific function. Honeybees emerge from their larval stage into one of three major preordained roles. One role is that of *queen*. Only one queen is tolerated in a hive at any time. Her functions are to produce eggs (up to 1,000 per day for a lifetime of 5 to 10 years) and regulate the lives of the workers. Male bees, called *drones*, serve merely as mates for the queen. Soon after the queen hatches, drones lured by her sex pheromones swarm around her, and she mates with as many as 15 of them. This relatively brief "orgy" supplies her with sperm that will last a lifetime, enough to fertilize more than 3 million eggs. Their sexual chore accomplished, the drones become superfluous and are eventually driven out of the hive or killed.

▲ **Figure 26-23 Cooperation in loosely organized social groups** A herd of musk oxen functions as a unit when threatened by predators such as wolves. Males form a circle, horns pointed outward, around the females and young.

Web Animation Communication in Honeybees

▲ **Figure 26-24 Bee language: the waggle dance** A forager, returning from a rich source of nectar, performs a waggle dance that communicates the distance and direction of the food source, as other foragers crowd around her, touching her with their antennae. The bee moves in a straight line while shaking her abdomen back and forth ("waggling") and buzzing her wings. She repeats this dance over and over in the same location, circling back in alternating directions.

Sex and Symmetry

Continued

Many animals pay attention to symmetry when choosing a mate, and some even use symmetry to choose food. For example, when nectar-feeding bees (such as honey-bees) locate a tempting patch of flowers to feed on and are faced with a multitude of seemingly identical flowers to choose from, they tend to select the most symmetrical flowers. Why would a bee favor symmetrical flowers? Because, among flowers of a given plant species, the most symmetrical flowers contain the most nectar. And why would flowers with more nectar tend to be more symmetrical? Read "Sex and Symmetry, *Revisited*" at the end of this chapter and then see if you can explain the connection between a flower's symmetry and its nectar-producing ability.

The hive is run by the third class of bees, sterile female *workers*. A worker's tasks are determined by her age and by conditions in the colony. A newly emerged worker starts life as a "waitress," carrying food such as honey and pollen to the queen, to other workers, and to developing larvae. As she matures, special glands begin wax production, and she becomes a builder, constructing hexagonal cells of wax in which the queen deposits her eggs and the larvae develop. She also takes shifts as a "maid," cleaning the hive and removing the dead, and as a guard, protecting the hive against intruders. Her final role in life is that of a forager, gathering pollen and nectar, food for the hive. She spends nearly half of her 2-month life in this role. Acting as a forager scout, she seeks new and rich sources of nectar and, if she finds one, returns to the hive and communicates its location to other foragers. She communicates by means of the **waggle dance,** an elegant form of symbolic expression (**Fig. 26-24**).

Pheromones play a major role in regulating the lives of social insects. Honeybee drones are drawn irresistibly to the queen's sex pheromone (*queen substance*), which she releases during her mating flights. Back at the hive, she uses the same substance to maintain her position as the only fertile female. The queen substance is licked off her body and passed among all the workers, rendering them sterile. The queen's presence and health are signaled by her continuing production of queen substance; a decrease in production alters the behavior of the workers. Almost immediately they begin building extra-large "royal cells." The workers feed the larvae that develop in these cells a special glandular secretion known as "royal jelly." This unique food alters the development of the growing larvae so that, instead of a worker, a new queen emerges from the royal cell. The old queen then leaves the hive, taking a swarm of workers with her to establish residence elsewhere. If more than one new queen emerges, a battle to the death ensues. The victorious queen takes over the hive.

Naked Mole Rats Form a Complex Society

The nervous systems of vertebrates are far more complex than those of insects, and we might therefore expect vertebrate societies to be proportionately more complex. With the exception of human society, however, they are not. Perhaps the most unusual society among nonhuman mammals is that of the naked mole rat (**Fig. 26-25**). These nearly blind, nearly hairless relatives of guinea pigs live in large underground colonies in southern Africa and have an antlike form of social organization not known to exist in any other mammalian society. The colony is dominated by the queen, a single reproducing female to whom all other members are subordinate.

The queen is the largest individual in the colony and maintains her status by aggressive behavior, particularly shoving. She prods and shoves lazy workers, stimulating them to become more active. As in honeybee hives, there is a division of labor among the workers, in this case based on size. Small, young rats clean the tunnels, gather food, and tunnel. Tunnelers line up head to tail and pass excavated dirt along the completed tunnel to an opening. Just below the opening, a larger mole rat flings the dirt into the air, adding it to a cone-shaped mound. Biologists observing this behavior from the surface dubbed it "volcanoing." In addition to volcanoing, large mole rats defend the colony against predators and members of other colonies.

If another female begins to become fertile, the queen apparently senses changes in the estrogen levels of the subordinate female's urine. The queen then selectively shoves the would-be breeder, causing stress that prevents her rival from ovulating. Although all adult males are fertile, large males are more likely to mate with the queen than are small ones. When the queen dies, a few of the females gain weight and begin shoving one another. The aggression may escalate until a rival is killed. Ultimately, a single female becomes dominant. Her body lengthens, and she assumes the queenship and begins to breed. Litters averaging 14 pups are produced about four times a year. During the first month, the queen nurses her pups, and the workers feed the queen. Then the workers begin feeding the pups solid food.

▲ **Figure 26-25 A naked mole rat queen rests atop a group of workers**

26.7 Can Biology Explain Human Behavior?

The behaviors of humans, like those of all other animals, have an evolutionary history. Thus, the techniques and concepts of ethology can help us understand and explain human behavior. Human ethology, however, will remain a less rigorous science than animal ethology, because we cannot treat people like laboratory animals, devising experiments that control and manipulate the factors that influence their attitudes and actions. In addition, some observers argue that human culture has been freed from the constraints of its evolutionary past for so long that we cannot explain our behavior in terms of biological evolution. Nevertheless, many scientists have taken an ethological, evolutionary approach to human behavior, and their work has had a major impact on our view of ourselves.

The Behavior of Newborn Infants Has a Large Innate Component

Because newborn infants have not had time to learn, we can assume that much of their behavior is innate. The rhythmic movement of an infant's head in search of its mother's breast is an innate behavior that is expressed in the first days after birth. Sucking, which can be observed even in a human fetus, is also innate (**Fig. 26-26**). Other behaviors seen in newborns and even premature infants include grasping with the hands and feet and making walking movements when the body is held upright and supported.

Another example is smiling, which can occur soon after birth. Initially, smiling can be induced by almost any object looming over the newborn. This initial indiscriminate response, however, is soon modified by experience. Infants up to 2 months old will smile in response to a stimulus consisting of two dark, eye-sized spots on a lighter background, which at that stage of development is a more potent stimulus for smiling than is an accurate representation of a human face. But as the child's development continues, learning and further development of the nervous system interact to limit the response to more correct representations of a face.

Newborns in their first 3 days of life can be conditioned to produce certain rhythms of sucking when their mother's voice is used as reinforcement. In experiments, infants preferred their own mothers' voices to other female voices,

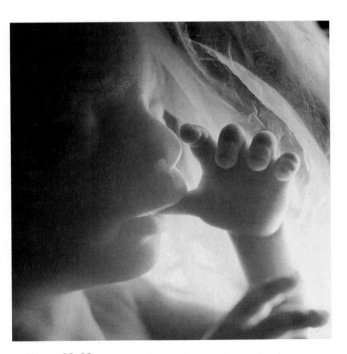

▲ **Figure 26-26 A human instinct** Thumb sucking is a difficult habit to discourage in young children, because sucking on appropriately sized objects is an instinctive, food-seeking behavior. This fetus sucks its thumb at about 4 months of development.

▲ **Figure 26-27 Newborns prefer their mother's voices** Using a nipple connected to a computer that plays audio tapes, researcher William Fifer demonstrated that newborns can be conditioned to suck at specific rates in order to listen to their own mothers' voices through headphones. For example, if the infant sucks faster than normal, her mother's voice is played; if she sucks more slowly, another woman's voice is played. Researchers found that infants easily learned and were willing to work hard at this task just to listen to their own mothers' voices, presumably because they had become used to her voice in the womb.

as indicated by their responses (Fig. 26-27). The infant's ability to learn his or her mother's voice and respond positively to it within days of birth has strong parallels to imprinting and may help initiate bonding with the mother.

Young Humans Acquire Language Easily

One of the most important insights from studies of animal learning is that animals tend to have an inborn predilection for specific types of learning that are important to their species' mode of life. In humans, one such inborn predilection seems to be for the acquisition of language. Young children are able to acquire language rapidly and nearly effortlessly; they typically acquire a vocabulary of 28,000 words before the age of 8. Research suggests that we are born with a brain that is already primed for this early facility with language. For example, a human fetus begins responding to sounds during the third trimester of pregnancy, and researchers have demonstrated that infants are able to distinguish among consonant sounds by 6 weeks after birth. In this experiment, infants sucked on a pacifier that contained a force transducer to record the sucking rate. The infants were conditioned to suck at a higher rate in response to playback of adult voices making various consonant sounds. When one sound (such as "ba") was presented repeatedly, the infant became habituated and decreased her sucking rate. But when a new sound (such as "pa") was presented, sucking rate increased, revealing that the infant perceived the new sound as different.

Behaviors Shared by Diverse Cultures May Be Innate

Another way to study the innate bases of human behavior is to compare simple acts performed by people from diverse cultures. This comparative approach, pioneered by ethologist Irenaus Eibl-Eibesfeldt, has revealed several gestures that seem to form a universal, and therefore probably innate, human signaling system. Such gestures include facial expressions for pleasure, rage, and disdain and greeting movements such as an upraised hand or the "eye flash" (in which the eyes are widely opened and the eyebrows rapidly elevated). The evolution of "hardwiring" for these gestures presumably depended on the advantages that accrued to both senders and receivers from sharing information about the emotional state and intentions of the sender. A species-wide method of communication was perhaps especially important before the advent of language and later remained useful during encounters between people who shared no common language.

Certain complex social behaviors are widespread among diverse cultures. For example, the incest taboo (avoidance of mating with close relatives) seems to be universal across human cultures (and even across many species of nonhuman primates). It seems unlikely, however, that a shared belief could be encoded in our genes. Some biologists have suggested that the taboo is instead a cultural expression of an evolved, adaptive behavior. According to this hypothesis, close contact among family members early in life suppresses sexual desire, an effect that arose because of the negative consequences of inbreeding (such as a higher incidence of genetic diseases). The hypothesis does not require us to assume an innate social belief, but rather proposes that we inherit a learning program that causes us to undergo a kind of imprinting early in life.

Humans May Respond to Pheromones

Although the main channels of human communication are through the eyes and ears, humans also seem to respond to chemical messages. The possible existence of human pheromones was hinted at in the early 1970s, when biologist Martha McClintock found that the menstrual cycles of roommates and close friends tended to become synchronized. McClintock suggested that the synchrony resulted from some chemical signal between the women, but almost 30 years passed before she and her colleagues uncovered more conclusive evidence that a pheromone was at work.

In 1998, McClintock's research group asked nine female volunteers to wear cotton pads in their armpits for 8 hours each day during their menstrual cycles. The pads were then disinfected with alcohol and swabbed above the upper lips of another set of 20 female subjects (who reported that they could detect no odors other than alcohol on the pads). The subjects were exposed to the pads in this way each day for 2 months, with half the group sniffing secretions from women in the early (preovulation) part of the menstrual cycle while the other half was exposed to secretions from later in the cycle (postovulation). Women exposed to early-cycle secretions had shorter-than-usual menstrual cycles, and women exposed to late-cycle secretions had delayed menstruation. It appears that women release different pheromones, with different effects on receivers, at different points in the menstrual cycle.

Although McClintock's experiment offers strong evidence for the existence of human pheromones, little else is known about chemical communication in humans. The actual molecules that caused the effects documented by McClintock remain unknown, as does their function. (What benefit would a woman gain by influencing the menstrual cycles of other women?) Receptors for chemical messages have not yet been found in humans, and we don't know if the "menstrual pheromones" are the first known example of an important communication system or merely an isolated case of a vestigial ability. Despite the hopeful advertisements for "sex attraction pheromones" on late-night television, chemical communication in humans is a scientific mystery awaiting a solution.

Studies of Twins Reveal Genetic Components of Behavior

Twins present an opportunity to examine the hypothesis that differences in human behavior are related to genetic differences. If a particular behavior is heavily influenced by genetic factors, we would expect to find that *identical twins* (which arise from a single fertilized egg and have identical genes) are more likely to share the behavior than are *fraternal twins* (which arise from two individual eggs fertilized by different sperm and are no more similar genetically than are other siblings). Data from twin studies, and from other within-family investigations, have tended to confirm the heritability of many human behavioral traits. These studies have documented a significant genetic component for traits such as activity level, alcoholism, sociability, anxiety, intelligence, dominance, and even political attitudes. On the basis of tests designed to measure many aspects of personality, identical twins are about two times more similar in personality than are fraternal twins.

The most fascinating twin findings come from observations of identical twins separated soon after birth, reared in different environments, and reunited for the first time as adults. Identical twins reared apart have been found to be as similar in personality as those reared together, indicating that the differences in their environments had little influence on their personality development. They have been found to share nearly identical taste in jewelry, clothing, humor, food, and names for children and pets. In some cases these separated twins share personal idiosyncrasies such as giggling, nail biting, drinking patterns, hypochondria, and mild phobias.

Biological Investigation of Human Behavior Is Controversial

The field of human behavioral genetics is controversial, especially among nonscientists, because it challenges the long-held belief that environment is the most important determinant of human behavior. As discussed earlier in this chapter, we now recognize that all behavior has some genetic basis and that complex behavior in nonhuman animals typically combines elements of both innate and learned behaviors. Thus, it seems certain that our own behavior is influenced by both our evolutionary history and our cultural heritage. The debate over the relative importance of heredity and environment in determining human behavior continues and is unlikely ever to be fully resolved. Human ethology is not yet recognized as a rigorous science, and it will always be hampered because we can neither view ourselves with detached objectivity nor experiment with people as if they were laboratory rats. Despite these limitations, there is much to be learned about the interaction of learning and innate tendencies in humans.

Sex and Symmetry Revisited

In the experiment described at the beginning of this chapter, women found males with the most symmetrical bodies to be the most attractive. But how did the women know which males were most symmetrical? After all, the researchers' measurement of male symmetry was based on small differences in the size of body parts that the female judges did not even see during the test.

Perhaps male body symmetry is reflected in facial symmetry, and females prefer symmetrical faces. To test this hypothesis, a group of researchers used computers to alter photos of male faces, either increasing or decreasing their symmetry (**Fig. 26-28**). Then female observers rated each face for attractiveness. The observers had a strong preference for more symmetrical faces.

Some evidence suggests that women need not even look at men to determine symmetry. In one study, researchers measured the body symmetry of 80 men and then issued a clean T-shirt to each one. Each subject wore his shirt to bed for two consecutive nights. A panel of 82 women sniffed the shirts and rated their scents for "pleasantness" and "sexiness." Which shirts had the sexiest, most pleasant scents? The ones worn by the most symmetrical men. The researchers concluded that women can identify symmetrical men by their scent.

Why would females prefer to mate with symmetrical males? The most likely explanation is that symmetry indicates good physical condition. Disruptions of normal embryological development can cause bodies to be asymmetrical, so a highly symmetrical body indicates healthy, normal development. Females that mate with individuals whose health and vitality are announced by their symmetrical bodies might have offspring that are similarly healthy and vital.

Consider This

Is our perception of human beauty determined by cultural standards, or is it part of our biological makeup, the product of our evolutionary heritage? What evidence would persuade you that beauty is a biological phenomenon, or that it is a cultural one?

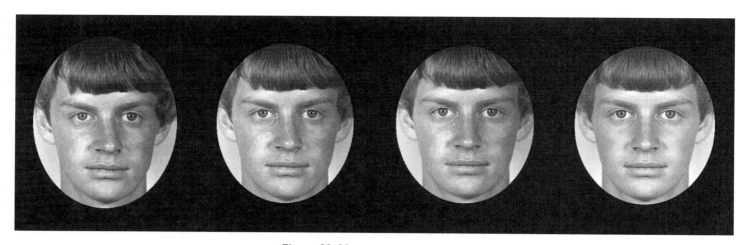

▲ **Figure 26-28 Faces of varying symmetry** Researchers used sophisticated software to modify facial symmetry. From left: a face modified to be less symmetrical; the original, unmodified face; a face modified to be more symmetrical; a perfectly symmetrical face.

Chapter Review

Summary of Key Concepts

For additional study help and activities, go to www.mybiology.com.

26.1 How Do Innate and Learned Behaviors Differ?

Although all animal behavior is influenced by both genetic and environmental factors, biologists distinguish between behaviors whose development is not highly dependent on external factors and behaviors that require more extensive environmental stimuli in order to develop. Behaviors in the first category are sometimes designated as innate and can be performed properly the first time an animal encounters the appropriate stimulus. Behavior that changes in response to input from an animal's social and physical environment is said to be learned. Learning can modify innate behavior to make it more appropriate.

Although the distinction between innate and learned behavior is conceptually useful, the distinction is not sharp in naturally occurring behaviors. Learning allows animals to modify innate responses so that they occur only with appropriate stimuli. Imprinting, a form of learning with innate constraints, is possible only at a certain time in an animal's development.

Web Animation Observing Behavior: Homing in Digger Wasps

26.2 How Do Animals Communicate?

Communication allows animals of the same species to interact effectively in their quest for mates, food, shelter, and other resources. Animals communicate through visual signals, sound, chemicals (pheromones), and touch. Visual communication is quiet and can convey rapidly changing information. Visual signals are active (body movements) or passive (body shape and color). Sound communication can also convey rapidly changing information, and it is effective when vision is impossible. Pheromones can be detected after the sender has departed, conveying a message over time. Physical contact reinforces social bonds and is a part of mating.

26.3 How Do Animals Compete for Resources?

Although many competitive interactions are resolved through aggression, serious injuries are rare. Most aggressive encounters are settled by displays that communicate the motivation, size, and strength of the combatants.

Some species establish dominance hierarchies that minimize aggression. On the basis of initial aggressive encounters, each animal acquires a status in which it defers to more dominant individuals and dominates subordinates. When resources are limited, dominant animals obtain the largest share and are most likely to reproduce.

Territoriality, a behavior in which animals defend areas where important resources are located, also minimizes aggressive encounters. In general, territorial boundaries are respected, and the best-adapted individuals defend the richest territories and produce the most offspring.

26.4 How Do Animals Find Mates?

Successful reproduction requires that animals recognize the species, sex, and sexual receptivity of potential mates. In many species, animals also assess the quality of potential mates. These requirements have contributed to the evolution of sexual displays that use all forms of communication.

26.5 Why Do Animals Play?

Animals of many species engage in seemingly wasteful (and sometimes dangerous) play behavior. Although engaging in play appears to provide no immediate benefit to an animal, play behavior in young animals may have been favored by natural selection because it provides opportunities to practice and perfect behaviors that will later be crucial for survival and mating.

26.6 What Kinds of Societies Do Animals Form?

Social living has both advantages and disadvantages, and species vary in the degree to which their members cooperate. Some species form cooperative societies. The most rigid and highly organized are those of the social insects such as the honeybee, in which the members follow rigidly defined roles throughout life. These roles are maintained through both genetic programming and the influence of certain pheromones. Naked mole rats exhibit the most complex and rigid vertebrate social interactions, resembling those of social insects.

Web Animation Communication in Honeybees

26.7 Can Biology Explain Human Behavior?

The degree to which human behavior is genetically influenced is highly controversial. Because we cannot freely experiment on humans, and because learning plays a major role in nearly all human behavior, investigators must rely on studies of newborn infants, comparative cultural studies, and studies of identical and fraternal twins. Evidence is mounting that our genetic heritage plays a role in personality, intelligence, simple universal gestures, our responses to certain stimuli, and our tendency to learn specific things such as language at particular stages of development.

aggression *p. 520*

altruism *p. 527*

behavior *p. 514*

communication *p. 518*

dominance hierarchy *p. 521*

imprinting *p. 517*

innate behavior *p. 514*

kin selection *p. 527*

learning *p. 515*

pheromone *p. 519*

territoriality *p. 522*

waggle dance *p. 528*

Thinking Through the Concepts

Suggested answers to end-of-chapter and figure-based questions can be found at the end of the text.

Fill-in-the-Blank

1. In general, animal behaviors arise from an interaction between the animal's _____ and its _____. Some behaviors are performed correctly the first time an animal encounters the proper _____. Such behaviors are described as _____.

2. Play is almost certainly an adaptive behavior because it uses considerable _____, it can distract the animal from watching for _____, and it may cause _____. The most likely explanation for why animals play is the "_____ hypothesis," which states that play teaches the young animal _____ that it can use when it becomes a(n) _____. This hypothesis is supported by the observation that animals that have larger _____ and are most capable of _____ are more likely to play.

3. One of the simplest forms of learning is _____, defined as a decline in response to a(n) _____, harmless stimulus. A different type of learning in which an animal's nervous system is rigidly programmed to learn a certain behavior during a certain period in its life is called _____. The time frame during which such learning occurs is called the _____.

4. Animals often deal with competition for resources through _____ behavior. Such conflicts are often resolved through _____, which allow the competing animals to assess each other without _____ each other. Animals that resolve conflicts this way often display their _____, and make their bodies appear _____.

5. The defense of an area where important resources are located is called _____. Examples of important resources that may be

defended include places to _____, _____, _____, and _____. Such resources are most commonly defended by which sex? _____ Are these spaces usually defended against members of the same or of different species? _____

6. After each form of communication, list a major advantage in the first blank, and a major disadvantage in the second blank. Pheromones: _____, _____; visual displays: _____, _____.

Review Questions

1. Explain why neither "innate" nor "learned" adequately describes the behavior of any given organism.

2. Explain why animals play. Include the features of play in your answer.

3. List four senses through which animals communicate, and give one example of each form of communication. After each, present both advantages and disadvantages of that form of communication.

4. A bird will ignore a squirrel in its territory but will act aggressively toward a member of its own species. Explain why.

5. Why are most aggressive encounters among members of the same species relatively harmless?

6. Discuss advantages and disadvantages of group living.

7. In what ways do naked mole rat societies resemble those of the honeybee?

Applying the Concepts

1. **IS THIS SCIENCE?** A company that markets a product it calls a "human sex pheromone" promises that men who apply this product to their skin will be "approached by women who will make eye contact, smile, initiate conversations, and compliment you." Further, the company cites a survey of its customers that found that 68% experienced an increase in the frequency of sexual intercourse after using the product. Do you find the company's claims plausible? Why or why not? Design a study to test the claims.

2. Male mosquitoes orient toward the high-pitched whine of the female, and female mosquitoes (only females suck blood) are

attracted to the warmth, humidity, and carbon dioxide exuded by their prey. Using this information, design a mosquito trap or killer that exploits a mosquito's innate behaviors. Now design one for moths.

3. Describe and give an example of a dominance hierarchy. What role does it play in social behavior? Give a human parallel, and describe its role in human society. Are the two roles similar? Why or why not? Repeat this exercise for territorial behavior in humans and in another animal.

For additional resources, go to www.mybiology.com.

unit six
Ecology

Human observers are captivated by the bright colors and ethereal beauty of coral reefs, which are among the most diverse, productive, and fragile of Earth's ecosystems.

chapter 27

Population Growth

If they could speak, would the mysterious statues of Easter Island tell the tale of a population that outgrew the capacity of the environment to sustain it?

Case Study   The Mystery of Easter Island

Why do civilizations vanish? Among those who have pondered this question were the first Europeans to reach Easter Island. When they arrived at the remote Pacific island in the 1700s, these voyagers were mystified by the enormous stone statues that dominate the island's barren landscape. The island's few inhabitants had no record or memory of the statues' creators and, in any case, did not possess the technology that would have been needed to transport and erect such huge and heavy structures. Moving such heavy objects over the 6 miles from the nearest stone quarry and then maneuvering them into an upright position would have required long ropes and strong timbers. Easter Island, however, was devoid of anything that could have furnished wood or fibers for rope. There were no trees at all on the island, and none of the few shrubs there grew higher than 10 feet.

An important clue to the mystery of Easter Island was revealed by scientists who studied pollen grains that they found in layers of ancient sediments. Because the age of each sediment layer can be determined, and because a plant species can be identified by the unique appearance of its pollen, pollen analysis can show how vegetation has changed over time.

The pollen record of Easter Island showed that, before the arrival of humans, the island supported a diverse forest, including plants that could supply fiber for rope and trees with long, straight trunks that would have been ideal for moving statues. But by A.D. 1400, about a thousand years after the arrival of humans, almost all of Easter Island's trees were gone. According to one hypothesis, a thousand years of clearing land for agriculture and cutting trees for firewood and construction materials had destroyed the forest.

Apparently, the culture responsible for the statues had disappeared along with the forest. Could there have been a connection between the two disappearances? ■

Pollen grains provided some of the key clues for solving the mystery of Easter Island.

This chapter begins our unit on **ecology,** the study of organisms' interactions with each other and with the nonliving part of their environment. These interactions can be examined at different levels of organization; different perspectives yield different insights. Ecologists work at three main levels of organization:

- A **population** consists of all the members of a particular species who live in a particular area.

- A **community** includes all the interacting populations of different species that coexist in a particular area.

- An **ecosystem** consists of all the communities within a defined area, along with their nonliving environment.

In Chapters 27 through 31, we will proceed through these levels of organization, starting with populations, moving to the communities that contain the populations, going on to the ecosystems that contain the communities, and then to the *biosphere*, the diverse collection of ecosystems that encompasses all life on Earth. We will close with an examination of human impacts on the biosphere and an overview of efforts to conserve biodiversity.

27.1 How Are Populations Distributed in Space and Time?

Organisms arrange themselves in space in many different ways. They might live in large flocks or herds, in small groups, as pairs, or as solitary individuals. They might spread themselves evenly through their habitat or cluster around a resource such as a water hole. Their spatial arrangement might vary with time, changing in the breeding season, for example. Ecologists recognize three major types of spatial distribution: clumped, uniform, and random.

Individuals in Many Populations Clump Together in Groups

The members of many populations live clumped together in groups. Examples of such **clumped distributions** include social groups such as elephant herds, wolf packs, lion prides, flocks of birds, and schools of fish (**Fig. 27-1a**). What are the advantages of clumping? In groups, there are many eyes that can search for localized food, such as a tree full of fruit. Membership in a large school of fish or flock of birds may reduce the odds that any one individual will be killed by an attacking predator. A temporary group that forms for mating may increase individuals' chances of finding a mate. Alternatively, plant or animal populations may cluster because resources are localized. Cottonwood trees, for example, cluster along streams and rivers in grasslands.

Some Individuals Disperse Themselves Evenly

Populations with **uniform distribution** maintain relatively even spacing between individuals. This type of distribution is most common among animals that defend territories to protect scarce resources. Male Galápagos iguanas, for example, establish regularly spaced breeding territories. Shorebirds are often found in evenly spaced nests during the breeding season. Other territorial species, such as the tawny owl, continuously occupy well-defined, relatively uniformly spaced territories. (For animals that remain together in pairs to raise their young, *uniform spacing* often refers to pairs rather than to individuals.) Some plants, such as creosote bushes, release chemicals into the soil around them that inhibit germination of other plants and thus space themselves relatively evenly (**Fig. 27-1b**).

In a Few Populations, Individuals Are Distributed at Random

In a population with **random distribution,** the distance between individuals varies unpredictably. Individuals in such populations do not form social groups. The resources they need are more or less equally available throughout the area they inhabit and are not scarce enough to require territorial spacing. Trees and other plants in rain forests may be randomly distributed (**Fig. 27-1c**).

27.2 How Do Populations Grow?

Studies of undisturbed ecosystems show that the size of some populations tends to remain relatively constant over time. In other populations, size fluctuates in a cyclical pattern, or varies sporadically in response to environmental changes.

(a) Clumped distribution

(b) Uniform distribution

(c) Random distribution

▲ **Figure 27-1 Population distributions** *(a)* Clumped: a school of fish. *(b)* Uniform: creosote bushes in the desert. *(c)* Random: trees and other plants in a rain forest.

Earth's human population, in contrast to most nonhuman populations, has grown steadily for centuries.

Births, Deaths, and Migration Determine Population Growth

A population's size remains stable if, on average, as many individuals join as leave. Individuals join a population through birth or immigration (migration in) and leave it through death or emigration (migration out). Thus, a population grows when the number of births plus immigrants exceeds the number of deaths plus emigrants. Populations shrink when the reverse occurs. A simple equation for the change in population size within a given time period is as follows:

(births – deaths) + (immigrants – emigrants) = change in population size

In many natural populations, organisms moving in and out contribute relatively little to population change, leaving birth and death rates as the primary factors that influence population growth.

Population Growth Can Be Expressed as a Rate

The **per capita growth rate** (r) of a population is a measure of how fast a population grows, expressed as a change in population size per individual per unit of time (that is, as a fraction). This value is determined by subtracting the per capita death rate (d) from the per capita birth rate (b):

$$b \quad - \quad d \quad = \quad r$$
$$\text{(birth rate)} \quad \text{(death rate)} \quad \text{(growth rate)}$$

For example, to calculate the annual growth rate of a human population of 10,000 in which there are 1,500 births and 500 deaths each year, we first calculate the annual per capita birth rate:

$$b = 1{,}500 \text{ births/10,000 people}$$
$$= 0.15 \text{ births per person per year}$$

Next, the annual per capita death rate:

$$d = 500 \text{ deaths/10,000 people}$$
$$= 0.05 \text{ deaths per person per year}$$

Finally, we calculate the per capita growth rate by subtracting the death rate from the birth rate:

$$r = 0.15 \text{ births per person per year} - 0.05 \text{ deaths per person per year}$$
$$= 0.10 \text{ (that is, 10\% increase per year)}$$

Note that if the death rate exceeds the birth rate, the growth rate will be negative and the population will shrink. In this chapter we focus on growing populations.

A Constant Growth Rate Increases Population Size Rapidly

To calculate the number of individuals added to a population in a year, multiply the annual per capita growth rate (r) by the original population size (N):

$$\text{population growth} = rN$$

In our example, population growth in the first year (rN) is $0.10 \times 10{,}000 = 1{,}000$ people. Thus, at the end of year 1, the population contains 11,000 people. If the per capita growth rate remains the same during year 2, then population growth in the second year is $0.10 \times 11{,}000 = 1{,}100$ and the population size at the end of year 2 is 12,100. In the third year, the population grows by 1,210 people, and so on in subsequent years. As you can see, if the per capita growth rate r remains constant, the number of people added to the population increases each year. This pattern of continuously accelerating increase in population size is **exponential growth.**

A Population's Growth Rate Depends on Patterns of Reproduction

A population grows during periods when births exceed deaths. Growth will persist if, on average, each individual produces more than one surviving offspring during its lifetime. Each individual, of course, has the potential to replace itself many times during its lifetime. An oyster may have millions of offspring, and even members of the most slowly reproducing species, such as elephants, may have several offspring over their lifetimes. So every species has a built-in capacity for population growth, but the speed of this potential growth varies among species, depending on several factors, including the following:

- The age at which each organism first reproduces.

- The frequency with which each organism reproduces.

- The average number of offspring produced each time.

- The length of each organism's reproductive life span.

- The death rate of the organisms under ideal conditions.

Let's look at some examples that illustrate how these factors affect population growth.

Some Species Produce Large Numbers of Offspring Quickly

The bacterium *Staphylococcus* is normally a harmless resident in and on the human body. If transferred to an ideal environment, such as warm custard, each bacterial cell can divide every 20 minutes, doubling the population three times each hour. The larger the population grows, the more cells there are to divide. The potential population growth of bacteria is so great that, hypothetically, the offspring of a single bacterium could swamp Earth in a layer 7 feet deep within 48 hours. (The fact that we are not buried in bacteria suggests that many of the offspring must die rather than divide.)

Other Species Produce Fewer but Longer-Lived Offspring

In contrast to *Staphylococcus*, the golden eagle is a relatively long-lived, rather slowly reproducing species. **Figure 27-2** compares the potential population growth of eagles with that of bacteria, assuming that there are no deaths in either population during the time graphed. Although both the timescale and the number of individuals differ tremendously for the two species, note that all three curves have the same shape—the J shape that represents exponential growth. That is, the bacteria and eagle populations grow at vastly different rates and reach vastly different sizes, but if both populations have constant growth rates, both grow exponentially.

Delayed Onset of Reproduction Slows Population Growth

Figure 27-2b also shows what happens if there is a difference in the age at which individuals in a population begin reproduction. The red line in the graph represents population growth if each golden eagle lives for 30 years, reaches sexual maturity at 4 years of age, and produces one offspring annually for its remaining 26 years (that is, each mated pair of eagles produces two offspring per year). The blue line shows the growth of another eagle population whose members also live for 30 years and produce one offspring each year, but birds in this population do not begin reproducing until they are 6 years old. Note that growth in both populations is exponential, but the population whose members begin reproduction later takes longer to reach a particular size.

This result has important implications for the human population: Delayed childbearing significantly slows population growth. For example, if each woman bears three children in her early teens, the population will grow much faster than if each woman bears five children but begins at age 30.

Death Rate Also Influences Growth Rate

So far we have looked mainly at the effects of birth rates. Even under ideal conditions, however, deaths are inevitable. To illustrate the effect of death rates, **Figure 27-3** compares three bacterial populations with different death rates. Again, the J shapes of the curves are the same. Note that the time required to reach any given population size is longer at higher death rates.

▼ **Figure 27-2 J-shaped exponential growth curves** All exponential growth curves share a similar J shape, although they may differ in the scales of the axes. **(a)** Growth of a population of bacteria, starting with a single individual and doubling every 20 minutes. **(b)** Growth of two populations of eagles (*i* and *ii*), each starting with a single pair. The age at first reproduction is 4 years for population *i* (red line) and 6 years for population *ii* (blue line). Notice that after 26 years, the population that began reproducing at age 4 is more than seven times the size of the population that began reproducing at age 6.

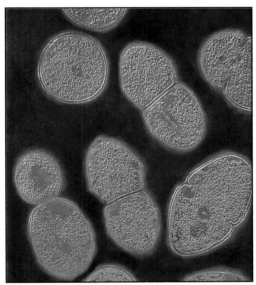

(a) Bacteria

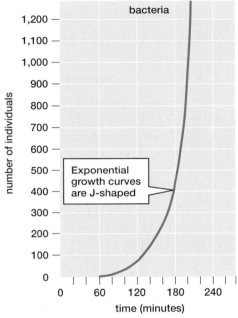

time (minutes)	number of bacteria
0	1
20	2
40	4
60	8
80	16
100	32
120	64
140	128
160	256
180	512
200	1,024
220	2,048

Exponential growth curves are J-shaped

(b) Eagles

Reproduction begins at 4 years

Reproduction begins at 6 years

time (years)	number of eagles (i)	number of eagles (ii)
0	2	2
2	2	2
4	4	2
6	8	4
8	14	8
10	28	12
12	52	18
14	100	32
16	190	54
18	362	86
20	630	142
22	1,314	238
24	2,504	392
26	4,770	644
28	9,088	1,066
30	17,314	1,764

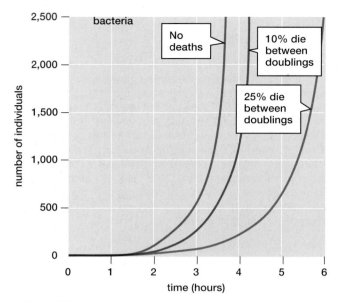

▲ **Figure 27-3 The effect of death rates on population growth** The graphs assume that the bacterial populations double every 20 minutes.

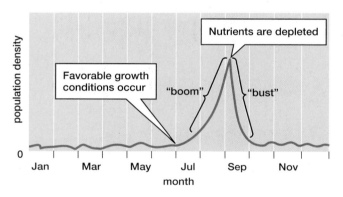

▲ **Figure 27-4 A boom-and-bust population cycle** In early July, conditions in a lake become favorable for bacterial growth, and an initially small population grows rapidly. The resulting large population soon depletes nutrients, and the population "goes bust."

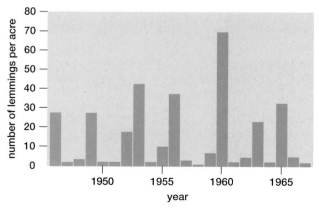

▲ **Figure 27-5 Lemming population cycles** The size of lemming populations cycles every 2 to 4 years (data from Point Barrow, Alaska). **QUESTION:** Why are the cycles shown by these data somewhat erratic and irregular?

27.3 How Is Population Growth Regulated?

A population's growth is influenced by its **biotic potential,** the maximum rate at which a population can increase, assuming ideal conditions that allow the highest possible birth rate and the lowest possible death rate. The ultimate size of a population, however, is also affected by limits that oppose this potential for growth. These limits, which are set by the living and nonliving environments, are collectively known as **environmental resistance.**

Environmental resistance is imposed by the availability of food and space; by interactions with competitors, predators, and disease-causing organisms; and by natural catastrophes such as storms, fires, freezing weather, floods, and droughts. These factors limit population size by increasing death rates, decreasing birth rates, or both. For example, a drought might restrict growth of an animal population both by increasing the number of deaths from starvation and by causing malnutrition that reduces the number of births.

Rapid Growth Cannot Continue Indefinitely

Populations grow rapidly only under certain circumstances and only for a limited time before environmental resistance brings their expansion to a halt.

Some Populations Fluctuate Cyclically

Short-term explosive growth occurs in populations that undergo regular cycles of rapid population growth followed by a sudden, massive die-off. These *boom-and-bust cycles* occur in a variety of organisms. Many short-lived, rapidly reproducing species, from bacteria to insects, have seasonal population cycles that are linked to predictable changes in rainfall, temperature, or nutrient availability (**Fig. 27-4**). For example, many insect populations in temperate climates grow rapidly during the spring and summer and then crash when the killing frosts of winter arrive.

Some populations experience boom-and-bust cycles in which the cycle time is longer than a year. For example, populations of lemmings, voles, and other small rodents grow in cycles in which population size peaks roughly every 4 years (**Fig. 27-5**). Hares, muskrats, and grouse experience longer boom and bust cycles.

A lemming population, for example, may grow until the animals overgraze their fragile arctic tundra ecosystem. Then, lack of food, increasing populations of predators, and social stress caused by overpopulation all contribute to a suddenly high death rate. Many lemmings die during dramatic mass emigrations away from regions of high population density. As they migrate, lemmings are easy targets for predators, and many drown trying to cross bodies of water. Eventually, the lemming population shrinks so much that number of predators declines (see "Scientific Inquiry: Cycles in Predator and Prey Populations") and the plants on which lemmings feed begin to recover. These responses, in turn, set the stage for another round of rapid growth in the lemming population.

Ecosystem Changes May Allow Temporary Rapid Growth

Even in populations that do not experience boom-and-bust cycles, temporary rapid growth may occur under special circumstances. For example, a population may grow rapidly if population-controlling factors, such as predators, are eliminated or if the food supply is increased. Growth can also be explosive when individuals invade a new habitat that has favorable conditions and few competitors. For example, a species' population may grow explosively when people introduce the species into an ecosystem. The introduction of foreign, or *exotic*, species into ecosystems, in many cases, has very damaging results (see "Earth Watch: Exotic Invaders" on pp. 560–561).

Cycles in Predator and Prey Populations

Are the sizes of predator and prey populations linked? If we assume that prey, such as hares, are eaten exclusively by a particular predator, such as lynx, it seems logical that both populations might show cyclic changes, with changes in the size of the predator population lagging behind changes in the size of the prey population. For example, a large hare population would provide abundant food for lynx and their offspring, which would then survive in large numbers. The increased lynx population would eat more hares, reducing the hare population. With fewer prey, fewer lynx would survive and reproduce, so the lynx population would decline a short time later.

Does this out-of-phase population cycling of predators and prey actually occur in nature? One example of such a cycle was revealed by examining the numbers of lynx and snowshoe hares purchased from trappers by the Hudson Bay Company in northern Canada between 1845 and 1935. The availability of pelts (which presumably reflects population size) showed dramatic, closely linked population cycles of

these predators and their prey (Fig. E27-1). Unfortunately, many uncontrolled variables could have influenced the relationship between the numbers of hares and lynx. One problem is that hare populations sometimes fluctuate even without lynx present, possibly because in the absence of predators, growing hare populations eat so much that they reduce their food supply. Further, lynx do not feed exclusively on hares but also eat other small mammals. Environmental variables such as exceptionally severe winters could have adversely affected both populations and produced similar cycles.

Recently, researchers tested the hare–predator relationship more rigorously by fencing off 1-kilometer-square areas in northern Canada to exclude predators. The crash of the hare population was lessened both by providing extra food and by excluding predators, but by far the greatest success in preventing the crash of the hare population was achieved when the researchers excluded predators *and* provided extra food.

To test the predator-prey cycle hypothesis in an even more controlled manner, investigators turned to laboratory studies on populations of small predators and their prey. One study used braconid wasps as predators and bean weevils as prey. The wasps lay their eggs on weevil larvae, which provide food for the newly hatched wasps. As predicted, the sizes of the two populations showed regular cycles, with the predator population increasing and decreasing slightly later than the prey population. A large weevil population ensures a high survival rate for wasp offspring, increasing the predator population. Then, as the predators feed, the weevil population plummets, reducing the population size of the next generation of wasps. Reduced predation then allows the weevil population to increase rapidly, and so on (Fig. E27-2).

It is highly unlikely that such a straightforward example will ever be found in nature, but this type of predator-prey interaction clearly can contribute to the fluctuations observed in many natural populations.

▶ **Figure E27-1 Population cycles of predators and prey** Numbers of snowshoe hares and their lynx predators are graphed on the basis of the number of pelts received by the Hudson Bay Company.

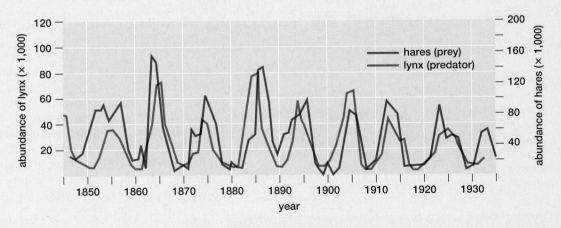

▶ **Figure E27-2 Experimental predator-prey cycles** Populations of bean weevils and their braconid wasp predators show out-of-phase but similar fluctuations in a laboratory experiment.

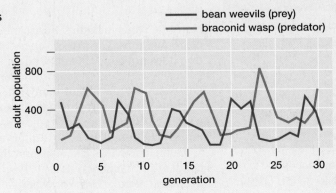

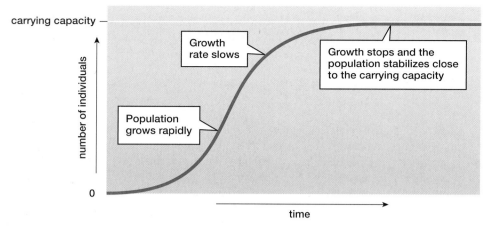

(a) An S-shaped growth curve stabilizes at carrying capacity

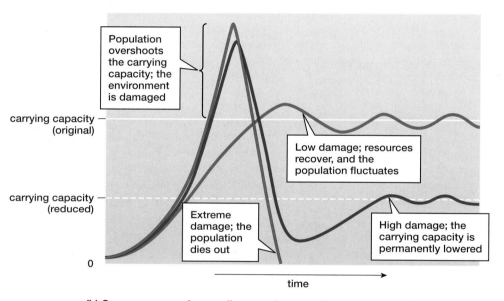

(b) Consequences of exceeding carrying capacity

▲ **Figure 27-6 Population growth is limited** *(a)* The population grows rapidly at first, but growth slows as populations' size approaches carrying capacity. Once the population is at or near the carrying capacity, it stops growing. *(b)* Populations can overshoot carrying capacity, but only for a limited time. Three possible results are illustrated.

The Mystery of Easter Island

Continued

Human activities can modify ecosystems in ways that affect carrying capacity. Some ecosystem modifications can increase carrying capacity (especially carrying capacity for human populations), but many human-caused changes decrease carrying capacity, both for humans and for other species. The ecological changes wrought by the early inhabitants of Easter Island may well have caused a dramatic reduction in the island's carrying capacity for human populations.

Environmental Resistance Limits Population Growth

Populations that grow rapidly must eventually stabilize or crash (decrease rapidly). They tend to stabilize at or below their ecosystem's **carrying capacity,** the maximum population of particular species that an ecosystem can support indefinitely. As a growing population approaches the carrying capacity of the environment, its growth rate gradually declines. Eventually, population growth stops, and population size remains in a long-term state of equilibrium, with a growth rate fluctuating around zero. In this equilibrium, the birth rate is balanced by the death rate, and population size is stable. This type of population growth is typical of long-lived organisms that colonize a new area. It is represented graphically by an S-shaped growth curve (**Fig. 27-6a**).

In some circumstances, a population may grow to a size larger than its ecosystem's carrying capacity (**Fig. 27-6b**; for a discussion of whether human population has reached this point, see "Earth Watch: Have We Exceeded Earth's Carrying Capacity?"). This overshooting of carrying capacity is necessarily temporary, because a population larger than carrying capacity is living on resources that cannot regenerate as fast as they are being depleted. A small overshoot of carrying capacity is likely to be followed by a decrease in population size until the resources recover and the original carrying capacity is restored. If, however, a population grows until it far exceeds the carrying capacity of its environment, consequences are more severe. The resulting excess demands on resources can severely and permanently reduce carrying capacity, causing the population to decline to a fraction of its former size or even to disappear entirely.

The factors that usually maintain populations at or below the carrying capacity of their environment can be classified into two broad categories. **Density-independent** factors limit population size regardless of the population density (number of individuals per given area). **Density-dependent** factors increase in effectiveness as the population density increases. In the following sections, we look more closely at these factors and how they control population growth.

Density-Independent Factors Limit Population Size Regardless of Population Density

Natural events including hurricanes, droughts, floods, and fire can reduce the size of populations. The effectiveness of such events in limiting the size of a population does not generally depend on the population's density.

Perhaps the most important natural density-independent factor is weather. For example, the size of many insect and annual plant populations is limited by the number of individuals that can be produced before the first hard freeze of winter. Such populations typically do not reach their carrying capacity, because density-independent factors intervene first.

Human activities can also limit the growth of populations in ways that are independent of population density. Pesticides and pollutants can cause drastic declines in natural populations. For example, the U.S. Environmental Protection Agency estimates that millions of birds are killed each year when they consume food tainted

Have We Exceeded Earth's Carrying Capacity?

In Côte d'Ivoire, a small country on the west African coast, the government is waging a battle to protect some of its rapidly dwindling tropical rain forest from thousands of illegal hunters, farmers, and loggers. Officials burn the homes of the squatters, who immediately return and rebuild. One illegal resident is Sep Djekoule. "I have ten children and we must eat," he explains. "The forest is where I can provide for my family, and everybody has that right."

Djekoule's words illustrate the conflict between population growth and environmental protection. A glance at the age structure of developing countries, where most of the world's population resides, shows a tremendous momentum for growth. The U.S. Census Bureau predicts that world population will reach almost 7 billion by 2010 and more than 9 billion by 2050. Even then, population will still be increasing.

How many people can Earth support? No one knows. Estimates of Earth's carrying capacity for humans have ranged from 3 billion to 44 billion people. Each of these estimates was based on different assumptions about future technological developments and lifestyles.

The upper limit of the planet's carrying capacity is determined by the ability of its photosynthetic organisms to harvest energy from sunlight and produce high-energy molecules that other organisms can use as food. As humans appropriate an ever-larger share of this captured energy, less is available for the rest of the planet's species. Worse, the total amount of captured energy may be declining. Stanford University biologist Peter Vitousek estimates that human activities have already reduced the productivity of Earth's forests and grasslands by 12%. Each year, overgrazing and deforestation further decrease the productivity of land, especially in developing countries (**Fig. E27-3**).

In a world where more than 800 million people are chronically undernourished (see Fig. E27-3, inset), the quest for more agricultural land leads to deforestation. More than 23,000 square miles of forest are destroyed each year. Yet, even as people in developing nations desperately clear land for farming, the United States loses nearly half a million acres of prime farmland each year to development of homes,

shopping malls, and roads. As our own growing population spreads onto our farmland, our ability to export grain to help sustain other nations will diminish, while their needs increase. Worldwide, the amount of cropland per person has diminished by half since the mid-1950s.

Although we take fresh water for granted, in many developing countries, water supplies are badly polluted and underground water supplies, called *aquifers*, are being depleted and not replaced. Both India and China are rapidly depleting aquifers to supply the needs of their growing cities and to irrigate their cropland, a policy that may result in reduced grain harvests in the near future.

The demand for wood in developing countries is far outstripping production, providing yet another source of deforestation. Deforestation in turn causes erosion of precious topsoil (two-thirds of the world's agricultural land is suffering moderate to severe erosion). Deforestation also leads to runoff of much-needed fresh water and to the spread of deserts.

The total world fish harvest peaked in the late 1980s and has been declining since, despite continual improvement in fish-finding and fish-catching technology. Almost all ocean fish populations have been overfished, and one-third of commercially exploited fish populations have collapsed (that is, the number of fish currently caught is less than 10% of the historical maximum for that fishery). A recent study of historical trends in fish harvests predicted that by 2050, all ocean fish populations will have collapsed and wild-caught seafood will no longer exist. We are clearly "overgrazing" the world's ocean ecosystems.

Taken together, the observations described above strongly suggest that we are severely stressing Earth's ability to support us. Furthermore, when we estimate how many people Earth can support, we must keep in mind that humans desire more out of life than a minimum caloric intake each day. Do we want to

be able to eat meat, drive a car, live in a single-family dwelling, walk in a wilderness, and know that somewhere eagles, pandas, and elephants are living free? For people living at Earth's maximum carrying capacity, these will probably be unattainable luxuries.

Hope for the future lies in using the intelligence that has allowed us to overcome environmental resistance to act before we have irrevocably damaged our biosphere. The explosive growth of the human population *will* ultimately stop. Either we will voluntarily reduce our birth rate, or sources of environmental resistance such as disease and starvation will increase human death rates. The choice is ours. Facing the problem of how to limit births is politically and emotionally difficult, but continued failure to do so could be disastrous. Our dignity, our intelligence, and our role as self-appointed stewards of life on Earth demand that we make the decision to halt population growth before we have permanently reduced Earth's ability to support life, including our own.

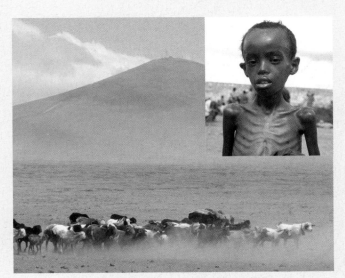

▲ **Figure E27-3 Deforestation can lead to the spread of deserts** Human activities, including overgrazing livestock, deforestation, and poor agricultural practices, may sometimes convert once-productive land into barren desert. (*Inset*) An expanding human population, coupled with a loss of productive land, can lead to widespread malnutrition.

▲ **Figure 27-7 Predators help control prey populations** Grey wolves, hunting in a pack, have brought down an elk, which may have been weakened by old age or parasites.

by toxic pesticides. Habitat destruction by humans also reduces populations. For example, the construction of farms, roads, and housing developments has pushed species such as the California condor and the giant panda to the brink of extinction.

Density-Dependent Factors Have Greater Effect as Population Density Increases

In many populations, the key population-limiting factors are density dependent. Density-dependent factors become increasingly effective as population density increases, thus exerting negative feedback that limits the size of populations. Conversely, density-dependent factors become less effective as population density decreases, allowing population size to stabilize or grow. The most important density-dependent factors are predation and competition.

In *predation*, one organism feeds on another, harming it in the process. Often, one organism (the **predator**) kills another (its **prey**) in order to eat it (**Fig. 27-7**), but the prey is not always killed. For example, when deer browse the buds of a maple tree, the tree is harmed but not killed. Prey also often survive the special form of predation known as *parasitism*. In parasitism, the predator (called the **parasite**) lives on or inside another organism (its **host,** usually a much larger organism) and feeds on the host's body without killing it—at least not immediately.

Predators Exert Density-Dependent Controls on Populations

Predation plays an increasingly important role in population control as prey populations increase. The more abundant the prey, the more often predators encounter them. The increased frequency of encounters with predators increases the prey population's death rate, because predators tend to eat larger numbers of whichever prey species is most abundant and easiest to find. For example, coyotes eat more mice when the mouse population is high. If, however, the mouse population shrinks because of the increased predation, the coyotes might switch to eating more ground squirrels, allowing the mouse population to recover.

Another reason for the increasing effect of predation as prey populations grow is that predator populations may also increase. For example, the reproductive output of predators that feed heavily on lemmings, such as the arctic fox and snowy owl, varies with the abundance of lemmings. Similarly, a pair of snowy owls might produce up to 13 chicks in a year when lemmings are abundant but none at all in years when lemmings are scarce.

Parasites Spread Faster When Host Population Density Is High

Like other forms of predation, parasitism is density dependent. Most parasites have limited ability to move, so they spread more readily from host to host at high host-population densities. For example, plant diseases spread readily through acres of densely planted crops, and childhood diseases spread rapidly through schools and day-care centers. (Many diseases are caused by parasitic bacteria, viruses, fungi, worms, or protists.)

Parasites can affect the death rate of a host population because the damage inflicted by a parasite on its host's body may kill the host. Parasites also influence population size by weakening their hosts and making them more susceptible to death from other causes, such as other predators. For example, a deer weakened by parasites may be easier prey for a mountain lion.

Populations Can Soar or Crash When Predator-Prey Relationships Are Disrupted

The population balance in ecosystems can be disrupted when predators are introduced to regions in which they were not previously present and where prey species have had no opportunity to evolve defenses against them. For example, the smallpox virus, inadvertently carried by traveling Europeans, ravaged the native populations of Hawaii, Argentina, and Australia. Similarly, rats, snakes, and mongooses were introduced in Hawaii and many other Pacific islands and have drastically reduced or exterminated many native bird populations.

Conversely, prey populations can grow out of control when introduced to areas where they have no predators. A dramatic example of this phenomenon is the prickly pear cactus, which was introduced into Australia from Latin America. Lacking natural predators, it spread uncontrollably, destroying millions of acres of valuable pasture and range land. Finally, in the 1920s, a cactus moth (a predator of the prickly pear cactus) was imported from Argentina and released to feed on the cacti. Within a few years, the cacti were almost eliminated. Today, the moth continues to keep the population density of its prey very low.

Competition for Resources Helps Curb Population Size

The resources that determine carrying capacity (space, nutrients, water, light) are limited. As a result, organisms must compete for access to resources. This **competition** among individuals limits population size in a density-dependent manner.

There are two major forms of competition: **interspecific competition,** among individuals of different species, and **intraspecific competition,** among individuals of the same species. Because the members of a species all have nearly identical resource requirements, intraspecific competition is more intense than interspecific competition.

Competition May Be Indirect or Direct

Some organisms, including most plants and many insects, engage in *scramble competition*, a kind of free-for-all in which individuals independently seek resources without interacting directly. For example, after gypsy moth caterpillars hatch from large masses of eggs laid on tree trunks, armies of caterpillars crawl up to the treetops to feed on leaves (**Fig. 27-8**). In large gypsy moth populations, competition among caterpillars for food may be so great that most die before they can metamorphose into moths. Similarly, when a plant disperses its seeds in a small area, hundreds may germinate. As they grow, however, some plants will grow taller and gain access to more sunlight, and some will have more extensive root systems and will absorb most of the water. Eventually, the less successful individuals will wither and die.

Many animals (and even a few plants) have evolved *contest competition*, in which individuals interact to contest access to important resources. For example, territorial species such as wolves, many fish, rabbits, and songbirds defend an area that contains important resources such as food or nesting sites. Only the best competitors are able to defend adequate territories. Poor competitors fail to secure territories and are unlikely to reproduce or perhaps even to survive.

Different Age Groups May Experience Different Death Rates

The effects of population-limiting factors do not necessarily fall equally on all members of a population. For example, the death rate in a population often differs among individuals of different ages. One way to describe such age-related differences in death rate is to say that *survivorship* varies with age. Researchers who study patterns of survivorship may track the fates, from birth to death, of the members of a group of individuals born at the same time. In each succeeding year (or other unit of time), the number of deaths in the group is recorded. These numbers can form the basis of a graph, called a *survivorship curve*, that shows the likelihood that an individual will survive to a given age.

Survivorship in Populations Follows Three Basic Patterns

The measured survivorship curve for a population will vary according to the population's particular characteristics and circumstances, but most curves fall into one of three general categories, described as "late loss," "early loss," and "constant loss" (**Fig. 27-9**).

▲ **Figure 27-8 Scramble competition** Gypsy moths gather on tree trunks to lay egg masses that each produce many hundreds of caterpillars (inset).

▼ **Figure 27-9 Survivorship curves** The horizontal axis shows percentage of life span (rather than age) so that species with different life spans can be shown on a single graph.

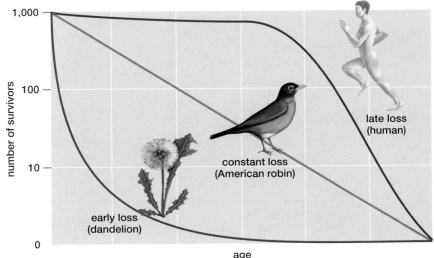

age
(in percentage of maximum life span)

Convex *late-loss* survivorship curves reflect populations in which the death rate is low for juveniles and in which most individuals survive to old age. Such curves are characteristic of humans and other large, long-lived animals such as elephants and mountain sheep. These species produce relatively few offspring that receive a great deal of parental care.

Early-loss survivorship produces a concave curve and is characteristic of organisms that produce large numbers of offspring that receive little or no parental care. In such organisms, the death rate among juveniles is very high, but individuals that manage to reach adulthood have a good chance of surviving to old age. Early-loss survivorship curves are common among invertebrates, plants, and fish.

Populations with straight-line *constant-loss* survivorship curves have a fairly constant death rate that does not vary much with age. In these populations, an individual has an equal chance of dying at any stage of its life. This pattern has been found in some bird and lizard species and in populations of some asexually reproducing invertebrate species.

Web Animation Human Population Growth and Regulation

27.4 How Is the Human Population Changing?

Compare the graph of human population growth in **Figure 27-10** with the exponential growth curves in Figure 27-2. The time spans are different, but each has the J shape characteristic of explosive growth. It took more than 100,000 years for the human population to reach 1 billion, but the second billion was added in just 100 years, the third billion in 30 years, the fourth billion in 15 years, the fifth billion in 12 years, and the sixth billion in another 12 years. In the year 1750, the human population was 800 million. It doubled in 150 years, reaching 1.6 billion in 1900. In contrast, the 1965 population of 3.5 billion will double to 7 billion in less than 50 years. According to the U.S. Census Bureau (the source of all the population statistics and projections included in this chapter), world population currently grows by about 1.2%, or almost 78 million people, yearly—more than 200,000 people are added every day, more than 1.4 million every week! Why hasn't environmental resistance put an end to our tremendous population growth?

▼ **Figure 27-10 Human population growth** The human population from the Stone Age to the present has shown rapid growth. Note the dip in the fourteenth century caused by the bubonic plague. **QUESTION:** Human population is still growing rapidly, but some evidence suggests we have already exceeded Earth's carrying capacity. What do you think this curve will look like in the year 2500? 3000? Explain.

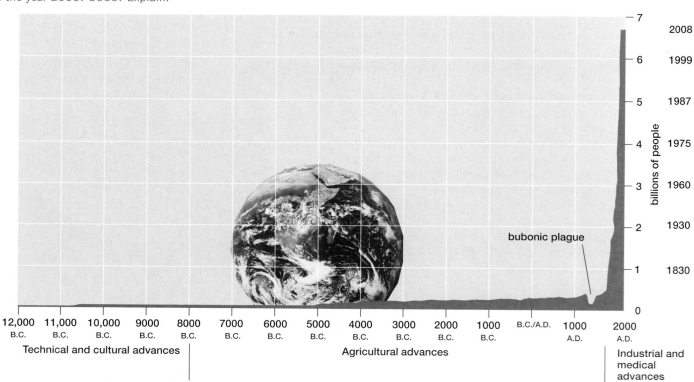

Technological Advances Have Increased Earth's Carrying Capacity for Humans

Like all populations, humans have encountered environmental resistance, but unlike other populations, we have responded to environmental resistance by devising ways to overcome it. As a result, the human population has grown rapidly for an unprecedented time span. Human population growth has been spurred by a series of cultural "revolutions" in which products of human culture conquered various aspects of environmental resistance and increased Earth's carrying capacity for people.

Early Tools Allowed Human Populations to Spread and Prosper

Prehistoric people produced a *tool-use revolution* when they discovered fire, invented tools and weapons, built shelters, and designed protective clothing. Tools and weapons allowed more effective hunting and thereby increased the supply of food. Shelter, fire, and clothing allowed humans to move into previously uninhabitable areas of the globe. Together, these developments set the stage for future population growth.

Agriculture Fostered Population Growth

By about 8000 B.C., cultivation of domesticated crops and animals had replaced hunting and gathering as humans' primary means of acquiring food. This *agricultural revolution* provided people with a larger, more dependable food supply and further increased Earth's carrying capacity for humans. Increased food resulted in a longer life span and more childbearing years, and the human population began to climb steadily.

Explosive Growth Began with the Industrial Revolution

Human population growth continued slowly for thousands of years until the *industrial-medical revolution* began in England in the mid-eighteenth century and ultimately spread to much of the rest of the world. This revolution marked the beginning of a period of extremely rapid population growth, as medical advances dramatically decreased the death rate. These advances included the discovery of bacteria and their role in infection, leading to the control of bacterial diseases through improved sanitation. Another key advance was the development of vaccines for diseases such as smallpox. The medical revolution continues today as research proceeds on vaccines against major killers such as malaria and AIDS. In addition, death rates are further reduced by continuing advances in sophisticated medical procedures that extend the human life span.

Population Growth Continues Today but Is Unevenly Distributed

In many developed countries, such as those of Western Europe, the population explosion fostered by the declining death rates of the industrial-medical revolution was eventually followed by a decline in birth rates. The tendency for birth rates to decline in affluent societies is not fully understood but has been attributed to many factors, including better education, increased availability of contraceptives, a shift to a primarily urban lifestyle, and more career options for women. In most developed countries, population sizes have stabilized or even declined, though in some developed countries immigration has offset or exceeded population declines.

In developing countries, such as those of Central and South America, Asia, and Africa, medical advances have decreased death rates and increased the life span (though the AIDS epidemic threatens this progress in some areas). Birth rates, however, remain high in many developing countries, which have not experienced the increased wealth associated with the birth rate decline in developed countries. In agricultural societies, children are an important source of labor and may be the only source of support for elderly parents. Social traditions may

The Mystery of Easter Island

Continued

Although agriculture can increase an environment's carrying capacity for humans, agricultural activity can also cause changes that have negative effects on carrying capacity. For example, when forests are cleared to make room for fields and pastures, life-sustaining resources provided by the forest are lost. Also, because other animals compete with humans for the plentiful food produced by farming, populations of these competitors may boom in agricultural landscapes. Rats, for example, can become abundant near farm fields and in crop storage areas. Thus, some rats accompanied the foodstuffs carried by the seafaring Polynesians who settled Easter Island. These stowaways gave rise to a thriving rat population on the island, which contained no large predatory animals to limit rat population growth. The rats survived in part by eating some of the food that the Easter Islanders grew, but the rats also consumed the seeds of the native palm trees, hastening the forest's demise.

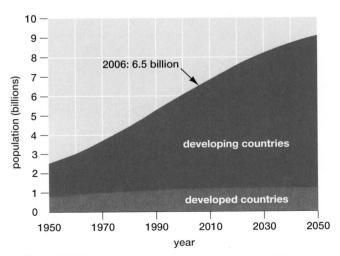

▲ Figure 27-11 Historical and projected world population

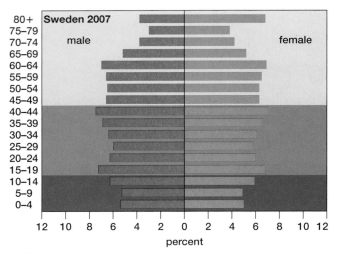

(a) Population age structure for Sweden

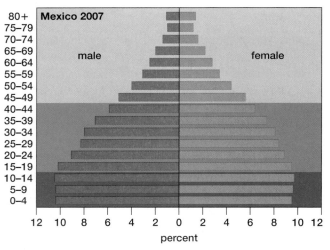

(b) Population age structure for Mexico

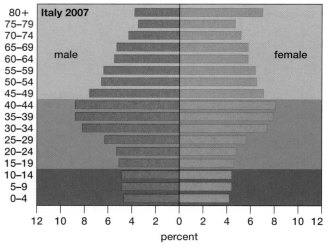

(c) Population age structure for Italy

▲ **Figure 27-12 Age-structure diagrams (a)** Sweden has a stable population. **(b)** Mexico's population is growing quite rapidly. **(c)** Italy's population is shrinking. The different shades of blue in the diagrams respresent (from bottom to top) prereproductive, reproductive, and postreproductive age groups.

offer prestige to men who father many children and to women who bear them. Although many developing nations have concluded that their population growth rates are too high, social traditions, lack of education, and lack of access to contraceptives impede progress in curbing population growth.

Of the more than 6.6 billion people on Earth today, more than 5.3 billion reside in developing countries. Of the more than 9 billion people expected to inhabit the planet in 2050, 8 billion will be living in the developing nations (**Fig. 27-11**).

The prospects for population stabilization—*zero population growth*—in the near future are nil. You can understand why by examining the age structures of the developing countries and comparing them with those of countries with stable populations, as we describe next.

The Age Structure of a Population Predicts Its Future Growth

The **age structure** of a population is determined by the numbers of males and females in each age group, and can be shown graphically in an age-structure diagram. The vertical axis of an age-structure diagram shows the age of people, and the horizontal axis shows numbers of individuals in each age group, with males and females graphed in opposite directions from the central axis. All age-structure diagrams rise to a peak at the top, because few people live into their nineties. The shape of the rest of the diagram, however, shows at a glance whether the population is expanding, stable, or shrinking.

The shape of an age-structure diagram is determined by the relative numbers of reproductive-age adults (ages 15 to 45) and their children (ages 0 to 14). If the adults of reproductive age are producing just enough children to replace themselves, the population is said to have **replacement level fertility (RLF).** The age-structure diagram of such a population has relatively straight sides (**Fig. 27-12a**). If children in the various age classes exceed the number of reproductive-age adults, the population is exceeding RLF and is expanding as a result. The age-structure diagrams of such expanding populations have a pyramidal shape (**Fig. 27-12b**). In shrinking populations, there are fewer children than reproducing adults and the age-structure diagram narrows at the base (**Fig. 27-12c**).

Almost All Future Population Growth Will Be in Developing Countries

Figure 27-13 shows the average age structures of the populations of developed and developing countries. The outermost boundaries represent the projected population structure for the year 2050 and the innermost green areas give actual values for 2006. The vertical axis of each graph has been divided into three parts to show individuals who are prereproductive (ages 0 to 14), reproductive (15 to 44), and postreproductive (45 and older). In 2006, the developing countries had an average annual *natural increase* (population growth based on the births minus deaths, not including migration) of 1.6% and a doubling time (the time required for the population to double) of 50 years. In contrast, developed countries showed an average annual natural increase of 0.07% and a doubling time of 985 years.

More than 97% of the growth of the human population occurs in the developing countries, where increasing numbers of people enter their reproductive years and give birth to an ever-increasing base of infants. Even if these countries were immediately to reach replacement-level fertility, their populations would continue to grow for decades. The growth would come as the children of the large families of the recent past entered their reproductive years and began having children—even if each couple were to have only two children. This entry into the reproductive age classes fuels China's population growth of 0.6% annually, even though China's fertility rate (the average number of children per woman) is below replacement level. The percentage growth per year caused by natural increases for various parts of the world is shown in **Figure 27-14**.

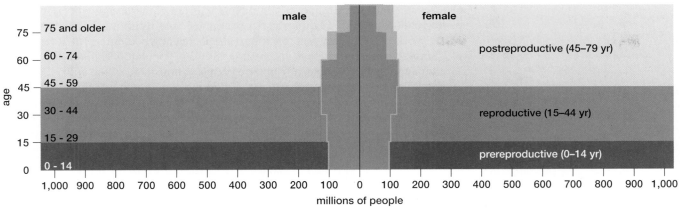

(a) Developed countries

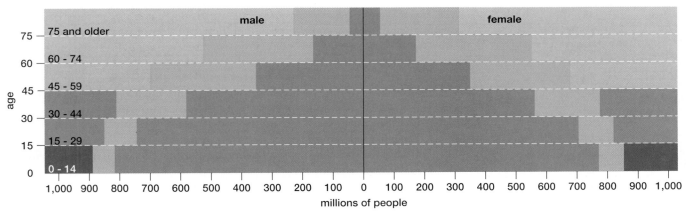

(b) Developing countries

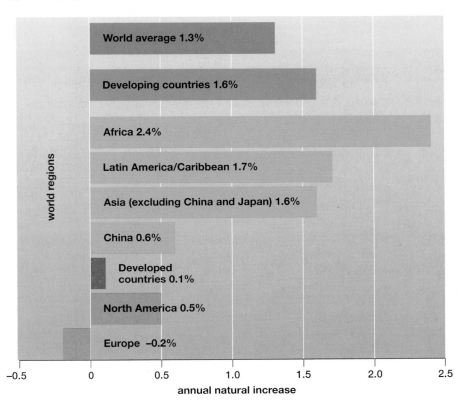

▲ Figure 27-13 Age structures compared The nearly vertical age structure of populations in (a) developed countries predicts a much slower population increase than the pyramid-shaped age structure of populations in (b) developing countries.

◄ Figure 27-14 Population growth by world regions Growth rates shown are due to natural increase (births – deaths) expressed as the percentage increase per year. These figures do not include immigration or emigration. QUESTION: Why are there such big differences between the growth rates of developed and developing countries?

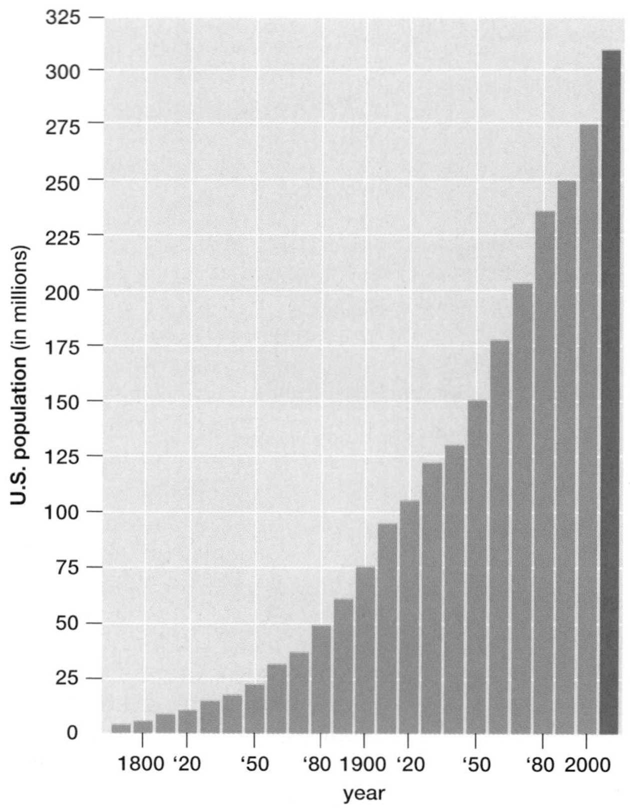

▲ Figure 27-15 **U.S. population growth** Since 1790, U.S. population growth has shown the J-shaped curve characteristic of rapid growth. The dark green bar at far right represents a projection for 2010. **QUESTION:** What factors do you think will cause U.S. population to stabilize, and when?

Fertility in Europe Is Below Replacement Level

In Europe, the fertility rate is well below RLF, and several European countries are experiencing a slight decline in population, as women increasingly delay or forgo having children. This situation raises economic concerns about the availability of future workers and taxpayers to support the increasing percentage of elderly people. Similar concerns prevail in Japan, a densely populated country the size of Montana that has more than 127 million people (about 42% of the entire U.S. population). The low Japanese fertility rate is causing concern about slowed economic growth and an aging population. The Japanese government provides a variety of subsidies and incentives to encourage people to have larger families.

The situations in Europe and Japan reveal one of the key barriers to controlling the world's population: Current economic structures are based on growing populations. Although a reduced and stable population may offer tremendous benefits in the future, the transition from rapid population growth is difficult.

The U.S. Population Is Growing Rapidly

The United States is the fastest growing industrialized country in the world. With a current population of more than 300 million and a natural increase of 0.6% annually (**Fig. 27-15**), the U.S. population grows at more than six times the average rate of developed countries. When immigration is added to this picture, the U.S. annual growth rate approaches 1%. During 2006, for example, the U.S. population grew by 2.7 million, an increase of more than 7,000 people daily. In 2050, there will be 420 million people in the United States and the population will still be growing. Even with RLF, immigration will guarantee continued U.S. population growth.

The rapid growth of the U.S. population has major environmental implications both for us and for the world. The average person in the United States uses five times as much energy as the average non-U.S. person, so, with less than 5% of the world's population, the United States accounts for 25% of world energy use. The ecological impact of the 2.7 million people added annually to the U.S. population is far greater than the total ecological impact of the 18 million people added annually to India's population.

The Mystery of Easter Island Revisited

The destruction of Easter Island's forest was paralleled by a wave of animal die-offs. Fossils show that Easter Island was once home to parrots, owls, herons, and numerous other bird species. At least 25 species of seabirds, including albatrosses, boobies, and frigate birds, once nested on the island. Today, Easter Island contains no native land birds or mammals. No native animal larger than an insect is a permanent resident there.

When the forest and its animals disappeared, the island's human population probably lacked adequate food. With no wood for canoes or boats, there was no way to leave the island. Ominously, garbage heaps from the post-forest period contain gnawed human bones, testimony to the cannibalism that emerged in this period. By the time Europeans arrived, the small number of people left on Easter Island lived impoverished lives.

The earliest human settlers of Easter Island found a forested isle that held abundant natural resources and a diverse array of plant and animal species useful to humans. Over time, however, the islanders' population grew until it exceeded the environment's capacity to support it. As a result, the island's human population declined dramatically, and its society came apart.

What can we learn from Easter Island? According to biologist and author Jared Diamond, "the meaning of Easter Island for us should be chillingly obvious. Easter Island is Earth writ small. Today, again, a rising population confronts shrinking resources. We too have no emigration valve, because all human societies are linked by international transport, and we can no more escape into space than the Easter Islanders could flee into the ocean. If we continue to follow our present course, we shall have exhausted the world's major fisheries, tropical rain forests, fossil fuels, and much of our soil by the time my sons reach my current age."

Consider This

Other inhabited Pacific islands did not share Easter Island's fate. What qualities might have made Easter Island susceptible to a population crash?

Chapter Review

Summary of Key Concepts

For additional study help and activities, go to www.mybiology.com.

27.1 How Are Populations Distributed in Space and Time?

The distribution of the members of a population may be clumped, uniform, or random. Distributions may be clumped for social reasons or around limited resources. Uniform distributions are typically the result of territorial spacing based on resources. Distributions are random only when individuals do not interact socially and when resources are abundant and evenly distributed.

27.2 How Do Populations Grow?

Individuals join populations through birth or immigration and leave through death or emigration. All organisms have the potential to more than replace themselves over their lifetimes, so all populations have potential for growth. A population's growth rate depends on the age at which individuals begin to reproduce, the number of offspring produced by each individual, and the life spans of individuals. Populations in which the growth rate remains constant over time grow exponentially, with increasing numbers of individuals added during each successive time period.

27.3 How Is Population Growth Regulated?

The ultimate size of a stable population is the result of interactions between biotic potential (the maximum possible growth rate) and environmental resistance (which limits population growth). Populations cannot grow indefinitely. They either stabilize or undergo periodic boom-and-bust cycles as a result of environmental resistance. Environmental resistance restrains population growth by increasing the death rate or decreasing the birth rate. The maximum size at which a population may be sustained indefinitely by an ecosystem is termed the carrying capacity, which is determined by limited resources such as space, nutrients, and light. Environmental resistance generally maintains populations at or below the carrying capacity. If a population growth overshoots carrying capacity, resources may be depleted more rapidly than they are replenished, ultimately resulting in environmental damage that reduces carrying capacity. In extreme cases, carrying capacity may become too low to support a population.

Population growth is restrained by density-independent forms of environmental resistance (such as weather) and density-dependent forms of resistance (including predation, parasitism, and competition).

27.4 How Is the Human Population Changing?

The human population has exhibited rapid growth for an unprecedented time by overcoming certain aspects of environmental resistance and thereby increasing Earth's carrying capacity for humans. These feats have been accomplished by the use of tools, agriculture, industry, and medical advances. Age-structure diagrams depict numbers of males and females in various age groups within a population. Expanding populations have pyramidal age structures; stable populations have straight-sided age structures; and shrinking populations have age structures that are constricted at the base.

Today, most of the world's people live in developing countries with rapidly expanding populations, where social and cultural conditions encourage people to have large families. The United States is the fastest growing of the developed countries, owing mainly to high immigration rates. Earth's carrying capacity for humans is unknown, but with a population of more than 6 billion, resources are already too limited for all to be supported at a high standard of living. A steady decline in productive land, fresh water, and in wood and fish harvests suggests that we are already damaging our world ecosystem and decreasing its ability to sustain us.

Web Animation Human Population Growth and Regulation

Key Terms

age structure *p. 550*
biotic potential *p. 542*
carrying capacity *p. 544*
clumped distribution *p. 538*
community *p. 537*
competition *p. 547*

density-dependent *p. 544*
density-independent *p. 544*
ecology *p. 537*
ecosystem *p. 537*
environmental resistance *p. 542*
exponential growth *p. 540*

host *p. 546*
interspecific competition *p. 547*
intraspecific competition *p. 547*
parasite *p. 546*
per capita growth rate *p. 539*
population *p. 537*

predator *p. 546*
prey *p. 546*
random distribution *p. 538*
replacement level fertility (RLF) *p. 550*
uniform distribution *p. 538*

Thinking Through the Concepts

Suggested answers to end-of-chapter and figure-based questions can be found at the end of the text.

Fill-in-the-Blank

1. Two examples of natural events that act as density-independent forms of environmental resistance are _____ and _____. The two most important density-dependent forms of environmental resistance are _____ and _____.

2. The general name for graphs that plot how the number of individuals born at the same time changes over time are _____ curves. The specific type of curve that describes a population of mosquitoes in which a million larvae hatch but most are consumed by fish and only 100 reach the adult stage is _____. The specific type of curve that describes the human population is _____.

3. The two fundamental types of competition that limit population size are _____ and _____. Of these, generally the most effective is _____. If an exotic species of grasshopper is introduced into an area and eats much of the vegetation, causing a reduction in the population of native grasshoppers, which type of competition is this? _____

4. The type of growth occurring in a population that increases by 0.1% per year is described as _____. Does this type of growth add the same number of individuals to the population each year? _____ Can such growth be sustained indefinitely? _____ The maximum population that can be sustained indefinitely without damaging its environment is called _____. If a population of long-lived organisms is introduced into a new, hospitable environment, and its numbers are graphed, the resulting curve is _____.

5. The type of spatial distribution most likely to occur when resources are localized is _____. The type of distribution that would result when pairs of animals defend territories used for nesting is _____.

6. An expanding population will have an age structure diagram shaped like a(n) _____. If the sides of an age structure diagram are roughly vertical, the population is _____. The shape of age-structure exhibited by developing countries considered as a group is _____.

7. A population grows whenever the number of _____ plus _____ exceeds the number of _____ plus _____. The growth rate of a population is increased when the age at which an organism first reproduces _____, when the frequency of reproduction _____, and when the length of the organism's reproductive life span _____.

Review Questions

1. Explain *biotic potential* and *environmental resistance*.

2. Draw the growth curve of a population before it encounters significant environmental resistance. What is the name of this type of growth, and what is its distinguishing characteristic?

3. Distinguish between density-independent and density-dependent environmental resistance.

4. Describe (or draw a graph illustrating) what is likely to happen to a population that far exceeds the carrying capacity of its ecosystem. Explain your answer.

5. List three density-dependent forms of environmental resistance, and explain why each is density dependent.

6. Discuss some reasons why making the transition from a growing to a stable population can be economically difficult.

Applying the Concepts

1. **IS THIS SCIENCE?** The makers of a popular "bug zapper" device claim that it will "reduce the mosquito population in your yard by 50%." Is this claim plausible? Why or why not? Devise a scientific procedure to test the claim.

2. What factors are responsible for rapid population growth in developing countries? What actions might slow or halt this growth?

3. Why is the concept of carrying capacity difficult to apply to human populations? In reference to human population, should the concept be modified to include quality of life?

For additional resources, go to www.mybiology.com.

chapter 28

Community Interactions

A past brush with extinction revealed the sea otter's key role in coastal ecological communities.

Case Study The Case of the Vanishing Sea Otters

Alaska's sea otters are disappearing. Again. What's more, their disappearance threatens the area's lush kelp forests. Why are the otters vanishing? And why are kelp forests affected? The story behind these events begins in the nineteenth century, when sea otters were abundant. But by the early 1900s, commercial hunting for fur had driven sea otters nearly to extinction. Before sea otters disappeared completely, however, new laws protected them, and their population began to rebound.

By the 1970s, sea otters were again abundant. The recovery, however, was uneven. Otter populations in some areas were large, but other areas held only small populations or no sea otters at all. This uneven distribution created a "natural experiment." Ecologists could compare similar habitats that held different numbers of sea otters. Researchers found that areas with many sea otters also had extensive kelp forests (see Fig. 16-9b), and areas without otters tended to lack kelp forests. Long blades of kelp, anchored to the seafloor and floating upward, play a role similar to that of trees in a terrestrial forest, providing food, shelter, and protection to a diverse array of other organisms. Clearly, otters and kelp forests are linked, but what is the connection?

The connection turns out to be sea urchins. These spiny animals eat underwater plants and algae, including kelp. They, in turn, are a favorite food of sea otters, so urchin populations were small where otters were common. But in areas with few or no otters, sea urchins were abundant and consumed kelp faster than it could grow. Kelp forests grew only where there were enough otters to limit the sea urchin population.

The crucial ecological role of sea otters shows that removing even a single species from a community can be profoundly disruptive. Therefore, ecologists were alarmed when, around 1990, sea otter populations in Alaska unexpectedly began to decline rapidly. Today, the number of sea otters in Alaskan waters is less than 10% of the number present in 1990. As you might expect, the area covered by kelp forests has shrunk dramatically. What has happened to Alaska's sea otters? Consider this question as you read about ecological communities in this chapter. ∎

Sea urchin populations grow explosively where sea otters are not present.

28.1 Why Are Interactions in Ecological Communities Important?

An ecological **community** consists of all the interacting populations in an ecosystem. The populations in a community interact in various ways, such as competition, predation (including parasitism), mutualism, and commensalism. The distinctions between these types of interactions are based on whether the interactions are harmful or beneficial to the participating organisms (**Table 28-1**).

Community Interactions Help Limit Population Size

The interacting web of life that forms a community tends to maintain a balance between resources and the numbers of individuals using them. For example, as you learned in Chapter 27, community interactions such as predation and com-

Table 28-1	Interactions within Communities	
Type of Interaction	**Effect on Organism A**	**Effect on Organism B**
Competition between A and B	Harms	Harms
Predation* by A on B	Benefits	Harms
Commensalism of A with B	Benefits	No effect
Mutualism between A and B	Benefits	Benefits

*Predation includes parasitism and herbivory.

petition help limit the size of populations. The finely tuned interactions that maintain balance among populations developed as a result of prolonged coexistence among populations, but the balance can be fragile and susceptible to disruption. Such disruption is especially likely when organisms are introduced into an ecosystem in which they did not evolve, as described in "Earth Watch: Exotic Invaders" on pp. 560–561.

Community Interactions Influence Evolutionary Change

When members of different populations interact with one another, they may influence each other's ability to survive and reproduce. Community interactions thus serve as agents of natural selection. Predators, for example, tend to harm the members of prey populations that are easiest to attack, thereby favoring individuals with better defenses against predation. These individuals therefore tend to survive longer and leave the most offspring, so that over time the inheritable characteristics that help them avoid predation become more common in the prey population. At the same time, the increasingly effective defenses of prey favor the survival and reproduction of those members of the predator population that are best able to overcome the defenses. Thus, even as predator-prey interactions limit population size, they also shape the bodies and behaviors of the interacting populations. This process, in which interacting species influence one another's evolution by acting as mutual agents of natural selection, is called **coevolution.** We discuss several kinds of interaction—competition, predation, and symbiosis—that can result in coevolution in sections 28.2, 28.3, and 28.4, respectively.

28.2 What Are the Effects of Competition Among Species?

During competition, an individual uses or defends a resource and thereby reduces the availability of that resource to other individuals. Organisms inevitably compete with members of their own species, but may also compete with members of other species. When different species compete, the interaction is called **interspecific competition.** In interspecific competition, each species is harmed, because access to resources is reduced for both (see Table 28-1). The intensity of interspecific competition depends on how similar the requirements of the competing species are. Ecologists express this idea by saying that the degree of competition depends on the degree to which the ecological niches of the competing species overlap.

Each Species Has a Unique Place in Its Ecosystem

Each species occupies a unique **ecological niche** that encompasses all aspects of its way of life. A species' niche includes the particular type of habitat in which it lives, the environmental factors necessary for its survival, and the methods by which it acquires its nutrients. Among the factors that may define a species' niche

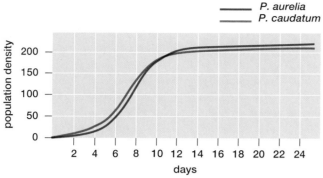

(a) Grown in separate flasks

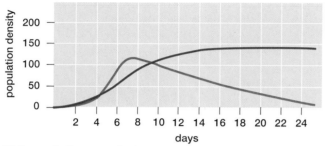

(b) Grown in the same flask

▲ **Figure 28-1** **Competitive exclusion** *(a)* Raised separately with a constant food supply, both *Paramecium aurelia* and *P. caudatum* show the S-curve typical of a population that initially grows rapidly, then stabilizes. *(b)* When the two species are raised together and forced to occupy the same niche, *P. aurelia* consistently outcompetes *P. caudatum* and causes that population to die off. *(Modified from G. F. Gause,* The Struggle for Existence. *Baltimore: Williams & Wilkins, 1934.)* **QUESTION:** How does competitive exclusion contribute to the threat that introduced exotic species can pose to native species?

▲ **Figure 28-2** **Interspecific competition affects spatial distribution** Competition from other species may prevent members of a barnacle species from living outside a particular portion of the intertidal zone.

are the range of temperatures under which it can survive, the amount of moisture it requires, and the types of nutrients needed to sustain it. For an animal species, we can add to that list items such as the types of food it eats, the time of day at which it forages for food, and the type of shelter it needs. A plant species' niche might include the range of soil pH it can tolerate, the amount of sunlight it needs, and the amount of soil nutrients it requires. A species' predators, prey, and competitors are also considered to be elements of its niche. In sum, a species' ecological niche is the combination of all factors that define how the species "makes a living" within its ecosystem.

The Ecological Niches of Coexisting Species Never Overlap Completely

Although different species may share some aspects of their niches, no two species ever occupy exactly the same ecological niche, just as no two organisms can occupy exactly the same physical space at the same time. This concept, called the **competitive exclusion principle,** was formulated in 1934 by the Russian microbiologist G. F. Gause.

Gause demonstrated competitive exclusion in an experiment with two species of the protist *Paramecium*: *P. aurelia* and *P. caudatum*. Grown separately in laboratory flasks and provided with bacteria to eat, each species flourished (**Fig. 28-1a**). But because both species used the same food source and sought food in the same part of the flask, Gause hypothesized that their niches were too similar for them to coexist. Sure enough, in trials in which Gause placed the two species together in a single flask, one species (usually *P. caudatum*) always died out (**Fig. 28-1b**). Gause then repeated the experiment, but replaced *P. caudatum* with a different species, *P. bursaria*, which tended to feed in a different part of the flask. In this case, the two species of *Paramecium* were able to coexist indefinitely, because their niches did not overlap completely.

Competitive Exclusion Helps Determine How Populations Are Distributed

A classic study of competitive exclusion under natural conditions was performed by the ecologist Joseph Connell. Connell experimented with barnacles of the genera *Chthamalus* and *Balanus*. (Barnacles are crustaceans that attach permanently to rocks and other surfaces; **Fig. 28-2**.) Both genera live in the *intertidal zone*, an area of the shore that is alternately covered and exposed by the tides. Connell found that *Chthamalus* dominates the upper shore and *Balanus* dominates the lower. When he scraped off *Balanus*, however, the *Chthamalus* population increased, spreading downward into the area its competitor had once inhabited. So *Chthamalus* is clearly able to thrive in the areas that are inhabited by *Balanus*. Connell concluded that *Balanus*, by virtue of growing faster, competitively excluded *Chthamalus* from the lower zone. *Chthamalus* survives in the drier upper zone, which is submerged only by the highest tides, because it is better than *Balanus* at tolerating the dry conditions there. As this example illustrates, interspecific competition can limit the distribution (and therefore the size) of populations.

Species Evolve in Ways That Reduce Niche Overlap

The ecologist Robert MacArthur documented the effects of competitive exclusion by observing five warbler species in the wild. All five species live in the same habitat, and all forage for insects and build their nests in the same spruce trees, so it appears at first glance that their niches overlap considerably. MacArthur, however, found that the different species search for food in different portions of the

Yellow-rumped warbler

Bay-breasted warbler

Cape May warbler

Black-throated green warbler

Blackburnian warbler

tree, employ different hunting tactics, and nest at slightly different times. The five species in effect divide the resource, thereby reducing niche overlap and reducing interspecific competition (Fig. 28-3).

As MacArthur discovered, when two species with similar requirements coexist, they typically occupy a smaller niche than either would if it were by itself. This phenomenon, called **resource partitioning,** is an outcome of the coevolution of species with extensive niche overlap. Because natural selection favors individuals with fewer competitors, competing species tend to evolve physical and behavioral characteristics that reduce their competitive interactions. A dramatic example of resource partitioning is found among the 13 species of Darwin's finches in the Galápagos Islands. The different finch species have evolved different bill sizes and shapes and different feeding behaviors that reduce competition among them (see Fig. 13-5).

28.3 What Are the Effects of Predator-Prey Interactions?

When a predator consumes its prey (Fig. 28-4), one species benefits at the expense of another (see Table 28-1). Many predators kill the organisms they eat, but some predators, known as **parasites,** live on or inside their prey, or **host,** and feed on its body without necessarily killing it. **Herbivores** (animals that eat plants) are also predators that do not necessarily kill the prey on which they feed.

▲ **Figure 28-3 Reducing niche overlap** Each of these five species of North American warblers searches for insects in a different region of spruce trees, highlighted in dark green.

Web Animation Competition

▼ **Figure 28-4 Predation** *(a)* A pika (a relative of rabbits) eats grass. *(b)* A long-eared bat uses echolocation to hunt a moth. *(c)* A red-tailed hawk prepares to eat a smaller bird. **QUESTION:** Can you describe some other examples of coevolution between predator and prey?

(a) Pika

(b) Long-eared bat

(c) Red-tailed hawk

Exotic Invaders

The residents of Monroe, Michigan, a town on the shores of Lake Erie, have a problem with zebra mussels. Since the late 1980s, when mussels growing within the intake pipe of a water-treatment plant repeatedly clogged the pipe and shut down the town's water supply, Monroe has spent hundreds of thousands of dollars on efforts to keep the pipes free of mussels. Monroe's problem is not unique—zebra mussels are abundant across the Great Lakes (**Fig. E28-1a**).

Where did the mussels come from? Sometime in the 1980s, a cargo ship from Europe discharged fresh water into Lake St. Clair, located between Lake Huron and Lake Erie. The water, used for ballast during the ship's transatlantic voyage, contained stowaways—millions of larvae of the zebra mussel. Although these mollusks are native to the Caspian and Black seas (large lakes in southeastern Europe), they found ideal conditions in Lake St. Clair. Now they inhabit all of the Great Lakes and the entire Mississippi and Ohio River drainage systems and are spreading throughout eastern waterways. The microscopic mussel larvae can be carried in currents for hundreds of miles. The inch-sized adults use sticky threads to attach themselves to nearly any underwater surface, including piers, pipes, machinery, underwater debris, boat hulls, and even sand and silt. Because they can survive out of water for days, mussels clinging to small boats may be transported to other lakes and rivers, where

they quickly move in. An adult female can produce 50,000 eggs each year, and the mussel menace has proven unstoppable.

The zebra mussel is an **exotic species,** meaning that it has been introduced into an ecosystem in which it did not evolve. Such species sometimes encounter no predators or parasites in their new environment and become *invasive*—their population grows unchecked. As their population explodes, the invaders may seriously damage the ecosystem as they displace, outcompete, or prey on native species. For example, starlings and house sparrows, deliberately introduced to the eastern United States in the 1890s, have displaced native songbirds, with which they compete for nesting sites. Similarly, red fire ants from South America, accidentally introduced into Alabama on shiploads of lumber in the 1930s, have spread throughout the southern United States. Fire ants displace and may kill other insects, birds, and even small mammals. Their mounds can ruin farm fields, and their fiery stings and aggressive temperament can make backyards uninhabitable. Another insect invader, the gypsy moth, was introduced from Europe in 1866 and still poses a serious threat to fruit and forest trees in North America. Now a new invader, the Asian long-horned beetle, is devouring hardwood trees in New York City and Chicago. The beetle, which officials believe may pose the biggest threat to forests since the gypsy moth, arrived around 1996 in wooden pallets and boxes shipped from China.

Many of the most persistent and destructive exotic species are plants. For example, a Japanese vine called kudzu was planted extensively in the 1940s to control erosion in the southern United States. Today kudzu is a major pest, overgrowing and killing trees and underbrush and occasionally engulfing small houses (**Fig. E28-1b**).

Some of the invasive plant species that crowd out native plants are quite beautiful. In light of this beauty, it is not surprising that many invasive species gained their foothold in new ecosystems by escaping from gardens. For example, the water hyacinth was introduced from South America as an ornamental plant and now clogs about 2 million acres of southern lakes and waterways, slowing boat traffic and displacing natural vegetation. Similarly, purple loosestrife, whose drifts of purple flowers now dominate many wetlands, was intentionally introduced by adventurous gardeners (**Fig. E28-1c**).

Once an invasive species becomes established, wildlife officials may attempt to control it by importing and releasing predators or parasites to attack the exotic species. However, this type of "biological control" is fraught with danger, because introducing exotic predators or parasites into an ecosystem can have unpredicted and possibly disastrous consequences for native species. For example, in 1958 a large predatory Florida snail, the rosy wolf snail (**Fig. E28-1d**), was imported into Hawaii to control another exotic, the giant African snail,

Predator and Prey Coevolve

To survive, predators must feed and prey must avoid becoming food. Therefore, predator and prey populations exert intense natural selection on one another. As prey become more difficult to catch, predators must become more adept at hunting them, and as predators become more adept, prey must get better at eluding them. Coevolution has thus endowed both the cheetah and its antelope prey with speed and camouflage. It has produced the keen eyesight of the hawk and the poisons and bright warning colors of the poison arrow frog and the coral snake.

Bats and moths provide an excellent example of how both physical structures and behaviors are molded by coevolution. Most bats are nighttime hunters that navigate and locate insect prey by echolocation. They emit pulses of extremely

► **Figure E28-1 Exotic species** *(a)* Workers blast jets of hot water at zebra mussels coating the interior of a water-treatment plant. *(b)* The Japanese vine kudzu rapidly covers entire trees and even houses. *(c)* Although its flowers are beautiful, purple loosestrife is an invasive species in North America. *(d)* Importing the rosy wolf snail proved disastrous for Hawaii's native snails, one of which is being attacked in this photo.

which was a menace to the native vegetation. The rosy wolf snail did not do much damage to the giant African snails, but instead preyed on many native snail species, threatening them with extinction.

Despite the risks of importing predators and parasites to control invasive species, there often seems to be little alternative, because poisons intended for exotics will kill natives as well. Having learned from past disasters, biologists now carefully screen proposed predator or parasite imports to make sure they attack only the intended invasive species. For example, a small fly from South America whose larvae feed selectively on fire ants is now being released in the southern United States.

Ecologists estimate that more than 6,500 exotic species have established themselves in the United States. The cost of attempting to control destructive invaders, when combined with the cost of the damage they do, is estimated at more than $120 billion annually in the United States. By evading the checks and balances imposed by millennia of coevolution, exotic species are wreaking havoc on natural ecosystems throughout the world.

(a) Zebra mussels

(c) Purple loosestrife

(d) Rosy wolf snail

(b) Kudzu

high-frequency sound and, by analyzing the returning echoes, create an "image" of their surroundings. In response to this highly effective prey-detection system, certain moth species (a favored prey of bats) have evolved ears that are particularly sensitive to the frequencies used by echolocating bats. When these moths hear a bat, they take evasive action, flying erratically or dropping to the ground. The bats may counter this defense by switching the frequency of their sound pulses away from the moth's sensitivity range. Some moth species have even evolved a way to interfere with the bats' echolocation by producing their own high-frequency clicks. In response, when hunting a clicking moth, a bat may turn off its own sound pulses temporarily and zero in on the moth by following the moth's clicks.

In the following sections, we examine some of the other evolutionary results of predator-prey interactions.

(a) A camouflaged fish

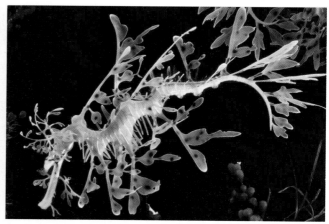

(b) A camouflaged bird

▲ **Figure 28-5 Camouflage by blending in** *(a)* The sand dab is a flattened, bottom-dwelling ocean fish with a mottled color that closely resembles the sand on which it rests. *(b)* This nightjar on its nest in Belize is barely visible among the surrounding leaf litter.

Camouflage Conceals Both Predators and Their Prey

Both predators and prey have evolved colors, patterns, and shapes that resemble their surroundings. Such disguises, called **camouflage,** render animals inconspicuous even when they are in plain sight (Fig. 28-5). Camouflaged prey animals may resemble specific objects such as leaves, twigs, seaweed, thorns, or even bird droppings (Fig. 28-6). Such animals tend to remain motionless when they need to be concealed; a moving bird dropping would be quite conspicuous.

(a) A camouflaged moth

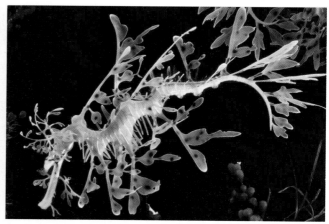

(b) A camouflaged fish

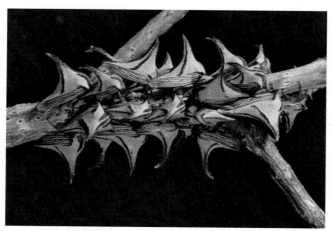

(c) Camouflaged treehoppers

▲ **Figure 28-6 Camouflage by resembling specific objects** *(a)* A moth whose color and shape resemble a bird dropping sits motionless on a leaf. *(b)* The leafy sea dragon (an Australian fish) has evolved extensions of its body that mimic the algae in which it often hides. *(c)* Florida treehopper insects avoid detection by resembling thorns on a branch.

Predators may also be camouflaged. For example, a spotted cheetah becomes inconspicuous in the grass as it watches for grazing mammals (Fig. 28-7a). The frogfish closely resembles the algae-covered rocks on which it sits motionless, dangling a small lure from a threadlike appendage on its upper lip (Fig. 28-7b). Small predatory fish notice only the lure and approach it, expecting to get a meal. Instead, they become a meal.

Bright Colors Often Warn of Danger

In contrast to camouflaged animals, some animal species have evolved bright, conspicuous **warning coloration.** Animals with warning coloration are usually distasteful, and many are poisonous (Fig. 28-8). Clearly, the defensive value of being poisonous is increased if potential predators can recognize dangerous prey *before* trying to eat them. Predators learn from their mistakes; a single unpleasant experience is often sufficient to teach a predator to avoid similarly colored prey in the future.

Some Organisms Gain Protection Through Mimicry

In **mimicry,** one species evolves to resemble another. Mimics are usually prey organisms that gain added protection from predators as a result of this adaptation. This benefit can be gained in several ways:

■ Harmless animals may evolve to resemble poisonous ones, thereby gaining some protection from predators that avoid the poisonous species. For example, the harmless mountain king snake avoids predation because it mimics the brilliant warning coloration of the deadly coral snake (Fig. 28-9a,b).

■ A distasteful species may evolve to resemble another distasteful species, such that each species benefits from predators' painful experience with the other. The shared color pattern leads to faster learning by predators and less predation on all similarly colored species. For example, predators rapidly learn to avoid insects with the conspicuous striped pattern shared by stinging bees, hornets, and yellow jackets. Similarly, the poisonous and distasteful monarch butterfly has wing patterns strikingly similar to those of the equally distasteful viceroy butterfly (Fig. 28-9c,d).

■ Prey species may evolve to mimic predators. A fascinating example of this type of mimicry is seen in snowberry flies, which are hunted by jumping spiders. When a fly detects an approaching spider, it spreads its wings, moving them back and forth in a jerky dance. A spider that sees this display is likely to flee from the harmless fly. Why? The markings on the fly's wings closely resemble the legs of another jumping spider. The jerky movements of the fly mimic those of a jumping spider when it drives another spider from its territory (Fig. 28-10). Natural selection has finely tuned both the behavior and the appearance of the fly to avoid predation by jumping spiders.

■ Prey species may evolve **startle coloration,** patterns that disrupt predation by distracting the predator. For example, some animal species, including several insects and a few vertebrates, have evolved patterns of color that closely resemble the eyes of a much larger animal (Fig. 28-11). If a predator gets close, the prey suddenly flashes its eyespots, startling the predator and allowing the prey to escape.

(a) A camouflaged cheetah

(b) A camouflaged frogfish

▲ **Figure 28-7 Camouflage assists predators** Prey may have difficulty detecting camouflaged predators such as the **(a)** cheetah and **(b)** frogfish.

▲ **Figure 28-8 Warning coloration** The South American poison arrow frog advertises its unpleasant taste and poisonous skin with bright, contrasting color patterns.

(a) Coral snake (venomous)

(b) Mountain king snake (non-venomous)

(c) Monarch (distasteful)

(d) Viceroy (distasteful)

▲ **Figure 28-9 Warning mimicry** *(a)* Dangerous animals may be mimicked by *(b)* harmless ones. Different species of distasteful insects *(c),(d)* may benefit from mutual mimicry.

(a) Jumping spider (predator)

(b) Snowberry fly (prey)

▲ **Figure 28-10 Visual and behavioral mimicry** *(a)* A predator's appearance and behavior may be mimicked by *(b)* its prey.

(a) False-eyed frog

(b) Peacock moth

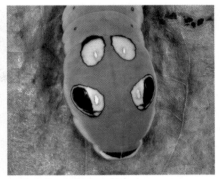

(c) Swallowtail caterpillar

Some Animal Predators and Prey Engage in Chemical Warfare

Some predators and prey have evolved toxic chemicals for attack and defense. Perhaps the best-known examples of such chemicals are the venoms that spiders and some snakes use to paralyze prey and deter predators. In addition, certain mollusks, including squid, octopus, and some sea slugs, emit clouds of ink when attacked. These chemical "smoke screens" confuse predators and mask the prey's escape.

A dramatic example of chemical defense is seen in the bombardier beetle. In response to the bite of an ant, the beetle releases secretions from special glands into an abdominal chamber. There, enzymes catalyze an explosive chemical reaction that shoots a toxic, boiling-hot spray onto the attacker (Fig. 28-12a).

Plants Have Defenses Against Herbivores

To an even greater extent than animals, plants have evolved chemical adaptations that deter their herbivorous predators. In particular, many plant species manufacture toxic and distasteful chemicals. Lupine plants, for example, produce chemicals that deter attack by the blue butterfly, whose larvae feed on the lupine's buds. Many of the chemicals that plants produce for defense have proved useful to people. Examples include stimulants such as caffeine, painkillers such as morphine, and medicines such as aspirin and the cancer drug Taxol. Fully half of today's medicinal drugs originated from plants.

▲ Figure 28-11 Startle coloration (a) When threatened, the false-eyed frog raises its rump, which resembles the eyes of a larger predator. (b) The peacock moth from Trinidad is well camouflaged, but should a predator approach, it opens its wings to reveal spots resembling large eyes. (c) Predators of this swallowtail butterfly caterpillar are deterred by its resemblance to a snake. Note that the caterpillar's head is the "snake's" nose.

(a) Bombadier beetle

(b) Monarch caterpillar

◀ Figure 28-12 Chemical warfare (a) The bombardier beetle sprays a hot, toxic brew in response to a leg pinch. (b) A monarch caterpillar feeds on milkweed that contains a powerful toxin. QUESTION: Why is the caterpillar so brightly colored?

The Case of the Vanishing Sea Otters

Continued

Chemical defense against predation is not limited to animals and plants; many fungi and protists have also evolved chemical defenses. Among the protists with chemical defenses are the seaweeds whose long blades, growing upward from the seafloor, make up kelp forests. The tissues of kelp blades contain chemicals that predators, including sea urchins, find distasteful.

Scientists have found that kelp in the waters off the coast of southern California contain higher concentrations of defensive chemicals than do kelp found in more northern waters. This difference may help explain why the kelp forests of southern California, unlike those farther north, did not undergo a catastrophic decline when sea otter populations were decimated. With their stronger chemical defenses, the southern kelp may have been better equipped to withstand predation by exploding sea urchin populations.

Plants may also defend themselves by producing compounds that toughen their edible parts. For example, many grasses have tough, glassy substances in their leaves, so grass blades are difficult for large herbivores to tear off and chew. Other plant species deter predation with structures such as thorns or thick bark.

Herbivores Have Adaptations for Overcoming Plant Defenses

Herbivorous predators have coevolved with the chemical defenses of plants. For example, the resistant leaves of grasses have fostered the evolution of strong, grinding teeth and powerful jaws in grazing animals. Similarly, some insect species have evolved increasingly efficient ways to detoxify or even use the toxic chemicals present in many plant tissues. In fact, nearly every toxic plant is eaten by at least one type of insect. For example, monarch butterflies lay their eggs on milkweed. When their larvae hatch, they consume the toxic plant (**Fig. 28-12b**). The caterpillars not only tolerate the milkweed poison but even store it in their tissues, where it serves as a defense against their own predators. The stored toxin is retained in the metamorphosed monarch butterfly.

28.4 What Is Symbiosis?

Symbiosis is an intimate, prolonged interaction between organisms of different species. In a symbiotic relationship, one species always benefits, but the second species may be unaffected, harmed, or helped by the relationship.

In Commensal Interactions, One Species Benefits and the Other Is Unaffected

A symbiotic relationship that benefits one species while neither helping nor harming the other is said to be *commensal*. For example, barnacles that attach themselves to the skin of a whale benefit by getting a free ride through nutrient-rich waters, and the whale does not seem to be affected. Similarly, many of the species of bacteria that inhabit the human gut benefit from living in a protected, nutrient-rich environment and seem to neither help nor harm their host. It is, however, often difficult to rule out the possibility that an apparently commensal relationship in fact slightly helps or harms a participant, and some biologists argue that truly commensal symbiosis is unlikely.

In Parasitic Interactions, One Species Benefits and the Other Is Harmed

In *parasitic* symbiosis, one organism benefits by feeding on another. Parasites live in or on their hosts, usually harming or weakening them but not immediately killing them. Parasites are generally much smaller and more numerous than their hosts. Symbiotic parasites include tapeworms, fleas, and numerous disease-causing protists, fungi, bacteria, and viruses.

Like other predator-prey interactions, parasite-host interactions are a powerful source of coevolutionary change. The effects of this mutual selection are evident in the precision and complexity of the human immune system (see Chapter 22) and in the adaptations that help disease-causing parasites overcome the immune response. Consider, for example, the protist that causes malaria. This parasite takes up residence inside human red blood cells. Ordinarily, red blood cells live for a few months before being destroyed in the spleen. Such destruction would also kill the malaria parasite, but the parasite resists this fate by producing a protein knob that extends from the surface of the infected red blood cell and firmly attaches the cell to the wall of a blood vessel so that it cannot be swept away to the spleen. The immune system in turn recognizes the foreign protein on the blood cell and sends its forces to attack. But the malaria parasite has evolved yet another defense: It can produce thousands of different versions of the sticky protein that anchors it to the blood vessel. By the time the immune system recognizes one version, the parasite has already switched to another.

In Mutualistic Interactions, Both Species Benefit

Many symbiotic relationships are *mutualistic*, that is, beneficial to both participants. For example, certain species of algae and fungi form mutualistic associations so intimate that the resulting structure appears to be a single organism: a lichen (Fig. 28-13a). The fungal body in the lichen provides support and protection for the photosynthetic algae, which in turn provide sugar molecules for the fungus.

Animals and Plants May Have Symbiotic Mutualistic Interactions with Microorganisms

Many animals benefit from mutualistic associations with microorganisms that inhabit their digestive systems. For example, even though animals lack the enzymes required to digest cellulose, some animals, including cows, horses, rabbits, and termites, can digest cellulose-containing plant tissues such as leaves or wood with the help of cellulose-digesting protists and bacteria that live in their guts. In these associations between animals and microorganisms, the animal provides a stable home and regular supply of food for the microorganisms, which return the favor by breaking cellulose into sugar molecules that the animal can absorb.

Plants may also benefit from mutualistic symbioses with microorganisms. For example, plants and certain fungal species may associate to form *mycorrhizae*, in which fungal bodies penetrate and become entwined with plant roots. The fungi acquire sugars that the plant has produced by photosynthesis, and the plant acquires mineral nutrients that the fungi extract from the soil. Another mutualistic symbiosis is the interaction between legume plants and the nitrogen-fixing bacteria that inhabit special chambers on their roots (see Fig. 17-18). These bacteria obtain food and shelter from the plant and in return convert soil nitrogen into a form that the plant can use.

Not All Mutualisms Are Symbiotic

Many mutualistic interactions between species do not involve the prolonged intimacy of symbiosis. For example, many flowering plants are pollinated by animals that visit only when the plant is in bloom. These animal pollinators—insects, hummingbirds, and bats—visit plants to eat the nutritious nectar produced in their flowers. As an animal probes a flower, grains of the flower's pollen may stick to the animal's body. On subsequent visits to other plants, the transported pollen (and the sperm cells it contains) may drop onto the female parts of a flower, fertilizing the egg contained within.

In the mutualistic relationship between plant and pollinator, the plant gains transportation of its pollen and the pollinator gains food. Such interactions have led to coevolution in which flowers have evolved colors, scents, and shapes that make them especially visible and attractive to pollinators, while pollinators evolved feeding structures and behaviors that are especially well suited to extracting nectar (Fig. 28-13b).

28.5 What Are Keystone Species?

In our descriptions of the various kinds of interactions between species, we have focused on interactions between two species. But it is important to keep in mind that in communities, each species participates in many interactions. A species

(a) Lichen

(b) Hawk moth

▲ Figure 28-13 Mutualism *(a)* Each of these brightly colored lichens is a symbiotic mutualistic association between an alga and a fungus. *(b)* The size and shape of a hawk moth's proboscis is often closely matched to the size and shape of the flowers it feeds on and pollinates.

Web Animation The Importance of Keystone Species

▲ **Figure 28-14 Keystone species** The starfish *Pisaster* is a keystone species along the rocky coast of the Pacific Northwest.

The Case of the Vanishing Sea Otters

Continued

In many cases, a species' keystone status is revealed only when the species disappears from a community. For example, ecologists were not aware that some kelp forests could not survive without sea otters until hunting decimated many otter populations.

might be a predator on multiple prey species, be preyed upon by several predator species, compete with still other species, and also participate in some mutualistic interactions. And each of the species with which our example species interacts itself interacts in multiple ways with a different set of other species. A community thus encompasses an intricate web of myriad community interactions. This network of interactions gives a community its distinctive properties and makes it far more than a simple collection of populations.

In some communities, a **keystone species** plays a major role in determining community structure—a role that is out of proportion to its abundance in the community. If a keystone species is eliminated, the community structure is dramatically altered. The sea otter, described in this chapter's opening essay, is an example of a keystone species; its removal caused the collapse of the kelp forest ecosystem. Another example was discovered by ecologist Robert Paine who performed an experiment in which he removed the predatory starfish *Pisaster* from sections of the rocky intertidal coast of Washington (**Fig. 28-14**). In areas lacking *Pisaster*, mussels (mollusks that are a favored prey of *Pisaster*) became so abundant that they outcompeted and displaced the other species that normally coexist in intertidal communities.

The importance of many keystone species, such as *Pisaster* starfish and sea otters, stems from their roles as predators. But keystone status can also arise from other kinds of interactions. For example, the feeding habits of the red-naped sapsucker (a woodpecker) make it a keystone species in mountain forest communities of the western United States. A sapsucker feeds mainly on tree sap that wells up from the many holes that it drills in living trees; the sap exposed by sapsuckers also serves as a key source of nourishment for many other bird, mammal, and insect species that would decline or disappear without access to sap. In cold Antarctic oceans, a small shrimp-like animal known as krill is a keystone species because of its role as prey; it is the primary food of many whales, seals, penguins, sea birds, fish, and squid. If krill were to disappear, the entire web of Antarctic animal life would likely disappear as well.

28.6 How Does a Community Change over Time?

The particular species and interactions that constitute a community vary from place to place and, in a given place, vary over time. As time passes, the makeup of a community may change. This gradual process of change is called **succession** because different community stages replace one another in a sequence that is somewhat predictable. For example, freshwater ponds and lakes tend to undergo a series of changes that, through time, transform them first into marshes and eventually to dry land. Coastal sand dunes tend to be stabilized by creeping plants and undergo changes that eventually lead to a forest. Different environments have different, characteristic patterns of succession.

We are able to observe succession because events frequently disrupt existing communities. Volcanic eruptions may create new islands ripe for colonization or leave behind a nutrient-rich environment that is rapidly invaded by new life (**Fig. 28-15**). Forest fires, while destroying an existing community, also release nutrients and create conditions that favor rapid succession.

The particular changes that take place during succession are as diverse as the environments in which succession occurs, but most succession follows the same overall pattern. It begins with a few hardy invaders called **pioneers** and, if undisturbed, progresses to a diverse and relatively stable **climax community,** the end point of succession.

Our discussion of succession in the following sections focuses largely on plant species. Keep in mind, however, that as plant species replace one another, they create food and habitats for a changing assemblage of animals, fungi, and microorganisms.

(a) Mt. St. Helens shortly after eruption

(b) Twenty years later

◀ **Figure 28-15 New life emerges after disruption** *(a)* On May 18, 1980, the explosion of Mount St. Helens in Washington State devastated the surrounding pine forest ecosystem. *(b)* Twenty years later, life abounds on the once-barren landscape.

There Are Two Main Types of Succession

Succession takes two major forms: primary and secondary. Succession that starts from scratch on bare rock, sand, a clear glacial pool, or other location where there is no trace of a previous community is **primary succession.** Succession that begins only after an existing ecosystem is disturbed, for example, by a forest fire or the abandonment of a farm field, is **secondary succession.**

Primary succession is slow; it typically requires thousands or even tens of thousands of years to generate a community in a previously lifeless location. Secondary succession happens more rapidly because the previous community has left its mark in the form of soil and seeds.

Primary Succession Can Begin on Bare Rock

Figure 28-16 illustrates primary succession on a rocky, northern island. Succession begins with bare rock, such as that exposed by a retreating glacier or formed when molten lava cooled. Over time, wind, rain, and cycles of freezing and thawing cause the rock to *weather*. Weathering causes cracks to form and also releases minerals that serve as nutrients for plants.

The weathered rock provides a place for lichens to settle. The lichens manufacture food by photosynthesis and obtain minerals by dissolving some of the rock with an acid they secrete. As the pioneering lichens spread over the rock, drought-resistant, sun-loving mosses begin growing in cracks. Fortified by nutrients liberated from the rock by the lichens, the moss forms a dense mat that traps

Web Animation Primary Succession: Glacier Bay Alaska

◀ **Figure 28-16 Primary succession** Changes over a thousand years, beginning with bare rock, in northern North America.

lichens and moss on bare rock

bluebell, yarrow

blueberry, juniper

jack pine, black spruce, aspen

fir-spruce-birch climax forest

time (years)

0

1000

plowed field

ragweed, crabgrass and other grasses

asters, goldenrod, broom sedge grass

blackberry

Virginia pine, tulip poplar, sweet gum

oak-hickory climax forest

time (years)

0 ———————————————————————→ 200

▲ **Figure 28-17** **Secondary succession** Changes over two hundred years, beginning with the abandonment of a southeastern farm field.

dust, tiny rock particles, and bits of organic debris. The moss mat also acts like a sponge, trapping moisture. Any moss body that dies decomposes, adding to a growing collection of nutrients.

Within the moss mat, seeds of larger plants, such as bluebell and yarrow, germinate. Later, these plants die and their bodies contribute to a growing layer of soil. Eventually, woody shrubs such as blueberry and juniper take advantage of the newly formed soil. As the shrubs grow, moss and lichens are shaded out and buried by decaying leaves and vegetation. Ultimately, trees such as jack pine, black spruce, and aspen take root in the deeper crevices, and the sun-loving shrubs are themselves shaded out. On the floor of the newly formed forest, shade-tolerant seedlings of taller or faster-growing trees, such as balsam fir, paper birch, and white spruce, thrive. In time they tower over and replace the original trees, which are intolerant of shade. After a thousand years or more, a tall climax forest thrives on what was once bare rock.

Abandoned Farmland May Undergo Secondary Succession

Figure 28-17 illustrates secondary succession on an abandoned farm in the southeastern United States. The pioneer species are fast-growing annual plants (species that live for a single growing season) such as crabgrass, ragweed, and sorrel, which sprout in the rich soil already present and thrive in direct sunlight. A few years later, longer-lived perennial plants such as asters, goldenrod, and broom sedge grass invade, along with woody shrubs such as blackberry. These plants multiply rapidly and dominate for the next few decades. Eventually, they are replaced by pines and fast-growing deciduous trees, such as tulip poplar and sweet gum, which sprout from windblown seeds. These trees become prominent after about 25 years, and a pine forest dominates the field for the rest of the first century. Meanwhile, shade-resistant, slow-growing hardwoods such as oak and hickory take root beneath the pines. After the first century, these hardwoods begin to tower over and shade the pines, which eventually die from lack of sun. A relatively stable climax forest dominated by oak and hickory is present by the end of the second century.

Succession Culminates in the Climax Community

Succession ends with a relatively stable climax community, which perpetuates itself as long as it is not disturbed by external forces (such as fire, invasion of an introduced species, or human activities). The species in a climax community have ecological niches that allow them to coexist without replacing one another. In general, climax communities have more species and more community interactions than do early stages of succession. The plant species that dominate climax communities are generally longer lived and larger than pioneer species. This trend is particularly evident in ecosystems where forest is the climax community.

In your travels, you have undoubtedly noticed that the type of climax community varies dramatically from one area to the next. For example, if you drive through Colorado, you will see a shortgrass prairie climax community on the eastern plains (in those areas where it has not been replaced by farms), pine-spruce forests in the mountains, tundra on their uppermost reaches, and a sagebrush-dominated climax community in the western valleys. The exact nature of a climax community is determined by numerous geological and climatic variables, including temperature, rainfall, elevation, latitude, type of rock (which determines the type of nutrients available), exposure to sun and wind, and many more.

Some Ecosystems Are Maintained in a Subclimax State

Some ecosystems are not allowed to reach the climax stage but are maintained in a **subclimax** stage. The tallgrass prairie that once covered northern Missouri and Illinois was a subclimax of a community whose climax would have been deciduous forest. The prairie was maintained by periodic fires, some set by lightning and others deliberately set by Native Americans to increase grazing land for buffalo. Forest now encroaches, and limited prairie preserves are maintained by carefully managed burning.

Agriculture also depends on the artificial maintenance of carefully selected subclimax communities. For example, a hayfield is a grass-dominated early stage of succession maintained by people to feed their livestock. Farmers must invest energy to prevent the growth of weeds and shrubs that represent the next stage of succession. Similarly, a suburban lawn is a painstakingly maintained subclimax ecosystem. Mowing destroys woody invaders, and the selective herbicides that some homeowners use kill pioneers such as crabgrass and dandelions.

The Case of the Vanishing Sea Otters Revisited

What caused the recent crash of the thriving Alaskan otter population? Dr. James Estes and his colleagues believe that otters are disappearing because killer whales (**Fig. 28-18**) are eating them. Although killer whales and sea otters have shared these waters for hundreds, if not thousands, of years, the researchers believe that whales did not begin to eat otters until the early 1990s.

The researchers first saw a whale kill and eat an otter in 1991, and more attacks have been observed since then. But is this new behavior responsible for declining otter populations? To test the hypothesis that predation by killer whales was reducing otter populations, the researchers performed a clever natural experiment. They tracked otters in two locations: Clam Lagoon, a cove accessible only via a shallow passage through which whales could not pass, and Kuluk Bay, a nearby open coastal area. After 1 year, the otter population in Kuluk Bay had declined by 65%. The population in Clam Lagoon, out of the reach of killer whales, remained stable. The researchers concluded that whales were capable of reducing otter populations.

Why have killer whales suddenly begun to eat otters? The researchers hypothesize that the whales have shifted their predation because of declining populations of their more typical prey species, especially sea lions and seals. Deprived of their preferred prey, the killer whales have turned to sea otters.

And why have the sea lions and seals declined? Most likely because of overfishing of the open-ocean fish on which they depend for food. Thus, human disruption of an oceanic community began a cascade of changes that ultimately resulted in vanishing sea otters and deteriorating nearshore kelp forests. Disturbance of a community can have unpredictable consequences in unexpected places.

Consider This

Another sea otter prey is the abalone, a mollusk that inhabits kelp forests and eats kelp. Abalone is also harvested for human consumption. Abalone fishermen blame sea otters for the recent dramatic decline in abalone populations. What data would you need to gather to determine whether sea otters are to blame?

▲ Figure 28-18 **Otter populations are again threatened** Are killer whales to blame?

Chapter Review

Summary of Key Concepts

For additional study help and activities, go to www.mybiology.com.

28.1 Why Are Interactions in Ecological Communities Important?

Community interactions influence population size, and the interacting populations within communities act on one another as agents of natural selection. Thus, community interactions also shape the evolution of the interacting populations.

28.2 What Are the Effects of Competition Among Species?

A species' ecological niche includes all aspects of the species' habitat and interactions with its living and nonliving environments. Each species occupies a unique ecological niche. When the niches of two populations within a community overlap, there is interspecific competition between the populations. When two species are forced under laboratory conditions to occupy the same ecological niche, one species always outcompetes the other. Species within natural communities have evolved in ways that reduce niche overlap, with adaptations that foster resource partitioning.

Web Animation   Competition

28.3 What Are the Effects of Predator-Prey Interactions?

Predators eat other organisms. Predators and prey act as strong agents of selection on one another (an example of coevolution). Some prey animals have evolved protective colorations that render them inconspicuous (camouflage) or startling (startle coloration) to their predators. Some prey are poisonous and exhibit warning coloration by which they are readily recognized and avoided. Some prey species have color patterns that mimic those of other species. Some predators and some prey have evolved toxic chemicals for attack and defense. Many plants have evolved chemical and physical defenses against predation. These defenses, in turn, have selected for predators that can overcome them.

28.4 What Is Symbiosis?

Two species in a symbiotic relationship interact closely over an extended time. Symbiosis includes parasitism, in which the parasite feeds on a larger, less abundant host, usually harming it but not killing it immediately. In commensal symbiotic relationships, one species benefits, typically by finding food more easily in the presence of the other species, which is not affected by the association. Mutualism benefits both symbiotic species.

28.5 What Are Keystone Species?

Keystone species have a greater influence on community structure than can be predicted by their numbers. Removal of a keystone species radically alters community structure to an extent that would not be predicted on the basis of the species' abundance.

Web Animation   The Importance of Keystone Species

28.6 How Does a Community Change over Time?

Over time, the species that compose the community in a particular place change, a process known as succession. Primary succession, which may take thousands of years, begins where there is no remnant of a previous community (such as on rock scraped bare by a glacier or cooled from molten lava, a sand dune, or in a newly formed glacial lake). Secondary succession proceeds more rapidly, because it builds on the remains of a disrupted community in, for example, an abandoned field or burned forest. Secondary succession is initiated by fast-growing, readily dispersing pioneer plants, which are eventually replaced by longer-lived, generally larger species. Uninterrupted succession ends with a climax community, which tends to be self-perpetuating unless acted on by outside forces, such as fire or human activities. Some ecosystems, including tallgrass prairie and farm fields, are maintained in relatively early stages of succession by periodic disruptions.

Web Animation   Primary Succession: Glacier Bay Alaska

Key Terms

Thinking Through the Concepts

Suggested answers to end-of-chapter and figure-based questions can be found at the end of the text.

Fill-in-the-Blank

1. Organisms that interact serve as agents of _____ selection on one another. This results in _____, which is the process by which species evolve adaptations to one another. The three major types of community interaction described in this chapter are _____, _____, and _____.

2. Two forms of predation in which the predator usually does not immediately kill its prey are _____ and _____. Both predators and their prey may evolve _____, which consists of colors, patterns, and shapes that blend into their surroundings. Some prey organisms are poisonous; many of these have evolved conspicuous _____.

3. Relationships in which members of two species interact in an intimate, prolonged manner are described as _____. Relationships of this type, in which both species benefit, are called _____. If one species benefits but the other is unharmed, the relationship is described as _____.

4. A species that plays a major role in determining community structure is called a(n) _____ species. For example, much of the Antarctic community ultimately depends on _____, which is a type of _____. In the scenario described in our chapter-opening essay, the _____ play a major role in determining the _____ community.

5. A somewhat predictable change in community structure over time is called _____. This process takes two basic forms: _____, which is the type of change that would start with bare rock, and _____, which is the type of change that would start after a forest fire. Species that enter the new habitat and begin the process of change are called _____. A relatively stable community that is the end product of this change is called a _____ community.

Review Questions

1. Define *ecological community*, and list three important types of community interactions.

2. Describe four different ways in which specific plants or animals protect themselves against being eaten. In each, describe an adaptation that might evolve in predators of these species that would overcome their defenses.

3. List two important types of symbiosis, and define and provide an example of each.

4. Which type of succession would occur on a clear-cut (a region in which all the trees have been removed by logging) in a national forest, and why?

5. List two subclimax and two climax communities. How do they differ?

6. Define *succession*, and explain why it occurs.

Applying the Concepts

1. IS THIS SCIENCE? Arguing against efforts to control the spread of invasive exotic species, an editorial writer claimed that "ecosystems and those of us who live in them benefit from change and competition, just like human societies and economies." Evaluate this claim from a scientific perspective. Can this hypothesis be tested using the scientific method?

2. An ecologist visiting an island finds two very closely related species of birds, one of which has a slightly larger bill than the other. Interpret this finding with respect to the competitive exclusion principle and the ecological niche, and explain both concepts.

3. Design an experiment to determine whether the kangaroo is a keystone species in the Australian outback.

For additional resources, go to www.mybiology.com.

chapter 29

How Do Ecosystems Work?

Can life be sustained in an enclosed, artificial ecosystem in the Arizona desert?

Case Study

A World in a Bubble

Is it possible for people to survive in an artificial ecosystem? In 1991, eight hardy individuals tried to answer this question by allowing themselves to be sealed inside Biosphere 2, a 3-acre glass and steel enclosure in the Arizona desert. Inside were 3,800 species in several miniature artificial ecosystems, including a rain forest, a tropical grassland, a desert, and a 900,000-gallon ocean. Electronic sensors, pumps, and fans controlled temperature and humidity within the giant greenhouse. The whole system could be isolated from the surrounding environment (except for the sunlight that entered though glass windows).

The eight "Biospherians" intended to remain sealed inside for 2 years, with nothing but energy coming from the outside. Biosphere 2's internal environment would, they predicted, supply their needs for food, air, and water. By demonstrating that an artificial ecosystem that includes humans could sustain itself, the creators of Biosphere 2 hoped to show that our knowledge of ecology is sufficient to prepare us for whatever ecological crises might lie ahead.

Unfortunately, it proved impossible to maintain livable conditions inside the enclosure. The oxygen concentration in the atmosphere dropped dramatically and the level of carbon dioxide soared, because respiration by microbes in the topsoil consumed the oxygen and released huge quantities of carbon dioxide. In addition, the internal atmosphere became polluted by high concentrations of nitrous oxide. To protect the safety of the researchers inside, Biosphere 2's seal was broken so that fresh air could be added.

Meanwhile, many of the Biosphere's species went extinct. Most vertebrate and insect species disappeared, including most of the pollinators, which doomed many plant species. Cockroaches and ants were among the few remaining insect species; their populations exploded and they swarmed over the Biosphere landscape. Food was in short supply and the human inhabitants lost weight. Water systems became polluted with excess nutrients. Overall, the closed ecosystem unraveled in fairly short order.

What lessons might be found in the outcome of the Biosphere 2 adventure? Consider that question as you review the principles of ecosystem ecology described in this chapter, and see if your new knowledge influences your answer. ■

A miniature ocean and a tropical rain forest are among the habitats inside Biosphere 2.

Web Animation   Ecology Models—Building a Food Web

29.1 How Do Ecosystems Obtain Energy and Nutrients?

The activities of life, from the flight of an eagle to the active transport of molecules across a cell membrane, are powered by the energy of sunlight. The solar energy that continuously bombards Earth is captured in ecosystems by photosynthetic organisms, transformed by the chemical reactions that power life, and ultimately converted to heat energy that radiates back into space. In this way, energy moves through ecosystems in a continuous one-way flow and is constantly replenished from an outside source, the sun.

In contrast to energy, nutrients are not replenished. **Nutrients** are elements and small molecules that form the chemical building blocks of life. Although these chemicals may be transported, redistributed, or converted to different molecular forms, they do not leave Earth, nor is their supply increased (except for small amounts in meteorites that reach Earth). Instead, nutrients are constantly recycled within and among ecosystems.

Recall that an *ecosystem* consists of all the communities within a defined area, along with their nonliving environment. Ecological study at the level of the ecosystem focuses in large measure on the flow of energy and nutrients. Ecosystem ecologists follow the pathways of energy and nutrients in order to understand the factors that shape the complex interactions within communities and between communities and their nonliving environment. **Figure 29-1** provides a preview of our study of ecosystems.

▶ **Figure 29-1 Energy flow, nutrient cycling, and feeding relationships in ecosystems** Note that nutrients (purple) neither enter nor leave the cycle. Energy (yellow), continuously supplied to producers as sunlight, is captured in chemical bonds and transferred through various levels of organisms (red), from primary producers to primary consumers and onward to higher-level consumers. Detritus feeders and decomposers gain energy from the wastes and dead bodies of both producers and consumers. At each transfer between levels, some energy is lost as heat (orange).

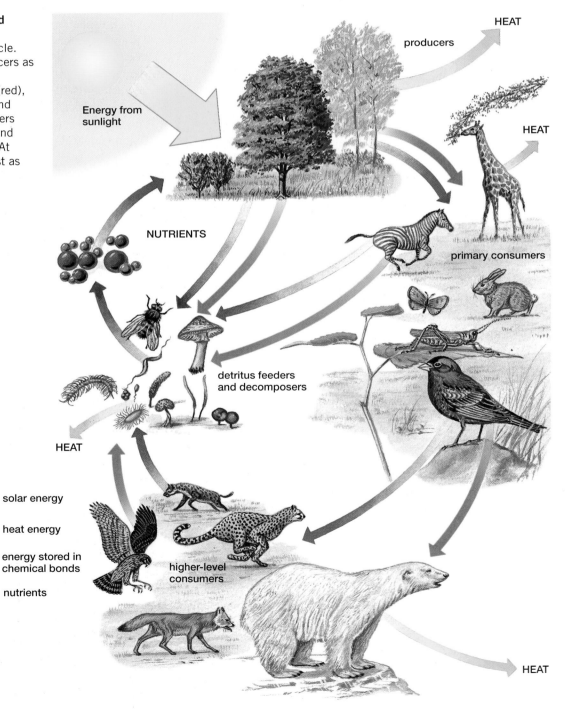

29.2 How Does Energy Flow Through Ecosystems?

Millions of miles from Earth, hydrogen atoms in the sun fuse to form helium atoms, releasing tremendous amounts of energy. A tiny fraction of this energy reaches Earth, arriving in the form of electromagnetic waves, including heat, light, and ultraviolet energy. Of the energy that reaches Earth, much is absorbed or reflected back into space by Earth's atmosphere, clouds, and surface. Overall, only about 1% of the solar energy that reaches Earth makes it to the surface as light that is available to power life. Of this 1%, green plants, algae, cyanobacteria, and other photosynthetic organisms capture 3% or less. Thus, the teeming life on this planet is supported by less than 0.03% of the energy that reaches Earth from the sun.

Energy Enters Ecosystems Mainly Through Photosynthesis

During photosynthesis, solar energy powers reactions that store energy in the chemical bonds of sugar and other high-energy molecules (**Fig. 29-2**). Photosynthetic organisms, from mighty oaks to single-celled diatoms in the ocean, produce food for themselves using nonliving nutrients and sunlight. Organisms that can manufacture their own food are called **autotrophs** or **producers.** In almost all of Earth's ecosystems, the producers are photosynthesizers that use solar energy to make food. However, in a few, such as the deep-sea ecosystems that flourish near hot-water-spewing vents in the ocean floor, the producers are non-photosynthetic prokaryotes that produce food using energy stored in the chemical bonds of inorganic compounds such as hydrogen sulfide.

As producers manufacture food for themselves, they directly or indirectly produce food for nearly all other organisms as well. Organisms that cannot produce their own food, called **heterotrophs** or **consumers,** must acquire energy and many of their nutrients from the bodies of other organisms.

Energy Captured by Producers Is Available to the Ecosystem

The amount of life that a particular ecosystem can support is determined by the amount of energy captured by the producers in that ecosystem. The energy that producers store and make available to other organisms is called **net primary productivity.** Net primary productivity is measured as the *biomass* (dry weight of living material) added by producers per unit area over a given time.

The net primary productivity of an ecosystem is influenced by many environmental variables, including the amount of nutrients available to the producers, the amount of sunlight reaching them, the availability of water, and the temperature. In the desert, for example, lack of water limits productivity. In the open ocean, light is limited in deep water and nutrients are limited in surface water. In ecosystems in which resources are abundant, productivity is high—for example, in tropical rain forests and in estuaries (where rivers meet the ocean, carrying nutrients washed from the land). Note that an ecosystem's overall contribution to Earth's productivity depends on both the ecosystem's net primary productivity per unit area and its prevalence. For example, tropical rain-forest ecosystems are far more productive per square meter than open ocean ecosystems (**Fig. 29-3**), but each ecosystem accounts for about a quarter of the planet's total productivity because open oceans cover about 65% of Earth, whereas tropical rain forests cover only about 4%.

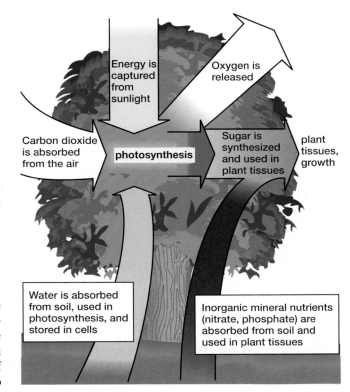

▲ **Figure 29-2 Photosynthesis** Photosynthetic organisms capture solar energy. They also remove inorganic nutrients from the nonliving parts of ecosystems and incorporate these nutrients into living tissue. They ultimately provide all the energy and most of the nutrients for other organisms.

▼ **Figure 29-3 Ecosystem productivities compared** The numbers give the average net primary productivity, in grams of organic material per square meter per year. Notice the enormous differences in productivity among the ecosystems. **QUESTION:** What factors are responsible for the differences in productivity among ecosystems?

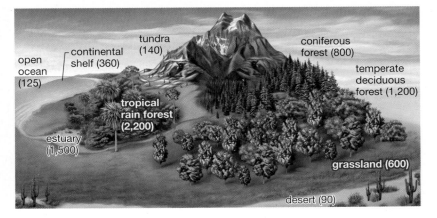

Energy Passes from One Trophic Level to Another

Energy flows through communities from producers through several levels of consumers. Each category of organism is called a **trophic level.** The producers are the first trophic level, generally obtaining their energy directly from sunlight (see Fig. 29-1). Consumers occupy several trophic levels. Some consumers feed directly on producers, the most abundant living energy source in any ecosystem. These **primary consumers,** or **herbivores** (plant eaters), ranging from grasshoppers to giraffes, form the second trophic level. The third trophic level consists of **secondary consumers** or **carnivores** (meat eaters), predators such as the spider, eagle, and wolf that feed mainly on primary consumers. Some carnivores occasionally eat other carnivores. When doing so, they occupy the fourth trophic level, **tertiary consumers.**

Feeding Relationships Within Ecosystems Form Chains and Webs

In illustrations of who feeds on whom in an ecosystem, it is common to identify a representative of each trophic level such that each representative species eats another on the level below it. This linear feeding relationship is called a **food chain.** As illustrated in **Figure 29-4,** different ecosystems have radically different food chains.

Natural communities rarely contain well-defined groups of primary, secondary, and tertiary consumers, however. Instead, the actual feeding relationships form a **food web** of many interconnecting food chains (**Fig. 29-5**). Some animals, such as raccoons, bears, rats, and humans, are **omnivores**—that is, at different times they act as primary, secondary, and tertiary consumers. Many carnivores eat both herbivores and other carnivores, thus acting as both secondary and tertiary consumers. An owl, for instance, is a secondary consumer when it eats a mouse, which feeds on plants, but a tertiary consumer when it eats a shrew, which feeds on insects. The owl becomes a quaternary (fourth-level) consumer when it eats a bird that is a tertiary consumer because it feeds on carnivorous spiders. A carnivorous plant, such as a Venus flytrap, that digests a fly is simultaneously a producer and a secondary consumer.

Detritus Feeders and Decomposers Release Nutrients for Reuse

Among the most important strands in a food web are the detritus feeders and decomposers. The **detritus feeders** are an army of mostly small and often unnoticed animals and protists that live on the refuse of life: molted

▼ **Figure 29-4 Food chains** The red arrows signify the transfer of energy, stored in chemical bonds, between trophic levels. In *(a)* terrestrial, *(b)* marine, and other ecosystems, energy is transferred from lower to higher trophic levels as organisms consume other organisms.

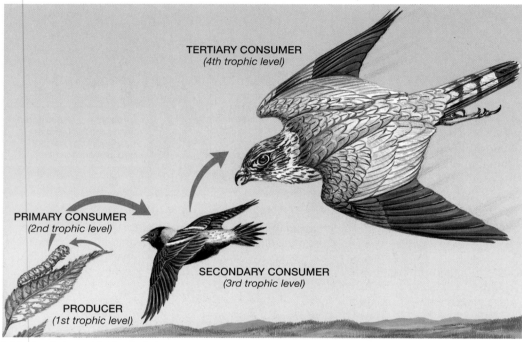

(a) A simple terrestrial food chain

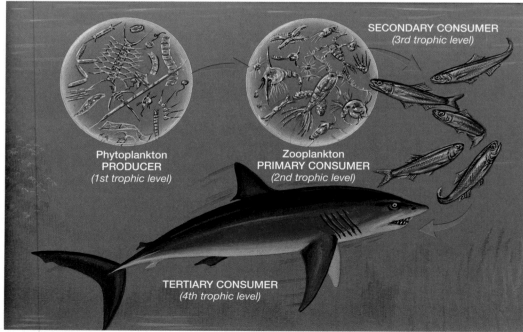

(b) A simple marine food chain

▲ **Figure 29-5 A food web in a shortgrass prairie** Paths of energy transfer, signified here by red arrows, can form complex webs. For example, primary consumers may consume multiple producer species and may be consumed by multiple secondary and tertiary consumers. Similarly, a secondary or tertiary consumer may consume multiple species from lower trophic levels. **QUESTION:** Point out the primary consumers in this food web.

exoskeletons, fallen leaves, wastes, and dead bodies. (Detritus means "debris.") The network of detritus feeders is complex. In terrestrial ecosystems it includes earthworms, mites, protists, centipedes, some insects, land-dwelling crustaceans called pillbugs, nematode worms, and even a few large vertebrates such as vultures. These organisms consume dead organic matter, extract some of the energy stored in it, and excrete it in a further decomposed state. Their excretory products serve as food for other detritus feeders and for decomposers.

The **decomposers** are primarily fungi and bacteria. For example, the patches of fuzz on an old piece of bread are the visible portions of the bodies of fungal decomposers, and the smelly slime on a decaying piece of meat marks the presence of bacterial decomposers. Decomposers digest food outside their bodies by secreting digestive enzymes into the environment. They then absorb the nutrients they need, and the remaining nutrients are released to the environment.

Detritus feeders and decomposers are absolutely essential to life on Earth. Their activities reduce the bodies and wastes of other organisms to simple molecules, such as carbon dioxide, water, minerals, and organic molecules that return to the atmosphere, soil, and water. By liberating nutrients for reuse, detritus feeders and decomposers perform a vital role in recycling nutrients. If detritus feeders and decomposers were to disappear, accumulating wastes and dead bodies would gradually smother Earth's ecosystems, and the nutrients stored in those bodies would remain locked up and unavailable for use by other organisms.

Energy Transfer Through Trophic Levels Is Inefficient

As we discussed in Chapter 5, a basic law of thermodynamics is that energy use is never completely efficient. For example, as your car burns gasoline, about 75% of the energy released is immediately lost as heat and cannot be used to do the work of moving the car. The same inefficiency also applies in living systems. For example, splitting the chemical bonds of ATP to cause muscular contraction produces heat as a by-product. This is why walking briskly on a cold day warms you. Waste heat is produced by all the biochemical reactions that keep cells alive.

Energy transfer from one trophic level to the next is also inefficient. When a caterpillar (a primary consumer) eats the leaves of a tomato plant (a producer), only a portion of the solar energy originally trapped by the plant is available to the insect. Some of the original energy was used by the plant for growth and maintenance, and some was converted into the chemical bonds of molecules such as cellulose, which the caterpillar cannot digest. Even more energy was lost as heat during these processes. Overall, only a fraction of the energy captured by the first trophic level is available to organisms in the second trophic level.

Additional energy is lost in each transfer to a higher trophic level. Some of the energy consumed by the caterpillar in the second trophic level is used to power crawling and chewing. Some is used to construct its indigestible exoskeleton, and much is given off as heat. All of this energy is unavailable to a songbird (in the third trophic level) that eats the caterpillar. The bird in turn loses some energy as body heat, uses some to fly, and converts a considerable amount into indigestible feathers, beak, and bone. All of this energy is unavailable to the hawk that catches the songbird. A simplified model of energy flow through the trophic levels in a deciduous forest ecosystem is illustrated in **Figure 29-6**.

Energy Pyramids Illustrate Energy Transfer Between Trophic Levels

Studies of ecosystems have shown that, in general, about 90% of the energy stored at a trophic level is lost in the transfer to the next level. Thus, the energy stored by an ecosystem's primary consumers (herbivores) is only about 10% of the energy stored in the bodies of producers. In turn, the bodies of secondary consumers contain roughly 10% of the energy stored in primary consumers. In other words, for every 1,000 calories of solar energy captured by grass, only about

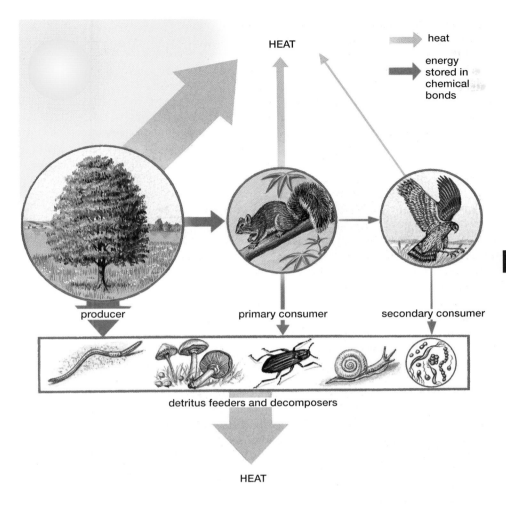

producer

primary consumer

secondary consumer

heat

energy stored in chemical bonds

HEAT

HEAT

detritus feeders and decomposers

Figure 29-6 caption:

◀Figure 29-6 **Energy transfer and loss in a forest** The width of the arrows is roughly proportional to the quantity of energy transferred (red) or lost as heat (orange). Note that, because energy is lost at each transfer between trophic levels, secondary consumers as a group contain less energy than primary consumers, and primary consumers contain less energy than producers. **QUESTION:** Why is so much energy lost as heat? (Hint: Review the laws of thermodynamics.)

A World in a Bubble

Continued

The energy pyramid helps explain the hunger and weight loss that plagued Biosphere 2's inhabitants. Ecosystems that provide food for people are affected by the inefficient energy transfer common to all ecosystems, and only a small portion of incoming energy is converted to edible biomass. One way to overcome this inherent inefficiency is to eat mainly foods drawn from the lowest trophic level, such as fruits, vegetables, grains, nuts, and other foods from plants. These foods contain a much higher proportion of an ecosystem's captured energy than do meat, milk, and other foods derived from animals. But even a diet rich in plant products did not save Biosphere 2's residents from hunger. The total amount of energy entering the sealed habitat was apparently too small to overcome the inherent inefficiency of the artificial ecosystems inside.

100 calories are converted into herbivores, and only 10 calories into carnivores. This inefficient energy transfer between trophic levels is called the "10% law."

An **energy pyramid,** which shows maximum energy at the base and steadily diminishing amounts at higher levels, illustrates the energy relationships between trophic levels (**Fig. 29-7**). Energy pyramids also show why primary producers are the most abundant organisms in an ecosystem, and carnivores the rarest. An unfortunate side effect of the inefficiency of energy transfer is that certain persistent toxic chemicals produced by human activities become concentrated in the bodies of carnivores, as described in "Earth Watch: Food Chains Magnify Toxic Substances" on p. 582.

29.3 How Do Nutrients Move Within and Among Ecosystems?

In contrast to the energy of sunlight, nutrients do not flow down onto Earth in a steady stream from above. The same pool of nutrients has been supporting life for more than 3 billion years. Some, called **macronutrients,** are required by organisms in large quantities. These include carbon, hydrogen, oxygen, nitrogen, phosphorus, sulfur, and calcium. **Micronutrients,** including zinc, molybdenum, iron, selenium, and iodine, are required only in trace quantities. **Nutrient cycles,** also called **biogeochemical cycles,** describe the pathways that substances follow as they move through communities to the nonliving portions of ecosystems and then back again to communities.

As nutrients move through biogeochemical cycles, they may accumulate in one portion of the cycle and remain there for long periods. These storage sites

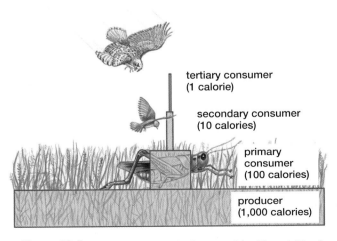

tertiary consumer (1 calorie)

secondary consumer (10 calories)

primary consumer (100 calories)

producer (1,000 calories)

▲ Figure 29-7 **An energy pyramid for a prairie** The width of each rectangle is proportional to the energy stored at that trophic level.

Food Chains Magnify Toxic Substances

In the 1940s, the new insecticide DDT seemed close to miraculous. In the tropics, DDT saved millions of lives by killing the mosquitoes that spread malaria. DDT also destroyed agricultural insect pests, increasing crop yields and saving millions from starvation. What's more, DDT is long lasting, so a single application keeps on killing. DDT's discoverer, the Swiss chemist Paul Müller, was awarded the 1948 Nobel Prize for Medicine and Physiology, and people looked forward to a new age of freedom from insect pests. Little did they realize that indiscriminate use of this pesticide was unraveling the complex web of life.

For example, in the mid-1950s the World Health Organization sprayed DDT on the island of Borneo to control malaria. One unintended casualty was a predatory wasp species that fed on caterpillars. The wasp population was wiped out, but its main prey, a caterpillar that fed on the thatched roofs of houses, was relatively unaffected. Within a short period, thatched roofs were collapsing, eaten by the uncontrolled population of caterpillars. At the same time, gecko lizards that ate poisoned insects accumulated high concentrations of DDT in their bodies. Both the geckos and the village cats that ate them died of DDT poisoning. With the cats eliminated, the rat population exploded. The human population was then threatened with an outbreak of plague, carried by the uncontrolled rat population. The outbreak was avoided only by airlifting new cats to the villages.

During the 1950s and 1960s in the United States, wildlife biologists witnessed an alarming decline in populations of several predatory birds, especially fish-eaters, such as bald eagles, cormorants, ospreys, and brown pelicans. These top predators are never abundant, and the decline pushed some, including the brown pelican and the bald eagle, close to extinction. The cause of the declining bird populations was eventually traced to DDT, which had been sprayed on the aquatic ecosystems supporting these birds. Scientists were amazed to find concentrations of DDT in the bodies of predatory birds that were up to *1 million* times the concentration present in the water.

This finding led to the discovery of **biological magnification,** the process by which toxic substances accumulate at increasingly higher concentrations in progressively higher trophic levels. Because the transfer of energy from lower to higher trophic levels is extremely inefficient, a pesticide that has been sprayed on plants is consumed in large amounts by herbivores, which must eat large quantities of plant material to survive. Carnivores, in turn, must eat many herbivores, along with the pesticide contained in the bodies.

Substances that undergo biological magnification have three properties that make them dangerous: First, they are not easily **biodegradable** (cannot readily be broken down into harmless substances by decomposer organisms). Because they are not biodegradable, they persist for a long time in the environment. Second, they are soluble in fat, and, finally, they are not soluble in water. Because they are fat soluble but not water soluble, they build up in the bodies of animals, particularly in fat, and are not readily broken down and excreted in watery urine. These substances are not flushed from animal bodies, so a predator accumulates poisons from its prey over many years.

Mercury is among the bioaccumulating toxins that threaten human health. A potent neurotoxin, mercury accumulates in predatory fish. When we eat these fish, we act as tertiary or even quaternary consumers, and hence run the risk of accumulating dangerous levels of mercury in our tissues. Because of this risk, the U.S. Food and Drug Administration advises young children and women of childbearing age to stop eating shark, swordfish, and king mackerel and to consume only limited amounts of other species, such as salmon, pollock, and canned light tuna. In the United States, coal-fired power plants are the largest single source of mercury pollution. Once in the atmosphere, mercury can be wafted thousands of miles, to be deposited in what should be pristine environments, such as the Arctic. As a result, even Inuit natives living north of the Arctic Circle have high levels of mercury and other bioaccumulating pollutants in their bodies.

Human health may also be at risk from a class of bioaccumulating chemicals called *endocrine disruptors*. Found in widely used pesticides, plastics, and flame retardants, these chemicals have become widespread in the environment. Endocrine disruptors accumulate in fat and either mimic or interfere with the actions of animal hormones. There is compelling evidence that these chemicals are interfering with the reproduction and development of fish, fish-eating birds such as cormorants (**Fig. E29-1**), frogs, salamanders, alligators, and many other animals. They are also suspected of causing reduced sperm counts in people.

▲ **Figure E29-1 The price of pollution** Deformities such as the twisted beak of this double-crested cormorant from Lake Michigan have been linked to bioaccumulating chemicals.

are called **reservoirs.** Most major reservoirs are in the nonliving part of an ecosystem. For example, carbon has several major reservoirs: It is stored as carbon dioxide gas in the atmosphere, in dissolved form in oceans and other waters, and as fossil fuels underground.

In the following sections, we briefly describe the cycles of carbon, nitrogen, phosphorus, and water.

Carbon Cycles Through the Atmosphere, Oceans, and Communities

Carbon enters the living part of the ecosystem when producers capture carbon dioxide (CO_2) during photosynthesis and incorporate its carbon atoms in organic molecules. On land, producers acquire CO_2 from the atmosphere, where it makes up 0.036% of the total gases. For producers in lakes, rivers, and oceans, there is abundant CO_2 dissolved in the water.

Some of the carbon taken up by producers is returned as CO_2 to the atmosphere and ocean during cellular respiration, and the rest is incorporated into the producers' bodies. Primary consumers, such as rabbits, shrimp, and grasshoppers, eat the producers and acquire the carbon stored in their tissues. These herbivores also release some carbon through respiration and store the rest, which may then be consumed by organisms in higher trophic levels. Both producers and consumers eventually die, and their bodies are broken down by detritus feeders and decomposers. Cellular respiration by these organisms returns CO_2 to the reservoirs in the atmosphere and oceans (**Fig. 29-8**).

The complementary processes of uptake by photosynthesis and release by respiration continually move carbon from the nonliving to the living portion of an ecosystem and back again. Some carbon, however, cycles much more slowly. For example, many mollusks and marine protists extract CO_2 dissolved in water and combine it with calcium to form calcium carbonate ($CaCO_3$), from which they

▼ **Figure 29-8 The carbon cycle** Arrows signify the movement of carbon through ecosystems. Carbon is extracted from its reservoirs in the atmosphere and oceans by photosynthesis by producers. The carbon captured by producers moves within ecosystems through consumption of carbon-containing bodies by consumers and decomposers. Carbon is ultimately returned to its reservoirs by respiration and by combustion of organic matter, including fossil fuels.

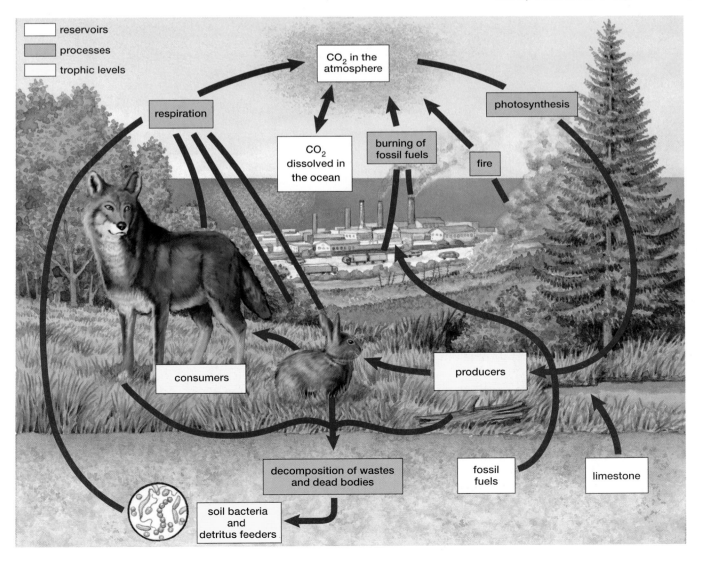

construct their shells. After these organisms die, their shells sink to the ocean floor, are buried, and may eventually be converted to limestone. Geological events may expose the limestone, which dissolves gradually as water runs over it, making the carbon available to living organisms once again, millions of years after it was incorporated into living biomass.

The Major Reservoir for Nitrogen Is the Atmosphere

The atmosphere contains about 79% nitrogen gas (N_2) and is thus the major reservoir for this important nutrient. Neither plants nor animals, however, can extract nitrogen from the atmosphere. Instead, nitrogen enters the food web mainly through certain bacteria in soil and water that engage in **nitrogen fixation,** a process in which nitrogen and hydrogen are combined to form ammonia (NH_3).

Nitrogen-fixing bacteria do not use all of the ammonia they produce, and the excess is excreted into the surrounding soil or water. Much of this ammonia is in turn converted by other bacteria to nitrate (NO_3^-). Nitrate also forms when atmospheric nitrogen and oxygen react during electrical storms and when fossil fuels are burned.

Plant roots can absorb ammonia and nitrate, and plants incorporate the nitrogen from these molecules into amino acids, proteins, nucleic acids, and some vitamins. These nitrogen-containing molecules in plants are eventually consumed by primary consumers, detritus feeders, or decomposers. As nitrogen passes through the food web, some is returned to soil or water when decomposer bacteria convert wastes and dead bodies back to nitrate and ammonia. Atmospheric nitrogen is replenished by **denitrifying bacteria,** which break down nitrate, releasing nitrogen gas back to the atmosphere (Fig. 29-9).

▼ **Figure 29-9 The nitrogen cycle** Arrows signify the movement of nitrogen through ecosystems. Nitrogen is extracted from its reservoir in the atmosphere by electrical storms, fertilizer production, and nitrogen fixation by bacteria. Some of the captured nitrogen remains as a reservoir in the soil, but a portion moves within ecosystems through uptake by producers followed by consumption of nitrogen-containing bodies by consumers and decomposers. Nitrogen is ultimately returned to the atmosphere by the metabolic activity of denitrifying bacteria. **QUESTION:** Why do humans augment the nitrogen cycle by capturing nitrogen from the air and transferring it to the soil?

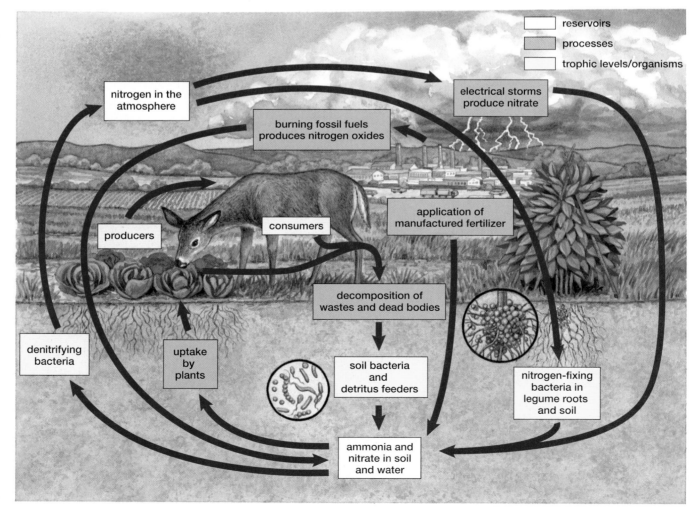

reservoirs
processes
trophic levels/organisms

nitrogen in the atmosphere

electrical storms produce nitrate

burning fossil fuels produces nitrogen oxides

consumers

application of manufactured fertilizer

producers

decomposition of wastes and dead bodies

denitrifying bacteria

uptake by plants

soil bacteria and detritus feeders

nitrogen-fixing bacteria in legume roots and soil

ammonia and nitrate in soil and water

The Major Reservoir for Phosphorus Is Rock

In contrast to the cycles of carbon and nitrogen, the phosphorus cycle does not include the atmosphere. The reservoir of phosphorus in ecosystems is rock, where phosphorus is bound to oxygen to form phosphate. As phosphate-rich rocks are exposed and eroded, rainwater dissolves the phosphate. Dissolved phosphate is readily absorbed through the roots of plants and by other autotrophs, such as photosynthetic protists and cyanobacteria. From these producers, phosphorus passes through food webs (Fig. 29-10). In both producers and consumers, phosphorus is a crucial component of biological molecules, including energy transfer molecules (ATP and NADP), nucleic acids, and the phospholipids of cell membranes. It is also a major component of vertebrate teeth and bones.

Phosphorus in the living parts of ecosystems is eventually returned to the nonliving parts. Organisms excrete excess phosphate into their surroundings, and decomposers ultimately return phosphate from dead bodies back to the soil and water. The phosphate may then be again absorbed by autotrophs, or it may become bound to sediment and eventually reincorporated into rock.

Some of the phosphate that dissolves in fresh water is carried to the oceans. Although much of this phosphate ends up in marine sediments, some is absorbed by marine producers and eventually incorporated into the bodies of invertebrates and fish. Some of these, in turn, are consumed by seabirds, which excrete large quantities of phosphorus back onto the land. At one time, the guano (droppings) deposited by seabirds along the western coast of South America was mined as a source of phosphorus for fertilizer and industrial use.

Water Remains Unchanged During the Water Cycle

The water cycle, or **hydrologic cycle** (Fig. 29-11), differs from most other biogeochemical cycles in that most water is not incorporated into other molecules during the cycle. The major reservoir of water is the ocean, which covers about three-quarters of Earth's surface and contains more than 97% of the available water.

▼ Figure 29-10 The phosphorus cycle Arrows signify the movement of phosphorus through ecosystems. Phosphorus, in the form of phosphate, dissolves from its largest reservoir in rocks and moves to the soil and water. Some of the dissolved phosphate settles into sediment reservoirs, which may ultimately be reincorporated in phosphate rock. Much of the dissolved phosphate, however, moves within ecosystems through uptake by producers followed by consumption of phosphorus-containing bodies by consumers and decomposers. This phosphorus is ultimately returned to soil, water, or sediments as dissolved phosphate.

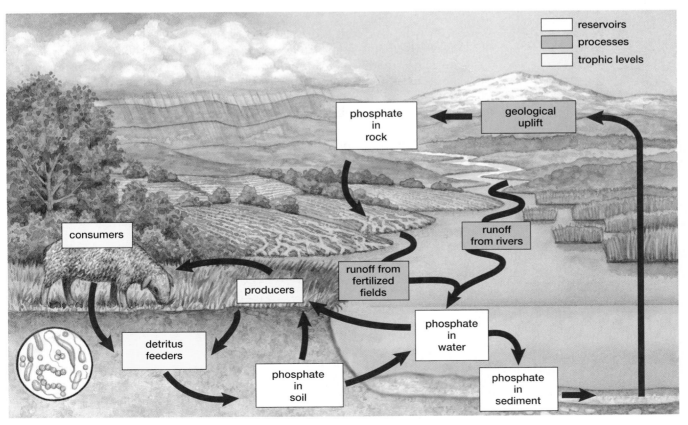

▶ Figure 29-11 **The hydrologic cycle** Water moves between atmospheric, aquatic, and terrestrial reservoirs by evaporation, transpiration, and precipitation.

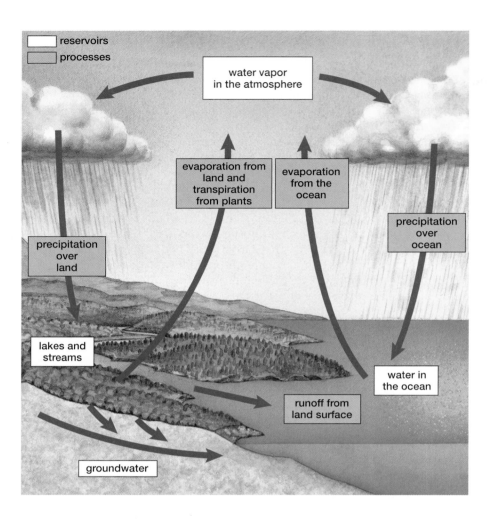

The hydrologic cycle is driven by solar energy, which evaporates water, and by gravity, which draws the water back to Earth in the form of precipitation (rain, snow, sleet, and dew). Most evaporated water comes from the oceans, and much water returns directly to them by precipitation. Water falling on land takes various paths. Some of it evaporates from soil, lakes, and streams. A portion runs off the land back to the oceans, and a small amount seeps down through the soil to underground reservoirs.

Some water enters the living communities of ecosystems. It is absorbed by the roots of plants; most of the absorbed water evaporates back to the atmosphere from leaves, but a small amount is combined with carbon dioxide during photosynthesis to produce high-energy molecules. Eventually these molecules are broken down during cellular respiration, releasing water back to the environment. Consumers get water from their food or by drinking.

29.4 What Happens When Humans Disrupt Nutrient Cycles?

Many of the environmental problems that plague modern society are caused by human disruption of geochemical cycles. In some cases, the effects of the disruption are fairly direct, as when industrial processes transfer toxic substances such as lead, arsenic, mercury, uranium, and oil from their normal reservoirs into the environment (Fig. 29-12). But other human-caused cycle disruptions do their damage in more indirect fashion. In the following sections, we describe some of the consequences of human impacts on nutrient cycles.

▲ Figure 29-12 **A natural substance out of place** This bald eagle was killed by an oil spill off the coast of Alaska.

Overloading the Phosphorus and Nitrogen Cycles Damages Aquatic Ecosystems

In human-dominated ecosystems—such as farm fields, gardens, and suburban lawns—ammonia, nitrate, and phosphate are supplied by chemical fertilizers. Each year, more than 100 million tons of phosphorus is removed from its reservoir in rocks, converted to phosphate fertilizer, and applied to cultivated land. Similarly, agricultural and industrial processes that artificially "fix" atmospheric nitrogen each year convert an estimated 150 million tons of nitrogen from its gaseous form to ammonia and nitrate fertilizers.

The soil that erodes from fertilized fields carries tremendous quantities of phosphorus- and nitrogen-containing chemicals into lakes, streams, and the ocean. In these waterways, the nutrients stimulate explosive growth of photosynthetic algae. As the algae die, their bodies are consumed by decomposer bacteria, whose respiration uses up most of the oxygen in the water, thereby killing other organisms. Often, the result is severe damage to aquatic and marine ecosystems. For example, agricultural runoff into the Mississippi River is carried to the Gulf of Mexico, where scientists believe it is the cause of a 7,000-square-mile "dead zone," nearly devoid of living organisms, that appears there each summer.

Overloading the Nitrogen and Sulfur Cycles Causes Acid Deposition

The combustion of fossil fuels in our vehicles, power plants, industrial boilers, smelters, and refineries releases huge amount of sulfur dioxide and nitrogen oxides into the atmosphere. Volcanoes, hot springs, and decomposer organisms also release sulfur dioxide, and decomposers, lightning, and fires release nitrogen oxides, but human industry and transportation account for most of the sulfur dioxide and nitrogen oxides in the atmosphere. The amounts present are far in excess of what can be absorbed and recycled by ecosystems and are the cause of a growing environmental threat: *acid rain,* more accurately called **acid deposition.**

Reactions in the Atmosphere Form Acid Deposition

When combined with water vapor in the atmosphere, nitrogen oxides and sulfur dioxide are converted to nitric acid and sulfuric acid. Days later, and often hundreds or thousands of miles from the source, the acids fall in rain or snow, eating away at statues and buildings and rendering lakes lifeless. In the Adirondack Mountains of New York, for example, acid rain has made about 25% of the lakes and ponds too acidic to support fish. But even before the fish die, much of the food web that sustains them is destroyed. Clams, snails, crayfish, and insect larvae die first, then amphibians, and finally fish. The result is a crystal-clear lake—beautiful but dead.

The impact of acid deposition is not limited to aquatic organisms. Acid rain also interferes with the growth and yield of many farm crops by leaching out essential nutrients such as calcium and potassium and killing decomposer microorganisms, thus preventing the return of nutrients to the soil. Crop plants, poisoned and deprived of nutrients, become weak and vulnerable to infection and insect attack. Trees and other plants in natural ecosystems are also at risk (**Fig. 29-13**).

Acid deposition also increases the exposure of organisms to toxic metals, including aluminum, lead, mercury, and cadmium, which are far more soluble in acidified water than in water of neutral pH. Aluminum dissolved from rock may inhibit plant growth and kill fish. Drinking water in some households has been found to be dangerously contaminated with lead, because acidic water dissolves the lead solder in old pipes.

A World in a Bubble

Continued

Like Earth, Biosphere 2 was a closed system with no new inputs of nutrients. Like Earth, Biosphere 2 therefore depended on nutrient cycling to maintain life-friendly levels of nutrients in the sealed environment's soil, water, and atmosphere. Like Earth, Biosphere 2 had nutrient cycles that were susceptible to disruption. But unlike Earth's nutrient cycles, those in Biosphere 2's artificial ecosystems seemed unable to compensate for even relatively minor perturbations. For example, carbon uptake did not keep pace with carbon released by normal respiration by Biosphere's organisms, and potential reservoirs such as the artificial "ocean" or increased wood growth failed to absorb the excess. Instead, carbon (in the form of carbon dioxide) accumulated in the atmosphere until it reached toxic concentrations.

▲ **Figure 29-13 Acid deposition can destroy forests** Acid rain and fog have destroyed this forest atop Mount Mitchell in North Carolina.

Web Animation The Global Carbon Cycle and Greenhouse Effect

Overloading the Carbon Cycle Contributes to Global Warming

Much of Earth's carbon is in long-term storage in reservoirs. One such reservoir is fossil fuels. **Fossil fuels** form from the buried remains of plants and animals. Over millions of years, the carbon in the organic molecules of these organisms is transformed by high temperatures and pressures into coal, oil, or natural gas. When people burn fossil fuels to supply energy for heat, light, transportation, manufacturing, and agriculture, CO_2 is released into the atmosphere. By freeing carbon from the fossil fuel reservoir, humans are increasing the amount of CO_2 in the atmosphere, as we describe later in this chapter.

Without human intervention, the carbon in fossil fuels would stay locked away. Since the Industrial Revolution, however, we have increasingly relied on the energy stored in these fuels. As a result of our prodigious consumption of fossil fuels, the amount of CO_2 in the atmosphere has increased by more than 36% since 1850, from 280 parts per million (ppm) to 381 ppm. Atmospheric CO_2 continues to increase at a rate of 1.5 ppm per year.

Carbon dioxide is also added to the atmosphere by **deforestation,** which destroys tens of millions of acres of forests each year. When forests are cut and burned, the carbon stored in the bodies of trees returns to the atmosphere. The rate of deforestation is especially rapid in the tropics, where rain forests are being converted to marginal agricultural land.

Altogether, human activities release almost 7 billion tons of carbon (in the form of CO_2) into the atmosphere each year. About half of this carbon is captured by the global carbon cycle—that is, it is absorbed into the plants, soil, and the oceans. The remaining 3.5 billion tons remains in the atmosphere, fueling global warming.

Greenhouse Gases Trap Heat in the Atmosphere

Atmospheric CO_2 acts something like the glass in a greenhouse: It allows solar energy to pass through and reach Earth's surface, but it absorbs and is heated by the longer-wavelength energy that then radiates from the surface back into the atmosphere (**Fig. 29-14**). This **greenhouse effect** traps some of the sun's energy as heat and keeps Earth's atmosphere warm enough to support life.

Most climate scientists have concluded that the greenhouse effect has been intensified by human activities that produce CO_2 and other **greenhouse gases** such as methane, chlorofluorocarbons (CFCs), water vapor, and nitrous oxide. Historical temperature records have revealed a **global warming** trend. The average global temperature has increased since 1860, paralleling the increase in atmospheric CO_2 (**Fig. 29-15**). Nineteen of the 20 hottest years on record have occurred since 1980, and the 6 hottest years were all after 1998.

A large, international group of climate scientists known as the Intergovernmental Panel on Climate Change (IPCC) predicts that, if greenhouse gas emissions are not curtailed, the average surface air temperature of Earth will increase by 3.6°F to 8.1°F (2.0°C to 4.5°C) by 2100. To put this change in perspective, average air temperatures during the peak of the last Ice Age (20,000 years ago), when much of North America was under a thick sheet of ice, were only about 9°F (5°C) lower than at present.

Global Warming Will Have Severe Consequences

Although the precise consequences of global warming are difficult to predict, disruptions that will affect living things are nearly certain. In fact, Earth has already begun to experience them. For example, scientists have documented effects of warming on glaciers and polar ice caps, on weather patterns, and on the living inhabitants of ecosystems. The speed and magnitude of these changes are expected to increase as warming continues.

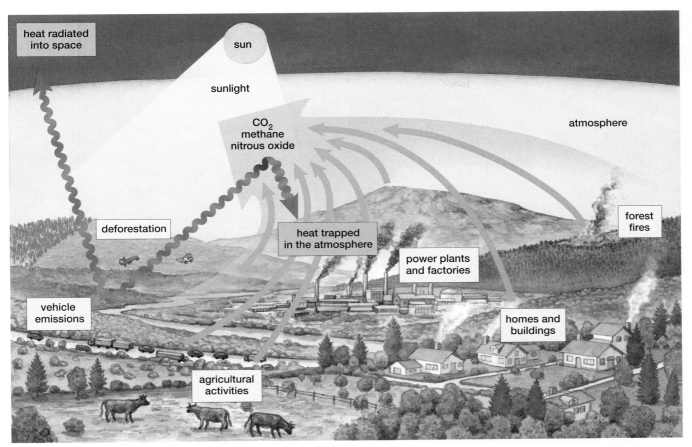

A Global Meltdown Is Under Way

Throughout the world, ice is melting (see "Earth Watch: Poles in Peril" on p. 591). Glaciers are retreating and disappearing (Fig. 29-16). In Glacier National Park, where 150 glaciers once graced the mountainsides, only 35 remain, and scientists estimate that none will remain 30 years from now. Greenland's ice sheet is melting at twice the rate of a decade ago, releasing 53 cubic miles (221 cubic kilometers) of water into the Atlantic annually. As glaciers and polar ice melt, global sea levels are rising. Increasing sea levels threaten to flood

▲ **Figure 29-14 Increases in greenhouse gas emissions contribute to global warming** Incoming sunlight warms Earth's surface and is radiated back to the atmosphere. Greenhouse gases absorb some of this radiated energy, trapping it as heat in the atmosphere. **QUESTION:** Why does the temperature rise in an actual greenhouse? Does this process provide a good analogy for heat trapping by greenhouse gases?

◀ **Figure 29-15 Global warming parallels CO_2 increases** The CO_2 concentration of the atmosphere (blue line) has increased steadily since 1860. Average global temperatures (red line) show a parallel increase.

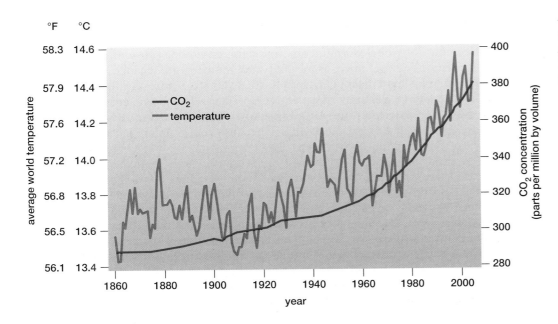

(a) Carroll Glacier, 1904

(b) Carroll Glacier, 2004

▲ **Figure 29-16 Glaciers are melting** Photos taken from the same vantage point in **(a)** 1904 and **(b)** 2004 document the retreat of Carroll Glacier in Glacier Bay, Alaska.

coastal cities, with island nations being especially vulnerable because inhabitants displaced by rising waters may have nowhere to go.

In addition to the meltdown of glaciers and ice caps, the permanently frozen soils of far northern lands are also melting. In Alaska, thawed permafrost allows tons of mud to wash into rivers and destroy salmon spawning grounds. Perversely, permafrost thawing due to global warming can cause even more global warming—the thawing releases additional CO_2 into the atmosphere as previously frozen organic matter decomposes. In Siberia, a region of frozen soil the size of France and Germany combined is melting, creating a gigantic bog that could release billions of tons of CO_2 and methane (an even more potent greenhouse gas) into the atmosphere.

Weather Is Growing More Extreme

Climate scientists predict that global warming will increase the incidence of extreme weather events. Increasing evidence suggests that the weather is already being affected. During the past 35 years, both the intensity and the duration of hurricanes has increased by 50%. The number of Category 4 and 5 hurricanes (the highest categories of wind speed and destruction) has doubled.

Even as some areas suffer from an increase in severe storms, other areas are getting drier. Scientists at the National Center for Atmospheric Research report that since the 1970s, the area of Earth impacted by severe drought has doubled from about 15% to about 30%. As the world warms further, droughts in these regions will last longer and be more severe. Agricultural disruption resulting from the newly emerging extremes in weather could be especially disastrous for poor nations that are already barely able to feed themselves.

Ecosystems Are Affected

The impact of global warming on ecosystems could be profound. On land, plant distributions will change as rainfall and temperature change. For example, sugar maples may disappear from northeastern U.S. forests, and southeastern forests could be partially replaced by grasslands. In the sea, coral reefs, highly productive ecosystems already stressed by human activities, are likely to suffer further damage from warmer waters, because reef-building corals are harmed by even small increases in water temperature.

Around the world, ecologists are discovering changes related to warming. The growing season in Europe has increased by more than 10 days during the past 28 years. In Europe and North America, spring flowers are blooming earlier. Birds in many temperate regions begin nesting a week or more earlier than they did a few decades ago. The geographic ranges of many species of butterflies and birds have shifted toward the poles. Among the tropical organisms with expanding ranges are disease-carrying organisms such as malaria-transmitting mosquitoes, suggesting that diseases currently restricted to the tropics will soon be present in temperate regions as well. Overall, the accumulated data from scientific investigations around the world provide strong evidence that warming-related biological changes are well under way.

Poles in Peril

Earth's polar ice is melting, in both the Arctic and the Antarctic. In Antarctica, the average annual temperature has increased by about 4.5°F (2.5°C) during the past 50 years, far more than the increase for the planet as a whole. The ice shelf off the Antarctic Peninsula, present there for thousands of years, is shrinking; more than 2,000 square miles of ice have disintegrated since 1995.

The loss of floating ice in the Antarctic has far-reaching consequences. Sea ice creates conditions that favor abundant growth of phytoplankton. These primary producers in turn provide food for krill, the shrimp-like crustaceans that are a keystone species in the Antarctic food web. Krill are the main food of seals, penguins, and several species of whales. During the past 30 years, however, krill populations in the southwest Atlantic have plummeted by about 80%, and researchers hypothesize that the decline is linked to the loss of sea ice. A likely scenario is that, as the ice shelves shrink, the algae that grow on the underside of the ice dwindle, and the krill that rely on the algae starve. Researchers are concerned that the impact of the krill loss may reverberate up the food chain, ultimately starving whales and seals and penguins as well. Ominously, the Adelie penguin population in the western Antarctic Peninsula has shrunk by 80% since 1975; only about 4,000 breeding pairs remain of the more than 20,000 pairs originally present. (Fortunately, most other Antarctic penguin populations remain healthy.)

At the other end of Earth, Arctic temperatures have risen almost twice as rapidly as the world average, causing a 20% to 30% decrease in late summer arctic sea ice over the past 30 years. Even larger changes are expected during the upcoming century, due to a predicted temperature increase of 7° to 12°F (4° to 7°C).

Arctic sea ice provides critical habitat for polar bears and for ringed seals, their major food source. Complete disappearance of sea ice, which some scientists predict will occur within the next century, would mean almost certain extinction for polar bears in the wild. In Canada's Hudson Bay, sea ice now breaks up 3 weeks earlier than it did 30 years ago, depriving the bears of a prime opportunity to hunt ringed seals on the ice (**Fig. E29-2**). As a result, Hudson Bay polar bears now start their summers 15% lighter. Leaner females produce fewer cubs with a lower survival rate. Shrinking habitat and food shortage has also induced dangerous changes in bear behavior. Hungry bears now venture further south than in the past, sometimes wandering into inhabited areas where they may be shot. Similarly, the

▲ **Figure E29-2 Polar bears on thin ice** The loss of arctic sea ice threatens polar bear populations.

bears, which can swim well, have been observed hunting up to 60 miles offshore, much further out than is usual for them. These seafaring bears risk drowning and several have been spotted floating dead after storms. As a result of decreased production of cubs and the increasingly risky behavior of displaced adults, the local polar bear population has declined by 22% since 1987.

Our Decisions Make a Difference

Americans make a disproportionately large contribution to environmental problems, such as global warming, that are caused by overloaded nutrient cycles. Despite accounting for less than 5% of the world's population, the United States generates about 25% of world's greenhouse gases, more than 6 tons per person each year. What can individuals do to help reduce this impact? Quite a lot; here are a few examples:

- **Use fuel-efficient vehicles, car pools, and public transportation.** A car with a fuel efficiency of 20 miles per gallon releases 1 pound of CO_2 per mile traveled. Therefore, we can substantially reduce emissions of CO_2 by using more efficient forms of transportation.

- **Conserve electricity.** Generating electricity in fossil fuel-fired power plants emits tremendous quantities of CO_2, sulfur dioxide, and nitrogen oxides. We can reduce these emissions by supporting the efforts of utility companies to use renewable energy sources such as wind and solar power. We can also conserve electricity by purchasing more efficient appliances, turning off unused computers and lights, and replacing incandescent with compact fluorescent light bulbs.

- **Improve the energy-efficiency of housing.** Insulating and weatherproofing our homes, incorporating solar energy features into new homes, and planting deciduous trees near our houses to provide for summer shade and winter sun will significantly reduce fuel consumption while also reducing heating and air conditioning costs.

- **Recycle.** Recycling is a tremendous energy saver. For example, 95% of the energy used to produce an aluminum can from raw materials is conserved when the can is recycled.

A World in a Bubble Revisited

According to the managers of Biosphere 2, the experiment with human inhabitants succeeded in demonstrating that "there is no demonstrated alternative to maintaining the viability of the Earth. . . . Earth remains the only known home that can sustain life." Nonetheless, the effort was widely perceived as a failure and, in 1996, Biosphere 2's owners turned management of the facility over to Columbia University. Since then, Biosphere 2 has served as a laboratory for more conventional ecological research.

Most recent research at Biosphere 2 has examined the effects of environmental changes that Earth is likely to face in the future, especially higher concentrations of atmospheric CO_2. For example, researchers tested the hypothesis that tropical rain forests can "soak up" excess CO_2 by increased photosynthesis. Using the instruments that precisely control the Biosphere's atmosphere, the researchers subjected the facility's rain forest to CO_2 levels of 400, 700, and 1,200 ppm (400 ppm is close to today's actual level). Under each condition, the researchers measured the forest's rate of CO_2 uptake (by photosynthesis) and release (by cellular respiration).

Results showed that the rain forest's ability to soak up CO_2 declined at an atmospheric concentration of 700 ppm and was severely reduced at 1,200 ppm. Uptake of CO_2 did increase at higher atmospheric concentrations, but respiration increased even more (especially respiration by soil microbes). The researchers concluded that rain forests are unlikely to soak up much of the additional carbon dioxide that human activities will add to the future atmosphere.

Biosphere 2 has served as a unique laboratory for ecological experiments that would be difficult or impossible to perform elsewhere. Although a dispute with the owners of Biosphere 2 ended Columbia University's arrangement to manage the facility and brought a temporary halt to research there, the University of Arizona recently agreed to assume managment. The new managers have begun a $3 million renovation project, and expect Biosphere 2 to soon resume its role as a site for ecological research.

Consider This

Can conclusions drawn from experiments in the closed, artificial system of Biosphere 2 be "scaled up" to apply to natural ecosystems? Can sound environmental policy be based on data from Biosphere 2?

Chapter Review

Summary of Key Concepts

For additional study help and activities, go to www.mybiology.com.

29.1 How Do Ecosystems Obtain Energy and Nutrients?
Ecosystems are sustained by a continuous flow of energy from sunlight and a constant recycling of nutrients.

> **Web Animation Ecology Models—Building a Food Web**

29.2 How Does Energy Flow Through Ecosystems?
Energy enters ecosystems when it is harnessed by producers during photosynthesis. The amount of energy that autotrophs store in a given unit of area during a given period is the ecosystem's net primary productivity.

Trophic levels describe feeding relationships in ecosystems. Autotrophs are the producers, the lowest trophic level. Herbivores occupy the second level as primary consumers. Carnivores act as secondary consumers when they prey on herbivores and as tertiary or higher-level consumers when they eat other carnivores.

Feeding relationships in which each trophic level is represented by one species are called food chains. In natural ecosystems, feeding relationships are more complex than in a food chain and are described as food webs. Detritus feeders and decomposers, which digest dead bodies and wastes, use some of the stored energy and free up nutrients for recycling. In general, only about 10% of the energy captured by organisms at one trophic level is converted to the bodies of organisms at the next level. The higher the trophic level, the less energy available to sustain it. As a result, plants are more abundant than herbivores, and herbivores are more abundant than carnivores. The storage of energy at successive trophic levels is illustrated graphically as an energy pyramid.

29.3 How Do Nutrients Move Within and Among Ecosystems?

A nutrient cycle depicts the movement of a particular nutrient from its reservoir (usually in the nonliving portion of the ecosystem) through the living portion of the ecosystem and back to its reservoir, where it is again available to producers. Carbon reservoirs include the oceans, the atmosphere, and fossil fuels. Carbon enters producers through photosynthesis. From producers, it is passed through the food web and released to the atmosphere as CO_2 during cellular respiration.

The major reservoir of nitrogen is the atmosphere. Bacteria and human industrial processes convert nitrogen gas into ammonia and nitrate, which plants can use. Nitrogen then passes from producers to consumers and is returned to the environment through excretion and the activities of detritus feeders and decomposers.

The reservoir of phosphorus is in rocks as phosphate, which dissolves in rainwater. Phosphate is absorbed by photosynthetic organisms, then passed through food webs. Some is excreted, and the rest is returned to the soil and water by decomposers. Some is carried to the oceans, where it is deposited in marine sediments. Humans mine phosphate-rich rock to produce fertilizer.

The major reservoir of water is the oceans. Water is evaporated by solar energy and returned to Earth as precipitation. The water flows into lakes and underground reservoirs and in rivers, which flow to the oceans. Water is absorbed directly by plants and animals and is also passed through food webs. A small amount is combined with CO_2 during photosynthesis to form high-energy molecules.

29.4 What Happens When Humans Disrupt Nutrient Cycles?

Environmental problems arise when human activities interfere with the natural functioning of ecosystems. Human industrial processes release toxic substances and produce more nutrients than nutrient cycles can efficiently process. Through massive consumption of fossil fuels, we have disrupted the natural cycles of carbon, sulfur, and nitrogen, causing acid deposition and global warming (an amplification of the greenhouse effect).

Web Animation The Global Carbon Cycle and Greenhouse Effect

Key Terms

acid deposition *p. 587*
autotroph *p. 577*
biodegradable *p. 582*
biogeochemical cycle *p. 581*
biological magnification *p. 582*
carnivore *p. 578*
consumer *p. 577*
decomposer *p. 580*
deforestation *p. 588*

denitrifying bacterium *p. 584*
detritus feeder *p. 578*
energy pyramid *p. 581*
food chain *p. 578*
food web *p. 578*
fossil fuel *p. 588*
global warming *p. 588*
greenhouse effect *p. 588*
greenhouse gas *p. 588*

herbivore *p. 578*
heterotroph *p. 577*
hydrologic cycle *p. 585*
macronutrient *p. 581*
micronutrient *p. 581*
net primary productivity *p. 577*
nitrogen fixation *p. 584*
nutrient *p. 575*

nutrient cycle *p. 581*
omnivore *p. 578*
primary consumer *p. 578*
producer *p. 577*
reservoir *p. 582*
secondary consumer *p. 578*
tertiary consumer *p. 578*
trophic level *p. 578*

Thinking Through the Concepts

Suggested answers to end-of-chapter and figure-based questions can be found at the end of the text.

Fill-in-the-Blank

1. Sunlight falls on Earth continuously and is captured by _____ organisms. In contrast, _____ are constantly recycled in processes called _____ cycles.

2. Photosynthetic organisms are called either _____ or _____. The energy that these organisms store and make available to other organisms is called _____. Photosynthetic organisms are consumed by organisms collectively called _____ or _____. Animals and protists that live on wastes and dead bodies are called _____. Organisms described as decomposers are primarily _____ and _____.

3. Feeding levels within ecosystems are also called _____ levels. In general, only about _____ percent of the energy in one such level is transferred to the organisms in the level above it. An animal that feeds on other animals is called a(n) _____. Animals that eat both plants and other animals are _____. Feeding relationships within ecosystems are most accurately depicted as _____.

4. As nutrients are recycled, they tend to accumulate in storage sites called _____. For carbon, the major storage sites are _____, _____, and _____. The major storage site for nitrogen is the _____. The organisms responsible for capturing nitrogen and making it available to plants are called _____.

5. Release of excessive amounts of nutrients into the Mississippi River from agricultural activities causes an enormous _____ in the Gulf of Mexico each summer. Excess nitrogen oxides and _____ released into the atmosphere from burning fossil fuels cause the phenomenon known as _____.

6. Carbon dioxide and methane are known as _____ gases. Human activities, primarily burning fossil fuels and _____ have increased the carbon dioxide content of the atmosphere by about _____ percent since 1850. Loss of the Antarctic ice floes threatens an important primary consumer, the shrimp-like _____ .

Review Questions

1. What makes the flow of energy through ecosystems fundamentally different from the flow of nutrients?

2. What is an autotroph? What trophic level does it occupy, and what is its importance in ecosystems?

3. Define *net primary productivity*. Would you predict higher net primary productivity in a farm pond or an alpine lake? Defend your answer.

4. List the first three trophic levels. Among the consumers, which are most abundant? Why would you predict that there will be a greater biomass of plants than herbivores in any ecosystem? Relate your answer to the "10% law."

5. How do food chains and food webs differ? Which is the more accurate representation of actual feeding relationships in ecosystems?

6. Define *detritus feeders* and *decomposers*, and explain their importance in ecosystems.

7. Trace the movement of carbon from its reservoir through the biological community and back to the reservoir. How have human activities altered the carbon cycle, and what are the implications for future climate?

8. Explain how nitrogen gets from the air to a plant.

9. Trace the movement of a phosphorus molecule from a phosphate-rich rock into the DNA of a carnivore. How does the phosphorus cycle differ from the carbon and nitrogen cycles?

10. Trace the movement of a water molecule from the moment it leaves the ocean until it eventually reaches a plant root, then a plant stoma, and then makes its way back to the ocean.

Applying the Concepts

1. **IS THIS SCIENCE?** Some skeptics argue that there is no scientific proof that increased fossil fuel use causes global warming, although fossil fuel use, atmospheric carbon dioxide, and average global temperature have all increased steadily for the past 150 years or so. Is their criticism valid? Explain your reasoning.

2. Define and give an example of biological magnification. What qualities are present in materials that undergo biological magnification? In which trophic level are the problems worst, and why?

3. Describe the contribution of population growth to (a) acid rain and to (b) the greenhouse effect.

For additional resources, go to www.mybiology.com.

chapter 30

Earth's Diverse Ecosystems

Do you know where your coffee comes from? Choosing the right beans could help protect rainforest biodiversity.

Case Study   Can Coffee Save Songbirds?

Can drinking coffee help preserve threatened species? Perhaps so, if you're willing to change the brand you brew. Many conservation organizations are urging coffee consumers to switch to "shade coffee." Shade coffee is grown beneath trees and was traditionally produced by farmers who started their crops by clearing the undergrowth from a patch of forest and planting coffee bushes in the shade of the remaining trees. In recent decades, however, many of these traditional plantings have been replaced with new varieties of "sun coffee" that thrive in bright sunlight. Sun coffee produces many more coffee beans per plant than shade coffee, so farmers have an economic incentive to make the switch. But growing sun coffee requires that the forest be completely cleared away.

In recent years, researchers have amassed considerable evidence that shade coffee plantations hold far more species than do sun coffee plantations. For example, researchers working in southern Mexico surveyed birds and butterflies in sun coffee plantings, shade coffee plantings, and undisturbed forest. For both kinds of animals, the number of species was similar in shade coffee and undisturbed forests, but far lower in sun coffee plantings. Studies in other tropical areas have yielded similar findings. Many of these studies have focused on birds, especially the songbirds that breed in temperate regions and then migrate to the Tropics.

Because ecological communities are so much more diverse and complex in shade coffee plantations than in sun coffee plantations, conservationists would like to slow or halt the conversion of traditional plantings to sun coffee. One strategy for achieving that goal is to enlist the help of consumers. If consumers were to insist on shade coffee and were willing to pay a small premium to get it, then perhaps coffee farmers would have an economic incentive to continue traditional growing methods.

Is promoting shade coffee production a good conservation strategy? Are there any drawbacks? See if you can think of any as you read this chapter's overview of Earth's ecosystems. ■

Plantings of shade-grown coffee provide habitat for rain-forest species.

Web Animation Tropical Atmospheric Circulation and World Biomes

30.1 What Factors Influence Earth's Climate?

The distribution of life, particularly on land, is strongly influenced by weather and climate. **Weather** refers to short-term fluctuations in temperature, humidity, cloud cover, wind, and precipitation over periods of hours or days. **Climate** refers to patterns of weather that prevail from year to year and even century to century in a particular region. A region's climate is determined by the range of temperatures and amount of sunlight and precipitation it receives.

Both Climate and Weather Are Driven by the Sun

Both climate and weather are powered by energy from the sun. This solar energy includes wavelengths that range from short, high-energy ultraviolet (UV) rays, through visible light, to longer infrared wavelengths.

Much of the solar energy that reaches Earth's atmosphere never hits the planet's surface. A large portion is reflected back into space by dust, water vapor, and clouds in the atmosphere. In addition, the atmosphere absorbs some solar

The Ozone Hole—A Puncture in Our Protective Shield

A small fraction of the radiant energy produced by the sun, called *ultraviolet*, or UV, radiation, is so highly energetic that it can damage biological molecules. In small quantities, UV radiation helps human skin produce vitamin D and causes tanning in fair-skinned people. But in larger doses, UV causes sunburn, premature aging of skin, skin cancer, and cataracts, a condition in which the lens of the eye becomes cloudy.

Fortunately, most UV radiation is filtered out by ozone in the stratosphere, a layer of atmosphere extending from 6 to 30 miles above Earth. In pure form, ozone (O_3) is a bluish, explosive, and highly poisonous gas. In the stratosphere, the normal concentration of ozone is about 0.1 part per million (ppm), compared with 0.02 ppm in the lower atmosphere. The ozone-enriched layer in the stratosphere is called the *ozone layer*. Ultraviolet radiation is converted to heat when it strikes the ozone layer. UV absorption destroys ozone, but UV radiation also converts some stratospheric oxygen gas to ozone, so the overall level of ozone remains reasonably constant—or did until humans intervened.

In 1985 British atmospheric scientists published a startling discovery: The springtime levels of stratospheric ozone over Antarctica had declined by more than 40% since 1977. A hole had been pierced in Earth's protective shield. In the *ozone hole* over Antarctica, ozone now dips during part of the year to about one-third of its original levels (**Fig. E30-1**). Although ozone layer depletion is most severe over Antarctica, the ozone layer is somewhat reduced over most of the world, including the continental United States.

Satellite data reveal that, since the early 1970s, UV radiation reaching Earth's surface has increased by nearly 7% per decade in the Northern Hemisphere and by 10% or more per decade in the Southern Hemisphere. This increase threatens human health.

Epidemiological studies indicate that for every 1% increase in lifetime exposure to UV radiation, the lifetime risk of melanoma (skin cancer) increases by about 1%. Human health effects are not the only cause for concern. Photosynthesis by phytoplankton, the producers for marine ecosystems, is reduced under the ozone hole above Antarctica. Some types of trees and crops are also harmed by increased UV radiation.

The thinning of the ozone layer is caused by human production and release of chlorofluoro-carbons (CFCs). Developed in 1928, these gases were widely used as coolants in refrigerators and air conditioners, aerosol spray propellants, in the production of foam plastic, and as cleansers for electronic parts. CFCs are very stable and were considered safe. Their stability, however, proved to be a major problem, because they remain chemically unchanged as they gradually rise into the stratosphere. There, under intense bombardment by UV light, the CFCs break down, releasing chlorine atoms. Chlorine catalyzes the breakdown of ozone to oxygen gas (O_2) while remaining unchanged itself.

Fortunately, we have taken the first steps toward "plugging" the ozone hole. In an almost unprecedented example of global concern and cooperation, industrialized nations throughout the world agreed, in a series of treaties beginning in 1987, to rapidly phase out ozone-depleting chemicals. These compounds are no longer used in the production of plastic foam, and CFC substitutes have been found for use in spray cans, refrigerators, and car air conditioners as well. Ground-level atmospheric chlorine levels (an indicator of CFC use) peaked in 1994. By 1999, scientists had detected chlorine reductions in the stratosphere as well. In 2005, the National Oceanic and Atmospheric Administration reported that ozone

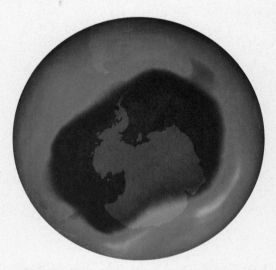

▲ **Figure E30-1 Satellite image of the Antarctic ozone hole** The ozone hole recorded in September 2006 is shown in blue and purple on this NASA satellite image. At 11.4 million square miles, it tied the previous record set in 2000. (*Image courtesy of NASA.*)

concentrations had leveled off between 1996 and 2002.

Chlorofluorocarbons can persist 50 to 100 years and take a decade or more to ascend into the stratosphere, so current release of CFCs by developing countries, along with the millions of tons already released, will continue to erode the protective ozone shield. Full recovery is decades away and will require further reductions in CFC release. The prospects for continued reduction received a boost in 2007, when China halted production of CFCs. And, in a spirit of continued cooperation, developed countries recently pledged funding to help developing countries devise alternatives to CFCs, providing more cause to believe that our shield will eventually be restored.

energy. Much of sunlight's high-energy UV radiation, which can damage biological molecules, is absorbed by the **ozone layer,** a region of the middle atmosphere that is relatively rich in ozone (see "Earth Watch: The Ozone Hole—A Puncture in Our Protective Shield"). Similarly, energy at infrared wavelengths is absorbed by carbon dioxide, water vapor, methane, and other *greenhouse gases* in the atmosphere.

So much solar energy is reflected or absorbed by the atmosphere that only about half the energy that reaches Earth actually strikes its surface. Of this amount, a small fraction is captured by photosynthetic organisms and used to power photosynthesis, and most of the rest is absorbed and converted to heat. The portion absorbed and temporarily stored as heat by Earth's surface and atmosphere keeps Earth warm enough to sustain life. Eventually, nearly all of the incoming solar energy returns to space, either as light or as heat.

Solar energy drives Earth's ocean currents, wind, and global water cycle. These large-scale movements of air and water interact with the physical features of Earth's surface to produce different climates in different places. Among the factors that affect climate are air currents, ocean currents, and the presence of mountains and irregularly shaped continents. A location's climate is also influenced by its *latitude* (a measure of the distance north or south of the equator, expressed in degrees).

Sunlight Strikes Earth at Various Angles

The amount of sunlight that strikes a given area of Earth's surface has a major effect on the average yearly temperatures of that area. At the equator, sunlight hits Earth's surface nearly at a right angle. At higher latitudes farther north or south, the sun's rays strike Earth's surface at a greater slant. This angle spreads the same amount of sunlight over a larger area, producing lower overall temperatures at higher latitudes than at the equator.

Because Earth is tilted on its axis, the angle of the sunlight at higher latitudes varies as the planet makes its yearly trip around the sun. This variation causes pronounced seasons. For example, when the Northern Hemisphere is tilted toward the sun, it receives sunlight more directly than when it is tilted away from the sun. The season we call "summer" results. Six months later, the Northern Hemisphere is tilted away from the sun and experiencing winter, but the Southern Hemisphere is now tilted toward the sun (**Fig. 30-1**). At the equator, Earth's tilt has only a small effect on the angle of the sun's rays, so there is little seasonal variation there.

Air Currents Produce Regional Climates

Air currents are generated by Earth's rotation and by differences in temperature between different air masses. As the sun's rays fall on the equator, the air there heats and rises, because warm air is less dense than cold air. The warm air

▶ **Figure 30-1 Earth's curvature and tilt produce seasons and climate** Near the equator, sunlight falls nearly perpendicularly to Earth's surface, so its warmth is concentrated on a relatively small area. Closer to the poles, the same amount of sunlight falls over a much larger surface area. Temperatures are thus highest at the equator and lowest at the poles. In addition, the tilt of Earth on its axis causes seasonal variations in the amount and directness of sunlight. The temperatures in each hemisphere are at their lowest when that hemisphere tilts away from the sun. (Arrows indicate the plane of Earth's orbit.) **QUESTION:** Describe the seasons and day length if Earth were not tilted on its axis. Would there still be a temperature gradient from the equator to the poles?

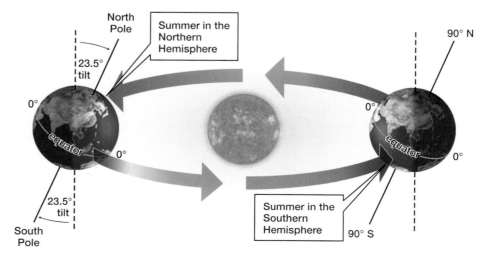

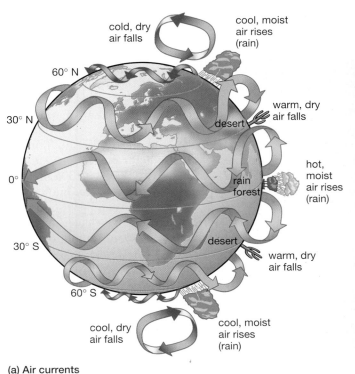

(a) Air currents

(b) Earth from space

near the equator is also laden with water evaporated by solar heat (**Fig. 30-2a**). As the water-saturated air rises, it cools somewhat. Cool air cannot hold as much moisture as can warm air, so water condenses from the rising air and falls as rain. This rainfall, along with the direct rays of the sun, creates a warm, wet zone around the equator, called the *tropics*.

The cooler, dry air that remains after the moisture has fallen from the equatorial air then flows north and south from the equator. Near 30° N and 30° S latitudes, the cooled air becomes dense enough to sink. As it sinks, it is warmed by heat radiated from Earth. By the time it reaches the surface, it is both warm and very dry. Not surprisingly, the major deserts of the world are found at these latitudes (**Fig. 30-2a,b**). This air then flows back toward the equator, completing a circular flow. Farther north and south, this general circulation pattern is repeated, dropping moisture at around 60° N and 60° S and creating extremely dry conditions at the North and South Poles.

Ocean Currents Moderate Nearshore Climates

Ocean currents are driven by Earth's rotation, winds, and the heating of water by the sun. Continents interrupt the currents, breaking them into roughly circular patterns that rotate clockwise in the Northern Hemisphere and counterclockwise in the Southern Hemisphere (**Fig. 30-3**).

Ocean currents influence the climates of coastal areas by transporting water over large distances. For example, a current in the Atlantic Ocean brings warm water (the Gulf Stream) from equatorial regions north along the eastern coast of North America, creating a warmer, moister climate than is found farther inland. It then carries the still-warm water farther north and east, warming the western coast of Europe before returning south.

In general, coastal areas tend to have less variable climates than do areas near the center of continents. Because water both heats and cools more slowly than land or air, ocean currents moderate temperature extremes.

▲ **Figure 30-2 Distribution of air currents and climatic regions** *(a)* Earth's rotation and the distribution of temperatures interact to create air currents that rise and fall predictably with latitude, producing broad climatic regions. *(b)* A photograph of the African continent taken from *Apollo 11*. Along the equator are heavy clouds that drop moisture on the central African rain forests. Note the lack of clouds over the Sahara and Arabian Deserts near 30° N and the South African Desert near 30° S.

▼ **Figure 30-3 Ocean circulation patterns** Currents run in circles that travel clockwise in the Northern Hemisphere and counterclockwise in the Southern Hemisphere, distributing warmth from the equator to northern and southern coastal areas.

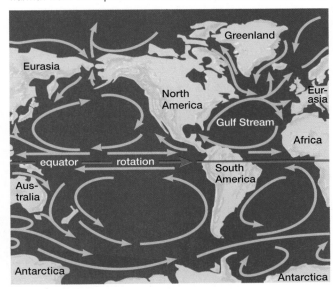

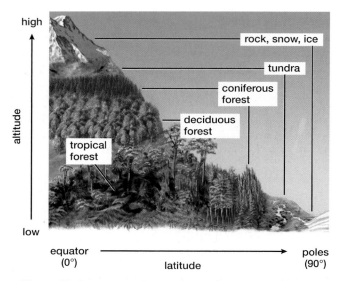

▲ **Figure 30-4 Effects of elevation on temperature** Climbing a mountain in some ways is like going toward one of Earth's poles. Increasingly cool temperatures produce a similar series of habitats in both cases.

Continents and Mountains Complicate Weather and Climate

If Earth's surface were uniform, climate would be determined entirely by latitude, and climate zones would occupy neat bands corresponding to latitude. However, the presence of irregularly shaped continents, which heat and cool more quickly than the oceans that surround them, alters the flow of wind and water and contributes to an irregular distribution of climate.

Variations in elevation within continents further complicate the situation. At higher elevations, the atmosphere becomes less dense and retains less heat. The temperature drops approximately 3.5°F (2°C) for every 1,000 feet (305 meters) of increased elevation. This phenomenon explains why snow-capped mountains are found even in the tropics (**Fig. 30-4**).

Mountains also modify rainfall patterns. When water-laden air meets a mountain and is forced to rise, the air cools. Cooling reduces the air's ability to retain water, and the water condenses as rain or snow on the windward (near) side of the mountain. The cool, now-dry air is warmed again as it travels down the far side of the mountain and absorbs water from the land, creating a local dry area called a **rain shadow.** For example, mountain ranges such as the Sierra Nevada of the western United States wring the moisture from the westerly winds that come off the Pacific Ocean, leaving deserts in the rain shadow on their eastern sides (**Fig. 30-5**).

30.2 What Conditions Does Life Require?

From bare Arctic rocks to steamy tropical rivers to the pressure-cooker conditions of a deep-sea vent, Earth's habitats teem with life. Underlying the diversity of habitats is a common ability to provide four fundamental resources required for life:

- nutrients from which to construct living tissue,

- energy to power that construction,

- liquid water in which metabolic reactions can take place,

- appropriate temperatures.

As we shall see in the following sections, these resources are very unevenly distributed over Earth. Their relative availability determines the types of organisms that live in Earth's various terrestrial and aquatic ecosystems.

The community that is characteristic of each ecosystem contains species that are adapted to particular environmental conditions. The desert community, for example, is dominated by plants with adaptations to hot, dry conditions. These adaptations tend to take similar forms wherever hot, dry conditions prevail. Thus, the cacti of the Mojave Desert of the American Southwest are strikingly similar to the euphorbia of the deserts of Africa, although these plants are only distantly related. Their spinelike leaves and thick, green, water-storing stems are adaptations for water conservation (**Fig. 30-6**). Likewise, the plants of the arctic tundra and those of the alpine tundra of the Rocky Mountains show growth patterns clearly recognizable as adaptations to a cold, dry, windy climate. Thus, in different regions where the environmental conditions are similar, organisms with similar adaptations are organized into similar types of communities.

▼ **Figure 30-5 Mountains create rain shadows**

▶Figure 30-6 **Environmental demands mold physical characteristics** Evolution in response to similar environments has molded the bodies of **(a)** American cacti and **(b)** African euphorbia into nearly identical shapes, although they are only distantly related.

(a) Cactus   **(b) Euphorbia**

30.3 How Is Life on Land Distributed?

The distribution of terrestrial organisms is determined largely by temperature and the availability of water (**Fig. 30-7**). Other resources tend be comparatively abundant. For example, terrestrial ecosystems receive plenty of light, even on an overcast day, and the soil provides ample nutrients. Water, however, is limited and very unevenly distributed, both in place and in time. Terrestrial organisms must be able to obtain water when it is available and to conserve it when it is scarce.

Like water, temperatures favorable to life are very unevenly distributed in place and time. At the South Pole, even in summer, the average temperature is well below freezing. Not surprisingly, life is scarce there. Conversely, the tropics have a uniformly warm climate, and life abounds there. Between these extremes are places that have favorable temperatures during only part of the year.

Rainfall and Temperature Determine the Vegetation an Area Can Support

The effects of temperature and rainfall interact. Temperature strongly influences the effectiveness of rainfall in providing soil moisture for plants and standing

▼Figure 30-7 **Rainfall and temperature influence the distribution of vegetation** Together, rainfall and temperature determine the soil moisture available for plant growth.

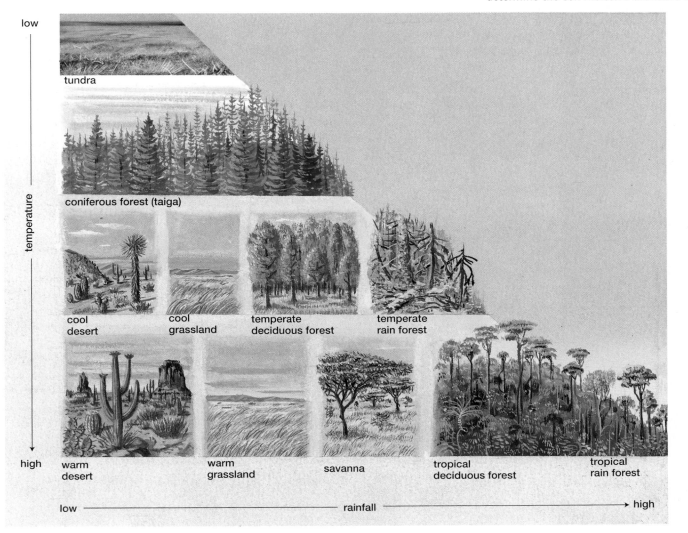

water for animals to drink. For example, the hotter it is, the more rapidly water evaporates, both from the ground and from plants. In winter, water may be abundant but frozen, and thus far less available to support life.

As a result of the interaction between temperature and rainfall and, to a lesser extent, of the distribution of rain throughout the year, different areas that receive almost exactly the same amount of rainfall can have startlingly different vegetation. To illustrate this phenomenon, let's take a trip from southern Arizona to central Alaska, visiting ecosystems that all receive about 12 inches (30 centimeters) of rain annually.

The Sonoran Desert near Tucson, Arizona, has an average annual temperature of 68°F (20°C). The landscape there is dominated by giant saguaro cacti and low-growing, drought-resistant bushes. Going north for 900 miles (1,500 kilometers) brings us into eastern Montana, where rainfall is about the same as in the desert, but the vegetation is mostly grass, largely because the average temperature is much lower, about 45°F (7°C). Much farther north, central Alaska also receives roughly the same 12-inch annual rainfall, but is covered with coniferous forest. Because of the low average annual temperature (about 25°F, or –4°C), permafrost underlies much of the ground there. During the summer thaw, the ground becomes wet and swampy, even though its annual rainfall is no different than that of the Sonoran Desert.

Terrestrial Biomes Have Characteristic Plant Communities

Terrestrial communities are dominated and defined by their plant life. Large land areas with similar environmental conditions and characteristic plant communities are called **biomes** (Fig. 30-8). Biomes are generally named after the major types of vegetation they contain.

▼ **Figure 30-8 The distribution of biomes** Although the distribution pattern of biomes is complex, note the overall consistencies. Tundras and coniferous forests are in the northernmost parts of the Northern Hemisphere, whereas the deserts of Mexico, the Sahara, Saudi Arabia, South Africa, and Australia are located about 20° to 30° N and S.

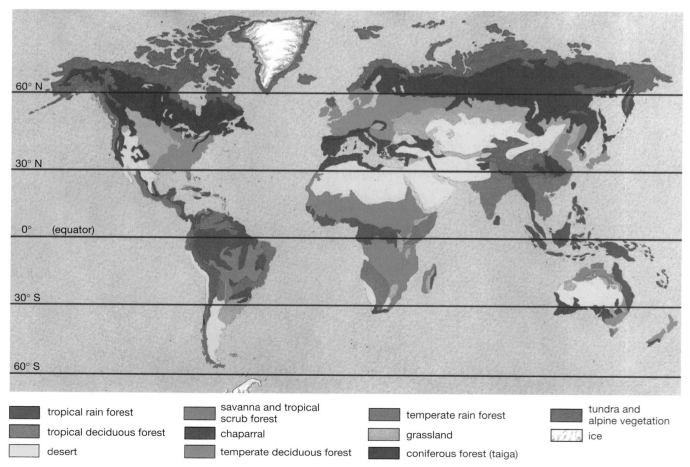

tropical rain forest	savanna and tropical scrub forest
tropical deciduous forest	chaparral
desert	temperate deciduous forest
	temperate rain forest
	grassland
	coniferous forest (taiga)
	tundra and alpine vegetation
	ice

In the following sections we discuss the major biomes, beginning at the equator and working our way poleward. We also discuss some of the effects of human activities on these biomes.

Tropical Rain Forests

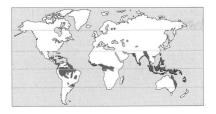

Near the equator, the temperature averages between 77°F and 86°F (25°C and 30°C) with little seasonal variation, and rainfall ranges from 100 to 160 inches (250 to 400 centimeters) annually. These evenly warm, evenly moist conditions create the most diverse biome on Earth, the **tropical rain forest,** dominated by huge broadleaf evergreen trees (**Fig. 30-9**). Large areas of rain forest are found in South America, Africa, and Southeast Asia.

▼ **Figure 30-9 The tropical rain-forest biome** Towering trees draped with vines reach for the light in the dense tropical rain forest. Amid their branches dwells the most diverse assortment of organisms on Earth, including (clockwise from top left) a fruit-eating toucan, a golden-eyed leaf frog, a howler monkey, and a tree-climbing orchid. **QUESTION:** Why does the tropical forest support the highest productivity (see Fig. 29-3) and greatest species diversity on Earth, even though its soils are very poor?

Rain forests have the highest biodiversity of any ecosystem on Earth. (**Biodiversity** refers to the total number of species within an ecosystem and the resulting complexity of interactions among them.) Ecologists estimate that rain forests contain from half to two-thirds of the world's species, even though rain forests cover only 6% of Earth's total land area. For example, a recent survey of a 3-square-mile tract of rain forest in Peru revealed more than 1,300 species of butterflies and 600 species of birds. In comparison, the entire United States is home to only 400 butterfly species and 700 bird species.

Tropical rain forests typically have several layers of vegetation. The tallest trees reach 150 feet (50 meters) and tower above the rest of the forest. Below is a fairly continuous canopy of treetops at about 90 to 120 feet (30 to 40 meters). Another layer of shorter trees typically stands below the canopy. Huge woody vines, commonly 325 feet (100 meters) or more in length, grow up the trees. These layers of vegetation block out most of the sunlight, and many of the plants that live in the dim light that filters through to the forest floor have enormous leaves to gather the small amount of available energy.

Because edible plant material close to the ground in tropical rain forests is relatively scarce, much of the animal life is *arboreal* (living in the trees). Competition for the nutrients that do reach the ground is intense among both animals and plants. Even such unlikely sources of food as the droppings of monkeys are in great demand. When ecologists attempted to collect droppings of the South American howler monkeys to find out what the monkeys had been eating, they found themselves in a race with the dung beetles that feed on and lay their eggs in monkey droppings. Hundreds of beetles would arrive within minutes after a dropping hit the ground.

Almost as soon as decomposer bacteria or fungi release nutrients from dead plants or animals into the soil, rain-forest trees and vines absorb the nutrients. This is one of the reasons why, despite the teeming vegetation, agriculture is risky and destructive in rain-forest soils. Because most of the nutrients in a rain forest are tied up in the vegetation, the soil is very infertile. If the trees are carried away for lumber, few nutrients remain to support crops. Further, even if the nutrients are released by burning the vegetation, the heavy year-round rainfall quickly washes them away, leaving the soil infertile after a few seasons of cultivation. The exposed soil, which is rich in iron and aluminum, then takes on an impenetrable, bricklike quality as it bakes in the tropical sun. As a result, secondary succession on cleared rain-forest land is slow. Even small forest cuttings take about 70 years to regenerate.

Human Impact   Rain forests are being destroyed at a rapid rate, felled for lumber or burned to clear land for ranching or farming. Estimates of current rain-forest destruction range from 22,000 square miles to 52,000 square miles per year. (For comparison, the state of Connecticut occupies about 5,000 square miles.) In many areas, the rain forest left standing consists only of fragments too small to allow reproduction of trees and provide adequate habitat for large animals. Further, as rain forests disappear, rainfall is reduced and the region becomes drier, more stressed, and more susceptible to fire. Rainfall decreases in deforested areas because much of the rain in a rain forest comes from water that evaporates from the forest's leaves.

At least 40% of the world's rain forests are now gone. Although disastrous forest losses continue, some areas have been set aside as protected preserves. Local residents in some places are becoming more involved in conservation efforts. These efforts are steps toward the ultimate solution, which is tragically slow in coming: sustainable use of all forests. Sustainable use means deriving benefits from an ecosystem, whether from tourism or harvesting products, in a way that can be sustained indefinitely without continuing to damage the ecosystem.

Can Coffee Save Songbirds?

Continued

Coffee is not the only tropical delicacy whose production threatens the inhabitants of rain-forest ecosystems. Cacao trees, the source of cocoa and chocolate, are also cultivated in equatorial regions throughout the world. As cacao growers try to increase production to meet increasing worldwide demand for chocolate, vast stretches of rain forest have been cleared to make room for large plantations.

Fortunately, some growers and governments have recognized the value of cultivating cacao under the conditions in which it evolved—beneath a dense rain-forest canopy. This environment provides cacao plants with natural protection from pests and diseases, and also provides habitat for other rain-forest species, some of them critically endangered. You can help support rain-forest cacao production by buying chocolate and cocoa that carry the Rainforest Alliance Certified label.

Tropical Deciduous Forests

Slightly farther from the equator, rainfall is not nearly as constant, and there are pronounced wet and dry seasons. In these areas, which include India, much of Southeast Asia, and parts of South and Central America, **tropical deciduous forests** grow. During the dry season, trees in these forests cannot get enough water from the soil to compensate for evaporation from their leaves. As a result, the plants cope with the dry season by shedding their leaves, thereby minimizing water loss. If the rains fail to return on schedule, the trees delay growth of new leaves until the drought passes.

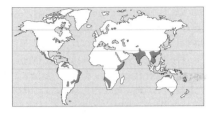

Savannas

Along the edges of tropical deciduous forests, trees gradually become more widely spaced, with grasses growing between them. Eventually, grasses become the dominant vegetation, with only scattered trees and thorny scrub forests here and there. This biome is the **savanna** (Fig. 30-10).

Savannas typically have a rainy season during which virtually all of the year's precipitation falls—12 inches (30 centimeters) or less. When the dry season arrives, it comes with a vengeance. Rain might not fall for months, and the soil becomes hard, dry, and dusty. Grasses are well adapted to this type of climate, growing very rapidly during the rainy season and dying back to drought-resistant

▼ **Figure 30-10 The African savanna** Beneath a rainbow, elephants roam the savanna. Elephants share this biome with other large herbivores, such as the white rhino (top left), seen here with a red oxpecker bird. Herds of grazing animals, such as zebras (bottom left), can still be seen on African preserves. The herds of herbivores provide food for the greatest assortment of large carnivores on Earth, including cheetahs (top right).

▲ **Figure 30-11 Illegal hunting threatens African wildlife**
Rhinoceros horns, believed by some people to have
aphrodisiac properties, fetch staggering prices and encourage
illegal hunting. The black rhino is now nearly extinct.

roots during dry periods. Only a few specialized trees, such as the thorny acacia
and the water-storing baobab, can survive the devastating dry seasons of the savannas (see Fig. 17-14b on p. 327).

The African savanna probably has the most diverse and impressive array of
large mammals on Earth. These mammals include numerous herbivores such as
antelope, wildebeest, water buffalo, elephants, and giraffes, and such carnivores
as the lion, leopard, hyena, and wild dog.

Human Impact   Africa's rapidly expanding human population threatens the
wildlife of the savanna. Illegal hunting has driven the black rhinoceros to the
brink of extinction (**Fig. 30-11**) and endangers the African elephant. The abundant
grasses that make the savanna a suitable habitat for wildlife also make it suitable
for grazing domestic cattle. As the human population of East Africa increases, so
does the population of cattle grazing on the savanna. Fences meant to restrain
the movements of cattle increasingly disrupt the migration of the great herds of
wild herbivores, which move in search of food and water.

Deserts

Even drought-resistant grasses need at least
10 to 20 inches (25 to 50 centimeters) of rain
a year. In areas with less than 10 inches of
annual rainfall, **deserts** are found (**Fig. 30-12**).
Although we tend to think of deserts as hot,
they are defined by lack of rain rather than
by temperature. In the Gobi Desert of Asia, for example, temperatures average
below freezing for half the year, while the average summer temperature may be
as high as 110°F (43°C).

The desert biome is found on every continent, mostly around 20° to 30° north
and south latitude but also in the rain shadows of major mountain ranges. The
desert category encompasses a range of environments. At one extreme are certain parts of the Sahara Desert and deserts in Chile, where it virtually never rains
and no vegetation grows (**Fig. 30-12a**). More commonly, deserts are characterized
by widely spaced vegetation and large areas of bare ground. Plants tend to be
spaced evenly, as if planted by hand (**Fig. 30-12b**).

In many cases, the perennial plants in deserts are bushes or cacti with large,
shallow root systems. The shallow roots quickly soak up the soil moisture after
the infrequent desert storms. The rest of the plant is typically covered with a waterproof, waxy coating to prevent evaporation of precious water. Water is stored
in the thick stems of cacti and other succulents. The spines of cacti are leaves
modified to protect the plant from herbivores and to conserve water, presenting
almost no surface area for evaporation. In many deserts, all of the rain falls in
just a few storms, and specialized annual wildflowers take advantage of the brief
period of moisture to race through germination, growth, flowering, and seed production in a month or less (**Fig. 30-13**).

The animals of deserts, like the plants, are adapted to survive on little water.
Most deserts seem devoid of animal life during summer days because the resident

▼ **Figure 30-12 The desert biome (a)** Under the most
extreme conditions of heat and drought, as in these sand
dunes of the Sahara Desert in Africa, deserts can be almost
devoid of life. **(b)** Throughout much of Utah and Nevada, the
Great Basin Desert presents a landscape of widely spaced
shrubs, such as sagebrush and greasewood. These shrubs
often secrete a growth inhibitor from their roots, preventing
germination of nearby plants and thus reducing competition
for water. **(c)** The kangaroo rat is an elusive inhabitant of the
deserts of North America.

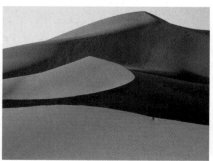

(a) Sahara dunes

(b) Utah desert

(c) A kangaroo rat

animals seek relief from the sun and heat in cool, underground burrows. After dark, when deserts cool down considerably, lizards, snakes, and other reptiles emerge to feed, as do mammals such as the kangaroo rat (**Fig. 30-12c**) and birds such as the burrowing owl. Most of the smaller animals survive without ever drinking, getting all the water they need from their food and from that produced during cellular respiration in their tissues. Larger animals, such as desert bighorn sheep, are dependent on permanent water holes during the driest times of the year.

Human Impact   Desert ecosystems are fragile. For example, the soil of the Mojave Desert in southern California is stabilized and enriched by bacteria whose filaments intertwine among sand grains. This crucial bacterial network is sensitive to disturbances such as those caused by the numerous off-road vehicles that careen about the desert for recreation. The vehicles destroy the bacterial community, allowing the soil to erode and reducing nutrients available to the desert's slow-growing plants. Ecologists estimate that the desert soil may require hundreds of years to fully recover from heavy vehicle use. Treadmarks made in the Mojave when General Patton trained tank crews there in 1940 remain visible today.

Chaparral

In many coastal regions that border on deserts, as in southern California and much of the Mediterranean region, we find a distinctive type of vegetation called **chaparral** (**Fig. 30-14**). Annual rainfall in these regions is as much as 30 inches (75 centimeters),  nearly all of which falls during cool, wet winters that alternate with hot, dry summers. The proximity of the sea lengthens the winter rainy season and causes frequent fog during the dry season, reducing evaporation. Chaparral vegetation consists of small trees or large bushes such as sages and evergreen oak, with thick, waxy or fuzzy leaves that conserve water. These hardy shrubs are also able to withstand the frequent summer fires started by lightning.

Grasslands

Most **grassland,** or **prairie,** biomes are located in the centers of continents, where they receive 10 to 30 inches (25 to 75 centimeters) of rain annually. Particularly large grasslands are found in the centers of the North American and Eurasian continents. In  general, grasslands have a continuous cover of grass and virtually no trees, except along rivers.

From the tallgrass prairies of Iowa, Missouri, and Illinois (**Fig. 30-15**) to the shortgrass prairies of eastern Colorado, Wyoming, and Montana (**Fig. 30-16**), the North American grassland once stretched across almost half the continent. These grasslands, on which grasses have grown and decomposed for thousands of years, contain what may be the most fertile soil in the world.

Water and fire are the key factors in the competition between grasses and trees. In the shortgrass prairies, grass can tolerate the hot, dry summers and frequent droughts, but trees cannot. In the more eastern tallgrass prairies, forests are actually the climax ecosystems but, historically, trees were destroyed by frequent fires, often started by lightning or by Native Americans to maintain grazing land for the bison. Trees are killed outright by fire, but the root systems of grasses usually survive, even if the tops of the grasses are destroyed.

▶ **Figure 30-14** **The chaparral biome** This biome is limited primarily to coastal mountains in dry regions, such as the San Gabriel Mountains in southern California. Chaparral is maintained by frequent fires set by summer lightning. Although the tops of the plants may be burned off, the roots send up new sprouts the following spring.

▲ **Figure 30-13** **The Sonoran Desert** In spring, this Arizona desert is carpeted with annual wildflowers. Through much of the year, the wildflower seeds lie dormant, waiting for the spring rains to fall.

◄Figure 30-15 **Tallgrass prairie** In the central United States, moisture-bearing winds out of the Gulf of Mexico produce summer rains, allowing a lush growth of tall grasses and wildflowers. Periodic fires, now carefully managed, prevent encroachment of forest. **QUESTION:** Why is tallgrass prairie one of Earth's most endangered biomes?

The grasslands of North America once supported huge herds of bison—as many as 60 million in the early nineteenth century. Pronghorn antelope can still be seen in some prairies of the western United States. Bobcats and coyotes are the major large predators there (a prairie food web is illustrated in Fig. 29-5 on p. 579).

Human Impact   When people developed plows that could break through dense grass turf, the stage was set for converting the prairies of the midwestern United States into the "breadbasket" of North America, so named because enormous quantities of grain are cultivated in its fertile soil. The tallgrass prairie has been converted to agricultural land, except for tiny protected remnants that are maintained by periodic controlled burning.

On the dry western shortgrass prairie, cattle have replaced the bison and pronghorn antelope. As a result of cattle overgrazing the prairie grasses, the

▼Figure 30-16 **Shortgrass prairie** The lands east of the Rocky Mountains receive relatively little rainfall and support shortgrass prairie, which is characterized by low-growing bunch grasses such as buffalo grass and grama grass. Among the organisms that inhabit this biome are (clockwise from top left) pronghorn antelope, prairie dogs, coneflowers, and bison.

boundary between the cool deserts and the grassland has been altered in favor of desert plants. Much of the sagebrush desert of the western United States is actually overgrazed shortgrass prairie (**Fig. 30-17**). Cattle prefer grass to sagebrush, so heavy grazing destroys the grass. Consequently, moisture that the grass would have absorbed is left in the soil, encouraging the growth of the woody sagebrush. Ultimately, prairie grasses are replaced by plants characteristic of the cool desert.

Temperate Deciduous Forests

At their eastern edge, the North American grasslands merge into the **temperate deciduous forest** biome, which is also found in Western Europe and East Asia (**Fig. 30-18**). Precipitation in this biome is higher than in grasslands (30 to 60 inches, or 75 to 150 centimeters). In particular, more rain falls during the summer. The soil retains enough moisture for trees to grow, and the resulting forest shades out grasses.

In contrast to tropical forests, the temperate deciduous forest biome has cold winters, usually with at least several hard frosts and often with long periods of below-freezing weather. Winter in this biome has an effect on trees similar to that of the dry season in the tropical deciduous forests: During periods of sub-freezing temperatures, liquid water is not available to the trees. To reduce evaporation when water is in short supply, the trees drop their leaves in the fall. They produce leaves again in the spring, when liquid water becomes available. During the brief time in spring when the ground has thawed but the leaves of trees have not yet blocked off all the sunlight, abundant wildflowers grace the forest floor.

▲ **Figure 30-17 Sagebrush desert or shortgrass prairie?** The shortgrass prairie field on the right has been overgrazed by cattle, causing the grasses to be replaced by sagebrush.

▼ **Figure 30-18 The temperate deciduous forest biome** In temperate deciduous forests of the eastern United States, the white-tailed deer (left) is the largest herbivore, and birds such as the blue jay (top right) are abundant. In spring, a profusion of woodland wildflowers (such as hepaticas, bottom right) blooms briefly before the trees produce leaves.

Insects and other arthropods are numerous and conspicuous in deciduous forests. The decaying leaf litter on the forest floor also provides food and habitat for bacteria, earthworms, fungi, and small plants. Many arthropods feed on these or on each other. Vertebrate animals, including mice, shrews, squirrels, raccoons, deer, bear, and many species of birds, dwell in the deciduous forests.

Human Impact Large predatory mammals such as black bear, wolves, bobcats, and mountain lions were formerly common in deciduous forests, but hunting and habitat loss have eliminated the wolves and severely reduced the populations of the others. In many deciduous forests, deer are abundant because few of their predators remain. Clearing for lumber, agriculture, and housing has dramatically reduced deciduous forests in the United States from their original extent, and virgin deciduous forests are now almost nonexistent. During the past 50 years, however, forest cover in the United States (both evergreen and deciduous) has increased due to regrowth of forests on abandoned farms, paper recycling that decreases demand for wood pulp, more efficient lumber milling and tree farming techniques, and the use of alternative building materials.

Temperate Rain Forests

On the U.S. Pacific Coast, from the lowlands of the Olympic Peninsula in Washington State to southeastern Alaska, lies the **temperate rain-forest** biome (**Fig. 30-19**). Temperate rain forests, which are relatively rare, are also located along the southeastern coast of Australia, the southwestern coast of New Zealand, and the southernmost portion of Chile and Argentina. As in tropical rain forests, there is ample liquid water year-round. Water is abundant for two reasons. First, there is a tremendous amount of rain. The Hoh River rain forest in Olympic National Park receives more than 160 inches (400 centimeters) of rain annually, in-

▼**Figure 30-19 The temperate rain-forest biome** In a temperate rain forest, ferns, mosses, and wildflowers grow in the pale green light of the forest floor. The dead feed the living, as new trees grow from the decay of a fallen giant (top right), called a "nurse log," and flowering foxglove (bottom left) and fungi (bottom right) find ideal conditions amid the moist, decaying vegetation.

cluding more than 24 inches (60 centimeters) in the month of December alone. Second, the moderating influence of the Pacific Ocean keeps temperatures mild along the coast, so the ground seldom freezes and liquid water remains available.

Because of the abundance of water, trees in a temperate rain forest have no need to shed their leaves in the fall, and almost all of the trees are evergreens. In contrast to the broadleaf evergreen trees of the tropics, temperate rain forests are dominated by conifers. The ground and the trunks of the trees are covered with mosses and ferns. As in tropical rain forests, so little light reaches the forest floor that tree seedlings usually cannot become established. When one of the forest giants falls, however, it opens up a patch of light, and new tree seedlings quickly sprout, commonly right on top of the fallen log. This event produces a "nurse log" (see Fig. 30-19).

Taiga

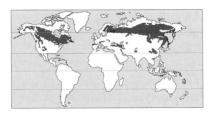

North of the grasslands and temperate forests, the **taiga,** also called the northern coniferous forest (**Fig. 30-20**), stretches across all of North America and Eurasia, including parts of the northern United States and much of southern Canada. Conditions in the taiga are harsher than those in the temperate deciduous forest. In the taiga, the winters are longer and colder, and the growing season is shorter. The few months of warm weather are too short to allow trees the luxury of regrowing leaves in the spring. As a result, the trees of the taiga are almost all evergreen conifers with narrow, waxy needles. The waxy coating and small surface area of the needles reduce water loss by evaporation during the cold months, and the leaves remain on the trees year-round. Thus, the trees are instantly ready to take advantage of good growing conditions when spring arrives, and they can continue slow growth late into the fall.

▼ **Figure 30-20 The taiga (or northern coniferous forest) biome** The small needles and pyramidal shape of conifers allow them to shed heavy snows. Winter is a challenge not only for the trees but also for animals such as the snowshoe hare and the bobcat that preys on it (bottom left). The hare is also prey for the great horned owl (top right).

▲ **Figure 30-21 Clear-cutting** The bare mountains of this Oregon forest have been clear-cut. Clear-cutting is relatively simple and cheap—but its environmental costs are high. Erosion will diminish the fertility of the soil, slowing new growth. Further, the dense stands of same-age trees that typically regrow are more vulnerable to attack by parasites than a natural stand of trees of various ages would be.

Because of the harsh climate in the taiga, the diversity of life there is lower than in many other biomes. Vast stretches of central Alaska, for example, are covered by a somber forest that consists almost exclusively of black spruce and an occasional birch. Large mammals such as the wood bison, grizzly bear, moose, and wolf, which have mostly been eradicated in the southern regions of their original range, still roam the taiga, as do smaller animals such as the wolverine, fox, snowshoe hare, and deer.

Human Impact   The taiga is a major source of lumber for construction. *Clear-cutting*, the removal of all the trees in a given area, has destroyed huge expanses of forest, in both Canada and the Pacific Northwest in the United States (**Fig. 30-21**). Owing to the remoteness of the northernmost taiga and the severity of its climate, a greater percentage of the taiga remains undisturbed than does any other North American biome except the tundra.

Tundra

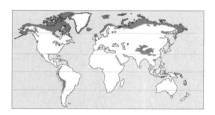

The last biome we encounter before reaching the polar ice cap is the arctic **tundra,** a vast treeless region bordering the Arctic Ocean (**Fig. 30-22**). Conditions in the tundra are severe. Winter temperatures often reach –40°F (–55°C) or below, winds howl at 30 to 60 miles (50 to 100 kilometers) per hour, and precipitation averages 10 inches (25 centimeters) or less each year, making the tundra a "freezing desert." Even during the summer, the temperature can drop to freezing, and the growing season may last only a few weeks before a hard frost occurs. Somewhat less cold but similar conditions produce alpine tundra on mountaintops above the elevation where trees can grow.

The cold climate of the arctic tundra results in **permafrost,** a permanently frozen layer of soil that lies no more than about 1.5 feet (45 centimeters) below the surface. As a result, root growth is limited to the topmost layer of soil, and trees cannot survive in the tundra. In summer, the water from melted snow and ice cannot soak into the ground, and the tundra becomes a huge marsh.

Nevertheless, the tundra supports a surprising abundance and variety of life. The ground is carpeted with small perennial plants and dwarf willows no more

▼ **Figure 30-22 The tundra biome** Life on the tundra is adapted to cold. Plants such as dwarf willows and perennial wildflowers such as dwarf clover (top left) grow low to the ground, escaping the chilling tundra wind. Tundra animals, such as caribou (bottom right) and arctic foxes (top right), can regulate blood flow in their legs, keeping them just warm enough to prevent frostbite, while preserving precious body heat for the brain and vital organs.

than a few inches tall and is often covered with a large lichen called "reindeer moss," a favorite food of caribou. The standing water provides a superb habitat for mosquitoes. The mosquitoes and other insects provide food for numerous birds, most of which migrate long distances to nest and raise their young during the brief, insect-filled summer. The tundra vegetation supports lemmings, which are eaten by wolves, snowy owls, arctic foxes, and even grizzly bears.

Human Impact   Tundra is among the most fragile of all the biomes because its short growing season allows only very slow recovery from damage or disturbance. It can take 50 years for a willow to reach a height of 4 inches (10 centimeters). Human activities in the tundra can leave scars that persist for centuries. Fortunately for the tundra inhabitants, the impact of civilization is localized around oil drilling sites, pipelines, mines, and military bases.

30.4 How Is Life in Water Distributed?

Saltwater oceans and seas are the largest ecosystems on Earth, covering about 71% of the planet's surface. Freshwater ecosystems, in contrast, cover less than 1%.

The special properties of water lend some common features to aquatic ecosystems. First, because water is slower to heat and cool than air, temperatures in aquatic ecosystems are generally more moderate than are those in terrestrial ecosystems. Second, although water may seem transparent, it absorbs a considerable amount of the light energy that sustains life. Even in the clearest water, the intensity of light decreases rapidly with depth. At depths of 650 feet (200 meters) or more, little light is left to power photosynthesis. If the water is at all cloudy—for example, because of suspended sediment or microorganisms—the depth to which light can penetrate is greatly reduced.

Of the four main requirements for life, aquatic ecosystems provide abundant water and appropriate temperatures. Thus, the quantity and type of life in aquatic ecosystems are determined by the remaining two factors: energy and nutrients. Nutrients in aquatic ecosystems tend to be concentrated near the bottom sediments, where light levels are often too low to support photosynthesis. This separation of energy and nutrients limits aquatic life.

There are many kinds of aquatic ecosystems. Freshwater ecosystems encompass rivers, streams, ponds, lakes, and marshes. Marine (saltwater) ecosystems include estuaries, tide pools, coral reefs, the open ocean, and hydrothermal vents. In the following sections, we look more closely at some of these aquatic ecosystems.

Freshwater Lakes Have Distinct Regions of Life

Freshwater lakes vary tremendously in size, depth, and nutrient content. Although each lake is unique, medium-sized to large lakes in temperate climates share some common features, including distinct zones of life.

Life Zones Are Determined by Light and Nutrients

The distribution of life in lakes depends largely on access to light, nutrients, and, in some cases, a place for attachment (the bottom). The life zones of a lake correspond to specific locations within it. We recognize three such zones: the littoral zone, the limnetic zone, and the profundal zone (**Fig. 30-23**).

Near the shore is the **littoral zone.** In this zone, the water is shallow, so plants have access to anchorage, abundant light, and nutrients from the bottom sediments. Not surprisingly, the littoral zone holds a lake's most diverse communities. Cattails and bulrushes abound in the area closest to shore; water lilies and entirely submerged plants and algae may flourish in the deepest part of the littoral zone.

The plants of the littoral zone trap sediments carried in by streams and runoff from the surrounding land, increasing the nutrient content in the zone. Living among the anchored plants are microscopic organisms called plankton. There are

▶ **Figure 30-23 Life zones of a lake** There are three life zones in a typical lake: a nearshore littoral zone with rooted plants, an open-water limnetic zone, and a deep, dark profundal zone.

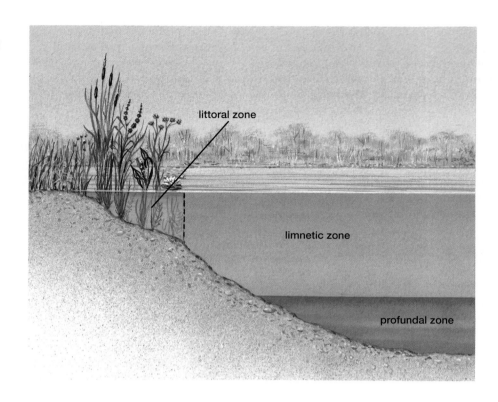

two categories of plankton: autotrophic, photosynthetic algae and bacteria called **phytoplankton;** and heterotrophic, nonphotosynthetic protists and tiny animals collectively known as **zooplankton.** Larger heterotrophic organisms in the littoral zone include insect larvae, crustaceans, snails, flatworms, frogs, minnows, snakes, and turtles.

Farther from shore, the water deepens. This open-water area is divided into two zones: the limnetic zone closer to the surface and the profundal zone below (see Fig. 30-23). In the **limnetic zone,** enough light penetrates to support photosynthesis. Here, drifting phytoplankton are the producers. They are eaten by zooplankton and small crustaceans, which are in turn consumed by fish.

In the lower **profundal zone,** there is not enough light to support photosynthesis, so plants and phytoplankton are unable to survive here. Organisms in the profundal are nourished mainly by detritus that falls from the littoral and limnetic zones and by incoming sediment. The profundal zone is inhabited by decomposers and detritus feeders, such as bacteria, snails, and insect larvae, and by fish that swim freely among the different zones.

Freshwater Lakes Are Classified by Nutrient Content

Although each lake is unique, freshwater lakes can be classified on the basis of their nutrient content as either oligotrophic or eutrophic.

Oligotrophic lakes are very low in nutrients. Many were formed by glaciers that scrape depressions in bare rock, and they are fed by mountain streams that carry little sediment. Because there is little sediment or microscopic life to cloud the water, oligotrophic lakes are clear, and light penetrates deeply. Therefore, photosynthesis is possible in comparatively deep water, and the limnetic zone may extend to the bottom of the lake. Because oxygen is a by-product of photosynthesis, oxygen-rich water also extends deeper in oligotrophic lakes, creating good conditions for oxygen-loving fish, such as trout.

Compared with oligotrophic lakes, **eutrophic lakes** receive larger inputs of sediments, organic material, and inorganic nutrients and can therefore support denser communities. Their water is murkier, because of suspended sediment and dense phytoplankton populations. As a result, the lighted limnetic zone is shallower.

Dense "blooms" of algae occur seasonally in the limnetic zone of a eutrophic lake. When these algae die, their bodies fall into the profundal zone, where they are consumed by decomposers. The metabolic activities of the decomposers use oxygen, reducing the oxygen content of the profundal zone.

Although very large lakes may persist for millions of years, lakes are transient ecosystems. Over time, lakes gradually fill with sediment and undergo succession to dry land (see pp. 569–570). As nutrient-rich sediment accumulates during this transition, oligotrophic lakes tend to become eutrophic, a process called *eutrophication*.

Human Impact   Human activities can greatly accelerate the process of eutrophication, because nutrients are carried into lakes from farms, feedlots, sewage, and even fertilized suburban lawns. Over-enriched lakes become clogged with microorganisms whose dead bodies are consumed by bacteria whose respiration depletes the water of oxygen. Normal community interactions are disrupted as organisms in higher trophic levels are smothered.

Lakes are the ultimate destination of many long-persisting toxic substances produced on land by humans. To take just one example, fish in the Great Lakes carry high levels of toxiphene, a pesticide used heavily in the past but banned in 1982 after the discovery that it could cause birth defects and cancer. Scientists hypothesize that air currents carry toxiphene to the Great Lakes from fields in the South, where it was heavily sprayed on cotton more than 30 years ago.

Marine Ecosystems Cover Much of Earth

In the oceans, the upper layer of water to a depth of about 650 feet (200 meters), where the light is strong enough to support photosynthesis, is called the **photic zone.** Below the photic zone lies the **aphotic zone,** where the only sources of energy are the wastes and bodies of organisms that live there or that sink down from the photic zone (**Fig. 30-24**).

As in lakes, most of the nutrients in the oceans are at or near the bottom, where there is not enough light for photosynthesis. Nutrients dissolved in the water of the photic zone are quickly incorporated into the bodies of living organisms. When these organisms die, some sink into the aphotic zone, providing the organisms there with energy and nutrients. This constant drain of nutrients from the photic zone would eventually end life there, if no additional nutrients entered the zone.

Fortunately, there are two sources of nutrients to the photic zone: the land, from which rivers constantly remove nutrients and carry them to the oceans, and **upwelling,** an upward flow that brings cold, nutrient-laden water from the ocean depths to the surface. Upwelling occurs along western coastlines, as in California, Peru, and West Africa, where prevailing winds displace surface water, causing it to be replaced by water from below. Upwelling also occurs around Antarctica. Not surprisingly, the major concentrations of life in the oceans are found where abundant light is combined with a source of nutrients, a combination most commonly found in regions of upwelling and in shallow coastal waters.

Coastal Waters

The greatest abundance of life in the oceans is found in a narrow zone surrounding Earth's landmasses, where the water is shallow and a steady flow of nutrients washes off the land. Coastal waters include two main types of habitat. The **intertidal zone** is the area that is alternately covered

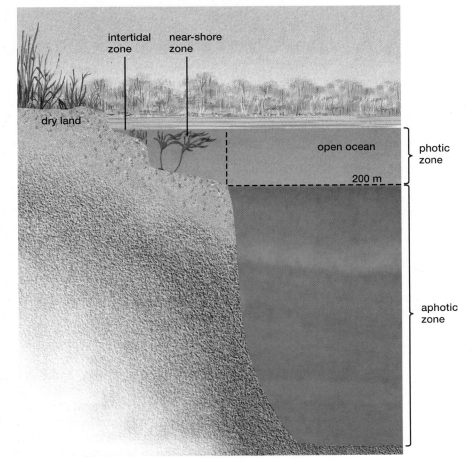

▼ **Figure 30-24 Ocean life zones** There is sufficient light for photosynthesis only in the photic zone, which includes the intertidal and nearshore zones and the upper waters of the open ocean. Life in the aphotic zone relies mainly on energy-rich material that drifts down from the photic zone.

(a) Salt marsh

(b) Beach

(c) Rocky intertidal zone

(d) Kelp forest

▲ **Figure 30-25 Nearshore ecosystems** *(a)* A salt marsh in the eastern United States. Expanses of shallow water fringed by marsh grass *(Spartina)* provide excellent habitat and breeding grounds for many marine organisms and shorebirds. *(b)* Although shifting sands in the intertidal zone present a challenge to life, grasses stabilize them, and animals such as (inset) this *Emerita* crab burrow in them. *(c)* A rocky intertidal shore in Oregon, where animals and algae grip the rock against the pounding waves and resist drying during low tide. (Inset) Colorful sea stars cling to the rocks, surrounded by a seaweed, *Fucus*. *(d)* Towering kelp sway through the clear water off southern California, providing the basis for a diverse community of invertebrates, fishes, and (inset) an occasional sea otter. **QUESTION:** Why do nearshore ecosystems have higher productivity than other ocean biomes? Of the ecosystems pictured here, which do you predict has the highest productivity? Why?

and uncovered by water with the rising and falling of the tides (see Fig. 30-24). To the seaward side of the intertidal zone is the **nearshore zone,** which includes relatively shallow but constantly submerged areas, including **estuaries,** which are bodies of water that form where rivers meet oceans. The nearshore zone also includes coastal wetlands such as salt marshes, which are grasslands that are periodically flooded by tides (**Fig. 30-25**).

The nearshore zone is the only part of the ocean where plants and seaweeds can grow anchored to the bottom. In addition, the abundance of nutrients and sunlight in this zone promotes the growth of photosynthetic phytoplankton. Associated with the plants and protists are animals from nearly every phylum: annelid worms, sea anemones, jellyfish, sea urchins, sea stars, mussels, snails, fish, and sea otters, to name just a few.

A large number and variety of organisms live permanently in coastal waters, but, in addition, many that spend most of their lives in the open ocean move into coastal waters to reproduce. Bays, salt marshes, and estuaries in particular are breeding grounds for crabs, shrimp, and fish, including most commercially important species.

Human Impact Coastal regions are of great importance not only to the organisms that live or breed there but also to humans, who use them for food sources, recreation, mineral and oil extraction, and places to live. As the human population has increased in coastal areas, the conflict between preservation and development of coastal wetlands has become increasingly intense. Estuaries are also threatened, especially by runoff from farming operations. For example, pig farms often collect the animals' wastes in large holding ponds, and when leaks or floods transport the waste material into rivers and estuaries, it can be fatal to the organisms of estuarine communities. Much of the life of the oceans depends on the well-being of coastal waters, so it is essential to protect these fragile, vital areas.

Coral Reefs

In warm, tropical waters with just the right combination of bottom depth, wave action, and nutrients, corals (types of cnidarians) build reefs from their own calcium carbonate skeletons. **Coral reefs** are most abundant in tropical waters of the Pacific and Indian Oceans, the Caribbean, and the Gulf of Mexico as far north as southern Florida, where the average water temperatures range between 72°F and 82°F (22°C and 28°C).

Reef-building corals thrive in the photic zone at depths of less than 130 feet (40 meters), where light penetrates the clear water and provides energy for photosynthesis. The corals require light because they harbor photosynthetic unicellular algae called *dinoflagellates* within their tissues. The relationship between the corals and the algae is mutualistic. The algae benefit from the high nitrogen, phosphorus, and carbon dioxide levels in coral tissues and, in return, provide food for the coral and help produce calcium carbonate, which forms the coral skeleton. The algae represent up to half the weight of the coral polyp and give the corals their diverse, brilliant colors (**Fig. 30-26**).

The skeletons of corals accumulate over thousands of years, providing anchoring sites for an array of algae species and supplying shelter and food for a diverse collection of invertebrates and fish. In some ways, coral reefs might be considered the marine equivalent of rain forests, as reefs are home to more than 93,000 known species, with probably 10 times that number yet to be identified. The

▼ **Figure 30-26 Coral reefs** Coral reefs, composed of the bodies of corals and algae, provide habitat for an extremely diverse community of extravagantly colored animals. Many fish, including blue tang (top right), feed on coral (note the bright yellow corals in background). A vast array of invertebrates such as sponges (bottom left) and blue-ringed octopus (bottom right) live among the corals of Australia's Great Barrier Reef. **QUESTION:** Why does "bleaching" threaten the health of a coral reef? What causes bleaching?

Great Barrier Reef in Australia supports more than 200 species of coral alone, and a single reef may harbor 3,000 species of fish, invertebrates, and algae.

Human Impact Coral reefs are extremely sensitive to disturbance. Anything that diminishes the clarity of the water harms the coral's photosynthetic partners and hinders coral growth. When sewage and agricultural runoff pollute coastal waters, eutrophication reduces both sunlight and oxygen. When people farm, log, and develop coastal land, erosion carries silt into the water. In addition to clouding the water, silt can settle directly onto the reef, smothering corals. In the Philippines and Indonesia, rain-forest logging has dramatically increased erosion. Destroying the rain forests is also destroying the local coral reefs.

Overfishing also threatens reef communities. In many tropical countries, mollusks, turtles, fish, crustaceans, and the corals themselves are being harvested from reefs faster than they can reproduce. In some areas, fish are harvested with dynamite, which kills indiscriminately and damages the reef. Most of the harvest is for food, but some is for sale to shell enthusiasts and aquarium owners in developed countries. Some collectors for the aquarium trade use poison to stun the fish before collecting them, leaving most dead. Removing fish and invertebrates from reefs may disrupt the ecological balance of the community, allowing populations of coral-eating sea urchins or sea stars to explode.

Throughout the tropics, researchers have observed a disturbing trend: Coral reefs are getting sick. Previously rare diseases are causing massive mortality of corals and their communities throughout the world. Researchers studying this trend suggest that the declining health of coral reefs stems from complex causes, including global warming and a variety of other human disturbances.

Bleaching is another symptom of poor health observed in corals worldwide. When waters become too warm, the corals expel their colorful symbiotic algae, leaving the corals a deathly white. Without their algal partners, the corals will eventually starve. Continued global warming poses a severe threat to the world's coral reefs.

The Open Ocean

Beyond the coastal regions lie vast areas of ocean. Most life in the open ocean (**Fig. 30-27**) is limited to the photic zone, where organisms are free swimming or floating for their entire lives. The food web of the open ocean depends on phytoplankton consisting of microscopic photosynthetic protists, mainly diatoms and dinoflagellates. These organisms are consumed by zooplankton, such as tiny crustaceans that are relatives of crabs and lobsters. Zooplankton in turn serve as food for larger invertebrates, small fish, and even marine mammals such as the humpback whale.

One challenge faced by inhabitants of the open ocean is to remain afloat in the photic zone, where sunlight and food are abundant. Many members of the plankton have elaborate flotation devices, such as oil droplets in their cells or long projections, to slow their rate of sinking. Most fish have a swim bladder that can be filled with gas to regulate their buoyancy. Some animals, and even some of the phytoplankton, actively swim to stay in the photic zone. Many small crustaceans migrate to the surface at night to feed, then sink into the dark depths during daylight, avoiding visual predators such as fish.

The amount of *pelagic* (open-ocean) life varies tremendously from place to place. The blue clarity of tropical waters is a result of a lack of nutrients, which limits the concentration of plankton in the water. Nutrient-rich waters that support a large plankton community are greenish and relatively murky.

Below the photic zone, the only available energy in most regions comes from the excrement and dead bodies that drift down from above. Nevertheless, a surprising quantity and variety of life exists in the aphotic zone, including fish with bizarre shapes, worms, sea cucumbers, sea stars, and mollusks.

Human Impact Two major threats to the open ocean are pollution and overfishing. Open-ocean pollution takes several forms. Oceangoing vessels such as

▲ **Figure 30-27 The open ocean** The open ocean supports abundant life in the photic zone, where light is available. Porpoises (top left), fish such as the blue jack (top right), and rare humpback whales (bottom left) swim in ocean waters. The photosynthetic phytoplankton (bottom center) are the producers on which most other marine life ultimately depends. Phytoplankton are eaten by zooplankton, which include microscopic crustaceans (bottom right).

cruise ships dump millions of plastic containers overboard daily, and plastic six-pack holders, foam cups, and packing material wash and blow off the land, collecting on parts of the ocean surface. The plastic looks like food to unsuspecting sea turtles, gulls, porpoises, seals, and whales, many of which die after trying to consume it. Until 1992, New York City placed its refuse and sewage sludge on barges and towed them out to sea, creating a heavily contaminated area covering 40 square miles of open-ocean floor. The open ocean has also served as a dumping ground for radioactive wastes. Oil contaminates the open ocean from many sources, including oil tanker spills, runoff from improper disposal on land, leakage from offshore oil wells, and natural seepage. In the Gulf of Mexico, the Mississippi River deposits nutrient-laden sediment from agricultural runoff onto the seafloor, creating an expanding "dead zone" where oxygen is depleted and the normal marine community has been eliminated. The dead zone now covers roughly 7,000 square miles during the warmer months of each year.

Increased demand for fish to feed a growing human population has depleted many commercially exploited fish populations. The rate of depletion has been accelerated by increasingly efficient fish-finding technologies including satellite technology, radar, and sonar. Most of the world's fisheries are harvesting fewer fish despite the advanced technology—evidence that fish are being taken in unsustainable numbers. Many governments heavily subsidize their fishing industries, a practice that promotes continued harvest of a diminishing resource. The cod fishery of the northeastern United States and eastern Canada has collapsed from overfishing. Natural populations of lobsters, salmon, haddock, king crab, swordfish, and many other types of seafood have also declined dramatically because of overfishing.

Can Coffee Save Songbirds?

Continued

Just as conservationists hope that consumer demand for shade-grown coffee and cocoa will help preserve the rain forest's endangered plants and animals, they also believe that informed seafood consumers can help lessen our impact on dwindling ocean fish populations. Among the fish species that we consume, not all are equally threatened; some populations have been decimated, but others are relatively healthy. Similarly, some species are harvested by ecologically destructive methods, while others are harvested by more benign techniques. Conservation-minded seafood consumers will certainly want to purchase fish of species that are in comparatively good shape, and avoid species whose populations are depleted.

How can you tell which species to favor? Conservation organizations, working closely with fisheries biologists, have done the research to help you decide. For example, the National Audubon Society encourages you to enjoy Pacific halibut, wild Alaskan salmon, sardines, and tilapia, but to avoid Chilean sea bass, cod, shark, and bluefin tuna. The Audubon Society provides its complete list in a handy, pocket-sized format that you can carry along to restaurants and stores. (To find it, do a web search for "Audubon seafood wallet card.")

Hydrothermal Vents

In 1977 a new and unusual source of nutrients, forming the basis of a spectacular undersea community, was discovered in the deep ocean. Geologists exploring the Galápagos Rift (an area of the Pacific floor where plates that form Earth's crust are separating) found vents spewing superheated water, black with sulfur and minerals. Surrounding these vents was a rich community of pink fish, blind white crabs, enormous mussels, giant white clams, sea anemones, and giant tube worms (**Fig. 30-28**). Hundreds of previously unknown species have been found in these **hydrothermal vent communities.** Scientists have now identified vent communities in many deep-sea areas where plates are separating and material from Earth's interior is spewing forth to form new crust.

In this ecosystem, sulfur bacteria serve as the primary producers. They harvest energy from an unlikely source that is deadly to most other forms of life—hydrogen sulfide released from cracks in Earth's crust. This process, called *chemosynthesis*, replaces photosynthesis in these vent communities, which flourish more than a mile below the ocean surface. Bacteria and archaea proliferate in the hot water surrounding the vents, covering nearby rocks with thick, matlike colonies. These colonies provide the food on which the animals of the vent community thrive. Many vent animals consume the microorganisms directly. Others, such as the giant tube worm (which lacks a digestive system), harbor the bacteria within special organs in their bodies. In this mutualistic association, the bacteria provide high-energy carbon compounds and the tube worm provides hydrogen sulfide. The worm, which can reach a length of 12 feet, derives its red color from a unique form of hemoglobin that transports hydrogen sulfide, rather than oxygen, to the symbiotic bacteria.

The bacteria and archaea that inhabit the vent communities hold the record for survival at high temperatures. One species stops reproducing at water temperatures below 194°F (90°C) because it gets too cold, and another can survive water temperatures of 248°F (106°C; water at this depth can reach temperatures much higher than boiling because of the tremendous pressure). Scientists are investigating how the enzymes and other proteins of these heat-loving microbes can function at temperatures that would destroy the proteins in our bodies.

The world still holds wonders and mysteries for those who seek them. We have only begun to explore the versatility and diversity of life on Earth.

▶ **Figure 30-28 Hydrothermal vent communities** Located in the ocean depths, vent communities include giant tube worms nearly 12 feet long. Parts of these worms are red with hemoglobin. They have no digestive tract but are host to sulfur bacteria, which provide the worms with energy by oxidizing hydrogen sulfide.

Can Coffee Save Songbirds? Revisited

Virtually every ecological study of coffee plantations has found evidence that shade coffee plantations are far superior to sun coffee plantations in terms of the abundance and diversity of organisms present. Nonetheless, not all conservationists are persuaded that an all-out effort to encourage the purchase of shade coffee is a good idea.

The biggest concern of those who urge caution is that farmers persuaded to grow shade coffee will do so not by converting sun coffee plantings but by converting undisturbed forest. Even though shade coffee plantings contain nearly as many species as forests, the identities of the species are not necessarily the same. For example, a comparison in Costa Rica of bird species present in shade coffee plantations and intact forest found that many of the species present in forests were missing from shade coffee plantations. Many species adapted for survival in the forest could not survive in a coffee plantation, no matter how shady. Only intact ecosystems fully protect the communities that inhabit them, so conversion of forest to shade coffee would reduce, rather than protect, biodiversity.

To ensure that promotion of shade coffee truly benefits tropical ecosystems, some conservation organizations have instituted systems to certify growers. Rigorous certification programs identify coffee that has been grown under a diverse tree canopy in plantings that were not carved out of previously intact forest. So if you decide to switch to shade coffee, stick to certified brands. Certified shade coffee will bear either the Rainforest Alliance's "Eco-OK" logo or the Smithsonian Migratory Bird Center's "Bird Friendly" logo.

Consider This

Why do conservation organizations devote attention to promoting certain agricultural practices instead of focusing all of their resources on preserving undisturbed habitat?

Chapter Review

Summary of Key Concepts

For additional study help and activities, go to www.mybiology.com.

30.1 What Factors Influence Earth's Climate?

Climate is determined by the availability and distribution of sunlight and water throughout the year. Each unit of solar energy is spread over a smaller area at the equator than farther north and south, making the equator relatively warm, whereas higher latitudes have lower overall temperatures. Earth's tilt on its axis causes dramatic seasonal variation at northern and southern latitudes.

Circular air currents consistently carry moist or dry air to particular latitudes, producing areas of low and high precipitation. These broad rainfall patterns are modified locally by the topography of continents and by ocean currents.

Web Animation Tropical Atmospheric Circulation and World Biomes

30.2 What Conditions Does Life Require?

The requirements for life include nutrients, energy, liquid water, and a suitable range of temperatures. The distribution (in time and space) and availability of these factors determine the type and abundance of living things in different geographical regions.

30.3 How Is Life on Land Distributed?

On land, the crucial limiting factors are temperature and liquid water. A biome comprises regions that have similar climates and vegetation owing to the interaction of temperature and rainfall or the availability of water.

Tropical forest biomes, located near the equator, vary in the amount of rainfall they receive. Tropical deciduous forests occur in drier climates and contain deciduous trees that lose their leaves seasonally, typically in the driest months. Tropical rain forests occur where rainfall is plentiful and are dominated by huge broadleaf evergreen trees. In rain forests, most nutrients are tied up in vegetation, and most animal life is arboreal. Rain forests, home to at least 50% of all species, are rapidly being cut for agriculture, although the soil is extremely poor.

The African savanna, a grassland studded with scattered trees, has pronounced wet and dry seasons. It is home to the world's most diverse and extensive herds of large mammals.

Most deserts are located between 20° and 30° north and south latitude or in the rain shadows of mountain ranges. In deserts, plants are often widely spaced and have adaptations to conserve water. Most animals are small and nocturnal.

Chaparral exists in desert-like conditions that are moderated by proximity to a coastline, allowing small trees and bushes to thrive. Grasslands, concentrated in the centers of continents, have a continuous grass cover without trees. They have the world's richest soils and have largely been converted to agriculture.

Temperate deciduous forests, whose broadleaf trees drop their leaves in winter to conserve moisture, dominate the eastern half of the United States and are also found in Western Europe and East Asia. Precipitation is higher than in grasslands. The wet temperate rain forests, dominated by evergreens, are on the northern Pacific Coast of the United States. The taiga, or northern coniferous forest, covers much of the northern United States, southern Canada, and northern Eurasia. It is dominated by conifers whose small, waxy needles are adapted for water conservation and year-round photosynthesis.

The tundra is a frozen desert where permafrost prevents the growth of trees, and bushes remain stunted. Nonetheless, diverse arrays of animal life and perennial plants flourish in this fragile biome, which is found on mountain peaks and in the Arctic.

30.4 How Is Life in Water Distributed?

Energy and nutrients are the major limiting factors in the distribution and abundance of life in aquatic ecosystems. Nutrients are found in bottom sediments and are washed in from surrounding land, concentrating them near shore and in deep water.

Freshwater lakes have three life zones. The littoral zone, near shore, is rich in energy and nutrients and supports the most diverse community. The limnetic zone is the lighted region of open water where photosynthesis can occur. The profundal zone is the deep water, where light is inadequate for photosynthesis and the community is dominated by heterotrophic organisms. Oligotrophic lakes are clear and low in nutrients, and they support sparse communities. Eutrophic lakes are rich in nutrients and support dense communities. During succession to dry land, lakes tend to go from an oligotrophic to a eutrophic condition.

Most life in the oceans is found in shallow water, where sunlight can penetrate, and is concentrated near the continents and in areas of upwelling, where nutrients are most plentiful. Coastal waters, consisting of the intertidal zone and the nearshore zone, contain the most abundant life. Producers consist of aquatic plants anchored to the bottom and photosynthetic protists called phytoplankton. Coral reefs are confined to warm, shallow seas. The calcium carbonate reefs form a complex habitat supporting the most diverse undersea ecosystem, threatened by silt, overfishing, and global warming.

In the open ocean, most life is found in the photic zone, where light supports phytoplankton. In the aphotic zone, life is supported by nutrients that drift down from the photic zone. Many ocean fisheries have been overexploited. Hydrothermal vent communities, supported by chemosynthetic bacteria, thrive at great depths in the superheated waters where Earth's crustal plates are separating.

Key Terms

aphotic zone *p. 615*
biodiversity *p. 604*
biome *p. 602*
chaparral *p. 607*
climate *p. 596*
coral reef *p. 617*
desert *p. 606*
estuary *p. 616*
eutrophic lake *p. 614*
grassland *p. 607*

hydrothermal vent community
    p. 620
intertidal zone *p. 615*
limnetic zone *p. 614*
littoral zone *p. 613*
nearshore zone *p. 616*
oligotrophic lake *p. 614*
ozone layer *p. 597*
permafrost *p. 612*

photic zone *p. 615*
phytoplankton *p. 614*
prairie *p. 607*
profundal zone *p. 614*
rain shadow *p. 600*
savanna *p. 605*
taiga *p. 611*
temperate deciduous forest
    p. 609

temperate rain forest *p. 603*
tropical deciduous forest *p. 605*
tropical rain forest *p. 603*
tundra *p. 612*
upwelling *p. 615*
weather *p. 596*
zooplankton *p. 614*

Thinking Through the Concepts

Suggested answers to end-of-chapter and figure-based questions can be found at the end of the text.

Fill-in-the-Blank

1. The tilt of Earth on its axis produces _____. Air currents are generated by _____ and _____. The warm, wet zone around the equator is called the _____. Nearshore climates are more moderate because of _____. A dry region on the side of a mountain range that faces away from the prevailing winds is called a(n) _____.

2. Of the four major requirements for life, which two are most limited on land? _____, _____ Which two are most limited in aquatic ecosystems? _____, _____ The most biologically diverse terrestrial ecosystems are _____. The most diverse aquatic ecosystems are _____.

3. The thorns of cacti are modified _____, molded by evolution to reduce _____ and provide _____. The desert growing season for annual plants is limited by the brief availability of _____. Many desert animals are most active during the _____.

4. In the _____ biome, most of the nutrients are found in the bodies of plants, rather than in the soil. In the _____ biome, there are pronounced seasons and trees drop their leaves in winter. In the _____ biome, there is too little rain to support forests, but the soil is so rich that much of the biome has been converted to farmland. The _____ biome is characterized by very low rainfall—fewer than 10 inches annually. The _____ biome, in which grasses are the dominant vegetation and trees are widely spaced, is found along the edges of tropical rain forests. Stunted vegetation thrives in the "freezing desert" biome known as _____.

5. The shallow portion of a freshwater lake is called the _____. Aquatic life is most diverse in this portion because two of the major requirements for life, _____ and _____, which are often lacking in aquatic ecosystems, are abundant here. Photosynthetic plankton is called _____; nonphotosynthetic plankton is called _____. The open-water portion of a lake is divided into two zones: the upper _____ and the lower _____. In this lower zone, organisms are nourished primarily by _____.

6. Lakes very low in nutrients are said to be _____. Lakes very high in nutrients are said to be _____. The ocean zone that is most comparable to the littoral zone of a lake is the _____ zone. A body of water formed where a rivers meets the ocean is called a(n) _____.

Review Questions

1. Explain how air currents contribute to the formation of the tropics and the large deserts.

2. What effect do the roughly circular ocean currents have on climate, and where is that effect strongest?

3. What are the four major requirements for life? Which two are most often limiting in terrestrial ecosystems? In ocean ecosystems?

4. Explain why traveling up a mountain takes you through biomes similar to those you would encounter traveling for a long distance toward one of Earth's poles.

5. Where are the nutrients of the tropical forest biome concentrated? Why is animal life in the tropical rain forest concentrated high above the ground?

6. Explain two undesirable effects of agriculture in the tropical rain-forest biome.

7. List some adaptations to heat and drought of (a) desert plants and (b) desert animals.

8. What human activities damage deserts?

9. How are trees of the taiga adapted to a lack of water and a short growing season?

10. How do deciduous and coniferous biomes differ?

11. What single environmental factor best explains why there is shortgrass prairie in Colorado, tallgrass prairie in Illinois, and deciduous forest in Ohio?

12. Where are the world's largest populations of large herbivores and carnivores located?

13. Where is life in the oceans most abundant, and why?

14. Why is the diversity of life so high in coral reefs? What human impacts threaten them?

15. Distinguish among the limnetic, littoral, and profundal zones of lakes in terms of their location and the communities they support.

16. Distinguish between oligotrophic and eutrophic lakes. Describe (a) a natural scenario and (b) a human-created scenario under which an oligotrophic lake might be converted to a eutrophic lake.

17. Distinguish between the photic and aphotic zones. How do organisms in the photic zone obtain nutrients? How are nutrients obtained in the aphotic zone?

18. What unusual primary producer forms the basis for hydrothermal vent communities?

19. On the basis of the location of the worst atmospheric ozone depletion, which biomes are likely to be most affected by increased UV penetration?

Applying the Concepts

1. **IS THIS SCIENCE?** Many observers have hypothesized that the hole in the ozone layer is the cause of recent increases in the rate of occurrence of skin cancer. In addition, the ozone hole is often cited as a cause of the worldwide decline in frog populations. Can you think of ways to test these two hypotheses?

2. Global warming is expected to change rainfall in ways that are hard to predict; some areas will get wetter, others drier. Would a decrease in rainfall be more likely to cause major changes in biomes that are adapted to low temperatures or to high temperatures? Expalin your answer.

3. More-northerly forests are better able to regenerate after logging than are tropical rain forests. Explain why. (*Hint:* The cold soils of northern climates greatly slow down decomposition rates.)

For additional resources, go to www.mybiology.com.

chapter 31

Conserving Earth's Biodiversity

Officially listed as endangered in the United States in 1967, the bald eagle was removed from the endangered species list in 2007.

🔍 At a Glance

Case Study   Back from the Brink

It's a story that has become sadly familiar: A wild species thrives in its natural habitat, living as it has lived for thousands of generations. But the species shares its habitat with humans, and as the human population grows, people make ever greater modifications to the environment. The wild species does not tolerate these changes very well. Its population begins to shrink, becoming smaller with each passing year. Eventually, the population dwindles to a tiny remnant, perhaps even to just a single individual. When the last individual dies, the species becomes extinct—permanently extinguished.

This scenario has been repeated many times over the past few centuries, and has become increasingly common. Today, thousands of species are in danger of extinction. But in at least a few cases, a species' seemingly inexorable slide into extinction has been halted. The American alligator (photo at left), peregrine falcon, brown pelican, and Robbins cinquefoil are among the once-threatened species that now seem safely recovered. Perhaps the highest-profile species to come back from the brink is the bald eagle.

When the United States was founded, hundreds of thousands of bald eagles patrolled the skies over North America. By the 1950s, however, eagles were in trouble. Only around 10,000 breeding pairs remained in the lower 48 states. Conservationists became alarmed, but the eagle decline continued. In 1963, only 417 pairs raised young in the lower 48. The eagle population's trend was steeply downward, and the species seemed to be facing certain doom.

The U.S. bald eagle population, however, did not disappear. Instead, it stopped shrinking and then began to grow. Today, the lower 48 states are home to nearly 10,000 breeding pairs. "It's one of the greatest wildlife success stories in the history of this country," says John Kostyack of the National Wildlife Federation.

The bald eagle's recovery was possible because concerned people took action. Of course, human actions were the reason the species became endangered in the first place. What factors caused the bald eagle's decline? And what steps brought the species back from the edge of oblivion? Read on to discover the answers to these questions, and to learn more about the scientific discipline devoted to conserving life's diversity. ■

The American alligator nearly disappeared from the swamps and waterways of the southern United States. Today, however, alligator populations are thriving.

31.1 What Is Conservation Biology?

Conservation biology is the branch of science that seeks to understand and conserve biological diversity. Biological diversity—more commonly called **biodiversity**—is simply the variety of life, the diversity of living organisms. Biologists analyze biodiversity at different levels:

- Genetic diversity is the variety and relative frequencies of different alleles in the gene pool of a species.

- Species diversity is the variety and relative abundance of different species.

- Ecological diversity is the variety of different ecosystems and of the community interactions within them.

Conservation biologists study and seek to preserve biodiversity at all levels. They apply methods drawn from ecology, genetics, and evolutionary biology to uncover information that will help protect and maintain the genetic, species, and ecological diversity of life. Although the different levels of biodiversity may be studied separately, they are intimately related. If the genetic diversity of a species is reduced, the species may lack the genetic variability necessary to evolve adaptations to changing environments. If the species diversity of an ecosystem is reduced, the intricate network of community interactions necessary to maintain a properly functioning ecosystem may be threatened. And if ecological diversity is reduced, the ability of the biosphere to support species may be compromised.

The knowledge gained by conservation biologists can help preserve biodiversity only if it is used to inform actions that prevent human-caused extinctions. Thus, conservation biologists work closely with social scientists, policymakers, lawyers, geographers, economists, ethicists, and historians. Conservation is necessarily a social enterprise.

31.2 Why Is Biodiversity Important?

Most of us live in cities or suburbs. Our food comes mostly in packages from a supermarket. We may spend months without glimpsing an ecosystem in its natural state. So why should we care about preserving biodiversity? Many people would say that species and ecosystems are worth preserving for their own sake. But even those who do not find inherent value in biodiversity benefit from its preservation. Diverse ecosystems provide concrete benefits to everyone (**Fig. 31-1**).

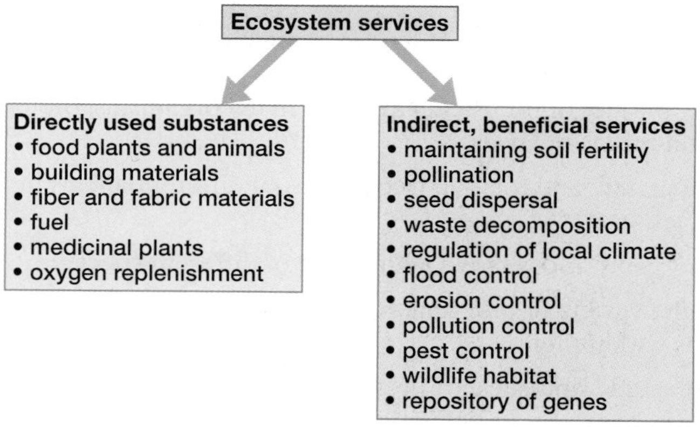

Ecosystem Services: Practical Uses for Biodiversity

In recent decades, scientists, economists, and policymakers have come to recognize that nature provides free but usually unrecognized benefits. These benefits, known as **ecosystem services,** sustain and enhance our lives and are provided by the normal activities of natural ecosystems and their living communities.

Ecosystem services include purifying air and water, replenishing atmospheric oxygen, pollinating plants and dispersing their seeds, providing wildlife habitat, decomposing wastes, limiting erosion and flooding, controlling pests, and providing recreational opportunities. These services are priceless; without them, human societies would collapse. But because we don't pay for ecosystem services, their monetary value is difficult to measure and is almost always ignored in economic transactions. When land is converted to housing, for example, there is usually no financial incentive for the developer to preserve ecosystems and their services, but there is often a considerable financial reward for destroying them. Governments, businesses, and individuals almost never attempt to weigh the true costs and benefits of altering the environment, and instead treat the loss of ecosystem services as a cost-free consequence of their actions.

Ecosystems Provide Goods Directly

Healthy ecosystems provide a variety of resources directly to people. For example, our diets are enriched by wild-caught fish and other seafood species that thrive only in healthy marine environments. Hunting for food and for sport in natural wildlife habitat is important to the economy of many rural areas. In Africa, most types of wild animals are harvested for food and provide an important source of inexpensive protein for a growing and often poorly nourished population (see "Earth Watch: Tangled Troubles—Logging and Bushmeat" on p. 631). Forests provide wood for housing and furniture and, especially in less-developed countries, supply firewood for heating and cooking. Wild plants are the main source of traditional medicines, used by about 80% of the world's people. Modern medicine also relies on plants; about 25% of prescription medications contain active ingredients that are now—or were originally—extracted from plants.

Ecosystems Also Benefit People Indirectly

The services provided indirectly by healthy, diverse ecosystems are far reaching and make a much greater contribution to human welfare than do

Ecosystem services

Directly used substances
• food plants and animals
• building materials
• fiber and fabric materials
• fuel
• medicinal plants
• oxygen replenishment

Indirect, beneficial services
• maintaining soil fertility
• pollination
• seed dispersal
• waste decomposition
• regulation of local climate
• flood control
• erosion control
• pollution control
• pest control
• wildlife habitat
• repository of genes

▲ **Figure 31-1 Ecosystem services**

goods harvested directly from nature. Here we describe just a few important examples.

Soil Formation   It can take hundreds of years to build up a single inch of soil. The rich soils of the U.S. Midwest accumulated under natural grasslands over thousands of years. Farming has converted these grasslands into one of the most productive agricultural regions in the world.

Soil, with its diverse community of decomposers and detritus feeders (bacteria, fungi, worms, insects, and others), plays a major role in breaking down wastes and recycling nutrients. We rely on soils to decompose waste products from industry, sewage, agriculture, and forestry. Thus, soil serves some of the same functions as a water purification plant.

Erosion and Flood Control   Grassland and forest plants block winds that would otherwise blow away the soil. The plants' roots stabilize the soil and increase its ability to hold water, reducing both soil erosion and flooding. The massive flooding and consequent erosion that took place along the Missouri River in 1993 was caused, in part, by conversion of riverside forests, marshes, and grasslands to farmland. In the wake of heavy rains, the converted land allowed greatly increased runoff and soil erosion (**Fig. 31-2a**).

Wetland ecosystems (marshes) reduce flooding because they act like enormous sponges that absorb storm water. They also cushion the impact of waves that batter the coastline. This protective function, and the consequences of destroying it, was illustrated in grim fashion by the catastrophic flooding of New Orleans during Hurricane Katrina in August 2005. Prior to the establishment and growth of New Orleans, the Mississippi River's naturally silt-laden waters replenished the area's wetlands with sediment and fortified outlying islands that served as a natural barrier to the force of incoming storms. By the time Katrina hit, however, the dredged, walled, and diverted waters of the Mississippi no longer sustained the wetlands and islands; southern Louisiana had lost 1,000 square miles of wetlands in the 50 years prior to Katrina. Huge levees (built at enormous cost to replace ecosystem services that were formerly provided for free) temporarily supplied some of the protection previously afforded by wetlands. But then Katrina struck, breaching the levees and flooding 80% of the city (**Fig. 31-2b**).

Climate Regulation   By providing shade, moderating temperature, and serving as windbreaks, plant communities have a major impact on local climates. Forests have dramatic effects on the water cycle, returning water to the atmosphere through transpiration (evaporation from leaves). For example, in the Amazon rain forest, one-third to one-half of the rain consists of water transpired by leaves. Clear-cutting of rain forests can cause the local climate to become hotter and drier, making it harder for the ecosystem to regenerate and damaging nearby intact forest as well.

Genetic Resources   Our food supply depends on a small number of domesticated species; 75% of human food is supplied by only 12 food crops. Such heavy dependence on just a few species leaves our food security vulnerable to pest or disease epidemics that attack the key crops. The best protection against such a potential catastrophe would be to increase food crop diversity by developing wild plant species into new food sources. In addition, researchers have identified genes in wild plants that might be transferred into existing crops to increase productivity and provide greater resistance to disease, drought, and salt accumulation in irrigated soil. The genetic treasure house of biodiversity may be the key to future food security and promises to become an increasingly important resource—but only if it is preserved.

Recreation   Many, perhaps most, people experience great pleasure in "returning to nature." Each year in the United States, about 350 million visitors flock to protected public lands such as national parks and wildlife refuges. In many rural

(a) Flooding along the Missouri River

(b) Flooding in New Orleans

▲ **Figure 31-2   Loss of flood control services (a)** Conversion of natural ecosystems to agriculture contributed to flooding of the Missouri River after unusually heavy rains in 1993. **(b)** New Orleans in the wake of Hurricane Katrina in 2005.

areas, the local economy depends on money spent by visitors who come to hike, camp, hunt, fish, or photograph nature.

Ecotourism, in which people travel to observe unique biological communities, is a rapidly growing industry worldwide. Examples of ecotourism destinations include tropical coral reefs and rain forests, the Galápagos Islands, the African savanna, and even Antarctica.

Ecological Economics Recognizes the Monetary Value of Ecosystem Services

The emerging discipline of *ecological economics* attempts to establish the monetary value of ecosystem services and to assess the economic trade-offs that occur when ecosystems are damaged to make way for profit-making activities. For example, a farmer planning to divert water from a wetland to irrigate a crop would typically weigh the monetary value of increased crop production against the monetary cost of the labor and machinery required to divert the wetland's water supply. If, however, the farmer also accounted for the cost of ecosystem services (neutralizing pollutants, controlling floods, providing breeding grounds for fish, birds, and many other animals) lost by destruction of the wetland, the decision on whether to proceed might be different.

In today's economy, the profits from damaging ecosystems usually go to an individual, while the costs are borne by society as a whole. Therefore, application of the principles of ecological economics often requires government involvement. In "Earth Watch: Restoring the Everglades," we describe a massive and costly project to undo human manipulation of the largest wetland ecosystem in the United States.

31.3 Is Earth's Biodiversity Diminishing?

We depend on ecosystem services, but the ability of ecosystems to provide those services declines when biodiversity declines. Conservation biologists thus devote significant effort to assessing the extent to which species extinctions are affecting biodiversity.

Extinction Is a Natural Process, but Rates Have Risen Dramatically

The fossil record indicates that, in the absence of cataclysmic events, extinctions occur naturally at a very low rate. However, the fossil record also provides evidence of occasional **mass extinctions,** in which many species were eradicated in a relatively short time. Five such mass extinction episodes have occurred; the most recent took place about 65 million years ago and abruptly ended the age of dinosaurs. The causes of mass extinctions are uncertain, but sudden changes in the environment (such as would be caused by enormous meteor impacts or rapid climate change) are the most likely explanations.

Most biologists have concluded that we are now in the midst of a sixth mass extinction, this one caused by human activities. Not all biologists, however, agree that current extinction rates are high enough to substantially reduce overall biodiversity. The lack of complete consensus with regard to current extinction rates reflects the difficulty of accurately measuring extinction rates. Because biologists have identified only a fraction of Earth's species, it is difficult to establish the proportion that has become extinct. The best estimates are for birds and mammals, fairly conspicuous organisms that have been studied for centuries.

Since the 1500s, about 2% of all known mammal species and 1.3% of all known bird species have become extinct. We have certainly also lost many undescribed bird and mammal species. Consider, for example, the African kipunji, a new monkey species discovered in 2005 (**Fig. 31-3**). There are only about 1,000

▲ Figure 31-3 **A recently discovered large mammal** The secretive African kipunji was discovered in rapidly disappearing African tropical rain forest.

earth watch

Restoring the Everglades

In 1948, the U.S. Congress authorized the Central and Southern Florida Project, creating an extensive series of canals, levees, and other structures to control flooding, irrigate farms, and provide drinking water for new Florida developments in the Everglades, the massive marshland that dominated central and southern Florida. The project also transformed the meandering, 103-mile-long Kissimmee River into a straightened, 56-mile-long channel, eliminating most of its surrounding wetlands (**Fig. E31-1**). As the Everglades and other wetlands in southern Florida diminished, so did the wildlife that depended on them. The natural water-purifying functions of the wetlands were also lost, compounding the pollution problem as new farms and cities sprang up. Native plants, birds, fish, and other species declined, while invasive species flourished.

Over the ensuing decades, many people began to regret the transformation of the Everglades. With half of the Everglades' original area converted to agriculture, housing, and industry, the diversity of species and richness of community interactions that had made the Everglades a unique ecosystem were rapidly being lost.

In 2000, Florida and the U.S. government launched the Comprehensive Everglades Restoration Plan. This 30-year plan is intended to restore 18,000 square miles of wetlands at a projected cost of $7.8 billion. One of the largest ecosystem restorations ever attempted, the plan will remove 240 miles of canals and levees, reestablish natural river flow, revive wetlands, and recycle some waste water. As a result of the efforts to date, Florida now has about 64 square miles of reconstructed wetlands. Work on the Kissimmee River will eventually restore 47 miles of river. Bird populations have already rebounded along portions of the river, and water quality has improved.

The Everglades restoration plan represents an expensive, long-term commitment to undo human damage to an ecosystem. The project provides evidence of a growing awareness of the economic and intrinsic values of diverse, healthy ecosystems.

(a) Pre-channelized Kissimmee River

(b) Channelized Kissimmee River

▶ **Figure E31-1** **Channelization of Florida's Kissimmee River**

kipunji remaining, and these are threatened by human activities; they could easily have become extinct before they were ever discovered, as many other species have undoubtedly done. Extinction of undiscovered species is probably even more prevalent among the many types of small, inconspicuous organisms that people tend to overlook. The World Conservation Union (IUCN) estimates that the current extinction rate is between 100 and 1,000 times greater than the background rate expected in the absence of people.

Increasing Numbers of Species Are Threatened with Extinction

The IUCN has established a "Red List" that classifies at-risk species. Species may be described as **critically endangered, endangered,** or **vulnerable,** depending on how likely they are to become extinct in the near future. Species that fall into any

▶ **Figure 31-4 Habitat destruction**
(a) Clear-cutting a rain forest.
(b) This satellite image from NASA shows new soybean plantations created in the Bolivian rain forest.

(a) Cutting down forest

(b) Plantations seen from space

of the three categories above are described as **threatened.** In 2007, the Red List showed 16,306 threatened species, including 12% of all birds, 20% of mammals, and 29% of amphibians. Many scientists fear that many if not most of the currently endangered species are on their way to extinction. Why is this happening?

31.4 What Are the Major Threats to Biodiversity?

The greatest hazards to biodiversity are posed by habitat destruction, overexploitation, invasive species, pollution, and global warming. Imperiled species usually face several or all of these threats simultaneously (see "Earth Watch: Tangled Troubles—Logging and Bushmeat"). For example, the steep decline in frog populations worldwide results from a combination of habitat destruction, invasive species, pollution, and a virulent fungal infection that many experts believe is linked to global warming (see "Earth Watch: Frogs in Peril" on p. 312). Similarly, coral reefs, home to about one-third of marine fish species, suffer from a combination of overharvesting of species, pollution, and global warming.

Habitat Destruction Is the Most Serious Threat to Biodiversity

The IUCN has identified habitat destruction as the leading threat to biodiversity worldwide. Loss of habitat affects the vast majority of threatened species. Habitat is destroyed as rivers are dammed, wetlands are drained, and grasslands and forests are converted to agriculture, roads, housing, and industry. For example, since people began to farm around 11,000 years ago, Earth has lost about half of its forests. More alarmingly, approximately half of all tropical rain forests have been cut down in just the past 50 years. This rapid destruction of rain forest is the result of intensive harvesting of wood for export and converting forest to pastures and farm fields that produce beef, coffee, soybeans, palm oil, sugarcane, and other crops (Fig. 31-4).

Even when a natural ecosystem is not completely destroyed, it may become split into small pieces surrounded by regions devoted to human activities (Fig. 31-5). This **habitat fragmentation** can be a serious threat to wildlife. For example, individuals of many bird species need large tracts of continuous forest to find food, mates, and breeding sites. Big cats are also threatened by habitat fragmentation. For example, forest reserves intended to protect the endangered Bengal tiger in India have be-

Web Animation Habitat Destruction and Fragmentation

▲ **Figure 31-5 Habitat fragmentation** Fields isolate forest patches in Paraguay. **QUESTION:** Which types of species do you think are most likely to disappear from small forest fragments?

earth watch

Tangled Troubles—Logging and Bushmeat

The bushmeat trade in Africa is a prime example of how threats to biodiversity interact and amplify one another. Historically, rural Africans have supplemented their diet by hunting a variety of animals, collectively called "bushmeat." Traditional subsistence hunting by small tribes using primitive weapons did not pose a serious threat to animal populations. Now, however, as logging roads penetrate deeper into rain forests, hunters follow, using shotguns and snares to kill any animal large enough to eat. Communities that spring up along logging roads develop a culture of hunting and selling bushmeat, and become dependent on this new and profitable industry. Logging trucks are sometimes used to carry the meat to urban markets. The World Conservation Society estimates that the bushmeat harvest in equatorial Africa exceeds a million tons annually. Because many of the hunted animals play an important role in dispersing tree seeds, loss of these animals reduces the ability of the logged forest to regenerate.

Because bushmeat hunters pay no attention to the sex, age, size, or rarity of the animal, many threatened species are declining rapidly.

For example, despite estimates that only about 2,000 to 3,000 pygmy hippos remain in the wild, pygmy hippo meat has been found in bushmeat markets, as has meat from elephants and rhinos.

The profitability of bushmeat has helped overcome traditional African taboos against eating primates. Although one-third of all primates (apes, monkeys, and lemurs) are threatened with extinction, in some bushmeat markets, 15% of the meat comes from primates (**Fig. E31-2**). In Cameroon, Africa, endangered gorillas are a favored target of poachers because of their large size. Even endangered chimpanzees and bonobos, our closest relatives, end up in cooking pots. Experts now believe that bushmeat hunting is an even greater threat than habitat loss for Africa's great apes, and that the combined threats of hunting and habitat loss make extinction in the wild a very real possibility for some of these magnificent, intelligent species.

Recognizing the threats to wildlife, several central African countries are working to reduce illegal timber harvesting and wildlife poaching. Together, they have established protected areas

▲ **Figure E31-2 Bushmeat** Primates are threatened by bushmeat hunters.

in the African rain forest of the Congo River Basin. Although enormous logging enterprises continue along the borders of these reserves, protecting them is a crucial step toward preserving some of Africa's rich natural heritage.

come islands in a sea of development, forcing the estimated 5,000 remaining tigers into about 160 isolated patches of woodland.

Habitat fragmentation may result in populations that are too small to survive. To be functional, a preserve must support a **minimum viable population (MVP).** This is the smallest isolated population that can persist despite the effects of inbreeding and potential disruptions from disease, fires, and floods. The MVP for a species is influenced by many factors including the quality of the environment, the species' average life span, its fertility, and how many young usually reach maturity. Some wildlife experts believe that a minimum viable population of Bengal tigers must include at least 50 females, more than are found in most of India's tiger reserves.

Overexploitation Threatens Many Species

Overexploitation refers to hunting or harvesting natural populations at a rate that exceeds their ability to replenish their numbers. Overexploitation has increased as growing demand is coupled with technological advances that greatly increase our efficiency at harvesting wild animals and plants. The IUCN estimates that overexploitation impacts about 30% of threatened mammals and birds.

Overfishing is the single greatest threat to marine life, causing dramatic declines of many species, including cod, sharks, red snapper, swordfish, and tuna. High-tech fishing fleets harvest huge numbers of commercially valuable fish and also unintentionally catch equally huge numbers of unwanted "bycatch" fish

Back from the Brink

Continued

Historically, many bald eagles were shot and killed by hunters. In addition, many eagles died when they ate poisoned meat that ranchers put out to kill coyotes and other mammalian predators. Far more harmful than this direct killing, however, were the effects of habitat loss. Nesting bald eagles require forested areas that are near fish-containing waters but not too near human habitations. As a growing human population steadily converted suitable eagle nesting habitat to human uses such as housing and agriculture, eagles found fewer and fewer usable nest sites. Many eagles failed to breed, and the eagle population began to shrink.

▲ **Figure 31-6 Overexploitation** Illegally harvested green sea turtle eggs for sale in a market in Borneo.

species, along with hundreds of thousands of whales, porpoises, dolphins, and seabirds. Most species of marine turtles are endangered as a result of overharvesting of both adults and their eggs (**Fig. 31-6**).

On land, rapidly growing populations in less-developed countries increase the demand for wild animal products, as hunger and poverty drive people to harvest all that can be eaten or sold, legally or illegally, without regard to its rarity. Compounding the problem, affluent consumers fuel the demand for endangered animals by paying high prices for illegal products such as elephant tusk ivory, rhinoceros horn, and exotic rain-forest birds. Similarly, demand for wood in developed countries encourages unsustainable logging in rain forests.

Invasive Species Displace Native Wildlife and Disrupt Community Interactions

Humans have transported a multitude of species around the world: eucalyptus trees from Australia to California, and redwood trees from California to England, for example. In many cases, these introduced species cause no great harm. Sometimes, however, non-native species become *invasive*: They increase in number at the expense of native species, competing for food or habitat, or preying on them directly. Introduced species often make native species more vulnerable to extinction from other causes, such as disease or habitat destruction. Approximately 7,000 invasive species have become established in the United States, and nearly half of threatened U.S. species suffer from competition with or predation by invasive species (see "Earth Watch: Exotic Invaders" on pp. 560–561).

Island species are particularly vulnerable to competition and predation from introduced species. Island populations are small, many island species are unique, and they have nowhere to go when conditions change. For example, 99% of Hawaii's 414 threatened plants and all but one of its 42 threatened bird species are endangered by invasive species. Among these invasive species are mongooses, which were deliberately imported in the 1800s to control accidentally introduced rats (**Fig. 31-7**). Now both mongooses and rats pose major threats to Hawaii's native ground-nesting birds. Wild pigs and goats, released by early Polynesian settlers to provide food, have decimated native Hawaiian plants.

Pollution Is a Multifaceted Threat to Biodiversity

Pollution takes many forms. Pollutants include synthetic chemicals such as plasticizers, flame retardants, and pesticides that enter the air, soil, and water, and then accumulate to toxic levels in animal tissues. Naturally occurring substances can also become pollutants when they are released in unnaturally high quantities. Some of these, such as mercury, lead, and arsenic, are directly toxic to both people and wildlife. Others become pollutants by disrupting biogeochemical cycles. For example, combustion of fossil fuels releases oxidized nitrogen and sulfur that disrupt the natural cycles of these nutrients, causing acid rain that threatens biodiversity in forests and lakes.

Global Warming Is an Emerging Threat to Biodiversity

The use of fossil fuels, coupled with deforestation, has substantially increased atmospheric carbon dioxide levels. As predicted by climatologists, the increase has been accompanied by increasing global temperatures. In response to this global warming, many species are shifting their ranges closer to the poles, plants and animals are beginning springtime activities earlier in the year, and glaciers, ice shelves, and ice caps are melting (see pp. 589–591). Some meteorologists hypothesize that global warming is also causing more extreme weather, such as heat waves, droughts, floods, and stronger hurricanes and other storms.

Back from the Brink

Continued

Although hunting, poisoning, and habitat loss set in motion the decline of the bald eagle, the truly crushing blow came in 1945, with the invention of the insecticide DDT. In the years following its introduction, DDT was sprayed regularly and repeatedly over huge areas all across the United States, to control mosquitoes and agricultural pests. Unfortunately, in addition to killing pests, the toxic pesticide also entered the food chain and accumulated in the tissues of organisms, especially in top-level predators such as the bald eagle. The DDT in the bodies of eagles interfered with the deposition of calcium in egg shells, leaving the shells weak and unable to support the weight of incubating mothers. The eggs cracked, killing the embryos within. Eagle reproductive output declined catastrophically, putting already stressed populations into a tailspin.

The rapid pace of human-induced climate change challenges the ability of species to evolve adaptations rapidly enough to keep up with their changing environment and may thereby make them vulnerable to extinction. Many conservation biologists have concluded that global warming is now as great a threat to biodiversity as is habitat destruction.

31.5 How Can Conservation Biology Help to Preserve Biodiversity?

Research in conservation biology yields knowledge that can be used to devise effective strategies for conserving biodiversity. In the sections that follow, we describe some of these strategies.

Conservation Biology Is an Integrated Science

Effective conservation depends on consensus and broad participation, so conservation biologists seek expertise and support from people outside of science. These people include government leaders, who establish environmental policy and laws; environmental lawyers and law enforcement officials, who help enforce laws that protect species and their habitats; and ecological economists, who help place a value on ecosystem services. In addition, conservation biologists collaborate with social scientists, who study how different cultural groups use and interact with their environments, and with educators, who help students and the public understand how ecosystems function, how they support human life, and how people can disrupt or preserve them. This educational effort is augmented by the outreach activities of conservation organizations, which also help organize grassroots support for conservation initiatives.

Conserving Wild Ecosystems

Because habitat destruction and habitat fragmentation are key factors threatening biodiversity, habitat preservation is a key strategy for preserving biodiversity. Protected reserves, connected by wildlife corridors, are essential to conserve natural ecosystems and their diverse communities.

Core Reserves Preserve All Levels of Biodiversity

Core reserves are natural areas protected from most human uses except very low-impact recreation. These reserves encompass enough space to preserve ecosystems and all their biodiversity. Because natural storms, fires, and floods are important to maintain ecosystems, core reserves must be large enough to sustain these events without loss of species. (See "Earth Watch: Restoring a Keystone Predator" on p. 634 for a description of how a core reserve has been managed to enhance populations and restore community interactions.)

To establish effective core reserves, conservationists must know the *minimum critical areas* required to sustain minimum viable populations of the species that require the most space. Minimum critical areas vary significantly among species, and also depend on the availability of food, water, and shelter in the habitat under consideration. Extensive research and analysis are typically required to determine the minimum critical area required to sustain a particular community.

Corridors Connect Critical Animal Habitats

In today's crowded world, it is seldom possible to establish a core reserve large enough to maintain biodiversity and complex community interactions. To compensate for limits on the size of core reserves, it is necessary to establish **wildlife corridors,** which are strips of protected land linking core reserves. Such corridors

▲ **Figure 31-7 Invasive species** The mongoose, imported from India to prey on rats, threatens native ground-nesting birds in Hawaii. **QUESTION:** Why are Hawaii's birds more vulnerable to predation by mongooses than are bird species in the mongoose's native habitat?

Restoring a Keystone Predator

A predator is described as a *keystone predator* when its hunting activities have major effects on the community structure of an ecosystem. Research in Yellowstone National Park in the western United States is documenting the complex interrelationships within this natural community and the critical role played by a keystone predator: the wolf.

Considered a threat to elk and bison herds, wolves were exterminated from Yellowstone by 1928. Unexpectedly, the decline and disappearance of wolves marked the beginning of the end for regeneration of aspen trees (Fig. E31-3). Aspen groves, which shelter a diverse community of plants and birds, have declined by more than 95% since the park was established in 1872. New research suggests that elk, the major prey of wolves, eat nearly all the young aspen, as well as young willow and cottonwood trees. In the absence of wolves, elk populations expanded, and the forest community was drastically altered.

In 1995 and 1996, after years of planning, study, and public comment, the U.S. Fish and Wildlife Service captured 21 wolves in Canada and released them in Yellowstone National Park (Fig. E31-3, inset). The wolves have thrived and now number about 1,000. Many ecologists are convinced that reintroduction of this keystone predator has had far-reaching, favorable impacts on the Yellowstone ecosystem.

A recent study suggests that wolf predation not only controls elk numbers, but also changes elk behavior. With wolves nearby, elk tend to avoid streamside aspen, willow, and cottonwood groves where wolves can hide, and these streamside communities are regenerating, providing more habitat for songbirds and better stream conditions for trout. As their favorite trees have rebounded, beaver have returned and built dams on the streams, creating marshlands that provide habitat for mink, muskrat, otter, ducks, and rare boreal toads. Succulent plants in beaver marshes are favored food for grizzly bears as they emerge from hibernation. Grizzlies also feed on elk carcasses left by wolves, as do bald and golden eagles. In a further twist of the tangled web of community interactions, wolves compete with and kill coyotes, which eat rodents. With rodent populations on the rise, red foxes that feed on rodents are proliferating, and biologists are anticipating a rebound in other small predators such as weasels and wolverines.

Some ecologists, noting the complexity of species interactions, correctly point out that the "wolf effect" needs more time and study before it is conclusively demonstrated. But the evidence so far suggests that, as the Wolf Restoration Project leader Douglas Smith puts it, "Wolves are to Yellowstone what water is to the Everglades."

▶ **Figure E31-3 Impact of a keystone predator** Remnants of a once-thriving aspen grove in Yellowstone National Park attest to the lack of aspen regeneration since the early 1900s. Wolves *(inset)* now roam Yellowstone, delighting visitors and exerting far-reaching positive effects on the natural community.

allow animals to move freely and safely between habitats that would otherwise be isolated by human activities (Fig. 31-8). Corridors effectively increase the size of core reserves by connecting them. Both core reserves and corridors should, ideally, be surrounded by *buffer zones* supporting low-impact human activities, such as small-scale agriculture to sustain local populations, that are compatible with wildlife. Buffer zones prevent high-impact uses such as clear-cutting, large-scale agriculture, mining, freeways, and housing from affecting wildlife in the core region.

A wildlife corridor can be as narrow as an underpass beneath a highway. For example, in densely populated southern California, plans for development of more than 1,000 new houses near San Diego were abandoned and the freeway exits closed after wildlife biologists tracked a cougar that was using an underpass to move between suitable habitats. Now an official wildlife corridor, the underpass and its surroundings are being restored to a more natural state, encouraging cougars and other wild animals to cross safely beneath the freeway (**Fig. 31-9**).

Larger scale establishment of wildlife corridors can have even greater impact. For example, in the northern Rocky Mountains, a coalition of conservation groups and scientists has proposed a series of corridors linking existing core reserves, such as Yellowstone National Park, with nearby ecosystems. These interconnected habitats would sustain populations of grizzly bears, elk, wolves, and mountain lions.

31.6 Why Is Sustainability the Key to Conservation?

Natural ecosystems share certain features that allow the ecosystems to persist and flourish. Among the most important characteristics of healthy ecosystems are:

- diverse communities with rich, complex community interactions;
- relatively stable population sizes that remain within the carrying capacity of the environment;
- recycling and efficient use of raw materials;
- reliance on renewable sources of energy.

Environments that have been modified by human development often do not possess these qualities. As a result, many human-modified ecosystems may not be sustainable over the long run, and persistence of current modes of development and land use could lead to loss of biodiversity and ecosystem services. It is therefore crucial that we learn to meet our material needs in ways that sustain the ecosystems on which we depend.

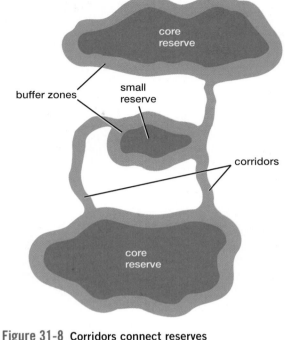

▲ **Figure 31-8** Corridors connect reserves

(a) Radio-tagging a cougar

(b) Underpass wildlife corridor

◄ **Figure 31-9** Wildlife corridors **(a)** National Park Service wildlife biologists identify and track cougars using ear tags and GPS radio collars, such as that worn by this tranquilized animal. **(b)** Asphalt has been removed and traffic barred from this underpass beneath the Riverside Freeway near San Diego to allow cougars to move safely between habitats on either side.

Sustainable Living and Sustainable Development Promote Long-Term Ecological and Human Well-Being

Respect for nature's operating principles is central to sustainability. In the landmark document *Caring for the Earth*, the IUCN states that **sustainable development** "meets the needs of the present without compromising the ability of future generations to meet their own needs." It explains that "(h)umanity must take no more from nature than nature can replenish. This in turn means adopting lifestyles and development paths that respect and work within nature's limits. It can be done without rejecting the many benefits that modern technology has brought, provided that technology also works within those limits."

Commercial fishing is a prime example of technology working outside nature's limits. Using sonar, enormous fishing nets, and trawls that may scrape entire communities from the ocean floor, commercial fishermen have harvested far more than can be replenished, endangering both commercial and noncommercial species in the process. Sustainable fishing requires that we preserve spawning grounds, limit fish catches, and improve technology to avoid unintended damage.

Unfortunately, in modern society, sustainable development is rarely achieved, because "development" typically means replacing natural ecosystems with human infrastructure such as housing and retail developments. People in developed countries have gained economic benefits and a high quality of life from such development. But they have done so by using large quantities of nonrenewable energy and by exploiting, in an unsustainable manner, the direct and indirect services provided by ecosystems.

Now, however, evidence from all parts of the world shows that such human growth activities are unraveling the complex tapestry of natural communities and undermining the ability of Earth to support life. As individuals and governments recognize the need to change, numerous projects have emerged to address the challenge of meeting human needs sustainably. We describe a few of these in the following sections.

Biosphere Reserves Provide Models for Conservation and Sustainable Development

A world network of **biosphere reserves** has been designated by the United Nations. The goal of biosphere reserves is to maintain biodiversity and evaluate techniques for sustainable human development while preserving local cultural values. Biosphere reserves consist of three regions. A central core reserve, while protected, allows research and sometimes tourism and some sustainable traditional uses. A surrounding buffer zone permits low-impact human activities and development. Outside the buffer zone is a *transition area* that supports settlements, tourism, fishing, and agriculture, all (in theory) operated sustainably. The first biosphere reserve was designated in the late 1970s, and there are now more than 480 sites worldwide.

Biosphere reserves are entirely voluntary and are managed by the countries and regional areas where they are located. This arrangement has greatly reduced local opposition to the reserves, but, as a result, few completely adhere to the full biosphere reserve model. In many cases, much of the land in the buffer and transition zones is privately owned, and some of the landowners may be unaware of its designation. Often, funding is inadequate to compensate them for restricting development and for promoting and coordinating sustainable development, particularly in the transition zones.

Biosphere reserves have met the same obstacles encountered by most efforts to preserve biodiversity and change the way people use natural resources, but they are gradually succeeding. The concept offers an elegant model for conservation in the context of sustainable development and provides a framework for both present and future efforts.

Sustainable Agriculture Helps Preserve Natural Communities

The greatest loss of habitat occurs when people convert natural ecosystems to monoculture farming, in which large expanses of land are devoted to single crops. For example, in the Midwestern United States, the original natural grasslands have been almost entirely converted to agriculture. Farming is necessary to feed humanity, and farmers face pressure to produce large amounts of food at the lowest possible cost. In some cases, this has led to unsustainable approaches to farming that interfere with ecosystem services. For example, the herbicides atrazine and 2, 4-D, widely used to kill weeds, have been found to be potent environmental toxins. Insecticides often indiscriminately kill many other species along with the pests that are the intended victims. In many regions throughout the world, irrigation for agriculture depletes underground water supplies faster than they can be replaced by natural processes.

Fortunately farmers are increasingly recognizing that sustainable agriculture ultimately saves money while preserving the land (Table 31-1). The **no-till** cropping technique, which leaves the remains of harvested crops in the fields to form mulch for the next year's crops, represents one component of sustainable agriculture (Fig. 31-10). In the United States, no-till methods are now used on about 20% of croplands, saving more than 300 million gallons of fuel annually and greatly reducing soil erosion.

Table 31-1 Agricultural Practices Affect Sustainability

	Unsustainable Agriculture	Sustainable Agriculture
Soil erosion	Allows soil to erode far faster than it can be replenished, because the remains of crops are plowed under, leaving the soil exposed until new crops grow.	Erosion is greatly reduced by no-till agriculture. Wind erosion is reduced by planting strips of trees as windbreaks around fields.
Pest control	Uses large amounts of pesticides to control crop pests.	Trees and shrubs near fields provide habitat for insect-eating birds and predatory insects. Reducing insecticide use helps to protect birds and insect predators.
Fertilizer use	Uses large amounts of synthetic fertilizer.	No-till agriculture retains nutrient-rich soil. Animal wastes are used as fertilizer. Legumes that replenish soil nitrogen (such as soybeans and alfalfa) are alternated with crops that deplete soil nitrogen (such as corn and wheat).
Water quality	Runoff from bare soil contaminates water with pesticides and fertilizers. Excessive amounts of animal wastes drain from feedlots.	Animal wastes are used to fertilize fields. Plant cover left by no-till agriculture reduces nutrient runoff.
Irrigation	May excessively irrigate crops, using groundwater pumped from natural underground storage at a rate faster than the water is replenished by rain or snow.	Modern irrigation technology reduces evaporation and delivers water only when and where it is needed. No-till agriculture reduces evaporation.
Crop diversity	Relies on a small number of high-profit crops, which encourages outbreaks of insects or plant diseases and leads to reliance on large quantities of pesticides.	Alternating crops and planting a wider variety of crops reduces the likelihood of major outbreaks of insects and diseases.
Fossil fuel use	Uses large amounts of nonrenewable fossil fuels to run farm equipment, produce fertilizer, and apply fertilizers and pesticides.	No-till agriculture reduces the need for plowing and fertilizing.

Most no-till farmers still use herbicides and pesticides. Organic farmers also often use no-till methods, but do not use synthetic herbicides, insecticides, or fertilizers. Organic farming relies on natural predators to control pests and relies on soil microorganisms to degrade animal and crop wastes, thereby recycling their nutrients. In contrast to conventional "factory farms," which may devote hundreds or even thousands of acres to a single crop, organic farms tend to be small.

Because the value of lost ecosystem services is not factored into the costs of unsustainable farming practices, food produced unsustainably tends to be cheaper, at least in the short term. Thus, consumers who wish to support sustainable agriculture must be prepared to pay a bit more for food produced this way. Although only about 0.5% of cropland in the United States is devoted to organic farming, consumer demand is driving steady growth of this sustainable approach.

Changes in Lifestyle and Use of Appropriate Technologies Are Also Essential

Government policies and institutional changes are vital for a sustainable future, but individual choices are also important. We can reduce our consumption of energy and nonrenewable fossil fuels by conserving and by using energy-saving technologies. In the absence of as-yet-unproven technologies such as nuclear fusion, sustainable living must ultimately rely on renewable energy sources (solar, wind, and wave energy, for example). We can emulate natural ecosystems by recycling nonrenewable resources. Our choices as consumers can provide markets for foods and durable goods that are produced sustainably.

This chapter has provided some examples of human activities moving in the right direction. You can also find some examples of steps that individuals can take at the end of Chapter 29 (see pp. 591–592). Look around your campus and community—what is being done sustainably? What isn't? What would it take to make the necessary changes?

Human Population Growth Is Unsustainable

Discussions of sustainability and preservation of biodiversity often skirt an inescapable fact. The root cause of environmental degradation is simple: too many people using too many resources and generating too much waste. As stated in the IUCN document *Who Will Care for the Earth?* a central issue is "... how to bring human populations into balance with the natural ecosystems that sustain them." Through technological advances and by depleting Earth's ecological capital, we have achieved a population that far exceeds this balance. Yet we currently add 75 to 80 million people to Earth every year. This is incompatible with a sustainable increase in the quality of life for the 6.5 billion already here, and with saving what is left of Earth's biodiversity for those who are to come.

(a) Soybean plants emerge in a no-till field

(b) The same field later in the season

▲ Figure 31-10 **No-till crops** *(a)* Soybeans can be seen growing up through the dead remains of an earlier wheat crop; the wheat stubble anchors soil and reduces evaporation. *(b)* Later in the season, the same field shows a healthy soybean crop mulched by the dead wheat. **QUESTION:** Make a list of environmental benefits gained from no-till farming (compared to farming with conventional tillage). Are any environmental costs associated with no-till farming?

Back from the Brink Revisited

Why were bald eagles in the lower 48 states able to bounce back from their brush with extinction? Because concerned citizens pushed for legislation to protect the eagles. In 1940, the Bald and Golden Eagle Protection Act (BGEPA) made it a crime to shoot or poison an eagle. This early law was an important first step, and made it plain that citizens and the federal government meant to protect eagles. But a straightforward ban on direct killing was not sufficient to counteract the devastating effects of habitat loss and DDT. Twenty years after BGEPA went into effect, the eagle population was lower than ever.

The tide finally turned in the early 1970s, when the U.S. government banned the use of DDT and passed the Endangered Species Act

(ESA). After the ban on DDT, its concentration in the environment and in eagle bodies began a gradual decline. The ESA authorized federal regulators to protect critical habitat and take other measures to protect and promote the recovery of all species officially designated as threatened or endangered.

With the two key laws on the books, the bald eagle gained a measure of protection from pollution and habitat loss, and its population began a period of steady growth. By the mid-1990s, most bald eagle populations in the lower 48 were no longer in imminent danger of extinction. In July 2007, the bald eagle was formally declared to have recovered, and the species was removed from the government's list of endangered species.

The "delisting" of the bald eagle stripped away the protections provided by the Endangered Species Act. Some conservationists argue that this change is too risky. Without ESA protection, the eagle is now protected only by the BGEPA, which remains in effect. We will now find out if this narrower legal protection and Americans' respect for their national symbol are sufficient to prevent the bald eagle from falling once again into a spiral toward extinction.

Consider This

The bald eagle is a majestic animal, but it is only a single species. Was saving it worth decades of effort, investment, and lost opportunities for development?

Chapter Review

Summary of Key Concepts

For additional study help and activities, go to www.mybiology.com.

31.1 What Is Conservation Biology

Conservation biology is the branch of science that seeks to understand and conserve biodiversity. Biodiversity includes genetic diversity, species diversity, and ecological diversity.

31.2 Why Is Biodiversity Important?

Biodiversity is a source of goods such as food, fuel, building materials, and medicines. Biodiversity provides ecosystem services such as forming soil, purifying water, controlling erosion and floods, moderating climate, and providing genetic reserves and recreational opportunities. The emerging discipline of ecological economics attempts to measure the contribution of ecosystem goods and services to the economy, and estimates the costs of losing them to unsustainable development.

31.3 Is Earth's Biodiversity Diminishing?

Natural communities have a low background extinction rate. Many biologists believe that human activities are currently causing a mass extinction, increasing extinction rates by a factor of 100 to 1,000. About 16,300 species are now known to be threatened with extinction.

31.4 What Are the Major Threats to Biodiversity?

Human use of natural resources has exceeded Earth's ability to replenish what we are taking from it. Major threats to biodiversity include habitat destruction and fragmentation as ecosystems are converted to human uses; overexploitation as wild animals and plants are harvested beyond their ability to regenerate; the introduction of non-native invasive species; pollution; and global warming.

Web Animation   Habitat Destruction and Fragmentation

31.5 How Can Conservation Biology Help to Preserve Biodiversity?

Conservation biology seeks to identify the diversity of life, explore the impact of human activities on natural ecosystems, and apply this knowledge to conserve species and foster the survival of healthy, self-sustaining communities. It is based on the premise that biodiversity has intrinsic value. Conservation biology integrates knowledge from many areas of science, and requires the efforts of government leaders, environmental lawyers, conservation organizations, and most importantly, individuals. Conservation efforts include establishing wildlife reserves connected by wildlife corridors, with the goal of preserving functional communities and self-sustaining populations.

31.6 Why Is Sustainability the Key to Conservation?

Sustainable development meets present needs without compromising the future. It requires that people maintain biodiversity, recycle raw materials, and rely on renewable resources. Biosphere reserves promote conservation and sustainable development. A shift to sustainable farming is crucial to conserving soil and water, reducing pollution and energy use, and preserving biodiversity.

Human population growth is unsustainable and is driving consumption of resources beyond nature's ability to replenish them. We must bring our population into line with Earth's ability to support us, leaving room and resources for all forms of life. Individuals can help by making responsible choices that reduce resource consumption so that it does not exceed what Earth can replenish.

Key Terms

biodiversity *p. 625*

biosphere reserves *p. 636*

conservation biology *p. 625*

core reserves *p. 633*

critically endangered species
  p. 629

ecosystem services *p. 626*

endangered species *p. 629*

habitat fragmentation *p. 630*

mass extinction *p. 628*

minimum viable population
  (MVP) *p. 631*

no-till *p. 637*

overexploitation *p. 631*

sustainable development *p. 636*

threatened species *p. 630*

vulnerable species *p. 629*

wildlife corridors *p. 633*

Thinking Through the Concepts

Suggested answers to end-of-chapter and figure-based questions can be found at the end of the text.

Fill-in-the-Blank

1. The three levels of biodiversity are _____, _____, and _____. If the population of a species becomes too small, it is likely to have lost much of its _____ diversity. A monoculture farm will have little _____ diversity.

2. Products or processes by which functioning ecosystems benefit humans are collectively called _____. Four important examples of these benefits include _____, _____, _____, and _____.

3. Many of the benefits that humans derive from functioning ecosystems, such as purifying water, have traditionally been considered to be free. The discipline of _____ tries to quantify the monetary value of these benefits.

4. The major threats to biodiversity include _____, _____, _____, and _____. For most endangered species, _____ is probably the major threat.

5. The smallest population of a species that is likely to be able to survive in the long term is called the _____. When suitable habitat for a given species is split up into areas that are too small to support a large enough population, this is called _____. One way in which conservation biologists seek to maintain large enough populations is to set up core reserves of suitable habitat, connected by _____.

6. A Native American saying tells us that "We do not inherit the Earth from our ancestors, we borrow it from our children." If this principle guided our activities, we would practice _____ development.

Review Questions

1. Define conservation biology. What are some of the disciplines it draws upon?

2. What are the three different levels of biodiversity, and why is each one important?

3. What is ecological economics? Why is it important?

4. List the types of goods and services that natural ecosystems provide.

5. What four specific threats to biodiversity are described in this chapter? Provide an example of each.

Applying the Concepts

1. What are the goals of conservation biology? Do you agree with them? Why or why not?

2. Search for and describe some examples of habitat destruction, pollution, and invasive species in the region around your home or campus. Predict how each of these might affect specific local populations of native animals and plants.

3. Identify a dense suburban development near your home or school. Redesign it in any ways necessary to make it into a sustainable development. (This would make a good group project.)

4. What economic arguments would conventional farmers be likely to raise against switching to organic farming techniques and other sustainable agricultural methods? What would be the advantages to farmers of switching to organic methods? How would widespread switching affect consumers?

5. Some U.S. government officials are encouraging the use of biofuels (gasoline supplemented with palm or soybean oil, or ethanol) to reduce reliance on oil imports. Discuss this from as many angles as possible.

For additional resources, go to www.mybiology.com.

Appendix I

Metric System Conversions

To Convert Metric Units:	Multiply by:	To Get English Equivalent:
Length		
Centimeters (cm)	0.3937	Inches (in)
Meters (m)	3.2808	Feet (ft)
Meters (m)	1.0936	Yards (yd)
Kilometers (km)	0.6214	Miles (mi)
Area		
Square centimeters (cm^2)	0.155	Square inches (in^2)
Square meters (m^2)	10.7639	Square feet (ft^2)
Square meters (m^2)	1.1960	Square yards (yd^2)
Square kilometers (km^2)	0.3831	Square miles (mi^2)
Hectares (ha) (10,000 m^2)	2.4710	Acres (a)
Volume		
Cubic centimeters (cm^3)	0.06	Cubic inches (in^3)
Cubic meters (m^3)	35.30	Cubic feet (ft^3)
Cubic meters (m^3)	1.3079	Cubic yards (yd^3)
Cubic kilometers (km^3)	0.24	Cubic miles (mi^3)
Liters (L)	1.0567	Quarts (qt), U.S.
Liters (L)	0.26	Gallons (gal), U.S.
Mass		
Grams (g)	0.03527	Ounces (oz)
Kilograms (kg)	2.2046	Pounds (lb)
Metric tons (tonne) (t)	1.10	Tons (tn), U.S.
Speed		
Meters/second (mps)	2.24	Miles/hour (mph)
Kilometers/hour (kmph)	0.62	Miles/hour (mph)

To Convert English Units:	Multiply by:	To Get Metric Equivalent:
Length		
Inches (in)	2.54	Centimeters (cm)
Feet (ft)	0.3048	Meters (m)
Yards (yd)	0.9144	Meters (m)
Miles (mi)	1.6094	Kilometers (km)
Area		
Square inches (in^2)	6.45	Square centimeters (cm^2)
Square feet (ft^2)	0.0929	Square meters (m^2)
Square yards (yd^2)	0.8361	Square meters (m^2)
Square miles (mi^2)	2.5900	Square kilometers (km^2)
Acres (a)	0.4047	Hectares (ha) (10,000m^2)
Volume		
Cubic inches (in^3)	16.39	Cubic centimeters (cm^3)
Cubic feet (ft^3)	0.028	Cubic meters (m^3)
Cubic yards (yd^3)	0.765	Cubic meters (m^3)
Cubic miles (mi^3)	4.17	Cubic kilometers (km^3)
Quarts (qt), U.S.	0.9463	Liters (L)
Gallons (gal), U.S.	3.8	Liters (L)
Mass		
Ounces (oz)	28.3495	Grams (g)
Pounds (lb)	0.4536	Kilograms (kg)
Tons (tn), U.S.	0.91	Metric tons (tonne) (t)
Speed		
Miles/hour (mph)	0.448	Meters/second (mps)
Miles/hour (mph)	1.6094	Kilometers/hour (kmph)

Metric Prefixes

Prefix			Meaning
giga-	G	$10^9 =$	1,000,000,000
mega-	M	$10^6 =$	1,000,000
kilo-	k	$10^3 =$	1,000
hecto-	h	$10^2 =$	100
deka-	da	$10^1 =$	10
		$10^0 =$	1
deci-	d	$10^{-1} =$	0.1
centi-	c	$10^{-2} =$	0.01
milli-	m	$10^{-3} =$	0.001
micro-	µ	$10^{-6} =$	0.000001
nano-	n	$10^{-9} =$	0.000000001

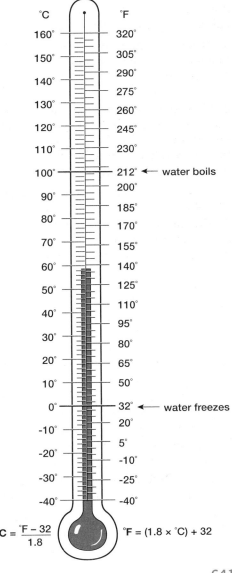

$$°C = \frac{°F - 32}{1.8} \qquad °F = (1.8 \times °C) + 32$$

Appendix II

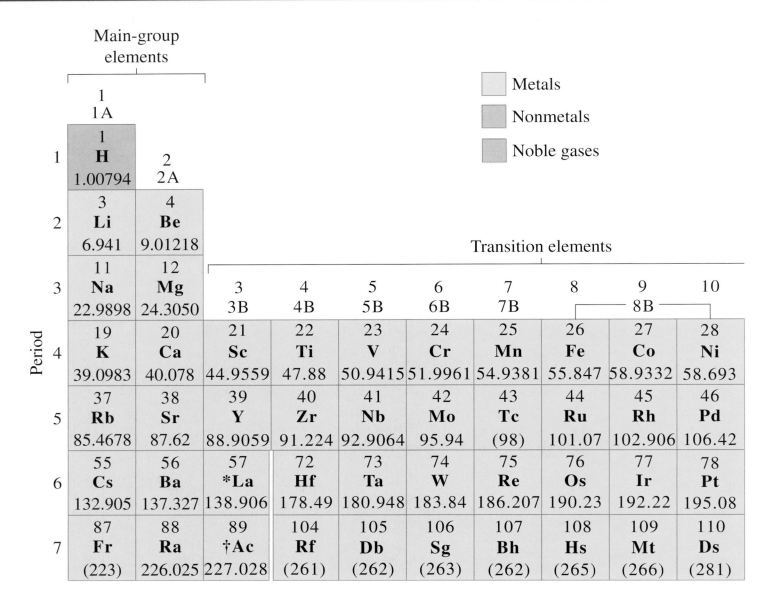

	Metals
	Nonmetals
	Noble gases

Main-group elements

Transition elements

Period	1 1A	2 2A								
1	1 **H** 1.00794									
2	3 **Li** 6.941	4 **Be** 9.01218								
			3 3B	4 4B	5 5B	6 6B	7 7B	8	9 8B	10
3	11 **Na** 22.9898	12 **Mg** 24.3050								
4	19 **K** 39.0983	20 **Ca** 40.078	21 **Sc** 44.9559	22 **Ti** 47.88	23 **V** 50.9415	24 **Cr** 51.9961	25 **Mn** 54.9381	26 **Fe** 55.847	27 **Co** 58.9332	28 **Ni** 58.693
5	37 **Rb** 85.4678	38 **Sr** 87.62	39 **Y** 88.9059	40 **Zr** 91.224	41 **Nb** 92.9064	42 **Mo** 95.94	43 **Tc** (98)	44 **Ru** 101.07	45 **Rh** 102.906	46 **Pd** 106.42
6	55 **Cs** 132.905	56 **Ba** 137.327	57 ***La** 138.906	72 **Hf** 178.49	73 **Ta** 180.948	74 **W** 183.84	75 **Re** 186.207	76 **Os** 190.23	77 **Ir** 192.22	78 **Pt** 195.08
7	87 **Fr** (223)	88 **Ra** 226.025	89 **†Ac** 227.028	104 **Rf** (261)	105 **Db** (262)	106 **Sg** (263)	107 **Bh** (262)	108 **Hs** (265)	109 **Mt** (266)	110 **Ds** (281)

	58 **Ce** 140.115	59 **Pr** 140.908	60 **Nd** 144.24	61 **Pm** (145)	62 **Sm** 150.36	63 **Eu** 151.965
*Lanthanide series						
†Actinide series	90 **Th** 232.038	91 **Pa** 231.036	92 **U** 238.029	93 **Np** 237.048	94 **Pu** (244)	95 **Am** (243)

** Not yet named

Notes: (1) Values in parentheses are the mass numbers of the most common or most stable isotopes of radioactive elements. (2) Some elements adjacent to the stair-step line between the metals and nonmetals have a metallic appearance but some nonmetallic properties. These elements are often called metalloids or semimetals. There is no general agreement on just which elements are so designated. Almost every list includes Si, Ge, As, Sb, and Te. Some also include B, At, and/or Po.

Main-group
elements

					18 8A
13 3A	14 4A	15 5A	16 6A	17 7A	2 **He** 4.00260
5 **B** 10.811	6 **C** 12.011	7 **N** 14.0067	8 **O** 15.9994	9 **F** 18.9984	10 **Ne** 20.1797

11 1B	12 2B	13 **Al** 26.9815	14 **Si** 28.0855	15 **P** 30.9738	16 **S** 32.066	17 **Cl** 35.4527	18 **Ar** 39.948
29 **Cu** 63.546	30 **Zn** 65.39	31 **Ga** 69.723	32 **Ge** 72.61	33 **As** 74.9216	34 **Se** 78.96	35 **Br** 79.904	36 **Kr** 83.80
47 **Ag** 107.868	48 **Cd** 112.411	49 **In** 114.818	50 **Sn** 118.710	51 **Sb** 121.76	52 **Te** 127.60	53 **I** 126.904	54 **Xe** 131.29
79 **Au** 196.967	80 **Hg** 200.59	81 **Tl** 204.383	82 **Pb** 207.2	83 **Bi** 208.980	84 **Po** (209)	85 **At** (210)	86 **Rn** (222)
111 ** (272)	112 ** (285)		114 ** (289)		116 ** (292)		

64 **Gd** 157.25	65 **Tb** 158.925	66 **Dy** 162.50	67 **Ho** 164.930	68 **Er** 167.26	69 **Tm** 168.934	70 **Yb** 173.04	71 **Lu** 174.967
96 **Cm** (247)	97 **Bk** (247)	98 **Cf** (251)	99 **Es** (252)	100 **Fm** (257)	101 **Md** (258)	102 **No** (259)	103 **Lr** (260)

Appendix III

Classification of Major Groups of Eukaryotic Organisms*

Kingdom	Phylum Class	Common Name
Chlorophyta		green algae
Amoebozoa	Tubulinea	amoebas
	Myxomycota	acellular slime molds
	Acrasiomycota	cellular slime molds
Cercozoa	Foraminifera	foraminiferans
Alveolata	Apicomplexa	sporozoans
	Pyrrophyta	dinoflagellates
	Ciliophora	ciliates
Stramenopila	Phaeophyta	brown algae
	Bacillariophyta	diatoms
Plantae (multicellular, photosynthetic)	Bryophyta	liverworts, mosses
	Pteridophyta	ferns
	Coniferophyta	evergreens
	Anthophyta	flowering plants
Fungi (multicellular, heterotrophic, absorb nutrients)	Chytridiomycota	chytrids
	Zygomycota	zygote fungi
	Ascomycota	sac fungi
	Basidiomycota	club fungi
Animalia (multicellular, heterotrophic, ingest nutrients)	Porifera	sponges
	Cnidaria	hydras, sea anemones, jellyfish, corals
	Annelida	segmented worms
	Oligochaeta	earthworms
	Polychaeta	tube worms
	Hirudinea	leeches
	Mollusca	mollusks
	Gastropoda	snails
	Pelecypoda	mussels, clams
	Cephalopoda	squid, octopuses
	Arthropoda	arthropods
	Insecta	insects
	Arachnida	spiders, ticks
	Crustacea	crabs, lobsters
	Chordata	chordates
	Urochordata	tunicates
	Cephalochordata	lancelets
	Myxini	hagfishes
	Petromyzontiformes	lampreys
	Chondrichthyes	sharks, rays
	Actinopterygii	ray-finned fishes
	Sarcopterygii	lobe-finned fishes
	Amphibia	frogs, salamanders
	Anapsida	turtles
	Diapsida	birds, crocodiles, lizards, snakes
	Mammalia	mammals

*This table lists only those taxonomic categories described in the textbook.

Glossary

abscisic acid (ab-sis´-ik): a plant hormone that generally inhibits the action of other hormones, enforcing dormancy in seeds and buds and causing the closing of stomata.

abscission layer: a layer of thin-walled cells, located at the base of the stalk of a leaf, that produces an enzyme that digests the cell wall holding leaf to stem, allowing the leaf to fall off.

absorption: the process by which nutrients are taken into cells.

acid: a substance that releases hydrogen ions (H^+) into solution; a solution with a pH of less than 7.

acid deposition: the deposition of nitric or sulfuric acid, either dissolved in rain (acid rain) or in the form of dry particles, as a result of the production of nitrogen oxides or sulfur dioxide through burning, primarily of fossil fuels.

acquired immune deficiency syndrome (AIDS): an infectious disease caused by the human immunodeficiency virus (HIV); attacks and destroys T cells, thus weakening the immune system.

action potential: a rapid change from a negative to a positive electrical potential in a nerve cell. This signal travels along an axon without a change in intensity.

activation energy: in a chemical reaction, the energy needed to force the electron shells of reactants together, before the formation of products.

active site: the region of an enzyme molecule that binds substrates and performs the catalytic function of the enzyme.

active transport: the movement of materials across a membrane through the use of cellular energy, normally against a concentration gradient.

adaptation: a trait that increases the ability of an individual to survive and reproduce compared to individuals without the trait.

adaptive radiation: the rise of many new species in a relatively short time as a result of a single species that invades different habitats and evolves under different environmental pressures in those habitats.

adenine (A): a nitrogenous base found in both DNA and RNA.

adenosine diphosphate (a-den´-ō-sēn-dī-fos´-fāt; ADP): a molecule composed of the sugar ribose, the base adenine, and two phosphate groups; a component of ATP.

adenosine triphosphate (a-den´-ō-sēn-trī-fos´-fāt; ATP): a molecule composed of the sugar ribose, the base adenine, and three phosphate groups; the major energy carrier in cells. The last two phosphate groups are attached by "high-energy" bonds.

ADP: see *adenosine diphosphate*.

adrenal gland: a mammalian endocrine gland, adjacent to the kidney; secretes hormones that function in water regulation and in the stress response.

age structure: the distribution of males and females in a population according to age groups.

aggression: antagonistic behavior, usually among individuals of the same species; often associated with competition for resources.

allantois (al-an-tō´-is): one of the embryonic membranes of reptiles, birds, and mammals; in reptiles and birds, serves as a waste-storage organ; in mammals, forms most of the umbilical cord.

allele (al-ēl´): one of several alternative forms of a particular gene.

allele frequency: for any given gene, the relative proportion of each allele of that gene in a population.

allergy: an inflammatory response produced by the body in response to invasion by foreign materials, such as pollen, that are themselves harmless.

allosteric regulation: the process by which enzyme action is enhanced or inhibited by small organic molecules that act as regulators by binding to the enzyme and altering its active site.

alternation of generations: a life cycle, typical of plants, in which a diploid sporophyte (spore-producing) generation alternates with a haploid gametophyte (gamete-producing) generation.

altruism: a type of behavior that may decrease the reproductive success of the individual performing it, but that increases the fitness of other individuals.

alveolus (al-vē´-ō-lus; pl., alveoli): a tiny air sac within the lungs, surrounded by capillaries, where gas exchange with the blood occurs.

amino acid: the individual subunit of which proteins are made, composed of a central carbon atom bonded to an amino group ($-NH_2$), a carboxyl group ($-COOH$), a hydrogen atom, and a variable group of atoms denoted by the letter *R*.

amino acid–derived hormone: a hormone consisting of a modified amino acid.

amnion (am´-nē-on): one of the embryonic membranes of reptiles, birds, and mammals; encloses a fluid-filled cavity that envelops the embryo.

amniotic egg (am-nē-ōt´-ik): the egg of reptiles and birds; contains an amnion that encloses the embryo in a watery environment, allowing the egg to be laid on dry land.

amphibian: a member of the chordate class Amphibia, which includes the frogs, toads, and salamanders, as well as the limbless caecilians.

anaerobe: an organism whose respiration does not require oxygen.

analogous structures: structures that have similar functions and superficially similar appearance but very different anatomies, such as the wings of insects and birds. The similarities are due to parallel evolution under similar environmental pressures rather than to common ancestry.

anaphase (an´-a-fāz): in mitosis, the stage in which the sister chromatids of each chromosome separate from one another and are moved to opposite poles of the cell; in meiosis I, the stage in which homologous chromosomes, consisting of two sister chromatids, are separated; in meiosis II, the stage in which the sister chromatids of each chromosome separate from one another and are moved to opposite poles of the cell.

angiosperm (an´-jē- ō-sperm): a flowering vascular plant.

anther (an´-ther): the uppermost part of the stamen, in which pollen develops.

antibiotic: a substance that kills or slows the reproduction of bacteria, fungi, and protists; may be used to treat infectious diseases.

antibody: a protein, produced by cells of the immune system, that combines with a specific antigen and normally facilitates the destruction of the antigen.

anticodon: a sequence of three bases in transfer RNA that is complementary to the three bases of a codon of messenger RNA.

antigen: a complex molecule, normally a protein or polysaccharide, that stimulates the production of a specific antibody.

aphotic zone: the region of the ocean deeper than 200 meters, where sunlight does not penetrate.

apical meristem (āp´-i-kul mer´-i-stem): the cluster of meristematic cells at the tip of a shoot or root (or one of their branches).

Archaea: one of life's three domains; consists of prokaryotes that are only distantly related to members of the domain Bacteria.

arteriole (ar-tēr´- ē- ōl): a small artery that empties into capillaries. Contraction of the arteriole regulates blood flow to various parts of the body.

artery (ar´-tuh-rē): a vessel with muscular, elastic walls that conducts blood away from the heart.

arthropod: a member of the animal phylum Arthropoda, which includes the insects, spiders, ticks, mites, scorpions, crustaceans, millipedes, and centipedes.

artificial selection: a selective breeding procedure in which only those individuals with particular traits are chosen as breeders; used mainly to enhance desirable traits in domestic plants and animals; may also be used in evolutionary biology experiments.

asexual reproduction: reproduction that does not involve the fusion of haploid sex cells. The parent body may divide and new parts regenerate, or a new, smaller individual may form as an attachment to the parent, to drop off when complete.

atom: the smallest particle of an element that retains the properties of the element.

atomic nucleus: the central region of an atom, consisting of protons and neutrons.

atomic number: the number of protons in the nuclei of all atoms of a particular element.

ATP: see *adenosine triphosphate*.

atrioventricular (AV) node (ā´trē- ō-ven-trik´- ū-lar nōd): a specialized mass of muscle at the base of the right atrium through which the electrical activity initiated in the sinoatrial node is transmitted to the ventricles.

atrioventricular valve: a heart valve that separates each atrium from each ventricle, preventing the backflow of blood into the atria during ventricular contraction.

atrium (ā´-trē-um): a chamber of the heart that receives venous blood and passes it to a ventricle.

autoimmune disease: a disorder in which the immune system produces antibodies against the body's own cells.

autonomic nervous system: the part of the peripheral nervous system of vertebrates that synapses on glands, internal organs, and smooth muscle and produces largely involuntary responses.

autosome (aw´-tō-sōm): a chromosome that occurs in homologous pairs in both males and females and that does not bear the genes determining sex.

autotroph (aw'-tō-trōf): literally, "self-feeder"; an organism that produces its own food; normally, a photosynthetic organism; a producer.

auxin (awk'-sin): a plant hormone that influences many plant functions, including phototropism, apical dominance, and root branching; generally stimulates cell elongation and, in some cases, cell division and differentiation.

axon: a long extension of a nerve cell, extending from the cell body to synaptic endings on other nerve cells or on muscles.

B cell: a type of lymphocyte that participates in humoral immunity; gives rise to plasma cells, which secrete antibodies into the circulatory system, and to memory cells.

Bacteria: one of life's three domains; consists of prokaryotes that are only distantly related to members of the domain Archaea.

base: (1) a substance capable of combining with and neutralizing H+ ions in a solution; a solution with a pH of more than 7; (2) in molecular genetics, one of the nitrogen-containing, single- or double-ringed structures that distinguish one nucleotide from another. In DNA, the bases are adenine, guanine, cytosine, and thymine.

behavior: any observable activity of a living animal.

bile (bīl): a liquid secretion, produced by the liver, that is stored in the gallbladder and released into the small intestine during digestion; a complex mixture of bile salts, water, other salts, and cholesterol.

binary fission: the process by which a single bacterium divides in half, producing two identical offspring.

biodegradable: able to be broken down into harmless substances by decomposers.

biodiversity: the diversity of living organisms; measured as the variety of different species, the variety of different alleles in a gene pool, or the variety of different community interactions in an ecosystem.

biogeochemical cycle: also called a *nutrient cycle*, the process by which a specific nutrient in an ecosystem is transferred between living organisms and the nutrient's reservoir in the nonliving environment.

biological magnification: the increasing accumulation of a toxic substance in progressively higher trophic levels.

biome (bī'-ōm): a terrestrial ecosystem that occupies an extensive geographical area and is characterized by a specific type of plant community; for example, a desert.

biosphere reserves: regions designated by the United Nations as reserves intended to maintain biodiversity while preserving local cultures and serving as models for sustainable economic development.

biotechnology: any industrial or commercial use or alteration of organisms, cells, or biological molecules to achieve specific practical goals.

biotic potential: the maximum rate at which a population could increase, assuming ideal conditions that allow a maximum birth rate and minimum death rate.

blastocyst (blas'-tō-sist): an early stage of human embryonic development, consisting of a hollow ball of cells, enclosing a mass of cells attached to its inner surface, which becomes the embryo.

blastula (blas'-tū-luh): in animals, the embryonic stage attained at the end of cleavage, in which the embryo normally consists of a hollow ball with a wall one or several cell layers thick.

blood: a fluid consisting of plasma in which blood cells are suspended; carried within the circulatory system.

blood clotting: a complex process by which platelets, the protein fibrin, and red blood cells block an irregular surface in or on the body, such as a damaged blood vessel, sealing the wound.

blood vessel: a channel that conducts blood throughout the body.

bone: a hard, mineralized connective tissue that is a major component of the vertebrate endoskeleton; provides support and sites for muscle attachment.

Bowman's capsule: the cup-shaped portion of the nephron in which blood filtrate is collected from the glomerulus.

brain: the part of the central nervous system of vertebrates that is enclosed within the skull.

bronchiole (bron'-kē- ōl): a narrow tube, formed by repeated branching of the bronchi, that conducts air into the alveoli.

bronchus (bron'-kus; pl., bronchi): a tube that conducts air from the trachea to each lung.

budding: asexual reproduction by the growth of a miniature copy, or bud, of the adult animal on the body of the parent. The bud breaks off to begin independent existence.

buffer: a compound that minimizes changes in pH by reversibly taking up or releasing H+ ions.

C₃ cycle: the cyclic series of reactions whereby carbon dioxide is fixed into carbohydrates during the light-independent reactions of photosynthesis; also called *Calvin-Benson cycle*.

C₄ pathway: the series of reactions in certain plants that fixes carbon dioxide into oxaloacetic acid, which is later broken down for use in the C₃ cycle of photosynthesis.

calorie (kal'-ō-rē): the amount of energy required to raise the temperature of 1 gram of water by 1 degree Celsius.

Calorie: a unit of energy, in which the energy content of foods is measured; the amount of energy required to raise the temperature of 1 liter of water 1 degree Celsius; also called a *kilocalorie*, equal to 1,000 calories.

CAM: abbreviation for crassulacean acid metabolism, a variant form of photosynthesis in which carbon dioxide is incorporated into organic molecules at night, when stomata are open, and then released to the C₃ cycle during the day, when the stomata are closed; used by plants in hot, dry environments.

cambium (kam'-bē-um; pl., cambia): a lateral meristem, parallel to the long axis of roots and stems, that causes secondary growth of woody plant stems and roots. See *cork cambium; vascular cambium*.

camouflage (cam'-a-flaj): coloration and/or shape that renders an organism inconspicuous in its environment.

cancer: a disease in which some of the body's cells escape from normal regulatory processes and divide without control.

capillary: the smallest type of blood vessel, connecting arterioles with venules. Capillary walls, through which the exchange of nutrients and wastes occurs, are only one cell thick.

carbohydrate: a compound composed of carbon, hydrogen, and oxygen, with the approximate chemical formula $(CH_2O)_n$; includes sugars and starches.

carbon fixation: the initial steps in the C₃ cycle, in which carbon dioxide reacts with ribulose bisphosphate to form a stable organic molecule.

carnivore (kar'-neh-vor): literally, "meat eater"; a predatory organism that feeds on herbivores or on other carnivores; a secondary (or higher) consumer.

carpel (kar'pel): the female reproductive structure of a flower, composed of stigma, style, and ovary.

carrier protein: a membrane protein that facilitates the diffusion of specific substances across the membrane. The molecule to be transported binds to the outer surface of the carrier protein; the protein then changes shape, allowing the molecule to move across the membrane through the protein.

carrying capacity: the maximum population size that an ecosystem can support indefinitely; determined primarily by the availability of space, nutrients, water, and light.

cartilage (kar'-teh-lij): a form of connective tissue that forms portions of the skeleton; consists of chondrocytes and their extracellular secretion of collagen; resembles flexible bone.

catalyst (kat'-uh-list): a substance that speeds up a chemical reaction without itself being permanently changed in the process; lowers the activation energy of a reaction.

cell: the smallest unit of life, consisting, at a minimum, of an outer membrane that encloses a watery medium containing organic molecules, including genetic material composed of DNA.

cell body: the part of a nerve cell in which most of the common cellular organelles are located; typically a site of integration of inputs to the nerve cell.

cell cycle: the sequence of events in the life of a cell, from one division to the next.

cell division: splitting of one cell into two; the process of cellular reproduction.

cell plate: a membrane-bound structure, formed in plant cells during cytokinesis, that becomes the cell wall separating two daughter cells.

cell wall: a layer of material, normally made up of cellulose or cellulose-like materials, that is outside the plasma membrane of plants, fungi, bacteria, and some protists.

cell-mediated immunity: an immune response in which foreign cells or substances are destroyed by contact with T cells.

cellular respiration: the oxygen-requiring reactions, occurring in mitochondria, that break down the end products of glycolysis into carbon dioxide and water while capturing large amounts of energy as ATP.

central nervous system (CNS): in vertebrates, the brain and spinal cord.

central vacuole: a large, fluid-filled vacuole occupying most of the volume of many plant cells; performs several functions, including maintaining turgor pressure.

centriole (sen'-trē- ōl): in animal cells, a short, barrel-shaped ring consisting of nine microtubule triplets; a microtubule-containing structure at the base of each cilium and flagellum; gives rise to the microtubules of cilia and flagella and is involved in spindle formation during cell division.

centromere (sen'-trō-mēr): the region of a replicated chromosome at which the sister chromatids are held together until they separate during cell division.

cervix (ser'-viks): a ring of connective tissue at the outer end of the uterus, leading into the vagina.

channel protein: a membrane protein that forms a channel or pore completely through the membrane and that is usually permeable to one or to a few water-soluble molecules, especially ions.

chaparral: a biome that is located in coastal regions but has very low annual rainfall.

chemical bond: the force of attraction between neighboring atoms that holds them together in a molecule.

chemical reaction: the process that forms and breaks chemical bonds that hold atoms together.

chemiosmosis (ke-mē-oz-mō′-sis): a process of adenosine triphosphate (ATP) generation in chloroplasts and mitochondria. The movement of electrons down an electron transport system is used to pump hydrogen ions across a membrane, thereby building up a concentration gradient of hydrogen ions across the membrane; the hydrogen ions diffuse back across the membrane through the pores of ATP-synthesizing enzymes; the energy of their movement down their concentration gradient drives ATP synthesis.

chiasma (kī-as′-muh; pl., chiasmata): a point at which a chromatid of one chromosome crosses with a chromatid of the homologous chromosome during prophase I of meiosis; the site of exchange of chromosomal material between chromosomes.

chlorophyll (klor′- ō-fil): a pigment found in chloroplasts that captures light energy during photosynthesis; absorbs violet, blue, and red light but reflects green light.

chloroplast (klor′- ō-plast): the organelle in plants and photosynthetic protists that is the site of photosynthesis; surrounded by a double membrane and containing an extensive internal membrane system that bears chlorophyll.

chorion (kor′-ē-on): the outermost embryonic membrane in reptiles, birds, and mammals; in birds and reptiles, functions mostly in gas exchange; in mammals, forms most of the embryonic part of the placenta.

chromatid (krō′ma-tid): one of the two identical strands of DNA and protein that forms a replicated chromosome. The two sister chromatids are joined at the centromere.

chromatin (krō′-ma-tin): the complex of DNA and proteins that makes up eukaryotic chromosomes.

chromosome (krō′-mō-sōm): a DNA double helix together with proteins that help to organize the DNA.

chyme (kīm): an acidic, souplike mixture of partially digested food, water, and digestive secretions that is released from the stomach into the small intestine.

cilium (sil′-ē-um; pl., cilia): a short, hairlike projection from the surface of certain eukaryotic cells that contains microtubules in a 9 + 2 arrangement. The movement of cilia may propel cells through a fluid medium or move fluids over a stationary surface layer of cells.

citric acid cycle: see *Krebs cycle*.

cleavage: the early cell divisions of embryos, in which little or no growth occurs between divisions; reduces cell size and distributes gene-regulating substances to the newly formed cells.

climate: patterns of weather that prevail from year to year and even from century to century in a given region.

climax community: a diverse and relatively stable community that forms the endpoint of succession.

clitoris: an external structure of the female reproductive system; composed of erectile tissue; a sensitive point of stimulation during sexual response.

clonal selection: the mechanism by which the immune response gains specificity; an invading antigen elicits a response from only a few lymphocytes, which proliferate to form a clone of cells that attack only the specific antigen that stimulated their production.

clone: offspring that are produced by mitosis and are therefore genetically identical to each other.

cloning: the process of producing many identical copies of a gene; also the production of many genetically identical copies of an organism.

clumped distribution: the distribution characteristic of populations in which individuals are clustered into groups; may be social or based on the need for a localized resource.

CNS: see *central nervous system*.

codominance: the relation between two alleles of a gene, such that both alleles are phenotypically expressed in heterozygous individuals.

codon: a sequence of three bases of messenger RNA that specifies a particular amino acid to be incorporated into a protein; certain codons also signal the beginning or end of protein synthesis.

coevolution: the evolution of adaptations in two species due to their extensive interactions with one another, in which each species is a source of natural selection on the other.

cohesion: the tendency of the molecules of a substance to stick together.

coleoptile (kō-lē-op′-tīl): a protective sheath surrounding the shoot in monocot seeds; allows the shoot to push aside soil particles as it grows.

collecting duct: a conducting tube, within the kidney, that collects urine from many nephrons and conducts it through the renal medulla into the renal pelvis. Urine may become concentrated in the collecting ducts if antidiuretic hormone (ADH) is present.

colon: the longest part of the large intestine, exclusive of the rectum.

communication: the act of producing a signal that causes another animal, normally of the same species, to change its behavior in a way that is, on average, beneficial to both individuals.

community: all of the interacting populations within an ecosystem.

competition: interaction among individuals who attempt to use a resource (for example, food or space) that is limited relative to the demand for it.

competitive exclusion principle: the concept that no two species can simultaneously and continuously occupy the same ecological niche.

competitive inhibition: the process by which two or more molecules that are somewhat similar in structure compete for the active site of an enzyme.

complementary base pair: in nucleic acids, bases that pair by hydrogen bonding. In DNA, adenine is complementary to thymine and guanine is complementary to cytosine; in RNA, adenine is complementary to uracil, and guanine to cytosine.

complete flower: a flower that has all four floral parts (sepals, petals, stamens, and carpels).

concentration: the number of particles of a dissolved substance in a given unit of volume.

concentration gradient: the difference in concentration of a substance between two parts of a fluid or across a barrier such as a membrane.

conclusion: the final operation in the scientific method; a decision made about the validity of a hypothesis on the basis of experimental evidence.

conifer (kon′-eh-fer): a member of a class of vascular plants (Coniferophyta) that reproduce by means of seeds formed inside cones and that retain their leaves throughout the year.

connective tissue: a tissue type that includes diverse tissues, such as bone, fat, and blood, that generally contain large amounts of extracellular material.

conservation biology: a scientific discipline that analyzes biological diversity and seeks to protect and preserve it.

constant region: the part of an antibody molecule that is similar in all antibodies.

consumer: an organism that eats other organisms; a heterotroph.

contraception: the prevention of pregnancy.

control: that portion of an experiment in which all possible variables are held constant; in contrast to the "experimental" portion, in which a particular variable is altered.

convergent evolution: the independent evolution of similar structures among distantly related organisms as a result of similar environmental pressures; see *analogous structures*.

copulation: reproductive behavior in which the penis of the male is inserted into the body of the female, where it releases sperm.

coral reef: a biome created by animals (reef-building corals) and plants in warm tropical waters.

core reserve: a natural area protected from most human uses and that encompasses an area large enough to preserve ecosystems and the species in them.

cork cambium: a lateral meristem in woody roots and stems that gives rise to cork cells.

corpus luteum (kor′-pus loo′-tē-um): in the mammalian ovary, a structure that is derived from the follicle after ovulation and that secretes the hormones estrogen and progesterone.

cortex: the part of a primary root or stem located between the epidermis and the vascular cylinder.

cotyledon (kot-ul-ē′don): a leaflike structure within a seed that absorbs food molecules from the endosperm and transfers them to the growing embryo; also called *seed leaf*.

coupled reaction: a pair of reactions, one exergonic and one endergonic, that are linked together such that the energy produced by the exergonic reaction provides the energy needed to drive the endergonic reaction.

covalent bond (kō-vā′-lent): a chemical bond between atoms in which electrons are shared.

critically endangered species: a species that faces an extremely high risk of extinction in the wild in the near future.

cross-fertilization: the union of sperm and egg from two individuals of the same species.

crossing over: the exchange of corresponding segments of the chromatids of two homologous chromosomes during meiosis.

cuticle (kū′-ti-kul): a waxy or fatty coating on the exposed surfaces of epidermal cells of many land plants, which aids in the retention of water.

cytokinesis (sī-tō-ki-nē′sis): the division of the cytoplasm and organelles into two daughter cells during cell division; normally occurs during telophase of mitosis.

cytokinin (sī-tō-kī′-nin): a plant hormone that promotes cell division, fruit growth, and the sprouting of lateral buds and prevents the aging of plant parts, especially leaves.

cytoplasm (sī′-tō-plaz-um): the material contained within the plasma membrane of a cell, exclusive of the nucleus.

cytosine (C): a nitrogenous base found in both DNA and RNA.

cytoskeleton: a network of protein fibers in the cytoplasm that gives shape to a cell, holds and moves organelles, and is typically involved in cell movement.

cytotoxic T cell: a type of T cell that, upon contacting foreign cells, directly destroys them.

daughter cell: a cell that is produced as the result of cell division.

decomposer: an organism, normally a fungus or bacterium, that digests organic material by secreting digestive enzymes into the environment, in the process liberating nutrients into the environment.

deforestation: the excessive cutting of forests, primarily rain forests in the tropics, to clear space for agriculture.

dehydration synthesis: a chemical reaction in which two molecules are joined by a covalent bond with the simultaneous removal of a hydrogen atom from one molecule and a hydroxyl group from the other, forming water; the reverse of hydrolysis.

deletion mutation: a mutation in which one or more pairs of nucleotides are removed from a gene.

denature: to disrupt the secondary and/or tertiary structure of a protein while leaving its amino acid sequence intact. Denatured proteins can no longer perform their biological functions.

dendrite (den′-drīt): a branched tendril that extends outward from the cell body of a neuron; specialized to respond to signals from the external environment or from other neurons.

denitrifying bacterium (dē-nī′-treh-fī-ing): a bacterium that breaks down nitrates, releasing nitrogen gas to the atmosphere.

density-dependent: referring to any factor, such as predation, that limits population size more effectively as the population density increases.

density-independent: referring to any factor that limits a population's size and growth regardless of its density.

deoxyribonucleic acid (dē-ox-ē-rī- bō-noo-klā′-ik; DNA): a molecule composed of deoxyribose nucleotides; contains the genetic information of all living cells.

dermal tissue: plant tissue that makes up the outer covering of the plant body.

desert: a biome in which less than 10 to 20 inches (25 to 50 centimeters) of rain falls each year.

detritus feeder (de-trī′-tus): one of a diverse group of organisms, ranging from worms to vultures, that lives off the wastes and dead remains of other organisms.

development: the process by which an organism proceeds from fertilized egg through adulthood to eventual death.

diaphragm (dī′-uh-fram): in the respiratory system, a dome-shaped muscle forming the floor of the chest cavity that, when it contracts, pulls itself downward, enlarging the chest cavity and causing air to be drawn into the lungs; in contraception, a rubber cap that fits snugly over the cervix, preventing the sperm from entering the uterus and thereby preventing pregnancy.

dicot (dī′-kaht): short for dicotyledon; a type of flowering plant characterized by embryos with two cotyledons, or seed leaves, modified for food storage.

differentiated cell: a mature cell specialized for a specific function; in plants, differentiated cells normally do not divide.

diffusion: the net movement of particles from a region of high concentration of that type of particle to a region of low concentration, driven by the concentration gradient; may occur entirely within a fluid or across a barrier such as a membrane.

digestion: the process by which food is physically and chemically broken down into molecules that can be absorbed by cells.

diploid (dip′-loid): referring to a cell with pairs of homologous chromosomes.

distal tubule: in the nephrons of the mammalian kidney, the last segment of the renal tubule through which the filtrate passes just before it empties into the collecting duct; a site of selective secretion and reabsorption as water and ions pass between the blood and the filtrate across the tubule membrane.

DNA: see *deoxyribonucleic acid.*

DNA helicase: an enzyme that helps unwind the DNA double helix during DNA replication.

DNA ligase: an enzyme that joins the sugars and phosphates in a DNA strand to create a continuous sugar-phosphate backbone.

DNA polymerase: an enzyme that bonds DNA nucleotides together into a continuous strand, using a preexisting DNA strand as a template.

DNA probe: a sequence of nucleotides that is complementary to the nucleotide sequence in a gene under study; used to locate a given gene within a DNA library.

DNA profile: a pattern determined by the number of short tandem repeats in particular DNA segments in an individual's genome. An individual is uniquely identified by a DNA profile that includes counts of short tandem repeats in several different DNA segments.

DNA replication: the copying of the double-stranded DNA molecule, producing two identical DNA double helices.

dominance hierarchy: a social arrangement in which a group of animals, usually through aggressive interactions, establishes a rank for some or all of the group members that determines access to resources.

dominant: an allele that can determine the phenotype of heterozygotes completely, such that they are indistinguishable from individuals homozygous for the allele; in the heterozygotes, the expression of the other (recessive) allele is completely masked.

dormancy: a state in which an organism does not grow or develop; usually marked by lowered metabolic activity and resistance to adverse environmental conditions.

double fertilization: in flowering plants, the fusion of two sperm nuclei with the nuclei of two cells of the female gametophyte. One sperm nucleus fuses with the egg to form the zygote; the second sperm nucleus fuses with the two haploid nuclei of the primary endosperm cell, forming a triploid endosperm cell.

double helix (hē′-liks): the shape of the two-stranded DNA molecule; like a ladder twisted lengthwise into a corkscrew shape.

ecological niche: the role of a particular species within an ecosystem, including all aspects of its interaction with the living and nonliving environments.

ecology (ē-kol′-uh-jē): the study of the interrelationships of organisms with each other and with their nonliving environment.

ecosystem (ē′kō-sis-tem): all of the organisms and their nonliving environment within a defined area.

ecosystem services: the effects of interactions among species in natural ecosystems that sustain and enrich human life.

ectoderm (ek′-tō-derm): the outermost embryonic tissue layer, which gives rise to structures such as hair, the epidermis of the skin, and the nervous system.

effector (ē-fek′-tor): a part of the body (normally a muscle or gland) that carries out responses as directed by the nervous system.

egg: the haploid female gamete, normally large and nonmotile, containing food reserves for the developing embryo.

electron: a subatomic particle, found in an electron shell outside the nucleus of an atom, that bears a unit of negative charge and very little mass.

electron carrier: a molecule that can reversibly gain or lose electrons. Electron carriers generally accept high-energy electrons produced during an exergonic reaction and donate the electrons to acceptor molecules that use the energy to drive endergonic reactions.

electron micrograph: a photographic image of an object viewed through an electron microscope.

electron shell: a region within which electrons orbit that corresponds to a fixed energy level at a given distance from the atomic nucleus of an atom.

electron transport chain (ETC): a series of electron carrier molecules, found in the thylakoid membranes of chloroplasts and the inner membrane of mitochondria, that extract energy from electrons and generate ATP or other energetic molecules.

element: a substance that cannot be broken down, or converted, to a simpler substance by ordinary chemical means.

embryo: in plants and animals, an organism in the earliest stages of development.

embryo sac: the haploid female gametophyte of flowering plants.

endangered species: a species that faces a high risk of extinction in the wild in the near future.

endergonic reaction: a chemical reaction that requires an input of energy to proceed.

endocrine gland: a ductless, hormone-producing gland consisting of cells that release their secretions into the extracellular fluid, from which the secretions diffuse into nearby capillaries.

endocrine system: an animal's organ system for cell-to-cell communication, composed of hormones and the cells that secrete them and receive them.

endocytosis (en-dō-sī-tō′-sis): the process in which the plasma membrane engulfs extracellular material, forming membrane-bound sacs that enter the cytoplasm and thereby move material into the cell.

endoderm (en′-dō-derm): the innermost embryonic tissue layer, which gives rise to structures such as the lining of the digestive and respiratory tracts.

endodermis (en-dō-der′-mis): the innermost layer of small, close-fitting cells of the cortex of a root that form a ring around the vascular cylinder.

endosperm: a triploid food-storage tissue in the seeds of flowering plants that nourishes the developing plant embryo.

endospore: a protective resting structure of some rod-shaped bacteria that withstands unfavorable environmental conditions.

endosymbiont hypothesis: the hypothesis that certain organelles, especially chloroplasts and mitochondria, arose as mutually beneficial associations between the ancestors of eukaryotic cells and captured bacteria that lived within the cytoplasm of the pre-eukaryotic cell.

energy: the capacity to do work.

energy-carrier molecule: a molecule that stores energy in "high-energy" chemical bonds and releases the energy to drive coupled endothermic reactions. In cells, adenosine triphosphate (ATP) is the most common energy-carrier molecule.

energy pyramid: a graphical representation of the energy contained in succeeding trophic levels, with maximum energy at the base (primary producers) and steadily diminishing amounts at higher levels.

entropy (en'trō-pē): a measure of the amount of randomness and disorder in a system.

environmental resistance: any factor that tends to counteract biotic potential, limiting population size.

enzyme (en'zīm): a protein catalyst that speeds up the rate of specific biological reactions.

epidermis (ep-uh-der'-mis): in animals, specialized epithelial tissue that forms the outer layer of skin; in plants, the outermost layer of cells of a leaf, young root, or young stem.

epididymis (e-pi-di'-di-mus): a series of tubes that connect with and receive sperm from the seminiferous tubules of the testis.

epithelial tissue (eh-puh-thē'-lē-ul): a tissue type that forms membranes that cover the body surface and line body cavities and that also gives rise to glands.

equilibrium population: a population in which allele frequencies and the distribution of genotypes do not change from generation to generation.

esophagus (eh-sof'-eh-gus): a muscular passageway that conducts food from the pharynx to the stomach in humans and other mammals.

essential amino acid: an amino acid that is a required nutrient; the body is unable to manufacture essential amino acids, so they must be supplied in the diet.

essential fatty acid: a fatty acid that is a required nutrient; the body is unable to manufacture essential fatty acids, so they must be supplied in the diet.

estuary: the area where a river meets the ocean; the salinity there is quite variable but lower than in seawater and higher than in fresh water.

ethylene: a plant hormone that promotes the ripening of fruits and the dropping of leaves and fruit.

Eukarya: one of life's three domains; consists of all eukaryotes (plants, animals, fungi, and protists).

eukaryote (ū-kar'-ē-ōt): an organism whose cells are eukaryotic; plants, animals, fungi, and protists are eukaryotes.

eukaryotic cell (ū-kar-ē-ot'-ik): referring to cells of organisms of the domain Eukarya (plants, animals, fungi, and protists). Eukaryotic cells have genetic material enclosed within a membrane-bound nucleus and contain other membrane-bound organelles.

eutrophic lake: a lake that receives sufficiently large inputs of sediments, organic material, and inorganic nutrients from its surroundings to support dense communities; murky with poor light penetration.

evolution: (1) the theory that all species are related by common ancestry and have changed over time; (2) any change in the proportions of different genotypes in a population from one generation to the next.

excretion: the elimination of waste substances from the body; can occur from the digestive system, skin glands, urinary system, or lungs.

exergonic reaction: a chemical reaction that liberates energy (either as heat or in the form of increased entropy).

exhalation: the act of releasing air from the lungs, which results from a relaxation of the respiratory muscles.

exocrine gland: a gland that releases its secretions into ducts that lead to the outside of the body or into the digestive tract.

exocytosis (ex-ō-sī-tō'-sis): the process in which intracellular material is enclosed within a membrane-bound sac that moves to the plasma membrane and fuses with it, releasing the material outside the cell.

exoskeleton (ex'-ō-skel'-uh-tun): a rigid external skeleton that supports the body, protects the internal organs, and has flexible joints that allow for movement.

exotic species: a foreign species introduced into an ecosystem where it did not evolve; such a species may flourish and outcompete native species.

experiment: the fifth operation in the scientific method; the testing, by further observations, of a prediction generated by a hypothesis, leading to a conclusion.

exponential growth: a continuously accelerating increase in population size.

extinction: the death of all members of a species.

extracellular fluid: fluid, similar in composition to plasma (except lacking large proteins), that leaks from capillaries and acts as a medium of exchange between the body cells and the capillaries.

extraembryonic membrane: in the embryonic development of reptiles, birds, and mammals, either the chorion, amnion, allantois, or yolk sac; functions in gas exchange, provision of the watery environment needed for development, waste storage, and storage of the yolk, respectively.

facilitated diffusion: the diffusion of molecules across a membrane, assisted by protein pores or carriers embedded in the membrane.

feces: semisolid waste material that remains in the intestine after absorption is complete and is voided through the anus. Feces consist of indigestible wastes and the dead bodies of bacteria.

fermentation: anaerobic reactions that convert the pyruvic acid produced by glycolysis into lactic acid or alcohol and CO_2.

fertilization: the fusion of male and female haploid gametes, forming a zygote.

fetus: a mamal in the later stages of embryonic development (after the second month for humans), when the developing animal has come to resemble the adult of the species.

fever: an elevation in body temperature caused by chemicals (pyrogens) that are released by white blood cells in response to infection.

fibrous roots: a root system, commonly found in monocots, characterized by many roots of approximately the same size arising from the base of the stem.

filtrate: the fluid produced by filtration; in the kidneys, the fluid produced by the filtration of blood through the glomerular capillaries.

filtration: within Bowman's capsule in each nephron of a kidney, the process by which blood is pumped under pressure through permeable capillaries of the glomerulus, forcing out water, dissolved wastes, and nutrients.

first law of thermodynamics: the principle of physics that states that within any isolated system, energy can

be neither created nor destroyed but can be converted from one form to another.

fission: asexual reproduction by dividing the body into two smaller, complete organisms.

fitness: the reproductive success of an organism, usually expressed in relation to the average reproductive success of all individuals in the same population.

flagellum (fla-jel'-um; pl., flagella): a long, hairlike extension of the plasma membrane; in eukaryotic cells, it contains microtubules arranged in a 9 + 2 pattern. The movement of flagella propel some cells through fluids.

flavin adenine dinucleotide (FADH₂): an electron carrier molecule produced in the mitochondrial matrix by the Krebs cycle; subsequently donates electrons to the electron transport chain.

flower: the reproductive structure of an angiosperm plant.

fluid mosaic: a model of membrane structure; according to this model, membranes are composed of a double layer of phospholipids in which various proteins are embedded. The phospholipid bilayer is a somewhat fluid matrix that allows the movement of proteins within it.

follicle: in the ovary of female mammals, the oocyte and its surrounding accessory cells.

food chain: a linear feeding relationship in a community, using a single representative from each of the trophic levels.

food web: a representation of the complex feeding relationships (in terms of interacting food chains) within a community, including many organisms at various trophic levels, with many of the consumers occupying more than one level simultaneously.

forebrain: during development, the anterior portion of the brain. In mammals, the forebrain differentiates into the thalamus, the limbic system, and the cerebrum. In humans, the cerebrum contains about half of all neurons in the brain.

fossil: the remains of a dead organism, normally preserved in rock—may be petrified bones or wood; shells; impressions of body forms, such as feathers, skin, or leaves; or markings made by organisms, such as footprints.

fossil fuel: a fuel, such as coal, oil, and natural gas, derived from the remains of ancient organisms.

founder effect: the result of an event in which an isolated population is founded by a small number of individuals; may result in genetic drift if allele frequencies in the founder population are by chance different from those of the parent population.

fruit: in flowering plants, the ripened ovary (plus, in some cases, other parts of the flower), which contains the seeds.

functional group: one of several groups of atoms commonly found in an organic molecule, including hydrogen, hydroxyl, amino, carboxyl, and phosphate groups, that determine the characteristics and chemical reactivity of the molecule.

gallbladder: a small sac, next to the liver, in which the bile secreted by the liver is stored and concentrated. Bile is released from the gallbladder to the small intestine through the bile duct.

gamete (gam'-ēt): a haploid sex cell formed in sexually reproducing organisms.

gametophyte (ga-mēt'-o-fīt): the multicellular haploid stage in the life cycle of plants.

gastrula (gas'-troo-luh): in animal development, a three-layered embryo with ectoderm, mesoderm, and endoderm cell layers. The endoderm layer normally encloses the primitive gut.

gastrulation (gas-troo-la′-shun): the process whereby a blastula develops into a gastrula, including the formation of endoderm, ectoderm, and mesoderm.

gel electrophoresis: a technique in which molecules (such as DNA fragments) are placed on restricted tracks in a thin sheet of gelatinous material and exposed to an electric field; the molecules then migrate at a rate determined by certain characteristics, such as length.

gene: a unit of heredity that encodes the information needed to specify the amino acid sequence of proteins and hence particular traits; a functional segment of DNA located at a particular place on a chromosome.

gene flow: the movement of alleles from one population to another owing to the movement of individual organisms or their gametes.

gene pool: the total of all alleles of all genes in a population; for a single gene, the total of all of the alleles of that gene that occur in a population.

genetic code: the collection of codons of messenger RNA (mRNA), each of which directs the incorporation of a particular amino acid into a protein during protein synthesis.

genetic drift: a change in the allele frequencies of a population that occurs purely by chance.

genetic engineering: the modification of genetic material to achieve specific goals.

genetically modified organism (GMO): an organism produced by genetic engineering.

genome (jē′-nōm): the entire set of genes carried by a member of any given species.

genotype (jēn′- ō-tīp): the genetic composition of an organism; the actual alleles of each gene carried by the organism.

germination: the growth and development of a seed, spore, or pollen grain.

gibberellin (jib-er-el′-in): a plant hormone that stimulates seed germination, fruit development, and cell division and elongation.

gill: in aquatic animals, a branched tissue richly supplied with capillaries around which water is circulated for gas exchange.

gland: a cluster of cells that are specialized to secrete substances such as sweat or hormones.

global warming: a gradual rise in global atmospheric temperature as a result of an amplification of the natural greenhouse effect due to human activities.

glomerulus (glō-mer′- ū-lus): a dense network of thin-walled capillaries, located within the Bowman's capsule of each nephron of the kidney, where blood pressure forces water and dissolved nutrients through capillary walls for filtration by the nephron.

glycogen (glī′-kō-jen): a long, branched polymer of glucose that is stored by animals in the muscles and liver and metabolized as a source of energy.

glycolysis (glī-kol′-i-sis): reactions, carried out in the cytoplasm, that break down glucose into two molecules of pyruvic acid, producing two adenosine triphosphate (ATP) molecules; does not require oxygen but can proceed when oxygen is present.

Golgi apparatus: a stack of membranous sacs, found in most eukaryotic cells, that is the site of processing and separation of membrane components and secretory materials.

gonad: an organ where reproductive cells are formed; in males, the testes, and in females, the ovaries.

grassland: a biome, located in the centers of continents, that supports grasses; also called *prairie*.

gravitropism: growth with respect to the direction of gravity.

greenhouse effect: the process in which certain gases such as carbon dioxide and methane trap sunlight energy in a planet's atmosphere as heat; the glass in a greenhouse does the same. The result, global warming, is being enhanced by the production of these gases by humans.

greenhouse gas: a gas, such as carbon dioxide or methane, that traps sunlight energy in a planet's atmosphere as heat; a gas that participates in the greenhouse effect.

ground tissue: plant tissue consisting of parenchyma, collenchyma, and sclerenchyma cells that makes up the bulk of a leaf or young stem, excluding vascular or dermal tissues. Most ground tissue cells function in photosynthesis, support, or carbohydrate storage.

guanine (G): a nitrogenous base found in both DNA and RNA.

guard cell: one of a pair of specialized epidermal cells surrounding the central opening of a stoma of a leaf, which regulates the size of the opening.

gymnosperm (jim′-nō-sperm): a nonflowering seed plant, such as a conifer, cycad, or gingko.

habitat fragmentation: the process by which human activities produce isolated patches of habitat that may not be large enough to sustain viable populations.

haploid (hap′-loid): referring to a cell that has only one member of each pair of homologous chromosomes.

Hardy-Weinberg principle: a mathematical model proposing that, under certain conditions, the allele frequencies and genotype frequencies in a sexually reproducing population will remain constant over generations.

heart: a muscular organ responsible for pumping blood within the circulatory system throughout the body.

helper T cell: a type of T cell that helps other immune cells recognize and act against antigens.

hemoglobin (hē′mō-glō-bin): the iron-containing protein that gives red blood cells their color; binds to oxygen in the lungs and releases it to the tissues.

herbivore (erb′-i-vor): literally, "plant-eater"; an organism that feeds directly and exclusively on producers; a primary consumer.

heterotroph (het′-er-ō-trōf′): literally, "other-feeder"; an organism that eats other organisms; a consumer.

heterozygous (het-er-ō-zī′-gus): carrying two different alleles of a given gene; also called *hybrid*.

hindbrain: the posterior portion of the brain, containing the medulla, pons, and cerebellum.

homeostasis (hōm-ē-ō-stā′sis): the maintenance of a relatively constant environment required for the optimal functioning of cells; maintained by the coordinated activity of numerous regulatory mechanisms, including the respiratory, endocrine, circulatory, and excretory systems.

hominid: a human or a prehistoric relative of humans, beginning with the Australopithecines, whose fossils date back at least 4.4 million years.

homologous chromosome: see *homologue*.

homologous structures: structures that may differ in function but that have similar anatomy, because the organisms that possess them have descended from common ancestors.

homologue (hō-′mō-log): a chromosome that is similar in appearance and genetic information to another chromosome with which it pairs during meiosis; also called *homologous chromosome*.

homozygous (hō-mō-zī′-gus): carrying two copies of the same allele of a given gene; also called *true-breeding*.

hormone: a chemical that is synthesized by one group of cells, secreted, and then carried in the bloodstream to other cells, whose activity is influenced by reception of the hormone.

host: the prey organism on or in which a parasite lives; is harmed by the relationship.

human immunodeficiency virus (HIV): a pathogenic retrovirus that causes acquired immune deficiency syndrome (AIDS) by attacking and destroying the immune system's T cells.

humoral immunity: an immune response in which foreign substances are inactivated or destroyed by antibodies that circulate in the blood.

hydrogen bond: the weak attraction between a hydrogen atom that bears a partial positive charge (due to polar covalent bonding with another atom) and another atom, normally oxygen or nitrogen, that bears a partial negative charge; hydrogen bonds may form between atoms of a single molecule or of different molecules.

hydrologic cycle: the water cycle, driven by solar energy; a nutrient cycle in which the main reservoir of water is the ocean and most of the water remains in the form of water throughout the cycle (rather than being used in the synthesis of new molecules).

hydrolysis (hī-drol′-i-sis): the chemical reaction that breaks a covalent bond by means of the addition of hydrogen to the atom on one side of the original bond and a hydroxyl group to the atom on the other side; the reverse of dehydration synthesis.

hydrophilic (hī-drō-fil′-ik): pertaining to a substance that dissolves readily in water or to parts of a large molecule that form hydrogen bonds with water.

hydrophobic (hī-drō-fō′-bik): pertaining to a substance that does not dissolve in water.

hydrothermal vent community: a community of unusual organisms, living in the deep ocean near hydrothermal vents, that depends on the chemosynthetic activities of sulfur bacteria.

hypha (hī′-fuh; pl., hyphae): a threadlike structure that consists of elongated cells, typically with many haploid nuclei; many hyphae make up the fungal body.

hypothalamus (hī-pō-thal′-a-mus): a region of the brain that controls the secretory activity of the pituitary gland; synthesizes, stores, and releases certain peptide hormones; directs autonomic nervous system responses.

hypothesis (hī-poth′-eh-sis): the third operation in the scientific method; a supposition based on previous observations that is offered as an explanation for the observed phenomenon and is used as the basis for further observations or experiments.

immune response: a specific response by the immune system to the invasion of the body by a particular foreign substance or microorganism, characterized by the recognition of the foreign substance by immune cells and its subsequent destruction by antibodies or by cellular attack.

immune system: cells such as macrophages, B cells, and T cells and molecules such as antibodies that work together to combat microbial invasion of the body.

implantation: the process whereby the early embryo embeds itself within the lining of the uterus.

imprinting: the process by which an animal forms an association with another animal or object in the environment during a sensitive period of development.

incomplete dominance: a pattern of inheritance in which the heterozygous phenotype is intermediate between the two homozygous phenotypes.

incomplete flower: a flower that is missing one of the four floral parts (sepals, petals, stamens, or carpels).

inflammatory response: a nonspecific, local response to injury to the body, characterized by the phagocytosis of foreign substances and tissue debris by white blood cells and by the walling off of the injury site by the clotting of fluids that escape from nearby blood vessels.

inhalation: the act of drawing air into the lungs by enlarging the chest cavity.

inheritance: the genetic transmission of characteristics from parent to offspring.

innate behavior (in-āt′): behavior that can be performed on the first attempt, without the need for experience or learning.

insertion mutation: a mutation in which one or more pairs of nucleotides are inserted into a gene.

intermembrane compartment: the fluid-filled space between the inner and outer membranes of a mitochondrion.

interneuron: a neuron that receives signals from hormones, sensory neurons, and other neurons and often activates motor neurons.

interphase: the stage of the cell cycle between cell divisions; the stage in which chromosomes are replicated and other cell functions occur, such as growth, movement, and acquisition of nutrients.

interspecific competition: competition among individuals of different species.

intertidal zone: an area of the ocean shore that is alternately covered and exposed by the tides.

intraspecific competition: competition among individuals of the same species.

invertebrate (in-vert′-uh-bret): an animal that does not possess a vertebral column at any stage of its life.

ion (ī′-on): a charged atom or molecule; an atom or molecule that either has an excess of electrons (and hence is negatively charged) or has lost electrons (and is positively charged).

ionic bond: a chemical bond formed by the electrical attraction between positively and negatively charged ions.

islet cell: a cluster of cells in the endocrine portion of the pancreas that produces insulin and glucagon.

isolating mechanism: a morphological, physiological, behavioral, or ecological difference that prevents members of two species from interbreeding.

isotope: one of several forms of a single element, the nuclei of which contain the same number of protons but different numbers of neutrons.

keystone species: a species whose influence on community structure is greater than its abundance would suggest.

kidney: one of a pair of organs of the excretory system that is located on either side of the spinal column and filters blood, removing wastes and regulating the composition and water content of the blood.

kin selection: a type of natural selection that favors alleles that increase the survival or reproductive success of relatives.

kinetic energy: the energy of movement; includes light, heat, mechanical movement, and electricity.

kinetochore (ki-net′-ō-kor): a protein structure that forms at the centromere regions of chromosomes; attaches the chromosomes to the spindle.

Krebs cycle: a cyclic series of reactions, occurring in the matrix of mitochondria, in which the acetyl groups from the pyruvic acids produced by glycolysis are broken down to CO_2, accompanied by the formation of adenosine triphosphate (ATP) and electron carriers; also called *citric acid cycle*.

labium (pl., labia): one of a pair of folds of skin of the external structures of the mammalian female reproductive system.

labor: a series of contractions of the uterus that result in birth.

large intestine: the final section of the digestive tract; consists of the colon and the rectum, where feces are formed and stored.

larva (lar′-vuh): an immature form of an organism with indirect development before metamorphosis into its adult form; includes the caterpillars of moths and butterflies and the maggots of flies.

larynx (lar′-inks): that portion of the air passage between the pharynx and the trachea; contains the vocal cords.

lateral bud: a cluster of meristematic cells at the node of a stem; under appropriate conditions, it grows into a branch.

lateral meristem: a meristematic tissue that forms cylinders parallel to the long axis of roots and stems; normally located between the primary xylem and primary phloem (vascular cambium) and just outside the phloem (cork cambium); also called *cambium*.

law of independent assortment: the independent inheritance of two or more distinct traits; states that the alleles for one trait may be distributed to the gametes independently of the alleles for other traits.

law of segregation: Gregor Mendel's conclusion that each gamete receives only one of each parent's pair of genes for each trait.

leaf: an outgrowth of a stem, normally flattened and photosynthetic.

learning: an adaptive change in behavior as a result of experience.

light-harvesting complex: in photosystems, the assembly of pigment molecules (chlorophyll and accessory pigments) that absorb light energy and transfer that energy to electrons.

limnetic zone: a lake zone in which enough light penetrates to support photosynthesis.

linkage: the inheritance of certain genes as a group because they are parts of the same chromosome. Linked genes do not show independent assortment.

lipid (li′-pid): one of a number of organic molecules containing large nonpolar regions composed solely of carbon and hydrogen, which make lipids hydrophobic and insoluble in water; includes oils, fats, waxes, phospholipids, and steroids.

littoral zone: a lake zone, near the shore, in which water is shallow and plants find abundant light, anchorage, and adequate nutrients.

liver: an organ with varied functions, including bile production, glycogen storage, and the detoxification of poisons.

lobefin: a member of the fish order Sarcopterygii, which includes coelacanths and lungfishes. Ancestors of today's lobefins gave rise to the first amphibians, and thus ultimately to all tetrapod vertebrates.

locus: the physical location of a gene on a chromosome.

loop of Henle (hen′-lē): a specialized portion of the tubule of the nephron in birds and mammals that creates an osmotic concentration gradient in the fluid immediately surrounding it. This gradient in turn makes possible the production of urine that is more osmotically concentrated than blood plasma.

lung: a paired respiratory organ consisting of inflatable chambers within the chest cavity in which gas exchange occurs.

lymph: a pale fluid, within the lymphatic system, that is composed primarily of interstitial fluid and lymphocytes.

lymph node: a small structure that filters lymph; contains lymphocytes and macrophages, which inactivate foreign particles such as bacteria.

lymphatic system: a system consisting of lymph vessels, lymph capillaries, lymph nodes, and the thymus and spleen; helps protect the body against infection, absorbs fats, and returns excess fluid and small proteins to the blood circulatory system.

lymphocyte (lim′-fō-sīt): a type of white blood cell important in the immune response.

lysosome (lī′-sō-sōm): a membrane-bound organelle containing intracellular digestive enzymes.

macronutrient: a nutrient needed in relatively large quantities (often defined as making up more than 0.1% of an organism's body).

macrophage (mak′-rō-fāj): a type of white blood cell that engulfs microbes and destroys them by phagocytosis; also presents microbial antigens to T cells, helping stimulate the immune response.

mammal: a member of the chordate class Mammalia, which includes vertebrates with hair and mammary glands.

mammary gland (mam′-uh-rē): a milk-producing gland used by female mammals to nourish their young.

mass extinction: the extinction of an extraordinarily large number of species in a short period of geologic time. Mass extinctions have recurred periodically throughout the history of life.

matrix: the fluid contained within the inner membrane of a mitochondrion.

megaspore: in plants, a haploid cell formed by meiosis from a diploid megaspore mother cell; through mitosis and differentiation, develops into the female gametophyte.

meiosis (mī-ō′-sis): in eukaryotes, a type of nuclear division in which a diploid nucleus divides twice to produce four haploid nuclei.

meiotic cell division: meiosis followed by cytokinesis.

memory cell: a type of white blood cell that is produced as a result of the binding of a receptor on a B cell or a T cell to an antigen on an invading microorganism. Memory cells persist in the bloodstream and provide future immunity to invaders bearing that antigen.

memory B cell: a memory cell that is descended from a B cell.

menstrual cycle: in human females, a complex 28-day cycle during which hormonal interactions among the hypothalamus, pituitary gland, and ovary coordinate ovulation and the preparation of the uterus to receive and nourish the fertilized egg. If pregnancy does not occur, the uterine lining is shed during menstruation.

menstruation: in human females, the monthly discharge of uterine tissue and blood from the uterus.

meristem cell (mer′-i-stem): an undifferentiated cell that remains capable of cell division throughout the life of a plant.

mesoderm (mēz'- ō-derm): the middle embryonic tissue layer, lying between the endoderm and ectoderm, and normally the last to develop; gives rise to structures such as muscle and skeleton.

mesophyll (mez'- ō-fil): loosely packed parenchyma cells beneath the epidermis of a leaf.

messenger RNA (mRNA): a strand of RNA, complementary to the DNA of a gene, that conveys the genetic information in DNA to the ribosomes to be used during protein synthesis; sequences of three bases (codons) in mRNA specify particular amino acids to be incorporated into a protein.

metabolic pathway: a sequence of chemical reactions within a cell, in which the products of one reaction are the reactants for the next reaction.

metabolism: the sum of all chemical reactions that occur within a single cell or within all of the cells of a multicellular organism.

metamorphosis (met-a-mor'-fō-sis): in animals with indirect development, a radical change in body form from larva to sexually mature adult, as seen in amphibians (tadpole to frog) and insects (caterpillar to butterfly).

metaphase (met'-a-fāz): the stage of mitosis in which the chromosomes, attached to spindle fibers at kinetochores, are lined up along the equator of the cell.

micronutrient: a nutrient needed only in small quantities (often defined as making up less than 0.01% of an organism's body).

microspore: in plants, a haploid cell formed by meiosis from a microspore mother cell; through mitosis and differentiation, develops into the male gametophyte.

microtubule: a hollow, cylindrical strand, found in eukaryotic cells, that is composed of the protein tubulin; part of the cytoskeleton used in the movement of organelles, cell growth, and the construction of cilia and flagella.

microvillus (mī-krō-vi'-lus; pl., microvilli): a microscopic projection of the plasma membrane of each villus; increases the surface area of the villus.

midbrain: during development, the central portion of the brain; contains an important relay center, the reticular formation.

mimicry (mim'ik-rē): the situation in which a species has evolved to resemble something else, typically another type of organism.

mineral: an inorganic substance, especially one in rocks or soil.

minimum viable population (MVP): the smallest isolated population that can persist indefinitely.

mitochondrion (mī-tō-kon'-drē-un): an organelle, bounded by two membranes, that is the site of the reactions of aerobic metabolism.

mitosis (mī-tō-sis): a type of nuclear division, used by eukaryotic cells, in which one copy of each chromosome (already duplicated during interphase before mitosis) moves into each of two daughter nuclei; the daughter nuclei are therefore genetically identical to each other.

mitotic cell division: mitosis followed by cytokinesis.

molecule (mol'-e-kūl): a particle composed of one or more atoms held together by chemical bonds; the smallest particle of a compound that displays all of the properties of that compound.

molt: to shed an external body covering, such as an exoskeleton, skin, feathers, or fur.

monocot: short for monocotyledon; a type of flowering plant characterized by embryos with one seed leaf, or cotyledon.

motor neuron: a neuron that receives instructions from the association neurons and activates effector organs, such as muscles or glands.

muscle tissue: tissue that consists mainly of muscle fibers and whose function is accomplished when the tissue contracts.

muscular dystrophy: a group of inherited disorders that result in degeneration of muscle tissue.

mutation: a change in the base sequence of DNA in a gene; normally refers to a genetic change significant enough to alter the appearance or function of the organism.

mycelium (mī sēl'-ē-um): the body of a fungus, consisting of a mass of hyphae.

mycorrhiza (mī-kō-rī'-zuh; pl., mycorrhizae): a symbiotic relationship between a fungus and the roots of a land plant that facilitates mineral extraction and absorption.

natural killer cell: a type of white blood cell that destroys some virus-infected cells and cancerous cells on contact; part of the immune system's nonspecific internal defense against disease.

natural selection: the process in which unequal survival and reproduction of organisms, favoring individuals with particular traits, causes those traits to become increasingly common in a population.

nearshore zone: the region of coastal water that is relatively shallow but constantly submerged; includes bays and coastal wetlands and can support large plants or seaweeds.

negative feedback: a process in which a change initiates a series of events that tend to counteract the change and restore the original state. Negative feedback in physiological systems maintains homeostasis.

nephron (nef'-ron): the functional unit of the kidney; where blood is filtered and urine formed.

nerve: a bundle of axons of nerve cells, bound together in a sheath.

nerve cord: in most animals, the nervous tissue running lengthwise from the head toward the tail, connecting the brain or principal ganglia of the head with the rest of the body. In chordates, a hollow structure lying just beneath the dorsal surface of the body, connecting the brain with the rest of the body. In vertebrates, this is the *spinal cord.*

nerve tissue: the tissue that makes up the brain, spinal cord, and nerves; consists of neurons and glial cells.

net primary productivity: the energy stored in the autotrophs of an ecosystem over a given time period.

neuron (noor'-on): a single nerve cell.

neurosecretory cell: a specialized nerve cell that synthesizes and releases hormones.

neurotransmitter: a chemical that is released by a nerve cell close to a second nerve cell, a muscle, or a gland cell and that influences the activity of the second cell.

neutron: a subatomic particle that is found in the nuclei of atoms, bears no charge, and has a mass approximately equal to that of a proton.

nicotinamide adenine dinucleotide (NADH): an electron carrier molecule produced in the cytoplasmic fluid by glycolysis and in the mitochondrial matrix by the Krebs cycle; subsequently donates electrons to the electron transport chain.

nicotinamide adenine dinucleotide phosphate (NADPH): an energy-carrier molecule produced by the light-dependent reactions of photosynthesis; transfers energy to the carbon-fixing (light-independent) reactions.

nitrogen fixation: the process that combines atmospheric nitrogen with hydrogen to form ammonium (NH_4^+).

nitrogen-fixing bacterium: a bacterium that possesses the ability to remove nitrogen from the atmosphere and combine it with hydrogen to produce ammonium (NH_4^+).

nondisjunction: an error in meiosis in which chromosomes fail to segregate properly into the daughter cells.

nonpolar molecule: a molecule bound by covalent bonds in which electrical charge is symmetrically distributed, so that no portion of the molecule is electrically charged relative to other portions.

nonvascular plant: a plant that lacks conducting vessels; a bryophyte; mosses, hornworts, and liverworts are nonvascular plants.

northern coniferous forest: see *taiga.*

no-till: a method of growing crops in which the soil is not plowed and the remains of harvested crops are left in place to serve as mulch for the following year's crops.

notochord (nōt'- ō-kord): a stiff but somewhat flexible, supportive rod found in all members of the phylum Chordata at some stage of development.

nuclear envelope: the double-membrane system surrounding the nucleus of eukaryotic cells; the outer membrane is typically continuous with the endoplasmic reticulum.

nucleic acid (noo-klā'-ik): an organic molecule composed of nucleotide subunits; the two common types of nucleic acids are ribonucleic acid (RNA) and deoxyribonucleic acid (DNA).

nucleoid (noo-klē-oid): the location of the genetic material in prokaryotic cells; not membrane-enclosed.

nucleolus (noo-klē'- ō-lus): the region of the eukaryotic nucleus that is engaged in ribosome synthesis; consists of the genes encoding ribosomal RNA, newly synthesized ribosomal RNA, and ribosomal proteins.

nucleotide: a subunit of which nucleic acids are composed; a phosphate group bonded to a sugar (deoxyribose in DNA), which is in turn bonded to a nitrogen-containing base (adenine, guanine, cytosine, or thymine in DNA). Nucleotides are linked together, forming a strand of nucleic acid, as follows: Bonds between the phosphate of one nucleotide link to the sugar of the next nucleotide.

nucleotide substitution: a mutation that replaces one nucleotide in a DNA molecule with another; for example, a change from an adenine to a guanine.

nucleus: (1) atomic nucleus: the central region of an atom, consisting of protons and neutrons; (2) cellular nucleus: the membrane-bound organelle of eukaryotic cells that contains the cell's genetic material.

nutrient: a substance acquired from the environment and needed for the survival, growth, and development of an organism.

nutrient cycle: a description of the pathways of a specific nutrient (such as carbon, nitrogen, phosphorus, or water) through the living and nonliving portions of an ecosystem; also called a *biogeochemical cycle.*

nutrition: the process of acquiring nutrients from the environment and, if necessary, processing them into a form that can be used by the body.

observation: the first operation in the scientific method; the noting of a specific phenomenon, leading to the formulation of a question.

oligotrophic lake: a lake that is very low in nutrients and hence clear, with extensive light penetration.

omnivore: an animal that consumes both plants and other animals.

oogenesis: the process by which egg cells are formed.

organ: a structure (such as the liver, kidney, or skin) composed of two or more distinct tissue types that function together.

organ system: two or more organs that work together to perform a specific function; for example, the digestive system.

organelle (or-guh-nel´): a structure, found in the cytoplasm of eukaryotic cells, that performs a specific function; sometimes refers specifically to membrane-bound structures, such as the nucleus or endoplasmic reticulum.

organic molecule: a molecule that contains both carbon and hydrogen.

organism (or´-guh-niz-um): an individual living thing.

organogenesis (or-gan-ō-jen´-uh-sis): the process by which the layers of the gastrula (endoderm, ectoderm, mesoderm) rearrange into organs.

osmosis (oz-mō´-sis): the diffusion of water across a differentially permeable membrane, normally down a concentration gradient of free water molecules. Water moves into the solution that has a lower concentration of free water from a solution with the higher concentration of free water.

ovary: in animals, the female gonad; in flowering plants, a structure at the base of the carpel that contains one or more ovules and develops into the fruit.

overexploitation: hunting or harvesting of a natural population at a rate that reduces population size more rapidly than reproduction can replace the lost individuals.

oviduct: see *uterine tube*.

ovulation: the release of a secondary oocyte, ready to be fertilized, from the ovary.

ovule: a structure within the ovary of a flower, inside which the female gametophyte develops; after fertilization, it develops into the seed.

ozone layer: the ozone-enriched layer of the upper atmosphere that filters out some of the sun's ultraviolet radiation.

pancreas (pan´-krē-us): a combined exocrine and endocrine gland located in the abdominal cavity next to the stomach. The endocrine portion secretes the hormones insulin and glucagon, which regulate glucose concentrations in the blood. The exocrine portion secretes enzymes for fat, carbohydrate, and protein digestion into the small intestine and neutralizes the acidic chyme.

pancreatic juice: a mixture of water, sodium bicarbonate, and enzymes released by the pancreas into the small intestine.

parasite (par´-uh-sīt): an organism that lives in or on a larger prey organism, called a *host*, weakening it.

passive transport: the movement of materials across a membrane down a gradient of concentration, pressure, or electrical charge without using cellular energy.

pathogenic (path´-ō-jen-ik): capable of producing disease; refers to an organism with such a capability (a pathogen).

pedigree: a diagram showing genetic relationships among a set of individuals, normally with respect to a specific genetic trait.

penis: an external structure of the male reproductive and urinary systems; serves to deposit sperm into the female reproductive system and delivers urine to the exterior.

peptide hormone: a hormone consisting of a chain of amino acids; includes small proteins that function as hormones.

per capita growth rate: the rate at which a population's size changes, expressed as the increase or decrease per individual per unit of time.

peripheral nervous system (PNS): in vertebrates, the part of the nervous system that connects the central nervous system to the rest of the body.

peristalsis: rhythmic coordinated contractions of the smooth muscles of the digestive tract that move substances through the digestive tract.

permafrost: a permanently frozen layer of soil in the arctic tundra that cannot support the growth of trees.

petal: part of a flower, typically brightly colored and fragrant, that attracts potential animal pollinators.

pH scale: a scale, with values from 0 to 14, used for measuring the relative acidity of a solution; at pH 7 a solution is neutral, pH 0 to 7 is acidic, and pH 7 to 14 is basic; each unit on the scale represents a tenfold change in H^+ concentration.

phagocytic cell (fa-gō-sit´-ik): a type of immune system cell that destroys invading microbes by using phagocytosis to engulf and digest the microbes.

pharyngeal gill slit (far-in´-jē-ul): an opening, located just posterior to the mouth, that connects the digestive tube to the outside environment; present (at some stage of life) in all chordates.

pharynx (far´-inks): in vertebrates, a chamber that is located at the back of the mouth and is shared by the digestive and respiratory systems; in some invertebrates, the portion of the digestive tube just posterior to the mouth.

phenotype (fēn´-ō-tīp): the physical characteristics of an organism; can be defined as outward appearance (such as flower color), as behavior, or in molecular terms (such as glycoproteins on red blood cells).

pheromone (fer´-uh-mōn): a chemical produced by an organism that alters the behavior or physiological state of another member of the same species.

phloem (flō´-um): a conducting tissue of vascular plants that transports a concentrated sugar solution up and down the plant.

phospholipid bilayer: a double layer of phospholipids that forms the basis of all cellular membranes. The phospholipid heads, which are hydrophilic, face the water of extracellular fluid or the cytoplasm; the tails, which are hydrophobic, are buried in the middle of the bilayer.

photic zone: the region of the ocean where light is strong enough to support photosynthesis.

photorespiration: a series of reactions in plants in which O_2 replaces CO_2 during the C_3 cycle, preventing carbon fixation; this wasteful process dominates when C_3 plants are forced to close their stomata to prevent water loss.

photosynthesis: the complete series of chemical reactions in which the energy of light is used to synthesize high-energy organic molecules, normally carbohydrates, from low-energy inorganic molecules, normally carbon dioxide and water.

photosystem: in thylakoid membranes, a light-harvesting complex and its associated electron transport system.

phototropism: growth with respect to the direction of light.

phytoplankton (fī´-tō-plank-ten): photosynthetic protists that are abundant in marine and freshwater environments.

pineal gland (pī-nē´-al): a small gland within the brain that secretes melatonin; controls the seasonal reproductive cycles of some mammals.

pioneer: an organism that is among the first to colonize an unoccupied habitat in the first stages of succession.

pituitary gland: an endocrine gland, located at the base of the brain, that produces several hormones, many of which influence the activity of other glands.

placenta (pluh-sen´-tuh): in mammals, a structure formed by a complex interweaving of the uterine lining and the embryonic membranes, especially the chorion; functions in gas, nutrient, and waste exchange between embryonic and maternal circulatory systems and secretes hormones.

plant hormone: the plant-regulating chemicals auxin, gibberellin, cytokinin, ethylene, and abscisic acid; somewhat resemble animal hormones in that they are chemicals produced by cells in one location that influence the growth or metabolic activity of other cells, typically some distance away in the plant body.

plasma: the fluid, noncellular portion of the blood.

plasma cell: an antibody-secreting descendant of a B cell.

plasma membrane: the outer membrane of a cell, composed of a bilayer of phospholipids in which proteins are embedded.

plasmid (plaz´-mid): a small, circular piece of DNA located in the cytoplasm of many bacteria; normally does not carry genes required for the normal functioning of the bacterium but may carry genes that assist bacterial survival in certain environments, such as a gene for antibiotic resistance.

plate tectonics: the theory that Earth's crust is divided into irregular plates that are converging, diverging, or slipping by one another; these motions cause continental drift, the movement of continents over Earth's surface.

platelet (plāt´-let): a cell fragment that is formed from megakaryocytes in bone marrow and lacks a nucleus; circulates in the blood and plays a role in blood clotting.

pleiotropy (plē´-ō-trō-pē): a situation in which a single gene influences more than one phenotypic characteristic.

point mutation: a mutation in which a single base pair in DNA has been changed.

polar molecule: a molecule bound by covalent bonds in which electrical charge is asymmetrically distributed, so that electrical charge differs in different portions of the molecule.

polar nucleus: in flowering plants, one of two nuclei in the primary endosperm cell of the female gametophyte; formed by the mitotic division of a megaspore.

pollen/pollen grain: the male gametophyte of a seed plant.

pollination: in flowering plants, when pollen grains land on the stigma of a flower of the same species; in conifers, when pollen grains land within the pollen chamber of a female cone of the same species.

polygenic inheritance: a pattern of inheritance in which the interactions of two or more functionally similar genes determine phenotype.

polymerase chain reaction (PCR): a method of producing virtually unlimited numbers of copies of a specific piece of DNA, starting with as little as one copy of the desired DNA.

population: all the members of a species that occupy a particular area at the same time.

population bottleneck: the result of an event that causes a population to become extremely small; may cause genetic drift that results in changed allele frequencies and loss of genetic variability.

positive feedback: a process in which a change initiates events that tend to amplify the original change.

post-anal tail: a tail that extends beyond the anus; exhibited by all chordates at some stage of development.

postmating isolating mechanism: any structure, physiological function, or developmental abnormality that prevents organisms of two different populations, once mating has occurred, from producing vigorous, fertile offspring.

postsynaptic neuron: at a synapse, the nerve cell that changes its electrical potential in response to a chemical (the neurotransmitter) released by another (presynaptic) cell.

postsynaptic potential (PSP): an electrical signal produced in a postsynaptic cell by transmission across the synapse; it may be excitatory (EPSP), making the cell more likely to produce an action potential, or inhibitory (IPSP), tending to inhibit an action potential.

potential energy: "stored" energy, normally chemical energy or energy of position within a gravitational field.

prairie: see *grassland*.

predation (pre-dā′-shun): the act of eating another living organism.

predator: an organism that eats other living organisms.

prediction: the fourth operation in the scientific method; a statement describing an observable outcome that would occur if a particular hypothesis were true.

premating isolating mechanism: any structure, physiological function, or behavior that prevents organisms of two different populations from interbreeding.

presynaptic neuron: a nerve cell that releases a chemical (the neurotransmitter) at a synapse, causing changes in the electrical activity of another (postsynaptic) cell.

prey: organisms that are eaten by another organism.

primary consumer: an organism that feeds on producers; an herbivore.

primary growth: growth in length and development of the initial structures of plant roots and shoots, due to the cell division of apical meristems and differentiation of the daughter cells.

primary succession: succession that occurs in an environment, such as bare rock, in which no trace of a previous community was present.

primate: a member of the mammal order Primates, characterized by the presence of an opposable thumb, forward-facing eyes, and a well-developed cerebral cortex; includes lemurs, monkeys, apes, and humans.

producer: a photosynthetic organism; an autotroph.

product: an atom or molecule that is formed from reactants in a chemical reaction.

profundal zone: a lake zone in which light is insufficient to support photosynthesis.

prokaryote (prō-kar′- ē- ōt′): an organism whose cells are prokaryotic; bacteria and archaea are prokaryotes.

prokaryotic cell (prō-kar-ê-ot′-ik): cells of the domains Bacteria and Archaea. Prokaryotic cells have genetic material that is not enclosed in a membrane-bound nucleus; they lack other membrane-bound organelles.

promoter: a specific sequence of DNA to which RNA polymerase binds, initiating gene transcription.

prophase (prō′-fāz): the first stage of mitosis, in which the chromosomes first become visible in the light microscope as thickened, condensed threads and the spindle begins to form; as the spindle is completed, the nuclear envelope breaks apart, and the spindle fibers invade the nuclear region and attach to the kinetochores of the chromosomes. Also, the first stage of meiosis: In meiosis I, the homologous chromosomes pair up and exchange parts at chiasmata; in meiosis II, the spindle re-forms and chromosomes attach to the microtubules.

protein: polymer of amino acids joined by peptide bonds.

protist (prō′-tist): a eukaryotic organism that is not a plant, animal, or fungus. The term encompasses a diverse array of organisms and does not represent a monophyletic group. Algae, amoebas, slime molds, and ciliates are examples of protists.

protocell: the hypothetical evolutionary precursor of living cells, consisting of a mixture of organic molecules within a membrane.

proton: a subatomic particle that is found in the nuclei of atoms, bears a unit of positive charge, and has a relatively large mass, roughly equal to the mass of the neutron.

proximal tubule: in nephrons of the mammalian kidney, the portion of the renal tubule just after the Bowman's capsule; receives filtrate from the capsule and is the site where selective secretion and reabsorption between the filtrate and the blood begin.

Punnett square method: an intuitive way to predict the genotypes and phenotypes of offspring in specific crosses.

pupa: a developmental stage in some insect species in which the organism stops moving and feeding and may be encased in a cocoon; occurs between the larval and the adult phases.

pyruvate: a three-carbon molecule that is formed by glycolysis and then used in fermentation or cellular respiration.

question: the second operation in the scientific method; a query that identifies a particular aspect of an observation that a scientist wishes to explain.

radioactive: pertaining to an atom with an unstable nucleus that spontaneously disintegrates, with the emission of radiation.

rain shadow: a local dry area created by the modification of rainfall patterns by a mountain range.

random distribution: distribution characteristic of populations in which the probability of finding an individual is equal in all parts of an area.

reactant: an atom or molecule that is used up in a chemical reaction to form a product.

reaction center: in the light-harvesting complex of a photosystem, the chlorophyll molecule to which light energy is transferred by the antenna molecules (light-absorbing pigments); the captured energy ejects an electron from the reaction center chlorophyll, and the electron is transferred to the electron transport system.

receptor: a cell that responds to an environmental stimulus (chemicals, sound, light, pH, and so on) by changing its electrical potential; also, a protein molecule in a plasma membrane that binds to another molecule (hormone, neurotransmitter), triggering metabolic or electrical changes in a cell.

receptor potential: an electrical potential change in a receptor cell, produced in response to the reception of an environmental stimulus (chemicals, sound, light, heat, and so on). The size of the receptor potential is proportional to the intensity of the stimulus.

receptor protein: a protein, located on a membrane (or in the cytoplasm), that recognizes and binds to specific molecules. Binding by receptor proteins typically triggers a response by a cell, such as endocytosis, increased metabolic rate, or cell division.

recessive: an allele that is expressed only in homozygotes and is completely masked in heterozygotes.

recognition protein: a protein or glycoprotein protruding from the outside surface of a plasma membrane that identifies a cell as belonging to a particular species, to a specific individual of that species, and in many cases to one specific organ within the individual.

recombinant DNA: DNA that has been altered by the recombination of genes from a different organism, typically from a different species.

recombination: the formation of new combinations of the different alleles of each gene on a chromosome; the result of crossing over.

rectum: the terminal portion of the vertebrate digestive tube, where feces are stored until they can be eliminated.

red blood cell: the most common type of cell in vertebrate blood; active in oxygen transport; contains the red pigment hemoglobin.

reflex: a simple, stereotyped movement of part of the body that occurs automatically in response to a stimulus.

replacement-level fertility (RLF): the birth rate at which a reproducing population exactly replaces itself during its lifetime.

reproductive isolation: the failure of organisms of one population to breed successfully with members of another; may be due to premating or postmating isolating mechanisms.

reptile: a member of the chordate group, which includes the snakes, lizards, turtles, alligators, and crocodiles; not a monophyletic group.

reservoir: the major source and storage site of a nutrient in an ecosystem, normally in the abiotic portion.

resource partitioning: the coexistence of two species with similar requirements, each occupying a smaller niche than either would if it were by itself; an evolutionary outcome that minimizes their competitive interactions.

respiration: the process by which an organism exchanges gases with the environment.

respiratory center: a cluster of neurons, located in the medulla of the brain, that sends rhythmic bursts of nerve impulses to the respiratory muscles, resulting in breathing.

resting potential: a negative electrical potential in unstimulated nerve cells.

restriction enzyme: an enzyme, normally isolated from bacteria, that cuts double-stranded DNA at a specific nucleotide sequence; the nucleotide sequence that is cut differs for different restriction enzymes.

ribonucleic acid (rī-bō-noo-klā′-ik; RNA): a molecule composed of ribose nucleotides, each of which consists of a phosphate group, the sugar ribose, and one of the bases adenine, cytosine, guanine, or uracil; transfers hereditary instructions from the nucleus to the cytoplasm; also the genetic material of some viruses.

ribosomal RNA (rRNA): a type of RNA that combines with proteins to form ribosomes.

ribosome: an organelle consisting of two subunits, each composed of ribosomal RNA and protein; the site of protein synthesis, during which the sequence of bases of messenger RNA is translated into the sequence of amino acids in a protein.

ribozyme: an RNA molecule that can catalyze chemical reactions, especially those involved in the synthesis and processing of RNA itself.

ribulose bisphosphate (RuBP): a six-carbon molecule that reacts with carbon dioxide in the carbon-fixing reaction of C_3 photosynthesis; an important participant in the Calvin-Benson cycle.

RNA: see *ribonucleic acid.*

RNA polymerase: in RNA synthesis, an enzyme that catalyzes the bonding of free RNA nucleotides into a continuous strand, using RNA nucleotides that are complementary to those of a strand of DNA.

root: the part of the plant body, normally underground, that provides anchorage, absorbs water and dissolved nutrients and transports them to the stem, produces some hormones, and in some plants serves as a storage site for carbohydrates.

root cap: a cluster of cells at the tip of a growing root, derived from the apical meristem; protects the growing tip from damage as it burrows through the soil.

root hair: a fine projection from an epidermal cell of a young root that increases the absorptive surface area of the root.

rough endoplasmic reticulum: endoplasmic reticulum lined on the outside with ribosomes.

savanna: a biome that is dominated by grasses and supports scattered trees and thorny scrub forests; typically has a rainy season in which all of the year's precipitation falls.

scientific method: a rigorous procedure for making observations of specific phenomena and searching for the order underlying those phenomena; consists of six operations: observation, question, hypothesis, prediction, experiment, and conclusion.

scientific theory: a general explanation of natural phenomena developed through extensive and reproducible observations; more general and reliable than a hypothesis.

scrotum (skrō´-tum): the pouch of skin containing the testes of male mammals.

second law of thermodynamics: the principle that any change in an isolated system causes the quantity of concentrated, useful energy to decrease and the amount of randomness and disorder (entropy) to increase.

second messenger: an intracellular chemical, such as cyclic AMP, that is synthesized or released within a cell in response to the binding of a hormone or neurotransmitter (the first messenger) to receptors on the cell surface; brings about specific changes in the metabolism of the cell.

secondary consumer: an organism that feeds on primary consumers; a carnivore.

secondary growth: growth in the diameter of a stem or root due to cell division in lateral meristems and differentiation of their daughter cells.

secondary succession: succession that occurs after an existing community is disturbed—for example, after a forest fire; much more rapid than primary succession.

seed: the reproductive structure of a seed plant; protected by a seed coat; contains an embryonic plant and a supply of food for it.

seed coat: the thin, tough, and waterproof outermost covering of a seed, formed from the integuments of the ovule.

selectively permeable: refers to membranes across which some substances may pass freely while other substances cannot pass.

self-fertilization: the union of sperm and egg from the same individual.

semiconservative replication: the process of replication of the DNA double helix; the two DNA strands separate, and each is used as a template for the synthesis of a complementary DNA strand. Consequently, each daughter double helix consists of one parental strand and one new strand.

semilunar valve: a paired valve between the ventricles of the heart and the pulmonary artery and aorta; prevents the backflow of blood into the ventricles when they relax.

seminiferous tubule (sem-i-ni´-fer-us): in the vertebrate testis, a series of tubes in which sperm are produced.

senescence: in plants, a specific aging process, typically including deterioration and the dropping of leaves and flowers.

sensory neuron: a nerve cell that responds to a stimulus from the internal or external environment.

sensory receptor: a cell (typically, a neuron) specialized to respond to particular internal or external environmental stimuli by producing an electrical potential.

sepal (sē´-pul): the set of modified leaves that surround and protect a flower bud, typically opening into green, leaflike structures when the flower blooms.

severe combined immune deficiency (SCID): a disorder in which no immune cells, or very few, are formed; the immune system is incapable of responding properly to invading disease organisms, and the individual is very vulnerable to common infections.

sex chromosome: one of the pair of chromosomes that usually determines the sex of an organism; for example, the X and Y chromosomes in mammals.

sex-linked: referring to a pattern of inheritance characteristic of genes located on one type of sex chromosome (for example, X) and not found on the other type (for example, Y); also called X-linked. In sex-linked inheritance, traits are controlled by genes carried on the X chromosome; females show the dominant trait unless they are homozygous recessive, whereas males express whichever allele is on their single X chromosome.

sexual reproduction: a form of reproduction in which genetic material from two parent organisms is combined in the offspring; normally, two haploid gametes fuse to form a diploid zygote.

sexual selection: a type of natural selection that acts on traits involved in finding and acquiring mates.

sexually transmitted disease (STD): a disease that is passed from person to person by sexual contact.

short tandem repeat: a short (two to five base pairs) DNA sequence that is repeated about 5 to 15 times; the repetitions of the sequence are adjacent.

simple diffusion: the diffusion of water, dissolved gases, or lipid-soluble molecules through the phospholipid bilayer of a cellular membrane.

sinoatrial (SA) node (sī´-nō- āt´-rē-ul): a small mass of specialized muscle in the wall of the right atrium of the heart; generates electrical signals rhythmically and spontaneously and serves as the heart's pacemaker.

small intestine: the portion of the digestive tract, located between the stomach and large intestine, in which most digestion and absorption of nutrients occur.

smooth endoplasmic reticulum: endoplasmic reticulum without ribosomes.

solvent: a liquid capable of dissolving (uniformly dispersing) other substances in itself.

somatic nervous system: that portion of the peripheral nervous system that controls voluntary movement by activating skeletal muscles.

speciation: the process of species formation, in which a single species splits into two or more species.

species (spē´-sēs): the basic unit of taxonomic classification, consisting of a group of populations that evolves independently. In sexually reproducing organisms, a species can be defined as a population or series of populations of organisms that interbreed freely with one another under natural conditions but that do not interbreed with members of other species.

sperm: the haploid male gamete, normally small, motile, and containing little cytoplasm.

spermatogenesis: the process by which sperm cells form.

spinal cord: the part of the central nervous system of vertebrates that extends from the base of the brain to the hips and is protected by the bones of the vertebral column; contains the cell bodies of motor neurons that form synapses with skeletal muscles, the circuitry for some simple reflex behaviors, and axons that communicate with the brain. See also *nerve cord.*

spindle microtubules: microtubules organized in a spindle shape that separate chromosomes during mitosis or meiosis.

spleen: an organ of the lymphatic system in which lymphocytes are produced and blood is filtered past lymphocytes and macrophages, which remove foreign particles and aged red blood cells.

spontaneous generation: the proposal that living organisms can arise from nonliving matter.

spore: a haploid reproductive cell capable of developing into an adult without fusing with another cell; in the alternation-of-generation life cycle of plants, a haploid cell that is produced by meiosis and then undergoes repeated mitotic divisions and differentiation of daughter cells to produce the gametophyte, a multicellular, haploid organism.

sporophyte (spor´- ō-fīt): the diploid stage of the plant life cycle produces haploid, asexual spores through meiosis.

stamen (stā´-men): the male reproductive structure of a flower, consisting of a filament and an anther, in which pollen grains develop.

start codon: the first AUG codon in a messenger RNA molecule.

startle coloration: a form of mimicry in which a color pattern (in many cases resembling large eyes) can be displayed suddenly by a prey organism when approached by a predator.

stem: the portion of the plant body, normally located aboveground, that bears leaves and reproductive structures such as flowers and fruit.

stem cell: an undifferentiated cell that is capable of dividing and giving rise to one or more types of differentiated cell.

steroid hormone: a class of hormone whose chemical structure (four fused carbon rings with various functional groups) resembles cholesterol; steroids, which are lipids, are secreted by the ovaries and placenta, the testes, and the adrenal cortex.

stigma (stig´-muh): the pollen-capturing tip of a carpel.

stoma (stō´-muh; pl., stomata): an adjustable opening in the epidermis of a leaf, surrounded by a pair of guard cells, that regulates the diffusion of carbon dioxide and water into and out of the leaf.

stomach: the muscular sac between the esophagus and small intestine where food is stored and mechanically broken down and in which protein digestion begins.

stop codon: a codon in messenger RNA that stops protein synthesis and causes the completed protein chain to be released from the ribosome.

strand: single polymer of nucleotides; DNA is composed of two strands.

stroma (strō´-muh): the semifluid material inside chloroplasts in which the grana are embedded.

style: a stalk connecting the stigma of a carpel with the ovary at its base.

subclimax: a community in which succession is stopped before the climax community is reached and is maintained by regular disturbances—for example, tallgrass prairie maintained by periodic fires.

substrate: the atoms or molecules that are the reactants for an enzyme-catalyzed chemical reaction.

succession (suk-seh´-shun): a structural change in a community and its nonliving environment over time. Community changes alter the ecosystem in ways that favor competitors, and species replace one another in a somewhat predictable manner until a stable, self-sustaining climax community is reached.

sugar-phosphate backbone: a major feature of DNA structure, formed by attaching the sugar of one nucleotide to the phosphate from the adjacent nucleotide in a DNA strand.

surface tension: the tendency of a liquid to resist penetration by objects at its interface with the air, due to cohesion between molecules of the liquid.

sustainable development: a pattern of economic development in which the economic needs of the current population are met by means of activities that minimize damage to ecosystems, so that the ecosystem services available to future generations are not reduced.

symbiosis (sim´-bī- ō´sis): an intimate association between organisms of different species over an extended period. Either or both species may benefit from the association, or (in the case of parasitism) one of the participants is harmed.

synapse (sin´-aps): the site of communication between nerve cells. At a synapse, one cell (presynaptic) normally releases a chemical (the neurotransmitter) that changes the electrical potential of the second (postsynaptic) cell.

synaptic terminal: a swelling at the branched ending of an axon; where the axon forms a synapse.

T cell: a type of lymphocyte that recognizes and destroys specific foreign cells or substances or that regulates other cells of the immune system.

T-cell receptor: a protein receptor, located on the surface of a T cell, that binds a specific antigen and triggers the immune response of the T cell.

taiga (tī´-guh): a biome with long, cold winters and only a few months of warm weather; populated almost entirely by evergreen coniferous trees; also called *northern coniferous forest.*

taproot: the long, thick, main root of a root system, commonly found in dicots, that also includes many smaller lateral roots, all of which grow from the main root.

target cell: a cell on which a particular hormone exerts its effect.

telophase (tēl´- ō-fāz): in mitosis, the final stage, in which a nuclear envelope re-forms around each new daughter nucleus, the spindle fibers disappear, and the chromosomes relax from their condensed form; in meiosis I, the stage during which the spindle fibers disappear and the chromosomes normally relax from their condensed form; in meiosis II, the stage during which chromosomes relax into their extended state, the nuclear envelopes re-form, and cytokinesis occurs.

temperate deciduous forest: a biome in which winters are cold and summer rainfall is sufficient to allow enough moisture for trees to grow and shade out grasses.

temperate rain forest: a biome in which there is no shortage of liquid water year-round and that is dominated by conifers.

template strand: the strand of the DNA double helix from which RNA is transcribed.

terminal bud: meristem tissue and surrounding leaf primordia that are located at the tip of a plant shoot.

territoriality: the defense of an area in which important resources are located.

tertiary consumer (ter´-shē-er-ē): a carnivore that feeds on other carnivores (secondary consumers).

testis (pl., testes): the male gonad in animals.

threatened species: umbrella term for any species classified as critically endangered, endangered, or vulnerable.

threshold: the electrical potential (less negative than the resting potential) at which an action potential is triggered.

thylakoid (thī´-luh-koid): a disk-shaped, membranous sac found in chloroplasts; thylakoid membranes contain the photosystems and ATP-synthesizing enzymes used in the light-dependent reactions of photosynthesis.

thymine (T): a nitrogenous base found only in DNA.

thymus (thī´-mus): an organ of the lymphatic system that is located in the upper chest in front of the heart and that secretes thymosin, which stimulates lymphocyte maturation; begins to degenerate at puberty and has little function in the adult.

thyroid gland: an endocrine gland, located in front of the larynx in the neck, that secretes the hormones thyroxine (affecting metabolic rate) and calcitonin (regulating calcium ion concentration in the blood).

tissue: a group of (normally similar) cells that together carry out a specific function; for example, muscle; may include extracellular material produced by its cells.

trachea (trā´-kē-uh): in birds and mammals, a rigid but flexible tube, supported by rings of cartilage, that conducts air between the larynx and the bronchi; in insects, an elaborately branching tube that carries air from openings called *spiracles* near each body cell.

transcription: the synthesis of an RNA molecule from a DNA template.

transfer RNA (tRNA): a type of RNA that binds to a specific amino acid by means of a set of three bases (the anticodon) on the tRNA that are complementary to the messenger RNA (mRNA) codon for that amino acid; carries its amino acid to a ribosome during protein synthesis, recognizes a codon of mRNA, and positions its amino acid for incorporation into the growing protein chain.

transformation: a method of acquiring new genes, whereby DNA from one bacterium (normally released after the death of the bacterium) becomes incorporated into the DNA of another, living, bacterium.

transgenic: referring to an animal or a plant that expresses DNA derived from another species.

translation: the process whereby the sequence of bases of messenger RNA (mRNA) is converted into the sequence of amino acids of a protein.

transpiration (trans´-per-ā-shun): the evaporation of water through the stomata of a leaf.

transport protein: a protein that regulates the movement of water-soluble molecules through the plasma membrane.

trophic level: literally, "feeding level"; the categories of organisms in a community, and the position of an organism in a food chain, defined by the organism's source of energy; includes producers, primary consumers, secondary consumers, and so on.

tropical deciduous forest: a biome with pronounced wet and dry seasons and plants that must shed their leaves during the dry season to minimize water loss.

tropical rain forest: a biome with evenly warm, evenly moist conditions; dominated by broadleaf evergreen trees; the most diverse biome.

tubular reabsorption: the process by which cells of the tubule of the nephron remove water and nutrients from the filtrate within the tubule and return those substances to the blood.

tubular secretion: the process by which cells of the tubule of the nephron remove additional wastes from the blood, actively secreting those wastes into the tubule.

tubule (toob´- ūl): the tubular portion of the nephron; includes a proximal portion, the loop of Henle, and a distal portion. Urine is formed from the blood filtrate as it passes through the tubule.

tundra: a biome with severe weather conditions (extreme cold and wind and little rainfall) that cannot support trees.

uniform distribution: the distribution characteristic of a population with a relatively regular spacing of individuals, commonly as a result of territorial behavior.

upwelling: an upward flow that brings cold, nutrient-laden water from the ocean depths to the surface; occurs along western coastlines.

ureter (ū´-re-ter): a tube that conducts urine from each kidney to the bladder.

urethra (ū-rē´-thruh): the tube leading from the urinary bladder to the outside of the body; in males, the urethra also receives sperm from the vas deferens and conducts both sperm and urine (at different times) to the tip of the penis.

urinary bladder: a hollow muscular storage organ for storing urine.

urine: the fluid produced and excreted by the urinary system of vertebrates; contains water and dissolved wastes, such as urea.

uterine tube: the tube leading out of the ovary to the uterus, into which the secondary oocyte (egg cell) is released; also called the *oviduct.*

uterus: in female mammals, the part of the reproductive tract that houses the embryo during pregnancy.

vaccine: a preventive treatment that contains antigens characteristic of a particular disease organism and that, when introduced into a body, stimulates an immune response that protects the body against the disease organism.

vagina: the passageway leading from the outside of a female mammal's body to the cervix of the uterus.

variable: a condition, particularly in a scientific experiment, that is subject to change.

variable region: the part of an antibody molecule that differs among antibodies; the ends of the variable regions of the light and heavy chains form the specific binding site for antigens.

vas deferens (vaz de′-fer-enz): the tube connecting the epididymis of the testis with the urethra.

vascular cambium: a lateral meristem that is located between the xylem and phloem of a woody root or stem and that gives rise to secondary xylem and phloem.

vascular cylinder: the centrally located conducting tissue of a young root, consisting of primary xylem and phloem.

vascular plant: a plant that has specialized structures (vessels) for transporting water and nutrients though its body. Horsetails, ferns, gymnosperms, and flowering plants are examples of vascular plants.

vascular tissue: plant tissue consisting of xylem (which transports water and minerals from root to shoot) and phloem (which transports water and sugars throughout the plant).

vein: in vertebrates, a large-diameter, thin-walled vessel that carries blood from venules back to the heart; in vascular plants, a vascular bundle, or a strand of xylem and phloem in leaves.

ventricle (ven′-tre-kul): the lower muscular chamber on each side of the heart that pumps blood out through the arteries. The right ventricle sends blood to the lungs; the left ventricle pumps blood to the rest of the body.

venule (ven′-ūl): a narrow vessel with thin walls that carries blood from capillaries to veins.

vertebral column (ver-tē′-brul): a column of serially arranged skeletal units (the vertebrae) that enclose the nerve cord in vertebrates; the backbone.

vertebrate: an animal that possesses a vertebral column.

vesicle (ves′-i-kul): (1) a small, membrane-bound sac within the cytoplasm; (2) a small, hollow ball composed of protein and lipid; forms spontaneously when a solution of protein and lipid is agitated.

vestigial structure (ves-tij′-ē-ul): a structure that has no apparent function, but is homologous to functional structures in related organisms.

villus (vi′-lus; pl., villi): a finger-like projection of the wall of the small intestine that increases the absorptive surface area.

virus (vī′-rus): a noncellular parasitic particle that consists of a protein coat surrounding a strand of genetic material; multiplies only within a cell of a living organism (the host).

vitamin: one of a group of diverse chemicals that must be present in trace amounts in the diet to maintain health; used by the body in conjunction with enzymes in a variety of metabolic reactions.

vulnerable species: a species that faces a high risk of extinction in the wild in the medium-term future.

waggle dance: a symbolic form of communication used by honeybee foragers to communicate the location of a food source to their hivemates.

warning coloration: bright coloration that warns predators that the potential prey is distasteful or even poisonous.

wax: a lipid composed of fatty acids covalently bonded to long-chain alcohols.

weather: short-term fluctuations in temperature, humidity, cloud cover, wind, and precipitation in a region over periods of hours to days.

white blood cell: cellular components of blood; not as numerous as red blood cells; most function as part of the immune system and help defend the body against invaders.

wildlife corridors: strips of protected land linking larger protected areas.

xylem (zī-lum): a conducting tissue of vascular plants that transports water and minerals from root to shoot.

yolk sac: one of the embryonic membranes of reptilian, bird, and mammalian embryos; in birds and reptiles, a membrane surrounding the yolk in the egg; in mammals, forms part of the umbilical cord and the digestive tract but is empty.

zooplankton: nonphotosynthetic protists and tiny animals that are abundant in marine and freshwater environments.

zygote (zī′-gōt): in sexual reproduction, a diploid cell (the fertilized egg) formed by the fusion of two haploid gametes.

Suggested Answers to Text Questions

CHAPTER 1

Figure-based Questions

Figure 1-5 Answerable at the cell level but not at the tissue level: How are signals transmitted in a neuron? How do white blood cells move to the site of wounds? How do chromosomes move during cell division? How do bacteria stick to surfaces?

Answerable at the tissue level but not at the cell level: Which part of the brain controls speech? How does the kidney help maintain the body's water balance? What are the functions of skin? How does water get from a plant's roots to its leaves?

Figure 1-7 The fungal antibacterial chemicals probably evolved because they kill bacteria that compete with fungi for food and space.

Figure E1-1 Redi's experiment demonstrated that the maggots were caused by something that was excluded by gauze, but some agent other than flies could have produced the maggots. Other maggot sources could be tested using a series of identical, gauze-covered, meat-containing jars, each with a single possible maggot-causing element added, such as one with flies, one with roaches, or one with dust or soot. And of course the control jar would have nothing added.

Fill-in-the-Blank

1. cells, tissues, organs, organisms, populations, communities
2. scientific theory; hypothesis; scientific method
3. question, hypothesis, prediction, experiment (or observation), conclusion
4. variable; controls
5. evolution; natural selection
6. deoxyribonucleic acid, DNA; genes
7. organized; reproduce, stimuli; materials and energy; DNA

Review Questions

1. The scientific method consists of six interrelated operations: (1) observation of a phenomenon; (2) formulation of a question about the observation; (3) development of a hypothesis that answers the question; (4) prediction of an outcome if the hypothesis is correct; (5) experimentation that either confirms or refutes the prediction; and (6) forming a conclusion about the validity of the hypothesis. The scientific method can be used to solve mechanical problems in everyday life, such as "Why won't my car start?" or "Why won't my computer monitor or cell phone come on?"
2. The cell is the smallest unit of life. In multicellular life-forms, cells of similar types combine to form tissues (e.g., muscle). Two or more tissue types combine to make an organ (e.g., heart). Organs functioning in harmony produce a multicellular organism. A group of organisms of the same type in the same place constitutes a population, and an interacting group of different populations makes up a community.
3. A scientific theory is a general explanation of natural phenomena, developed through extensive and reproducible observations. A hypothesis is an educated guess that a certain preceding cause produces the observable phenomenon. Scientists use hypotheses as proposed explanations that are then repeatedly tested to either support them or

prove them false. If a hypothesis is repeatedly confirmed, it may be given the status of a scientific theory, which is used as a fundamental principle that has never been proven false.
4. A salt crystal is well organized and can grow; however, because it does not possess all seven attributes of living things, it is nonliving. A tree does possess all seven attributes and is therefore alive. By carefully observing each object under a microscope, you could determine that the tree is composed of cells and the salt crystal isn't.
5. Evolution is the scientific theory that modern organisms descended, with modification, from preexisting life-forms. Evolution occurs as a result of (1) genetic variation among members of a population, caused by mutation; (2) inheritance of those variations by offspring; and (3) natural selection of the variations that best adapt an organism to its environment.

Applying the Concepts

Hints

1. Include a clear statement of your hypothesis and at least one testable prediction. Include control groups fed other diets. Specify the number of subjects in each group, the variables, and how you will measure them and compare them to controls. What results would support your hypothesis?
2. A question for the caterpillars and milkweeds might be "Do these caterpillars prefer milkweed leaves?" A hypothesis might be "These caterpillars prefer milkweed to all other nearby plants." What would you predict if the caterpillars were presented with a choice of milkweed and leaves from different nearby plants? Think about a good sample size of caterpillars. What kind of containers would you use for your tests? How would you measure your results? What results would support your hypothesis? The variables would be different leaves; what would be the control in each experimental setup?

CHAPTER 2

Figure-based Questions

Figure 2-2 Atoms with partially empty outer shells have a strong tendency to react with other atoms in ways that gain or share electrons to fill their outer shells, or that give up electrons to empty their outer shells.

Figure 2-6 Oxygen's nucleus has eight protons, whereas hydrogen's has only a single proton. So the positive charge of the oxygen nucleus attracts electrons far more strongly than that of the hydrogen nucleus.

Figure 2-8 Cellular energy is derived from the chemical bonds in sugar molecules. Blood, the fluid surrounding each cell, and cytoplasm consist largely of water, so the solubility of sugar in water allows sugar molecules to be transported to and into all of the body's cells.

Figure 2-11 The pH of lemon juice is lower than the pH of tea, so the hydrogen ion concentration would increase.

Figure 2-15 Starch is easily digested by decomposer microbes, so starch-based plastics would be far more biodegradable than plastics made from more microbe-resistant molecules such as cellulose.

Figure 2-24 The hydrogen bonds that account for secondary and higher level protein structure can be broken by heat. Because a protein's function generally depends on its shape, breaking shape-controlling bonds disrupts function.

Consider This

Hints: Research the risks of obesity and of food additives that reduce calories. Research the effectiveness of the additives in helping people lose weight. Also consider whether it is possible to promote behavior that avoids both kinds of risks.

Fill-in-the-Blank

1. ion; positive; negative; ionic
2. neutrons; isotopes; energy; radioactive
3. inert; reactive, chemical bonds
4. polar; hydrogen; cohesion
5. dehydration synthesis; hydrolysis; glucose, amino acids
6. carbohydrates, lipids, proteins, nucleic acids; carbohydrate; lipid; protein; lipid; carbohydrate; nucleic acid; carbohydrate

Review Questions

1. An atom is the smallest particle of an element that still retains the properties of that element. A molecule consists of two or more atoms chemically bonded together. Protons are positively charged subatomic particles found in the atomic nucleus. Neutrons are uncharged subatomic particles found in the atomic nucleus. Electrons are negatively charged subatomic particles found in electron shells around the atomic nucleus.
2. Covalent bonds are formed by sharing electrons between two atoms. Ionic bonds result when oppositely charged ions attract one another. The charges on the atoms result from the transfer of one or more electrons from one atom to another to fill or empty the outer electron shell.
3. The polar water molecules are attracted to the charged ions. The water molecules surround and insulate the ions from one another, allowing the ions to separate.
4. An acid is a substance that releases hydrogen ions when dissolved in water. A base is a substance that combines with hydrogen ions in water. A buffer is a compound that tends to maintain a solution at a constant pH by accepting or releasing hydrogen ions in response to small changes in hydrogen ion concentration. Small changes in pH from normal levels found in the body can interfere with the structure and proper functioning of biological molecules.
5. The three most abundant atoms are carbon (C), hydrogen (H), and oxygen (O); nitrogen (N) is also relatively common (found in all amino acids and nucleic acids), and phosphorus (P) is present in nucleic acids and ATP.
6. Carbohydrates—glucose, sucrose, starch, glycogen, cellulose, chitin; Lipids—oils, fats, waxes, cholesterol; Proteins—keratin, silk, hemoglobin; Nucleic acids—DNA, RNA
7. Nucleotides are subunit monomers in DNA and RNA; ATP serves as a short-term energy storage molecule; cyclic AMP is an intracellular messenger.
8. A monosaccharide is a single sugar molecule; glucose and ribose are examples.

Monosaccharides provide energy and serve as building blocks for polysaccharides. A disaccharide is two monosaccharides bonded together; sucrose and lactose are examples. Disaccharides are used for short-term energy storage. A polysaccharide is a long chain of bonded monosaccharides; cellulose and glycogen are examples. Polysaccharides are used for long-term energy storage, and many are structural components of cells.

9. Proteins are produced by dehydration synthesis, which joins amino acids together with peptide bonds, forming polypeptides.

Applying the Concepts

Hints

1. Fat tissue consists mostly of fat molecules, and muscle tissue consists mostly of protein molecules. Do they contain all of the same kinds of atoms?

2. Antacids may work by being basic and neutralizing stomach acid, or by buffering the stomach contents at a pH that is normal for the stomach. Experiments might involve mixing fixed amounts of different antacids into solutions with acidity in the range found in stomachs and seeing how much the pH changes. Trials with different starting pH would help make the experiments more revealing.

3. If water molecules were nonpolar, there would be no hydrogen bonds between water molecules. As a result, we might hypothesize that water would be less effective as a solvent, and that it would have less internal cohesion. Think about how such changes would affect such biological processes as transport of substances in blood and movement of water and nutrients in plants.

CHAPTER 3

Figure-based Questions

Figure 3-6   Both water and glucose will diffuse from compartment B to compartment A.

Figure 3-7   For simple diffusion, the greater the concentration of the solution, the faster the initial diffusion rate. For facilitated diffusion, the greater the concentration of the solution, the faster the initial diffusion rate until it reaches an upper limit and levels off when the carrier proteins are saturated.

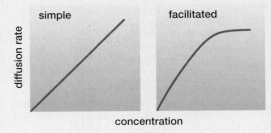

Figure 3-8   Freshwater fishes have physiological mechanisms that constantly export water to the environment to compensate for the water that flows into their bodies by osmosis.

Consider This

Hints: Think about the food that animals consume. Consider how a substance that is toxic in one setting might be beneficial in another. Look ahead to Chapter 21 and find out how the digestive tract is protected from enzymes.

Fill-in-the-Blank

1. passive transport; active transport
2. fluid mosaic; phospholipids, proteins; phospholipid bilayer; proteins

3. selectively permeable; swell
4. phospholipids; they repel water (or are "water-fearing"); inside
5. simple diffusion; simple diffusion; facilitated diffusion; facilitated diffusion, osmosis; facilitated diffusion
6. endocytosis; yes; pinocytosis; phagocytosis; vesicles

Review Questions

1. The plasma membrane is formed by a double layer of phospholipids with hydrophilic heads on the outside and hydrophobic tails inside. Within this fluid bilayer float various proteins.

2. The three types of proteins found in the plasma membrane are (1) transport proteins used to regulate the movement of water-soluble molecules through the plasma membrane, (2) receptor proteins used to signal cellular responses when particular extracellular molecules bind to them, and (3) recognition proteins that serve as identification and cell-surface attachment sites.

3. Diffusion is the net movement of molecules in a fluid from regions of high concentration of those molecules to regions of low concentration. Osmosis is the diffusion of water though a differentially permeable membrane. Plants remain firm as plant cell vacuoles take in water by osmosis; this creates turgor pressure that "inflates" and supports the plant cells.

4. A hypotonic solution is one in which the concentration of solutes is lower than that of a second solution separated from the first by a water-permeable membrane. Water will enter an animal cell placed in a solution that is hypotonic to the cell's cytoplasm, causing it to burst. A hypertonic solution has a higher concentration of solutes than the cell cytoplasm and would cause water to leave the cell, which would shrivel. An isotonic solution has the same concentration of solutes, so there would be no net loss or gain of water by the cell.

5. Simple diffusion is the diffusion of water, dissolved gases, or lipid-soluble molecules through the phospholipid bilayer of a membrane. Facilitated diffusion is the diffusion of molecules through a membrane, driven by concentration gradients and assisted by membrane transport proteins. Active transport is the movement of small molecules and ions through membrane proteins against their concentration gradients, using cellular energy. Pinocytosis draws extracellular fluid into a cell by an infolding of the plasma membrane that then pinches off. During receptor-mediated endocytosis, specific molecules bind to membrane receptor proteins and the membrane then folds inward, forming a sac containing the molecules. Phagocytosis moves large food particles into a cell by surrounding them with pseudopodia that merge to encase the particle in a vesicle. In exocytosis, vesicles formed by the Golgi fuse with the plasma membrane and expel their contents outside the cell.

Applying the Concepts

Hints

1. Research whether cholesterol in the diet contributes to cholesterol in the blood. Research where cholesterol is absorbed along the digestive tract. Use this information in your answer.

2. Is the water in ponds fresh or salty? Red blood cells are surrounded by fluid similar in salt concentration to the red blood cell. How would osmosis affect cells in these different environments? Which cell type is likely to have more aquaporin in its membrane? In designing tests of your hypothesis, assume that you could identify these water channels under a microscope.

3. Virtually all cells, including the cells of the immune system, can manufacture and secrete proteins. Some proteins form pores in membranes.

4. Think about the concentration gradient of minerals in soil vs. root cell. Would diffusion suffice, or would active transport be needed to move minerals into root cells?

CHAPTER 4

Figure-based Questions

Figure 4-4   Of these, only the ribosome is found in bacteria, archaea, and all eukaryotes. Thus, ribosomes must have been present in the common ancestor of all cells, and nuclei, mitochondria, and chloroplasts must have arisen later.

Figure 4-6   Key processes such as DNA replication and transcription require that enzyme molecules have access to the DNA strand. Condensation restricts that access because it leaves little space around individual strands.

Consider This

Hints: Could patients' own stem cells correct a genetic disorder? Would the body's immune system be likely to accept stem cells from an outside donor?

Fill-in-the-Blank

1. cell wall; vacuole; chloroplast
2. nuclear envelope, nuclear pores, water, ions, many small molecules
3. DNA, RNA, protein, ribosomes under construction
4. ribosome, endoplasmic reticulum; vesicles, Golgi apparatus
5. lysosomes, Golgi; enzymes; membranes, organelles
6. cell shape, cell movement, organelle movement, cell division
7. microtubules; cilia are shorter, more numerous, and exert force parallel to the plasma membrane

Review Questions

1. *Observation*: Members of some animal species can regenerate lost limbs. *Question*: What is the difference between animals that can regenerate limbs and those that cannot? *Hypothesis*: Animals that can regenerate limbs produce substances that cause cells at a wound site to reproduce instead of forming scar tissue. *Prediction*: If the hypothesis is true, then applying an extract of tissue from a regenerating animal (salamander) to tissue of a nonregenerating animal (mouse) will cause the mouse cells to divide and reproduce. *Experiment*: Apply tissue extract from a regenerating salamander to cultured mouse cells (of a type that do not normally divide) and observe whether cell division is stimulated. If yes, draw the following conclusion: *Conclusion*: Some substance in the tissue of a regenerating salamander limb causes nondividing mouse cells to change to a growing state.

2. Prokaryotic and eukaryotic cells both possess a plasma membrane surrounding the cytoplasm. Prokaryotic cells are smaller and simpler than eukaryotic cells. Eukaryotic cells possess several membrane-enclosed organelles in the cytoplasm, and a membrane-bound nucleus. See Figures 4-3, 4-4, and 4-11.

3. Organelles common to plant and animal cells: nucleus, mitochondria, ribosomes, endoplasmic reticulum, Golgi, lysosomes, cytoskeleton. Organelles unique to plant cells: chloroplasts, plastids, central vacuole. Organelles unique to animal cells: centrioles.

4. The nucleus houses the DNA in eukaryotic organisms. It is surrounded by the nuclear

envelope, which consists of two membranes containing numerous pores that allow various molecules to pass between the nucleus and the cytoplasm. DNA exists in the nucleus associated with several proteins and condensed tightly in a form called chromatin. A dark-staining region of the nucleus, called the nucleolus, is responsible for ribosome synthesis.

5. Mitochondria extract energy from food molecules and store it in the bonds of ATP; chloroplasts are the sites of photosynthesis.

6. Ribosomes are responsible for protein synthesis and are found embedded in the endoplasmic reticulum.

7. The endoplasmic reticulum forms membrane-enclosed channels in the cytoplasm. Proteins synthesized on ribosomes enter these channels and accumulate in pockets at their ends. The ends pinch off, forming membrane-enclosed vesicles that transport the proteins to the Golgi. The Golgi apparatus consists of specialized membranous sacs derived from the endoplasmic reticulum. It sorts, chemically alters, and packages proteins and other biological molecules into membrane-enclosed vesicles. These are transported within the cell or exported.

Applying the Concepts

Hints

1. Could something other than a living cell leave the kinds of traces found on the Martian meteor? Is there evidence that Mars has ever had conditions that could support life? What features of living things (see Chapter 1) might be preserved in fossils that you could look for?

2. Which of these individuals' muscles would require large amounts of energy on a sustained basis?

3. Think about how organelles and other materials are transported within cells.

4. Think about how substances enter and exit cells and how far they would need to travel in an increasingly large cell. Consider how a spherical cell's surface area changes relative to its volume as its size increases.

CHAPTER 5

Figure-based Questions

Figure 5-6 Other possibilities include mechanical energy (e.g., shaking), electricity, and radiation.
Figure 5-9 Breaking a phosphate bond allows the energy stored in that bond to be transferred to another molecule involved in a reaction.
Figure 5-13 No. A catalyst lowers activation energy, but does not eliminate it. The reaction energy must still be overcome for the reaction to proceed.
Figure 5-14 Increase the concentration of enzyme, because the reaction rate can be limited by the available enzyme molecules; increase the reaction temperature (but not so much as to harm the enzyme); adjust the pH to the optimum for the enzyme's activity.

Consider This

Hint: Consider how common a disorder is, how serious it is, and the cost of testing for it.

Fill-in-the-Blank

1. constant; kinetic, potential
2. more, less; organized; entropy
3. activation; electron shells; heat
4. exergonic; endergonic; exergonic; endergonic; coupled
5. adenosine triphosphate; energy-carrier; adenosine diphosphate, phosphate; energy

6. proteins; activation energy; active site; shape, electrical charges

Review Questions

1. Although living organisms increase in complexity, order, and amount of concentrated energy over time, life does not violate the second law of thermodynamics because Earth is not an isolated system, but instead receives an enormous amount of energy from sunlight, the ultimate source of energy for most life-forms.

2. Metabolism is the sum total of all chemical reactions in a cell. Reactions are coupled so that one reaction releases energy (exergonic) to drive another reaction (endergonic).

3. Activation energy is the initial input of energy required to start any chemical reaction. Catalysts reduce the amount of activation energy required to start a reaction, which results in a faster rate of reaction.

4. The reactions of photosynthesis and cellular respiration both involve coupled exergonic and endergonic reactions. Dehydration synthesis (endergonic) and hydrolysis (exergonic) reactions are also examples.

5. Enzymes are proteins with complex three-dimensional shapes that catalyze biochemical reactions involving one or two specific substrates.

Applying the Concepts

Hints

1. The second law describes conditions inside a closed system. Consider outside inputs of energy that life may be using to offset a tendency to increase disorder.

2. Consider the fate of most enzymes in the acidic environment of the stomach. Would the body likely produce more enzymes than were needed? What purpose could they serve? What differences would you look for between subjects who had or had not consumed enzyme supplements, and how would you measure the differences?

3. What do the laws of thermodynamics tell us about the efficiency of energy conversions? Apply this to animals who obtain energy from other animals.

4. How does the shape of an enzyme determine its function? Are all organic molecules shaped the same? Would the same shape of enzyme work on all of them?

CHAPTER 6

Figure-based Questions

Figure 6-3 Carotenoids are yellow, orange, or red.
Figure 6-5 Most of the ATP and NADPH produced in the chloroplast is used to produce glucose in the C_3 cycle. Mitochondria are needed to extract the energy stored in the sugar molecules.
Figure 6-7 No. Although some oxygen produced by photosynthesis might be used for cellular respiration, much of it would diffuse into the atmosphere and flow out through open stomata, leaving inadequate amounts for cellular respiration.
Figure 6-9 The C_4 pathway uses one more ATP per CO_2 molecule than does the C_3 pathway, so the C_3 pathway is more efficient if water is adequate to allow stomata to remain open.

Consider This

Hints: Can you envision a sudden decline in photosynthesis that only lasts for months, but starts a series of events leading to extinctions tens of thousands or hundreds of thousands of years later? Or might an event that blocked photosynthesis have recurred many times over 300,000 years?

Fill-in-the-Blank

1. stomata, carbon dioxide (CO_2), oxygen (O_2); chloroplasts, mesophyll
2. violet, blue, red; green; carotenoids; green, blue; orange, yellow
3. photosystem II; reaction center; electron transport chain; energy, ATP
4. water (H_2O), carbon dioxide (CO_2); independent, C_3 cycle
5. carbon dioxide (CO_2); photorespiration; C_4 or CAM; dry
6. water (H_2O), oxygen (O_2); ATP, NADPH; carbon dioxide (CO_2), glucose

Review Questions

1. The photosynthetic equation is the same for both C_3 and C_4 plants:
$$6\ CO_2 + 6\ H_2O + light \rightarrow C_6H_{12}O_6 + 6\ O_2.$$

2. Refer to Figure 6-1c for a diagram of a chloroplast. The extensive thylakoid membranes allow large numbers of molecules, including light-harvesting photopigments and the electron transport chain, to be arranged in the appropriate sequence to capture energy in high-energy molecules such as ATP.

3. The light-dependent reaction converts light energy to chemical energy and occurs in the thylakoid membranes. The light-independent reaction utilizes the chemical energy harnessed in the light-dependent reaction to produce high-energy glucose molecules from carbon dioxide. The light-independent reaction occurs in the stroma of the chloroplast.

4. C_4 photosynthesis has a two-stage carbon fixation pathway involving the production of the four-carbon compound oxaloacetic acid from the reaction of PEP with CO_2. C_4 plants possess chloroplasts in their bundle sheath cells to accommodate this two-stage reaction, whereas C_3 plants do not contain chloroplasts in their bundle sheath cells. C_4 photosynthesis is most efficient under conditions of abundant light and little water, whereas C_3 photosynthesis is most efficient when water is plentiful.

Applying the Concepts

Hints

1. Breath and body odors are generally caused by bacteria in the mouth and on the body surface. Can you find any evidence that chlorophyll would kill such bacteria or absorb their odors? Consider how you might objectively assess the breath and body odors of experimental subjects who had and had not eaten chlorophyll tablets.

2. What molecules are broken down to produce the oxygen gas released during photosynthesis? From what molecules do the oxygen atoms in the sugar produced during photosynthesis originate?

3. Consider the benefits that could be gained if photorespiration were prevented and plants could use C_3 photosynthesis even in hot, dry conditions.

CHAPTER 7

Figure-based Questions

Figure 7-5 Without oxygen, ATP production halts. Oxygen is the final acceptor in the electron transport chain, and if it is not present, electrons cannot proceed along the chain and production of ATP by chemiosmosis ceases.
Figure 7-7 In oxygen-rich environments, both types of bacteria can survive, but aerobic bacteria prevail because their respiration produces far more ATP per glucose molecule. In oxygen-poor environments, aerobic bacteria are limited by the oxygen shortage

(unless they can switch to fermentation), and anaerobes prevail.

Consider This

Hints: How do you define "cheating"? Is there a difference between an athlete injecting a substance vs. having it produced by his or her genetically modified cells? If the person was genetically modified for the purpose of gaining a sports advantage, would this be different than a case in which the modification was made for other reasons and happened to provide a sports advantage? Is this a fundamentally different scenario from people with naturally occurring mutations or gene combinations that give them competitive advantages?

Fill-in-the-Blank

1. glycolysis, cellular respiration; cytoplasmic fluid; mitochondria; cellular respiration
2. fermentation, two; fermentation, NAD+
3. carbon dioxide (CO_2), alcohol
4. oxygen; glycolysis, lactic acid, lactate; pyruvate
5. matrix, intermembrane compartment, concentration gradient; chemiosmosis; ATP
6. Krebs cycle, citric acid cycle; acetyl CoA, citric acid; $FADH_2$, NADH

Review Questions

1. *(a)* $C_6H_{12}O_6 + 6 O_2 \rightarrow 6 CO_2 + 6 H_2O + 36$ ATP. (Note: This assumes 34 ATP from aerobic cellular respiration, and includes 2 ATP derived from glycolysis.) *(b)* $C_6H_{12}O_6 \rightarrow 2 CO_2 + 2 C_2H_6O$ (ethanol) $+ 2$ ATP. (Note: The 2 ATP are generated by glycolysis.)
2. See Figure 7-3 for mitochondrial structure. In the mitochondrial matrix, pyruvic acid is broken down in a series of reactions, including those of the Krebs cycle. The inner mitochondrial membrane contains an electron transport system that accepts electrons from electron carriers and drives the formation of a hydrogen ion gradient. It also contains ATP-synthesizing enzymes. The intermembrane compartment stores H^+ ions.
3. *(a)* In glycolysis, the six-carbon glucose is split into two three-carbon pyruvates. *(b)* During chemiosmosis, hydrogen ions flow down their concentration gradient through ATP-synthesizing enzymes that use the energy released to synthesize ATP. *(c)* Fermentation is a metabolic pathway used to regenerate NAD^+ in the absence of oxygen. *(d)* NAD^+ accepts two high-energy electrons and one hydrogen ion, becoming NADH, and carries them to the inner mitochondrial membrane for use in the electron transport system.
4. *(a)* In aerobic respiration, glycolysis splits glucose into two pyruvates, yielding two ATP and two NADH, which eventually yield an additional two ATP each from chemiosmosis. The two pyruvates are converted into two-carbon acetyl groups, and in the process two NADH are created, which results in the production of three ATP, each though chemiosmosis. Each acetyl group enters the Krebs cycle, which generates an ATP, three NADH (producing three ATP each), and one $FADH_2$ (two ATP each). One molecule of glucose yields a total of 36 ATP. *(b)* In anaerobic respiration, glucose is broken down into two pyruvates during glycolysis. Fermentation then converts pyruvate into lactic acid or ethanol. Only two ATP are produced.
5. In the Krebs cycle, an acetyl group (two carbons) derived originally from glucose reacts with a molecule of oxaloacetic acid (four carbons) to form citric acid (six carbons). Citric acid is rearranged and undergoes several reactions, releasing two carbons as CO_2 and eventually

regenerating oxaloacetic acid, which completes the cycle. Most of the energy released is harvested in the form of one ATP and four high-energy electron carriers.

6. The electron transport chain is a series of molecules within the inner mitochondrial membrane. High-energy electron carriers formed during glycolysis and the citric acid cycle donate their electrons to the first molecule, which passes them along the chain, liberating energy in a stepwise process. Some of the energy is used to pump hydrogen ions across the inner membrane from the matrix into the intermembrane space, creating a hydrogen ion gradient. During chemiosmosis, the hydrogen ions move down their gradient through ATP-generating channel proteins in the inner membrane. Movement of hydrogen ions down their gradient provides energy that drives ATP synthesis.
7. Oxygen is the final electron acceptor for electrons that have passed through the electron transport chain. Without oxygen, energy-depleted electrons could not leave the chain, and so there would be no room for more high-energy electrons to enter. Hydrogen ion pumping, chemiosmosis, and ATP production would cease and the organism would die.

Applying the Concepts

Hints

1. If cells are not generating enough ATP, will they burn more or less glucose to try to compensate? Will this cause weight loss? What will happen to any living cell if too little ATP is produced to maintain itself?
2. How do yeast generate ATP when oxygen is present? How do they generate ATP when oxygen is absent? What substance is produced as a by-product? Which process is more efficient?
3. Consider what is required for glucose activation during glycolysis.

CHAPTER 8

Figure-based Questions

Figure 8-9 One daughter nucleus would receive a replicated chromosome (if the chromatids ever did separate, this nucleus would then have an extra copy of that chromosome), while the other daughter nucleus would be missing that chromosome.
Figure 8-12 One of the resulting daughter cells (and the gametes produced from it) would have both homologues and the other daughter cell (and the gametes produced from it) would not have any copies of that homologue.

Consider This

Hints: You are on your own here!

Fill-in-the-Blank

1. binary fission
2. mitotic, differentiation
3. homologues (or homologous chromosomes); autosomes; sex chromosomes
4. prophase, metaphase, anaphase, telophase; cytokinesis; telophase
5. four; meiosis I; gametes (or sperm and eggs)
6. prophase, chiasmata; crossing over
7. shuffling of homologues, crossing over, fusion of gametes

Review Questions

1. Mitosis is division of the nucleus. Cytokinesis is division of the cytoplasm. Although these two processes usually proceed together, some cells undergo mitosis without cytokinesis, resulting in cells with multiple nuclei.

2. Refer to Figure 8-9 for a diagram of the stages of mitosis. During normal mitosis, each of the two sister chromatids of each replicated chromosome become attached to spindle microtubules leading to opposite poles of the cell. Therefore, during anaphase, when the chromatids separate and become independent chromosomes, one copy of each chromosome moves to each pole of the cell.
3. Homologous chromosomes (homologues) are paired eukaryotic chromosomes that contain similar genetic information. A centromere is a specialized region of a chromosome where sister chromatids are attached to one another and to which the spindle fibers attach during cell division. A kinetochore is a structure located at the centromere that becomes attached to spindle microtubules during mitosis. A chromatid consists of a replicated DNA molecule; the replicated chromosome consists of two identical sister chromatids attached to one another at the centromere. "Diploid" refers to cells that contain pairs of homologous chromosomes; "haploid" to cells that contain only one (unpaired) copy of each chromosome.
4. Cytokinesis divides the cytoplasm into approximately equal halves at the end of cell division. In animal cells, the plasma membrane is pinched along the equator by a ring of microfilaments, resulting in two daughter cells. In plant cells, plasma membrane forms along the equator by fusion of vesicles produced by the Golgi apparatus. The contents of the fused vesicles become the new cell wall between the two daughter cells.
5. Refer to Figure 8-12 for a diagram of meiosis. Homologous chromosomes separate during anaphase I, forming haploid daughter cells.
6. *Similarities:* Both require DNA replication prior to division, and both involve the same sequence of steps using many of the same cellular components. *Differences:* Mitosis produces two daughter cells with the same complement of chromosomes; they are genetically identical to each other and to the parent cell. Meiosis separates homologous chromosomes, and produces four haploid daughter cells that are genetically different.
7. Meiosis generates genetic variability in two ways. First, crossing over mixes segments of homologous pairs, creating new combinations of genes. Second, each homologous pair aligns independently of the other pairs in metaphase I, resulting in haploid daughter cells with a random assortment of maternal and paternal homologues. If we ignore crossing over, an animal with a haploid number of 2 can produce $2^2 = 4$ genetically different gametes; with a haploid number of 5, the animal can produce $2^5 = 32$ genetically different gametes.

Applying the Concepts

Hints

1. If nondividing brain or heart muscle cells die, will the organs be repaired? Because the intestinal lining is under constant attack from digestive enzymes and its cells have a short life span, would it survive long without cell division?
2. We lose hairs every day, and these are usually replaced by dividing cells within the hair follicle. Nausea can result from damage to the lining of the stomach and intestines if this everyday damage is not repaired by cell division. Why are cancerous cells so destructive? What type of cells are likely targeted by most cancer therapies? If cancers could be cured with relatively harmless herbs, would doctors routinely prescribe radiation and chemotherapy? How might you test claims using cancerous cells grown in dishes? How would you

test the claims in laboratory animals? Would it be feasible to test them in people?

3. Are asexually reproduced offspring genetically identical to or different from their parents? If the parents are well adapted to an unchanging environment, would it be more adaptive to produce identical or different offspring? In an unstable environment, or an environment to which an organism is not well adapted, would sexual or asexual reproduction be most likely to produce some well-adapted offspring? Is an organism that can't reproduce sexually likely to leave many offspring if the environment changes and the organism is no longer well adapted to it? Is the capacity to use sexual reproduction important over evolutionary time?

CHAPTER 9

Figure-based Questions

Figure 9-8   A heterozygous Pp plant will produce half P gametes and half p gametes. The homozygous recessive pp plant will produce only p gametes. A Punnett square will show that half the offspring are expected to be Pp (purple) and half pp (white). See Figure 9-9.

Figure 9-11   A plant with wrinkled green seeds always has the homozygous recessive genotype $ssyy$. A plant with smooth yellow seeds could be $SSYY, SsYY, SSYy,$ or $SsYy$. Set up four Punnett squares (one for each of the four possible smooth yellow genotypes crossed with the wrinkled green genotype). (1) If the smooth yellow parent is $SSYY$, then all of the offspring of the test cross will be $SsYy$, or smooth yellow. (2) If it is $SsYY$, then one-half of the offspring will be $SsYy$ (smooth yellow) and one-half will be $ssYy$ (wrinkled yellow). (3) If it is $SSYy$, then one-half of the offspring will be $SsYy$ (smooth yellow) and one-half will be $Ssyy$ (smooth green). (4) If it is $SsYy$, then one-quarter of the offspring will be $SsYy$ (smooth yellow), one-quarter will be $Ssyy$ (smooth green), one-quarter will be $ssYy$ (wrinkled yellow), and one-quarter will be $ssyy$ (wrinkled green). Therefore, the test cross will allow you to determine the genotype of the smooth yellow parent.

Figure 9-12   As the figure shows, chromosomes are the structures that assort independently. Genes will assort independently only if they are on different chromosomes.

Figure 9-15   Because human female gametes all carry an X chromosome, the sex of offspring is determined by which sex chromosome is in the sperm that fertilizes the egg. Therefore, a man passes his X chromosome only to his daughters, and his Y chromosomes only to his sons.

Figure 9-20   In principle, you could determine if a phenotypically dominant person is homozygous or heterozygous by using a test cross with a homozygous recessive person. Of course, in addition to ethical considerations, you would need *lots* of children to be sure of the answer.

Consider This:

Hints: You are on your own here, but people's opinions about this issue might be influenced by the costs of these procedures, their potential dangers to the embryo, and feelings about adoption and abortion.

Fill-in-the-Blank

1. locus; alleles; mutations
2. genotype, phenotype; heterozygous
3. independently; as a group; linked
4. an X and a Y; two X; sperm
5. sex-linked
6. incomplete dominance; codominance; polygenic inheritance
7. nondisjunction; Turner syndrome (XO), trisomy X (XXX), Klinefelter syndrome (XXY), Jacob

syndrome (XYY) (order not important); Down syndrome, 21

Review Questions

1. A gene is a segment of DNA coding for a specific protein. An allele is any of the alternative forms of the same gene. "Dominant" is a term that describes alleles that are expressed when a single copy is present. "Recessive" is a term that describes alleles that are masked in the presence of a dominant allele. True-breeding refers to organisms in which all of the offspring produced by self-fertilization are genetically identical to the parent. Homozygous means having a pair of alleles that are the same for a particular gene. Heterozygous means having two different alleles for a particular gene. Cross-fertilization is the fertilization of an egg produced by one organism by a sperm produced by a different organism. Self-fertilization is fertilization of an egg by a sperm produced by the same organism.

2. Independent assortment occurs during meiosis I as pairs of homologous chromosomes are randomly separated during anaphase. Therefore, individual genes assort independently only if they are on different chromosomes. Genes on the same chromosome tend to be moved as a unit into the same daughter cell and tend to be inherited together, or linked. Crossing over may separate alleles of linked genes during meiosis.

3. Genes that are close together would be more tightly linked. Crossing over is roughly equally likely to occur at any place along a chromosome. If genes are far apart on a chromosome, then there is a long section of chromosome between the genes, providing many places for crossing over between the genes. If the genes are close together, then there is only a very short section of chromosome between them, so crossing over will seldom happen to occur between the genes.

4. Polygenic inheritance occurs when two or more genes influence a particular trait. These traits often involve genes that show incomplete dominance. The result is a gradient of different phenotypes depending on the numbers of particular alleles present. Two parents may have "medium" skin color, but be heterozygous at several gene loci that together influence skin color. Because of independent assortment of alleles during meiosis, each parent will produce a wide range of allele combinations in gametes. Random joining of sperm and egg could then produce offspring whose skin color was much paler or darker than that of either parent.

5. Sex-linked genes are found on only one type of sex chromosome (for example, the X chromosome in mammals). In mammals, the X chromosome contains many more genes than the Y chromosome does, and most X chromosome genes have no corresponding gene on the Y chromosome. Therefore, males are most likely to exhibit recessive sex-linked traits, because they possess only one copy of the X chromosome, and their Y chromosome does not have a corresponding gene that could mask recessive genes present on the X chromosome.

6. An organism's phenotype is its physical characteristics (for example, outward appearance, behaviors, or biochemical composition). An organism's genotype is the actual combination of alleles in the organism's genome. Knowing the phenotype of an organism does not always allow you to determine the genotype because an organism with the dominant phenotype for a particular trait might be homozygous or heterozygous for the gene that controls that trait. Examining the offspring of a test cross (mating

the phenotypically dominant individual with a homozygous recessive individual) allows the genotype of the phenotypically dominant individual to be determined.

7. Nondisjunction is an error in meiosis in which a chromosome fails to segregate properly into daughter cells, resulting in some cells that have too few and some that have too many copies of some chromosomes. Common syndromes resulting from nondisjunction include Turner syndrome, trisomy X, Klinefelter syndrome, Jacob syndrome, and trisomy 21, or Down syndrome. See the text for descriptions of these conditions.

Applying the Concepts

Hints

1. Which term is more general? More specific? Can a gene have more than one allele? If you are describing differences in a trait between individuals, should you say that the individuals have different genes for that trait or different alleles for that trait?

2. There are no "correct" answers. Scientific misconduct will be discovered as other scientists try to repeat or build on earlier work. Most scientific research is funded by government agencies, using tax money. Therefore, time, effort, and money are wasted uncovering fraud. Currently, the most common punishments for scientific misconduct are public humiliation, loss of funding, and loss of jobs.

3. Assortative mating tends to keep multiple, distinctly different phenotypes (including what are often called "races") in the human population. What changes would random mating produce? Do you think this would be preferable?

4. There are no "correct" answers. Some factors to consider include personal freedom to choose; the lack of a readily available medical test for salt sensitivity; most people's generally poor understanding of biology, genetics, and medicine; and the fact that public funds pay for a large proportion of medical care. Should the dietary guidelines seek to protect the most sensitive people, or should they only seek to protect the majority? Are there alternative ways to present the guidelines?

Answers to Genetics Problems

1. (a) A red bull (R_1R_1) is mated to a white cow (R_2R_2). The bull will produce all R_1 sperm; the cow will produce all R_2 eggs. All the offspring will be R_1R_2 and will have roan hair (codominance).

   (b) A roan bull (R_1R_2) is mated to a white cow (R_2R_2). The bull produces half R_1 and half R_2 sperm; the cow produces R_2 eggs. Using the Punnett square method:

   eggs

	R_2
sperm R_1	R_1R_2
R_2	R_2R_2

   The predicted offspring will be one-half R_1R_2 (roan) and one-half R_2R_2 (white).

2. The offspring occur in three types, classifiable as dark (chestnut), light (cream), and intermediate (palomino). This distribution suggests incomplete dominance, with the alleles for chestnut (C_1) combining with the allele for cream (C_2) to produce palomino heterozygotes (C_1C_2). We can test this hypothesis by examining the offspring numbers. There are approximately one-quarter chestnut (C_1C_1), one-half palomino (C_1C_2), and one-quarter cream (C_2C_2). If palominos are

heterozygotes, we would expect the cross $C_1C_2 \times C_1C_2$ to yield one-quarter C_1C_1, one-half C_1C_2, and one-quarter C_2C_2. Our hypothesis is supported.

3. (a) $TtGg \times TtGg$. This is a "standard" cross of F_1 heterozygotes whose parents were true breeding for different phenotypes of two traits. Both parents produce TG, Tg, tG, and tg gametes. The expected proportions of offspring are 9/16 tall green, 3/16 tall yellow, 3/16 short green, 1/16 short yellow.
(b) $TtGg \times TTGG$. In this cross, the heterozygous parent produces TG, Tg, tG, and tg gametes. However, the homozygous dominant parent produces only TG gametes. Therefore, all offspring receive at least one T allele for tallness and one G allele for green pods and are tall with green pods.
(c) $TtGg \times Ttgg$. The second parent produces two types of gametes, Tg and tg. Using a Punnett square:

eggs

	Tg	tg
TG	$TTGg$	$TtGg$
Tg	$TTgg$	$Ttgg$
tG	$TtGg$	$ttGg$
tg	$Ttgg$	$ttgg$

(sperm)

The expected proportions of offspring are 3/8 tall green, 3/8 tall yellow, 1/8 short green, 1/8 short yellow.

4. If the genes were on separate chromosomes, then this would be a cross of individuals heterozygous for two independently assorting traits with expected offspring in the proportions 9/16 round smooth, 3/16 round fuzzy, 3/16 long smooth, and 1/16 long fuzzy. However, only the parental combinations show up in the offspring, indicating that the genes are on the same chromosome.

5. The genes are on the same chromosome and are quite close together. On rare occasions, crossing over occurs between the two genes, recombining the alleles.

6. (a) $BBMM$ (brown) $\times BbMm$ (brown). The first parent can produce only BM gametes, so all offspring will receive at least one dominant allele for each gene. Therefore, all offspring will have brown hair.
(b) $BbMm$ (brown) $\times BbMm$ (brown). Both parents can produce four types of gametes: BM, Bm, bM, and bm. Filling in the Punnett square:

eggs

	BM	Bm	bM	bm
BM	$BBMM$	$BBMm$	$BbMM$	$BbMm$
Bm	$BBMm$	$BBmm$	$BbMm$	$Bbmm$
bM	$BbMM$	$BbMm$	$bbMM$	$bbMm$
bm	$BbMm$	$Bbmm$	$bbMm$	$bbmm$

(sperm)

All mm offspring are albino, so we get the expected proportions 9/16 brown-haired, 3/16 blond-haired, 4/16 albino.
(c) $BbMm$ (brown) $\times bbmm$ (albino).

eggs

	bm
BM	$BbMm$
Bm	$Bbmm$
bM	$bbMm$
bm	$bbmm$

(sperm)

The expected proportions of offspring are one-quarter brown-haired, one-quarter blond-haired, one-half albino.

7. A man with normal color vision is CY (the Y chromosome does not have the gene for color vision). His color-blind wife is cc. Their expected offspring will be:
We therefore expect that all of the sons will be color-blind and all of the daughters will have normal color vision.

eggs

	c
C	Cc
Y	cY

(sperm)

8. The husband should win his case. All of his daughters must receive one X chromosome, with the C allele, from him and therefore should have normal color vision. If his wife gives birth to a color-blind daughter, her husband cannot be the father (unless there was a new mutation for color-blindness in his sperm line, which is very unlikely).

CHAPTER 10

Figure-based Questions

Figure 10-2 It takes more energy to break apart a C-G base-pair, because these are held together by three hydrogen bonds, compared with the two hydrogen bonds that bind A to T.

Figure 10-5 DNA polymerase always moves toward the free phosphate end of a parental strand. Because the two strands of a DNA double helix are oriented in opposite directions, the free phosphate end on one strand points toward the replication fork and the free phosphate end on the other strand points away from the fork. Therefore, DNA polymerase must move in opposite directions on the two strands.

Figure E10-1 If the diameter is constant, then every *pair* of bases must be approximately the same total width. Because adenine and guanine are so much larger than thymine and cytosine, the base-pairs could be (1) A-A, A-G, or G-G (all very wide); (2) T-T, T-C, or C-C (all very thin); or (3) A-T, A-C, G-T, or G-C (all intermediate in width). Only the third set uses all of the bases, so the first and second sets must not be correct. Then, by examining the possible hydrogen bonds that each pair could form, Watson and Crick deduced that the pairs must be A-T and G-C.

Consider This

Hints: If the sequence and function of all human alleles were known, and athletes' genomes were mapped, do you think they would all be genetically equal in genes conferring athletic ability except for a few rare mutations such as in the myostatin gene?

Fill-in-the-Blank

1. nucleotides; sugar (deoxyribose), phosphate, base (order not important)
2. phosphate, sugar (order not important); strand; double helix
3. thymine, cytosine; complementary
4. semiconservative
5. DNA helicase; DNA polymerase
6. mutations

Review Questions

1. Refer to Figure 10-1 for a diagram of a nucleotide. Each nucleotide consists of a phosphate group, a sugar (deoxyribose, for DNA), and a nitrogenous base (any of adenine, guanine, cytosine, or thymine). For DNA, all nucleotides have the same phosphate group and sugar but vary in the type of nitrogenous base.
2. The nitrogenous bases found in DNA are adenine, guanine, cytosine, and thymine.

3. Adenine is complementary to thymine and guanine is complementary to cytosine. Complementary bases are held together in DNA by hydrogen bonds between the nitrogenous bases.
4. The DNA of a chromosome is composed of two strands of nucleotides wound around each other in a double helix. In each strand, the sugar of one nucleotide is covalently bonded to the phosphate of the next nucleotide in the strand, forming the backbone on each side of the double helix. The bases from each strand pair up in the middle of the helix and are held together by hydrogen bonds. Adenine is held to thymine by two hydrogen bonds, while cytosine is held to guanine by three hydrogen bonds.
5. The two DNA strands of the double helix unwind. DNA polymerase enzymes move along each strand, linking up free nucleotides into new DNA strands. The sequence of nucleotides in each newly formed strand is complementary to the sequence on the parent strand. As a result, two double helices are synthesized, each consisting of one parental DNA strand plus one newly synthesized complementary strand that is an exact copy of the other parental strand. This type of replication is called semiconservative replication.

Applying the Concepts

Hints

1. The hair that protrudes from your scalp is dead, so—can it make use of DNA? The cells in the hair follicle (from which the hair grows) and the scalp are living, but do you think human cells are likely to pick up and use arbitrary strands of DNA from the environment? Would this be adaptive?
2. There are no "correct" answers. What are some benefits of competition? What are some problems with keeping information secret? How does collaboration avoid these problems? What factors can cause individuals or small groups to avoid collaborating with "outsiders"?
3. For an analogy, suppose that English were a "complementary language," with letters at opposite ends of the alphabet complementary to one another (that is, A complementary to Z, B to Y, C to X, etc.). Would a sentence complementary to "To be or not to be" make sense? How does this correspond to a protein encoded in the nucleotide sequence of the "other" strand of the DNA coding for the hemoglobin molecule? How does the double-stranded structure of DNA provide the information needed to replicate DNA during cell division?

CHAPTER 11

Figure-based Questions

Figure 11-3 No, because the "nontemplate" strand is unlikely to code for a functioning protein.

Figure 11-4 Cells produce far more of some proteins than others. For example, cells that produce protein hormones often secrete large quantities into the bloodstream, producing effects throughout the body. A cell that needs to produce large quantities of a protein will probably synthesize more mRNA that will be translated into that protein.

Figure 11-5 Grouped in codons, the original mRNA sequence is CGA AUC UAG UAA. Changing all G to U would produce the sequence CUA AUC UAU UAA. Refer to Table 11-2. CGA encodes arginine, while CUA encodes leucine, so the first G→U change would substitute leucine for arginine in the protein. Second, UAG is a stop codon, but UAU encodes tyrosine. Therefore, the second G→U change would add tyrosine to the protein instead of stopping translation. The final codon in the illustration, UAA, is also a stop codon, so the new protein would end with tyrosine.

Consider This

Hints: There are no "correct" answers here.

Fill-in-the-Blank

1. transcription; translation; ribosome
2. messenger RNA, transfer RNA, ribosomal RNA (order not important)
3. three; codon; anticodon
4. RNA polymerase; template; promoter; termination signal
5. start, stop; transfer; peptide
6. point; insertion; deletion

Review Questions

1. RNA is usually single stranded; RNA has a different type of sugar in its backbone (ribose instead of deoxyribose); and the base thymine in DNA is replaced by uracil in RNA.
2. Messenger RNA (mRNA) carries a copy of genetic information (to make a protein) from the nucleus into the cytoplasm. Transfer RNA (tRNA) carries amino acids to the site of protein synthesis in the cytoplasm (the ribosome). Ribosomal RNA (rRNA) is a structural component of the ribosomes that "read" the code of nucleotides of mRNA to synthesize an amino acid sequence, using the amino acids carried by the tRNAs.
3. The genetic code is the set of three-base sequences (codons in messenger RNA) corresponding to the amino acids that will be incorporated into a protein. A codon is a specific three-base sequence in mRNA that specifies a particular amino acid, or the start or termination of protein synthesis. An anticodon is a triplet of tRNA nucleotides that is complementary to one of the mRNA codons. The relationship between bases in DNA, codons, and anticodons is based on complementary base-pairing: mRNA codons are complementary to three-base sequences in DNA, and tRNA anticodons are complementary to mRNA codons.
4. Refer to Figure 11-5 for a diagram and description of protein synthesis.
5. In transcription, complementary base-pairing is used to copy the nucleotide sequence of a gene in DNA into a sequence of mRNA nucleotides. In translation, complementary base-pairing is used to match each mRNA codon of nucleotides to a complementary anticodon of a tRNA molecule carrying a specific amino acid.
6. Cells regulate gene expression in many ways, including control of the rate of transcription of mRNA; the rate at which different mRNA molecules are broken down after they are transcribed; how fast the mRNA molecule is translated into protein; and how long the protein lasts before its broken down.
7. A mutation is a change in the nucleotide sequence of a DNA molecule. Mutations can occur through mistakes in DNA replication or through damage to DNA caused by ultraviolet light, ionizing radiation, or chemicals. Most mutations are not beneficial. Some mutations occur in non-protein-coding portions of DNA; others do not alter the activity of the protein coded from the mutated DNA sequence, and some reduce or destroy protein function. A small percentage of mutations are beneficial to the organism, such as when altered protein function produces resistance in bacteria to the antibiotics that people use against them.

Applying the Concepts

Hints

1. RNA synthesis is required for learning, in that proteins synthesized from the instructions in messenger RNA are involved in the changes in connections between neurons that produce learning. Synthesis of a specific RNA molecule is not likely to be required for learning a specific fact or a specific behavior. In proposing a hypothesis for the *Planaria* results, consider what nutrients might have been transferred when *Planaria* were fed trained flatworms. Because proteins are involved in learning, what types of nutrients are required to code for and build proteins? To test the hypothesis that a specific behavior could be transferred, what would you predict would happen if the experimental group were fed trained flatworms, and control worms were fed untrained flatworms? Would they be receiving the same nutrients? What would you learn by comparing learning between "control" flatworms and flatworms that have been fed worms that were trained for other tasks?
2. There are no "correct" answers. Factors to consider include (1) the possible health risks of hormone treatment (for example, anabolic steroids often cause the testes to shrink, change natural hormone levels, and may decrease fertility); (2) the "fairness" of allowing people taking hormone treatments to compete against people not taking hormones; and (3) a person's perception of the relative benefits vs. risks of hormone treatments.
3. The gene for the androgen receptor is on the X chromosome. Can a man develop a normal male phenotype (testes, penis, sperm production) if he has a defective androgen receptor allele on his single X chromosome? Could a woman, with two X chromosomes, have a normal androgen receptor allele on one X chromosome and a defective allele on the other? Could androgen insensitivity potentially be passed from heterozygous female to heterozygous female indefinitely? Could an XY child inherit androgen insensitivity from the mother, making the child genetically male but phenotypically female? Can a genetic male with androgen insensitivity produce sperm and sire children?

CHAPTER 12

Figure-based Questions

Figure 12-3 *Thermus aquaticus* lives in hot springs, and its DNA polymerase has evolved to be efficient at high temperatures. The DNA polymerase would probably work more slowly at both higher and lower temperatures than the springs in which *T. aquaticus* resides; 72°C is close to optimal.

Figure 12-7 Each person normally has two copies of each STR gene, one on each homologous chromosome. A person may be homozygous or heterozygous for each STR gene. The bands on the gel represent individual alleles of an STR gene. Therefore, a single person can have one band (if homozygous) or two bands (if heterozygous). If a person is homozygous for an STR allele, that (single) band will have twice as much DNA as each of the two bands of DNA from a heterozygote. The more DNA, the brighter the band.

Figure 12-11 A heterozygote has one normal globin allele and one sickle-cell allele. Therefore, MstII would produce two DNA bands from the normal allele and one from the sickle-cell allele, for a total of three bands of different sizes. The amount of DNA in each of the bands from heterozygotes would be half as much as from the homozygotes, because DNA from heterozygotes contains only one copy of each type of allele, whereas homozygous normal DNA has two

copies of the normal allele, and homozygous sickle-cell DNA has two copies of the sickle-cell allele. Therefore, bands from heterozygotes should be about half as bright as bands from homozygotes.

Consider This

Hints: There are no correct answers here. Keep in mind that many, if not most, scientific breakthroughs relied in part on discoveries made during studies that had no obvious or direct applicability to human welfare.

Fill-in-the-Blank

1. genetically modified organisms
2. transformation; plasmids
3. polymerase chain reaction
4. short tandem repeats; repeats; DNA profile
5. electrophoresis, or gel electrophoresis; DNA probe, base-pairing or hydrogen bonds

Review Questions

1. Three natural forms of genetic recombination include sexual reproduction, bacterial transformation, and viral infection. In sexual reproduction, haploid gametes combine to form a diploid zygote. Genetic recombination occurs because of crossing over and independent assortment during meiosis prior to gamete formation, and by combining differing genetic material from the two parents. Bacterial transformation is the acquisition of DNA by bacteria from other dead bacteria, sometimes of a different species. During viral infection, newly formed viruses acquire some of the host's DNA and may transfer host DNA when they infect a second host of the same or different species. Both natural genetic recombination and recombinant DNA technology involve exchange of DNA between organisms, and sometimes between species. They are different, however, in that natural recombination is relatively random and its usefulness is driven by natural selection, whereas recombinant DNA technology is deliberate and directed toward a specified use.
2. A plasmid is a small, circular DNA molecule within the cytoplasm of bacteria. Living bacteria exchange plasmids, or bacteria may acquire plasmids when a bacterium dies and its cellular contents are released.
3. A restriction enzyme cuts a DNA molecule at specific short nucleotide sequences. Some make a "staggered" cut across the DNA, leaving complementary single-stranded ends protruding. If both human and plasmid DNA are cut with the same restriction enzyme and then mixed together, their complementary single-stranded ends base-pair with one another. DNA ligase then splices the cut ends of the human and plasmid DNA together.
4. The polymerase chain reaction amplifies specific segments of DNA. Primer segments are synthesized, complementary to the beginning of each of the two strands of DNA to be copied. The DNA sample, the primers, free nucleotides, and a special DNA polymerase that is active at very high temperatures are mixed together in a tube. The tube is heated to 194°F (90°C) to separate the DNA strands, then cooled to 122°F (50°C) to allow the primers and DNA polymerase to bind to the separated DNA strands. The tube is then heated to 161°F (72°C), allowing the DNA polymerase to copy the DNA. Repeated heating and cooling cycles copy the DNA, which doubles with each cycle.
5. A short tandem repeat (STR) is a sequence of usually 2 to 5 nucleotides in DNA repeated many times (usually 5 to 15 times) head-to-tail.

Different people have the same sets of STRs, but differ in the number of repeats. By analyzing the numbers of repeats in a set of 10 to 13 different, standard STRs, forensic scientists can identify any individual person's DNA with a probability of misidentification of less than one in a trillion.

6. Gel electrophoresis separates DNA pieces according to size. The gel has tiny holes of various sizes. The smaller the molecule, the more rapidly it can move through the gel. DNA samples are applied to one end of the gel, and the negatively charged DNA pieces are driven through the gel by an electric field. The smaller the DNA segment, the farther it moves through the gel in any given time period, so segments of different sizes end up at different locations along the gel.

7. DNA probes are pieces of DNA with a sequence of nucleotides complementary to the DNA to be identified and labeled with radioactive or colored molecules. DNA samples are bathed in DNA probes, which only bind to pieces of sample DNA with the complementary nucleotide sequence. DNA probes for defective human alleles that cause genetic diseases can be spotted in an orderly array on special paper. A patient's DNA is cut into small pieces and labeled with colored molecules. The probes in the array are bathed with this labeled DNA, which binds only to the complementary probes, showing the physician which defective alleles the patient possesses.

8. Genetic engineering in agriculture usually adds genes from other species to crop plants or livestock. These added genes may produce crop plants that resist insects (for example, by expressing a protein from *Bacillus thuringiensis* that damages insects' digestive tracts), resist herbicides (so that herbicides kill weeds without harming crops), or resist disease or environmental stresses such as drought or salty soil. In animals, added genes may increase growth rate and lean muscle mass and may cause the animal to produce useful proteins such as spider silk or antibodies.

9. Restriction enzymes can be used to diagnose diseases such as sickle-cell anemia, because certain restriction enzymes cut the normal allele but not the disease-causing, mutant allele (or vice versa). DNA sequencing or DNA probes can be used to diagnose diseases such as cystic fibrosis. Recombinant DNA technology is used to produce vaccines by synthesizing DNA or pieces of protein from disease-causing organisms. Injected into people, these cause an immune response and produce immunity without any danger of causing the disease from the vaccine. Recombinant DNA techniques are used to make insulin for diabetics and growth hormone for unusually short children. In severe combined immune deficiency, transplanted bone marrow cells merged with disabled viruses containing the normal, functional allele have allowed some children to develop functional immune systems.

10. In amniocentesis, amniotic fluid is removed after 16 weeks of pregnancy. This fluid contains some fetal cells, which are cultured for at least a week, providing enough cells to be analyzed for genetic disorders such as trisomy 21. Amniotic fluid can also be analyzed for molecules that indicate fetal disorders such as spina bifida. In chorionic villus sampling, a few villi are removed from the fetal chorion and analyzed. This can be performed in the eighth week of pregnancy, and the process provides enough cells for immediate genetic analysis. There is, however, a slightly higher risk of abortion or damage to the fetus. Chorion cells from healthy fetuses are more likely to have abnormal numbers of chromosomes than cells obtained from amniocentesis. Chorionic villus sampling does not obtain amniotic fluid, so certain types of fetal defects cannot be detected.

Applying the Concepts

Hints

1. A gene is a sequence of nucleotides. Were people reacting to the *source* of the vitamin C gene? Is this scientifically valid? Or were they reacting to the greater value of the vaccine? Was this a well-designed question to test people's perceptions of genetically modified food? If not, how would you reword it?

2. *Bacillus thuringiensis* in crops acts as an agent of natural selection. Can you think of ways to reduce or delay the development of resistance? Does *Bt* have advantages over the alternative ways to control insects?

3. Genetically modified plants often contain genes that confer insect or herbicide resistance. Might viruses or pollen transfer these genes to nearby wild relatives? If so, what might be the consequences? Are there similar dangers with genetically modified livestock? What about fish or invertebrates? Could they be sterilized? Whether the benefits of GMOs justify the risks is a social, not a biological, question.

4. These are personal, ethical decisions; biotechnology has given us the tools to make informed choices.

CHAPTER 13

Figure-based Questions

Figure 13-6   No. Mutations, the ultimate source of the phenotypic variation on which selection acts, occur in all organisms, including those that reproduce asexually.

Figure 13-7   No. Evolution can include changes in nonstructural elements such as physiological systems and metabolic pathways. Evolution—changes in a species' gene pool over time—is inevitable.

Figure 13-10   Homologous. Vertebrates share common ancestry, and both birds and dogs have vertebrae that support the tail. These structures did not arise by convergent evolution, nor do they serve a similar function, as do analogous structures.

Consider This

Hints: Try to describe a scenario in which natural selection causes the evolution of a structure that is not essential at the time it first evolves. Is the scenario consistent with how natural selection works?

Fill-in-the-Blank

1. wing, arm; analogous, convergent; vestigial
2. common ancestor; amino acids, ATP
3. catastrophism; uniformitarianism; old
4. evolution; mutations, DNA
5. natural selection; artificial selection
6. many traits are inherited; Gregor Mendel

Review Questions

1. Evolution is defined as a change in the genetic makeup of a population over time.

2. Catastrophism hypothesized that all species were created at the same time, and that successive catastrophes killed off certain species and fossilized them in successive rock layers. Uniformitarianism hypothesized that ordinary natural processes that occurred repeatedly and continuously over long periods of time produced the layers of rock. Catastrophism attempted to explain fossils in the context of creationism. Uniformitarianism is the accepted scientific theory and explains that the rock strata were laid down over many millions of years, providing ample time for species to evolve and to become extinct.

3. Lamarck's hypothesis states that organisms modify their bodies through use or disuse of parts, and these modifications can be inherited by their offspring. This is invalid because acquired characteristics are not inherited.

4. Natural selection is the process by which the environment selects for those individuals whose inherited traits best adapt them to a particular environment by allowing the best adapted to leave more reproducing offspring. Barracuda whose genetic traits allow them to swim faster can catch more prey. They would leave more healthy offspring inheriting the genes that confer speed than would slower individuals.

5. Natural populations have the potential to increase rapidly, because all organisms can produce far more offspring than are required to replace the parents. Nevertheless, the sizes of most natural populations and the resources available to maintain them remain relatively constant over time. Therefore, many offspring do not survive and some fail to reproduce. Genetic differences make some individuals better adapted to their environment, so they produce more offspring that inherit these adaptations; this is natural selection. Over many generations, unequal reproduction among individuals with different genetic makeup changes the overall genetic composition of the population, resulting in evolution.

6. Convergent evolution leads to similar structures in unrelated organisms exposed to similar environmental demands. An example is the wings of flies and birds.

7. Biochemistry and molecular genetics reveal that all cells use DNA as hereditary material. All cells use ribosomes, RNA, and the same genetic code to transcribe proteins from roughly the same set of amino acids. This provides strong evidence that all cells alive today descended from a common ancestor.

Applying the Concepts

Hints

1. At rest, the human brain consumes about 20% of the body's energy. Consider the likelihood that natural selection would cause the evolution and persistence of a large, complex, energy-guzzling structure that went largely unused.

2. Review the description of science in Chapter 1. Think about whether the theory of evolution and/or the concept of special creation make predictions that can be tested with objective observations or experimentation.

3. Consider how the "best" traits might change when environmental conditions change.

CHAPTER 14

Figure-based Questions

Figure 14-3   A random selection of colonies would have contained individual bacteria that mutated in response to the antibiotic, resulting in different distributions of surviving colonies on each of the three test plates. Another possibility would be that the antibiotic *always* caused mutations and then all the colonies would survive on the test plates.

Figure 14-4   For a locus with two alleles, one dominant and one recessive, there are two possible phenotypes. A mating between a heterozygote and a homozygote-recessive yields offspring with a 50:50 ratio of the two phenotypes.

Figure 14-5   Allele *A* should behave roughly as it does in the size 4 population, its frequency drifting to fixation or loss in almost all cases. But, in the larger population, the allele should, on average, take more generations to reach fixation or loss. The longer period of drift should also allow for more reversals of direction (e.g., frequency drifting down, then up, then down again, etc.) than in the size 4 population.

Figure 14-6 Mutations inevitably and continually add variability to a population and, after the population becomes larger, the counteracting, diversity-reducing effects of drift decrease. The net result is an increase in genetic diversity.

Figure 14-9 Greater for males. A female's reproductive success is limited by her maximum litter size, but a male's potential reproductive success is limited only by the number of available females. When, as in bighorn sheep, males battle for access to females, the most successful males can impregnate many females, while unsuccessful males may not mate at all. Thus, the difference between the most and least successful male can be very large.

Consider This

Hints: Some species of bacteria and fungi secrete chemicals that help them compete for space and food. Are most natural populations of bacteria exposed to antibiotics? Would natural selection favor production of proteins that cause antibiotic resistance if the proteins were not used?

Fill-in-the-Blank

1. Hardy-Weinberg principle, equilibrium, allele; no
2. alleles; nucleotides; mutations; homozygous, heterozygous
3. genotype; phenotype; phenotype
4. genetic drift; small; founder effect, population bottleneck; founder effect
5. the same species; natural selection, coevolution; adaptations
6. reproducing; environment
7. behavioral isolation; hybrid inviability; temporal isolation; gametic incompatibility; mechanical incompatibility

Review Questions

1. A gene pool is all the genes that are present in a population. Allelic frequencies are determined by adding all of the alleles for a trait in a population and determining their relative proportions.
2. An equilibrium population is one in which the allelic frequencies and the distribution of genotypes remain constant in succeeding generations. This can occur only if (1) there are no mutations, (2) there is no gene flow between populations, (3) the population is very large, (4) all mating is random, and (5) there is no natural selection.
3. Chance alone can change allelic frequencies in small populations, but the effect is weak in very large populations. Yes, genetic drift can alter allelic frequencies, especially in small populations.
4. This would not prove that natural selection was occurring because there are other factors such as migration, mutation, nonrandom mating, or genetic drift that could cause changes in allelic frequencies.
5. This experiment does not prove that bacteria don't develop resistance in response to the antibiotic. It is possible that only those four colonies contained individuals that were capable of responding to the antibiotic by developing resistance. Testing the colonies in the same four locations on the original dish for antibiotic resistance would support the hypothesis that antibiotic resistance was present before exposure to the antibiotic.
6. Sexual selection is the process by which females act as agents of natural selection on males of their species, by choosing to mate with males that exhibit particular characteristics, such as brighter coloration. This is a form of natural selection because the brightly colored males will produce

more offspring, and the genetic trait will increase among males of the species. Sexual selection differs from other forms of natural selection in that the traits selected by the female may be disadvantageous—in this case making the males more vulnerable to predation.

7. By the definition of a species given in the text, naturally interbreeding species are not technically "true species."
8. Premating isolating mechanisms include geographical isolation (related species may not be able to reach each other; for example, if one is restricted to an island), ecological isolation (related species may inhabit different parts of a general area; for example, thicket-dwelling vs. field-dwelling sparrows), temporal isolation (related species may breed at different times of the year; for example, Bishop pines and Monterey pines), behavioral isolation (related species may have different courtship rituals, such as different courtship calls produced by related species of frogs), and mechanical incompatibility (the reproductive structures of the two species may be incompatible, such as flower structures that attract different pollinators, so cross-pollination cannot occur). Postmating isolating mechanisms include gametic incompatibility (in which the sperm fails to fertilize the egg, common in different species of plants); hybrid inviability (the hybrid embryo aborts or the hybrid offspring is unable to reproduce, such as hybrid lovebirds who are unable to build a nest); and hybrid infertility (the hybrid is infertile, such as with the mule, which is a cross between a donkey and a horse).

Applying the Concepts

Hints

1. When one species gives rise to another, does the first species disappear?
2. Think about how one might determine whether two populations "interbreed freely under natural conditions."
3. Gene flow introduces new alleles to populations. Consider how this result might benefit an endangered population.

CHAPTER 15

Figure-based Questions

Figure 15-2 The presence of oxygen would prevent the accumulation of organic compounds by quickly oxidizing them or their precursors. All successful abiotic synthesis experiments used oxygen-free "atmospheres."

Figure 15-5 The bacterial sequence would be most similar to that of the plant mitochondrion (the bacterium shares a more recent common ancestor with the mitochondrion than with the chloroplast or the nucleus).

Figure 15-9 No. The mudskipper merely demonstrates the plausibility of a hypothetical intermediate step in the proposed scenario for the origin of land-dwelling tetrapods. But the existence of a modern example similar in form to the hypothetical intermediate form does not provide information about the actual identity of that intermediate form.

Figure E15-1 356.5 million years old. (3:1 ratio means that three-quarters of the original uranium-235 is left, so one-half of its half-life has passed.)

Consider This

Hints: Is it likely that the fine details of a complex structure could evolve twice independently? Is this kind of evolutionary pattern absolutely impossible? Think about how you would interpret a fossil bird that was older than the oldest dinosaur fossil.

Fill-in-the-Blank

1. anaerobic; photosynthesize; poisonous (toxic, harmful), aerobic (cellular), energy
2. RNA, enzymes (catalysts), ribozymes
3. eukaryotic; endosymbiotic; aerobic; DNA
4. swimming, moist/wet; pollen
5. conifers; wind; flowers, insects; efficient
6. arthropods, exoskeletons, drying
7. reptiles; eggs; internal; skin; lungs

Review Questions

1. Evidence for life arising from nonliving matter is that organic molecules can form spontaneously under hypothetical prebiotic conditions.
2. Many habitats remained anaerobic and anaerobic bacteria were best adapted to these habitats.
3. The endosymbiont hypothesis states that bacterial cells acquired the precursors of mitochondria and chloroplasts by engulfing, and entering into a symbiotic relationship with, certain other types of bacteria. Mitochondria may have evolved from engulfed aerobic bacteria, and chloroplasts from engulfed photosynthetic cyanobacteria.
4. Multicellularity allowed early algae to become larger and thus more resistant to predation, and allowed them to evolve specialized rootlike structures for anchoring and leaflike structures to capture solar energy.
5. The advantages of terrestrial existence for early plants included abundant light, nutrients, and few predators. Disadvantages include limited water (for metabolic processes and for sperm to swim through during reproduction) and a need to support body weight. The advantages of terrestrial existence for the first animals included few predators and abundant food from newly evolved land plants. Disadvantages included the need to support body weight, to waterproof the body, to breathe air without drying out their respiratory structures, and to keep their embryos in a watery environment during reproduction.
6. Amphibians evolved legs and simple lungs, which helped them to move on land and to breathe the air, but their skin needed to be kept moist, and they deposited their sperm in water. Reptiles evolved internal fertilization, waterproof eggs, waterproof skin, and better lungs. Birds evolved warm body temperature and insulating feathers that protected them from the cold temperatures on land and allowed them to exploit the air for flight. Mammals evolved insulating fur rather than feathers and internal development of their embryos, which freed them from guarding eggs and protected the embryo from the cold.
7. Primate evolution began with tree shrews that had grasping hands, which allowed them to perform powerful and precise manipulations. Early primates evolved binocular vision, which provided accurate depth perception for moving from tree to tree. A large brain facilitated hand-eye coordination as well as complex social interactions. *Australopithecus* evolved bipedalism, which freed the hands from the task of walking. Free hands and further brain enlargement in the genus *Homo* allowed them to fashion and manipulate tools.

Applying the Concepts

Hints

1. Some meteors and comets contain organic molecules. Liquid water was probably present on Mars in the past and may currently be present elsewhere in our solar system. Could a living cell survive space travel in an asteroid? Is the idea of life having originated elsewhere more likely than life originating here on Earth?

2. Can you think of adaptations to the harsh conditions that early hominids faced (need to obtain food, high infant mortality, need to avoid predation) that persist in people today?

3. What was required for the protocell to evolve to meet the criteria of life? How might a fundamental new type of cell have evolved? What advances helped cells obtain energy? What advance allowed cells to specialize? What advances allowed organisms to invade new habitats? Be sure to explain why each was important.

CHAPTER 16

Figure-based Questions

Figure 16-7 Compared to other environments inhabited by bacteria, soils are especially vulnerable to drying out, which can be fatal to unprotected bacteria. Bacteria that form protective endospores that resist long dry periods would gain an evolutionary advantage in soil.

Figure 16-17 Nonvascular plants remain small because they lack any stiffening agent, which is needed to support a larger body, and they lack the conducting vessels necessary to supply nutrients to a larger plant.

Figure 16-20 Animal pollination requires far less pollen, and increases the chances of successful fertilization because pollen is delivered directly to the flower rather than being dispersed randomly by wind.

Figure 16-21 Filaments allow the fungal body to penetrate into its food sources, and provide a large surface area for absorbing nutrients.

Figure 16-25 Although sponges arose early in the evolutionary history of animals and their body plan is comparatively simple, it allows for effective survival and reproduction in many aquatic habitats.

Figure 16-36 One advantage is that adults and juveniles occupy different habitats and therefore do not compete with one another for resources.

Figure 16-38 Flight consumes a lot of energy, so in circumstances in which the benefits of flight are low, such as in habitats without predators or in species whose size is very large, natural selection may favor flightlessness.

Consider This

Hints: Think of some economic benefits of biodiversity (look ahead to Chapter 31). How might you estimate the dollar value of those benefits? Think also of some noneconomic benefits of biodiversity and how the value of such benefits might be assessed.

Fill-in-the-Blank

1. domains; Bacteria, Archaea; Eukarya; Bacteria; toxins
2. diatoms; dinoflagellates; ciliates; foraminiferans; amoebozoans
3. green algae; nonvascular plants (bryophytes), vascular plants (tracheophytes); alternation of generations
4. swim to the egg; seeds, pollen; gymnosperms, angiosperms; attract pollinators
5. reproduction, spores; mycelium, hyphae; yeasts
6. invertebrates, vertebrates; invertebrates; sponges (or phylum Porifera), protists; cnidarians (or phylum Cnidaria); annelids (or phylum Annelida)
7. bivalves, gastropods, cephalopods; arthropods; insects, arachnids, crustaceans
8. mammary glands; monotremes, marsupials, placental; rodents

Review Questions

1. Dolphins breathe with lungs rather than gills, and maintain a relatively constant body temperature. They nourish their young with milk from mammary glands, and their DNA sequence is more like a bear than a fish. Dolphins, like bears, are mammals.

2. There are many inconspicuous species, living in areas that have not been well characterized, including tropical and deep ocean habitats.

3. Some bacteria acquire energy by photosynthesizing, some by metabolizing hydrogen sulfide, and most by breaking down organic molecules synthesized by other organisms. Bacteria can obtain nutrients from almost any organic or inorganic molecules, some can capture (fix) atmospheric nitrogen.

4. Nitrogen-fixing bacteria inhabit soil and the roots of legume plants and capture nitrogen gas (N_2), releasing ammonium (NH_4^+), which is important in plant nutrition and ultimately our own nutrition.

5. Unicellular algae account for nearly 70% of the photosynthetic activity on Earth. They provide food for small invertebrates, and release oxygen into the atmosphere.

6. Mushrooms and puffballs are spore-producing reproductive structures. They are elevated above the ground to disperse their spores in the wind.

7. Adaptations to dry land include the development of roots that anchor the plant and absorb water and nutrients from the soil, conducting tissues that transport water and minerals up from the roots and move photosynthetic products from the leaves throughout the rest of the plant, lignin-impregnated cell walls that support the plant body, a waxy cuticle that limits evaporation of water, stomata that allow gas exchange and limit water loss when water is scarce, and pollen that allows sperm to be transported without drying out.

8. Success of angiosperms can be attributed to flowers, fruit, and broad leaves. Flowers attract pollinators, fruits assist in seed dispersal, and broad leaves capture more sunlight for photosynthesis.

9. Vertebrates have a backbone (vertebral column); invertebrates do not.

10. Four distinguishing features of the chordates (during some stage) are (1) a notochord; (2) a dorsal, hollow nerve cord; (3) pharyngeal gill slits; and (4) a post-anal tail.

11. Most amphibian adults have lungs and legs that adapt them to life on land. They remain tied to water because their skin (a respiratory organ) must be kept moist. Amphibian eggs are fertilized in water and must remain moist during development. Most amphibian larvae are aquatic.

12. Reptiles evolved a tough, scaly skin that resists water loss and protects the body; reptiles evolved internal fertilization and a shelled egg that encloses the embryo in a watery environment, although the egg itself may exist in very dry surroundings.

Applying the Concepts

Hints

1. Have all or even most of Earth's species been discovered and identified?
2. What type of understanding is required to produce the best classification schemes? How does knowledge of species influence efforts to protect them?
3. Do cells in multicellular organisms need to perform all of the cellular processes necessary to keep the organism alive and functioning?

CHAPTER 17

Figure-based Questions

Figure 17-3 Primary growth occurs at the tip of the primary root, at the tips of all lateral roots, at the terminal bud, at all lateral buds, and at the tips of lateral branches. Secondary growth will occur all along the margins of the primary and lateral roots and shoots.

Figure 17-8 Cortex and endodermis are ground tissues; xylem and phloem are vascular tissues. None of the cells shown are part of epidermal tissue (the root epidermis lies outside the area shown in the photo).

Figure 17-13 As can be seen in the figure, the phloem layer is part of the bark. Removing a strip of bark entirely around the trunk creates a complete break in the phloem vessels connecting the roots to the rest of the plant. Because phloem carries nutrients from their point of acquisition to their point of use, the plant cannot survive without a phloem connection between (for example) its sugar sources (e.g., leaves) and its sugar sinks (e.g., growing root tips).

Consider This

Hints: Try to think of some possible "costs" (disadvantages) of producing anthocyanins and some circumstances in which the benefits of anthocyanins would be small or nonexistent. Also, What if the series of mutations that led to anthocyanin production failed to occur in the ancestors of some modern genera of trees? Is it possible to determine whether fossil trees produced anthocyanins?

Fill-in-the-Blank

1. mycorrhizae; fungi, roots (plant); bacteria; nodules, ammonium
2. xylem, vascular; hydrogen bonds; transpiration
3. carbon dioxide, oxygen, diffusion; minerals; active transport
4. fibrous, parallel, three, cotyledon
5. taproot, netlike, five, two
6. one-half; stomata; humidity

Review Questions

1. Dermal tissues cover the plant body, protecting it and regulating water loss. Ground tissues make up the internal tissue of the plant body and, in various types of plants, may photosynthesize, provide support, and store food. Vascular tissue is used to transport water and dissolved minerals and sugars.

2. Primary growth is cell division in apical meristems, increasing the length of roots and stems or producing flowers and leaves. Secondary growth is cell division of the lateral meristems, resulting in an increase in the diameter of the plant.

3. Meristem cells are capable of cell division, whereas differentiated cells are specialized in structure and function and do not divide. Apical meristems, located at the tips of stems and roots, are responsible for primary growth; lateral meristems, located internally as cylinders that run parallel to the long axis of stems and roots, are responsible for secondary growth.

4. See Figure 17-8. The epidermis provides a large surface for water and mineral absorption. The cortex provides support and regulates the movement of substances into the vascular cylinder. Xylem transports water and dissolved minerals to upper parts of plant. Phloem transports water, sugars, amino acids, and hormones throughout the plant body.

5. Xylem consists of tracheids and vessel elements that die at maturity. Xylem conducts water and dissolved minerals up from the roots. Phloem consists of sieve tubes and companion cells that are alive at maturity. Phloem conducts water, sugars, amino acids, and hormones throughout the plant body.

6. Root functions include anchorage, absorption of water and minerals, and storage of sugars and

starches. Stems support leaves and conduct minerals and water up through the plant. Leaves perform photosynthesis and transpiration, which pulls water and dissolved minerals up through the plant body.

7. Epidermal cells form root hairs. Root hairs provide increased surface area through which water and minerals can be absorbed.

8. See Figure 17-15. Stomata regulate water loss and gas exchange.

Applying the Concepts

Hints

1. What aspects of plant growth or reproduction might differ in response to different types of music? Include appropriate controls and specify the variables you will measure to test your hypotheses.

2. Many desert grasses and herbs die or become dormant during the hot, dry summer, while some remain alive and growing. Think about the advantages and disadvantages of deep versus shallow roots for tapping permanent groundwater or brief spring showers.

3. Consider the role of primary meristems in plant growth and the likely consequences if they are removed.

CHAPTER 18

Figure-based Questions

Figure 18-4   Separate bloom times reduce the chances of inbreeding through self-fertilization. Because inbreeding increases the number of individuals homozygous for recessive alleles and deleterious traits they may produce, plants that do not inbreed have a greater chance to survive and reproduce.

Figure 18-9   Double fertilization ensures that no energy will be wasted producing endosperm unless the pollen tube reaches the ovule and both the egg and the polar nuclei are fertilized.

Consider This

Hint: Consider the role of ethylene in the plant life cycle, and the problems with plants that lack ethylene receptors.

Fill-in-the-Blank

1. anther, male gametophyte; stigma; style; ovule, female gametophyte; sperm
2. egg, polar nuclei, endosperm; food (nourishment)
3. meiosis; haploid; sporophyte, mitosis; diploid; alternation of generations
4. seed coat; fertilized egg (zygote); shoot, root, cotyledons (seed leaves); cotyledons (seed leaves)
5. germination; root, root cap; coleoptile, hook
6. abscisic acid; gibberellin; auxin; roots; ethylene

Review Questions

1. Refer to Figures 18-1 and 18-2. These life cycles differ in that ferns have an independent gametophyte generation, while flowering plants have a reduced gametophyte generation that is contained within, and dependent on, the sporophyte. For both, the sporophyte is diploid, while the gametophyte is haploid. Gametes are formed by mitosis within gametophytes.

2. Reduced gametophytes of flowering plants are protected within the sporophyte and draw nourishment from it. They do not require water for sperm to reach the eggs as do fern gametophytes, and so are not restricted to moist habitats.

3. Refer to Figure 18-3. Male gametophytes are formed in the anther, and female gametophytes in the ovary. Male gametophytes are pollen grains, and female gametophytes are embryo sacs.

4. A meiotic division of the diploid megaspore mother cell produces four haploid megaspores. Three degenerate and the fourth divides mitotically, producing eight haploid nuclei. The female gametophyte forms when these nuclei are partitioned into one egg cell, five small cells, and a large central cell with two nuclei. Double fertilization occurs when one sperm fuses with the two nuclei of the central cell, and a second sperm fuses with the egg, forming a zygote.

5. Pollen is formed by meiotic division of a microspore mother cell. This produces microspores that divide by mitosis to produce the pollen grain (male gametophyte), consisting of two haploid sperm cells and a tube cell surrounded by a tough coat.

6. Endosperm is a food source contained in a seed that is produced after the two polar nuclei in the central cell are fertilized by a sperm and undergo repeated mitotic divisions.

7. The Darwins determined that the coleoptile tip detects light direction and transmits this information farther down the coleoptile, which bends in response. Boysen-Jensen concluded that a chemical is produced in the side of the shoot tip exposed to light and moves down the stem, causing elongation. Went isolated this elongation chemical and named it auxin. A useful experiment would be to study an auxin-deficient mutant to determine if the plant still exhibits phototropism.

8. High auxin concentrations promote cell elongation in shoots but inhibit elongation in roots.

9. Gibberellin and cytokinin are involved in stimulating fruit development. Cytokinin comes from the roots, and gibberellin from the seeds. Ripening is often stimulated by ethylene, which is produced in response to auxin.

10. Ethylene is involved in leaf and fruit drop. Bud dormancy is controlled by abscisic acid.

Applying the Concepts

Hints

1. Does the same hormone have the same effects on all plants? What hormone hastens apple ripening and softening? Might the same hormone inhibit potato sprouting? Be sure to include appropriate controls. This is one you could do at home!

2. Consider the consequences to a plant with a single, specialized pollinator species if the pollinator's population declines. Also consider the likelihood that members of the pollinator species will feed on other plant species.

3. Hormones in animals and plants are normally present and effective at very low concentrations. What do you think would happen if excessive quantities of hormones were used on plants?

CHAPTER 19

Figure-based Questions

Figure 19-5   Because they have a larger ratio of surface area (through which heat is lost) to body volume, smaller individuals require additional insulation to maintain body temperature in cold climates.

Figure 19-9   Skin is composed of several tissue types (connective, epithelial, muscle, nerve), whereas blood is composed of fairly similar cells and their surrounding matrix. One cannot identify two tissue types in blood.

Figure 19-10   If the mammal's heat-sensing nerve endings were destroyed, they could no longer signal the hypothalamus when the body reached or exceeded its set-point temperature. Consequently, the hypothalamus would send continuous "turn on" signals to the body's heat-generating and heat-retention mechanisms,

causing the body to overheat and possibly die. Failure of the sensor removes the negative feedback control, so the system remains "on."

Consider This

Hint: Consider the kinds of physiological responses to submersion in water that would result in a survival advantage to a mammal.

Fill-in-the-Blank

1. homeostasis; negative feedback
2. tissues, organs, organ systems
3. connective tissue; epithelial tissue; connective tissue; connective tissue; epithelial tissue; muscle tissue; nerve tissue; connective tissue
4. exocrine; any three of the following: sweat, sebaceous, salivary, gland cells in the stomach lining; endocrine; hormones
5. cardiac; skeletal; cardiac, skeletal; smooth, cardiac; smooth; skeletal

Review Questions

1. Homeostasis is the maintenance of physiological conditions within the range required for optimal functioning of the organism. Homeostasis is maintained by negative feedback, in which a change in the internal environment triggers a response that counteracts the change, maintaining the original conditions. For example, a change in body temperature is detected by temperature receptors that signal the hypothalamus. The hypothalamus activates mechanisms that restore normal body temperature, and turns off these responses when normal body temperature is restored.

2. Positive feedback occurs when a change in the internal environment initiates a response that intensifies the original change. An example occurs during childbirth, when labor contractions press the baby against the cervix, stretching it. Stretch-receptive neurons in the cervix signal the hypothalamus, which triggers oxytocin release, intensifying uterine contractions. This creates more pressure on the cervix, triggering more oxytocin release. This positive feedback is terminated by birth. Positive feedback is rare and must be self-limiting because it results in intense reactions that proceed away from the initial set point.

3. Body temperature cannot be precisely maintained because in order for feedback mechanisms to be initiated, a deviation from the set point must occur and be detected.

4. Epithelial tissue is composed of cells in sheets called membranes that cover the body and line body cavities. Membranes regulate movement of substances across them, depending on their location and function (those in the capillaries allow significant flow of water, nutrients, and cellular wastes; those lining the bladder do not). Some epithelial tissues fold inward during development, creating endocrine and exocrine glands that secrete substances inside and outside the body.

5. Connective tissues (except blood) contain collagen. Dermis supports the epidermis and surrounds the capillaries, muscles, glands, and receptors of the skin; tendons attach muscles to bones; ligaments attach bones to bones; cartilage covers the ends of bones at joints, forms pads between vertebrae, and supports the respiratory passages, ears, and nose; bone provides support for the body; fat is used for long-term energy storage; blood carries dissolved oxygen and nutrients to cells and removes carbon dioxide and cellular wastes.

6. The epidermis is specialized epithelial tissue that is covered by a protective layer of dead cells packed with keratin protein that keeps the skin

airtight and waterproof. Under the epidermis lies the dermis, which consists of loosely packed cells permeated with capillaries that nourish the epithelial and dermal tissues. Neural tissue in skin includes receptors for touch, temperature, and pain. Arterioles in the skin may constrict or dilate, conserving or releasing heat. The dermis of skin contains glands derived from epithelial tissue, including sweat and sebaceous glands and hair follicles. Tiny muscles attached to hair follicles erect the hairs to help retain heat.

Applying the Concepts

Hints
1. Consider the substances that are used up during vigorous activity. Sweat contains sodium, chloride, and small amounts of potassium. What would you need to know about the conditions and the amount to exercise to determine whether "water is enough" or not?
2. In the figure, note the locations of nerve cells, hair follicles and sweat glands, and epithelial cells.
3. Think about what physiological components could be modeled by electrical currents, motors, and sensors. What kind of feedback loops should your device include?

CHAPTER 20

Figure-based Questions

Figure 20-4 When the right ventricle contracted, some blood would flow back into the right atrium; less blood would reach the lungs. Flow from the vena cava into the right atrium would be impeded because the atrium would be partly full of backwashed blood. Overall, each round of the cardiac cycle would circulate a smaller volume of blood.

Figure 20-5

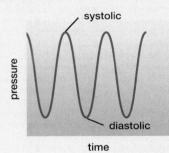

Figure 20-14 Molecules diffuse down gradients from high to low concentration. Capillaries, carrying blood that has released carbon dioxide and picked up oxygen in the lungs, pass through tissues whose cells are consuming oxygen and generating carbon dioxide. This creates gradients that cause oxygen to diffuse out of capillary blood and carbon dioxide to diffuse in.

Figure 20-20 Forcing air to travel over a large area of membrane helps conserve water and heat because as warm moist air is exhaled, water condenses on the nasal passages and heat is transferred to them. Conversely, inhaled air is warmed and moistened before it reaches the lungs. The membrane contains odor receptors and it is lined with cilia and mucus that trap foreign particles, preventing them from entering the lungs.

Consider This

Hints: This is a complex issue about the balance between individual rights and responsibilities and social well-being. There are no correct (or easy) answers.

Fill-in-the-Blank

1. capillaries; arterioles, venules
2. ventricles; diastolic; left ventricle; right ventricle

3. plasma; contractions of nearby muscles; blood (circulatory system); lymph nodes; thymus, spleen
4. pharynx; epiglottis; trachea, bronchi, bronchioles, alveoli
5. capillaries; hemoglobin; bicarbonate ions; cellular respiration
6. any three of the following: cancer, chronic bronchitis, emphysema, cough, female reproductive problems, heart attacks, atherosclerosis; any three of the following: bronchitis, pneumonia, ear infections, coughs, colds, allergies, asthma, decreased lung capacity

Review Questions

1. Deoxygenated blood in the right atrium is pumped to the right ventricle, which pumps it to the lungs via the pulmonary artery. Blood travels to arterioles, then to lung capillaries (where it picks up oxygen), and into venules that merge to form the pulmonary vein, which empties into the left atrium. From the left atrium, blood is pumped into the left ventricle, which pumps it to the rest of the body through arteries to arterioles to capillaries, which release oxygen to tissues and pick up carbon dioxide. Deoxygenated blood flows into venules and then into veins, which return it to the right atrium.
2. The vertebrate circulatory system (any of the following) (1) transports oxygen from the lungs to the tissues; (2) transports carbon dioxide from the tissues to the lungs; (3) distributes nutrients from the digestive system to all body cells; (4) transports waste products and toxic substances to the liver for detoxification, and to the kidney for excretion; (5) distributes hormones from endocrine glands to the tissues on which they act; (6) helps regulate body temperature by adjusting blood flow; (7) helps prevent blood loss by clotting; and (8) helps protect the body from bacteria and viruses by the actions of white blood cells and circulating antibodies.
3. The lymphatic system (1) returns excess fluid and its dissolved proteins and other substances to the blood, (2) transports fats from the intestines to the blood, and (3) assists in defending the body against infectious disease.
4. Blood plasma is primarily water in which proteins, salts, nutrients, and wastes are dissolved. Extracellular fluid is derived from plasma that has filtered through capillary walls into spaces surrounding capillaries and tissues. It consists of water and dissolved nutrients, hormones, gases, wastes, and small (but not large) plasma proteins. Lymph is extracellular fluid that has moved into lymph capillaries to be returned to the circulatory system.
5. Veins carry blood back to the heart. Vein walls are thinner and more expandable than arteries. Both veins and arteries contain a layer of smooth muscle. Arteries are large, thick-walled blood vessels that carry blood away from the heart. Capillaries, microscopic blood vessels with walls only one cell thick, are the sites of exchange between the blood and tissues (via the extracellular fluid).
6. The contraction of the heart is initiated and coordinated by regular, spontaneous, electrical signals produced by the sinoatrial (SA) node, a pacemaker consisting of a cluster of specialized muscle cells in the right atrium. Heart muscle cells communicate directly with one another through gap junctions, transmitting signals from the SA node rapidly through the two atria, which contract synchronously in response. A barrier of unexcitable tissue separates the atria from the ventricles, channeling the electrical signal

through the atrioventricular (AV) node in the floor of the right atrium. From here, the signal to contract spreads along tracts of excitable fibers to the bases of both ventricles, which contract synchronously from the bottom up, forcing blood out through arteries to the lungs and body.
7. Atherosclerotic plaques, composed of cholesterol, other fatty substances, calcium, and fibrin, are deposited within the artery wall. Plaques narrow the artery and may rupture into the artery through its inner lining, stimulating blood clotting. Blood clots may break loose and be forced into smaller arteries, which they block, causing strokes (in brain) or heart attacks (in coronary arteries).
8. Air enters the nose or mouth and passes through the pharynx and then the larynx. Air then enters the trachea, which splits into two bronchi, one to each lung. Inside the lung, each bronchus branches repeatedly into smaller tubes called bronchioles. These lead to microscopic alveoli, tiny air sacs where gas exchange occurs.
9. Inhalation occurs when the diaphragm and rib muscles contract, enlarging the chest cavity by drawing the diaphragm downward and the ribs up and outward. Because there is no air space between the lungs and chest cavity, they expand together, drawing air into the lungs. Exhalation occurs when the muscles causing inhalation are relaxed, the chest cavity decreases in size, and air is forced out of the lungs. Inhalation is always an active process.
10. In the alveoli, oxygen diffuses from air into lung capillary blood, where it binds to hemoglobin. Oxygenated blood returns to the heart and is pumped to capillaries within body tissues. Here, oxygen concentration is lower, and the oxygen diffuses from capillaries into the cells.
11. Toxic substances in smoke paralyze the cilia lining the respiratory tract, allowing carcinogen-laden carbon particles to stick to the walls of the respiratory tract or enter the lungs, causing most cases of lung cancer. Smoke increases mucous production, causing smokers' cough, and can lead to the chronic infection and inflammation of bronchitis. Emphysema caused by smoking ruptures alveoli; both bronchitis and emphysema reduce lung function. Smoking promotes atherosclerosis, and carbon monoxide in smoke reduces the oxygen-carrying capacity of the blood. Together, these greatly increase the risk of heart disease in smokers.
12. Each lung contains about 2 million alveoli enmeshed in capillaries, providing an enormous surface area for diffusion. The alveoli are coated with watery fluid in which gases dissolve. Gases diffuse through the alveolar and capillary walls, which are each only one cell thick, bringing alveolar air extremely close to the blood in the capillaries.

Applying the Concepts

Hints
1. Will increasing the number of red blood cells in the circulatory system increase the total amount of oxygen it can carry? What will excess blood do to blood pressure? Could this be dangerous if the heart is already working at near-maximum capacity?
2. Nicotine is addictive, so are reduced levels of nicotine likely to cause smokers to smoke less—or more? Are most of the harmful effects of smoking caused by nicotine?
3. Consider the feedback mechanisms that determine breathing rate. Could an unconscious person exert control over these mechanisms?

CHAPTER 21

Figure-based Questions

Figure 21-5   Yes. The label shows a product relatively low in calories, with low fat and no cholesterol. It has lots of fiber and a substantial percentage of the recommended vitamins.

Figure 21-9   The muscular walls churn food, breaking it into smaller pieces. An expandable stomach conserves space but allows periodic consumption of large amounts of food when it becomes available.

Figure 21-12   The absorptive surface area would have been achieved by other adaptations, such as a much longer intestine.

Figure 21-15   Ammonia is excreted by aquatic animals, which can excrete ammonia more or less continuously and thus escape its toxic effects. This is especially common in freshwater animals, which must expel a lot of water to maintain osmotic balance.

Figure 21-16   The kidneys regulate blood composition, which must be tightly controlled, by filtering blood continuously and rapidly.

Figure 21-20   Alcohol consumption tends to dehydrate the body by reducing negative feedback that causes ADH to be released in response to increased blood osmolarity.

Consider This

Hints: To what extent do fashion shows and advertising feed the perception that extreme slimness is important? Should government agencies get involved?

Fill-in-the-Blank

1. nutrient; digestion; excretion, digestive tract, urinary tract, respiratory tract, skin (order not important)
2. carbohydrates, lipids; vitamins; minerals
3. essential (essential amino acids; essential fatty acids)
4. ingestion, mechanical breakdown, chemical breakdown, absorption, elimination
5. mouth (oral cavity), pharynx, esophagus, small intestine, large intestine
6. (hydrochloric) acid, pepsinogen (or pepsin), mucus
7. pancreas, bicarbonate; bile, lipids (fats, triglycerides)
8. nephrons; Bowman's capsule; collecting ducts; (urinary) bladder, urethra
9. pituitary; antidiuretic hormone (ADH)

Review Questions

1. Categories are lipids, carbohydrates, proteins, minerals, vitamins, and water. Lipids are energy sources, components of cell membranes, and certain hormones. Carbohydrates are energy sources. Proteins are sources of amino acids for synthesis of new proteins. Some minerals are structural components of bones and teeth; others are essential for muscle contraction and conduction of nerve impulses. Vitamins often work in conjunction with enzymes to promote metabolic reactions. Water provides a medium for many metabolic reactions and participates in breakdown of proteins, carbohydrates, and fats. It serves in temperature regulation, acts as a lubricant, and is responsible for the fluidity of saliva, blood, lymph, extracellular fluid, and cytoplasm.
2. The primary secretions of the stomach are hydrochloric acid, pepsinogen, and mucus. Hydrochloric acid converts inactive pepsinogen to active pepsin, a protein-digesting enzyme. Mucus coats the stomach lining, serving as a barrier to self-digestion.
3. Bile is produced in the liver, concentrated in the gallbladder, and released in the small intestine. Bile disperses fat into microscopic particles. Pancreatic

juice is secreted by the pancreas into the small intestine. Its ingredients include bicarbonate to neutralize acidic chyme, and several enzymes that digest carbohydrates, lipids, and proteins.

4. The small intestine has many folds that are covered with tiny, finger-like villi, whose cells each bear a fringe of microvilli, all of which greatly enhance the absorptive surface area of the intestine. Circular smooth muscles create rhythmic, unsynchronized contractions that bring nutrients into contact with the absorptive surfaces. Each villus has numerous blood capillaries that carry off absorbed nutrients, and a lacteal that carries fats. Cells of the intestinal wall contain enzymes that complete digestion

5. Seeing, smelling, tasting, and chewing send nervous signals to the stomach, stimulating gastrin and acid secretion. The acidity of the stomach converts the pepsinogen to pepsin, which begins protein digestion. As liquid chyme is released into the small intestine, its acidity stimulates the release of the hormone secretin, which causes the pancreas to release sodium bicarbonate into the small intestine to neutralize the acid chyme. The hormone cholecystokinin is produced by the small intestine in response to chyme and stimulates pancreatic release of digestive enzymes (including protein-digesting enzymes) into the small intestine where the final digestion of small peptides occurs through enzymes on microvilli of intestinal cells.

6. First, urinary systems play a crucial role in homeostasis, maintaining a stable internal environment. The kidneys of mammalian urinary systems regulate blood pH, ions, and water content. Second, urinary systems eliminate cellular wastes, as well as excess and toxic substances.

7. Filtration is the process by which blood is forced through the glomerular capillaries, driving water and small dissolved molecules and ions (but not cells or proteins) into Bowman's capsule. During tubular reabsorption, proximal tubule cells actively transport nutrients such as salts, amino acids, and glucose out of the tubule, where they diffuse back into the surrounding capillaries; water follows by osmosis. During tubular secretion, distal tubule cells actively transport wastes and excess substances that diffuse from the surrounding capillaries (such as hydrogen and potassium ions, ammonia, and many drugs) into the tubule.

Applying the Concepts

Hints
1. Most animals, including humans, evolved under conditions in which food was limited. What food preferences and eating tendencies are likely to have evolved?
2. Magnesium salts in the large intestine cause water to move in by osmosis, stretching the large intestine, which promotes peristalsis, and moistening the feces, making them easier to expel.
3. Based on your knowledge of the role of the kidneys, what would you expect to find in urine after a person took drugs?

CHAPTER 22

Figure-based Questions

Figure 22-4   Some fluid always escapes through small gaps in capillary walls. If more blood flows through leakier capillaries, then more fluid will escape into surrounding tissues, causing swelling.

Figure 22-5   The variable region allows the antibody to recognize and bind to a specific antigen, and the constant region allows the antibody to carry out functions common to all antibodies in its class.

Figure 22-16   Cancer cell membranes bear distinctive, abnormal molecules that are not recognized as "self" by the immune system.

Consider This

Hint: Research scientists don't always agree on the relative benefits and risks of dual-use research. Should scientists and governments try to estimate the relative risks and benefits? Do we know enough about the likely outcomes of research to be able to predict the risk/benefit ratio?

Fill-in-the-Blank

1. skin; digestive, respiratory, urogenital (order not important)
2. phagocytic cells (or macrophages); natural killer cells; inflammatory response; fever
3. antigens; antibodies, T-cell receptors
4. light, heavy (order not important); variable, constant (order not important); variable
5. humoral, plasma cells; cell-mediated; cytotoxic; helper; memory
6. vaccine
7. allergy; immune deficiency disease; autoimmune disease

Review Questions

1. The body's three lines of defense against invading microbes are (1) nonspecific external barriers that keep microbes out of the body, (2) nonspecific internal defenses that combat all invading microbes, and (3) the immune system, which attacks specific microbes.
2. Both secrete enzymes that damage the target cell, and pore-forming proteins that create large holes in the target cell's plasma membrane.
3. Humoral immunity is produced by antibodies in the blood. B cells that bind the antigen divide rapidly, producing many more B cells; some of them become memory B cells and others become plasma cells. Plasma cells release specific antibodies into the blood or interstitial fluid. Antibodies may inactivate toxic molecules or bind to microbes, causing the microbes to be recognized and engulfed by macrophages. Cell-mediated immunity defends against the body's own cells when they become cancerous or infected with a virus. Cytotoxic T cells bind to antigens on the surface of the infected cell and release proteins that destroy the infected cell. Finally, helper T cells stimulate both humoral and cell-mediated immunity. Helper T cells have receptors that bind to antigens on target cells. They release chemicals that stimulate the production of B cells and cytotoxic T cells.
4. In B cells, segments of DNA that encode the parts of antibodies are spliced together in random ways to create a multitude of antibodies, each with a different pair of variable regions, capable of binding different antigens.
5. Immature immune system cells that bind "self antigens" are destroyed. Only cells that do not respond to "self" are retained and mature into functional B and T cells.
6. See Figure 22-5 for antibody structure. The variable regions of the light chains bind to specific antigens because their binding sites have specific sizes, shapes, and electrical charges that interact only with specific antigens.
7. Memory cells are offspring of activated B cells, cytotoxic T cells, and helper T cells, which survive for many years, retaining the capacity to respond to the original antigen. Large numbers of fast-responding memory cells allow a much more effective attack on the antigen, often destroying the invaders before they cause disease symptoms.

8. Vaccination introduces weakened or dead microbes, or genetically engineered antigens that are found on a microbe, into the body to stimulate the development of memory cells, conferring immunity against subsequent exposure to living, dangerous microbes.

9. Both wounds and allergies provoke inflammatory responses. In a wound, chemicals released from damaged cells cause mast cells to release histamine, which increases blood flow, causes capillaries to become leaky, and causes injured tissues to become warm, red, swollen, and painful. This inflammatory response attracts phagocytic cells to the area to engulf pathogens and debris, promotes blood clotting, and causes pain that stimulates protective behaviors. In an allergy, "allergy antibodies" are present on mast cells of the respiratory or digestive tract. The allergens bind these antibodies, causing the mast cells to release histamine, producing inflammation similar to that of a wound. The congestion of "hay fever" is mostly caused by increased blood flow through leaky capillaries in the nasal passages.

10. An autoimmune disease is an immune response against some of the body's own molecules and cells. Examples are some types of anemia and juvenile-onset diabetes. An immune deficiency disease results from the inability to mount an effective immune response to infection. Examples are severe combined immune deficiency (SCID) and acquired immune deficiency syndrome (AIDS).

11. HIV viruses cause AIDS by infecting and destroying helper T cells. As the helper T-cell population declines, the AIDS victim becomes increasingly susceptible to other diseases, eventually dying of infection. AIDS treatments interfere with the HIV replication. AIDS is spread by direct contact of either mucous membranes or broken skin with body fluids such as semen, blood, vaginal secretions, and breast milk.

12. Some cancer cells evade detection because they do not bear antigens that allow the immune system to recognize them as foreign. Some cancers, such as leukemia, suppress the immune system; other cancerous cells multiply so rapidly that the immune response cannot kill them as rapidly as they are produced.

Applying the Concepts

Hints

1. Refer to the discussion of the scientific method in Chapter 1. Which steps were probably followed by ancient people? How much of the difference between ancient and modern vaccinations might be caused by differences in the people's approaches to inquiring about their world and how much by technology?

2. Think about how one might evaluate potential dangers vs. potential benefits of maintaining the viruses. How secure do you think the storage facilities should be, and how stable (i.e., permanent) should their governments be?

3. Refer back to this chapter for a discussion of how the body defends against cancers. What would happen if these defenses were inhibited?

CHAPTER 23

Figure-based Questions

Figure 23-6   Dwarfism can be treated by administering growth hormone, but it is much more difficult to halt the production of excess hormone. Disruption of the pituitary is likely to disrupt the release of many other hormones as well. (Some cases of gigantism are caused by pituitary tumors, which can be treated by surgery.)

Figure 23-7   Suckling is controlled by negative feedback, because a change (being hungry) causes a response (suckling) that counteracts the change and restores the original condition (not hungry). The mother's response is also negative feedback, in which a change (suckling) causes a response (oxytocin release, causing milk letdown) that counteracts the change and restores the original condition (no oxytocin release after suckling stops).

Consider This

Hints: This are no correct answers. Consider the effects of anabolic steroids on the body, how likely it is that high school students will be fully informed about these effects, pressures for success in athletics, and individual rights and responsibilities.

Fill-in-the-Blank

1. endocrine system; target cells, receptors
2. peptide, amino acid-derived, steroid (order not important); peptide hormones, amino acid-based hormones; second messengers; steroid; changes in gene transcription (or synthesis of messenger RNA)
3. negative feedback; positive feedback; negative feedback
4. hypothalamus; neurosecretory cells, oxytocin, antidiuretic hormone (order of hormones not important); releasing or inhibiting hormones
5. adrenocorticotropic hormone; follicle-stimulating hormone; luteinizing hormone; growth hormone; prolactin; thyroid-stimulating hormone (order not important)
6. insulin, glucagon (order not important); insulin; diabetes mellitus; glucagon, glycogen
7. testes, testosterone; ovaries, estrogen, progesterone; steroid
8. cortex, medulla; glucocorticoids, mineralocorticoids, testosterone (order of hormones not important); norepinephrine, epinephrine (order not important)

Review Questions

1. The binding of a peptide hormone to a receptor on the plasma membrane triggers the release of a chemical second messenger in the cell that initiates a cascade of biochemical reactions.
2. Lipid-soluble steroid hormones can enter the cytoplasm or nucleus, where they bind receptors. The receptor-hormone complex binds to DNA and stimulates particular genes to be transcribed into mRNA and finally translated into protein.
3. Refer to Figures 23-7 and 23-10 for diagrams involving negative feedback of hormone secretion and action. Other examples include control of water in blood and extracellular fluid. Water loss causes the pituitary to release antidiuretic hormone (ADH), causing the kidneys to reabsorb water and produce concentrated urine. If you restore normal blood water by drinking, negative feedback turns off ADH secretion.
4. See Figure 23-4.
5. The hypothalamus contains clusters of specialized neurosecretory cells that control release of hormones from both the anterior and posterior lobes of the pituitary. The posterior pituitary secretes two hormones synthesized in the hypothalamus: antidiuretic hormone, which causes water to be reabsorbed from the urine and retained in the body, and oxytocin, which triggers milk secretion in nursing mothers.

   Releasing hormones are synthesized in the hypothalamus and are carried in the blood to the endocrine cells of the anterior pituitary, where they control release of the six peptide hormones secreted by the anterior pituitary. Follicle-stimulating hormone and luteinizing hormone

stimulate the production of sperm and testosterone in males, and eggs and estrogen in females. Thyroid-stimulating hormone stimulates the thyroid to release its hormones. Adrenocorticotropic hormone causes release of hormones from the adrenal cortex. Prolactin stimulates development of mammary glands during pregnancy. Growth hormone regulates body growth.

6. When blood glucose rises, insulin is released, causing body cells to take up glucose. When glucose levels drop, insulin release is inhibited, and glucagon is released. Glucagon activates a liver enzyme that breaks down glycogen, releasing glucose into the blood. When glucose levels rise, glucagon release is inhibited. If blood glucose rises too far, insulin is released again.

7. The adrenal medulla, in the center of each adrenal gland, produces epinephrine and norepinephrine in response to stress. These hormones increase the heart and respiratory rates, increase blood glucose levels, and direct blood flow away from the digestive tract and toward the brain and muscles. The adrenal cortex forms the outer layer of the adrenal gland and secretes glucocorticoids that help control glucose metabolism and mineralocorticoids (such as aldosterone) that regulate the sodium content of the blood. Aldosterone causes the kidneys and sweat glands to retain sodium.

Applying the Concepts

Hints

1. On the Internet, use reliable sites; addresses that end with *.edu* (usually universities) or *.gov* (government agencies, such as the National Institutes of Health) usually provide accurate information.
2. Research the effects of growth hormone including "side effects." What is your opinion about artificially increasing the size of one's body? Is this more or less fair than the "genetic lottery" that produces many thousands of physically ordinary specimens for every Shaquille O'Neal or Michael Jordan?
3. There is no correct answer. How would you balance costs vs. benefits? How much effort and money should go into finding replacements with less estrogenic activity?
4. Review the section in this chapter about the pineal gland.

CHAPTER 24

Figure-based Questions

Figure 24-3   The toxin must be blocking the postsynaptic neurotransmitter receptors.

Figure 24-4   Sensitive areas have a higher density of touch-sensitive sensory neurons, so touching any spot on the skin area would be more likely to stimulate at least one touch receptor, and a touch of a particular force would stimulate a larger number of receptors than in lower-density areas.

Figure 24-9   A damaged spinal cord prevents the sensation of pain from being relayed to the brain but does not disrupt the reflex circuit, which is contained within a small portion of the spinal cord.

Figure 24-19   For nearsightedness, the cornea is flattened (made less convex), which causes the light rays to converge less before they reach the lens. For farsightedness, the edges of the cornea are reduced, creating a more rounded, convex shape that causes the light rays to converge more before they reach the lens.

Consider This

Hints: Have you ever felt this way? This song was written long before neurobiologists documented the links between love and addiction.

Fill-in-the-Blank

1. neuron; dendrite; cell body; axon, synaptic terminal
2. negative; threshold; positive
3. neurotransmitter; receptors; excitatory postsynaptic potential; inhibitory postsynaptic potential
4. frequency; number
5. somatic nervous system; autonomic nervous system, sympathetic, parasympathetic
6. cerebellum, pons, medulla (order not important); cerebellum
7. thalamus; limbic system
8. frontal, parietal, temporal, occipital (order not important); occipital; frontal
9. mechanoreceptors; photoreceptors; chemoreceptors
10. pinna; eardrum, or tympanic membrane; hammer (malleus), anvil (incus), stirrup (stapes) (order IS important); oval window; hair cells
11. rods, cones (order not important); rods; cones, fovea

Review Questions

1. Dendrites are branched tendrils extending from the nerve cell body that respond to signals from other neurons or the external environment. The cell body maintains the neuron and integrates electrical signals from the dendrites, deciding whether to produce an action potential. The axon is a long, thin fiber that carries action potentials from the cell body to their destination. Synaptic terminals are swellings at the branched ends of the axons that communicate with other neurons, muscles, or glands by releasing neurotransmitters in response to an action potential traveling down the axon.
2. Refer to Figure 24-3. Signals are transmitted by release of neurotransmitter molecules from the presynaptic neuron into the synaptic cleft. Neurotransmitter diffuses across the synaptic cleft and binds to receptors on the postsynaptic neuron dendrite. Receptor activation may cause an excitatory postsynaptic potential (making the postsynaptic neuron less negative and more likely to produce an action potential) or an inhibitory postsynaptic potential (more negative or less likely to produce an action potential).
3. Stimulus intensity is signaled by the frequency of action potentials in a single neuron and by the number of neurons firing in response to the stimulus. The type of stimulus is determined by the brain region to which the sensory signals travel (e.g., action potentials in the optic nerve travel to visual areas of the brain and are always perceived as light).
4. The four elements of a neural pathway include sensory neurons, interneurons, motor neurons, and effectors. In the pain-withdrawal reflex, a painful stimulus is received by the sensory neuron, which transmits the signal to an interneuron in the spinal cord, which signals a motor neuron to contract the effector (a muscle), which causes withdrawal from the painful stimulus.
5. Refer to Figure 24-8. The spinal cord contains cell bodies of motor neurons that control voluntary muscles and involuntary (autonomic) responses, synaptic terminals of sensory neurons, axons and synaptic terminals of some neurons whose cell bodies are in the brain, and many interneurons. Severing the spinal cord severs axons from the brain that command movements, preventing them from stimulating spinal cord interneurons or motor neurons below the cut.
6. The medulla controls several automatic functions, such as swallowing, breathing, heart rate, and blood pressure. The cerebellum coordinates body movements and is important in learning motor skills such as writing or typing. The reticular formation is a relay center that receives input from most senses and helps filter the sensory inputs that reach consciousness. The thalamus relays sensory information from nearly all parts of the body and from the cerebellum to the cerebral cortex. The limbic system produces our most basic emotions, drives, and behaviors, including fear, rage, tranquility, hunger, thirst, pleasure, and sexual responses. The cerebral cortex receives and interprets sensory information, stores memories, plans and weighs consequences of actions, is responsible for thought and reasoning, and directs all voluntary movements.
7. The corpus callosum connects the hemispheres. In most right-handed people, the left hemisphere controls the right side of the body and is principally responsible for speech, reading, writing, language comprehension, mathematical ability, and logical problem solving. The right side of the brain is superior in musical skills, artistic ability, recognizing faces, spatial visualization, and recognizing and expressing emotions.
8. Long-term memory is semipermanent, perhaps due to changes in expression of certain genes and the formation of new (or strengthened) permanent synaptic connections between specific neurons. Working memory is very brief and either requires repeated activation of electrical circuits or involves temporary biochemical changes within neurons in a circuit.
9. The senses of taste and smell are provided by chemoreceptors in the tongue and nasal epithelium, respectively. Photoreceptors respond to light. Mechanoreceptors are the receptors for touch and hearing.
10. We perceive a great variety of odors. In food, odors add a component to the basic taste that we perceive as a unique flavor. Much of what we call "taste" is actually "smell."
11. Different odors are distinguished by about 500 different types of receptor proteins in the olfactory dendrites. Each receptor binds to a different type of odor molecule.
12. Sound waves enter the auditory canal and vibrate the tympanic membrane, which vibrates the hammer, anvil, and stirrup bones of the middle ear. These bones transmit the vibrations to the oval window membrane, which covers an opening into the cochlea. The fluid in the cochlea is vibrated, shifting the basilar membrane up and down and bending the hairs of its embedded hair cells. Bending the hairs causes receptor potentials in the hair cells, which then release neurotransmitter onto neurons whose axons form the auditory nerve.
13. The perception of the pitch occurs because different frequencies of sound vibrate different regions of the basilar membrane and thus stimulate hair cells preferentially in these locations. These hair cells in turn stimulate different axons in the auditory nerve, which connect with different auditory neurons in the brain. Sound intensity is perceived by the amplitude of vibrations of the basilar membrane. Small vibrations cause slight bending of the hairs of hair cells, producing small receptor potentials and a low frequency of action potentials in the auditory nerve axons. Loud sounds deflect the hairs further, producing larger receptor potentials and a higher frequency of action potentials in the auditory nerve.
14. Refer to Figure 24-18 for a diagram of the structure of the human eye. The cornea is the transparent covering of the front of the eye that begins bending incoming light for focusing on the retina. The iris is muscular tissue that regulates the amount of light entering the eye through the pupil. The lens is composed of transparent protein fibers, and its flexible shape is responsible for bending light and for final fine focusing of images on the retina. The sclera is tough connective tissue surrounding the outer portion of the eye. The retina is a multilayered sheet of photoreceptors and neurons where light is converted into electrical impulses that are transmitted to the brain. The choroid is darkly pigmented tissue located behind the retina, whose blood supply helps nourish the cells of the retina. Its dark pigment also absorbs stray light, whose reflection inside the eyeball would interfere with clear vision.
15. A circular muscle surrounding the lens changes its shape. To focus on a distant object, the muscle relaxes, which pulls on the lens, stretching and flattening it. A relatively flat lens does not bend light much, which allows light from distant objects to be focused on the retina. Nearsightedness does not allow the focusing of distant objects, because the eyeball is too long or the cornea is too curved and so the light is focused in front of the retina. A concave lens is used to diverge light rays so that they can be properly focused on the retina.
16. Rods and cones are both photoreceptors located in the retina. Both use the same general molecules and mechanisms to absorb and respond to light, although their photopigments differ. In humans, rods are far more numerous than cones and dominate in the periphery of the retina. Rods have much deeper stacks of pigment-bearing membrane and are much more sensitive to light, but they cannot distinguish colors. Cones are concentrated in the fovea and respond to different wavelengths of light, giving us color vision.
17. When cells are damaged, they leak potassium ions and they release enzymes that produce bradykinin from a protein in blood that leaks from damaged capillaries. Both of these substances stimulate pain receptors.

Applying the Concepts

Hints

1. Look for Web sites that end in *.edu, .gov,* or *.org* (generally educational institutions such as colleges, government agencies such as the National Institutes of Health, or nonprofit organizations such as those devoted to seeking cures for Alzheimer's or Parkinson's disease, respectively).
2. Consider the difficulty and possible ethical objections to using aborted fetuses, the large number of people with Parkinson's, the high cost of the procedure, and the uncertain results.
3. Sensory perceptions all arise from electrical signals that travel to specific parts of the brain. Might different brains perceive stimuli entirely differently? Do people's perceptions differ? Can you assume that animals perceive the world as we do? How about "aliens"?

CHAPTER 25

Figure-based Questions

Figure 25-1 Mitotic cell division.
Figure 25-5 Courtship rituals are reproductive isolating mechanisms that help ensure that animals mate with other individuals of the same species. Courtship activities in some species stimulate ovulation.
Figure 25-11 The head contains the nucleus with chromosomes from the male parent and the acrosome, which releases enzymes that digest through the barriers surrounding the egg. The midpiece contains mitochondria that provide ATP to power sperm swimming by the flagellum, or tail.

Figure 25-17 The barriers surrounding the egg cannot be breached using the enzymes from the acrosome of a single sperm, but require the enzymes from many sperm.

Figure 25-24 Because there is no placenta to exchange nutrients, wastes, and gases, the embryo is retained in the uterus for a much shorter time. Offspring are born in a much less developed state (kangaroos) or even leave the mother as eggs (platypuses).

Consider This

Hints: This complex question touches on the nature of addiction, societal obligations to protect its present and future citizens, individual responsibility, and the obligations of friendship. Good luck!

Fill-in-the-Blank

1. asexual; bud; fission
2. testis; testosterone (or androgen); seminiferous tubules
3. nucleus, acrosome; mitochondria
4. epididymis, vas deferens; seminal vesicles, prostate gland, bulbourethral gland (order not important); urethra
5. ovary; estrogen, progesterone (order not important); secondary oocyte; polar body; oviduct, or uterine tube; uterus
6. corpus luteum; chorionic gonadotropin
7. cleavage, morula; blastula; ectoderm, mesoderm, endoderm (order not important)
8. amnion; chorion; umbilical cord; yolk sac

Review Questions

1. Asexual reproduction is more efficient because it does not require a mate, and no gametes are produced. Offspring are genetically identical to the parent, an advantage if the parent is well adapted to its environment, but lack of genetic diversity is a serious disadvantage in changing environments. Sponges can reproduce asexually. Sexual reproduction produces greater genetic diversity, which facilitates survival under a variety of environmental conditions. Disadvantages include finding and competing for a suitable mate. Most animals reproduce sexually at least some of the time. Mammals always reproduce sexually. An advantage of external fertilization is that it avoids the need for copulation. Disadvantages include the necessity to synchronize release of eggs and sperm, and to produce large numbers of gametes to guarantee successful fertilization. The developing young are exposed to the dangers of the external environment, and most usually perish. Many aquatic species, including most fish, use external fertilization. Internal fertilization reduces the number of gametes produced, increases the likelihood that most eggs will be fertilized, and affords some protection to the developing embryo. Mammals use internal fertilization.
2. The egg is surrounded by two barriers: the jelly-like zona pellucida, and an outer layer of follicle cells, the corona radiata.
3. The corpus luteum secretes estrogen and progesterone, which prevent the development of additional follicles after ovulation, and stimulate growth of the endometrial lining to prepare for embryo implantation. The corpus luteum disintegrates if fertilization does not occur, but during pregnancy, the embryo secretes chorionic gonadotropin, which maintains the corpus luteum during early pregnancy.
4. Sperm travel through the epididymis, vas deferens, and urethra as they are ejaculated from the penis. The sperm travel through the vagina, enter the uterus via the opening in the cervix,

and then enter the uterine tube of the female, where fertilization usually takes place.
5. Refer to Figure 25-15. Under stimulation by spontaneous release of GnRH from the hypothalamus, the anterior pituitary gland produces FSH and LH, which signal the ovaries to produce estrogen. FSH, LH, and estrogen cause development of follicles and increased estrogen production. High levels of estrogen trigger a surge of GnRH, which causes a surge in LH and FSH release by the anterior pituitary. The surge in LH and FSH causes ovulation. After ovulation, follicle cells in the ovary develop into the corpus luteum, which produces estrogen and progesterone, which suppresses production of LH and FSH. Without LH, the corpus luteum dies. Therefore, no more progesterone is produced. Freed from inhibition by progesterone, the hypothalamus resumes production of GnRH, and the next month's menstrual cycle begins.
6. See Table 25-3. Endoderm-derived tissues include the lining of the digestive and respiratory tracts, the liver, and pancreas. Ectoderm-derived tissues include the epidermis and glands of the skin, lining of the mouth and nose, hair, nervous system, lens of the eye, and the inner ear. Mesoderm-derived tissues include the dermis of the skin, muscle, skeleton, circulatory system, gonads, kidneys, and the outer layers of the digestive and respiratory tracts.
7. The chorionic villi contain a dense network of fetal capillaries that are bathed in pools of maternal blood. Many small molecules including oxygen, carbon dioxide, nutrients, and urea diffuse through the capillary membranes. These same membranes act as barriers to the passage of large molecules and most cells.
8. When pregnancy occurs, large quantities of estrogen and progesterone stimulate mammary glands to grow, branch, and develop the capacity to secrete milk. Lactation (milk secretion) is promoted by the pituitary hormone prolactin,

Applying the Concepts

Hints

1. Have the results been published in reputable scientific journals? Have they been reproduced by others? The problems inherent in animal cloning provide the strongest arguments against human cloning at our present level of knowledge. Future human cloning has no "correct" answer.
2. The woman should consult an obstetrician about the types of drugs she already takes or expects to take. Uncertain balancing of risks and benefits to mother vs. child can force difficult choices that will ultimately rest with the woman, in consultation with her doctor.
3. Factors to consider: the importance of individual choice, the drive to reproduce, the likelihood of high medical costs being borne both by the individual and by society, and the likelihood of embryonic developmental defects without intervention.

CHAPTER 26

Figure-based Questions

Figure 26-1 The cuckoo egg and chick resemble the foster parents' eggs and chicks, so parents cannot distinguish them. The cuckoo's large size may make the foster parents more likely to feed it, because it resembles the largest and healthiest of their own offspring. Any mutations that reduced the tendency to feed cuckoo chicks might reduce feeding of the foster parents' own offspring, and would be strongly selected against.

Figure 26-6 Because orientation is a genetic trait, birds from the western population should orient in a

southwesterly direction, regardless of the environment in which they are raised.

Figure 26-18 Females are most likely to reproduce successfully if they mate with the fittest males. If a male's fitness is reflected in his ability to build and decorate a bower, females benefit by choosing mates that build especially good bowers.

Figure 26-21 Dogs search for food mainly by smell, which is far more acute in dogs than in apes, which are mostly visual foragers. Sexual signaling evolves in a way that reflects the sensory capabilities of the species.

Consider This

Hints: Consider how comparisons of criteria for beauty in different cultures might be used to distinguish innate from learned preferences. Think about how to ensure that different cultures represent independent data points (i.e., have not influenced one another).

Fill-in-the-Blank

1. genes, environment; stimulus; innate
2. energy, predators, injury; practice, skills, adult; brains, learning
3. habituation, repeated; imprinting; sensitive period
4. aggressive; displays, injuring; weapons (such as fangs, claws), larger
5. territoriality (territorial behavior); mate, raise young, feed, store food (order not important); male; same species
6. Pheromones: any of the following advantages: long-lasting, use little energy to produce, species specific, don't attract predators, can convey messages over long distances; any of the following disadvantages: cannot convey changing information, conveys fewer different types of information. Visual displays: any of the following advantages: instantaneous, silent, can change rapidly, many different messages can be sent; any of the following disadvantages: makes animals conspicuous to predators, generally ineffective in darkness or dense vegetation, requires that recipient be in visual range (close by)

Review Questions

1. Every behavior is a mixture of innate and learned. In some cases, the nature and timing of learning is so rigidly programmed by genes that it can be considered as innate learning. Genes influence all learning, and learning can modify innate behaviors.
2. Play has survival value, providing young animals with experience in a variety of behaviors that they will use as adults in hunting, fleeing, or social interactions. (1) Play seems to lack any clear goal. (2) Play is abandoned in favor of escaping from danger, feeding, and courtship. (3) Play seems to involve feelings of pleasure. (4) Young animals play more frequently than adults. (5) Play often involves movements borrowed from other behaviors. (6) Play uses considerable energy. (7) Play is potentially dangerous.
3. Animals communicate through (1) sight (e.g., female mandrils communicate sexual receptivity by a brightly colored swelling of their buttocks)—visual communication is instantaneous and conveys a great amount of information, but it can be used only over relatively short distances and is ineffective in the dark, and may make the animal more visible to predators; (2) sound (e.g., monkeys use different calls to warn others of different types of predators)—sound communication is instantaneous and can convey a great number of different messages over rather long distances, it is effective in darkness, but it requires a great deal of energy and may attract predators; (3) smell (e.g., insects use pheromones that are sensed through smell as sex attractants)—chemical communication is silent and doesn't attract predators, uses little

energy, and is active over long distances, but cannot convey many different or rapidly changing signals; and (4) touch (e.g., primates use touch to establish social bonds)—touch is especially effective in social bonding and mating behaviors but can also spread infectious diseases.

4. Territories are usually defended against members of the same species who compete most directly for the resources being protected.

5. Natural selection favors the evolution of symbolic displays or rituals for resolving conflicts, because even the fittest animals could be seriously harmed in direct conflict.

6. Advantages of group living include (1) increased ability to detect, repel, and confuse predators; (2) increased hunting efficiency or increased ability to spot localized food resources; (3) potential for division of labor; (4) conservation of energy; and (5) increased likelihood of finding mates. Disadvantages include (1) increased competition within the group for limited resources; (2) increased risk of infection from contagious diseases; (3) increased risk that offspring will be killed by other members of the group; and (4) increased risk of being spotted by predators.

7. Naked mole rat societies are dominated by a single reproducing female, called the queen. There is division of labor among the workers. When a queen dies, there is a fight among rival females to determine the new queen.

Applying the Concepts

Hints

1. No human sex-attractant pheromones have been identified, though they have been found in other mammals. Did the company provide any research data? Be sure to include appropriate controls in your experimental design.

2. Consider the type of lure(s) most likely to attract mosquitoes and moths. Incorporate practical lures into your traps.

3. Think about examples of human dominance and territorial behavior in universities, businesses, and homes.

CHAPTER 27

Figure-based Questions

Figure 27-5 Many variables (including weather, diseases, and the abundance of other predators and prey) interact in complex ways to produce real population cycles, so they will not be completely regular.
Figure 27-10 A sustainable population cannot exceed carrying capacity, so hundreds of years in the future, the population would need to be smaller than its present size, or carrying capacity would need to have been increased to support more people sustainably.
Figure 27-14 High birthrates in developing countries are sustained by cultural expectations, lack of health education, and lack of access to contraception. Lower birthrates in developed countries are encouraged by easy access to contraception, the relatively high cost of raising children, and more varied career opportunities for women. (Students should be able to expand on, or add to, these factors.)
Figure 27-15 U.S. population growth is in the rapidly rising "exponential" phase of the S-curve. Stabilization will require some combination of reduction in immigration rates and birthrates. An increase in death rates is less likely, but not impossible. Speculate about the timing and the reason for various time frames.

Consider This

Hints: Consider that emigration is a typical "safety valve" for overcrowded populations, and consider factors that might restrict this option for Easter Island's humans.

Fill-in-the-Blank

1. any of the following: freezing, drought, flood, fire, hurricane; predation, competition
2. survivorship; early loss; late loss
3. interspecific competition, intraspecific competition; intraspecific competition; interspecific competition
4. logarithmic; no; no; carrying capacity; S-shaped
5. clumped; even
6. pyramid; roughly stable; pyramidal
7. births and immigrants, deaths and emigrants; decreases, increases, increases

Review Questions

1. Biotic potential is the maximum rate of population growth assuming ideal conditions that allow a maximum birthrate and a minimum death rate. Environmental resistance is limits on population growth that are set by the living and nonliving environment, including food and space availability, competition with other organisms, predation, and parasitism.

2. See Figure 27-2 for the growth curve of an exponentially growing population. An exponentially growing population adds an increasing number of individuals to the population in each succeeding time period.

3. Density-independent factors limit population size regardless of population density. Density-dependent factors increase in effectiveness as the population density increases.

4. Exceeding carrying capacity can damage an ecosystem, reducing its ability to support a population. Population growth can be exponential until the ecosystem is severely damaged, and then population size declines dramatically, and may remain at a smaller size if the ecosystem's carrying capacity has been reduced.

5. Density-dependent factors include (1) predation, (2) parasitism, and (3) competition. Predators tend to encounter, and prey on, the prey species that are most abundant. Most parasites have limited mobility and spread more readily among hosts at higher population densities. Competition for limited resources increases with increases in population size.

6. The transition from a growing to a stable population can be economically difficult because current economic structures are based on growing populations. Also, it reduces the availability of future workers and taxpayers.

Applying the Concepts

Hints

1. Consider the likely outcome of a high death rate in a small portion of a very large population of mobile organisms. In designing your experiment, be sure to include appropriate controls and specify how you will measure changes in population size.

2. Consider the outcome expected when death rates decline but birthrates do not. Consider the social and economic factors that affect birthrates in a country. Consider possible explanations of the link between affluence and birthrate.

3. Consider the ways in which human activities and technology can change the carrying capacity of an environment. Consider what humans require for a good quality of life and whether this could be achieved with the maximum population that Earth is capable of supporting.

CHAPTER 28

Figure-based Questions

Figure 28-1 Exotic species, because they did not evolve in the habitat to which they were introduced, may occupy a niche that is nearly identical to a native

species (for example, zebra mussels compete with and may exclude other freshwater mussel and clam species). A successful exotic species may have adaptations (e.g., a higher growth rate or reproductive rate) that allow it to outcompete native species.
Figure 28-4 Examples include keen eyesight of hawk and camouflage color of mouse; toxic substances in milkweed evolved under pressure from herbivores, but monarch larvae have evolved immunity to the toxin; grasses have evolved tough silicon embedded in their leaves, and grazing herbivores have teeth that grow continuously throughout life so that they are not completely worn down by the abrasive grasses.
Figure 28-12 Poisonous prey, such as the monarch caterpillar, frequently advertise their presence with bright, warning colors that make it easy for predators to learn to avoid the toxic prey.

Consider This

Hint: First, formulate a simple hypothesis: "Sea otter predation causes a decline in abalone populations." What would be ideal sites to monitor and compare abalone populations? Can you find (or create) an area with only otter predation? With human abalone harvesting but no otter predation? With both? What data would you collect in each site? What other factor(s) could cause changes in abalone populations? How would you control for these?

Fill-in-the-Blank

1. natural; coevolution; competition, predation, mutualism
2. parasitism, herbivory; camouflage; warning coloration
3. symbiotic; mutualistic; commensal
4. keystone; krill, shrimp (or crustacean); sea otters, kelp forest
5. succession; primary succession, secondary succession; pioneers; climax

Review Questions

1. An ecological community is all the interacting populations within an ecosystem. Community interactions are competition, predation, and symbiosis.

2. Milkweed evolved toxins, and the monarch evolved resistance to them; grasses evolved abrasive silica granules, and herbivores evolved strong jaws and grinding teeth; moths evolved an erratic flight behavior in response to bats' echolocating sounds, and bats switch to a frequency that moths don't hear; mice have evolved gray coats for camouflage against the ground, and hawks have evolved superb eyesight.

3. Types of symbiosis include parasitism, commensalism, and mutualism. In parasitism, one species benefits while the other is harmed; an example is the malarial parasite in humans. In commensalism, one species benefits while the other is neither harmed nor benefits; an example is barnacles that attach to the skin of whales, gaining exposure to food-rich water without harming the whale. Mutualism is a relationship that benefits both species involved; an example is bees that obtain nectar while pollinating flowers.

4. A clear-cut forest would experience secondary succession, because a previous ecosystem had already been established there.

5. Subclimax communities: suburban lawns, tallgrass prairies. Climax communities: deciduous hardwood forests, tundra. Climax communities are relatively stable, whereas subclimax communities progress to their climax communities unless they are maintained by human-caused or natural events (such as mowing or fire).

6. Succession is a change in a community and its nonliving environment over time. Successional

changes alter the ecosystem in ways that favor new communities, causing species to replace one another somewhat predictably, until a relatively stable climax community is formed.

Applying the Concepts

Hints

1. How do you define "benefit" for an ecosystem? If scientists could agree on this definition, should it be possible to study comparable ecosystems with and without the invader?
2. Did competitive exclusion occur? Think about possible evolutionary outcomes of competition, such as character displacement.
3. Consider how you might manipulate field plots so that some contain kangaroos and others do not. What variables would you measure? How long would the experiment last?

CHAPTER 29

Figure-based Questions

Figure 29-3 Ecosystems with the highest productivity have adequate nutrients, light energy, and optimal temperatures to support primary producers. As one or more of these factors becomes inadequate, productivity declines.

Figure 29-5 The primary consumers shown are a mouse, a ground squirrel, a bison, a prairie dog, a quail, a pheasant, a pronghorn, a rabbit, and a grasshopper.

Figure 29-6 The second law of thermodynamics tells us that when energy is converted from one form to another, the amount of useful energy decreases. Much of this energy is lost as heat. Because relatively little energy is captured in chemical bonds and thus available to maintain life, animals must consume a large number of Calories from the trophic level below them to obtain the useful energy they need.

Figure 29-9 Growth of many crop plants is limited by a lack of nitrogen in a form that plants can absorb. Humanity's need to grow crops to feed our growing population has led us to invent industrial processes for fixing nitrogen, which is used as fertilizer.

Figure 29-14 In a greenhouse, solar energy enters as light that hits surfaces within the greenhouse and is converted to heat, which is trapped inside by the glass. In much the same way, the greenhouse effect occurs when greenhouse gases (including CO_2 and methane) absorb sunlight and convert it to heat, holding this heat in the atmosphere.

Consider This

Hints: How does Biosphere 2 compare to most natural ecosystems in size? In complexity? In what ways are the differences between Biosphere 2 and natural ecosystems an advantage for controlled studies? Can experiments performed in Biosphere 2 be replicated elsewhere?

Fill-in-the-Blank

1. photosynthetic; nutrients, biogeochemical (nutrient)
2. autotrophs, producers; net primary productivity; heterotrophs, consumers; detritus feeders; fungi, bacteria
3. trophic; 10; carnivore; omnivores; food webs
4. reservoirs; atmosphere, oceans, fossil fuels; atmosphere; nitrogen-fixing bacteria
5. dead zone; sulfur dioxide, acid deposition
6. greenhouse; deforestation (cutting forests), 36; krill

Review Questions

1. Energy continuously enters ecosystems as light, flows through the trophic levels and is progressively lost as heat. Nutrients flow through

trophic levels, and often through the abiotic environment, but are recycled.
2. An autotroph (producer) captures energy from sunlight and produces its own food through photosynthesis. Autotrophs occupy the first trophic level, and provide the energy utilized by all the organisms in higher trophic levels.
3. Primary productivity is the amount of light energy captured and converted to chemical energy (in bonds of organic molecules) by the producers in an ecosystem. Net primary productivity is the portion of primary productivity left over after the metabolic needs of the producers have been met. Net primary productivity will be higher in a farm pond because, although light and water are abundant in both ecosystems, nutrients and favorable temperatures are less available in the alpine lake.
4. The first three trophic levels are (1) producers, (2) primary consumers or herbivores, and (3) secondary consumers or carnivores. Because herbivores acquire their energy from plants, and only a fraction of the energy contained in the plants is available to the herbivores for use in building their own biomass (the 10% law), plants must have a higher biomass than their consumers.
5. A food chain describes a linear feeding relationship among organisms in an ecosystem, whereas a food web depicts the complex interconnecting feeding relationships in an ecosystem far more accurately.
6. Detritus feeders are animals and protists that live on wastes and dead bodies, including earthworms, some insects and crustaceans, nematode worms, and vultures. Decomposers are primarily fungi and bacteria that digest food outside their bodies and absorb some of the liberated nutrients. Both groups of organisms are necessary for recycling nutrients.
7. Carbon from the atmosphere enters producers during photosynthesis. Consumers acquire carbon by consuming photosynthetic organisms or other consumers. Detritus feeders and decomposers acquire carbon from dead organisms and wastes. Carbon is returned to the atmosphere during aerobic respiration, which releases CO_2. Human activities have added CO_2 (a greenhouse gas) to the atmosphere by burning fossil fuels and forests, causing global warming.
8. Nitrogen gas in the atmosphere is combined with hydrogen by nitrogen-fixing bacteria, producing ammonia. Nitrogen gas is combined with oxygen by lightning or by burning fossil fuels. This produces nitrates that are carried to Earth dissolved in rain. Nitrates and ammonia are absorbed by plant roots.
9. Phosphate dissolved in water from phosphate-rich rocks enters plant roots and becomes incorporated into biological molecules. These molecules pass through from plants to herbivores, which are eaten by carnivores. The phosphorus cycle lacks an atmospheric component.
10. Water evaporates from the ocean and is carried inland as water vapor until it condenses and falls as rain. The rain enters the soil and is absorbed by plant roots. This water is moved up through the plant body by transpiration through the stomata. Transpired water vapor returns to the atmosphere, eventually condenses and falls as rain, entering the ocean either directly or as runoff from land.

Applying the Concepts

Hints

1. Correlation between two trends does not prove that one caused the other. Consider methods and approaches that might be used to supplement data on recent worldwide trends in temperature

and carbon dioxide levels. Possibilities include controlled experiments (what kinds?) and temperature and CO_2 data from the more distant past when the two were not rising in tandem (how might one get such data?).
2. Examples are in the essay "Earth Watch: Food Chains Magnify Toxic Substances" on p. 582, but you might also look up additional examples on the Internet. In which tissue do bioaccumulating substances build up? Think about energy pyramids and how they affect the amount of bioaccumulating chemicals consumed by organisms at different trophic levels.
3. Consider the contribution of each additional person and the relative impact of births in different parts of the world.

CHAPTER 30

Figure-based Questions

Figure 30-1 The equator-to-pole temperature gradient would still remain, but there would be no change in day length or seasons throughout the year.

Figure 30-9 Near optimal growing conditions of temperature, light, and moisture support such a huge number and diversity of plants that these, in turn, provide a wealth of habitats and food sources for diverse animals. Nutrients are abundant, but are primarily stored in plant bodies, not in the soil. When plants die, the optimal conditions for decomposers and detritus feeders allow nutrients to be recycled almost immediately back into the bodies of plants.

Figure 30-15 First, tallgrass prairie is a subclimax ecosystem that relies on fire. If people suppress fires, tallgrass prairie is replaced by forest. Second, these biomes provide fertile soils, and farming has now displaced the native vegetation over most former tallgrass prairies.

Figure 30-25 Nearshore ecosystems have an abundance of the two limiting factors for life in water: nutrients and light to support photosynthetic organisms. Both upwelling from ocean depths and runoff from the land provide nutrients. The shallow water allows adequate light to penetrate to support rooted plants and/or anchored algae, which in turn provide food and shelter for a wealth of marine animals. The salt marsh would have the highest productivity because nutrients from land can accumulate in its calm shallow water, and many animals breed here.

Figure 30-26 Bleaching refers to the loss of symbiotic algae that normally inhabit the corals' tissues, providing them with food energy captured during photosynthesis and calcium carbonate used in coral skeletons. Loss of these resources can eventually kill the coral. Bleaching is a common response to water that is excessively warm, and thus global warming may contribute to the demise of coral reefs.

Consider This

Hints: Will people with inadequate food respect preserved habitat? Can sustainable agriculture benefit both people and wildlife?

Fill-in-the-Blank

1. seasons; Earth's rotation, temperature differences; tropics; ocean currents; rain shadow
2. temperature, water; nutrients, light; tropical rain forests; coral reefs
3. leaves, evaporation, protection; water (rain); night
4. tropical rain forest; deciduous forest; shortgrass prairie; desert; savanna; tundra
5. littoral zone; nutrients, light; phytoplankton, zooplankton; limnetic zone, profundal zone; detritus
6. oligotrophic; eutrophic; nearshore zone; estuary or salt marsh

Review Questions

1. Warm air saturated with water vapor rises over the equator, and as the air cools, it rains over the tropics. The cooler, dry air then flows north and south from the equator and sinks but is warmed by heat radiated from Earth, so by the time it reaches Earth, it is both warm and dry and influences the formation of deserts 30° north and south of the equator.

2. Ocean circulation patterns distribute warmth from the equator to northern and southern coastal areas.

3. The four requirements for life are nutrients, energy, water, and appropriate temperatures. Water and temperature are the most limiting in terrestrial ecosystems. Nutrients and energy are the most limiting in aquatic ecosystems.

4. In both cases, increasingly cold temperature is limiting primary productivity.

5. Nutrients are concentrated in the vegetation. Much animal life is high in trees because most of the vegetation is in the canopy where light levels are higher than on the forest floor.

6. Limited nutrients are rapidly washed away and the exposed soil hardens in the intense sunlight.

7. (a) Desert plants often have extensive shallow root systems to trap water from brief showers, quick reproductive cycles, and thickened stems and sometimes leaves that store water. Leaves are often small, with waxy coatings, sometimes modified into spines. (b) Many animals are nocturnal, hiding in cool moist burrows during the day; some derive all of their water from food.

8. Offroad vehicles disrupt delicate bacterial networks that sustain desert soils.

9. Taiga trees are evergreens with short, waxy needles that resist water loss and can take advantage of the short growing season.

10. In deciduous forests, trees shed their leaves when water is limiting, whereas the evergreens in coniferous forests retain their leaves the entire year. Coniferous biomes have a lower productivity and are found in areas of lower temperature and rainfall.

11. Precipitation levels explain the differences among the three biomes.

12. The African savanna

13. In the coastal waters because their shallow depths allow algae and plants to anchor and get enough light for photosynthesis, and there is a steady flow of nutrients from the land.

14. Coral reefs have a high productivity due to the optimal combinations of adequate light, temperature, and nutrients. Human development creates silt that clouds the water, reducing the photosynthetic rates of coral's symbiotic algae. Warming water, likely caused by global warming, results in the loss of these algae. Overfishing, damage from tourist boats, and sewage pollution add to the destruction.

15. The nearshore littoral zone supports the most diverse community because there is abundant light and nutrients, supporting anchored algae and aquatic plants. The limnetic zone is the open water below the surface, with sufficient light to support phytoplankton, which support zooplankton, eaten by small crustaceans and fish. The profundal zone, where light is too dim for photosynthesis, lies between the limnetic zone and the lake floor; life consists mainly of detritus feeders and decomposers, and some fish.

16. Oligotrophic lakes are low in nutrients and clear, due to low sediment and few microscopic organisms. Eutrophic lakes are high in nutrients and support dense communities. (a) An oligotrophic lake may undergo slow eutrophication naturally as runoff from land brings in nutrient-laden sediment over time. (b) People greatly speed eutrophication by allowing runoff into lakes from fertilized farm fields, feedlots, and sewage.

17. The photic zone is the upper ocean water, where light is sufficient to support photosynthesis. Its inhabitants obtain nutrients that run off the land or that upwelling carries into the photic zone. The aphotic zone has inadequate light for photosynthesis; both energy and nutrients are derived from the bodies and excrement of organisms that rely on the photic zone for energy.

18. Sulfur bacteria, which derive energy from chemosynthesis, are the primary producers of vent communities.

19. Temperate rain forests in the Southern Hemisphere, as well as open-ocean ecosystems.

Applying the Concepts

Hints

1. Think of ways to identify and compare otherwise similar populations of frogs and people in places that differ in the extent of ozone layer decay.

2. Refer to Figure 30-8. Look at the biomes adapted to low vs. high temperatures, and think about how much each would be affected if rainfall were increased or decreased.

3. Think about how the rate of decomposition in a biome affects soil fertility. Which of these biomes is likely to have richer soil? How does clear-cutting affect rain-forest soil?

CHAPTER 31

Figure-based Questions

Figure 31-5 Large carnivores are especially likely to disappear from small forest fragments. Due to the energy pyramid phenomenon (see Chapter 29), supporting a breeding population of top-level predators requires very large populations of prey species, which in turn require a large area of suitable habitat. Large herbivores are vulnerable for similar reasons; for example, it takes a huge amount of vegetation to support a population of forest elephants.

Figure 31-7 Hawaii lacks any predatory native mammals or reptiles, so many native birds nest on the ground because they evolved when it was safe to do so. These birds are easy prey for the mongoose. Bird species that evolved in continental, predator-rich environments possess robust defenses against predation.

Figure 31-10 Benefits include reduced soil erosion; less need for irrigation (intact soil retains water better); less airborne dust (a pollutant); better soil ecology (more beneficial microbes, insects, and worms) in undisturbed soil; and reduced carbon emissions (soil releases CO_2 when tilled, as does the farm machinery used for tilling). The main environmental disadvantage is increased application of toxic herbicides (except by organic no-till farming).

Consider This

Hint: Aside from the inherent value of preserving eagles, what additional benefits might arise from efforts to protect a large, aerial predator?

Fill-in-the-Blank

1. genetic diversity, species diversity, ecological diversity (order not important); genetic; species

2. ecosystem services; any four of the following: purifying air and water, replenishing atmospheric oxygen, pollinating plants and dispersing their seeds, providing wildlife habitat, decomposing wastes, limiting erosion and flooding, controlling pests, and providing recreational opportunities

3. ecological economics

4. any four of the following: habitat destruction, overexploitation, invasive species, pollution, global warming; habitat destruction

5. minimum viable population; habitat fragmentation; wildlife corridors

6. sustainable

Review Questions

1. Conservation biology is the branch of science that seeks to understand and conserve biological diversity. It draws from ecology, genetics, and evolutionary biology, and also involves social scientists, policymakers, lawyers, geographers, economists, ethicists, and historians.

2. (1) Genetic diversity is the variety and relative frequencies of different alleles in the gene pool of a species. If genetic diversity is reduced, the species may lack the genetic variability necessary to evolve adaptations to changing environments. (2) Species diversity is the variety and relative abundance of different species. The intricate network of community interactions necessary to maintain a properly functioning ecosystem relies on species diversity. (3) Ecological diversity is the variety of different ecosystems and of the community interactions within them. Loss of ecological diversity reduces the ability of the biosphere to support its current wealth of species.

3. Ecological economics attempts to value ecosystem services and assess the monetary trade-offs that occur when natural ecosystems are damaged to make way for human profit-making activities. It is important because ecosystem services sustain people and all other forms of life, and if they are not appropriately valued, they are often damaged or destroyed by human activities.

4. Directly used goods include food, building materials, oxygen, and medicinal plants. Indirect services include maintenance of soil fertility, controlling erosion and flooding, neutralizing pollutants, disposing of organic wastes, recreational opportunities, and providing wildlife habitat.

5. Threats to biodiversity include habitat destruction, such as cutting tropical rain forests; harmful interactions with invasive species, such as the mongoose, which threatens native Hawaiian ground-nesting birds; overexploitation, such as overfishing of sharks, bluefin tuna, and turtles; pollutants, such as pesticides; and global warming from excess production of greenhouse gases, which is disrupting community interactions by altering breeding seasons.

Applying the Concepts

Hints

1. Why should we understand and preserve species?

2. The Internet, government agencies such as your local division of wildlife, and your own powers of observation are the best resources here.

3. Think about how people use space and energy and how this could be made more efficient. How could you preserve maximum habitat for wildlife and ecosystem services, while still meeting human needs?

4. Think about, and perhaps use the Internet to learn more about, organic farming. What attributes would make it unattractive to farmers accustomed to planting several hundred acres of corn? What are its advantages? Think about the cost of organic produce. Why is it sometimes more expensive? What are you paying for?

5. Where do biofuels come from? Are all sources equally good for preserving ecosystems? What are the costs vs. the benefits of this type of fuel?

Photo Credits

Chapter 1
Opener 1a   NASA Headquarters
Opener 1b   Yorgos Nikas/Photo Researchers, Inc.
1-1   AP Photo
1-4   Corbis/Bettmann
1-6   E.H. Newcomb & W.P. Wergin/Biological Photo Service
1-7   Christine Case
1-8a   Andrew Syred/Science Photo Library/Photo Researchers, Inc.
1-8b   Craig Tuttle/Corbis/Stock Market
1-8c   Kim Taylor/Bruce Coleman Inc.
1-9   Johnny Johnson/DRK Photo
1-10   Kim Taylor/Bruce Coleman Inc.
1-11   Lawrence Livermore National Laboratory/Photo Researchers, Inc.
E1-2   Nigel J. Dennis/Photo Researchers, Inc.
E1-4   Franz Lanting/Minden Pictures

Unit 1
Opener 1a   Manfred Kage/Peter Arnold, Inc.
Opener 1b   Lester V. Bergman/CORBIS

Chapter 2
Opener 2a   Roberto Brosan/Time Life Pictures/Getty Images/Time Life Pictures
Opener 2b   Digital Vision/SuperStock
2-5cL   Arnold Fisher/Photo Researchers, Inc
2-9a   Stephen Dalton/Photo Researchers, Inc.
2-9b   Teresa and Gerald Audesirk
2-15L   Larry Ulrich Stock Photography, Inc./DRK Photo
2-15M   Jeremy Burgess/Science Photo Library/Photo Researchers, Inc.
2-15R   Biophoto Associates/Photo Researchers, Inc.
2-17a   Jean-Michel Labat/Auscape International Proprietary Ltd.
2-17b   Donald Specker/Animals Animals/Earth Scenes
2-21a   Robert Pearcy/Animals Animals/Earth Scenes
2-21b   Jeff Foott/DRK Photo
2-21c   Nuridsany et Perennou/Photo Researchers, Inc.
E2-1   Don Mason/Brand X Pictures/Jupiter Images
E2-2   Photo Researchers, Inc.

Chapter 3
Opener 3a   Daniel J. Lyons/Bruce Coleman USA
Opener 3b   Don W. Fawcett/Photo Researchers, Inc.
3-8aT   Joseph Kurantsin-Mills, The George Washington University Medical Center
3-8bT   Joseph Kurantsin-Mills, The George Washington University Medical Center
3-8cT   Joseph Kurantsin-Mills, The George Washington University Medical Center
3-11R   L.A. Hufnagel, Ultrastructural Aspects of Chemoreception in Ciliated Protists (Ciliophora), *Journal of Electron Microscopy Technique*, 1991. Photomicrograph by Jurgen Bohmer and Linda Hufnagel, University of Rhode Island
3-12a   American College of Physicians
3-12b   Scott Camazine/Phototake NYC

Chapter 4
Opener 4a   David Young-Wolff/PhotoEdit Inc.
Opener 4b   John P. Clare (www.caudata.org)
4-5b   E. Guth, T. Hashimoto, and S.F. Conti
4-6   Lester V. Bergman/OpenerRBIS
4-7b   R. Bolender and Donald Fawcett/Visuals Unlimited
4-8M   Don W. Fawcett/Photo Researchers, Inc.

4-9b   Molecular Probes, Inc.
4-10aR   Ellen R. Dirksen/Visuals Unlimited
4-10bR   Yorgos Nikas/Getty Images Inc. - Stone Allstock
E4-1aT   Biophoto Associates/Photo Researchers, Inc.
E4-1aB   Cecil Fox/Science Source/Photo Researchers, Inc.
E4-1bT   Brian J. Ford
E4-1bB   Brian J. Ford
E4-1c   Jean-Claude Revy/Phototake NYC
E4-2a   M.I. Walker/Photo Researchers, Inc.
E4-2b   David M. Phillips/Visuals Unlimited
E4-2c   Manfred Kage/Peter Arnold, Inc.
E4-2d   National Library of Medicine
E4-3   BBC Photo Library
E4-4   AP Photo

Chapter 5
Opener 5a   Phil Cole/Getty Images
Opener 5b   BananaStock/SuperStock
5-1   Tim Davis/OpenerRBIS
5-2   SOHO
E5-1   Baumgartner Olivia/Corbis Sygma

Chapter 6
Opener 6a   Joe Tucciarone
Opener 6b   BIOS/Peter Arnold, Inc.
6-1a   Ken W. Davis/Tom Stack & Associates, Inc.
6-8a   Jeremy Burgess/Science Photo Library/Photo Researchers, Inc.
6-8b   Jeremy Burgess/Photo Researchers, Inc.
E6-1   Frans Lanting/Minden Pictures

Chapter 7
Opener 7a   AP Photo
Opener 7b   Agence Zoom/Stringer/Getty Images, Inc.
7-7   Jesse Ceballos/SuperStock

Unit 2
Opener 2a   GK Hart/Vicky Hart/Getty Images
Opener 2b   Eric Isselée/Shutterstock

Chapter 8
Opener 8a   Young Ho/SIPA
Opener 8b   Viagen. Photo Candace Dobson-Gunslinger LLC
8-2a   David Scharf/Science Photo Library/Photo Researchers, Inc.
8-2b   Michael Abbey/Photo Researchers, Inc.
8-2c   Carolina Biological Supply Company/Phototake NYC
8-2d   Teresa and Gerald Audesirk
8-4a   Professor Ulrich K. Laemmli
8-4b   Andrew Syred/Photo Researchers, Inc.
8-6   CNRI/Science Photo Library/Photo Researchers, Inc.
8-9aT   Michael W. Davidson/The Florida State University
8-9bT   Michael W. Davidson/The Florida State University
8-9cT   Michael W. Davidson/The Florida State University
8-9dT   Michael W. Davidson/The Florida State University
8-9eT   Michael W. Davidson/The Florida State University
8-9fT   Michael W. Davidson/The Florida State University
8-9gT   Michael W. Davidson/The Florida State University
8-9hT   Michael W. Davidson/The Florida State University
8-10b   T.E. Schroeder/Biological Photo Service
E8-2BR   Photograph courtesy of The Roslin Institute

Chapter 9
Opener 9a   Ronald Modra/Sports Illustrated
Opener 9b   Corbis/Bettmann
9-2   Archive/Photo Researchers, Inc.
9-14   Biophoto Associates/Photo Researchers, Inc.
9-16   Hart-Davis/Science Photo Library/Photo Researchers, Inc.
9-18   Sarah Leen/National Geographic Image Collection
9-19   Jane Burton/Bruce Coleman Inc.
9-21a   Dennis Kunkel/Phototake NYC
9-21b   Walter Reinhart/Phototake NYC
9-22a   CNRI/Science Photo Library/Photo Researchers, Inc.
9-22b   Lawrence Migdale/Pix
E9-1   Ezra Shaw/Getty Images
E9-2a   Custom Medical Stock Photo/Newscom.com
E9-2b   Custom Medical Stock Photo/Newscom.com

Chapter 10
Opener 10aT Yann Arthus-Bertrand/OpenerRBIS-NY
Opener 10aB Yann Arthus-Bertrand/OpenerRBIS-NY
Opener 10b   Stuart Isett/Polaris
10-2c   Michael Freeman/Phototake NYC
10-5   Gopal Murti/Science Photo Library/Photo Researchers, Inc.
E10-1   A. Barrington Brown/Photo Researchers, Inc.

Chapter 11
Opener 11a   Pete Moss by permission of Sony BMG and Luke Martineau
Opener 11b   Delly Carr. Reprinted from the UW Magazine Office of Alumni Affairs, University of Waterloo
11-4   Oscar L. Miller, Jr.
11-7   Nature Genetics, Volume 30, pp. 167-174 (2002), Figure 3A (top). Chromosomal silencing and localization are mediated by different domains of Xist RNA Anton Wutz, Theodore P. Rasmussen & Rudolf Jaenisch
11-8   Frederic Jacana/Photo Researchers, Inc.
E11-1   Howard W. Jones, Jr.

Chapter 12
Opener 12a   Virginia Cronis/The Virginian Pilot
Opener 12b   Uli Holz/The Innocence Project <www.innocenceproject.org>
12-1aR   Stanley N. Cohen/Science Photo Library
12-7   Courtesy of Orchid Cellmark, Inc., Germantown, Maryland
12-10   Monsanto Company
12-13   Photo courtesy of Promega Corporation
E12-1   Janet Chapple/Granite Peak Publications
E12-2   KEREN SU/DanitaDelimont.com
E12-3   Courtesy Syngenta

Unit 3
Opener 3a   Francois Gohier/Photo Researchers, Inc.
Opener 3b   SMC Images/Getty Images

Chapter 13
Opener 13a   Peter Lillie/Photolibrary
Opener 13b   Vem/Photo Researchers, Inc.
13-4a   John Cancalosi/DRK PHOTO
13-4b   David M. Dennis/Tom Stack & Associates, Inc.
13-4c   Chip Clark
13-6   James Carmichael/NHPA/Photo Researchers, Inc.
13-10a   Stephen Dalton/Photo Researchers, Inc.
13-10b   Stephen Dalton/Photo Researchers, Inc.

13-11a   Photo Lennart Nilsson/Albert Bonniers Forlag AB
13-11b   Photo Lennart Nilsson/Albert Bonniers Forlag AB
13-11c   Photo Lennart Nilsson/Albert Bonniers Forlag AB
13-13a   Stephen J. Krasemann/DRK Photo
13-13b   Timothy O'Keefe/Tom Stack & Associates, Inc.
13-14   Courtesy of Sean Earnshaw & Anne Magurran/University of St. Andrews, Scotland
13-15   Patti Murray/Animals Animals/Earth Scenes
E13-1   Corbis/Bettmann
E13-2   Don & Pat Valenti/DRK Photo
E13-3   Yva Momatiuk/John Eastcott/Minden

Chapter 14
Opener 14a   Richard Price/Getty Images, Inc. - Taxi
Opener 14b   David Scharf/Peter Arnold, Inc.
14-7a   Alan Mason Chesney Medical Archives of the John Hopkins Institutions
14-7b   SPL Science Photo Library/Photo Researchers, Inc.
14-8a   M.P.L. Fogden/Bruce Coleman Inc.
14-8b   Tim Davis/Photo Researchers, Inc.
14-9   W. Perry Conway/Tom Stack & Associates, Inc.
14-10   D. Cavagnaro/DRK Photo
14-11a   Rick & Nora Bowers/VIREO
14-11b   Bob Steele/VIREO
14-12a   Wayne Lankinen/Valan Photos
14-12b   Edgar T. Jones/Bruce Coleman Inc.
14-13   Tim Laman/National Geographic Image Collection
14-14   Loic Degen
14-15   Gerard Lacz/Animals Animals/Earth Scenes
14-16a   Dani/Jeske/Animals Animals/Earth Scenes
14-16b   Jeff Jeffrey/Photo Resource Hawaii/Alamy Images
14-16c   A.C. Medeiros/Gerald D. Carr
14-16d   Gerald D. Carr
E14-1   Andrew J. Martinez/Photo Researchers, Inc.
E14-2   Frans Lanting/Minden Pictures

Chapter 15
Opener 15a   O. Louis Mazzatenta/National Geographic Image Collection
Opener 15b   O. Louis Mazzatenta/National Geographic/Getty Images
15-4   Hybrid painting, 204, for Scientific American. ©2006 by Don Dixon/cosmographica.com
15-6   Michael Abbey/Visuals Unlimited
15-7a   Milwaukee Public Museum
15-7b   James L. Amos/Photo Researchers, Inc.
15-7c   Carolina Biological Supply Company/Phototake NYC
15-7d   Douglas Faulkner/Photo Researchers, Inc.
15-8   Science Photo Library/Photo Researchers, Inc.
15-9   Terry Whittaker/Photo Researchers, Inc.
15-10   Science Photo Library/Photo Researchers, Inc.
15-13a   Tom McHugh/Chicago Zoological Park/Photo Researchers, Inc.
15-13b   Frans Lanting/Minden Pictures
15-13c   Nancy Adams/Tom Stack & Associates, Inc.
15-14   Brunet Michel
15-16   Prof. David W. Frayer
15-17   Jerome Chatin/Gamma Press USA, Inc.

Chapter 16
Opener 16a   Claro Cortes IV/Reuters Limited
Opener 16b   AP Photo

16-1a   Wayne Lankinen/Bruce Coleman, Inc.
16-1b   M.C. Chamberlain/DRK Photo
16-1c   Maslowski/Photo Researchers, Inc.
16-2a   C. Steven Murphree/Biological Photo Service
16-2b   Greg Rouse, Department of Invertebrate Zoology, National Museum of Natural History, Smithsonian Institution
16-2c   Jeremy Burgess/Science Photo Library/Photo Researchers, Inc.
16-3a   Hans Gelderblom/Getty Images Inc. - Stone Allstock
16-3b   W. Jack Jones/Springer-Verlag GmbH & Co KG
16-5a   David M. Phillips/Visuals Unlimited
16-5b   Karl O. Stetter, University of Regensburg, Germany
16-5c   CNRI/Science Photo Library/Photo Researchers, Inc.
16-6   Karl O. Stetter, University of Regensburg, Germany
16-7   George Chapman/Visuals Unlimited
16-8   Manfred Kage/Peter Arnold, Inc.
16-9a   D.P. Wilson/Eric and David Hosking/Photo Researchers, Inc.
16-9b   Lawrence E. Naylor/Photo Researchers, Inc.
16-10   David M. Phillips/Visuals Unlimited
16-11   Kevin Schafer/Peter Arnold, Inc.
16-12   Oliver Meckes & Nicole Ottawa/Eye of Science/Photo Researchers, Inc.
16-13   Ed Degginger/Color-Pic, Inc.
16-14   Dennis Kunkel/Phototake NYC
16-15a   P.W. Grace/Science Source/Photo Researchers, Inc.
16-15b   Cabisco/Visuals Unlimited
16-16   Ray Simons/Photo Researchers, Inc.
16-17a   Lee W. Wilcox
16-17b   John Gerlach/Tom Stack & Associates, Inc.
16-17c   John Shaw/Tom Stack & Associates, Inc.
16-18a   Dwight R. Kuhn/Dwight R. Kuhn Photography
16-18b   Milton Rand/Tom Stack & Associates, Inc.
16-18c   Larry Ulrich/DRK Photo
16-19a   Image 100/Alamy Images
16-19b   Maurice Nimmo/A-Z Botanical Collection, Ltd.
16-19c   Teresa and Gerald Audesirk
16-20a   Dwight R. Kuhn/Dwight R. Kuhn Photography
16-20b   David Dare Parker
16-20b Inset   Matt Jones/Auscape International Pty. Ltd.
16-20c   Dwight R. Kuhn/Dwight R. Kuhn Photography
16-20d   Larry West/Photo Researchers, Inc.
16-21a   Robert & Linda Mitchell Photography
16-21b   Elmer Koneman/Visuals Unlimited
16-22   Jeff Lepore/Photo Researchers, Inc.
16-23   David Dvorak, Jr./AAAAJPF0
16-24   David M. Dennis/Tom Stack & Associates, Inc.
16-25a   Larry Lipsky/DRK Photo
16-25b   Brian Parker/Tom Stack & Associates, Inc.
16-26a   Gregory Ochocki/Photo Researchers, Inc.
16-26b   Mark Webster/Photolibrary.com
16-26c   Teresa and Gerald Audesirk
16-27a   Kjell B. Sandved/Butterfly Alphabet, Inc.
16-27b   Peter Batson/Image Quest Marine
16-27c   J.H. Robinson/Photo Researchers, Inc.
16-28a   Ray Coleman/Photo Researchers, Inc.
16-28b   Alex Kerstitch/Estate of Alex Kerstitch
16-29a   Fred Bavendam/Peter Arnold, Inc.
16-29b   Ed Reschke/Peter Arnold, Inc.
16-30a   Fred Bavendam/Peter Arnold, Inc.
16-30b   Kjell B. Sandved/Photo Researchers, Inc.

16-30c   Alex Kerstitch/Estate of Alex Kerstitch
16-31a   Carolina Biological Supply Company/Phototake NYC
16-31b   Peter J. Bryant/Biological Photo Service
16-31c   Stephen Dalton/Photo Researchers, Inc.
16-31d   Stanley Breeden/DRK Photo
16-32a   Dwight Kuhn/Dwight R. Kuhn Photography
16-32b   Tim Flach/Getty Images Inc. - Stone Allstock
16-32c   Teresa and Gerald Audesirk/Teresa and Gerald Audesirk
16-33a   Tom Branch/Photo Researchers, Inc.
16-33b   Peter J. Bryant/Biological Photo Service
16-33c   Carolina Biological Supply Company/Phototake NYC
16-33d   Alex Kerstitch/Estate of Alex Kerstitch
16-34   Tom McHugh/Photo Researchers, Inc.
16-35a   Peter David/Getty Images Inc.-Hulton Archive Photos
16-35b   Mike Neumann/Photo Researchers, Inc.
16-35c   Stephen Frink/Getty Images Inc.-Stone Allstock
16-36a   Breck P. Kent/Animals Animals/Earth Scenes
16-36b   Joe McDonald/Tom Stack & Associates, Inc.
16-36c   Cosmos Blank/National Audubon Society/Photo Researchers, Inc.
16-37a   David G. Barker/Tom Stack & Associates, Inc.
16-37b   Roger K. Burnard/Biological Photo Service
16-37c   Frans Lanting/Minden Pictures
16-38a   Walter E. Harvey/Photo Researchers, Inc.
16-38b   Carolina Biological Supply Company/Phototake NYC
16-38c   Ray Ellis/Photo Researchers, Inc.
16-39a   Craig Watts/Nature Picture Library
16-39b   Mark Newman/Index Stock Imagery, Inc.
16-39b Inset   D. Parer and E. Parer-Cook/Auscape International Proprietary Ltd.
16-39c   Flip Nicklin/Minden Pictures
16-39d   Jonathan Watts/Science Photo Library/Photo Researchers, Inc.
E16-1   Stanley Breeden/DRK Photo

Unit 4
Opener 4a   Purestock/Getty Images
Opener 4b   Tatiana53/Shutterstock

Chapter 17
Opener 17a   Chris Cheadle/Alamy
Opener 17b   Peter E. Smith/The Natural Sciences Image Library (NSIL)
17-4   Jeremy Burgess/Science Photo Library/Photo Researchers, Inc.
17-5b   Richard Kessel & Gene Shih/Visuals Unlimited/Getty Images, Inc.
17-6b   Lee W. Wilcox
17-7a   Lynwood M. Chace/Photo Researchers, Inc.
17-7b   Dwight R. Kuhn Photography
17-8RT   Ed Reschke/Peter Arnold, Inc.
17-8RB   E.R. Degginger/Animals Animals/Earth Scenes
17-9   Teresa and Gerald Audesirk
17-10   J. Robert Waaland, University of Washington/Biological Photo Service
17-11RT   Ed Reschke/Peter Arnold, Inc.
17-12   Max Westby
17-14a   Kenneth W. Fink/Photo Researchers, Inc.
17-14b   Belinda Wright/DRK Photo
17-15b   Lee W. Wilcox
17-16   Kim Taylor/Bruce Coleman Inc.
17-17   Biophoto Associates/Science Source/Photo Researchers, Inc.
17-18R   Hugh Spencer/National Audubon Society/Photo Researchers, Inc.
E17-1   Michael Fogden/DRK Photo

Chapter 18
Opener 18a Maximilian Stock Ltd/Animals Animals/Earth Scenes
Opener 18b Getty Images Inc. - Stone Allstock
18-1TL Jeff Foott/Bruce Coleman Inc.
18-3b Heather Audesirk
18-4 Teresa and Gerald Audesirk
18-5 Dennis Kunkel/Dennis Kunkel Microscopy, Inc.
18-6 Teresa and Gerald Audesirk
18-7TL Thomas Eisner, Cornell University
18-7TR Thomas Eisner, Cornell University
18-8 Lucy Nicholson/Agence France Presse/Getty Images
18-9L Carolina Biological Supply Company/Phototake NYC
18-11L David Rhodes/Purdue University/Dept of Horticulture and Landscape Architecture
18-11R Teresa Audesirk
18-12a Teresa and Gerald Audesirk
18-12b Carolina Biological Supply Company/Phototake NYC
18-13 Teresa and Gerald Audesirk
18-14 Scott Camazine/Photo Researchers, Inc.
18-16 Teresa and Gerald Audesirk
18-17RT Lee W. Wilcox
18-17RB Malcolm B. Wilkins
18-19 John Cancalosi/DRK Photo
18-20 Biophoto Associates/Photo Researchers, Inc.
E18-1 Myrleen Ferguson Cate/PhotoEdit Inc.

Unit 5
Opener 5a Alonzo King's Lines Ballet - Drew Jacoby in The Moroccan Project - costume design Colleen Quen & Robert Rosenwasser. ©Marty Sohl
Opener 5b Dee Conway/Lebrecht Music & Arts/Lebrecht Music & Arts

Chapter 19
Opener 19a Steve May
Opener 19b Norbert Wu/Peter Arnold, Inc.
19-3 Robert Brons/Biological Photo Service
19-4 Manfred Kage/Peter Arnold, Inc.
19-5a Susumu Nishinaga/SPL/Photo Researchers, Inc.
19-5b Fred Bruemmer/OKAPIA/Photo Researchers, Inc.
19-6 Yorgos Nikas/Science Photo Library/Photo Researchers, Inc.
19-7 James Cavallini/Photo Researchers, Inc.
19-8 James Cavallini/Photo Researchers, Inc.

Chapter 20
Opener 20a Studio MPM/Stone/Getty Images
Opener 20b Shaker Mousa, AP World Wide Photos
20-9a Copyright Dennis Kunkel Microscopy, Inc.
20-9b Don Rubbelke
20-9c Kenneth Kaushansky, University of Washington Medical Center
20-10 David M. Phillips/Visuals Unlimited/Getty Images, Inc - Stock Image
20-11 CNRI/Photo Researchers, Inc.
20-14 Photo Lennart Nilsson/Albert Bonniers Forlag
20-18 Watkins, Bruce/Animals Animals/Earth Scenes
E20-4a Matt Meadows/Peter Arnold, Inc.
E20-4b Southern Illinois University/Photo Researchers, Inc.
E20-4c Bill Travis/National Cancer Institute

Chapter 21
Opener 21a Schito/Olympia/SIPA
Opener 21b REUTERS/L'Equipe Agence/HO/Landov
21-1 Robert Lubeck/Animals Animals - Earth Scenes
21-2 General Board of Global Ministries
21-3 A.D.A.M., Inc.
21-4 Biophoto Associates/Photo Researchers, Inc.
21-5 Robert and Beth Plowes Photography
21-6a Therisa Stack/Tom Stack & Associates, Inc.
21-7a Roland Birke/Okapia/Photo Researchers, Inc.
21-13LB Tom Adams/Visuals Unlimited

Chapter 22
Opener 22a James Cavallini/Photo Researchers, Inc.
Opener 22b Stu Forster/Getty Images
22-1T David Scharf/Peter Arnold, Inc.
22-2 Juergen Berger/Photo Researchers, Inc.
22-3a NIBSC/Science Photo Library/Photo Researchers, Inc.
22-3b S.H.E. Kaufmann & J.R. Golecki/Science Photo Library/Photo Researchers, Inc.
22-9bT Secchi-Lecaque/Roussel/CNRI/Science Source/Photo Researchers, Inc.
22-9bB CNRI/Science Photo Library/Photo Researchers, Inc.
22-15 National Institute for Biological Standards and Control (U.K.)/Science Photo Library/Photo Researchers, Inc.
22-16 Oliver Meckes & Nicole Ottawa/Photo Researchers, Inc.
E22-1 Corbis/Bettmann
E22-2 Dennis Kunkel/Visuals Unlimited/Getty Images

Chapter 23
Opener 23a Gabriel Moisa/Shutterstock
Opener 23b Jeff Minton/Jeff Minton Photography
23-6 Ringling Bros. and Barnum & Bailey Circus Feld Entertainment
23-9 Biophoto Associates/Photo Researchers, Inc.
23-12 Johanna Watson, School of Veterinary Medicine, University of California, Davis
23-13 Photo courtesy of The Jackson Laboratory, Bar Harbor, Maine
E23-1 Novocell

Chapter 24
Opener 24a Miramar Films, Universal Pictures/Laurie Sparham/Picture Desk, Inc./Kobal Collection
Opener 24b Michael R. Jeffords
24-14 From H. Damasio, T. Grabowski, R. Frank, A.M. Galaburda, A.R. Damasio, The Return of Phineas Gage: Clues about the brain from the skull of a famous patient. Science 264:112-115, 1994. Department of Neurology and Image Analysis Facility, University of Iowa.
24-17a Robert S. Preston/Joseph E. Hawkins Jr.
24-21B Susan Eichorst
E24-1 Ken Graham/Bruce Coleman Inc.
E24-2a Reprinted by permission from Nature V2; May 2001, P357; Fig 3a (left), Andrew J. Calder, et al, Neuropsychology of Fear and Loathing. Copyright ©2001 Macmillan Magazines Limited
E24-2b From NeuroReport Vol II, #17,27 November 2000 Fig 3a (middle and right). Andreas Bartels & Zemi Scki, The Neural Basis of Romantic Love.

Chapter 25
Opener 25a Mary Chind/Tucson Citizen
Opener 25b Mary Chind/Tucson Citizen
25-1 David Wrobel/Visuals Unlimited/Getty Images, Inc - Stock Image
25-2 Photograph courtesy of Alexa Bely/University of Maryland
25-3 Peter J. Bryant/Biological Photo Service
25-4a Teresa and Gerald Audesirk
25-4b Peter Harrison
25-5 Anne et Jacques Six
25-6 Michael Fogden/Oxford Scientific Films/Animals Animals/Earth Scenes
25-7a Stephan J. Krasemann/DRK Photo
25-7b Frans Lanting/Minden Pictures
25-14L C. Edelmann/La Villette/Photo Researchers, Inc.
25-17a Photo Lennart Nilsson/Albert Bonniers Forlag
25-17b Y. Nikas/Photo Researchers, Inc.
25-20b Photo Lennart Nilsson/Albert Bonniers Forlag AB
25-23a Photo Lennart Nilsson/Albert Bonniers Forlag AB
25-23b Petit Format/Nestle/Science Source/Photo Researchers, Inc.
25-23c Photo Lennart Nilsson/Albert Bonniers Forlag AB
E25-1a Linda Stannard, UCT/Photo Researchers, Inc.
E25-1b Eye of Science/Photo Researchers, Inc.
E25-2 Haag and Kropp Mauritius, GMBH/Phototake NYC

Chapter 26
Opener 26a Jennifer Graylock/AP Wide World Photos
Opener 26b Joe McDonald/DRK Photo
26-1a Eric and David Hosking/Frank Lane Picture Agency Limited
26-1b Eric and David Hosking/Frank Lane Picture Agency Limited
26-2TL Boltin Picture Library
26-2TM Boltin Picture Library
26-2TR Boltin Picture Library
26-2BL Boltin Picture Library
26-2BR Boltin Picture Library
26-3 Frans Lanting/Minden Pictures
26-5 Thomas McAvoy/Getty Images/Time Life Pictures
26-7 Renee Lynn/Photo Researchers, Inc.
26-8 Ken Cole/Animals Animals/Earth Scenes
26-9 Richard K. LaVal/Animals Animals/Earth Scenes
26-10 Ingo Arndt/Minden Pictures
26-11 Robert & Linda Mitchell/Robert & Linda Mitchell Photography
26-12a Stephen J. Krasemann/Photo Researchers, Inc.
26-12b Hans Pfletschinger/Peter Arnold, Inc.
26-13a M.P. Kahl/DRK Photo
26-13b Marc Chamberlain
26-14 Ray Dove/Visuals Unlimited
26-15 William Ervin/Natural Imagery
26-16 Michael K. Nichols/National Geographic Image Collection
26-17 John D. Cunningham/Visuals Unlimited
26-18a Konrad Wothe/Minden Pictures
26-18b Fans Lanting/Minden Pictures
26-19 Anne et Jacques Six
26-21a Dwight R. Kuhn Photography
26-21b Teresa and Gerald Audesirk
26-22a Frans Lanting/Minden Pictures
26-22b Daniel J. Cox/Getty Images, Inc. - Liaison
26-22c Michio Hoshino/Minden Pictures
26-23 Fred Bruemmer/Peter Arnold, Inc.
26-25 Raymond Mendez/Animals Animals/Earth Scenes
26-26 Lennart Nilsson/Albert Bonniers Forlag AB
26-27 William P. Fifer, New York State Psychiatric Institute, Columbia University
26-28 Reproduced by permission from Low-, normal-, high- and perfect-symmetry versions of a male face From Fig 1 page 235 in Animal Behaviour, 202, 64, 233-238 by Koehler N, Rhodes, G, & Simmons L.W. Copyright © 2006 by Elsevier Science Ltd. Image courtesy of Animal Behavior.

Index